The Practice of Medicinal Chemistry

Second edition

Books are to be returned on or before
the last date below.

The Practice of Medicinal Chemistry

Second edition

Edited by

Camille G. Wermuth

Laboratoire de Pharmacochimie Moléculaire, Faculté de Pharmacie,
Université Louis Pasteur, Illkirch, France

ELSEVIER
ACADEMIC
PRESS

AMSTERDAM • BOSTON • HEIDELBERG • LONDON • NEW YORK • OXFORD
PARIS • SAN DIEGO • SAN FRANCISCO • SINGAPORE • SYDNEY • TOKYO

First published 1996
Reprinted 2001
Second edition 2003
Reprinted 2004

Academic Press
An Imprint of Elsevier
84 Theobald's Road, London WC1X 8RR, UK
http://www.academicpress.com

Academic Press
An Imprint of Elsevier
525 B Street, Suite 1900 San Diego, California 92101-4495, USA
http://www.academicpress.com

ISBN 0-12-744481-5

Library of Congress Catalog Number: 2003104296

A catalogue record for this book is available from the British Library

Typeset by Alden Dataset
Printed and bound in Great Britain by the Bath Press, Avon

Contents

Part III Primary Exploration of Structure–Activity Relationships

Part IV Substituents and Functions: Qualitative and Quantitative Aspects of Structure–Activity Relationships

Part V Spatial Organization, Receptor Mapping and Molecular Modeling.

Biography

Camille-Georges Wermuth PhD, Prof. and Founder of Prestwick Chemical, was Professor of Organic Chemistry and Medicinal Chemistry at the Faculty of Pharmacy, Louis Pasteur University, Strasbourg, France from 1969 to 2002. He became interested in Medicinal Chemistry during his two years of military service in the French Navy at the "Centre d'Etudes Physio-biologiques Appliquées à la Marine" in Toulon. During this time he worked under the supervision of Dr Henri Laborit, the scientist who invented artificial hibernation and discovered chlorpromazine.

Professor Wermuths' main research themes focus on the chemistry and the pharmacology of pyridazine derivatives. The 3-aminopyridazine pharmacophore, in particular, allowed him to accede to an impressive variety of biological activities, including antidepressant and anticonvulsant molecules; inhibitors of enzymes such as mono-amine-oxidases, phosphodi-esterases and acetylcholinesterase; ligands for neuro-receptors: GABA-A receptor antagonists, serotonine 5-HT$_3$ receptor antagonists, dopaminergic and muscarinic agonists. More recently, in collaboration with the scientists of the Sanofi Company, he developed potent antagonists of the 41 amino-acid neuropeptide CRF (corticotrophin-releasing factor) which regulates the release of ACTH and thus the synthesis of corticoids in the adrenal glands. Professor Wermuth has also, in collaboration with Professor Jean-Charles Schwartz and Doctor Pierre Sokoloff (INSERM, Paris), developed selective ligands of the newly discovered dopamine D$_3$ receptor. After a three-year exploratory phase, this research has led to nanomolar partial agonists which may prove useful in the treatment of the cocaine-withdrawal syndrome.

Besides about 300 scientific papers and about 60 patents, Professor Wermuth is co-author or editor of several books including; *Pharmacologie Moléculaire*, Masson & Cie, Paris; *Médicaments Organiques de Synthèse*, Masson & Cie, Paris; *Medicinal Chemistry for the Twenty-first Century*, Blackwell Scientific Publications, Oxford; *Trends in QSAR and Molecular Modeling*, ESCOM, Leyden, two editions of *The Practice of Medicinal Chemistry*, Academic Press, London and *The Handbook of Pharmaceutical Salts, Properties Selection and Use*, Wiley-VCH.

Professor Wermuth was awarded the Charles Mentzer Prize of the Société Française de Chimie Thérapeutique in 1984, the Léon Velluz Prize of the French Academy of Science in 1995, the Prix de l'Ordre des Pharmaciens 1998 by the French Academy of Pharmacy and the Carl Mannich Prize of the German Pharmaceutical Society in 2000. He is Corresponding Member of the German Pharmaceutical Society and was nominated Commandeur des Palmes Académiques in 1995. He has been President of the Medicinal Chemistry Section of the International Union of Pure and Applied Chemistry (IUPAC) from 1988 to 1992 and from January 1998 to January 2000 was President of the IUPAC Division on Chemistry and Human Health.

Contributors

Andrews, Peter, Center for Drug Design and Development, The University of Queensland, Brisbane, Queensland 4072, Australia

Belpaire, Franz, JF&C Heymans Institute for Pharmacology, University of Gent Medical School, De Pintelaan 185, B-9000 Gent, Belgium

Blundell, Tom, Department of Biochemistry, University of Cambridge, 80 Tennis Court Road, Cambridge, CB2 1GA, UK

Bogaert, Marc G., JF&C Heymans Institute for Pharmacology, University of Gent Medical School, De Pintelaan 185, B-9000 Gebt, Belgium

Bourguignon, Jean-Jacques, Laboratoire de Pharmacochimie de la Communication Cellulaire, Faculté de Pharmacie, 74 Route du Rhin, 67401 ILLKIRCH-Cedex, France

Cavalla, David, Arachnova, St John's Innovation Centre, Cambridge CB4 4WS, UK

Chast, François, Hôpital de l'Hôtel-Dieu, 1 Place du Parvis de Notre-Dame, 75004, Paris, France

Contreras, Jean-Marie, Prestwick Chemical Inc., Rue Tobias Stimmer, Bâtiment Tycho, Brahé, 67400 Illkirch, France

Cragg, Gordon M., Natural Products Branch, Developmental Therapeutics Program, Division of Cancer Treatment and Diagnosis, National Cancer Institute, Frederick Cancer Research and Development Center, P.O. Box B, Frederick, MD 21701-1013, USA

Cruciani, Gabriele, Laboratory for Chemometrics, Chemistry Department, University of Perugia, Via Elce di Sotto 10, Perugia I-06123, Italy

Dansette, Patrick, Laboratoire de chimie et Biochimie, Pharmacologiques et toxicologiques, CNRS UMR 8601, 45 rue des Saints-Pères, 75006 Paris, France

Deane, Charlotte, Department of Biochemistry, University of Cambridge, 80 Tennis Court Road, Cambridge, CB2 1GA, UK

Faller, Bernard, Novartis Pharma AG, Werkklybeck, Klybeckstrasse 141, CH-4057, Basel, Switzerland

Fournel-Gigleux, Sylvie, UMR 7561 CNRS/UMP, Faculté de Médicine, Boite Postale 184, 54505 Nancy, France

Galli, Bruno, Novartis Pharma AG, Postfach, Ch-4002, Basel, Switzerland

Gasteiger, Johann, Computer-Chemie-Centrum, Institut für Organische Chemie, Universität Erlangen-Nürnberg, Naegelbachstrasse 25, D-91052 Erlangen Germany

Gies, Jean-Pierre, Laboratoire de Pathologie des Communications entre Cellules Nerveuses et Musculaires, Université Louis Pasteur, Faculté de Pharmacie, 74 Route du Rhin, 67401 ILLKIRCH-Cedex, France

Hirayama, Fumitoshi, Faculty of Pharmaceutical Sciences, Kumamoto University, 5-1 Oe Honmachi, Kumamoto 862, Japan

Hirschmann, Ralph S., University of Pennsylvania, Philadelphia, USA

Hobden, Adrian N., Myriad Genetics, Inc., 320 Wakara Way, Salt Lake City, Utah 84108, USA

Höltje, Hans-Dieter, Institut für Pharmazeutische Chemie, Heirich Heine Universität, Universitätstrassse 1, Düsseldorf 40225, Germany

Hoste, Katty, University of Gent, Department of Organic Chemistry, Polymer Materials Research Group, Krijgslaan 281 54 Bis, Ghent, Belgium

Kahn, Michaël, Department of Pathobiology, Box 35 72 38, University of Washington, Seattle, WA 98195, USA

Kan, Jean-Paul, Sanofi-Synthelabo Recherche, Département Système Nerveux Central, 371 Rue du Professeur J. Blayac, 34082 Montpellier-Cedex 04, France

Kingston, David G.I., Department of Chemistry College of Arts and Sciences, Virginia Polytechnic Institute and State University, Blacksburg, Virginia 24061-0212, USA

Kopp-Kubel, Sabine, DMP-RGS World Health Organization, 20 Avenue Appia, 1211 Genève-27, Switzerland

Landry, Yves, Laboratoire de Neuroimmuno-Pharmacologie Pulmonaire INSERM U425, Université Louis Pasteur, Faculté de Pharmacie, 74 Route du Rhin, 67401 ILLKIRCH-Cedex, France

Lipinski, Christopher A., Pfizer Global Research and Development, Mail Stop 8118W/220, Eastern Point Road, Groton, Connecticut 06340/0177, USA

Macherey, Anne-Christine, UPS 831-Institut de Chimie des Substances Naturelles, Avenue de la Terasse, 91198 Gif sur Yvette-Cedex, France

Magdalou, Jacques, UMR 7561 GNRS/UMP, Faculté de Médecine, Boite Postale 184, 22 Rue de Vandoeuvre, 54505 Nancy, France

Mann, André, Laboratoire de Pharmacochimie de la Communication Cellulaire, Faculté de Pharmacie, 74 Route du Rhin, 67401 ILLKIRCH-Cedex, France

Nakanishi, Hiroshi, Department of Pathobiology, Box 35 72 38, University of Washington, Seattle, WA 98195, USA

Newman, David, Natural Products Branch, Developmental Therapeutics Program, Division of Cancer Treatment and Diagnosis, National Cancer Institute, Frederick Cancer Research and Development Center, P.O. Box B, Frederick, MD 21701-1013, USA

Ouzzine, Mohamed, UMR 7561 CNRS/UMP, Faculté de Médecine, Boite Postale 184, 22 Rue de Vandoeuvre, 54505 Nancy, France

Picot, André, UPS 831-Institut de Chimie des Substances Naturelles, Avenue de la Terasse, 91198 Gif sur Yvette-Cedex, France

Polinsky, Alexander, Pfizer Global R&D, La Jolla Laboratories, 10770 Science Center Drive, San Diego, CA 92121, USA

Reuben, Bryan G., 7 Clarence Avenue, London SW4 8LA, UK

Rondeau, Jean-Michel, Novartis Pharma Research, WSJ-88.9.09A, 4002 BASEL, Switzerland

Rose, Sally, BioFocus plc., Sittingbourne Research Centre, Sittingbourne, Kent ME9 8AZ, UK

Rich, Daniel H., School of Pharmacy, 425 North Charter Street, Madison, WI 53706-1508, USA

Schacht, Etienne H., University of Gent, Department of Organic Chemistry, Polymer Materials Research Group, Krijgslaan 281 54 Bis, Ghent, Belgium

Schreuder, Herman, Aventis Pharma Deutschland GmbH, Building G 865 A, 65926 Frankfurt am Main, Germany

Seymour, Len, University of Birmingham Clinical Oncology, Queen Elizabeth Hospital, Edgbaston Birmingham, B15 2TH UK

Souleau, Maria, Sanofi-Synthélabo, Departement Brevets, 174 Avenue de France, 75 013-Paris, France

Stahl, P. Heinrich, Cosmas Consult, Lerchenstrasse No. 28, D-79104 Freiburg im Breisgau, Germany

Steenkiste, Stefan van, University of Gent, Department of Organic Chemistry, Polymer Materials Research Group, Krijgslaan 281 54 Bis, Ghent, Belgium

Terfloth, Lothar, Computer-Chemie-Centrum, Institut für Organische Chemie, Universität Erlangen-Nürnberg, Nägelbach-strasse 25, D-91052 Erlangen, Germany

Testa, Bernard, Université de Lausanne, École de Pharmacie, B.E.P., CH-1015 Lausanne-Dorigny, Switzerland

Triggle, David J., State University of New York, School of Pharmacy, Amherst Campus, Buffalo, NY 14260, USA

Uekama, Kaneto, Faculty of Pharmaceutical Sciences, Kumamoto University, 5-1 Oe Honmachi, Kumamoto, 862, Japan

Vanier, Cécile, Graffinity Pharmaceuticals AG Im, Neuenheimer Feld 518-519 69120, Heidelberg, Germany

Wagner, Alain, Laboratoire de Synthèse Bioorganique, UMR 7514 du CNRS, Université Louis Pasteur, Faculté de Pharmacie, 74 Route du Rhin, 67401 ILLKIRCH-Cedex, France

Waterbeemd, Han van de, Pfizer Global Research and Development, PDM, Department of Drug Metabolism, Director and Head of Discovery and Preclinical Development, IPC 331, Sandwich, Kent CT13 9NJ, UK

Wermuth, Camille G., Laboratoire de Pharmacochimie de la Communication Cellulaire, Université Louis Pasteur, Faculté de Pharmacie, 74 Route du Rhin, 67401 Illkirch-Cedex, France

Winne, Katleen de, University of Gent, Department of Organic Chemistry, Polymer Materials Research Group, Krijgslaan 281 54 Bis, Ghent, Belgium

Zavitz, Kenton H., Myriad Pharmaceuticals Inc., Salt Lake City, UT 84108, USA

Foreword

It is a privilege to write a Foreword to *The Practice of Medicinal Chemistry* written by a distinguished group of contributors. It is also a privilege because this book will be read by medicinal chemists from diverse backgrounds, interests and expertise. What these scientists share is a chosen career in medicinal chemistry, surely one of the most satisfying, because it is dedicated to improving mankind's quality of life and also because it provides an intellectually satisfying environment.

The eight general topics discussed insightfully in forty-three chapters of this textbook were wisely selected and do indeed describe medicinal chemistry as it is practiced today. This Foreword does not analyze or summarize these chapters, but offers instead some personal reflections which, it is hoped, have some relevance to the volume.

Modern medicinal chemistry began in the 1950s when organic chemists began to apply newly developed steric and electronic concepts to an understanding of the structure-activity relationships of the steroids. During the second half of the twentieth century, chemistry and biology made possible the discovery of a steady stream of important new medicines. Chemistry contributed to these discoveries through impactful advances in both theory and practice of this art/science. Notable examples include invaluable advances in physical measurements, computational techniques, inorganic catalysis, stereochemical control of synthesis and the application of physical organic chemical concepts, typified by the transition state analog principle, to enzyme inhibitor design. At the same time, biology continued to contribute through the discoveries of new concepts and understanding at a rate that may well be termed explosive.

Specifically, in 1953 Watson and Crick had proposed the double-helical structure of DNA, and suggested that the sequence of nucleotide units in DNA carries encoded genetic information that determines the amino acid sequence of proteins. These discoveries proved to be a revolutionary event in biology with a profound impact on the vitality of all biomedical research. In the early 1970s, biochemical research made possible the application by Herbert W. Boyer of this new understanding to the introduction of recombinant DNA research, a new technology. This in turn led to the Boyer and Cohen collaboration on prokaryotic and eukaryotic DNA research, early cloning successes, and the founding of Genentech, Inc. in 1976. The beginning of the successful commercialization of the recombinant DNA technology by Boyer and Robert A. Swanson thus occurred twenty-three years after the discovery of the double-helix. The early (and subsequent) successes of Genentech, Inc. amply validated DNA technology as an industrial enterprise. Biotechnology became a household word and with it the search for new biotechnology came to be accepted as a desirable end in itself.

During the 1970s, target validation became an important consideration in the selection of therapeutic programs explored by the pharmaceutical industry. In the strictest sense this strategy holds that intervention in any particular biochemical or pharmacological pathway has been fully validated only if it has been shown to work in human subjects. Any research program that does not pass this definitive test is therefore thought to be a 'long shot.' In practice, this leads to the conclusion that a conservative portfolio of an organization's research programs should strike some appropriate balance between 'validated' and 'long shot' targets. In recent years successful use of antibodies in neutralizing a target protein or other substance has come to be accepted as adequate validation; this is also the case for another validated technology, the use of 'knock-out' or 'knock-in' mice.

At the end of the twentieth century there was every reason to expect that the flow of new drugs from the laboratory stages, *via* clinical trials, to the expected regulatory approval would continue to accelerate. As this book goes to press, the pipelines of both the pharmaceutical industry and of the biotechnology companies are, however, relatively dry. This is at the very time when spectacular advances in biology such as genomics and proteomics seemed to have laid the basis for the discovery of many new breakthrough drugs.

Surprisingly, there has been relatively little discussion of the causes of this paradoxical state of affairs. It is believed by some that the growing impact of marketing departments on the choice of clinical targets may have played a role in bringing about this disappointing state of affairs. Perhaps so, but it seems prudent to suggest that scientific decisions originating within the research organizations themselves may also deserve scrutiny (see below).

The unexpectedly dry pipelines of the industry raise the interesting question of whether it might be wise for any company to also 'validate' *new technologies* on a modest scale before they are extensively embraced by any organization. Two examples serve to illustrate this concern. In the 1980s the recognition that 'rational design' of enzyme inhibitors is a fruitful approach to drug discovery, led to the belief that knowing the tertiary structure of the active sites of such enzyme targets would greatly facilitate the discovery process. Significant time and effort was invested in this approach by several companies before it was recognized that the X-ray structure of the *uninhibited* enzyme is likely to be misleading, because the important role of water molecules and of conformational effects on molecular recognition was not appreciated.

More importantly, the extensive early commitment by the pharmaceutical industry to the use of combinatorial chemistry as the principal source of new chemical entities for lead discovery may be an even more serious issue. Combinatorial chemistry has proven to be an effective tool for lead optimization, but its use as the principal source of compounds for screening for lead discovery has been problematic. At the present time, very few – if any – compounds that are approved drugs or that are in Phase III clinical trials, are thought to owe their existence solely to leads generated by combinatorial chemistry. If this assessment is indeed a valid one, the extensive reliance by the industry on this unvalidated technology may well have contributed to the current disappointing status of the pipelines. For combinatorial chemistry to become a useful source of compounds for lead discovery, two requirements must be met. (1) The successive reactions must proceed sufficiently well to afford the expected final product in good yield. This requirement has generally been met. (2) It is equally important, however, that the chosen synthetic targets incorporate sufficient complexity to have a good chance that some of them will become true leads. The bar for this second objective may often have been lowered too much in order to achieve the first requirement – the desired purity. There is reason to hope that now-a-days both requirements are being met, but only time will tell.

On the other hand, a change in tactics in lead optimization has allowed for huge advances, especially in the discovery of ligands for G-protein coupled receptors. As their first objective, medicinal chemists used to seek to optimize *in vitro* potency. Given the hydrophobic nature of many of these receptors, it is not surprising that the most potent compounds to emerge from this tactic proved to be equally hydrophobic, and thus to display poor pharmacokinetic properties. Attempts to deal with this matter in the so-called 'endgame' more often than not proved to be futile. The program had thus become a victim of the 'hydrophobic trap.' By the end of the twentieth century we had learned to appreciate that measurements or calculations that relate to solubility, and thus to oral bioavailability such as log P and polar molecular surface properties, should guide the synthetic program from the very beginning.

I shall close this Foreword by returning to the theme of medicinal chemists in the service of humanity and illustrate it with two examples: the first relates to AIDS. In the mid-1980s, the human immunodeficiency virus (HIV) had been identified as the cause of AIDS. By 1996 medicinal chemists, in collaboration with biologists, had discovered reverse transcriptase inhibitors and later HIV protease inhibitors, which were combined to provide cocktails of therapy. Various regimes of what was termed 'highly active antiretroviral therapy' (HAART) reduced the viral load below detection levels — an enormous advance. Today there are sixteen approved drugs against AIDS which include seven nucleoside reverse transcriptase inhibitors (NRTIs), three non-nucleoside inhibitors of these enzymes (NNRTIs) and six protease inhibitors (PIs). These medicines thus represent an enormous step forward, although they have failed to eliminate the virus entirely from patients, resulting in the need to continue these therapies, along with their considerable side effects, for life. Resistance to these anti-AIDS medications represents an even more serious challenge, as is our inability to make anti-AIDS therapy readily available to patients in developing nations.

In contrast, the drug Mectizan has all but eliminated another dreaded disease, river blindness, in both Latin America and in West Africa. In other parts of Africa where progress varies on a country-by-country basis, success has been less dramatic. Surely there is every reason for medicinal chemists to be proud of this drug, first discovered and developed as an animal health anthelmintic, and which ultimately proved to have a far more glorious role to play in human medicine. Almost miraculously, oral administration of this drug is required only three times a year, an enormous plus in an environment where patient compliance is so poor. It should be a source of great pride and satisfaction to all medicinal chemists, whether working in the pharmaceutical industry or in other organizations, that at the time of writing Merck & Co., Inc. continues to donate 150 million tablets annually to patients in the developing world.

Looking to the future, great challenges remain including cancer, AIDS and — as life expectancy increases — Alzheimer's disease. It is exciting to contemplate that in the twenty-first century medicinal chemists will, in addition, become increasingly successful in using small molecules to block undesired protein–protein interactions. It is surely the hope of the scientists who have contributed to this book that *The Practice of Medicinal Chemistry* will become a much-consulted adjunct to the medicinal chemists in their search for the drugs of the future.

Ralph Hirschmann
Philadelphia, Pennsylvania

Preface to the First Edition

The role of chemistry in the manufacture of new drugs, and also of cosmetics and agrochemicals, is essential. It is doubtful, however, whether chemists have been properly trained to design and synthesize new drugs or other bioactive compounds. The majority of medicinal chemists working in the pharmaceutical industry are organic synthetic chemists with little or no background in medicinal chemistry who have to acquire the specific aspects of medicinal chemistry during their early years in the pharmaceutical industry. This book is precisely aimed to be their 'bedside book' at the beginning of their career.

After a concise introduction covering background subject matter, such as the definition and history of medicinal chemistry, the measurement of biological activities and the three main phases of drug activity, the second part of the book discusses the most appropriate approach to *finding a new lead compound or an original working hypothesis*. This most uncertain stage in the development of a new drug is nowadays characterized by high-throughput screening methods, synthesis of combinatorial libraries, data base mining and a return to natural product screening. The core of the book (Parts III to V) considers the *optimization of the lead in terms of potency, selectivity, and safety*. In 'Primary Exploration of Structure-Activity Relationships', the most common operational stratagems are discussed, allowing identification of the portions of the molecule that are important for potency. 'Substituents and functions' deals with the rapid and systematic optimization of the lead compound. 'Spatial Organization, Receptor Mapping and Molecular Modelling' considers the three-dimensional aspects of drug-receptor interactions, giving particular emphasis to the design of peptidomimetic drugs and to the control of the agonist–antagonist transition. Parts VI and VII concentrate on the definition of satisfactory drug-delivery conditions, i.e. means to ensure that the molecule reaches its target organ. Pharmacokinetic properties are improved through adequate chemical modifications, notably prodrug design, obtaining suitable water solubility (of utmost importance in medical practice) and improving organoleptic properties (and thus rendering the drug administration acceptable to the patient). Part VIII, 'Development of New Drugs: Legal and Economic Aspects', constitutes an important area in which chemists are almost wholly self taught following their entry into industry.

This book fills a gap in the available bibliography of medicinal chemistry texts. There is not, to the author-editor's knowledge, any other current work in print which deals with the practical aspects of medicinal chemistry, from conception of molecules to their marketing. In this single volume, all the disparate bits of information which medicinal chemists gather over a career, and generally share by word-of-mouth with their colleagues, but which have never been organized and presented in coherent form in print, are brought together. Traditional approaches are not neglected and are illustrated by modern examples and, conversely, the most recent discovery and development technologies are presented and discussed by specialists. Therefore, *The Practice of Medicinal Chemistry* is exactly the type of book to be recommended as a text or as first reading to a synthetic chemist beginning a career in medicinal chemistry. And, even if primarily aimed at organic chemists entering into pharmaceutical research, all medicinal chemists will derive a great deal from reading the book.

The involvement of a large number of authors presents the risk of a certain lack of cohesiveness and of some overlaps, especially as each chapter is written as an autonomic piece of information. Such a situation was anticipated and accepted, especially for a first edition. It can be defended because each contributor is an expert in his/her field and many of them are 'heavyweights' in medicinal chemistry. In editing the book I have tried to ensure a balanced content and a more-or-less consistent style. However, the temptation to influence the personal views of the authors has been resisted. On the contrary, my objective was to combine a plurality of opinions, and to present and discuss a given topic from different angles. Such as it is, this first edition can still be improved and I am grateful in advance to all colleagues for comments and suggestions for future editions.

Special care has been taken to give complete references and, in general, each compound described has been identified by at least one reference. *For compounds for which no specific literature indication is given, the reader is referred to the Merck Index.*

The cover picture of the book is a reproduction of a copperplate engraving designed for me by the late Charles Gutknecht, who was my secondary school chemistry teacher in Mulhouse. It represents an extract of Brueghel's engraving *The alchemist ruining his family in pursuing his chimera*, surmounted by the aquarius symbol. Represented on the left-hand side is my lucky charm caster oil plant (*Ricinus communis* L., *Euphorbiaceae*), which was the starting point of the pyridazine chemistry in my laboratory. The historical cascade of events was as follows: cracking of caster oil produces n-heptanal and aldolization of n-heptanal – and, more generally, of any enolisable aldehyde or ketone – with pyruvic acid leads to a-hydroxy-γ-ketonic acids. Finally, the condensation of these keto acids with hydrazine yields pyrodazones. Thus, all our present research on pyridazine derivatives originates from my schoolboy chemistry, when I prepared in my home in Mulhouse n-heptanal and undecylenic acid by cracking caster oil!

Preparing this book was a collective adventure and I am most grateful to all authors for their cooperation and for the time and the effort they spent to write their respective contributions. I appreciate also their patience, especially as the editing process took much more time than initially expected.

I am very grateful to Brad Anderson (University of Utah, Salt Lake city), Jean-Jacques André (Marion Merrell Dow, Strasbourg), Richard Baker (Eli Lilly, Erl Wood, UK), Thomas C. Jones (Sandoz, Basle), Isabelle Morin (Servier, Paris), Bryan Reuben (South Bank University, London) and John Topliss (University of Michigan, Ann Arbor) for their invaluable assistance, comments and contributions.

My thanks go also to the editorial staff of Academic Press in London, Particularly to Susan Lord, Nicola Linton and Fran Kingston, to the two copy editors Len Cegielka and Peter Cross, and finally, to the two secretaries of our laboratory, François Herth and Marylse Wernert.

Last but not least, I want to thank my wife Renée for all her encouragement and for sacrificing evenings an Saturday family life over the past year and a half, to allow me to sit before my computer for about 2500 hours!

Camille G. Wermuth

Preface to the Second Edition

Like the first edition of *The Practice of Medicinal Chemistry* (nicknamed 'The Bible' by medicinal chemists) the second edition is intended primarily for organic chemists beginning a career in drug research. Furthermore, it is a valuable reference source for academic, as well as industrial, medicinal chemists. The general philosophy of the book is to complete the biological progress — Intellectualization at the level of function — using the chemical progress — Intellectualization at the level of structure (Professor Samuel J. Danishevsky, *Studies in the chemistry and biology of the epothilones and eleutherobins*, Conference given at the XXXIVémes Rencontres Internationales de Chimie Thérapeutique, Faculté de Pharmacie, Nantes, 8–10 July, 1998).

The recent results from genomic research have allowed for the identification of a great number of new targets, corresponding to hitherto unknown receptors or to new subtypes of already existing receptors. The massive use of combinatorial chemistry, associated with high throughput screening technologies, has identified thousands of hits for these targets. The present challenge is to develop these hits into usable and useful drug candidates. This book is, therefore, particularly timely as it covers abundantly the subject of drug optimization.

The new edition of the book has been updated, expanded and refocused to reflect developments over the nine years since the first edition was published. Experts in the field have provided personal accounts of both traditional methodologies, and the newest discovery and development technologies, giving us an insight into diverse aspects of medicinal chemistry, usually only gained from years of practical experience.

Like the previous edition, this edition includes a concise introduction covering the definition and history of medicinal chemistry, the measurement of biological activities and the three main phases of drug activity. This is followed by detailed discussions on the discovery of new lead compounds including automated, high throughput screening techniques, combinatorial chemistry and the use of the internet, all of which serve to reduce pre-clinical development times and, thus, the cost of drugs. Further chapters discuss the optimization of lead compounds in terms of potency, selectivity, and safety; the contribution of genomics; molecular biology and X-ray crystallization to drug discovery and development, including the design of peptidomimetic drugs; and the development of drug-delivery systems, including organ targeting and the preparation of pharmaceutically acceptable salts. The final section covers legal and economic aspects of drug discovery and production, including drug sources, good manufacturing practices, drug nomenclature, patent protection, social-economic implications and the future of the pharmaceutical industry.

I am deeply indebted to all co-authors for their cooperation, for the time they spent writing their respective contributions and for their patience during the editing process. I am very grateful to Didier Rognan, Paola Ciapetti, Bruno Giethlen, Annie Marcincal, Marie-Louise Jung, Jean-Marie Contreras and Patrick Bazzini for their helpful comments.

My thanks go also to the editorial staff of *Academic Press* in London, particularly to Margaret Macdonald and Jacqueline Read.

Last but not least, I want to express my gratitude to my wife Renée for all her encouragements and for her comprehensiveness.

Camille G. Wermuth

PART I

GENERAL ASPECTS OF MEDICINAL CHEMISTRY

1

A BRIEF HISTORY OF DRUGS: FROM PLANT EXTRACTS TO DNA TECHNOLOGY

François Chast*

Le médicament place l'organisme dans des conditions particulières qui en modifient heureusement les procédés physiques et chimiques lorsqu'ils ont été troublés.

Claude Bernard[**]

I THE ANCIENT LINK BETWEEN MEDICINE AND RELIGION

The earliest written records of therapeutic practices are to be found in the Ebers Papyrus, dating from the sixteenth century BC. This is historically of value, since by itself, it represents a compilation of earlier works that contain a large number (877) of prescriptions and recipes. Many plants are mentioned, including opium, cannabis, myrrh, frankincense, fennel, cassia, senna, thyme, henna, juniper, linseed, aloe, castor oil and garlic.[1] Cloves of garlic have been found in Egyptian burials, including the tomb of Tutankhamun and in the sacred underground temple of the bulls at Saqqara.[2] The Ebers Papyrus describes several charms and invocations that were used to encourage healing. The Egyptians were also well known for other healing techniques: spiritual healing, massage and surgery, as well as the extensive use of therapeutic herbs and foods. The Egyptian Shaman physician had to discover the nature of the particular entity possessing the person and then attack, drive it out, or otherwise destroy it. This was done by some powerful magic for which rituals, spells, incantations, talismans and amulets were used.

The art of divination is first known to be used in Babylonian-Assyrian medicine along with the use of astrology to determine the influence of the stellar constellations on human welfare and medical ethics. Two others aspects were usually outlined: besides divination, exorcism and medical treatment were blended together to form a composite picture. Two hundred and fifty vegetable drugs and 120 mineral drugs were identified in the clay tablets from the library of King Assurbanipal. Excrements likewise played an important part in the therapy. They were supposed to throw out the evil spirit that had invaded the body of the patient.[3]

Ancient civilizations tended to borrow and adopt the skills and knowledge of medicine and healing of various cultures to their own. When Alexander the Great conquered and encompassed virtually the known world, he did so with the intention of extolling the humanizing Greek culture.

*Adresse de l'auteur: 1, place du Parvis Notre-Dame, F-75181, PARIS Cedex 04

** Leçons sur les Effets des Substances Toxiques et Médicamenteuses

All the nations brought under the wing of Greece, however, brought with them their own traditions and customs including their healing knowledge.[4]

Hippocrates (approximately 460–377 BC) is considered as the father of medicine through a major, but anonymous writing called *Corpus Hippocraticum*. The regulation of diet occupied the most important place in therapeutics. At his time, drugs were mainly from vegetal origin: juice of the poppy, henbane and mandrake are cited side by side with castor oil, fennel plant, linseed, juniper, saffron, etc. Aspects of the theoretical basis for their use and application were also adopted. Purgatives, sudorifics and emetics were frequently used in order to purify sick organisms. Just as the Greek universe was ordered according to the principles of four dynamic elements: fire, water, air and earth, Hippocrates saw the body as governed by four correspond-ing 'humors', consisting of: sanguine, melancholic, phleg-matic and choleric. Such theories, common to most ancient civilizations, outline essential differences between holistic objectives of traditional medicine in contrast to that of contemporary medicine.[5] These principles, formulated 2400 years ago, attempted to weed out various aspects of superstition which dominated people's minds at the time, in favour of applied logic and reason. Health and disease were seen as a question of humoral balance or imbalance with foods and herbs classified according to their ability to affect natural homeostasis. Of his many aphorisms the most memorable are: 'above all else, do no harm', or ' let your medicine be your food and your food, medicine'. The classification of herbs as 'hot, 'cold', 'wet', 'dry' for instance was not thought to represent absolutes in the scientific sense, but rather aspects to be utilized as part of the art of medicine.[6]

Ancient times were a period when poisoning was raised to a high art, and in turn spurred on dazzling efforts to discover or create effective antidotes. Thus the art of Greek pharmacy was strongly supported and encouraged by the wealthy. *Mithridaticum* was an antidote containing no less than 54 ingredients, developed for Mithridates, king of Pontus during the first century BC. The remedy consisted of small amounts of various poisons which taken over a period of time are supposed to make one immune to their fatal effects.[7] The Romans, famous for incorporating the best of their Greek forbears, attempted through the efforts of Andromachus, Nero's physician, to improve or at least enlarge upon Mithridatesshotgun anti-poison by increasing the number of toxic ingredients from 54 to 70. Under the name *Theriac*, it was described in pharmacopoeias for centuries, through the European Renaissance to the modern pharmacopoeias, at the end of the nineteenth century.

One of the most significant virtues of the Romans, responsible for the long-lasting success of their civilization, was their ability to adopt local customs, religions and cultural mores, along with incorporating the accumulated knowledge and wisdom of foreign cultures under Roman dominion.

The two most important medical figures of Rome, whose contributions remain the uncontested 'standard' for botany and medicine are Dioscorides and Galen.[8] Dioscorides was born in Turkey, in the first century. His most significant contribution was the five botanical books entitled *De Materia Medica*,[9] forming the basis for all subsequent *Materia Medica* for the next 1600 years throughout Europe.

Most of Dioscorides' *Materia Medica* consists of plant medicines, while the remainder is divided more or less 10% mineral and 10% animal. If we consider that many chemically synthesized drugs were once derived from plant products, the percentages of Dioscorides' work is remarkably similar to today's. Dioscorides sought to classify drugs according to broad physiological categories of action, including: warming, mollifying and softening, astringent, bitter, or binding, diuretics, drying, etc. He raised herbal medicine beyond the purely empirical principle of finding a specific herb for a specific disease and presupposed a corresponding system of diagnosis for which the above physiologic actions will be useful. His work became the primary source of future herbalists for over 1500 years.

Galen, born in Sicily, lived around 130 AD, learned anatomy at the Greek School in Alexandria, and was the last of the important Greek herbalists, writing over 400 works, of which 83 are extant. His major herbal, *De Simplicibus* represents the fruits of his extensive travel and research. Drugs supposed to have only one quality were classified as 'simples', while those with more qualities were considered 'composites'. He described 473 drugs from vegetable, animal and mineral origin. Galen, as a continuator of Hippocrates, kept the humoral pathology scheme, which was to rule western medicine throughout the Middle Ages.

Many other Roman medical authors may be cited: Celsus, with his book *De Medicina*, was very influential, as was Scribonius with *Compositiones* and Plinius with *Historia Naturalis*.

During the Middle Ages, which lasted from AD 400 to the 1500s, the Muslim Empire of Southwest and Central Asia made significant contributions to medicine. Rhazes, a Persian-born physician of the late 800s and early 900s wrote the first accurate descriptions of measles and smallpox, Avicenna, an Arab physician of the late 900s and early 1000s, wrote a vast medical encyclopedia called *Canon of Medicine*. It represented a summation of medical knowledge of the time and influenced medical education for more than 600 years.[10]

The primary medical advance of the Middle Ages was the founding of many hospitals and university medical schools. Christian religious groups maintained hundreds of charitable hospitals for victims of leprosy. In the 900s a medical school established in Salerno (Italy), became the primary centre of medical learning in Europe during the 1000s and 1100s.

The *Leech Book*, the oldest known Anglo-Saxon herbarium, probably written in Winchester, circa AD 920, by Cyril Bald or at his special request, is the oldest book written in the vernacular and the first medical treatise of Western Europe.

During the twelfth century, pharmaceutical history was dominated by the high personality of Hildegard of Bingen (St. Hildegard). Through devotion and mysticism, she uses the Hippocratic four-humor system and integrated body-mind and spirit with specific descriptions of diet, herbs and gems. She recommended the use of various plants: psyllium, aloe, horehound, galangal, geranium, fennel, parsley, nettles and spices, and prepared wines, infusions, syrups, oils, salves, powders and smoking mixtures. Later, in the fifteenth century, the *Herbarius* and in 1491, *Hortus Sanitatis* both have some of the best woodcuts prior to the new period of botanical illustration beginning in 1530. In France, Le '*Grant Herbier*' was important because of its later English translation in 1526.[11]

Paracelsus (1493–1541), more properly Theophrastus Phillippus Aureolus Bombastus Von Hohenheim, was born in Einsiedeln (Switzerland), in 1493. He was a phenomenon in the history of medicine, who tried to substitute something better for what seemed to him antiquated and erroneous in therapeutics, thus falling into the mistake of other radical reformers, who, during the process of rebuilding, underestimated the work of their contemporaries. Like Hippocrates, he prescribed the observation of nature and dietetic directions, but attached too great a value to experience (empiricism). In nature, all substances have two kinds of influences, helpful (*essentia*) and harmful (*venena*), which were separated by means of alchemy. It required experience to recognize essences as such and to employ them at the proper moment. His aim was to discover a specific remedy (*arcanum*) for every disease. It was precisely here, however, that he fell into error, since not infrequently he drew conclusions as to the availability of certain remedies from purely external signs, e.g. when he taught that the pricking of thistles cured internal inflammation. This untrustworthy 'doctrine of signatures' was developed at a later date by Rademacher, and also to a certain extent also by Hahnemann.[12] Although the theories of Paracelsus, as contrasted with the Galeno-Arabic system, indicate no advance inasmuch as they ignore entirely the study of anatomy, his reputation as a reformer of therapeutics is still justified in that he broke new paths in the science. He may be taken as the founder of modern *Materia Medica*, and pioneer of scientific chemistry, since before his time medical science received no assistance from alchemy. To Paracelsus is due the use of mercury for syphilis as well as a number of other metallic remedies. He was the first to point out the value of mineral waters. He recognized the tincture of gallnut as a reagent for the iron properties of mineral water and showed

a particular preference for native herbs, from which he obtained 'essences' and 'tinctures', the use of which was to replace the composite medicines so popular at the time.

Robert Boyle (1627–1691) is noted for his pioneer experiments on the properties of gases and his espousal of a corpuscular view of matter that was a forerunner of the modern theory of chemical elements and atomic theory. Boyle conducted pioneering experiments in which he demonstrated the physical characteristics of air and the necessity of air for combustion and respiration.[13] In 1661, he described, in the second edition of his work, *New Experiments Physio-Mechanical*, the relationship, known as Boyle's Law, of the volume of gases and pressure. Attacking the Aristotelian theory of the four elements (earth, air, fire, and water) and the three principles (salt, sulphur and mercury), proposed by Paracelsus, in *The Skeptical Chymist*, he can be considered as the founder of modern chemistry.[14]

In 1676, the British physician Thomas Sydenham published *Observationes Medicae* as a standard textbook for two centuries noted for its detailed observations and the accuracy of its records. His treatise on gout (1683) is considered his masterpiece. Sydenham was among the first to use iron to treat iron-deficiency anaemia and used laudanum (a solution of opium in alcohol) as a medication, and helped popularize the use of quinquina for malaria.

Despite these prestigious glories in medicine or pharmacotherapy, pharmacy remained an empiric science until the end of the eighteenth century, guided by ancient medicine, inherited from Hippocrates or Galen.

II MODERN CHEMISTRY AS THE BASIS OF THE CONCEPT OF MODERN DRUGS

The eighteenth century concluded its progress in chemistry with an enthusiastic environment. Joseph Priestley (1733–1804) in the United Kingdom, Carl Wilhelm Scheele (1742–1786) in Sweden, Antoine Augustin Lavoisier (1743–1794) in France,[15] pulled down alchemist practice by propounding a precise signification to the chemical reactivity and giving to a large number of substances the statute of chemical reagents. They overthrew the 'phlogiston' doctrine, which holds that a component of matter (phlogiston) is given off by a substance in the process of combustion. That theory had held sway for a century.

Scheele prepared and studied oxygen, but his account in *Chemical Observations and Experiments on Air and Fire* appeared after the publication of Joseph Priestley's studies: *Observations on the Different Kinds of Air*. He discovered nitrogen to be a constituent of air. His treatise on manganese was influential, as well as the discovery of barium and chlorine. He also isolates glycerin and many acids, including tartaric, lactic, uric, prussic, citric, and gallic. Priestley is also considered the discoverer of nitrogen, carbon monoxide, ammonia, and several other gases, and in 1774 he became the first to identify oxygen. His report led Lavoisier to repeat the experiment, deduce oxygen's nature and role, and name it. Lavoisier is generally considered as the founder of modern chemistry. He should be known as one of the most astonishing eighteenth century 'men of the Enlightenment', the founder of modern scientific experimental methodology. As he worked on combustion, Lavoisier observed the oxidation caused by a gas contained in the air. He formulated the principle of the conservation of mass (the weights of the reactants must add up to the weights of the products) in chemical reactions, being the first to use quantitative procedures in chemical investigations. He gave a clear differentiation between elements and compounds, something so important for pharmaceutical chemistry. He devised the modern system of chemical nomenclature, naming oxygen, hydrogen and carbon. These preliminary works formed the basis of future preparations or synthesis.

The works performed by Antoine François de Fourcroy (1755–1809), Louis Nicolas Vauquelin (1763–1829), Joseph Louis Proust (1754–1826) and Jöns Jakob Berzelius (1779–1848) introduced new concepts in chemistry. Gay-Lussac published his *Law of Combining Volumes* in 1809, the year after John Dalton (1766–1844) had proposed his *Atomic Theory of Matter* around 1803. It was left to Amedeo Avogadro (1776–1856) to take the first major step in rationalizing Gay-Lussac's results two years later.

At the same time, Louis-Joseph Gay-Lussac (1778–1850) made many less celebrated, but perhaps more important, contributions to chemistry. Along with his great rival, Humphrey Davy (1778–1829), Gay-Lussac established the elemental nature of chlorine, iodine and boron. He prepared pure sodium and potassium in large quantities. Within few years, those scientists integrated the practical advancements of a new generation of experimenters.

All these industrial innovations would have their own impact on other developments in industrial and then medicinal chemistry.[16] At the beginning of the nineteenth century, as the result of a scientific approach, drugs were becoming an industrial item. Claude Louis Berthollet (1748–1822) begins the industrial exploitation of chlorine (circa 1785). Nicolas Leblanc (1742–1806) prepared sodium hydroxide (circa 1789) and then bleach (circa 1796). Davy performed electrolysis and distinguished between acids and anhydrides. Louis Jacques Thénard (1777–1857) prepared hydrogen peroxide and Antoine Jérôme Balard (1802–1876) discovered bromide (1826).

The increase in therapeutic resources was mainly due to the mastery of chemical or physico-chemical principles proposed by Gay-Lussac, Justus Von Liebig (1803–1876),[17] and the new rules given to biology through the works of Claude Bernard (1813–1878),[18] Rudolph Virchow (1821–1902)[19] and Louis Pasteur (1822–1895).[20] Besides these fundamental sciences, physiology, biochemistry or microbiology were becoming natural tributaries of the outbreak of pharmacology. Thus, rational treatments were being designed, not on the Hippocratic basis of regulating humors, but on the purpose of new knowledge in various clinical or fundamental fields.

After a period of extraction and purification from nature (mainly plants), drugs were synthesized in factories or prepared through biotechnology (fermentation or gene technology) before being rationally designed in research laboratories. When the purpose was to isolate active molecules from plants during the first half of the nineteenth century, the birth of organic chemistry following the charcoal and oil industries, progressively lead pharmacists towards organic synthesis. It is precisely in the new concept of the laboratory that this research was performed.

Even when those laboratories host discoveries such as active principles extracted from plants, progress in drug compounding and packaging make the industrialization process irreversible. Gradually, this made the traditional apothecary less and less linked to the manufacture of drugs. Nevertheless, if chemical industries (dyes) gave birth to pharmaceutical companies in Great Britain or Germany, traditional pharmacies remain the origin of such companies in France or the United States. At the same time, the economic dimension of the growing pharmaceutical industry made drugs strategic items, mainly when they were involved with military processes.

During the first half of the nineteenth century, the study of drugs was included within the *Materia Medica*, the old traditional term through which pharmacists are considered the description of drugs and the way to obtain them. On the other hand, the 'modern' word *pharmacology* was more and more often used by physicians. Gradually a clear dichotomy took place between those two entities. *Materia Medica* was a static view of drugs, their production and the compounding of medicines, somewhere within the natural history of drugs, whereas pharmacology considers drugs from a more dynamic point of view. Pharmacology is the study of drugs considering their site or mechanism of action.

III THE BIRTH OF ORGANIC CHEMISTRY

A radical turn in the development of new chemicals occurred when coal, and then oil distillation offered so many opportunities. After the first chemical revolution, the birth of organic chemistry was also a leap forward in chemical industry developments. After the extract of paraffin, carbon derivatives chemistry was developed with many industrial consequences during the second third of the century. Michael Faraday (1791–1867), the British physicist, discovered benzene in 1825, but the first organic molecules used for their therapeutic properties were acyclic. Chloroform was discovered in 1831 by three chemists, each working independently of the others: Eugene Soubeiran (1793–1858) in France (1831),[21] Justus Von Liebig (1803–1873) of Germany,[22] and Samuel Guthrie (1782–1848) in the United States (1832).[23] Sir James Simpson (1811–1870), in Scotland, publicly demonstrated chloroform as an anaesthetic in 1847.

Jean-Baptiste Dumas (1800–1884) proposed the ether theory, the theory of substitution with the 'radicals theory', the measurement of vapor densities, the determination of nitrogen in organic compounds, and the isolation of anthracene from tar, chloral, iodoform, bromoform and picric acid.

Von Liebig came to Paris to study with Louis-Jacques Thénard (1777–1857), Gay-Lussac, Michel-Eugène Chevreul (1783–1886) and Nicolas Vauquelin (1763–1829). Returning to the University of Giessen, in Germany, he became a university professor at the age of 21, something quite unique in history.

One of Liebig's greatest contributions to pure chemistry is his reformation of the methods for teaching the subject including teaching books like *Organic Chemistry and its Application to Agriculture and Physiology* (1840), and *Organic Chemistry in its Application to Physiology and Pathology* (1842).[24] From Giessen, he also edited the journal that was to become the pre-eminent publication in chemistry—*Annalen der Chemie und Pharmazie*. Friedrich Wöhler (1800–1882), after having studied at the University of Heidelberg went to Sweden to study with J.J. Berzelius before settling for nearly 50 years at the University of Göttingen.

In 1825, Liebig and Wöhler began various studies on two substances that apparently had the same composition—cyanic acid and fulminic acid—but very different characteristics. The silver compound of fulminic acid, investigated by Liebig is explosive, whereas silver cyanate, as Wöhler discovered, is not. These substances, called 'isomers' by Berzelius, lead chemists to suspect that substances are defined not simply by the number and kind of atoms in the molecule, but also by the arrangement of those atoms. The most famous creation of an isomeric compound is Wöhler's 'accidental' synthesis of urea (1828), when failing to prepare ammonium cyanate. For the first time someone prepared an organic compound by means of an inorganic one.[25] This 'incident' resulted in Wöhler saying: 'I can no longer, so to speak, hold my chemical water and must tell you that I can make urea without needing a kidney, whether of man or dog; the ammonium salt of cyanic acid is urea'.[26] Liebig and Wöhler discovered certain stable groupings of atoms in organic compounds that retain their identity, even when those compounds were transformed into others. The first to be identified is the 'benzoyl radical', found in 1832 during a study of oil of bitter almonds (benzaldehyde) and its derivatives. Their original objective was to interpret radicals as organic chemical equivalents of inorganic atoms. Their identification of radicals can be seen as an early step along the path to structural chemistry.

Those timid approaches took a precise shape when chemistry precipitously entered the medicinal arena in 1856 when William Perkin, in an unsuccessful attempt to synthesize quinine, stumbled upon mauvein, the first synthetic coal tar dye. This discovery led to the development of many synthetic dyes. The industrial world also understood that some of these dyes could have therapeutic effects. Synthetic dyes, and especially their medicinal 'side-effects,' helped put Germany and Switzerland in the forefront of both organic chemistry and synthesized drugs. The dye—drug connection began to be a very prolific way to discover drugs. Acetanilide was derived from aniline dye in 1886.

IV THE EXTRACTION OF ALKALOIDS FROM PLANTS

Besides this conceptual progress, another evolution in the concept of medicines formed the basis of another revolution. Many pharmacists, mainly in France and Germany, encouraged by an improved knowledge in extraction procedures, tried to isolate the substances responsible for drug action. The 'polypharmacy' was to be abolished. One of the theorists of this trend was the French pharmacist Charles Louis Cadet de Gassicourt.[27] In the inaugural issue of the *Bulletin de Pharmacie* (1809), he reported that the use of complex preparations must be withdrawn in favour of pure substances. It was necessary to study and classify them.[28] As Carl Von Linné did with plants when he published *Species plantarum* in 1753, which is still considered as the starting point for modern botanical nomenclature, pharmacist and physicians tried to classify drugs and their use. This trend was much more convenient with pure substances.

It was between the years 1815 and 1820 that the first active principles were isolated from plants. The French apothecary Jean François Derosne (1780–1846) probably isolated the alkaloid later known as narcotine in 1803, and the German apothecary Friedrich Serturner (1783–1841) further investigated opium and isolated a new compound, 'morphium' (1805), later named morphine. After administration to dogs, solutions of the white powder induced sedation and sleep in the dogs. His work was completed and published in 1817.[29]

This discovery was received with great perplexity: morphine has an alkaline reaction towards litmus paper. Up to that time, chemicals found in plants (in this case the poppy) had exhibited acidic reaction, and the scientific world was doubtful. Pierre Jean Robiquet (1780–1840) performed new experiments in Paris in order to check Serturner's results. Gay-Lussac accepted the revolutionary idea, following which alkaline drugs could be found in plants. All alkaline substances isolated in plants were to be given a name with the suffix '-ine' (Wilhelm Meissner, 1818) in order to recall the basic reaction of all these drugs. At that time, a whole new era in pharmaceutical chemistry was beginning.

Following Serturner's works, Pierre Joseph Pelletier (1788–1842) and François Magendie (1783–1855) found the first alkaloid ever isolated in the traditional ipecac. The pharmacist and the physician succeeded in the purification of emetine from *Ipecacuanha* (1817).[30] The same year, Joseph Pelletier and Joseph Bienaymé Caventou (1795–1877) extracted strychnine, a powerful neurostimulating agent, from *Strychnos*. Three years later (1820) the same extract quinine was derived from various *Cinchona* species.[31] Many attempts had previously been made to purify *Cinchona* bark. Pelletier and Caventou began the industrialization of quinine production, the drug being more and more popular as a tonic and anti-fever drug (before being recognized as a treatment of choice for malaria). No-one was aware of its parasiticide action, although its favourable effect on spleen congestion had already been described.

Other alkaloids were extracted soon after. Brucine (1819), piperine (1819), caffeine (1819), colchicine (1820) and coniine (1826), codeine (1832),[32] atropine (1833),[33] papaverine (1848),[34] were subsequently obtained. These first isolations were coincidental with the advent of the percolation process for the extraction of drugs. Coniine was the first alkaloid to have its structure established (Schiff, 1870) and to be synthesized,[35] but for others, such as colchicine, it was well over a century before the structures were finally elucidated.

Between the years 1817 and 1850, a new generation of scientists gave rise to a new relationship between medicine and new therapeutic tools. Drug formulation became more rational. Hereafter, drug activity would not depend on the concentrations of extracts or tinctures in active principles. The only variable would be the patient himself.

Nevertheless, in the first two-thirds of the nineteenth century, pure alkaloids were seldom used. For instance, even if recommended by a hospital physician, morphine was barely prescribed, as most physicians remained faithful to Sydenham's laudanum (opium tincture) to treat pain. There was greater curiosity from chemists than from physicians or pharmacists because it was impossible to find alkaline

components in plants, where an alternative source for therapeutic strategies could be found.

The first medical textbook including alkaloids as a source of drugs was the *Formulaire des médicaments* by François Magendie (1822), where he tried to popularize the use of morphine, and fought against old formulas.[36]

V SALICYLATES AND ASPIRIN: THE FIRST BEST-SELLER

Another active principle soon extracted from plants was salicylic acid. Willow and salicin, in extensive competition with *Cinchona* bark and quinine, never became a very popular treatment for fever or rheumatic symptoms due to a Scottish physician, Thomas John MacLagan (1838–1934), who launched salicin in 1876.[37] His rationale was inspired by the Paracelse signature's theory: willow is growing in moisture along rivers and lakes, an ideal place to be crippled with rheumatism. Raffaele Piria (1815–1865), after isolation of salicylaldehyde (1839),[38] in *Spireae* species, prepared salicylic acid from salicin. This acid was easier to use and was an ideal step before future synthesis. His structure was closely related to that of benzoic acid, an effective preservative useful as an intestinal antiseptic, for instance in typhoid fever. Patients treated with salicylic acid were dying as frequently as untreated patients, but without fever.

Salol, a condensed product of salicylic acid and carbolic acid, described by Joseph Lister as a precious antibacterial drug, was more palatable and gained huge popularity.

Acetylsalicylic acid was first synthesized by Charles Frederic Gerhardt (1816–1856) in 1853[39] and then, in a purer form, by Johann Kraut (1869). Hermann Kolbe (1818–1884) improved acetylsalicylic acid synthesis with carbolic acid and carbonic anhydride, in 1874, but in fact nobody registered any pharmacological interest. It was almost a century later that Bayer began work on the topic. During the 1880s and 1890s, physicians became intensely interested in the possible adverse effects of fever on the human body and the use of antipyretics became one of the hottest fields in therapeutic medicine. The name of Arthur Eichengrün (1867–1949), who ran the research and development-based pharmaceutical division where Felix Hoffmann (1868–1946) worked, and Heinrich Dreser (1860–1924) who was in charge of testing the drug with Kurt Wotthauer and Julius Wohlgemuth should be remembered for this historical discovery (1897). It is likely that acetylsalicylic acid was synthesized under Arthur Eichengrün's direction and that it would not have been introduced in 1899 without his intervention.[40] Dreser carried out comparative studies of aspirin and other salicylates to demonstrate that the former was less noxious and more beneficial than the latter.[41]

Bayer built his fortune upon this drug, which was given the name of 'aspirin', the most well-known and familiar drug name. Few groups of drugs have provided the manufacturers with such fortunes, physicians with such therapeutic resources, and the laity with so many semi-proprietary remedies, as have the so-called antipyretic or analgesic derivatives of coal tar. Nor is there any industrial group, which illustrates so well the close relationship between chemistry and practical therapeutics, and the relationship between chemical constitution and physiological action.

VI FIRST DRUGS FOR THE HEART

The fact that leaves from the foxglove contain a substance which increases the ability to pump blood round the weakened heart has been known by old wives, priests and botanical experts for several hundred years.

William Withering (1741–1799), an English doctor, learned that the local population was able to cure dropsy using a decoction of 20 different plants, one of which was the leaf from the foxglove. After having tested the various herbs on dropsy, the digitalis leaf remained the most active. In 1775, William Withering published a pamphlet in which he reported his discoveries about the way in which the foxglove can be used in medicine. He meticulously described his discovery, gave an account of how the extract of the digitalis should be prepared, and gave precise instructions on dosage including warnings about side-effects and overdose that he documented through the results obtained in 163 patients.[42] The only, but not least problem is a dreadful continuous vomiting and diarrhoea during the treatment. It is caused by the fact that the boundary between the therapeutic dose and poisoning is exceedingly fine. It was therefore evident and absolutely necessary to purify the active substance in order to fix the effective and non-toxic dosage.

Despite dropsy, caused by a deficiency in heart function, being one of the biggest scourges and the most common cause of death, Withering's discovery was forgotten at the beginning of the new century.

However, after decades of works, Augustin Eugène Homolle (1808–1875) and Théodore Quevenne (1806–1855) obtained an amorphous substance from foxglove leaves which they called 'digitaline', as they were sure that it was another alkaloid. In fact, it was a complex substance containing a specific sugar. It was not until 1867 that another French pharmacist, Claude Adolphe Nativelle was able to purify the leaves of the foxglove and produce the effective substance in the form of white crystals.[43] The Frenchman called the substance 'digitaline

cristallisée'. A few years later the German, Oswald Schmiedberg (1838–1921), managed to produce digitoxin (1875),[44] and Alphonse Adrian (1832–1911) found a convenient way to prepare indictable digitalis preparations. In 1905, James Mackenzie (1853–1925) found a new justification for digitalis use: it was not only effective on cardiac load, facilitating the myocardial work, but it was also a drug for decreasing cardiac rate, making digitalis a drug of choice in atrial tachycardia or flutter.[45] Shortly thereafter, reports began to appear about other medicinal herbs which had the same effect on the heart as the foxglove products. Ethnopharmacology gave birth to ouabain, extracted by Albert Arnaud (1853–1915) from *Acocanthera* roots and bark, and strophantin, extracted from *Strophantus*. Arrow hunters in Equatorial Africa had previously used both of these drugs.

Antoine Jérôme Balard (1802–1876) synthesized nitroglycerine in 1844. He observed that when the drug was administered to animals they collapsed in a few minutes. Two years later, Ascani Sobrero (1812–1888) observed that a small quantity of the oily substance placed on the tongue elicited a severe headache. Konstantin Hering (1834–1918) in 1847 developed the sublingual dosage form of nitroglycerine, which he advocated for a number of diseases. Johann Friedrich Albers (1805–1867) had previously developed the cardiac properties of this yellow liquid, whose explosive property had been discovered by Alfred Nobel (1833–1896).[46] The English physician Thomas Lauder Brunton (1844–1916) was unable to relieve severe recurrent anginal pain, except when he bled his patient. He believed that phlebotomy provided relief by lowering arterial blood pressure.

The concept that reduced cardiac afterload and work are beneficial continues to the present day. In 1867, Brunton administered amyl nitrite, a potent vasodilator, by inhalation.[47] He notes that coronary pain was relieved within 30 to 60 seconds after administration. However, the action of amyl nitrite was transitory. The dosage was difficult to adjust. In 1879, William Murrell (1853–1912), proved that the action of nitroglycerine mimicked that of amyl nitrite, and he establishes the use of sublingual nitroglycerin for relief of acute angina attacks and as a prophylactic agent to be taken prior to physical exercise. The empirical observation that organic nitrates could be used safely for the rapid, dramatic alleviation of the symptoms of angina pectoris led to their widespread acceptance by the medical profession.

VII TREATMENT FOR HYPERTENSION AS A DISEASE

In the 1930s and 1940s, only few antihypertensive treatments were available: sympathectomy,[48] very-low-sodium diets,[49] thiocyanates,[50] and pyrogen therapy.[51] Sympathectomy, which involved cutting nerves to blood vessels, lowered blood pressure in some patients, but it required more than ordinary surgical skill, often produced life-threatening complications, and has unpleasant side effects.[52] Rigid low sodium diets were also unpleasant because they limited food choice, but they were effective in lowering blood pressure. Pyrogen therapy (intravenous infusion of bacterial products) was based on the observation that fever lowered blood pressure.

The first successful drug treatments for hypertension were introduced after World War II. By that time, researchers had learnt that blocking the sympathetic nervous system could lower blood pressure. In 1946, tetraethylammonium, a drug known for 30 years to block nerve impulses, was introduced as a treatment for hypertension. Hexamethonium, an improved version of tetraethylammonium, was available for use by 1951.[53] Another effective blood pressure-lowering drug, hydralazine,[54] resulting from the search for antimalarial compounds, was diverted to the treatment of hypertension when it was found to have no antimalarial activity, but to lower blood pressure and increase kidney blood flow.

For a few years, hexamethonium and hydralazine were mainstays in the treatment of severe hypertension. They were reasonably effective in lowering blood pressure, but often caused severe side-effects. The final drug developed in those early days, reserpine,[55] was the product of more than two decades of research into compounds derived from *Rauwolfia serpentina*, a plant used for centuries by physicians and herbalists on the Indian subcontinent.[56] The quality of the result obtained with the various drugs used in mono- or combined therapy to treat hypertension proved clearly that fatal outcomes associated with this disease are caused by high blood pressure.[57]

Relevant clinical trials were conducted, using the three drugs then available: hydrochlorothiazide, hydralazine and reserpine. In one of them, males with elevated blood pressure were randomly divided into two groups. One group received antihypertensive drugs; the other group received a placebo. The study was planned to last for approximately 5 years, but was stopped after 18 months. The men with severe hypertension and receiving the placebo were dying at a greater rate than those receiving the antihypertensive drugs.[58] The clinical interest of treating hypertension was definitively proven.

Among recent discoveries, the research pointing out the role of converting enzyme is crucial. Advances leading to recognition of the relationship of the renin-angiotensin system to aldosterone includes the measurement of aldosterone plasma levels, the discovery of an aldosterone-stimulating factor in plasma, the finding that a potent aldosterone-stimulating factor is secreted by the kidney, and the evidence that synthetic angiotensin II increases aldosterone secretion. The fractionation of crude kidney extracts allowed Robert Tigerstedt (1853–1923) to find that aldosterone-stimulating factor was a peptide: renin.[59] The renin-angiotensin-aldosterone system plays an important role in congestive heart failure and in renovascular and malignant hypertension. The early use of blocking agents for the renin-angiotensin system had been proposed because arterial pressure decreased in experimental renovascular hypertension.[60] Major steps in the initial development of angiotensin I conversion inhibitors include the discovery of the *Bothrops* peptides (bradykinin potentiating factor) and the demonstration of its therapeutic potential. It is a history where chance, serendipity and clear scientific reasoning weave together the work of several scientists. It is also a classical example of drug development for which the initial basic research was made at university level, but the useful product is achieved by industry.[61]

The renin-angiotensin system was a key element in blood pressure regulation and fluid volume homeostasis. Since angiotensin II (AII) is the effector molecule of the RAS, the most direct approach to block this system was to antagonize AII at the level of its receptor. Therefore, at Du Pont Merck, the working hypothesis was that the identification of metabolically stable and orally effective AII-receptor antagonists would constitute a new and superior class of agents, useful in treating hypertension and congestive heart failure. The program began with a detailed pharmacological evaluation of some simple *N*-benzylimidazoles, originally described by Takeda Chemical Industries in Osaka, Japan. Potent and orally effective nonpeptide antagonists were found. The first major breakthrough, to increase the potency of the compounds, came with the development of a series of *N*-benzylimidazole phthalamic acid derivatives and the discovery of losartan, a highly potent angiotensin type 1 (AT$_1$) selective receptor antagonist with a long duration of action.[62]

VIII GENERAL AND LOCAL ANAESTHESIA

One of the greatest therapeutic revolutions during the nineteenth century was the introduction of general anaesthesia in the practice of surgery. As early as 1776, Joseph Priestley discovered laughing gas (nitrous oxide), but analgesia seemed to be beyond reach. Priestley and Humphrey Davy commented in 1796: 'it may probably be used with advantage during surgical operations in which no great effusion of blood takes place'. Michael Faraday (1791–1867) proposed the use of diethyl ether to induce similar action. However, their inhalation was proposed during exhibitions for shows named 'ether frolics'. Neither diethyl ether nor nitrous oxide was clinically used before 1846. Surgery was so difficult before that it was very uncommon until the middle of the century: the pain and the infection risk resulting from the surgical procedure were very discouraging.

Dentists set the pace in the field of analgesia. They became familiar with both diethyl ether (sulphuric ether) and nitrous oxide. They were in permanent contact with pain in complaining patients. They also produced pain through unfair or badly controlled operations. Horace Wells (1815–1848), a dentist, asked a colleague to extract his own teeth while under the influence of nitrous oxide.[63] This trial, held in 1844, was successful and painless. Shortly thereafter, in 1845, he attempted to demonstrate his discovery at the Massachusetts General Hospital in Boston. His first attempt was a total failure. Another Bostonian dentist, William T.G. Morton (1819–1868), familiar with the use of nitrous oxide from his friendship with Wells, asked the surgeons of the Massachusetts General Hospital to demonstrate his technique after many attempts on animals, himself and friends. The first patient, Gilbert Abbott was to be operated on by the chief surgeon Dr. John Collins Warren (1778–1856). Morton arrived with special apparatus with which to administer the ether and only a few minutes of ether inhalation were necessary to make the patient unconscious.[64] According to the eminent surgeon, Henry J. Bigelow (1818–1890), who noted 'I have seen something today that will go around the world', a new era in the history of medicine had begun.[65]

Techniques and safety of anaesthesia will continue to improve. Even though ether was an interesting agent, other drugs were rapidly being tested, among which was chloroform, introduced into the surgery by the Scottish obstetrician James Simpson (1811–1870) in 1847. As ether is flammable, chloroform is safer from this point of view.[66] Unfortunately, chloroform is a severe hepatotoxic drug and cardiovascular depressant. Despite the relatively high incidence of deaths associated with the use of chloroform, it was the anaesthetic of choice for nearly 100 years. Many other halogenoalkanes had been synthesized, among which ethylene chloride and more recently halothane, a non-flammable anaesthetic, was introduced into clinical practice in 1956, after its preparation at Imperial Chemical Industries. It revolutionized anaesthesia.

In the 1860s, the introduction of the hypodermic syringe presented new opportunities for the use of drugs for anaesthesia. Injectable anaesthetics were introduced after the works of Eugene Baumann (1846–1896) and Alfred Kast (1856–1903) who, in 1887, introduced a major advance with sulfones, mainly Sulfonal®, a long-acting sedative drug.[67]

Prepared by Adolf Von Baeyer (1835–1917) as early as 1864, barbituric acid was not used before 1903, at that time Emil Fischer (1852–1919) had already prepared the first derivative of barbituric acid, one of which, diethylmalonylurea, would be marketed under the trade

name Veronal, also known as Barbital.[68] Barbiturates, mainly used as sleep inducers, are also very useful in the operating room, especially when Thiopental, a very rapid and short-acting derivative was launched in 1935, after the work of the American anaesthetist John S. Lundy (born 1894). Thiopental had been enthusiastically accepted as an agent for the rapid induction of general anaesthesia. Barbiturates enabled the patient to go to sleep quickly, smoothly and pleasantly contrary to inhaled agents.

In the 1940s and early 1950s, muscle relaxants were introduced, firstly with curare (derived from the original South American Indian poison studied by Claude Bernard 100 years before) and then over subsequent decades a whole series of other agents.[69] Curare, in the form of tubocurarine, was first used in clinical anaesthesia in Montreal in 1943 by Harold Griffith (born 1894) and Enid Johnson.[70] In 1946, T.C. Gray first used Curare in Liverpool in the UK: 'The road lies open before us and ... we venture to say we have passed yet another milestone, and the distance to our goal is considerably shortened'.[71]

Local anaesthesia began in Vienna (1884) when Carl Koller (1857–1944)[72] and Sigmund Freud (1856–1939)[73] administered cocaine locally over the cornea, in order to anaesthetize the eye before cataract surgery. They noticed that the drug was able to prevent the oculomotor reflex in frogs. Cocaine had been previously isolated from coca leaves by Albert Niemann (1834–1861) in 1860.[74] Before this founding step, local sensitivity could be abolished by the dermal administration of organic derivatives like diethyl ether or ethylene chloride on the skin. A few years later, William Halsted (1852–1922) in the United States used cocaine to block nerves. Paul Reclus (1847–1914) in France and August Bier (1861–1949) used it for locoregional anaesthesia.[75] Unfortunately, cocaine is an addictive drug and between the years 1890s to 1910s, it became a pillar of drug addiction.[76] Cocaine was to be completely eradicated from clinical use in the years 1914–1916 with restrictive law in the USA, as well as in Europe. Fortunately, synthetic local anaesthetics would appear, thanks to the works of Alfred Einhorn (1856–1939)[77] and Wilhem Filehne (1844–1927),[78] in Germany, and Ernest Fourneau (1872–1949)[79] in France.

IX ANTIEPILEPTIC DRUGS

Soon after its introduction as a hypnotic drug, phenobarbital was found to be an excellent antiepileptic drug. Historically, agents introduced for the treatment of epilepsy are also turned to for psychiatric indications. The original 'first generation' antiepileptic drug, a bromide salt, which appears in 1857,[80] was also known for its tranquilizing properties. After phenobarbital came into use for epilepsy in 1912, reports of its psychopharmacologic application soon followed. Tracy J. Putnam (born 1894) and Houston H. Merritt[81] (1902–1979) introduced phenytoin as an antiepileptic drug in 1938 and immediately described its psychotropic advantages. This is generally considered as the beginning of the modern psycho-pharmacological usage of antiepileptic drugs (1942), although relatively few psychiatric uses for antiepileptic drugs were reported over the next decade.

Major events preceding this work are the fortuitous discovery of phenobarbital as an anticonvulsant agent, structure/hypnotic activity studies with barbiturates and hydantoins in the early 1920s by A.W. Dox in the Parke Davis laboratories, and the development of anticonvulsivant assay techniques in animals, by a number of laboratories. Phenytoin was the first item on the list of compounds sent to Putnam by Dox and W.G. Bywater in April 1936. It was found to have anticonvulsivants properties in animals late in 1936, but no public reports were issued until the following year. Clinical efficacy was established in 1937, but again no public reports were issued until 1938. Dilantin® sodium capsules were prepared by Parke, Davis & Co. and were ready for marketing the same year.[82]

In the early 1960s, there was a near-simultaneous introduction of carbamazepine and valproic acid and its derivatives, as new treatments for epilepsy. Although in 1882 Beverly S. Burton, an American working in Europe, had prepared valproic acid,[83] its antiepileptic utility was not appreciated until this was serendipitously discovered 65 years later by Meunier in France.[84] Carbamazepine was first synthesized in 1960, in the United States by Schindler — who, a decade earlier, had patented the structurally closely related imipramine and it was found to have antiepileptic properties.[85] When concurrent remedial effects on mood and behaviour were noted with both carbamazepine and valproic acid in the very early epilepsy trials, both drugs were soon appropriated by psychiatrists, first by Lambert[86] in France (1966), using the amide derivative of valproic acid.

It has only been since the mid-1990s that a series of novel antiepileptic drug has been approved. There are currently five of these agents available, which might then be termed the 'third generation'. These are felbamate, lamotrigine,[87] gabapentin,[88] topiramate and tiagabine.[89]

X FIGHT AGAINST MICROBES

After the initial developments in organic chemistry during the first half of the twentieth century, the question of the chemical basis of life was clearly put in the forefront of the scientific debate. Since Wöhler's works, it was clear that chemistry is a unique science, with the same rules governing reactions, kinetics and atomic, radical or molecular arrangements. The advent of scientific cooperation using multidisciplinary approaches led to a greater understanding of natural and experimental phenomena. A typical example of this approach is Louis Pasteur. This leading physicist began his career as a specialist in crystallography. He studied the impact of bacteria on stereochemical properties of tartaric acid crystals. Later, his interest remained fixed on alcoholic and acetic fermentations. Afterwards, he pulled the concept of spontaneous generation to pieces. As microorganisms reacted on organic substances, he presumed that they could also be active on living beings, which is why he began his research on animal epidemics, culminating in the discovery of the rabies vaccine.

Carl Wilhelm Scheele inaugurated the practical use of disinfectant fumigation. It preceded 'Guytonian's fumigations', which was based on chlorine activity. As early as in 1785, a solution of chlorine gas in water was used to bleach textiles. Potassium hypochlorite (Eau de Javel) was prepared by Berthollet in 1789. After its first use as a means of discolouration, bleach was revealed to be a really good antibacterial activity, used by Antoine Germain Labarraque (1777–1850).[90] In 1820, he replaced potash liquor by the cheaper caustic soda liquor, and thus sodium hypochlorite was invented. At the end of the 1820s, Robert Collins, followed by Oliver Wendell Holmes (1809–1894), showed that puerperal fever frequency decreased when midwives washed their hands in chlorinated water.[91] A few decades later (1861) Ignaz Philip Semmelweis (1818–1865) published his research on the transmissible nature of puerperal (childbed) fever.[92] But he failed to convince physicians either in Vienna or in Budapest that this was the cause of infection in pregnant women.

Thomas Alcock published his *Essay on the Use of Chlorites of Oxide of Sodium and Lime* in 1827 (sodium and calcium hypochlorites). Recommended for disinfecting and deodorizing a wide range of environments, e.g. hospitals, workshops, stables, toilets, reservoirs, sewers and areas contaminated with blood or other body fluids. At about the same time a health commission in Marseilles recommended hypochlorite for disinfecting hands, clothes and drinking water. In 1881, Robert Koch (1843–1910) demonstrated the lethal effect of hypochlorites on pure cultures of bacteria. A few years later, Isidore Traube (1860–1943) established the purifying and disinfecting properties of hypochlorites in water treatment.[93] During the First World War, much progress was made: Dakin's solution (0.5% sodium hypochlorite) for disinfection of open and infected wounds was widely used in 1915. Milton fluid (containing 1% sodium hypochlorite and 16.5% sodium chloride) was marketed in the UK in 1916 used as a general disinfectant and antiseptic in paediatrics and childcare. In 1917, Halazone® tablets were introduced and provided a dose-controlled method of disinfecting small volumes of drinking water.

In 1881, Bernard Courtois (1777–1838) introduced another halogen, iodine, extracting the element from wracks at the seashore. William Wallace proposed iodine tincture in 1835 to disinfect wounds. It was superseded by iodoform, which was less of an irritant, invented by Georges Simon Serullas (1774–1832). Structurally, it was very comparable to chloroform, the chlorine atom being substituted by an iodine one. Aqueous iodine solutions were proposed by Casimir Davaine (Lugol's solution) as antiseptics.

Joseph Lister (1827–1912) instituted a revolutionary change in hospital hygiene[94] by introducing carbolic acid, prepared by distillation of coal tar.[95] The disinfectant was used for surgical ligatures and dressings. Sprays of carbolic acid, improved by the French surgeon, Just Lucas-Championnière (1843–1913) were used in operating rooms around 1860.[96] All these procedures were deeply contested, but the final proof was obtained from the works of prestigious biologists, like Robert Koch. He performed laboratory tests demonstrating the bactericidal activity of carbolic acid.[97] Increasingly, the experimental proof confirmed empirical behaviour.

In this environment, the microbial theory many diseases constituted the hallmark of nineteenth-century medicine. The theory that infectious diseases were caused by invisible agents provided an opportunity for much progress. The laboratory took its rightful place when microscopes, staining of preparations and sterilization became available for new discoveries. For example, *Escherichia coli*, discovered in 1879, became the perfect example of an easily grown, 'safe' bacteria for laboratory practice. Working with pure cultures of the diphtheria bacillus in the Pasteur Institute, in Paris, Emile Roux (1853–1933) and Alexandre Yersin (1863–1943) first isolated, in 1888, the deadly toxin that causes most of diphtheria's lethal effects.[98] One by one over the next several decades, various diseases revealed their microbial origins.

XI SULFONAMIDES

Up to the advent of the twentieth century, the fight against microbes was devoted to disinfecting external wounds and to sanitizing drinking water. Since Pasteur's works, the objective was to treat infectious diseases: cholera, tuberculosis, diphtheria, etc. Some vaccines were already available, but only for smallpox and rabies.

The breakthrough came from an unexpected side of the scientific field: the dye industry. In 1865, Friedrich Engelhorn (1821–1902) founded Badische Anilin und Soda-Fabrik AG (BASF). The company produced coal tar dyes and precursors, gaining a leading position in the world dye market within only a few decades. The demand for dyes was strong, reflecting soaring population growth, matched by the textile industry. BASF's first products included aniline dyes. In 1871, the company marketed the red dye alizarin, and other new dyestuffs follow: eosin, auramine and methylene blue, together with the azo dyes, which were eventually developed into the largest group of synthetic dyestuffs.

Around the 1880s, German chemists, following in Paul Ehrlich's wake, discovered the fact that living cells absorb dyes in a different way to dead cells.[99] If a microorganism could be coloured, vital properties of the bacteria or the parasite could also be transformed.[100] What conclusions can be drawn from this new information on the viability of coloured microorganisms? Ehrlich refined the use of methylene blue in bacteriological staining and used it to stain the tubercle bacillus, showing that the dye binds to the bacterium and resists discolouration with an acid alcohol wash.[101] Following this hypothesis, Paul Ehrlich (1854–1915) administered methylene blue to patients suffering from malaria.[102]

Ehrlich is looking for a cure or treatment for 'sleeping sickness', a disease caused by a microbe.[103] He found that a chemical called Atoxyl® worked well, but was a fairly strong poisonous arsenic compound. Ehrlich began an exhaustive search for an arsenic compound that would be a 'magic bullet', capable of killing the microbe but not the patient. In 1909, after testing over 900 different compounds on mice, Ehrlich's new colleague Sahachiro Hata went back to No. 606: doxydiaminoarsenobenzol, dihydrochloride. Although unsuccessful against the sleeping sickness microbe, it seemed to kill another (recently discovered) trypanosoma, which caused syphilis. At that time, syphilis was a disabling and prevalent disease. Ehrlich and Hata tested No. 606 repeatedly on mice, guinea pigs, and then rabbits contaminated with syphilis. They achieved complete cures within three weeks, with no mortalities.[104] Production of the first batch of Salvarsan® at Hoechst started in July 1910. It was an almost immediate success, being sold all over the world. It spurred Germany to become a leader in chemical and drug production, and made syphilis a curable disease. The concept of the 'magic bullet' was born simultaneously with the concept of chemotherapy.

The following year, Julius Morgenroth (1871–1924) worked on experimental trypanosomiasis. He had just discovered that not only was quinine the drug of choice for malaria, but that it was also very active against the parasites *Trypanosoma* spp. In the same laboratory, other work was performed on *Pneumococcus* and particularly on the nature of the external capsule of the microorganism. The bacterium *Streptococcus pneumoniae* is the most common cause of severe pneumonia. Morgenroth noticed that biliary salts could dissolve *Trypanosomia* structures as well as *Pneumococcus* ones. Another concept concerning unspecific targets for drugs in infectious diseases was founded, which also explained the activity of various isoquinolines derivatives for treating different infectious diseases.[105]

The influenza pandemic of 1918–1920 clearly demonstrated the inability of medical science to stand up to disease. More than 20 million people worldwide were killed by flu that attacked not the old and frail, but the young and strong. This was a disease that no magic bullet could cure and no government could stamp out. Chemotherapy research had to be improved and continued.

In 1927, Gerhardt Domagk (1895–1964), who had been promoted in Bayer's research department, aimed to find a drug capable of destroying microorganisms after oral administration. The internal route is imperative to treat most infectious diseases. The experimental model he used was the streptococci infection of mice. This model allowed the study of the effect of a large number of drugs. Among a large number of candidates, Domagk turned his attention to azo dyes, so-called because the two major parts of the molecule are linked by a double bond between two nitrogen atoms. Some of these dyes attach strongly to protein in fibres or leather, so that they hold fast against fading or cleaning. Domagk reasoned that they might also attach themselves to the protein in bacteria, inhibiting if not killing microorganisms.

Chrysoidine, which was a deep-red dye, had to be grafted to a sulfonylurea derivative (sulfamidochrysoidine) in order to be active. In 1932, it was studied by two chemists, Fritz Mietzsch and Josef Klarer. Testing the new dye on laboratory rats and rabbits infected with streptococci bacteria, Domagk found that it was highly antibacterial but not toxic. It was called Streptozan®, but its name soon changed to Prontosil®.

It gave birth to the new era of antimicrobial chemotherapy. The first cure occurred in 1932. At least two versions of the same story coexist. It is still not clear whether it was administered in an act of desperation, to a 10-month-old boy who was dying of staphylococcal septicaemia; the baby made an unexpectedly rapid recovery. Another account is that Domagk himself used Prontosil® to treat his own daughter, who was deathly ill from a streptococcal infection following a pin prick.

Domagk did not immediately publish his remarkable results. His landmark paper of February 1935 was edited shortly after having taken a patent on the product and won wide acclaim in Europe.[106] Domagk was awarded the Nobel Prize in Physiology or Medicine in 1939, but due to the Nazi veto, did not receive his medal until 1947, after World War II. However, although Domagk discovered sulfonamides, he did not discover the way in which they were active.

A French research team with Ernest Fourneau (1872–1949), Jacques Trefouel (1897–1977) and Thérèse Trefouel (born 1892), Federico Nitti (1905–1947) and Daniel Bovet (1907–1992) at the Pasteur Institute in Paris did the work. Prontosil® was inactive on bacilli cultures because it needed the presence of an esterase to split the molecule. The active part was the sulfonamide (amino-4-benzene sulfonamide) itself and not the dye.[107]

Doctors in Europe in 1936 had excellent results using the new drug to treat childbed fever and meningitis. Tests in the USA in 1936, initially at Johns Hopkins Hospital in Baltimore and Western Pennsylvania Hospital in Pittsburgh, showed that it was also effective against various streptococci infections and pneumonia. Prontosil® won wide publicity in the USA in 1936 when it was used to treat President Franklin D. Roosevelt's son, Franklin, Jr., who was severely ill from a streptococcal infection.

More than 5000 sulphur drugs were prepared in the late 1930s and early 1940s. Among them, sulfapyridine was used against pneumonia (it was used to treat Winston Churchill when he came down with pneumonia in 1943 just before the Casablanca Conference). Sulfathiazole was used against both pneumonia and staphylococci infections; sulfadiazine was used against pneumonococci, streptococci and staphylococci; and sulfaguanadine against dysentery.

XII ANTIBIOTICS

The 1930s were also the beginning of another chapter of this new era: the birth of antibiotic treatments.[108] The first antibiotic ever used was Gramicidin, the first natural antibiotic extracted from soil bacteria. It was prepared By René Dubos (1901–1982),[109] in 1939, and showed an interesting capacity to arrest the growth of staphylococcus, limited by its high toxicity. The works led by Alexander Fleming (1881–1955) are symbolic of the history of twentieth century's drug development. It was the desire to find an internal antiseptic that drove Scottish-born doctor Alexander Fleming in his pioneering work in London in the 1920s. In 1922 Fleming made the amazing observation that the human teardrop contains a chemical capable of destroying bacteria—and at an alarming rate. However, the excitement at this discovery was soon dashed. While the new discovery, which Fleming called lysozyme, was effective at dissolving harmless microbes it proved ineffective at negating those that cause disease.

Fleming, however, did not give up. In 1928 his diligence was rewarded. In his laboratory Fleming was in the process of developing staphylococci. Removing the lid from one of these cultures, Fleming was surprised to see that around the mould of *Penicillium*, the colonies of staphylococci had been dissolved. Something produced by the mould has dissolved the bacteria. After further testing, Fleming was able to isolate the essence of the mould and it was this that he named penicillin.[110] Bacteria which cause diseases such as gonorrhea, meningitis, diphtheria and pneumonia were about to be dramatically affected by this new breakthrough.

Best of all, although very toxic for microorganisms, it was not poisonous to humans. At the time, the medical community reacted coldly to this new discovery, however: everyone thought that once a bacteria entered the body, nothing could be done, and penicillin was seen as a non-event.

Twelve years later, the overwhelming casualties on the battlefield during the Second World War led two medical researchers, Howard Florey (1898–1968) and Boris Ernst Chain (1906–1979), to look into resurrecting Fleming's work with penicillin. After much refinement they were able to develop a powdered form of penicillin and experimented with mice.[111] In 1941 the first human was successfully treated. Before long, penicillin was in full production. Fleming, Florey and Chain were awarded the Nobel Prize for Medicine in 1945.[112]

As early as 1945, in an interview with The New York Times, Fleming warned that the misuse of penicillin might lead to resistant forms of bacteria.[113] In fact, Fleming had already experimentally derived such strains by varying the dosage and conditions upon which he added the antibiotic to bacterial cultures. As a result, Fleming warned that the drug carried great potential for misuse, especially with patients taking it orally at home, and that inadequate treatments would likely lead to mutant forms. Fleming stated that resistance to penicillin can be conferred in two ways—either through the strengthening of the bacterial cell wall, which the drug destroyed, or through the selection of bacteria expressing mutant proteins capable of degrading penicillin. The study was performed by B.E. Chain and coworkers.[114] Nevertheless, until the mid-1950's, penicillin is available orally to the public without prescription. During this period, the drug was indeed sometimes used inappropriately.

In 1942, full-scale production for therapeutic use in World War II began. Many factors were responsible for the delay between Fleming's discovery and the industrial step. A scientific explanation of Fleming's 'phenomenon' was needed, the finding of a classification of the fungus secreting the active substance, source of the mould, difficulties of bacteriologists in reproducing Fleming's discovery, identifying the chemical make-up of penicillin, search for other penicillin-producing organisms to enhance production of penicillin, purification and crystallization of penicillin, experiments on animals (chiefly mice) to determine toxicity, hesitancy to administer the drug to humans, standardization of an effective dosage for humans, and search for equipment and financial resources to enhance full-scale production. The adjunctive role of serendipity (chance, happenstance, improbability and luck) in overcoming these obstacles and in contributing to the successful, scientific conclusion of the penicillin project is an unusual story.[115]

By 1946, one hospital reported that 14% of the strains of *Staphylococcus* isolated from sick patients were penicillin resistant. By the end of the decade, the same hospital reported that resistance had been conferred to 59% of the strains *Staphylococcus* studied.

In 1940, Selman Waksman (1888–1973) isolated and purified actinomycin from *Actinomyces griseus* (later named *Streptomyces griseus*), which led to the discovery of many other antibiotics from that same group of microorganisms. Actinomycin attacks Gram-negative bacteria responsible for diseases like typhoid, dysentery, cholera and undulant fever and was the first antibiotic purified from an actinomycete. Considered too toxic for the treatment of diseases in animals or humans, actinomycin is primarily used as an investigative tool in cell biology. Waksman, with Albert Schatz (born 1920) and Elizabeth Bugie isolated the first aminoglycoside, streptomycin, from *S. griseus*.[116] Like penicillin, aminoglycosides decrease protein synthesis in bacterial cells, except that streptomycin targets Gram-positive organisms instead of Gram-negative ones. Waksman studied the value of streptomycin in treating bacterial infections, especially tuberculosis. In 1942, several hundred thousand deaths resulted from tuberculosis in Europe and another 5 to 10 million people suffered from the disease.

Merck immediately started manufacturing streptomycin. Simultaneously, studies by William H. Feldman (born 1892) and H. Corwin Hinshaw (1902–2000) at the Mayo Clinic confirm streptomycin's efficacy and relatively low toxicity against tuberculosis in guinea pigs.[117] On 20 November 1944, doctors administered streptomycin for the first time to a seriously ill tuberculosis patient and observed a rapid and impressive recovery.[118] His advanced disease was visibly arrested, the bacteria disappeared from his sputum, and he made a rapid recovery. The only problem was that the new drug made the patient deaf. Streptomycin was particularly toxic on the inner ear. No longer unconquerable, tuberculosis could be tamed and beaten into retreat. In 1952, Waksman was awarded the Nobel Prize in Physiology or Medicine for his discovery of streptomycin. During the following years, a succession of tuberculicid drugs appeared. These were important because with streptomycin monotherapy, resistant mutants began to appear within a few years, *p*-aminosalicylic acid (1949), isoniazid (1952),[119] pyrazinamide (1954), cycloserine (1955), ethambutol (1962),[120] ethionamid (1959)[121] were introduced as anti-tuberculosis drugs.

The discovery of rifampicin in 1967 is considered one of the greatest achievements in the history of chemotherapy against tuberculosis. Rifampin was developed in the Lepetit Research Laboratories (Italy) as part of an extensive program of chemical modification of the rifamycins, the natural metabolites of *Nocardia mediterranei*. All of the studies leading to highly active derivatives were performed on a molecule (rifamycin B) that was itself practically inactive. Systematic structural modifications of most of the functional groups of the rifamycin molecule were performed with the objective of finding a derivative that was active when administered orally. The understanding of structure–activity relations in the rifamycins led to the synthesis of several hydrazones of 3-formylrifamycin SV. Among them, the hydrazone with *N*-amino-*N'*-methylpiperazine (rifampin) was found to be the most active in the oral treatment of infections in animals and, after successful clinical trials, was introduced into therapeutic use in 1968.[122]

Aminoglycosides such as capreomycin, viomycin, kanamycin and amikacin and the newer quinolones (e.g. pefloxacin, ofloxacin and ciprofloxacin) were only used in drug resistance situations. Tuberculosis in particular experienced a resurgence. In the mid-1980s, the worldwide decline in tuberculosis cases levelled off and then began to rise.

In 1948, Benjamin M. Duggar (1872–1956), a professor at the University of Wisconsin and a consultant to Lederle, isolated chlortetracycline from *Streptomyces aureofaciens*. Chlortetracycline, also called aureomycin, was the first tetracycline antibiotic and the first broad-spectrum antibiotic. Active against various organisms, aureomycin works by inhibiting protein synthesis. The discovery of the tetracycline ring system also enabled further development of other important antibiotics.[123] Since that time more than a hundred molecules active against a wide range of bacteria have been discovered.

XIII AIDS: AN EMERGING DISEASE

The first published reports of the new disease seemed at first to be no more than medical curiosities. On 5 June 1981, the Atlanta-based Centers for Disease Control and Prevention (CDC), the US agency charged with keeping tabs on disease, published an unusual notice in its *Morbidity and Mortality Weekly Report*: the occurrence of *Pneumocystis carinii* pneumonia among gay men.[124] In New York, a dermatologist encountered cases of a rare cancer, Kaposi's sarcoma,[125] a disease so obscure he recognized it only from descriptions in old textbooks.[126]

By the end of 1981, those symptoms were recognized as harbingers of a new and deadly disease.[127] The disease is initially called 'Gay Related Immune Deficiency'. Within a year, similar symptoms appeared in other demographic groups, primarily haemophiliacs and users of intravenous drugs. The CDC renamed the disease Acquired Immune Deficiency Syndrome (AIDS). By the end of 1983, the CDC had recorded some 3000 cases of this new plague. The prospects for AIDS patients were not good: almost half had already died. Twenty years after, more than 30 million cases are estimated worldwide.

In 1984, Luc Montagnier (born 1932) of the Pasteur Institute[128] and Robert Gallo (born 1937) of the National Cancer Institute (NCI)[129] proved that AIDS was caused by a retrovirus (whose replication is linked to a key enzyme, reverse transcriptase). There is still a controversy over priority of discovery. A number of therapeutic strategies can now be used in the treatment of AIDS. Many of these are suitable for immediate application in clinical trials and have already yielded positive results in many patients.[130]

The first drug introduced to treat the disease was AZT (azidothymidine, zidovudine), a thymidine analogue. AZT had been developed in 1964 as an anticancer drug by Jerome Horowitz of the Michigan Cancer Foundation (Detroit). Because AZT was ineffective against cancer, Horowitz never filed a patent. Nevertheless, *in vitro* studies showed some activity of this supposed anticancer drug against the AIDS virus.[131] In a six-week clinical trial 4 dose regimens of 3'-azido-3'-deoxythymidine, with potent anti-viral activity against HTLV-III *in vitro*, were examined in 19 patients with the acquired immunodeficiency syndrome (AIDS) or AIDS-related complex (ARC). Fifteen of the 19 patients had increases in their numbers of circulating helper-inducer T lymphocytes during therapy. Six who were anergic at entry showed positive delayed-type hypersensitivity skin test reactions during treatment, two had clearance of chronic fungal nailbed infections without specific anti-fungal therapy, six had other evidence of clinical improvement, and the group as a whole had a weight gain of 2.2 kg. This was the first clinical trial for an anti-HIV drug.[132]

The Delta trial was a major clinical trial of combination antiretroviral therapy. In September 1995 the results of this trial showed that combining AZT with ddI (didanosine) or ddC (zalcitabine), provided a major improvement in treatment compared with AZT on its own.[133] Of the different steps of the HIV replicative cycle, the reverse transcription step has received most attention as a target for chemotherapeutic intervention. Reverse transcriptase (RT) can be blocked by both nucleoside (nucleotide) and non-nucleoside types of inhibitors. Whereas the former act as competitive inhibitors with respect to the natural substrates or alternative substrates (chain terminators), the latter act allosterically with a nonsubstrate binding site of the enzyme.[134] Other nucleotide analogues including stavudine, lamivudine, abacavir and tenofovir, began to be used in 1994 while non-nucleoside reverse transcriptase inhibitors, nevirapine, delavirdine, efavirenz came to the market in 1996. The third subclass of antiretroviral drugs was introduced at the same time: protease inhibitors such as saquinavir, ritonavir, indinavir, nelfinavir, amprenavir and lopinavir. Highly active antiretroviral therapies usually consisting of two nucleoside reverse transcriptase inhibitors plus an HIV protease inhibitor, have been widely used since 1996. They produce durable suppression of viral replication with undetectable plasma levels of HIV-RNA in more than half of patients. Immunity recovers, and morbidity and mortality fall by more than 80%. Treatment was thought to be particularly effective when started early. Despite these successes, however, antiretroviral therapies also produce numerous side-effects.

The use of chemotherapy to suppress replication of the human immunodeficiency virus (HIV) has transformed the face of AIDS in the developed world. Pronounced reductions in illness and death have been achieved and healthcare utilization has diminished. HIV therapy has also provided many new insights into the pathogenesis and the viral and cellular dynamics of HIV infection. However, challenges remain. Treatment does not suppress HIV replication in all patients, and the emergence of drug-resistant virus hinders subsequent treatment. Highly active antiretroviral therapy has revolutionized the treatment of human immunodeficiency virus (HIV) infection, which can now be viewed as a chronic and manageable disease. However, HIV infection differs from other chronic diseases in that early treatment decisions can irrevocably alter the patient's response to future therapy.[135] Chronic therapy can also result in toxicity. These challenges prompt the search for new drugs and new therapeutic strategies to control chronic viral replication.[136] The preparation of an effective vaccine is probably the only way to eradicate the disease.[137]

XIV DRUGS FOR ENDOCRINE DISORDERS

The flowering of biochemistry in the early part of the new century is key, especially as it relates to human nutrition, anatomy and disease. Some critical breakthroughs in metabolic medicine were made in the 1890s, but these were exceptions rather than regular occurrences. In 1891, myxedema was treated with sheep thyroid injections. This event is the first proof that animal gland solutions could benefit humans. In 1896, Addison's disease was treated with minced adrenal glands from a pig. These test treatments provide the starting point for all hormone research. From the 1920s to the 1940s, new treatments for physiological disorders were discovered, among them insulin for diabetes mellitus and cortisone for inflammatory diseases. Throughout human history, the condition diabetes mellitus meant certain death. Since the late nineteenth century, scientists have attempted to isolate the essential hormone and inject it into the body to control this disorder. Using dogs, numerous researchers had tried and failed, but in 1921, Canadian physician Frederick Banting (1891–1941) realized that if he tied off the duct to the pancreas of a living dog to atrophy the gland before removing it, there would be no digestive juices left to dissolve the hormone. It was first called 'iletin'.

Beginning in the late spring of 1921, Banting worked on his project in Toronto with his assistant, medical student Charles Best (1899–1978). After many failures, one of the dogs whose pancreas had been tied off showed signs of diabetes. Banting and Best removed the pancreas, ground it up, and dissolved it in a salt solution to create the long-sought extract. They injected the extract into the diabetic dog, and within a few hours the canine's blood sugar returned to normal. The scientists had created the first effective treatment for diabetes.[138] John Macleod (1876–1935), physiologist at the University of Toronto provided facilities for Banting's work, biochemists James Collip (1892–1965) and E.C. Noble joined the research team to help purify and standardize the hormone, which was renamed insulin.[139] It was Connaught and Lilly in Northern America and Novo in Europe (Denmark) who performed the technical developments that enabled large-scale collection of raw material, extraction and purification of insulin, and supplied the drug in a state suitable for clinical use. Only after proper bulk production and/or synthesis techniques were established did insulin and many other hormones discovered in the 1920s and 1930s become useful and readily available to the public. This would continue to be the case with most pharmaceutical breakthroughs throughout the century. During the 1960s new developments in peptide engineering led to synthetic insulin and in the 1970s, with biotechnology developments genetically manufactured insulin.

If insulin revolutionized diabetes mellitus treatment, cortisone discovery was another revolution in inflammatory and arthritis management. The discovery of corticosteroids as a therapeutic can be linked to Thomas Addison (1793–1860), who made the connection between the adrenal glands and the rare Addison's disease in 1855.[140] Edward C. Kendall (1886–1972)[141] at the Mayo Clinic (Rochester, USA) and Tadeus Reichstein (1897–1996)[142] at the University of Basel (Switzerland) independently isolated several hormones from the adrenal cortex. In 1948, Kendall and Philip S. Hench (1896–1965) demonstrated the successful treatment of patients with rheumatoid arthritis using cortisone.[143] Kendall, Reichstein, and Hench were awarded the 1950 Nobel Prize in Physiology or Medicine for determining the structure and biological effects of adrenal cortex hormones.

The birth of steroid chemistry meant that the female hormonal cycle was being controllable.[144] The modern knowledge of the menstrual cycle began when Edgar Allen (1892–1943) and Edward Doisy (1893–1986) showed that uterine bleeding occurs as a withdrawal effect when estrogen ceases to act on the endometrium.[145] At the same time, the chemistry of steroids becomes clearer with the works of Adolf Butenandt (1903–1995),[146] John Browne,[147] Leopold Ruzicka (1887–1976),[148] etc. Perhaps no contribution of chemistry in the second half of the twentieth century has a greater impact on social customs than the development of oral contraceptives. Several people were important in its development — among them Margaret Sanger (1879–1966), Katherine MacCormick (1875–1967), advocates of birth control as a means of solving the world's overpopulation,[149] Russell Marker (1903–1995), Gregory Pincus (1903–1967), and Carl Djerassi (born 1923). Pincus agreed with the project when he was asked by the feminist leaders to produce a physiological contraceptive. The key to the problem was the use of a female sex hormone, such as progesterone. This hormone prevents physiological ovulation and could be considered as a pregnancy-preventing hormone. The first difficulty to be overcome was to find a suitable, inexpensive source of the scarce compound to do the necessary research.[150]

The chemist Russell Marker who converted sapogenin steroids, extracted from dioscoreas, into progesterone has performed the task. Until 1970, his source for the sapogenins was a yam grown in Mexico. Marker created a new company, Syntex, in Mexico to produce progesterone. In 1949, he left the company to a young scientist hired that same year by Syntex and who ultimately figured prominently in the further development of 'the Pill'. The period from late 1949 through 1951 is an extraordinarily productive one in steroid chemistry and especially so at Syntex in Mexico City.[151] Two of the most important Syntex contributions — the synthesis of 19-nor,17-alpha-ethynyltestosterone (norethindrone) and of cortisone from diosgenin. Djerassi first worked on the synthesis of cortisone from diosgenin. He later turned his attention to synthesizing an 'improved' progesterone, one to be taken orally. Progesterone is able to be absorbed by the oral route after a minimal change in the carbone 19 of the steroid, the withdrawal of a methyl group. Those derivatives are called '19-nor'. In 1951, his group developed a progesterone-like compound called norethindrone.

Shortly thereafter, G.D. Searle & Co. initiated a major effort in steroid research, the objective of which was to discover better steroid drugs than those available at that time or steroids that could be used for conditions for which no compounds were previously available. This effort was remarkably successful and resulted in the introduction of several important pioneering drugs. These included norethandrolone, marketed in 1956 as Nilevar®, the first anabolic agent with a favourable separation between protein building and virilization, and spironolactone, introduced in 1959 as Aldactone®, the first steroid antialdosterone antihypertensive agent. Of special importance was the research that culminated in the discovery of Enovid®. This substance, a combination of the progestin norethynodrel and the estrogen mestranol is first approved in 1957 for the treatment of a variety of disorders associated with the menstrual cycle. The era of oral contraception began in May 1960, when Enovid® was approved by the Food and Drug Administration for ovulation inhibition.[152]

XV DRUGS OF THE MIND

The field of psychiatry is so complex that until the middle of the twentieth century, it is clear that the only a behaviourist approach could represent a means to explore and treat mental disorders. This idea proved more popular than 'biological' explanations, which stressed that the cause of schizophrenia, depression or other mental illnesses was an imbalance in the chemicals of the brain. Severe mental illness had been increasing since the beginning of the century. In 1904, 0.2% of people were hospitalized in mental hospitals; by 1955, the number had doubled. Psychiatrists argued whether it was the result of biology or of experience, but there was nothing to help the chronically mentally ill. They were usually housed in state institutions.

If psychoanalysis was extremely popular across Europe and in the USA, science lagged behind. Early treatments for depression at the beginning of the twentieth century involved dosing patients with barbiturates, keeping them unconscious for several days, in the hope that sleep would restore them to a healthier frame of mind.[153]

It was then discovered that, in certain cases, patients who experienced epileptic fits also experienced less severe symptoms of mental illness. By causing a person to have a controlled fit (first by dosing the patients with camphor then, in 1938, by the use of electricity), doctors found they could lessen the effects of depression. Nowadays, electroconvulsive therapy (ECT) is still used as a treatment for severe depression.

The understanding of depression depends on the understanding of the brain itself. This took a leap forward in 1928, when Austrian scientist Otto Loewi (1873–1961) discovered the first neurotransmitter in the brain, acetylcholine.[154] He concluded that this substance was necessary to help electrical messages pass through the brain, from one nerve ending (neuron) to the next. It was to be another 24 years before scientists would discover the presence of other neurotransmitter substances in the brain, such as serotonin, norepinephrine and dopamine. By the 1980s, scientists isolated 40 different neurotransmitter substances in the brain.[155]

Modern psychiatric treatments were introduced in 1948, when lithium carbonate was discovered as a treatment for mania by Australian psychiatrist John F. Cade.[156] After Cade's initial report, lithium treatment was principally developed in Denmark by Mogens Schou (1918–), beginning in 1954.[157] After a decade of trials by these and other groups in the USA and abroad, the Psychiatric Association and the Lithium Task Force recommended lithium to the Food and Drug Administration for therapy of mania in 1969, 20 years after its discovery by Cade. In 1970, the FDA approved the prescription drug. A breakthrough had finally been achieved in the treatment and prevention of one of the world's major mental health problems in the form of manic depression, and the genetically related forms of recurrent depression.

In 1952, reserpine had been isolated from *Rauwolfia* and eventually was used for treating essential hypertension, and a year later, reserpine, a *Rauwolfia* alkaloid is used as a first tranquillizer drug.[158] The source plant comes from India, where it has long been used as a folk medicine.

A potent type of antidepressant, monoamine oxidase inhibitor (MAOI), was developed in the 1950s. This works by blocking the action of certain enzymes in the brain (oxidases) which break down neurotransmitters.[159] The brain thus remains 'bathed' in extra quantities of neurotransmitters, and is able to fight off the depression. Though an effective remedy, this drug could have unpleasant reactions when taken alongside certain foods and drink. Patients taking MAOIs have to observe quite strict dietary rules because the side-effects could be fatal. Iproniazid, the first modern antidepressant, was originally developed as a drug for the chemotherapy of tuberculosis in the early 1950s. Nathan Kline (1916–1984) observed that iproniazid, in addition to its ability to treat tuberculosis, could elevate mood and stimulate activity in many patients.[160] These effects led researchers to investigate the ability of iproniazid to treat the symptoms of depression. After promising preliminary findings reported in 1957, iproniazid was widely prescribed to patients with major depression.[161]

The first tricyclic antidepressant, imipramine, was originally developed in the search for drugs useful in the treatment of schizophrenia. Although clinical trials demonstrate a lack of effect in treating schizophrenia, an astute clinician decided to examine its effectiveness in depressed patients. Early studies in 1958, performed by Roland Kuhn (born 1912), published in the *American Journal of Psychiatry*, led to the first antidepressant to be introduced. Imipramine is reported to significantly alleviate symptoms in patients with major depression.[162] Interestingly, although imipramine elevated mood and increased energy in depressed patients, the drug proved to be sedating in individuals without major depression. These effects led to the idea that imipramine selectively reversed the depression, rather than simply producing a general activating effect. Subsequent biochemical studies on imipramine demonstrate that this drug increases the activity of the monoamine neurotransmitters, norepinephrine and serotonin, by inhibiting their reuptake into neurons.[163]

Most of the early antidepressants work by affecting several different neurotransmitter chemicals at the same time. Scientists began to work on drugs that would target one specific neurotransmitter, while leaving others unaffected. In 1968, a Swedish scientist Arvid Carlsson (born 1923), made discoveries that would eventually lead to the creation of the drug Prozac. He found that when an electrical impulse passed from one neuron to another, serotonin was released into the space between the neurons—the synapse—to help the 'message' be transmitted. Once the job was done, the serotonin was reabsorbed by the neuron. But antidepressants prevented the neurons from taking the serotonin. It remained in the synapse, where its presence seemed

to help the patient recover from depression. Carlsson was the 2000 Nobel Laureate in Physiology or Medicine. By 1974, American scientists were testing a drug which prevented the neurons from reabsorbing serotonin, while not preventing the absorption of other brain chemicals, such as noradrenaline. They called this drug a specific serotonin re-uptake inhibitor (SSRI). Its name is fluoxetine. In tests, they discover that it provided rapid relief from the symptoms of depression, without any of the unpleasant side-effects associated with the older, tricyclic antidepressants or the dietary restrictions that were necessary with MAOI drugs. By 1987, the drug was marketed as Prozac®. By 1994, it was the number two best-selling drug in the world.[164]

In 1952, Henri Laborit (1914–1995), a naval surgeon in Paris, was looking for a way to reduce surgical shock in his patients. Much of the shock came from the anaesthesia, and if he would find a way to use less, his patients could recover quicker. He knew that shock was the result of certain brain chemicals and looked for a chemical that might counteract these. He tried antihistamines, usually used to fight allergies. He noticed when he gave a strong dose to his patients, their mental state changed. They did not seem anxious about their forthcoming surgery, in fact, they were rather indifferent. Laborit was able to operate using much lower doses of anaesthetic. He was so struck by the effect on his patients, especially with a drug called chlorpromazine (Largactil® in Europe; Thorazine® in the USA), he thought the drug must have some use in psychiatry.[165] A chemistry team at Rhône-Poulenc had prepared this new drug.[166] The French psychiatrist Pierre Deniker (1907–1987) was interested in these results and ordered some chlorpromazine to try on his most agitated, uncontrollable patients. The results were stunning. Patients who had stood in one spot without moving for weeks and patients who had to be restrained because of violent behaviour, could now make contact with others and be left without supervision.[167] Another psychiatrist reported, 'For the first time we could see that they were sick individuals to whom we could now talk'.

Meanwhile, Smith Kline purchased the rights to chlorpromazine from Rhône-Poulenc in 1952. Smith Kline put it on the market as an anti-vomiting treatment but tried to convince American medical schools and hospitals to test the drug as an antipsychotic drug. Unfortunately, the academics saw it as just another sedative and were more interested in psychoanalysis and behaviourism. Smith Kline invited Pierre Deniker to help them convince US practitioners to try the drug. Their success came by way of the state institutions seeing the drug as a money saver. Testing began at these institutions, where the most hopeless cases were housed. The results are convincing, even miraculous. In 1954 the US Food and Drug Administration approved chlorpromazine. By 1964, some 50 million people around the world had taken the drug.

The demand for sedatives was profound, and the drug marketplace responded rapidly. Although meprobamate, the first of the major tranche, discovered by Frank M. Berger is called the Wonder Drug of 1954,[168] sedatives were not widely used until 1961, when Librium® (chlordiazepoxide) was discovered by Leo Sternbach (born 1908) and marketed as a treatment for tension. Librium® was a phenomenal success. Then Valium® (diazepam), discovered in 1960, was marketed by Roche in 1963 and rapidly become the most widely prescribed drug in history. These drugs were touted to the general population, mass-marketed and prescribed by doctors with what many claimed was blithe abandon. While the youth of occidental world were smoking joints and tripping on 'acid', their parents' generation of businessmen and housewives were downing an unprecedented number of sedatives. For many, physical and psychological addiction followed.

XVI DRUGS FOR IMMUNOSUPPRESSION

Over the past 50 years, many immunosuppressive drugs have been described. Often their mechanisms of action were established long after their discovery. Eventually these mechanisms were found to fall into five groups: regulators of gene expression; alkylating agents; inhibitors of purine synthesis; inhibitors of pyrimidine synthesis and inhibitors of kinases or phosphatases. Glucocorticoids exert immunosuppressive and anti-inflammatory activity mainly by inhibiting the expression of genes for interleukin-2 and other mediators. After the 1950s and 1960s when corticosteroids were proposed to prevent organ rejection in renal transplantation, surgery enters a new era of optimism, characterized by improving allograft survival rates when tissue typing and new immunosuppressive drugs such as cyclosporin A were introduced. Revolutionary methods of rejection treatment have been responsible for this new era,[169] and a few years later in 1967, the first heart transplant was performed in Capetown by Christiaan Barnard.[170]

A few determined individuals in the medical and research community spent the next two decades attempting to solve the organ rejection puzzle. One of these scientists was Jean-François Borel (1933–), who worked in Switzerland for Sandoz Pharmaceuticals. He discovered the immunosuppressant agent that ultimately moved transplantation from the realm of curiosity into routine therapy. Both J. Borel and Hartmann Stahelin markedly contributed to the discovery and characterization of the biological profile of the drug.[171] In its subsequent exploitation, Borel played a leading role.[172] He chose to examine a weak compound that

was isolated from the soil fungus *Tolypocladium inflatum* Gams (subsequently renamed *Beauveria nivea*). The compound was thought to have little practical value, yet Sandoz chemists continued to study and purify the compound because of its 'interesting' chemical properties. Borel discovered that this compound selectively suppressed the T-cells of the immune system. Excited by these characteristics, Borel continued his study and, in 1973, purified a compound called cyclosporine A.[173] Cyclosporine acts in two ways. First, it impedes the production and release of interleukin-2 by T-helper white blood cells. Secondly, it inhibits interleukin-2 receptor expression on both T-helper and T-cytotoxic white blood cells.

Other immunosuppressants are available. Cyclophosphamide, a nitrogen mustard,[174] alkylates DNA bases and preferentially suppresses immune responses mediated by B-lymphocytes. Methotrexate and its polyglutamate derivatives suppress inflammatory response through release of adenosine; they suppress immune response by inducing the apoptosis of activated T-lymphocytes and inhibiting the synthesis of both purines and pyrimidines.[175] Azathioprine metabolites, studied by Roy Calne (1930–), inhibit several enzymes of purine synthesis.[176] Mycophenolic acid inhibits inosine monophosphate dehydrogenase, thereby depleting guanosine nucleotides and induces apoptosis of activated T-lymphocytes.[177] Like cyclosporine, tacrolimus suppresses the production of IL-2 and other cytokines. In addition, these compounds have recently been found to block signalling pathways triggered by antigen recognition in T-cells.[178] In contrast, rapamycin inhibits kinases required for cell cycling and responses to IL-2. Rapamycin also induces apoptosis of activated T-lymphocytes. Immunosuppressive and anti-inflammatory compounds in development include inhibitors of p38 kinase and of the type IV isoform of cyclic AMP phosphodiesterase, which is expressed in lymphocytes and monocytes. A promising future application of immunosuppressive drugs is their use in a regime to induce tolerance to allografts.[179]

XVII CHEMISTRY AGAINST CANCER

Many anticancer drugs are extracted from plants.[180] Galen proposed the juice expressed from woody nightshade (*Solanum dulcamara*) to treat tumours and warts, which has been demonstrated to exert anti-inflammatory properties.[181] More than 1600 genera have been examined in recent decades.[182]

Since the use of *Podophyllum* in ancient China, many vegetal derivatives from this plant are being used in cancer chemotherapy. Two glycosides were extracted from *Podophyllum* to prepare semisynthetic derivatives of podophyllotoxin, etoposide and teniposide.[183] In folklore medicine, extracts of the leaves of the subtropical plant *Catharanthus roseus* (L.) (Madagascar periwinkle) were reputed to be useful in the treatment of diabetes. The attempt to verify the antidiabetic properties of the extracts led instead to the discovery and isolation of two complex indole alkaloids, vinblastine and vincristine, which are used in the clinical treatment of a variety of lymphomas, leukaemias and various cancers such as small cell lung or cervical and breast cancer. The two alkaloids, although structurally almost identical, nevertheless differ markedly in the type of tumours that they affect and in their toxic properties.

As the 1970s opened, new chemistries and the war on cancer seized centre stage. US President Richard Nixon established the National Cancer Program, popularly known as the war on cancer, with an initial half-billion dollars of new funding. This explains why in the recent years, many new compounds with antineoplastic properties were isolated in plants. Among them, the pyridocarbazole alkaloids ellipticine and 9-methoxyellipticine from *Ochrosia ellip-tica*, that intercalate between the base pairs of DNA.[184] Camptothecin and its derivatives, alkaloids from Chinese tree *Camptotheca acuminata*, showed a broad-spectrum activity. 10-Hydroxycamptothecin and moreover, 9-amino-camptothecin which was more active, gave birth to topotecan and irinotecan.[185] New alkaloid esters from *Cephalotoaxus* species are currently being isolated for experimental and clinical studies. If the parent alkaloid is inactive, but the esters harringtonine and homoharringto-nine form *Cephalotaxus harringtonia*, give new hopes in the cure of solid tumours or leukaemias.

The most enthusiastic reports concern the diterpenoids Taxol® and Taxotere® having unique tri- or tetracyclic 20-carbon skeletons. Taxol has been extracted from the bark of the Pacific yew (*Taxus brevifolia*), a slow-growing tree found in the virgin rain forests of the Pacific Northwest United States. Yew was known as a toxic plant for animals and humans for centuries.[186] Monroe E. Wall (1917–2002) and Mansukh C. Wani at Research Triangle Park, identified the active principle of the yew tree in 1971.[187] In 1979, Susan Horwitz of the Department of Molecular Pharma-cology, Albert Einstein College of Medicine, New York, suggested that Taxol's® mechanism of action is different from that of any previously known cytotoxic agent. She observed an increase in the mitotic index of P388 cells and an inhibition of human HeLa and mouse fibroblast cells in the G2 and M phases of the cell cycle.[188] It was suggested that Taxol® exerted its activity by preventing depolymer-ization of the microtubule skeleton.

Clinical use of Taxol® included many solid tumours with best results in ovarian and breast cancers. Extraction of Taxol® (paclitaxel) from the yew bark is quite difficult: three trees were needed for 1 gram of drug (one cycle of chemotherapy). This difficulty has encouraged the pursuit of semisynthetic production. The strategy included immediately increasing the amount of Taxol® derived from

yew bark and establishing a broad research program to evaluate alternative sourcing options and their commercial feasibility.[189] Taxol® introduced into the marketplace in January 1993 by Bristol-Myers Squibb reached worldwide sales of $1.2 billion in 1998. The prospects for finding a solution to the Taxol® supply problem through semisynthe-sis using a naturally occurring taxane as a starting material were considerable. This approach was pioneered by Pierre Potier (born 1935) in Gif-sur-Yvette (France). He found in the early 1980s that a naturally occurring taxane containing the Taxol® core, 10-deacetyl baccatin III, was twenty times more abundant than Taxol® and was primarily contained in the needles of the abundant English Yew (*Taxus baccata*). Potier succeeded in the difficult conversion of 10-DAB into Taxol®, in 1988 using only four steps with an overall yield being only 35%.[190] Pierre Potier and coworkers discovered a paclitaxel semisynthetic analogue, docetaxel (Taxotere®), which represented a significant advance in the treatment of various malignancies. Although paclitaxel and docetaxel have a similar chemical root, extensive research and clinical experience indicate that important biological and clinical differences exist between the two compounds. Although the mechanism by which they disrupt mitosis and cell replication is novel and unique to this class of compounds, there are small but important differences in the formation of the stable, nonfunctional microtubule bundles and in the affinity of the two compounds for binding sites.[191] These differences may explain the lack of complete cross-resistance observed between docetaxel and paclitaxel in clinical studies.[192]

Besides natural products, synthetic anticancer drugs flourished in various directions. The first agents were nitrogen mustards (halogenated alkyl amine hydrochlo-rides) among which 2-2′-2′-trichlotriethylamine was the prototype first studied by two prestigious pharmacologists from Yale University, Louis Goodman (1906–2000) and Alfred Gilman (1908–1984).

Louis Sanford Goodman, Maxwell M. Wintrobe, William Dameshek, Morton Goodman, Alfred Gilman and Margaret MacLennan[193] performed studies in 1943, but only presented the salutary results obtained in patients treated for Hodgkin's disease, lymphosarcoma and leukaemia in 1946. Indeed, in the first two disorders dramatic improvement was observed in an impressive proportion of terminal and so-called radiation-resistant cases.[193] First constant successes in haematological malignancies were obtained in 1970 with the 'MOPP' therapy (mechlorethamine, vincristine, procar-bazine, prednisone). This protocol was superior to that previously reported with the use of single drugs with 35 of 43, or 81% of the patients achieving a complete remission, defined as the complete disappearance of all tumour and return to normal performance status.[194]

Antimetabolites in cancer treatment was discovered by George Hitchings (1905–1998) and Gertrude Elion

(1918–1999), utilizing what today is termed 'rational drug design'. They methodically investigated areas where they could see cellular and molecular targets for the development of useful drugs.[195] During their long collaboration, they produced a number of effective drugs to treat a variety of illnesses including leukaemia, malaria, herpes, and gout.

Hitchings and Elion began examining the nucleic acids, particularly purines, including adenine and guanine, two of DNA's building blocks at Wellcome Research Laboratories. They discovered that bacteria could not produce nucleic acids without the presence of certain purines, and set to work on antimetabolite compounds, which locked up the enzymes necessary for the incorporation of these purines into nucleic acids. They synthesized two substances: diaminopurine and thioguanine, which the enzymes apparently latched on to, instead of adenine and guanine. These new substances proved to be effective treatments for leukaemia. Elion later substituted an oxygen atom with a sulphur atom on a purine molecule, thereby creating 6-mercaptopurine used to treat leukaemia.[196] After this success, Elion and Hitchings developed a number of additional drugs using the same principle. Later, these related drugs were found to not only interfere with the multiplication of white blood cells, but also suppress the immune system. This latter discovery led to a new drug, Imuran® (azathioprine), and a new application—organ transplants. Imuran suppresses the immune system that would otherwise reject newly transplanted organs. For the first time, patients can receive organ transplants without their bodies rejecting the new organs. The team also develops allopurinol, a drug successful in reducing the body's production of uric acid, thereby treating gout. They discovered pyrimethamine, used to treat malaria, and trimethoprim used to treat meningitis, septicaemia, and bacterial infections of the urinary and respiratory tracts. With Howard Schaeffer, Elion is also at the origin of acyclovir, marketed as Zovirax®, which interferes with the replication process of the herpes virus.[197]

Anthracyclines may be listed among the main anticancer drugs. Daunomycin (also called daunorubicin) was isolated from *Streptomyces peucetius* in 1962 by Aurelio Di Marco (Farmitalia, Milano).[198] With Adriamycin, it is the prototypical member in the anthracycline antitumour antibiotic family. Adriamycin, a 14-hydroxy derivative of daunorubicin, was isolated from the same microorganism, in 1967. Despite their severe cardiotoxicity and other side-effects, these drugs have been widely used as dose-limited chemotherapeutic agents for the treatment of human solid cancers or leukaemias since their discovery.[199] These antibiotics contain a quinone-containing chromophore and an aminoglycoside sugar. The antineoplastic activity of these drugs has been mainly attributed to their strong interactions with DNA in the target cells. While anthracyclines can be very effective against breast and other cancers,

they pose a risk of cardiotoxicity and therefore, they are typically used in limited doses. Doxorubicin and epirubicin are examples of anthracyclines used to treat breast cancer, commonly used in combination with other chemotherapy drugs to help decrease the risk of side-effects.[200]

Cisplatin was discovered serendipitously in 1965 while Rosenberg's team was studying the effect of an electric current on *E. coli* cultures. It was found that cell division was inhibited by the production of *cis*-diamminedichloroplatinum from the platinum electrodes rather than by the method expected.[201] Further studies on the drug indicated that it possessed antitumour activity. In 1972 the National Cancer Institute introduced cisplatin into clinical trials. It now has a major role in the treatment of several human malignancies, including testicular, ovarian, head and neck, bladder, esophageal and small cell lung cancers. Cisplatin is a square planer compound containing a central platinum atom surrounded by two chloride atoms and two molecules of ammonia moieties. The antitumour activity has been shown to be much greater when the chloride and ammonia moieties are in the *cis* configuration as opposed to the *trans* configuration. The cytotoxicity of cisplatin is due to its ability to form DNA adducts which include DNA–protein cross-links, DNA monoadducts, and inter/intra DNA cross-links.[202] The drug is able to enter the cell freely in its neutral form, yet once in the cell the chloride ions are displaced to allow the formation of a more reactive, aquated compound. In 1975, Memorial Sloan-Kettering Cancer Center initiated trials of cisplatin alone and later in combination with cyclophosphamide and/or adriamycin in patients with urothelial tract cancer. The results were not as positive as those seen in the testicular cancer studies, but they were favourable. Combination of cisplatin and adriamycin also showed noteworthy improvements in tumour cells. Studies by Holland using cisplatin alone and in combination with adriamycin to fight ovarian cancer gave substantial improvements.[203]

Due to the extreme toxicity of cisplatin, as well as resistance against it, there has been a need for the development of analogues which are just as potent, but not as toxic. One of the most widely known analogues of cisplatin is carboplatin.[204] Carboplatin has the same two amine groups that cisplatin does, but rather than chloride it contains two cyclobutanedicarboxylated groups. These groups are much less labile, thus the reactions in water to activate it are much slower. Thus, carboplatin is more stable and less reactive than cisplatin. There has been less testing and fewer trials with carboplatin than with cisplatin, but it has been shown that the neurotoxicity is no longer a limiting side-effect. This does not mean it does not occur, simply it means that the effects are not as drastic. It has been suggested however, that in higher doses carboplatin may have similar effects, although no studies have been performed as yet.

XVIII FROM GENETICS TO DNA TECHNOLOGY

As chemists and pharmacists joined together in finding new drugs, a second revolution in biology (the first came from Claude Bernard's and Louis Pasteur's generation) took place when advances in experimental genetics, biology, and virology happened in the middle of the century.

In 1935, George W. Beadle (1903–1989), before collaborating with Edward L. Tatum (1909–1975), began studying the development of eye pigment in *Drosophila* with Boris Ephrussi (1901–1979). After producing mutants of *Neurospora crassa*, a bread mould by irradiation and searching for interesting phenotypes, they concluded in a 1940 report that each gene produced a single enzyme, also called the 'single gene–single enzyme' concept. The two scientists shared the Nobel Prize in Physiology or Medicine in 1958 for discovering that genes regulate the function of enzymes and that each gene controls a specific enzyme.[205]

Eminent scientific research was carried out by Joshua Lederberg (born 1925) on plasmid concept, John F. Enders (1897–1985), Thomas H. Weller (born 1915), and Frederick C. Robbins (born 1916) on virus cultures, Salvador Luria (1912–1991), and Alfred D. Hershey (1908–1997) on the bacteriophage role. James Watson (born 1928) and Francis Crick (born 1916) determined the structure of genetic material in 1953.[206] In 1955, Severo Ochoa (1905–1993) at New York University School of Medicine discovered polynucleotide phosphorylase, an RNA-degrading enzyme. In 1956, Arthur Kornberg (born 1918) at Washington University Medical School (St. Louis, MO) discovered DNA polymerase. In 1960, François Jacob (born 1920) and Jacques Monod (1910–1976) and André Lwoff (1902–1994) proposed their operon model. It was the birth of gene regulation models which launched a continuing quest for gene promoters and triggering agents. In 1964, Bruce Merrifield (born 1921) invented a simplified technique for protein and nucleic acid synthesis. These discoveries were made outside the pharmaceutical industry, but gave enormous contributions to the understanding of the mechanisms of diseases and on the discovery of new drugs.

The next following step was the manufacture of therapeutic proteins. In 1970, two years before the birth of recombinant DNA (rDNA) technology, cytogeneticist Robert J. Harris coined the term 'genetic engineering'. However, DNA recombinant technology needed the discovery by Werner Arber, in 1970, of restriction enzymes which cut DNA in the middle of a specific symmetrical sequence. Modern genetic engineering began in 1973 when Herbert Boyer (born 1928) and Stanley Cohen (born 1922) used enzymes to cut a bacteria plasmid and insert another strand of DNA in the gap. The invention of recombinant DNA technology offered a window into the previously impossible: the mixing of traits between totally dissimilar organisms. To prove it was possible, Cohen and Boyer used the same process to put frog DNA into a bacteria. Since 1973, this technology has been made more controllable by the discovery of new enzymes to cut the DNA differently and by mapping the genetic code of different organisms.

The year 1975 heralded DNA sequencing. Walter Gilbert (born 1932) and Allan Maxam and Frederick Sanger (born 1918) simultaneously develop different methods for determining the sequence of bases in DNA with relative ease and efficiency.

Genentech's goal of cloning human insulin in *Escherichia coli* was achieved in 1978, and the technology was licensed to Eli Lilly. The recombinant DNA era grew from these beginnings and have a major impact on pharmaceutical production and research in the 1980s and 1990s. Since that time, dozens of protein drugs, have been marketed: growth hormone, colony-stimulating factors, erythropoietin, tissue plasminogen activator, antihaemophilic factors, interferons, etc.

XIX CONCLUSION

Modern medicinal chemistry is more and more focused on the interactions of small molecules with proteins than with genes, which code for the synthesis of those proteins. Many of the medically relevant proteins have already been identified and will continue to be important targets for modern therapies even after the human genome is fully sequenced. The human genome sequence may help scientists finish the task of identifying these proteins. It looks likely that protein drugs may be a prominent part of the pharmaceutical library of the future.

Pharmaceuticals will be more personalized, thanks to a growing field known as pharmacogenomics, which focuses on polymorphism in drug-metabolizing enzymes and the resulting differences in drug effects. Due to slight genetic differences between humans in drug absorption, metabolism and excretion, pharmacogenomics will identify the patient population most likely to benefit from a given medication. Greater integration of chemistry into biological research will allow biology to be studied in a less reductionist way. This seems vital, given that tissue engineering taking cells (stem cells) to restructure or rebuild damaged or congenitally defective tissues will probably be the next step towards an effective cure.

REFERENCES

1. Nunn, J.F. (1996) *Ancient Egypt Medicine*. University of Oklahoma Press, Norman, Oklahoma.
2. Frey, E.F. (1985–86) The earliest medical texts. *Clio. Med.* **20**: 79–90.

3. Sigerist, H. (1951) *History of Medicine*, vol. 1, Primitive and archaic medicine. Oxford University Press, New York.

4. Sabbah, G., Corsetti, P.P. and Fischer, K.D. (1987) *Bibliographie des Textes Médicaux Latins, Antiquité et Haut Moyen Age*. Saint Etienne.

5. Grmek, M.D. (1995) *Histoire de la Pensée Médicale en Occident*, vol. 1. Antiquité et Moyen Age, Le Seuil, Paris. Especially following chapters: Introduction by Mirko D. Grmek; La naissance de l'art médical occidental, by Jacques Jouanna; Entre le savoir et la pratique: la médecine hellénistique, by Mario Vegetti; Les voies de la connaissance: la médecine dans le monde romain by Danielle Gourevitch; Le concept de maladie, by Mirko D. Grmek; Stratégies thérapeutiques: les médicaments, by Alain Touwaide.

6. Bariety, M. and Coury, C. (1963) *Histoire de la Médecine*. Fayard, Paris.

7. Griffin, J.P. (1995) Famous names in toxicology. Mithridates VI of Pontus, the first experimental toxicologist. *Adverse Drug React. Toxicol. Rev.* **14**: 1–6.

8. Hankinson, R.J. (1991) *Galen: On the Therapeutic Method*, I and II. Oxford.

9. Dioscorides, P.A. (1906–14) *De Materia Medica*. Berolini, Weidmann.

10. Nurhussein, M.A. (1989) Rhazes and Avicenna. *Ann. Intern. Med.* **111**: 691–692.

11. Nyverd, G. (1520) *Le Grant Herbier en Francoys, Contenant les Qualitez, Vertus et Proprietez des Herbes, Arbres, Gommes et Semences*. Paris.

12. Court, W.E. (1999) The ancient doctrine of signatures or similitudes. *Pharm. Hist. (Lond.)*, **29**: 41–48.

13. Oster, M. (1993) Biography, culture, and science: the formative years of Robert Boyle. *Hist. Sci.* **31**: 177–226.

14. Reti, L. and Gibson, W.C. (1968) *Some Aspects of Seventeenth-Century Medicine and Science*. University of California Press, Los Angeles.

15. Lavoisier, A.A. (1789) *Traité Elémentaire de Chimie Présenté dans un Ordre Nouveau et d'Après les Découvertes Récentes, Avec Figures*, Cuchet, Paris.

16. Aftalion, F. (1988) *Histoire de la Chimie*, Masson, Paris.

17. Bensaude-Vincent, B. and Stengers, I. (1993) *Histoire de la Chimie*, La Découverte, Paris.

18. Bernard, C. (1855–56) *Leçons de Physiologie Expérimentale Appliquée à la Médecine*. 2 vols. J.-B. Baillière, Paris.

19. Virchow, R.L. (1858) *Die Cellularpathologie in ihrer Begründung auf Physiologische und Pathologische Gewebelehre*. A. Hirschwald, Berlin.

20. Pasteur, L. (1922–39) Œuvres complètes réunies par Pasteur Vallery-Radot, 7 vols., Masson, Paris.

21. Soubeiran, E. (1831) Recherches sur quelques combinaisons du chlore. *Ann. Chim. (Paris)*, **48**: 113–157.

22. Von Liebig, J. (1832) Ueber die Verbindungen, welche durch die Einwirkung des Chlors auf Alkohol, Aether, ölbildenes Gas und Essiggeist entstehen. *Ann. Pharm.* **1**: 182–230.

23. Guthrie, S. (1832) New mode of preparing a spirituous solution of chloric ether. *Am J. Sci. Arts* **21**: 64–65 and **22**: 105–106.

24. Von Liebig, J. (1842) *Die organische Chemie in ihrer Anwendung auf Physiologie und Pathologie*. Braunschweig, F. Vieweg.

25. Wohler, F. (1828) Ueber kunstliche Bildung des Harnstoffs. *Ann. Phys. Chem.* (Leipzig). (French translation of this article appears in *Ann. Chim (Pans)* **37**: 33–34.)

26. Müller-Jahncke, W.D. and Friedrich, C. (1996) *Geschichte der Arzneimitteltherapie*, Deutscher Apotheker Verlag, Stuttgart.

27. Cadetde Gassicourt, C.L. (1809) Considérations sur l'etat actuel de la pharmacie. *Bull. Pharm.* **1**: 5–12.

28. Chast, F. (1995) *Histoire Contemporaine des Médicaments*, La Découverte Ed., Paris.

29. Sertürner, F. (1805) Auszuge aus briefen an den Herausgeber. (a) Säure im opium. (b) Ein deres Schreiben von Ebendenselben. Nachtrag zur Charakteristik der Saüre im Opium. *J. Pharmazie für Artze, Apotheker, und Chemisten von D.J.B. Trommsdorff* **13**: 29–30.

30. Pelletier, P.J. and Magendie, F. (1817) Recherches chimiques et physiologiques sur l'ipecacuanha. *Ann. Chim. Phys.* **4**: 172–185.

31. Pelletier, P.J. and Caventou, J.-B. (1820) Recherches chimiques sur les quinquinas. *Ann. Chim. Phys.* **15**: 289–318, 337–365.

32. Robiquet, P.J. (1832) Nouvelles observations sur les principaux produits de l'opium. *Ann. Chim.* **51**: 225–267.

33. Mein, H.F. (1833) Ueber die Darstellung des Atropins in weissen Krystallen. *Ann. Chem. Pharm.* **6**: 67–72.

34. Merck, G.F. (1848) Vorläufige Notiz über eine neue organische Base im Opium. *Ann. Phys. Chem. (Lpz.)* **66**: 125–128.

35. Ladenburg, A. (1881) Die Natürlich vorkommenden mydriatische wirkenden Alkaloïde. *Ann. Chem. Pharm.* **206**: 274–307.

36. Magendie, F. (1822) *Formulaire pour la préparation et l'emploi de plusieurs nouveaux médicamens, tels que la noix vomique, la morphine, etc.* Méquignon-Marvis, Paris.

37. McLagan, T.J. (1876) The treatment of acute rheumatism with salicin. *Lancet*, **1**: 342–343, 383–384.

38. Piria, R. (1839) Recherches sur la salicine et les produits qui en dérivent. *CR Acad. Sci. (Paris)*, **8**: 479–485.

39. Gerhardt, C.F. (1853) Untersuchungen über die wasserfrei organischen Saüren. *Ann. Chem. Pharm.* **87**: 149–179.

40. Sneader, W. (2000) The discovery of aspirin: a reappraisal. *Br. Med. J.* **321**: 1591–1594.

41. Dreser, H. (1899) Pharmakologisches über Aspirin (Acetylsalicylsaüre). *Pflüg. Arch. Ges. Physiol.* **76**: 306–318.

42. Withering, W. (1785) *An account of the foxglove, and some of its medical uses*, G.G.J. & J. Robinson, Birmingham.

43. Nativelle, C.A. (1872) De la digitaline et de la digitale. *Bull. Acad. Méd.* Paris i: 201.

44. Schmiedeberg, J.E.O. (1875) Untersuchungen uber die pharmakologische wirksamen Bestandteile der *Digitalis purpurea*. *L. Arch. Exp. Path. Pharmak.* **3**: 16–43.

45. Mackenzie, J. (1905) New methods of studying affections of the heart. *Br. Med. J.* **1**: 519–521, 587–589, 702–505, 759–562, 812–515.

46. Albers, J.F.H. (1864) Die Physiologische und Therapeutische wirkung. *Dtsch. Klin.* **42**: 405.

47. Brunton, T.L. (1867) On the use of nitrite of amyl in angina pectoris. *Lancet* **2**: 97–98.

48. Smithwick, R.H. (1947) Surgery of the autonomic nervous system. *N. Engl. J. Med.* **236**: 662–669.

49. Kempner, W. (1948) Treatment of hypertensive vascular disease with rice diet. *Am. J. Med.* **4**: 545–577.

50. Barker, M.H. (1936) Blood cyanates in the treatment of hypertension. *J. Am. Med. Ass.* **106**: 762–767.

51. Page, I.H. and Corcoran, A.C. (1949) *Arterial Hypertension: Its Diagnosis and Treatment*. Year Book Medical Publishers, Chicago.

52. Dustan, H.P., Rocella, E.J. and Garrison, H.H. (1996) Controlling hypertension: a research success story. *Arch. Intern. Med.* **17**: 1926–1935.

53. Freis, E.D., Finnerty, F.A., Schnaper, H.W. and Johnson, R.L. (1952) The treatment of hypertension with hexamethonium. *Circulation* **5**: 20–27.

54. Schroeder, H.A. (1952) The effect of 1-hydrazinophthalazine in hypertension. *Circulation* **5**: 28–37.

55. Wilkins, R.W. and Judson, W.E. (1953) Use of *Rauwolfia serpentina* in hypertensive patients. *N. Engl. J. Med.* **248**: 48–53.

56. Vakil, R.J. (1949) A clinical trial of *Rauwolfia serpentina* in essential hypertension. *Br. Heart J.* **11**: 350–355.

57. Perry, H.M. and Schroeder, H.A. (1958) The effect of treatment on mortality ratio in severe hypertension. *Arch. Intern. Med.* **102**: 418–425.

58. Veterans Administration Cooperative Study Group on Antihypertensive Agents (1967) Effects of treatment on morbidity in hypertension, I: results in patients. with diastolic blood pressure averaging 115 through 129 mm Hg. *J. Am. Med. Assoc.* **202**: 1028–1034.

59. Tigerstedt, R.A. and Bergman, P.G. (1898) Niere und Kreislauf. *Skand. Arch. Physiol.* **8**: 223–271.

60. Davis, J.O. and Freeman, R.H. (1982) Historical perspectives on the renin-angiotensin-aldosterone system and angiotensin blockade. *Am. J. Cardiol.* **49**: 1385–1389.

61. Ferreira, S.H. (1985) History of the development of inhibitors of angiotensin I conversion. *Drugs* **30** (Suppl. 1): 1–5.

62. Wexler, R.R., Carini, D.J., Duncia, J.V., Johnson, A.L., Wells, G.J., Chiu, A.T., Wong, P.C. and Timmermans, P.B. (1992) Rationale for the chemical development of angiotensin II receptor antagonists. *Am. J. Hypertens* **5**: 209S–220S.

63. Wells, J. G. (1847) *A History of the Discovery of the Applications of Nitrous Oxyde Gas, Ether and Other Vapours to Surgical Operations.* Hartford.

64. Morton, W.T.G. (1847) *Remarks on the Proper Mode of Administering Sulfuric Ether by Inhalation.* Dutton & Wentworth, Boston.

65. Bigelow, H. J. (1846) Insensibility during surgical operations produced by inhalation. *Boston Med. Surg. J.* **35**: 309–317, 379–382.

66. Snow, J. (1848) On the inhalation of chloroform and ether. With description of an apparatus. *Lancet* **1**: 177–180.

67. Baumann, E. (1886) Ueber Disulfone. *Ber. Dtsch. Chem. Ges.* **19**: 2806–2814.

68. Fischer, E. and Mering, J. Von (1903) Ueber eine neue Klasse von Schlafmitteln. *Ther. Gegenw.* **44**: 97–101.

69. Bernard, C. (1883) Cours de Médecine du Collège de France. Première leçon (29 Février 1856), In *Leçons sur les Effets des Substances Toxiques et Médicamenteuses.* J.-B. Baillière, Paris.

70. Grifith, H.R. and Johnson, E. (1942) The use of curare in general anaesthesia. *Anesthesiology* **3**: 418–420.

71. Gray, T.C. and Halton, J. (1946) Technique for the use of d-tubocurarine chloride with balanced anaesthesia. *Br. Med. J.* **2**: 29329–29335.

72. Koller, C. (1884) Vorläufige Mittheilung über locale Anästhesirung am Auge. *Klin. Mbl. Augenheilk.* **22**: 60–63.

73. Freud, S. (1885) *Über Coca.* Verlag Von Moritz Perles, Wien.

74. Niemann, A. (1860) *Ueber eine neue organische Base in den Cocablättern.* E.A. Huth, Göttingen.

75. Bier, A. K. G. (1899) Versuche über Cocainisirung des Rückenmarkes. *Dtsch. Z. Chir.* **51**: 361–369.

76. Musto, D.F. (1992) Cocaine's history, especially the American experience. *Ciba Found. Symp.*, **166**: 7–14; discussion, 14–19.

77. Einhorn, A. (1899) Ueber die Chemie der localen Anesthaestica. *Munch. Med. Wschr.* **46**: 1218–1220, 1254–1256.

78. Filehne, W. (1887) Die local Anästhesirende Wirkung von Benzoylderivaten. *Berliner Klinische Wochenschrift* **24**: 107–108.

79. Fourneau, E. (1904) Stovaine, anesthésique local. *Bull. Soc. Pharmacol.* **10**: 141–148.

80. Dreifuss, F.E. (1995) Other antiepileptic drugs. In Levy, R., Mattson, R.H. and Meldrum, B.S. (eds). *Antiepileptic Drugs*, 4th edn., pp. 949–962. Raven Press, New York.

81. Merritt, H.H. and Putnam, T.J. (1938) Sodium diphenylhydantoinate in the treatment of convulsive disorders. *J. Am. Med. Ass.* **111**: 1068.

82. Glazko, A.J. (1986) Discovery of phenytoin. *Ther. Drug Monit.* **4**: 490–497.

83. Burton, B.S. (1882) On the propyl derivatives and decomposition products of ethyl acetoacetate. *Am. Chem. J.* **3**: 385–395.

84. McElroy, S.L. and Keck, P.E. (1995) Antiepileptic drugs. In Schatzberg, A. and Nemeroff, C.B. (eds). *Textbook of Psychopharmacology*, Vol. 17, pp. 351–375. American Psychiatric Press, Washington, DC.

85. Schindler, W. (1960) 5H-Dibenz[b,f]azepines, US patent 2,948, 718 (1960 to Geigy). *Chem. Abstr.* **55**: 1671c.

86. Lambert, P.A., Carraz, G. and Carbel, S. (1966) Action neuropsychotrope d'un nouvel anti-épileptique: le dépamide. *Ann. Méd. Psychol.* **124**: 707–710.

87. Cohen, A.F., Ashby, L., Crowley, D., Land, G., Peck, A.W. and Miller, A.A. (1985) Lamotrigine (BW 430C), a potential anticonvulsant. Effects on the central nervous system in comparison with phenytoin and diazepam. *Br. J. Clin. Pharmacol.* **6**: 619–629.

88. Saletu, B., Grunberger, J. and Linzmayer, L. (1986) Evaluation of encephalotropic and psychotropic properties of gabapentin in man by pharmaco-EEG and psychometry. *Int. J. Clin. Pharmacol. Ther. Toxicol.* **24**: 362–373.

89. Pierce, M.W., Suzdak, P.D., Gustavson, L.E., Mengel, H.B., McKelvy, J.F. and Mant, T. (1991) Tiagabine. *Epilepsy Res. Suppl.* **3**: 157–160.

90. Labarraque, A.G. (1825) *De l'Emploi des Chlorures d'Oxide de Sodium et de Chaux.* Mme Hazard, Paris.

91. Holmes, O.W. (1855) *Puerperal Fever, as a Private Pestilence.* Ticknor & Fields, Boston.

92. Semmelweis, I.P. (1861) *Die Aetiologie, der Begriff und die Prophylaxis des Kindbettfiebers.* C. A. Hartleben, Vienna.

93. Traube, I. (1910, 1911) Die Theorie des Haftdrucks (Oberflächendrucks) und ihre Bedeutung fur die Physiologie. *Pfldg. Arch. Ges. Physiol.* **132**: 511–538; **140**: 109–134.

94. Lister, J. (1867) On a new method of treating compound fracture, abscess, etc., with observations on the conditions of suppuration. *Lancet* **1**: 326–329, 357–359, 387–389, 507–509; **2**: 95–96.

95. Runge, F. F. (1834) Ueber einige Producte den Steinkohlendestillation. *Ann. Phys. Chem. (Lpz.)* **31**: 65–77, 308–332, 513–532.

96. Lucas-Championnière, J.M.M (1876) *Chirurgie Antiseptique.* J. B. Baillière, Paris.

97. Koch, R. (1881) Ueber Desinfection. *Mitt. k. GesandhAmte* **1**: 234–282.

98. Roux, P.P.E. and Yersin, A.E.J. (1888, 1889, 1890) Contribution à l'étude de la diphtérie. *Ann. Inst. Pasteur* **2**: 629–661; **3**: 273–288; **4**: 385–426.

99. Ehrlich, P. (1881) Ueber das Methylenblau und seine klinisch-bakterioskopische Verwerthung. *Z. Klin. Med.* **2**: 710–713.

100. Ehrlich, P. (1877) Beitrag zur Kenntnis der Anilinfärbungen und ihrer Verwendung in der mikroskopischen Technik. *Arch. Mikr. Anat.* **13**: 263–277.

101. Ehrlich, P. (1881) Ueber das Methyleneblau und seine klinisch-bakterioskopische Verwerthung. *Ztschr. F. Klin. Med.* **ii**: 710–713.

102. Guttmann, P. and Ehrlich, P. (1891) Ueber die Wirkung des Methylenblau bei Malaria. *Berl. Klin. Wschr.* **28**: 953–956.

103. Ehrlich, P. (1907) Chemotherapeutische Trypanosomen-Studien. *Berl. Klin. Wschr.* **44**: 233–236, 280–283, 310–314, 341–344.

104. Ehrlich, P. (1912) Ueber Laboratoriumsversuche und Klinische Erprobung von Heistoffen. *Chem. Ztg.* **36**: 637–638.

105. Parascandola, J. (1997) Alkaloids to arsenicals: systematic drug discovery before the First World War. *Publ. Am. Inst. Hist. Pharm.* **16**: 77–91.

106. Domagk, G. (1935) Ein Beitrag zur Chemotherapie der bakteriellen Infektionen. *Dtsch. Med. Wschr.* **C1**: 250–253.

107. Tréfouël, J., Tréfouël, T., Nitti, F. and Bovet, D. (1935) Activité du p-aminophénylsulfamide sur les infections streptococciques expérimentales de la souris et du lapin. *CR Soc. Biol. (Paris)* **120**: 756–758.

108. *The Pharmaceutical Century. Ten Decades of Drug Discoveries.* The American Chemical Society Web Site.

109. Dubos, R.J. (1939) Bactericidal effect of an extract of a soil bacillus on gram-positive cocci. *Proc. Soc. Exp. Biol.* (NY) **40**: 311–312.

110. Fleming, A. (1929) On the antibacterial action of cultures of a penicillium, with special reference to their use in the isolation of *B. influenzae. Br. J. Exp. Path.* **10**: 226–236.

111. Chain, E.B., Florey, H.W., Gardner, A.D., Heatley, N.G., Jennings, M.A., Orr-Ewing, J. and Sanders, A.G. (1940) Penicillin as a chemotherapeutic agent. *Lancet* **2**: 226–228.

112. Wainwright, M. (1990) *Miracle Cure: The Story of Penicillin and the Golden Age of Antibiotics.* Basil Blackwell, Oxford.

113. Levy, S.B. (1992) *The Antibiotic Paradox.* Plenum Press, New York.

114. Abraham, E.P. and Chain, E.B. (1940) An enzyme from bacteria able to destroy penicillin. *Nature (Lond.)* **146**: 837.

115. Henderson, J.W. (1997) The yellow brick road to penicillin: a story of serendipity. *Mayo Clin. Proc.* **7**: 683–687.

116. Schatz, A., Bugie, E. and Waksman, S.A. (1944) Streptomycin, a substance exhibiting antibiotic activity against gram-positive and gram-negative bacteria. *Proc. Soc. Exptl. Biol. Med.* **55**: 66–69.

117. Hinshaw, H.C. and Feldman, W.H. (1945) Streptomycin in treatment of clinical tuberculosis: a preliminary report. *Proc. Mayo Clin.* **20**: 313–318.

118. Hinshaw, H.C., Feldman, W.H. and Pfuetze, K.H. (1946) Streptomycin in treatment of clinical tuberculosis. *Am. Rev. Tuberc.* **54**: 191–203.

119. Robitzek, E.H., Selikoff, J. and Ornstein, G.G. (1952) Chemotherapy of human tuberculosis with hydrazine derivatives of isonicotinic acid, (preliminary report of representative cases). *Quart. Bull. Sea View Hosp.* **13**: 27–51.

120. Thomas, J.P., Baughn, C.O., Wilkinson, R.G. and Shepherd, R.G. (1961) A new synthetic compound with antituberculous activity in mice: ethambutol (dextro-2,2′-(ethylenediimino)-di-l-butanol). *Am. Rev. Resp. Dis.* **83**: 891–893.

121. Rist, N., Grumbach, F. and Liberman, D. (1959) Experiments on the antituberculous activity of alpha-ethyl-thioisonicotinamide. *Am. Rev. Tuberc.* **79**: 1–5.

122. Sensi, P. (1983) History of the development of rifampin. *Rev. Infect. Dis.* **5**(Suppl. 3): S402–S406.

123. Finland, M. (1974) Twenty-fifth anniversary of the discovery of Aureomycin: the place of the tetracyclines in antimicrobial therapy. *Clin. Pharmacol. Ther.* **1**: 3–8.

124. Anonymous (1981) *Pneumocystis* pneumonia — Los Angeles. *MMWR Morb. Mortal. Wkly. Rep.* **21**: 250–252.

125. Anonymous (1981) Kaposi's sarcoma and Pneumocystis pneumonia among homosexual men — New York City and California. *MMWR Morb. Mortal. Wkly. Rep.* **30**: 305–308.

126. Hymes, K.B., Cheung, T., Greene, J.B., Prose, N.S., Marcus, A., Ballard, H., William, D.C. and Laubenstein, L.J. (1981) Kaposi's sarcoma in homosexual men — a report of eight cases. *Lancet* **2**: 598–600.

127. Gottlieb, M.S., Schroff, R., Schanker, H.M., Weisman, J.D., Fan, P.T., Wolf, R.A. and Saxon, A. (1981) *Pneumocystis carinii* pneumonia and mucosal candidiasis in previously healthy homosexual men: evidence of a new acquired cellular immunodeficiency. *N. Engl. J. Med.* **305**: 1425–1431.

128. Barre-Sinoussi, F., Chermann, J.C., Rey, F., Nugeyre, M.T., Chamaret, S., Gruest, J., Dauguet, C., Axler-Blin, C., Vezinet-Brun, F., Rouzioux, C., Rozenbaum, W. and Montagnier, L. (1983) Isolation of a T-lymphotropic retrovirus from a patient at risk for acquired immune deficiency syndrome (AIDS). *Science* **4599**: 868–871.

129. Gallo, R.C., Sarin, P.S., Gelmann, E.P., Robert-Guroff, M., Richardson, E., Kalyanaraman, V.S., Mann, D., Sidhu, G.D., Stahl, R.E., Zolla-Pazner, S., Leibowitch, J. and Popovic, M. (1983) Isolation of human T-cell leukemia virus in acquired immune deficiency syndrome (AIDS). *Science* **4599**: 865–867.

130. Mitsuya, H. and Broder, S. (1987) Strategies for antiviral therapy in AIDS. *Nature* **6107**: 773–778.

131. Mitsuya, H., Weinhold, K.J., Furman, P.A., St. Clair, M.H., Lehrman, S.N., Gallo, R.C., Bolognesi, D., Barry, D.W. and Broder, S. (1985) 3′-Azido-3′-deoxythymidine (BW A509U): an antiviral agent that inhibits the infectivity and cytopathic effect of human T-lymphotropic virus type III/lymphadenopathy-associated virus *in vitro. Proc. Natl. Acad. Sci. USA* **82**: 7096–7100.

132. Yarchoan, R., Klecker, R.W., Weinhold, K.J., Markham, P.D., Lyerly, H.K., Durack, D.T., Gelmann, E., Lehrman, S.N., Blum, R.M. and Barry, D.W. (1986) Administration of 3′-azido-3′-deoxythymidine, an inhibitor of HTLV-III/LAV replication, to patients with AIDS or AIDS-related complex. *Lancet* **8481**: 575–580.

133. Delta Coordinating Committee. (1996) Delta: a randomised double-blind controlled trial comparing combinations of zidovudine plus didanosine or zalcitabine with zidovudine alone in HIV-infected individuals. *Lancet* **348**: 283–291.

134. De Clercq, E. (1994) New developments in the chemotherapy of lentivirus (human immunodeficiency virus) infections: sensitivity/resistance of HIV-1 to non-nucleoside HIV-1-specific inhibitors. *Ann. NY Acad. Sci.* **724**: 438–456.

135. Gallant, J.E. (2000) Strategies for long-term success in the treatment of HIV infection. *J. Am. Med. Assoc.* **283**: 1329–1334.

136. Richman, D.D. (2001) HIV chemotherapy. *Nature* **6831**: 995–1001.

137. Moore, J.P. (2002) AIDS vaccines: on the trail of two trials. *Nature* **6870**: 365–366.

138. Banting, F.G., Best, C.H. and MacLeod, J.J.R. (1922) The internal secretion of the pancreas. *Am. J. Physiol.* **59**: 479.

139. Banting, F.G., Best, C.H., Collip, J.B., Campbell, W.R. and Fletcher, A.A. (1922) Pancreatic extracts in the treatment of diabetes mellitus. *J. Can. Med. Assoc. J.* **12**: 141–146.

140. Addison, T. (1855) *On the Constitutional and Local Effects of Disease of the Suprarenal Capsules.* S. Highley, London.

141. Kendall, E.C., Mason, H.L., Mackenzie, B.F., Myers, C.S. and Koelsche, G.A. (1934) Isolation in crystalline form of the hormone essential to life from the suprarenal cortex; its chemical nature and physiologic properties. *Proc. Mayo Clin.* **9**: 245–250.

142. Reichstein, T. (1936) Über Bestandteile der Nebennieren-Rinde VI. Trennungsmethoden sowie Isolierung der Substanzen Fa, H, und J. *Helv. Chim. Acta* **19**: 1107–1126.

143. Hench, P.S., Kendall, E.C., Slocumb, C.H. and Polley, H.F. (1949) The effect of a hormone of the adrenal cortex (17-hydroxy-11-dehydrocorticosterone: compound E) and of pituitary adrenocorticotropic hormone on rheumatoid arthritis. *Proc. Mayo Clin.* **24**: 181–197.

144. Medvei, V.C. (1993) *The History of Clinical Endocrinology.* Parthenon, New York.

145. Allen, E. and Doisy, E.A. (1927) The menstrual cycle of the monkey, *Macacus rhesus*: Observations on normal animals, the effects of removal of the ovaries and the effects of injection of ovarian and placental extracts into the spayed animals. *Contr. Embryol. Carneg. Instn* **19**: 1–44.

146. Butenandt, A.F.J. (1931) Ueber die chemische Untersuchung der Sexualhormone. *Z. Angew. Chem.* **44**: 905–908.

147. Browne, J.S.L. (1933) Chemical and physiological properties of crystalline oestrogenic hormones. *Canad. J. Res.* **8**: 180–197.

148. Ruzicka, L., Goldberg, M.W., Meyer, J., Brüngger, H. and Eichenberger, E. (1934) Über die Synthese des Testikelhormons (Androsteron) und Stereoisomerer desselben durch Abbau hydrierter Sterine. *Helv. Chim. Acta* **17**: 1395–1406.

149. Chast, F. (1997) Histoire de la Contraception. *Pour la Science* **233**: 14–17.

150. Pincus, G.G. and Chang, M.C. (1953) The effects of progesterone and related compounds on ovulation early development in the rabbit. *Acta Physiol. Latinoamer.* **3**: 177–183.

151. Djerassi, C. (1992) Steroid research at Syntex: 'the pill' and cortisone. *Steroids* **12**: 631–641.

152. Colton, F.B. (1992) Steroids and 'the pill': early steroid research at Searle. *Steroids* **12**: 624–630.

153. Persidis, A. and Copen, R.M. (1999) Mental disorder drug discovery. *Nat. Biotechnol.* **17**: 307–309.

154. Loewi, O. (1921, 1922, 1924) Ueber humorale Uebertragbarkeit der Herznervenwirkung. *Pflüg. Arch. Ges. Physiol.* **189**: 239–242; **193**: 201–213; **203**: 408–412.

155. Triggle, D.J. (2000) Pharmacological receptors: a century of discovery — and more. *Pharm. Acta Helv.* **74**: 79–84.

156. Cade, J.F.J. (1949) Lithium salts in the treatment of psychotic excitement. *Med. J. Aust.* **36**: 349–352.

157. Schou, M. (1993) *Lithium Treatment of Manic Depressive Illness.* Karger, Basel.

158. Muller, J.M., Schlittler, E. and Bein, H.J. (1952) Reserpin, der sedative Wirkstoff aus *Rauwolfia serpentina* Benth. *Experientia (Basel)* **8**: 338.

159. Youdim, M.B. (1975) Monoamine oxidase. Its inhibition. *Mod. Probl. Pharmacopsyc.* **10**: 65–88.

160. Kline, N.S. (1958) Clinical experience with iproniazid (Marsilid). *J. Clin. Exp. Psychopathol.* **19** (Suppl.): 72–78.

161. Healy, D. (1998) Pioneers in psychopharmacology. *Int. J. Neuropsychopharmcol.* **1**: 191–194.

162. Kuhn, R. (1958) The treatment of depressive states with G22355 (Imipramine hydrochloride). *Am. J. Psych.* **115**: 459–464.

163. Iversen, L.L. (1999) The discovery of monoamine transporters and their role in CNS drug discovery. *Brain Res. Bull.* **50**: 379.

164. Shorter, E. (1997) *A History of Psychiatry — From the Era of the Asylum to the Age of Prozac.* John Wiley FRSC, University of Toronto.

165. Laborit, H., Huguenard, P. and Alluaume, R. (1952) Un nouveau stabilisateur végétatif, le 4560 RP. *Presse Méd.* **60**: 206–208.

166. Courvoisier, S., Fournel, J., Ducrot, R., Kolsky, M. and Koetschet, P. (1953) Propriétés pharmacodynamiques du chlorhydrate de chloro 3-(diméthylamino-3′propyl) 10-phénothiazine (4560 RP). *Arch. Int. Pharmacodyn.* **92**: 305–361.

167. Delay, J. and Deniker, P. (1952) Trente-huit cas de psychoses traitées par la cure prolongée et continue de 4560 RP. *Le Congrès des Aliénistes et Neurologues de Langue Française.* In *Compte Rendu du Congrès.* Masson et Cie, Paris.

168. Berger, F.M. (1954) The pharmacological properties of 2-methyl 2 *m*-propyl-1,3-propane-diol dicarbamate (Miltown), a new interneuronal blocking agent. *J. Pharmacol.* **112**: 413–423.

169. Helderman, J.H. and Gailiunas, P., Jr. (1983) Transplantation. *Am. J. Kidney Dis.* **3**: 194–198.

170. Barnard, C.N. (1967) The operation. A human cardiac transplant: an interim report of a successful operation performed at Groote Schuur Hospital, Cape Town. *S. Afr. Med. J.* **48**: 1271–1274.

171. Stahelin, H.F. (1996) The history of cyclosporin A (Sandimmune) revisited: another point of view. *Experientia* **52**: 5–13.

172. Heusler, K. and Pletscher, A. (2001) The controversial early history of cyclosporin. *Swiss Med. Wkly.* **131**: 299–302.

173. Borel, J.F. and Kis, Z.L. (1991) The discovery and development of cyclosporine (Sandimmune). *Transplant. Proc.* **23**: 1867–1874.

174. Gilman, A. and Philips, F.S. (1946) The biological actions and therapeutic applications of the chlorethylamines and sulfides. *Science* **103**: 409–415.

175. Jukes, T.H. (1978) The history of methotrexate. *Cutis* **3**: 396–398.

176. Calne, R.Y., Alexander, G. P. J. and Murray, J.E. (1962) A study of drugs in prolonging survival of homologous renal transplant in dogs. *Ann. NY Acad. Sci.* **99**: 743–761.

177. Taylor, D.O., Ensley, R.D., Olsen, S.L., Dunn, D. and Renlund, D.G. (1994) Mycophenolate mofetil (RS-61443): preclinical, clinical, and three-year experience in heart transplantation. *J. Heart Lung Transplant.* **13**: 571–582.

178. Borel, J.F., Feurer, C., Gubler, H.U. and Stahelin, H. (1976) Biological effects of cyclosporin A: a new antilymphocytic agent. *Agents Actions* **6**: 478–475.

179. Allison, A.C. (2000) Immunosuppressive drugs: the first 50 years and a glance forward. *Immunopharmacology* **47**: 63–83.

180. Abelson, P.H. (1990) Medicine from plants. *Science* **4942**: 513.

181. Tunon, H., Olavsdotter, C. and Bohlin, L. (1995) Evaluation of anti-inflammatory activity of some Swedish medicinal plants. Inhibition of prostaglandin biosynthesis and PAF-induced exocytosis. *J. Ethnopharmacol.* **2**: 61–76.

182. Dewick, P.M. (1996) Tumor inhibitors from plants. In Evans, W.C. (ed.). *Trease and Evans' Pharmacognosy*, 14th edn. Saunders, London.

183. Stahelin, H.F. and Von Wartburg, A. (1991) The chemical and biological route from podophyllotoxin glucoside to etoposide: Ninth Cain Memorial Award Lecture. *Cancer Res.* **1**: 5–15.

184. Kouadio, K., Chenieux, J.C., Rideau, M. and Viel, C. (1984) Antitumor alkaloids in callus cultures of *Ochrosia elliptica. J. Nat. Prod.* **47**: 872–874.

185. Wall, M.E. and Wani, M.C. (1996) Camptothecin. Discovery to clinic. *Ann. NY Acad. Sci.* **803**: 1–12.

186. Langreth, R. (1991) Whom do yew trust? *Science* **5014**: 1780.

187. Wani, M.C., Taylor, H.L., Wall, M.E., Coggon, P. and MacPhail, A.T. (1971) Plant antitumor agents. VI. The isolation and structure of Taxol, a novel antileukemic and antitumor agent from *Taxus brevifolia. J. Am. Chem. Soc.* **93**: 2325–2327.

188. Schiff, P.B., Fant, J. and Horwitz, S.B. (1979) Promotion of microtubule assembly *in vitro* by Taxol. *Nature* **5698**: 665–667.

189. Defuria, M.D. and Horovitz, Z. (1993) Taxol commercial supply strategy. *J. Natl Cancer Inst. Monogr.* **15**: 195–198.

190. Gueritte-Voegelein, F., Guenard, D., Lavelle, F., Le Goff, M.T., Mangatal, L. and Potier, P. (1991) Relationships between the structure of Taxol analogues and their antimitotic activity. *J. Med. Chem.* **34**: 992–998.

191. Guenard, D., Gueritte-Voegelein, F., Dubois, J. and Potier, P. (1993) Structure–activity relationships of Taxol and Taxotere analogues. *J. Natl. Cancer Inst. Monogr.* **15**: 79–82.

192. Von Hoff, D.D. (1997) The taxoids: same roots, different drugs. *Semin. Oncol.* **24** (Suppl. 13): S13-3–S13-10.

193. Goodman, L.S., Wintrobe, M.M., Dameshek, W., Goodman, M.J., Gilman, A. and McLennan, M.T. (1946) Nitrogen mustard therapy. Use of methyl-*bis*(beta-chlorethylamine hydrochloride and *tris*(beta-chlorcethyl)amine hydrochloride for Hodgkin's disease, lymphosarcoma, leukemia and certain allied miscellaneous disorders. *J. Am. Med. Assoc.* **132**: 126–132.

194. Devita, V.T., Serpick, A.A. and Carbone, P.P. (1970) Combination chemotherapy in the treatment of advanced Hodgkin's disease. *Ann. Intern. Med.* **73**: 881–895.

195. Elion, G.B. and Hitchings, G.H. (1965) Metabolic basis for the actions of analogs of purines and pyrimidines. *Adv. Chemother.* **2**: 91–177.

196. Elion, G.B. (1969) Actions of purine analogs: enzyme specificity studies as a basis for interpretation and design. *Cancer Res.* **12**: 2448–2453.

197. Bowden, M.E. (2001) *Pharmaceutical Achievers.* Chemical Heritage Foundation, Philadelphia.

198. Di Marco, A., Cassinelli, G. and Arcamone, F. (1981) The discovery of daunorubicin. *Cancer Treat. Rep.* **65** (Suppl. 4): 3–8.

199. Weiss, R.B., Sarosy, G., Clagett-Carr, K., Russo, M. and Leyland-Jones, B. (1986) Anthracycline analogs: the past, present, and future. *Cancer Chemother. Pharmacol.* **18**: 185.

200. Weiss, R.B. (1992) The anthracyclines: will we ever find a better doxorubicin? *Semin. Oncol.* **19**: 670–686.

201. Rosenberg, B., Van Camp, L. and Krigas, T. (1965) Inhibition of cell division in *Escherichia coli* by electrolysis products from a platinum electrode. *Nature* **205**: 698–699.

202. Bellon, S.F., Coleman, J.H. and Lippard, S.J. (1991) DNA unwinding produced by site-specific intrastrand cross-links of the antitumor drug *cis*-diammine-dichloroplatinum(II). *Biochemistry* **30**: 8026–8035.

203. Bruckner, H.W., Ratner, L.H., Cohen, C.J., Wallach, R., Kabakow, B., Greenspan, E.M. and Holland, J.F. (1978) Combination chemotherapy for ovarian carcinoma with cyclophosphamide, adriamycin, and cis-dichlorodiammineplatinum (II) after failure of initial chemotherapy. *Cancer Treat. Rep.* **62**: 1021–1023.

204. Knox, R., Friedlos, F., Lydall, D. and Roberts, J. (1986) Mechanism of cytotoxicity of anticancer platinum drugs: evidence that *cis*-diammine-dichloroplatinum (II) and *cis*-diammine-(1,1-cyclobutane-dicarboxylato)platinum (II) differ only in the kinetics of their interaction with DNA. *Cancer Res.* **46**: 1972–1979.

205. Raju, T.N. (1999) The Nobel chronicles, 1958: George Wells Beadle, Edward Lawrie Tatum and Joshua Lederberg. *Lancet* **353**: 2082.

206. Watson, J.D. and Crick, F. (1953) Molecular structure of nucleic acids a structure for deoxyribose nucleic acid. *Nature* **171**: 724–738.

2

MEDICINAL CHEMISTRY: DEFINITION AND OBJECTIVES, THE THREE MAIN PHASES OF DRUG ACTIVITY, DRUG AND DISEASE CLASSIFICATIONS

Camille G. Wermuth

Medicinal chemistry remains a challenging science, which provides profound satisfaction to its practitioners. It intrigues those of us who like to solve problems posed by nature. It verges increasingly on biochemistry and on all the physical, genetic and chemical riddles in animal physiology which bear on medicine. Medicinal chemists have a chance to participate in the fundamentals of prevention, therapy and understanding of diseases and thereby to contribute to a healthier and happier life.

A. Burger[1]

I DEFINITION AND OBJECTIVES

A Medicinal chemistry

Taken in the *prospective sense* the objective of medicinal chemistry is the design and the production of compounds that can be used in medicine for the prevention, treatment and cure of human or animal diseases. Thus medicinal chemistry is a part of pharmacology, this latter being taken in its etymological sense ('pharmakon' + 'logos': study of drugs; Fig. 2.1).

Taken in the *retrospective sense*, medicinal chemistry also includes the study of already existing drugs, of their pharmacological properties, and their structure–activity relationships. An official definition of medicinal chemistry was given by an IUPAC specialized commission.[2]

Medicinal chemistry concerns the discovery, the development, the identification and the interpretation of the mode of action of biologically active compounds at the molecular level. Emphasis is put on drugs, but the interest of the medicinal chemistry is also concerned with the study, identification, and synthesis of the metabolic products of drugs and related compounds.

The main activities of medicinal chemists appear clearly from the analysis of their most important scientific journals (*Journal of Medicinal Chemistry, European Journal of Medicinal Chemistry, Bioorg MedChem, Il Farmaco, Archiv der Pharmazie, Arzneimittelforschung, Chemical and Pharmaceutical Bulletin*, etc.). Thus, medicinal chemistry covers three critical steps:[3]

- A **discovery step**, consisting of the *choice of the therapeutic target* (receptor, enzyme, transport group, cellular or *in vivo* model) and the *identification* (or *discovery*) *and production* of new active substances interacting with the selected target. Such compounds are usually called lead compounds, they can originate from

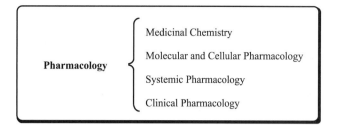

Fig. 2.1 The domains of pharmacology.

synthetic organic chemistry, from natural sources, or from biotechnological processes.

- An **optimization step**, that deals with the improvement of the lead structure. The optimization process takes primarily into account the increase in potency, selectivity and toxicity. Its characteristics are the establishment and analysis of *structure–activity relationships*, in an ideal context to enable the understanding of the molecular mode of action. However, an assessment of the pharmacokinetic parameters such as absorption, distribution, metabolism, excretion and oral bioavailability is almost systematically practised at an early stage of the development in order to eliminate unsatisfactory candidates.

- A **development step**, whose purpose is the continuation of the improvement of the pharmacokinetic properties and the fine tuning of the pharmaceutic properties (*chemical formulation*) of the active substances in order to render them suitable for clinical use. This chemical formulation process can consist in the preparation of better absorbed compounds, of sustained release formulations, of water-soluble derivatives or in the elimination of properties related to the patient's compliance (causticity, irritation, painful injections, undesirable organoleptic properties).

Medicinal chemistry is an interdisciplinary science covering a particularly wide domain situated at the interface of organic chemistry with life sciences, such as biochemistry, pharmacology, molecular biology, immunology, pharmacokinetics and toxicology on one side, and chemistry-based disciplines such as physical chemistry, crystallography, spectroscopy and computer-based information technologies on the other.

The knowledge of the molecular targets (enzymes, receptors, nucleic acids), has benefitted from the progress made in molecular biology, genetic engineering and structural biology. For an increasing number of targets the three-dimensional structure and the precise location of the active site are known. The design of new active substances is therefore more and more based on results obtained from ligand–receptor modelling studies. One can actually consider the existence of a molecular pharmaco*chemistry* making a pair with molecular pharmaco*logy*.

If the main objective is the discovery of new drug candidates, medicinal chemistry is also concerned with the fate of drugs in living organisms ('ADME' studies: absorption, distribution, metabolism, excretion), and with the study of bioactive compounds not related to medicine (agrochemicals, food additives, etc.).

A certain number of terms more or less synonymous with medicinal chemistry are used: pharmacochemistry, molecular pharmacochemistry, drug design, selective toxicity. The French equivalent to medicinal chemistry is 'Chimie Thérapeutique' and the German one is 'Arzneimittelforschung'.

B Molecular and cellular pharmacology

This is the study of the pharmacological action of drug at the molecular or at the cellular level. The first objective is to identify the cellular levels of action. Three levels, important for drug activity can be distinguished: (1) the plastic membrane which is very rich in potential targets, notably in receptors; (2) the cytosol with its enzymatic equipment and the organelle membranes with their particular ion transporters; and (3) the nucleus which notably responds to the steroid hormones, to anticancer drugs and to gene therapy. The second objective is to elucidate the precise biochemical and biophysical sequence of events that result from the drug–target interaction. All these studies are performed *in vitro*, therefore they yield generally rather reliable quantitative data. They are also free from pharmacokinetic factors such as the peregrination and the metabolism of the drug between the site of administration and the site of action. Finally they save animals and are thus better accepted by the animal protection leagues.

C Systemic pharmacology

The systemic pharmacology considers the effects of biologically active substances in integrated systems (cardiovascular, skeletal, central nervous, gastrointestinal, pulmonary, etc.). The experimentation is performed in intact animals or in isolated organs (isolated heart, isolated arteria, perfused kidney, etc.). The main difficulty resides in the design of animal experimental models that are predictive of an activity in a human disease. As many pharmacological experiments are still performed on healthy animals or on disease-simulating paradigms, their extrapolation to clinical situations is questionable. The availability of transgenic mice, in which the genes of a human disease were introduced, represent an interesting progress. However in all animal models, intra- and interspecific physiological variations account for rather imprecise results the margins being often as elevated as $\pm 50\%$.

D Clinical pharmacology

Clinical pharmacology deals with the examination in humans of the effects of a new drug candidate. The tests are performed under the responsibility of the clinical pharmacologist who is generally a medical doctor and who has to report to an ethical committee. Phase I tests take place in healthy volunteers. They aim to assess the level of dosing and the tolerance ('dose ranging') and to initiate metabolic studies in humans. Once the safety margin has been determined, phase II, III and IV studies examine successively the beneficial effects in patients, the possible side-effects, the comparison of the drug with reference drugs and the emergence of new therapeutic indications.

II THE THREE MAIN PHASES OF DRUG ACTIVITY

The activity of a given drug depends on a sequence of physico-chemical events that begin when the active molecule penetrates into the living organism and which culminates when the active molecule reaches its target and elicits the appropriate biological response. Classically it is admitted that three characteristic phases govern the biological activity of a drug in a living organism[4,5] the *pharmaceutical phase*, the *pharmacokinetic phase*, and the *pharmacodynamic phase* (Table 2.1).

A The pharmaceutical phase

The *pharmaceutical phase*, sometimes also called *biopharmaceutical phase*, deals with the choice of the appropriate *route of administration* and with the choice of the *pharmaceutical formulation* most suited to the desired medical treatment.

Routes of administration

Possible routes of administration are divided into two major classes: *enteral*, whereby drugs are absorbed from the alimentary canal and *parenteral*, in which drugs enter the bloodstream directly (intravenous injection) or by some other non-enteral absorptive route (intramuscular or subcutaneous injections, transdermal delivery systems, nasal sprays, etc.). Below we describe in brief the intravenous and the oral route; other routes of administration are considered in Chapter 30.

Intravenous injections. The *intravenous injection* is the route of administration leading to the fastest effects. The drug preparation is directly injected into the bloodstream and from there the active principle is carried along to its site of action. The intravenous route shunts the natural barriers of the body to absorption, and therapeutic blood levels are reached almost instantaneously. The drug solution must be completely clear with no particulate matter present. On the other hand, injection solutions should be sterile to avoid any infections, also isotonic and at pH values close to that of the plasma (pH = 7.4) to avoid local pain and tissue necrosis.[6-8] Once arrived at the target, the drug can trigger its receptor mechanism and induce the awaited biological response. Actually the situation is not so simple and many additional and sometimes unwanted events can occur (Fig. 2.2):

(1) In the bloodstream the drug can bind to the plasmatic proteins or to the blood cells or the platelets and never reach the target organs with a sufficient concentration.
(2) Due to its ionized character or to its inadequate partition coefficient, the drug may be unable to cross the lipidic biomembranes.
(3) Instead of being carried to the biophase, the drug can be concentrated in the fat storage compartment.
(4) The drug can also be rapidly altered by metabolic processes. Drug metabolism usually yields more water-soluble, less active and much less toxic derivatives of the parent drug. However, sometimes metabolic processes can generate active or toxic molecules.

Table 2.1 *The three phases that govern the activity of a drug*

Phase	Concerned events	Objectives
Pharmaceutical phase	Selection of the administration route	Optimize the distribution
	Preparation of the most appropriate pharmaceutical formulation	Facilitate the absorption
		Eliminate unwanted organoleptic properties
Pharmacokinetic phase	Fate of the drug in the organism: absorption, distribution, metabolism, excretion ('ADME')	Control the bioavailability, i.e. the ratio of the administered dose over the concentration at the site of action, in function of the time
Pharmacodynamic phase	Quality of the drug–receptor interaction	Maximal activity
	Nature and intensity of the biological response	Maximal selectivity
		Minimal toxicity

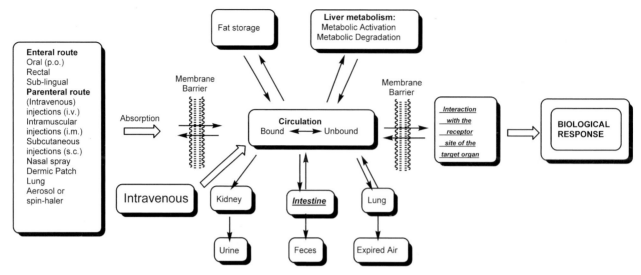

Fig. 2.2 Schematic diagram of *in vivo* events and compartments (after Kier[9] and Ariëns[10]).

(5) Due to exchanges with the intestine, the kidneys and the lungs, the parent drug or its metabolites can be too quickly removed from the organism.

(6) On a practical side, only water-soluble drugs can easily be administrated by the intravenous route and the injections have to be done very slowly to avoid excessively high concentrations (as much as 400 times the final blood level) of the drug in the heart tissue.

Oral route. The most common route of administration is the *oral route.* In this case additional peregrinations of the drug are involved to ensure its passage from the gastrointestinal tract to the bloodstream. The drug preparation is swallowed and the active principle is absorbed through the mucous membrane of the small intestine or, to a limited extent, from the stomach. As absorption is maximal for unionized drugs, acidic drugs are rather well absorbed through the stomachal epithelium (pH of the stomachal juice = 2–3) and weakly basic drugs through the small intestine epithelium (pH varying progressively from 5 to 7). Once absorbed, drugs do not directly reach the sites of action, but are carried to the portal vein and from there into the liver where they are subjected to chemical attacks (oxidations, reductions, hydrolyses, coupling with solubilizing moieties) before being released into the bloodstream. These metabolic attacks, taking place before the drug reach the general circulation, are called 'first pass effects'. They are avoided when using intravenous injections and thus represent the major difference between the intravenous and the oral route.

For oral administration, the active compound is usually integrated in tablets, soft or hard gelatine capsules or coated tablets. As a rule, a tablet is a compressed preparation that contains approximately 5–10% of the active principle, 80% of various excipients (fillers, disintegrants, lubricants, glidants and binders) and about 10% of compounds ensuring an easy disintegration, deaggregation and dissolution of the tablet in the stomach or the intestine. Tablets are relatively simple to manufacture and to use. They represent the most current presentation.

Thanks to appropriate *pharmaceutical formulation techniques,* the disintegration time can be modified so that a fast effect or a sustained release is achieved. Special coatings can also render the tablet gastro-resistant, the disintegration taking place only in the duodenum, under the combined action of the intestinal enzymes and of the pH change. In the family of tablets one finds equally the dragées. These are tablets covered with a colored sugar layer and a fine layer of varnish or wax. Recently films have replaced the sugar and the dyestuffs.

Capsules are constituted of a gelatinous envelope that contains the active substance in the form of powder or granules. The most used form are capsules in hard or tender gelatine, capsules to chew and capsules for rectal use.

Other pharmaceutical preparations (see also Chapter 30). Suppositories are composed from an excipient that melts at body temperature. It can be a natural fat (cocoa butter) or a polyethylene-glycol (Carbowax). They are exclusively destined to be introduced in the anus. They allow a rapid action because the rectum is richly irrigated, moreover, they avoid loading the digestive system.

Ovules are destined to be introduced in the vagina, so as to exert a local action. They are usually constituted of a dissolution of the active principle in a soft gelatine. *Ointments* are coatings that one spreads on the skin or on mucus. They are generally used for the treatment of cutaneous or subcutaneous lesions. *Aerosols* are sprays for local applications. Their therapeutic advantages are as follows: possibility to process large surfaces, good resorption, simple and easy utilization. Aerosols are of particular importance for inhalation therapies of the respiratory system. *Liquid medicines*, sterile for most, are composed of active substances in solution. Besides the intravenous injection, other liquid medicines are perfusions destined for parenteral nutrition after a surgical intervention or a traumatic coma, or solutions for stomach washings after an intoxication. Finally drinkable ampoules also belong to the group of liquid medicines.

The bioequivalence problem. A given formulation procedure of an active principle ensures the corresponding bioavailability in patients (bioavailability = the fractional part of a drug that reaches the general circulation in a given time span). A slight modification in the galenic procedure (change of one excipient, changes in the granulation process before tabletting, changes in the drying process, modification of the aging or storage conditions) can sometimes dramatically influence the drug release of the final product. The same is true for changes in the final purification process of the active principle. Thus re-crystallization from another solvent system or under other temperature and/or concentration conditions can produce mesomorphic crystalline forms, presenting other solubilities and as a consequence a change in bioavailability (see Chapter 39). As quoted by Kellaway:[11] 'When a patient is successfully treated or stabilized on a branded product, it is therefore undesirable to change to a chemically equivalent product from an alternative manufacturer unless bioequivalence has been proven. Economic pressures advocating change of product should be resisted, at least until bioequivalence data are presented'.

When the pharmaceutical formulation of an active compound is ineffective, slight chemical modifications or formation of bioreversible derivatives (esters, amides, peptides) can improve its physico-chemical properties (lipophilicity, pK, polarity) and optimize the dissolution rates in the biological fluids and the passage through the very first biological membranes (cutaneous, intestinal, etc.). The global result is better penetration into the organism. Compared with the pharmaceutical formulation mentioned above, this process can be considered as a chemical formulation and will be considered in Chapter 39.

B The pharmacokinetic phase

The *pharmacokinetic phase* controls the different parameters that govern the random walk of the drug between its application point and its final site of action and which ensure the destruction and/or the elimination once the effect is produced. The site of action is often separated in space and

time from the administration or penetration place. In a chronological order the events of the pharmacokinetic phase are as follows: *Absorption:* The absorption processes through the different biological membranes and compartments, they are highly dependent on the physico-chemical properties of the drug (ionized or unionized state, partition coefficient, size) and can proceed simply through passive diffusion or to more sophisticated physiological transport mechanisms (Chapter 30). *Distribution:* The distribution of the drug substance into the various compartments is ensured by the blood, and to a minor extent by the lymphatic circulation. Blood plasma, contains the suspended blood cells and platelets and is essentially a solution of 70 g per litre of proteins (albumin and globulins), of 9 g‰ of mineral salts (essentially sodium chloride) and of ∼ 1 g‰ of glucose (exact composition see Table 2.2, from Rettig[12]). The proteins, especially albumin, are able to bind to various drugs and thus to temporarily subtract them from their pharmacological destination. Albumin has a molecular weight of 69 000 and is mainly negatively charged at the physiological pH of the blood (pH = 7.4). At pH = 5, its isoelectric point, it has 100 negative and 100 positive charges, which explains its important role also as physiological buffer molecule. *Metabolism:* The apparent finality of metabolism is to chemically transform drugs or any other substances that are foreign to the organism (xenobiotics) into water-soluble derivatives to facilitate their urinary elimination. This change normally produces a diminution, or even suppression, of the pharmacological activity and of the eventual toxicity. However it can happen that the metabolism activates the parent molecule (see Chapters 31 and 32) or even generates highly reactive intermediates (mostly electrophiles) that induce toxicity mechanisms (see Chapter 32). If metabolic activation is implied, drugs inactive *in vitro* may be found to be active *in vivo*.

Table 2.2 *Constitution of blood*[12]

Constituent	%	% of the total
Blood cells (hematocrit)		45
Erythrocytes	44.6	
Leukocytes	0.15	
Thrombocytes	0.25	
Plasma		55
Fibrinogen	0.3	
Serum	54.7	
Proteins	3.5	
Electrolytes	∼0.4	
Carbohydrates	∼0.05	
Hormones	Trace	
Enzymes	Trace	
Vitamins	Trace	
Antibodies, gases, dyestuffs	Trace	

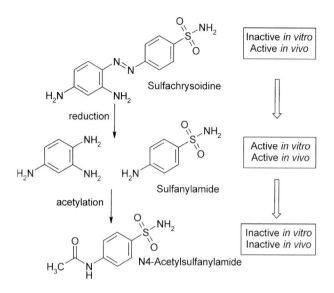

| | Inactive *in vitro* Active *in vivo* |

Sulfachrysoidine

reduction

Sulfanylamide

| | Active *in vitro* Active *in vivo* |

acetylation

N4-Acetylsulfanylamide

| | Inactive *in vitro* Inactive *in vivo* |

Fig. 2.3 Metabolic activation and inactivation.

Table 2.3 *Examples of drugs possessing very different elimination half-times (taken from references 13–15)*

Compound	$t_{\frac{1}{2}}$ (hours)
Pyridostigmine	0.1
Fentanyl	0.2
Morphine	2–3
Clonidine	8
Adamantane	10
Griseofulvin	20
Phenobarbital	35–96
Chloroquine	48
Reserpine	150
Sulfamethoxine	200
Bromide ion	280–340

Sulfamidochrysoidine ('Prontosil rubrum'), converted *in vivo* to the anti-infectious agent sulphanilamide, is the historical example. In its turn, the active metabolite sulphonamide is inactivated through acetylation (Fig. 2.3). Metabolic reactions take place in majority in the liver, but other organs such as the kidneys, the lungs or the brain can also ensure drug transformations. *Elimination:* Once the awaited pharmacological effect is produced, drugs and metabolites have to be eliminated from the organism with adequate kinetics. A too slow elimination process produces a progressive accumulation of the drug and appearance of toxic effects. Conversely too fast elimination leads to repeated daily administrations and low patient acceptance. The main elimination routes are renal (urine) and rectal (faeces). They can occasionally be pulmonary (expired air), oral (salivary) and cutaneous (sweat). The elimination kinetics are very seldom of order zero (Fig. 2.4a). One of the best known examples is the linear elimination of ethanol (which allows the calculation of the ethanol blood level at the moment of a car accident even when the blood sample was taken some time after the accident).

The usual elimination kinetic of a drug from the circulating blood is of first order (Fig. 2.4b). In this case

the time at which the drug is completely cleared from the blood is relatively difficult to determine, since the curve does not intersect with the x axis, but only approaches it in an asymptotic way. Much easier is the determination of the biological half-time ($Et_{\frac{1}{2}}$), i.e. the time after which the blood level has reached half of the original level. Table 2.3 gives some examples of elimination half-times.

The four above-mentioned pharmacokinetic processes ('ADME': absorption, distribution, metabolism, excretion) account for the bioavailability of a given drug. As the bioavailability expresses the percentage of a given drug that reaches the general circulation in a given time span, an intravenous administration represents therefore, by definition, a 100% bioavailability. After an oral dosage, a 100% bioavailability would imply a complete absorption and no first pass destruction; such a situation is rather improbable.

C The pharmacodynamic phase

The *pharmacodynamic phase* is the phase of the greatest interest to the medicinal chemist as it deals directly with the nature and the quality of the interaction of the drug with its biological target. Starting often with a relatively weak and non-selective compound, the challenge is to maximize the potency and to minimize the deleterious or undesired effects of the molecule. The biological response obtained is maximal when the active principle shows the precise stereoelectronic complementarity with the target structure. Ideally the medicinal chemist, from the knowledge of the characteristics of the target tissue (enzyme, receptor, transport protein, nucleic acid), tries to design drugs with the optimal size, shape, hydrophilic–lipophilic ratio, disposition of functional groups. The sharper is the obtained fit between the receptor site and the molecule, the more selective will be the drug in eliciting only the expected biological response.

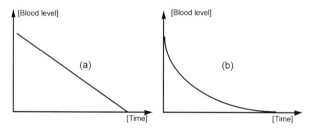

Fig. 2.4 Elimination kinetics of order zero (a) and of order 1 (b).

Dose–effect relationships

During the search of relationships between the concentration of an active compound and its effects it is essential to accede to quantitative data describing the potency of the compound and allowing comparisons with other substances. In classical pharmacological studies, the activity of the compound is often examined towards isolated organs maintained at an adequate temperature in nutritive and oxygenated solutions. In cardiovascular pharmacology for example, the changes in the diameter of an isolated blood vessel (portal or mesenteric vein, coronary artery) represent a means of measurement of vasodilating or vasoconstrictive properties. Experimental research conducted on isolated organs presents many *advantages,* such as the precise knowledge of the concentration of the active substance in the nutritive solution, easy observation conditions, elimination of the compensation reactions found in whole animals through feed-back reactions, and the possibility to study a given compound until its maximum effect. The *disadvantages* are unavoidable lesions caused to the tissue during the preparation, loss of the physiological control of function of the isolated organ, and the artificial character of the environment.

Dose–response curves

In increasing the concentration stepwise, the increase of the effect is constant at the beginning and tends progressively to zero when approaching the maximal active concentration (parabolic branch). However the concentration triggering the maximal effect is difficult to measure precisely, whereas the concentration producing half of the maximal effect can be determined exactly. This value (EC_{50}; EC = *effective concentration*) corresponds to the inflexion point of the sigmoid curve obtained in plotting the effect/concentration data in semi-logarithmic coordinates (see Chapter 3). To fully characterize a dose–response curve, the E_{max} (maximal possible effect) and the slope of the curve (concentration range used to establish the curve) have also to be indicated.

Binding curves

In molecular pharmacology, biochemical tests performed on tissue homogenates or partially purified preparations, but also on isolated enzymes or on cloned receptors, the binding curves are based on the action mass law. This law describes the hyperbolic function (B) that links the binding to the ligand concentration (c). This curve is characterized by the affinity $1/K_D$ and the maximal binding B_{max} (which corresponds to the total number of binding sites contained in a given weight unit of the preparation):

$$B = B_{max} \frac{c}{c + K_D}$$

In this equation, K_D is the equilibrium dissociation constant and corresponds to the ligand concentration for which 50% of the binding sites are occupied by the ligand. As discussed in Chapter 3, the biological response can be an activation mechanism (post-synaptic agonists, presynaptic antagonists) or an inhibition (post-synaptic antagonists, enzyme inhibitors). Therefore compounds having the same affinity for a receptor site do not necessarily produce the same response, one possessing an agonist and the other one an antagonist profile. Moreover, two agonists with similar affinities can induce quantitatively different biological responses depending on their intrinsic activity (also called efficacy).[16]

Criteria for the expression of pharmacological results

Quantitative dose–effect relationships alone do not provide enough information. Ideally a well-conceived pharmacological study should comply with the following criteria.[17]

(1) There should be a relationship between the utilized dose and the observed effect. Increasing the concentration of the compound must be accompanied by an increase in activity.
(2) The results should always be presented with the confidence limits (example: $ED_{50} = 5.8 \pm 1.7$ mg kg^{-1}). This avoids considering as different two values that in fact have overlapping intervals.
(3) The results should always be published in comparison with one or more reference substances serving as internal standards.
(4) For tests performed in whole animals, an activity kinetic study is highly recommended. It will allow determination of the time of the peak action and to compare the different molecules at the optimal time.

D   Medicinal chemistry approaches as a function of the therapeutic classes

The approaches utilized by medicinal chemists do not follow strict rules and in daily practice they vary with the therapeutic classes that are being targetted:

Agents acting on the central nervous system: psychotropic and neurological drugs

Regulations at the level of the central nervous systems are ensured by neurotransmitters. Such substances can be simple molecules (acetylcholine, γ-aminobutyric acid, norepinephrine, etc.) or neuropeptides (substance P, vasopressin, corticotrophin releasing factor, neurotensin, etc.). Pharmacological interventions aim then to correct dysregulations, either in reducing an excessive neuronal activity or in stimulating a weak working system. The easiest ways for medicinal chemists to intervene are at the enzymatic level (inhibition of biosynthesis or biodegradative enzymes), at the receptor level (activation or blockade by a synthetic analogue of the endogenous neurotransmitter) and at the transport systems (blockade of the neurotransmitter re-uptake by a structural analogue). As these systems are well characterized, most of the initial testing can be performed *in vitro* on a practically unlimited number of compounds, thus rendering the search and the discovery of new CNS agents rather easy. The real difficulties in the development of CNS drugs reside in the pharmacokinetics. A satisfying CNS drug must be able to enter into the central nervous system, in other words to cross the so-called blood–brain barrier (BBB). On the other hand, the molecule has to resist to premature metabolic degradation and it should not be mainly distributed to peripheral action sites.

Pharmacodynamic agents

Despite the increasing possibility of biochemical approaches, the design of pharmacodynamic agents in the cardiovascular, anti-asthmatic and anti-allergic domains is still heavily dependent on animal models or animal-derived models, such as isolated organs, perfused hearts or kidneys. Mass screening can therefore be somewhat limited. Conversely, access to the target organs is facilitated by their peripheral location. In some rare instances pharmacodynamic agents can be used in a preventive way (e.g. aspirin to diminish the risks of heart infarcts).

Chemotherapy

Chemotherapeutic treatments rely dominantly on *selective toxicity* approaches targetted at killing the invader without killing the host. Albert distinguishes three selectivity principles:[18] favourable differences in drug distribution, favourable differences in biochemistry and favourable differences in cell structure. Certainly the most seducing of them are the differences in biochemistry. A practical application is the inhibition of the bacterial cell wall construction by β-lactam antibiotics, the activity is absolutely restricted to the bacteria's, just because similar cell walls do not exist in mammals.

Agents acting on metabolic or immunologic diseases and on endocrine functions

In this group we find deficiency diseases (vitamin deficiencies, diabetes, Addison's disease, etc.). The logical treatment is to compensate these deficiencies by an outside supply of the missing molecule (vitamin or hormone). In hormonology often modified versions of the original endogenous hormone were developed. Examples are modified vitamins (menadione, benfotiamine), steroids (fluocinolone acetonide, flumethasone) and modified peptides (modified TRH, angiotensine, somatostatine, etc.).

Immunological diseases have benefited recently from progress in the knowledge of protein structure and protein chemistry. Immunostimulants and immunosupressing drugs are now available and immunological approaches to rheumatoid arthritis and to some auto-immune diseases are underway. Vaccines and serums represent other aspects of immunological therapies, vaccination acting in a preventive manner.

III DRUG CLASSIFICATIONS

All attempts to establish clear-cut and well-balanced drug classifications lead to failures, due to the complex nature of the medicaments, which do not fit into simplified systems. The best way to illustrate how drugs are situated with regard to each other, is to use several classification systems, based on different criteria.

A Classification systems

The most common classifications take into account:[5,19–21]

The origin of the drug

Drugs from *natural origin* can come from three sources. They can originate from minerals. Various simple inorganic substances are still in use in medicine: sulphur, iodine, phosphates and arseniates, calcium, sodium, magnesium, iron and bismuth salts, etc. From the animal kingdom some hormones (e.g. insulin) and fish liver oils (vitamins A and E) were extracted for a long time. Biliary salts yield precursors for steroid hemisynthesis (corticoids, sexual hormones). However, the majority of natural compounds are produced by vegetals (alkaloids, cardiac glycosides, antibiotics, anti-cancer drugs).

Drugs from *synthetic origin* relay the natural compounds in providing improved and or simplified synthetic analogues, the production of which are not dependent on hazardous agricultural supplies.

In an intermediary position between natural and synthetic compounds are found the various *fermentation products* (vitamins, amino acids) and the products issued from *genetic engineering* (e.g. recombinant insulin).

The mode of action

One can distinguish between medicaments which really treat the cause of the disease, medicaments which compensate for the deficiency of a given substance and medicaments which only aim to alleviate the symptoms of the disease.

The *drugs acting directly on the causal agent* of the disease are called *etiological* drugs and represent 'true' medicaments. Presently most of them belong to the class of chimiotherapeutic drugs, i.e. to compounds used to treat infectious diseases (antibacterials and antivirals) and parasitic diseases. The principle of their activity resides in their *selective toxicity*: destroy the invader without distroying the host.

Logically can be added to this group a certain number of drugs used by healthy persons in a preventive way in order to protect them from a future illness (vaccinotherapy, aspirin and anticoagulants to prevent cardiac infarcts, vitamins and antioxidants against neurodegenerative disorders). Other drugs can temporarily modify a physiological process (steroidal contraceptives).

Substitutive drugs take the place of a *missing substance*: the lack can be due to dietary reasons (vitamin deficiency) or to a physiological disturbance (insulin in diabetes, estrogens in menopause). The substitutive treatment can cover a very short period (intravenous rehydration in case of haemorrhages and diarrhoea), or can last the whole life (hormonal treatment in Addison's disease).

Symptomatic treatments are given in order to attenuate or to neutralize the disorders which result from a pathological state. They abolish 'general' symptoms such as fever, pain or insomnia. However, their activity can be much more specific and targeted to a particular system: thus symptomatic drugs are available for cardio-vascular, for neuro-psychiatric, for respiratory, for digestive diseases, etc. As a rule, a symptomatic treatment is not supposed to cure the patient, but rather to render his all-day life more comfortable and to prolong his life. In fact the distinction between substitutive, preventive and symptomatic treatments is not always easy. Antihypertensive drugs for example abolish, or at least diminish, the symptoms associated with arterial hypertension, but they play also a preventive role against the cardio-vascular complications of hypertension (notably myocardial infarct).

The nature of the illness

The so-called *physiological classification* was adopted by the World Health Organization (WHO) in 1968. It classifies the drugs by the body system on which they act (drugs affecting the central nervous system, the genito-urinary tract, the musculoskeletal system, for example). The WHO requests member countries who report to it to classify diseases according to the scheme laid down in the Manual of the International Statistical Classification of Diseases, 9th revision (1975), WHO, Geneva, 1977. There are 17 main categories (see Table 2.4) of drugs and each individual disease is represented by a three-digit number. For example cholera is quoted as 001, and mental disorders such as schizophrenic psychoses are found under the entries 295–299.

Table 2.4   *International classification of diseases*

1 Infectious and parasitic diseases
2 Neoplasms (cancers)
3 Endocrine, nutritional and metabolic diseases, and immunity disorders
4 Diseases of blood and blood-forming organs
5 Mental disorders
6 Diseases of the nervous system and sense organs
7 Diseases of the circulatory system
8 Diseases of the respiratory system
9 Diseases of the digestive system
10 Diseases of the genito-urinary system
11 Complications of pregnancy, childbirth and puerperium
12 Diseases of the skin and subcutaneous tissue
13 Diseases of the musculoskeletal system and connective tissue
14 Congenital anomalies
15 Conditions originating in the prenatal period
16 Symptoms, signs, and ill-defined conditions
17 Injury and poisoning

The chemical structure

This classification is important to people involved in pharmaceutical research. An expert in peptide or prostaglandin chemistry will be primarily concerned with the various chemical manipulations that can be performed on these molecules and will rely on someone else to screen them for effects against the various illnesses susceptible to peptide or prostaglandin therapy. On the other hand, the chemical classification allows an excellent overview of all the congeners and analogues derived from an initial lead and thus facilitates structure–activity considerations.

B   Practical classifications

In practise the most powerful and useful system developed so far is a compromise between the methods, known as the anatomical-therapeutic-chemical system (ATC). The system divides products into 13 general groups (Table 2.5) according to the body system on which they act: A, alimentary system; B, blood and blood-forming organs, and so on. This is the usually followed by the name of the disease they cure and finally by a description of the chemical classes involved.

An even simpler classification is usual among the medicinal chemist community, it distinguishes between four major classes of drugs:

Agents acting on the central nervous system: psychotropic and neurological drugs

In man, the central nervous system (CNS) comprises the brain and the spinal cord and it controls the thoughts, emotions, sensations and motor functions. The CNS-active

Table 2.5 *Drug classification by general anatomical groups*[5]

Code letter	Code heading	Examples
A	Alimentary tract and metabolism	Anti-peptic ulcerants, anticholinergics antidiarrheals, anti-emetics, vitamins, anorectics, hypoglycemics
B	Blood and blood-forming organs	Anticoagulants, thrombolytics, hypolipemics
C	Cardiovascular system	Cardiovascular drugs
D	Dermatologicals	Antifungals, antibiotics, corticosteroids, anti-acne
G	Genito-urinary system and sex hormones	Antibacterials, corticosteroids, sex hormones
H	Other systemic hormonal preparations	Glucocorticoids, thyroid therapy
J	General systemic anti-infectives	Antibacterials and antibiotics, antivirals
L	Antineoplastic and immunosuppressive drugs	Antineoplastics
M	Musculoskeletal system	Corticosteroids, antigout agents, non-steroid anti-inflammatory agents, muscle relaxants
N	Central nervous system	Psychotropic drugs, neurological drugs (ex: anti-Parkinson drugs) analgesics
P	Antiparasitic products	Drugs for tropical diseases
R	Respiratory system	Antihistamines, anti-asthmatics cough and cold preparations
S	Sensory organs	
V	Various	

drugs are: The *psychotropic drugs* consist of the antidepressants, the antipsychotics, the anxiolytics and the psychomimetics which all do *affect the mood or mental functioning*. The *neurological drugs* such as: (1) the *anticonvulsants* intended for the treatment of epilepsy; (2) the *sedatives* and *hypnotics* used for sleep disorders; (3) the *analgesics* such as painkillers; and (4) *anti-Parkinson* drugs, etc.

Pharmacodynamic agents

The pharmacodynamic agents are drugs affecting the normal dynamic processes of the body, and particularly the cardio-vascular domain. This group is composed of the *anti-arrythmics*, the *anti-anginals*, the *vasodilators*, the *anti-hypertensives*, the *diuretics*, and the *anti-thrombotics* which all, directly or indirectly, concern the heart or the blood circulation. Traditionally the anti-allergic drugs and the drugs acting on the gastrointestinal tract are also included in the class of pharmacodynamic agents.

Chemotherapeutic agents

Initially, the term chemotherapy referred to the treatment by means of drugs preventing selectively the development of various kinds of infesting hosts: protozoa (amoebas, leishmania, hematozoa, treponema, trypanosoma), microbes, viruses and as a rule all parasites which propagate infectious diseases. In the same context of selective toxicity, the anti-cancer treatments also belong to the class of chemotherapeutic agents.

Agents acting on metabolic diseases and endocrine functions

This category of drugs is made up of a collection of agents which do not easily fit in the previous classes. It consists of the anti-inflammatory drugs, the anti-arthritics, the anti-diabetics, the hypolipemic agents, the anorectics and most of the peptidic and steroidal hormones.

IV CONCLUSION

About 6000 chemical entities can be used, in various pharmaceutical formulations, to treat human or animal diseases; all attempts towards their classification represent arbitrary procedures. The first reason for that is that not a single drug has ever been encountered which exhibits only one biological activity. The antimalarial drug chloroquine is also active on some inflammatory processes, the anxiolytic benzodiazepines possess antiepileptic properties, etc. On the other hand, the scientific communities have different needs: a chemical classification may sometimes be very useful to medicinal chemists but strictly of no interest to a social security employee. Conversely pharmacologists and physicians will probably prefer the physiological classification.

REFERENCES

1. Burger, A. (1990) Preface. In Hansch, C., Sammes, P.G. and Taylor, J.B. (eds). *Comprehensive Medicinal Chemistry*, p. 1. Pergamon Press, Oxford.
2. Wermuth, C.G., Ganellin, C.R., Lindberg, P. and Mitscher, L.A. (1998) Glossary of terms used in medicinal chemistry (IUPAC Recommendations, 1997). *Annual Reports in Medicinal Chemistry*, pp. 385–395. Academic Press, San Diego.

3. Wermuth, C.G. (1993) Preface. *Trends in QSAR and Molecular Modeling '92. Strasbourg (France), September*, pp. 7–11. ESCOM, Leiden.

4. Ariëns, E.J. (1974) Drug design and cancer. *27th Annual Symposium on Fundamental Cancer Research*, pp. 127–152. The Williams and Wilkins Company, Baltimore, and M.D. Anderson Hospital and Tumor Institute, Houston.

5. Taylor, J.B. and Kenewell, P.D. (1993) *Modern Medicinal Chemistry*. Ellis Horwood, London.

6. Taylor, J.B., Kennewell, P.D. (eds). (1993) The pharmaceutical phase. *Modern Medicinal Chemistry*, pp. 49–71. Ellis Horwood, New York.

7. Trissel, L.A. (1986) *Handbook on Injectable Drugs*, 4th edn., American Society of Hospital Pharmacists, Bethesda.

8. Sinkula, A.A. and Yalkowsky, S.H. (1975) Rationale for design of biologically reversible drug derivatives: Prodrugs. *J. Pharm. Sci.* **64**: 181–210.

9. Kier, L.B. (1971) *Molecular Orbital Theory in Drug Research. Medicinal Chemistry*. Academic Press, New York.

10. Ariëns, E.J. (1966) Some of the principal processes that take place in drug action. In *Progress in Drug Research*, pp. 429–529. Karger Verlag, Basel.

11. Kellaway, I.W. (1983) The influence of formulation on drug availability. In *Introduction to the Principles of Drug Design*, pp. 39–51. Wright. PSG, Bristol.

12. Rettig, H. (1981) Physiologische Transportvorgänge. *Biopharmazie-Theorie und Praxis der Pharmakokinetik*, pp. 93–124. Georg Thieme Verlag, Stuttgart.

13. Forth, W., Henschler, D., Rummel, W., *et al.* (1987) *Pharmakologie und Toxicologie*. B.I. Wissenschaftsverlag, Mannheim.

14. Bennet, W.M., Singer, I. and Coggins, C.H. (1970) A practical guide to drug usage in adult patients with impaired renal function. *J. Am. Med. Assoc.* **214**: 1468–1475.

15. Bennet, W.M., Singer, I. and Coggins, C.H. (1973) Guide to drug usage in adult patients with impaired renal function. *J. Am. Med. Assoc.* **223**: 991–997.

16. Ariëns, E.J. and Simonis, A.M. (1964) A molecular basis for drug action. The interaction of one or more drugs with different receptors. *J. Pharm. Pharmacol.* **16**: 289–312.

17. Bizière., K. (1984) The biological activity. Lecture given at the Louis Pasteur University: 8 March, 1984.

18. Albert, A. (1975) The selectivity of drugs. In *Outline Studies in Biology*. Chapman and Hall, London.

19. Burger, A. (1990) Classification of drugs. In Hansch, C., Sammes, P.G. and Taylor, J.B. (eds). *Comprehensive Medicinal Chemistry*, pp. 249–260. Pergamon Press, Oxford.

20. Reuben, B.G. and Wittkoff, H.A. (1989) *Pharmaceutical Chemicals in Perspective*. John Wiley, Chichester.

21. Pradal, H. (1975) *Les Grands Médicaments*. Editions du Seuil, Paris.

3

MEASUREMENT AND EXPRESSION OF DRUG EFFECTS

Jean-Paul Kan

To most of the modern pharmacologists the receptor is like a beautiful but remote lady. He has written her many a letter and quite often she has answered the letters. From these answers the pharmacologist has built himself an image of this fair lady. He cannot, however, truly claim ever to have seen her, although one day he may do so.

D. K. de Jongh (1964)[1]

I INTRODUCTION

The knowledge of the full sequence of the biochemical and pharmacological effects of a given drug and the understanding of its mechanism of action, are of basic utility for the chemist, the pharmacologist and the clinician. The identification, description and quantification of each step leading from the primary action of a drug to the resulting effects, provides the basis for a rational therapeutic use and the design of improved or original chemical agents. In addition, the study of the effects of drugs and the analysis of their modes of action also provides essential clues for the discovery and elucidation of biochemical and physiological mechanisms. The aim of this chapter is to describe the main strategies used to assess drug effects at various levels of complexity of organization: cell, tissue, organ and living animal. Kinetic analysis of drug–receptor interactions and drug–receptor effects will be briefly presented in a comprehensive manner. The main classical terms and expressions commonly used in describing drug effects will also be defined. For further information, a dictionary of pharmacology may be consulted.[2]

II MECHANISMS OF DRUG EFFECTS

Since the situation picturesquely described by de Jongh,[1] not only has the concept of receptor first proposed by Langley[3] and Ehrlich[4] at the turn of the nineteenth century been universally accepted, but also the powerful techniques used in modern pharmacology have allowed us to truly see 'the fair lady'.[5]

Thus, the effects of drugs basically result from their interaction with functional macromolecules — generally cellular proteins termed receptive substances or, more simply, receptors[3,4] — which normally serve as natural targets for endogenous regulatory ligands. The latter are mostly cell-to-cell information-transferring molecules (e.g. hormones, neurotransmitters) acting on specific receptors confined to the cell surface (e.g. membrane-bound receptors, ion channels, transport sites). However, other drug targets also reside in the cytosol (soluble and particulate enzymes) and the cell nucleus (DNA, RNA). Receptor activation induces a cascade of biochemical responses that depends, both on the type of the receptor and on the type of the effector cell. These unitary responses are then integrated by the cell to yield a specific cellular response (e.g. release of Ca^{2+} from the sarcoplasmic reticulum of the myocyte and contraction of the myofibrils). Finally, the individual cellular responses are summed up by the functional tissue or organ to generate the physiological response (e.g. contraction of the muscle) (Fig. 3.1). In theory, a drug can interfere with any step of this process. In this respect, it does not create effects, but merely modulates ongoing cellular functions.

It must be noted that the term *ligand* concerns any molecule, endogenous or exogenous, able to bind to a

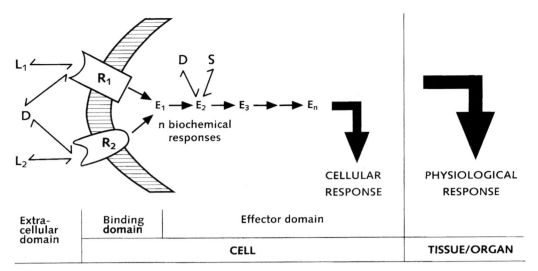

Fig. 3.1 Schematic representation of the series of biochemical steps leading from ligand–receptor interaction to the cellular and physiological responses. L1, L2 and D stand for ligand and drug, respectively. R1 and R2 represent specific membrane receptors that recognize L1 and L2. E1 → n are effector proteins, generally enzymes, modulated by receptor activation or by drugs acting as inhibitors. S is a specific substrate.

receptor. Any drug usually refers to exogenous chemicals, however. In addition, a ligand can bind to acceptor sites that are not coupled to effector systems. They are often circulating proteins (e.g. plasma albumin) that trap the ligand thus preventing its effects.

For a given effector cell, the drug–receptor interaction initiates a series of biochemical and physiological changes which are characteristic of the effect of, or response to, the drug. If most of these changes are identical to those induced by the endogenous ligand, the drug is termed *agonist*. A compound that is itself devoid of intrinsic activity, but causes effects by inhibition of the action of the endogenous ligand or a specific agonist is designated as an *antagonist*. Finally, if a compound elicits by itself, effects opposite to those induced by the agonist, it is termed *inverse agonist* or *negative antagonist*. Generally, during interaction with the receptor, the chemical structures of both the regulatory ligand and the drug remain unchanged.

Agonist-induced effects can also be defined with reference to those typically caused by the endogenous ligand itself. For example, noradrenergic (from Greek: ergon, work) drugs are compounds that initiate the typical effects induced by the endogenous neurotransmitter noradrenaline (norepinephrine). They are also termed sympathomimetics since noradrenaline triggers part of its effects on peripheral organs (e.g. heart, vascular, visceral and bronchial smooth muscles, secretory glands) by stimulating the sympathetic peripheral nerves (autonomic nervous system). Corresponding antagonists may be termed adrenolytics or sympatholytics. Pharmacological effects can also be classified according to the specific activation of receptor

subtypes. For example, the effects of acetylcholine which are mimicked by the alkaloid muscarine and selectively antagonized by atropine are termed *muscarinic effects*. Other effects of acetylcholine that are mimicked by nicotine and selectively blocked by D-tubocurarine are described as *nicotinic effects*. These effects are actually mediated by activation of muscarinic and nicotinic cholinergic receptors that are two distinct cases of molecular entities buried in the surface of the effector cell.

The suffix 'ergic' is also used to characterize nerve endings that work by releasing a particular neurotransmitter (e.g. neurons which release noradrenaline are termed noradrenergic, serotonin/serotonergic, acetylcholine/cholinergic, dopamine/dopaminergic, and so on). Finally, the expression *synergistic action* is employed when a drug action increases the effect produced by another drug. A synergistic action may be simply an additive effect or may involve potentiation. Enzymes of critical metabolic or regulatory pathways of the cell (e.g. acetylcholinesterase, monoamine oxidase) are also receptors for many drugs. These compounds interact with the endogenous substrate of the enzyme to inhibit the catalytic activity of the protein. Conversely to the ligand, the substrate combines with the active site of the enzyme during the course of the reaction (enzyme–substrate complex) to produce a new, chemically distinct compound.

Of equal interest are proteins of the cytoskeleton that serve structural roles (e.g. tubulin), or those involved in transport processes. For example, many drugs act indirectly by releasing, or preventing the release, of pharmacologically active molecules (e.g. neurotransmitters) concentrated in cytoplasmic vesicles. Such compounds can also interact with autoreceptors that respond to the neurotransmitter by modulating the release of the neurotransmitter (e.g. noradrenergic nerve endings possess α2-adrenergic receptors mediating suppression of noradrenaline release).

Nucleic acids like DNA are important drug receptors, e.g. for chemicals controlling malignancy such as intercolator agents. Finally, general anaesthetics interact with and alter the structure and function of the lipids of the cellular membrane.

A Sites of drug effects

Drug effects can be measured at various levels of complexity: from the cell to the living animal a wide spectrum of biological preparations are accessible to the pharmacologist. As already pointed out, most drugs interfere at the membrane receptors with physiological regulatory molecules, to modulate the response of the cell and, ultimately, that of a particular tissue (Fig. 3.1). Complex pathways involving integrated control systems, i.e. homeostasis, are often essential in order to lead from receptor activation to the observable effect (e.g. drug–receptor interaction → *n* integrated steps → alteration of cardiac contractility and blood pressure). Thus, the more complex the biological preparation is (e.g. isolated organ → *in situ* organ), the more complicated is the stimulus–effect relationship (see below). It is therefore necessary to have access, at the molecular level, to the actual primary effects of a drug by measuring its interaction with the receptor and the direct consequences on cellular regulatory effectors.

The receptor–effector complex of the cell

The eukaryotic plasma membrane is the cellular switchboard responsible for receiving all sorts of extracellular messengers mainly hormones and neurotransmitters, but also growth factors, pheromones, odours and light. These messengers must be detected, decoded, amplified, integrated with each other and with cellular metabolism, and conveyed to the cytoplasm as intracellular chemical messengers. Thus, for a given cell, a particular receptor–effector complex (Fig. 3.1) regulates information received from other neighbouring and distant cells. Despite the diversity of extracellular signals, cell surface receptor proteins utilize only a few biochemical mechanisms for these processes. Actually, two main transducing mechanisms of the signal can be considered: (1) the same molecular entity recognizes the extracellular messenger and transduces the signal (e.g. transmitter-gated ion channels and receptor tyrosine kinases); (2) distinct molecular entities coupled by regulatory guanine (G)-proteins, recognize the extracellular messenger and transduce the signal. These G-coupled membrane-bound receptors are of particular interest since

in the clinic the actions of numerous important drugs are mediated via occupancy of these receptors.[6] All these receptors consist of a single polypeptidic chain forming seven transmembrane (7TM) domains joined by intracellular and extracellular loops. G-proteins that link cell-surface receptors to effector proteins of the plasma membrane (e.g. the enzyme adenylyl-cyclase which catalyses the synthesis of cAMP), are membrane proteins able to strongly bind G-nucleotides. Upon activation by the ligand, the receptor interacts with the G-protein and triggers a complex regulatory cycle leading to the activation of effector proteins that mobilize chemical second messengers (e.g. cAMP) easily detectable by *in vitro* techniques (for further details see Chapter 5 and below).

The ability of membrane receptors to be regulated in response to their level of activation is well documented. For example, sustained stimulation of cells with agonists generally results in a state of desensitization such that the effect which follows subsequent exposure to the same concentration of drug is diminished, i.e. seems to result from an apparent loss of intrinsic efficacy or efficiency coupling (see below). This can be very important in therapeutic situations (e.g. treatment of asthma by the β-adrenergic agonist isoproterenol). Finally, receptor desensitization may lead to tachyphylaxis, i.e. a rapid diminution in response to each of a succession of repeated doses of a drug. Conversely, supersensitivity to receptor agonist, or up-regulation, is also frequently observed following a reduction in the chronic level of receptor stimulation. Such a situation is observed after long-term treatment with certain antagonists (e.g. propranolol) or after chronic denervation.

The use of specific markers (e.g. selected radiolabelled ligands or substrates, specific antibodies, hybridization probes) associated with powerful analytical techniques (separation by chromatography or electrophoresis, blotting, highly sensitive immuno-assays, etc), are utilized to determine the cellular sites of action of endogenous ligands and drugs at the molecular level. These methods are applied on various biological preparations, including subcellular structures, cell membrane suspensions and tissue homogenates. Cell and tissue culture preparations are also widely used to assess biological processes and drug actions. They have a number of general advantages: cells are more accessible for study, diffusion delays and barriers to applied substances are minimized, the humoral and cellular components of the culture environment can be controlled and progressive changes in intracellular and intercellular events can be directly monitored. Primary cell cultures from various embryonic tissues (e.g. brain structures) can be achieved. A number of transformed cell lines derived from tumoral tissues are also available. Cells (e.g. Chinese hamster ovary (CHO) cells) transfected with the cloned gene of a particular receptor, including human clones, generally express a large amount of this receptor on their cell surface. These cells, all identical, are widely and routinely used to assess the transducing pathways linked to the receptor and to screen specific drugs *in vitro*.

From the cell to the living animal

Although many drugs are known to act at well-defined cellular receptor–effector systems, more complex biological preparations are needed to reveal their subsequent physiological effects. Isolated tissues or organs, generally smooth muscles, are routinely used *in vitro* to classify drugs as agonists or antagonists. For example, cholinergic agonists (e.g. arecaidine) acting at peripheral muscarinic receptors, induce a concentration-dependent contraction of the uterine horn of the guinea pig. Muscarinic antagonists antagonize this response.[7] Once isolated, such organs are free from the nervous and regulatory influences which complicate their functions *in vivo*. Using standardized test conditions, typical reactions can be reliably reproduced.

In addition, there are still many examples of drug action where the knowledge is less precise (e.g. unknown receptor–effector complex), and the action is recognized as a response involving a tissue, an organ or even a living animal. The action of the drug may be on many cells rather than on a particular type of cell. For example, a drug that acts as a diuretic can only act in that way on an organized tissue, namely the kidney, although the mechanism of action may be known at the molecular or cellular level. Although apparently simple, isolated preparations have a relative degree of local regulatory control systems that plays a role in the understanding of drug action.

Living animals are also needed to assess secondary effects of drugs resulting from homeostatic mechanisms. For example, a drug that relaxes smooth muscle in the walls of blood vessels and so causes vasodilatation may thereby elicit a reflex acceleration of the heart rate. Thus, the direct action of acetylcholine or of a muscarinic agonist, on the heart is to decrease the rate of beating (bradycardia), yet the intravenous injection of small amounts produces the opposite effect, providing that cardiovascular reflexes are operating.

Finally, chemicals aimed at relieving symptoms observed in psychiatric illnesses (e.g. anxiety, depression, schizophrenia) are studied in various animal models that reveal typical behaviours claimed to mimic those exhibited by patients. Animal tests are also needed to discover and develop drugs to treat cognitive decline. Recent studies of the effects of certain brain lesions on learning in rats or monkeys as animal models of Alzheimer's disease may go some way towards developing an experimental model for this disorder.

III. MEASUREMENT OF DRUG EFFECTS

A Kinetic analysis of ligand–receptor interactions

Theoretical background

Considerable evidence suggests that 7TM receptors can exist in two different coexisting conformational states: an 'inactive' state R and an 'active' state R*. The relative proportion of R and R* will depend on both the actual receptor–effector system and the presence of ligands (L) having affinity for the receptor, i.e. able to form complexes with R (LR) and R* (LR*). Only R* and AR* can couple with G-proteins to induce a cellular response (e.g. increase in cAMP levels). If, to simplify, the reversible interaction between the receptor and the G-protein is omitted, the minimal scheme depicting the relationship between A, R, R*, LR and LR* can be written as follows:
where:

(1) γ is an intrinsic allosteric constant ($\gamma = R^*/R$), reflecting the relative proportions of R and R* forms which coexist in the absence of any ligand L.

(2) Ka and α Ka are the association constants of L for R and R*, respectively. These constants actually measure the affinity of the ligand L for the R and R* forms of the receptor. The factor α denotes the differential affinity of L for R*. This means that L, along with γ, can modify the relative proportions of R, R*, LR and LR* (see Scheme 1).

Two main experimental situations can be observed (Fig. 3.2):

(1) In the absence of any ligand L, a cellular response can be easily measured. In such constitutively active systems (e.g. native or mutated receptors, cells highly transfected with a particular cloned receptor) a sufficient proportion of R*, spontaneously present, binds to the G-protein to induce a cellular response (high basal level). In this case,

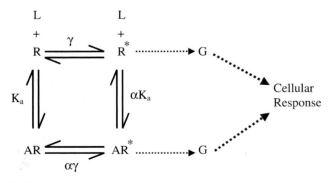

Scheme 1

A

B

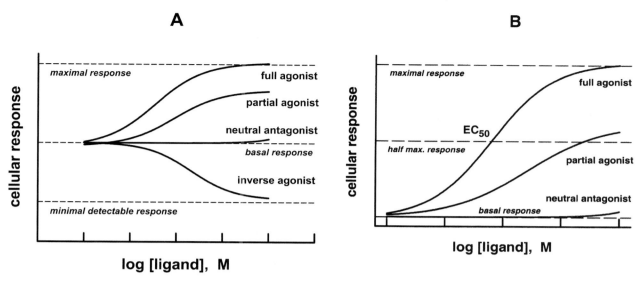

Fig. 3.2 Evolution of the cellular response as a function of the ligand and the receptor–effector complex. (A) Constitutively active systems. (B) Quiescent systems.

ligands L are classified according to the value of the α constant, as full agonists and partial agonists ($\alpha > 0$), neutral antagonists ($\alpha = 0$) and inverse agonists ($\alpha < 0$) (Fig. 3.2A). Thus, the higher the absolute value of the α constant, the higher the cellular response. The α constant actually measures the intrinsic efficacy — positive, null or negative — of a particular ligand L for a given receptor–effector system. In this respect, a neutral antagonist with zero efficacy, does nothing to the receptor but bind to it and therefore preclude activation of the receptor by an agonist. As pointed out by Kenakin[8] efficacy is 'the property of a drug that modifies subsequent interaction of the 7TM receptor with other membrane proteins' (see also below). The inverse agonism patterns depicted in Fig. 3.2A have been observed with β-adrenergic ligands on basal adenylyl-cyclase activity mediated by a constitutively active mutant of the β2-adrenergic receptor[9] or on Sf9 cells overexpressed with β2 wild-type receptors as well.[10]

(2) A low basal level of activity is measured in the absence of any ligand L. In such quiescent systems (e.g. in general native wild-type receptors present on cellular or tissue membranes) inverse agonists cannot be revealed since they behave as neutral antagonists. In contrast, as in the former situation, full and partial agonists will induce a cellular response the magnitude of which depends on their efficacy (Fig. 3.2B) (see also below).

Extensive critical reviews aimed at analysing the numerous mathematical methods describing the complex interaction between ligand and G-coupled receptors, have been discussed in the literature.[8,11–13]

Quantitative measurement

As far as affinity and efficacy of drugs are concerned in their therapeutic activity in humans, the correct measurement of these parameters using well-chosen and reliable methods is highly important for drug discovery. These methods mainly, but not exclusively, concern radioligand-binding techniques and specific functional assays.

Radioligand binding. The simplest assumption about the reversible formation of a ligand–receptor complex (see left part of Scheme **1**) is that it may be expressed according to the mass-action law, as the following chemical reaction:

$$[L] + [R] \underset{k_{-1}}{\overset{k_1}{\rightleftharpoons}} [LR] \tag{1}$$

where L is the ligand, k_1 and k_{-1} are the rate constants for association and dissociation of the LR complex. Square brackets signify the concentration of the entity they enclose.

At equilibrium (or steady state), the rates of association and dissociation of the complex LR are equal, and:

$$k_1[L][R] = k_{-1}[LR] \tag{2}$$

The equilibrium binding constant may then be defined as dissociation constant (k_d) such as:

$$k_d = \frac{k_{-1}}{k_1} = \frac{[L][R]}{[LR]} \tag{3}$$

The total number of receptors (RT) is the maximum number of receptors on which the drug is able to bind specifically. This finite number, also termed B_{max}, is the sum of the receptors engaged in forming the complex (LR), plus the free receptors (R), such as:

$$[R_T] = [LR] + [R] = B_{max} \tag{4}$$

From equations (3) and (4) after rearranging:

$$[LR] = \frac{B_{max}[L]}{[L] + k_d} \qquad (5)$$

If we now define LR as bound ligand = B, and L as free ligand = F, from equation (5):

$$B = \frac{B_{max}F}{F + k_d} \qquad (6)$$

and

$$B/F = \frac{B_{max} - B}{k_d} \qquad (7)$$

Equation (7) is the Scatchard equation.[14]

Receptor labelling studies involve binding of a radioactive form of the ligand to receptor preparations of target tissues. Radiolabelled ligands — ideally neutral antagonists that are devoid of efficacy for the receptor used (see above) — must be pure, stable and have a high radiochemical specific activity. The most commonly used isotopes are tritium [^{3}H] and radioactive iodine [^{125}I]. Radioactive ligand and receptor preparations are incubated in appropriate conditions: receptor concentration, temperature, pH and ionic strength. At the end of the incubation period, while the binding reaction is at equilibrium, bound and free ligands are promptly separated using filtration or centrifugation.

Experimental determination of k_d and B_{max} for a given radioligand and receptor preparation requires incubation of various concentrations of the radioligand with a fixed concentration of receptors: this is the so-called saturation experiment. Such experiments can be performed in the absence (control) or presence of various concentrations of non-radioactive ('cold') ligand or drug — also termed displacer — able to compete with the radioligand for the same receptor. Finally, interaction of displacers with a given receptor can also be studied by incubating various concentrations of the displacer with fixed concentrations of radioligand and receptors: this is the so-called displacement experiment (Fig. 3.3B). In all cases, the non-specific binding is measured by the bound radioactivity remaining in the presence of a large excess of cold ligand.

In most cases, data generated from a saturation experiment are analysed according to equation (7). When the radioactive ligand interacts with a single population of non-interacting receptors, the Scatchard-plot,[14] i.e. B/F vs B, leads to a linear curve. K_d is estimated as the negative reciprocal of the slope of the line of best fit, and B_{max} by the abscissa intercept of the line (Fig. 3.3A). The reciprocal of K_d measures the *association constant* K_a (see Scheme 1) of the radioligand for the receptor. Thus, for a given ligand–receptor pair, the smaller the K_d (generally 0.1–10 nmol l^{-1} or nM) the higher is its *affinity*. B_{max} is expressed as pmol or fmol per mg tissue or protein. A downward curvilinear Scatchard plot may indicate that there is more than one type of binding site, and that the different types differ in their dissociation constants for binding with the radiolabelled ligand under investigation. Alternatively, the binding of ligand to its receptors may inhibit further binding and hasten dissociation from the receptor (e.g. insulin receptor). This latter phenomenon is called *negative cooperativity. Positive cooperativity* also occurs, i.e. the binding of ligand facilitates the binding of subsequent molecules to the receptors (e.g. nicotinic receptor). In this case, the Scatchard plot is upward concave. Cooperativity generally implies that the ligand behaves as an allosteric modifier that induces a conformational change in the structure of the receptor protein.

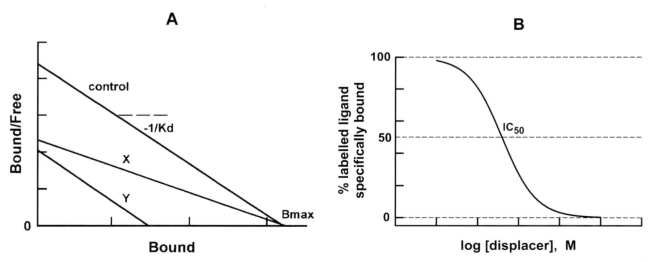

Fig. 3.3 Analysis of ligand–receptor interactions. (A) Scatchard plot in absence (control) or presence of two different displacers. X and Y are competitive and non-competitive displacers, respectively. (B) Displacement curve.

When the saturation experiment is performed in the presence of a displacer, the line of best fit of the Scatchard plot can be modified in a manner that depends on the type of receptor interaction exhibited by the displacer. Two main cases exist: (1) decreased slope and unchanged B_{max}, the displacement is of the competitive type; (2) unchanged slope and decreased B_{max}, the displacement is of the non-competitive type (Fig. 3.3A). Intermediate cases where both the slope and B_{max} are modified also exist.

Data generated from a displacement experiment are generally fitted by a sigmoid curve termed *displacement* or *inhibition curve*, i.e. percentage radiolabelled ligand specifically bound vs log [Displacer] (in mol 1^{-1} or M) (Fig. 3.3B). The abscissa of the inflexion point of the curve gives the IC_{50} value, i.e. the concentration of displacer that displaces or inhibits 50% of the radioactive ligand specifically bound. IC_{50} is a measure of the *inhibitory* or *affinity constant* (K_i) of the displacer for the receptor. IC_{50} and K_i are linked as follows: if the displacement is of the competitive type, then,

$$K_i = \frac{IC_{50}}{1 + [L]/K_d}$$

This is the Cheng–Prusoff[15] equation, where [L] is the concentration of radioactive ligand used in the experiment and K_d its dissociation constant determined from the Scatchard plot. Finally, if the displacement is of the non-competitive type, then, $K_i = IC_{50}$.

Functional assays of drugs. For a given relevant receptor–effector complex (e.g. cells expressing a particular cloned receptor, isolated organ) there exists a complex function (f) that determines the magnitude of the observable response (e.g. increase in cAMP levels, contraction of the isolated ileum) induced by a drug D,[16]

$$\text{Response} = f\frac{([D].\alpha.[R_T]}{[D] + K_d)} \qquad (8)$$

If the response is plotted as a function of [D], a symmetrical sigmoid curve is generally obtained (Figs 3.2 and 3.4A). The abscissa of the inflexion point is termed EC_{50}, i.e. the concentration that causes 50% of the maximum possible effect induced by the drug (Fig. 3.4A). The EC_{50} parameter actually reflects the potency of the drug for a given receptor–effector system.

More precisely, equation (8) signifies that the response is proportional to the intrinsic efficacy α of the drug and the total number of viable specific receptors RT, but inversely proportional to the equilibrium dissociation constant K_d of the drug for the drug–receptor complex. This equation contains two parameters related solely to the drug (α and K_d), and two others completely related to the receptor–effector system (RT and f). In other words, both the type and the magnitude of the apparent response will completely depend on the couple drug/receptor–effector system, i.e. on the efficiency coupling. For example, the low efficacy β1-adrenoceptor agonist prenalterol[17] can be nearly a full agonist in the thyroxine-treated guinea-pig atria, a partial agonist in the rat atria, and a neutral antagonist in the canine coronary artery, indicating that efficiency coupling increases from dog coronaries to rat atria and guinea-pig atria. In comparison, isoproterenol behaves as a full agonist

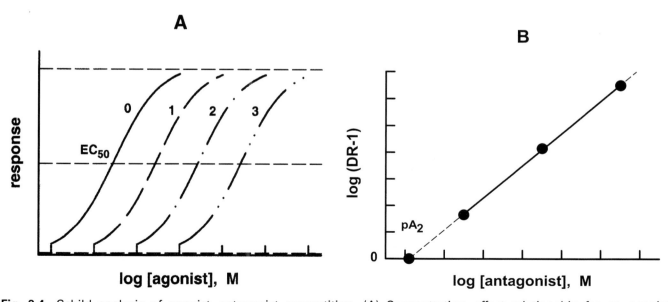

Fig. 3.4 Schild analysis of agonist–antagonist competition. (A) Concentration–effect relationship for an agonist in the absence (curve 0) and the presence of three increasing concentrations of antagonist (curves 1, 2 and 3). (B) Schild plot generated from curve A (see text for explanations).

in the three tissues with increasing potency from dog coronaries to rat atria and guinea-pig atria. Therefore, the lack of observation of an agonist response does not necessarily involve the absence of drug efficacy, only that the system used is inadequate to make it observable. The comparison between binding (K_d) and potency (EC_{50}) also reveals the efficiency of receptor coupling. For example, in intact S49 lymphoma cells, the stimulation of adenylyl-cyclase activity by the full agonist adrenaline through activation of β-adrenoceptors, exhibits a wide gap between the K_d (2 μM) and the EC_{50} (10 nM). The existence of this apparent discrepancy actually reveals an amplification step between adrenaline binding and the observed response. In this case, the amplification effect seems to be mainly due to the mobility of the β-receptor and of the G-protein that makes it possible for one receptor to activate numerous G-protein/adenylyl-cyclase complexes.[18]

Interaction between agonist and antagonist: the Schild analysis. As indicated above, a competitive antagonist may be regarded as a drug which interacts reversibly with a set of receptors to form a complex which does not elicit any response. The antagonist–receptor interaction is characterized by dissociation constant, but the intrinsic efficacy is zero. Schild analysis[19] allows the determination of the dissociation constant (K_d) of a competitive antagonist. Practically, several concentration–response curves are generated for a given agonist–tissue pair in the absence or presence of increasing concentrations of the antagonist (Fig. 3.4A). Then a Schild plot (Fig. 3.4B) is constructed by plotting log (dose ratio − 1) or log (DR − 1) vs log [antagonist], where 'dose-ratio' is the ratio of equiactive agonist concentrations measured as EC_{50}, in the absence and presence of antagonist concentrations[20] (Fig. 3.4A). The Schild plot is analysed by linear regression. This line meets the abscissa at a point that provides an estimate of pA_2 (Fig. 3.4B). Assuming that $pA_2 = -\log K_d$, Schild analysis is therefore a useful method to calculate the dissociation constant of a given antagonist in fixed conditions (agonist/receptor–effector complex).

B   Relationship between dose and effect

When drugs are administered *in vivo* to living animals or humans, there is no single characteristic relationship between intensity of drug effect and drug dosage. A dose–effect curve may be linear, concave upward or downward, or sigmoid. In addition, differences appear in the magnitude of response among individuals in the same population for the same dose of a given drug, i.e. drug effects are never identical in all animals or all patients. In other words, a range of doses is required to produce a specified intensity of effect in all individuals. Alternatively,

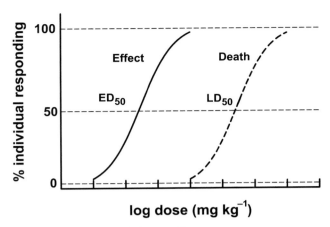

Fig. 3.5   Examples of dose–effect curves.

a range of effects will be produced if a given dose is administered to a group of individuals. It is possible to determine the dose of a drug required to produce a specified effect in an individual, i.e. the individual effective dose for which the specified intensity of effect is present (e.g. animals injected with a hypnotic, sleep or do not sleep). The percentage of individuals that exhibits the specified effect plotted as a function of log dose (mg kg^{-1}), gives the so-called sigmoid dose–effect curve (Fig. 3.5).

The dose required to produce a specified intensity of effect in 50% of individuals is known as the median effective dose or ED_{50} (Fig. 3.5). If death is the specified effect, the median effective dose is termed the median lethal dose or LD_{50}. The ratio LD_{50}/ED_{50} leads to the therapeutic index (or therapeutic ratio). This index has limited usefulness since it cannot be calculated for man and data relating to one species cannot reliably be transferred to another.

REFERENCES

1. de Jongh, D.K. (1964) Introductory remarks. In Ariëns, E.J. (ed.). *Molecular Pharmacology*. Academic Press, New York.
2. Bowman, W.C., Bowman, A. and Bowman, A. (1986) *Dictionary of Pharmacology*, 234 pp. Blackwell Scientific Publications, Oxford.
3. Langley, J.N. (1878) On the physiology of the salivary secretion. *J. Physiol. (Lond.)* **1**: 339–369.
4. Ehrlich, P. (1913) Chemotherapeutics: scientific principles, methods and results. *Lancet* **2**: 445–451.
5. Giraudat, J. and Changeux, J.P. (1981) The acetylcholine receptor. In Lamble, J.W. (ed.). *Towards Understanding Receptors; Current Reviews in Biomedicine 1*, pp. 34–41. Elsevier, North Holland Biomedical Press, Amsterdam.
6. Herz, J.M., Thomsen, W.J. and Yarbrough, G.G. (1997) Molecular approaches to receptors as targets for drug discovery. *J. Recep. Signal Trans. Res.* **17**: 671–776.
7. Dörje, F., Friebe, T., Tacke, R., Mutschler, E. and Lambrecht, G. (1990) Novel pharmacological profile of muscarinic receptors mediating contraction of the guinea-pig uterus. *Naunyn-Schmiedeberg's Arch. Pharmacol.* **342**: 284–289.

8. Kenakin, T. (1996) The classification of seven transmembrane receptors in recombinant expression systems. *Pharmacol. Rev.* **48**: 413–463.

9. Chidiac, P., Hebert, T.E., Valiquette, M., Dennis, M. and Bouvier, M. (1994) Inverse agonist activity of β-adrenergic antagonists. *Mol. Pharmacol.* **45**: 490–499.

10. Samama, P., Pei, G., Costa, T., Cotecchia, S. and Lefkowitz, R.J. (1994) Negative antagonists promote an inactive conformation of the β2-adrenergic receptor. *Mol. Pharmacol.* **45**: 390–394.

11. Costa, T., Ogino, Y., Munson, P.J., Onaran, H.O. and Rodbard, D. (1992) Drug efficacy at guanine nucleotide-binding regulatory protein-linked receptors: thermodynamic interpretation of negative antagonism and receptor activity in the absence of ligand. *Mol. Pharmacol.* **41**: 549–560.

12. Colquhoun, D. (1998) Binding, gating, affinity and efficacy. *Br. J. Pharmacol.* **125**: 923–947.

13. Defflaer, L. and Landry, Y. (2000) Inverse agonism at heleptical receptors: concept, experimental approach and therapeutic potential. *Fund. Clin. Pharmacol.* **14**: 73–87.

14. Scatchard, G. (1949) The attraction of proteins for small molecules and ions. *Ann. NY Acad. Sci.* **51**: 660–672.

15. Cheng, Y.C. and Prusoff, W.H. (1973) Relationship between the inhibition constant (K_i) and the concentration of inhibitor which causes 50 percent inhibition (I_{50}) of an enzymatic reaction. *Biochem. Pharmacol.* **22**: 3099–3108.

16. Furchgott, R.F. (1966) The use of β-haloalkylamines in the differentiation and determination of dissociation conatants of receptor agonist complexes. *Adv. Drug Res.* **3**: 21–55.

17. Kenakin, T. (1990) Drugs and receptors, an overview of the current state of knowledge. *Drugs* **40**: 666–687.

18. Stickle, D. and Barber, R. (1989) Evidence for the role of epinephrine binding frequency in activation of adenylate cyclase. *J. Pharmacol. Exp. Ther.* **36**: 437–445.

19. Schild, H.O. (1957) Drug antagonism and pA2. *Pharmacol. Rev.* **9**: 242–246.

20. Lazareno, S. and Birdsall, N.J.M. (1993) Estimation of competitive antagonist affinity from functional inhibition curves using the Gaddum, Schild and Cheng-Prusoff equations. *Br. J. Pharmacol.* **109**: 1110–1119.

4

DRUG TARGETS: MOLECULAR MECHANISMS OF DRUG ACTION

Jean-Pierre Gies and Yves Landry

Only such substances can be anchored at any particular part of the organism, as fit into the molecules of the recipient complex like a piece of mosaic finds its place in a pattern.
Paul Ehrlich, 1956[1]

I INTRODUCTION

To understand drug actions, it is necessary to consider the effects produced by the drug on the biological system at various levels of complexity of organization. The main levels, from the most complex to the most simple, could be designated as follows: intact organism, organized cells (tissues and organs), cells, subcellular structures and biological molecules. The effects produced by a drug can be recognized only as an alteration in a function or process that maintains the existence of the living organism, since all drugs act by producing changes in some known physiologi-

cal function or process. Drugs may increase or decrease the normal function of tissues or organs, but they do not endow them with new functions. Thus the particular effect of a drug is always expressed in relative terms, relative to the physiological conditions at the time of administration of the drug.

The action of the drug may be specific, i.e. aimed directly at the agent responsible for the disease, or non-specific, i.e. ameliorating a symptom of the disease (fever, for example) without getting to the cause of the disorder. Clearly, the distinction between what is produced by an agent, its effect, and where and how the effect is produced, its action, becomes of consequence in determining the purpose for which the drug may be used.[2]

How can the properties of drugs be discriminated at a molecular level? To even begin to answer this question, manipulations of the structure of the ligand (drug, hormone, neurotransmitter, etc.) have been often performed by chemists. Moreover, a comprehensive understanding of pharmacological selectivity and responses cannot be dissociated from a knowledge of the fundamentals of the molecular structure of the drug binding sites.

The functional macromolecular structures which are targets for drug action could be arbitrarily divided into receptors, enzymes, proteins involved in the transport, and nucleic acids. A number of other molecular interactions between drugs and the components of the biological system may occur, such as the binding of drugs to plasma albumin or other constituents of the tissues. Serum albumin can transport drugs in the circulation to various organs, and it can hold drugs up, preventing them from binding to their site of action. Those interactions have secondary rather than primary consequences for pharmacodynamic actions, since the duration of the drug action or its rate of actions is affected. Albumin might then be considered as an acceptor site for the drug rather than a receptor. In this chapter we do not deal with these acceptor or binding sites, but we focus interest on the molecular levels of the drug actions on cell receptors.

A   Induction of drug responses

The common event in the induction of pharmacologic responses is the formation of a complex between the ligand, or the drug, molecule and its site of action. The component of the organisms with which the drug interacts is termed *receptor*. Such interaction alters the function of the pertinent component and thereby initiates biochemical and physiological changes that are characteristic of the response to the drug. The statement that the receptor for a drug can be any functional macromolecular component of the organism has several fundamental corollaries. One is that a drug is potentially capable of altering the rate at which any bodily function proceeds. Another is that drugs do not create effects, but merely modulate ongoing functions; a drug does not impart a new function to a cell. Although gene therapy may soon challenge this principle, it remains valid for the immediate future.

The vast majority of drugs produce their effects by interacting with cells, either with components of the interior of the cell or with those on the surface of the cell comprising the cellular membrane. Some others act extracellularly at noncellular constituents of the body without involving a drug–receptor interaction.[3] The simplest example is that of the neutralization of gastric acid by antacid drugs. In this reaction the excess of gastric acid is neutralized by the base sodium bicarbonate. The base reacts chemically with the hydrochloric acid and removes the acid by inducing the formation of salt, water and carbon dioxide. This reaction is not considered as a drug–receptor interaction, since no macromolecular cell component is involved. Other types of extracellular mechanisms, where the action of the drug is the result of a chemical reaction, can be illustrated, for example, by the action of heparin, which prevents blood coagulation. There are still other mechanisms of drug action which are not mediated by receptors. These actions may occur at cellular sites and may involve macromolecular components, but the biological effects produced are nonspecific consequences of the chemical properties of the drugs. Detergents, alcohol, oxidizing agents, phenol derivatives act by destroying the integrity of the cell through disrupting the cellular constituents, such as proteins or nucleic acids. These actions are not treated in the present chapter, which gives emphasis to the drug actions on cell receptors.

B   Chemistry of the drug binding to receptors

The receptor concept was formulated by Langley and the term 'receptor' was proposed by Ehrlich.[4] The concept of receptor binding, *Corpora non agunt nisi fixata* (compounds do not act unless bound), has been subject to refinement but is still valid.

According to the receptor theory of drug action, the common event in the initiation of pharmacological responses is the formation of a complex between the drug molecule and its site of action. Since most pharmacological responses are mediated through receptors, recognition of the more mobile drug molecules by the cellular receptor is the critical element determining the specificity of the response. There must be some forces that not only attract the drug to its receptor, but also hold it in combination with the receptor long enough to initiate the chain of events leading to the effect. Thus the combination of various chemical bonds is of great interest to drug potency.

- Hydrophobic binding plays an important role in stabilizing the conformation of proteins and in the association of hydrophobic structure between the drug and the receptor.
- Hydrogen bonding, which is strongly directional, has considerable importance in stabilizing structures by intramolecular bond formation. The formation of such bonds between a drug and a receptor can result in a relatively stable and reversible interaction. Such bonds are also involved in the maintaining of the tertiary structure of the receptor macromolecule and are thought to be involved in the specificity and selectivity of drug–receptor interaction.
- Drug–receptor interaction often involves charge transfer complexes formed between electron-rich donor molecules and electron-deficient acceptors.
- Ionic bonds which are very ubiquitous are of importance in the actions of ionizable drugs since they act across long distances. The formation of an ionic bond results from the electrostatic attraction that occurs between oppositely charged ions. Most receptors have a number of ionizable groups (COO^-, OH^-, NH_3^+) at physiological pH that are available for the binding with charged drugs.
- The covalent bonds are less important in drug–receptor interaction. Since bonds of this type are so stable at physiological temperature, the binding of a drug to a receptor through covalent bond formation would result in the formation of a long-lasting complex. Although most drug–receptor interactions are readily reversible, some drugs, such as anticancer nitrogen mustards and alkylating compounds form reactive cationic intermediates (i.e. aziridinium ion) that can react with electron donor groups on the receptor. Covalent bonds are also seen for example in the case of penicillin, which acylates a transpeptidase enzyme that is essential to bacterial

cell-wall synthesis. In this case a long-lasting inhibition of bacteria replication is needed. Most drugs, however, induce a brief formation of a reversible drug–receptor complex.

C Drug receptors and physiological mediators

The drug–receptor complex formed at a first stage, triggers, in a second stage, the formation of intracellular messengers or modulates the opening of ion channels. At a third stage, other members of molecular cascades such as protein kinases can be activated, resulting in the physiological changes attributed to the drug. Since most drugs have a considerable degree of selectivity or specificity in their actions, it follows that the receptors with which they interact must be equally unique. It is generally accepted that endogenous or exogenous agents interact specially with a specific receptor site. The cascades of events initiated by drugs binding also occurs with endogenous agents, such as hormones and neurotransmitters.[5]

Many drugs act on such physiological receptors. Those that mimic the effects of the endogenous regulatory compound are termed *agonists*. Other compounds may bind to the receptor, but have no intrinsic regulatory activity. These drugs usually prevent the neurotransmitter from reaching the receptor, by competition at the agonist binding site, and are termed *antagonists* (Fig. 4.1).

The differential capacity to induce responses is attributed either to conformational changes in the receptor or to different states of association of the receptor with active complexes of coupling proteins.

II RECEPTORS AS PRIMARY TARGETS FOR DRUG ACTION

Receptors are commonly classified according to the mediator to which they respond and hence according to their chemical specificity. The name given to a receptor is often derived from this classification (i.e. cholinergic receptors, glutamic receptors). In some cases, when the endogenous mediator is unknown, the receptor is named according to an exogenous compound to which the receptor responds (this was the case for the opioid receptors).

At present, a large number of cell surface receptors for transmitters have been identified, and corresponding amino acid sequences have been determined. These sequences have been used to design synthetic oligonucleotides in order to screen cDNA libraries by conventional gene cloning methods. Molecular pharmacology has allowed detailed analysis of structure–function relationships of signalling proteins through site-directed mutagenesis. The classification inherent in the primary amino acid sequence of the receptor shows a small number of receptor subfamilies that share homologous structures and common mechanisms of action (Fig. 4.2). These different subfamilies correspond to (1) the receptors which gated ion channels; (2) the receptors containing enzymatic activity; (3) the receptors tightly coupled to cytosolic enzymes; (4) the G-protein-coupled receptors; and (5) the nuclear receptors which regulated the transcription of different genes. These receptors subfamilies are coupled to different effectors, yielding different cellular effects. The effects produced by the ionic receptors are very fast (milliseconds), those produced by steroids and thyroids hormones very slow (several minutes to several hours), and those induced by the G coupling receptors occupy an intermediate time scale (seconds to minutes).

A Ligand-gated ion channels

Receptors for several neurotransmitters form ion-selective channels in the plasma membrane and diffuse their signals by altering the cell's membrane potential or the cytoplasmic ionic composition. This subfamily of receptors included the well-characterized nicotinic acetylcholine receptor from electric organ, muscle and brain, the receptors for the excitatory amino acids (aspartate and glutamate), the inhibitory amino acids (γ-aminobutyrate (GABA), glycine),

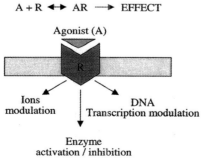
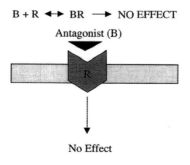

Fig. 4.1 Mode of action of agonists and antagonists.

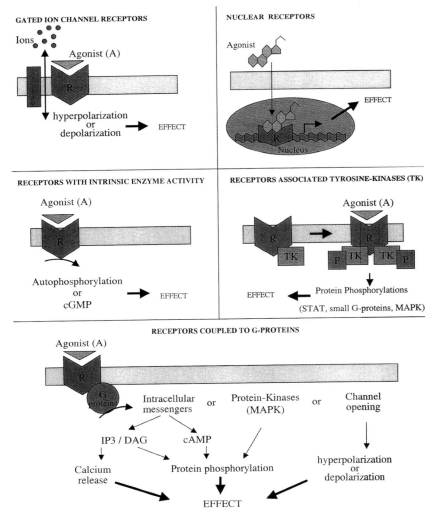

Fig. 4.2 Different subfamilies of receptors.

and the serotoninergic 5HT3 receptors.[6-8] Structural and functional evidence support the view that these membrane-bound allosteric proteins are heteropentameric oligomers. The nicotinic receptor of the electric organ and vertebrate skeletal muscle is a pentamer composed of four distinct subunits (α, β, γ, δ) in the stoichiometric ratio of 2:1:1:1, respectively. In the mature muscle end plates, the gamma subunit is replaced by epsilon. The neuronal type is a pentamer α_2-β_3. The distinct pharmacological and electro-physiological properties are associated with defined combinations of subunits. Chemical labelling and site-directed mutagenesis show that the various members of the superfamily are composed of subunits sharing the same conformational pattern and showing great sequence homology with the nicotinic acetylcholine receptor. The receptors for these rapid-acting neurotransmitters, both excitatory (glutamate, aspartate) and inhibitory (glycine, GABA), appear at a superfamily which has probably arisen by

duplication and divergence of an ancestral neurotransmitter receptor. The significance of this is that the knowledge obtained about the function of any one of these receptors can also be applied to all the receptors belonging to the same family.

The neuromuscular nicotinic acetylcholine receptor is a pentamere in which five structurally related subunits are assembled to form an integral ion channel.[9] Each subunit is believed to contain four transmembrane domains inserted into the plasma membrane. The second transmembrane domain of each subunit forms the ion channel lining. This is also true for the glycine and glutamate receptors. Thus, unlike the G-protein-coupled receptors, the ligand-gated ion channel receptors do not appear to form the ligand-binding site. The channel selectivity among monovalent cations may be confined to a layer located at the cytoplasmic end of the channel domain, as well as to the geometry of the quaternary structure.

The receptors of this family control the fastest synaptic events in the nervous system by increasing transient permeability to particular ions. Excitatory neurotransmitters, such as acetylcholine at the neuromuscular junction or glutamate in the central nervous system, induce an opening of cation channels. These channels are relatively unselective for cations, but because of the electrochemical gradient across the plasma membrane, the major effect of channel opening is an increase in Na^+ and K^+ permeability. This results in a net Na^+ inward current, which depolarizes the cell and increases the generation of an action potential. The action of the transmitter occurs in a fraction of a millisecond and decays within a few milliseconds. In this way the receptor converts a chemical signal (neurotransmitter) into an electrical signal (depolarization). The firing of the action potential is inhibited by other receptor-gated channels as $GABA_A$ or glycine. The GABA receptors exist as a family of subtypes with their pharmacology determined by their composition. The majority of these receptors are potentiated by volatile anaesthetics and alcohols, hence inhibiting levels of synaptic activity in the brain. In addition, the anaesthetic potency correlates very well with their activity at $GABA_A$ receptor subtypes.[10,11]

The current belief is that these receptors are pentameric and the different subunits assemble to form a chloride-sensitive pore.[12,13] Stimulation of these receptors induces the opening of anionic channels (Cl^-) which results in an inward influx of Cl^- (Fig. 4.3). The opening of these channels cause slight hyperpolarization and will resist depolarization induced by excitatory ligands.

B Receptors with intrinsic enzyme activity

The receptors of this family possess an intrinsic enzymatic activity (tyrosine-kinase or guanylate-cyclase) activated following the agonist–receptor interactions.

Receptors with intrinsic tyrosine-kinase activity

These receptors mediate the actions of several hormonal agonists, such as insulin, insulin-like growth factor, epidermal growth factor, and platelet-derived growth

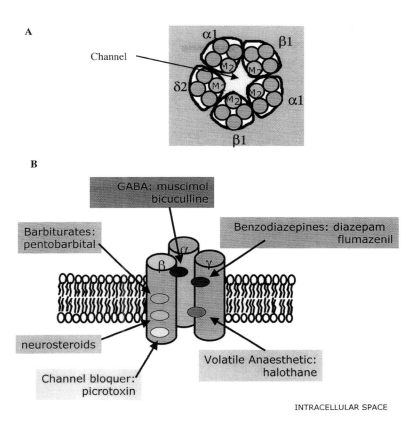

Fig. 4.3 Proposed topography of the γ-amino acid receptor, GABA-A. (A) Cross-section in the plane of the membrane. The receptor complex contained 20 membrane-spanning segments surrounding the central ion channel. Only the M2 membrane-spanning helices form the inner wall of the central channel. (B) The different subunits with their ligand-binding sites are represented.

factor.[14] These receptors contain a very large extracellular domain, rich in cysteines residues, and a cytoplasmic domain containing the tyrosine kinase activity, as well as the sites of autophosphorylation. There are about 400–700 amino-acyls residues in each domain. These receptors span the plasma membrane only once. While certain of these receptors possess a single polypeptide chain (epidermal growth factor receptors, platelet-derived growth factor receptors), others like the insulin receptor possess two chains (α-β) linked as a dimer by disulphide bounds. In this case the α chains possess the ligand-binding site and the β chains, the tyrosyl kinase activity.[15] The tyrosine kinase domain seems to be similar among these receptors, while the ligand-binding domain shows very little sequence homology between the members of this family. The autophosphorylation on tyrosine residues allows association of various proteins characterized by SH2 domains: PLC-γ, adaptors such as Grb2 proteins. These first events initiate cascades of reactions, including small G-proteins, tyrosine-kinases and mitogen-activated protein kinases (MAPKinases) and PI3K. These receptors are able to produce both rapid and slower effects on target cells.

Receptors with intrinsic guanylyl-cyclase activity

Some receptors appear to be linked to the stimulation of guanylyl-cyclase. These membrane receptors mediate the action of the atrial natriuretic peptide (ANP), the heat-stable enterotoxin of *E. coli* and the receptors involved in fertilization in some species. These membrane receptors are glycoproteins of about 130–160 kDa spanning the membrane only once.[16] The binding of the peptides occurs at the extracellular domain, while the intracellular regions contain both cyclase catalytic domains, converting GTP to cGMP, and a protein kinase-like domain.

A second family of guanylyl-cyclases is found in the cytosol. These enzymes are activated by NO, now considered as an endogenous mediator. Thus cytosolic guanylyl-cyclase can be considered as a receptor.

As with to cAMP, cGMP mediates most of its intracellular effects through the activation of specific cGMP-dependent protein kinases (PKG)[17] or by the modulation of ion channel opening.

C Receptors linked to cytosolic tyrosine-kinases

These receptors are quite similar to the above receptors with intrinsic tyrosine-kinase activity, but the receptor and the enzyme entities are two separate proteins.

A large number of cytokines receptor superfamily[18] and growth factors serve as signal carriers in a dynamic cellular communication network. These pleiotropic mediators act synergistically or antagonistically to orchestrate the proliferation and death of cells by acting directly and/or by regulating the expression of other cytokines. Cytokines or growth factors bind to the extracellular domain of the transmembrane receptors and induce intracellular responses through associated cytosolic tyrosine-kinase. Ligand-induced dimerization of the receptor induces the reciprocal tyrosine phosphorylation of the associated proteins, which, in turn, phosphorylates tyrosine residues on the cytoplasmic tail of the receptor.[19] These phosphorylated tyrosines serve as docking sites for the Src homology-2 (SH-2) domain of several endogenous proteins, initiating the regulation of gene transcription.[20]

D G-Protein-coupled receptors

A wide range of neurotransmitters, polypeptides, inflammatory mediators transduce their signal into the interior of the cell by a specific interaction with receptors coupled to G-proteins.[21] The most familiar are the muscarinic acetylcholine receptors, the adrenergic receptors, the dopaminergic receptors, the opioids receptors, as well as many peptides receptors. However, the rapid acquisition of structural information on the members of this subfamily (about 1000 genes) indicates a much greater diversity than previously considered[22] (Table 4.1). This family of receptors also includes receptors for calcium, sphingosine, photons, taste and odorant molecules. The characterization of novel molecular receptor subtypes and their corresponding pharmacological properties extended the existing definitions on the receptor diversity based on pharmacological criteria.

The receptors of this subfamily are characterized by a common topology and by varying degrees of primary sequences similarities. Most of the receptors are formed by a single polypeptide chain of 350–800 residues. Hydropathy

Table 4.1 *Illustration of the diversity of the G-protein-coupled receptor family*

Neurotransmitters receptors	Peptides/Peptide hormones
Adenosine (4)	Angiotensin (2)
Adrenergic alpha-1 (3)	Bombesin (3)
Adrenergic alpha-2 (3)	Bradykinin (2)
Adrenergic-beta (3)	Chemokine-alpha (4)
Dopamine (5)	Chemokine-beta (5)
Glutamate metabotropic (7)	Cholecystokinin (2)
Histamine (3)	Endothelin (2)
Muscarinic acetylcholine (5)	Neuropeptide Y (5)
Serotonine (14)	Somatostatin (5)
Opioid (3)	Tachykinin (3)

The number in brackets refers to the number of receptor subtypes identifed to date.

plots revealed seven hydrophobic regions. They are likely to correspond to transmembrane α-helices which are membrane-spanning domains. This topology is highly conserved between the members of the family.[23] The amino terminal extracellular domain contains potential N-linked glycosylation sites in most receptors. The carboxyl-terminal cytoplasmic end is involved in the coupling to G-proteins and contains a palmitoylation site (Cys) and phosphorylation sites (Thr, Ser), both involved in the receptor desensitization. The three cytoplasmic loops are implicated in the coupling with G-proteins and the third confers the specificity of the coupling to different G-proteins.

Ligand-binding domains

The receptors for the glycoprotein hormones follicle-stimulating hormone (FSH), luteinizing hormone/chorionic gonadotropin (LH/CG) and thyroid stimulating hormone (TSH), show a very large extracellular extremity NH_2 terminal (> 300 amino acids). The construction of chimeric receptors of this subfamily has established that this extremity represented the ligand-binding site.[24,25]

The ligand-binding sites for the neurotransmitter receptors are lacking the significant amino acid extracellular domain. Chimeric α_2-β_2 adrenoceptor constructs reported that switching of the transmembrane domains indicated that those appear to be important in determining ligand-binding specificity. Consistent with findings obtained with the adrenergic receptors,[26] ligand-binding to muscarinic receptors, as well as the other biogenic amine receptors occur in a pocket formed by a ring-like arrangement of seven transmembrane α-helices. Site-directed mutagenesis studies and construction of chimeric adrenoceptors have pointed to the importance of negatively charged aspartic acid residues in transmembrane segments III[27] in the ligand-binding. The role of these negatively charged residues appears to be in binding positively charged ligands (e.g. muscarinic, adrenergic) into the centre of the receptor. Interestingly, these residues seem to be conserved among all other biogenic amine receptors. However, additional molecular interactions can be required to determine the specificity of the binding of a given amine for a particular receptor.[28–30]

Coupling of the receptors to G-proteins

Several lines of evidence suggest that multiple intracellular receptor domains (particularly the cytoplasmic loop between transmembrane segment III–IV and V–VI) are involved in G-protein coupling. In order to delineate receptor domains responsible for the G-protein coupling selectivity displayed by the individual muscarinic receptor subtypes, chimeric m1/m2 and m2/m3 muscarinic receptors have been constructed and functionally characterized in different expression systems. Biochemical and electro-

physiological studies have shown that the exchange of the third intracellular loop between two functionally different muscarinic receptor subtypes results in mutant receptors that display qualitatively the same coupling properties as the wild-type receptor from which the third intracellular loop is derived. These data indicate that the third intracellular loop represents the primary structural determinant dictating G-protein coupling selectivity.[31,32] Similar findings obtained with the chimaeric human β_2-α_{2A} adrenoceptor, where segments of the β_2 adrenoceptor, which normally couples to Gs (stimulation of adenylyl-cyclase), have been replaced with the corresponding segments of the α_{2A} adrenoceptor, coupled to Gi (inhibition of adenylyl-cyclase), indicate that the carboxyl-terminal segment of the third cytoplasmic loop is involved in the coupling with Gs.[33] Interestingly, synthetic peptides as short as 15 amino acids in length that mimic those sequences can activate purified G-proteins *in vitro*.[34] Thus one function of the intact receptor structure must be to prevent these small activating domains from interacting with G-proteins until an agonist binds to the receptor.

Mechanism of multifunctional signalling by G-proteins

Addition of a drug to a cell results frequently in a regulation of multiple cellular signalling cascades. G-proteins represent the level of middle management in the organizational hierarchy able to communicate between the receptors and the effectors (enzymes or ion channels). These G-proteins are heterotrimeric and are composed of α, β, and γ subunits.[35] Activated cell surface receptors initiate G-protein signalling, promoting the exchange of GDP, inducing the dissociation of the α-subunit from a high stable β-γ dimer. In this dissociated state both α- and β-γ subunits modulate the activity of effector molecules.[36] Slow hydrolysis of bound GTP by the GTPase intrinsic to the (α-subunit leads to reassociation of the oligomer and cessation of the signal (Fig. 4.4).

The diversity of heterotrimeric G-proteins was demonstrated around 1980, with the purification of Gs (s = stimulatory for adenylyl-cyclase), Gi (i = inhibitory for adenylyl-cyclase), and Gt (t = transducine which activate cGMP phosphodiesterase in retinal cells). To date, at least 20 G-α (39 to 52 kDa in size), 5 G-β (36 kDa) and 13 G-γ (7–8 kDa) subunits have been identified in mammalian system. The usual classification of G-proteins remains based on their α-subunits with four subfamilies: Gs including α-s and α-olf (olf = olfactive); Gi including α-i1, i2, i3, o1, o2, t1, t2, z and α-gust (gust = gustducin); Gq including α-q, 11, 14, 15, 16; and G12 including α_{12} and α_{13}.

G-protein subunits are highly homologous in both primary sequence and tertiary structure, but the effectors they regulate are extremely diverse.[37,38]

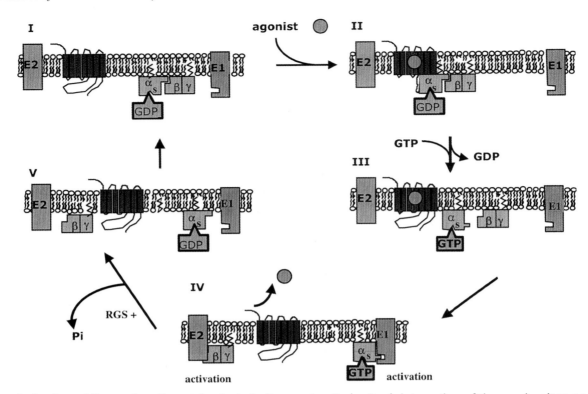

Fig. 4.4 Activation of G-proteins. G-proteins include three subunits (α, β, γ). Interaction of the α-subunit to an agonist stimulated receptor (I–II) causes the exchange of the bound GDP with GTP (II–III). The α-GTP complex and the dimer β-γ dissociated. The α-GTP complex interacted with an effector (E1) and the dimer β-γ with another effector (E2) (IV). The α-subunit catalyses hydrolysis of the bound GTP to GDP (V) and reassociated with the dimer β-γ (I). This deactivation of signalling can be accelerated by proteins termed regulators of G-protein signalling (RGS) which have been shown to directly bind to the α-subunit of G-proteins. (Adapted from Gies, J. -P. (1993) *Bases de Pharmacologie Moléculaire*, Published by Ellipses-Edition Marketing, Paris).

The key aspect in the function of G-proteins is that they can interact with several different receptors and effectors in a promiscuous fashion. A number of G-protein-linked receptors have recently been shown to regulate multiple effector pathways both when expressed endogenously and, more frequently, following transfection into heterologous systems. Direct evidence for the interaction of a single receptor with multiple classes of G-proteins has been obtained. The thyrotrophin-releasing hormone (THR) G-protein-linked receptor has, for example, been shown to be able to activate both adenylyl-cyclase and phosphoinositidase C, probably by interacting with two different G-proteins.[39] In another case, a single receptor can modulate two distinct effectors by interacting with a single G-protein. Particularly the α-2-adrenoceptor, when expressed at high level in fibroblasts, has been shown to inhibit adenylyl cyclase and to stimulate phosphoinositidase C. This may reflect of Gi-α-mediated inhibition of adenylyl-cyclase and a Gi-β-γ-mediated activation of phosphoinositidase C.[40] Many other receptors have been

shown to have multifunctional signalling potential which vary with the cell type.[41]

Thus, the receptor–G-protein–effector systems are complex networks of convergent and divergent interactions that permit extraordinarily versatile regulation of cell function.

Effectors modulated by the G-protein-coupled receptors

This family of membrane receptors regulated distinct effector proteins through the activation of GTP-binding proteins, known as G-proteins. The targets of G-proteins include enzymes such as adenylyl-cyclase, phospholipases C and PI3K, channels that are specific for Ca^{++} or K^{+}.

Adenylyl-cyclase/cAMP system. Sutherland's studies revealed that the intracellular messenger cyclic $3',5'$-adenosine monophosphate (cyclic-AMP) is a nucleotide synthesized within the cell from ATP by the action of adenylyl-cyclase. Receptors linked to Gs activate adenylyl-cyclase and those linked to Gi inhibit the enzyme and reduce the c-AMP formation.[42,43]

The cloning and the modelled topography revealed nine cloned isoforms of mammalian adenylyl-cyclases. This enzyme contained a pair of six membrane-spanning segments separated by a large cytoplasmic loop and one major extracellular loop.[44] Two similar cytoplasmic domains around 250 amino acids should be the nucleotide-binding sites. The adenylyl-cyclase bear a striking resemblance to proteins of markedly different function such as transporters, channels (dihydropyridine sensitive Ca^{++} channel) and the drug efflux pump (P-glycoprotein), whose synthesis is enhanced in multidrug resistant cells.[45]

The c-AMP is broken down by phosphodiesterases which hydrolyse the $3'$-phosphate ester to give the common inactive metabolite, $5'$-AMP. Given that cyclic nucleotide signalling regulates a wide variety of cellular functions, it is not surprising that cyclic nucleotide phosphodiesterases (PDEs) are represented by a large superfamily of enzymes.[46] The PDEs superfamily currently includes 19 different genes subgrouped into different PDE families, and it is likely that more will be added in the coming years. PDEs 5, 6, 9 and 11 are selective for the hydrolysis of cGMP, others families are selective for cAMP.

There are also several pharmacological agents which elevate c-AMP itself inside the cell, such as caffeine and theophilline. Isobutylmethylxanthine is another phosphodiesterase inhibitor which increases the c-AMP content of the cells. The plant terpenoid forskolin stimulates c-AMP formation by acting directly on the adenylyl-cyclase.

A large number of drugs and neurotransmitters exert their effects by increasing or decreasing the catalytic activity of adenylyl-cyclase and thus raising or lowering the c-AMP content in the cell. There are numerous and varied regulatory effects of c-AMP on cellular function, including, for example, enzymes involved in energy metabolism, ion transport, ion channel function leading to changes in neuronal excitability, cell differentiation, or contractile proteins. The biological effects can be as diverse as gluconeogenesis (glucagon in liver), lipolysis (adrenaline on β_1 receptors in adipocytes), Na^+/water reabsorption (vasopressin on V2 receptors in kidney), contraction (adrenaline on α_2 vascular smooth muscle receptors), relaxation (acetycholine on muscarinic M_2 receptors in the heart). These various effects are however, all brought about a common mechanism, namely the activation of protein kinase A by the intracellular mediator, cAMP.

Phospholipase C/DAG/IP3. The phosphoinositide system is an important intracellular messenger system. Two enzymes families are targets of G-proteins, phosphoinositide-specific phospholipase C (PLC), and phosphoinositide 3-kinase (PI3K). It is well established that receptor-dependent activation of phospholipase C (PLC) results in the hydrolysis of phosphatidylinositol 4,5-bisphosphate (PiP_2) and the formation of two intracellular mediators: inositol (1,4,5)-triphosphate (IP_3) and diacylglycerol (DAG).

Three mammalian PLC subtypes (β: four isoforms, γ: two isoforms, δ: four isoforms) have been isolated, but only the PLC-β family appears to be regulated directly by G-proteins.[47] IP_3 binds to endoplasmic membrane receptors and liberates the calcium from the sequestered stores (endoplasmic reticulum) inducing an increase of cytoplasmic calcium. An elevation of free calcium in the cell can induce, for example, a smooth muscle contraction, secretion from exocrine glands and transmitter release.

DAG is the principal endogenous regulator of membrane-bound protein kinases C. So far, ten different PKC isoenzymes have been identified in different species, tissues and cell lines and more are being identified by reverse transcription of RNA.[48] It is now known to consist of a family of isoenzymes that differ in their structures, cofactor requirements and substrate specificities. These enzymes control phosphorylation of serine and threonine residues of a variety of intracellular proteins, also inducing a wide range of pharmacological effects: tumour propagation, inflammatory responses, contraction or relaxation of smooth muscle, increase or decrease of neurotransmitter release, increase or decrease of neuronal excitability (by phosphorylating ion channels) and receptor desensitization. There are other types of kinases involved in these pharmacological effects such as those regulated by cAMP or cGMP, those controlled directly by the family of receptor-protein kinases (discussed latter). This multitude of pharmacological effects underline the importance of proteins phosphorylation regulation.

The phosphoinositide 3-kinase (PI3K). Phosphoinositide 3-kinases (PI3Ks) are a large family of intracellular signal transducers that have attracted much attention over the past 10 years. PI3Ks phosphorylate inositol lipids at the $3'$ position of the inositol ring to generate the 3-phosphoinositides PI(3)P, PI(3,4)P2 and PI(3,4,5)P3. Three PI3K classes have been defined on the basis of their primary structure and *in vitro* lipid substrate specificity.[49] PI3Ks of class 1B are directly activated by G$\beta\gamma$. PI3K lipid products are not substrates for PLC, and thus initiate original signal pathways. PIP3 remains associated to membranes and is recognized by PH domains of various proteins leading to their recruitment to the plasma membrane and to their activation. The PH domain was originally described as a novel protein motif of about 100 amino acid residues, repeated twice in the protein plekstrin. These motifs have now been identified in more than 100 proteins. PH domains function as adaptors or tethers linking their host proteins to the plasma membrane inner surface. Principal membrane-binding partners are G$\beta\gamma$ and phosphoinositides, such as PIP3.

Also in 1998, PTEN (a major tumour suppressor in human cancer) was also shown to antagonize PI3K signalling by removing the 3-phosphate from 3-phosphoinositides. This 3-phosphoinositide-phosphatase activity clearly establishes the importance of PI3Ks in cancer and provides an excellent rationale for the development of PI3K inhibitors for therapeutic application.[50]

Regulation of ion channels. G-protein-coupled receptors can control ion channel function by interacting directly with the channel. It has been reported that G-protein $\beta\gamma$ dimers may directly modulate the opening of the voltage-gated calcium[51] and potassium[52] channels. For example, stimulation of M2 muscarinic receptors by acetylcholine in pacemaker cells (sinoatrial, atrioventricular and atrial) activates a K^+ current (I_{KAch}.). Heart rate is thus slowed by hyperpolarization of the pacemaker depolarization potential, as well as by blocking tonic β-adrenergic stimulation of depolarizing pacemaking channels.

E Intracellular receptors

Intracellular receptors are a class of ligand-dependent transcription factors that include receptors for both steroid and non-steroid hormones. In the non-liganded state, these receptors reside in the cytoplasm and/or in the nucleus. Upon binding to the hormone, these receptors homo- or heterodimerize, then regulate, positively or negatively, gene expression by binding to specific sequences in the chromatin, termed HRE (hormone responsive element).

Steroid receptors include receptors for glucocorticoids, progestins, estrogens, androgens and mineralocorticoids. The non-steroidal compounds include retinoids, thyroid hormones and vitamin D, and a large set of exogenous compounds, such as dioxin. These receptors are organized into three major domains:[53] a variable amino-terminal domain responsible for antigenic properties of the receptors; a relatively well-conserved carboxyterminal domain which represents the hormone binding domain, a well-conserved, cystein-rich central domain containing two Zn^{++}-stabilized fingers, which mediates binding to specific sites on nuclear DNA to activate or inhibit transcription of the nearby gene.

It is therefore not surprising that the steroids are identified in many quarters of modern biology as agents that regulate gene expression. Another area of steroid action is related to their lipophilic character and their effect on cell surface events.

Certain steroids positively modulate GABA-induced Cl^- flux in a manner that resembles that of the barbiturates, although the steroid modulatory site is believed to differ from that of barbiturates. Steroids are also able to activate different genes and thus activate different patterns of protein synthesis. For instance, glucocorticoids enhance the production of anti-inflammatory compounds like annexins (lipocortin); mineralocorticoids stimulate the production of transport proteins (sodium channels) involved in renal tubular function.

There is also an impressive array of genomic and non-genomic effects of steroids on neural activity.[54]

III DRUG ACTING THROUGH VOLTAGE-SENSITIVE IONIC CHANNELS

Ion channels are essential for a wide range of functions such as muscle contraction, sensory transduction, endocrine and exocrine secretions. The ion channels mediate Na^+, Ca^{++}, K^+ and Cl^- conductance induced by membrane potential changes. These channels propagate action potentials in excitable cells and are also involved in the regulation of membrane potential and intracellular Ca^{++} transients in most eukaryotic cells. The molecular structure of the ion channels differ from those of the ligand-gated ion channels (or neurotransmitter channels), in which the ligand receptor and the ion channel form a single functional entity. Moreover, the voltage-gated channels reveal some common structural and functional features with the ligand-gated channels (Fig. 4.5). In this chapter we consider the main types of ion channels.[55]

A The Na^+ channels

These play a critical role in initiating action potential. The activation of the channels allows for the inward movement of Na^+ from the extracellular space of the cell. The Na^+ channels from brain and skeletal muscle are heterooligomeric and composed of α- and β-subunits. The α-subunit determines the major functional characteristics of Na^+ channels. Nine related functional α-subunits have been described.[56] Each subunit consists of a 1800–2000 residue polypeptide composed of four repeats, each containing six putative transmembrane helices.

The α-subunit shares significant homology with other voltage-gated ion channels, particularly the α_1 subunit of the Ca^{++} channel.

Site-directed mutagenesis studies suggest that: (1) the fourth membrane segment may act as the voltage sensor; (2) the cytoplasmic loop linking repeats III and IV is essential for inactivation and that the region between segment V and VI forms the permeation pathway.

B The Ca^{++} channels

In addition to their normal physiological functions (muscle contraction, neurotransmitter release, gene transcription, etc.), calcium channels are also implicated in human disorders (cerebellar ataxia, epilepsy, hypertension, arrhyth-

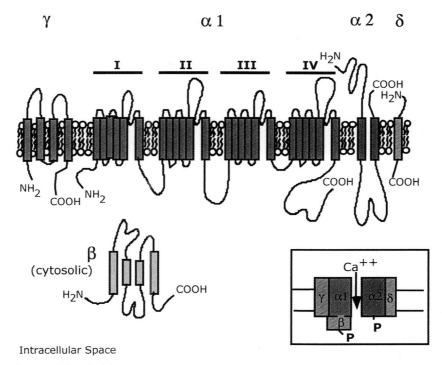

Fig. 4.5 Putative topology of the Ca^{++} voltage-gated channel. The α-subunits of the calcium and sodium channels share similar presumptive six-transmembrane structure, repeated four times. The calcium channel also requires auxiliary proteins such as α_2, β, γ and δ.

mias, etc.). The treatment of some of these disorders has been aided by the development of therapeutic calcium channel antagonists.

A number of calcium channel subtypes have been identified and are classified by their distinct pharmacological and electrophysiological properties into T-, N-, L-, P-, Q- and R-type.[57]

The L-type Ca^{++} channels are the best characterized of the voltage-gated Ca^{++} channels. They are heterooligomeric and composed of α_1-, α_2-, β-, δ- and γ-subunits. The α_1-subunit provides the binding site for the L channel blockers, such as nifedipine. Different types of the α_1-subunit, present in skeletal and cardiac muscle and brain, are encoded by distinct genes; alternative splicing may also occur. Each α_1-subunit includes four homologous repeats containing six transmembrane segments, and shares significant homology with other voltage-gated ion channels, particularly the Na^+ channel.

The Ca^{++} channels of intracellular membranes, which regulate Ca^{++} release from internal stores, comprise the ryanodine and inositol trisphosphate receptor families.[58] The sarcoplasmic reticulum Ca^{++}-release channel of the skeletal muscle is the best ryanodin receptor that has been studied so far. It occupies a central position in excitation–contraction coupling in skeletal muscle by

linking T-tubule depolarization to sarcoplasmic reticulum Ca^{++} release.

C The K^+ channels

The potassium channels are highly heterogeneous.[59] The K^+ channels gene only encodes a single repeat and four polypeptides come together to form the channel. Four types can be distinguished: (1) the voltage-dependent K^+ channels; (2) the Ca^{++}-activated K^+ channels; (3) the receptor-coupled K^+ channels, and other K^+ channels (ATP-sensitive K^+ channels, Na^+-activated K^+ channels, etc.).

D The Cl^- channels

Chloride channels, with the exception of the ligand-gated γ-aminobutiric (GABA) and glycine receptors, have been grouped into four main families on the basis of their molecular properties: the CFTR (cystic fibrosis transmembrane conductance regulator) channel,[60] the calcium-activated chloride channel family,[61] the voltage-dependent anion-selective channels (VDAC),[62] and the ClC family of voltage-gated chloride channels.[63]

ClC channels have emerged as an influential family, members of which play important cellular functions, ranging from the control of membrane excitability in neurons and muscle to salt movement through epithelia.[64]

IV DRUG ACTING THROUGH TRANSPORTERS

For many years, there was little information about these integral membrane proteins. Cloning and sequencing have considerably increased knowledge in this field. Transporter genes encode proteins, generally constituted by 12 transmembrane spanning regions. These mediated Na^+- or H^+-dependent accumulation of small molecules such as neurotransmitters, antibiotics and ions into the cells or organelles. The transport is performed by different mechanisms: uniport, substrate–ion symport, substrate–ion antiport, substrate–substrate antiport or ATP-dependent translocation. The transporters can be classified into families and subfamilies based on ion dependence. The family that is of key importance for neurotransmitter uptake is the Na^+/Cl^--dependent neurotransmitter transporters. This contained monoamine-recognizing (dopamine, noradrenaline, 5HT) and amino acid-like (GABA, glycine, betaine, taurine, proline) subfamilies. So far there are other families which have been described such as the Na^+/K^+/glutamate transporter, the Na^+-dependent glucose transporter, the H^+-dependent vesicular monoamine transporters. These transporters, which are transmembrane proteins, represent sites of action of several drugs, such as antidepressants, or psychostimulants like cocaine.[65]

V ENZYMES AS DRUG TARGETS

Several families of drugs do not act on receptors and their therapeutic properties are attributed to inhibition or activation of enzyme activities. A number of drugs in clinical use exert their effect by inhibiting a specific enzyme present either in tissues of an individual under treatment or in those of an invading organism. The basis of using enzyme inhibitors as drugs is that inhibition of a suitable selected target enzyme leads to a build-up in concentration of substrate and a corresponding decrease in concentration of the metabolite, which leads to a useful clinical response. The type of inhibitor selected for a particular target enzyme may be important in producing a useful clinical effect. Enzyme inhibiting processes may be divided into two main classes, reversible and irreversible, depending upon the manner in which the inhibitor is attached to the enzyme.

Reversible inhibition occurs when the inhibitor is bound to the enzyme through a suitable combination of Van der Vaals', electrostatic, hydrogen bonding, and hydrophobic attractive forces. Reversible inhibitors may be competitive, noncompetitive, uncompetitive, or of mixed type.

During irreversible inhibition, after initial binding of the inhibitor to the enzyme, covalent bonds are formed between a functional group on the enzyme and the inhibitor. This is the case, for example, for the active-site-directed inhibitors (affinity labelling).

The inhibitors used in therapy must possess a high specificity towards the target enzyme, since inhibition of closely related enzymes with different biological functions could lead to a range of side-effects.

There is a very large area of enzyme targets as illustrated in Table 4.2. For example, dihydrofolate reductase (DHFR) catalyses the NADPH-linked reduction of dihydrofolate to tetrahydrofolate. The tetrahydrofolate are cofactors for the biosynthesis of nucleic acids and aminoacids. The reduction of their level induces a limitation of cell growth.[66]

Thymidylate synthase methylated the deoxyuridylate into thymidylate using 5,10-methylenetetrahydrofolate as a cofactor. This reaction is the rate-limiting step in the *de novo* synthetic pathway to thymidine nucleotide. Antitumoral effects are obtained with antifolate compounds.[67]

Table 4.2 *Selective illustration of enzymes inhibitors*

Enzymes	Inhibitors	Diseases
Dihydrofolate reductase	Methotrexate	Cancer
Thymidylate synthase	Fluorouracyl	Cancer
HIV reverse transcriptase	Zidovudine (AZT)	AIDS
HIV protease	U75875	AIDS
Angiotensin converting enzyme	Captopril, enalapril	Hypertension
Cyclooxygenase	Aspirin, indomethacin	Inflammation, pain
Catechol-O-methyltransferase	Ro41-0960	Parkinson's Disease
Acetylcholinesterase	Organophosphorus	Myasthenia gravis, glaucoma, Alzheimer disease
H^+/K^+ ATPase 'proton pump'	Omeprazol	Gastric secretion, ulcers

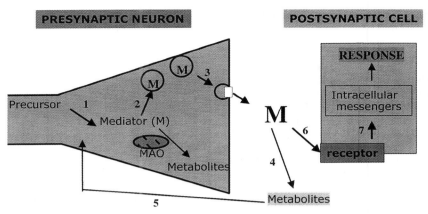

Fig. 4.6 Drugs influencing synaptic transmission by: (1) inhibiting enzymes that synthesize neurotransmitters; (2) preventing neurotransmitter storage in synaptic vesicles; (3) blocking the release of neurotransmitter into the synapse; (4) blocking enzymes that degrade neurotransmitters; (5) blocking neurotransmitter (or metabolite) re-uptake; (6) binding to the receptor and either mimicking or blocking neurotransmitters; and (7) interfering with second-messenger activity. (Adapted from Gies, J. P. (1993) *Bases de Pharmacologie Moléculaire*. Published by Ellipses-Edition Marketing, Paris.)

A functional HIV protease is required for the production of infective virions; this key role of the protease in the viral life cycle makes the inhibition of this enzyme a potential way for therapeutic intervention in the treatment of AIDS. New strategies for the development of bifunctional inhibitors, which combine the protease inhibitor and another enzyme inhibitor in one molecule are under investigation.[68]

Angiotensin-converting enzyme (ACE) inhibitors are used for the treatment of high blood pressure,[69] and were designated using the carboxypeptidase structure as a model for Zn^{++} protease action.[70] Captopril is a small, potent, orally available, dipeptidyl inhibitor of ACE.

Acetylcholinesterase (AchE) hydrolyses the neurotransmitter acetylcholine and yields acetic acid and choline. AchE is a serine hydrolase inhibited by organophosphorus poisons, as well as by carbamates and sulfonyl halides which form a covalent bond to a serine residue in the active site.[71] AchE inhibitors are used in the treatment of various disorders.[72]

In conclusion, the modification by drugs of a precise function can be achieved in several different ways, acting on the receptors of mediators, on enzymes or transmembrane exchange processes. This view is illustrated in Fig. 4.6 showing a synapse. Future development of drug targets will be based on the diversity of biological targets, according to the definition of the human genome, increasing the selectivity of drugs. Recent development in gene drug delivery systems[73] and in antisense oligonucleotide technology[74,75] might also be crucial factors in drug development.

REFERENCES

1. Ehrlich, P. (1956) The relation existing between chemical constitution, distribution and pharmacological action. In *The Collected Papers of Paul Ehrlich*, vol. 1. Pergamon Press, London.

2. Levine, R.R. (1990) *Pharmacology: Drug Actions and Reactions*, 4th edn. Little Brown and Company, Boston.

3. Ross, E.M. (1996) Pharmacodynamics: mechanisms of drug action and the relationship between drug concentration and effect. In Hardman, J.G., Limbird, L.E., Molinoff, P.B., Ruddon, R.W. and Gilman, A.G. (eds). *The Pharmacological Basis of Therapeutics*, pp. 29–41. McGraw-Hill, New York.

4. Ehrlich, P. (1913) Chemotherapeutics: scientific principles, methods and results. *Lancet* **ii**: 445–451.

5. Rang, H.P., Dale, M.M. and Ritter, J.M. (1995) How drugs act: General principles. In Rang, H.P., Dale, M.M. and Ritter, J.M. (eds). *Pharmacology*, pp. 3–21. Churchill Livingstone, London.

6. Lena, C., de Kerchove D'Exaerde, A. M., Cordero-Erausquin, M., Le Novere, N., del Mar Arroyo-Jimenez, M. and Changeux, J.-P. (1999) Diversity and distribution of nicotinic acetylcholine receptors in the locus ceruleus neutrons. *Proc. Natl Acad. Sci. USA* **96**: 12126–12131.

7. Bormann, J. (2000) The 'ABC' of GABA receptors. *Trends Pharmacol. Sci.* **21**: 16–19.

8. Ortells, M.O. and Lunt, G.G. (1995) Evolutionary history of the ligand-gated ion-channel superfamily of receptors. *Trends Neurosci.* **18**: 121–127.

9. Dani, J.A. (2001) Overview of nicotinic receptors and their roles in the central nervous system. *Biol. Psych.* **49**: 166–174.

10. Brussaard, A.B., Devay, P., Leyting-Vermeulen, J.L. and Kits, K.S. (1999) Changes in properties and neurosteroid regulation of GABA-ergic synapses in the supraoptic nucleus during the mammalian female reproductive cycle. *J. Physiol.* **516**: 513–524.

11. Thompson, S.A. and Wafford, K. (2001) Mechanism of action of general anaesthetics-new information from molecular pharmacology. *Curr. Opin. Pharmacol.* **1**: 78–83.

12. Chebib, M. and Johnston, G.A. (2000) GABA-activated ligand gated ion channels: medicinal chemistry and molecular biology. *J. Med. Chem.* **43**: 1427–1447.

13. Zhang, D., Pan, Z.H., Awobuluyi, M. and Lipton, S.A. (2001) Structure and function of GABA-C receptors: a comparison of native versus recombinant receptors. *Trends Pharmacol. Sci.* **22**: 121–132.

14. Sclessinger, J. (2000) Cell signalling by receptor tyrosine kinases. *Cell* **103**: 211–225.

15. Whitehead, J.P., Clark, S.F., Urso, B. and James, D.E. (2000) Signalling through the insulin receptor. *Curr. Opin. Cel. Biol.* **12**: 222–228.

16. Potter, L.R. and Hunter, T. (2001) Guanylcyclase-linked natriuretic peptide receptors: structure and regulation. *J. Biol. Chem.* **276**: 6057–6060.

17. Francis, S.H. and Corbin, J.D. (1999) Cyclic nucleotide-dependent protein kinases: intracellular receptors for cAMP and cGMP action. *Crit. Rev. Clin. Lab. Sci.* **36**: 275–328.

18. Touw, I.P., De Koning, J.P., Ward, A.C. and Hermans, M.H. (2000) Signaling mechanisms of cytokine receptors and their perturbances in disease. *Mol. Cell. Endocrinol.* **160**: 1–9.

19. Schindler, C. (1999) Cytokines and JAK-STAT signalling. *Exp. Cell. Res.* **253**: 7–14.

20. Horvath, C.M. (2000) STAT proteins and transcriptional responses to extracellular signals. *Trends Biochem. Sci.* **25**: 496–502.

21. Wess, J. (1998) Molecular basis of receptor/G-protein-coupling selectivity. *Pharmacol. Ther.* **80**: 231–264.

22. Bockaert, J. and Pin, J.-P. (1998) Use of a G-protein-coupled receptor to communicate. A success during evolution. *CR Acad. Sci. Paris* **321**: 529–551.

23. Zeng, F.Y. and Wess, J. (2000) Molecular aspects of muscarinic receptor dimerization. *Neuropsychopharmacol.* **23**: S19–S31.

24. Nagayama, Y., Wadsworth, H.L., Chazenbalk, G.D., Russo, D., Seto, P. and Rapoport, B. (1991) Thyrotropin-luteinizing hormone/chorionic gonadotropin receptor extracellular domain chimeras as probes for thyrotropin receptor function. *Proc. Natl Acad. Sci. USA* **88**: 902–905.

25. Braun, T., Schofield, P.R. and Sprengel, R. (1991) Amino-terminal leucine-rich repeats in gonadotropin receptors determine hormone selectivity. *EMBO J.* **10**: 1885–1890.

26. Dohlman, H.G., Thorner, J., Caron, M.G. and Lefkowitz, R.J. (1991) Model systems for the study of seven-transmembrane-segment receptors. *Annu. Rev. Biochem.* **60**: 653–688.

27. Kobilka, B.K., Kobilka, T.S., Daniel, K., Regan, J.W., Caron, M.G. and Lefkowitz, R.J. (1988) Chimeric alpha-2 beta-2 adrenergic receptors: delineation of domains involved in effector coupling and ligand binding specificity. *Science* **240**: 1310–1316.

28. Wess, J. (1993) Mutational analysis of muscarinic acetylcholine receptors: structural basis of ligand/receptor/G protein interactions. *Life Sci.* **53**: 1447–1463.

29. Heitz, F., Holzwarth, J.A., Gies, J.-P., Pruss, R.M., Trumpp-Kallmeyer, S., Hibert, M. and Guenet, C. (1999) Site directed mutagenesis of the putative human muscarinic M2 receptor binding site. *Eur. J. Pharmacol.* **380**: 183–195.

30. Haddad, E.B., Landry, Y. and Gies, J.P. (1990) Sialic acid residues as catalysts for M2-muscarinic agonist receptor interactions. *Mol. Pharmacol.* **37**: 682–688.

31. Lechleiter, J., Hellmiss, R., Duerson, K., Ennulat, D., David, N., Clapham, D. and Peralta, E. (1990) Distinct sequence elements control the specificity of G-protein activation by muscarinic acetylcholine receptor subtypes. *EMBO J.* **9**: 4381–4390.

32. Wess, J., Bonner, T.I., Doürje, F. and Brann, M.R. (1990) Delineation of muscarinic receptor domains conferring selectivity of coupling to guanine nucleotide-binding proteins and second messengers. *Mol. Pharmacol.* **38**: 517–523.

33. Ligget, S.B., Caron, M.G., Lefkowitz, R.J. and Hnatowich, M. (1991) Coupling of a mutated form of the human beta-2 adrenergic receptor to Gi and Gs; requirement for multiple cytoplasmic domains in the coupling process. *J. Biol. Chem.* **266**: 4816–4821.

34. Okamoto, T., Murayama, Y., Hayashi, Y., Inagaki, M., Ogata, E. and Nishimoto, I. (1991) Identification of a Gs activator region of the beta 2-adrenergic receptor that is autoregulated via protein kinase A-dependent phosphorylation. *Cell* **67**: 723–730.

35. Hamm, H.E. (1998) The many faces of G-protein signalling. *J. Biol. Chem.* **273**: 669–672.

36. Ivanova-Nikolova, T.T. and Breitwieser, G.E. (1997) Effector contributions to G beta gamma-mediated signaling as revealed by muscarinic potassium channel gating. *J. Gen. Physiol.* **109**: 245–253.

37. Clapham, D.E. and Neer, E.J. (1997) G protein beta-gamma subunits. *Annu. Rev. Pharmacol. Toxicol.* **37**: 167–203.

38. Watson, A.J., Aragay, A.M., Slepak, V.Z. and Simon, M.I. (1996) A novel form of the G protein beta-subunit G-beta5 is specifically expressed in the vertebrate retina. *J. Biol. Chem.* **271**: 28154–28160.

39. Paulssen, R.H., Paulssen, E.J., Gautvik, K.M. and Gordeladze, J.O. (1992) The thyroliberin receptor interacts directly with a stimulatory guanine-nucleotide-binding protein in the activation of adenylyl cyclase in GH_3 rat pituitary tumor cells. Evidence obtained by the use of antisense RNA inhibition and immunoblocking of the stimulatory guanine-nucleotide-binding protein. *Eur. J. Biochem.* **204**: 413–418.

40. Cotecchia, S., Kobilka, B.K., Daniel, K.W., Nolan, R.D., Lapetina, E.Y., Caron, M.G., Lefkowitz, R.J. and Regan, J.W. (1990) Multiple second messenger pathways of alpha-adrenergic receptor subtypes expressed in eukaryotic cells. *J. Biol. Chem.* **265**: 63–69.

41. Milligan, G. (1993) Mechanisms of multifunctional signalling by G-protein-linked receptors. *Trends Pharmacol. Sci.* **14**: 239–244.

42. Hurley, J.H. (1999) Structure, mechanism, and regulation of mammalian adenylyl cyclase. *J. Biol. Chem.* **274**: 7599–7602.

43. Chern, Y. (2000) Regulation of adenylyl cyclase in the central nervous system. *Cell Signal* **12**: 195–204.

44. Dessauer, C.W., Tesmer, J.J., Sprang, S.R. and Gilman, A.G. (1998) Identification of a Gialpha binding site on type V adenylyl cyclase. *J. Biol. Chem.* **273**: 25831–25839.

45. Gottesman, M.M. and Pastan, I. (1988) Resistance to multiple chemotherapeutic agents in human cancer cells. *Trends Pharmacol. Sci.* **9**: 54–58.

46. Soderling, S.H. and Beavo, J.A. (2000) Regulation of cAMP and cGMP signalling: new phosphodiesterases and new functions. *Curr. Opin. Cell Biol.* **12**: 174–179.

47. Rebecchi, M.J. and Pentyala, S.N. (2000) Structure, function, and control of phosphoinositide-specific phospholipase C. *Physiol. Rev.* **80**: 1291–1335.

48. Chang, J.D., Xu, Y., Raychowdhury, M.K. and Ware, J.A. (1993) Molecular cloning and expression of a cDNA encoding a novel isoenzyme of protein kinase C (nPKC). *J. Biol. Chem.* **268**: 14208–14214.

49. Vanhaesebroeck, B., Leevers, S.J., Panayotou, G. and Waterfield, M.D. (1997) Phosphoinositide 3-kinases; a conserved family of signal transducers. *Trends Biochem. Sci.* **22**: 267–272.

50. Leevers, S.J., Vanhaesebroeck, B. and Waterfield, M.D. (1999) Signalling through phosphoinositide 3-kinases: the lipids take centre stage. *Curr. Opin. Cell Biol.* **11**: 219–225.

51. Herlitze, S., Zhong, H., Scheur, T. and Catterall, W.A. (2001) Allosteric modulation of Ca^{2+} channels by G-proteins, voltage-dependent facilitation, protein kinase C, and Ca(v)beta subunits. *Proc. Natl Acad. Sci. USA* **98**: 4699–4704.

52. Mark, M.D. and Herlitze, S. (2000) G-protein mediated gating of inward-rectifier K^+ channels. *Eur. J. Biochem.* **267**: 5830–5836.

53. Kumar, R. and Thompson, E.B. (1999) The structure of the nuclear hormone receptors. *Steroids* **64**: 310–319.

54. Falkenstein, E., Tillmann, H. -C., Christ, M., Feuring, M. and Wehling, M. (2000) Multiple actions of steroid hormones — A focus on rapid, nongenomic effects. *Pharmacol. Rev.* **52**: 513–555.

55. Watson, S. and Girdlestone, D. (1994) Receptor and ion channel nomenclature. *Trends Pharmacol. Sci.* (Suppl.) 43–48.

56. Catterall, W.A. (2000) From ionic currents to molecular mechanisms: the structure and function of voltage-gated sodium channels. *Neuron* **26**: 13–25.

57. Catterall, W.A. (2000) Structure and regulation of voltage-gated calcium channels. *Annu. Rev. Cell Dev. Biol.* **16**: 521–555.

58. Ehrlich, B.E., Kaftan, E., Bezprozvannaya, S. and Bezprozvanny, L. (1994) The pharmacology of intracellular Ca^{++} release channels. *Trends Pharmacol. Sci.* **15**: 145–149.

59. Bloom, F.E. (1996) Neurotransmission and the central nervous system. In Hardman, J.G., Limbird, L.E., Molinoff, P.B., Ruddon, R.W. and Gilman, A.G. (eds). *The Pharmacological Basis of Therapeutics*, pp. 267–293. McGraw-Hill, New York.

60. Sheppard, D.N. and Welsh, M.J. (1999) Structure and function of the CFTR chloride channel. *Physiol. Rev.* **79**: S23–S45.

61. Romio, L., Musante, L., Cinti, R., Seri, M., Moran, O., Zegara-Moran, O. and Galietta, U.V. (1999) Characterization of a murine gene homologous to the bovine CaCC chloride channel. *Gene* **228**: 181–188.

62. Reymann, S., Florke, H., Heiden, M., Jakob, C., Stadtmuller, U., Steinacker, P., Lalk, V.E., Pardowitz, I. and Thinnes, F. (1995) Further evidence for multitopological localization of mammalian porin (VDAC) in the plasmalemma forming part of a chloride channel complex affected in cystic fibrosis and encephalomyopathy. *Biochem. Mol. Med.* **54**: 75–87.

63. Jentsch, T.J., Friedrich, T., Schriever, A. and Yamada, H. (1999) The CIC chloride channel family. *Eur. J. Physiol.* **437**: 783–795.

64. Valverde, M.A. (1999) CIC channels: leaving the dark ages on the verge of a new millennium. *Curr. Opin. Cell Biol.* **11**: 509–516.

65. Henderson, P. J. F. (1993) The transmembrane helix transporteurs. *Curr. Opin. Cell. Biol.* **5**: 708–721.

66. Gangjee, A., Yu, J., McGuire, J.J., Cody, V., Galitsky, N., Kisliuk, R.L. and Queener, S.F. (2000) Design, synthesis, and X-ray crystal structure of a potent dual inhibitor of thymidylate synthase and dihydrofolate reductase as an antitumor agent. *J. Med. Chem.* **43**: 3837–3851.

67. Gangjee, A., Vidwans, A., Elzein, E., McGuire, J.J., Queener, S.F. and Kisliuk, R.L. (2001) Synthesis, antifolate, and antitumor activities of classical and nonclassical 2-amino-4-oxo-5-substituted-pyrrolo. *J. Med. Chem.* **44**: 1993–2003.

68. Le, V., Mak, C.C., Lin, Y., Elder, J.H. and Wong, C. (2001) Structure–activity studies of FIV and HIV protease inhibitors containing allophenylnorstatine. *Biorg. Med. Chem.* **9**: 1185–1195.

69. Kaplan, N.M. (2001) Low-dose combination therapy: the rationalization for an ACE inhibitor and a calcium channel blocker in higher risk patients. *Am. J. Hypertens.* **14**: 8S–11S.

70. Hooper, N.M. (1991) Angiotensin converting enzyme: implications from molecular biology for its physiological functions. *Int. J. Biochem.* **23**: 641–647.

71. Quinn, D.M. (1987) Acetylcholinesterase: enzyme, structure, reaction dynamics and virtual transition states. *Chem. Rev.* **87**: 955–975.

72. Taylor, P. (1996) Anticholinesterase agents. In Hardman, J.G., Limbird, L.E., Molinoff, P.B., Ruddon, R.W. and Gilman, A.G. (eds). The Pharmacological Basis of Therapeutics, pp. 161–176. McGraw-Hill, New York.

73. Takakura, Y., Nishikawa, M., Yamashita, F. and Hashida, M. (2001) Development of gene drug delivery systems based on pharmacokinetic studies. *Eur. J. Pharm. Sci.* **13**: 71–76.

74. Stepkowski, S.M. (2000) Development of antisense oligodeoxynucleotides for transplantation. *Curr. Opin. Mol. Ther.* **2**: 304–317.

75. Giles, R.V. (2000) Antisense oligonucleotide technology: from EST to therapeutics. *Curr. Opin. Mol. Ther.* **2**: 238–252.

PART II

LEAD COMPOUND DISCOVERY STRATEGIES

5

STRATEGIES IN THE SEARCH FOR NEW LEAD COMPOUNDS OR ORIGINAL WORKING HYPOTHESES

Camille G. Wermuth

So ist denn in der Strategie alles sehr einfach, aber darum nicht auch alles sehr leicht. (Thus in the strategy everything is very simple, but not necessarily very easy)

Carl von Clausewitz[1]

I INTRODUCTION

Medicinal chemists have efficient methods for optimizing the potency and the profile of a given active substance.

These methods may consist of more or less intuitive approaches such as the synthesis of analogues and isomers and isosteres, or the modification of ring systems, etc. They may also rest on computer-assisted design, such as identifying pharmacophores by molecular modelling or optimizing activity by means of quantitative structure–activity relationships. Finally, structural biology studies yield more and more X-rays or high-field NMR descriptions of drug–receptor interactions (see Chapter 26). In each case, at the start of the process whether one is to identify a new chemical structure or a new mechanism of action, the medicinal chemist is responsible for developing as rapidly as possible more active molecules that are also more selective and less toxic.

The real challenge is the absolute requirement to discover or identify an original research track. Indeed, for this major step, no codified receipt exists and the creativity of a laboratory cannot be planned. As a result, the discovery of a new lead substance represents the most uncertain stage in a drug development programme. Until the 1970s, the discovery of lead compounds depended essentially upon randomly occurring parameters such as accidental observations, fortuitous findings, hearsay or laborious screening of a large number of molecules. Since then, more rational approaches have become available, based on the knowledge of structures of the endogenous metabolites, the enzymes and the receptors, or on the nature of the biochemical disorder implied in the disease. Today, the decryption of the human genome represents a mine of potentially useful targets for which ligands have to be found (see Chapter 8).

A retrospective analysis of the ways leading to the discovery of new drugs allows one to distinguish essentially four types of strategies giving rise to new lead compounds. The first strategy is based on the modification

The Practice of Medicinal Chemistry
ISBN 0-12-744481-5

and improvement of already existing active molecules. The second one consists of the systematic screening of sets of arbitrarily chosen compounds on selected biological assays. The third approach resides in the retroactive exploitation of various pieces of biological information which result sometimes from new discoveries made in biology and medicine, and sometimes are just the fruits of more or less fortuitous observations. Finally, the fourth route to new active compounds is a rational design based on the knowledge of the molecular cause of the pathological dysfunction.

II FIRST STRATEGY: IMPROVEMENT OF EXISTING DRUGS

The objective of this strategy is, starting from already known active principles, to prepare, by various chemical transformations new molecules for which can be claimed an increase in potency, a better specific activity profile, improved safety, or a formulation more easily handled by physicians and nurses, or more acceptable to the patient.

In the pharmaceutical industry motivations for this approach are often driven by competitive and economic factors. Indeed, if the sales of a given medicine are high and the firm is found in a monopolistic situation, protected by patents and trade marks, other companies will want to produce similar medicines, if possible with some therapeutic improvements. They will therefore use the already commercialized drug as a lead compound and search for

ways to modify its structure and some of its physical and chemical properties while retaining or improving its therapeutic properties. This approach lacks originality and has often been a source of criticism of the pharmaceutical industry. Each laboratory wants to have its own antiulcer drug, its own antihypertensive, and so on. These drug copies are referred to 'me-too', products. Generally, the owner firm of the original drug continues to prepare new analogues, both to ensure a maximum perimeter of protection of its patents and to remain the leader in a given area. For these reasons, the chemical transformation of known active molecules constitutes the most widespread practice in pharmaceutical research. In an interesting review article, Fischer and Gere[2] define two subclasses of analogues. *Early-phase analogues* are defined as structurally similar drugs discovered before the original drug is launched (Fig. 5.1). As a result of early-phase parallel research, the discovery dates of such derivatives are often very close to each other. *Drug analogues* are structurally similar drugs that were discovered later (much later) as a successful analogue research (me-too copy) of a pioneer drug (Fig. 5.2).

A Pros and cons of therapeutic copies

A reassuring aspect of making a therapeutic copy resides in the certainty of ending with an active drug in the desired therapeutic area. It is indeed extremely rare, and practically improbable, that a given biological activity is unique to a single molecule. Molecular modifications allow the preparation of additional products for which one can

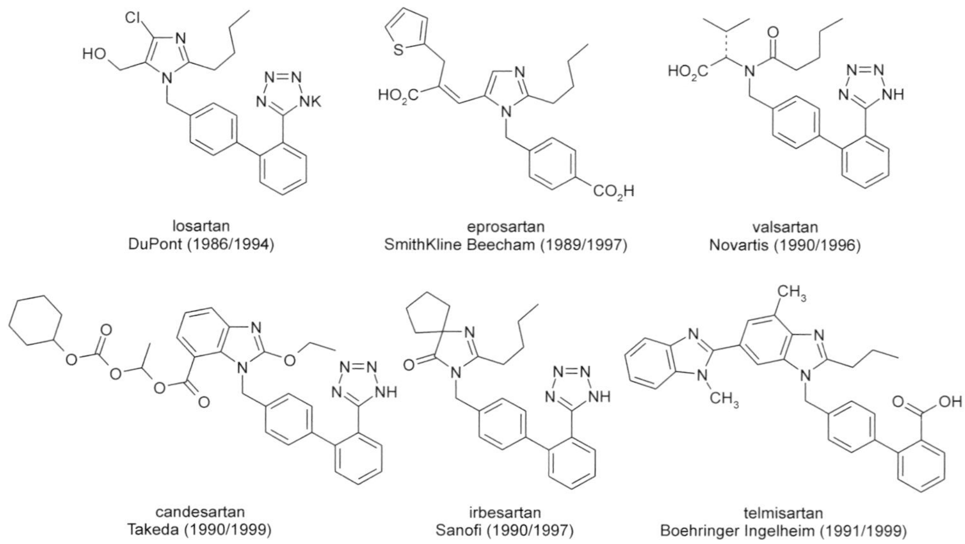

Fig. 5.1 Angiotensin AT1 receptor antagonists derived from losartan are *early-phase analogues*. Despite the similarity of the structures, it can be assumed that the corresponding discoveries were made independently. The first year in parentheses is the basic patent year, the second one is the year of the first launch.

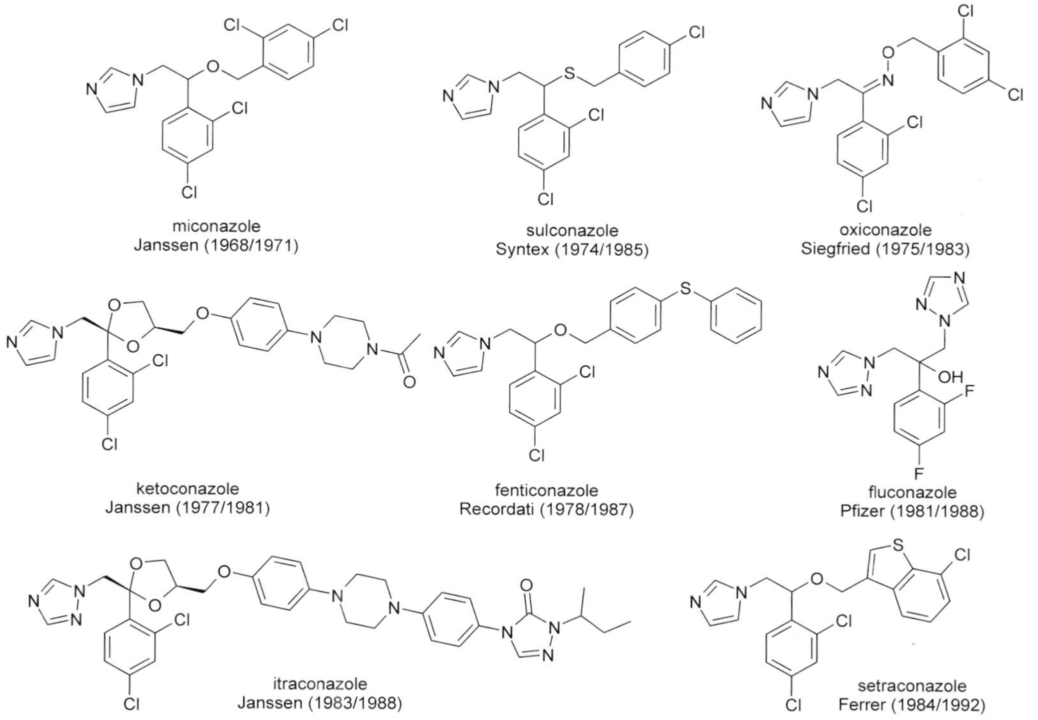

Fig. 5.2 An example of me-too compounds (classical *drug analogues*) is given by micoconazole-derived fungistatics which act by inhibition of the ergosterol biosynthesis. The first year under parentheses is the basic patent year, the second one is the year of the first launch.

expect, if the investigation has been sufficiently prolonged, a comparable activity to that of the copied model, perhaps even a better one. This is comforting for the copier as well as for the financiers who subsidize him. It is necessary, however, to keep in mind that the original inventor of a new medicine possesses a technological and scientific advantage over the copier and that, he too, has been able to design a certain number of copies of his own compound before he selected the molecule ensuring the best compromises between activity, secondary effects, toxicity and invested money.

A second element that favours the copy derives from the information already gained, which then facilitates subsequent pharmacological and clinical studies. As soon as the pharmacological models that serve to identify the activity profile of a new prototype are known, it suffices to apply them to the therapeutic copies. In other terms, the pharmacologist will know in advance what kind of activity is desired and which tests he will have to apply to select the desired activity profile. In addition, during clinical studies, the original research, undertaken with the lead compound will serve as a reference and can be adopted unchanged to the evaluation of the copy. Criticism of this approach is a result of the obvious fact that, in selecting a new active molecule by means of the same pharmacological models as were used for the original compound, one will inevitably end with a compound presenting an identical activity profile and thus the innovative character of such a research is practically nil.

Finally, financial arguments may favour the therapeutic copy. Thus it may be important, and even vital, for a pharmaceutical company or for a national industry, to have its own drugs rather than to subcontract a licence. Indeed, in paying licence dues, an industry impoverishes its own research. Moreover, the financial profitability of research based on me-too drugs can appear to be higher, because no investment in fundamental research is required. The counterpart is that the placement on the market of the copy will naturally occur later than that of the original drug and thus it will make it more difficult to achieve a high sales ranking, all the more so because the me-too drug will be in competition with other copies targeting a similar market.

In reality the situation is more subtle because very often the synthesis of me-too drugs is justified by a desire to improve the existing drug. Thus for penicillins, the chemical structure that surrounds the β-lactamic cycle is still being modified. Current antibiotics that have been derived from this research (the cephalosporins for example) are more

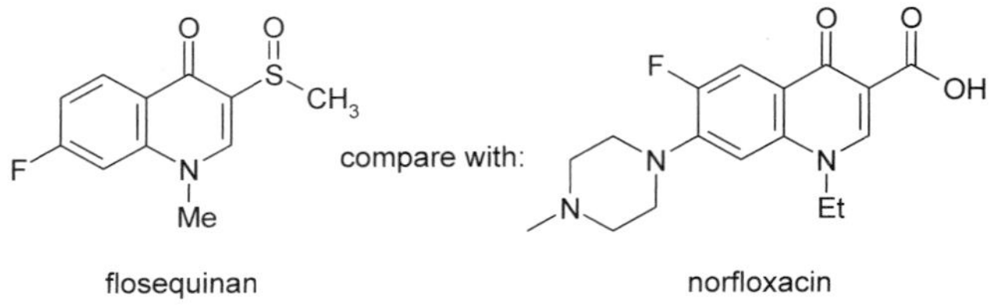

Fig. 5.3 The striking analogy between the vasodilator drug flosequinan and the quinolone antibiotic norfloxacin.

selective, more active on resistant strains and can be administered by the oral route. They are as different from the parent molecule as a recent car compared with a forty-year-old model. In other terms, innovation can result from the sum of a great number of stepwise improvements as well as from a major breakthrough.

It can also happen that during the pharmacological or clinical studies of a me-too compound a totally new property, not present in the original molecule, appears unexpectedly. Thanks to the emergence of such a new activity, the therapeutic copy becomes in turn a new lead structure. This was the case for imipramine, initially synthesized as an analogue of chlorpromazine and presented to clinical investigators for study of its antipsychotic profile.[3] During its clinical evaluation this substance demonstrated much more activity against depressive states than against psychoses. Imipramine has truly opened, since 1954, a therapeutic avenue for the pharmacological treatment of depression. On its way to becoming Viagra, the compound UK-92,480, prepared in 1989 by the Pfizer scientists in Sandwich, UK went first from a drug for hypertension to a drug for angina. Then it changed again when a 10-day toleration study in Wales revealed an unusual side-effect: penile erections.[4] It seems probable that a similar emergence of a new activity occurred with flosequinan, which is a sulfoxide bioisostere of the quinolone antibiotics (Fig. 5.3). This compound proved to be a vasodilator and cardiotonic drug having totally lost any antibiotic activity.[5]

III SECOND STRATEGY: SYSTEMATIC SCREENING

This method consists in screening new molecules, whether they are synthetic or of natural origin, on animals or in any biological test without regard to hypotheses on its pharmacological or therapeutic potential. It rests on the systematic use of selective batteries of experimental models destined to mimic closely the pathological events. The trend is to undertake *in vitro* rather than *in vivo* tests: binding

assays, enzyme inhibition measurements, activity on isolated organs or cell cultures, and so on. In practice, systematic screening can be achieved in two different manners. The first one is to apply to a small number of chemically sophisticated and original molecules, a very exhaustive pharmacological investigation: this is called 'extensive screening'. The second method, in contrast, strives to find, among a great number of molecules (several hundreds or thousands), one that could be active in a given indication: this is 'random screening'.

A Extensive screening

Extensive screening is generally applied to totally new chemical entities coming from an original effort of chemical research or from a laborious extraction from a natural source. For such molecules, the high investment in synthetic or extractive chemistry justifies an extensive pharmacological study (central nervous, cardiovascular, pulmonary and digestive systems, antiviral, antibacterial or chemotherapeutic properties, etc.) to detect whether there exists an interesting potential linked to these new structures. In summary, a limited number of molecules is studied in a thorough manner (vertical screening). It is by such an approach that the antihistaminic, and later on the neuroleptic properties of the amines derived from phenothiazine, were identified. Initially these compounds had been submitted, with negative results, to a limited screening study only directed towards possible chemotherapeutic, antimalarial, trypanocidal and anthelmintic activities.

Original chemical research is also at the origin of the discovery of the benzodiazepines by Sternbach.[6] This author specified that the class of compounds he was seeking had to fulfil the following criteria: (1) the chemical series had to be relatively unexplored; (2) it had to be easily accessible; (3) it had to allow a great number of variations and transformations; (4) it had to offer some challenging chemical problems; (5) it had to 'look' as though it could lead to biologically active products.

The extensive screening approach has often led to original molecules; it is, however, highly dependent on the skill and the intuition of the medicinal chemist, and even more on the talents of the pharmacologist who has to be able to adapt and to orient his tests as soon as his findings evolve to reveal the real therapeutic potential of the molecule under study.

More recent examples are seen by the discovery, thanks to systematic screening programme, of the cyclopyrrolones, e.g. zopiclone (Fig. 5.4), as ligands for the central benzodiazepine receptor,[7,8] or of taxol as an original and potent anticancer drug (for a review see Suffness[9]).

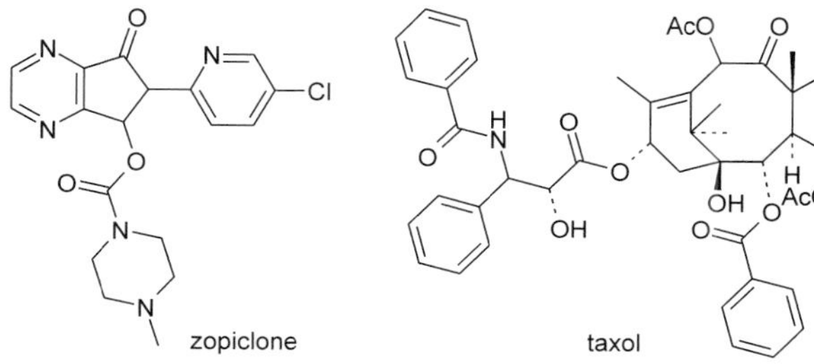

Fig. 5.4 Drugs discovered by random screening.

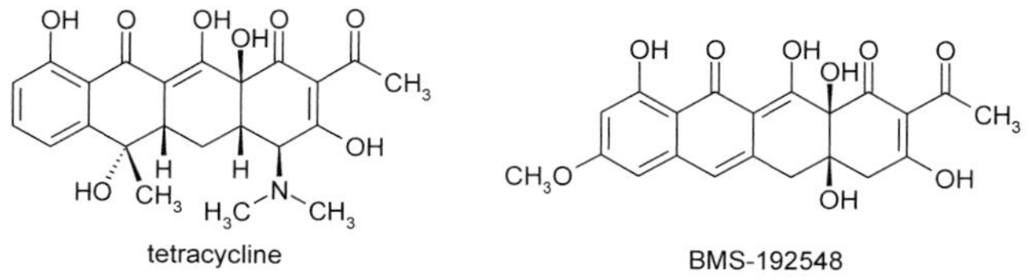

B Random screening

In this case the therapeutic objective is fixed in advance and, in contrast to the preceding case, a great number (several thousands) of molecules is tested, but on a limited number of experimental models only. With this method one practices so-called random screening. This method has been used for the discovery of new antibiotics. By submitting samples of earth collected in countries from all over the world to a selective antibacterial and antifungal screening, the rich arsenal of anti-infectious drugs which are presently at the disposal of clinicians, was developed. During the Second World War, an avaian model in chickens infected with *Plasmodium gallinaceum*, was used for the massive screening of thousands of potential antimalarials. The objective was to solve, by finding a synthetic antimalarial, the problem of the shortage of quinine. Unfortunately, no satisfactory drug was found. Massive screening was implemented in Europe and the United States to discover new anticancer[10] and antiepileptic drugs. Here again the problem is to select some predictive, but cheap cellular or animal model. A common criticism of these methods is that they constitute, by the absence of a rational

R1 = R2 = H :   compactin (mevastatin)
R1 = CH3; R2 = H : lovastatin (mevinolin)
R1 = R2 = CH3 :   simvastatin

pravastatin

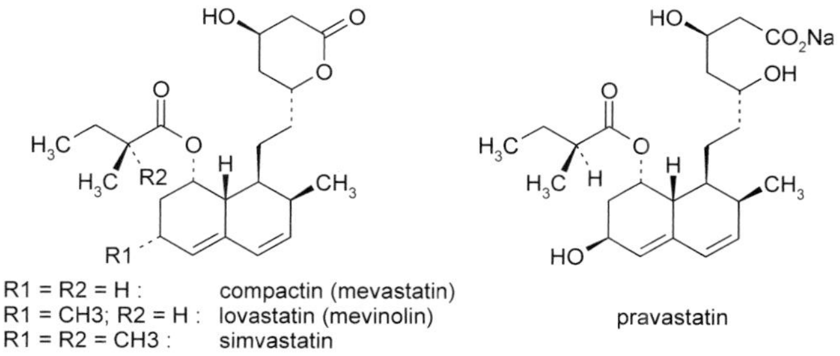

Fig. 5.5 The natural compounds compactin (mevastatin) and lovastatin block the cholesterol biosynthesis in inhibiting the enzyme hydroxymethyl-glutaryl-CoA reductase (HMG-CoA reductase). The later developed compounds, simvastatin and pravastatin, are semisynthetic analogues. The open-ring derivative pravastatin is less lipophilic and therefore presents less central side-effects. For all these compounds the ring-opened form is the actual active form *in vivo*.

Fig. 5.6 Unexpected CNS activity of the tetracycline analogue BMS-192548.

lead, a sort of fishing. Besides, the results are very variable: nil for the discovery of new antimalarials, rather weak for the anticancer drugs but excellent, in their time, for the discovery of antibiotics.

Among the recent successes of this approach the discovery of lovastatin, also called mevinolin, should be mentioned (Fig. 5.5),[11,12] which was the basis of a new generation of hypocholesterolemic agents, acting by inhibition of hydroxymethyl-glutaryl-CoA reductase (HMG-CoA reductase).

Sometimes unexpected findings result from systematic screening applied in an unprejudiced manner. A good example is found in the tetracyclic compound BMS-192548 extracted from *Aspergillus niger* WB2346 (Fig. 5.6). For any medicinal chemist or pharmacologist the similarity of this compound with the antibiotic tetracycline is striking. However, none of them would a priori forecast that BMS-192548 exhibits CNS activities. In fact, the compound is a ligand for the neuropeptide Y receptor preparation.[13]

C High-throughput screening

Since the 1980s, with the arrival of robotics and with the miniaturization of *in vitro* testing methods, it has become possible to combine the two preceding approaches; in other words, to screen thousands of compounds on a large number of biological targets. This high-throughput screening is usually applied to the displacement of radioligands and to the inhibition of enzymes. The present trend is to replace radioligand-based assays by fluorescence-based measurements. As it is now possible for a pharmaceutical company to screen several thousand molecules simultaneously in 30 to 50 different biochemical tests, the problem becomes one of feeding the robots with interesting molecules. Primary sources are chemicals coming from in-house libraries or from commercial collections, but the samples can also be crudely purified vegetal extracts or fermentation fluids. In this latter case, one proceeds to the isolation and to the identification of the responsible active principle[14,15] only when an interesting activity, after the screening, is observed.

Many success stories based on this approach are found in the literature, among them, the discovery of insulin mimetics or of ORL1 receptor agonists.

Insulin mimetics: Considerable evidence suggests that insulin receptor tyrosine kinase (IRTK) activity is essential

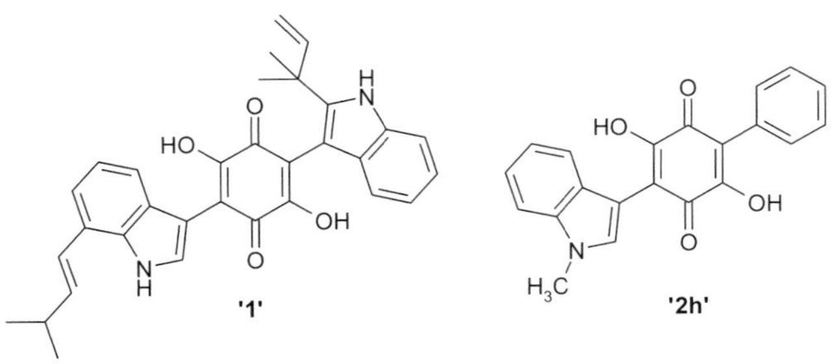

Fig. 5.7 Insulin mimicking tyrosine-kinase inhibitors resulting from systematic screening.[16]

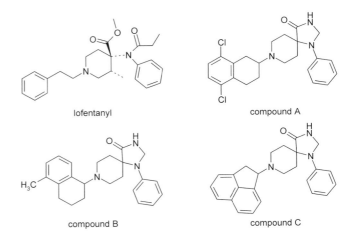

Fig. 5.8 Structures of lofentanyl and analogous ORL1 receptor agonists.[17]

for many, if not all, of the biological effects of insulin. Thus, IRTK activation may be useful to treat patients suffering from insulin resistance, a characteristic feature of noninsulin-dependent diabetes mellitus (NIDDM). In an extensive screening effort for small molecules that activate human IRTK compound '1' (Fig. 5.7), a fungal metabolite isolated from a culture broth of *Pseudomassaria* was identified. This compound acted as an insulin mimetic in several biochemical and cellular assays. Oral administration of '1' to two mouse models of diabetes resulted in a significant decrease in blood glucose levels. An initial optimization study[16] led to the simplified and about 15 times more potent compound '2h'.

ORL1 receptor agonists: The cloning of G-protein-coupled receptors selective for δ, μ and κ opioids led several research groups (see the first reference in reference 17) to the discovery of a fourth member of the family by homology cloning. This fourth receptor was named ORL1 (orphanin FQ/nociceptin) receptor. High-throughput screening identified 8-(5,8-dichloro-1,2,3,4-tetrahydro-naphtalen-2-yl)-1-phenyl-1,3,8-triazaspiro[4.5]decan-4-one (compound **A,** Fig. 5.8) as an initial nanomolar, but unselective, hit.[17] This compound shows a relatively close resemblance to lofentanyl, a high-affinity μ opioid receptor agonist having appreciable affinity for this new opioid receptor. The synthesis of a series of optimized ligands led to the two high-affinity, potent ORL1 receptor agonists **B** and **C** which exhibit moderate to good selectivity versus the other opioid receptors.

Lamellarin-20-sulfate: The screening of diverse marine natural products for compounds active against integrase

in vitro identified a series of ascidian alkaloids, the lamellarins. One new member of the family, named lamellarin-20-sulfate (Fig. 5.9) displayed the most favourable therapeutic index.[18]

Active site directed thrombin inhibitors: A broad screening approach for thrombin inhibitors led to the identification of the 2,3-disubstituted benzo[19] thiophene derivative '1a' (Fig. 5.9). A systematic structure–activity relationship study led to '31c' which is 1271 times more potent.

Selective neuropeptide Y5 receptor antagonists: Screening of an in-house chemical library identified the novel aminotetralin '2' as having low micromolar binding affinity for the human Y5 receptor.[20] Optimization yielded the nanomolar antagonist '8i'(Fig. 5.10).

Selective COX-2 inhibitors: Mass screening of the Parke-Davis library for compounds that would inhibit COX-2 identified original hits derived from 2,6-di-*tert*-butylphenol (generic structures '**I**' and '**II**', Fig. 5.11).[21] These compounds are related to probucol and are totally different from the previously reported selective COX-2 inhibitors.[21]

In parallel with high-throughput screening approaches, an extraordinary acceleration of synthesis technologies has occurred. Combinatorial chemistry (see Chapter 7) now renders possible the simultaneous synthesis of thousands of

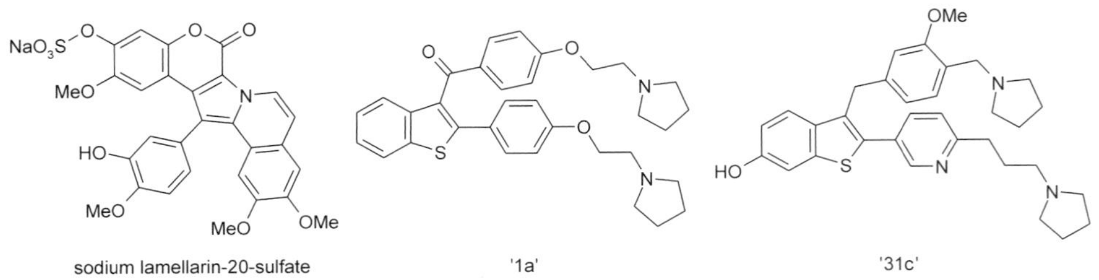

sodium lamellarin-20-sulfate   '1a'   '31c'

Fig. 5.9 Integrase and thrombin inhibitors.

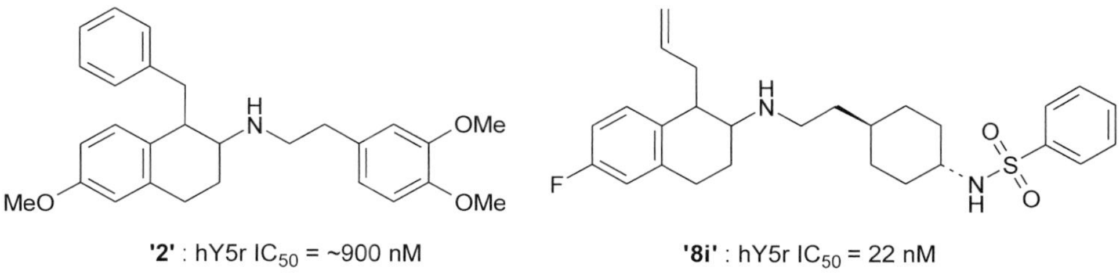

Fig. 5.10 Aminotetralin-derived hit '2' yields the optimized Y5 receptor antagonist '8i'. Note the replacement of the benzyl side chain by an allyl side chain.

diverse molecules that are necessary to supply enough samples to the biochemical screening robots. As quoted by Moos *et al.*,[22] 'The power of multiple compound synthesis methodologies suggests that more compounds have been synthesized and screened in the 1990s than in the combined histories of the pharmaceutical and biotechnology industries pre-1990.' Generation and screening of molecular diversity has become a major tool in the search for novel lead structures. Besides chemical laboratory synthesis, molecular diversity may derive from biological systems. The exploitation of biological systems is beyond the scope of this book, however, references can be found in the review of Bull *et al.*[23] On the other hand, biodiversity resulting from natural sources is treated in Chapter 6 and in references 14 and 15. More chemistry-oriented approaches such as synthesis of peptide and of nonpeptide combinatorial libraries[22,24] for lead structure screening are detailed in Chapter 7. Finally, so-called electronic screening, which means finding the lead from database mining, is described in Chapters 9 and 10.

D Limitations of the high-throughput screening approach

Experience gathered over the past few years has confirmed that high-throughput screening allows for the rapid identification of many hits. However, this strategy for drug discovery has several limitations that are mainly due to the nature of the chemical libraries fed to the robots. These libraries are usually massive and typically contain 100 000 to 500 000 compounds. They are most often assembled by combinatorial chemistry. Limitations of these libraries include inadequate diversity leading to decreased chances of success, very moderate or even low hit rates, and finally low 'biopharmaceutical' quality of the hits.

Inadequate diversity: Initially, libraries were assembled from available in-house compound collections. The disadvantage of such collections is that they often consist of very large series of similar compounds. Commercial catalogues are another source of compounds, but they contain compounds or reagents essentially made for chemical purposes. Combinatorial chemistry is by far the

most productive route for the creation of large series of molecules. This approach, however, is limited by the restricted number of synthetic reactions that are useable for combinatorial chemistry. Often badly designed combinatorial syntheses lead to more or less homologous series (e.g. amides and ureas) that show no real chemical diversity and thus decrease the chances of 'hitting'. In addition, the simplicity of many high-throughput syntheses has often excluded access to pharmacophores of sophisticated molecules such as natural compounds (asperlicin, cyclosporin, mevinolin, taxol, neocarzinostatine, epothilones, etc.).

Low hit rates: The yield of screening massive libraries is often low, approximating a 0.01% hit rate.[25] Many experts agree that the discovery of new drug entities has quantitatively not increased, compared with traditional medicinal chemistry approaches, since the systematic utilization of high-throughput screening.[26]

Low-quality hits: Furthermore, many of the 'hits' obtained from massive libraries are 'low-quality hits', i.e. their bioavailability or their toxicity in humans is largely unknown. This can be a serious issue. Since the compounds that constitute massive libraries derive from arbitrarily chosen chemical structures, little is known about their bioavailability or their toxicity *in man*. In other words, there is a real risk that many hits will be compounds that are not bioavailable and/or are toxic, making it more than likely that only a small number of hits will be suitable for further development. As a consequence, many hits from currently available massive libraries never make it past initial pharmacokinetic and toxicological screening.

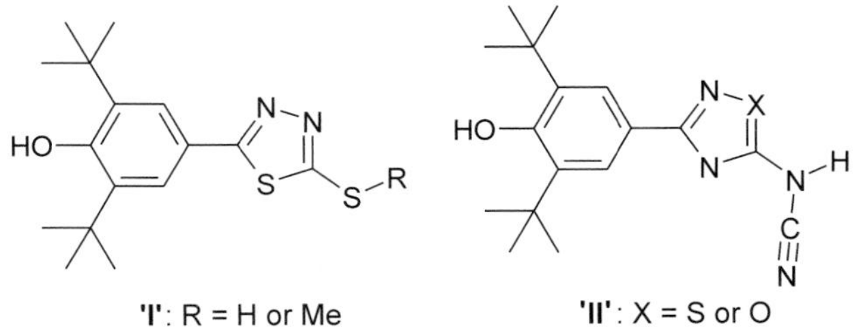

Fig. 5.11 Original COX-2 inhibitors discovered through a mass screening process.

Cost: In addition to the above limitations, massive libraries are expensive to assemble and to screen. Time is also an issue. Often the time needed to optimize the chemical syntheses and to assemble all the required starting reagents is even greater than the time needed for the final automated synthesis procedure. Moreover, screening several thousand compounds can be expensive. For example, consider a low average cost of US$2 per assay (which does not include the cost of the library or of the equipment to run the assays), and consider screening 150 000 compounds at a single concentration, the total cost will be US$300 000. To this cost must be added the cost of data management and data analysis. In the example just given, 150 000 data points will be generated. Since these assays can be run on average once or twice per week, the systematic use of this approach may cost millions of dollars per year.

In conclusion, high-throughput screening of massive libraries is expensive, time-consuming, diversity is inadequate, discovery is often limited by monotony, yield is low, and there is a risk of low-quality hits. Not surprisingly, the present trend is to use smaller libraries, and, especially, libraries with increased 'drug-likeness'.

E New trends in high-throughput screening

The present trends in high-throughput screening focus on reducing the size of libraries, on making them more effective and on excluding as early as possible ADME- and toxicological-inadequate candidates.

Computational chemists can help reduce the number of compounds to be selected for high-throughput screening. Instead of performing random screening to find leads, they propose a set of compounds presenting diversity in their structures and their physicochemical properties. The aim of computational chemistry is to select a smaller number of compounds to be tested, while gaining as much information as possible about the data set. Any reduction of the number of compounds to be tested, even if it reduces only the amount of redundancy within a database without introducing any voids, will have an important impact on research efficiency and associated costs. Among the techniques recently published one can cite hierarchical clustering and maximum dissimilarity methods,[27] and neural network techniques.[28]

Other solutions include the synthesis of targeted libraries and computational-driven compound purchase ('targeted drug acquisition'). Alternative approaches rely on drug design by means of virtual screening and database mining ('*in silico* design') or on computational assessment of drug-likeness. Taken together, all these procedures pursue the same objective: isolate from a pool of possible hit molecules those which present the most credible drug-like properties.

Once hits are identified, it becomes of utmost importance to eliminate the unwanted candidates using data visualization, data mining, chemoinformatics and computational chemistry. Emerging technologies in ADME and predictive toxicology make use of models for the estimation, on one hand of oral bioavailability of drugs in humans and other species from their *in vitro* ADME screens, and of predictive toxicity testing by transcription profiling on the other hand. Taking into account these considerations, it is clear that profitability of high-throughput screening approaches is lower than had initially been expected.

F Particular cases

Screening of synthesis intermediates

As synthesis intermediates are chemically connected to final products, and as they often present some common groupings with them, it is not inconceivable that they share some pharmacological properties. For this reason, it is always prudent also to submit these compounds to a pharmacological evaluation. Among drugs discovered in this way are the tuberculostatic semicarbazones: they were initially used in the synthesis of antibacterial sulfathiazoles. Subsequent testing of isonicotinic acid hydrazide, destined for the synthesis of a particular thiosemicarbazone, revealed the powerful tuberculostatic activity of the precursor which has since become a major antitubercular drug (isoniazide).

Inhibitors of the enzyme dihydrofolate-reductase such as methotrexate (Fig. 5.12) are used in the treatment of leukaemia. During the search for methotrexate analogues a very simple intermediate, mercaptopurine, was also submitted to testing. It proved to be active but relatively toxic.

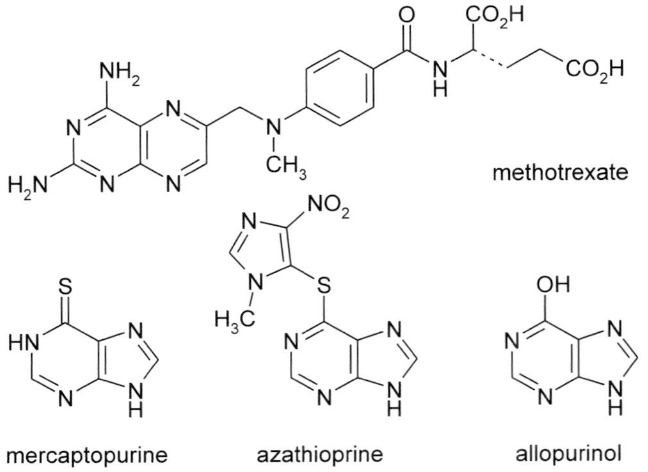

Fig. 5.12 Departing from methotrexate, simple intermediates led to new drugs. Mercaptopurine and azathioprine are immunosuppressants and allopurinol is used in the treatment of gout.

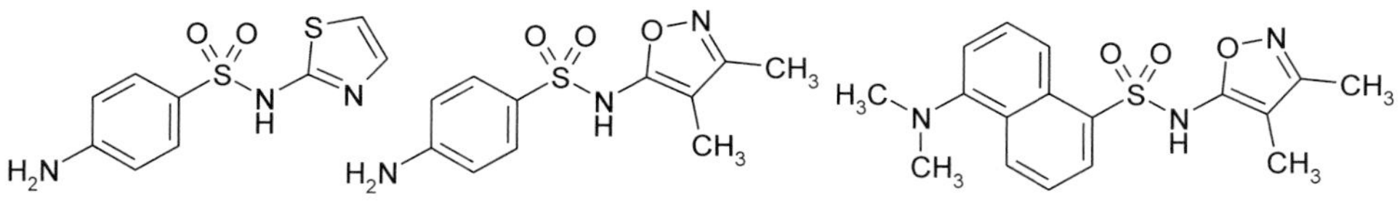

Fig. 5.13 A successful SOSA approach allowed the identification of the antibacterial sulfonamide sulfathiazole as a ligand of the endothelin ET_A receptor and its optimization to the selective and potent compound BMS-182874.[34]

Subsequent optimization led to azathioprine, a prodrug releasing mercaptopurine *in vivo*. Azathioprine was found to be more potent as an immunosuppressive agent than previously used corticoids and was systematically used in all organ transplantations until the advent of cyclosporine. Another intermediate in this series, allopurinol, inhibits xanthine-oxidase and is therefore used in the treatment of gout.[29]

New leads from old drugs: the SOSA approach

The SOSA approach (SOSA = selective optimization of side activities) represents an original alternative to high-throughput screening.[30–33] It consists of screening on newly identified pharmacological targets only a limited set (less than 1000 compounds) of well-known drug molecules for which bioavailability and toxicity studies have already been performed and which have proven useful in human therapy. By definition, in using such a library, all hits that are found are drug-like!

Once a hit is observed with a given drug molecule, the objective is to prepare analogues of this molecule in order to transform the observed 'side activity' into the main effect and to strongly reduce or abolish the initial pharmacological activity. The rationale behind the SOSA approach lies in the fact that, in addition to their main activity, almost all drugs used in human therapy show one or several side-effects. In other words, if they are able to exert a strong interaction with the main target, they also exert less strong interactions with some other biological targets. Most of these targets are unrelated to the primary therapeutic activity of the compound. The objective of medicinal chemists is then to proceed to a reversal of the affinities, the identified side-effect becoming the main effect and vice versa. Many cases of activity profile reversals by means of the SOSA approach have been published.

Among them, a typical illustration of the SOSA approach is shown by the development of a selective ligand for the endotheline ET_A receptors by scientists from Bristol-Myers-Squibb.[34] Starting from an in-house library, the antibacterial compound sulfathiazole (Fig. 5.13) was an initial, but weak, hit (IC_{50} = 69 μM). Testing of related sulfonamides identified the more potent sulfisoxazole (IC_{50} = 0.78 μM). Systematic variations led finally to the potent and selective

ligand BMS-182874. *In vivo*, this compound was orally active and produced a long-lasting hypotensive effect.

Another example is the antidepressant minaprine (Fig. 5.14). In addition to reinforcing serotoninergic and dopaminergic transmission, this amino-pyridazine possesses weak affinity for muscarinic M_1 receptors (K_i = 17 μM). Simple chemical variations allow the dopaminergic and serotoninergic activities to be abolished, and the cholinergic activity up to nanomolar concentrations to be boosted.[35–37] Similarly, chemical variations of the D_2/D_3 nonselective benzamide sulpiride (Fig. 5.15) led to compound Do 897, a selective and potent D_3 receptor partial agonist.[38]

As mentioned above, a differentiating peculiarity of this type of library is that it is constituted by compounds that have already been safely given to humans. Thus, if a compound were to 'hit' with sufficient potency on an orphan target, there is a high chance that it could quickly be tested in patients for Proof of Principle. Alternatively, if one or more compounds hit, but with insufficient potency, optimized analogues can be synthesized, and the chances that these analogues will be good candidate drugs for further development are much higher than if the initial lead is toxic or not bioavailable. One of these 'new-type' of chemical libraries, the Prestwick Chemical Library, has recently become available.[39] It contains 880 biologically active compounds with high chemical and pharmacological diversity as well as known bioavailability and safety in

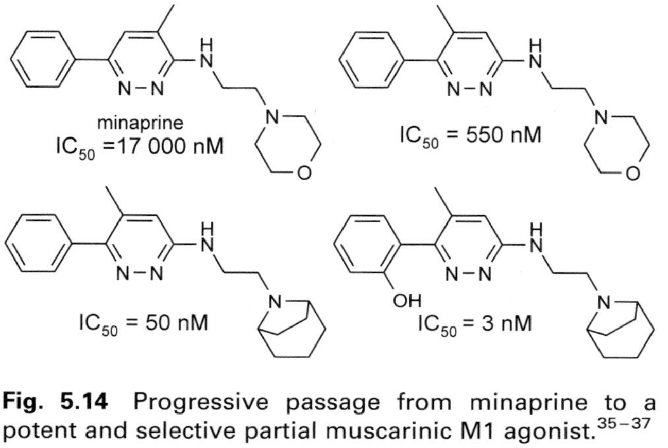

Fig. 5.14 Progressive passage from minaprine to a potent and selective partial muscarinic M1 agonist.[35–37]

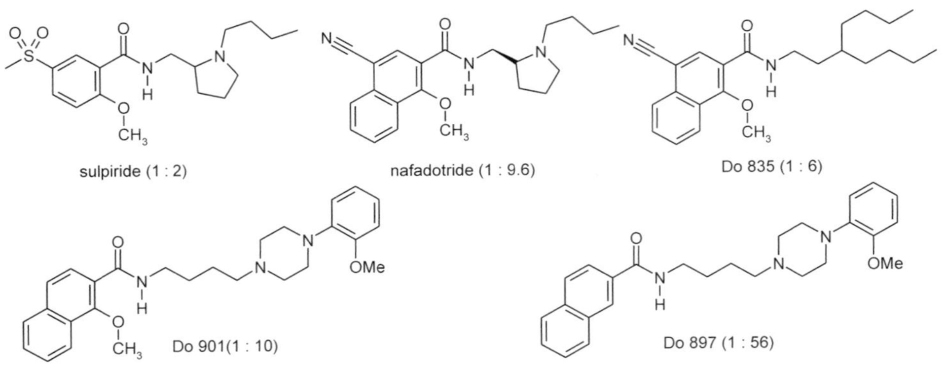

Fig. 5.15 The progressive change from the D_2/D_3 receptor nonselective antagonists to the highly D_3-selective compound Do 897.[38] The numbers between parentheses indicate the D_2/D_3 affinity ratio.

humans. Over 85% of the compounds are well-established drugs, and 15% are bioactive alkaloids. For scientists interested in drug-likeness such a library fulfils, in the most convincing way, the quest for 'drug-like' leads!

IV THIRD STRATEGY: EXPLOITATION OF BIOLOGICAL INFORMATION

A major contribution to the discovery of new active principles comes from the exploitation of biological information. By this is meant information that relates to a given biological effect (fortuitous or voluntary) provoked by some substances in man, in animals or even in plants or bacteria. When such information becomes accessible to the medicinal chemist, it can serve to initiate a specific line of therapeutic research. Originally, the observed biological effect can simply be noticed without any rational knowledge on how it works.

A Exploitation of observations made in man

The activity of exogenous chemical substances on the human organism can be observed under various circumstances: ethnopharmacology, popular medicines, clinical observation of secondary effects or adverse events, fortuitous observation of activities of industrial chemical products, and so on. Since in all cases the information harvested is observed directly in man, this approach represents a notable advantage.

Study of indigenous medicines (ethnopharmacology)

Natural substances were for a long time the unique source of medicines. At present, they constitute 30% of the active principles used and probably more (approximately 50%) if one considers the number of prescriptions that utilize them, particularly since use of antibiotics plays a major role.[40]

Behind most of these substances one finds indigenous medicines. As a consequence, ethnopharmacology represents a useful source of lead compounds. Historically, we are indebted to this approach for the identification of the cardiotonic digitalis glucosides of the digital, the opiates and the cinchona alkaloids. Curare was obtained from a South American plant long used by natives to make arrow poisons. The cardiotonic glucosides of *Strophantus* seeds and the alkaloid eserine from Calabar beans are other examples of active drugs originally used by natives as poisons. The *Rauwolfia serpentina* was used for centuries in India before Western medicine became interested in its tranquillizing properties and extracted reserpine from it. Atropine, pilocarpine, nicotine, ephedrine, cocaine, theophylline and innumerable other medicines have thus been extracted from plants to which the popular medicine attributed therapeutic virtues.

Despite its extremely useful contributions to the modern pharmacopoeia such as artemisin and huperzine, folk medicine is a rather unreliable guide in the search for new medicines. This is illustrated by the example of antifertility agents: according to natives of some islands of the Pacific, approximately 200 plants would be efficient in reducing male or female fertility. Extracts have been prepared from 80 of these plants and have been administered at high dose to rats for periods of 4 weeks and more, without observing the slightest effect upon pregnancies or litter sizes.[41] When ethnopharmacology and the natural substance chemistry end in the discovery of a new active substance, this latter is first reproduced by total synthesis. It is then the object of systematic modifications and simplifications that aim to recognize by trial and error the minimal requirements that are responsible for the biological activity.

Clinical observation of side-effects of medicines

The clinical observation of entirely unexpected side-effects constitutes a quasi-inexhaustible source of tracks in the search for lead compounds. Indeed, besides the desired therapeutic action, most drugs possess side-effects. These are accepted either from the beginning as a necessary evil, or recognized only after some years of use. When side-effects present a medical interest in themselves, a planned objective can be the dissociation of the primary from the side-effect activities: enhance the activity originally considered as secondary and diminish or cancel the activity that was initially dominant. Promethazine, for example, an antihistaminic derivative of phenothiazine, has important sedative effects. Like Laborit *et al.*,[42] a clinician might promote the utilization of this side-effect and direct research towards better-profiled analogues. This impulse was the origin of chlorpromazine, the prototype of a new therapeutic series, the neuroleptics, whose existence was previously unsuspected and which has revolutionized the practice of psychiatry.[3,43] Innumerable other examples can be found in the literature, such as the hypoglycaemic effect of some antibacterial sulfamides, the uricosuric effect of the coronary-dilating drug benziodarone, the antidepressant effect of isoniazid, an antitubercular drug, and the hypotensive effect of β-blocking agents.

This last example is beautifully illustrated by the discovery of the potassium channel activator cromakalim.[44] Cromakalim is the first antihypertensive agent shown to act exclusively through potassium channel activation.[45] This novel mechanism of action involves an increase in the outward movement of potassium ions through channels in the membranes of vascular smooth muscle cells, leading to relaxation of the smooth muscle. The discovery of this compound can be summarized as follows: β-Adrenergic receptor blocking drugs were not thought to have antihypertensive effects when they were first investigated. However, pronethalol, a drug that was never marketed, was found to reduce arterial blood pressure in hypertensive patients with angina pectoris. This antihypertensive effect was subsequently demonstrated for propranolol and all other β-adrenergic antagonists.[46] Later there were some doubts that blockade of the β-adrenergic receptors was responsible for the hypotensive activity and attempts were made, in the classical β-blocking molecules, to dissociate the β-blockade from the antihypertensive activity. Among the various conceivable molecular variations which are possible for the flexible β-blockers, it was found that conformational

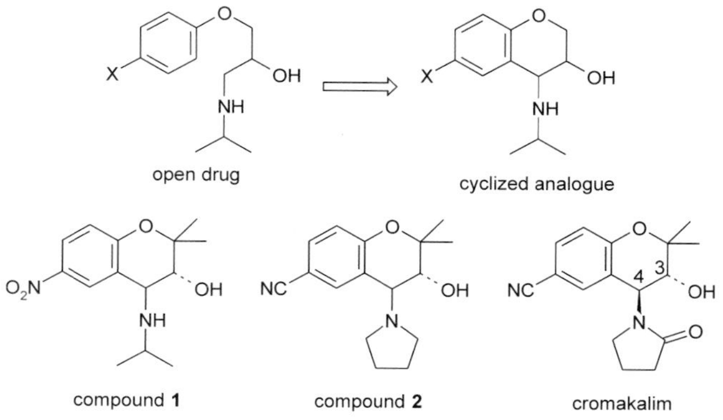

Fig. 5.16 The clinical observation of the hypotensive activity of the 'open' (and therefore flexible) β-blocking agents was the initial lead to cyclized analogues devoid of β-blocking activity, but retaining the antihypertensive activity.[44]

restriction obtained in cyclizing the carbon atom bearing the terminal amino group on to the aromatic ring yielded derivatives devoid of β-blocking activity, but retaining the antihypertensive activity (Fig. 5.16). One of the first compounds prepared (compound **1**, Fig. 5.16) was indeed found to lower blood pressure in hypertensive rats by a direct peripheral vasodilator mechanism; no β-blocking activity was observed. Optimization of the activity led to the 6-cyano-4-pyrrolidinylbenzopyran (compound **2**), which was more than 100-fold more potent than the nitro derivative. The replacement of the pyrrolidine by a pyrrolidinone (which is the active metabolite) produced a three-fold increase in activity and the optical resolution led to the (−)-3R, 4S enantiomer of cromakalim (BRL 38227), which concentrates almost exclusively the hypotensive activity.[44,47,48]

New uses for old drugs

In some cases a new clinical activity observed for an old drug is sufficiently potent and interesting to justify the immediate use of the drug in the new indication, this is illustrated hereafter.

Amiodarone, for example (Fig. 5.17), was introduced as a coronary dilator for angina. Concern about corneal deposits, discolouration of skin exposed to sunlight and thyroid

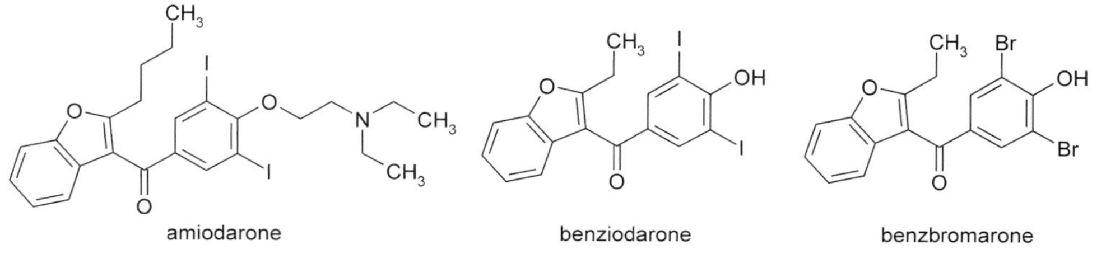

Fig. 5.17 Structures of the arones.

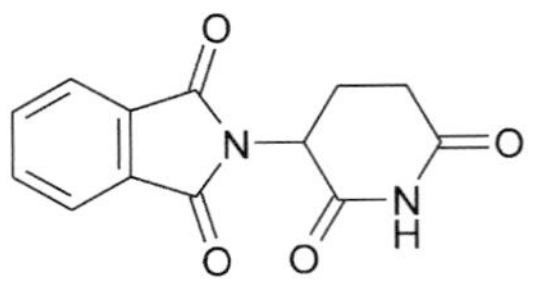

Fig. 5.18 Structure of thalidomide. The marketed compound is the racemate.

disorders led to the withdrawal of the drug in 1967. However, in 1974 it was discovered that amiodarone was highly effective in the treatment of a rare type of arrhythmia known as the Wolff-Parkinson-White syndrome. Accordingly, amiodarone was reintroduced specifically for that purpose.[49]

Benziodarone, initially used in Europe as a coronary dilator, proved later to be a useful uricosuric agent. At the present time it is withdrawn from the market due to several cases of jaundice associated with its use.[49] The corresponding brominated analogue, benzbromarone, was specifically marketed for its uricosuric properties.

Thalidomide, was initially launched as a sedative/hypnotic drug (Fig. 5.18), but withdrawn because of its extreme teratogenicity. Under restricted conditions (no administration during pregnancy or to any woman of childbearing age), it found a new use as an immunomodulator. Particularly, it seems efficacious for the treatment of *erythema nodosum leprosum*, a possible complication of the chemotherapy of leprosy.[50]

In 1978 the synthesis of the indenoisoquinoline NSC 314622 (Fig. 5.19) was reported as the result of an unexpected transformation during synthesis of nitidine chloride. Given its weak antitumour activity, it was not investigated further. Twenty years later, NSC 314622 resurfaced as a potential topoisomerase I (top 1) inhibitor and served as lead structure for the design of cytotoxic noncamptothecin topoisomerase I inhibitors such as the compound '19a'.[51]

In 2001, the antimalarial drug quinacrine and the antipsychotic drug chlorpromazine (Fig. 5.20) were shown

to inhibit prion infection in cells. Prusiner and coworkers identified the drugs independently and found that they inhibit conversion of normal prion protein into infectious prions and clear prions from infected cells.[52] Both drugs can crossover from the bloodstream to the brain, where prion diseases are localized.

A more recent example is provided by the discovery of the use of sildenafil (Viagra®, Fig. 5.21), a phosphodiesterase type 5 (PDE5) inhibitor, as an efficacious, orally active agent for the treatment of male erectile dysfunction.[53,54] Initially this compound was brought to the clinic as a hypotensive and cardiotonic substance and its usefulness in male erectile dysfunction resulted from clinical observations.

In many therapeutic families each generation of compounds induces the birth of the following one. This happened in the past for the sulfamides, penicillins, steroids, prostaglandins and tricyclic psychotropics families, and real genealogical trees representing the progeny of the discoveries can be drawn. More recent examples are found in the domain of ACE inhibitors and in the family of histaminergic H$_2$ antagonists.

Research programmes based on the exploitation of side-effects are of great interest in the discovery of new tracks in so far as they depend on information about activities *observed directly in man* and not in animals. On the other hand, they allow detection of new therapeutic activities *even when no pharmacological models in animals exist*.

The fortuitous discovery of activities of industrial chemical products

During the industrial manufacture of nitroglycerin, toxic manifestations due to this compound, i.e. particularly strong vasodilating properties, were observed in workers. From this observation came the utilization of this substance, and later of other nitric esters of aliphatic alcohols, in angina pectoris and as cerebral vasodilators. In an analogous manner it was observed during the manufacture of the sulfa drug sulfathiazole that 2-amino-thiazole, one of the starting materials, was endowed with antithyroidal properties. This

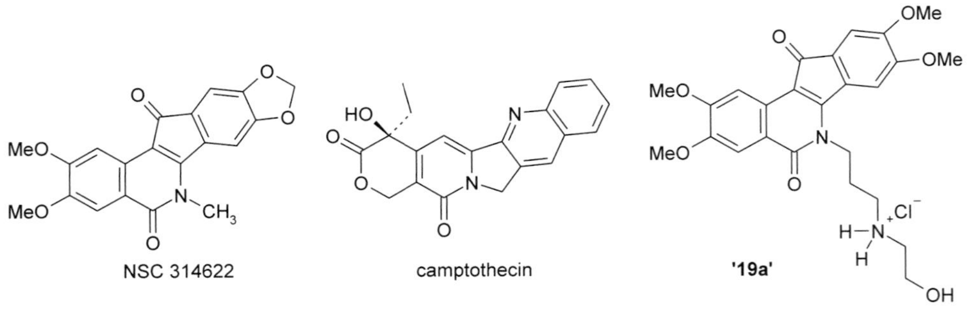

Fig. 5.19 The indenoisoquinoline NSC 314622 resurfaced 20 years after its first testing as an antitumour agent.[51]

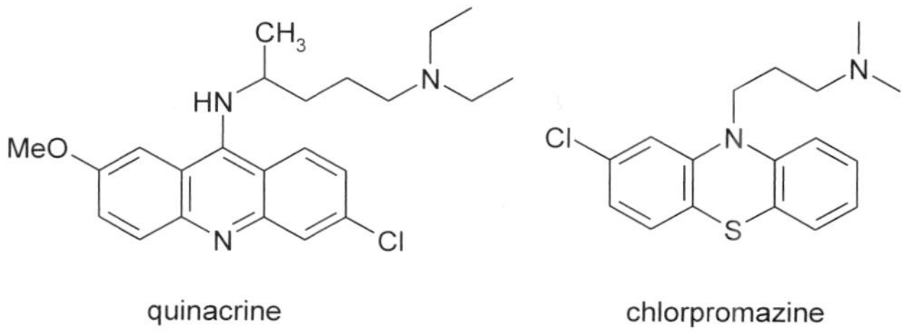

Fig. 5.20 Old drugs, new use. The antimalarial drug quinacrine and the antipsychotic drug chlorpromazine are able to inhibit prion infection.[52]

observation fostered the use of this compound, and of aminothiazoles in general, for the treatment of thyroid gland hyperactivity. Tetraethylthiurame disulfide was originally used as an antioxidant in the rubber industry. After having handled this product, workers felt an intolerance to alcohol. Therefore this product was proposed for ethylic alcohol withdrawal cures (disulfiram). On the molecular level, the mode of action of disulfiram rests on the inhibition of the enzyme aldehyde dehydrogenase that normally ensures the oxidation of acetaldehyde into acetic acid. The intake of alcohol under disulfiram provokes an accumulation of acetaldehyde that achieves intoxication of the patient. Another example of a fortuitous discovery is probucol. This antihyperlipoproteinemic compound was originally synthesized as an antioxidant for plastics and rubber.[55,56]

B Exploitation of observations made in animals

Here we find all the research done by physiologists which has been the basis of the discovery of vitamins, hormones and neurotransmitters and the fall-out of various pharmacological studies, when they were performed *in vivo*. Other observations made on animals, often in a more or less fortuitous manner, have led to useful discoveries. An example is provided by the dicoumarol-derived anticoagulants.

The discovery of the anticancer properties of the alkaloids of *Vinca rosea* constitutes a particularly beautiful example of pharmacological feedback. Preparations from this plant had the reputation in some popular medicines of possessing antidiabetic virtues. During a controlled pharmacological test, these extracts were proved to be devoid of hypoglycaemic activity. On the other hand, it was frequently observed that the treated rats died from acute septicaemia. A study of this phenomenon showed that it was due to massive leukopenia. Taking the leukocyte count as the activity end-point criterion, it became possible to isolate the main alkaloid, vinblastine.[57] At the same time, in another laboratory, routine anticancer screening had revealed the activity of the crude extract on murine leukaemia.[58] Subsequently, the antileukaemic activity became a screening tool. Out of 30 alkaloids isolated from various periwinkles, four (vinblastine, vinleurosine, vincristine and vinrosidine) were found active in human leukaemias.[59]

Analogues of L-arginine with modifications at the terminal guanidino nitrogen and/or the carboxyl terminus of the molecule have been widely used for their ability to inhibit the production of nitric oxide and are thought to be competitive antagonists of nitric synthase. In studies designed to elucidate the role of nitric oxide in the gastrointestinal tract, an inhibitory effect of N^G-nitro-L-arginine methyl ester (L-NAME) on cholinergic neural responses was sometimes observed. This inhibitory effect was shown to be consistent with a blockade of the muscarinic receptors.[60]

Remember too that it was the research of insecticides that led to the discovery of the organophosphorus acetylcholinesterase inhibitors by Schrader at the Bayer laboratory.[61] The study of their mechanism of action has shown that they act by acylation of a serine hydroxyl in the catalytic site of the enzyme. This was one of the first examples to describe a molecular mechanism for an enzymatic inhibition.

Replacement by Janssen *et al.*[62] of the N-methyl group of pethidine by various propio- and butyrophenones led to potent analgesics such as R951 and R1187 (Fig. 5.22). During their pharmacological study it was noted that mice that had been injected with these drugs became progressively calm and sedated. The resemblance of the sedation

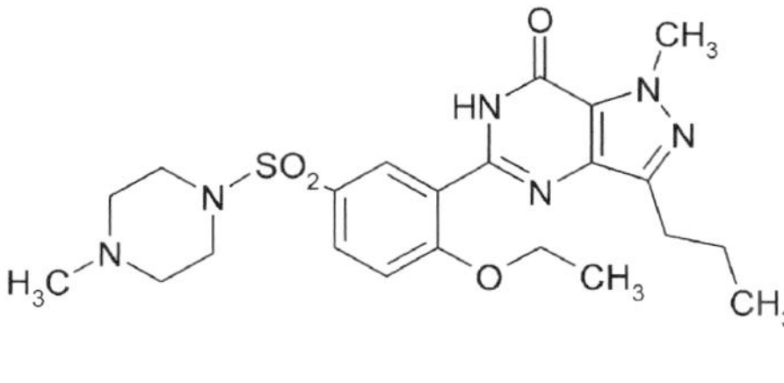

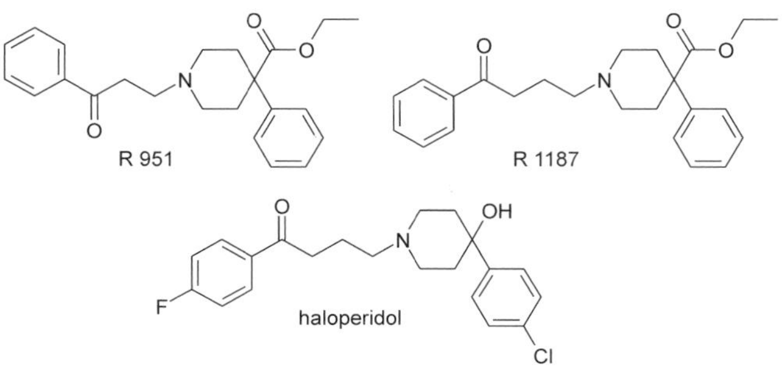

Fig. 5.22 The passage from pethidine-related opiate analgesics to the dopaminergic antagonist haloperidol.

Fig. 5.21 Structure of the phosphodiesterase type 5 (PDE5) inhibitor sildenafil.[53,54]

to that produced by chlorpromazine encouraged Janssen to synthesize analogues of R1187 in the hope that one might be devoid of analgesic activity while retaining tranquillizing activity. From this effort, haloperidol emerged in 1958 as the most potent tranquillizer yet to be discovered. It is 50–100 times as potent as chlorpromazine, with fewer side-effects.[62,63]

C Exploitation of observations made in the plant kingdom and in microbiology

Among the numerous discoveries that we owe to botanists and pharmacognosts are the development of tryptophan metabolites, and especially indolylacetic acid.[64] This compound acts as growth hormone in plants. *Para*-chlorinated phenoxyacetic acids (MCPA or methoxone; 2,4-D or chloroxone) are mimics of indolylacetic acid (bioisostery) and show similar phytohormonal properties: at high doses they serve as weeders. Ring-chlorinated phenoxyacetic acids were later introduced in molecules as varied as meclofenoxate (cerebral metabolism), clofibrate (lipid metabolism) and ethacrynic acid (diuretic).

The 5-hydroxylated analogue of indolacetic acid is the principal urinary metabolite of serotonin. On the basis of two biochemical observations — the possible role of serotonin in inflammatory processes and the increase of urinary metabolites of tryptophan in rheumatic patients — Shen, from the Merck Laboratories, designed anti-inflammatory compounds derived from indolacetic acid. Among them in 1963 he found indomethacin, one of the most powerful nonsteroidal anti-inflammatory drugs currently known.[65]

A particularly rich contribution of this approach in the therapeutic area has been the discovery and the development of penicillin (see Chapter 1). It initiated the discovery of many other major antibiotics such as chloramphenicol, streptomycin, tetracyclines, rifampicine, etc.

In conclusion, whatever its origin, the use of biological information constitutes a preferential source for original molecule research. It offers creative approaches that do not rest on the exploitation of routine pharmacological models. Once the lead molecule is identified, it will immediately be the object of thorough studies to elucidate its molecular mechanism of action. Simultaneously, one will proceed to the synthesis of structural analogues, as well as to the establishment of structure–activity relationships, and to the optimization of all the important parameters for its development: potency, selectivity, metabolism, bioavailability, toxicity, cost price, etc. In other terms, even if the initial discovery was purely fortuitous, subsequent research must be marked by a very important effort of rationalization.

V FOURTH STRATEGY: PLANNED RESEARCH AND RATIONAL APPROACHES

The approaches described so far leave too much to chance (screening, fortuitous discoveries) or they lack originality (therapeutic copies). A more scientific approach is based on the knowledge of the incriminated molecular target: enzyme, receptor, ion channel, signalling protein, transport protein or DNA. The progress in molecular and structural biology has allowed the identification and characterization of several hundreds of new molecular targets and made it possible to envisage the design of drugs on a more scientific basis.

A L-Dopa and parkinsonism

A historical example in which the key information, which rendered possible a rational approach to drug design, is the discovery of the usefulness of L-dopa in the treatment of Parkinson's disease. Thus, since in patients suffering from parkinsonism it was observed that the dopamine levels in the basal ganglions were much lower than those found in the brains of healthy persons,[66] a symptomatic, but rational, therapy became possible. This therapy consists of administering to patients the L-3,4-dihydroxy-phenylalanine (L-dopa); this amino acid is able to cross the blood–brain barrier, and is then decarboxylated into dopamine by brain dopa-decarboxylase. Initial clinical studies were undertaken by Cotzias, Van Woert and Schiffer.[67] Several hundred thousand patients have benefited from this treatment. However, 95% of the dopa administered by the oral route is decarboxylated in the periphery before having crossed the blood–brain barrier. To preserve the peripheral dopa from this unwanted precocious degradation, a peripheral inhibitor of dopa-decarboxylase is usually added to the treatment. An additional improvement of the treatment is the simultaneous addition of an inhibitor of catechol *O*-methyltransferase such as tolcapone or entacapone (see the section 'Vinylogy' in Chapter 12).

Other examples of the rational approach in pharmacology are the discovery of the inhibitors of the angiotensin-converting enzyme or of antagonists of histaminergic H_2-receptors.

B Inhibitors of the angiotensin-converting enzyme

The angiotensin-converting enzyme catalyses two reactions which are supposed to play an important role in the regulation of arterial pressure: (1) conversion of angiotensin I, which is an inactive decapeptide, into angiotensin II, an octapeptide with a very potent vasoconstrictor activity;

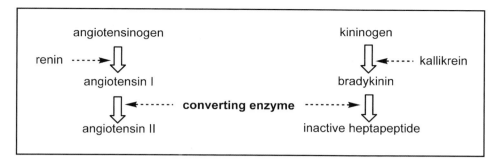

Fig. 5.23 Scheme of the renin–angiotensin and of the kallikrein–kinin systems. The converting enzyme (a carboxy-dipeptidyl-hydrolase) is common to the two systems.

pyro-Glu-Trp-Pro-Arg-Pro-Glu-Ile-Pro-Pro-OH

Fig. 5.24 The structure of the nonapeptide teprotide.

(2) inactivation of the nonapeptide, bradykinin, which is a potent vasodilator (Fig. 5.23).

An inhibitor of the converting enzyme would therefore constitute a good candidate for the treatment of patients suffering from hypertension. The first substance developed in this sense was teprotide, a nonapeptide presenting an identical sequence to that of some peptides isolated in 1965 by Ferreira from the venom of *Bothrops jararaca*, a Brazilian viper (Fig. 5.24). Teprotide inhibits in a competitive manner the degradation of angiotensin I by the converting enzyme. The presence of four prolines and a pyroglutamate renders this peptide relatively resistant to hydrolysis, but not sufficiently to allow its oral administration. In the search for a molecule offering better bioavailability, the reasoning of the Squibb scientists rested on the analogy of the angiotensin-converting enzyme to the bovine carboxypeptidase A.[68] In fact, both enzymes are carboxypeptidases; carboxypeptidase A detaches only one

C-terminal amino acid while the converting enzyme detaches two. Furthermore, it was known that the active site of carboxypeptidase A comprises three important elements for the interaction with the substrate (Fig. 5.25): an electrophilic centre, establishing an ionic bond with a carboxylic function, a site capable of establishing a hydrogen bond with a peptidic C-terminal function, and an atom of zinc, solidly fixed on the enzyme and serving to form a coordinating bond with the carbonyl group of the penultimate (the scissile) peptidic function.

By identifying that the conversion enzyme had a similar function, although altered by one amino acid unit (cleavage of the second peptidic bond instead of the first, departing from the terminal carboxyl group), scientists of the Squibb company have created the model drawn in Fig. 5.26. According to this model, *N*-succinyl amino acids such as the succinyl prolines shown in Fig. 5.26 (right) should be able to interact with each of the above-mentioned sites based on first their proline carboxyl (ionic bond), their amide function (hydrogen bond) and then on the carboxyl of the succinyl moiety (coordination with the zinc atom). These compounds should then be able to act as competitive and specific inhibitors of the converting enzyme. Therefore a series of

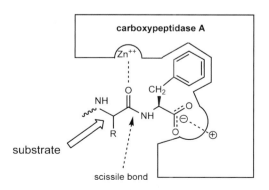
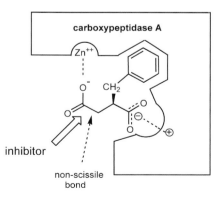

Fig. 5.25 Interactions between carboxypeptidase A and a substrate (left) or an inhibitor (right). Adapted after Cushman *et al.*[68]

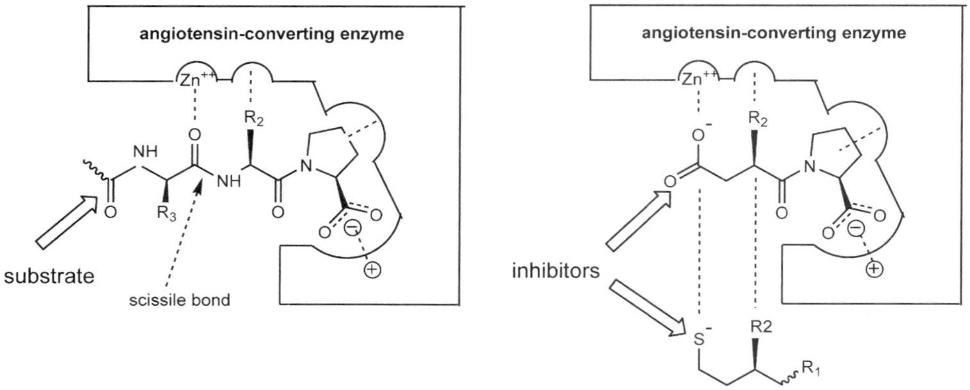

Fig. 5.26 Interactions between angiotensin-converting enzyme (a dipeptidyl carboxypeptidase) and a substrate (left) or inhibitors (right). Adapted after Cushman *et al.*[68]

N-succinyl amino acids was prepared and the *N*-L-proline derivative **1** (Fig. 5.27) showed some activity (IC_{50} = 330 μM). Amino acids other than L-proline lead to less active succinyl derivatives; this result is in agreement with the fact that several peptidic inhibitors (notably teprotide) also possess a proline in the C-terminal position. In the present example, *N*-succinyl-L-proline was selected as lead compound. The next task was to optimize its activity and this was done by researching the best interaction with the active site of the enzyme. Two steps were decisive in this quest: 'the fishing for hydrophobic pockets' and the research for a better coordinant for the zinc atom (Fig. 5.27). The exploration of hydrophobic pockets was achieved by substituting the succinyl moiety with methyl groups (four possibilities taking account of the regio- and the stereoisomers). Structure **2**, methylated at position β to the amide clearly appeared more active than **1** (IC_{50} = 22 instead of 330 μM). In this process, one observes an important stereoselectivity, since the IC_{50} value of epimer **3** of the compound **2** drops to 1480 μM. The best coordination with the zinc was achieved in replacing the carboxyl function by a mercapto group. The gain resulting from this modification has been extremely important as shown by the comparison of compounds **1** and **4** or also **2** and **5**. Compound **5** (SQ 14225) with an IC_{50} of 0.023 μM, and a K_i of 0.0017 μM is active by the oral route and has been introduced into therapeutic use under the designation of captopril.

It is interesting to observe that the loss in affinity caused by the replacement of the mercapto function by a carboxyl rest can be compensated with the help of an additional hydrophobic interaction. Thus, scientists from Merck developed enalaprilat **(6a)**, a compound of comparable

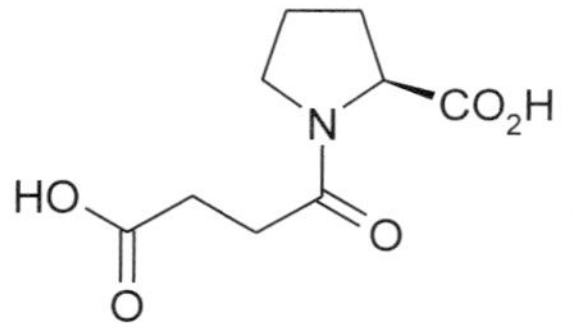

1 (330)
N-succinyl-proline

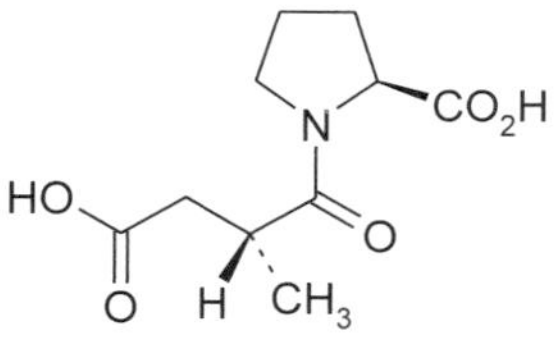

2 (22)

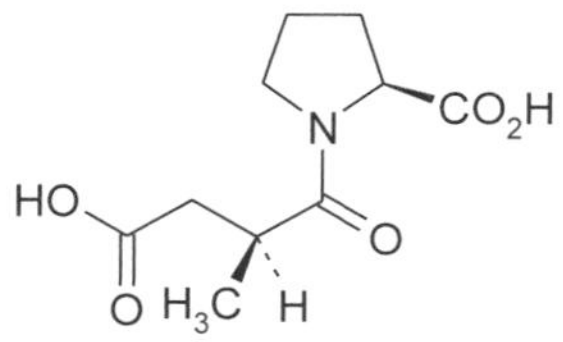

3 (1480)

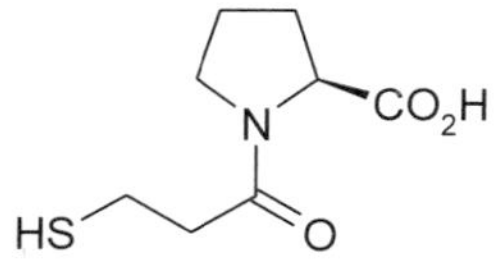

4 (0.20)

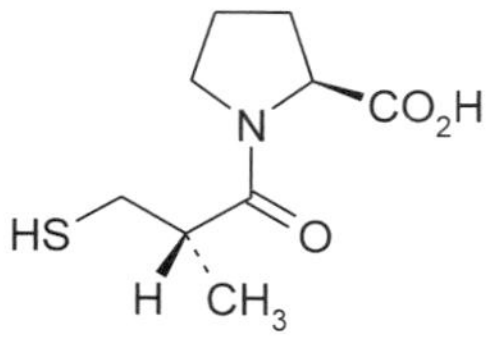

5 (0.023)
captopril

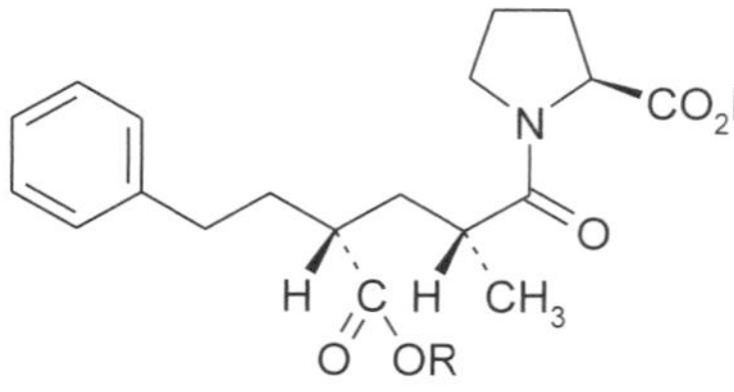

6a R = H: enalaprilat
6b R = C_2H_5: enalapril

Fig. 5.27 Structures of some key compounds in the development of captopril and enalapril.[68]

effectiveness and for which the additional hydrophobic interaction is provided by a phenethyl substituent. However, enalaprilat is poorly absorbed orally, therefore the commercial compound is enalapril (**6b**), the corresponding ethyl ester.

C Example of the H$_2$-receptor antagonists

Research to develop specific antagonists of the H$_2$ histamine receptor with a view to the treatment of gastric ulcers has also proceeded through a rational process.[69,70] Starting from the observation that the antihistaminic compounds known at that time (antagonists of H$_1$-receptors) were not capable of antagonizing the gastric secretion provoked by histamine, Black and his collaborators envisaged the existence of an unknown subclass of the histamine receptor (the future H$_2$-receptor). From 1964 onwards, they initiated a programme of systematic research of specific antagonists for this receptor.

The starting point was guanylhistamine (Fig. 5.28), which possesses weak antagonistic properties against the gastric secretion induced by histamine. The lengthening of the side chain of this compound clearly increased the H$_2$-antagonistic activity, but a residual agonist effect remained. In replacing the strongly basic guanidino function by a neutral thiourea, burimamide was obtained. Although very active, this compound was rejected for its low oral bioavailability. The addition of a methyl group in position 4 of the imidazolic ring, followed by the introduction of an electron-withdrawing sulfur atom in the side chain, finally led to a compound that was both very active and less ionized, properties which improved its absorption by the oral route. The derivative thus obtained, metiamide, was excellent and, moreover, 10 times more potent than burimamide. However, metiamide, because of its thiourea grouping, was tainted with side-effects (agranulocytosis, nephrotoxicity), that would limit its clinical use. The replacement of the thiourea by an isosteric grouping having the same pK_a (*N*-cyanoguanidine) led finally to cimetidine, which became a medicine of choice in the treatment of gastric ulcers. Later, it appeared that the imidazolic ring, present in histamine and in all H$_2$-antagonists discussed hitherto, was not indispensable to the H$_2$-antagonistic activity. Thus, ranitidine, which possesses a furan ring, has appeared to be even more active than cimetidine. The same proved to be true of famotidine and roxatidine.

D Recent molecular targets

Currently, hundreds of potential targets are available which are all starting points for the inventiveness of medicinal chemists:

Farnesyltransferase inhibitors as potential anticancer drugs: In 1989, it was discovered that farnesylation of a cysteine residue near the *ras* gene protein's C-terminus is required for its biological activity.[71] Interfering with Ras function by blocking the enzyme responsible for its

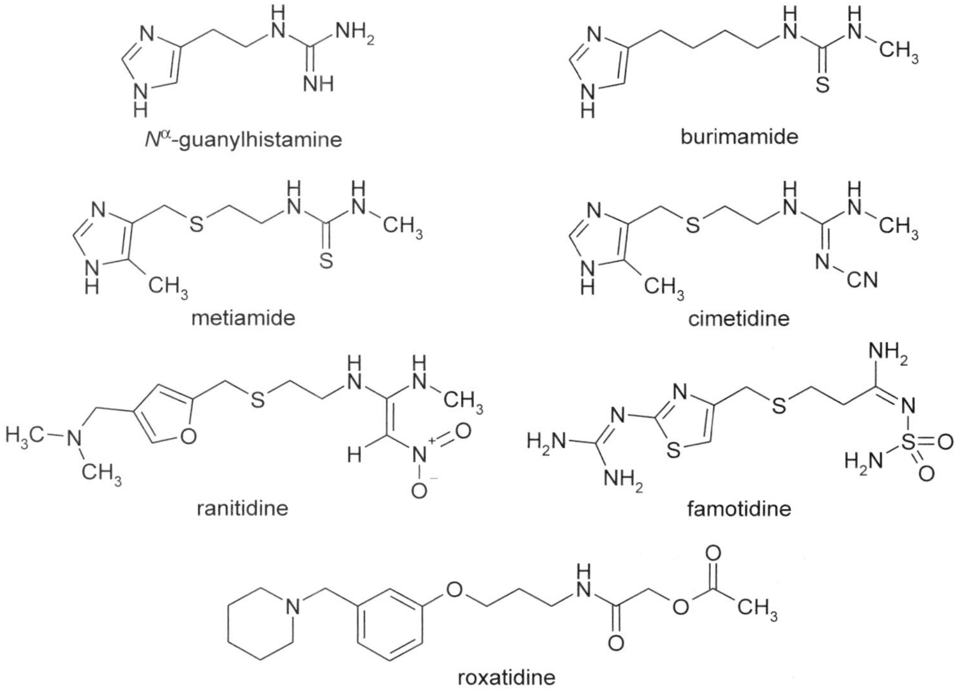

Fig. 5.28 Structures of some key compounds in the development of H$_2$-receptor antagonists.

farnesylation thus became a pharmacological target to suppress *ras*-dependent tumour growth.[72]

Disease-modifying antirheumatic drugs: Quinoline and quinazoline derivatives were found to inhibit preferentially IFNγ production by T_{H1}-type clones over IL-4 production by T_{H2}-type clones. This preferential suppression of T_{H1} cytokine production is considered essential for immunomodulating activity.[73]

Orexins and orexin receptors: Orexin A and orexin B are hypothalamic neuropeptides acting on a G-protein-coupled receptor that regulates feeding behaviour.[74]

Corticotropin-releasing factor (CRF) receptor modulators: CRF antagonists may act on depression, various other neuropsychiatric disorders and immune diseases.[75]

PPARs, an essential group of nuclear receptors: PPARα receptors are involved in dyslipemias, atherosclerosis and obesity; PPARγ receptors are mainly involved in diabetes but also in dyslipemias, hypertension, inflammation and cancer; PPARδ receptors, in addition to dyslipemia control, are involved in fertility and cancer.[76]

Type I fatty acid synthase (FAS) inhibitors as a new class of antituberculosis agents: While the synthesis of fatty acids occurs in all living organisms, mycobacteria possess accessory fatty acid synthase (FAS) enzymes with specialized substrate and product specificities that are attractive targets for drug development. Among a series of 3-sulfonyl-alcanamides examined, 3-sulfonyl-tridecanamide exhibited a minimum inhibitory concentration (MIC) of $0.75-1.5 \, \mu g \, mL^{-1}$, comparable with first-line antituberculosis drugs.[77]

Caspases as targets for anti-inflammatory and anti-apoptotic drug discovery: The biochemical mechanisms underlying proinflammatory cytokine maturation and cellular apoptosis are linked by the family of cysteine proteases known as caspases. These enzymes have necessary roles in both processes and are widely considered as promising targets for drug discovery.[78]

E Computer-assisted drug design

Like many other research areas, computational chemistry has evolved dramatically over the last 5 years in order to mirror changes occurring in the experimental disciplines on which it relies. It is now commonplace to address the study of hundreds of thousands of molecules for either physico-chemical descriptors, compatibility to a known pharmacophore or three-dimensional coordinates of a macromolecular target.[79] This tendency is likely to be strengthened by at least four concomitant trends:

(1) The full sequencing of the human genome[80] and its future annotation are going to provide medicinal chemists with an overwhelming number of potential targets for which very little data will be available. Electronic screening of chemical libraries[79] is thus going to play a major role for preselecting the very first ligands for interesting genomic targets.

(2) The protein data base (PDB)[81] which stores three-dimensional (3-D) coordinates of macromolecules (proteins, nucleic acids) is growing at an incredible pace, moving from 2241 entries in 1992 to 17 300 entries. This number is planned to increase significantly in the future with the advent of high-throughput X-ray crystallography.[82] High-throughput homology modelling of entire genome products has already been reported.[83] As a consequence, protein structure-based drug design will no longer be an exceptional situation.

(3) The constant increase of computing power at a reasonable price allows now complex simulations of protein behaviour[84] on massively parallel computing architecture. It will thus be conceivable to study the dynamic properties of supramolecular assemblies in the near future.

(4) Significant advances in software development are now emerging for predicting target-ligand 3D structures,[85] absolute binding-free energies (affinities)[86] and physicochemical descriptors able to predict ADME (absorption, distribution, metabolism, excretion) properties.[87]

However, the considerable gain in the throughput of computational chemistry should not be obtained at the detriment of the quality of predicted properties. There are still some hurdles of which to be aware. First, bigger is not always better.[19,31] Increasing the size of any library (ligands, targets) has no value if diversity is not created. Second, chemical similarity does not necessarily mean biological similarity. For example, steroid hormones (testosterone, progesterone) although chemically very similar have totally different biological functions. Third, one should avoid comparing molecules (full agonist, partial agonist, inverse agonist, antagonist, allosteric modulator) which have different pharmacological responses. Being aware of these limitations, it is now possible to rationalize as much as possible the identification and optimization of novel ligands (Fig. 5.29).

Starting from a 3-D structure of a target of interest obtained by either experimental techniques (X-ray diffraction or NMR) of molecular modelling, 3-D libraries can be rapidly screened (up to ca. 100 000 ligands/day) for virtual hits showing the best complementarity to a defined active site.[88] Reducing the number of molecules to be experimentally screened is not only of economic importance, especially for academia and small-sized companies, but allows the virtual hit list in true

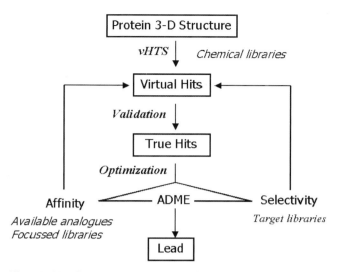

Fig. 5.29 Rational scheme of computer-based hit identification and optimization (vHTS: virtual high-throughput screening).

ligands to be enriched. Generally, validated hit rates of about 20–30% can be expected when screening a protein X-ray structure.[89] Screening a 3-D model is obviously less successful (hit rates of about 10–15% can be expected) (D. Rognan, personal communication) but is still significantly better than random screening of a full library. Once a hit has been identified, three different properties generally need to be optimized. First, the affinity for the respective target should usually be brought to the low nanomolar range. This can be done by looking at commercially (or in-house) available analogues of the hits for deriving structure–activity relationships, and then enabling the design of focused libraries (database of ligands sharing the same chemical scaffold) for prioritizing the synthesis of the most promising analogues.[90] Second, one should not forget to optimize the selectivity of the hit/lead versus other potential targets. A rational way to address the selectivity issue is to screen a 3-D library of defined active sites for complementarity to the known ligand.[91] One could even imagine a multidimensional screening of ligand libraries against target libraries for picking up the required ligand profile (selective for a single or several targets). Last, some pharmacokinetic properties (solubility, membrane permeation) can be predicted early from physicochemical descriptors (free energy of solvation, polar surface area)[87] while optimizing both affinity and selectivity, to focus synthesis on the most potent, selective and bioavailable ligands. Taking these three parameters (potency, selectivity, bioavailability) simultaneously into account to design 'drug-like' molecules[92] is believed to hasten lead optimization and reduce failure rates in clinical development.

VI CONCLUSION

The means leading to the discovery of new lead compounds, and possibly to new drugs, can be schematically classified into four approaches. These consist of the improvement of already existing drugs, of systematic screening, of retro-active exploitation of biological information, and of attempts towards rational design. Depending on which of these four strategies they apply, medicinal chemists can be seen as copiers, industrious, intuitive and deductive. It would be imprudent to compare the merit of each of these characteristics hastily. Indeed 'poor' research can end with a universally recognized medicine and, conversely, a brilliant rational demonstration can remain sterile. It is therefore of the highest importance, given the random character of discovery and the virtual impossibility of planned invention of new active principles, that decision-makers in the pharmaceutical industry appeal to all the four strategies that have been described and that they realize that they are not mutually exclusive. On the other hand, it would be inappropriate, once a lead compound is discovered and characterized, not to study its molecular mechanism of action. Every possible effort should be made in this direction. In conclusion, all strategies resulting in identification of lead compounds are a priori equally good and advisable, provided that the research they subsequently induce is done in a rational manner.

REFERENCES

1. Clausewitz von, C. (1973) *Vom Kriege, Drittes Buch, 1. Kapitel Strategie*, 18th edn. Ferdinand Dümmlers Verlag, Bonn, 1973.
2. Fischer, J. and Gere, A. (2001) Timing of analog research in the medicinal chemistry. *Pharmazie* **56**: 675–682.
3. Thuilier, J. (1981) *Les dix ans qui ont changé la folie*, pp. 253–257. Robert Laffont, Paris.
4. Kling, J. (1988) From hypertension to angina to Viagra. *Mod. Drug Discovery* 31–38.
5. Mannhold, R. (1986) BTS-49465. *Drug Fut.* **11**: 177–178.
6. Sternbach, L.H. (1979) The Benzodiazepine story. *J. Med. Chem.* **22**: 1–7.
7. Jeanmart, C. and Cotrel, C. (1978) Synthèse de (chloro-5 pyridyl-2)-6 méthyl-4 pipérazinyl-1)carbonyloxy-5-oxo-7 dihydro-6,7 5*H*-pyrrolo [3,4-*b*] pyrazine. *Compt. Rend. Acad. Sci. Paris Ser. C* **287**: 377–378.
8. Blanchard, J.C., Boireau, A., Garret, C. and Julou, L. (1979) *In vitro* and *in vivo* inhibition by zopiclone of benzodiazepine binding to rodent brain receptors. *Life Sci.* **24**: 2417–2420.
9. Suffness, M. (1993) Taxol: from discovery to therapeutic use. In Bristol, J.A. (ed.). *Annual Reports in Medicinal Chemistry*, pp. 305–314. Academic Press, San Diego.
10. Boyd, M.R. and Paull, K.D. (1995) Some practical consideration and applications of the NCI *in vitro* drug discover screen. *Drug Dev. Res.* **34**: 91–109.
11. Endo, A. (1985) Compactin (ML-236B) and related compounds as potential cholesterol-lowering agents that inhibit HMG-CoA reductase. *J. Med. Chem.* **28**: 401–405.
12. Lee, T.J. (1987) Synthesis, SARs and therapeutic potential of

HMG-CoA reductase inhibitors. *Trends Pharmacol. Sci.* **8**: 206–208.

13. Shu, Y.Z., Cutrone, J.Q., Klohr, S.E. and Huang, S. (1995) BMS-192548, a tetracyclic binding inhibitor of neuropeptide Y receptors, from Aspergillus niger WB2346. II. Physico-chemical properties and structural characterization. *J. Antibiot.* **48**: 1060–1065.

14. Nisbet, L.J. and Westley, J.W. (1986) Developments in microbial products screening. In Bristol, J.A. (ed.). *Annual Reports in Medicinal Chemistry*, pp. 149–157. Academic Press, San Diego.

15. Hylands, P.J. and Nisbet, L.J. (1991) The search for molecular diversity (I). In Bristol, J.A. (ed.). *Annual Reports in Medicinal Chemistry*, pp. 259–269. Academic Press, San Diego.

16. Liu, K., Xu, L., Szalkowski, D., Li, Z., Ding, V., *et al.* (2000) Discovery of a potent, highly selective, and orally efficacious small-molecule activator of the insulin receptor. *J. Med. Chem.* **43**: 3487–3494.

17. Roever, S., Adam, G., Cesura, A.M., Galley, G., Jenck, F., *et al.* (2000) High-affinity, non-peptide agonists for the ORL1 (Orphanin FQ/Nociceptin) receptor. *J. Med. Chem.* **43**: 1329–1338.

18. Reddy, M. V. R., Rao, M.R., Rhodes, D., Hansen, M. S. T., Rubins, K., *et al.* (1999) Lamellarin a 20-sulfate, an inhibitor of HIV-I integrase active against HIV-I virus in cell culture. *J. Med. Chem.* **42**: 1901–1907.

19. Mander, T. (2000) Beyond uHTS: ridiculously HTS? *Drug Discov. Today* **3**: 223–225.

20. Youngman, M.A., McNally, J.J., Lovenberg, T.W., Reitz, A.B., Willard, N.M., *et al.* (2000) N-(Sulfonamido)alkyl-β-amonotetralins: Potent and selective neuropeptide Y Y5 receptor antagonists. *J. Med. Chem.* **43**: 346–350.

21. Song, Y., Connor, D.T., Sercel, A.D., Doubleday, R., *et al.* (1999) Synthesis, structure–activity relationships and *in vivo* evaluations of substituted Di-*tert*-butylphenols as a novel class of potent, selective, and orally active cyclooxygenase-2 inhibitors. 2. 1,3,4- and 1,2,4-thiadiazole series. *J. Med. Chem.* **42**: 1161–1169.

22. Moos, W.H., Green, G.D. and Pavia, M.R. (1993) Recent advances in the generation of molecular diversity. In Bristol, J.A. (ed.). *Annual Reports Medicinal Chemistry*, pp. 315–324. Academic Press, San Diego.

23. Bull, A.T., Goodfellow, M. and Slater, J.H. (1992) Biodiversity as a source of innovation in biotechnology. *Annu. Rev. Microbiol.* **46**: 219–252.

24. Dower, W.J. and Fodor, A.P. (1991) The search of molecular diversity (II): recombinant and synthetic randomized peptide libraries. In Bristol, J.A. (ed.). *Annual Reports Medicinal Chemistry*, pp. 271–280. Academic Press, San Diego.

25. Thorpe, D.S., Chan, A. W. E., Binnie, A., *et al.* (1999) Efficient discovery of inhibitory ligands for diverse targets from a small combinatorial chemical library of chimeric molecules. *Biochem. Biophys. Res. Commun.* **266**: 62–65.

26. Lahana, R. (1999) How many leads from HTS? *Drug Discov. Today* **4**: 447–448.

27. Pötter, T. and Matter, H. (1998) Random or rational design? Evaluation of diverse compound subsets from chemical structure databases. *J. Med. Chem.* **41**: 478–488.

28. Frimurer, T.M., Bywater, R., Naerum, L., Lauritsen, L.N. and Brunak, S. (2000) Improving the odds in discriminating 'drug-like' from 'non-drug-like' compounds. *J. Chem. Inf. Comput. Sci.* **40**: 1315–1324.

29. Böhm, H.-J., Klebe, G. and Kubinyi, H. (1996) *Wirkstoffdesign*. Spektrum Akademischer Verlag, Heidelberg.

30. Wermuth, C.G. (1998) Search for new lead compounds: the example of the chemical and pharmacological dissection of aminopyridazines. *J. Heterocyclic Chem.* **35**: 1091–1100.

31. Wermuth, C.G. and Clarence-Smith, K. (2000) 'Drug-like' leads: bigger is not always better. *Pharmaceutical News* **7**: 53–57.

32. Poroikov, V.V., Akimov, D.I., Shabelnikova, E. and Filimonov, D.

(2001) Top 200 medicines: can new actions be discovered through computer-aided prediction? *SAR QSAR Environ. Res.* **12**: 327–344.

33. Wermuth, C.G. (2001) The 'SOSA' approach: an alternative to high-throughput screening. *Med. Chem. Res.* **10**: 431–439.

34. Riechers, H., Albrecht, H.-P., Amberg, W., Baumann, E., Bernard, H., *et al.* (1996) Discovery and optimization of a novel class of orally active non-peptide endothelin-A receptor antagonists. *J. Med. Chem.* **39**: 2123–2128.

35. Wermuth, C.G., Schlewer, G., Bourguignon, J.-J., Maghioros, G., Bouchet, M.J., *et al.* (1989) 3-Aminopyridazine derivatives with atypical antidepressant, serotonergic and dopaminergic activities. *J. Med. Chem.* **32**: 528–537.

36. Wermuth, C.G. (1993) Aminopyridazines — an alternative route to potent muscarinic agonists with no cholinergic syndrome. *Il Farmaco* **48**: 253–274.

37. Wermuth, C.G., Bourguignon, J.-J., Hoffmann, R., Boigegrain, R., Brodin, R., *et al.* (1992) SR 46559 and related aminopyridazines are potent muscarinic agonist with no cholinergic syndrome. *Biorg Med. Chem Lett.* **2**: 833–836.

38. Pilla, M., Perachon, S., Sautel, F., Garrido, F., Mann, A., *et al.* (1999) Selective inhibition of cocaine-seeking behaviour by a partial dopamine D_3 agonist. *Nature* **400**: 371–375.

39. Prestwick Chemical Inc., 1825 K Street, N.W., Suite 1475,Washington DC 20006-1202; www.prestwickchemical.com.

40. Kleemann, A. and Engel, J. (1982) *Pharmazeutische Wirkstoffe-Synthese, Patente, Anwendungen*, (Preface). Georg Thieme Verlag, Stuttgart.

41. Price, J.R. (1965) *Antifertility Agents of Plant Origin. A Symposium on Agents Affecting Fertility*, pp. 3–17. Little, Brown and Co, Boston.

42. Laborit, H., Huguenard, P. and Alluaume, R. (1952) Un nouveau stabilisateur végétatif, le 4560 R.P. *Presse Méd.* **60**: 206–208.

43. Maxwell, R.A. and Eckhardt, S.B. (1990) *Drug Discovery — A Casebook and Analysis*. Humana Press, Clifton.

44. Stemp, G. and Evans, J.M. (1993) Discovery and development of cromakalim and related potassium channel activators. *Medicinal Chemistry, the Role of Organic Chemistry in Drug Research*, 2nd edn., pp. 141–162. Academic Press, London.

45. Hamilton, T.C., Weir, S.W. and Weston, A.H. (1986) Comparison of the effects of BRL 34915 and verapamil on electrical and mechanical activity in rat portal vein. *Br. J. Pharmacol.* **88**: 103–111.

46. Gerber, J.G. and Nies, A.S. (1990) Antihypertensive agents and the drug therapy of hypertension. In Gilman, A., Rall, T.W., Nies, A.S. and Taylor, P. (eds). *Goodman and Gilman's The Pharmacological Basis of Therapeutics*, 8th edn., pp. 784–813. Pergamon Press, New York.

47. Evans, J.M., Fake, C.S., Hamilton, T.C., Poyser, R.H. and Watts, E.A. (1983) Synthesis and antihypertensive activity of substituted *trans*-4-amino-3,4-dihydro-2,2-dimethyl-2*H*-1-benzopyran-3-ols. *J. Med. Chem.* **26**: 1582–1589.

48. Evans, J.M., Fake, C.S., Hamilton, T.C., Poyser, R.H. and Showell, G.A. (1984) Synthesis and antihypertensive activity of 6,7-disubstituted trans-4-amino-3,4-dihydro-2,2-dimethyl-2H-1-benzopyran-3-ols. *J. Med. Chem.* **27**: 1127–1131.

49. Sneader, W. (1996) *Drug Prototypes and Their Exploitation*. John Wiley & Sons, Chichester.

50. Iyer, C. G. S., Languillon, J. and Ramanujam, K. (1971) WHO coordinated short-term double-blind trial with thalidomide in the treatment of acute lepra reactions in male lepromatus patients. *Bull. WHO* **45**: 719–732.

51. Cushman, M., Jayaraman, M., Vroman, J.A., Fukunaga, A.K., Fox, B.M., *et al.* (2000) Synthesis of new indeno[1,2-*c*]isoquinolines: cytotoxic non-camptothecin topoisomerase I inhibitors. *J. Med. Chem.* **43**: 3688–3698.

52. Korth, C., May, B.C., Cohen, F.E. and Prusiner, S.B. (2001) Acridine

and phenothiazine derivatives as pharmacotherapeutics for prior disease. *Proc. Natl Acad. Sci. USA* **98**: 9836–9841.

53. Terret, N.K., Bell, A.S., Brown, D. and Ellis, P. (1996) Sildenafil (Viagra), a potent and selective inhibitor of type 5 cGMP phosphodiesterase with utility for the treatment of male erectile dysfunction. *Bioorg. Med. Chem. Lett.* **6**: 1819–1824.

54. Boolell, M., Allen, M.J., Ballard, S.A., Gepi-Attee, S., Muirhead, G.J., *et al.* (1996) Sildenafil: an orally active type 5 cyclic GMP-specific phosphodiesterase inhibitor for the treatment of penile erectile dysfunction. *Int. J. Urol. Res.* **8**: 47–52.

55. Foye, W.O., Lemke, T.L. and Williams, D.A. (1995) *Principles of Medicinal Chemistry*. Lea & Febiger, Baltimore.

56. Neuworth, M.B., Laufer, R.J., Barnhart, J.W., Sefranka, J.A. and McIntosh, D.D. (1970) Synthesis and hypocholesterolemic activity of alkylidenedithio bisphenols. *J. Med. Chem.* **13**: 722–725.

57. Noble, R.L., Beer, C.T. and Cutts, J.H. (1958) Role of chance-observations in chemotherapy. *Vinca rosea. Ann. N.Y. Acad. Sci.* **76**: 882–894.

58. Johnson, I.S., Wright, H.F., Svoboda, G.H. and Vlantis, J. (1960) Antitumor principles derived from *Vinca rosea* Linn. I. Vincaleukoblastine and leurosine. *Cancer Res.* **20**: 1016–1022.

59. Johnson, I.S., Armstrong, J.G., Gorman, M. and Burnett, J.P., Jr. (1963) The Vinca alkaloids: a new class of oncolytic agents. *Cancer Res.* **23**: 1390–1427.

60. Buxton, I.L.O., Cheek, D.J., Eckman, D., Westfall, D.P., Sanders, K.M., *et al.* (1993) N^G-Nitro-L-arginine methyl ester and other alkyl esters of arginine are muscarinic antagonists. *Circulation. Res.* **72**: 387–395.

61. Schrader, G. (1952) *Die Entwicklung neuer Insektizide auf Grundlage von Organischen Fluor und Phosphorverbindungen*. Monographie No. 62, Verlag Chemie, Weinheim.

62. Janssen, P.A.J., Van de Westeringh, C., Jagneau, A. W. M., Demoen, P. J. A., Hermans, B. K. F., *et al.* (1959) Chemistry and pharmacology of CNS depressants related to 4-(4-hydroxy-4-phenylpiperidino)butyrophenone. Part I. Synthesis and screening data in mice. *J. Med. Pharm. Chem.* **1**: 281–297.

63. Janssen, P.A.J. (1965) The evolution of the butyrophenones, haloperidol and trifluperidol, from meperidine-like 4-phenylpiperidines. *Int. Rev. Neurobiol.* **8**: 221–263.

64. Albert, A. (1979) *Selective Toxicity*, 6th edn. Chapman and Hall, London.

65. Shen, T.Y. (1972) Perspectives in non-steroidal anti-inflammatory agents. *Angew. Chem. Int. Ed.* **11**: 460–472.

66. Ehringer, H. and Hornykiewicz, O. (1960) Verteilung von Noradrenalin und Dopamin (3-hydroxytyramin) im Gehirn des Menschen und ihr Verhalten bei Erkrankungen des Extrapyramidalen Systems. *Klin. Wochenschr.* **38**: 1236–1239.

67. Cotzias, G.E., Van Woert, M.H. and Schiffer, L.M. (1967) Aromatic amino acids and modification of Parkinsonism. *New. Engl. J. Med.* **276**: 374–379.

68. Cushman, D.W., Cheung, H.S., Sabo, E.F. and Ondetti, M.A. (1977) Design of potent competitive inhibitors of angiotensin-converting enzyme. Carboxyalkanoyl and mercaptoalkanoyl amino acids. *Biochemistry* **16**: 5485–5491.

69. Black, J.W., Duncan, W. A. M., Durant, J.C., Ganellin, C.R. and Parsons, M.E. (1972) Definition and antagonism of histamine H2-receptors. *Nature* **236**: 385–390.

70. Ganellin, C.R. (1982) Cimetidine. *Chronicles of Drug Discovery*, pp. 1–38. John Wiley and Sons, New York.

71. Leonard, D.M. (1997) Ras farnesyltransferase: a new therapeutic target. *J. Med. Chem.* **40**: 2971–2990.

72. Hunt, J.T., Ding, C.Z., Batorsky, R., Bednarz, M., Bhide, R., *et al.*

(2000) Discovery of (R)-7-cyano-2,3,4,5-tetrahydro-1-(1*H*-imidazol-4ylmethyl)-3-(phenylmethyl)-4-(2-thienylsulfonyl)-1*H*-1,4-benzodiazepine (BMS-214662), a farnesyltransferase inhibitor with potent preclinical antitumor activity. *J. Med. Chem.* **43**: 3587.

73. Baba, A., Kawamura, N., Makino, H., Ohta, Y., Taketomi, S., *et al.* (1996) Studies on disease-modifying antirheumatic drugs: synthesis of novel quinoline and quinazoline derivatives and their anti-inflammatory effect. *J. Med. Chem.* **39**: 5176–5182.

74. Sakurai, T., Amemiya, A., Ishii, M., Matsuzaki, I., Chemelli, R.M., *et al.* (1998) Orexins and orexin receptors: a family of hypothalamic neuropeptides and G protein-coupled receptors that regulate feeding behavior. *Cell* **92**: 573–585.

75. Gilligan, P.J., Robertson, D.W. and Zaczek, R. (2000) Corticotropin releasing factor (CRF) receptor modulators: progress and opportunities for new therapeutic agents. *J. Med. Chem.* **43**: 1641–1660.

76. Willson, T.M., Brown, P.J., Sternbach, D.D. and Henke, B.R. (2000) The PPARs: from orphan receptors to drug discovery. *J. Med. Chem.* **43**: 527–550.

77. Jones, P.B., Parrish, N., Houston, T.A., Stapon, A., Bansal, N.P., *et al.* (2000) A new class of antituberculosis agents. *J. Med. Chem.* **43**: 3304–3314.

78. Talanian, R.V., Brady, K.D. and Cryns, V.L. (2000) Caspases as targets for anti-inflammatory and anti-apoptotic drug discovery. *J. Med. Chem.* **43**: 3351–3371.

79. Walters, W.P., Stahl, M.T. and Murcko, M.A. (1998) Virtual screening — an overview. *Drug Discov. Today* **3**: 160–178.

80. Wenter, J.C. (2001) The sequence of the human genome. *Science* **291**: 1304–1335.

81. Berman, H.M., Westbrook, J., Feng, Z., Gilliland, G., Bhat, T.N., *et al.* (2000) The protein data bank. *Nucleic Acids Res.* **28**: 235–242.

82. Stevens, R.C., Yokohama, S. and Wilson, I.A. (2001) Global efforts in structural genomics. *Science* **294**: 89–92.

83. Sanchez, R. and Sali, A. (1998) Large-scale protein structure modeling of the *Saccharomyces cerevisiae* genome. *Proc. Natl Acad. Sci. USA* **95**: 13597–13602.

84. Duan, Y. and Kollman, P.A. (1998) Pathways to a protein folding intermediate observed in a 1-microsecond simulation in aqueous solution. *Science* **282**: 740–744.

85. Abagyan, R. and Totrov, M. (2000) High-throughput docking and lead generation. *Curr. Opin. Struct. Biol.* **5**: 375–382.

86. Gohlke, H. and Klebe, G. (2001) Statistical potentials and scoring functions applied to protein-ligand binding. *Curr. Opin. Struct. Biol.* **11**: 231–235.

87. Lipinski, C.A., Lombardo, F., Dominy, B.W. and Feeney, P.J. (2001) Experimental and computational approaches to estimate solubility and permeability in drug discovery and development settings. *Adv. Drug Deliv. Rev.* 2001 **46**: 3–26.

88. Schneider, G. and Böhm, H.J. (2001) Virtual screening and fast automated docking methods. *Drug Discovery Today* **7**: 64–70.

89. Bissantz, C., Folkers, G. and Rognan, D. (2000) Protein-based virtual screening of chemical databases 1. Evaluation of different docking/scoring combinations. *J. Med. Chem.* **43**: 4759–4767.

90. Tondi, D., Slomcczynska, U., Costi, M.P., Watterson, D.M., Ghelli, S., *et al.* (1999) Structure-based discovery and in-parallel optimisation of non competitive inhibitors of thymidylate kinase. *Chem. Biol.* **6**: 319–331.

91. Chen, Y. and Zhi, D.G. (2001) Ligand–protein inverse docking and its potential use in the computer search of protein targets of a small molecule. *Proteins* **43**: 217–226.

92. Clark, D.E. and Pickett, S.D. (2000) Computational methods for the prediction of 'drug-likeness'. *Drug Discov. Today* **5**: 49–58.

6

NATURAL PRODUCTS AS PHARMACEUTICALS AND SOURCES FOR LEAD STRUCTURES

David J. Newman, Gordon M. Cragg and David G.I. Kingston

Accuse not Nature, she hath done her part; do thou but thine   **Milton,** *Paradise Lost*

I INTRODUCTION

This urgent need for the discovery and development of new pharmaceuticals for the treatment of cancer, AIDS, infectious diseases, as well as a host of other diseases demands that all approaches to drug discovery be exploited aggressively. Among the possible approaches, the one from natural products has made many unique and vital contributions to drug discovery, and this chapter will address these contributions and will offer some perspective on the potential of the natural products approach for future drug discovery and development.

II THE IMPORTANCE OF NATURAL PRODUCTS IN DRUG DISCOVERY AND DEVELOPMENT

The existence of bioactive compounds in plants and other natural sources has been known for millennia, with history recording the use of such poisons as hemlock, used by Socrates in his "court-ordered suicide" and the infusion of yew used by a Gallic chieftain to avoid capture by Julius Caesar. The medicinal use of natural products, particularly those from plants, is also very ancient.[1] The first known records, written in cuneiform on hundreds of clay tablets, are from Mesopotamia and date from about 2600 BC. Amongst the approximately 1000 plant-derived substances that were used are some that are still in use today for the treatment of ailments, ranging from coughs and colds to parasitic infections and inflammation. Many "codifications" or what may well be called "Pharmacopoeias" were compiled in the next 4000 years, culminating in the publication in

The Practice of Medicinal Chemistry
ISBN 0-12-744481-5

1618 of the *London Pharmacopoeia*. The concept of pure compounds as drugs may be traced to the isolation of the active principles of commonly used plants such as strychnine, morphine, atropine and colchicines in the early 1800s. These isolations were then followed by what may be considered to be the first commercial pure natural product, morphine, produced by E. Merck in 1826, and the first semi-synthetic pure drug based on a natural product, aspirin, produced by Bayer in 1899.[2]

Given such a history of plant-derived pharmaceuticals (and until the second third of the twentieth century, almost all pharmaceuticals were either natural products or based thereon), at least three questions cry out for answers. Firstly, why do plants and other organisms produce compounds that are effective in human medicine? Secondly, why did natural products fall out of favor in the later part of the twentieth century? Thirdly, what factors have brought about the present interest in natural products as "leads" in the pharmaceutical industry? Each of these will be discussed in turn.

A   The origin of natural products

The question of the origins of secondary metabolites (all non-proteinaceous natural products would fall under this term) has long intrigued chemists and biochemists. Six major hypotheses have been proposed, and these have been well summarized by Haslam.[3] (1) Secondary metabolites are simply waste products with no particular physiological role. (2) Secondary metabolites are compounds that at one time had a functional metabolic role, which has now been lost. (3) Secondary metabolites are products of random mutations, and have no real function in the organism. (4) Secondary metabolites are an example of "evolution in progress", and provide a pool of compounds out of which new biochemical processes can emerge. (5) Secondary metabolism provides a way of enabling the enzymes of primary metabolism to function when they are not needed for their primary purpose. "It is the *processes* of secondary metabolism, rather than the *products* (secondary metabolites) which are important". (6) Secondary metabolites play a key role in the organism's survival, providing defensive substances or other physiologically important compounds.

Although each of the above has (or has had) its supporters, Williams[4] and Harborne[5] amongst others, argue convincingly that the weight of the evidence is behind the sixth hypothesis. This hypothesis is consistent with the fact that most secondary metabolites are produced by (relatively) immobile organisms (plants, sessile marine invertebrates, microbes in soil environments) and thus must rely on "chemical warfare" for defense. Consistent with this hypothesis is the finding that annual plants like grasses that can renew themselves each year, tend to have fewer secondary metabolites than the more permanent shrubs and trees. A further consideration is that plants and marine invertebrates lack the sophisticated immune systems of vertebrates, and must thus rely on chemical defense against viruses. In addition, this hypothesis is consistent with the fact that the biosynthesis of secondary metabolites requires much metabolic energy and the storage of significant amounts of genetic information. It thus makes intuitive sense that all of this energy and information should be used for some specific purpose. Put in simpler terms, there are many easier pathways for organisms to dispose of "products of waste metabolism" or to "keep the enzymes of primary metabolism functional", without invoking the synthesis of very complex chemicals.

Finally, however, and most importantly in the present context, many secondary metabolites do indeed "trigger very specific physiological responses … in many cases by binding to receptors with a remarkable complementarity".[4] One very specific example would be the peptidic toxins elaborated by cone snails and originally identified by Olivera,[6] that have led to the development of Ziconotide®, a non-narcotic analgesic that is currently awaiting final FDA approval in the USA. This finding and many others provide powerful incentives to search for bioactive compounds from the microbial, plant and animal kingdoms, since it is assured that such compounds do exist. Some of these compounds will be effective in directly treating human disease or just as importantly, will become leads to pharmacophores that will help develop novel agents for human use.

B   The decline of natural products research

In spite of the obvious and continuing need for new drugs described in the opening section, the search for new natural product drugs fell into disfavor in many pharmaceutical companies during the 1980s and most of the 1990s. The reasons for this were two-fold. In the first place, the rapid development of combinatorial chemistry since its discovery in the late 1970s enabled chemists to prepare tens of thousands of compounds rapidly and efficiently. Secondly, and coupled with it, the development of high-throughput screening with automated screening systems capable of assaying up to 50 000 samples per day provided a rapid way of screening the products of combinatorial chemistry. Given this high throughput, a screen might run only a few weeks before it is replaced with another screen, and this time period is too short for bioactive natural products to be isolated by classical bioassay-directed fractionation. In addition to this incompatibility between modern high-throughput screening and bioassay-directed fractionation, the natural products approach was seen as slow, expensive and inefficient. Added to these factors, the problems of scale-up of the isolation process and of the acquisition of adequate amounts of biomass undoubtedly served as additional deterrents to work in this area. Many pharmaceutical companies thus dropped research programs on natural products in favor of chemical synthesis programs using various combinatorial chemistry strategies, where compounds could be produced much more readily than by the natural products approach. The only exception to this was with fermentation programs, which were continued in some cases as the logistics were well worked out. Even the US National Cancer Institute, long a bulwark of natural products research in the anticancer area, discontinued its plant and fractionation contracts in 1979. Thus the prospects for natural products research in the pharmaceutical area appeared to be particularly gloomy in the 1980s.

C   The revival of natural products research

The success of a natural products approach

As examples of the continued influence of natural products in areas ranging from all diseases to certain specific areas such as cancer, the reader can consult the papers by Cragg *et al.*[7,8] and Newman *et al.*[9] for background. Specifically, in Table 6.1 the source of all drugs that sold US$1 billion (or more) in 1997 is shown, using the conventions established by Cragg *et al.*[7] to denote the source of the pharmacophore. Further details as to sales of other drugs and their sources for the then top 12 pharmaceutical houses are given in a recent article by Newman and Laird.[10]

Inspection of Table 6.1 shows that at the end of 1997, the second best-selling series of drugs were the natural product/natural product-derived cholesterol lowering agents derived from the fungi *Monascus ruber*, *Aspergillus terreus* and *Penicillium citrinum*.[11] These are followed by the hypertensive agents, one of which, enalapril maleate, was derived from a natural product pharmacophore.[7] The next class of natural product/derived drugs were those in the antimicrobial area. One can also look at the influence of natural products in another way from the 1997 sales data, which is "sales by source", and is shown in Table 6.2 where over the 11 therapeutic areas used in Table 6.1, 43% of all the drugs were natural products or derived in some way from natural products.

Although not listed in the tables, at least four other natural product drugs have given yeoman service in the antitumor area. The first of these is paclitaxel (Taxol®) which sold US$1.6 billion in 2000; this is followed by the *vinca* alkaloids vinblastine and vincristine. Completing the quartet are the natural product-derived *epi*-podophyllotoxin derivatives teniposide and etoposide and the materials derived from camptothecin, topotecan and CPT-11. These will probably not be the only natural product drugs in the antitumor field, as can be seen by inspection of Table 6.3, where Cragg and Newman recently reported on the source of compounds currently under clinical trials as antitumor agents.[8] Inspection of this Table shows that of the 294 clinical candidates identified, 60% are *other than synthetic*.

The uniqueness of the natural products approach

Natural products are generally complex chemical structures, whether they are cyclic peptides like cyclosporin A, or complex diterpenes like paclitaxel. Inspection of the structures that are discussed in Section IV is usually enough to convince any skeptic that few of them would have been discovered without application of natural products chemistry. Recognition of this structural diversity has certainly made an impact on the design of combinatorial strategies, as exemplified by the 2 million plus compound library that was recently assembled by the Schreiber group using natural product-like structures as the initial scaffolds.[12]

Structural diversity is not the only reason why natural products are of interest to drug development, since they often provide highly selective and specific biological activities based on mechanisms of action. Two very good examples of this are the HMG-CoA reductase inhibition exhibited by lovastatin and the tubulin-assembly promotion activity of paclitaxel. These activities would not have been discovered without the natural product lead and investigation of their mechanisms of action.

The bioactivity of natural products stems from the previously discussed hypothesis that essentially all natural products have some receptor-binding binding activity; the problem is to find to which receptor a given natural product is binding. Viewed another way, a given organism provides the investigator with a complex library of unique bioactive constituents, analogous to the library of crude synthetic products initially produced by combinatorial chemistry techniques. The natural products approach can thus be seen as complementary to the synthetic approach, each providing access to (initially) different lead structures. In addition, as will be demonstrated in a later section (c.f. discussion of eleutherobin and the sarcodictyins), development of an active natural product structure by combinatorially directed synthesis is an extremely powerful tool. The task of the natural products researcher is thus to select those compounds of pharmacological interest from the "natural combinatorial libraries" produced by extraction of organisms. Fortunately, the means to do this are now at hand.

The impact of new screening methods

In the early days of natural products research, new compounds were simply isolated at random, or at best by the use of simple broad-based bioactivity screens based on antimicrobial or cytotoxicity activities. Although these screens did result in the isolation of bioactive compounds, including many of those described earlier, they have been

Table 6.1 *The world's best selling pharmaceuticals (sales > US$ 1 billion in 1997)*

Ranking	US brand name	Generic name	Marketed by	Global sales US$ billion	Source
Anti-ulcer					
2	Losec	Omeprazole	Astra AB	2.82	S
6	Prilosec	Omeprazole	Astra Merck	2.24	S
5	Zantac	Ranitidine	Glaxo Wellcome	2.26	S
17	Pepcid	Famotidine	Merck	1.18	S
Cholesterol-lowering					
1	Zocor	Simivastin	Merck	3.58	ND
13	Pravachol	Pravastatin	BristolMyers Squibb	1.44	N
14	Mevalotin	Pravastatin	Sankyo	1.41	N
22	Mevacor	Lovastatin	Merck	1.10	N
Hypertension					
4	Vasotec	Enalapril	Merck	2.50	S*
7	Norvasc	Amilopidine	Pfizer	2.22	S
21	Adalat	Nifedipine	Bayer	1.10	S
Antidepressant					
3	Prozac	Flupxetine	Lilly	2.56	S
10	Zoloft	Sertraline	Pfizer	1.51	S
11	Paxil	Paroxetine	SmithKline Beecham	1.47	S
Hematologic					
18	Procrit	Epoetin alfa	Johnson and Johnson	1.17	B
19	Epogen	Epoetin alfa	Amgen	1.16	B
24	Neupogen	Filigastrim	Amgen	1.06	B
Antibacterial					
9	Augmentin	Amoxicillin and Clavulanate	SmithKline Beecham	1.52	ND
15	Biaxin	Clarithromycin	Abbott	1.30	ND
12	Cipro	Ciprofloxacin	Bayer	1.44	S
Antihistamine (H1 antagonist)					
8	Claritan	Loratidine	Schering-Plough	1.73	S
Immunosuppressant					
16	Sandimmune and Neoral	Cyclosporine	Novartis	1.25	N
NSAID					
20	Voltaren-XR	Diclofenac	Novartis	1.11	S
Antimigraine					
23	Imitrex	Sumatriptan	Glaxo Wellcome	1.09	S
Heartburn					
25	Propulsid	Cisapride	Johnson and Johnson	1.05	S

B = biological; N = natural product; ND = derived from natural product; S = synthetic; S* = synthetic, but natural product pharmacophore.

increasingly seen as too nonspecific for the next generation of drugs. Fortunately, a large number of robust and specific biochemical and genetics-based screens, particularly those that are or will be derived as a result of the decoding and subsequent comparisons of the human and other genomes in the late 1990s to early 2000s, are now or will soon be in routine use. These screens will permit the detection of bioactive compounds in the complex matrices that are

Table 6.2 *The top 25 drugs in 1997: Sales by source*

Source of agent	Gross sales (US$ billion)	% of top 25
Biologicals (B)	3.39	8.2
Natural products or derived from (N/ND)	11.59	28.1
Synthetic (S)	23.75	57.6
NP pharmacophore (S*)	2.50	6.1

Table 6.3 *The source of compounds currently in clinical trial as antitumor agents, worldwide in first quarter of 2000*

Source of material	Number	Percentage of total
Natural product	34	11.6
Derived from natural product	73	24.8
Natural product pharmacophore	49	16.7
Synthetic chemical	115	39.1
NP-linked delivery system	6	2.0
"Biologicals"	17	5.8
Total	294	100.0

natural product extracts with great precision. The targets of the screens may be transformed cells, a key regulatory intermediate in a biochemical or genetic pathway or a receptor–ligand interaction, to name but a few now in use.

One interesting feature of such screens has increased the attractiveness of natural products to the pharmaceutical industry. The screens themselves are all highly automated and high throughput (upwards of 50 000 assay points per day in a number of cases). Because of this, the screening capacity at many companies is significantly larger than the potential input from in-house chemical libraries. Since screening capacity is no longer the rate-limiting step, many major pharmaceutical companies are becoming very interested in screening natural products (either as crude extracts or as prefractionated "peak libraries") as a low-cost means of discovering novel lead compounds.

III THE DESIGN OF AN EFFECTIVE NATURAL PRODUCTS-BASED APPROACH TO DRUG DISCOVERY

There are four major elements in the design of any successful natural products-based drug discovery program: acquisition of biomass, effective screening, bioactivity-driven fractionation and rapid and effective structure elucidation (which includes dereplication). Although some of these have been mentioned earlier, it is instructive to bring them together here.

A Acquisition and extraction of biomass

The acquisition of biomass has undergone a very significant transition from the days when drug companies and others routinely collected organisms without any thought of ownership by or reimbursement to the country of origin. Today, thanks to the Convention on Biodiversity (or CBD) and similar documents/agreements such as the US National Cancer Institute's Letter of Collection (LOC), all ethical biomass acquisitions now include provisions for the country of origin to be recompensed in some way for the use of its biomass. It should be noted that the LOC predated the CBD by three years; its tenets, as a minimum, must be adhered to by any investigator who has his or her collections funded by the NCI/NIH. Such recompense to the country of origin is best provided through formal agreements with government organizations and collectors in the host country, with such agreements providing not only for reimbursement of collecting expenses, but also for further benefits (often in the form of milestone and/or royalty payments) in the event that a drug is developed from a collected sample. Agreements often include terms related to the training of host country scientists and transfer of technologies involved in the early drug discovery process. Recognition of the role played by indigenous peoples through the stewardship of resources in their region and/or the sharing of their ethno-pharmacological information in guiding the selection of materials for collection is important in determining the distribution of such benefits. There have been sample legal agreements[13] and discussions as to methods used by various groups published in the last few years.[14,15] A recently published supplement to the journal, *Pharmaceutical Biology*, has brought together discussions covering these areas in much greater detail.[16]

It is axiomatic that all samples collected, irrespective of type of source, must if at all possible be fully identified to genus and species. Such identification is usually possible for all plant species, but it is not always possible for microbes and marine organisms. Voucher specimens should be provided to an appropriate depository in the host country, as well as to a similar operation in the home country of the collector.

In the case of microbes and marine organisms, extraction is normally carried out on the whole organism (though now some groups are isolating the commensal/associated microbes from marine invertebrates before a formal extraction). With plants however, which may be large and have well-differentiated parts, it is common to take multiple samples from one organism and to extract them separately.

The procedures used for extraction vary with the nature of the sample, and in some cases are dependent upon the nature of the ultimate assay. Thus a number of screens are sensitive to the tannins and complex carbohydrates that are extractable from a variety of organisms, and systems have been developed that permit easy removal of such "nuisance" compounds before assay. In the simplest cases, however, extraction with a lower alcohol (methanol to isopropanol) will bring out compounds of interest, though in most cases a sequential extraction system is utilized with compounds being extracted with solvents of ever-increasing polarity.

The selection of plant samples often raises the question of the ethnobotanical/ethnopharmacological approach versus a random approach. The former method, which usually involves the selection of plants that have a documented (written or oral) use by native healers, is attractive in that it can tap into the empirical knowledge developed over centuries of use by large numbers of people. In addition, the bioactive constituents have had a form of continuing clinical trial in man. The benefits of this approach have been extolled in several relatively recent articles,[17-19] and one author provides personal experience of the effectiveness of some jungle medicines.[20] The weakness of the ethnobotanical approach has always been that it is slow, requiring careful interviewing of native healers by skilled scientists, including ethnobotanists, anthropologists, western-trained physicians and pharmacologists. In addition, the quoted folkloric activity in the collected plant(s) may not be detectable, given the particular screens in use by the screening laboratory. Where ethnobotanical approaches have the highest possibility of success is in studies related to overt diseases/conditions, such as parasitic infections, fungal sores, and contraception/conception to name a few. In such cases, there are adequate controls, even on the same patient. Where there does not yet appear to be any successful relationship is in diseases such as cancer and AIDS-related conditions, where extensive testing of the patient is required for an accurate diagnosis.

A thorough retrospective analysis of the NCI's collections for utility of ethnobotanical information was carried out by Cragg *et al.* and no successful linkage was found irrespective of prior claims, once nuisance compounds were removed.[21] What is important to remember however, is that ethnobotanical information of any type is indicative of biological activity in a particular sample and thus priorities may be altered if resources for work-up are limited. A preliminary answer to these questions has come from a study funded jointly by NIH and NSF in Suriname. In this study, two collection teams operated independently, one using ethnobotany, the other random collections. Extracts from both sets were assayed by scientists at Virginia Polytechnic Institute and State University in a consistent set of yeast-based screens for DNA-damaging activity. A simple comparison of "hit rates" in these bioassays between "ethnobotanical extracts" and "random extracts" indicated that the ethnobotanical extracts had a slightly higher hit rate of 3.8% versus 2.8% for the random extracts. This difference between hit rates is not statistically significant, which is perhaps not surprising since there was no real connection between the disease usage reported by the tribal healers and the bioassay used.[22] As noted above, the value of the ethnobotanical collecting is greatly enhanced where there is a clear connection between an overt disease condition and the bioassay used. As one example of this, a collection of nine plants collected in Suriname on the basis of their ethnomedical use as antimalarial agents yielded five extracts with strong activity in a screen for activity against malarial parasites; an additional two extracts had weak activity. These hit rates are much higher than would have been expected from collections made on a random basis (Kingston *et al.*, unpublished work).

B Screening methods

As mentioned earlier, the advent of new and robust high-throughput screens has had, and continues to have, a major impact on natural products research in the pharmaceutical industry. Most of the screens used are proprietary and published information is rare, although general summaries of this approach have been published, albeit in proceedings from commercially organized meetings.[23] One screen that has been described in detail is the National Cancer Institute's 60 cell-line cytotoxicity screen for anticancer agents.[24] Although this is not a true receptor-based screen it has now been developed into a system whereby a large number of molecular targets within the cell lines may be identified by informatics techniques, and refinements are continuing. Information can be obtained from the following URL: http://dtp.nci.nih.gov as to the current status of the screens involved. An assay based on differential susceptibility to genetically modified yeast strains has been described,[25] and in addition there are simple but robust assays that can be utilized by workers in academia that do not have access to or may not need high throughput screens. Examples would be brine shrimp and potato disc assays[26,27] or the still useful disc-based antimicrobial assays.

In the next few years, assays based upon gene-chip technologies, where large numbers of samples can be screened in a short period of time, will become less expensive, and it is foreseeable that such assays will move from the industrial or industrial-academic consortium-based groups to academia in general, with specific expression systems being employed as targets for natural product lead discovery.

C Isolation of active compounds

The isolation of the bioactive constituent(s) from a given biomass can be a challenging task, particularly if the active constituent of interest is present in very low amounts. The actual procedure will depend to a large extent on the nature of the sample extract: a marine sample,[28] for example, may well require a somewhat different extraction and purification process from that derived from a plant sample.[29] Nevertheless, the essential feature in all of these methods is the use of an appropriate and reproducible bioassay to guide the isolation of the active compound. It is also extremely important that compounds that are known to inhibit a particular assay, or those that are nuisance compounds be dereplicated (identified and eliminated) as early in the process as possible. Procedures for doing this have been discussed,[30] and various new approaches to isolation and structure elucidation have been reviewed.[31]

D Structure elucidation

Structure elucidation of the bioactive constituent depends almost exclusively on the application of modern instrumental methods, particularly high-field NMR and MS. These powerful techniques, coupled in some cases with selective chemical manipulations, are usually adequate to solve the structures of most secondary metabolites up to 2 kDa molecular weight. X-ray crystallography is also a valuable tool if crystallization of the material can be induced, and in some cases, it is the only method to unambiguously assign absolute configurations. Nowadays, the determination of the amino acid sequences of polypeptides or peptide-containing natural products up to 10–12 kDa is a relatively straightforward task, requiring less than 5 mg of a polypeptide. In addition, MS techniques have developed to the stage where polypeptides containing unusual amino acids not recognized by conventional sequence techniques can be sequenced entirely by MS.

E Further biological assessment

Once the bioactive component has been obtained in pure form and shown to be either novel in structure or to exhibit a previously unknown function (if it is a compound that is in the literature), then it must undergo a series of biological assays to determine its efficacy, potency, toxicity and pharmacokinetics. These will help to position the new compound's spectrum of activity within the portfolio of compounds that a group may be judging for their utility as either drug candidates or leads thereto. If an idea can be gained as to its putative mechanism of action (MOA) (assuming that the screening techniques used to discover it were not MOA-driven) at this stage, then it too can help as a discriminator in the prioritizing process.

F Procurement of large-scale supplies

Once a compound successfully completes evaluation in the secondary and tertiary assays described above, then large amounts of material will be required for the further studies necessary to convert what is initially a "hit" to a "drug lead" and then to a "clinical candidate". The large supplies could be made available by cultivation of the plant or marine starting material, or by large-scale fermentation in the case of a microbial product. Chemical synthesis or partial synthesis may also be possible if the structure of the active complex is not too complex. The example of paclitaxel is instructive here: after initial large-scale production by direct extraction from *Taxus brevifolia* bark, it is now also produced by a semisynthetic procedure starting from the more readily available precursor 10-deacetylbaccatin III.[32]

Another method of obtaining adequate supplies of a natural plant product is by plant tissue culture methods. Although there are a few examples of the commercial production of secondary metabolites by plant cell culture (shikonin being perhaps the best known one[33]), the application of this technique to commercial production of pharmaceuticals has yet to find general acceptance, primarily for economic reasons.[34] It is probable however, that this approach will find wider acceptance in the future as yields are improved and costs reduced. There are credible reports that Phyton Catalytic has produced paclitaxel in adequate yields for commercialization using plant cell tissue culture and a licensed technology from the USDA (Venkat R, personal communication), and at least one other company (Phytera, Inc.) is actively involved in this type of technology for pharmaceutical production.

G Determination of structure–activity relationships

The initial hit isolated from the biomass, irrespective of source, is not necessarily the lead required for further development into a drug. It may be too insoluble, not potent enough, or be broadly, rather than specifically, active. Once the structure has been determined, then synthetic chemistry, both the conventional type or combinatorially related methods may be used in order to generate derivatives/analogs that have the more desirable characteristics of a potential drug. Several examples of these types of processes are shown below.

IV EXAMPLES OF NATURAL PRODUCTS OR ANALOGS AS PHARMACEUTICALS

A Captopril

This synthetic product was derived from studies reported in 1965 by Ferreira,[35] where he demonstrated that there were principles in the venom of the pit viper, *Bothrops jararaca*, which inhibited the degradation of the nonapeptide, bradykinin. Subsequently, workers at Squibb[36,37]

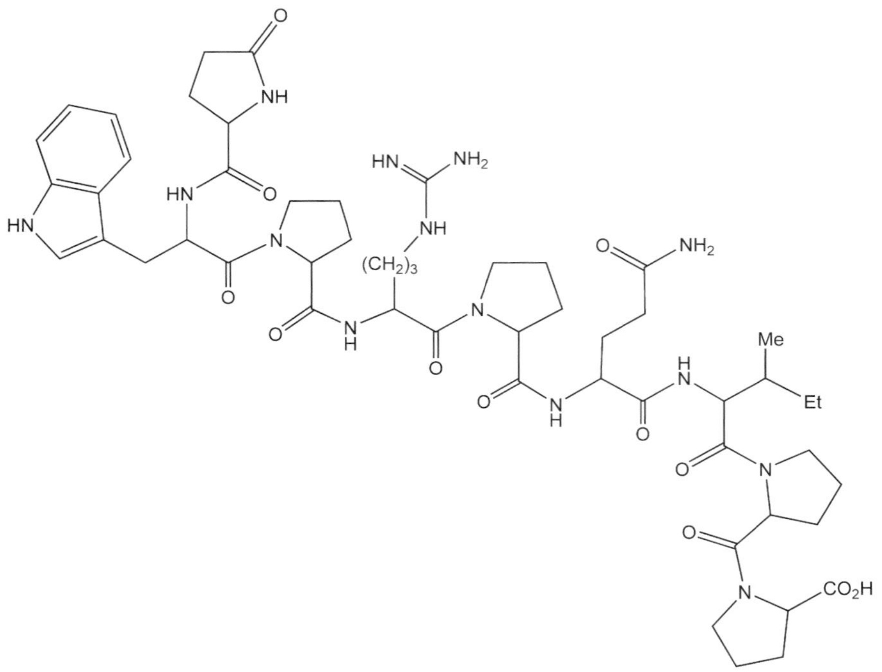

(1) Teprotide

demonstrated that the active principle was a simple nonapeptide, teprotide (**1**) and that this peptide had a specific activity as an angiotensin converting enzyme (ACE) inhibitor and also had hypotensive activity in clinical trials. However, due to the lack of oral activity, it was not a good drug candidate. With the recognition that ACE (a dipetidyl carboxypeptidase) was a metallo-enzyme came the utilization of a mono-carboxypeptidase, carboxypeptidase A (CpdA) as a surrogate model. Prior work had shown that all of the peptidic inhibitors of ACE had a C-terminal proline and this fact, plus the information that benzyl-succinic acid was a specific inhibitor of CpdA, was used to derive a series of carboxy- and mercapto-alkanoyl esters of proline that demonstrated good to excellent inhibition of ACE. One of the compounds (SQ14225) (**2**) subsequently became the prototypical ACE drug, Captopril[®].[36,37] Subsequent development of the concepts used in this work has led to more active agents such as Enalapril[®] (**3a**) and Quinapril[®] (**3b**), which can be considered prodrugs releasing the active metabolite upon ester cleavage.

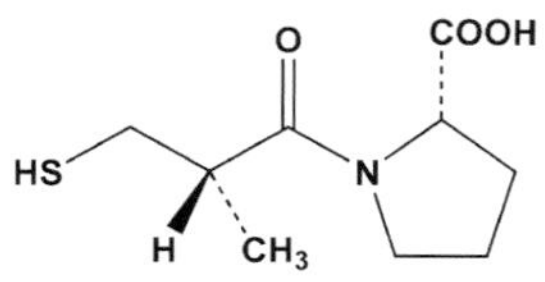

(2) Captopril

B Lovastatin

An elevated serum cholesterol level is an important risk factor in cardiac disease (and in hypertension), and thus a drug which could lower this level would be an important prophylactic against cardiovascular diseases in general. Humans synthesize about 50% of their cholesterol requirement, with the rest coming from diet. A potential site for inhibition of cholesterol biosynthesis is at the rate-limiting step in the system, the reduction of hydroxymethyl-glutaryl

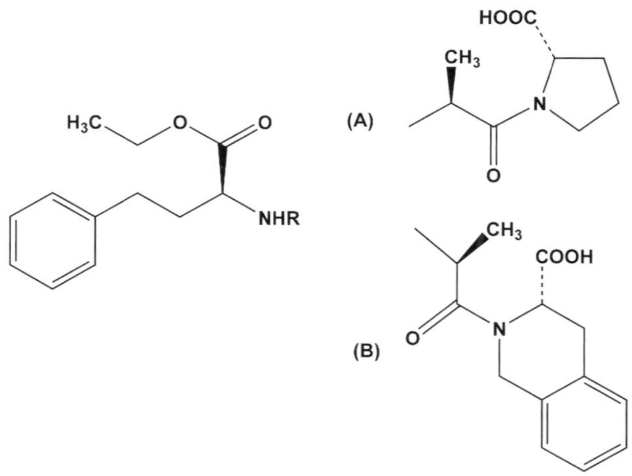

3a Enalapril R = A
3b Quinapril R = B

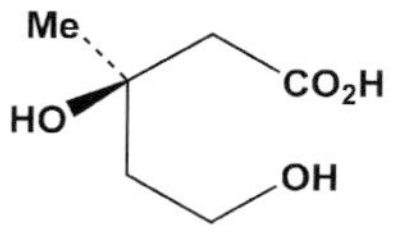

(4) Mevalonic Acid

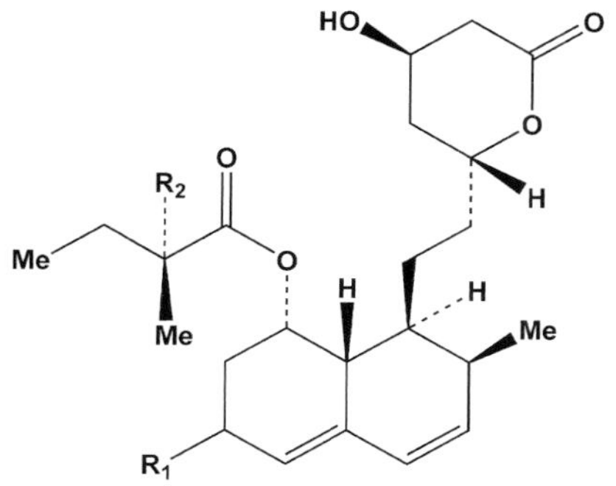

5a Compactin R$_1$ = R$_2$ = H
5b Mevinolin R$_1$ = CH$_3$; R$_2$ = H
5c Simvastin R$_1$ = CH$_3$; R$_2$ = CH$_3$

(7) Fluvastatin

coenzyme A by HMG-CoA reductase to produce mevalonic acid (**4**). Following the original identification of compactin (**5a**) from a fermentation beer of *Penicillium brevicompactum* as an inhibitor of HMG-CoA reductase by Sankyo,[38,39] it was also reported as an antifungal agent the next year by Brown *et al.*[40] Using the inhibitor assay, Endo isolated the 7-methyl derivative of compactin, mevinolin (**5b**), from *Monascus ruber* and submitted a patent for its biological activity to the Japanese patent office under the name Monacolin K[41,42] without a structure. Concomitantly, Merck discovered the same material using a similar assay from an extract of *Aspergillus terreus*. It was reported in 1980[43] and a US patent was issued in the same year.[44] Following a significant amount of development work, mevinolin (Lovastatin®) became the first commercialized HMG-CoA-reductase inhibitor in 1987.[45,46] Further work using either chemical modification of the basic structure or

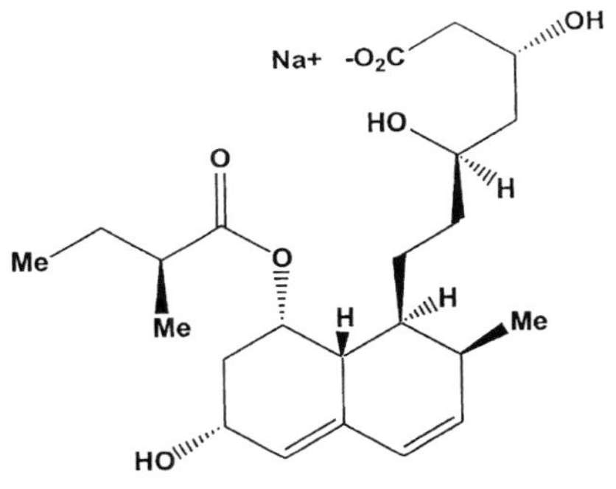

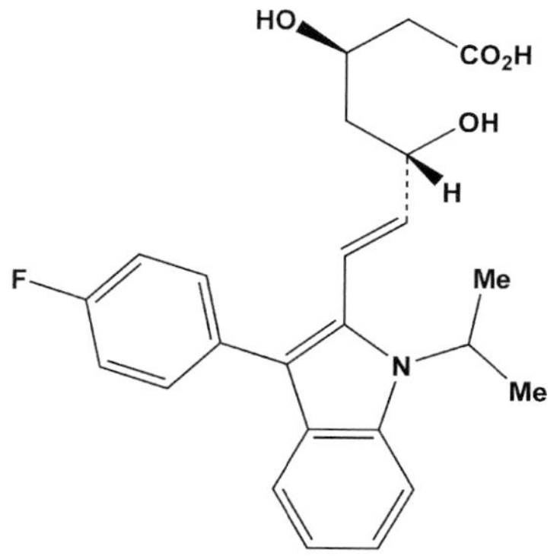

by use of biotransformation techniques led to two further compounds by converting the 2-methylbutanoate side-chain into 2,2-dimethylbutanoate (Simvastin®) (**5c**) or opening of the exocyclic lactone to give the free hydroxy acid (Pravastatin®) (**6**). Data from 1997 show that at that time, these three "statins" were the second best selling drugs in the world with combined sales of US$7.53 billion.[10] Further development of these compounds has led to totally synthetic "statins", three of which (fluvastatin (**7**), cerivastatin (**8**) and atorvastin (**9**)) are shown below. What is significant about these last three compounds is that in the case of the first two, their "operative ends" are the dihydroxy-heptenoic acid side chain of the fungal products linked to a lipophilic ring structure whilst the third, compound (**9**), uses the reduced form of the acid. These three compounds demonstrate the intrinsic value of natural products as the source of the pharmacophore and compound (**9**), Lipitor®, grossed over US$ 1 billion in sales in 2000. It is significant that evidence has been reported that this class of compounds has immunomodulatory activity,[47] as well as reducing the risk of developing Alzheimer's disease,[48,49] and may also protect against osteoporosis.[50]

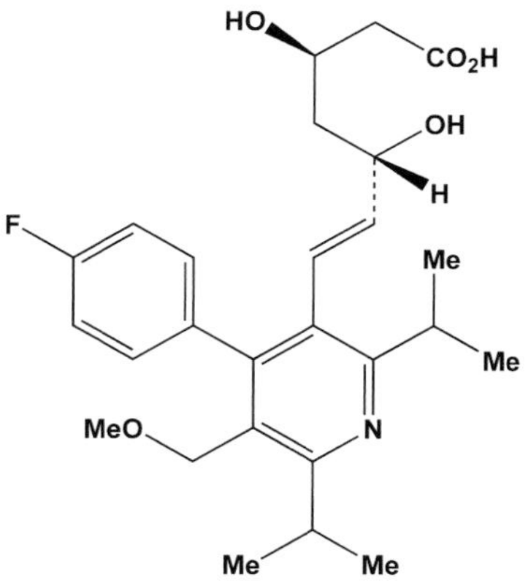

(6) Pravastatin

(8) Cerivastin

(9) Atorvastin

C   Cyclosporin A

This compound (**10**), originally isolated from the fungus *Trichoderma polysporum*[51] revolutionized transplant surgery from the early 1980s as the quintessential immunosuppressant. It is still in use, though other naturally occurring compounds have been found (particularly the FK506 series from microbes) that work by a similar mechanism. The identification of the cyclosporin-binding proteins in eukaryotic cells and the subsequent realization that subtle modifications in their subunit interactions on binding of these microbial metabolites appears to alter the immune response to foreign proteins has opened up a vast arena that companies are now investigating in order to find simpler molecules that perform the same function.

D   Avermectin and doramectin

The avermectins are a family of broad-spectrum antiparasitic compounds with avermectin A (**11a**) being an example. These compounds were originally sold by Merck (with a slight chemical modification) as Ivermectins®, initially as veterinary agents. Subsequently, it was discovered that they had excellent activity against some of the West African parasites that caused river blindness and they moved into human medicine. These molecules are polyketides, and over the last three or four years groups at Pfizer and elsewhere have been working on ways to modify the keto-synthases that would permit novel agents to be expressed by suitable microbial hosts. These efforts have led to the modified agents known collectively as the doramectins (**11b**) which have improved activities.[52] These genetic techniques and their applicability will be briefly discussed later.

E   Azithromicin

The 14-ring macrolides of the erythromycin family have been extremely important antibacterial agents since the late 1950s, and over the subsequent years thousands, if not millions of kilograms of erythromycin in the form of various salts have been prescribed worldwide for Gram-positive infections. As with all other antibiotics in current widespread use, microbial resistance to the erythromycin class arose fairly rapidly, and for a significant number of infective agents erythromycins were no longer usable as first-line therapy. In the 1980s, the Croatian chemical company Pliva managed to generate a 15-ring macrolide (an azalide) by reducing the ring ketone of erythromycin and inserting

(10)   Cyclosporin A

11a Avermectin R = (A)
11b Doramectin R = (B)

a methylated nitrogen alongside it. This compound was commercialized as Zithromax® (Azithromycin, (**12**)) by Pfizer and has been very successful, particularly with Gram-positive infections.

F Paclitaxel

The most exciting new drug in the anticancer area in the last few years is the plant-derived compound paclitaxel (Taxol® (**13a**)) which along with several key precursors (the baccatins), occurs in the leaves of various *Taxus* species. Paclitaxel was first isolated from the bark of *T. brevifolia* collected in the north-west of the USA as part of a random collection for the NCI by the US Department of

Agriculture. Although isolated as a cytotoxic agent, it languished for many years due to problems with availability and solubility. It was rescued as a result of its potent activity in various new *in vivo* bioassays and by the work of Horwitz' group, who discovered that it had a previously unknown mechanism of action, the stabilization of tubulin assemblies.[32,53] Paclitaxel and its close chemical relative, docetaxel[54] (**13b**), are now well-established antitumor

(12) Azithromycin

13a Taxol R = (A)
13b Docetaxel R = (B)

14a Podophyllotoxin $R_1 = H$; $R_4 = OCH_3$
14b Etoposide $R_1 = $ (A) with $R_2 = OH$; $R_3 = CH_3$; $R_4 = H$
14c Teniposide As **14b** with $R_3 = $ 2-thiophenyl
14d NK-611 As **14b** with $R_2 = N(CH_3)_2$

agents, primarily used for breast and ovarian carcinoma treatments, but being studied by many groups for their efficacy as components of multi-agent therapies in many types of carcinoma. Chemists have not been idle since the discovery of this class of compounds and many hundreds of variations on the taxane skeleton have been synthesized in efforts to produce agents with better solubilities and pharmacokinetic distributions. Two relatively recent papers give some idea of the breadth of such studies, one by Kingston *et al.*[55] that covers 2-acyl analogs and a more

general review of tubulin-active agents by von Angerer,[56] which includes taxanes.

G Podophyllotoxin, etoposide and teniposide

Podophyllotoxin (**14a**), originally isolated as the major cytotoxic component of the May apple, *Podophyllum pelatatum*,[57] was placed into clinical trials, but was shelved due to intractable toxicities. Elegant work by Sandoz, prior to its merger with Ciba-Geigy using the naturally occurring isomer *epi*-podophyllotoxin, led to the semisynthetic compounds etoposide (**14b**) and teniposide (**14c**), which are currently in clinical use worldwide. Fairly recently, work by Nippon Kayaku has led to a water-soluble derivative of etoposide, NK-611 (**14d**) by including a dimethylamino functionality in the sugar ring; this compound is currently in clinical trials. An interesting comment is that although podophyllotoxin is a tubulin polymerization inhibitor, etoposide and teniposide appear to function as DNA topoisomerase II inhibitors.[58]

H The Vinca alkaloids, vinblastine, vincristine and related compounds

Vinblastine (**15a**) and vincristine (**15b**) were isolated from the Madagascan periwinkle as a result of a serendipitous

15a Vinblastine $R_1 = CH_3$; $R_2 = OCH_3$; $R_3 = COCH_3$
15b Vincristine $R_1 = CHO$; $R_2 = OCH_3$; $R_3 = COCH_3$
15d Vindesine $R_1 = CH_3$; $R_2 = NH_2$; $R_3 = H$

15c Vinorelbine

observation whilst studying extracts of the plant as a source of hypoglycemic agents.[21,59] Two other alkaloids, vinleurosine and vinrosidine, were also isolated at around the same time,[60] though these did not become antitumor agents. The two clinically active agents are tubulin polymerization inhibitors, and their efficacy is therefore probably due to their antimitotic activity, but at a different site on the tubulin molecule to that of the taxanes. Many variations on the basic structure of the *Vinca* alkaloids have been synthesized and currently vinorelbine (**15c**) has been approved in Europe and the USA for treatment of lung cancer, with at least one more analog, vindesine (**15d**) in clinical trials.[61]

I Camptothecin and analogs

The alkaloid camptothecin (**16a**) was first isolated from the Chinese tree *Camptotheca acuminata* by Wani and Wall contemporaneously with their discovery of Taxol®.[62] It demonstrated encouraging results in animal models and commenced clinical trials at the NCI in the 1970s as the sodium salt of the ring-opened lactone. However, it demonstrated severe bladder toxicity and was dropped. It was resurrected as a result of its very specific biochemical activity as an inhibitor of topoisomerase I, and a series of semisynthetic derivatives has led to the approval of two, topotecan (Hycamptin®, (**16b**)) and irinotecan (Camptosar®, (**16c**)) for ovarian and other carcinomas. Two others, 9-amino- (**16d**) and 9-nitro-camptothecin (**16e**) are awaiting approval. The interesting finding with camptothecin and its analogs is that they function as topoisomerase I inhibitors, thus differentiating them from the podophyllotoxin derivatives and the anthracyclines, such as adriamycin, which function as topoisomerase II inhibitors. At the time of writing, no other topoisomerase I inhibitors have yet been approved for clinical use.

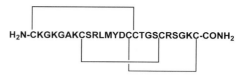

(17) Ziconotide

J Peptides

The magainins illustrate the truth that bioactive compounds can be obtained from animals, as well as plants and microorganisms. They are a family of cationic peptides isolated from the skin of the African clawed frog, *Xenopus laevis*, and demonstrate excellent antibiotic and also wound-healing activities. They were discovered after a researcher noted that the frogs seldom developed infections after experimental surgery in spite of being placed into a totally non-sterile environment, an aquarium. Magainins 1 and 2 are 23-residue peptides, differing only at positions 10 and 22.[63] Various derivatives of these peptidic agents are currently in clinical trials as topical antibiotics or for treatment of diabetic ulcers.

Perhaps the most interesting current natural product-derived peptide is the compound known as Ziconotide® (**17**). This is one of the class of peptides known as conotoxins which are produced as offensive agents by marine cone snails of the genus *Conus*.[6] Although a very significant number of peptides were synthesized by the original biotech company (Neurex) which developed the cone snail toxins, the best one from an analgesic perspective was the original compound isolated from the cone snail. To give some idea of the potential of this compound, the Irish company Elan Pharmaceuticals purchased Neurex in early 1999 for US\$750 million, whilst the New Drug Application (NDA) was still before the US FDA in order to obtain this compound and others in the pipeline. Many other groups are also working on the exquisitely potent toxins produced by

16a	Camptothecin	$R_1 = R_1 = R_3 = H$
16b	Topotecan	$R_1 = H$; $R_2 = CH_2N(CH_3)_2$; $R_3 = OH$
16c	Irinotecan	$R_1 = CH_2CH_3$; $R_2 = H$; $R_3 = (A)$
16d	9-Aminocamptothecin	$R_1 = R_3 = H$; $R_2 = NH_2$
16e	9-Nitrocamptothecin	$R_1 = R_3 = H$; $R_2 = NO_2$

18a Penicillin G, R = (A)
18b 6-Aminopenicillanic Acid, R = H

this genus and we expect to see many more compounds of this type approved in the relatively near future for treatment of a variety of diseases.

K The β-Lactams

It is probable that the largest number of medicinally directed semi-synthetic agents ever produced containing a specific pharmacophore, the β-lactam ring, are the penicillins, cephalosporins and mono-bactams. To these must also be added the synthetic cephems and cephams and the many variations upon those basic themes. Although current figures are not available, it was estimated in 1980 that over 10 000 variations on the basic penicillin structure had been made and assayed for their ability to inhibit microbial growth (Hoover, J.R.E., personal communication).

The serendipitous discovery by Fleming in 1929 of the activity of a *Penicillium notatum* strain upon the growth of a Gram-positive *Staphylococcus* is too well known to repeat.

What is significant however, is that by the early 1940s, only a very few years after the introduction of Penicillin G (**18a**) into clinical practice in the US and UK, bacterial resistance to the molecule was recognized in patient isolates and, in addition, it almost alone amongst the antibiotics, can produce an anaphylactic response in a subset of humans. Thus the search for modified penicillins began within a few years of its introduction into clinical practice. Initially the methods used supplemented fermentations, whereby precursors were added to broths and the biosynthetically derived materials isolated. Then came a whole series of reports that the probable precursor to all of the then known penicillins, 6-aminopenicillanic acid (6-APA) (**18b**) could be made by enzymatic cleavage of penicillin G. This work culminated in the synthesis of 6-APA by Sheehan in 1958 and its isolation from fermentation broths by Batchelor *et al.* working at Beecham in the UK. This discovery led to immense numbers of semi-synthetic penicillins being made by academic and industrial researchers and the introduction into clinical practice of materials, such as ampicillin, methicillin, oxacillin, etc.

In 1948, Brotzu[64] reported on the biological activity of a crude extract from a *Cephalosporium* sp. that subsequently led to the identification in the late 1950s of another β-lactam series, the basis for which was cephalosporin C (**19a**). Very soon after the identification of this molecule, workers at Lilly were able to convert methyl penicillin V via its sulphoxide into a cephalosporin. Further work then led to

19a Cephalosporin C,   R₁ = (A); R₂ = (B)
19b 7-Aminocephalosporanic Acid, R₁ = CH₃; R₂ = H
19c Cefactor,   R₁ = Cl; R₂ = (C)
19d Ceftazidime,   R₁ = (E); R₂ = (D)

20a Sulfazecin $R_1 = H$; $R_1 = OCH_3$; $R_3 = (A)$
20b Aztreonam $R_1 = CH_3$; $R_2 = H$; $R_3 = (B)$

21a Morphine, R = H
21b Codeine, R = CH$_3$

the cephalosporin equivalent of 6-APA, 7-amino-cephalosporanic acid (7-ACA) (**19b**). With this precursor in hand, medicinal chemists proceeded to generate the cephalosporin equivalents of penicillins and also to make molecules that contained other substituents not seen in the penicillin series, such as the 3-chloro derivative Cefaclor® (**19c**) and the fairly lactamase-resistant Ceftazidime® (**19d**). A recent reprint of an old report gives the history of penicillin discovery together with some early information on cephalosporins, and should be consulted for background information.[65] For many years in the 1970s, medicinal chemists had attempted to synthesize the base pharmacophore of these antibiotics, a cyclic 4-membered β-lactam ring. Many man-years were spent in attempting to do this, with the closest stable ring system being a β-lactol, rather than the lactam. Again, Mother Nature came to the rescue with the 1981 report by Imada[66] of the monobactams known as sulfazecins (**20a**) and within two months, Sykes *et al.* at Squibb reported that after Herculean efforts, involving over 1 000 000 small-scale fermentations, they successfully isolated monobactams from aquatic microorganisms collected near their laboratories in New Jersey.[67] With the identification of the naturally occurring monobactam compounds, it became "chemically obvious" why Mother Nature was the best chemist. Of all of the synthetic modifications made by medicinal chemists from Squibb, SmithKline and other pharmaceutical houses, none had included the use of an N-sulphonic acid as a stabilizing group. In a clever move, the Squibb group then proceeded to synthesize and patent all of the side-chain modifications that had been reported to give improved stabilities and activities in the penicillin and cephalosporin series; an early compound in this series, aztreonam (Azactam®) (**20b**) is in clinical use.

That the battle between microbes producing β-lactamases (degrading enzymes against specific classes of β-lactams) and medicinal chemists is still continuing, can

be seen by the continued reports of novel structures based on β-lactams that appear in many journals and in reports from annual meetings such as the Interscience Conferences on Antimicrobial Agents and Chemotherapy.[68]

L The morphine alkaloids

Morphine (**21a**) was first isolated by Serturner in 1806, followed by codeine (**21b**) in 1832 by Robiquet and then the non-morphine alkaloid, papaverine by Merck in 1848, all from crude opium preparations. With the invention of the hypodermic needle and the availability of the purified alkaloids, the benefits and problems associated with these compounds (and their synthetic derivative, heroin) rapidly became apparent,[69,70] leading to the search for potent drugs without the abuse potential of the morphinoids. With the exception of the semi-synthetic compound, buprenorphine (**22**), which is approximately 25–50 times more potent than morphine and has a lower addiction potential, none of the compounds made to date from modifications around the phenanthrene structure of morphine have exceeded the pain control properties without a concomitant addiction potential. An interesting compound, whose structure is based on that of morphine is pentazocine (Talwin®) (**23**). This has about 30% of the efficacy of morphine, but has a much lower incidence of abuse and does in fact cause withdrawal

(22) Buprenorphine

(23) Pentazocine

24a Purvalanol A;	$R_1 = (A; R = H);$ $R_2 = CH(CH_3)_2; R_3 = (B)$
24b Purvalanol B; As **24a** with $R = CO_2H$	
24c Olomucine;	$R_1 = CH_2C_6H_5;$ $R_2 = CH(CH_3)_2;$ $R_3 = CH_2CH_2OH$
24d Roscovitine; As **24c** with	$R_3 = (C)$

symptoms in addicts due to antagonist activity at the μ-opioid receptor.

M   Other natural products

The examples given above are simply a selection of the natural products and natural product analogs that have, in general, entered clinical use. There are recent reviews that cover natural products as drugs and sources of structures[7-9] and these should be consulted for further examples of where natural products have led to novel drugs in a multiplicity of diseases.

V   FUTURE DIRECTIONS

The probability that a directly isolated natural product (e.g. adriamycin or vincristine) will be the drug for a given disease in the future is low, but what is highly probable is that a natural product structure will give the initial lead to an active agent in a screen and the material will then be optimized for activity using combinatorial chemistry techniques. A subtle variation of this would be the design of the synthesis of a novel active natural product so that combinatorial chemistry could be utilized to produce many variants of a known active structure. In a similar fashion, combinatorial biosyntheses can be utilized to produce what are now being called "unnatural natural products" where the biosynthetic machinery of a microbial cell is dissected and the relevant genes are "mixed and matched" followed by expression in a suitable heterologous host. That these ideas are not just pipe dreams can be seen in the following examples.

The combinatorial chemistry optimization can be seen in the production of the purvalanols (**24a**,**24b**) by Schultz' group at the Scripps Institute from original studies with olomucine (**24c**) and roscovitine (**24d**) as cyclin-dependent kinase inhibitors.[71-73] As a result of these studies a thousand-fold increase in binding and specificity was achieved as a result of minor changes in the side chains.

The synthetic approach is best observed in the papers by Nicolaou's group dealing with the syntheses of the sarcodictyins,[74] where they were able to produce a molecule that inhibited tubulin at close to the activity of the marine gorgonian-derived product, eleutherobin A.[75] In another example from the same group, only this time dealing with the synthesis of the microbial anti-mitotic agent, epothilone, Nicolaou was able to produce hundreds of similar molecules that may well lead to novel antitumor agents.[76]

The production of "unnatural natural products" was pioneered by the work of Hopwood at the John Innes Institute in the UK in his work with *Streptomyces* and is now coming to fruition with many groups in the UK and the USA working on production of molecules derived from polyketides and peptide synthases in heterologous hosts. An example would be the combinatorial biosynthesis of a pikromycin aglycone coupled to a calicheamicin-derived sugar,[77] while Hutchinson has provided an excellent short review of the processes involved.[78]

Finally, the knowledge of a structure of an active natural product and its possible binding site on an individual enzyme may well enable a chemist to make compounds that will bind more tightly (and/or specifically) to that specific site. Such compounds could be significantly less complicated than the natural product, as demonstrated by Wender in his production of simpler variants of bryostatin that still give comparable activities to the base molecule.[79-83] That Mother Nature can produce the bryostatins as efficiently has been shown by the work of Haygood at The Scripps Institute of Oceanography, who has demonstrated that the bryostatins are probably produced by a commensal microbe that contains polyketide synthases coding for the basic structure.[84,85] A variety of syntheses of this agent have been reported in the literature and the reader should consult the recent review by Mutter and Wills for a thorough overview.[86]

VI SUMMARY

The preceding pages have given just a glimpse of the importance of natural products as both pharmaceutical agents and/or as leads to active molecules. With the advent of novel screening systems related to the explosion of genetic information now becoming available, it will be necessary to rapidly identify novel lead structures. Our belief is that a very significant proportion of these will continue to be natural product derived. It should be remembered that Mother Nature has had three billion years to refine her chemistry and we are only now scratching the surface of the potential structures that are there.

Due to ease of access, plant-derived materials have been in the majority as far as sources are concerned. Now, with the advent of genetic techniques that permit the isolation and expression of biosynthetic cassettes, microbes and their marine invertebrate hosts may well be the new frontier for natural products.

REFERENCES

1. Anon (1998) A pictorial history of herbs in medicine and pharmacy. *Herbalgram* 33–47.
2. Grabley, S. and Thiericke, R. (1999) Bioactive agents from natural sources: trends in discovery and application. *Adv. Biochem. Eng./ Biotech.* 64: 104–154.
3. Haslam, E. (1986) Secondary metabolism — fact and fiction. *Nat. Prod. Rep.* 3: 217–249.
4. Williams, D.H., Stone, M.J., Hauck, P.R. and Rahman, S.K. (1989) Why are secondary metabolites (natural products) biosynthesized? *J. Nat. Prod.* 52: 1189–1208.
5. Harborne, J.B. (1990) Role of secondary metabolites in chemical defence mechanisms in plants. In Chadwick, D.J. and Marsh, J. (eds). *Bioactive Compounds from Plants*, pp. 126–139. Wiley, Chichester.
6. Olivera, B.M. (1997) Conus venom peptides, receptor and ion channel targets, and drug design: 50 million years of neuropharmacology. *Mol. Biol. Cell* 8: 2101–2109.
7. Cragg, G.M., Newman, D.J. and Snader, K.M. (1997) Natural products in drug discovery and development. *J. Nat. Prod.* 60: 52–60.
8. Cragg, G.M. and Newman, D.J. (2000) Antineoplastic agents from natural sources: achievements and future directions. *Exp. Opin. Invest. Drugs* 9: 2783–2797.
9. Newman, D.J., Cragg, G.M. and Snader, K.M. (2000) The influence of natural products on drug discovery. *Nat. Prod. Rep.* 17: 215–234.
10. Newman, D.J. and Laird, S.A. (1999) The influence of natural products on 1997 pharmaceutical sales figures. In ten Kate, K. and Laird, S.A. (eds). *The Commercial Use of Biodiversity*, pp. 333–335. Earthscan Publications, London.
11. Vagelos, P.R. (1991) Are prescription drug prices high? *Science* 252: 1080–1084.
12. Schreiber, S.L., Tan, D.S., Foley, M.A. and Shair, M.D. (1998) Stereoselective synthesis of over two million compounds having structural features both reminiscent of natural products and compatible with miniaturized cell-based assays. *J. Am. Chem. Soc.* 120: 8565–8566.
13. Downes, D., Laird, S.A., Klein, C. and Carney, B.K. (1993) Biodiversity prospecting contract. In Reid, W.V., Laird, S.A., Meyer, C.A., Gamez, R., Sittenfeld, A., Janzen, D.H., Gollin, M.A. and Juma, C. (eds). *Biodiversity Prospecting: Using Genetic Resources for Sustainable Development*, pp. 255–287. World Resources Institute, Washington, DC.
14. Baker, J.T., Borris, R.P., Carte, B., Cordell, G.A., Soejarto, D.D., Cragg, G.M., Gupta, M.P., Iwu, M.M., Madulid, D.R. and Tyler, V.E. (1995) Natural product drug discovery and development: new perspectives on international collaboration. *J. Nat. Prod.* 58: 1325–1357.
15. ten Kate, K., Wells, A. (1998) *Benefit-Sharing Case Study. The Access and Benefit-Sharing Policies of the United States National Cancer Institute: A Comparative Account of the Discovery and Development of the Drugs Calanolide and Topotecan. Submission to the Executive Secretary of the Convention on Biological Diversity*, The Royal Botanic Gardens, Kew, UK.
16. Rosenthal, J.P. (1999) Drug discovery, economic development and conservation: the International Cooperative Diodiversity Groups. *Pharm. Biol.* 37: 5.
17. Farnsworth, N.R. (1990) The role of ethnopharmacology in drug development. In Chadwick, D.J. and Marsh, J. (eds). *Bioactive Compounds from Plants*, pp. 2–21. Wiley, Chichester.
18. Cox, P.A. (1990) Ethnopharmacology and the search for new drugs. In Chadwick, D.J. and Marsh, J. (eds). *Bioactive Compounds from Plants*, pp. 40–55. Wiley, Chichester.
19. Balick, M.J. (1990) Ethnobotany and the identification of therapeutic agents from the rainforest. In Chadwick, D.J. and Marsh, J. (eds). *Bioactive Compounds from Plants*, pp. 22–39. Wiley, Chichester.
20. Plotkin, M.J. (1988) Conservation, ethnobotany and the search for new jungle medicines: pharmacognosy comes of age … Again. *Pharmacotherapy* 8: 257–262.
21. Cragg, G.M., Boyd, M.R., Cardellina II, J.H., Newman, D.J., Snader, K.M. and McCloud, T.G. (1994) Ethnobotany and drug discovery: the experience of the US National Cancer Institute. In Chadwick, D.J. and Marsh, J. (eds). *Ethnobotany and the Search for New Drugs, Ciba Foundation Symposium*, vol. 185, pp. 178–196. Wiley, Chichester.
22. Kingston, D.G.I., Abdel-Kader, M., Zhou, B.-N., Yang, S.-W., Berger, J.M., van der Werff, H., Miller, J.S., Evans, R., Mittermeier, R., Famolare, L., Guerin-McManus, M., Malone, S., Nelson, R., Moniz, E., Wisse, J.H., Vyas, D.M., Wright, J.J.K. and Aboikonie, S. (1999) The Suriname International Cooperative Biodiversity Group Program: Lessons from the first five years. *Pharm. Biol.* 37(suppl.): 22–34.
23. Coombes, J.D. (1992) *New Drugs from Natural Sources*, IBC Technical Services, London, UK.
24. Boyd, M.R. and Paull, K.D. (1995) Some practical considerations and applications of the National Cancer Institute *In Vitro* Anticancer Drug Discovery Screen. *Drug Dev. Res.* 34: 91–109.
25. Johnson, R.K., Bartus, H.F., Hofmann, G.A., Bartus, J.O., Mong, S.-M., Faucette, L.F., McCabe, F.L., Chan, J.A. and Mirabelli, C.K. (1986) Discovery of new DNA-reactive drugs. In Hanka, L.J., Kondo, T. and White, R.J. (eds). *In Vitro and In Vivo Models for Detection of New Antitumor Drugs*, pp. 15–26. Organizing Committee of the 14th International Congress of Chemotherapy.
26. Meyer, B.N., Ferrigni, N.R., Putnam, J.E., Jacobsen, L.B., Nichols, D.E. and McLaughlin, J.L. (1982) Brine shrimp: a convenient general bioassay for active plant constituents. *Planta Med.* 45: 31–34.
27. McLaughlin, J.L. (1991) Crown gall tumours on potato discs and brine shrimp lethality: two simple bioassays for higher plant screening and fractionation. In Hostettmann, K. (ed.). *Methods in Plant Biochemistry: Assays for Bioactivity*, vol. 6. pp. 1–32. Academic Press, San Diego.
28. Shimizu, Y. (1985) Bioactive marine natural products with emphasis on handling of water-soluble compounds. *J. Nat. Prod.* 48: 223–235.
29. Marston, A. and Hostettmann, K. (1991) Modern separation methods. *Nat. Prod. Rep.* 8: 391–413.

30. Hostettmann, K., Wolfender, J.-L. and Rodriguez, S. (1997) Rapid detection and subsequent isolation of bioactive constituents of crude plant extracts. *Planta Med.* **63**: 2–10.

31. Cordell, G.A. (1995) Changing strategies in natural products chemistry. *Phytochemistry* **40**: 1612–1685.

32. Cragg, G.M., Schepartz, S.A., Suffness, M. and Grever, M.R. (1993) The taxol supply crisis. New NCI policies for handling the large-scale production of novel natural product anticancer and anti-HIV agents. *J. Nat. Prod.* **56**: 1657–1668.

33. Fujita, Y. (1988) Industrial production of shikonin and berberine. In Bock, G. and Marsh, J. (eds). *Applications of Plant Cell and Tissue Culture*, pp. 157–174. Wiley, Chichester.

34. Fowler, M.W., Cresswell, R.C. and Stafford, A.M. (1990) An economic and technical assessment of the use of plant cell cultures for natural product synthesis on an industrial scale. In Chadwick, D.J. and Marsh, J. (eds). *Bioactive Compounds from Plants*, pp. 228–238. Wiley, Chichester.

35. Ferreira, S.H. (1965) A bradykinin-potentiating factor (BPF) present in the venom of *Bothrops jararaca*. *Br. J. Pharmacol.* **24**: 163–169.

36. Ondetti, M.A., Rubin, B. and Cushman, D.W. (1977) Design of specific inhibitors of angiotensin-converting enzyme: new class of orally active hypertensive agents. *Science* **196**: 441–444.

37. Cushman, D.W., Cheung, H.S., Sabo, E.F. and Ondetti, M.A. (1977) Design of potent competitive inhibitors of angiotensin-converting enzyme. Carboxyalkanoyl and mercaptoalkanoyl amino acids. *Biochemistry* **16**: 5484–5491.

38. Endo, A., Kuroda, M., Tsujita, Y., Terahara, A. and Tamura, C. (1975) Physiologically Active Compounds. German Patent No. 2524355.

39. Endo, A. (1985) Compactin (ML-236B) and related compounds as potential cholesterol-lowering agents that inhibit HMG-CoA reductase. *J. Med. Chem.* **28**: 401–405.

40. Brown, A.G., Smale, T.C., King, T.J., Hasenkamp, R. and Thompson, R.H. (1976) Crystal and molecular structure of compactin, a new antifungal metabolite from *Penicillium brevicompactum*. *J. Chem. Soc. Perkin Trans.* **1**: 1165–1170.

41. Endo, A. (1979) Monacolin K, a new hypocholesterolemic agent produced by a *Monascus* species. *J. Antibiot.* **32**: 852–854.

42. Endo, A. (1980) Monacolin K, a new hypocholesterolemic agent that specifically inhibits 3-hydroxy-3-methylglutaryl coenzyme A reductase. *J. Antibiot.* **33**: 334–336.

43. Alberts, A.W., Chen, J., Kuron, G., Hunt, V., Huff, J., Hoffman, C., Rothrock, J., Lopez, M., Joshua, H., Harris, E., Patchett, A., Monaghan, R., Currie, S., Stapley, E., Albers-Shonberg, G., Hensens, O., Hirshfield, J., Hoogsteen, K., Liesch, J. and Springer, J. (1980) Mevinolin: a highly potent competitive inhibitor of hydroxymethyl-glutaryl-coenzyme A reductase and a cholesterol lowering agent. *Proc. Natl Acad. Sci. USA* **77**: 3957–3961.

44. Monaghan, R. L., Alberts, A. W., Hoffman, C. G. and Albers-Schonberg, G. (1980) Hypocholesteremic fermentation products and process of preparation USA Patent No. 4231938.

45. Alberts, A.W., MacDonald, J.S., Till, A.E. and Tobert, J.A. (1989) Lovastatin. *Cardiol. Drug. Dev.* **7**: 89–109.

46. Vagelos, P.R. (1991) Are prescription drug prices high? *Science* **252**: 1080–1084.

47. Kwak, B., Mulhaupt, F., Myit, S. and Mach, F. (2000) Statins as a newly recognized type of immunomodulator. *Nat. Med.* **6**: 1399–1402.

48. Jick, H., Zornberg, G.L., Jick, S.S., Seshadri, S. and Drachman, D.A. (2000) Statins and the risk of dementia. *Lancet* **356**: 1627–1631.

49. Wolozin, B., Kellman, W., Ruosseau, P., Celesia, G.G. and Siegel, G. (2000) Decreased prevalence of Alzheimer disease associated with 3-hydroxy-3-methylglutaryl coenzyme A reductase inhibitors. *Arch. Neurol.* **57**: 1439–1443.

50. Meyer, C.R., Schlienger, R.G., Kraenzlin, M.E., Schlegel, B. and Jick, H. (2000) HMG-CoA reductase inhibitors and the risk of fractures. *J. Am. Med. Assoc.* **283**: 3255–3257.

51. Borel, J.F. (ed.). In *Progress in Allergy*, vol. 38. *Cyclosporin*, pp. 1–465. Karger, Basel.

52. Cropp, T.A., Wilson, D.J. and Reynolds, K.A. (2000) Identification of a cyclohexylcarbonyl CoA biosynthetic gene cluster and application in the production of doramectin. *Nat. Biotechnol.* **18**: 980–983.

53. Suffness, M. and Wall, M.E. (1995) Discovery and development of taxol. In Suffness, M. (ed.). *Taxol: Science and Applications*, pp. 3–25. CRC Press, Boca Raton, FL.

54. Cortes, J.E. and Pazdur, R. (1995) Docetaxel. *J. Clin. Oncol.* **13**: 2643–2655.

55. Kingston, D.G.I., Chaudhary, A.G., Chordia, M.D., Gharpure, M., Gunatilaka, A.A.L., Higgs, P.I., Rimoldi, J.M., Samala, L., Jaktap, P.G., Giannakakou, P., Jiang, Y.Q., Lin, C.M., Hamel, E., Long, B.H., Fairchild, C.R. and Johnston, K.A. (1998) Synthesis and biological evaluation of 2-acyl analogs of paclitaxel (Taxol). *J. Med. Chem.* **41**: 3715–3726.

56. von Angerer, E. (2000) Tubulin as a target for anticancer drugs. *Curr. Opin. Drug Disc. Dev.* **3**: 575–584.

57. Hartwell, J.L. and Schrecker, A.W. (1958) Chemistry of podophyllum. In Zechmeister, L. (ed.). *Fortschritte Chemie der Organische Naturstoffe*, vol. 15, pp. 83–166. Springer-Verlag, Vienna.

58. Chen, G.L., Yang, L., Rowe, T.C., Halligan, B.D., Tewey, K. and Liu, L. (1984) Nonintercalative antitumor drugs interfere with the breakage-reunion of mammalian DNA topoisomerase II. *J. Biol. Chem.* **259**: 13560–13566.

59. Noble, R.L., Beer, C.T. and Cutts, J.H. (1958) Role of chance observations in chemotherapy: *Vinca rosea*. *Ann. NY Acad. Sci.* **76**: 882–894.

60. Johnson, I.S., Wright, H.F., Suoboda, G.H. and Lantis, J. (1960) Antitumor principles derived from *Vinca rosea* Linn. I. Vincaleuko-blastine and leurosine. *Cancer Res.* **40**: 1016–1022.

61. Pezzuto, J.M. (1997) Plant-derived anticancer agents. *Biochem. Pharmacol.* **53**: 121–133.

62. Potmeisel, M. and Pinedo, H. (1995) *Camptothecins: New Anticancer Agents*. CRC Press, Boca Raton.

63. Zasloff, M. (1987) Magainins, a class of antimicrobial peptides from *Xenopus* skin: isolation and characterization of two active forms, and partial cDNA sequence of a precursor. *Proc. Natl Acad. Sci. USA* **84**: 5449–5453.

64. Brotzu, G. (1948) Research as a new antibiotic. *Lav. Ist. Igiene Cagliari*.

65. Mateles, R.I. (1998) *Penicillin: A Paradigm for Biotechnology*. Candida Corporation, Chicago.

66. Imada, A., Kitano, K., Kintaka, K., Muroi, M. and Asai, M. (1981) Sulfazecin and isosulfazecin, novel beta-lactam antibiotics of bacterial origin. *Nature* **289**: 590–591.

67. Sykes, R.B., Cimarusti, C.M., Bonner, D.P., Bush, K., Floyd, D.M., Georgopapadakou, N.H., Koster, W.H., Liu, W.C., Parker, W.L., Principe, P.A., Rathnum, M.L., Slusarchyk, W.A., Trejo, W.H. and Wells, J.S. (1981) Monocyclic beta-lactam antibiotics produced by bacteria. *Nature* **291**: 489–491.

68. Ryan, B., Ho, H. -T., Wu, P., Frosco, M. -B., Dougherty, T. and Barrett, J.F. (2000) 40th Interscience Conference on Antimicrobial Agents and Chemotherapy (ICAAC). *Exp. Opin. Invest. Drugs* **9**: 2945–2972.

69. Terry, C.E. and Pellens, M. (1928) *The Opium Problem*. Bureau of Social Hygiene, New York.

70. Musto, D.F. (1973) *The American Disease*. Yale University Press, New Haven.

71. Gray, N.S., Wodicka, L., Thunnissen, A.-M.W.H., Norman, T.C., Kwon, S., Espinosa, F.H., Morgan, D.O., Barnes, G., LeClerc, S., Meijer, L., Kim, S.-H., Lockhart, D.J. and Schultz, P.G. (1998)

Exploiting chemical libraries, structure and genomics in the search for kinase inhibitors. *Science* **281**: 533–538.

72. Gray, N., Detivaud, L., Doerig, C. and Meijer, L. (1999) ATP-site directed inhibitors of cyclin-dependent kinases. *Curr. Med. Chem.* **6**: 859–875.

73. Chang, Y.T., Gray, N.S., Rosania, G.R., Sutherlin, D.P., Kwon, S., Norman, T.C., Sarohia, R., Leost, M., Meijer, L. and Schultz, P.G. (1999) Synthesis and application of functionally diverse 2,6, 9-trisubstituted purine libraries as CDK inhibitors. *Chem. Biol.* **6**: 361–375.

74. Nicolaou, K.C., Winssinger, N., Vourloumis, D., Ohshima, T., Kim, S., Pfefferkorn, J., Xu, J.-Y. and Li, T. (1998) Solid and solution phase synthesis and biological evaluation of combinatorial sarcodictyin libraries. *J. Am. Chem. Soc.* **120**: 10814–10826.

75. Hamel, E., Sackett, D.L., Vourloumis, D. and Nicolaou, K.C. (1999) The coral-derived natural products eleutherobin and sarcodictyins A and B: effects on the assembly of purified tubulin with and without microtubule-associated proteins and binding at the polymer taxoid site. *Biochemistry* **38**: 5490–5498.

76. Nicolaou, K.C., Roschangar, F. and Vourloumis, D. (1998) Chemical biology of epothilones. *Angew. Chem. Int. Ed. Engl.* **37**: 2014–2045.

77. Zhao, L., Ahlert, J., Xue, Y., Thorson, J.S., Sherman, D.H. and Liu, H.-W. (1999) Engineering a methymycin/pikromycin-calicheamicin hybrid: construction of two new macrolides carrying a designed sugar moiety. *J. Am. Chem. Soc.* **121**: 9881–9882.

78. Hutchinson, C.R. (1999) Microbial polyketide synthases: more and more prolific. *Proc. Natl Acad. Sci. USA* **96**: 3336–3338.

79. Wender, P.A., Hinkle, K.W., Koehler, M.F. and Lippa, B. (1999) The rational design of potential therapeutic agents: synthesis of bryostatin analogs. *Med. Res. Rev.* **19**: 388–407.

80. Wender, P.A., De Brabander, J., Harran, P.G., Jimenez, J.-M., Koehler, M.F.T., Lippa, B., Park, C.-M. and Shiozaki, M. (1998) Synthesis of the first members of a new class of biologically active bryostatin analogs. *J. Am. Chem. Soc.* **120**: 4534–4535.

81. Wender, P.A., De Brabander, J., Harran, P.G., Jimenez, J.-M., Koehler, M.F.T., Lippa, B., Park, C.-M., Siedenbiedel, C. and Pettit, G.R. (1998) The design, computer modeling, solution structure, and biological evaluation of synthetic analogs of bryostatin 1. *Proc. Natl Acad. Sci. USA* **95**: 6624–6629.

82. Wender, P.A., De Brabander, J., Harran, P.G., Hinkle, K.W., Lippa, B. and Pettit, G.R. (1998) Synthesis and biological evaluation of fully synthetic bryostatin analogs. *Tetrahedron Lett.* **39**: 8625–8628.

83. Wender, P.A., Lippa, B., Park, C.-M., Irie, K., Nakahara, A. and Ohigashi, H. (1999) Selective binding of bryostatin analogs to the cysteine rich domains of protein kinase C isozymes. *Bioorg. Med. Chem. Lett.* **9**: 1687–1690.

84. Haygood, M.G. and Davidson, S.K. (1997) Small-subunit rRNA genes and *in situ* hybridization with oligonucleotides specific for the bacterial symbionts in the larvae of the bryozoan *Bugula neritina* and proposal of "Candidatus endobugula sertula". *Appl. Environ. Microbiol.* **63**: 4612–4616.

85. Haygood, M.G., Schmidt, E.W., Davidson, S.K. and Faulkner, D.J. (1999) Microbial symbionts of marine invertebrates: opportunities for microbial biotechnology. *J. Mol. Microbiol. Biotechnol.* **1**: 33–43.

86. Mutter, R. and Wills, M. (2000) Chemistry and clinical biology of the bryostatins. *Bioorg. Med. Chem.* **8**: 1841–1860.

7

BASICS OF COMBINATORIAL CHEMISTRY

Cécile Vanier and Alain Wagner

Perfection is attained by slow degrees; it requires the hand of time   **Voltaire**

I INTRODUCTION

As new tools such as high-throughput biological screening became routinely involved in the process of drug finding, testing thousands of compounds within a couple of days was no longer a challenge. As a consequence, the demand for new molecules dramatically increased, leading medicinal chemists to rethink the way they provide biologists with chemical compounds. The so-called combinatorial chemistry emerged as a powerful tool in drug discovery. It is now of prime interest not only in the field of drug discovery but also in many others domains such as material science and asymmetric catalysis. Literally, combinatorial chemistry means 'a combinatorial process to prepare sets of compounds from sets of building blocks',[1] but for many chemists it is still difficult to get a clear view of this, no longer novel,[2] quite heterogeneous domain. It might even be more difficult to decide when and how to use combinatorial chemistry methods. The purpose of this chapter is not to give an extensive review of combinatorial chemistry, many excellent papers and books address this topic,[3] but rather to give an overview to help the novice to tackle this domain.

II PRINCIPLES OF COMBINATORIAL CHEMISTRY

The field of 'combinatorial chemistry' covers a wide range of interdependent concepts and techniques as diverse as solid-phase synthesis, supported reagents, parallel homogeneous-phase chemistry, solid-phase extractions and adapted analytical methods. They were designed and developed both to address specific problems, libraries synthesis, and to help chemists in their everyday work.

High-throughput synthesis is part of a global drug discovery strategy that usually involves dedicated, often expensive, automated synthesizers, on-line purification, computational analysis, data handling as well as high-throughput screening. It relies on the synthesis and screening of large numbers of compounds and involves heavy technological investment. These are usually carried out in dedicated laboratories and performed by chemists well-trained in the field. It seems, however, that people tend to prepare smaller libraries and privilege quality rather than quantity.

Small-scale parallel-synthesis methods can also speed up and significantly simplify everyday chemistry. Parallel synthesis can indeed be implemented in cheap, readily available apparatus in a routine manner. It is, however, often under used or misused by organic chemists not accustomed to these techniques. There is thus a need to demonstrate that any organic synthetic chemist can benefit from combinatorial chemistry techniques, to prepare more analogues or save bench time by improving the efficacy of the experimental work.

Another mapping that could help to gain an overview on combinatorial chemistry is based on the philosophical approach, combinatorial synthesis dealing with the preparation of compound mixtures and parallel synthesis dealing with the synthesis of separate single compounds.

ISBN 0-12-744481-5

If identifying an active pool of molecule is easy, identifying the active component within this pool in a reliable and meaningful way is very difficult. Thus, interest in combinatorial synthesis has gradually decreased because of the lengthy and uncertain hit identification process. Even though this trend might be reverted in the future, currently most of the combinatorial approaches focus on rapid parallel synthesis of individual compounds.

III SYNTHETIC METHODOLOGIES

Combinatorial chemistry can be carried out both in the solution phase or on a solid support. These two complementary approaches offer valuable solutions to the chemist either on a high throughput automated method or on a laboratory scale. Both solution- and solid-phase approaches present characteristics and requirements that have to be considered before deciding whether the planned synthesis should be carried out with one or the other technique.

A Solid-phase synthesis (SPS)

As illustrated in Scheme 1, an organic synthesis on the solid phase consists of a sequence where the starting material is covalently bound to a swollen insoluble polymeric support through an anchor and/or a linker. During all synthetic steps, the compound under construction remains linked to the support. The most obvious advantage of this method is the possibility of using any excess of reagents to complete the reaction, since the removal of unreacted reagents can be achieved by simple filtration. Despite the significant ease in handling the reaction, two more steps (linkage/cleavage) have to be added to every synthesis.

The early days of solid-phase chemistry were mainly dedicated to the field of peptide synthesis. Peptide chemists have now fully incorporated these techniques in the way they carry out chemistry and very few of them nowadays would prepare a tetrapeptide in solution on a laboratory scale. As combinatorial chemistry has turned to the development of smaller libraries of low molecular weight compounds, almost all kinds of reactions can be carried out

on solid supports. It is to be expected that organic chemists will follow the same path, and use combinatorial tools whenever possible.

Chemical structures of common polymeric supports

The most commonly used polymeric support is a copolymer styrene/divinylbenzene functionalized with various reactive groups such as chloromethyl (Merrifield resin) and various spacers such as ethylene glycol (Wang-type resin). The supports now available vary one from another in many aspects and properties: functionalization, cross-linking, porosity, loading, bead size. All these variations generate different swelling behaviours, different hydrophilicity, and different chemical stability.

Cross-linked polystyrene allows the use of a broad variety of reaction parameters, as diverse as the use of oxidants (ozone, DDQ), various acids (TfOH, TFA, HBr), strong bases (LDA), from low to relatively high temperatures ($-20°C$ to $150°C$).

Anchors and linkers

Many linkers, reactions and synthetic transformations have been documented and exemplified. Thus without much study, many basic or complex chemical transformations can be carried out on solid supports in a quite straightforward manner. Even a novice can achieve this ambitious goal, as long as basic reactivity rules are followed.

- Structural modifications and cleavage have to proceed under orthogonal conditions.
- As for any synthetic schemes, reactions conditions involved have to be chemo-selective towards the various functionalities introduced and towards the anchor.
- Protecting groups' removal and cleavage reactions have to be compatible with the target structure. Hence it is primordial to select an appropriate anchoring/releasing strategy.

Since they are the most versatile and documented, peptide chemistry-type linkers are the most widely used. Reactivity charts are found in many catalogues and SPS books.[4] Inherent to these linkers, however, the whole library

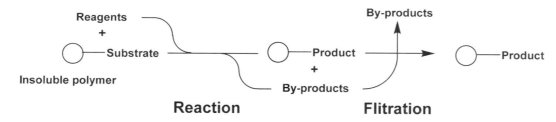

Scheme 1

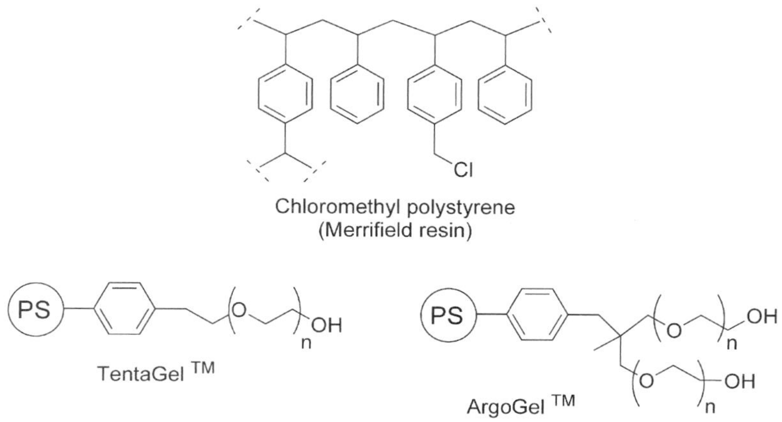

Chloromethyl polystyrene
(Merrifield resin)

TentaGel ™

ArgoGel ™

Scheme 2

bears a common functional group or structural motif at the position to which the molecule is grafted on to the polymeric support. When designing a SP synthetic scheme the choice of the attachment point is generally closely related to the chemistry or known structure–activity relationship. Advantageously the building block available with the poorest diversity, or generally speaking the part of the molecule that is not or cannot be varied, serves as the attachment point to the support. The type of linker will be chosen for its compatibility with the considered reaction sequence. However, if rather straightforward for peptide or peptide-like syntheses, the choice can become tricky for more complex and varied SPOS.

Some data about linker reactivity towards organic reagents are, however, still missing, such as the stability of a Wang linker in the presence of Lewis acids or oxidizing agents. Interestingly, a resin-bound version of most of the protecting groups routinely used in organic chemistry is now commercially available, i.e. resin-bound DHP, chlorosilane, *tert*-butylchlorocarbamate, 1,2-diol, thus enlarging and facilitating the choice of an anchoring strategy for nonpeptidic chemistry.[5]

In addition, novel linkers have recently been developed,[6] enabling new types of cleavage. The first strategy allows 'traceless synthesis', generally meaning that a carbon–hydrogen bond is formed upon cleavage of the molecule–resin bond. The second strategy called 'functional cleavage' involves an advantageous release step that proceeds via functional group interconversion with concomitant introduction of an additional diversity source. These latest strategies are worth considering if the target library holds the adequate functional group or residue, i.e. among others, an aromatic for traceless cleavage, and amide, urea, thiourea, guanidine for functional cleavage.

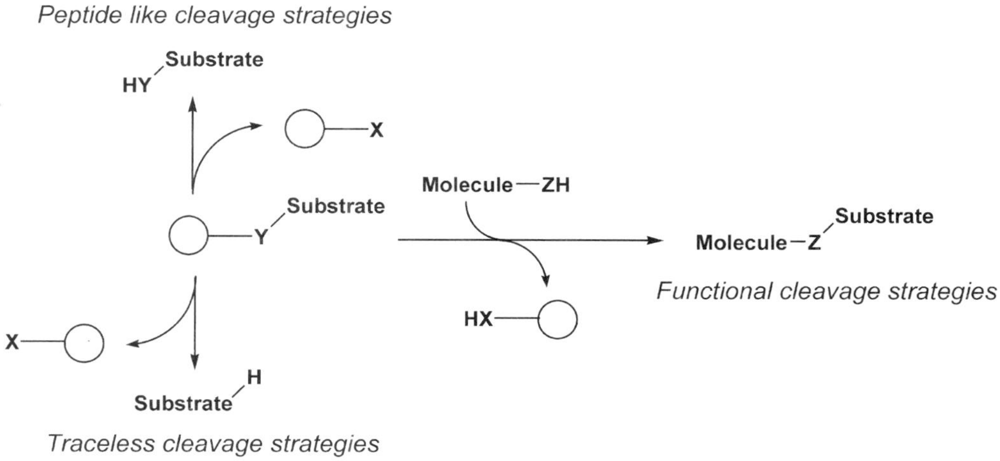

Peptide like cleavage strategies

Functional cleavage strategies

Traceless cleavage strategies

Scheme 3

Synthetic repertoire

From a retro-synthetic point of view, one can understand that the wider the chemical diversity of the attempted library is, the more chemo-selective the reactions have to be. In addition, since all reactions will be run in parallel under the same conditions, the conditions have to be as robust as possible. Thus, contrary to a widespread idea, SPS requires the experimenter to have a good knowledge and understanding of organic chemistry.

It appears also that, for the synthesis of large libraries, there are few reactions that fulfil these prerequisites. Among the reactions that proved to have wide applicability, amide bond formations and peptide-like protecting group strategies can be found. In a general way all amine functionalization reactions are usually effective and lead to stable structures[7] (i.e. sulfonamide, carbamate, urea, thiourea formation, reductive amination and nucleophilic substitution). Moreover, for any of these reactions the suitable linker can be easily selected from various literature sources.

To illustrate the scope of the reactions that can be carried out in the solid phase, we will focus on one type of transformation — reductive amination — and see how versatile SPS can be. Reductive amination is indeed very well suited to SPS, and has already been widely described in the literature. The reaction can be carried out with the same success with either resin-bound amine or resin-bound carbonyl compounds.

The sequence can proceed in a two- or in a one-step process. The first strategy affords clean monoalkylation of the bound primary amines. To help the formation of the intermediate Schiff base, catalytic amounts of acids or dehydrating agents can be added to the reaction mixture. The reduction of the imine is then easily achieved by treatment with, for instance, $LiBH_4$ or $NaBH(OAc)_3NaBH_4$.

The reductive amination of resin-bound aldehydes by amino acid esters has been successfully achieved in the presence of $NaBH(OAc)_3$ and acetic acid without racemization.[8]

Another interesting reducing agent is BAP ($= BH_3-Py$), which has been found to be the best reagent for less stable iminium ion intermediates formed with *sec*-amines and aldehydes or ketones.[9] It represents then the best protocol for the reductive amination of supported *sec*-amines. Thus, using successive reductive amination as the key reaction, complex libraries can easily be synthesized.[10]

For the synthesis of a targeted library, since building blocks will be structurally related and their number lower, complex reactions can be used to build-up the desired structure. In recent years many of these reactions have been reported and have proved to be particularly successful, e.g. Mitsunobu,[11] cyclodehydration,[12] Pd coupling.[13] However, one has to keep in mind that these reactions can present a moderate chemo-selectivity, or require either sensitive or harsh reaction conditions, that could, to a certain extend, hamper their use, especially in the preparation of large libraries.

With the aim of preparing elaborate molecules in the solid phase, chemists are searching for robust and versatile methods to assemble complex carbon frameworks in the insoluble support. One of the predominant sequences for creating carbon–carbon bonds in the solid phase is Pd-catalysed coupling reactions. The Heck reaction, for instance, is a convenient way to diversify activated alkenes (Scheme **5a**).[14] This coupling involves an arene with iodide, bromide, triflate or diazonium on one hand, and an alkene (generally electron poor) on the other, in the presence of a catalyst and a base.

Resin-bound biaryl structure can also conveniently be synthesized using a Pd-coupling approach. Interestingly, special linkers have been designed to allow concomitant cleavage and carbon–carbon bond formation, thus

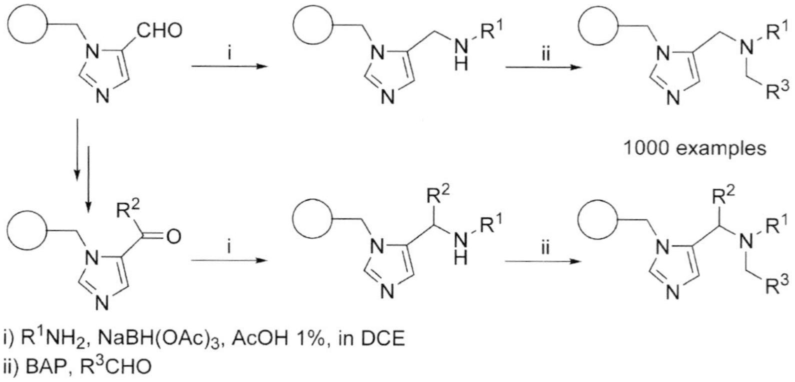

i) R^1NH_2, $NaBH(OAc)_3$, AcOH 1%, in DCE
ii) BAP, R^3CHO

Scheme 4

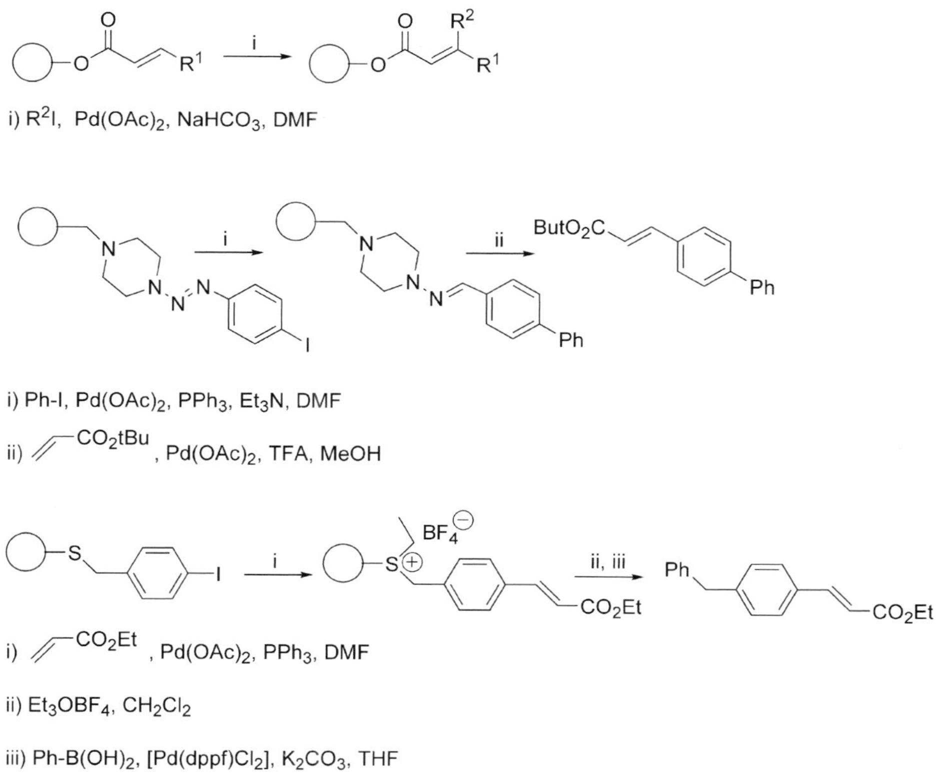

i) R²I, Pd(OAc)₂, NaHCO₃, DMF

i) Ph-I, Pd(OAc)₂, PPh₃, Et₃N, DMF

ii) $\diagup$—CO₂tBu , Pd(OAc)₂, TFA, MeOH

i) $\diagup$—CO₂Et , Pd(OAc)₂, PPh₃, DMF

ii) Et₃OBF₄, CH₂Cl₂

iii) Ph-B(OH)₂, [Pd(dppf)Cl₂], K₂CO₃, THF

Scheme 5

producing the target structure in a traceless manner Scheme **5c, d**).[15]

Among other reactions which have proved to be useful for the synthesis of targeted libraries are the Wittig and the closely related Horner–Emmons couplings. These powerful reactions can be carried out with either polymer-bound aldehydes/ketones or polymer-bound ylides.

The Wittig olefination involves non-stabilized ylides, which are generated upon treatment of the corresponding resin-bound alkylphosphonium salts with strong bases (NaHMDS, LiHMDS, KOᵗBu, nBuLi). Supported phosponium salts can be prepared from the corresponding bromoalkene by refluxing with a resin-bound phosphine.

The Horner–Wadsworth–Emmons reaction is also a useful option for producing olefins in the solid phase. Supported stabilized phosphonoactates, as they are more acidic than phosphonium salts, can readily give the corresponding ylides with weak bases such as DBU, DIPEA, Et₃N. One still has to make sure that these reaction conditions are chemoselective. Both types of reactions, Wittig[16] and Horner–Wadsworth–Emmons,[17] have been successfully used in cyclative and/or functionalizing cleavage.

Many other reactions, which were developed recently and were hardly manageable in the solution phase, are now routinely carried out in the solid phase — metathesis, stereoselective reductions. Although solid-phase organic synthesis is in its infancy the number of described and available reactions as well as the global comprehension of the problems are rapidly growing.

B Solution-phase chemistry

It is commonly argued that the major impediment to parallel solution-phase synthesis of large numbers of individual organic molecules is the time and effort required for purification of the reaction mixtures at each synthetic step. Nevertheless, the recent improvements to bypass this difficulty should be stressed.[18] Various alternatives based on different although complementary concepts have been developed: scavenger resins,[19] ion exchange resins, tagging reagents, solid-phase extraction (SPE),[20] solid supported liquid–liquid extraction (SLE)[21] and supported reagents and catalysts.

As some of these techniques can also, and mainly, be applied to the purification of all libraries, and not only the ones built-up in solution, they will be addressed later. In this section only the methods developed to speed up and facilitate liquid-phase synthesis of libraries, i.e. capture-release methods, supported reagents and supported

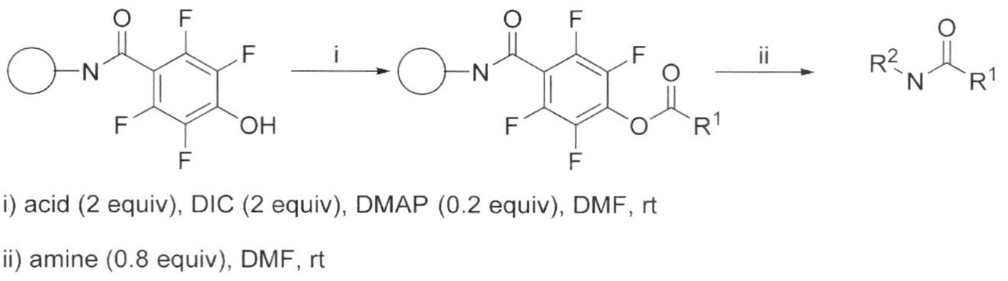

i) acid (2 equiv), DIC (2 equiv), DMAP (0.2 equiv), DMF, rt

ii) amine (0.8 equiv), DMF, rt

Scheme 6

catalysts, will be discussed. Thus, chemistry in solution is often the first choice, especially in the preparation of relatively small libraries (up to 1000 compounds), while for large libraries, SPS is often preferred, since it allows easy product handling and tagging.

Capture-release method

A basic example to illustrate the capture-release strategy is the use of a supported tetrafluorophenol moiety for the synthesis of amide libraries in solution.[22] As shown in Scheme **6**, a carboxylic acid is first activated with the corresponding resin-bound tetrafluorophenol ester (capture step), which in a second step reacts with nucleophilic amine (release step).

Other advantageous reaction sequences enable the preparation of complex libraries of thiourea or guanidine via an analogous catch and release process.[23] This strategy, although very efficient, is nevertheless closely related to SPS, and limited to only one step.

Supported reagents and catalysts

Supported reagent philosophy is the blueprint of SPS. In this approach the substrate is kept in solution while reagents are bound to resins. Hence, purification also consists of a simple filtration but the reaction product is recovered in the filtrate.[24] As a consequence, it can be analysed by classical means and reactions optimized very easily. More and more supported reagents and catalysts are now commercially available.

Contrary to SPS that usually requires long optimizations, supported reagents can quite simply be introduced into reaction sequences which were designed for classical homogeneous chemistry.[25] Many of them have to be used in stoichiometric amounts, or even in excess (reagents), others can simply be used in catalytic amounts (catalysts).

Resin-bound triphenylphosphine, hydrogenation catalysts, halogenation reagents, bases, oxidizing and reducing agents and chiral co-reagents or catalysts are among others commercially available.[26] These polymer-bound reagents not only facilitate the work-up of reaction mixtures, but also allow better yields and/or better selectivity to be attained. Another advantage of polymer-bound catalysts is that reactions can be carried out on a relatively large scale (100 mg) which could be ideal for building blocks or focused library preparation with the aim of providing biologists with larger quantities.

An aspect that should not be neglected is that chemistry in solution is familiar to any chemist, and many more reaction types are exemplified and illustrated in the literature (even if this difference tends to become smaller). It does not require (or few and affordable) specific apparatus and can be easily set up in any laboratory. Moreover, monitoring the reactions as well as analysis of the final product, are quick and easy. There is no need to adapt the reaction conditions to the solid support requirements, and obviously, the anchoring and the cleavage steps can be omitted.

Once the library is built-up, whatever technology has been implemented to achieve the synthesis, the final

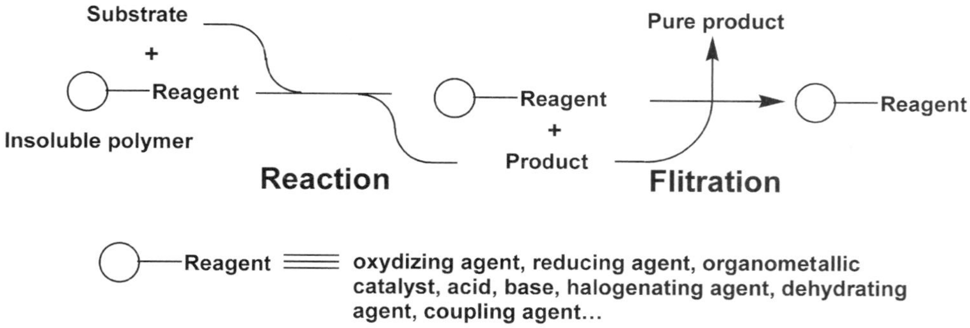

Scheme 7

products may require purification. Here again, many tools and techniques have been developed and optimized to allow fast and efficient parallel work-up of hundreds and thousands of compounds.

IV LIBRARY PURIFICATION

The level of purity as well as the required amount of compound is generally imposed by the targeted biological evaluation. The crude cleavage product, however, often does not meet the purity requirement. To allow quick parallel purification and structural assessment an increasing amount of apparatus, softwares and techniques are required in laboratories. The most efficient is probably fast LC/MS and its multicolumn evolution that enables purification of hundreds of compounds a day. Similarly, parallel chromatography devices and TLC spotting apparatus can be found on the market.[27]

Here again, without resorting to these expensive machines, alternative techniques like scavenger resins, solid-phase extraction (SPE) and solid-supported liquid–liquid extraction can help the chemist to simplify repetitive operations.

Scavenger resins are designed to selectively trap a molecule according to a defined criteria of reactivity or physical property. They can thus be classified according to their reactivity,[28] i.e. acidic, basic, nucleophilic, electrophilic (Scheme **8**).

This technique is particularly efficient in removing from the reaction media, acids (using basic resin), bases (using acidic resin), aldehydes (using tosylhydrazyne resin), nucleophiles (using electrophilic resin). Likewise, a wide range of extraction cartridges that allow rapid purification of non-similar compounds exists such as polar, non-polar, hydrophilic, ionic phases.

Traditional aqueous extraction is obviously cumbersome as the number of reaction mixtures to work-up increases. It would, however, be efficient to separate the desired products from impurities in many cases. Solid-liquid extraction (SLE) might be an answer. It enables parallel aqueous extractions to be run by efficiently trapping the aqueous layer. A number of scavenger resins, SPE and SLE devices, with clear application notes,[29] are available.

Scavenger resins have proved to be quite successful, especially when combined with supported reagents.[30] They quickly emerged as versatile and complementary tools to facilitate the set-up of repetitive sequences by avoiding tedious work-up and purification steps. By experimenting with heterogeneous reagents, scavenger resins and SPE, it is possible to produce multistep synthesis in the homogeneous phase without classical chromatographic purification and to recover a reasonably pure material.[31] Of course, these rough separation processes require the target molecule to be different from the reagent to be separated. Hence, in a way, this drawback limits SPE to the purification of targeted libraries or sub-libraries bearing common structural features.

V ANALYTIC TOOLS IN SPS

From a practical point of view, if running parallel reactions is quite straightforward in SPS, the analysis of resin-bound substrates and resin-bound products is not. For solution-phase synthesis the experimenter has at his disposal a wide panel of routine analytical techniques that give rapid and readable information. In contrast, for solid-phase reactions, analysis is not routine and chemists often do not have the know-how and the experimental background to analyse reactions in solid support.

One can distinguish on-bead methods from analysis of the crude mixture after cleavage.[32] IR proved to be the most

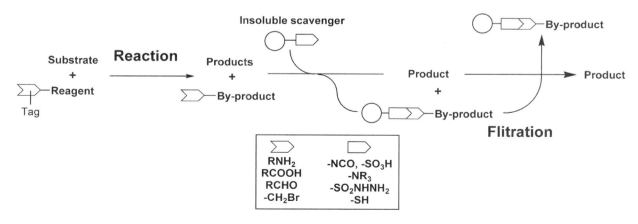

Scheme 8

useful and versatile tool to analyse resin-grafted molecules.[33] More particularly, IR microscope and micro-ATR techniques allow high quality spectra from very small amounts of resin (few beads) to be recorded. Interestingly, these techniques produce sharp signals even in the NH and OH region which are involved in many reactions. On-bead IR analysis can give a good idea of the reaction progress, and can now be considered as the SPS 'TLC'. When searching for optimum reaction conditions it is often advantageous to put on each reaction partner an IR tag, i.e. nitro-, carbonyl-, trifluoromethyl- or sulfonyl-groups.

Elemental analysis also proved to be a helpful routine method for determining reaction progress and calculate resin loading, especially when bromine or fluorine tags are present.

NMR techniques adapted for SPS produce useful information but are often not routine techniques.[34] Standard NMR spectra of polymers usually results in significant line broadening. The resolution may, however, be improved when the mobility of the support-bound molecules increases. Here again, the use of fluoride or phosphorus as tracers might be of interest.[35] A technique especially adapted for recording NMR spectra of support-bound compounds is MAS (magic angle spinning) NMR. Although efficient, this method requires a specialized expensive accessory.

Cleavage of the molecule from the support gives access to the use of classical analytic techniques i.e. NMR, MS.[36] It is the most widely used procedure, it is, however, only permitted for stable structures and will only result in information about side reactions when the side product does not decompose during the cleavage.

VI CONCLUSION

Combinatorial chemistry and biology developed hand in hand, in turn one challenging the other. The way that both chemists and biologists work is changing, from a practical and a philosophical point of view. It has resulted in the introduction of automated apparatus, novel experimental methodologies and informatics into research laboratories. One should, however, maintain a pragmatic approach and not forget that without the right reaction, reaction sequence, proteins, enzymes, all machines are useless and will not solve any problem.

REFERENCES

1. Maclean, D., Baldwinn, J.J., Ivanov, V.T., Kato, Y., Shaw, A., Schneider, P. and Gordon, E.M. (1999) Glossary of terms used in combinatorial chemistry. *Pure Appl. Chem.* **71**: 2349–2365.

2. Crowley, J.I. and Rapoport, H. (1976) Solid-phase synthesis: novelty or fundamental concept. *Acc. Chem. Res.* **9**: 135–144.

3. Sucholeiki, I. (ed.) (2001) *High-throughput Synthesis.* Marcel Decker, New York, NY.; and Jung, G. (ed.) (2001) *Combinatorial Chemistry: Synthesis, Analysis and Screening.* Wiley-VCH, Weinheim.

4. www.peptide.com Advance ChemTech; www.nova.ch NovaBiochem; www.argotech.com Argonaut Technology.

5. Barany, G. and Kempe, M. (1997) The context of solid-phase synthesis. In Czarnik, A.W. and DeWitt, S.H. (eds). *A Practical Guide to Combinatorial Chemistry*, pp. 51–97. American Chemical Society.

6. Guillier, F., Orain, D. and Bradley, M. (2000) Linkers strategies in solid-phase synthesis and combinatorial chemistry. *Chem. Rev.* **100**: 2091–2157.

7. Akritopoulou-Zanke, I. and Sowin, T.J. (2001) Solid-phase synthesis of macrolide analogues. *J. Comb. Chem.* **3**: 301–311.

8. Boojamara, C.G., Burow, K.M., Thompson, L.A. and Ellmann, J.A. (1997) Solid-phase synthesis of 1,4-benzodiazepine-2,5-diones. Library preparation and demonstration of synthesis generality. *J. Org. Chem.* **62**: 1240–1256.

9. Kahn, N.M., Arumugam, V. and Balasubramanian, S. (1996) Solid phase reductive alkylation of secondary amines. *Tetrahedron Lett.* **37**: 4819–4822.

10. Saha, A.K., Liu, L., Marichal, P. and Odds, F. (2000) Novel antifungals based on 4-substituted imidazole: solid-phase synthesis of substituted aryl sulfonamides towards optimization of *in vitro* activity. *Bioorg. Med. Chem. Lett.* **10**: 2735–2739. Dolle, R. E. (2001) Comprehensive survey of combinatorial library synthesis: 2000. *J. Comb. Chem.* **3**: 1–41.

11. Wang, F.J. and Hauske, R. (1997) Solid-phase synthesis of benzoxazoles via Mitsunobu reaction. *Tetrahedron Lett.* **38**: 6529–6532.

12. Chauhan, P.M. (2001) Recent development in the combinatorial synthesis of nitrogen heterocycles using solid phase technology. *Combi. Chem. High Throughput Screening* **4**: 35–51.

13. Franzen, R. (2000) The Suzuki, the Heck, the Stille reactions; three versatile methods for introduction of C-C bond on solid support. *Can. J. Chem.* **78**: 957–962. Ruhland, B., Bombrun, A. and Gallop, M.A. (1997) Solid-phase synthesis of 4-arylazetidin-2-ones via Suzuki and Heck cross-coupling reactions. *J. Org. Chem.* **62**: 7820–7826.

14. Blettner, C.G., König, W.A., Stenzel, W. and Schtten, T. (1999) Highly stereoselective preparation of 3,3-disubstituted acrylates on polyethylene glycol. *Tetrahedron Lett.* **40**: 2101–2102.

15. Bräse, S., Enders, D., Köbberling, J. and Avemaria, F. (1998) A surprising solid-phase effect: development of a recyclable traceless linker system for reactions on solid support. *Angew. Chem. Int. Ed. Engl.* **37**: 3413–3415. Vanier, C., Lorgé, F., Wagner, A. and Mioskowski, C. (2000) Traceless solid phase synthesis of biarylmethane structures through Pd-catalyzed release of supported benzylsulfonium salts. *Angew. Chem. Intl. Ed. Engl.* **39**: 1679–1683.

16. Bolli, M.H. and Ley, S.V. (1998) Development of a polymer bound Wittig reaction and use in multi-step organic synthesis for the overall conversion of alcohols to hydroxyamines. *J. Chem. Soc. Perkin Trans.* **1**: 2243–2246.

17. Nicolaou, K. C., Pastor, J., Wissinger, N. and Murphy, F. (1998) Solid phase synthesis of macrocycles by an intramolecular ketophosphonate reaction. Synthesis of a (dl)-muscone library. *J. Am. Chem. Soc.* **120**: 5132–5133.

18. Hayun, A. and Cook, P.D. (2000) Methodologies for generating solution-phase combinatorial libraries. *Chem. Rev.* **100**: 3311–3340.

19. Booth, R.J. and Hodges, J.C. (1997) Polymer-supported quenching reagents for parallel purification. *J. Am. Chem. Soc.* **119**: 4882–4886.

20. Flynn, D.L., Crich, J.Z., Devraj, R.V., Hockerman, S.L., Parlow, J.J., South, M.S. and Woodard, J. (1997) Chemical library purification strategies based on principles of complementary molecular reactivity

and molecular recognition. *J. Am. Chem. Soc.* **119**: 4874–4881. Kaldor, S. W., Siegel, M. G., Fritz, J. E., Dressman, B. A. and Hahn, P. J. (1996) Use of solid supported nucleophiles and electrophiles for the purification of non-peptide small molecule libraries. *Tetrahedron Lett.* **37**: 7193–7196.

21. Breitenbucher, J.G., Arienti, K.L. and McLure, K.J. (2001) Scope and limitations of solid-supported liquid–liquid extraction for the high-throughput purification of compound libraries. *J. Comb. Chem.* **3**: 528–533.

22. Salvino, J.M., Kumar, N.V., Orton, E., Airey, J., Kiesow, T., Crawford, K., Mathew, R., Krowlikowski, P., Drew, M., Engers, D., Krolikowski, D., Herpin, T., Gardyan, M., McGeehan, G. and Labaudinière, R. (2000) Polymer-supported tetrafluorophenol: a new activated resin for chemical library synthesis. *J. Comb. Chem.* **2**: 691–697.

23. Gomez, L., Gellibert, F., Wagner, A. and Mioskowski, C. (2000) An efficient procedure for traceless solid-phase synthesis of *N,N'*-substituted thioureas by thermolytic cleavage of resin-bound dithiocarbamate. *J. Comb. Chem.* **2**: 75–79. Gomez, L., Gellibert, F., Wagner, A. and Mioskowski, C. (2000) An original traceless linker strategy for solid phase synthesis of *N,N',N''*-substituted guanidines. *Chem. Eur. J.* **6**: 4016–4020.

24. Sherington, J.C. (2001) Polymer-supported reagents, catalysts, and sorbents: evolution and exploitation — a personalized view. *J. Polym. Sci. Part A: Polym. Chem.* **9**: 2364–2376.

25. Ley, S.V. and Massi, A. (2000) Polymer supported reagents in synthesis: preparation of bicyclo[2.2.2]octane derivatives via tandem Michael addition reactions and subsequent combinatorial decoration. *J. Comb. Chem.* **2**: 104–107.

26. Kobayashi, S. and Akiyama, R. (2001) New methods for high-throughput synthesis. *Pure Appl. Chem.* **73**: 1103–1111.

27. Nilsson, U.J. (2000) Solid-phase extraction for combinatorial libraries. *J. Chromatogr. A* **885**: 305–319. Takahashi, T. (2001) New tools for isolation and analysis in combinatorial chemistry. *Chromatography* **22**: 45–48. Schultz, L., Garr, C.D., Cameron, L.M. and Bukowski, J. (1998) High throughput purification of combinatorial libraries. *Bioorg. Med. Chem. Lett.* **8**: 2409–2414.

28. Eames, J. and Watkinson, M. (2001) Polymeric scavenger reagents in organic synthesis. *Eur. J. Org. Chem.* 1213–1224. Flynn, D. L., Crich, J. Z., Devraj, R. V., Hockerman, S. L., Parlow, J. J., South, M. S. and Woodard, S. (1997) Chemical purification strategies based on principles of complementary molecular reactivity and molecular recognition. *J. Am. Chem. Soc.* **119**: 4874–4881.

29. For example, see: www.alltechweb.com; www.macherey-nagel.com.

30. Ley, S.V., Baxendale, I.R., Bream, R.N., Jackson, P.S., Leach, A.G., Longbottom, D.A., Nesi, M., Scott, J.S., Storer, R.I. and Taylor, S.J. (2000) Multi-step organic synthesis using solid-supported reagents and scavengers: a new paradigm in chemical library generation. *J. Chem. Soc. Perkin Trans. I* **23**: 3815–4195.

31. Kirschning, A., Monenschein, H. and Wittenberg, R. (2001) Functionalized polymer-emerging versatile tools for solution-phase chemistry and automated parallel synthesis. *Angew. Chem. Int. Ed. Engl.* **40**: 651–679.

32. Schatz, M.E. (ed.) (2000) *Analytical Techniques in Combinatorial Chemistry.* Dekker, New York, NY.

33. De Miguel, Y., Shearer, A.R. and Alison, S. (2001) Infrared spectroscopy in solid-phase synthesis. *Biotechnol. Bioeng.* **71**: 119–129. Yan, B., Fell, J. B. and Kumaravel, G (1996) Progression of organic reactions on resin supports monitored by single bead FTIR microscopy. *J. Org. Chem.* **61**: 7467–7472.

34. Keifer, P.A., Baltusis, L., Rice, D.M., Tymiak, A.A. and Shoolery, J.N. (1996) A comparison of NMR spectra obtained for solid-phase synthesis resins using conventional high-resolution, magic-angle-spinning, and high resolution magic-angle-spinning probes. *J. Magnet. Reson. serie A* **119**: 65–75.

35. Johnson, C.R. and Zhang, B. (1995) Solid phase synthesis of alkene using Horner-Wadsworth-Emmens reaction and monitoring by gel phase ^{31}P NMR. *Tetrahedron Lett.* **36**: 9253–9256. Svensson, A., Fex, T. and Kihlberg, J. (1996) Use of ^{19}F to evaluate reaction in solid phase organic synthesis. *Tetrahedron Lett.* **37**: 7649–7652. Swayze, E. E. (1997) The solid phase synthesis of trisubstituted 11,4-diazabicyclo[4.3.0]nonnan-2-one scaffolds: on bead monitoring of heterocycle forming reaction using ^{15}N NMR. *Tetrathedron Lett.* **38**: 7649–7652.

36. Swali, V., Langley, G.J. and Bradley, M. (1999) Mass spectrometry analysis in combinatorial chemistry. *Curr. Opin. Chem. Biol.* **3**: 337–341.

8

THE CONTRIBUTION OF MOLECULAR BIOLOGY TO DRUG DISCOVERY

Kenton H. Zavitz and Adrian N. Hobden

Molecular biology will continue to provide drug research, with extraordinary analytical methods and lend a richer texture to our imaginations

James W. Black, *Nobel Lecture*, December 8, 1988

I INTRODUCTION

Over the past fifty years, the pharmaceutical industry has been extremely successful in its search for new and improved medicines. However, a quick survey of the world's best-selling drugs rapidly reveals that the majority are small molecules that were discovered by using natural product screening, medicinal chemistry and animal testing but without the aid of molecular biology. If that technology existed and was so successful, why do we need molecular biology? Of course, we should not forget that genetic engineering is a relatively new science dating only from 1975[1] and the process of drug discovery, refinement and testing can take a long time. It is, therefore, not surprising that the current drugs are just beginning to reflect the revolution that has occurred in the pharmaceutical and biotechnology industries. It is very unlikely that any of tomorrow's drugs will not have benefitted from molecular biology at some stage in their discovery. Indeed, for most,

molecular biology will have been used, directly or indirectly, at all stages in the drug discovery process.

In its infancy, genetic engineering was considered to be useful only for the production of therapeutic proteins. Many companies, e.g. Genentech and Biogen, were founded solely with that objective in mind. However, proteins do not make ideal drugs, being difficult to administer, rapidly cleared and potentially immunogenic. Despite these disadvantages, a rapidly increasing number of "biopharmaceuticals", including recombinant proteins, therapeutic monoclonal antibodies, polyclonal antibodies and even antisense oligonucleotides (e.g. Vitravene for CMV retenitis), have been approved by the FDA (26 new products between 1995 and 1999) for indications ranging from metastatic breast cancer (Herceptin) to rheumatoid arthritis (Remicade, Enbrel).[2] These biopharmaceutical therapies have been made possible by advances in molecular biology that allow the routine cloning of genes, expression of the corresponding proteins and the purification of the resulting product in commercially viable quantities, as well as a favorable regulatory environment. Nonetheless, the pharmaceutical industry has begun to recognize a much bigger role for molecular biology than simply expressing therapeutic proteins. There cannot now be a pharmaceutical company large or small that does not employ considerable numbers of staff with expertise in these diverse techniques.

It is not the intention of this chapter to describe, in great detail, the techniques of molecular biology. There are numerous specialized textbooks available to those who wish to learn them. Nor do we want to describe the process of drug discovery. That is covered elsewhere within this book. Rather, this chapter will illustrate the various uses of molecular biology in the various stages of the drug

The Practice of Medicinal Chemistry
ISBN 0-12-744481-5

Table 8.1 *Uses of molecular biology in drug discovery*

Dissection of disease etiology
Molecular diagnostics
Target identification and validation
Therapeutic proteins and monoclonal antibodies
Protein structure determination
Provision of reagents for screening
Screening organisms
Transgenic animals
Drug metabolism
Toxicology

discovery process (Table 8.1). Some of these applications are well established, almost mature for such a young science, others are just now being applied and still more applications will be conceived of and brought to fruition in the future. The essence of pharmaceutical research is innovative thought and competition. The winners will be those who have the best ideas and can most rapidly exploit them by bringing a drug to market. Molecular biology has become critical to that process.

II DISSECTION OF DISEASE ETIOLOGY

It is self-evident that all drug discovery programs require a disease and a therapeutic target. In the past, that target did not need to be defined too closely. Antibacterial therapy required the discovery of a compound, often a natural product, that killed the organism. The exact molecular target did not need to be known. However, the limitations of this approach have become apparent as pathogenic strains of bacteria remained immune to the best cephalosporin or as previously sensitive strains became resistant. It has become obvious that an understanding of antibiotic resistance is required in order to overcome it and that new antibacterial targets are required. Equally, many of today's drugs effectively combat the symptoms of disease, e.g. acetyl-cholinesterase inhibitors for Alzheimer's disease, but do nothing to modify the causes leading to the development of the condition. In other diseases (e.g. certain forms of cancer), knowledge of the precise cause of an individual's tumor at the molecular level can be used to determine the optimal course of treatment (choice of drug therapy, surgery, etc.). For example, the monoclonal antibody therapy Herceptin is recommended for use in the approximately 25% of breast cancer patients who overexpress the HER2 receptor, as determined by immunohistochemical analysis. Thus, the technologies that are driving the drug discovery process are also providing molecular diagnostics to identify the patients who will benefit most from the use of these new medicines.

Molecular biology has enormously expanded our ability to explore disease processes, to dissect the etiology of these diseases, to diagnose individual patients with unprecedented precision and, ultimately, to identify new molecular targets for drug discovery.

There are some 5000 known inherited disorders in man[3] leading to a wide range of diseases. In general, these diseases are so rare that a drug discovery program to treat these specific conditions cannot be commercially viable. However, identification of the gene causing a rare familial disorder can provide a starting point for the discovery of the biochemical pathway involved in the more common nonfamilial (sporadic) forms of the disease and thereby identify novel drug targets of great commercial value. For example, familial adenomatous polyposis (FAP) is a rare dominantly inherited syndrome that accounts for less than 1% of colorectal cancers in the USA. When it was discovered that FAP is caused by inherited mutations in the APC gene,[4] it quickly became apparent that somatic mutations of APC occur in the vast majority of sporadic colorectal tumors. Furthermore, identification of the proteins that bind to APC provided direct links to the Wnt signaling pathway and ultimately to the regulation of transcription of known oncogenes via β-catenin/Tcf transcription complexes.[5] Many components of this biochemical pathway are now actively being targetted by the pharmaceutical industry for the development of new colon cancer drug therapies.

In addition to the rare genetic disorders, it is well known that many common diseases show familial predispositions. These are the so-called polygenic diseases where one cannot point to a simple Mendelian inheritance of the disease. Rather, inheritance of a specific mutation in a particular gene leads to a predisposition to the disease. A simple example is loss of the retinoblastoma gene, which as the name implies, predisposes to cancer—often first presenting in the eye.[6,7] Since cancer is believed to be a multistep process resulting from a series of somatic mutations and leading to uncontrolled cell division, then any individual who is born without one gene required to regulate the process of cell division, presumably needs to accumulate a fewer number of mutations. Hence, the disease can appear earlier than it might in a normal individual. Equally, we know that many genes are involved in the control of blood pressure and that particular mutations in some of these genes predispose the individual to hypertension. A greater knowledge of all the genes involved in a disease and how they interact will allow us to define the critical new drug targets of the future.

III DRUG TARGET IDENTIFICATION

Perhaps the greatest impact of the molecular biology revolution is in the realm of drug target identification.

With the completion of the sequencing of the human genome, as well as the genomes of numerous human pathogens and important biological model systems (e.g. mouse, fruit fly (*Drosophila*) and worm (*C. elegans*)), the explosion of potential drug targets represents an unprecedented challenge and opportunity to the pharmaceutical industry. As with the process of DNA sequencing itself, a vast array of molecular biological techniques have been automated and industrialized to generate enormous amounts of data detailing, for example, in which tissue and to what extent each gene is expressed (expression profiling using DNA microarray hybridization),[8] comprehensive maps of protein–protein interactions (yeast two-hybrid systems, described below) and the effects of systematically over-expressing or reducing the expression (e.g. via antisense) of hundreds or even thousands of individual genes in a single experiment (microarrays of transfected cells).[9] Pharmaceutical companies, large and small, currently apply enormous resources to the generation and interpretation of these types of data in an effort to identify and prioritize the most promising of these new drug targets for development. It will be several years before we truly see if these efforts have paid off.

As an example of the application of these new approaches, we will examine the yeast two-hybrid system for the identification of protein–protein interactions within a cell. This system consists of a yeast genetic assay in which the physical association of two proteins is measured by the reconstitution of a functional transcriptional activator to drive a reporter gene such as β-galactosidase or an auxotrophic marker for selection[10,11] (Fig. 8.1). In general, a protein of known function, such as a disease causing protein from an inherited syndrome, is fused to the DNA-binding domain of a transcription factor, for example, GAL4. This hybrid "bait" protein is introduced into a yeast strain along with a library of human "prey" proteins fused to a transcriptional activation domain. Activation of the reporter gene indicates a direct interaction has occurred between the "bait" protein and the selected "prey" protein which is then easily recovered and identified by DNA sequence. Since its introduction in 1989, a huge number of interacting proteins from a vast variety of studies have been used by small and large laboratories alike to piece together previously undiscovered biological pathways. For example, this methodology was instrumental in tying the APC colon cancer gene, described above, into the Wnt/β-catenin signal transduction pathway. The advantages of the yeast two-hybrid system include its rapidity, low cost, robustness and applicability to virtually any protein of interest, and its adaptability to automation and high-throughput methodologies. The difficulty comes in attempting to determine which of this vast array of interactions is of biological relevance and, furthermore, which interactions can be linked to the disease state of interest. The necessary target

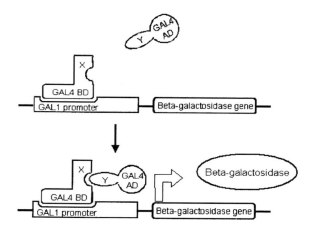

Fig. 8.1 The yeast two-hybrid system. Two hybrid proteins are expressed in yeast. The "bait" protein consists of protein X fused to the DNA-binding domain of the GAL4 transcription factor (BD). The "prey" protein consists of protein Y fused to the GAL4 transcriptional activation domain (AD). Neither hybrid alone is able to activate transcription of the reporter gene (β-galactosidase in this example). However, if a protein–protein interaction occurs between protein X and protein Y, a functional transcriptional activator is generated resulting in expression of β-galactosidase which is detected as a color change phenotype of the host yeast cells growing in the presence of a suitable substrate (e.g. X-gal).

validation steps developed thus far are of greatly lower-throughput. Current efforts to improve this situation include the development of human cell-based assays which provide a suitable biological readout (e.g. apoptosis or cell adhesion for carcinogenesis) coupled to automated methodologies to express full-length or smaller domains of the candidate proteins of interest.[9] These approaches will allow for the functional analysis of many candidate drug targets in parallel.

An additional development of enormous potential is that variations of the yeast two-hybrid system are now themselves being used as drug screens.[12] Such a screen is designed to detect small molecules capable of specifically affecting the association of the two target proteins in yeast. Recently, a high-throughput yeast two-hybrid screen successfully identified small molecules which disrupt the interaction between the α1B and β3 subunits of the N-type calcium channel. These compounds were subsequently found to inhibit selectively the activity of the N-type calcium channel in neurons in culture and may thus serve as the basis for a structure–activity synthetic chemistry program.[13] The success of this approach raises the exciting and daunting possibility that, in addition to the traditional drug targets of G-protein-coupled receptors (60% of drugs on the market), ion channels, nuclear hormone receptors and enzymes (proteases, kinases), virtually any protein, even those of unknown biochemical function, may serve as a viable target for therapeutic intervention. These novel classes of drug targets will certainly present new challenges to the practice of medicinal chemistry in the future.

IV PRODUCTION OF PROTEIN

As mentioned in the introduction, the ability to move DNA from man to bacteria, or indeed, from bacteria to man, made it possible, suddenly, to do what had previously been impossible. Human proteins could be produced in sufficient quantities for it to be possible to use them as drugs. The first commercial example was human insulin, which has now taken over from porcine insulin as the drug of choice for type 1 diabetics.

The techniques of molecular biology started to reveal a whole range of proteins that could be used as drugs. However, their structure did not always allow successful production in bacteria. As a general rule, *Escherichia coli* do not readily secrete proteins nor will they glycosylate them. As a consequence, if the protein required a large number of specific disulphide bonds or glycosylation for activity, *E. coli* were unsuitable hosts for their production. Although it was possible to produce the protein, it was unfolded and usually precipitated within the cell. No amount of protein re-folding *in vitro* could produce reasonable quantities of active product. It became necessary to use other protein expression systems. Nowadays, we have a vast array of systems from which to choose, each with its own advantages and disadvantages. For example, the yeast *Saccharomyces cerevisiae* is easy to grow and to manipulate genetically and will secrete proteins. However, quantities of secreted protein tend to be low and the glycosylation profile of proteins secreted from yeast is distinct from that of mammalian cells. Most of the therapeutic proteins currently on the market, e.g. erythropoietin, G-CSF and tPA, are produced in mammalian cell expression systems. Obviously these cells will secrete and glycosylate the protein in a manner similar to the natural protein. However, the cells are harder to manipulate and much more expensive to grow than their microbial counterparts. Furthermore, the expression levels have until recently been relatively low. Newer expression systems based upon viruses have started to make expression in complex eukaryotic cells much more straightforward due to the ease of getting foreign DNA into the cells and the high level of expression of recombinant protein following infection of the cells. Particular systems of great merit are baculovirus,[14] which will infect certain insect cells and the semliki forest virus system,[15] which has a very broad host range allowing a large number of different cell lines to be used.

While therapeutic proteins were an obvious use for the technology, it is evident that any protein can be produced provided the right system is chosen. Drug discovery requires that, if small molecules are the objective, they should work against the correct target, i.e. the human protein or specific viral enzyme. Genetic engineering often provides the only mechanism to acquire sufficient protein for high-throughput screening or X-ray crystallography to facilitate rational drug design. The technology is in routine use to supply proteins for these purposes. However, expression of recombinant proteins is not always a neutral event for the host cell. As an example, some years ago we attempted to express HIV protease in *E. coli* in order to acquire sufficient material for high-throughput screening. It proved impossible, however, to express large quantities since the moment the cell was induced to make the HIV protease it stopped growing (Fig. 8.2). If the active site aspartic acid was mutated to asparagine (making the protease inactive), then large quantities of protein were produced by actively growing cells. It became apparent that production of active protein prevented cell growth, presumably because of the protease activity of the recombinant product and it occurred to us and to others[16] that it was unnecessary to purify large quantities of HIV protease to use in some biochemical screen for inhibitors of the enzyme. The recombinant *E. coli* could act as the screen. The bacteria would grow while expressing HIV protease provided an inhibitor of the enzyme was present.

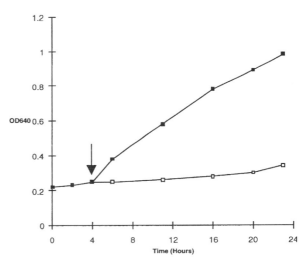

Fig. 8.2 Growth response of *E. coli* to the expression of HIV protease. *E. coli* containing the HIV protease gene were either induced (□) or uninduced (■) to express HIV protease at 4 hours (arrow) and their subsequent growth measured. The growth phenotype of such an *E. coli* strain could be used as a screen for inhibitors of the HIV protease.

The use of molecular biology to alter the phenotype of a host cell and, thus, permit the use of that cell in high-throughput screening is the subject of the next section.

V   WHOLE CELL SCREENS

In the process of drug discovery, we can envisage two phases of compound screening. The first phase is the screening of vast numbers of randomly selected compounds, whether as single molecular entities, e.g. from companies' compound libraries, or as mixtures of compounds such as may be present in multisynthesis chemical libraries or in microbial broths. In either case, the primary objective is to run through as many compounds, broths, etc., in as short a period as possible to identify a few molecules that may act as leads for further chemical synthesis. It does not matter, at this stage, whether the compounds are particularly potent or selective. Medicinal chemistry will address these issues. However, if possible, the screen should avoid identifying too many false positives. They can be time-consuming to identify and eliminate.

In the second phase, the lead compound has been identified and the medicinal chemist is seeking, in collaboration with pharmacologists, to identify a potent, selective compound to take forward into development. At this stage, it is important that the assay is predictive of what will be seen when the compound ultimately is tested in man.

Molecular biology has been used extensively for both these activities. Its utility appears limited only by the imagination and inventiveness of the molecular biologists. Below are a few examples of its use. They are but the tip of a large iceberg of work, most of which goes unreported by the pharmaceutical industry.

VI   INTRACELLULAR RECEPTORS

We have come to recognize that the intracellular receptor gene family is both large and diverse. Its best-characterized members are the sex hormone receptors, estrogen, progesterone and testosterone, but also included are receptors for corticosteroids, vitamin D_3, thyroxine and retinoids. In addition, molecular biology has revealed a number of "orphan" receptors, i.e. proteins known to be produced that carry a sequence motif suggestive of the ability to bind a small molecule, but for which the ligand is currently unknown.

This family of receptors expressed, as their name suggests, within the cell are already the targets for many drug discovery programs. For example, tamoxifen is widely used for the treatment of breast cancer and is an antagonist of the estrogen receptor. Many synthetic analogs of corticosteroids are used in asthma treatment. The estrogen receptor is present in the cytoplasm in association with HSP90, a heat-shock protein. Upon binding estrogen, this complex dissociates and the receptor enters the nucleus where it binds to specific DNA sequences and activates transcription of certain genes. This chain of events has been reconstructed in the yeast, *S. cerevisiae*.[17] The estrogen-responsive DNA sequence was inserted into a yeast promoter upstream of a reporter gene. The reporter, in this case β-galactosidase, is usually an enzyme whose presence can be detected simply by a colorimetric indicator. The effect of inserting the DNA sequence into the yeast promoter is to render it inactive until bound by an estrogen receptor/estradiol complex. Obviously, therefore, the yeast must also express the receptor. With this combination of receptor, responsive element and indicator (Fig. 8.3), the yeast is ready to be used as a screen for estrogen agonists or antagonists. Similar systems have been reported for corticosteroids[18] and androgens.[19]

Of course, the above is a rather simplistic description of the screen. In reality, the screener is seeking to achieve stability and sensitivity in the screen. The recombinant yeast must, therefore, be "fine-tuned" to ensure that the "foreign" DNA is not lost upon frequent growth of the cells and the concentration of estrogen receptor is sufficient, but not too high, so as to detect small quantities of active material. Once a therapeutic opportunity has been defined for the growing number of newly discovered orphan receptors, it is likely that agonists and antagonists will be sought using this technology.

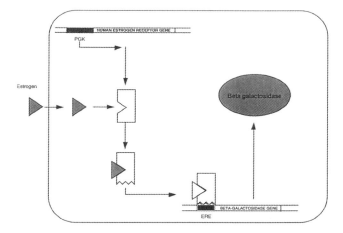

Fig. 8.3 Yeast-based screen for agonists or antagonists of the human estrogen receptor. The hormone estrogen (estradiol) binds to the estrogen receptor which is expressed from a gene driven by the PGK promoter. The hormone–receptor complex binds to an estrogen-responsive element (ERE) that controls the expression of the β-galactosidase reporter gene. The assay measures the activity of the enzyme using a substrate that forms a colored product on conversion.

VII INTRACELLULAR ENZYMES

It has been previously mentioned that the expression of HIV protease within *E. coli* gives rise to a phenotype. In a similar fashion, it has been observed that phosphodiesterases (PDEs) when expressed in yeast affect the cells. These enzymes function to modulate intracellular concentrations of the cyclic mononucleotides cAMP or cGMP. Yeast has two endogenous genes encoding PDEs which, when deleted, lead to elevated levels of cAMP within the cell. The consequence, to the yeast, of elevated cAMP is increased sensitivity to heat shock and inability to utilize acetate as sole carbon source. These yeast mutants may be complemented by the human PDE gene and the phenotype reversed (Fig. 8.4). The use of such yeast in the search for inhibitors of PDEs with utility in, for example, asthma has been proposed[20] and certainly works with the known type IV PDE inhibitor, rolipram.

In a similar fashion, it is evident that the estrogen screen described above, could be modified to include enzymes required for the synthesis or degradation of estradiol. An alternative therapeutic objective for estradiol inhibition might be to prevent its synthesis. Thus, a yeast strain already built to be sensitive to estradiol could be supplied instead with the precursor to estradiol, 19-nortestosterone, and the enzyme, aromatase, required for its conversion to estradiol. An inhibitor of the enzyme would, therefore, lead to the inability to synthesize estradiol and the loss of production of β-galactosidase.

A major potential objection to the above approaches is that the compound is required to cross the yeast cell wall and membrane. Failure of a compound to do so would lead it to not being identified in this type of screen. Obviously, an *in vitro* biochemical screen does not suffer from this constraint. There is no simple argument to counter this objection but a series of observations should allow the reader to make some judgement on the relative merits of the two approaches. Firstly, biochemical assays can be expensive and complicated preventing their use in high-throughput screening. Secondly, screening of random compounds rarely results in a complete failure to identify leads. Rather, it is often difficult to decide which, of a series of structurally diverse but relatively inactive leads, should progress into medicinal chemistry. The mechanism by which compounds enter cells is poorly defined but there is considerable overlap in their ability to cross microbial and mammalian cell membranes. Starting with a compound already able to cross the membrane may well be advantageous to the medicinal chemist.

There are, of course, a number of targets for drug discovery that are not located within the cell. Rather, they are located within the cytoplasmic membrane where they serve to tell the cell about its environment. They are the cell surface receptors and have been, over the years, the targets of many of the world's best-selling drugs.

VIII G-PROTEIN-COUPLED RECEPTORS

The G-protein-coupled receptors (GPCRs) are a superfamily of structurally related proteins, located in the cell membrane and consisting of seven transmembrane

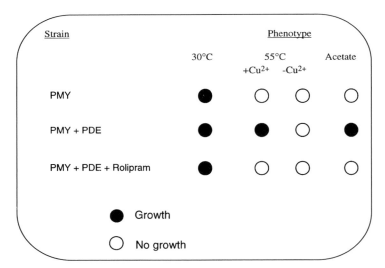

Fig. 8.4 Yeast-based screen for inhibitors of human phosphodiesterase IV. A phosphodiesterase (PDE)-deficient yeast (PMY) will not utilize acetate as a sole carbon source and is sensitive to heat shock (55°C). Complementation with a human type IV PDE (PMY + PDE) expressed from a copper-dependent (CUP1) promoter reverses the mutant phenotype. Addition of type IV PDE inhibitor (rolipram) to the complemented yeast restores the mutant phenotype (PMY + PDE + rolipram).

segments. Their primary amino acid sequence, however, can be quite diverse. Agonists or antagonists acting at these receptors constitute a large number of today's best-selling pharmaceuticals. Examples include the H_2 antagonists for ulcer therapy, β-blockers for hypertension, β-agonists for asthma and serotonin agonists for migraine.

In addition to the extensive families of these receptors that have small molecules as their agonists (e.g. histamine, prostaglandins, acetylcholine), many have peptides or even proteins as their ligand (e.g. angiotensin II, gastrin, luteinizing hormone). There is, in addition, an extensive collection of "orphan" 7-transmembrane receptors, identified by molecular biology techniques but for which a ligand has not yet been identified. Table 8.2 gives an impression of the diversity of this family.

There is enormous activity worldwide seeking to identify nonpeptide agonists or antagonists for both the peptide receptors and the orphans, since it is expected that this will be a fruitful area for drug discovery – witness the success of the nonpeptide angiotensin II receptor antagonists (collectively known as the "sartans") approved several years ago for the treatment of hypertension.

The standard approach to finding such molecules has been to express the cloned human receptor in mammalian cells and look for molecules able to inhibit ligand binding. This method can be successful, as with the angiotensin II receptor antagonists, however, it is most useful for identifying antagonists (rather than agonists) and requires both the ligand to be known and for a radio-labeled ligand

derivative to be available. Recently, two novel approaches have been reported, which potentially, should facilitate the whole process.

The first system again makes use of yeast. It has been known for some time that *S. cerevisiae* can exist as two sexual types, a cells and α cells, which communicate with each other via sex pheromones, a-factor and α-factor. The receptors for these two pheromones are members of the 7-transmembrane family, although their amino acid sequences are quite distinct from their mammalian counterparts. The consequence of the binding of the pheromone to its receptor is to set in motion a complex set of biochemical events that lead, ultimately, to mating of the two opposite cell types. However, there are two principal events that can readily be detected. The cells undergo rapid, but transient, cell cycle arrest and express on their cell surface a variety of proteins that aid in fusion of the mating types. Unlike their mammalian counterparts, the intracellular signal is transmitted via the β- and γ-subunits of the trimeric G-protein complex and not by the α-subunit. A detailed description of this pathway can be found elsewhere.[21] More importantly, from the point of view of this chapter, the system has been engineered so that the yeast expresses the human $β_2$-adrenergic receptor and its cognate Gsα-subunit instead of the yeast homologs.[22] The yeast responds to the presence of a $β_2$-agonist by inducing the FUS1 promoter which, in turn, has been connected to a β-galactosidase reporter gene.

Table 8.2 *Diversity of ligands for cloned G-protein-coupled receptors*

Ligand	Example
Amines	Adrenalin
	Histamine
	Dopamine
Protein hormones	Luteinizing hormone
	Follicle-stimulating hormone
Peptide hormones	Angiotensin
	Bradykinin
	Substance P
	Endothelin
	Gastrin
Sensory stimuli	Light
	Odorants
	Calcium ions
Others	Thrombin
	Low-density lipoprotein
	cAMP

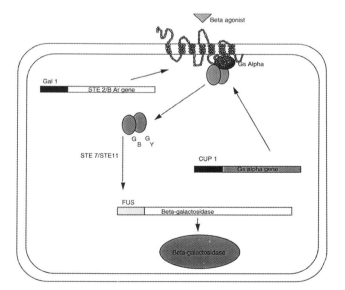

Fig. 8.5 Yeast-based screen for $β_2$-agonists. The $β_2$-adrenergic receptor (B Ar) expressed from a GAL1 promoter links to the mating type response via the human Gsα-subunit expressed off the CUP1 promoter, by complementation of GPA-1. The detection of signaling is by induction of the FUS promoter linked to a β-galactosidase reporter gene.

The yeast, therefore, turn blue in the presence of the indicator 5-bromo-4-chloro-3-indolyl-β-D-galactopyranoside (X-gal) (Fig. 8.5).

It is evident that this yeast strain could be used to look for agonists or antagonists of this receptor and, because yeast cells can be grown rapidly and inexpensively, such a system has the potential to be used for very high-throughput screens. Indeed, a series of novel and selective peptide agonists of an orphan GPCR were identified in a screen conducted in a similar yeast system designed to couple receptor activation to histidine prototrophy as a selectable marker.[23] Many companies are seeking to exploit similar technology for their favorite 7-transmembrane receptors, however, it is worth pointing out that these are very complex yeast strains to construct and not all receptors will be amenable to this approach.

As an alternative approach, the second system uses frog melanophores. This is an immortalized cell line derived from frog melanophores that responds to melanocyte-stimulating hormone or melatonin by, respectively, a dispersal or aggregation of pigment.[24] The consequence of addition of these ligands, which act at 7-transmembrane receptors, is that the cells change color within 30 minutes. Indeed, a dose–response curve for agonists or antagonists can be constructed using these cells in a 96-well plate in combination with an ELISA plate reader. Human GPCRs, such as the β_2-adrenergic receptor, as well as chemokine receptors and tyrosine kinase receptors, have been expressed in these cells to generate a screenable bioassay for the identification and characterization of novel agonists and antagonists of these receptors.[25]

The system is especially powerful when one wishes to determine structure–activity relationships for compounds derived in a medicinal chemistry program. Dose–response curves can be constructed versus the human receptor in less than 1 hour. However, the system is not without its problems. Firstly, the cells need special conditions to grow. They are amphibian and, therefore, need lower temperatures and a frog-derived growth factor supplement. Secondly, they have an endogenous background of a variety of frog receptors that may complicate analysis. Thirdly, they are difficult to transfect with exogenous DNA. Nonetheless, the system has great promise and may even have application for high-throughput screening of random compounds.

In the process of drug discovery, one can imagine running the initial lead discovery part of the program in yeast, then switching to frog melanophores for the lead refinement stage. There is still, of course, a requirement that the compound will work *in vivo*. This is a combination of drug absorption, excretion and metabolism activity of the compound at its target *in vivo* and a lack of other activities, i.e. toxicity. Molecular biology techniques are starting to address all these issues.

IX TRANSGENIC ANIMALS

It has become possible to manipulate the genetic material of mice so that extra genes can be expressed or mouse genes deleted. Obviously, this also allows for mouse genes to be replaced by their human homologs. It is only a matter of time before it will be possible to do this for other animal species. Such manipulations make it possible to screen *in vivo* against the human target or predispose the animal to a particular disease. For example, a mouse has been constructed which develops breast cancer at a high frequency.[26] Because the change that has induced the phenotype is known, it allows drugs directed against the tumor-promoting gene to be tested. Equally, it is no longer necessary to treat animals with carcinogens in order to induce tumors.

Mice and other animal models of disease are often poor mimics of the human condition. However, the expression of human genes in these animals can initiate development of the disease. In mice, the distribution of cholesterol between low-density lipoproteins (LDL) and high-density lipoproteins (HDL) is quite distinct from that in humans. However, if the human enzyme, cholesterol ester transfer protein (CETP) is expressed in mice, the ratio of LDL:HDL becomes more human in profile.[27] Inhibition of CETP is a target for antiatherosclerotic drugs and there now exists an animal model in which to test them.

Rodents are used extensively in toxicity evaluation of drugs and transgenic mice are now being considered for this use. This area is addressed later.

X DRUG METABOLISM

A major problem in all drug discovery programs is to discover compounds with good pharmacokinetics. Although it is possible to examine the metabolism of the drug in animals, it has often been difficult to predict what would happen in man. The obvious implications of drug metabolism are an effect on half-life *in vivo* and the production of toxic metabolic products.

In seeking to establish an effective dose for a new drug, the clinician needs to know what ranges of abilities humans will have to metabolize the drug and what effect the drug will have on the metabolizing enzymes. Failure to metabolize the drug may lead to overdose, whereas rapid metabolism could lead to lack of clinical benefit. Equally, inhibition of the metabolism of another drug could cause problems in a patient receiving several medications.

A large proportion of the metabolizing enzymes are members of the P450 superfamily[28] and a large number of these genes have now been cloned and their metabolic potential determined. Increasingly, the enzymes are being expressed in microbial systems, e.g. yeast, where their ability to metabolize the drug can be evaluated. In a few years, it would not be surprising if all new drugs were "typed" for their complete P450 metabolism profile. Equally, their metabolic products can be identified and their biological activity and toxicity determined.

An additional application is likely to be the P450 genotyping of patients. As "poor metabolizers" become recognized in the population, the problem is often found to be mutations in one or more of their P450 genes. Once identified, such mutations are easily screened for and it is entirely likely that some degree of P450 "profiling" will take place for patients in the future. Armed with knowledge on the metabolic fate of new drugs, the physician will then be able to prescribe the best drug for an individual depending on their P450 profile. This individualization of drug therapy based on genetic information is known as pharmacogenomics. There is a massive effort currently underway to identify and characterize polymorphisms in a wide variety of genes, including drug receptors and effectors, in addition to drug metabolizing enzymes. It is hoped that the correlation of these polymorphisms with clinical outcomes and drug effects across a population will allow for the prediction of the safety, efficacy and toxicity of both established drugs and new drugs in development and, thereby, a reduction in the size and expense of clinical trials.[29,30]

XI TOXICOLOGY

Toxicity testing for new drugs is a legal requirement of drug discovery and, of course, reflects our ignorance of biological processes. Toxicity constitutes the unwanted effects of the molecule. While it is hard to imagine that long-term testing of compounds in animals will not always need to be performed, molecular biology is starting to impact on genetic toxicology, i.e. the ability of compounds to induce mutations in DNA and, thus, to act as potential carcinogens. Systems have been constructed that permit identification of genetic mutation *in vitro* and *in vivo* extremely rapidly, therefore, a compound's potential as a carcinogen can be identified with concomitant savings in numbers of animals, human effort and the supply of compound needed for the larger-scale animal studies.

Most systems reported so far depend upon the detection of mutations either in an indicator gene, e.g. β-galactosidase, or a gene controlling the expression of the indicator gene.[31,32] The bacteriophage λ, which normally infects and lyses the bacteria *E. coli*, has been altered genetically such that the β-galactosidase gene is contained within its DNA. This gene will only be expressed when the phage infects *E. coli*. The phage λ DNA is then incorporated into the mouse genome such that it is inherited in subsequent generations of mice. Since the phage λ DNA is not capable of expressing any proteins in the mouse, it is effectively neutral in the mouse's growth and development. However, the λ DNA may be rescued from mouse by extracting total DNA and adding a "λ packaging extract" *in vitro*. This complex, which is commercially available, finds and extracts the λ DNA and packages it into infectious phage particles. The phages are then used to infect *E. coli* where they replicate and lyse the bacteria. The lysed "plaques" can be stained for β-galactosidase. Mutations within the λ DNA are scored by the proportion of plaques scoring negative for β-galactosidase.

In practice, the transgenic mice are given the new drug over a period of a few days before sacrifice. DNA is then extracted from a variety of organs and the mutation frequency scored by counting λ plaques. It therefore becomes possible to demonstrate the ability of the drug to induce mutations and determine whether it shows any tissue selectivity. Furthermore, since the λ packaging reaction can be repeated several times for each DNA sample, far fewer animals are required to obtain a statistically significant result.

It can be expected that several refinements to this system will be developed with time allowing, for example, the scoring of mutation frequency to be automated.

XII CONCLUSIONS

Molecular biology has already had an enormous impact on the process of drug discovery and its influence will certainly increase in the future. The power of these technologies will facilitate the discovery and development of novel pharmaceuticals and shorten the time and cost from idea to market. The pharmaceutical industry is very competitive and companies will prosper through a combination of hard work, innovation and serendipity. The companies that fully adopt the diverse techniques of molecular biology as part of their drug discovery strategy and use them in imaginative and inventive ways will ultimately find that serendipity has a less important role to play.

REFERENCES

1. Cohen, S.N. and Boyer, H.W. (1980) Process for producing biologically functional molecular chimeras. US Patent 4 237 224 (Stanford University).
2. Reichert, J.M. (2000) New biopharmaceuticals in the USA: trends in development and marketing approvals 1995–1999. *Trends Biotechnol.* **18**: 364–369.
3. McKusick, V.A. (1990) *Mendelian Inheritance in Man. Catalogs of Autosomal Dominant, Autosomal Recessive and x-linked Phenotypes*, 9th edn. The Johns Hopkins Press, Baltimore.

4. Groden, J., Thliveris, A., Samowitz, W., Carlson, M., Gelbert, L., Albertsen, H., Joslyn, G., Stevens, J., Spirio, L. and Robertson, M. (1991) Identification and characterization of the Familial Adenomatous Polyposis Coli gene. *Cell* **66**: 589–600.

5. Kinzler, K.W. and Vogelstein, B. (1996) Lessons from hereditary colorectal cancer. *Cell* **87**: 159–160.

6. Lee, W.H., Bookstein, R., Hong, F., Young, L.J., Shaw, J.Y. and Lee, E.Y. (1987) Human retinoblastoma susceptibility gene: cloning, identification and sequence. *Science* **235**: 1394–1399.

7. Ewen, M.E. (1994) The cell cycle and the retinoblastoma protein family. *Cancer Metastasis Rev.* **13**: 45–66.

8. Brown, P.O. and Botstein, D. (1999) Exploring the new world of the genome with DNA microarrays. *Nat. Genet.* **21**: 33–37.

9. Ziauddin, J. and Sabatini, D.M. (2001) Microarrays of cells expressing defined cDNAs. *Nature* **411**: 107–110.

10. Fields, S. and Song, O. (1989) A novel genetic system to detect protein–protein interactions. *Nature* **340**: 245–246.

11. Bartel, P.L. and Fields, S. (1997) *The Yeast Two-hybrid System.* Oxford University Press, New York, NY.

12. Vidal, M. and Endoh, H. (1999) Prospects for drug screening using the reverse two-hybrid system. *Trends Biotechnol.* **19**: 374–381.

13. Young, K., Lin, S., Sun, L., Lee, E., Modi, M., Hellings, S., Husbands, M., Ozenberger, B. and Franco, R. (1999) Identification of a calcium channel modulator using a high-throughput yeast two-hybrid screen. *Nat. Biotechnol.* **16**: 946–950.

14. Miller, L.K. (1989) Insect baculoviruses: powerful gene expression vectors. *Bioessays* **11**: 91–95.

15. Liljestrom, P. and Gargoof, H. (1991) A new generation of animal cell expression vectors based on the semliki forest virus replicon. *Bio/Technology* **9**: 1356–1361.

16. Baum, E.Z., Bebernitz, G.A. and Gluzman, Y. (1990) Isolation of mutants of human immunodeficiency virus protease based on the toxicity of the enzyme in *Escherichia coli. Proc. Natl Acad. Sci. USA* **87**: 5573–5577.

17. Metzger, D., White, J.H. and Chambon, P. (1980) The human oestrogen receptor functions in yeast. *Nature* **334**: 31–36.

18. Schena, M. and Yamamoto, K.R. (1988) Mammalian glucorticoid receptor derivatives enhance transcription in yeast. *Science* **241**: 965–967.

19. Purvis, I.J., Chotai, D., Dykes, C.W., Lubatin, D.B., French, F.S., Wilson, E.M. and Hobden, A.N. (1991) An androgen inducible expression system for *Saccharomyces cerevisiae. Gene* **106**: 35–42.

20. McHale, M.M., Cieslinski, L.B., Eng, W.-K., Johnson, R.K., Trophy, J.J. and Livi, G.P. (1991) Expression of human recombinant cAMP phosphodiesterase isozyme reverses growth arrest phenotypes in phosphodiesterase-deficient yeast. *Mol. Pharmacol.* **39**: 109–113.

21. Konopka, J.B. and Fields, S. (1992) The pheromone signal pathway in *Saccharomyces cerevisiae. Antonie Van Leeuwenhoek* **62**: 95–108.

22. King, K., Dohlman, H.G., Thorner, J., Caron, M.G. and Lefkowitz, R.J. (1990) Control of yeast mating signal transduction by a mammalian β2-adrenergic receptor and Gsα subunit. *Science* **250**: 121–123.

23. Klein, C., Paul, J.I., Sauve, K., Schmidt, M.M., Arcangeli, L., Ransom, J., Trueheart, J., Manfredi, J.P., Broach, J.R. and Schlatter, C. (1998) Identification of surrogate agonists for the human FPRL-1 receptor by autocrine selection in yeast. *Nat. Biotechnol.* **16**: 1334–1337.

24. Putenza, M.N. and Lerner, M.R. (1992) A rapid quantitative bioassay for evaluating the effects of ligands upon receptors that modulate cAMP levels in a melanophore cell line. *Pigment Cell Res.* **5**: 372–378.

25. Carrithers, M.D., Marotti, L.A., Yoshimura, A. and Lerner, M.R. (1999) A melanophore-based screening assay for erythropoietin receptors. *J. Biomol. Screen.* **4**: 9–14.

26. Muller, W.J., Sinn, E., Pattengale, P.K., Wallace, R. and Leder, P. (1988) Single step induction of mammary adenocarcinoma in transgenic mice bearing the activated *c-neu* oncogene. *Cell* **54**: 105–115.

27. Hayek, T., Azrolan, N., Verdeny, R.B., Walsh, A., Shajek-Shaul, J., Agellon, L.B., Jall, A.R. and Breslow, J.L. (1993) Hypertension and cholesteryl ester transfer protein interact to dramatically alter high density lipoprotein levels, particle sizes and metabolism. Studies in transgenic mice. *J. Clin. Invest.* **92**: 1143–1152.

28. Nebert, D.W. (1991) Proposed role of drug-metabolising enzymes: regulation of steady state levels of the ligands that affect growth, homeostasis, differentiation and neuroendocrine functions. *Mol. Endocrinol.* **5**: 1203–1214.

29. Norton, R.M. (2001) Clinical pharmacogenomics: applications in pharmaceutical R&D. *Drug Dev. Today* **4**: 180–185.

30. Phillips, K.A., Veenstra, D.L., Oren, E., Lee, J.K. and Sadee, W. (2001) Potential role of pharmacogenomics in reducing adverse drug reactions: a systematic review. *J. Am. Med. Assoc.* **286**: 2270–2279.

31. Myhr, B.C. (1991) Validation studies with Muta Mouse: a transgenic mouse model for detecting mutations *in vivo. Environ. Mol. Mutatag.* **18**: 308–315.

32. Shepherd, S.E., Sengstag, C., Lutz, W.K. and Schlatter, C. (1993) Mutations in liver DNA of *lac I* transgenic mice (Big Blue) following subchronic exposure to z-acetylaminofluorene. *Mutat. Res.* **302**: 91–96.

9

ELECTRONIC SCREENING: LEAD FINDING FROM DATABASE MINING

Lothar Terfloth and Johann Gasteiger

In silico screening of compound databases is currently one of the most popular chemoinformatics applications in pharmaceutical research. A wide spectrum of methods has been, and continues to be, developed.

J. Bajorath, *Current Drug Discovery* March 2002

I INTRODUCTION

Many drugs such as the sulphonamides introduced by Domagk or penicillin by Fleming were discovered by serendipity and not as a result of rational drug design.[1]

Up to the 1970s hypothetical acitivity models dominated the syntheses of new compounds in drug research. The biological activity of these compounds was verified by experiments with isolated organs or animals. Accordingly, the throughput was limited by the speed of the biological tests. From about 1980 on, molecular modelling and the development of *in vitro* models for enzyme inhibition and receptor-binding studies attained a growing impact on drug research. In these years, the syntheses of compounds became the time-limiting factor. Based on the progress achieved in gene technology, combinatorial chemistry and high-throughput test models, it became feasible to produce the proteins of interest and to conduct a structure-based design of ligands. Nevertheless, it turned out that many compounds designed by structure-based drug design showed inappropriate ADMET (absorption, distribution, metabolism, excretion and toxicity) properties and consequently caused their attrition in the preclinical or clinical phase.

Remarkable progress was achieved in the different disciplines involved in pharmaceutical research in the last five decades. It culminated in the elucidation of the sequence of the human genome. Today genomics, proteomics, bioinformatics, combinatorial chemistry, and ultra high-throughput screening (u-HTS) provide an enormous number of targets and data. Hence, the application of data mining methods and virtual screening obtained a growing impact on the validation of 'druggable' targets, lead finding and the prediction of suitable ADMET profiles. In this chapter, we give a brief overview on the application of *in silico* screening and database mining for the search and optimization of leads in drug design, and provide a concise insight into the methods used for this purpose.

A The role of chemoinformatics in drug development

Some recent publications from consulting firms singled out the discrepancy between technological progress in drug design and rising costs.[2,3] The long development periods until a new compound comes to the market are joined with a high financial effort. Despite the introduction of combinatorial chemistry and the establishment of HTS, the number of new chemical entities (NCEs) introduced into the world market was about 37 NCEs in the last decade (average for the period 1991–1999; 1999, 35; 1998, 27; 1997, 39; 1996, 38; 1995, 35; 1994, 44; 1993, 43; 1992, 36; 1991, 36).[4,5] In their report, the Boston Consulting Group came to the conclusion that genomics could reduce the cost of producing a new drug from $880 million by about $335 million in an ideal case.[2] The average expected savings across the research and development pipeline under consideration of scientific and market limitations are estimated to be about $80 million. In particular, *in silico* methods are expected to speed up the drug discovery process, to provide a quicker and cheaper alternative to *in vitro* tests, and to reduce the number of compounds with unfavourable pharmacological properties at an early stage of drug development. Bad ADMET profiles are the reason for attrition of the majority of new drug candidates during the development process. Therefore, the prediction of pharmacological properties is a central task of chemoinformatics in drug development in addition to lead finding. The drug discovery process comprises the following steps:

(1) target identification
(2) target validation
(3) lead finding (including design and synthesis of compound libraries, as well as screening of compound libraries)
(4) lead optimization (acceptable pharmacokinetic profile, toxicity and mutagenicity)
(5) preclinical studies
(6) clinical studies
(7) FDA approval.

B Definition of the term lead structure

Within this chapter we use Valler's and Green's definition of lead structure.[6] According to this definition a lead structure is 'a representative of a compound series with sufficient potential (as measured by potency, selectivity, pharmacokinetics, physicochemical properties, absence of toxicity and novelty) to progress to a full drug development programme'.[6]

C Definition of the term data mining and its impact on chemoinformatics

A central problem in chemoinformatics is the establishment of a relationship between a chemical structure and its biological activity. Huge amounts of data are gathered, particularly through the synthesis of combinatorial libraries and subsequent high-throughput screening. With these massive amounts of data, tools become necessary which can navigate through these data to extract the necessary information, to search, as they say, for a needle in a haystack.

A variety of methods has been developed by mathematicians and computer scientists to address this task and this has become known as data mining. Terms and expressions closely associated with this area of data mining are data warehousing, knowledge acquisition, knowledge discovery, data harvesting, fuzzy modelling, machine learning, web farming, etc. All these words might seem equally as fuzzy as the expression 'data mining' itself. Before a definition of the term 'data mining' is given Scheme 1 shows how one can come from data through information to knowledge and illustrates this process with an example.

Fayyad defined and described the term 'data mining' as the 'nontrivial extraction of implicit, previously unknown and potentially useful information from data, or the search for relationships and global patterns that exist in databases'.[7] In order to extract information from huge quantities of data and to gain knowledge from this information, analysis and exploration has to be performed by automatic or semi-automatic methods. The data mining process can be divided into the following steps: selection, preprocessing, transformation, interpretation and evaluation. In particular, for the final two steps, the visual representation of data plays a pivotal role. Here, we will not discuss the selection and preprocessing step.

II STRUCTURE REPRESENTATION

The task of finding a lead by database mining requires the analysis of the relationships between the structure of potential new drugs and their biological activity. Due to

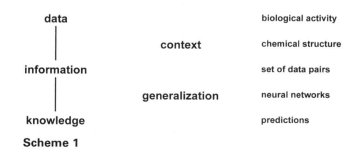

Scheme 1

the amount of data to be processed it is advisable to use a hierarchical representation of the chemical structures starting from 1D fingerprints, going further to topological descriptors such as 2D autocorrelation, and finally considering 3D structures and molecular surface properties. In this context we have to mention linear notations that convert the connection table of a molecule into a sequence of alphanumeric symbols by a set of rules. The more prominent examples are the Wiswesser Line Notation (WLN)[8] and the simplified molecular input line system (SMILES).[9,10] A structure can be searched for in a database by string matching, as long as each compound has a unique WLN or a unique SMILES[10] string. The SMILES arbitrary target specification (SMARTS) is based on the SMILES notation and is used to encode a query for substructure searches.

A Structural keys and 1D fingerprints

Structural keys describe the chemical composition and eventually structural motifs of molecules represented as a boolean array. If a certain structural feature is present in a molecule or a substructure, a particular bit is set to 1 (true) and otherwise to 0 (false). A bit in this array may encode a particular functional group (such as a carboxylic acid or an amid linkage), a structural element (e.g. a substituted cyclohexane), or at least *n* occurrences of a particular element (e.g. a nitrogen atom). Alternatively, the structural key can be defined as an array of integers, where the elements of this array contain the frequency of how often a specific feature occurs in the molecule.

In order to perform a database search the structural key of the query molecule or substructure is compared with the stored structural keys of the database entries. This implies that each array element in the structural key has to be initially defined. Therefore, this key is inflexible and can become extremely long. The choice and number of patterns included in the key effects the search speed across the database. Long keys slow down searching, whereas short keys may screen out a lot of structures that are of no interest. On the other hand, this structural key can be optimized for the compounds present in the database to be investigated.

Hence, fingerprints were developed to overcome the shortcomings of structural keys. A fingerprint is a boolean array, but in contrast to a structural key the meaning of any particular bit is not predefined. Initially all bits of a fingerprint with a fixed size are set to 0. In the second step a list of patterns is generated for each atom (0-bond path), for each pair of adjacent atoms and the bond connecting them (1-bond path), and for each group of atoms joined by longer pathways (usually paths of up to seven or eight bonds). Each pattern of the compound is assigned a unique set of bits (typically 4 or 5 bits per pattern) along the fingerprint by a

hash coding algorithm. The set of bits thus obtained is added to the fingerprint with a logical OR. Assuming a pattern is a substructure of a molecule, each bit in the pattern's fingerprint will be set in the molecule's fingerprint.

Using fingerprints, a database can be screened in the same way as with structural keys by simple boolean operations. Compared with structural keys fingerprints have a higher information density without losing specificity. Further on, the concept of folding a fingerprint was developed whereby the fingerprint is split into two equal halves and then the two halves are connected by a logical OR.

In MDL's structure database systems (ISIS and MACCS) 166 search keys, and 960 extended search keys are available.[11,12] Several types of molecular fingerprints exist depending on the substructures used to generate the fingerprints. MDL fingerprints represent the substructures defined by the MACCS search keys. Daylight fingerprints can be calculated with the Daylight Fingerprint Toolkit.[13]

B Topological descriptors

2D autocorrelation

The structural diagram of molecules can be considered as a mathematical graph. Each atom is represented by a vertex (or node) in the graph. Accordingly, the bonds are described by the edges. Problems like substructure searching[14] or the search for the maximum common substructure[15,16] of a set of ligands can be solved with algorithms developed in graph theory. Graph theory therefore influences the computer handling of structure information.

Both statistical methods and artifical neural networks need a fixed number of descriptors for the analysis of a set of molecules independent of their size and number of atoms. In order to transform the information from the structural diagram into a representation with a fixed number of components an autocorrelation function can be used:[17]

$$A(d) = \sum_{j=i}^{N} \sum_{i=1}^{N} \delta(d_{ij} - d)p_j p_i \quad \delta = \begin{cases} 1 & \forall \quad d_{ij} = d \\ 0 & \forall \quad d_{ij} \neq d \end{cases} \quad (1)$$

In this equation $A(d)$ is the component of the autocorrelation vector for the topological distance d. The number of atoms in the molecule is given by N. We denote the topological distance between atoms i and j (i.e. the number of bonds for the shortest path in the structure diagram) d_{ij}, and the properties for atom i and j are refered to as p_i and p_j, respectively. The value of the autocorrelation function $A(d)$ for a certain topological distance d results from the summation over all products of a property p of atoms i and j having the required distance d.

A range of physicochemical properties such as partial atomic charges[18] or measures of the polarizability[19] can be calculated for example with the program package PETRA.[20]

The topological autocorrelation vector is invariant with respect to translation, rotation and the conformer of the molecule considered. An alignment of molecules is not necessary for the calculation of their autocorrelation vectors.

Feature trees

The concept of feature trees as molecular descriptors was introduced by Rarey and Dixon.[21] A similarity value for two molecules can be calculated based on molecular profiles and rough mapping. In this paragraph only the basic concepts are described. For more detailed information the reader is referred to reference 21.

Molecules are represented by a tree which Rarey and Dixon called a feature tree. Within this feature tree the nodes are fragments of the molecule. The atoms belonging to one node are connected in the molecular graph. A node consists of at least one atom. Edges in the feature tree connect two nodes which have atoms in common or which have atoms connected in the molecular graph. Rings are collapsed to single nodes.

It is possible to represent molecules with feature trees at various levels of resolution. The maximum simplification of a molecule is its representation as a feature tree with a single node. On the other hand each acyclic atom forms a node at the highest level. Due to the hierarchical nature of feature trees all levels of resolution can be derived from the highest level. A subtree is replaced by a single node which represents the union of the atom sets of the nodes belonging to this subtree.

Two different kinds of features can be assigned to a node: steric or chemical features. The steric features are the number of atoms in the fragment and an estimate for the van der Waals volume. For atoms belonging to several fragments only the relevant fractional amount is taken into account. The chemical features are stored in an array and denote that the fragment has the ability to form interactions. The atom type profile considers the number of carbon or nitrogen atoms in different hybridization states, as well as the number of oxygen, phosphorus, sulphur, fluorine, chlorine, bromine, iodine or other non-hydrogen atoms. The FlexX interaction profile comprises hydrogen donors, hydrogen acceptors, aromatic ring centres, aromatic ring atoms and a hydrophobic interaction. For a pair of feature values a similarity value within the range from 0 (dissimilar) to 1 (identical) is calculated. For the comparison of two feature trees, the trees have to be matched to each other. The similarity value of the feature trees results from a weighted average of the similarity values of all matches within the two feature trees to be compared.

Further topological descriptors

In 1947, Wiener introduced a topological index.[22-24] The Wiener index or path number is calculated as the sum of all topological distances wherein the hydrogen atoms can be omitted from the molecular graph. Further topological descriptors are the Randic connectivity index,[25] the Kier-Hall connectivity indices,[26] the electrotopological state index (or *E*-state index);[27] and eigenvalue-based descriptors like Burden eigenvalues.[28]

C 3D descriptors

3D structure generation

Physical, chemical, and biological properties are related to the 3D structure of a molecule. In essence, the experimental sources of 3D structure information are X-ray crystallography, electron diffraction, and NMR spectroscopy. The Cambridge Structural Database (CSD)[29] currently comprises about 257 000 experimentally determined molecular structures of organic and coordination compounds. Nevertheless, the number of compounds with known 3D structure is small in comparison to the number of over 20 million known organic and inorganic compounds.[30] Furthermore, in drug design hypothetical molecules which still have to be synthesized, are also investigated. In particular, huge virtual libraries are generated by combinatorial chemistry. Although quantum or molecular mechanics calculation produce 3D molecular models of good accuracy and allow the calculation of a number of molecular properties, they are, up to now, too slow to process millions of compounds in a reasonable period of time. In addition, they require at least a reasonable 3D geometry of the molecule as a starting point. Therefore, automatic methods for the conversion of the 2D connectivity information into a 3D model are required. This field has been reviewed in detail by Sadowski.[31]

Two widely used programs for the generation of 3D structures are CONCORD and CORINA. CONCORD was developed by Pearlman and coworkers[32,33] and is distributed by Tripos.[34] The 3D structure generator CORINA originates from research groups,[35-38] can be tested via the internet, and is available from Molecular Networks.[39]

In general, these programs generate one low-energy conformation for each molecule. Ligands binding to a receptor may adopt conformations which differ from the calculated ones. In order to tackle the problem of conformational flexibility the conformational space has to be screened yielding a diverse set of conformations. A range of different approaches starting from a systematic grid search,[40] going further to stochastic methods,[41,42] the application of genetic algorithms,[43,44] and finally simulation methods such as molecular dynamics,[45] Monte Carlo,[46] or simulated annealing[47] address this topic. ROTATE is one program amongst others which generates a diverse and user-controlled set of conformations starting from a given 3D model.[48]

3D autocorrelation

In the calculation of a 3D autocorrelation vector the spatial distance is used as given by the following equation:

$$A(d_l, d_u) = \sum_{j=i}^{N} \sum_{i=1}^{N} \delta(d_{ij}, d_l, d_u) p_j p_i$$

$$\delta(d_{ij}, d_l, d_u) = \begin{cases} 1 & \forall \quad d_l < d_{ij} \leq d_u \\ 0 & \forall \quad d_{ij} \leq d_l \vee d_{ij} > d_u \end{cases} \tag{2}$$

Here, the component of the autocorrelation vector A for the distance interval between the boundaries d_l (lower) and d_u (upper) is the sum of the products of property p for atom i and j, respectively, having an Euclidian distance d within this interval.

In contrast to the topological autocorrelation vector it is possible to distinguish between different conformations of a molecule using 3D autocorrelation vectors.

The calculation of autocorrelation vectors of surface properties[49] is similar:

$$A(d) = \frac{1}{L} \sum_{x} p(x) \cdot p(x + d) \tag{3}$$

with the distance d within the interval $d_l < d \leq d_u$, a certain property $p(x)$, e.g. the electrostatic potential (ESP) at a point x on the molecular surface, and the number of distance intervals L. The component of the autocorrelation vector for a certain distance d within the interval $d_l < d \leq d_u$ is the sum of the product of the property $p(x)$ at a point x on the molecular surface, with the same property $p(x + d)$ at a certain distance d normalized by the number of distance intervals L. All pairs of points on the surface are considered only once.

3D molecule representation of structures based on electron diffraction code (3D MoRSE code)

In order to search for a novel code for the representation of the 3D structure of a molecule by a fixed number of variables irrespective of the number of atoms in the molecule, experimental methods used for 3D structure determination were surveyed. One of the methods employed is electron diffraction. The theoretical foundation of electron diffraction as given by Wierl[50] and some modification of this equation partly following suggestions made by Soltzberg and Wilkins[51] gives

$$I(s) = \sum_{i=2}^{N} \sum_{j=1}^{i-1} p_i p_j \frac{\sin sr_{ij}}{sr_{ij}} \tag{4}$$

with the atomic property p for the atoms i and j, a reciprocal distance s, and the distance r_{ij} between the atoms i and j. The reciprocal distance s depends on the scattering angle ϑ and

the wavelength λ following equation 5:

$$s = 4\pi\sin(\vartheta/2)/\lambda \tag{5}$$

This structure-encoding method has been applied both for the classification of a dataset comprising 31 corticosteroids for which affinity data were available in the literature binding to the corticosteroid binding globulin (CBG) receptor and the simulation of infrared spectra.[52,53]

Radial distribution function code

Closely related to the 3D-MoRSE code is the radial distribution function code (RDF code) which is calculated by equation 6:

$$g(r) = f \sum_{i=1}^{N-1} \sum_{j=i+1}^{N} p_i p_j e^{-B(r - r_{ij})^2} \tag{6}$$

where f is a scaling factor, N is the number of atoms in the molecule, p_i and p_j are properties of the atoms i and j, B a smoothing parameter, and r_{ij} the distance between the atoms i and j. $g(r)$ is calculated at a number of discrete points with defined intervals.[54,55] A slightly simplified interpretation of the radial distribution function for an ensemble of atoms is a kind of probability distribution of the individual interatomic distances. The smoothing parameter B can be regarded as a temperature factor that defines the movement of the atoms.

By including characteristic atomic properties p_i and p_j, the RDF code can be adapted to the requirements of information to be represented in different tasks. In the simplest case, solely the structure of a molecule is considered by choosing $p_i = p_j = 1$. For the study of a set of ligands binding to a receptor it may be advantageous to use properties describing the atomic partial charges or their capability to act as an electron donor or acceptor. The length or dimension of the RDF code is independent of the number of atoms and the size of a molecule, unambiguous regarding the three-dimensional arrangement of the atoms, and invariant against translation and rotation of the entire molecule.

D   Further descriptors

Besides the aforementioned descriptors, grid-based methods are frequently used in the field of QSAR (quantitative structure–activity relationships).[56] A molecule is placed in a box and the interaction energy values between this molecule and another small molecule, such as water, is calculated for an orthogonal grid of points. The grid map thus obtained characterizes the molecular shape, charge distribution and hydrophobicity.

After an alignment of a set of molecules known to bind to the same receptor, a comparative molecular field analysis (CoMFA) allows determination and visualization of molecular interaction regions involved in ligand-receptor binding.[57] Further, statistical methods such as partial least squares regression (PLS) are applied to search for a correlation between CoMFA descriptors and biological activity.

Also commonly used descriptors are some eigenvalue-based descriptors including 3D information, such as BCUT descriptors (Burden, CAS, University of Texas eigenvalues),[58] EVA,[59,60] and WHIM descriptors (weighted holistic invariant molecular descriptors).[61,62] Hopfinger *et al.*[63,64] construct 3D-QSAR models with the 4D-QSAR analysis formalism. This formalism allows both conformational flexibility and alignment freedom by ensemble averaging, i.e. the fourth dimension is the 'dimension' of ensemble sampling. A comprehensive overview about molecular descriptors is given by Todeschini and Consonni.[65]

III   DATA MINING METHODS

Data mining in essence combines methods and tools from three areas: statistics, machine learning and databases. It would be beyond the scope of this chapter to explain the methods originating from these fields. Therefore, we restrict ourselves to enumerating only the most important methods from statistics and machine learning.

Statistical methods

- Correlation analysis
- Factor analysis
- Multi-linear regression analysis (MLRA)
- Principal component analysis (PCA)
- Partial least squares regression (PLS)
- Principal component regression (PCR)

Machine learning

- Decision-tree learning or rule induction (ID3, C5)
- K-nearest neighbour analysis (KNN)
- Clustering algorithms (hierarchical and non-hierarchical clustering; k-means, Ward, Jarvis-Patrick)
- Artifical neural networks (feed-forward neural networks, self-organizing neural networks, counterpropagation neural networks, Bayesian neural networks)
- Genetic algorithms.

The learning algorithm can be supervised or unsupervised. It is reasonable to start the investigation of datasets with unsupervised learning methods.

IV   DATABASE SEARCHES

For the search of chemical structures in a database, the structural data have to be stored in searchable format.

In general, the encoding scheme can be a reversible or irreversible representation. Reversible or two-way encoding schemes — such as the systematic nomenclature, the SMILES notation, and connectivity tables — allow conversion of the original chemical structure diagram into the representation and vice versa. A one-way encoding is irreversible and it is not possible to regenerate the structural graph from the representation. Therefore, a one-way encoding may only be used as an index in the database search. The representation and search of chemical structures are reviewed by Barnard.[66] Database searches are useful tools both in ligand- and structure-based drug design. Lead finding from database mining in structure-based drug design is discussed in detail by Weber.[67]

In order to reduce the number of structures in a first step, it became common practice to perform pre-screening. For example the 'rule of five' from Lipinski and coworkers is applied to discard compounds possessing poor absorption and permeation properties with a high probability.[68] Other filters used for pre-screening account for lead-[69,70] or drug-likeness,[71–73] an appropriate ADMET profile,[74–77] or favourable properties concerning receptor-binding.[78]

Examples of chemical structure databases range from the comprehensive collection of small organic molecules in the CAS registry file (Chemical Abstracts Service) and the Beilstein online database, to specialized collections of experimentally determined 3D structures in CSD (Cambridge Structural Database) and PDB (Brookhaven Protein Data Bank), to commercial catalogues (ACD), and to pharmaceutical compounds (MDDR, NCI).

The 'similar property principle' states that high-ranked structures in a similarity search are likely to have similar physicochemical and biological properties to that of the target structure. Accordingly similarity searches play a pivotal role in database searches related to drug design. Frequently used similarity measures are illustrated in the next section.

A   Distance and similarity measures

The similarity between compounds is estimated in terms of a distance measure d_{st} between two different objects s and t. The objects s and t are described by the vectors $X_s = (x_{s1}, x_{s2}, ..., x_{sm})$ and $X_t = (x_{t1}, x_{t2}, ..., x_{tm})$, where m denotes the number of real variables and x_{sj} and x_{tj} are the jth element of the corresponding vectors. A distance measure should be defined in such a way that the conditions

$$d_{st} \geq 0$$
$$d_{ss} = 0$$
$$d_{st} = d_{ts} \tag{7}$$
$$d_{st} \leq d_{sz} + d_{zt}$$

are fulfilled. Furthermore, the variables x_j should have a comparable magnitude. Otherwise scaling or normalization of the variables has to be performed. In the following, the most prominent distance measures are listed:

$$\text{Euclidean distance:} \quad d_{st} = \sqrt{\sum_{j=1}^{m}(X_{sj} - X_{tj})^2} \quad (8)$$

$$\text{Average Euclidean distance:} \quad d_{st} = \sqrt{\frac{\sum_{j=1}^{m}(X_{sj} - X_{tj})^2}{m}} \quad (9)$$

$$\text{Manhattan distance:} \quad d_{st} = \sum_{j=1}^{m}\left|X_{sj} - X_{tj}\right| \quad (10)$$

A similarity measure s_{st} ($0 \leq S_{st} \leq 1$; $S_{ss} = 0$; $S_{st} = S_{ts}$) can be calculated from the distance measure d_{st} by the functions

$$S_{st} = \frac{1}{1 + d_{st}} \quad \text{or} \quad S_{st} = 1 - \frac{d_{st}}{d^{\max}} \quad (11, 12)$$

with the distance measure d_{st} for the objects s and t and $d^{\max}$ being the maximum distance between a pair of objects from the dataset. As the scalar product of two vectors is related to the cosine of the angle included by these vectors by the equation

$$\boldsymbol{X_s} \cdot \boldsymbol{X_t} = |\boldsymbol{X_s}||\boldsymbol{X_t}|\cos(\angle \boldsymbol{X_s}, \boldsymbol{X_t}) \quad (13)$$

a frequently used similarity measure is the cosine coefficient.

$$\text{cosine coefficient:} \quad S_{st} = \cos(\angle \boldsymbol{X_s}, \boldsymbol{X_t}) = \frac{\boldsymbol{X_s} \cdot \boldsymbol{X_t}}{|\boldsymbol{X_s}||\boldsymbol{X_t}|}$$

$$= \frac{\sum_{j=1}^{m} X_{sj} \cdot X_{tj}}{\sqrt{\sum_{j=1}^{m} X_{sj}^2 \cdot \sum_{j=1}^{m} X_{tj}^2}} \quad (14)$$

The calculation of a distance measure for two objects s and t represented by binary descriptors $\boldsymbol{X_s}$ and $\boldsymbol{X_t}$ with m binary values is based on the frequencies of common and different components. For this purpose we define the frequencies a, b, c and d as follows:

a: number of components with $x_{sj} = x_{tj} = 1$
b: number of components with $x_{sj} = 1$ and $x_{tj} = 0$
c: number of components with $x_{sj} = 0$ and $x_{tj} = 1$
d: number of components with $x_{sj} = 0$ and $x_{tj} = 0$

The frequencies a and d reflect the similarity between two object s and t, whereas b and c provide information about their dissimilarity.

$$\text{Hamming distance:} \quad d_{st} = b + c \quad (15)$$

$$\text{normalized Hamming distance:} \quad d_{st} = \frac{b+c}{m} \quad (16)$$

$$\text{Tanimoto coefficient:} \quad d_{st} = \frac{a}{a+b+c} \quad (17)$$

$$\text{Cosine coefficient:} \quad S_{st} = \frac{a}{\sqrt{(a+b)\cdot(a+c)}} \quad (18)$$

If the binary descriptors for the objects s and t are substructure keys, the Hamming distance gives the number of different substructures in s and t (components that are 1 in either s or t, but not in both). On the other hand, the Tanimoto coefficient is a measure of the number of substructures s and t have in common (i.e. the frequency a) relative to the total number of substructures they could share (given by the number of components that are 1 in either s or t).

Besides similarity searches similarity measures are frequently applied in the analysis and design of combinatorial libraries.[79–83]

B 2D database searches

Four different, general types of 2D database searches can be distinguished:

- *Exact 2D structure search:* All entries in the database that match exactly and completely to a unique query structure have to be retrieved. The structures can be retrieved rapidly by using hashed lexicographic codes.[84]
- *Exact 2D substructure search:* All entries in the database containing the user-defined partial structure are searched. The query defines the substructure exactly. The results from substructure searching are equal in size or larger than the query. This kind of search proceeds usually in two steps. First the database is screened based on 2D chemical fragment indices, such as molecular fingerprints. All records passing this key filtering are compared by atom-by-atom mapping with the query substructure, i.e. the subgraphs are compared. The final retrieval list comprises all records passing both test.
- *R-Group and Markush searching:* These search modes are extensions of an exact 2D substructure search. The substructure is defined as a partial structure with a substitution pattern and a list of substituents. Markush structure searching is even more flexible and is used in particular for chemical patent searching. The retrieval of compounds by R-Group and Markush searching results in the same way as substructure searching. Before the key screening, the features common to the substitution pattern elements are merged. The atom-by-atom mapping is extended in order to consider the different substitution patterns.

- *2D similarity searching:* The motivation to perform 2D similarity searches follows from the similarity property principle. All compounds having a similarity above a certain threshold are retrieved from the database by 2D similarity searching.[85-87] In most cases the similarity is calculated by one of the aforementioned similarity measures based on molecular fingerprints. Therefore, in the first step the fingerprint for the query structure is calculated.

The foundations of graph theory and algorithms required for structure and substructure searching are reviewed by Hopkinson.[88]

C 3D database searches

Searching in 3D databases falls into the catagories pharmacophore-based and shape- or volume-based methods, respectively. Queries for the first category give topological and geometric constraints for the search. The latter uses a query with a description of the shape of the ligand or of the shape of a binding pocket.

Ehrlich introduced the term *pharmacophore* about 90 years ago in the early twentieth century. A pharmacophore refers to the molecular framework that carries the essential features which are responsible for a drug's biological activity. Instead of the structural graph, the query of pharmacophore-based searches defines the relative position of atoms in 3D space or derived features anchored to these atoms or pseudo-centres. In many cases the pharmacophore is simply defined by three atoms and three distances (three-point pharmacophore).[89] Usually, pharmacophores have a ligand-based definition enabling an easier search. If the structure of the receptor is completely unknown, a pharmacophore model can be defined by pharmacophore perception from 3D structure alignment of ligands. 3D searches with both rigid and flexible pharmacophores are possible.

In contrast to a 2D search, the structural graph of the molecules does not account for a match of two molecules in 3D pharmacophore searching.

Docking programs (e.g. *DOCK*) take the 3D shape of a ligand and a receptor and their complementarity into account.[90-99] Solely the steric requirements are considered by shape similarity searches (volume-based searching).

The retrieval of all 3D structures from a database considered to be similar to a given target structure is comparable with 2D similarity searching. 3D similarity searching raises the problem of conformational flexibility. Schuffenhauer *et al.*[100] analysed the BIOSTER database by similarity search using 2D fingerprints and molecular field descriptors. A comprehensive overview on pharmacophore perception and 3D database searches is given in references 101 and 102.

V EXAMPLES FOR LIGAND-BASED DRUG DESIGN

A Lead optimization from the analysis of HTS data of a combinatorial library of hydantoins

In this section we report on a study of a dataset using a Kohonen network. The dataset investigated comprises a combinatorial library of hydantoins. Starting from 18 amino acids, 24 aldehydes and 24 isocyanates a maximum number of $18 \times 24 \times 24 = 10\,368$ compounds is accessible (Fig. 9.1). We obtained data from Boehringer for the activity of 5513 compounds of this set in a specific assay (Teckentrup, A., Briem, H. and Gasteiger, J., manuscript in preparation). This assay provided 185 compounds which were identified as hits, i.e. a hit fraction of 3.4%.

The aim of these studies of HTS data was to derive a classification rule for the biological activity of hydantoins for the specific target investigated in the assay. Based on this classification rule other hydantoins, not included in this study, should be selected for the screening of smaller focused libraries in lead optimization.

A variety of different structure representations was examined. Apart from 2D and 3D representations of the molecules representations of molecular surface properties were chosen. In Plate 1 (see plate section), the results obtained from the training of a Kohonen network using Daylight 2D fingerprints as representation of the molecules are depicted.[13] Fingerprints with three different lengths were investigated: 256-, 512-, and 1024-dimensional fingerprints.

After training the Kohonen network, the entire dataset was sent through the Kohonen network marking the neurons containing a hit in magenta and those populated with inactive compounds with red colour. Whenever both hits and inactive compounds were assigned to the same neuron the most frequently observed compound class in this particular neuron determines its colour. The Kohonen maps shown in Plate 1 are the top view on the Kohonen network. Each neuron is represented by a little square.

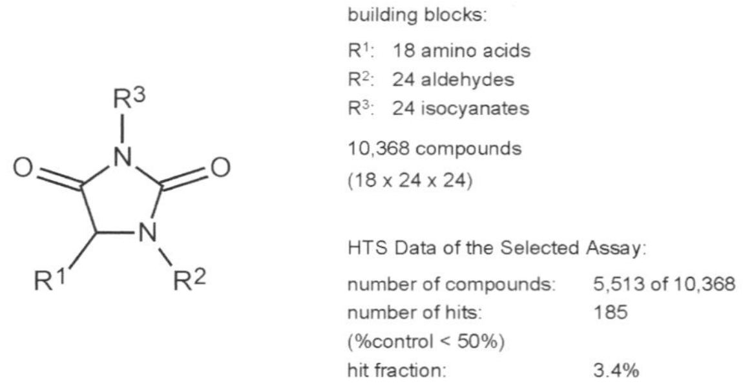

building blocks:

R^1: 18 amino acids
R^2: 24 aldehydes
R^3: 24 isocyanates

10,368 compounds
(18 x 24 x 24)

HTS Data of the Selected Assay:

number of compounds:	5,513 of 10,368
number of hits: (%control < 50%)	185
hit fraction:	3.4%

Fig. 9.1 Dataset from high-throughput screening (a hydantoin library).

256-dim.   512-dim.   1024-dim.

Plate 1 Kohonen maps of the entire dataset of 5513 hydantoins, represented by Daylight 2D fingerprints of three different lengths. Magenta indicates neurons that contain hits; red neurons that have only received non-hits.

No clear-cut separation of the hits from the non-hits can be perceived. Obviously, this structure representation in conjunction with a Kohonen network as a classification method is not suitable for the solution of the given problem.

As receptor binding is related to the complementarity between the ligands and the binding site in a protein, representations of three different molecular surface properties were explored: the molecular electrostatic potential (ESP), the hydrogen-bonding potential (HBP), and the hydrophobicity potential (HYP). As the molecules of the dataset have a different surface area, the number of points on their surface at which the potential is calculated differs if they are calculated with a constant point density (number of points per Å^2). However, pattern recognition methods or neural networks need the same number of descriptors for the objects to be studied. Thus, the data describing the surface properties of molecules have to be transformed resulting in a fixed number of descriptors irrespective of the size of the molecule. The calculation of an autocorrelation vector[17] of molecular surface properties[50] as described in Chapter 2 was used for this purpose.

Choosing autocorrelation of the electrostatic (ESP), the hydrogen bonding (HBP), and the hydrophobicity potential (HYP) for the representation of the molecules in this dataset results in the Kohonen maps illustrated in Plate 2 (see plate section). Neurons with hits are coloured in magenta, whereas neurons containing only non-hits are in red.

Again, no satisfying separation of hits from non-hits can be observed. Nevertheless, the Kohonen map in the center of Plate 2 showing the results of the representation of the molecules by autocorrelation of the hydrogen bonding potential looks promising because the hits cluster in the centre of the Kohonen map, albeit they are mixed with non-hits.

Therefore, the representation by autocorrelation of the hydrogen bonding potential is used in order to develop a classification model capable of filtering the hits from the dataset as well as possible, while accepting retrieval of some non-hits. New compounds presented to the network having the biological activity tested in the assay are expected to be found in neurons that are already occupied by hits. To make sure that all hits are gathered, compounds assigned to neurons directly adjacent to a neuron containing a hit are also considered to potentially receive hits.

The first step to test this hypothesis is the division of the dataset into a training and a test dataset. Two-thirds (3685 compounds) of the entire dataset were assigned to the training dataset and the remaining third (1828 compounds) to the test dataset by a random selection with the restriction that the ratio of hits to non-hits in the entire dataset is preserved in both the training and test dataset.

In the second step, a Kohonen network was trained with the training data. A rectangular topology of the network with a size of $48 \times 38 = 1824$ neurons was chosen, i.e. the ratio of data points to the number of neurons is about 2:1. The resulting Kohonen map for the training set is depicted in Plate 3 (see plate section).

Furthermore, the classification map was derived as the third step. Each neuron occupied by a hit and its first sphere of neighbours is coloured in magenta (Plate 3b). The magenta area in Plate 3b acts as filter for the prediction of hits.

Finally, sending the test set through the Kohonen network obtained from the training set 66 molecules (96%) of the 67 hits of the test set are assigned to neurons within the area selected as filter. Only 142 molecules (8%) of the non-hits are gathered as false-positive hits. Additionally, 1619 (92%) molecules are correctly predicted as non-hits. Based on this analysis the number of compounds which would have to be run through the assay was reduced from 1828 compounds to about a ninth (208 compounds) losing only one out of 67 active compounds. These findings are summed up in Table 9.1.

This study clearly illustrates the power of the unsupervised learning method comprised in a Kohonen network as

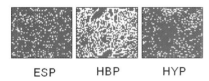

ESP   HBP   HYP

Plate 2 Kohonen maps of the entire dataset of 5513 hydantoins, represented by autocorrelation of the electrostatic (ESP), the hydrogen bonding (HBP), and the hydrophobicity potential (HYP) on the molecular surface. Magenta indicates hits, red non-hits.

a)   b)   c)

training data: classification map   test data:
118 hits, 3,567 non-hits   67 hits, 1,761 non-hits

Plate 3 Kohonen maps obtained by representing molecules by autocorrelation of the hydrogen bonding potential. (a) Training set; (b) filter obtained by recolouring (see text); (c) test set.

Table 9.1 *Classifications results on HTS data*

	Total dataset	Training set	Test set	
			Contained	Classified
Hits	185	118	67	66 (96%)
Non-hits	5328	3567	1761	1619 (92%)

a data mining tool. The establishment of a relationship between the biological activity of molecules with their structure requires appropriate structure-encoding schemes. In previous studies the applicability of a Kohonen network in locating biologically active compounds in an excess of inactive ones with a large structural variety,[103] in the separation of compounds with different biological activities,[103] and for the comparison of the similarity and diversity of combinatorial libraries[104] has been demonstrated. The first and second example will be described in more detail in the following section.

B Distinguishing molecules of different biological activity and finding new lead structures

The separation of a dataset with 172 molecules into benzodiazepine agonists (60 compounds) and dopamine agonist (112) compounds by a Kohonen network was the aim of the study presented here.[103] As no information about the steric and electronic requirements of the receptor was available a rather broad structure representation seemed to be a good starting point. Therefore, a variety of physico-chemical effects were considered for the calculation of a suitable descriptor. A topological autocorrelation vector for seven topological distances was calculated for each of the following properties: σ, atomic charges, $\sigma + \pi$, atomic charges, σ, electronegativity, π, electronegativity, lone pair electronegativity, atomic polarizability, and an atomic property of 1 (to simply represent the molecular graph). Concatenation of these seven autocorrelation vectors resulted in a 49-dimensional representation of the 172 molecules in the dataset. Training a 10×7 Kohonen

network with the entire dataset gave a map that was coloured according to the neuron's content (Fig. 9.2). Neurons containing a dopamine agonist are coloured dark grey and those occupied by benzodiazepine agonists light grey. Unoccupied neurons remain white.

It can be seen that the two different sets of molecules separate quite well, however class membership was not used in training the network. Class membership served only for the visualization of the self-organizing map after training. Thus, effects that are responsible for the different binding of dopamine and benzodiazepine agonists are reproduced by the chosen structure representation.

After it has been shown that these compounds with different biological activities can be distinguished using this structure representation in conjunction with a Kohonen neural network we turn our attention to another question. If the two sets of molecules are buried in a large dataset of diverse structures will we still see a separation between the different classes? For this purpose the dataset of 172 molecules and the entire catalogue of 8223 compounds available from a chemical supplier (Janssen Chimica) were investigated together. Due to the larger size of the dataset the size of the network had to be increased. A network of 40×30 neurons was chosen. Training this network with the previously described structure representation for all 8395 structures in the merged dataset provided the map depicted in Fig. 9.3.

Again the dopamine and benzodiazepine agonists could be separated quite well even in this diverse dataset of structures. Collisions between these two types of compounds occur only in two neurons. In addition, the self-organizing map guides our search for new lead structures for dopamine or for benzodiazepine agonists through the chemical space. As topology of the input space is preserved by the projection with self-organizing neural networks, similar compounds are nearby in the resulting self-organizing map. Structures which are considered to be similar to the dopamine or benzodiazepine agonists are likely to have a similar biological activity. To illustrate this point, Fig. 9.4 shows the contents of the neuron at position 3.9. In this neuron are two dopamine agonists and three compounds of the Janssen Chimica catalogue with an unknown biological activity.

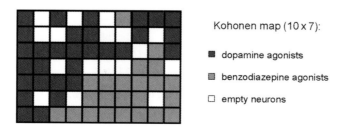

Kohonen map (10 x 7):

■ dopamine agonists

▨ benzodiazepine agonists

□ empty neurons

Fig. 9.2 Kohonen map obtained from a dataset of 112 dopamine and 60 benzodiazepine agonists.

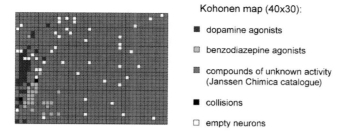

Kohonen map (40x30):

■ dopamine agonists

▨ benzodiazepine agonists

▦ compounds of unknown activity
(Janssen Chimica catalogue)

■ collisions

□ empty neurons

Fig. 9.3 Kohonen map of a dataset consisting of the dopamine and benzodiazepine agonists of Plate 3 and 8323 compounds of a chemical supplier catalogue.

These three structures might be taken as starting points for the search and optimization of a new lead structure for developing dopamine agonists.

The results outlined in this section show that this approach can be applied to compare two different libraries and to determine the degree of overlap between the compounds in these two libraries.

VI EXAMPLES FOR STRUCTURE-BASED DRUG DESIGN

Virtual screening of chemical databases has been successfully applied to find new leads in structure-based drug design.[105–107] Wang and coworkers[108] found a novel, potent dopamine transporter inhibitor by 3D-database pharmacophore searching and subsequent chemical modification. The analogue has a high affinity and is a promising lead for the treatement of cocaine abuse.

Another approach followed by Wang *et al.* led to an inhibitor of *Tritrichomonas foetus* hypoxanthine-guanine-xanthine phosphoribosyltransferase (HGXPRT).[109] The X-ray structure of this enzyme is available and it was used in a search for novel scaffolds. The Available Chemicals Directory (ACD) was screened with the program DOCK. Isatin and phthalic anhydride were capable of mimicking the substrate purine base and acting as competitive inhibitors. A virtual library of substituted

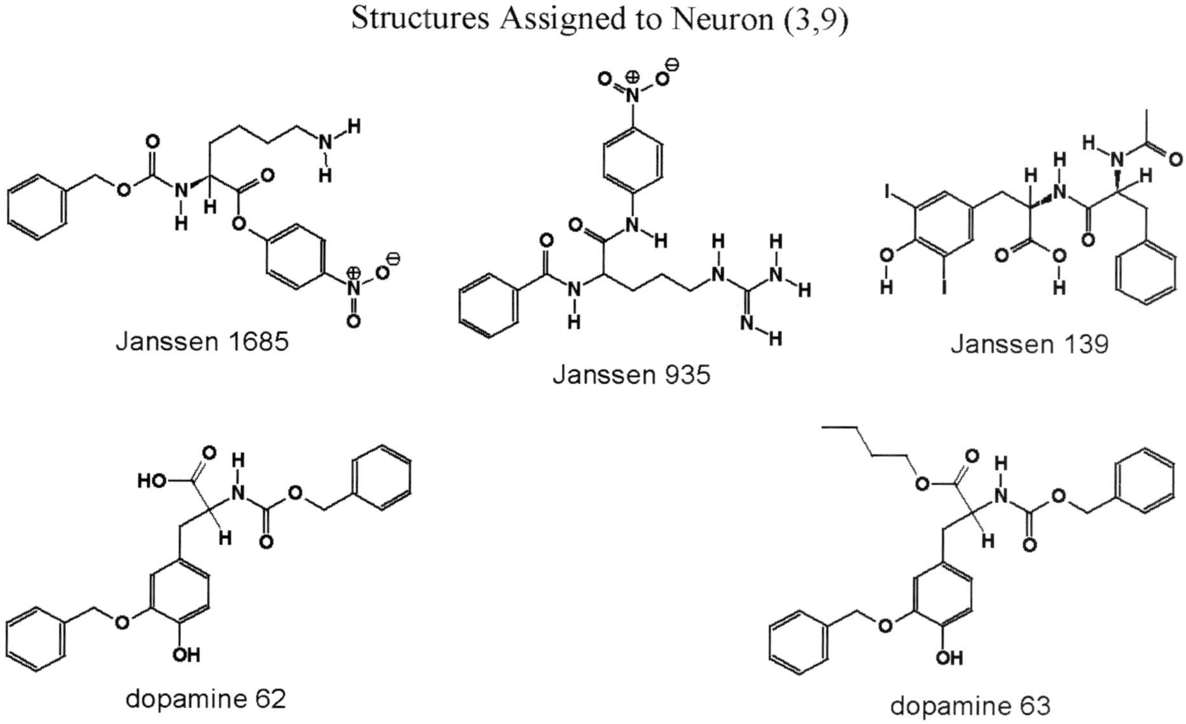

Structures Assigned to Neuron (3,9)

Janssen 1685

Janssen 935

Janssen 139

dopamine 62

dopamine 63

Fig. 9.4 Structures that were mapped into neuron at position 3.9 of the Kohonen map of Fig. 9.3.

4-phthalimidocarboxanilides was constructed and synthesized in solid support. The most active compound was used as lead for the further optimization providing [(4'-phthalimido)carboxamido-3-(4-bromobenzyloxy)benzene] as selective submicromolar inhibitor for *T. foetus* HPRT. Zhang and coworkers worked on the structure-based, computer-assisted search for low molecular weight, non-peptidic protein tyrosine phosphate 1B (PTP1B) inhibitors also using the DOCK methodology.[110] They identified several potent and selective PTP1B inhibitors by screening the ACD.

Researchers from Hoffman-La Roche discovered new inhibitors of DNA gyrase by combining several key techniques.[111] Based on the 3D structural information of the binding site they started with an *in silico* screening for potential low molecular weight inhibitors, followed by a biased high-throughput DNA gyrase screen. The hits of the screening were validated by biophysical methods and finally optimized in a 3D-guided optimization process. The lead optimization of an indazole derivative resulted in a ten times more potent DNA gyrase inhibitor than novobiocin.

Pang *et al.* also performed a virtual screening of the ACD with a newly developed docking program (EUDOC).[112] Their interest was the search for new farnesyltransferase inhibitor leads. Among the compounds identified by the virtual screening, four showed IC_{50} values in the range from 25 to 100 μM.

Docking also provided insight into the three-dimensional hydrophobic requirements for binding of nonpeptide molecules at the SH2 domain of pp60src at Ariad Pharmaceuticals.[113]

VII FUTURE DIRECTIONS

Nowadays a broad range of methods is available in the field of chemoinformatics. These methods will have a growing impact on drug design. In particular, the discovery of new lead structures and their optimization will profit from virtual screening.[105,114-119]

The vast amount of data produced by HTS and combinatorial chemistry enforces the use of database and data mining techniques.

Integrated systems suitable for processing the sometimes quite complex workflows more easily or automatically and to optimize new compounds in parallel for their potency, selectivity, and ADMET profile have to be developed. The reliability of the *in silico* models will be improved and their scope for predictions will be broader as soon as more reliable experimental data are available. However, there is the paradox of predictivity versus diversity. The greater the chemical diversity in a dataset, the more difficult is the establishment of a predictive structure—activity relationship. Otherwise, a model developed based on compounds representing only a small subspace of the chemical space has no predictivity for compounds beyond its boundaries.

REFERENCES

1. Kubinyi, H. (1999) Chance favors the prepared mind. From serendipity to rational drug design. *J. Receptor Signal Trans. Res.* **19**: 15–39.
2. Tollman, P., Guy, P., Althuler, J., Flanagan, A. and Steiner, M. (2001) A revolution in R&D: How genomics and genetics are transforming the biopharmaceutical industry. *Boston Consulting Group Report* 1–60.
3. Veverka, M.J. (2000) Speed to value. Delivering on the quest for better medicines. *Accenture Report* **23**: 1–53.
4. Olsson, T. and Oprea, T.I. (2001) Cheminformatics: A tool for decision-makers in drug discovery. *Curr. Opin. Drug Discov. Dev.* **4**: 308–313.
5. Gaudillière, B. and Berna, P. (2000) To market, to market — 1999. *Annu. Rep. Med. Chem.* **35**: 331–356.
6. Valler, M.J. and Green, D. (2000) Diversity screening versus focussed screening in drug discovery. *Drug Disc. Today* **5**: 286–293.
7. Fayyad, U.M., Piatetsky-Shapiro, G. and Smyth, P. (1996) From data mining to knowledge discovery: An overview. In Fayyad, U.M., Piatetsky-Shapiro, G., Smyth, P. and Uthurusamy, R. (eds). *Advances in Knowledge Discovery and Data Mining*, pp. 1–37. AAAI Press, Menlo Park.
8. Wiswesser, W.J. (1985) Historical development of chemical notations. *J. Chem. Inf. Comput. Sci.* **25**: 258–263.
9. Weininger, D. (1988) SMILES, a chemical language and information system. 1. Introduction to methodology and encoding rules. *J. Chem. Inf. Comp. Sci.* **28**: 31–36.
10. Weininger, D., Weininger, A. and Weininger, J.L. (1989) SMILES. 2. Algorithm for generation of unique SMILES notation. *J. Chem. Inf. Comp. Sci.* **29**: 97–101.
11. MDL Information Systems, Inc. 14600 Catalina Street, San Leandro, CA 94577, USA http://www.mdli.com.
12. The definitions of MDL's 166 MACCS search keys can be found from the ISIS/Base help file at Section 49.2.4, specifying searchable keys as a query.
13. Daylight Chemical Information Systems, Inc., 7401 Los Altos, Suite 360, Mission Viejo, CA 92691, USA http://www.daylight.com.
14. Hopkinson, G.A. (1998) Structure and substructure searching. In Von Ragué Schleyer, P., Allinger, N.L., Clark, T., Gasteiger, J., Kollman, P.A., Schaefer III, H.F. and Schreiner, P.R. (eds). *Encyclopedia of Computational Chemistry*, vol. 4, pp. 2764–2771. John Wiley, New York.
15. Bron, C. and Kerbosch, J. (1973) Algorithm 457: Finding all cliques of an undirected graph. 4*Commun. ACM.* **16**: 575–577.
16. Carraghan, R. and Pardalos, P.M. (1990) Exact algorithm for the maximum clique problem. *Op. Res. Lett.* **9**: 375.
17. Moreau, G. and Broto, P. (1980) The autocorrelation of a topological structure: A new molecular descriptor. *Nouv. J. Chim.* **4**: 359–360.
18. Gasteiger, J. and Marsili, M. (1980) Iterative partial equalization of orbital electronegativity — A rapid access to atomic charges. *Tetrahedron* **36**: 3219–3228.

19. Gasteiger, J. and Hutchings, M.G. (1984) Quantification of effective polarisability. Applications to studies of X-ray photoelectron spectroscopy and alkylamine protonation. *J. Chem. Soc. Perkin* 2: 559–564.

20. PETRA Version 3 is available from Molecular Networks GmbH, Nägelsbachstr. 25, 91052 Erlangen, Germany (http://www.mol-net.de) and can be tested via the internet at http://www2.chemie.uni-erlangen.de/software/petra.

21. Rarey, M. and Dixon, J.S. (1998) Feature trees: A new molecular similarity measure based on tree matching. *J. Comp.-Aided Mol. Design* 12: 471–490.

22. Wiener, H. (1947) Structural determination of paraffin boiling points. *J. Am. Chem. Soc.* 69: 17–20.

23. Wiener, H. (1947) Correlation of heat of isomerization, and differences in heats of vaporization of isomers, among the paraffin hydrocarbons. *J. Am. Chem. Soc.* 69: 2636–2638.

24. Wiener, H. (1947) Influence of interatomic forces on paraffin properties. *J. Chem. Phys.* 15: 766.

25. Randic, M. (1975) On characterization of molecular branching. *J. Am. Chem. Soc.* 97: 6609–6615.

26. Kier, L.B. and Hall, L.H. (1977) The nature of structure–activity relationships and their relation to molecular connectivity. *Eur. J. Med. Chem.* 12: 307–312.

27. Kier, L.B. and Hall, L.H. (1990) An electrotopological-state index for atoms in molecules. *Pharm. Res.* 7: 801–807.

28. Burden, F.R. (1989) Molecular identification number for substructure searches. *J. Chem. Inf. Comput. Sci.* 29: 225–228.

29. Cambridge Crystallographic Data Center, 12 Union Road, Cambridge, CB2 1EZ, UK, http://www.ccdc.cam.uk.

30. Chemical Abstracts Service, 2540 Olentangy River Road, P.O. Box 3012, Columbus, Ohio 43210, USA. http://www.cas.org/cgi-bin/regreport.pl.

31. Sadowski, J. (1998) Three-dimensional structure generation: Automation. In Von Ragué Schleyer, P., Allinger, N.L., Clark, T., Gasteiger, J., Kollman, P.A., Schaefer III, H.F. and Schreiner, P.R. (eds). *Encyclopedia of Computational Chemistry*, vol. 5, pp. 2976–2988. John Wiley, New York.

32. Pearlman, R.S. (1987) Rapid generation of high quality approximate 3D molecular structures. *Chem. Design Auto. News* 2: 1–7.

33. Pearlman, R.S. (1993) 3D Molecular structures: Generation and use in 3D-searching. In Kubinyi, H. (ed.). *3D QSAR in Drug Design: Theory, Methods and Applications*, pp. 21–58. ESCOM Science Publishers, Leiden.

34. Tripos, Inc. 1699 South Hanley Road, St. Louis, Missouri 63144, USA, http://www.tripos.com.

35. Sadowski, J. and Gasteiger, J. (1993) From atoms and bonds to three-dimensional atomic coordinates: Automatic model builders. *Chem. Rev.* 93: 2567–2581.

36. Sadowski, J., Gasteiger, J. and Klebe, G. (1994) Comparison of automatic three-dimensional model builders using 639 X-ray structures. *J. Chem. Inf. Comp. Sci.* 34: 1000–1008.

37. Gasteiger, J., Rudolph, C. and Gasteiger, J. (1990) Automatic generation of 3D-atomic coordinates for organic molecules. *Tetrahedron Comp. Meth.* 3: 537–547.

38. Hiller, C. and Gasteiger, J. (1987) Ein automatisierter Molekülbaukasten. In Gasteiger, J. (ed.). *Software-Entwicklung in der Chemie*, vol. 1, pp. 53–66. Springer, Berlin.

39. CORINA is available from Molecular Networks GmbH, Nägelsbachstr. 25, 91052 Erlangen, Germany, (http://www.mol-net.de) and can be tested via the internet at http://www2.chemie.uni-erlangen.de/software/corina.

40. Leach, A.R. (1991) A survey of methods for searching the conformational space of small and medium-sized molecules. In Lipkowitz, K.B. and Boyd, D.B. (eds). *Reviews in Computational Chemistry*, vol. 2, pp. 1–55. VCH Verlagsgesellschaft, New York.

41. Saunders, M. (1987) Stochastic exploration of molecular mechanics energy surfaces. Hunting for the global minimum. *J. Am. Chem. Soc.* 109: 3150–3152.

42. Saunders, M. (1989) Stochastic search for the conformations of bicyclic hydrocarbons. *J. Comput. Chem.* 10: 203–208.

43. McGarrah, D.B. and Judson, R.S. (1993) Analysis of the genetic algorithm method of molecular conformation determination. *J. Comp. Chem.* 14: 1385–1395.

44. Judson, R.S., Jaeger, E.P., Treasurywala, A.M. and Peterson, M.L. (1993) Conformational searching methods. II. Genetic algorithm approach. *J. Comput. Chem.* 14: 1407–1414.

45. Lybrand, T.P. (1990) Computer simulations of biomolecular systems using molecular dynamics and free energy perturbation methods. In Lipkowitz, K.B. and Boyd, D.B. (eds). *Reviews in Computational Chemistry*, vol. 1, pp. 295–320. VCH Verlagsgesellschaft, New York.

46. Chang, G., Guida, W.C. and Still, W.C. (1989) An internal coordinate Monte Carlo method for searching conformational space. *J. Am. Chem. Soc.* 111: 4379–4386.

47. Schönberger, H., Schwab, C.H., Hirsch, A. and Gasteiger, J. (2000) Molecular modelling of fullerene dendrimers. *J. Mol. Model.* 6: 379–395.

48. ROTATE is available from Molecular Networks GmbH, Nägelsbachstr. 25, 91052 Erlangen, Germany (http://www.mol-net.de).

49. Wagener, M., Sadowski, J. and Gasteiger, J. (1995) Autocorrelation of molecular surface properties for modeling corticosteroid binding globulin and cytosolic Ah receptor activity by neural networks. *J. Am. Chem. Soc.* 117: 7769–7775.

50. Wierl, R. (1931) Elektronenbeugung und Molekülbau. *Ann. Phys. (Leipzig)* 8: 521–564.

51. Soltzberg, L.J. and Wilkins, C.L. (1977) Molecular transforms: A potential tool for structure activity studies. *J. Am. Chem. Soc.* 99: 439–443.

52. Schuur, J.H., Selzer, P. and Gasteiger, J. (1996) The coding of the three-dimensional structure of molecules by molecular transforms and its application to structure–spectra correlations and studies of biological activity. *J. Chem. Inf. Comp. Sci.* 36: 334–344.

53. Gasteiger, J., Sadowski, J., Schuur, J., Selzer, P., Steinhauer, L. and Steinhauer, V. (1996) Chemical information in 3D space. *J. Chem. Inf. Comp. Sci.* 36: 1030–1037.

54. Gasteiger, J., Schuur, J., Selzer, P., Steinhauer, L. and Steinhauer, V. (1997) Finding the 3D structure of a molecule in its IR spectrum. *Fresenius J. Anal. Chem.* 359: 50–55.

55. Hemmer, M.C., Steinhauer, V. and Gasteiger, J. (1999) Deriving the 3D structure of organic molecules from their infrared spectra. *Vibrat. Spectrosc.* 19: 151–164.

56. Goodford, P.J. (1985) A computational procedure for determining energetically favorable binding sites on biologically important macromolecules. *J. Med. Chem.* 28: 849–857.

57. Cramer III, R.D., Patterson, D.E. and Bunce, J.D. (1988) Comparative molecular field analysis (CoMFA). 1. Effect of shape on binding of steroids to carrier proteins. *J. Am. Chem. Soc.* 110: 5959–5967.

58. Pearlman, R.S. and Smith, K.M. (1998) Novel software tools for addressing chemical diversity. In Kubinyi, H., Folkers, G. and Martin, Y.C. (eds). *3D QSAR in drug design*, vol. 2, pp. 339–353. Kluwer/ESCOM, Dordrecht.

59. Ferguson, A.M., Heritage, T.W., Jonathon, P., Pack, S.E., Phillips, L., Rogan, J. and Snaith, P.J. (1997) EVA: A new theoretically based molecular descriptor for use in QSAR/QSPR analysis. *J. Comp.-Aided Mol. Design* 11: 143–152.

60. Turner, D.B., Willett, P., Ferguson, A.M. and Heritage, T.W. (1997) Evaluation of a novel infrared range vibration-based descriptor (EVA) for QSAR studies. 1. General application. *J. Comp.-Aided Mol. Design* 11: 409–422.

61. Todeschini, R., Lasagni, M. and Marengo, E. (1994) New molecular descriptors for 2D- and 3D-structures. *Theory. J. Chemo.* **8**: 263–273.

62. Todeschini, R. and Gramatica, P. (1997) 3D-modelling and prediction by WHIM descriptors. Part 5. Theory development and chemical meaning of WHIM descriptors. *Quant. Struct.-Act. Relat.* **16**: 113–119.

63. Hopfinger, A.J., Wang, S., Tokarski, J.S., Jin, B., Albuquerque, M., Madhav, P.J. and Duraiswami, C. (1997) Construction of 3D-QSAR models using the 4D-QSAR analysis formalism. *J. Am. Chem. Soc.* **119**: 10509–10524.

64. Hopfinger, A.J., Reaka, A., Venkatarangan, P., Duca, J.S. and Wang, S. (1999) Construction of a virtual high throughput screen by 4D-QSAR analysis: Application to a combinatorial library of glucose inhibitors of glycogen phosphorylase *b*. *J. Chem. Inf. Comp. Sci.* **39**: 1151–1160.

65. Todeschini, R. and Consonni, V. (2000) The handbook of molecular descriptors. In Mannhold, R., Kubinyi, H. and Timmerman, H. (eds). *Methods and Principles in Medicinal Chemistry*, vol. 115, pp. 1–667. Wiley-VCH, Weinheim.

66. Barnard, J.M. (1991) Structure representation and searching. In Ash, J.E., Warr, W. and Willett, P. (eds). *Chemical Structure Systems: Computational Techniques for Representation, Searching, and Processing of Structural Information*, pp. 9–56. Ellis Horwood, Chichester.

67. Weber, H.P. (1996) Electronic screening: Lead finding from database mining. In Wermuth, C.G. (ed.). *Practice of Medicinal Chemistry*, pp. 167–178. Academic Press, London.

68. Lipinski, C.A., Lombardo, F., Dominy, B.W. and Feeney, P.J. (1997) Experimental and computational approaches to estimate solubility and permeability in drug discovery and development settings. *Adv. Drug Del. Rev.* **23**: 3–25.

69. Teague, S.J., Davis, A.M., Leeson, P.D. and Oprea, T. (1999) The design of leadlike combinatorial libraries. *Angew. Chem. Int. Ed. Engl.* **38**: 3743–3747.

70. Oprea, T.I., Davis, A.M., Teague, S.J. and Leeson, P.D. (2001) Is there a difference between leads and drugs? A historical perspective. *J. Chem. Inf. Comp. Sci.* **41**: 1308–1315.

71. Clark, D.E. and Pickett, S.D. (2000) Computational methods for the prediction of 'drug-likeness'. *Drug Disc. Today* **5**: 49–58.

72. Ajay, A., Walters, W.P. and Murcko, M.A. (1998) Can we learn to distinguish between 'drug-like' and 'nondrug-like' molecules? *J. Med. Chem.* **41**: 3314–3324.

73. Blake, J.F. (2000) Chemoinformatics — predicting the physicochemical properties of 'drug-like' molecules. *Curr. Opin. Biotechnol.* **11**: 104–107.

74. Li, A.P. and Segall, M. (2002) Early ADME/Tox studies and *in silico* screening. *Drug Disc. Today* **7**: 25–27.

75. Li, A.P. (2001) Screening for human ADME/Tox drug properties in drug discovery. *Drug Disc. Today* **6**: 357–366.

76. Thompson, T.N. (2000) Early ADME in support of drug discovery: The role of metabolic stability studies. *Curr. Drug Metabol.* **1**: 215–241.

77. Keserü, G.M. and Molnár, L. (2002) METAPRINT: A metabolic fingerprint. Application to cassette design for high-throughput ADME screening. *J. Chem. Inf. Comp. Sci.* **42**: 437–444.

78. Andrews, P.R., Craik, D.J. and Martin, J.L. (1984) Functional group contributions to drug–receptor interactions. *J. Med. Chem.* **27**: 1648–1657.

79. Agrafiotis, D.K. (1998) Diversity of chemical libraries. In Von Ragué Schleyer, P., Allinger, N.L., Clark, T., Gasteiger, J., Kollman, P.A., Schaefer III, H.F. and Schreiner, P.R. (eds) *Encyclopedia of Computational Chemistry*, vol. 1, pp. 742–761. John Wiley, New York.

80. Agrafiotis, D.K. (2001) Multiobjective optimization of combinatorial libraries. *IBM J. Res. Dev.* **45**: 545–566.

81. Lewis, R.A., Pickett, S.D. and Clark, D.E. (2000) Computer-aided molecular diversity analysis and combinatorial library design. In Lipkowitz, K.B. and Boyd, D.B. (eds). *Reviews in Computational Chemistry*, vol. 16, pp. 1–51. Wiley-VCH, New York.

82. Warr, W.A. (1997) Combinatorial chemistry and molecular diversity. An overview. *J. Chem. Inf. Comp. Sci.* **37**: 134–140.

83. Blaney, J.M. and Martin, E.J. (1997) Computational approaches for combinatorial library design and molecular diversity analysis. *Curr. Opin. Chem. Biol.* **1**: 54–59.

84. Wipke, W.T. and Dyott, T.M. (1974) Stereochemically unique naming algorithm. *J. Am. Chem. Soc.* **96**: 4834–4842.

85. Willett, P. (1987) Similarity and clustering techniques in chemical information systems. In Bawden, D. (ed.). *Chemometrics Series*, vol. 18, pp. 1–254. Research Studies Press, Letchworth.

86. Willett, P. (1998) Structural similarity measures for database searching. In Von Ragué Schleyer, P., Allinger, N.L., Clark, T., Gasteiger, J., Kollman, P.A., Schaefer III, H.F. and Schreiner, P.R. (eds) *Encyclopedia of Computational Chemistry*, vol. 4, pp. 2748–2756. John Wiley, New York.

87. Willett, P. (2000) Chemoinformatics — similarity and diversity in chemical libraries. *Curr. Opin. Biotechnol.* **11**: 85–88.

88. Hopkinson, G.A. (1998) Structure and substructure searching. In Von Ragué Schleyer, P., Allinger, N.L., Clark, T., Gasteiger, J., Kollman, P.A., Schaefer III, H.F. and Schreiner, P.R. (eds) *Encyclopedia of Computational Chemistry*, vol. 4, pp. 2764–2771. John Wiley, New York.

89. Güner, O.F., Hughes, D.W. and Dumont, L.M. (1991) An integrated approach to three-dimensional information management with MACCS-3D. *J. Chem. Inf. Comp. Sci.* **31**: 408–414.

90. Kuntz, I.D., Blaney, J.M., Oatley, S.J., Langridge, R. and Ferrin, T.E. (1982) A geometric approach to macromolecule–ligand interactions. *J. Mol. Biol.* **161**: 269–288.

91. Jones, G. and Willett, P. (1995) Docking small-molecule ligands into active sites. *Curr. Opin. Biotechnol.* **6**: 652–656.

92. Jones, G., Willett, P., Glen, R.C., Leach, A.R. and Taylor, R. (1997) Development and validation of a genetic algorithm for flexible docking. *J. Mol. Biol.* **267**: 727–748.

93. Shoichet, B.K., Bodian, D.L. and Kuntz, I.D. (1992) Molecular docking using shape descriptors. *J. Comp. Chem.* **13**: 380–397.

94. Meng, E.C., Shoichet, B.K. and Kuntz, I.D. (1992) Automated docking with grid-based energy evaluation. *J. Comp. Chem.* **13**: 505–524.

95. Sun, Y., Ewing, T.J.A., Skillman, A.G. and Kuntz, I.D. (1998) CombiDOCK: Structure-based combinatorial docking and library design. *J. Comp.-Aided Mol. Design* **12**: 597–604.

96. Rarey, M., Kramer, B., Lengauer, T. and Klebe, G. (1996) A fast flexible docking method using an incremental construction algorithm. *J. Mol. Biol.* **261**: 470–489.

97. Rarey, M., Kramer, B. and Lengauer, T. (1997) Multiple automatic base selection: Protein-ligand docking based on incremental construction without manual intervention. *J. Comp.-Aided Mol. Design* **11**: 369–384.

98. Yang, J.-M. and Kao, C.-Y. (2000) Flexible ligand docking using a robust evolutionary algorithm. *J. Comp. Chem.* **21**: 988–998.

99. Morris, G.M., Goodsell, D.S., Halliday, R.S., Huey, R., Hart, W.E., Belew, R.K. and Olson, A.J. (1998) Automated docking using a lamarckian genetic algorithm and an empirical binding free energy function. *J. Comp. Chem.* **19**: 1639–1662.

100. Schuffenhauer, A., Gillet, V.J. and Willet, P. (2000) Similarity searching in files of three-dimensional chemical structures: Analysis of the BIOSTER database using two-dimensional fingerprints and molecular field descriptors. *J. Chem. Inf. Comp. Sci.* **40**: 295–307.

101. Güner, O.F. (ed.). (2000) *Pharmacophore Perception, Development, and Use in Drug Design*, pp. 1–537. International University Line, La Jolla, CA, USA.

102. Paris, C.G. (1998) Structure databases. In Von Ragué Schleyer, P., Allinger, N.L., Clark, T., Gasteiger, J., Kollman, P.A., Schaefer III, H.F. and Schreiner, P.R. (eds). *Encyclopedia of Computational Chemistry*, vol. 4, pp. 2771–2785. John Wiley, New York.

103. Bauknecht, H., Zell, A., Bayer, H., Levi, P., Wagener, M., Sadowski, J. and Gasteiger, J. (1996) Locating biologically active compounds in medium-sized heterogeneous datasets by topological autocorrelation vectors: Dopamine and benzodiazepine agonists. *J. Chem. Inf. Comp. Sci.* **36**: 1205–1213.

104. Sadowski, J., Wagener, M. and Gasteiger, J. (1995) Assessing similarity and diversity of combinatorial libraries by spatial autocorrelation functions and neural networks. *Angew. Chem. Int. Ed. Engl.* **34**: 2674–2677.

105. Klebe, G. (ed.) (2000) Virtual screening: an alternative or complement to high throughput screening. *Persp. Drug Design Disc.* **20**: 1–287.

106. Bissantz, C., Folkers, G. and Rognan, D. (2000) Protein-based virtual screening of chemical databases. 1. Evaluation of different docking/scoring combinations. *J. Med. Chem.* **43**: 4759–4767.

107. Böhm, H.-J. and Schneider, G. (2000) Virtual screening for bioactive molecules. In Mannhold, R., Kubinyi, H. and Timmerman, H. (eds). *Methods and Principles in Medicinal Chemistry*, vol. 10, pp. 1–307. Wiley-VCH, Weinheim.

108. Wang, S., Sakamuri, S., Enyedy, I.J., Kozikowski, A.P., Deschaux, O., Bandyopadhyay, B.C., Tella, S.R., Zaman, W.A. and Johnson, K.M. (2000) Discovery of a novel dopamine transporter inhibitor, 4-hydroy-1-methyl-4-(4-methylphenyl)-3-piperidyl 4-methylphenyl ketone, as a potential cocaine antagonist through 3D-database pharmacophore searching. Molecular modeling. Structure–activity relationships, and behavioral pharmacological studies. *J. Med. Chem.* **43**: 351–360.

109. Aronov, A.M., Munagala, N.R., Ortiz de Montellano, P.R., Kuntz, I.D. and Wang, C. (2000) Rational design of selective submicromolar inhibitors of *Tritrichomonas foetus* hypoxanthine-guanine-xanthine phosphoribosyltransferase. *Biochemistry* **39**: 4684–4691.

110. Sarmiento, M., Wu, L., Keng, Y.-F., Song, L., Luo, Z., Huang, Z., Wu, G.-Z., Yuan, A.-K. and Zhang, Z.-Y. (2000) Structure-based discovery of small molecule inhibitors targeted to protein phosphatase 1B. *J. Med. Chem.* **43**: 146–155.

111. Böhm, H., Böhringer, M., Bur, D., Gmünder, H., Huber, W., Klaus, W., Kostrewa, D., Kühne, H., Lübbers, T., Meunier-Keller, N. and Müller, F. (2000) Novel inhibitors of DNA gyrase: 3D structure based biased needle screening. Hit validation by biophysical methods, and 3D guided optimization. A promising alternative to random screening. *J. Med. Chem.* **43**: 2664–2674.

112. Perola, E., Xu, K., Kollmeyer, T.M., Kaufmann, S.H., Prendergast, F.G. and Pang, Y.-P. (2000) Successful virtual screening of a chemical database for farnesyltransferase inhibitor leads. *J. Med. Chem.* **43**: 401–408.

113. Metcalf III C.A., Eyermann, C.J., Bohacek, R.S., Haraldson, C.A., Varkhedkar, V.M., Lynch, B.A., Bartlett, C., Violette, S.M. and Sawyer, T.K. (2000) Structure-based design and solid-phase parallel synthesis of phosphorylated nonpeptides to explore hydrophobic binding at the Src SH2 Domain. *J. Comb. Chem.* **2**: 305–313.

114. Walters, P.W., Stahl, M.T. and Murcko, M.A. (1998) High-throughput 'virtual' chemistry. In Ragué Schleyer, P., Allinger, N.L., Clark, T., Gasteiger, J., Kollman, P.A., Schaefer III, H.F. and Schreiner, P.R. (eds). *Encyclopedia of Computational Chemistry*, vol. 2, pp. 1225–1237. John Wiley, New York.

115. Good, A.C., Krystek, S.R. and Mason, J.S. (2000) High-throughput and virtual screening: core lead discovery technologies move towards integration. *Drug Disc. Today* **5** (suppl.): S61–S69.

116. Schneider, G. and Böhm, H.-J. (2002) Virtual screening and fast automated docking methods. *Drug Disc. Today* **7**: 64–70.

117. Waszkowycz, B., Perkins, T.D.J., Sykes, R.A. and Li, J. (2001) Large-scale virtual screening for discovering leads in the postgenomic era. *IBM Syst. J.* **40**: 360–376.

118. Walters, W.P., Stahl, M.T. and Murcko, M.A. (1998) Virtual screening — an overview. *Drug Disc. Today* **3**: 160–178.

119. Oprea, T.I. (2002) Virtual screening in lead discovery: A viewpoint. *Molecules* **7**: 51–62.

10

HIGH-SPEED CHEMISTRY LIBRARIES: ASSESSMENT OF DRUG-LIKENESS

Alex Polinsky

The golden rule is that there are no golden rules   **George Bernard Shaw**

I INTRODUCTION

A High-speed chemistry in drug discovery

Over the last decade, combinatorial chemistry has become an essential tool in lead generation and lead optimization. Originally conceived as a technology to produce very large numbers of peptidic compounds in mixtures for lead discovery, it has evolved over time in many respects. Main drive for change has come from the demand from 'customers' — medicinal chemists — that they produce leads that are chemically tractable and have the ADME-related properties such as solubility, permeability, metabolic stability and clearance, that are similar to what they seek in a drug, thus the term 'drug-like' molecules. To address this need, synthetic targets have changed from peptides, peptidomimetics and oligonucleotides to small molecules, while synthetic technology has become focused on *high-speed parallel chemistry*, developing chemical versatility needed to make diverse chemical classes of compounds. Parallel chemistry technology quickly found its

way to the medicinal chemist's bench where it is being used for producing relatively small lead optimization libraries. At the same time, methods for designing libraries with desirable properties started to be developed, forming the new field of ChemInformatics,[1,2] and these methods can be applied for both large, 'exploratory' and smaller, 'focused' or 'targeted' libraries.

B What is a 'drug-like' molecule?

'Drug-likeness' literally means similarity to known drugs. Considering the diversity of mechanisms of action and properties the known drugs have, and the variety of methods by which these drugs are administered, it is not surprising that 'drug-likeness' is difficult to define precisely and that there are multiple approaches to its assessment.[3-6] First of all though the question arises as to why anybody would want to come up with formal algorithms of assessing 'drug-likeness'.

When working on a specific drug discovery project, medicinal chemists are guided by property measurements and experience to achieve a desired drug property profile. Since not all properties can be measured during lead optimization (human PK and toxicity, for example), even compounds that fit the desired property profile often fail in the development stages. It is this uncertainty which drives our efforts to understand what structural features or molecular properties constitute the liabilities that might cause candidates to fail. For this reason, 'drug-likeness' is most commonly thought of in terms of those liabilities, and it is assessed to alert chemists about them.

The ability to assess 'drug-likeness' is even more important in the design of libraries made using high-speed chemistry. If library design is driven primarily by the ease-of-synthesis and structural diversity of an array of compounds, it is likely that significant portions of the library will have undesirable properties, for example, high molecular weight or low solubility. For libraries containing many thousands of compounds, computational methods for assessing 'drug-likeness' need to be applied to filter out molecules with potential liabilities before resources are spent on synthesis, characterization and high-throughput screening.

II ASSESSING 'DRUG-LIKENESS'

A Avoiding known threats

One approach to defining a space of 'drug-like' molecules is to do so by elimination. There are molecular features that from industry experience are known to cause toxicity, lack of bioavailability or other undesirable effects. Based on these features, a set of 'filters' can be defined and applied to libraries of molecules—real or virtual—to eliminate the ones possessing those features.

Exclusion of known undesirable functionalities

If a molecule has functionalities that are known to cause problems in screening, or to make it metabolically unstable or toxic, such a molecule should be classified as 'nondrug-like' and be removed from the library at the design stage. This filter does not require any special computational algorithms, just a substructure search against a list of known undesirable functionalities. There is no universal exclusion list, and the lists in general use are those that combine general medicinal chemistry experience with specific corporate or personal experience.

Compounds that have reactive functional groups can form covalent bonds with proteins and will generate false positives in screening or may appear to be toxic. An example of an exclusion list of reactive groups is shown in Fig. 10.1.[7]

Structural alerts have long been used to detect compounds with potential mutagenicity and carcinogenicity.[8] The following groups are defined as the causative substructure of one or more chemical carcinogens: arylamine function, ring epoxides, alkane sulphonate groups, arylnitro functions, azo groups, ring N-oxides and NMe_2 groups, methylols and aliphatic aldehydes, vinyl groups attached to aromatic rings, aziridines, nitrogen mustards and chloramines, benzyl halides, alkylnitrosamines and alkylurethanes. It should be noted that analysis based on these structural alerts is oversensitive: most carcinogens possess these features, yet there are many benign, noncarcinogenic molecules that have these same functional groups. This is why the exclusion lists applied in practice are subject to corporate or personal choice, especially when general libraries intended for lead discovery are designed. There

is a significant overlap with the reactive functionalities in Fig. 10.1, though, so the reactive groups that appear on both lists should be avoided under all circumstances.

Rule-of-Five

In order to be bioavailable, a drug molecule must be transported through biological membranes. Therefore, molecular properties that correlate with poor membrane permeability (in the absence of active transport) can be used to filter out undesirable molecules.

The 'Rule-of-Five' (Ro5)[9] is one of the best known and widely accepted 'drug-likeness' filters. The Ro5 was derived from an analysis of 2245 drug candidates from WDI[10] with assigned United States Adopted Name (USAN) or International Proprietary Name (INN) that are believed to have reached phase II trials. It states that a compound violating any two of the following rules is likely to be poorly absorbed: (1) molecular weight less than 500 Da; (2) number of hydrogen bond donors (OH or NH groups) equal to or less than 5; (3) number of hydrogen bond acceptors less than 10; and (4) calculated $LogP$ is less than 5.0 (by $ClogP$) or 4.15 (by $MlogP$).

The Ro5 is intended to filter out molecules with potential absorption problems and it does so very well using simple terms that are easily implemented and used by medicinal chemists. This rule should not be expected, however, to discriminate well between drugs and nondrugs. Neither can it provide a quantitative estimate of compound absorption or take into account active transport mechanisms.

Other rule-based filters

A consensus definition of a 'drug-like' molecule has been derived[11] from the analysis of the CMC database[12] by defining qualifying (covering 80% of drug molecules) ranges of calculated physical properties such as molecular weight, $logP$, molar refractivity and number of atoms:

- Molecular weight between 160 and 480, average 357
- Calculated $logP$ between -0.4 and 5.6, average 2.52
- Molar refractivity between 40 and 130, average 97
- Total number of atoms between 20 and 70, average 48

In addition to these property ranges, the consensus definition requires molecules to have a combination of several of the following functional groups: a benzene ring, a heterocyclic ring (both aliphatic and aromatic), an aliphatic amine (preferably tertiary), a carboxamide group, an alcoholic hydroxyl group, a carboxy ester, and a keto group.

Another study[13] analysed property distribution in drug databases and found that 70% of drugs are found within the following ranges of properties: number of H-bond donors between 0 and 2; number of H-bond acceptors between 2 and 9; number of rotatable bonds between 2 and 8; number of rings between 1 and 4.

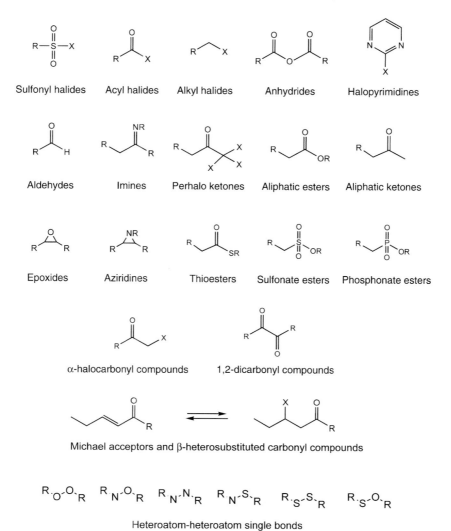

Fig. 10.1 Reactive functional groups responsible for *in vitro* false positives (X = F, Cl, Br, I, tosyl, mesyl, etc.; R = alkyl, aryl, heteroalkyl, heteroaryl, etc.). These reactive functional groups are generally prone to decomposition under hydrolytic conditions (i.e. aqueous Na_2CO_3/methanol). They are reactive towards protein and biological nucleophils (e.g. glutathione, dithiothreitol), and they exhibit poor stability in serum.[7]

An important parameter that strongly correlates with membrane permeability is polar surface area (PSA) — a sum of Van-der-Waals surface areas of polar atoms (oxygens and nitrogens).[14,15] Multiple studies[14,16] have found an upper PSA threshold value of $\sim$ 140–150 Å^2 for oral absorption. Similarly, for blood–brain barrier penetration, the upper PSA threshold is $\sim$90 Å^2 (Fig. 10.2). Recently, fast algorithms for PSA calculation have been suggested that do not need a 3D structure and use molecular topology instead (topological PSA or TPSA).[17] Using these algorithms, a PSA filter can be applied to large collections of real or virtual compounds to eliminate the ones with high PSA.

An even more refined PSA-based filter,[18] which also takes into account molecular flexibility, is based on the analysis of oral bioavailability measurements in rats for over 1100 drug candidates. The following two criteria were suggested for achieving bioavailability above 20–40% in rats: (1) PSA equal to or less than 140 Å^2 (or 12 or fewer H-bond donors and acceptors); (2) 10 or fewer rotatable bonds. The first criterion is related to the desolvation energy of polar groups which is required for permeation through membrane. The second criterion probably reflects the entropic cost of conformational changes required to present an appropriate exterior to the hydrocarbon interior of the membrane.[18]

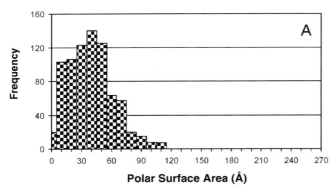

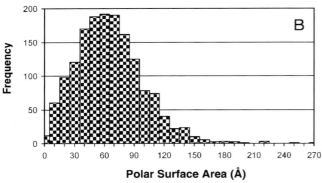

Fig. 10.2 Distribution of the polar surface area (Å) for drugs which have reached at least phase II efficacy studies. (A) 776 orally administered CNS drugs; (B) 1590 orally administered nonCNS drugs.[16]

It is noteworthy that these two criteria worked well independently of molecular weight, a property which has been perceived as very important: consider a 500 Da cutoff that is commonly used to define the 'drug-like' molecular weight range. The perception that this property is important probably arises from the apparent dependence of oral bioavailability on properties which correlate with molecular weight, which certainly include the number of rotatable bonds and PSA or H-bond donor count. As such, the value of a hard molecular weight cutoff may need to be re-examined.

A Pharmacophore Point Filter[19] contains a set of rules based on medicinal chemistry experience and on the observation that nondrug molecules are often underfunctionalized. Fundamental to this system is the evaluation of molecules for the occurrence of pharmacophore points, functional groups that potentially provide key interactions with the receptor or enzyme. The following rules constitute this filter:

(1) Molecules with less than two pharmacophore points are not 'drug-like'.
(2) Molecules with more than seven pharmacophore points are not 'drug-like'.
(3) The basic set of pharmacophore points includes: amine, amide, alcohol, ketone, sulfone, sulfonamide, car-

boxylic acid, carbamate, guanidine, amidine urea, and ester.
(4) Pharmacophore points are fused and counted as one if their heteroatoms are not separated by more than one carbon atom.
(5) Intracyclic amines that occur in the same ring (e.g. piperazine) are counted as one pharmacophore point.
(6) Primary, secondary and tertiary amines are considered pharmacophore points, but not pyrrole, indole, thiazole, isoxazole, other azoles and diazines.
(7) Compounds with more than one carboxylic acid are not 'drug-like'.
(8) Compounds without a ring structure are not 'drug-like'.

When applied to the MDDR[20] and CMC databases, this filter classified almost 70% of molecules as 'drug-like', while in the ACD database only 36% of molecules were found to be 'drug-like'. The advantage of this filter is that it provides a detailed structural reason for classifying molecules as drugs or nondrugs.

The filters described above or others described in the literature[21,22] should not be used indiscriminately. It is critical to consider the context in which they are applied. For example, different property filters should be considered when looking for an oral CNS agent for chronic use and for a chemotherapeutic agent administered intravenously for a short period of time.

B Mimicking known drugs

The introduction of desirable features found in known drugs into the structures of library compounds is an alternative to the use of negative filters.

Privileged structures

The concept of privileged structures has been introduced[23] to describe select structural types capable of binding to multiple, unrelated classes of receptors or enzymes with high affinity. Privileged structures are typically constrained, heterocyclic multiring systems capable of orienting varied substituent patterns in a well-defined 3D space.[24] In addition to known drugs, privileged structures can be derived from natural products.[25] Examples of privileged structures are shown in Fig. 10.3.

Using privileged structures as scaffolds is a powerful approach to making 'drug-like' libraries.[24,26-29] One has to keep in mind, though, that having these scaffolds does not guarantee that all compounds in the library will be bioavailable and nontoxic. Proper choice of substituents is required to avoid structural alerts and keep physico-chemical properties such as PSA or solubility within the desired range.

In recent years, the meaning of the term 'privileged structures' has been expanded to include all structural fragments, not just scaffolds, that are frequently found in drugs. Sets of such fragments are offered commercially as building blocks for libraries.[30,31]

Extraction of privileged substructures from drug databases

Databases of known drugs can be examined to identify any preferred substructures or combinations thereof that could be used in the design of 'drug-like' libraries. An evaluation of 5120 structures from the CMC database identified a diverse set of 1179 different topological frameworks.[32] Only 32 frameworks, however, were required to describe 50% of the drugs in the set (Fig. 10.4). This indicates that a surprisingly small number of common privileged 'shapes' can be used to design a variety of drug molecules with a wide variety of physical properties, binding to different classes of receptors. A six-member ring is the most commonly used framework, with 23 of the above-mentioned 32 frameworks containing at least two linked or fused six-member rings. Privileged frameworks tend to be relatively rigid, with only three out of 32 having more than five rotatable bonds. Only 6% of analysed drugs had an acyclic framework.

An analysis of side chains in drug molecules, performed by the same authors,[33] showed that there are only 1246 different side chains among molecules in the CMC database. The average number of side chains among the molecules is four, and the average number of heavy atoms in each side chain is two, although 66% of all side chains have only one heavy atom. Similar analysis of the same database[11] found that the most abundant functional groups are tertiary amines, alcohols and carboxamides.

RECAP (retrosynthetic combinatorial analysis procedure[34]) is an automated method for extracting privileged structural fragments in a form that makes them usable in combinatorial synthesis. The RECAP procedure starts with collecting a set of structures with activity at a common target or target class — known drugs, for example. Each structure is then 'cleft' into fragments. Eleven chemical bond types at which molecules are cleft have been chosen: amide, ester, amine, urea, ether, olefin, quarternary N, aromatic N– aliphatic C, lactam N–aliphatic C, aromatic C–aromatic C, and sulphonamide. All these bonds are formed using common chemical reactions accessible in a combinatorial format. A collection of obtained fragments is then analysed, and the most frequently occurring representative fragments are incorporated into libraries as building blocks.

Artificial intelligence-based classifiers discriminating drugs from nondrugs

Utilizing privileged structures to mimic known drugs has the advantage of providing chemists with very specific suggestions about what molecular fragments to include in their synthesis. However, this concept does not cover all molecular properties that might make a molecule 'drug-like'. Because of the general vagueness of the term 'drug-like', it is difficult to even define what those properties or combinations of properties are.

Meanwhile, there are several public drug databases[10,12,20] that lend themselves to examination for possible property patterns intrinsic to drugs. Several pattern recognition approaches have been tried[35–39] in which a computer is presented with examples of drugs (molecules from drug databases) and nondrugs (typically, ACD database cleaned

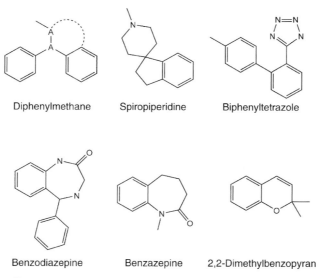

Diphenylmethane   Spiropiperidine   Biphenyltetrazole

Benzodiazepine   Benzazepine   2,2-Dimethylbenzopyran

Fig. 10.3 Examples of 'privileged' structures.

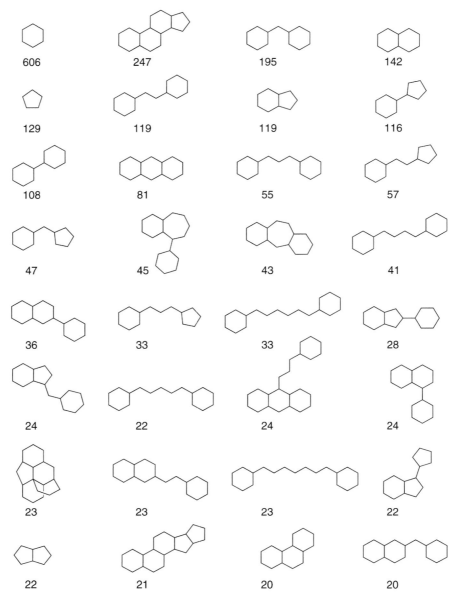

Fig. 10.4 Topological frameworks for compounds in the CMC database (numbers indicate frequency of occurrence).[32] Drug molecules are first reduced to a combination of rings and linkers forming the framework of the molecule, then atom and bond type identities are removed.

of reactive molecules) and is trained to discriminate between them. These classifiers differ from each other in what molecular descriptors or properties are used to present molecules to the computer, and in the type of learning algorithms applied to train the computer. Without describing details of each classifier, it is nonetheless instructive to review these applications as they represent an active area of research; in the future, similar applications might be routinely used by medicinal chemists.

Several types of *artificial neural networks* have been used as pattern recognition engines.[35–37] In general, neural networks are mathematical models that emulate some of the observed properties of biological nervous systems, in particular the ability to learn complex relationships. Examples of 'drug-like' molecules were taken from the WDI, CMC or MDDR databases, while compounds from the ACD database exemplified nondrugs. Molecules were represented by a large number of descriptors, such as ISIS substructure keys,[35] atom types,[36,37] as well as calculated log$P$.[35] On the completion of training, neural networks successfully recognized 80–90% of the molecules in the drug databases as drugs, and 70–90% of the molecules in the ACD database as nondrugs. These numbers are surprisingly high considering the multitude of factors other than chemical structure which contribute to a particular compound becoming a drug: patent protection, complexity and cost of synthesis, competitive situation, government regulations and drug approval policy, etc. A serious drawback of neural network-based classifiers is that they do not provide a clear analysis of molecular properties and structural features that can be used as guidelines for medicinal chemists in designing new molecules.

Genetic algorithms can also be used to distinguish 'drug-like' molecules from 'nondrug-like' ones based on the differential occurrence of several calculated molecular descriptors such as the number of H-bond donors and acceptors, number of rings, molecular weight and number of rotatable bonds.[38] Genetic algorithms are inspired by the Darwinian theory of evolution. Simply put, potential solutions to a problem (in this case molecular property profiles represented by a string of ones and zeros) evolve to arrive at the best one. The algorithm is seeded with a set of possible solutions (chromosomes) called a population. The solutions from one population are mated to produce a new population of solutions (offspring) containing features of their parents but in different combinations. Solutions that are selected for reproduction are selected according to their fitness (in this case ability to discriminate between drugs and nondrugs or between active and inactive compounds); the more suitable they are the likelier they are to be allowed to reproduce. The best solutions found by this classifier provided a six-fold enhancement in recognizing drugs over random selection. Drugs were represented by molecules from the WDI database, and nondrugs by a subset of the SPRESI database.[40] Importantly, an effective discrimination was achieved using only a few simple descriptors. As with neural network-based classifiers, no clear guidelines for medicinal chemists can be derived, although the same authors demonstrate how genetic algorithms can be used to select subsets of a virtual library with optimized diversity and physicochemical properties.[41]

Another classifier that has been used is based on the *decision tree method*.[39] Classification using decision trees is done using a set of tests (or decisions) that are arranged in the form of a tree (hence the name). This classifier was capable of discriminating between potential drugs and nondrugs and showed an efficiency similar to that of neural network-based classifiers (82% of drugs recognized). Compounds from the WDI and ACD databases were used as a training set, and atom types developed for estimating log$P$ were used as molecular descriptors.[42] The main advantage of this type of classifier is that it provides comprehensible models that can be visualized, inspected and interpreted. The authors found that just by testing for the presence of hydroxyl, tertiary and secondary amino, carboxyl, phenol or enol groups, 75% of all the drugs could be correctly recognized. The nondrugs are more aromatic in nature and have fewer functional groups (except halogens).

A concern intrinsic to all these classifiers is that they merely reflect the characteristics of existing drugs, and that using these classifiers as filters would impede the discovery of novel structural classes of drugs in the future. Furthermore, since these classifiers are trained on specific, carefully processed databases, the results could be strongly database-dependent.

C Direct property prediction

Prediction of general 'drug-likeness' based on chemical structure is an ambitious goal considering how vague the concept of 'drug-likeness' is. Many factors determine whether a compound can become a drug or not, and only some of them are structure-dependent. This is why most general 'drug-likeness' classifiers derived from analysing databases of known drugs cannot be expected to be precise and unambiguous. When large general libraries are designed for lead generation, these classifiers are instrumental in filtering out compounds with liabilities from huge virtual libraries. However, at the lead optimization stage of drug discovery projects, when specific issues need to be addressed for specific chemical series, more reliable predictions are desirable.

The issues chemists deal with during lead optimization are the same that contribute to general 'drug-likeness': the need to reduce toxicity or to improve solubility, oral absorption, blood–brain barrier penetration and metabolic stability. Computational methods that can directly predict these properties of compounds would allow for the design of high-speed chemistry libraries which are instrumental in solving these specific needs.

Solubility

Solubility plays an important role at all stages of drug discovery. Insoluble compounds cause problems in screening. Compounds with low solubility are poorly absorbed in the intestine. Predicting solubility is a challenge because it is a complex phenomenon. It depends on compound lipophilicity, the number of H-bonds that can be formed with the solvent, the ability to form intramolecular H-bonds, the ionization state of functional groups, and the properties of the crystalline form, in particular, the nature and energy of crystal packing.

Of particular interest are methods that do not require any measured parameters (for example, melting point) as an input. In one example of such an approach,[43] standard regression statistics has been applied to a series of whole-molecule descriptors that have a straightforward physical interpretation (as opposed to fragment-based descriptors): Lennard-Jones interaction energy, solvent accessible surface area, number of H-bond donors and acceptors, and a count of amino and nitro groups. QikProp,[44] a commercially available program implements this algorithm to calculate solubility and other physicochemical properties. The reader is referred to a more comprehensive account of solubility prediction methods[45] for further details.

Oral bioavailability

Oral bioavailability, a fraction of the oral dose that reaches blood circulation, is also a multimechanism phenomenon. It depends on the drug's solubility, chemical and metabolic stability, and membrane permeability. Most compounds are absorbed through passive diffusion across the membrane, while some are transferred by a carrier or by transporter molecules residing in the membrane.

The key factors determining membrane permeability (for passive transport) include polar surface area and lipophilicity ($\log P$), discussed earlier in this chapter. Many statistical models for predicting permeability have been described, for example in reference 46, and reviewed in detail in reference 47.

Blood–brain barrier penetration

The determinants of blood–brain barrier penetration are similar to the determinants of membrane permeability. They include $\log P$, H-bonding capacity, ionization profile, size and flexibility. A simple QSAR equation has been described[48] to calculate the ratio of the steady-state concentration of the drug molecule in the brain and in the blood:

$$\log(C_{\text{brain}}/C_{\text{blood}}) = -0.0148 * \text{PSA} + 0.152 * C \log P + 0.139$$

Metabolic stability

Currently, there are no reliable methods for the quantitative prediction of metabolic stability. However, over the years researchers have accumulated significant knowledge of which chemical groups might be metabolically labile and which are the resulting metabolites. Two software packages based on knowledge bases are available — MetabolExpert[49] and META.[50] Both provide alerts about potential metabolic pathways but do not indicate how probable those pathways are.

Understanding the structural requirements of specific metabolic enzymes (e.g. various isoforms of P450) is an active area of research, and in the future it might provide quantitative predictive models of metabolic stability.

Toxicity

The main causes of chemical toxicity are side reactions of drug molecules with DNA or proteins, as well as interference with enzymatic systems. There are two databases containing factual toxicological information. One is the RTECS (Registry of Toxic Effects of Chemical Substances), with data on 100 000 compounds.[51] The other is the TOXSYS, with data on 240 000 compounds.[52]

Computational approaches to predicting toxicity are based on expert systems containing empirical or computationally derived rules, typically encoded via structural fragments that could potentially cause toxicity. Several databases are available commercially: rule-based DEREK[53] and HazardExpert,[54] as well as systems based on automated QSAR — TOPKAT[55] and CASETOX.[56] Computational methods for predicting drug toxicity are reviewed in reference 57.

III DESIGNING 'DRUG-LIKE' LIBRARIES

Medicinal chemists design chemical structures with desirable properties, and then choose appropriate synthetic routes utilizing available synthetic precursors. The synthetic repertoire of high-speed chemistry is much more limited because there are fewer applicable synthetic reactions, a lower number of synthetic steps, and the need to execute reactions under the same conditions for diverse sets of starting materials. This explains why library design begins with the definition of a set of potentially synthesizable molecules — a virtual library (VL). The VL is determined by a combination of usable synthetic protocols and available building blocks compatible with those protocols. The molecules that will actually be made are then selected from the VL based on the design purpose. General libraries for screening aim at high structural diversity while lead optimization libraries could focus on a specific structural feature. In each case, there are two ways of ensuring the 'drug-like' character of the library. First is to design a library based on diversity (or similarity) considerations, and then to apply any of the 'drug-likeness' filters described earlier in this chapter. Second is to optimize diversity/similarity, 'drug-likeness' and any other relevant properties simultaneously using an appropriate optimization technique, for example, simulated annealing[58] or genetic algorithms.[41]

A Full combinatorial arrays or cherry-picking?

The original concept of combinatorial chemistry implied making molecules with all possible combinations of building blocks, although one could never be sure that all molecules in the array were actually made, especially when libraries were synthesized as mixtures. With the development of parallel chemistry and high-speed purification technology, making full combinatorial arrays is still an option, although no longer a necessity. In fact, the application of 'drug-likeness' filters almost guarantees that molecules that pass them will not form the full combinatorial array and, therefore, will have to be 'cherry-picked' from the VL and synthesized individually. An implementation of such a noncombinatorial approach as an automated system for the synthesis and purification of library compounds, including the design software, has recently been described.[59] The advantage of the 'cherry-picking' approach is that it gives medicinal chemists complete freedom to choose any selection of single compounds from a given VL that fit the desired physical property constraints and/or uses any available structural information (for example, protein X-ray structures or computationally derived pharmacophores).

B 'Drug-like' or 'lead-like'?

FLAThe various 'drug-likeness' assessment tools and filters described above have been developed and applied to combinatorial libraries with the intention of obtaining 'high-quality' lead compounds. Such an approach is based on an implicit assumption that any 'drug-like' molecule is in fact a 'high-quality' lead. For some this reflects an original naive hope from the early days of combinatorial chemistry that the generation and screening of large libraries could lead directly to a drug. Others observed that in many projects the structure of a drug candidate sent to clinic was very close to the structure of the original lead, so it made sense to start with a lead that resembled a drug most closely.

Certainly, a 'drug-like' lead is more desirable than some carcinogenic or insoluble or too polar compound. However, it has recently been questioned[60,61] whether 'drug-like' lead compounds are indeed the best starting points for optimization. Evaluation of 470 pairs of drugs with their prototype leads indicated that on average leads tend to have lower molecular weight, lower logP, fewer aromatic rings and fewer hydrogen bond acceptors. In other words, during lead optimization, molecular weight and lipophilicity tend to increase. This happens because, in the absence of structural information, potency is more likely to be gained by adding more nondirectional hydrophobic interactions than through forming a new, highly direction-sensitive, hydrogen bond. Another possible reason is that changing the bulk properties

of a molecule is often used as a means of improving the PK properties.

If an increase in molecular weight and lipophilicity is to be expected during optimization, it has been suggested[60] that screening libraries should consist of 'lead-like' compounds with molecular weight in the 100–350 range and ClogP between 1 and 3. Leads from these libraries are not expected to have very high affinity, but they have the potential for introducing additional drug–receptor interactions without increasing molecular weight or lipophilicity beyond 'drug-like' ranges. Owing to their smaller size, 'lead-like' molecules are more likely to find a binding mode than larger molecules containing the same 'lead-like' fragment.[61] To achieve a level of affinity measurable in HTS (10–20 μM), 'lead-like' libraries should be enriched with functionalities capable of forming strong interactions at the receptor site, for example, ionizable functional groups. Special attention has to be paid when following up on these 'weak' HTS hits in order to separate them from the numerous false positives usually found at this low level of potency.

REFERENCES

1. Blake, J. F. (2000) Chemoinformatics—predicting the physicochemical properties of 'drug-like' molecules. *Curr. Opin. Biotechnol.* **11**: 104–107.
2. Polinsky, A. (1999) Combichem and cheminformatics. *Curr. Opin. Drug Discov. Devel.* **2**: 197–203.
3. Matter, H., Baringhaus, K. -H., Naumann, T., Klabunde, T. and Pirard, B. (2001) Computational approaches towards the rational design of drug-like compound libraries. *Comb. Chem. High Throughput Screen* **4**: 453–475.
4. Walters, W.P., Ajay, A. and Murcko, M.A. (1999) Recognizing molecules with drug-like properties. *Curr. Opin. Chem. Biol.* **3**: 384–387.
5. Mitchell, T. and Showell, G.A. (2001) Design strategies for building drug-like chemical libraries. *Curr. Opin. Drug Discov. Devel.* **4**: 314–318.
6. Podlogar, B.L., Muegge, I. and Brice, J.L. (2001) Computational methods to estimate drug development parameters. *Curr. Opin. Drug Discov. Devel.* **4**: 102–109.
7. Rishton, G.M. (1997) Reactive compounds and *in vitro* false positives in HTS. *Drug Discov. Today* **2**: 382–384.
8. Ashby, J. (1985) Fundamental structural alerts to potential carcinogenicity or noncarcinogenicity. *Environ. Mutagen.* **7**: 919–921.
9. Lipinski, C.A., Lombardo, F., Dominy, B.W. and Feeney, P.J. (1997) Experimental and computational approaches to estimate solubility and permeability in drug discovery and development settings. *Adv. Drug Deliv. Rev.* **23**: 3–25.
10. World Drug Index. Derwent Information, Alexandria, VA www.derwent.com.
11. Ghose, A.K., Viswanadham, V.N. and Wendoloski, J.J. (1999) A knowledge-based approach in designing combinatorial or medicinal chemistry libraries for drug discovery. 1. A qualitative and quantitative® characterization of known drug databases. *J. Comb. Chem.* **1**: 55–68.

12. Comprehensive Medicinal Chemistry Database. Distributed by MDL Information Systems, San Leandro, CA.

13. Oprea, T.I. (2000) Property distribution of drug-related chemical databases. *J. Comput.-Aided Mol. Des.* **14**: 251–264.

14. Palm, K., Stenberg, P., Luthman, K. and Artursson, P. (1997) Polar molecular surface properties predict the intestinal absorption of drugs in humans. *Pharm. Res.* **14**: 568–571.

15. Clark, D.E. (1999) Rapid calculation of polar molecular surface area and its application to the prediction of transport phenomena. 1. Prediction of intestinal absorption. *J. Pharm. Sci.* **88**: 807–814.

16. Kelder, J., Grootenhuis, P. D. J., Bayada, D.M., Delbressine, L. P. C. and Ploemen, J. -P. (1999) Polar molecular surface as a dominating determinant for oral absorption and brain penetration of drugs. *Pharm. Res.* **16**: 1514–1519.

17. Ertl, P., Rohde, B. and Selzer, P. (2000) Fast calculation of molecular polar surface area as a sum of fragment-based contributions and its application to the prediction of drug transport properties. *J. Med. Chem.* **43**: 3714–3717.

18. Veber, D.F., Johnson, S.R., Cheng, H.-Y., Smith, B.R., Ward, K.W. and Kopple, K.D. (2002) Molecular properties that influence the oral bioavailability of drug candidates. *J. Med. Chem.* **45**: 2615–2623.

19. Muegge, I., Heald, S.L. and Brittelli, D. (2001) Simple selection criteria for drug-like chemical matter. *J. Med. Chem.* **44**: 1841–1846.

20. MACCS-II Drug Data Report Database. Distributed by MDL Information Systems, Inc., San Leandro, CA.

21. Wang, J. and Ramnarayan, K. (1999) Toward designing drug-like libraries: a novel computational approach for prediction of drug feasibility of compounds. *J. Chem. Inf. Comput. Sci.* **1**: 524–533.

22. Xu, J. and Stevenson, J. (2000) Drug-like index: a new approach to measure drug-like compounds and their diversity. *J. Chem. Inf. Comput. Sci.* **40**: 1177–1187.

23. Evans, B.E., Rittle, K.E., Bock, M.G., DiPardo, R.M., Freidinger, R.M., Whitter, W.L., Lundell, G.F., Veber, D.F., Anderson, P.S., Chang, R. S. L., Lotti, V.J., Cerino, D.J., Chem, T.B., Kling, P.J., Kunkel, K.A., Springer, J.P. and Hirshfield, J. (1988) Methods for drug discovery: development of potent, selective, orally effective chole-cystokinin antagonists. *J. Med. Chem.* **31**: 2235–2246.

24. Mason, J.S., Morize, I., Menard, P.R., Cheney, D.L., Hulme, C. and Labaudiniere, R.F. (1999) New 4-point pharmacophore method for molecular similarity and diversity applications: overview of the method and applications, including a novel approach to the design of combinatorial libraries containing privileged substructures. *J. Med. Chem.* **42**: 3251–3264.

25. Nicolaou, K.C., Pfefferkorn, J.A., Roecker, A.J., Cao, G.-Q., Barluenga, S. and Mitchell, H.J. (2000) Natural product-like combinatorial libraries based on privileged structures. 1. General principles and solid-phase synthesis of benzopyrans. *J. Am. Chem. Soc.* **122**: 9939–9953.

26. Dolle, R.E. (1998) Comprehensive survey of chemical libraries yielding enzyme inhibitors, receptor agonists and antagonists, and other biologically active agents: 1992 through 1997. *Mol. Divers.* **3**: 199–233.

27. Dolle, R.E. and Nelson, K.H., Jr. (1999) Comprehensive survey of combinatorial library synthesis. (1998) *J. Comb. Chem.* **1**: 235–282.

28. Andres, C.J., Denhart, D.J., Deshpande, M.S. and Gillman, K.W. (1999) Recent advances in the solid phase synthesis of drug-like heterocyclic small molecules. *Comb. Chem. High Throughput Screen* **2**: 191–210.

29. Bunin, B.A., Plunkett, M.J. and Ellman, J.A. (1996) Synthesis and evaluation of three 1,4-benzodiazepine libraries. In Jung, G. (ed.). *Combinatorial Peptide and Nonpeptide Libraries. A Handbook*, pp. 405–424. Weinheim, New York.

30. PHARMABlock® Proprietary Combinatorial Building Blocks. Chem-Bridge Corporation, San Diego, CA, www.chembridge.com.

31. Optimer Building Blocks. Array Biopharma, Bolder, CO, www.arraybiopharma.com.

32. Bemis, G.W. and Murcko, M.A. (1996) The properties of known drugs. 1. Molecular frameworks. *J. Med. Chem.* **39**: 2887–2893.

33. Bemis, G.W. and Murcko, M.A. (1999) Properties of known drugs. 2. Side chains. *J. Med. Chem.* **42**: 5095–5099.

34. Lewell, X.Q., Judd, D.B., Watson, S.P. and Hann, M.M. (1998) RECAP-retrosynthetic combinatorial analysis procedure: a powerful new technique for identifying privileged molecular fragments with useful applications in combinatorial chemistry. *J. Chem. Inf. Comput. Sci.* **38**: 511–522.

35. Ajay, A., Walters, W.P. and Murcko, M.A. (1998) Can we learn to distinguish between 'drug-like' and 'nondrug-like' molecules? *J. Med. Chem.* **41**: 3314–3324.

36. Sadowski, J. and Kubinyi, H. (1998) A scoring scheme for discriminating between drugs and nondrugs. *J. Med. Chem.* **41**: 3325–3329.

37. Frimurer, T.M., Bywater, R., Naerum, L., Lauritsen, L. and Brunak, S. (2000) Improving the odds in discriminating 'drug-like' from 'nondrug-like' compounds. *J. Chem. Inf. Comput. Sci.* **40**: 1315–1324.

38. Gillet, V.J., Willett, P. and Bradshaw, J. (1998) Identification of biological activity profiles using substructural analysis and genetic algorithms. *J. Chem. Inf. Comput. Sci.* **38**: 165–179.

39. Wagener, M. and van Geerestein, V.J. (2000) Potential drugs and nondrugs: prediction and identification of important structural features. *J. Chem. Inf. Comput. Sci.* **40**: 280–292.

40. The SPRESI database. Daylight Chemical Information Systems, Inc., Mission Vieho, CA www.daylight.com.

41. Gillet, V.J., Willett, P. and Bradshaw, J. (1999) Selecting combinatorial libraries to optimize diversity and physical properties. *J. Chem. Inf. Comput. Sci.* **39**: 167–177.

42. Ghose, A.K. and Crippen, G.M. (1986) Atomic physicochemical parameters for three-dimensional structure-directed quantitative structure-activity relationships I. Partition coefficients as a measure of hydrophobicity. *J. Comput. Chem.* **7**: 565–577.

43. Jorgensen, W.L. and Duffy, E.M. (2000) Prediction of drug solubility from Monte Carlo simulations. *Bioorg. Med. Chem. Lett.* **10**: 1155–1158.

44. QikProp. Schrodinger, Inc., Portland, OR www.schrodinger.com.

45. Taskinen, J. (2000) Prediction of aqueous solubility in drug design. *Curr. Opin. Drug Discov. Devel.* **3**: 102–107.

46. Egan, W.J., Merz, K.M., Jr. and Baldwin, J.J. (2000) Prediction of drug absorption using multivariate statistics. *J. Med. Chem.* **43**: 3867–3877.

47. Pagliara, A., Reist, M., Geinoz, S., Carrupt, P.A. and Testa, B. (1999) Evaluation and prediction of drug permeation. *J. Pharm. Pharmacol.* **51**: 1339–1357.

48. Clark, D.E. (1999) Rapid calculation of polar molecular surface area and its application to the prediction of transport phenomena. 2. Prediction of blood–brain barrier penetration. *J. Pharm. Sci.* **88**: 815–821.

49. MetabolExpert. CompuDrug, South San Francisco, CA www.compudrug.com.

50. META. Charles River Laboratories, Wilmington, MA www.criver.com.

51. Registry of Toxic Effects of Chemical Substances. National Institute for Occupational Safety and Health (NIOSH). www.cas.org/ONLINE/DBSS/rtecsss.html.

52. TOXSYS. SciVision, Burlington, MA www.scivision.com.

53. DEREK. LHASA group at Harvard University. Cambridge, MA www.lhasa.harvard.edu.

54. HazardExpert. CompuDrug, South San Francisco, CA www.compudrug.com.

55. TOPKAT. Accelrys, San Diego, CA www.accelrys.com.

56. CASETOX. Charles River Laboratories, Wilmington, MA www.criver.com.

57. Cronin, M.T. (2000) Computational methods for the predicition of drug toxicity. *Curr. Opin. Drug Discov. Devel.* **3**: 292–297.

58. Brown, R.D., Hassan, M. and Waldman, M. (2000) Combinatorial library design for diversity, cost efficiency, and drug-like character. *J. Mol. Graph. Model.* **18**: 427–437.

59. Everett, J., Gardner, M., Pullen, F., Smith, G.F., Snarey, M. and Terrett, N. (2001) The application of non-combinatorial chemistry to lead discovery. *Drug Discov. Today* **6**: 779–785.

60. Teague, S.J., Davis, A.M., Leeson, P.D. and Oprea, T. (1999) The design of leadlike combinatorial libraries. *Angew. Chem., Int. Ed. Engl.* **38**: 3743–3748.

61. Hann, M.M., Leach, A.R. and Harper, G. (2001) Molecular complexity and its impact on the probability of finding leads for drug discovery. *J. Chem. Inf. Comput. Sci.* **41**: 856–864.

11

WEB ALERT — USING THE INTERNET FOR MEDICINAL CHEMISTRY

David Cavalla

In five years time (i.e., in 50 Internet years time) we won't be having another ChemInt ('Chemistry and the Internet') meeting. We might have "Molecular Sciences on the Internet" but we won't call the scientists "chemists". The metaphor we call the Net will change. You won't need a wire or optic fiber to connect your computers: you will use wireless transmission.
Alan Arnold

I INTRODUCTION

Over the past three years the landscape of scientific information has exploded, and the expectation is for substantial future change. This article can only describe the situation as it currently stands and gives some predictions as to the future. It is in the nature of such reviews that they can never be complete and up to date, moreover they deteriorate rapidly with time.

This review is intended to provide resources for the tools that medicinal chemists generally use in the work they do, which necessarily involves a variety of tasks, from drug design to chemical synthesis. Sites for prediction of physical

activity are just as relevant as those for prediction of biological activity and patents. The chapter is divided into sections, but besides the meta-search sites (see below) already there are some information providers and community sites for chemists that fit into more than one section. Perhaps the best known of these is ChemWeb (http://www.chemweb.com). This site provides a range of services to visitors, from a newsletter to detailed access to databases. The information is sometimes free, sometimes for a price. ChemWeb acts as a gateway to about 35 different databases (currently), some of them without restrictions. The databases described on the ChemWeb site tend to change regularly.

The statistics of use of ChemWeb provide an inkling as to the exponential rise in access and use of the web for news and information. These figures are likely to increase substantially. For instance, the usage of ChemWeb has risen from 13 147 in March 1998 to 83 229 in March 2001, an increase of over six-fold in three years. The statistics for the University of Cambridge Chemistry Department server are even more invigorating: from under 1000 in June 1994 to over 100 000 by May 2001, and the rate of increase is still growing. Similarly for the nomenclature website hosted by Queen Mary College, London, usage has grown from around 400 over a 4-week period at the beginning of 1996 to nearly 35 000 for a similar period at the end of 2000. These figures suggest, in the UK at least a growth in usage of approximately two orders of magnitude over seven years.

In the future, the prognosis is for a number of changes. The present fragmented situation is likely to consolidate, and prediction of which sites will remain active in the next decade is hazardous. Currently, much of the information is available without monetary compensation, although site owners are finding out about their users by requiring them to donate personal information, which is used to contact them when enhancements occur and/or the sites assumes a more commercial basis of operation. Though there is a gathering momentum for commercialization (such as, for instance the recent adoption of a subscription-based price model for the patent information site Delphion (http://www.delphion. com), there are also new sites springing up which offer similar information access for nothing. Commercial prospects are hemmed in by not-for-profit organizations putting information out for free public access. In the case of Delphion, its commercial stance is in the framework of a market bounded by the publicly accessible and free US Patent and Trademark Office (http://www.uspto.gov) and the European Patent Agency (http://ep.espacenet.com/). In other cases, pricing models requiring the provider of the information to pay a fee (such as Pharmalicensing, http://www.pharmalicensing.com) permit the user to gain information without paying.

II COMPOUND INFORMATION

Scientific Technical Network (STN) (http://stnweb.cas.org/) is probably the most useful and extensive structure-searchable resource for chemistry worldwide. More than 200 scientific databases are available via STN (this service is not free). Several related databases can be searched simultaneously ('cluster search'). Cluster presearch is free and very useful, as the number of relevant hits per database can be visualized in this way. The most important files on STN are: the CA Registry, CA Plus, Beilstein and the INPADOC (patent data) file.

Beilstein Abstracts (http://www.chemweb.com/data bases/belabs) is one of the most important databases in organic chemistry due to its completeness (although not for patent data). It is clearly data-oriented. About 300 different physical properties can be searched and retrieved for each compound, if they are reported in the literature. Full access to the database is restricted, but Beilstein Abstracts is available free through ChemWeb (author, title or abstracts are searchable).

ChemIDplus (http://chem.sis.nlm.nih.gov/chemidplus) is a search engine which allows retrieval of about 350 000 chemical substance files. Files may include structure (65 000 structures available), official name, systematic name, other names, classification code (therapeutic use), molecular formula, STN locator code and CAS registry number.

ChemFinder (http://chemfinder.camsoft.com/) is a very large and specific chemical substances search engine, which provides basic information about chemicals, such as physical property data and 2D chemical structures. Obvious spelling errors and invalid CAS registry numbers are corrected. Chemicals and pharmaceuticals can be searched by chemical name, CAS registry number, molecular formula or molecular weight. About 75 000 compounds are registered to date.

III NAMING OF CHEMICALS

There are a number of chemical naming programs on the market, and competition is intensifying. Perhaps as a result, Beilstein's version can now be used free of charge to ChemWeb.com members at http://ChemWeb.com/ autonom. The program works well with Netscape Navigator requiring Isis Draw or Chime to draw structures for input, which are then named according to the IUPAC convention, although it needs careful handling with Internet Explorer 5.0. Two other products are available at present; one is Nomenclator, from Cheminnovation (http://www. cheminnovation.com/nomencla.html), the other from ACDLabs who have two products at http://www.acdlabs. com/products/name_lab/, which can produce names according to the IUPAC or the CAS systems. Not all of these programs produce the correct results under all test conditions, and a degree of caution is suggested. ACDLabs have recently produced some software that can work from the chemical structural image to the name.

The International Union of Pure and Applied Chemistry recommendations on organic & biochemical nomenclature, symbols & terminology, etc. are obtainable at http://www. chem.qmw.ac.uk/iupac/

IV CHEMICAL SUPPLIERS

There is currently a very large amount of information on available chemicals on the web. This information is relevant for both laboratory-scale synthesis and for larger scale preparations, however, it is more easily searched for as laboratory synthesis. A useful reference is to a mega-search site which incorporates the websites of on-line searchable chemicals and suppliers; for example http://www.mdpi.org/ forum.htm#chemicals offers a range of options, which are also accessible through its European mirror site at http:// www.unibas.ch/mdpi/forum.htm#chemicals

Examples of meta-sites for chemical suppliers on-line are provided in Table 11.1. The site http://www.chemexper. com also allows access to Expereact™ WEB, a laboratory management program that helps to keep stock control, order products, add reactions (electronic laboratory journal), export all the information to a word processor, etc.

Another site providing software for inventory management is ChemSW, at http://www.chemsw.com. Products include the CIS Inventory system Pro 2000, and a digital 'MSDS digital filing cabinet' (very useful for managing data

Table 11.1 *Meta-sites for chemical suppliers on-line*

Website address	Comments
http://pubs.acs.org/chemcy/	Chemcyclopedia online. Also access to Chemcenter for registered ACS members
http://www.chemacx.com	Available Chemicals Xchange features the complete catalogues of over 200 vendors
http://www.buyersguidechem.de/	Excellent site with wide variety of chemicals, no prices; useful for bulk
http://www.chemexper.be/	Excellent search capability on a wide variety of research chemicals, and information that includes the exact chemical name as well as formula, melting point and other physical properties. Searching can be conducted by CAS number, molecular formula, substructure, name and a range of other terms. Hot links allow the user to directly go to the individual supplier
http://www.molmall.org	MolMall features the Rare Chemical Samples ExchangeCenter. Compounds are made available from small samples provided by individual researchers. Full structure search or substructure searches are permitted on the website, as well as for the name of the submitter and several other very useful searching functions. Links to Molecules MolBank (http://www.molbank.org) papers if the compounds are published there. There are plans to allow the sample submitters to add additional information to the data sheet, such as the literature where the compound was published
http://www.chemexpo.com/	Fairly wide variety of chemicals but no prices
http://www.chemicalonline.com/	202 companies listed under 'Custom Synthesis and Manufacturing'
http://www.chem-edata.com	Text search capability. Fairly limited selection
http://www.chemsources.com/chemonline.html	The database includes the products of more than 7200 chemical firms spanning 128 countries. The Chemical Section lists approximately 200 000 chemical compounds. Charge $3.95 per supplier hit
http://www.sourcerer.co.uk/asp/english/onlineen.asp	Relatively useful, includes more than 100 000 products and services with comprehensive information on the supplying companies

sheets as they go out of date), as well as more conventional chemical drawing and molecular modelling programs.

Many suppliers offer database searching capabilities themselves. Large companies such as Sigma-Aldrich have managed to offer a complete searchable database of their products, by name, structure and CAS number (http://www.sigma-aldrich.com). They also feature on-line ordering via a secure interface. The smaller suppliers have been later in arriving at an on-line database with searching and secure ordering. Until recently however Maybridge (http://www.maybridge.co.uk) and another small chemical supplier, Specs (http://www.specs.net), which acts as a broker of academically derived molecules, supplied regular updates of their catalogue on CD-ROM. In the future more business will be delivered through on-line services and it is possible that the CD-ROM versions will disappear, although the paper-based catalogues still fulfil a function that will not be completely replaced by computerized versions.

There are some commercial database products in addition to the sources listed above, such as ChemSources International, which includes the products of more than 8000 chemical companies worldwide. The Chemical Section lists approximately 250 000 chemical compounds and provides contact data necessary for making direct enquiries to each chemical firm. The product is available at a cost of approximately $1000 for a single user CD-ROM license.

CHEMCATS is an on-line database accessible through Chemical Abstracts that contains over 1.3 million unique substances with CAS Registry Numbers, including pricing information when available from suppliers and for many, direct hyperlinks to suppliers' sites as well. CHEMCATS is routinely updated with new information provided by suppliers already in the database and with new suppliers and/or catalogues. This is another commercial product, but there is no up-front fee: pricing is based on pay as you go and can be accessed through STN Easy.

The Available Chemicals Directory (ACD) is available through http://www.chemweb.com (see above) along with a wide range of other useful databases for a commercial price.

V CHEMICAL SYNTHESIS

WebReactions (http://www.WebReactions.net) is a new, unique reaction search system offering direct retrieval of reaction precedents through the Internet. For a limited time the 40 0000-reaction archive from InfoChem GmbH, Munich, is available for searching without cost. The WebReactions system is easy to learn and use, though it works only with Netscape Navigator, not Internet Explorer 5.0. It is virtually instantaneous in displaying matches, not just for the input reaction itself, but for as broad a range of analogues as desired.

The complete Organic Synthesis (OS) is now available free on-line at http://www.orgsyn.org/. Exact and substructure searches are supported following download of a free ChemDraw plug-in, as well as chemical name, formula, OS reference, keyword index searches. This site is available free of charge to all chemists and contains all of the nine collectives, as well as annual volumes and indices. Organic Syntheses (OS) is a compilation of 77 annual volumes containing selected and independently checked procedures and new reactions in the field of organic synthesis. Since the 1920s, volumes of OS consisting of synthetic procedures have been published annually. The first six collective volumes were published every ten years, and the last three at five-year intervals.

Two other sites related to chemical synthesis include http://orgchem.chem.uconn.edu/namereact/named.html which includes details of about 100 named reactions (including a recent example and literature references); and the reaction index at http://www.pmf.ukim.edu.mk/PMF/Chemistry/reactions/rindex.htm, which contains a very extensive list of named reactions in organic chemistry.

For biotechnological synthesis, there is a superb database containing information on microbial biocatalytic reactions and biodegradation pathways for primarily xenobiotic, chemical compounds. It is called the University of Minnesota Biocatalysis/Biodegradation Database (UM-BBD) and can be found at http://umbbd.ahc.umn.edu/search/index.html. The goal of the UM-BBD is to provide information on microbial enzyme-catalysed reactions which are important for biotechnology.

VI COMBINATORIAL CHEMISTRY

An attempt at amalgamation of the disparate set of information providers in combinatorial chemistry has been made through the establishment of the Combi-web consortium (http://www.combi-web.com/), which links to Combichem.net (http://www.combichem.net), Combichemlab.com (http://www.combichemlab.com), 5z.com (http://www.5z.com) and Combinatorial.com (http://www.combinatorial.com). The consortium is useful in identifying the offerings of these similarly named providers, and helping the user to identify the various strengths and weaknesses of each. Presently, the most complete and relevant site devoted to combinatorial chemistry is http://www.combichem.net. This site has news items from both the business and technical communities in this field. For instance, press releases of the latest alliances between companies in the start-up sector with their large pharmaceutical cousins are reported, along with new products from the combinatorial supplier companies. Meetings, chemicals,

libraries, methods, are all there. Combichem's discussion group at http://www.groups.yahoo.com/group/combiChem/ has forums on automation, compounds and resins, informatics and materials technology, and should also be considered in comparison with the existing Laboratory Robotics Interest Group, which is focused on robotics applications in the laboratory. This group's membership consists of over 5000 scientists and engineers from a wide range of areas in the USA; further details can be found at http://lab-robotics.org/conferen.htm, and there is a useful list of moderated newsgroups related to laboratory automation at http://lab-robotics.org/newsgrou.htm. The 5z site (http://www.5z.com) is home to the publication *Molecular Diversity* and includes an archive of this journal's articles, as well as a list of companies active in the combinatorial chemistry area. Another of the links is to the seemingly more authoritatively named http://www.combinatorial.com, which offers a much more basic set of information.

A site worth watching for the future is http://www.combichemlab.com. This academic-oriented site is intended to provide free on-line information in the areas of combinatorial chemistry, high-throughput screening, and other related topics. Refreshingly, these days, there are no specials, offers or advertisements of any kind.

Finally, there are references to companies in the combinatorial and allied laboratory automation fields at http://www.warr.com/ombichem.html and in databases of pharmaceutical technology providers (see below).

VII PHYSICAL CHEMICAL PREDICTION

Syracuse Research Corporation, experts and consultants in environmental chemistry and toxicology (SRC) (http://esc.syrres.com/interkow/onlinedb.htm) offers free on-line searches of a number of physical property databases as a public service, including on-line $\log P$ measurements (octanol-water partition coefficient), environmental fate and physical properties databases. The NIST Chemistry WebBook is another useful source of chemical property information, at http://webbook.nist.gov/chemistry/. ChemFinder (http://www.chemfinder.com) also offers general information and many links concerning chemicals. Each search result gives structure, synonyms and physical data for the compound in question and then a list of links.

In addition to measurement, there are several methods to calculate $\log P$, not all giving the same answer, and with varying exactitude compared with the experimental figure. The SRC site mentioned above links to their proprietary KowWin software (http://esc.syrres.com/interkow/logkow.htm); BioByte's clogP software works for large datasets and has become the *de facto* industry standard. It can be tried for free for individual structures at http://www.daylight.com.

Another program is XLOGP v2.0, which can be downloaded free via anonymous FTP to ftp2.ipc.pku.edu.cn/pub/software/xlogp/. (Unfortunately it requires registration for an encryption key to be revealed.) XLOGP is an atom-additive method for calculating the log P value for a given compound by summing the contributions from component atoms and correction factors. It is claimed that XLOGP v2.0 gives better results than other atom-additive approaches and is at least comparable to some popular fragmental approaches. XLOGP v2.0 is written in C++ and has been tested on UNIX, LINUX and DOS platforms.

The Interactive Laboratory (I-Lab: http://www.acdlabs.com/ilab) available from Advanced Chemistry Development (ACD) provides on-line computation of molecular physical properties according to ACD's industry algorithms for Log P, pKa, Log D and aqueous solubility. These properties are key for making good decisions in medicinal chemistry, and this software has been licensed to a number of pharmaceutical and biotechnology companies. I-Lab also includes database searching of ACD's compilations of Spectra and Physical Properties. These are available at the URL above and also via http://www.chemweb.com. The ACD/log P calculator has been compared with competitive products at: http://www.acdlabs.com/products/phys_chem_lab/logp/competit.html. It is claimed to calculate an accurate Log P derived from an internal ACD/Log P database containing over 5000 experimental Log P values.

Finally, a very versatile program for calculating multiple physical properties of molecules is available from Schrodinger (http://www.schrodinger.com/Products/qikprop.html). This software offers the capability to calculate the solubility and permeability across gut and brain membranes, as well as the Log P of a molecule.

VIII CHEMICAL STRUCTURE DRAWING AND VIEWING PROGRAMS

There are now a number of 2D and 3D molecule viewers available for free download, which work to make chemical structures visible on web pages (Table 11.2). A summary of some of the available software is also held at http://www.indiana.edu/~cheminfo/mvts.html.

Unfortunately, the range of possibilities for conversion from paper-based graphical format into chemical structures is not so wide. This subject comes close to the graphic-to-text conversion that underpins optical character recognition of a scanned image, and has been termed 'chemical OCR'. The only known programme for enabling this conversion is Kekulé supplied by Oxford Molecular (http://www.oxmol.com/software/kekule/spec.shtml), now part of Pharmacopeia (http://www.pcop.com).

IX ANALYSIS

The Analytical Forum on ChemWeb (http://analytical.chemweb.com/search/search.exe) gives access to probably one of the most interesting sites of analytical chemistry on the web. Extensive search options in the 'Analytical Abstracts Database (AA)', as well as in a literature database (collection of analytical journals) are provided after registration (free). Articles of AA can be purchased on-line; references from the literature database generally also available as abstracts without cost.

The Analytical Abstracts database (http://www.rsc.org/CFAA/AASearchPage.cfm) can be searched according to analyte name, matrix name and analytical technique. CAS registry numbers are also accepted and searches can be restricted to publication year and other parameters. Hits give access to the title, the author and the publication month. The full abstract can be purchased.

Analytik (http://www.analytik.de/) is a very comprehensive German information site for analytical chemists. It relates discussions of analytical problems, contains a small but excellent link collection to chemical databases and literature (with an application database) and press releases from the German Chemical Society etc.

Spectra Online (http://spectra.galactic.com/) provides about 6000 IR, MS, NMR, UV/VIS and NIR spectra which can be consulted free on this site. Retrieval of information is possible by entering the compound name, CAS registry number, molecular formula etc. Requires registration (free).

Macherey-Nagel (http://www.macherey-nagel.com/mnappli.nsf/) provides GC, HPLC, SFC, SPE and TLC methods with details and literature references listed on this site. A classification is made according to categories of products (e.g. HPLC analysis of amides, amines, etc.).

The Buyer's Guide Company chromatography search engine at (http://www.bygd.com/search/Chromatography/) provides information on a collection of suppliers of analytical instruments and columns covers the following areas: CE, electrophoresis, LC, supercritical LC and TLC.

ChemicalAnalysis.com (http://www.chemicalanalysis.com) matches organizations and individuals in need of chemical analysis with a database of laboratories and consultants from around the world. This service is free to organizations and individuals requesting chemical analysis. Laboratories and consultants only incur a nominal fee after an organization or individual has accepted their bid.

X CHEMICAL JOURNALS

Nearly all journals have a web presence, and an increasing majority have electronic versions of their publications (including archives) available through the website.

Table 11.2 Chemical structure drawing and viewing programs

Viewer	Description	URL
Chime	The Chime plug-in displays 2D and 3D molecules directly within a web page and works with both Netscape and Microsoft browsers. The molecules in the web page are 'live', meaning they are not just pictures, but chemical structures that can be rotated, reformatted and saved in various file formats for use in modelling or database software	http://www.mdli.com
JMol	JMol, initiated by Dan Gezelter at Columbia University is a free, open source molecule viewer and editor. It is a collaboratively developed visualization and measurement tool for chemical scientists	http://jmol.sourceforge.net/
ChemDraw plug-in	This is claimed as being more than a mere structure viewer or a slow Java applet, rather it runs as fast and is as familiar as the regular ChemDraw application. Available without charge, it enables searching of web databases by structure or substructure, and viewing of ChemDraw documents that others have placed on the web	http://www.camsoft.com
JME molecular editor	JME Molecular Editor is a Java program (free for non-commercial use), which enables drawing and editing of molecules and reactions in a stand-alone mode or as an applet directly within an HTML page. The Editor can generate SMILES of created structure, so it can be used as a front-end to molecular and reaction databases or property calculation services	http://www.ch.ic.ac.uk/vchemlib/mol/search/spurt/
MarvinSketch and MarvinView	Marvin* is a Java-based chemistry software that is available in various forms. Marvin Applets are created for building chemical Internet/Intranet sites. The package contains MarvinSketch, an applet for editing and visualizing molecules on a web page, and MarvinView, an applet for viewing molecules on a web page. MarvinView can display a 2D or 3D molecule, or many molecules in a table	http://www.chemaxon/marvin
WebMolecules	Web visualization of molecules in 3D — in real-time — has now been achieved. Over 150 000 molecular models are available on-site, which may be searched by CAS number or exact formula. In addition partial formula searching is permitted to look into the Top 2000 molecules, which includes molecules of commercial value, educational importance, and of topical interest. Thousands of common molecules are organized into over 30 categories	http://www.webmolecules.com
Waltz and CSD	ChemViz (short for chemistry visualization) is a set of web-based applications created by NCSA designed to catalyse a better understanding of chemical processes through visualization. Two tools are currently supported: Waltz, which generates images and animations of desired molecules and ions; and CSD, which presents a three-dimensional model of complex organic compounds	http://chemviz.ncsa.uiuc.edu/
Depict	Daylight's Depict service, accepts a SMILES string as input and returns an HTML page with an embedded image. Unfortunately, there is no control on the output style and image size	http://www.daylight.com/daycgi/depict
GIF/PNG-Creator	Converts chemical structures into.gif format (increasingly important as a portable encoding standard for image data). Among other formats, it will accept a SMILES string as input and generate a.gif or.png image	http://www2.ccc.uni-erlangen.de/services/gif.html
Corina	CORINA (COoRdINAtes) is a rule and data based system that automatically generates three-dimensional atomic coordinates from the constitution of a molecule as expressed by a connection table or linear code, and can convert large databases of compounds	http://zabib.chemie.uni-erlangen.de/software/corina/corina.html
ACD/3D Viewer	Converts two-dimensional structures from ACD/ChemSketch into their three-dimensional counterparts and views, measures and handles them in virtual 3D, including setting up a rotational video	http://www.acdlabs.com

Table 11.3 *Journals related to medicinal chemistry*

Publisher	Journal title	URL
American Chemical Society	Bioconjugate Chemistry	http://pubs.acs.org/journals/bcches
American Chemical Society	Journal of Natural Products	http://pubs.acs.org/journals/jnprdf
American Chemical Society	Journal of Pharmaceutical Sciences	http://pubs.acs.org/journals/jpmsae
American Chemical Society	Journal of Medicinal Chemistry	http://pubs.acs.org/journals/jmcmar/
American Chemical Society	Modern Drug Discovery	http://pubs.acs.org/journals/mdd/index.html
American Chemical Society	Organic Process Research and Development	http://pubs.acs.org/journals/oprdfk
Bentham Scientific Publishers	Current Medicinal Chemistry	http://www.bscipubl.demon.co.uk/cmc
Bentham Scientific Publishers	Current Pharmaceutical Design	http://www.bscipubl.demon.co.uk/cpd
Current Drugs	Current Drug Discovery	http://currentdrugdiscovery.com/
Current Drugs	Current Opinion in Drug Discovery and Development	http://www.current-drugs.com/products/coddd/coddd.htm
Elsevier	Bioorganic and Medicinal Chemistry	http://www.elsevier.com/inca/publications/store/1/2/9/
Elsevier	Bioorganic and Medicinal Chemistry Letters	http://www.elsevier.com/inca/publications/store/9/7/2/
Elsevier	Drug Discovery Today	http://www.drugdiscoverytoday.com
Elsevier	European Journal of Medicinal Chemistry	http://www.elsevier.nl/inca/publications/store/5/0/5/8/1/3/
Wiley-Interscience	Chirality	http://www.interscience.wiley.com/jpages/0899-0042

A convenient listing of them is available in the chemistry section of the WWW virtual library at http://www.liv.ac.uk/Chemistry/Links/journals.html. Salient journals related to medicinal chemistry include those shown in Table 11.3.

XI PATENT INFORMATION

Delphion (http://www.delphion.com/) permits text and number searching of patents registered in the USA, Europe, Japan and by the World Intellectual Property Organization (WIPO). The patent references are hyperlinked, allowing users to locate and view the cited and citing patents. A separate gateway gives access to the INPADOC database. For US patents, front pages, claims or the full text can be visualized. Scanned images are available for patents issued from 1987 to 1997, although patents are missing for some time periods. There is an order function for ordering complete patent documents for a fee. Until June 2001, the bulk of the excellent searching capabilities were free. Since the introduction of a subscriber system, there has been a greater impetus for alternative websites.

Esp@cenet at the European Patent Office (EPO) databases (http://www.european-patent-office.org/index.htm) allows free on-line patent searching in over 30 million documents (in EPO member states and worldwide) by entering keywords, patent numbers, institute names, etc.

The US Patent and Trademark Office web patent database (http://www.uspto.gov/patft/) provides access to both the US Patent Bibliographic database, which includes bibliographic data from 1976 to the present and the AIDS Patent database, which includes the full text and images of AIDS-related patents issued by the US, European and Japanese patent offices. There is a patent number search page, as well as Boolean and advanced search pages for text field searching. Both cited and citing patents are hyperlinked to each patent. There are hyperlinks between the classification numbers and their definitions and good help pages for each search type.

The US Patents Citation database from Community of Science (http://patents.cos.com/) provides fully searchable bibliographic file containing all of the approximately 1.7 million US patents issued since 1975. All important fields in the database are fully indexed and can be searched using standard Boolean query mechanisms. The database tracks the 'lineage' of each patent — how each patent cites previous patents or is cited by subsequent ones. The database is also searchable using the US Patent Manual of Classification approach, which permits searchers to identify specific areas of interest through a meta-database query and to navigate across the hierarchy of the classification using browse-and-search combinations. Search pages are table-rich and the display includes the first claim.

A related database is provided through Chemweb: the IFI Claims database (http://www.chemweb.com/databases/claims) provides text-searchable information on claims in US patents from 1950 (chemical patents) and 1963 (utilities patents) to the present day. As well as non-specific search methods, the interface on Chemweb permits searching by Boolean combinations of title, inventor, exemplary claim(s), assignee, patent number, application number, US patent classifications or international patent classifications (IPCs). Search terms may be stemmed for additional flexibility.

US Patent Front Pages at QPat (http://www.qpat.com/) offers full text of all US patents issued since 1 January 1974 to subscribers only. Other users may register and get free search and display access to all of the patent abstracts contained in the database. The search system supports the use of Boolean operators, as well as proximity operators. Field searching can be performed using preselected supergroups or limited to one or more fields using suffix codes. There are help screens and sample searches for each kind of search. Information given includes inventors, assignees, foreign application data, references cited (hyperlinked for US, but not foreign patent documents), and the abstract. Each abstract is linked to similar abstracts by a clickable button.

MicroPatent (http://www.micropat.com/) provides some of the capabilities of its PatentWEB site free, such as searching of the US Online Gazette, browsing of the exemplary drawings and the most recent four weeks of US patents viewed by US Major Classification. Also free is a search of full text of all US patents issued for the current and past week. For a small charge ASCII full text of any US patent since 1974 is provided via instant e-mail or full document download of a patent as published by the Patent Office.

Lexis-Nexis Academic Universe (http://web.lexis-nexis.com/universe/) is a useful site for legal and government information on the internet. A separate section under 'Legal Research' leads to 'Patent Research'. Keyword searching of the 'Patents' section is available, with an option to narrow fields by type of patent, etc. The 'Class' section is searchable by patent number. The database covers patents back to 1971. Results are returned in HTML format, and pictures are (currently) not available.

There are various general sites for patents, such as that from the US Patent and Trademarks Office about patents and patenting procedures (http://www.uspto.gov/web/offices/pac/doc/general/). Another site is http://users.pandora.be/synthesis/wpatent.html, which contains general information on patents and a list of useful hypertext references. Various other useful bits of information about patent terms and procedures in other countries are available from Derwent (http://www.derwent.com/). A comparison table of web patent databases from Duke University is presented by the university library to help users compare the various resources available and assess which is best for each individual's needs (http://www.lib.duke.edu/chem/patcomp.htm and at http://www.lib.duke.edu/reference/subjects/patents.htm).

XII BIOINFORMATICS: GENOMICS AND PROTEOMICS

For an easy introduction to bioinformatics, http://www.sequenceanalysis.com is a guide to molecular sequence analysis. The site tells how to use freely available software to compare an unknown sequence to all of the sequences in a database. The guide provides an interactive working introduction, for scientists with no working knowledge of molecular sequence analysis to the techniques of *in silico* biology.

One of the sources of such freely available analysis tools is at http://www.up.univ-mrs.fr/~wabim/english/logligne.html, produced by the Atelier Bioinformatique at the Université of Aix-Marseilles. There is a truly vast array of links here to worldwide scientific research into bioinformatics, and direct, free links to sequence analysis programs, like BLAST and BioSCAN.

The Genome Database (http://www.gdb.org/) is the 'official central repository for genomic mapping data resulting from the Human Genome Initiative'. Users can search by gene, sequence, or for alleles, etc. In a similar vein, the relationship of DNA sequence in a gene to amino acid in a protein is derivable from the table of standard genetic code available at http://molbio.info.nih.gov/molbio/gcode.html, and it is worthwhile making a bookmark of this page.

Much more complicated than sequence is the relationship of genetic sequence to disease, which is now known for a small number of single-gene defect diseases. A useful site for this purpose is at http://www.ncbi.nlm.nih.gov/disease, which provides an introduction to this subject on the 'front end' NCBI site, from which the visitor is led through subsequent layers of detail eventually to specific PubMed searches on the disease in question. In total over 75 associations between disease and genetic defects are shown. They vary in strength of association, from Huntington's diseases (in which 3 bp on chromosome 4 are incorrectly repeated on the gene for the brain protein *huntingtin*), to the much weaker association between chromosome 5 variation and asthma.

The Yale Molecular Biophysics and Biochemistry Center Structural Biology at http://www.csb.yale.edu details masses of information related to biological structures; the section on databases includes an atlas that depicts how amino acid side-chains pack against one another within the known protein structures. This is based on the printed atlas of Singh and Thornton (http://www.biochem.ucl.ac.uk/bsm/sidechains/reference.html). It presents all 400 possible pair-wise interactions between the 20 amino-acid side-chains. There is a link too to the Dali server at the European Bioinformatics Institute, (http://www2.ebi.ac.uk/dali/) a network service for comparing protein structures in 3D. The user can submit the coordinates of a query protein structure and Dali compares them against those in the Protein Data Bank. A multiple alignment of structural neighbours is returned by e-mail. In favourable cases, comparing 3D structures may reveal biologically interesting similarities that are not detectable by comparing sequences.

The Atlas of Nucleic Acid-Containing Structures from the Nucleic Acid Database Project http://202.213.175.11/NDB/NDBATLAS/index.html offers a similar range of information and services, but particularly geared to nucleotides rather than proteins. The Nucleic Acid Database Project (NDB) assembles and distributes structural information about nucleic acids, and the Atlas of Nucleic Acid Structures was developed to highlight the special aspects of each structure in the NDB. Each page formats the data contained in each NDB file and generates pictures of the biological unit and the crystal packing for each structure. The index page for each category lists each structure by the NDB identifying number, first author and compound name. The Atlas pages aid in checking the structures contained within the NDB and also serve as a teaching tool.

Beyond nucleic acids, prediction of protein structure is becoming a major effort for drug hunters in the pharmaceutical and biotechnology industries. There are a series of links to research into protein folding from various academic groups as shown in Table 11.4.

On a more basic note in relation to proteins, the Amino Acid Information site at http://prowl.rockefeller.edu/aainfo/contents.htm is very useful. There is a wide range of information here. The index page divides the site into structural, genetic and chemical properties, as well as general information and an area for bioinformaticians and peptide mimetic design called 'Suggested amino acid substitutions'. This latter deals with similarity indices between amino acid residues, based on various proposed scoring mechanisms. Amongst many other features, elsewhere the site includes information on the universal genetic code (three-letter DNA codons for protein translation), one-letter abbreviations, hydrophobicity scales and bond geometry.

XIII PREDICTION OF BIOLOGICAL ACTIVITY

With the advent of more extensive late discovery and preclinical testing for development candidates, the importance of these disciplines for medicinal chemistry has become more important. A useful introduction to the subject (albeit rather old) called *The Emerging Role of A.D.M.E. in Optimising Drug Discovery and Design* by Robert J. Guttendorf, Ph.D. can be found at http://www.netsci.org/Science/Special/feature06.html.

Using *in silico* prediction of biological activity, various attempts have been made to estimate pharmacological properties of molecules based on their structure. PASS (Prediction of Activity Spectra for Substance) (http://www.ibmh.msk.su/PASS) estimates the probabilities for more than 400 pharmacological effects, mechanisms of action, and such specific toxicity or mutagenicity, carcinogenicity, teratogenicity and embryotoxicity. Based only on structural formula, an estimate is made of which kinds of activity are

the most probable for a drug-like organic compounds. Molecular Design Ltd's (http://www.mdli.com) structure data SDfiles or molfiles are the preferred format for inputting compounds.

The National Cancer Institute (NCI) database at http://129.43.27.140/ncidb2/ contains over 250 000 entries, which are searchable by structure, formula, CAS registry number or name. The results are analysed in terms of physical molecular descriptors (such as $\log P$) and predicted biological activities against around 100 biological activities.

For metabolism data, there is a superb database in the University of Minnesota Biocatalysis/Biodegradation Database at http://www.labmed.umn.edu/umbbd/index.html. The database includes a search capability for compound, enzyme, or micro-organism name; chemical formula; CAS Registry Number; or EC (enzyme classification) code. It also has lists of reaction pathways, enzymes, micro-organism entries, and organic functional groups. It specifically includes a large number of reactions of naphthalene 1,2-dioxygenase and of toluene dioxygenase. There is a paper describing the database, which was published in Nucleic Acids Research in January 2001, which can also be downloaded in full text or in pdf format from the site.

XIV TOXICOLOGY

There is a good introduction to the subject, which should be interesting to students, called 'Toxicology Tutor' at http://sis.nlm.nih.gov/toxtutor.htm.

There are a number of toxicology databases available on the internet, and recently there has been an amalgamation of the best in the form of TOXNET (http://sis.nlm.nih.gov/sis1/), a cluster of databases on toxicology, hazardous chemicals, and related areas. The old TOXNET Web search engine is still available through the National Library of Medicine (http://toxnet.nlm.nih.gov/servlets/simplesearch), as is the interface through Internet Grateful Med at http://igm.nlm.nih.gov/. The website provides access to an impressive array of files containing factual information related to the toxicity and other hazards of chemicals. Users can readily extract toxicology data and literature references, as well as toxic release information, on particular chemicals. Alternatively, one can perform a search with subject terms to identify chemicals that cause certain effects. A variety of display and sorting options are available. A summary of further resources in this area is provided in Table 11.5.

Fee-based resources include the updated US EPA Toxic Substance Control Act (TSCA) chemical inventory of 62 000 chemicals, which is available cross-referenced with Superfund Amendments and Reauthorization Act (SARA) Title III RCRA reporting requirements on CD-ROM.

Table 11.4 *Links to research into protein folding*

Centre	Main investigator	Comments	Link
National Center for Biotechnology Information (NCBI)	Various	The Basic Local Alignment Search Tool (BLAST), for comparing gene and protein sequences against others in public databases	http://www.ncbi.nlm.nih.gov/Tools/index.html
University of California, Santa Cruz	Kevin Karplus	Primary (sequence), secondary (folding), and tertiary (3-dimensional) structures of DNA, RNA and protein sequences	http://www.cse.ucsc.edu/research/compbio
Brunel University, London	David Jones	On-line structure prediction service allows the user to submit a protein sequence, perform a prediction method of choice and receive the results of the prediction via e-mail. The user is permitted to select one of three prediction methods to apply to the sequence: PSIPRED, MEMSAT 2 or GenTHREADER (http://insulin.brunel.ac.uk/threader/threader.html), a new sequence profile-based fold recognition method	http://insulin.brunel.ac.uk/psipred/
Washington University	David Baker	Combined molecular biological, biophysical and computational approach to protein folding	http://depts.washington.edu/bakerpg
Rockefeller University	André Sali	Methods for protein structure modelling, primarily in the area of homology or comparative modelling	http://guitar.rockefeller.edu/
National Center for Biotechnology Information (NCBI)	Stephen Bryant	Emphasis on conserved structure domains with linked database (Conserved Domain Database; CDD) and search service	http://www.ncbi.nlm.nih.gov/Structure/RESEARCH/res.html
Stanford University	Michael Levitt	Specialized in protein folding	http://csb.stanford.edu/levitt
Cornell University	Harold Scheraga	Theoretical and practical aspects of polypeptide folding	http://www.chem.cornell.edu/department/Faculty/Scheraga/scheraga.html

Table 11.5 *A summary of the further resources on toxicology*

Type of database	Database name	URL	Description
Toxicology data files	HSDB (Hazardous Substances Data Bank)	http://toxnet.nlm.nih.gov/cgi-bin/sis/htmlgen?HSDB	Factual data bank of over 4500 potentially hazardous chemicals. In addition to toxicity data, the file covers emergency handling procedures, environmental fate, human exposure, detection methods, and regulatory requirements. The data are fully referenced and peer-reviewed by expert toxicologists and other scientists
	IRIS (Integrated Risk Information System)	http://toxnet.nlm.nih.gov/cgi-bin/sis/htmlgen?IRIS.htm	On-line database of the Environmental Protection Agency (EPA) containing carcinogenic and non-carcinogenic health risk information on over 500 chemicals. Data have been scientifically reviewed and represent EPA consensus
	CCRIS (Chemical Carcinogenesis Research Information System)	http://toxnet.nlm.nih.gov/cgi-bin/sis/htmlgen?CCRIS	Sponsored by the National Cancer Institute (NCI), CCRIS contains scientifically evaluated data derived from carcinogenicity, mutagenicity, tumour promotion and tumour inhibition tests on about 8000 chemicals
	GENE-TOX (Genetic Toxicology)	http://toxnet.nlm.nih.gov/cgi-bin/sis/htmlgen?GENETOX	Another EPA database, contains genetic toxicology test results on over 3000 chemicals. Selectively reviewed for each of the test systems under evaluation. The GENE-TOX data bank is the product of these data review activities
Toxicology literature files	TOXLINE	http://toxnet.nlm.nih.gov/cgi-bin/sis/htmlgen?TOXLINE	Bibliographic database covering the biochemical, pharmacological, physiological, and toxicological effects of drugs and other chemicals; over 2.5 million citations, almost all with abstracts and/or index terms and CAS Registry Numbers
	DART/ETIC (Development and Reproductive Toxicology/Environmental Teratology Information Center)	http://toxnet.nlm.nih.gov/cgi-bin/sis/htmlgen?DARTETIC.htm	Bibliographic database covering literature on teratology and other aspects of developmental toxicology, contains over 90 000 references to teratology literature published since 1965
	EMIC (Environmental Mutagen Information Center)	http://toxnet.nlm.nih.gov/cgi-bin/sis/htmlgen?EMIC	Bibliographic database containing over 100 000 references on chemical, biological, and physical agents that have been tested for genotoxicity, covers literature published since 1965
Toxic releases files (TRI)	TRI (Toxic Chemical Release Inventory Files)	http://toxnet.nlm.nih.gov/cgi-bin/sis/htmlgen?TRI	Contains information on the annual estimated releases of toxic chemicals to the environment for 1995–1997. It is based upon data submitted to the EPA from industrial facilities throughout the US and includes the amounts of certain toxic chemicals released into the environment on over 650 chemicals and chemical categories. Pollution prevention data are also reported
Chemical information	ChemIDplus	http://chem.sis.nlm.nih.gov/chemidplus/	Provides structure and nomenclature data for the identification of chemical substances cited in NLM databases, contains over 349 000 records on over 56 000 include chemical structures, searchable by name, synonym, CAS registry number, molecular formula, and structure
	HSDB Structures	http://chem.sis.nlm.nih.gov/hsdb/	A database of over 4500 substances with 2D structural information, which was compiled from the HSDB database on TOXNET
	NCI-3D	http://chem.sis.nlm.nih.gov/nci3d/	A database of over 100 000 substances with 2D and 3D information compiled from the National Cancer Institute, and augmented by MDL

Table 11.6 *Meta-sites and technology service provider databases*

Title	URL	Comments
Chemfinder site	http://chemfinder.camsoft.com/siteslist.asp	The following categories of information are available among others: biochemistry, health, medications, MSDS, pesticides/herbicides, physical properties, regulations, structures and trading
Chemistry section of the WWW Virtual Library	http://www.liv.ac.uk/Chemistry/Links/links.html	Thorough, up-to-date and accurate listings of a large number of chemistry sites. The chemical database section at http://www.liv.ac.uk/Chemistry/Links/refdatabases.html gives details of a collection of about 80 chemical databases (among them: Analytical Abstracts, Beilstein, CCDC, CA Selects Plus, ChemFinder, DrugDB, FT-IR Library, STN, etc)
Organic Chemistry Resources Worldwide	http://www.organicworldwide.net	Created by Koen Van Aken, a Belgian chemist, a well-organized and highly useful site for all engaged in synthetic organic chemistry research
Competia	http://www.competia.com/express	A collection of web resources dealing with chemical industry
	http://www.diax.ch/users/herbert_benz/	A collection of commented websites, which are of interest for work in pharmaceutical chemistry
Caltech	http://library.caltech.edu/collections/chemistry.htm	Practically an indispensible point of call for databases and search engines for chemistry
University of Liège, Belgium	http://www.ulg.ac.be/libnet/ud18.htm	An initiative from Simone Jérôme, chemistry librarian at the University of Liège, Belgium
The Chemical Database Service (CDS)	http://cds3.dl.ac.uk/cds/cds.html	The CDS provides online access to numerous chemical databases, which are available free of charge to academics at UK universities. The chemistry links cover a large variety of topics (among them general information sites, reference databases, chemical sources, chemical WWW sites, UK universities, chemistry FTP sites). New entries to be cited are the solid-phase synthesis database (over 15 000 reactions described) and the available chemicals directory
University of Cincinnati Online	http://www.engrlib.uc.edu/selfhelp/alphlist.htm	Links to engineering, biology and chemistry databases, etc., are listed on this important site
Database Collection Chemiedatenbanken	http://www.chemie-datenbanken.de/	An excellent collection of German and international chemical databases, e.g. free resources, general collections and commercial database providers

It features SARA III fields integrated with TSCA information and Adobe™ Acrobat™ (PDF) format for instant search/retrieval. For details see http://www.env-sol.com/solutions/TSCASARA.HTML.

For further information and internet links in toxicology, try http://www.links2go.com/topic/Toxicology.

XV DRUG INFORMATION

While the FDA has a very good searchable website of approved drugs at http://www.fda.gov/cder/ob/ (and FDA-approved biologics and other biopharmaceutical products are at http://www.biopharma.com) this is not structure-searchable and does not contain information on compounds in development. More complete database products, like PJB Publications' Pharmaprojects® (http://www.pjbpubs.co.uk), Prous' Ensemble (http://www.dailydrugnews.com) and Current Drugs' IDdb (http://www.current-drugs.com) are only available for a substantial price. An alternative, mid-priced option is to use the Negwer (Organic-Chemical Drugs and Their Synonyms) Database, which is structure-searchable and describes over 13 000 synthetic drugs with more than 120 000 synonyms. CAS Registry Numbers, molecular formulae and CA Index Names, trivial/trade names and non-protected names suggested by the World Health Organization are given for each pharmaceutical compound, together with its use and associated reference information.

There are multiple other sources of information on marketed compounds, similar to that which is conventionally available in pharmacopoeias, indeed the names of these sites often reflects that connection. The Internet Drug Index (http://www.rxlist.com) is a prescription drug database, which provides good basic information about products on the market, searchable by keyword, brand or interaction. RxList is a trove of pharmaceutical knowledge with more than 4500 medications on file, a pharmaceutical discussion board, and an on-line dictionary of medical jargon. It provides useful basic information about conventional drugs and a handful of herbal remedies as well, in the form of drug FAQs (frequently asked questions) and patient monographs.

Another source is the electronic Medicines Compendium (eMC; http://emc.vhn.net/), with electronic versions of data sheets and Summaries of Product Characteristics (SPCs, sometimes also called SmPCs to differentiate them from Supplementary Patent Certificates) for medicines. It provides the same information as that contained in the latest edition of the Compendium of Data Sheets and Summaries of Product Characteristics, which covers 2500 medicines licensed in the UK. As an ongoing process, the eMC is also incorporating the SPCs of several thousand other medicines approved by the licensing authorities.

The eMC ultimately aims to provide information on every licensed prescription, pharmacy and general sale medicine in the UK, including generics. As well as SPCs, the eMC will eventually include all Patient Information Leaflets (PILs), and will also be enhanced with dynamic updating and on-line links to complementary sources of medicines information.

Finally, Clinical Pharmacology 2000 at http://cp.gsm.com/ is part of the Gold Standard Multimedia group, providing information on all FDA-approved drugs and OTC pharmaceuticals. Herbal products and nutraceutical products are included when it appears that a reasonable amount of scientific data are available for review. Investigational drugs are included in the database when they have reached phase III clinical investigation. It is an excellent resource; like the Electronic Medicines Compendium, it requires registration; but unlike the other site, which is free, it is available only by subscription.

XVI META-SITES AND TECHNOLOGY SERVICE PROVIDER DATABASES

Meta-sites providing access to a range of resources devoted to chemistry are listed in Table 11.6.

For specific resources related to the pharmaceutical industry, there are a number of databases providing information on technology and service providers of various parts of the discovery and developments functions required for entry of a new molecular entity on to the market. A recently expanded database is found at http://www.arachnova.com/cro_guide.php which is searchable by contract research organization (CRO) name, country of origin or speciality of service, and includes over 1300 companies across the entire service range, both research and development in origin. A large number of chemical providers are featured on this site. Similar databases are provided by Pharmalicensing (http://www.pharmalicensing.com), which also provides a service as a broker of intellectual property and product licensing in the healthcare field. Another is provided by Soteros Consultants, another consultancy and training service for regulatory toxicology and related disciplines (http://www.soteros.co.uk/). In addition to databases of CROs and consultants, information provided here also includes toxicology meetings and salient literature references. Presently, these sites are freely accessible. A very complete and up to date database is also provided commercially by the Technomark CRO Gateway http://www.technomarkregisters.com/technomark/, which is available by subscription to provide on-line data that is updated twice a month.

PART III

PRIMARY EXPLORATION OF STRUCTURE–ACTIVITY
RELATIONSHIPS

12

MOLECULAR VARIATIONS IN HOMOLOGOUS SERIES: VINYLOGUES AND BENZOLOGUES

Camille G. Wermuth

Methyl, ethyl, propyl, butyl … futile **Old adage**

I HOMOLOGOUS SERIES

The concept of a homologous series was introduced into organic chemistry by Gerhardt.[1] In medicinal chemistry the term has the same meaning, namely molecules differing one from another by only a methylene group.

II CLASSIFICATION OF THE HOMOLOGOUS SERIES

The most frequently encountered homologous series in medicinal chemistry are monoalkylated derivatives, cyclopolymethylenic compounds, straight chain difunctional systems, polymethylenic compounds and substituted cationic heads.

A Monoalkylated derivatives

$$R—X \rightarrow R—CH_2X \rightarrow R—CH_2—CH_2—X, \text{ etc…}$$

An example is provided by a series of neuraminidase inhibitors with a cyclohexene scaffold containing lipophilic side chains.[2]

As shown in Fig. 12.1, a 6300-fold increase in potency is observed when the hydroxylic hydrogen is replaced by a diethyl-methyl side chain. A comparable increase in potency is observed in a series of 1-methyl-1,2,3,4-tetrahydro-pyridyl-pyrazines described by Ward *et al.*[3] exhibiting M_1 muscarinic agonists. In changing from *O*-methyl to *O*-butyl, the affinity for the M_1 receptor varies from 850 nM to 17 nM. Another example is found in a series of 2-pyrone-derived elastase inhibitors.[4]

B Cyclopolymethylenic compounds

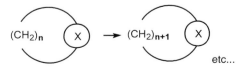

etc…

Examples of such structures with regularly increasing ring sizes are found for guanethidine (see Chapter 14), or for enalaprilat analogues (Fig. 12.2).[5] In the latter example, a

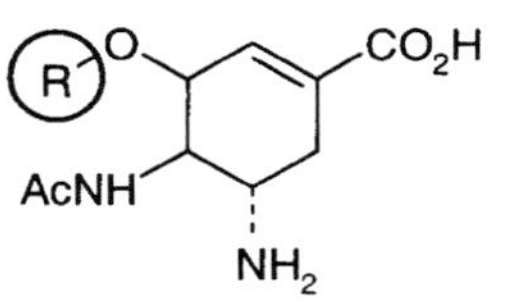

Neuraminidase inhibition

R =	IC50 (nM)
H	6.300
CH_3 -	3.700
CH_3 - CH_2 -	2.000
CH_3 - CH_2 - CH_2 -	180
CH_3 - CH_2 - CH_2 - CH_2 -	300
$(CH_3)_2$ -CH_2 - CH_2-	200
CH_3 - CH_2 - $CH(CH_3)$ -	10
$(CH_3 - CH_2)_2$ - CH -	1
$(CH_3 - CH_2 - CH_2)$ CH -	16
Cyclopentyl	22
Cyclohexyl	60
Phenyl	530

Fig. 12.1 Monoalkylated, cyclohexene-derived, neuraminidase inhibitors.

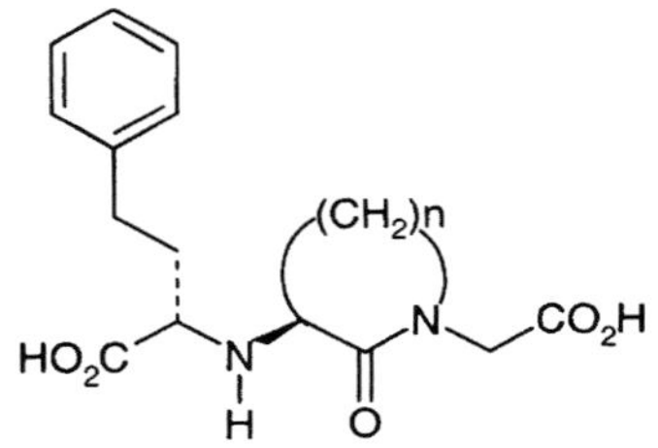

Size	IC_{50} (nM)
n = 2	19.000
n = 3	1.700
n = 4	19
n = 5	4.8
n = 6	8.1

Fig. 12.2 Angiotensine-convertase inhibiting potency of enalaprilat analogues.[5]

4000-fold increase in inhibition of angiotensin converting enzyme is obtained when changing from the five-membered ring ($n = 2$) to the eight-membered ring ($n = 5$).

Another example, published by scientists from Parke-Davis, relates to a series of 'dipeptoid' analogues of cholecystokinin.[6] These compounds are α-methyl-tryptophan derivatives, *N*-substituted by carbamic esters of cyclanols with ring sizes increasing from cyclobutyl to cyclododecyl (Fig. 12.3). Here again an optimal size was found (cyclononyl).

C Open, difunctional, polymethylenic series

$$X—(CH_2)_n—Y \rightarrow X—(CH_2)_{n+1}—Y$$

In the above general formula, X and Y can represent very diverse functional elements. The compounds can be symmetrical (X = Y; 'dimers') or non-symmetrical (X ≠ Y); see Chapter 15. Usually, X and Y represent *polar functions* or *functionalized cyclic systems*.

When X and Y are polar functions, they are made essentially from functional groups such as alcohols, amines, acids, amides, amidines or guanidines (Fig. 12.4). A classical representative of a difunctionalized, symmetrical compound is decamethonium.

When X and Y are functionalized cyclic systems, they can be alicyclic or aromatic, as well as homocyclic or heterocyclic (Fig. 12.5). In any case they bear some polar function or polar element. An example of this type of compound is pentamidine.

Other examples are symmetrical bradykinine antagonists[7] and symmetrical lexitropsines (netropsine, distamycine), active against HIV-I viruses.[8] Non-symmetrical polymethylenic thromboxane synthetase inhibitors are described by Press *et al.*[9] The compounds contain a thiophen-2-carboxamide moiety, separated from an imidazole ring by 3 to 8 methylene units. Surprisingly, whereas most of the compounds show similar thromboxane-synthetase inhibiting activities, only the two medium-sized ones ($n = 3$ and $n = 4$) showed hypotensive effects in spontaneously hypertensive rats (Fig. 12.6).

In a series of benzimidazole-derived thromboxane A_2 receptor antagonists described by Nicolaï *et al.*,[10] the crucial element is the distance between the carboxylic group and the benzimidazole ring. A 200-fold increase in affinity was observed when changing from propionic to a butyric side chain (Fig. 12.7).

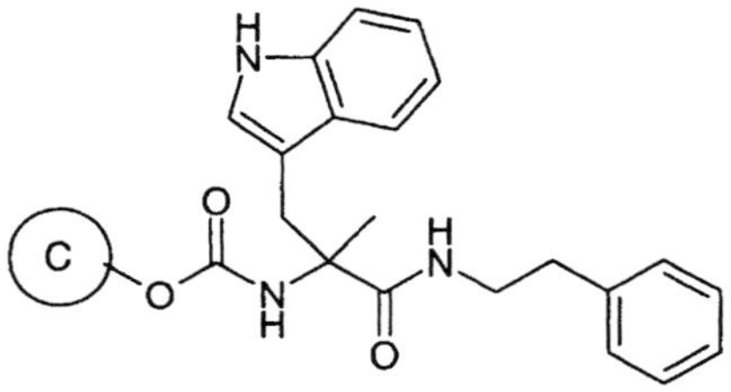

C	log P	IC_{50} (nM)
cyclobutyl	3.88	12100
cyclopentyl	4.44	5170
cyclohexyl	5.00	520
cycloheptyl	5.55	190
cyclooctyl	6.11	125
cyclononyl	6.67	85
cyclodecyl	7.23	247
cyclododecyl	8.34	1437

Fig. 12.3 Optimal ring size for a series of cyclanol carbamates.[6]

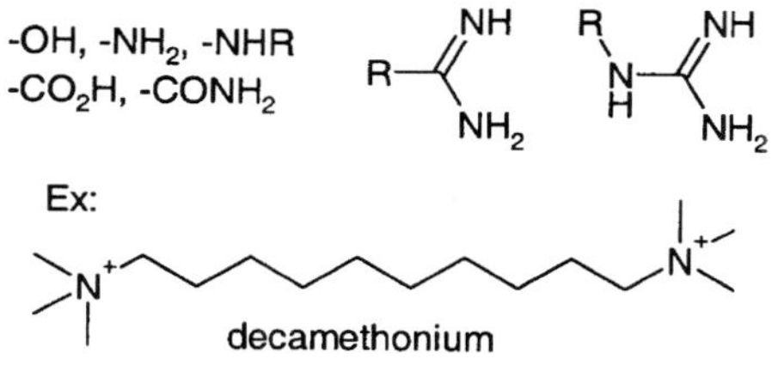

-OH, -NH$_2$, -NHR
-CO$_2$H, -CONH$_2$

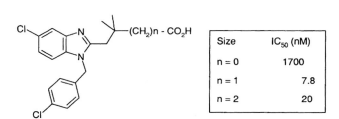

decamethonium

Fig. 12.4 Examples of functional groups encountered in open, difunctional, polymethylenic compounds and structure of decamethonium.

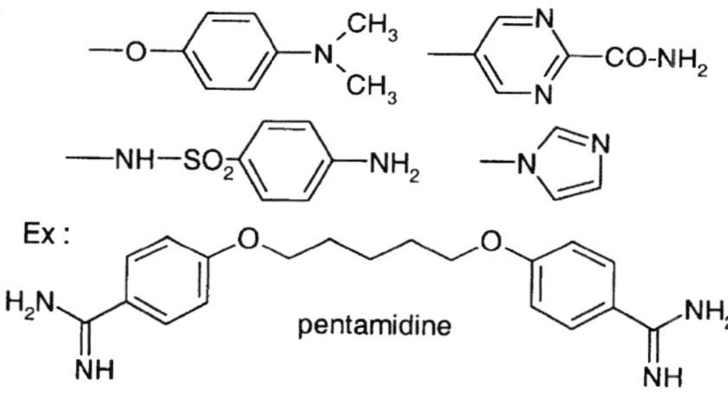

Ex :

pentamidine

Fig. 12.5 Examples of functionalized rings found in straight chain, difunctional, polymethylenic compounds. Structure of pentamidine.

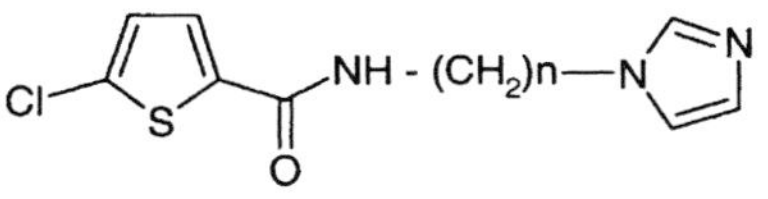

Size	IC$_{50 (nM)}$
n = 3	600
n = 4	200
n = 5	200
n = 6	200
n = 8	70

Fig. 12.6 Thromboxane synthetase inhibiting activity of a series of *N*-(imidazolyl-alkyl)-thiophene-5-carboxamide.[9]

Size	IC$_{50}$ (nM)
n = 0	1700
n = 1	7.8
n = 2	20

Fig. 12.7 Affinity for the thromboxane A$_2$ receptor.[10]

D Substituted cationic heads

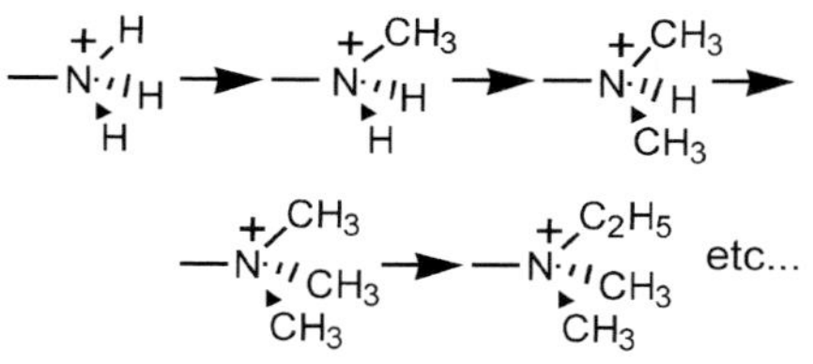

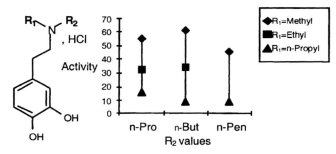

, HCl

	R$_1$=Methyl
	R$_1$=Ethyl
	R$_1$=n-Propyl

Fig. 12.8 Anticataleptic activity of substituted dopamines.[11]

With cationic head groups, homology simultaneously achieves a progressive increase in bulkiness and in lipophilicity. Figure 12.8 illustrates the influence of increasing bulkiness around the dopamine nitrogen in the antagonism of reserpine-induced catalepsy in mice.[11]

III SHAPES OF THE BIOLOGICAL RESPONSE CURVES

The most common curves are bell-shaped, the peak activity corresponding to a given value of the number *n* of carbon atoms (curve A, Fig. 12.9). However, many other relationships were found among homologous series:

(1) The activity can increase, without any particular rule, with the number of carbon atoms (curve B).
(2) The biological activity can alternate with the number of carbon atoms, resulting in a zigzag pattern (curve C).
(3) In other series the activity increases first with the number of carbon atoms and then reaches a plateau (curve D).
(4) The activity can also decrease regularly, starting with the first member of the series (curve E). This was found for the toxicity of aliphatic nitriles or for the antiseptic properties of aliphatic aldehydes.
(5) A last possibility resides in inversion of the pharmacological activity accompanying the increase in the number of carbon atoms (not shown in Fig. 12.9; this will be discussed below).

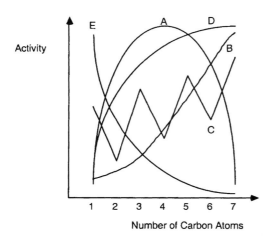

Fig. 12.9 Shapes of the biological response curves in homologous series.

IV RESULTS AND INTERPRETATION

A Curves with a maximum activity peak (bell-shaped curves) and curves with a continuous increase of activity

In such series the continuous growth of an alkyl chain or of methylene units increases the hydrophobic part of the molecule. Various physicochemical parameters, such as solubility in water, partition coefficients, chromatographic R_f values, and critical micellar concentration, are precisely governed by the same fundamental property: the hydrophobic character.

Bell-shaped curves

Curves with an activity maximum are the most common ones, it is currently admitted that they reflect the existence of an optimal partition coefficient associated with the easiest crossing of the biological membranes. The relationship between the biological response and the partition coefficient is then illustrated by a parabolic curve. An example is found in structural analogues of PAF-acether.[12] In varying the length of the alkoxy chain from *n*-butyl to *n*-eicosanyl, the authors observed the peak activity for the *n*-hexadecyl chain, with a 1200-fold interval between the most and the least active compounds (Fig. 12.10). The drop in activity observed for the descending branch, is usually attributed to insufficient solubility in water (incapability to cross the aqueous biophases), but can also be due to the formation of micelles. In this case, the concentration of the free drug, which represents the directly available form, lies under the critical threshold level. Bell-shaped curves are also seen when using isolated cells for which it can be demonstrated that the receptor is outside the membrane. In this case the dominant factor is probably not the crossing of biological membranes. Changes in critical micellar concentration with increasing chain length could explain the effect in some cases, however the curve is often too steep for this to be an acceptable explanation. Another possibility is that there is a lipophilic pocket of finite size. In many cases this pocket is not actually in the receptor protein. An argument in favour of this explanation is that the top of the bell is at C_{16} or C_{18} which fits with the length of the alkyl chains making up part of the bilayer, examples being PAF-acether analogues[12] and leucotriene D_4 agonists/antagonists. Another bit of evidence that supports this idea is the observation that the position of the peak of the curve can vary depending on which cell type is expressing the same receptor protein.

The study of the activities of *some* homologous compounds, can, through interpolation, identify which term is associated the highest potency. The optimization method proposed by Bustard,[13] makes use of the Fibonacci numbers, and allows the identification of the most active compound (presumed to exist in a given interval) with the smallest possible number of syntheses (see also References 14 and 15).

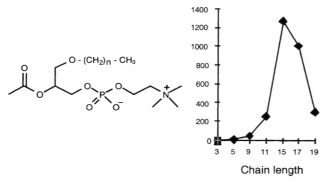

Fig. 12.10 Antiaggregant activity of structural analogues of PAF-acether.[12]

Apparently continuous increase

Actually, an apparently continuous increase of activity may correspond simply to the ascendant branch of the parabola (see the two curves in Fig. 12.11). The observed 'pseudolinear' curve usually occurs in an insufficiently explored series. A true linear correlation would imply the existence of compounds of infinite potency!

B Non-symmetrical curves with a maximum activity peak

In some instances curves with maximum activity peaks are not symmetrical and one side shows very sharp activity

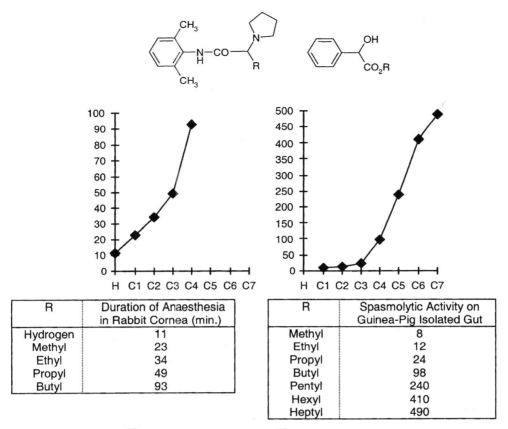

R	Duration of Anaesthesia in Rabbit Cornea (min.)
Hydrogen	11
Methyl	23
Ethyl	34
Propyl	49
Butyl	93

R	Spasmolytic Activity on Guinea-Pig Isolated Gut
Methyl	8
Ethyl	12
Propyl	24
Butyl	98
Pentyl	240
Hexyl	410
Heptyl	490

Fig. 12.11 Local anaesthetic activity[16] and spasmolytic activity[17] in homologous series.

variations, whereas the other one corresponds to a progressive variation. The shape of such a curve is represented on Fig. 12.12.

For the GABA$_A$ antagonists represented in Fig. 12.13, the peak activity corresponds to the branching of a *butyric* side chain on the aminopyridazine system. The affinity diminishes drastically for shorter chains, but very progressively for longer chains.[18]

The particular case of polymethylenic bisammonium compounds

Compounds having the general formula $(CH_3)_3N^+$ $—(CH_2)_n—N^+(CH_3)_3$ usually have high affinity for the cholinergic receptors. When the values of n are intermediary ($n = 5$ or 6: penta- or hexamethonium), such compounds behave like cholinergic *agonists* (towards the sympathic ganglions). For higher values ($n = 10$: decamethonium)

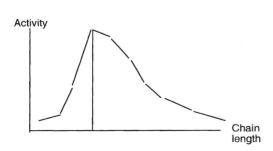

Fig. 12.12 Non-symmetrical curve with a maximum activity peak.

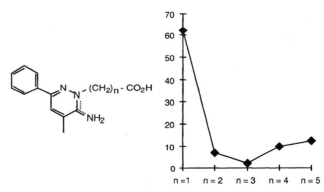

Fig. 12.13 Affinity of GABA$_A$ antagonists for the GABA$_A$ receptor site.[18]

the compounds become *antagonists* of acetylcholine (at the muscular end plate). In both cases increasing acetylcholine levels displace them from their binding sites. When considering the neuro-muscular blockade, one observes again a curve with an asymmetric profile: sudden changes between $n = 6$ and $n = 8$, and then progressive diminution between $n = 9$ and $n = 12$.

To explain this, if we suppose that compounds of general structure X—$(CH_2)_n$—Y interact by means of their polar groups X and Y with complementary groups at the receptor, four interaction schemes can be envisaged, depending on the value of n (Fig. 12.14: (1–5)):

(1) *n is small:* The molecule is too short and only one of its polar ends can establish an interaction with the complementary sites of the receptor (Fig. 12.14: (1)). The molecule will be inactive or poorly active. This is the case for the pyridazinyl-glycine of Fig. 12.13.

(2) *n possesses sufficient length:* A good interaction can be established with complementary sites of the receptor and trigger off the biological response (Fig. 12.14: (2)). This represents the optimal case.

(3) *n is too great:* Two situations are foreseeable. If the molecule is rigid or if there is steric hindrance, the interaction is not possible for Y (Fig. 12.14: (3)). If the molecule is flexible and if the steric tolerance is sufficient, the fit can be entirely satisfactory (Fig. 12.14:(4)).

(4) *n is very great:* In this case (Fig. 12.14: (5)), the fit with the receptor is again very good, but with a further located subsite Y″ instead of Y′, the substance can then

behave as an *antagonist* (this was the case discussed above of decamethonium).

C Serrated variations

One sometimes observes alternating (serrated) variations of activity (zig-zag curves) according to whether the number of atoms of carbon is even or odd. Such an example is found for antimalarials derived from methoxy-6-amino-8-quinoline (Fig. 12.15).

For these derivatives the antimalarial activity is greater if n represents an odd number (for studied values that vary from $n = 4$ to $n = 10$).[19] Another example is provided by leukotriene B_4 antagonists derived from hydroxyacetophenones[20] (Table 12.1).

For both cases, the findings are a reflection of the rotational energy curves for adjacent CH_2 groups. Similar observations were made in a series of 4,4′-dimethylaminodiphenoxyalkanes tested as potential schistosomicides.[21] For diamines where $n = 4$ to $n = 10$, the activity on the schistosomes varies in alternate manner (Fig. 12.16). Alkyl-linked bis(amidinobenzimidazoles) with an even number of methylenes connecting the benzimidazole rings have a higher affinity for the minor groove of DNA than those with an odd number of methylenes (Fig. 12.16).[22] Serrated variations of acetylcholinesterase inhibiting activity were also observed for donepezil analogues (Fig. 12.16).[23]

Zig-zag variations are well known in homologous series for physical properties such as melting points and solubilities. Thus, propane, with an odd number of carbon atoms, melts at $-189.9°C$ whereas butane, with an even number, melts $51.6°C$ higher at $-138.3°C$. However, odd-numbered pentane melts at $-129.7°C$, only $8.6°C$ higher than butane. Boese[24] studying X-ray structures of *n*-propane to *n*-nonane at $-90°C$ indicate that the methyl groups on chains lying end-to-end are the culprits. In even-numbered

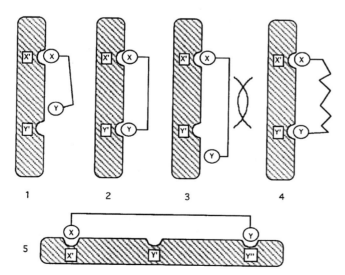

Fig. 12.14 Different modes of interaction of bifunctional molecules according to their length.

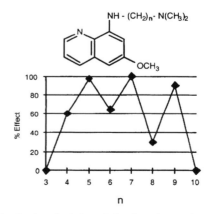

Fig. 12.15 Antimalarial activity in a homologous series of bifunctional methoxy-6-amino-8-quinolines (after Magidson and Strukov[19]).

Table 12.1 *Zig-zag variations of the affinity of hydroxyacetophenone derivatives for the human peripheral neutrophils. Inhibition of [³H] LTB₄ binding at 0.1mM*[20]

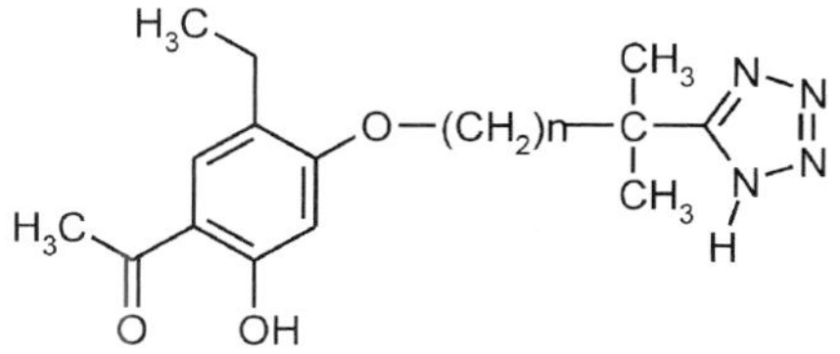

n = length of the methylene chain	% inhibition of [³H] LTB₄ binding at 0.1 μM
3	28
4	17
5	56
6	13
7	49

chains the methyl groups dovetail nicely and stay out of one another's way. However in odd-numbered chains methyl groups on one end can only avoid one another by increasing the distance between chain ends. This less-than-tight packing in odd-numbered chains apparently results in their anomalous melting points. In the examples above, the alkyl chain represents a spacer group between two binding groups. In some cases it can be shown that the energy required to fold the molecule to obtain the required separation should change in a zig-zag manner with increasing chain length.

In the biological domain, variations of activity are not necessarily linked to the induction of effects at the level of a given receptor, but could have come from a pharmacokinetic factor (urinary or biliary excretion, plasma protein binding, differential metabolism). A case of differential metabolism is illustrated by the comparison of the toxicities of odd and even ω-fluoro acids.[25] The β-oxidation of odd chain length compounds leads to the extremely toxic fluoroacetic acid, while that of the acids with even numbers of carbon atoms generates β-fluoropropionic acid which is clearly less toxic (Table 12.2).

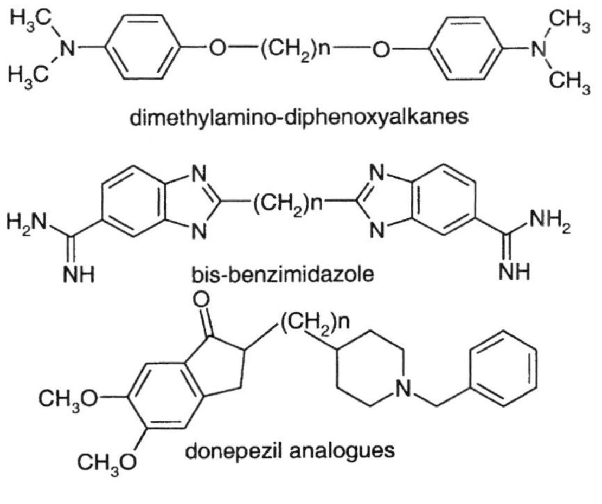

D Inversion of the activity

It can happen that the lower members of a series possess one activity profile and that the higher terms possess a different activity, which contrasts with that of the lower members.

This phenomenon is particularly observed when the bulkiness of cationic heads is progressively increased. In *N*-alkylated derivatives of norepinephrine,[26] progressive alkylation reduces the hypertensive activity according to the sequence: —NH₂, —NHMe, —NHEt, —NH-NPro. Finally, the molecules become hypotensive for the values: —NH-IsoPro, —NH-nBu and —NH-IsoBu (Table 12.3).

This anomaly is explained by the fact that norepinephrine can interact with two subclasses of receptors (α- and β-adrenergic receptors). The less hindered derivatives are able to bind to both α- and β-receptors, hindered ones solely to β-receptors. A similar inversion of properties is observed when the cholinergic agonist carbachol is modified by dibutylation at the carbamate function and exchange of one of its methyl groups for an ethyl group (Fig. 12.17). The analogue, dibutoline, is a powerful cholinolytic.

In morphine (agonist), the replacement of the *N*-methyl group by a more bulky radical such as *N*-allyl, *N*-cyclopropyl-methyl or *N*-cyclobutyl-methyl leads to powerful antagonists of the opiate receptors (see Chapter 19: Unsaturated Groups).

Introduction of bulkiness in a cationic head does not always cause a change from agonist to antagonist.

Fig. 12.16 4,4′-Dimethylamino-diphenoxyalkanes,[21] alkyl-linked bis(amidinobenzimidazoles),[22] and donepezil analogues.[23]

Table 12.2 *Zig-zag variations of the toxicity of aliphatic ω-fluoro derivatives (LD_{50} for mice in mg kg^{-1} intraperitoneally)*[25]

n	$F(CH_2)_nCOOH$ Odd	Even	$F(CH_2)_nCO$ Odd	Even	$F(CH_2)_nCOH$ Odd	Even	$F(CH_2)_nCH_3$ Odd	Even
1	6.6		6		10			
2		60				46.5		
3	0.65		2		0.9			
4		>100		81		>100		
5	1.35		0.58		1.2		1.7	
6		40		>100		80		35
7	0.64		2		0.6		2.7	
8		>100		53		32		21.7
9	1.5		1.9		1		1.7	
10		57		>40		>100		15.5
11	1.25				1.5		2.5	

Table 12.3 *Gradual inversion of the activity in a homologous series*[26]

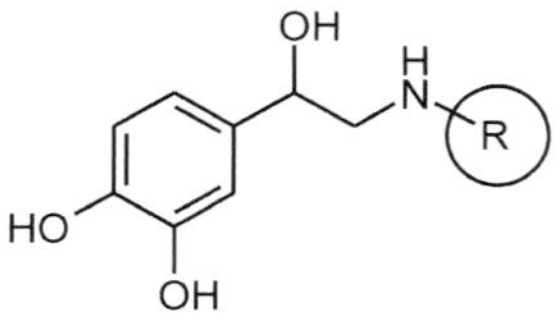

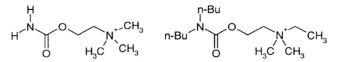

R	Blood pressure of the cat Hypertensive	Hypotensive
Hydrogen	++	−
Methyl	++	−
Ethyl	+	+
Propyl	−	+
Isopropyl	−	++
Butyl	−	++
Isobutyl	−	++

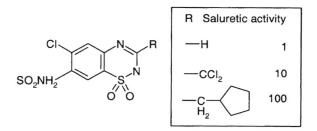

Fig. 12.17 Carbachol (left) and dibutoline (right).

R	Saluretic activity
—H	1
—CCl$_2$	10
—CH$_2$-cyclopentyl	100

Fig. 12.18 Tolerance to bulkiness.[27]

Thus the analogue *N*-propyl-apomorphine is a more powerful dopaminergic agonist than the apomorphine itself. The creation of bulkiness is obviously not limited to cationic head groups and lipophilic groups can be attached to any other part of the molecule (Fig. 12.18).[27]

E Conclusion

Variations in homologous series generally relate to the search for optimal lipophilicity. In the cyclo-polymethylenic series, conformational problems may be added. For difunctional polymethylenic derivatives, intercharge distances and, possibly, elements of symmetry (see Chapters 16 and 28) can take over. Whereas the activity profile is generally preserved during homology changes, very large differences in potency can be found, that confound the old adage 'methyl, ethyl, propyl, butyl ... futile'.

V VINYLOGUES AND BENZOLOGUES

The vinylogy principle was first formulated by Claisen in 1926,[28] who observed for formylacetone acidic properties similar to that of acetic acid. The vinyl group plays the role of an electron-conducting channel between the carbonyl and the hydroxyl group. The same effect explains the acidity of ascorbic acid (Fig. 12.19).

Today the vinylogy principle is explained by the mesomeric effect and it applies to all conjugated systems: imine and ethynyl groups, phenyl rings, aromatic heterocycles (Fig. 12.20). For a review of the chemical aspects of the vinylogy principle see Krishnamurthy.[29]

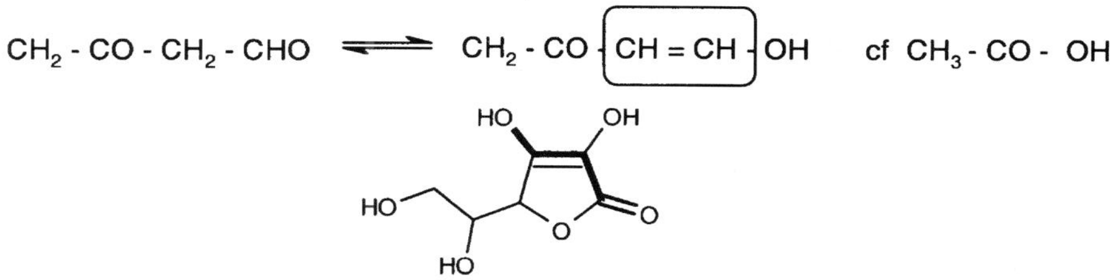

CH$_2$ - CO - CH$_2$ - CHO $\rightleftharpoons$ CH$_2$ - CO $\boxed{CH = CH}$ OH   cf CH$_3$ - CO - OH

Fig. 12.19 Formylacetone (enolic form) and vitamin C are comparable in acidity to carboxylic acids.

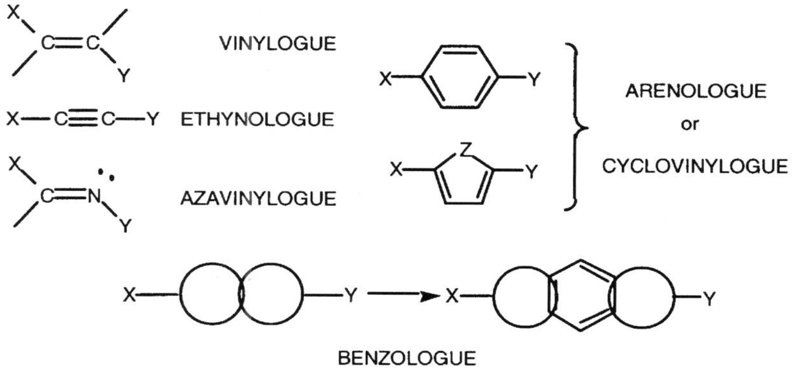

VINYLOGUE

X — C≡C — Y   ETHYNOLOGUE

AZAVINYLOGUE

ARENOLOGUE
or
CYCLOVINYLOGUE

BENZOLOGUE

Fig. 12.20 Vinylogy and its extensions.

A Applications of the vinylogy principle

Although numerous applications of the vinylogy concept are found in the medicinal chemistry literature, only a very few of them are of practical interest, mainly because the preparation of vinylogues usually leads to compounds which are more sensitive to metabolic degradation and more toxic (reactivity of the conjugated double bond) than the parent drug, without being more active.

Authentic vinylogues

The vinylogues of phenylbutazone,[30] and of pethidine (C. G. Wermuth, unpublished results) have the same type of activity as the parent drug, but the duration of action, especially for the pethidine analogue, is notably shorter than that of the initial molecule (Fig. 12.21). This is probably due to the easier metabolic degradation of the styryl double bond.

In preparing the vinylogues of acetylcholine (Fig. 12.21), Tenconi and Barzaghi[31] succeeded in separating the nicotinic from the muscarinic activity (Table 12.4).

Tolcapone (Fig. 12.22) was designed as an inhibitor of the enzyme catechol *O*-methyltransferase useful in the L-DOPA treatment of Parkinson's disease.[32] In avoiding the methylation of L-DOPA, as well as that of dopamine it prolongs the beneficial activities of these molecules.

Catechol *O*-methyltransferase inhibition therefore represents a valuable adjuvant to L-DOPA decarboxylase

inhibition. Unfortunately, tolcapone exhibited severe liver damage and had to be removed from the market. The corresponding vinylogue entacapone is devoid of these side-effects.[33]

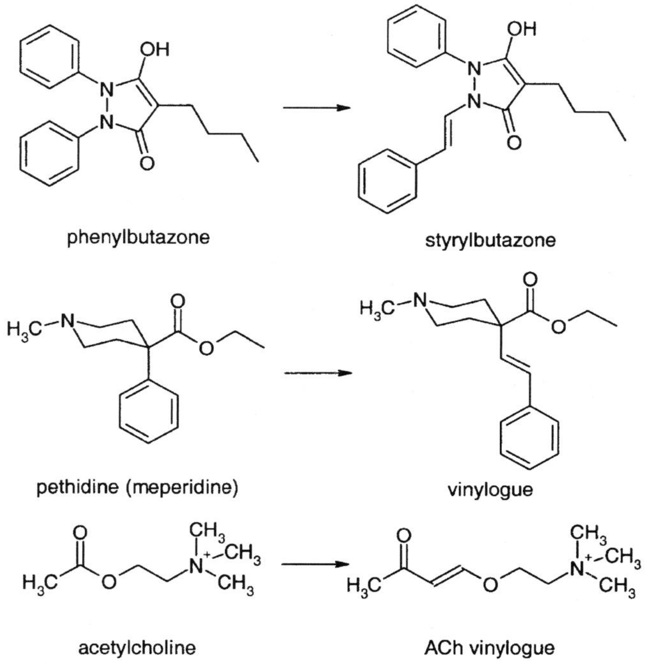

phenylbutazone

styrylbutazone

pethidine (meperidine)

vinylogue

acetylcholine

ACh vinylogue

Fig. 12.21 Vinylogues of phenylbutazone, pethidine and acetylcholine.

Table 12.4 *Cholinergic profile of the vinylogue of acetylcholine*[31]

Compound	Nicotinic activity	Muscarinic activity	Sensitivity to ACh-esterase
ACh	+	+	Sensitive
Vinylogue	+	Insensitive	Insensitive

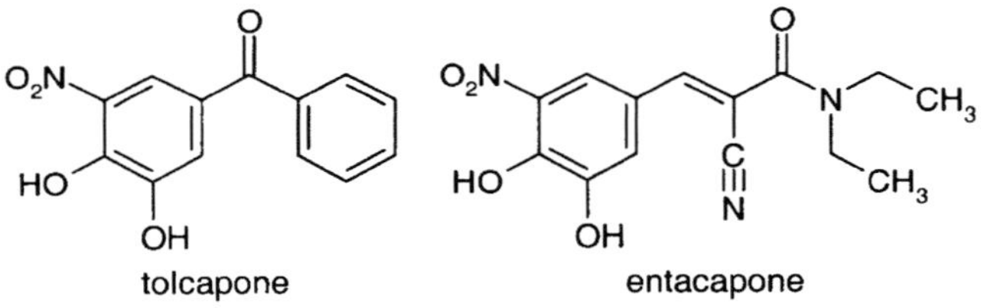

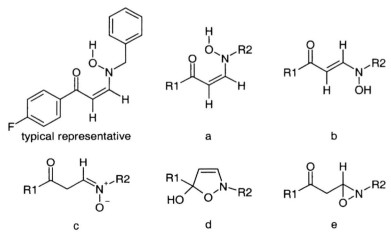

tolcapone   entacapone

Fig. 12.22 The vinylogy principle applied to the catechol *O*-methyltransferase inhibitor tolcapone.

Wright *et al.*[34] describe a series of hydroxamic acid vinylogues acting as dual inhibitors of 5-lipoxygenase (5-LO; $IC_{50} = 0.15 \times 10^{-6}$ M) and of interleukin-1β (IL-1β) biosynthesis ($IC_{50} = 2.8 \times 10^{-6}$ M) which might be useful as anti-inflammatory drugs (Fig. 12.23).

For such compounds several possible isomeric and tautomeric forms can be considered. These include the (E) and (Z) geometrical isomers (a and b), as well as the tautomers nitrone (c), 5-hydroxy-isoxazolidine (d), and oxaziridine (e). Examination of the ¹H and ¹³C NMR spectra of the vinylogues revealed that each of the tautomeric possibilities are present in solution in varying proportions. The relative proportion of each isomer was found to be dependent upon the solvent, the pH, and its chemical structure.

In order to design compounds able to react covalently with the nucleophilic cysteine of human 3C rhinovirus protease, scientists from Agouron designed vinylogous derivatives of the prototype inhibitor 3-carbamoyl-benzaldehyde (Fig. 12.24). In this particular case Michael acceptor reactivity is eventually the wanted feature.[35]

Ethynologues

Some ethynologues of biologically active compounds were prepared by Dunoguès and his group, unfortunately they did not describe their biological activity: aspirin ethynologue,[36] nicotinamide and isoniazide ethynologues,[37] chalcones ethynologues.[38] In some way the cholinergic antagonist oxotremorine can be considered as an ethynologue of the acetylcholine pharmacophore (Fig. 12.25).

Azavinylogues

As a rule, simple azavinylogues are unstable compounds, due to the easy hydrolysis of the imino bond. However the particular case of *O*-alkylated oximes (X—CH=N—O—Y; with Y=R or Ar) can be interesting insofar as the oximic imino bond was shown to be biostable.[39] Preparing azavinylogues of β-blocking agents (Fig. 12.26) led to some active compounds.[40-42] The proposal was made that the stable oxime C—NOCH₂ could mimic a portion of an aromatic ring, thus simulating an aryl or an aryloxymethylene group.[43]

Reduction of the imino bond results in a decrease but not a loss of activity and ether derivatives retain activity.[42] Tricyclic oxime β-blockers showed high selectivity for β₂ receptors.[40,41] Noxiptyline (Agedal®, Bayer) is the oximic equivalent of amitriptyline (for review see Hoffmeister;[39,44] for crystal structure see Bandoli[45]).

typical representative   a   b

c   d   e

Fig. 12.23 Hydroxamic acid vinylogues and their various isomeric forms.[34]

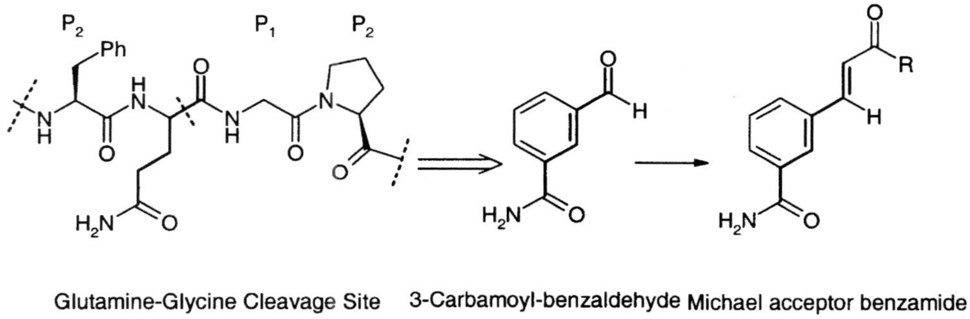

P2
Ph
P1
P2
H2N
Glutamine-Glycine Cleavage Site
3-Carbamoyl-benzaldehyde
Michael acceptor benzamide

Fig. 12.24 Cinnamic derivatives as vinylogues of benzaldehyde.[35]

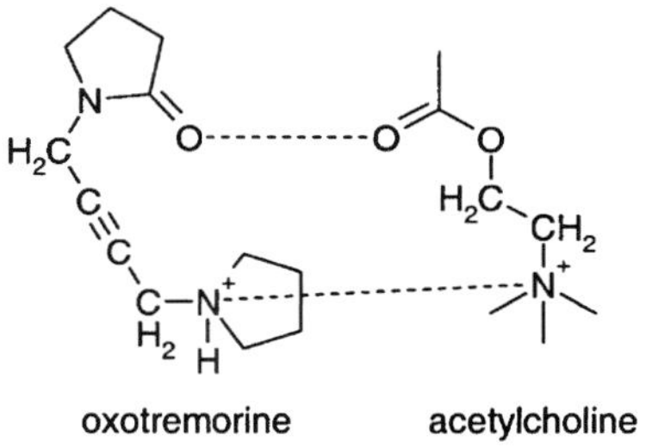

oxotremorine
acetylcholine

Fig. 12.25 Oxotremorine is an ethynologue of the acetylcholine pharmacophore.

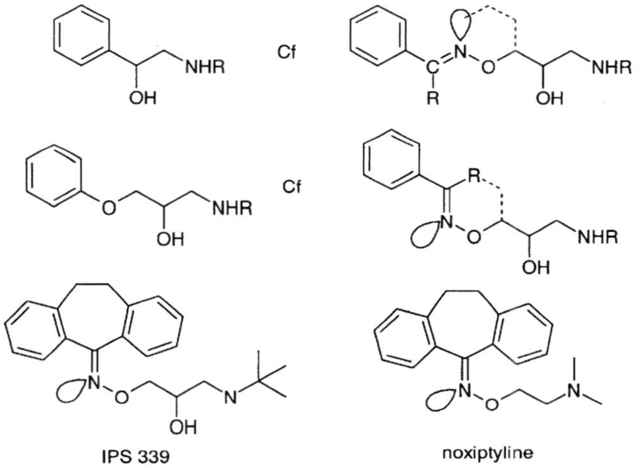

NHR
OH
Cf
IPS 339
noxiptyline

Fig. 12.26 Oxime ethers as azavinylogues.

Cyclovinylogues

These vinylogues have the advantage of being more stable towards *in vivo* metabolism. In addition they allow molecular variations with *ortho*, *meta* and *para* positional isomers. Thus, for cyclovinylogues of procainamide, the highest local anaesthetic activity was found with the *meta* derivative, which also showed the best dissociation between local anaesthetic and antiarrhytmic activity (Table 12.5).[46]

Similar results were observed with cyclovinylogues of lidocaine.[47] For other references, see Valenti *et al.*[48] Compound TA-1801 (Ethyl 2-(4-chlorophenyl)-5-(2-furyl)-4-oxazoleacetate),[49] can to some extent be considered as an arenologue of clofibrate (Fig. 12.27).

Pyrroline-3-ones were used as peptidic bond surrogates by Hirshman and his group.[50] In such compounds (Fig. 12.28), thanks to vinylogy, the carbonyl and the

Table 12.5 *Cyclovinylogues of procainamide, relative activity with regard to procainamide*[46]

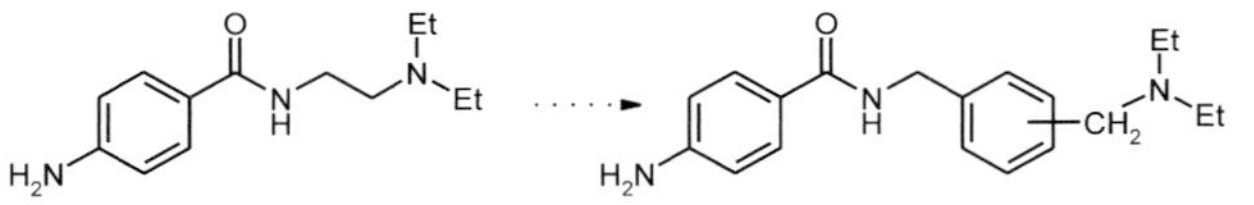

Compound	Local anaesthetic power	Antiarrhytmic activity
Procainamide	1	1
Ortho-cyclovinylogue	~0	0.17
Meta-cyclovinylogue	47	0
Para-cyclovinylogue	35	0

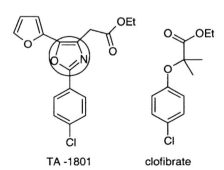

TA-1801
clofibrate

Fig. 12.27 TA-1801, an arenologue of clofibrate.[49]

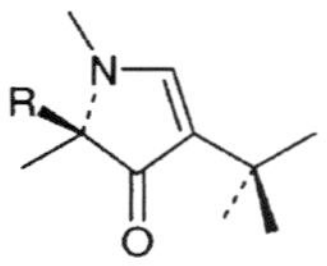

Fig. 12.28 The vinylogous relationship between the carbonyl and the amino group in pyrroline-3-ones gives them the reactivity of secondary amides.[50]

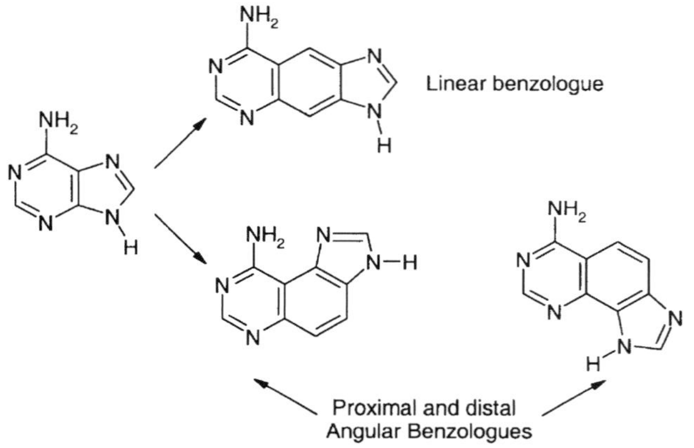

Fig. 12.29 Linear and angular adenine benzologues.

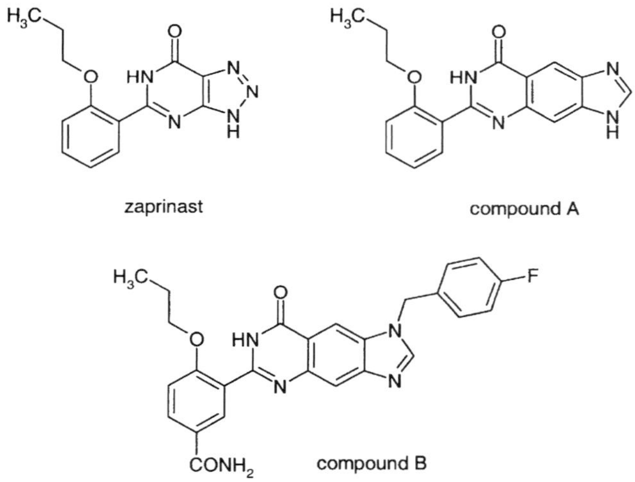

Fig. 12.30 Linear benzologues derived from zaprinast.[58]

amino group show the same chemical reactivity to that of a secondary amide.

Benzologues

Linear and angular benzologues of guanine[51] and adenine,[52-54] were published without any indication of biological activity (Fig. 12.29). A review article on chemistry and biochemistry of benzologues was published by Leonard and Hiremath.[55] More recently, linear and angular benzologues of xanthines showed submicromolar affinities for rat brain A_1 and A_2 adenosine receptors[56] and benzologues of quinolone antibacterials maintained high antimicrobial activity.[57]

A very convincing example of the usefulness of benzologues is provided by the synthesis of compound 'A', a linear benzologue of the prototypical PDE-5 inhibitor zaprinast and its optimization to potent and selective PDE5 inhibitors such as 'B'(Fig. 12.30).[58]

B Comments

Due to important changes in geometry, vinylogues often have unpredictable activity. For this reason vinylogues play a minor role in medicinal chemistry. In addition, their metabolic vulnerability or their increased toxicity may represent a significant drawback.

However, the vinylogy principle is sometimes applied to the design of bioisosteres. Thus the guanidinic group of the benzimidazole fenbendazole[59] can be compared with its vinylogue[60] in the corresponding imidazo[1,2-*a*]pyridine (Fig. 12.31). Both compounds are anthelminthics of similar potency.

The vinylogy principle can account for unexpected chemical reactivity that is not always recognized at first glance (Fig. 12.32). So, for example the basicity of the N1 nitrogen is strengthened in compound CGS 8216 thanks to the vinylogous influence of the quinoline nitrogen. For a similar reason, the carbonyl group of benzopiperidones or of 3-acyl-indoles behaves chemically more like an amidic carbonyl than a ketonic one. In 2-methoxy-*para*-benzoquinone the reactivity of the methoxy group is that of a carboxylic ester, rendering it susceptible to attack by secondary amines.

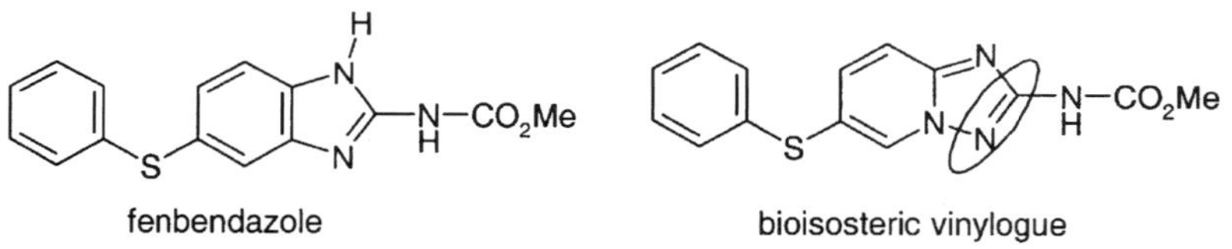

Fig. 12.31 Application of the vinylogy principle to the design of a fenbendazole bioisostere.[60]

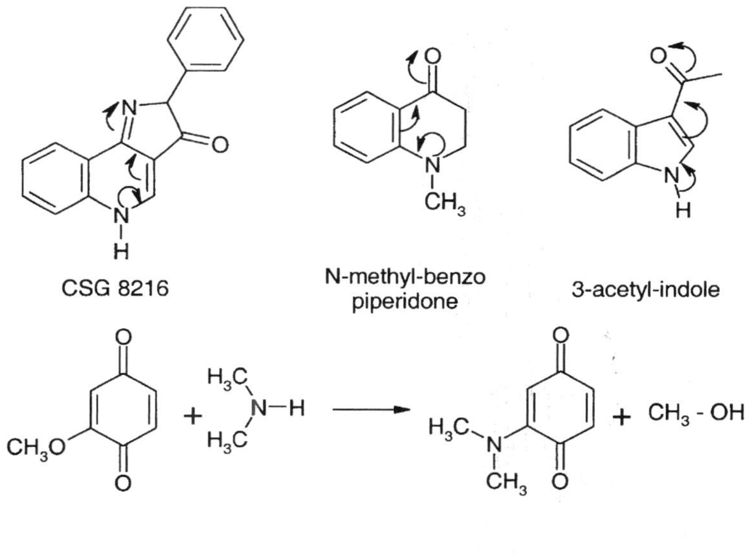

Fig. 12.32 Unexpected chemical reactivities attributable to vinylogy.

REFERENCES

1. Gerhardt, C. (1853) Principes de la classification sériaire. *Traité de Chimie Organique*, pp. 121–142. Firmin Didot Frères, Paris.
2. Kim, C.U., Lew, W., Williams, M.A., Wu, H., Zhang, L., *et al.* (1998) Structure–activity relationship studies of novel carbocyclic influenza neuraminidase inhibitors. *J. Med. Chem.* **41**: 2451–2460.
3. Ward, J.S., Merritt, L., Klimkowski, V.J., Lamb, M.L., Mitch, C.H., *et al.* (1992) Novel functional M_1 selective muscarinic agonists. 2. Synthesis and structure–activity relationships of 3-pyrazinyl-1,2,5,6-tetrahydro-1-methylpyridines. Construction of a molecular model for the M_1 pharmacophore. *J. Med. Chem.* **35**: 4011–4019.
4. Cook, L., Ternai, B. and Ghosh, P. (1987) Inhibition of human sputum elastase by substituted 2-pyrones. *J. Med. Chem.* **30**: 1017–1023.
5. Thorsett, E.D. (1986) Conformationally restricted inhibitors of angiotensin converting enzyme. In Combet-Farnoux, C. (ed.). *Actualités de Chimie Thérapeutique*, pp. 257–268. Société de Chimie Thérapeutique, Chatenay-Malabry.
6. Eden, J.M., Higginbottom, M., Hill, D.R., Horwell, D.C., Hunter, J.C., *et al.* (1993) Rationally designed 'dipeptoid' analogues of cholecystokinin (CCK): N-terminal structure–affinity relationships of a-methyltryptophan derivatives. *Eur. J. Med. Chem.* **28**: 37–45.
7. Cheronis, J.C., Whalley, E.T., Nguyen, K.T., Eubanks, S.R., Allen, L.G., *et al.* (1992) A new class of bradykinin antagonists: synthesis and *in vitro* activity of bissuccinimidoalkane peptide dimers. *J. Med. Chem.* **35**: 1563–1572.
8. Wang, W. and Lown, J.W. (1992) Anti-HIV-I activity of linked lexitropsins. *J. Med. Chem.* **35**: 2890–2897.
9. Press, J.B., Wright, W.B., Jr., Chan, P.S., Haug, M.F., Marsico, J.W., *et al.* (1987) Thromboxane synthetase inhibitors and antihypertensive agents. 3. N-[(1H-imidazol-1-yl)alkyl]heteroaryl amides as potent enzyme inhibitors. *J. Med. Chem.* **30**: 1036–1040.
10. Nicolaï, E., Goyard, J., Benchetrit, T., Teulon, J.M., Caussade, F., *et al.* (1992) Synthesis and structure–activity relationships of novel benzimidazole and imidazo[4,5-b]pyridine acid derivatives as thromboxane A_2 receptor antagonists. *J. Med. Chem.* **36**: 1175–1187.
11. Ginos, J.Z., Stevens, J.M. and Nichols, D.E. (1979) Structure–activity relationships of N-substituted dopamine and 2-amino-6,7-dihydroxy-1,

2,3,4-tetrahydronaphtalene analogues: behavioral effects in lesioned and reserpinized mice. *J. Med. Chem.* **22**: 1323–1329.
12. Godfroid, J.-J., Broquet, C., Jouquey, S., Lebbar, M., Heymans, F., *et al.* (1987) Structure–activity relationship in PAF-acether. 3. Hydrophobic contribution to agonistic activity. *J. Med. Chem.* **30**: 792–797.
13. Bustard, T.M. (1974) Optimization of alkyl modifications by Fibonacci search. *J. Med. Chem.* **17**: 777–778.
14. Santora, N.J. and Auyang, K. (1975) Non-computer approach to structure–activity study. An expanded Fibonacci search applied to structurally diverse types of compounds. *J. Med. Chem.* **18**: 959–963.
15. Martin, Y.C. (1987) *Quantitative Drug Design, a Critical Introduction*, pp. 257–261. Marcel Dekker, New York.
16. Koelzer, P.P. and Wehr, K.H. (1958) Beziehungen zwischen chemischer Konstitution un pharmakologischer Wirkung bei mehreren Klassen neure Lokalanaesthetica. *Arzneimittel-Forsch.* **8**: 544–550.
17. Funcke, A.B.H., Ernsting, M.J.E., Rekker, R.F. and Nauta, W.T. (1953) Untersuchungen über Spasmolytica. 1. Mandelsäureester. *Arzneimitt.-Forsch.* **3**: 503–506.
18. Wermuth, C.G., Bourguignon, J.-J., Schlewer, G., Gies, J.P., Schoenfelder, A., *et al.* (1987) Synthesis and structure–activity relationships of a series of aminopyridazine derivatives of g-aminobutyric acid acting as selective $GABA_A$ antagonists. *J. Med. Chem.* **30**: 239–249.
19. Magidson, O.J. and Strukow, I.T. (1933) Die derivate des 8-Aminochinolins als Antimalariapräparate. Mitteilung II: Der Einfluß der Länge der Kette in Stellung 8. *Arch. Pharm.* **271**: 569–580.
20. Herron, D.K., Goodson, T., Bollinger, N.G., Swanson-Bean, D., Wright, I.G., *et al.* (1992) Leukotriene B4 receptor antagonists: The LY 255 283 series of hydroxyacetophenones. *J. Med. Chem.* **35**: 1818–1828.
21. Raison, C.G. and Standen, O.D. (1955) The schistosomicidal activity of symmetrical diaminodiphenoxyalkanes. *Br. J. Pharmacol.* **10**: 191–199.
22. Fairley, T.A., Tidwell, R.R., Donkor, I., Naiman, N.A., Ohemeng, K.A., *et al.* (1993) Structure, DNA minor groove binding, and base pair specificity of alkyl- and aryl-linked bis(amidinobenzimidazoles) and bis(amidinoindoles). *J. Med. Chem.* **36**: 1746–1753.
23. Sugimoto, H., Iimura, Y., Yamanishi, Y. and Yamatsu, K. (1996) Synthesis and structure–activity relationships of acetylcholinesterase

inhibitors: 1-benzyl-4-[(5,6-dimethoxy-1-oxoindan-2-yl)methyl]piper-idine hydrochloride and related compounds. *J. Med. Chem.* **38**: 4821–4829.

24. Boese, R., Weiss, H.-C. and Bläser, D. (1999) The melting point alternation in the short claim *n*-alkanes: Single crystal X-ray analyses of propane at 30 K and of *n*-butane to *n*-nonane at 90 K. *Angew. Chem. Int. Ed.* **38**: 988–992.

25. Pattison, F.L.M. (1959) *Toxic Aliphatic Fluorine Compounds*. Elsevier, Amsterdam.

26. Ariëns, E.J. (1964) *Molecular Pharmacology*. Academic Press, New York.

27. Beyer, K.H. and Baer, J.E. (1961) Physiological basis for the action of newer diuretic agents. *Pharmacol. Rev.* 517–562.

28. Claisen, L. (1926) Zu den O-alkylderivaten des benzoyl-acetons und den aus ihnen entstehenden isooxazolen. *Ber. Dtsch. Chem. Ges.* **59**: 144–153.

29. Krishnamurthy, S. (1982) The principle of vinylogy. *J. Chem. Educ.* **59**: 543–547.

30. Yamamoto, H. and Kaneko, S.-I. (1970) Synthesis of 1-phenyl-2-styryl-3,5-dioxopyrazolidines as antiinflammatory agents. *J. Med. Chem.* **13**: 292–295.

31. Tenconi, F. and Barzaghi, F. (1964) Attivita nicotinica di vinil-analoghi di esteri della colina. *Boll. Chim. Pharmaceut.* **103**: 569–575.

32. Zürcher, G., Keller, H.H., Kettler, R., Borgulya, J., Bonetti, E.P., *et al.* (1990) Ro 40-7592, a novel, very potent, and orally active inhibitor of catechol-*O*-methyltransferase: a pharmacological study in rats. *Adv. Neurol.* **53**: 497–503.

33. Nissinen, E. and Linden, I.B. (1992) Biochemical and pharmacological properties of a peripherally acing catechol-*O*-methyltransferase inhibitor: entacapone. *Naunyn Schmiedebergs Arch. Pharmacol.* **346**: 262–266.

34. Wright, S.W., Harris, R.R., Kerr, J.S., Green, A.M., Pinto, D.J., *et al.* (1962) Synthesis, chemical, and biological properties of vinylogous hydroxamic acids: dual inhibitors of 5-lipoxygenase and IL-1 biosynthesis. *J. Med. Chem.* **35**: 4061–4068.

35. Reich, S.H., Johnson, T., Wallace, M.B., Kephart, S.E., Fuhrman, S.A., *et al.* (2000) Substituted benzamide inhibitors of human rhinovirus 3C protease: structure-based design, synthesis, and biological evaluation. *J. Med. Chem.* **43**: 1670–1683.

36. Babin, P., Bourgeois, P. and Dunoguès, J. (1976) Synthèse de l'éthynologue de l'acide acétylsalicylique. *CR Acad. Sci. (Paris)* **283**: 149–152.

37. Babin, P., Cassagne, A., Dunoguès, J., Duboudin, F. and Lapouyade, P. (1981) Ethynologues du nicotinamide et de l'isoniazide. *J. Heterocyclic Chem.* **18**: 519–523.

38. Babin, P., Lapouyade, P. and Dunoguès, J. (1982) Synthesis of chalcone ethynologues with a pharmacological objective. *Can. J. Chem.* **60**: 379–382.

39. Hoffmeister, F. (1969) Zur Frage pharmakologisch-klinischer Wirkungsbeziehungen bei Antidepressiva, dargestellt am Beispiel von Noxiptilin. *Arzneimitt.-Forsch.* **19**: 458–467.

40. Leclerc, G., Mann, A., Wermuth, C.G., Bieth, N. and Schwartz, J. (1977) Synthesis and β-adrenergic blocking activity of a novel class of aromatic oxime ethers. *J. Med. Chem.* **20**: 1657–1662.

41. Imbs, J.L., Miesch, F., Schwartz, J., Velly, J., Leclerc, G., *et al.* (1977) A potent new β2-adrenoceptor blocking agent. *Br. J. Pharmacol.* 357–362.

42. Leclerc, G., Bieth, N. and Schwartz, J. (1980) Synthesis and β-adrenergic blocking activity of new aliphatic oxime ethers. *J. Med. Chem.* **23**: 620–624.

43. Macchia, B., Balsamo, A., Lapucci, A., Martinelli, A., Macchia, F., *et al.* (1985) An interdisciplinary approach to the design of new structures active at the β-adrenergic receptor. Aliphatic oxime ether derivatives. *J. Med. Chem.* **28**: 153–160.

44. Aichinger, G., Behner, O., Hoffmeister, F. and Schütz, S. (1969) Basische tricyclische oximinoäther und ihre pharmakologischen eigenschaften. *Arzneimitt.-Forsch.* **19**: 838–845.

45. Bandoli, G. and Nicolini, M. (1983) Crystal structure of the anti-depressant noxiptyline hydrochloride (5-dimethylaminoethyloximino-5H-dibenzo[a,d]-cyclohepta-1,4-diene hydrochloride). *J. Crystallogr. Spectrosc. Res.* **13**: 191–199.

46. Valenti, P., Mazzotti, M., Rampa, A. and Magistretti, M.J. (1982) Cyclovinylogues of procainamide. *Arch. Pharm.* **315**: 1003–1007.

47. Valenti, P., Montanari, P., Da Re, P., Soldani, G. and Bertelli, A. (1980) Synthesis and pharmacological properties of three lidocaine cyclovinylogues. *Arch. Pharm.* **313**: 280–284.

48. Valenti, P., Montanari, P., Fabbri, G., Giovannini, L. and Giacomelli, A. (1985) Cyclo-vinylogues of some antimuscarinic drugs. *Arch. Pharm.* **318**: 222–224.

49. Mooriya, T., Seki, M., Takabe, S., Matsumoto, K., Takashima, K., *et al.* (1987) Compound TA-1801 [ethyl 2-(4-chlorophenyl)-5-(2-furyl)-4-oxazoleacetate]. *J. Pharm. Sci.* **76**: S164.

50. Smith, A.B., III, Keenan, T.P., Holcomb, R.C., Sprengeler, P.A., Guzman, M.C., *et al.* (1992) Design, synthesis and crystal structure of a pyrrolinone-based peptidomimetic possessing the conformation of a b-strand: potential application to the design of novel inhibitors of proteolytic enzymes. *J. Am. Chem. Soc.* **114**: 10672–10674.

51. Cottis, S.G., Clarke, P.B. and Tieckelmann, H. (1965) Pyrazolo[3,4-b]pyridines and pyrazolo[3',4'6,5]pyrido[2,3-d]pyrimidines. *J. Heterocycl. Chem.* **2**: 192–201.

52. Leonard, N.J., Morrice, A.G. and Sprecker, M.A. (1975) Linear benzoadenine. A stretched-out analog of adenine. *J. Org. Chem.* **40**: 356–363.

53. Morrice, A.G., Sprecker, M.A. and Leonard, N.J. (1975) The angular benzoadenines. 9-Aminoimidazo [4,5-f] quinazoline and 6-aminoimidazo [4,5-h] quinazoline. *J. Org. Chem.* **40**: 363–366.

54. Leonard, N.J., Sprecker, M.A. and Morrice, A.G. (1976) Defined dimensional changes in enzyme substrates and cofactors. Synthesis of lin-benzoadenosine and enzymatic evaluation of derivatives of the benzopurines. *J. Am. Chem. Soc.* **98**: 3987–3994.

55. Leonard, N.J. and Hiremath, S.P. (1986) Dimensional probes of binding and activity. *Tetrahedron* **42**: 1917–1961.

56. Schneller, S.W., Ibay, A.C., Christ, W.J. and Bruns, R.F. (1989) Linear and proximal benzo-separated alkylated xanthines as adenosine–receptor antagonists. *J. Med. Chem.* **32**: 2247–2254.

57. Jordis, U., Sauter, F., Rudolf, M. and Cai, G. (1988) Synthesen neuer chinolon-chemotherapeutika 1: pyridochinoline und pyridophenanthroline als 'lin-benzo-nalidixinsäure'-derivate. *Monatsh. Chem.* **119**: 761–780.

58. Rotella, D.P., Sun, Z., Zhu, Y., Krupinski, J., Pongrac, R., *et al.* (2000) N-3-Substituted imidazoquinazolinones: potent and selective PDE5 inhibitors as potential agents for treatment of erectile dysfunction. *J. Med. Chem.* **43**: 1257–1263.

59. Averkin, E.A., Beard, C.C., Dvorak, C.A., Edwards, J.A., Fried, J.H., *et al.* (1972) Methyl 5(6)-phenylsulfinyl-2-benzimidazolecarbamate, a new, potent anthelmintic. *J. Med. Chem.* **15**: 1164–1166.

60. Bochis, R.J., Dybas, R.A., Eskola, P., Kulsa, P., Linn, B.O., *et al.* (1978) Methyl 6-(phenylsulfinyl) imidazo [1,2-a] pyridine-2-carbamate, a potent, new anthelmintic. *J. Med. Chem.* **21**: 235–237.

13

MOLECULAR VARIATIONS BASED ON ISOSTERIC REPLACEMENTS

Camille G. Wermuth

Si ce n'est toi, c'est donc ton frère
If it isn't you, then it's your brother

Jean de La Fontaine (1621–1695) *Le loup et l'agneau*

The replacement in an active molecule of an atom or a group of atoms by another one, presenting a comparable electronic and steric arrangement is based on the concept of *isosterism*. The term isosterism was introduced in 1919 by the physicist Langmuir[1] who was mainly interested in the physico-chemical relationships of isosteric molecules. When, in addition to their physicochemical analogy, compounds share some common biological properties, the term *bioisosterism*, introduced by Friedman in 1951[2] is used, even if the physicochemical resemblance is only vague.

I HISTORY: DEVELOPMENT OF THE ISOSTERISM CONCEPT

The development of the concept of isosterism has its roots in the attempts to extend to entire molecules the knowledge acquired for elements, namely that two elements possessing an identical peripheral electronic distribution also possess similar chemical properties.

A The molecular number

Allen, in 1918 defined the *molecular number* of a compound in a similar way to the atomic number:

$$N = aN_1 + bN_2 + cN_3 + \cdots + zN_i$$

where

N = molecular number,
$N_1, N_2, N_3, \ldots N_i$ = respective atomic numbers of each element of the molecule,
$a, b, \ldots z$ = number of atoms of each element present in the molecule.

The Practice of Medicinal Chemistry
ISBN 0-12-744481-5

189

Compare ammonium and sodium cations as an example. The atomic number of nitrogen is 7 and that of hydrogen is 1. Thus the molecular number of the ammonium cation can be calculated and compared with that of the sodium ion:

	Atomic number	*Molecular number*
NH_4^+	$7 + (4 \times 1)$	11
Na^+	11	11

Possessing the same molecular number, the ammonium cation should resemble the sodium cation. This is roughly true. More generally, two compounds with identical molecular numbers present at least some similar physical properties (e.g. specific heat).

B The isosterism concept

Independently from Allen, Langmuir in 1919[1] defined the concept of isosterism

Comolecules are thus isosteric if they contain the same number and arrangement of electrons. The comolecules of isosteres must, therefore, contain the same number of atoms. The essential differences between isosteres are confined to the charges on the nuclei of the constituent atoms.

Langmuir cites a list of 21 kinds of isosteres, such as:

$$O^=, F^-, Ne^+, Na^+, Mg^{++}, Al^{+++} \text{ or } ClO_4^-, SO_4^{2-}, PO_4^{3-}$$

The first example clearly demonstrates that isosterism does not inevitably imply 'isoelectric' structures (having the same total electric charge), but it becomes evident that isoelectronic isosteres show the closest analogies:

$$C=O \text{ and } N=N \quad CO_2 \text{ and } NO_2 \quad N=N=N^- \text{ and } N=C=O^-$$

In the field of organic chemistry, Langmuir predicted the analogy between diazomethane and ketene, which was only discovered later.

$$\overset{\backslash}{\underset{/}{C}}{=}\overset{+}{N}{=}\overset{..}{\underset{..}{N}} \quad cf. \quad \overset{\backslash}{\underset{/}{C}}{=}C{=}\overset{..}{\underset{..}{O}}$$

C The notion of pseudoatoms and Grimm's hydride displacement law

Later, in 1925, Grimm formulated the 'hydride displacement law',[3-5] according to which the addition of hydrogen to an atom confers on the aggregate the properties of the atom of the next highest atomic number. An isoelectronic relationship exists among such aggregates which were named *pseudoatoms*. Thus, when a proton is 'added' to the $O^=$ ion in the nuclear sense, an isotope of fluorine is obtained (Fig. 13.1).

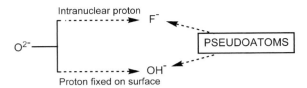

Fig. 13.1 The notion of pseudoatoms.

When the same proton is introduced at the peripheral electronic level, a 'pseudo-F', in other words an OH^-, is created. In this context, the H^+ ion having penetrated into the electronic shell of the oxygen, is assumed to be masked by the greater atom and to exert only negligible effect towards the outside. The fluoride anion F^- and the hydroxyl anion OH^- therefore show some analogies. The generalization of the pseudoatom concept represents the so-called 'hydride displacement law' proposed in tabular form by Grimm.[3,4] In each vertical column (Table 13.1), the original atom is followed by its isosteric pseudoatoms.

D Erlenmeyer's expansion of the isosterism concept

Starting in 1932, Erlenmeyer published a series of detailed studies on the isosterism concept, and particularly about its first applications to biological problems.[6] Erlenmeyer proposed his own definition of isosteres as elements, molecules or ions in which the peripheral layers of electrons may be considered identical.[7] Erlenmeyer also proposed three expansions of the isosterism concept:

(1) To the whole group of elements present in a given column of the periodic table. Thus silicon becomes isosteric to carbon, sulphur to oxygen, etc.
(2) To the pseudoatoms, with the aim of including groups which at a first glance seem totally different, but which, in practice, possess rather similar properties. This is

Table 13.1 *Hydride displacement law: in each vertical column the atom is followed by its pseudoatoms*

			Number of electrons			
6	7	8	9	10	11
C^{4-}	N^{3-}	$-O-$	$-F$	Ne	Na^+
	CH^{3-}	$-NH-$	$-OH$	FH	
		$-CH_2-$	$-NH_2$	OH_2	
			$-CH_3$	NH_3	OH_3^+
				CH_4	NH_4^+

Table 13.2 *The sulphur atom is approximately equivalent to an ethylenic group (size, mass, capacity to provide an aromatic lone pair)*

$$M_{(-CH = CH-)} = 26 \qquad M_{(S)} = 32$$

Compound	E°C	Isostere	E°C
Benzene	80°	Thiophene	84°
Methylbenzene	110°	2-Methyl-thiophene	113°
Chlorobenzene	132°	2-Chloro-thiophene	130°
Acetylbenzene	200°	2-Acetyl-thiophene	214°

the case for the pseudohalogens, for example (Cl ≅ CN ≅ SCN, etc.)

(3) To the ring equivalents: The equivalence between —CH=CH– and –S– explaining the well-known analogy between benzene and thiophene (Table 13.2).

E Isosterism criteria, present conceptions

The main criteria for isosterism is that two isosteric molecules must present similar, if not identical, volumes and shapes. Ideally, isosteric compounds should be isomorphic and able to cocrystallize. Among the other physical properties that isosteric compounds usually share one can cite: boiling point, density, viscosity and thermal conductivity. However, certain properties must be different: dipolar moments, polarity, polarization, size and shape (e.g. in comparing F^- and OH^-, the size and the shape of H cannot be totally neglected). After all the external orbitals may be hybridized differently.

In conclusion, it became evident to the physicists that the concept of isosterism, developed before quantum-mechanical theories, could not provide at the molecular level the same results as those that the periodic classification had provided for the elements, namely a correlation between electronic structure and physical and chemical properties. In the field of medicinal chemistry the isosterism concept, taken in its broadest sense, has proved to be a research tool of the utmost importance. The main reason for this is because isosteres are often much more alike in their biological than in their physical and chemical properties. An illustrative example is found in the comparison of oxazolidine-diones and hydantoins which possess different chemical reactivities, but present a similar antiepileptic profile (Fig. 13.2).

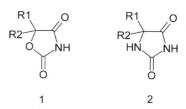

Fig. 13.2 5,5′-Disubstituted oxazolidine-diones (1) and hydantoins (2) show similar antiepileptic profiles.

F The bioisosterism concept—Friedman's and Thornber's definitions

Recognizing the usefulness of the isosterism concept in the design of biologically active molecules, Friedman[2] proposed to call *bio*isosteres compounds 'which fit the broadest definition of isosteres and have the same type of biological activity'. This definition received rapid acceptance and is now commonly used. Moreover, Friedman considers that isosteres that exhibit opposite properties (antagonists) also have to be considered as bioisosteres, since they usually interact with the same recognition site. This is the case for *p*-aminobenzoic acid and *p*-aminobenzene sulfonamide and also for glutamic acid and its phosphonic analogues.

The use of the word isosterism has largely been taken beyond its original meaning when employed in medicinal chemistry, and Thornber[8] proposes a loose and flexible definition of the term bioisostere: 'Bioisosteres are groups or molecules which have chemical and physical similarities producing broadly similar biological effects'.

The term non-classical isosterism is often used, interchangeably with the term bioisosterism, for example, when one has to deal with isosteres that do not possess the same number of atoms, but which have in common some key parameter of importance for the activity in a given series. Thus, the two GABAergic agonists isoguvacine and THIP (Fig. 13.3) possess similar pharmacological properties to GABA itself. The key parameters in these compounds are

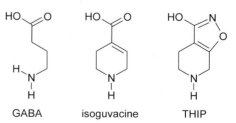

GABA   isoguvacine   THIP

Fig. 13.3 An example of bioisosterism, or nonclassical isosterism: GABA, isoguvacine and THIP are all agonists at the GABA$_A$ receptor. The 3-hydroxy-isoxazole ring has a comparable acidity to that of a carboxylic acid function.[9]

the acidic ($pK_A \approx 4$) and the basic (protonated nitrogen) functions with a ≈ 5.1 Å intercharge distance.[9]

II CURRENTLY ENCOUNTERED ISOSTERIC AND BIOISOSTERIC MODIFICATIONS

As the distinction between isosteres and bioisosteres is rather of academic interest, it is preferred, in this chapter to treat both categories together. Consequently, divalent series such as O=, HN=, and H_2C= can be discussed together with S=, for example. However, the correct nomenclature will be used as much as possible, bearing in mind nevertheless that 'isosteric replacement' embraces both true isosteres and bioisosteres.

A Replacement of univalent atoms or groups

Halogens (particularly chlorine) can be replaced by other electron-attracting functions, such as trifluoromethyl or cyano groups. In the antibiotic chloramphenicol, both the chlorine atoms of the dichloroacetic moiety and of the *p*-nitrophenyl group yielded productive isosteric replacements (Table 13.3). Many other examples of univalent atom or group replacements are found in the chapters dealing with substituent effects (Chapter 17) and with quantitative structure–activity relationships (Chapter 19).

B Interchange of divalent atoms and groups

A first series of frequently interchanged divalent atoms or groups is represented by O, S, NH and CH_2 and many interesting examples are found in the literature. In a study on meperidine analogues (Table 13.4) potent analgesic compounds were found for X=O, NH and CH_2.[10] Surprisingly, the sulphur analogue showed only moderate activity. As an

Table 13.3 *Isosteric replacements in the amphenicol family*

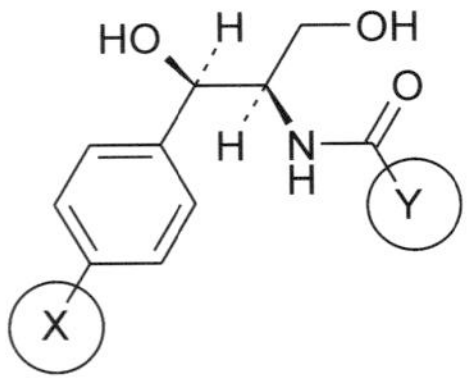

Compound	X	Y
Chloramphenicol	—NO₂	—CH—Cl₂
Thiamphenicol	CH₃—SO₂—	—CH—Cl₂
Cetophenicol	CH₃—CO—	—CH—Cl₂
Azidamphenicol	—NO₂	—CH—N₃

Table 13.4 *Meperidine analogues*[10]

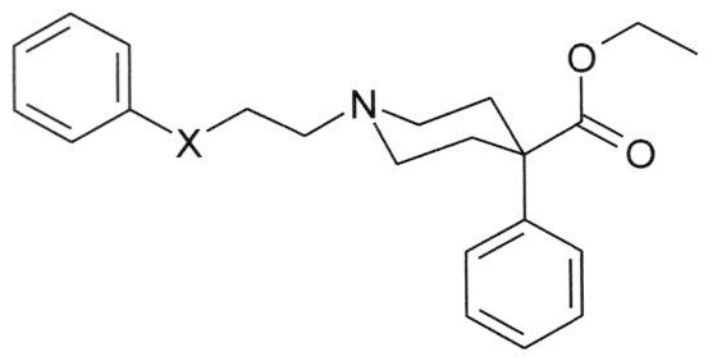

X	Analgesic potency (meperidine = 1)
O	12
NH	80
CH₂	20
S	1,5

in vivo test was used to assess the activity, the weaker effect may be attributable to a faster metabolism (sulfoxide or sulfone formation?).

Similar changes can be applied to cyclic series, for example to a series such as piperidine–morpholine–thiomorpholine–piperazine or in introducing oxygen or sulphur atoms into cyclic ketoprofen analogues.[11]

Isosteric variations were observed in a series of thermolysin inhibitors.[12] For these isosteres the replacement of the phosphonamide (X=NH) function by a phosphonate (X=O) or a phosphinate (X=CH_2) function demonstrated clearly that the maximal activity was associated with the phosphonamide which is able to establish an hydrogen bond with alanine 113 (Fig. 13.4).

With the objective of designing a more stable analogue of the acetylcholinesterase inhibitor alkaloid physostigmine, Chen *et al.*[13] prepared some 8-carbaisosteres of physostigmine (Table 13.5). The authors envisaged that replacing the *N*-methyl group at N_8 of the physostigmine nucleus by a methylene group would increase its chemical and metabolic stability, thanks to the change of the less stable aminal group to a more stable amino group.

Other interesting bioisosteric replacements of oxygen atoms were devised during the search of nonhydrolysable phophotyrosyl (pTyr) mimetics. The phosphoryl ester

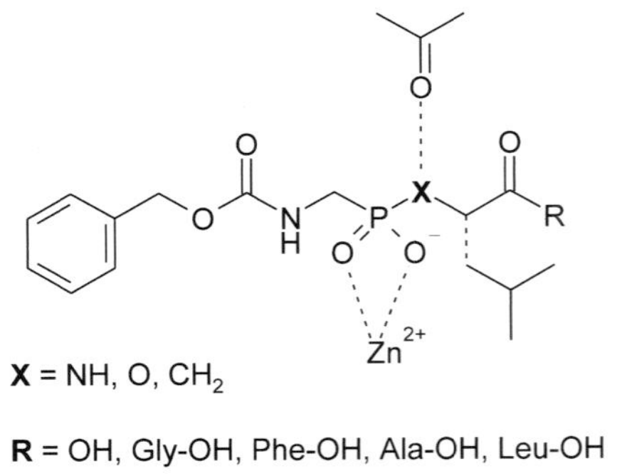

X = NH, O, CH₂

R = OH, Gly-OH, Phe-OH, Ala-OH, Leu-OH

Fig. 13.4 Isostery in thermolysin inhibitors.[12]

Table 13.5 *Physostigmines (left) and carbaisosteres (right)*[13]

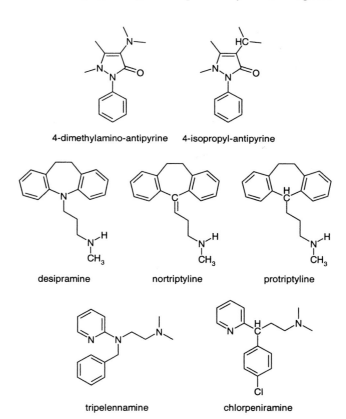

Compound	R1	R2	IC$_{50}$ (nM)	LD$_{50}$ (mg/kg)
(−)-Physostigmine	CH$_3$	CH$_3$	128	0.88
(−)-Heptyl physostigmine	n-C$_7$H$_{13}$	CH$_3$	110	24
(±)-Carba-isostere 1	n-C$_7$H$_{13}$	CH$_3$	114	21
(−)-Carba-isostere 2a	n-C$_7$H$_{13}$	C$_2$H$_5$	36	6
(+)-Carba-isostere 2b	n-C$_7$H$_{13}$	C$_2$H$_5$	211	18

oxygen of pTyr has been replaced either by a methylene (Pmp)[14] or by a difluoromethylene (F$_2$Pmp).[15]

The carbaisosteres are as potent or more potent than the corresponding physostigmines. In addition, the (−)-enantiomers which possess the same absolute configuration at C$_{3a}$ and C$_{8a}$ as that of physostigmine are generally six to 12 times more potent in inhibiting acetylcholinesterase than the corresponding (+)-enantiomers.

C Interchange of trivalent atoms and groups

The substitution of —CH= by —N= in aromatic rings has been one of the most successful applications of classical isosterism (see following paragraph on ring equivalents). Aminopyrine and its isostere are about equally active as antipyretics[16] (Fig. 13.5).

Similar interchanges are found in proceeding from desipramine to nortriptyline and protriptyline (Fig. 13.5) or among the antihistaminics, when comparing ethylene-diamine derived compounds with the diaryl-propylamines (Fig. 13.5).

D Ring equivalents

The substitution of —CH= by —N= or —CH=CH— by —S— in aromatic rings has been one of the most successful applications of classical isosterism. Early examples are found in the sulphonamide antibacterials with the development of sulphapyridine, sulphapyrimidine, sulphathiazole, etc. (Fig. 13.6).

Other examples are found in the neuroleptic or antidepressant tricyclics, in the benzodiazepine tranquillizers and antiepileptics, and in the development of semi-synthetic penicillins and cephalosporins with broader spectra of activity and greater stability towards β-lactamases. In some instances an aromatic ring can be replaced by an ethynyl group. Such a replacement was recently described by

Wallace *et al.*[17] in a series of inhibitors of endothelin-converting enzyme (ECE-1) derived from the biphenyl compound CGS 26 303 and its analogues.[18]

In all these cases no essential activity difference is found between the original drug and its isostere. However, it can happen that the procedure fails. Binder *et al.*[19] for example, reported that thieno[2,3-*d*]isoxazole-3-methanesulfonamide, the thiophene analogue of the anticonvulsant drug zonisamide (Fig. 13.7),[20] was practically inactive against

Fig. 13.5 Interchange of trivalent atoms and groups.

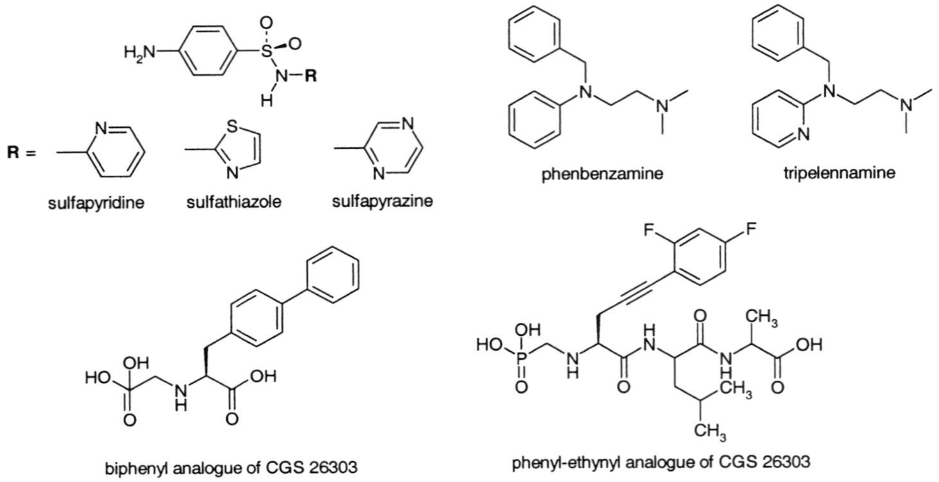

Fig. 13.6 Classical ring equivalents.

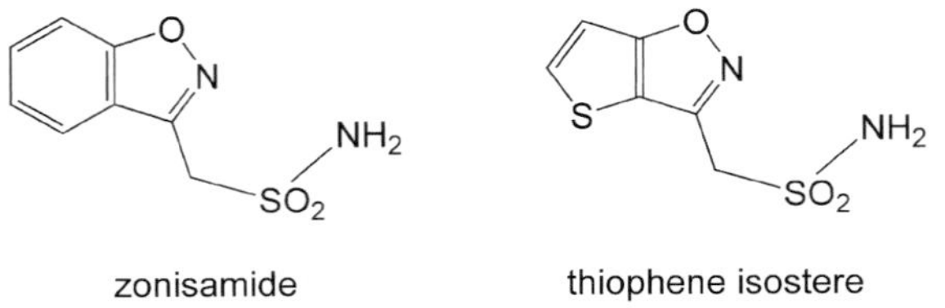

Fig. 13.7 The thiophene isostere of zonisamide is practically inactive as an anticonvulsant.

pentetrazole- or electroshock-induced convulsions in mice, even at high doses.

The concept of ring equivalents has been generalized to any possible heterocyclic system and represents a huge number of possible variations. Table 13.6 lists some less well-known studies on ring equivalents in aromatic series.

Another particularly impressive example of ring bioisosterism is found in the development of the antiulcer H$_2$-receptor histamine antagonists in which the initial imidazole ring was changed to various other 'equivalents'

such as a furan, a thiazole and finally a phenyl ring (Fig. 13.8). A detailed and very interesting account of the discovery and the development of these compounds is found in Ganellin and Roberts' book.[30]

One of the major problems when dealing with isosteric or bioisosteric replacements in *heterocyclic systems* is the selection of the a priori most promising candidate among several dozens of possible rings. A simple clue can be the knowledge and the comparison of the *boiling points* of the basic heterocycles. Thus, in the search for an ideal surrogate of the pyridazine ring, the comparison of the boiling points of seven possible ring candidates (Fig. 13.9), led us to reject the isomeric pyrimidine ring and to select the 1,2,4-triazine and the 1,3,4-thiadiazole rings.[31]

Effectively the observed biological activities were at least partially in accordance with the boiling point selection criteria (Table 13.7). On the reserpine ptosis and the 5-HT potentiation tests, the closest activities result from the replacement of the pyridazine ring (BP = 208°C) by the 1,3,4,-thiadiazole (BP = 204–205°C) or the 1,2,4-triazine

Table 13.6 *Ring equivalents*

Original ring	Bioisostere	Activity	Reference
1-Phenyl-pyrazolone	1-Phenyl-triazolone	Analgesic	Gold-Aubert et al.[21]
1,2,4-Triazole	1,3-Thiazole imidazole	Antiviral	Alonzo et al.[22]
Pyridine	Isoxazole or isothiazole	Nicotinic agonist	Garvey et al.[23]
Indole	Indazole	5-HT3 antagonists	Fludzinski et al.[24]
3,4-Dialkoxy-phenyl	Indole	Phospho diesterase inhibitors	Blascó et al.[25]
3,4-Dialkoxy-phenyl	Indole	GABA uptake inhibitors	Kardos et al.[26]
Quinoline-2-carboxylate	Indole-2-carboxylate	Glycine antagonists	Salituro et al.[27]
o-Nitro-phenyl	Furoxane	Calcium antagonist	Calvino et al.[28]
Spiro-hydantoin	Spiro hydroxyacetic acid unit	Aldose reductase inhibitor	Lipinski et al.[29]

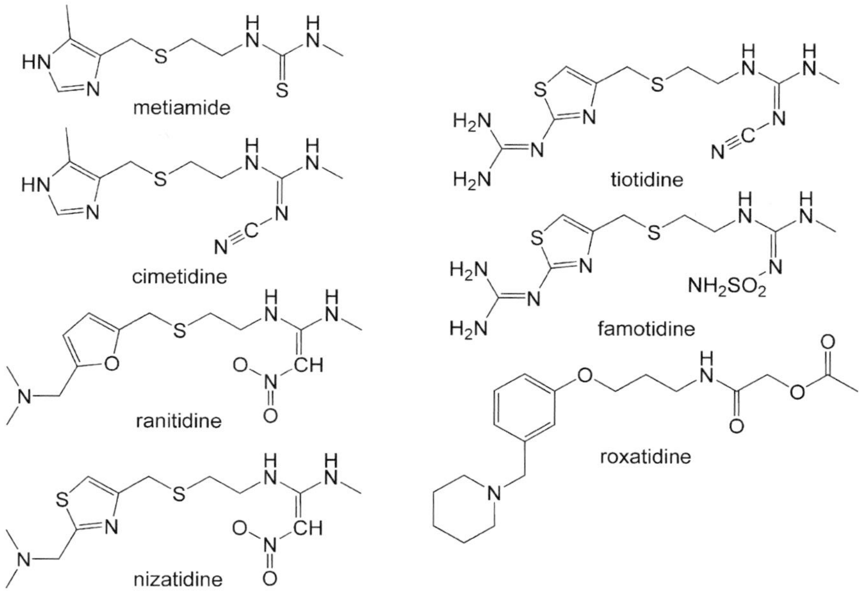

Fig. 13.8 Antiulcer H_2-receptor histamine antagonists: evolution of structures in the course of the time. Note the progressive use of a furan, a thiazole, and finally a phenyl ring in place of the original imidazole ring.

(BP = 200°C) ring. The pyrazine- and the pyrimidine-derived analogues are clearly less active on these two tests. The attenuation of the turning behaviour, after unilateral intrastriatal injection of the compounds in 6-hydroxydopamine-lesioned mice, reflects the dopaminergic properties of the molecules. Apparently these properties are insensitive to the bioisosteric variations.

A possible interpretation of these results can be the fact that in the heterocyclic series, the boiling point is correlated to the dipolar moment of the molecule and that, for two heterocyclic rings having the same aromatic geometry, the similarity of the dipolar moments may represent the dominant feature.

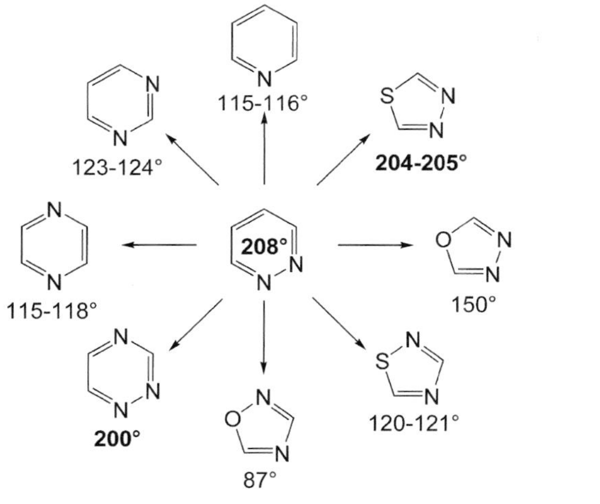

Table 13.7 *Cyclic equivalents of the pyridazine ring*[31]

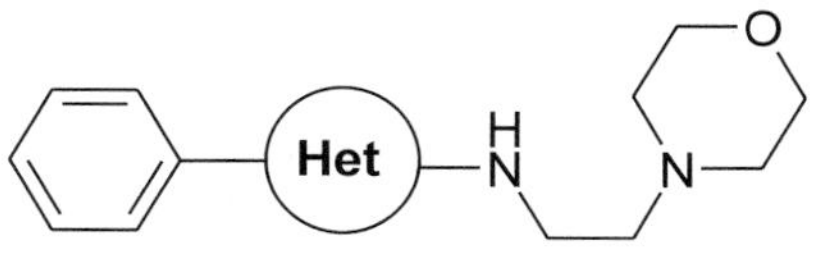

Central heterocycle	Reserpine ptosis (ED$_{50}$)	5-HT potentiation (ED$_{50}$)	Turning (minimal effective dose)
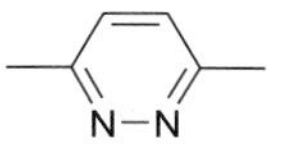	6	3.7	0.5
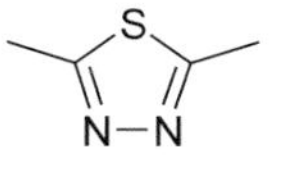	4.5	6	0.1
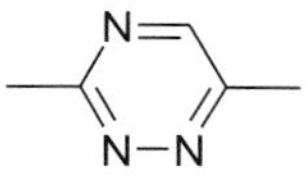	>10	6	2
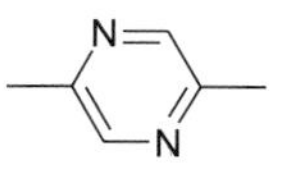	24	30	0.1
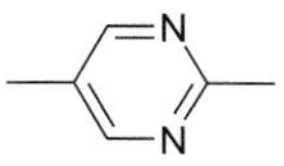	>100	>50	2

Fig. 13.9 Structure and boiling points of pyridazine isosteres.

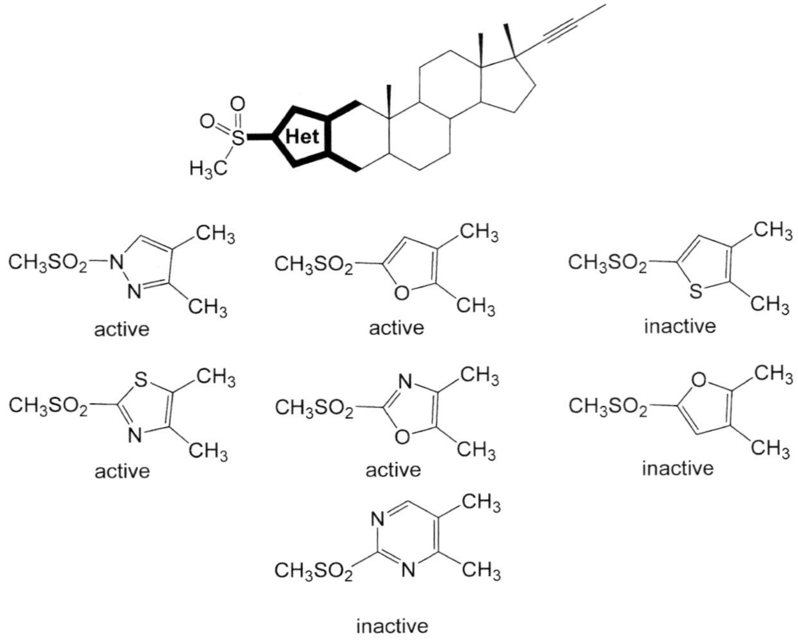

Fig. 13.10 Zanoterone isosteres.[32]

Better bioisosteric design possibilities are provided by quantum chemical calculations. Mallamo *et al.*[32] made use of electrostatic potential surface map complementarity in defining sulfonyl heterocycles bioisosteric to the steroidal antiandrogenic drug zanoterone (Fig. 13.10). Striking differences in the electrostatic potential surfaces accounted for the observed variability in the furan (active) and the thiophene (inactive) analogues of zanoterone (which contains a pyrazole ring). Good androgen receptor affinity was then anticipated, and effectively found, for the oxazole and the thiazole analogues of zanoterone.

The apparent failure of the isosterism concept for the inactive thiophene, inversed furan and pyrimidine is thus interpretable on a rational basis.

QSAR in heterocycles

Not many QSAR articles deal with the selection of heterocycles. Clementi *et al.*[33] describe the heterocycles by means of a set of principal properties. A collection of 40 heteroaromatic systems was submitted to a multivariate analysis using 13 descriptors derived by GRID. From the data matrix obtained a second generation of principal properties for heteroaromatics was derived. Such principal properties are claimed to be suitable for designing series of molecules of biological interest containing heteroaromatic moieties. Gibson *et al.*[34] have calculated ten physicochemical variables for each of 100 different aromatic rings. These variables were selected because of their potential involvement in the molecular recognition of drug–receptor binding interactions, and they include size, lipophilicity, dipole magnitude and orientation, HOMO and LUMO energies, and electronic point charges. A total of 59 different aromatic ring systems were studied including monocyclics and [5.5]-, [6.5]- and [6.6]-fused bicyclics. A principal components analysis of these results generated four principal components which account for 84% of the total variance in the data. These principal components provide a quantitative measure of molecular diversity, and their relevance for structure–activity relationships is discussed. The principal components correlate with the *in vitro* biological activity of heterocyclic aromatic fragments within a series of previously reported HIV-1 reverse transcriptase inhibitors.[35] Langer and Hoffmann[36] report on the determination of new descriptors for heteroaromatic ring fragments which take into account differences in anchoring positions of such substructures in a given molecule. For a set of 72 aromatic moieties (five- and six-membered monocyclic and benzo-fused heteroaromatics containing one or two heteroatoms) parameters including 12 2D physicochemical descriptors, six receptor surface models descriptors, and three 3D variables have been calculated and new sets of QSAR parameters were generated from principal components analyses of the data matrix this obtained. Hierarchical clustering of the principal components, as well as applying both D-optimal design and a new function for molecular estimation allowed partitioning of the heteroaromatic substituents into similar families and permitted a classification of the structures according to their molecular similarity. The descriptors obtained were finally tested

Table 13.8 *Carboxylic acid isosteres*

	hydroxamic acids	High chelating power	Almquist et al.[37]
	acyl-cyanamides	Mainly academic interest	von Kohler et al.[38] Kwon et al.[39]
	tetrazoles	Very popular Great number of publications. Recent in use. $pK_a = 6.6$ to 7.2	Bovy et al.[40] Marshall et al.[41]
	oxo-oxadiazoles	Lipophilic bioisosteric replacement for tetrazoles	Kohara et al.[42]
	oxo-thiadiazole	Lipophilic bioisosteric replacement for tetrazoles	Kohara et al.[42]
	mercaptoazoles + sulfinylazoles + sulfonylazoles	Phosphonate isosteres pK_a mercapto: 8.2–11.5 pK_a sulfinyl: 5.2–9.8 pK_a sulfonyl: 4.8–8.7	Chen et al.[13]
	isoxazoles isothiazoles	GABA and glutamic acid analogues	Krogsgaard-Larsen et al.[9] Krogsgaard-Larsen[43]
	hydroxy-thiadiazole	Isoxazole isostere pK_a # 5	Lunn et al.[44]
	hydroxy-chromones	Kojic acid derivatives: As GABA agonists	Atkinson et al.[45]
	phosphinates phosphonates phosphonamides	Many examples in the glutamate antagonist series and in the GABA$_B$ antagonists	Froestl et al.[46]
	sulphonates	Sulphonic analogues of GABA and glutamic acid	Rosowsky et al.[47]
	sulphonamides	Weak acids, used rather as equivalents of phenolic hydroxyls: catecholamine analogues	von Kohler et al.[38]
	acyl-sulphonamides	Glycine GABA β-alanine antiatherosclerotics pK_a # 4.5	Drummond and Johnson[48] Albright et al.[49]

for their prediction capacity using a previously reported dataset of biologically active compounds. The results obtained hereby clearly demonstrate the usefulness of the new principal properties.

E   Groups with similar polar effects

Surrogates of the carboxylic acid function

The carboxylic function of active compounds has been changed to direct derivatives such as hydroxamic acids: R—CO—NH—OH, acyl-cyanamides: R—CO—NH—CN and acyl-sulfonamides: R—CO—NH—SO$_2$—R', to planar acidic heterocycles such as tetrazoles, hydroxy-isoxazoles, etc., or even to non-planar sulfur- or phosphorus-derived acidic functions (Table 13.8).

Direct derivatives comprise hydroxamic acids,[50–54] acyl-cyanamides[38,39] and acyl-sulfonamides,[48] in which an acidic NH group replaces the acidic OH group. These bioisosteres are mainly of academic interest. Exceptions are the anti-inflammatory hydroxamates bufexamac[50] ibuproxam[51] and oxametacin[52,55] (Fig. 13.11). While ibuproxam is metabolized to ibuprofen (CONHOH → COOH) in man,[56] oxametacin is metabolically stable in man, and is a true bioisostere rather than a prodrug.[57,58]

Among the *planar acidic heterocycles*, the main representatives are tetrazoles and 3-hydroxy-isoxazoles. The medicinal chemistry of tetrazoles has been reviewed[59] and recent examples are found in various domains.[27,40,41,60] Tetrazole surrogates have the broadest field of applications, they can increase potency,[40,41] improve bioavailability[41,60] or bring some selectivity (the GABA tetrazole analogue inhibits GABA-transaminase, but not succinic semialdehyde dehydrogenase.[61] However, in some instances tetrazole analogues are poorly active.[62]

Hydroxy-isoxazoles and other cognate heterocyclic phenols encompassing an acidity range from 3.0 to 7.1 were incorporated in GABA agonists, antagonists and uptake inhibitors.[63,64] The experience gained with 3-hydroxy-isoxazoles in the GABA field was also transferable to glutamate receptor ligands and led to selective antagonists for glutamic acid receptor subtypes.[65]

Other interesting, but less studied, heterocyclic surrogates are: 3,5-dioxo-1,2,4-oxadiazolidine,[66] 3-hydroxy-1,2,5-thiadiazoles,[44] and 3-hydroxy-γ-pyrones.[45,67]

Non-planar sulfur or phosphorous-derived acidic functions: The most extensive use of phosphonates was made in the design of amino acid neurotransmitter antagonists such as glutamate[68] and GABA$_B$ antagonists.[46]

In a series of CCK antagonists derived from the non-peptide CCK-B selective antagonist CI-988, Drysdale *et al.*[69] prepared a series of carboxylate surrogates spanning a pK_a range of <1 (sulfonic acid) to >9.5 (thio-1,2,4-triazole). The affinity and the selectivity of the compounds were rationalized by consideration of the pK_a values, charge distribution, and geometry of the respective acid mimics (Table 13.9).

Diamino-cyclobutene-dione, was proposed by Kinney *et al.*[70] as an original surrogate of the α-amino carboxylic acid function (Fig. 13.12).

Carboxylic functions as the surrogates of phosphonates: Non-hydrolysable phosphotyrosyl (pTyr) mimetics serve as important components of many competitive protein-tyrosine kinases inhibitors. To date, the most potent of these inhibitors have relied on phosphonate-based structures to replace the 4-phosphoryl group of the parent pTyr residue (Fig. 13.13). Interestingly it was found that carboxy-based pTyr analogues can be utilized to introduce the anionic oxygen functionality of the parent phosphate. Particularly, when *p*-(2-malonyl)phenylalanine (Pmf) was incorporated as pTyr replacement in the high affinity Grb2 SH2 domain binding sequence, potencies approaching that of phosphonate mimetics were obtained.[71]

The above example is an elegant illustration of the possibility of mimicking the pyramidal structure of phosphates or phosphonates by means of two planar carboxylic groups, the three-dimensionality originating from the malonic methylenic carbon atom.

Surrogates of the ester function

The change from ester to amide (procaine → procainamide) has already been illustrated above as an example of classical isosterism. Similarly the lactone ring of the muscarinic agonist pilocarpine was changed into various, still active

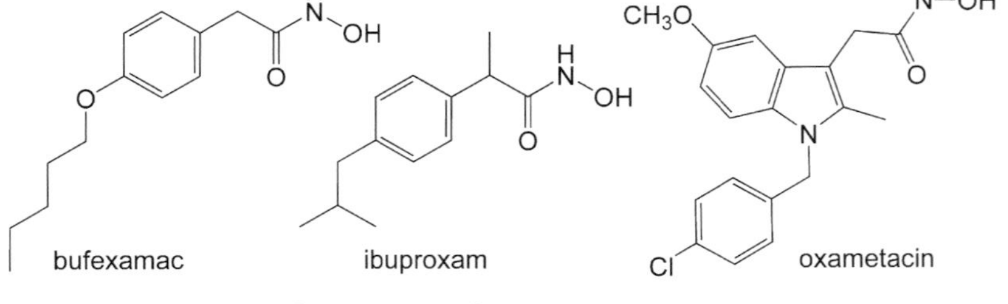

Fig. 13.11   Hydroxamate isosteres of anti-inflammatory drugs.

Table 13.9 *Exploration of the carboxyl isosterism possibilities in a series of CCK antagonists*[69]

R	IC$_{50}$ (nM) CCK-B	IC$_{50}$ (nM) CCK-A	A/B ratio	pK$_a$
—R—CH$_2$—COOH	1.7	4500	2500	5.6
Charge distributed monoanionic acid mimics				
	6.0	970	160	5.4
	2.6	1700	650	6.5
	2.4	620	260	4.3
	2.5	680	270	>9.5
	16	850	53	>9.5
	4.3	660	150	7.7
	1.7	940	550	7.0
	6.3	1300	200	5.2
	18	600	33	>8.2
	14	1300	93	>9.5

(*continued on next page*)

Table 13.9 (*continued*)

R	IC$_{50}$ (nM) CCK-B	IC$_{50}$ (nM) CCK-A	A/B ratio	pK$_a$
Point charge monoanionic acid mimics				
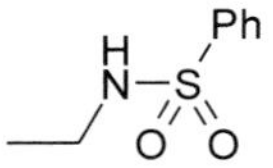	70	300	4.3	>9.5
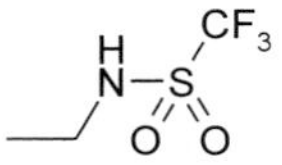	77	680	9	7.9
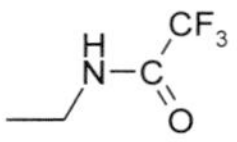	110	790	7	>9.5
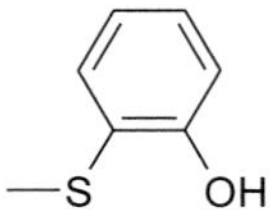	80	510	6.4	>9.5
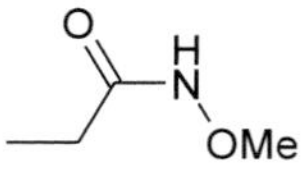	21	1500	71	>9.5
Tetrahedral acid mimics				
P(O)(OH)$_2$	27	5200	190	3.4; 7.7
CH$_2$–P(O)(OH)$_2$	23	2700	120	3.4; 7.8
P(O)(OH)(OEt)	12	480	40	6.5
P(O)(OH)Me	12	1700	140	3.8
CH$_2$—P(O)(OH)Me	23	4400	190	3.7
CH$_2$—SO$_3$Na	1.3	1010	780	–

isosteres such as the corresponding thiolactone, lactam, lactol, and thiolactol.[72] A series of aspirin isosteres was prepared by replacing the carboxylic ether oxygen successively by a nitrogen, sulfur or carbon isosteric equivalent.[73] None of the isosteric compounds showed any activity. This result is readily understood since the particular role of aspirin as an acylating agent of the enzyme cyclooxygenase has been demonstrated.[74]

In addition to these classical changes, much use was made of 1,2,4-oxadiazoles or 1,2,4,-thiadiazoles as carboxylic ester surrogates in series of benzodiazepine and muscarinic[75,76] receptor ligands (Fig. 13.14). For muscarinic agonists, numerous successful attempts to replace the oxadiazole ring by other heterocyclic ring

systems were published.[77-79] By substituting in pilocarpine the lactonic ester function by its carbamate equivalent, a much more stable analogue was obtained (see Chapter 33).

The change in (−)-cocaine of the carbomethoxy substituent into carbethoxyisoxazole (Fig. 13.15) doubles

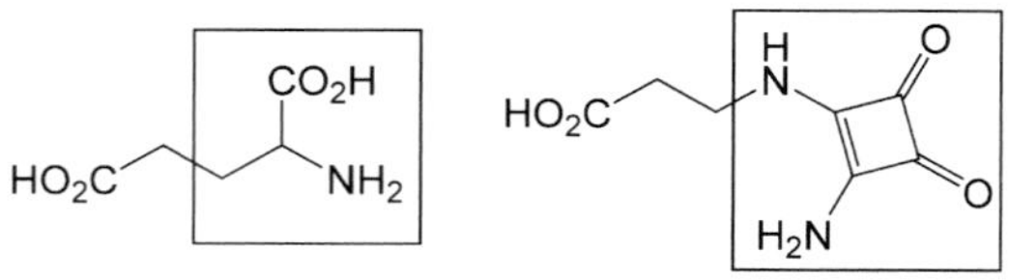

Fig. 13.12 3,4-Diamino-3-cyclobutene-1,2-dione as surrogate of the α-amino carboxylic acid function.[70]

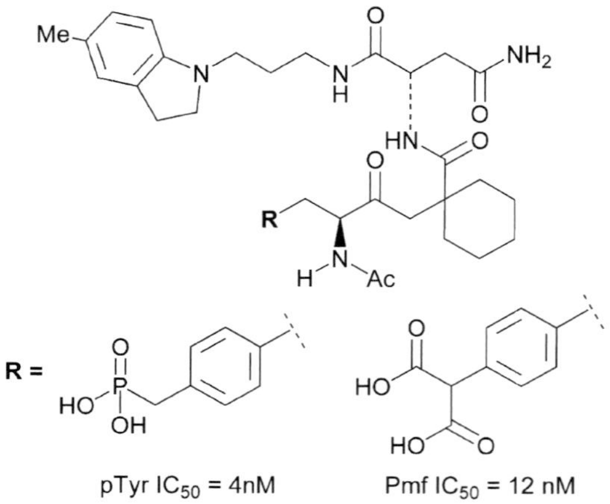

Fig. 13.13 Malonates as surrogates of phosphonates.

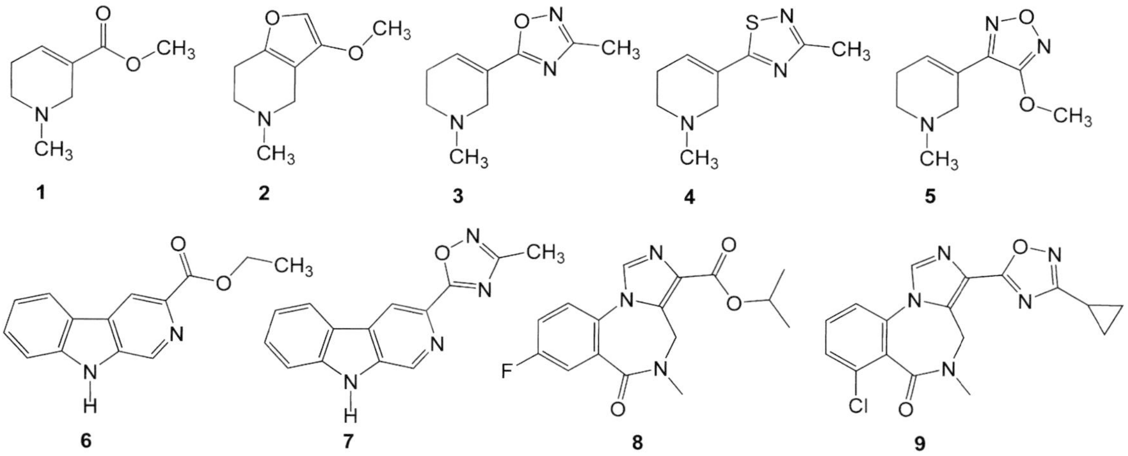

Fig. 13.14 1,2,4-Oxadiazoles and related five-membered heterocycles as ester surrogates.

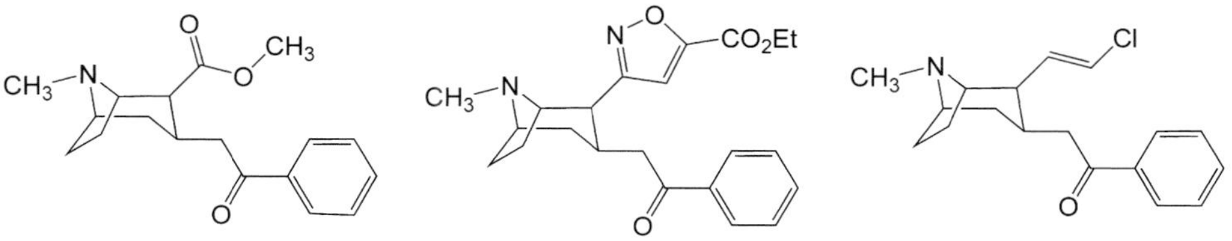

Fig. 13.15 Replacement in (−)-cocaine of the carbomethoxy group by a carbethoxyisoxazole and a chlorovinyl moiety.

the potency in [³H] mazindol binding and [³H] dopamine uptake. Astonishingly the replacement of the carbomethoxy group by a chlorovinyl moiety produces a comparable gain in potency arguing thus against the involvement of the carbomethoxy group in H-bonding.[80]

Another rather unusual example of ester isosterism is the replacement of the ether oxygen by a fluoro-nitrogen

(Fig. 13.16a) as mentioned by Lipinski.[81] Other uncommon examples are found in the replacement of the ester function of acetylcholine by exo-endo amidinic functions of 3-aminopyridazines in muscarinic agonists (Fig. 13.16b)[82] and of the carbomethoxy group of α-yohimbine (rauwolscine) by a *N*-methylsulfonamide function (Fig. 13.16c).[83]

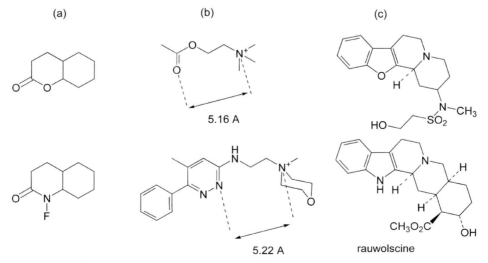

Fig. 13.16 (a) Replacement of ester ether oxygen by a fluoro-nitrogen; (b) exo-endo amidine in place of a carboxylic ester functionality; (c) *N*-methylsulfonamide analogue of α-yohimbine (rauwolscine).

Amides and peptides

Carboxamides are usually converted to sulfonamides as illustrated by the synthesis of the hypoglycemic sulfonyl isostere of glybenclamide.[84] The isosteric replacements for peptidic bonds have been summarized by Spatola[85] and by Fauchère.[86] The most used and well-established modifications are *N*-methylation; configuration change (D-configuration at Cα); formation of a retroamide or an α-azapeptide; use of aminoisobutyric or dehydroamino acids; replacement of the amidic bond by an ester (depsipeptide), ketomethylene, hydroxyethylene or thio-amide functional group; carba replacement of the amidic carbonyl; and use of an olefinic double bond (Fig. 13.17).

More unusual isosteric replacements for the peptidic bond were recently proposed (Fig. 13.18). Among them hydroxyethylureas served in the design of a novel class of potent HIV-1 protease inhibitors, diacylcyclopropanes in the design of novel renin inhibitors, and pyrroline-3-ones for various proteolytic enzyme inhibitors.[87,88] Vinyl fluorides can probably be considered as representing the closest possible bioisosteres of the peptide bond. The synthetic methods available allow, by an appropriate selection of the precursors, the preparation of analogues of dipeptidic combinations of amino acids bearing no other functionalities in their side chains, e.g. Gly, Ala, Val, Phe, Pro.[89] Vinyl fluorides have been used in the design of bioisosteres of Substance P[87] and of the analgesic dipeptide 2,6,-dimethyl-L-tyrosyl-D-alanine-phenylpropionamide.[90]

Urea and thiourea equivalents

In the histaminic H$_2$ receptor antagonist series, the classical urea–thiourea–guanidine progression was successfully completed by the use of the *N*-nitro and *N*-cyanoguanidines and, later on, by 1,1-diamino-2-nitroethylene groups[30] (Fig. 13.19). Cyano amidines and carbamoyl amidines were also used,[91] and structure–activity relationship patterns were rationalized in terms of dipole moment orientation of related bioisosteric groups.[92]

Among more exotic surrogates, the 3,4-diamino thiadia-zole dioxide moiety was proposed as a weakly acidic urea equivalent.[93] The similar thiatriazole dioxide is found in the H2 antagonist tuvatidine (HUK 978). Other bioisosteres are exo-endo amidinic heterocyles bearing an electron-attract-ing function in the α position[94,95] (Fig. 13.19).

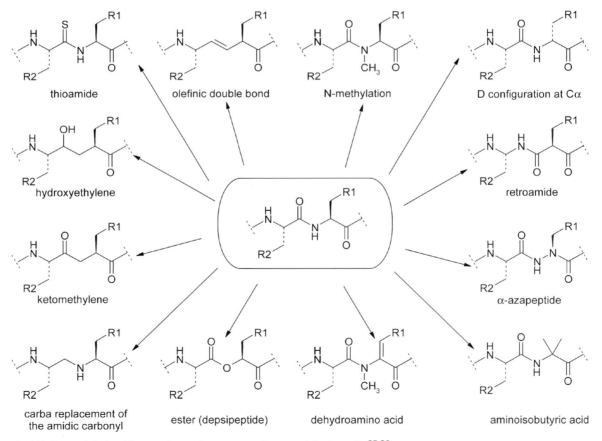

Fig. 13.17 Well-established isosteric replacements for peptidic bonds.[85,86]

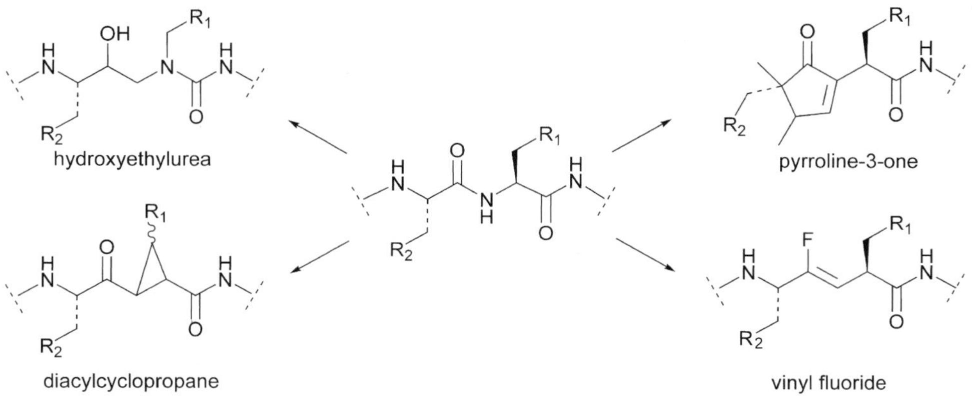

Fig. 13.18 Unusual isosteric replacements for the peptidic bonds.

Surrogates for phenolic groups

The most popular surrogates for phenolic functions are NH groups rendered acidic through the presence of an electron-attracting group. Table 13.10 shows an application of this bioisostery in the design of *N*-methyl-D-aspartate (NMDA) receptor antagonists.[96] In this bioisostery the replacement of the phenol by heterocyclic NH-containing rings were expected to slow metabolism and hence to improve oral bioavailability. The potent and NR1A/2B-receptor selective benzimidazolone bioisostere demonstrated oral activity in a rodent model of Parkinson's disease at 10 and 30 mg kg^{-1}.

F   Reversal of functional groups

The reversal of the peptidic functional groups is often used in peptide chemistry. The retropeptides obtained are generally more resistant to enzymatic attacks (Fig. 13.17).[86,97] For thiorphan and retro-thiorphan an identical binding mode to the zinc protease thermolysin was demonstrated.[98] Similar inhibition values for thermolysin and neutral endopeptidase were observed, whereas, for another zinc protease, angiotensin converting enzyme (ACE), noticeable differences for inhibition were found (Fig. 13.20).

However, the strategy of functional inversion can also be applied to non-peptidic compounds. A historical example is

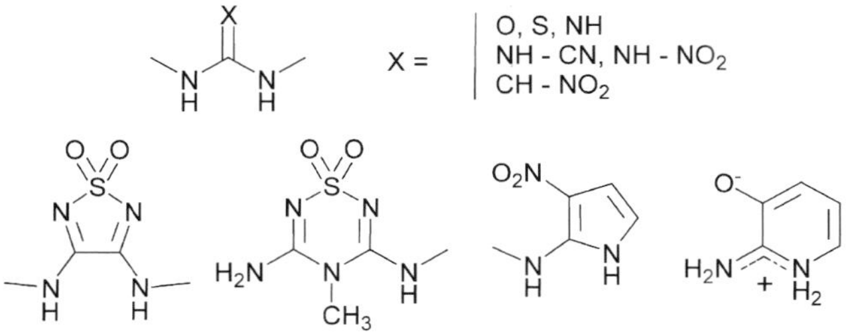

Fig. 13.19 Urea and thiourea equivalents.

the change from orthoform to neo-orthoform (orthocaine; Fig. 13.21). The unwanted side-effects, often encountered with aromatic *p*-amino substituted compounds ('para effects', essentially of allergic origin) are abolished in the *m*-amino isomer, whereas the local anaesthetic activity is maintained. Similarly the '*meta*' isomer of benoxinate has a local anaesthetic activity identical to that benoxinate itself.[99]

The β-blocking agent practolol was one of the first cardioselective β-blockers. It was rapidly replaced by its isomeric analogue atenolol which presents fewer side-effects (Fig. 13.22).

The inversion of the ester function of *meperidine* leads to 1-methyl-4-phenyl-4-propionoxy piperidine (Fig. 13.23) which is five times more potent as an analgesic drug than meperidine and represents the model compound of the series of inverted esters.[10]

The change from *indomethacin* to *clometacin*, although representing a clean example of functional group reversal, causes more profound alterations than that shown in the previous examples (Fig. 13.24). At a first glance, this change can even seem too drastic, however, in turning the molecule of clometacin by 180°, the resemblance with the parent molecule becomes evident. Indomethacin is mainly used as a non-steroidal anti-inflammatory agent and occasionally as an analgesic, clometacin on the other hand is usually recommended as an analgesic and shows weak anti-inflammatory properties. Applied to serotonin, a similar reversal of a functional moiety yielded 5HT$_{2C}$ receptor selective agonists[100] (Fig. 13.25).

III   ANALYSIS OF THE MODIFICATIONS RESULTING FROM ISOSTERISM

It is rare that the replacement of a part of a molecule by an isosteric or bioisosteric group leads to a *strictly identical*

Table 13.10 Bioisosteric replacements of the phenolic hydroxylic function

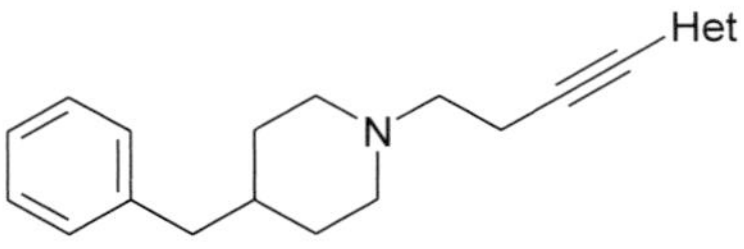

Name of Het	Structure of Het	IC50 values (μM) for the NMDA receptors NR1A/2B
4-phenol	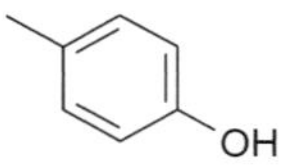	0.17
5-indole	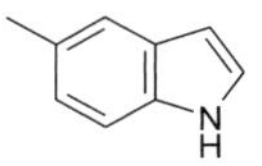	0.63
5-indazole	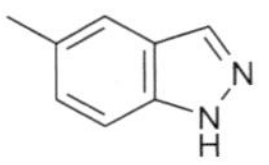	0.25
5-benzotriazole	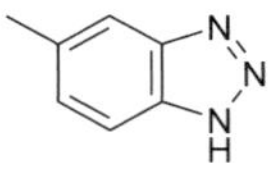	0.22
5-indolone	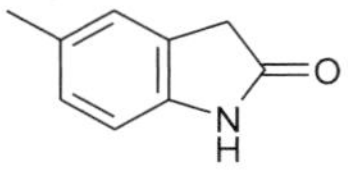	0.32
5-imidazolone	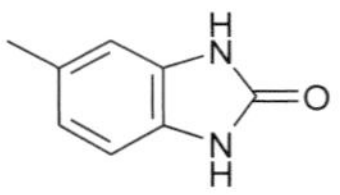	0.09
5-imidazole-thione	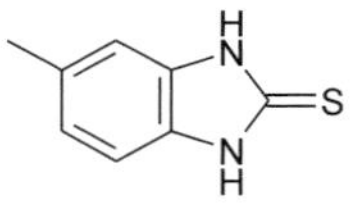	0.18
5-benzoxazolone	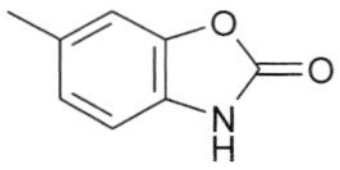	0.12

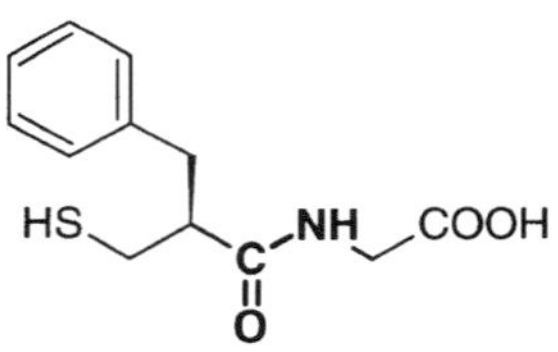

thiorphan

Enzyme	Ki value in μM
NEP 24.11	0.0019
Thermolysin	1.8
ACE	0.14

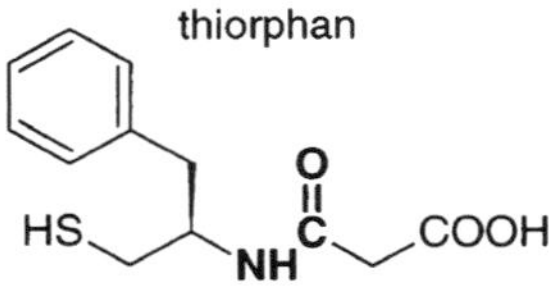

retro-thiorphan

Enzyme	Ki value in μM
NEP 24.11	0.0023
Thermolysin	2.3
ACE	>10

Fig. 13.20 Inhibition values of thiorphan and retro-thiorphan for three zinc proteases.

active principle. In practice, that is not even sought, and one prefers that the new compound produces a change as compared with the parent molecule. In general the isosteric replacement, even though it represents a subtle structural change, results in a modified profile: some properties of the parent molecule will remain unaltered, others will be changed. Bioisosterism will be productive if it increases the potency, the selectivity and the bioavailability or decreases the toxicity and undesirable effects of the compound. In proceeding to isosteric modifications one will focus predominantly on a given parameter: structural, electronic, hydrophilic, but it is all but impossible not to alter several parameters simultaneously.

A Structural parameters

These will be important when the portion of the molecule involved in the isosteric change serves to maintain other functions in a particular geometry. That is the case for tricyclic psychotropic drugs (Fig. 13.26).

In the two antidepressants (imipramine and maprotiline), the bioisosterism is geometrical insofar that the dihedral angle α formed by the two benzo rings is comparable: $\alpha = 65°$ for the dibenzazepine and $\alpha = 55°$ for the dibenzo-cycloheptadiene.[101] This same angle is only 25° for the neuroleptic phenothiazines and the thioxanthenes. In these examples the part of the molecule modified by isosterism is not involved in the interaction with the receptor. It serves only to position correctly the other elements of the molecule.

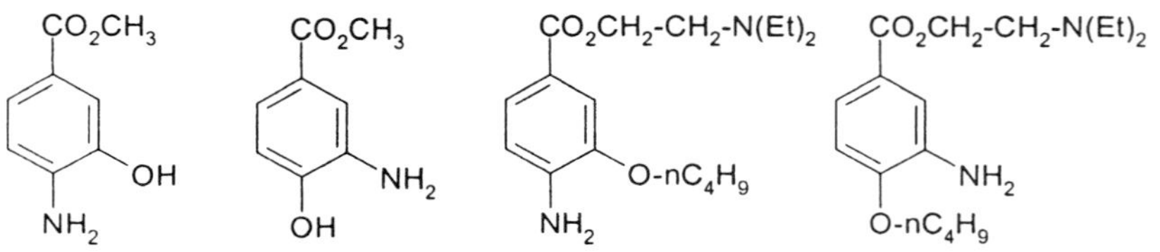

orthoform "old" orthoform "new" benoxinate benoxinate isomer

Fig. 13.21 Positional isomery in local anaesthetics.[99]

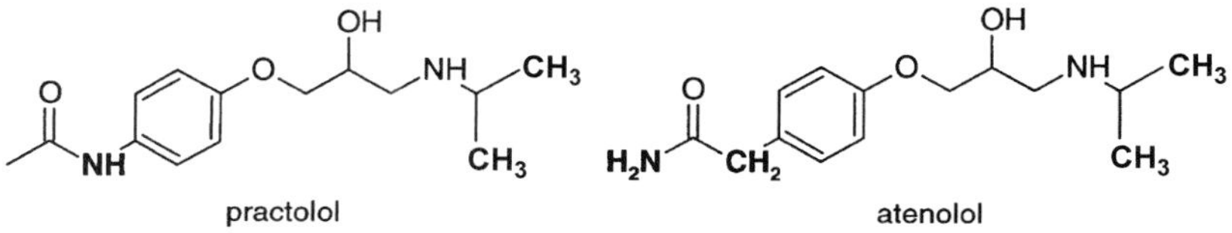

Fig. 13.22 Reversal of functional groups in practolol yields atenolol.

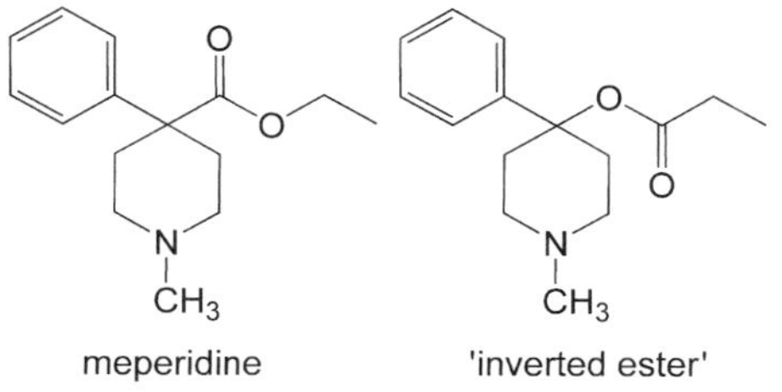

meperidine   'inverted ester'

Fig. 13.23 Meperidine and the corresponding inverted ester.[10]

The structure of various bioisosteric retinoic acid receptor agonists highlights the dominantly geometric parameter of this bioisostery (Fig. 13.27).[102]

B   Electronic parameters

Electronic parameters govern the nature and the quality of ligand–receptor or ligand–enzyme interactions. The relevant parameters will be inductive or mesomeric effects, polarizability, pK_a, capacity to form hydrogen bonds, etc. Despite their very different substituents in the *meta* position, the two epinephrine analogues (Fig. 13.28) exert comparable biological effects: they are both β-adrenergic agonists. In fact the key parameter resides in the very close pK_a values.[97]

C   Solubility parameters

When the functional group involved in the isosteric change plays a role in the absorption, the distribution or the excretion of the active molecule, the hydrophilic-lipophilic parameters become important. Imagine in an

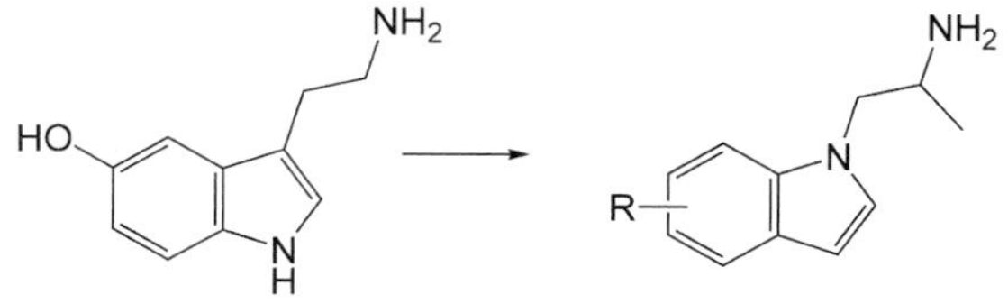

Fig. 13.25 Serotonin analogues resulting from a functional reversal.[100]

active molecule the replacement of —CF_3 by —CN (Fig. 13.29). The electron-attracting effect of the two groups will be comparable, but the molecule with the cyano function will be clearly more hydrophilic. This loss in lipophilicity can then be corrected in attaching elsewhere on the molecule a propyl, isopropyl or cyclopropyl group.

IV   ANOMALIES IN ISOSTERISM

In this section, two applications of the bioisosterism concept that imply unusual behaviours of commonly encountered atoms or groups are discussed.

A   Fluorine–hydrogen isosterism

There is an anomaly residing in the fact that fluorine does not resemble other halogens, notably chlorine, and that, on the other hand, it often mimics an atom of hydrogen.[103]

Steric aspects

The fluorine atom is considerably smaller than the rest of the halogen atoms. Seen from the steric point of view it resembles

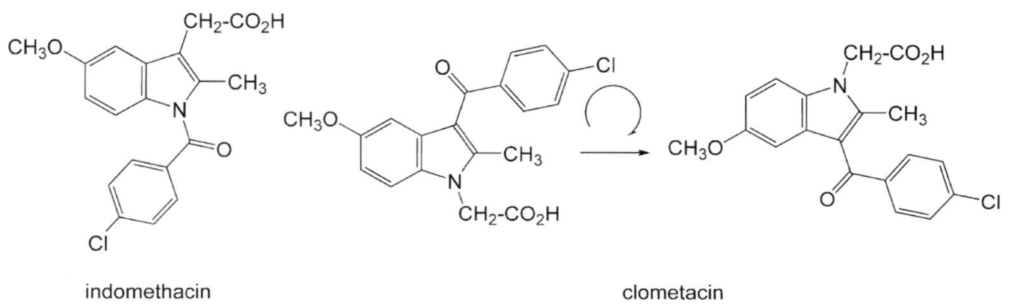

indomethacin   clometacin

Fig. 13.24 Functional inversion applied to indomethacin.

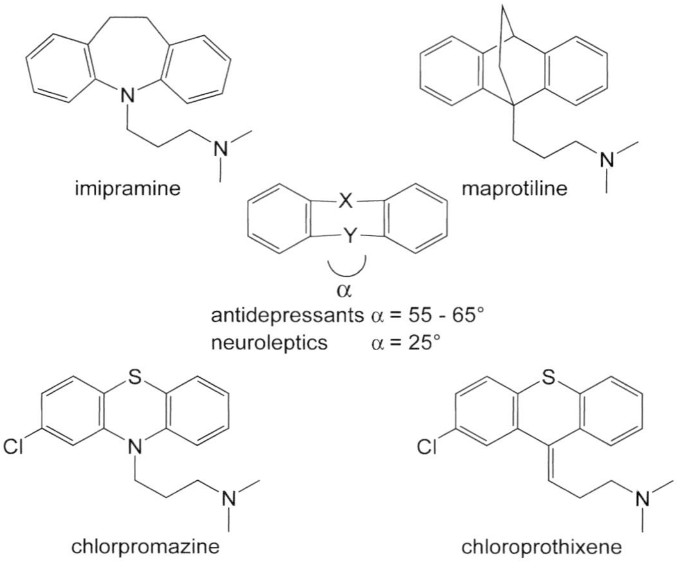

imipramine

maprotiline

α

antidepressants α = 55 - 65°
neuroleptics   α = 25°

chlorpromazine

chloroprothixene

Fig. 13.26   The tricyclic antidepressants (imipramine and maprotiline) are characterized by a dihedral angle of 55° to 65° between the two benzo rings, this angle is only 25° for the tricyclic neuroleptics (chlorpromazine, chlorprothixene).[101]

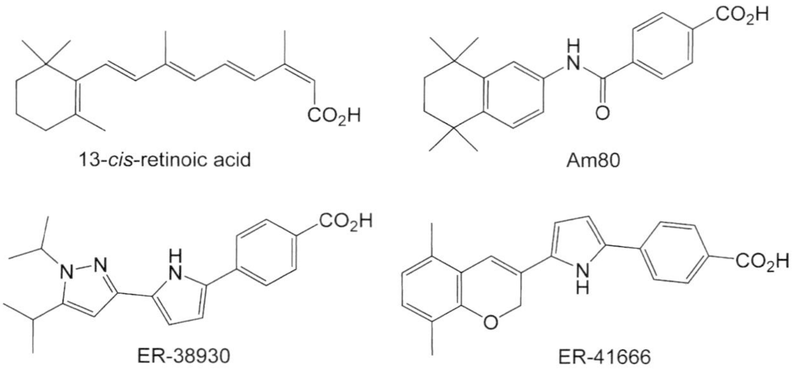

13-*cis*-retinoic acid

Am80

ER-38930

ER-41666

Fig. 13.27   Bioisostery in its broadest sense.

more hydrogen than chlorine (Table 13.11). Effectively fluoro-derivatives differ from the other halogenated derivatives because with carbon fluorine forms particularly stable bonds and, in contrast to other halogens, is only rarely ionized

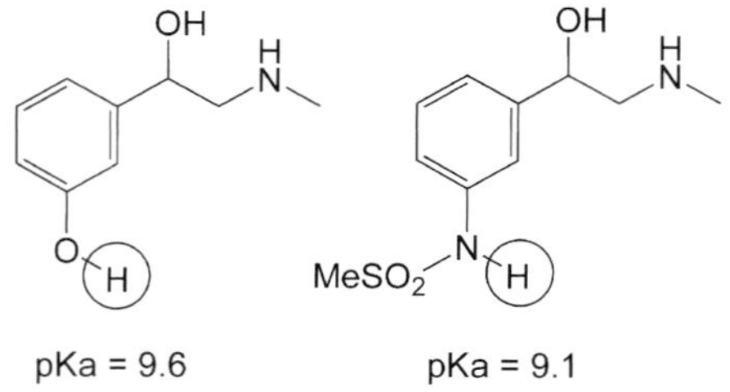

or displaced. Because it is both chemically inert and of small size, organic fluorine is often compared with hydrogen.

This relates particularly to the incorporation by living organisms of fluoroacetic acid in the place of acetic[104] acid or of 5-fluoro-nicotinic acid and 5-fluoro-uracil as antimetabolites (see Chapter 17, the section on halogens). This 'fraudulent' incorporation leads to lethal syntheses.[105] This is generally not the case with the corresponding chlorinated, brominated or iodinated analogues.

Electronic aspects

Fluorine is the most electronegative of the halogens (Table 13.10) and forms particularly stable bonds with carbon atoms. This chemical inertia explains why fluoroderivatives are more resistant to metabolic degradation (Fig. 13.30). Thus for the β-haloalkylamines (nitrogen

pKa = 9.6

pKa = 9.1

Fig. 13.28   An example of bioisosterism, or non classical isosterism, the methylsulfonamide substituent has comparable acidity to the phenolic hydroxyl group.[97]

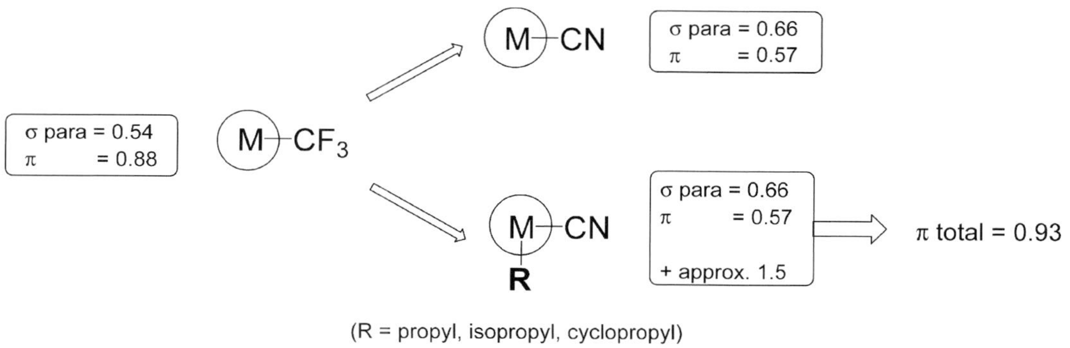

(R = propyl, isopropyl, cyclopropyl)

Fig. 13.29 The loss in lipophilicity resulting from the bioisosteric exchange of a CF_3 for a CN has to be compensated by the equivalent of a three carbon residue.

Table 13.11 *Fluorine-hydrogen isosterism. Observe the comparable sizes of the two atoms, whereas chlorine is close to the methyl and trifluoromethyl*

Parameter	H	F	Cl	CH_3	CF_3
Atomic radius	0.29	0.64	0.99	–	–
Van der Waals radius	1.2	1.35	1.80	≈ 2	≈ 2
Molecular refractivity	1.03	0.92	6.03	5.65	5.02
Electronic effect (*para* σ)[a]	0.00	0.06	0.23	−0.17	0.54
Resonance effect (R)[a]	0.00	−0.34	−0.15	−0.13	0.19
Electronic effect (σ^*)[b]	–	3.08	2.68	0.00	2.85

[a] For aromatic systems.
[b] For aliphatic systems.

mustards), the alkylating activity is lost when chlorine or bromine are replaced by fluorine or by hydrogen.[106]

H $\leftrightarrow$ F isosterism will therefore often serve to give analogues that are more resistant to metabolic degradation (obstructive halogenation: flunarizine and in flufenisal; Fig. 13.30). Similarly the CF_3 group is biostable whereas CH_3 is easily oxidized.[103]

Absence of d orbitals

Another difference between fluorine and the other halogens comes from the absence of a d orbital for fluorine, and thus

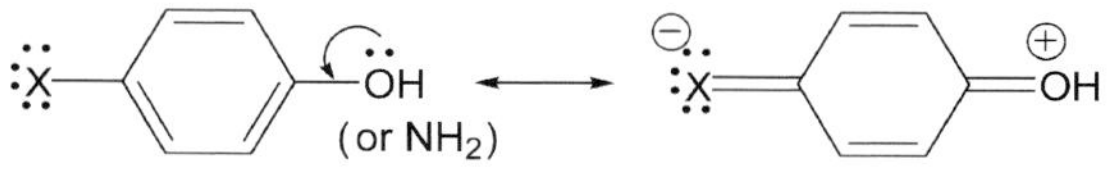

Fig. 13.31 The resonance between the OH lone pair and the X group is not possible if X=F.

its incapacity to participate in resonance effects with a donor of π electrons (Fig. 13.31).

This explains among why *p*-fluorophenol is slightly less acidic than phenol, while for other *p*-halogenated phenols the acidity changes in parallel with the atomic number (Table 13.12).[103]

Case study

A good example of continuous variation of activity in halogenated compounds is provided by a series of antihistaminic drugs related to tripelennamine (Fig. 13.32, X=H).

Apparently we are dealing here with a classical isosteric series: F, Cl, Br, I, but sensitive to steric hindrance in the *para* position. Probably what happens *in vivo*, is *p*-hydroxylation of the benzene ring. The best candidate becomes then the *p*-fluoro compound, since it is not bulkier than the unsubstituted compound, while being biostable.

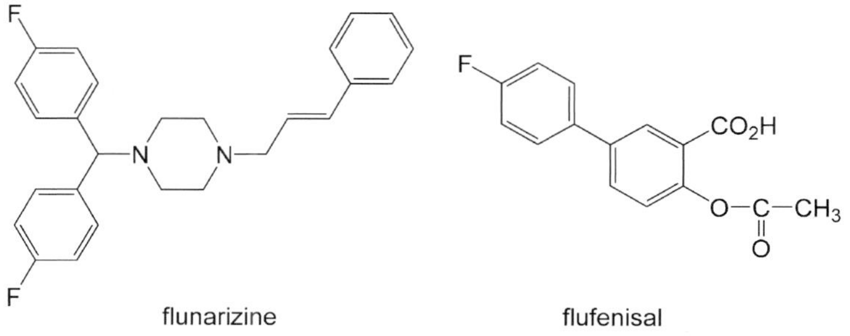

flunarizine   flufenisal

Fig. 13.30 In flunarizine and in flufenisal fluorine atoms in para position prevent metabolic hydroxylation.

Table 13.12 *Dissociation constants of p-halogenated phenols*[103]

Compound	Dissociation constant $K_a \times 10^{-10}$
Phenol	0.32
p-Fluorophenol	0.26
p-Chlorophenol	1.32
p-Bromophenol	1.55
p-Iodophenol	2.19

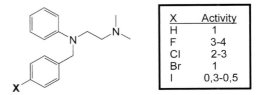

X	Activity
H	1
F	3-4
Cl	2-3
Br	1
I	0,3-0,5

Fig. 13.32 Variation of activity in a series of antihistaminic compounds as a function of the halogated para-substituent.[107]

B   Exchange of ether oxygen and methylene group

Ether oxygen atoms and methylene groups possess a similar tetrahedral structure and should normally be isosteric. In fact the $O \leftrightarrow CH_2$ isosterism yields very often anomalous results and brought Friedman[2] to the interesting observation 'that the omission of the ether oxygen changes biological activity much less in some cases than the replacement by the isosteric methylene group' (Fig. 13.33). In the meperidine series for example, the change from the *N*-phenoxypropyl derivative to the isosteric phenoxybutyl decreases the analgesic potency by a factor of 10, whereas the omission of the ether oxygen yields a slightly more potent compound.[10] A list of seven other examples is given by Schatz in the second edition of Burger's Medicinal Chemistry.[108]

The explanation for this anomalous behaviour may be that the omission of the ether oxygen yields a closer compound in terms of lipophilicity than its replacement by a methylene. An example which can be compared with Friedman's paradox is found in the resemblance of the phenylethyl type β-blockers (e.g. dichloroisoprenaline, sotalol) with the phenoxypropanol type (e.g. practolol, acebutolol).

V   MINOR METALLOIDS—TOXIC ISOSTERES

In this section we describe some 'exotic' applications of the bioisostery concept implying the utilization of unusual elements, such as silicon, boron, selenium, arsenic and antimony.

A   Carbon–silicon bioisosterism

Silicon is directly below carbon in the periodic table and the incorporation of silicon in place of carbon in biologically active substances has been a temptation for many organic chemists. The extent of this isosterism remains however limited. For reviews on the subject, see Fessenden and Fessenden,[109] Tacke and Zilch[110,111] and Ricci *et al.*[112]

Silicon is more electropositive than carbon (and even more if compared with oxygen and nitrogen) and the covalent silicon–carbon bonds in the sp3 hybridization state, are 20% longer than the corresponding carbon–carbon bond. Compared with their carbon bioisosteres, silicon-containing molecules are more sensitive to hydrolysis and to nucleophilic attack in general. Given the chemical

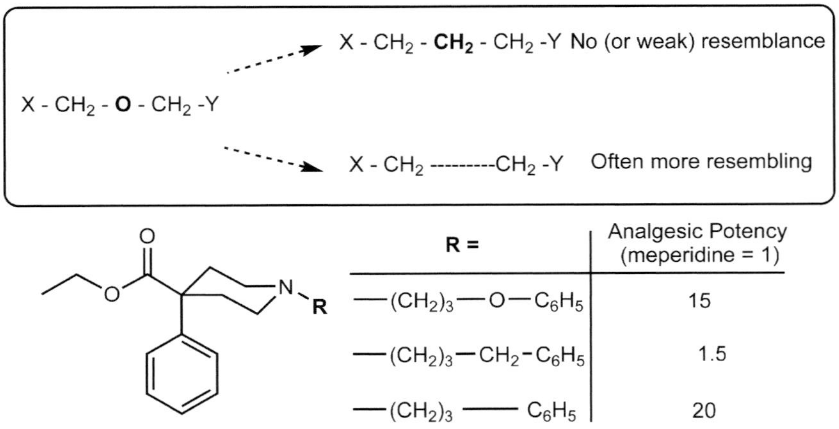

X - CH₂ - **O** - CH₂ -Y
→ X - CH₂ - **CH₂** - CH₂ -Y   No (or weak) resemblance
→ X - CH₂ ---------CH₂ -Y   Often more resembling

R =	Analgesic Potency (meperidine = 1)
—(CH₂)₃—O—C₆H₅	15
—(CH₂)₃—CH₂-C₆H₅	1.5
—(CH₂)₃——— C₆H₅	20

Fig. 13.33 Friedman's ether oxygen-methylene group paradox.[2]

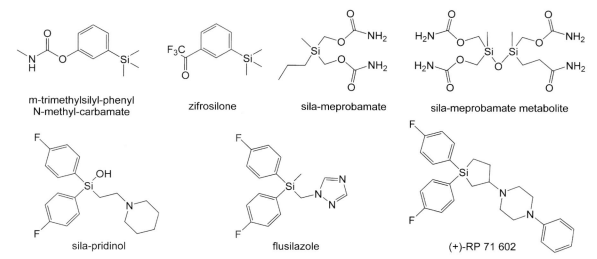

Fig. 13.34 Organosilicon active substances.

reactivity of silicon, carbon–silicon isosterism is generally practised only if the silicon is present in the centre of a quaternary structure as is the case for substances collected in Fig. 13.34. Among these, *m*-trimethylsilyl-phenyl *N*-methylcarbamate and *m*-trimethylsilyl-α-trifluoroaceto-phenone (zifrosilone) are acetylcholinesterase inhibitors,[113–115] sila-meprobamate is a CNS depressant,[116] sila-pridinol is an anticholinergic,[117] flusilazole is a fungicide for agricultural use,[118] and (+)-RP 71,602 is a potent and selective 5-HT$_{2A}$ antagonist.[119]

More recently a silicon-containing hypocholesterolemic squalene epoxidase inhibitor,[120,121] the silicon analogue of the α$_2$-adrenergic antagonist atipamezole,[122] and a highly potent, stable and CNS-penetrating silatecan[123] were published. As for the preceding molecules, the silicon atom of these compounds is quaternary and thus expected to be less sensitive to metabolic degradation (Fig. 13.35).

However, even when located in the centre of a quaternary structure, the risk exists that the silicon atom will be attacked. Thus, 1-chloro-1-sila-bicyclo-(2,2,1)-heptane can still be hydrolysed by an attack on the vacant d orbital;[124] this attack is *lateral* and therefore possible even in the cases where the corresponding carbon derivative would have been inert towards a SN$_2$ reaction (Fig. 13.36).

This sensitivity towards lateral attacks explains the four times shorter duration of action of sila-meprobamate compared with its carbon isostere on a model of tranquilliz-ing activity in mice (rotarod test, potentiation of hexobarbi-tal-induced sleep; intraperitoneal injection).[116] On the other hand, when given orally, sila-meprobamate is practically inactive. One of the first metabolites formed has been characterized as being a di-siloxane[125] (Fig. 13.34). For the two phenyl-trimethylsilyl-derived AChE inhibitors, the rather positively charged trimethyl-silyl group mimics the trimethyl-ammonium function present in acetylcholine.

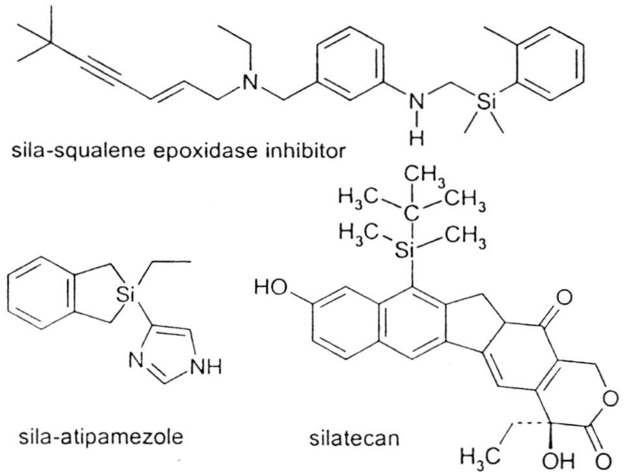

Fig. 13.35 Structures of a hypocholesterolemic squa-lene epoxidase inhibitor, an α$_2$-adrenergic antagonist and an anticancer camptothecin analogue containing silicon in a quaternary carbon environment.

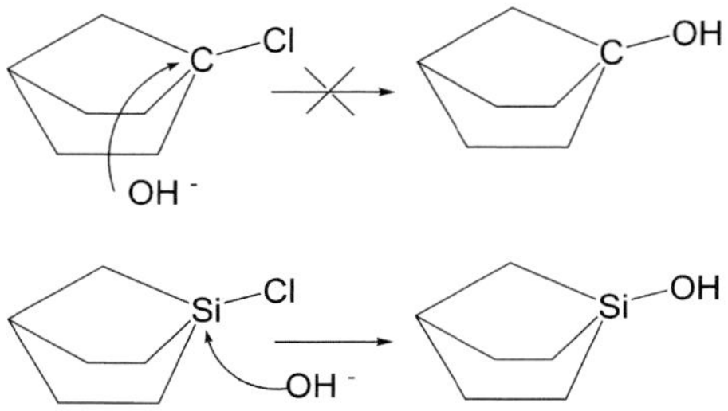

Fig. 13.36 Due to the presence of a vacant d orbital, a lateral attack can substitute for dorsal attacks in organo-silicon derivatives.[124]

For these compounds, metabolic oxidation does not take place on the silicon, but on one of methyl groups $(Si-CH_3 \rightarrow Si-CH_2-OH)$.[114]

B   Carbon–boron isosterism

Organoboron derivatives, even more than organosilicon compounds, are sensitive to hydrolytic degradation that always leads to the final formation of boric acid. However, boric acid has teratogenic properties in chickens. It produces the same malformations as those produced by a riboflavine (vitamin B_2) deficiency and the administration of riboflavine prevents these toxic effects.[126,127] The mechanism by which boric acid produces a deficiency in riboflavine is not known. In man the chronic utilization of boron derivatives results in cases of borism (dry skin, cutaneous eruptions, gastric troubles).[128]

Few medicines based on boron are known, in general boric acid or a boronic acid serve to esterify an α-diol or an *o*-diphenol. This is the case for the emetic antimony borotartrates of the ancient pharmacopoeias, for the injectable catecholamine solutions, for tolboxane,[129] which is close to meprobamate and which was commercially available as a tranquillizer some decades ago, and also for the phenylboronic esters of chloramphenicol.[130] Boromycine was the first natural product containing boron. It is a complex between boric acid and a polyhydroxylated tetradentate macrocycle.[131] Some boronic analogues of amino acids were prepared as chymotrypsine and elastase inhibitors.[132] The most important medical use of derivatives of boron derivatives is the treatment of some tumours by neutron capture therapy,[133–135] the problem here being to ensure a sufficient concentration of the product in the tumour being treated.

C   Bioisosterism involving selenium

Selenium and its derivatives are highly toxic and with the exception of ^{75}Se derivatives which serve diagnostic purposes (e.g. ^{75}Se-selenomethionine, used as a radioactive imaging agent in pancreatic scanning), there is no chemically defined seleno-organic drug on the market. Klayman reviewed a large number of selenium derivatives as chemotherapeutic agents in 1973.[136] Selenium bioisosteres of sulphur compounds are mainly used as research tools (e.g. bis[2-chloroethyl] selenide as selenium bioisostere of the classical sulphur mustards[137]). Selenocysteine is present in the catalytic site of mammalian glutathione-peroxidase and this explains the importance of selenium as an essential trace element.

The only selenium-containing drug candidate is ebselen (Fig. 13.37) which owes its antioxidant and anti-inflammatory properties to its interference with the selenoenzyme

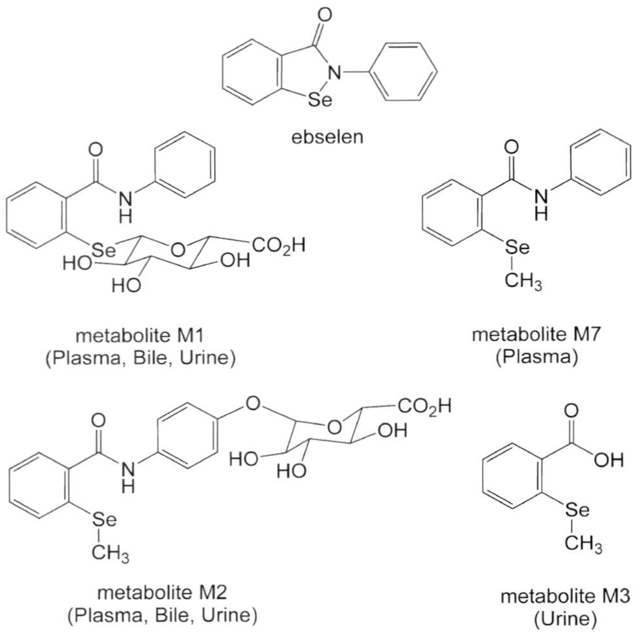

Fig. 13.37   Ebselen and its main metabolites.[138]

glutathione-peroxidase.[139] Because of its strongly bound selenium moiety, only metabolites of low toxicity are formed.[138]

REFERENCES

1. Langmuir, I. (1919) Isomorphism, isosterism and covalence. *J. Am. Chem. Soc.* **41**: 1543–1559.
2. Friedman, H.L. (1951) Influence of isosteric replacements upon biological activity, *Symposium on Chemical-Biological Correlation*, National Research Council Publication, Washington, DC.
3. Grimm, H.G. (1925) Über bau und grosse der nichtmetallhydride. *Z. Elektrochem* **31**: 474–480.
4. Grimm, H.G. (1929) The system of chemical compounds from the viewpoint of atom research, several problems of experimental research. Part I. *Naturwissenschaften* **17**: 535–540.
5. Grimm, H.G. (1929) The system of chemical compounds from the viewpoint of atom research, several problems of experimental research. Part II. *Naturwissenschaften* **17**: 557–564.
6. Erlenmeyer, H. and Leo, M. (1932) Über pseudoatome. *Helv. Chim. Acta* **15**: 1171–1186.
7. Erlenmeyer, H. (1948) Les composés isostères et le problème de la resemblance en chimie. *Bull. Soc. Chim. Biol.* **30**: 792–805.
8. Thornber, C.W. (1979) Isosterism and molecular modification in drug design. *Chem. Soc. Rev.* **8**: 563–580.
9. Krogsgaard-Larsen, P., Hjeds, H., Falch, E., Jørgensen, F.S. and Nielsen, L. (1988) Recent advances in GABA agonists, antagonists and uptake inhibitors: structure–activity relationships and therapeutic potential. In Testa, B. (ed.). *Advances in Drug Research*, pp. 381–456. Academic Press, London.
10. Janssen, P.A.J. and Van der Eycken, C.A.M. The chemical anatomy of potent morphine-like analgesics. Drugs Affecting the Central Nervous System, pp. 25–60. Marcel Dekker, New York.

11. Boyle, E.A., Mangan, F.R., Markwell, R.E., Smith, S.A., Thomson, M.J., *et al.* (1986) 7-Aroyl-2,3-dihydrobenzo[*b*]furan-3-carboxylic acids and 7-benzoyl-2,3-dihydrobenzo[*b*]thiophene-3-carboxylic acids as analgesic agents. *J. Med. Chem.* **29**: 894–898.

12. Morgan, B.P., Scholtz, J.M., Ballinger, M.D., Zipkin, I.D. and Bartlett, P.A. (1991) Differential binding energy: a detailed evaluation of the influence of hydrogen bonding and hydrophobic groups on the inhibition of thermolysin by phosphorous-containing inhibitors. *J. Am. Chem. Soc.* **113**: 297–307.

13. Chen, Y.L., Nielsen, J., Hedberg, K.D.A., Jones, S., Russo, L., *et al.* (1992) Syntheses, resolution, and structure–activity relationships of potent acetylcholinesterase inhibitors: 898-carbaphysostigmine analogues. *J. Med. Chem.* **35**: 1429–1434.

14. Marseigne, I. and Roques, B.P. Synthesis of new amino-acids mimicking sulfated and phosphorylated tyrosine residues. *J. Org. Chem.* **53**: 3621–3624.

15. Burke, T.R. Jr., Smyth, M., Nomizu, M., Otaka, A. and Roller, P.P. (1993) Preparation of fluoro- and hydroxy-4-phosphonomethyl-D,L-phenylalanine suitably protected for solid phase synthesis of peptides containing hydrolytically stable analogues of *O*-phosphotyrosine. *J. Org. Chem.* **58**: 1336–1340.

16. Erlenmeyer, H. and Willi, E. (1935) Zusammenhänge zwischen Konstitution und Wirkung bei Pyrazolonderivaten. *Helv. Chim. Acta* **18**: 740–743.

17. Wallace, E.M., Moliterni, J.A., Moskal, M.A., Neubert, A.D., Marcopoulos, N., *et al.* (1998) Design and synthesis of potent, selective inhibitors of endothelin-converting enzyme. *J. Med. Chem.* **41**: 1513–1523.

18. De Lombaert, S., Stamford, L.B., Blanchard, L., Tan, J., Hoyer, D., *et al.* (1997) Potent non-peptidic dual inhibitors of endothelin-converting enzyme and neutral endopeptidase. 22.11. *Bioorg. Med. Chem. Lett.* **8**: 1059–1064.

19. Binder, D., Noe, C.R., Holzer, W. and Baumann, K. (1987) Thiophen als strukturelement physiologisch aktiver substanzen,16. Thienoisoxazole durch substitution am oximstickstoff. *Arch. Pharm.* **320**: 837–843.

20. Uno, H., Kurokawa, M., Masuda, Y. and Nishimura, H. (1979) Studies on 3-substituted 1,2-benzisoxazole derivatives. 6. Syntheses of 3-(sulfamoylmethyl)-1,2-benzisoxazole derivatives and their anticonvulsant activities. *J. Med. Chem.* **22**: 180–183.

21. Gold-Aubert, P., Melkonian, D. and Toribio, L. (1964) Synthèses de nouvelles phényl-1-triazoline-1,2,4-ones-5 substituées en 3 et 4. *Helv. Chim. Acta* **47**: 2068–2071.

22. Alonzo, R., Andrès, J.I., Garcia-Lopez, M.-T., de las Heras, F.G., Herranz, R., *et al.* (1985) Synthesis and antiviral evaluation of nucleosides of 5-methylimidazole-4-carboxamide. *J. Med. Chem.* **28**: 834–838.

23. Garvey, D.S., Wasicak, J.T., Elliott, R.L., Lebold, S.A., Hettinger, A.-M., *et al.* (1994) Ligands for brain cholinergic channel receptors: synthesis and *in vitro* characterization of novel isoxazoles and isothiazoles as bioisosteric replacements for the pyridine ring in nicotine. *J. Med. Chem.* **37**: 4455–4463.

24. Fludzinski, P., Evrard, D.A., Bloomquist, W.E. and Lacefield, W.B. (1987) Indazoles as indole bioisosteres: synthesis and evaluation of the tropanyl ester and amide of indazole-3-carboxylate as antagonists to the serotonin 5HT$_3$ receptor. *J. Med. Chem.* **30**: 1535–1537.

25. Blaskó, G., Major, E., Blaskó, G., Rózsa, I. and Szántay, C. (1986) Pyrimido[1,6-*a*]pyrido[3,4-*b*] indoles as new platelet inhibiting agents. *Eur. J. Med. Chem.* **21**: 91–95.

26. Kardos, J., Blaskó, G., Simonyi, M. and Szántay, C. (1985) Octahydroindolo[2,3-*a*]quinolizin-2-one, a novel structure for γ-aminobutyric acid (GABA) uptake inhibition. *Eur. J. Med. Chem.* **21**: 151–154.

27. Salituro, F.G., Harrison, B.L., Baron, B.M., Nyce, P.L., Stewart, K.T., *et al.* (1992) 3-(2-Carboxyindol-3-yl)propionic acid-based antagonists of the *N*-methyl-D-aspartic receptor associated glycine binding site. *J. Med. Chem.* **35**: 1791–1799.

28. Calvino, R., Stilo, A.D., Fruttero, R., Gasco, A.M., Sorba, G., *et al.* (1993) Pharmacochemistry of the furoxan ring: recent developments. *Il Farmaco* **48**: 321–334.

29. Lipinski, C.A., Aldinger, C.E., Beyer, T.A., Bordner, J., Burdi, D.F., *et al.* (1992) Hydantoin isosteres. *In vivo* active spiro hydroxy acetic aldose reductase inhibitors. *J. Med. Chem.* **35**: 2169–2177.

30. Ganellin, C.R. (1993) Discovery of cimetidine, ranitidine and other H2-receptor histamine antagonists. *Medicinal Chemistry—The Role of Organic Chemistry in Drug Research*, pp. 227–255. Academic Press, London.

31. Morin, I. (1991) 3-Aryl as-triazines: Bioisostérie avec les 6-aryl pyridazines; Thesis, Université Louis Pasteur, Strasbourg, 17 January.

32. Mallamo, J.P., Pilling, G.M., Wetzel, J.R., Kowalczik, P.J., Bell, M.R., *et al.* (1992) Antiandrogenic steroidal sulfonyl heterocycles. Utility of electrostatic complementarity in defining bioisosteric sulfonyl heterocycles. *J. Med. Chem.* **35**: 1663–1670.

33. Clementi, S. and Cruciani, G. (1996) New set of principal properties for heteroaromatics obtained by grid. *Quant. Struct. Act. Rel.* **15**: 108–120.

34. Gibson, S., McGuire, R. and Rees, D.C. (1996) Principal components describing biological activities and molecular diversity of heterocyclic aromatic ring fragments. *J. Med. Chem.* **39**: 4065–4072.

35. Saari, W.S., Wai, J.S., Fisher, T.E., Thomas, C.M., Hoffman, J.M., *et al.* (1992) Synthesis and evaluation of 2-pyridone derivatives as HIV-1-specific everse transcriptase inhibitors. 2. Analogues of 3-aminopyridin-2-(1*H*)-one. *J. Med. Chem.* **35**: 3792–3802.

36. Langer, T. and Hoffmann, R.D. (1998) New principal components derived parameters describing molecular diversity of heteroaromatic residues. *Quant. Struct. Act. Rel.* **17**: 211–223.

37. Almquist, R.G., Chao, W.R. and Jennings-White, C. (1985) Synthesis and biological activity of carboxylic acid replacement analogues of the potent angiotensin converting enzyme inhibitor 5(S)-benzamido-4-oxo-6-phenylhexanoyl-L-proline. *J. Med. Chem.* **28**: 1067–1071.

38. Kohler, H.v., Eichler, B. and Salewski, R. (1970) Untersuchungen zum sauerstoffanalogen Charakter der C(CN)$_2$- und NCN- Gruppen. *Z. Anorg. Chem.* **379**: 183–192.

39. Kwon, C.-H., Nagasawa, H.T., DeMaster, E.G. and Shirota, F.N. (1986) Acyl, *N*-protected α-aminoacyl, and peptidyl derivatives as prodrug forms of the alcohol deterrent agent cyanamide. *J. Med. Chem.* **29**: 1922–1929.

40. Bovy, P.R., Reitz, D.B., Collins, J.T., Chamberlain, T.S., Olins, G.M., *et al.* (1993) Nonpeptide angiotensin II antagonists: *N*-phenyl-1-*H*-pyrrole derivatives are angiotensin II receptor antagonists. *J. Med. Chem.* **36**: 101–110.

41. Marshall, W.S., Goodson, T., Cullinan, G.J., Swanson-Bean, D., Haisch, K.D., *et al.* (1987) Leukotriene receptor antagonists. 1. Synthesis and structure-activity relationships of alkoxyacetophenone derivatives. *J. Med. Chem.* **30**: 682–689.

42. Kohara, Y., Kuba, K., Imamiya, E., Wada, T., Inada, Y., *et al.* (1996) Synthesis and angiotensin II receptor antagonistic activities of benzimidazole derivatives bearing acidic heterocycles as novel tetrazole nioisosteres. *J. Med. Chem.* **39**: 5228–5235.

43. Krogsgaard-Larsen, P. (1990) *Comprehensive Medicinal Chemistry*, pp. 493–537. Pergamon Press, Oxford.

44. Lunn, W. H. W., Schoepp, D. D., Lodge, D., True, R. A. and Millar, J. D (1992) Poster No. P-041.A. LY262466, DL-2-amino-3-(4-hydroxy-1, 2,5-thiazol-3-yl) propanoic acid hydrochloride, a novel and selective agonist at the AMPA excitatory amino acid receptor. XIIth International Symposium on Medicinal Chemistry, pp. 133–137. Base1: Switzerland, 13–17 September.

45. Atkinson, J.G., Girard, Y., Rokach, J., Rooney, C.S., McFarlane, C.S., *et al.* (1979) Kojic amine-A novel γ-aminobutyric acid analogue. *J. Med. Chem.* **22**: 90–106.

46. Froestl, W., Furet, P., Hall, R.G., Mickel, S.J., Strub, D., *et al.* (1993) GABA_B antagonists: novel CNS-active compounds. *Perspectives in Medicinal Chemistry*, pp. 259–272. VHC, Weinheim.

47. Rosowsky, A., Forsch, R.A., Freisheim, J.H., Moran, R.G. and Wick, M. (1984) Methotrexate analogues. 19. Replacement of the glutamate side chain in classical antifolates by L-homocysteic acid and L-cysteic acid: effect on enzyme inhibition and antitumor activity. *J. Med. Chem.* **27**: 600–604.

48. Drummond, J.T. and Johnson, G. (1988) Convenient procedure for the preparation of alkyl and aryl substituted *N*-(aminoalkylacyl)sulfonamides. *Tetrahedron Lett.* **29**: 1653–1656.

49. Albright, J.D., DeVries, V.G., Du, M.D., Largis, E.E., Miner, T.G., *et al.* (1983) Potential antiatherosclerotic agents. 3. Substituted benzoic and non-benzoic analogues of cetabon. *J. Med. Chem.* **26**: 1393–1411.

50. Buu-Hoï, N.P., Lambelin, G., Lepoivre, C., Gillet, C., Gautier, M., *et al.* (1965) Un nouvel agent anti-inflammatoire de structure non stéroïdique: l'acide p-butoxyphénylacéthydroxamique. *CR Acad. Sci. (Paris)* **261**: 2259–2262.

51. Orzalesi, G., Selleri, R (1974) Pharmaceutical 2-(4-isobutylphenyl) propionohydroxamic acid. German Patent 2 400 531 (24 July 1974, to Societa Italo-Britannica L. Manetti & H. Roberts e C.). *Chem. Abst.* **81**: 120272i.

52. De Martiis, F., Corsico, N., Franzone, J.S. and Tamietto, T. (1975) Valutazione farmaco-tossicologica di un nuovo agente antifiammatorio non-steroideo: l'acido indoxamico. *Boll. Chim. Farm.* **114**: 319–333.

53. Summers, J.B., Masdiyasni, H., Holms, J.H., Ratajczik, J.D., Dyer, R.D., *et al.* (1987) Hydroxamic acid inhibitors of 5-lipoxygenase. *J. Med. Chem.* **30**: 574–580.

54. Bergeron, R.J., Liu, Z.-R., McManis, J.S. and Wiegand, J. (1992) Structural alterations in desferioxamine compatible with iron clearance in animals. *J. Med. Chem.* **35**: 4739–4744.

55. De Martiis, F., Franzone, J.S. and Tamietto, T. (1975) Sintesi e proprieta antiflogistiche di alcuni acidici indolil-acetoidrossammici. *Bol. Chim. Farm.* **114**: 309–318.

56. Orzalesi, G., Mari, F., Bertol, E., Selleri, R. and Pisaturo, G. (1980) Anti-inflammatory agents: determination of ibuproxam and its metabolite in humans. *Arzneim.-Forsch.* **30**: 1607–1609.

57. Demay, F. and De Sy, J. (1982) A new non-steroidal anti-inflammatory drug (NSAID) in current rheumatologic practice (oxamethacin). *Curr. Ther. Res.* **31**: 113–118.

58. Vergin, H., Ferber, H., Brunner, F. and Kukovetz, W.R. (1981) Pharmakokinetik und biotransformation von Oxametacin bei gesunden probanden. *Arzneim.-Forsch.* **31**: 513–518.

59. Singh, H., Chawla, A.S., Kapoor, V.K., Paul, D. and Malhotra, R.K. (1980) Medicinal chemistry of tetrazoles. In Ellis, G.P. and West, G.B. (eds). *Progress in Medicinal Chemistry*, pp. 151–183. Elsevier, Amsterdam.

60. Ashton, W.T., Cantone, C.L., Chang, L.L., Hutchins, S.M., Strelitz, R., *et al.* (1993) Non-peptide angiotensin II antagonists derived from 4*H*-1, 2,4-triazoles and 3*H*-imidazo[1,2-*b*][1,2,4]triazoles. *J. Med. Chem.* **36**: 591–609.

61. Kraus, J.L. (1983) Isoterism and molecular modification in drug design: tetrazole analogue of GABA: effects on enzymes of the gamma-aminobutyrate system. *Pharmacol. Res. Commun.* **15**: 183–189.

62. Schlewer, G., Wermuth, C.G. and Chambon, J.-P. (1984) Analogues tétrazoliques d'agents GABA-mimétiques. *Eur. J. Med. Chem.* **19**: 181–186.

63. Krogsgaard-Larsen, P. and Rodolskov-Christiansen, T. (1979) GABA agonists. Synthesis and structure-activity studies on analogues of isoguvacine and THIP. *Eur. J. Med. Chem.* **14**: 157–164.

64. Krogsgaard-Larsen, P. (1981) γ-Aminobutyric acid agonists, antagonists, and uptake inhibitors. Design and therapeutic aspects. *J. Med. Chem.* **24**: 1377–1383.

65. Krogsgaard-Larsen, P., Ferkany, J.W., Nielsen, E.O., Madsen, U., Ebert, B., *et al.* (1991) Novel class of amino acid antagonists at non-*N*-methyl-D-aspartic acid excitatory amino acid receptors. Synthesis, *in vitro* and *in vivo* pharmacology, and neuroprotection. *J. Med. Chem.* **34**: 123–130.

66. Kraus, J.L. (1983) Isoterism and molecular modification in drug design: new *n*-dipropylacetate analogs as inhibitors of succinic semi aldehyde dehydrogenase. *Pharmacol Res. Commun.* **15**: 119–129.

67. Lichtenthaler, F.W. and Heidel, P. (1969) Intermediates in the formation of γ-pyrones from hexose derivatives: a simple synthesis of Kojic acid and hydroxymaltol. *Angew. Chem. Int. Ed.* **8**: 978–979.

68. Watkins, J.C., Krogsgaard-Larsen, P. and Honoré, T. (1990) Structure–activity relationships in the development of excitatory amino acid receptor agonists and competitive antagonists. *Trends Pharm. Sci.* **11**: 25–33.

69. Drysdale, M.J., Pritchard, M.C. and Horwell, D.C. (1992) Rationally designed 'dipeptoid' analogues of CCK. Acid mimics of the potent and selective non-peptide CCK-B receptor antagonist CI-988. *J. Med. Chem.* **35**: 2573–2581.

70. Kinney, W.A., Lee, N.E., Garrison, D.T., Podlesny, E.J. Jr., Simmonds, J.T., *et al.* (1992) Bioisosteric replacement of the α-amino carboxylic functionality in 2-amino-5-phosphonopentanoic acid yields unique 3,4-diamino-3-cyclobutene-1,2-dione containing NMDA antagonists. *J. Med. Chem.* **35**: 4720–4726.

71. Gao, Y., Luo, J., Yao, Z.J., Guo, R., Zou, H., *et al.* (2000) Inhibition of Grb2 SH2 domain binding by non-phosphate-containing ligands. 2. 4-(2-Malonyl)phenylalanine as a potent phosphotyrosyl mimetic. *J. Med. Chem.* **43**: 911–920.

72. Shapiro, G., Floersheim, P., Boelsterli, J., Amstutz, R., Bolliger, G., *et al.* (1992) Muscarinic activity of the thiolactone, lactam, lactol and thiolactol analogues of pilocarpine and a hypothetical model for the binding of agonists to the m_1 receptor. *J. Med. Chem.* **35**: 15–27.

73. Thompkins, L. and Lee, K.H. (1975) Comparison of analgesic effects of isosteric variations of salicylic acid and aspirin (acetylsalicylic acid). *J. Pharm. Sci.* **64**: 760–763.

74. Roth, G.J., Stanford, N. and Majerus, P.W. (1975) Acetylation of prostaglandine synthase by aspirin. *Proc. Nat. Acad. Sci. USA* **72**: 3073–3076.

75. Saunders, J., Cassidy, M., Freedman, S.B., Harley, E.A., Iversen, L.L., *et al.* (1990) Novel quinuclidine-based ligands for the muscarinic cholinergic receptor. *J. Med. Chem.* **33**: 1128–1138.

76. Sauerberg, P., Kindtler, J.W., Nielsen, L., Sheardown, M.J. and Honoré, T. (1991) Muscarinic cholinergic agonists and antagonists of the 3-(3-alkyl-1,2,4-oxadiazol-5-yl)1,2,5,6-tetrahydropyridine type. Synthesis and structure-activity relationships. *J. Med. Chem.* **34**: 687–692.

77. Sauerberg, P., Olesen, P.H., Nielsen, S., Treppendahl, S., Sheardown, M.J., *et al.* (1992) Novel functional M_1 selective muscarinic agonists. Synthesis and structure-activity relationships of 3-(1,2,5-thiadiazolyl)-1,2,5,6-tetrahydro-1-methylpyridines. *J. Med. Chem.* **35**: 2263–2274.

78. Wadsworth, H.J., Jenkins, S.M., Orlek, B.S., Cassidy, F., Clark, M.S., *et al.* (1992) Synthesis and muscarinic activities of quinuclidin-3-yltriazole and -tetrazole derivatives. *J. Med. Chem.* **35**: 1280–1290.

79. Street, L.J., Baker, R., Book, T., Reeve, A.J., Saunders, J., *et al.* (1992) Synthesis and muscarinic activity of quinuclidinyl- and (1-azanorbornyl)pyrazine derivatives. *J. Med. Chem.* **35**: 295–305.

80. Kozikowski, A.P., Roberti, M., Xiang, L., Bergmann, J.S., Callahan, P.M., *et al.* (1992) Structure–activity relationship studies of cocaine: replacement of the C-2 ester group by vinyl argues against H-bonding and provides an esterase-resistant, high-affinity cocaine analogue. *J. Med. Chem.* **35**: 4764–4766.

81. Lipinski, C.A. (1986) Bioisosterism in drug design. In Bailey, D.M. (ed.). *Annual Reports in Medicinal Chemistry*, pp. 283–291. Academic Press, San Diego.

82. Wermuth, C.G. (1993) Aminopyridazines—an alternate route to potent muscarinic agonists with no cholinergic syndrome. *Il Farmaco* **48**: 253–274.

83. Huff, J.R., Anderson, P.S., Baldwin, J.J., Clineschmidt, B.V., Guare, J.P., et al. (1985) N-(1,3,4,6,7,12b-hexahydro-2H-benzo[b]furo[2,3-a] quinolizin-2-yl)-N-methyl-2-hydroxyethane-sulfonamide: a potent and selective α₂-adrenoreceptor antagonist. *J. Med. Chem.* **28**: 1756–1759.

84. Fournier, J.-P., Moreau, R.C., Narcisse, G. and Choay, P. (1982) Synthèse et propriétés pharmacologiques de sulfonylurées isostères du glibenclamide. *Eur. J. Med. Chem.* **17**: 81–84.

85. Spatola, A.F. (1983) Peptide backbone modifications: structure-activity analysis of peptides containing amide bond surrogates. In Weinstein, B. (ed.). *Chemistry and Biochemistry of Amino Acids, Peptides and Proteins*, pp. 267–357. Marcel Dekker, New York.

86. Fauchère, J.-L. (1986) Elements for the rational design of peptide drugs. In Testa, B. (ed.). *Advances in Drug Research*, pp. 29–69. Academic Press, London.

87. Smith, A.B., III, Keenan, T.P., Holcomb, R.C., Sprengeler, P.A., Guzman, M.C., et al. (1992) Design, synthesis, and crystal structure of a pyrrolinone-based peptidomimetic possessing the conformation of a β-strand: potential application to the design of novel inhibitors of proteolytic enzymes. *J. Am. Chem. Soc.* **114**: 10672–10674.

88. Smith, A.B., III, Holcomb, R.C., Guzman, M.C., Keenan, T.P., Sprengeler, P.A., et al. (1993) An effective synthesis of scalemic 3,5, 5-trisubstituted pyrrolin-4-ones. *Tetrahedron Lett.* **34**: 63–66.

89. Allmendinger, T., Felder, E. and Hungerbuehler, E. (1991) Fluoro-olefin dipeptide isosteres. In Weldi, J.T. (ed.). *Selective Fluorination in Organic and Bioorganic Chemistry*, pp. 186–195. American Chemical Society, Washington.

90. Chandrakumar, N.S., Yonan, P.K., Stapelfeld, A., Svage, M., Rorbacher, E., et al. (1992) Preparation and opioid activity of analogues of the analgesic dipeptide 2,6-dimethyl-L-tyrosyl-N-(3-phenylpropyl)-D-alanylamide. *J. Med. Chem.* **35**: 223–233.

91. Yanagisawa, I., Hirata, Y. and Ishii, Y. (1984) Histamine H₂ receptor antagonists. 1. Synthesis of N-cyano and N-carbamoyl amidine derivatives and their biological activities. *J. Med. Chem.* **27**: 849–857.

92. Young, R.C., Durant, G.J., Emmet, J.C., Ganellin, C.R., Graham, M.J., et al. (1986) Dipole moment in relation to H₂ receptor antagonist activity for cimetidine analogues. *J. Med. Chem.* **29**: 44–49.

93. Lumma, W.C., Jr., Anderson, P.S., Baldwin, J.J., Bolhofer, W.A., Habecker, C.N., et al. (1982) Inhibitors of gastric acid secretion: 3,4-diamino-1,2,5-thiadiazole 1-oxides and 1,1-dioxides as urea equivalents in a series of histamine H₂-receptor antagonists. *J. Med. Chem.* **25**: 207–210.

94. Young, R.C., Ganellin, C.R., Graham, M.J. and Grant, E.H. (1982) The dipole moments of 1,3-dimethylthiourea, 1,3-dimethyl-2-cyanoguanidine and 1,1-bis-methylamino-2-nitroethene in aqueous solution. *Tetrahedron* **38**: 1493–1497.

95. Young, R.C., Ganellin, C.R., Graham, M.J., Roantree, M.J. and Grant, E.H. (1985) The dielectric properties of seven polar amidine-containing compounds of biological interest. *Tetrahedron Lett.* **26**: 1897–1900.

96. Wright, J.L., Gregory, T.F., Kesten, S.R., Boxer, P.A., Serpa, K.A., et al. (2000) Subtype-selective N-methyl-D-aspartate receptor antagonists: synthesis and biological evaluation of 1-(heteroarylalkynyl)-4-benzylpiperidines. *J. Med. Chem.* **43**: 3408–3419.

97. Larson, A.A. and Lish, P.M. (1964) A new bio-isostere: alkylsulphonamidophenethanolamines. *Nature (London)* **203**: 1283–1285.

98. Roderick, S.L., Fournié-Zaluski, M.C., Roques, B.P. and Matthews, B.W. (1989) Thiorphan and *retro*-thiorphan display equivalent interactions when bound to crystalline thermolysin. *Biochemistry* **28**: 1493–1497.

99. Büchi, J., Stünzi, E., Flury, M., Hirt, R., Labhart, P., et al. (1951) Über lokalanästhetisch wirksame basische ester und amide verschiedener alkoxy-amino-benzoesäuren. *Helv. Chim. Acta* **34**: 1002–1013.

100. Bös, M., Jenck, F., Martin, J.R., Moreau, J.L., Sleight, A.J., et al. (1997) Novel agonists of 5HT2C receptors. Synthesis and biological evaluation of substituted 2-(indol-1-yl)-1-methylethylamines and 2-(indeno[1,2-b]pyrrol-1-yl)-1-methylethylamines. Improved therapeutics for obsessive compulsive disorder. *J. Med. Chem.* **40**: 2762–2769.

101. Wilhelm, M. (1975) The chemistry of polycyclic psycho-active drugs: serendipity or systematic investigation? *Pharm. J.* **214**: 414–416.

102. Yoshimura, H., Kikuchi, K., Hibi, S., Tagami, K., Satoh, T., et al. (2000) Discovery of novel and potent retinoic acid receptor α agonists: syntheses and evaluation of benzofuranyl-pyrrole and benzothiophenyl-pyrrole. *J. Med. Chem.* **43**: 2929–2937.

103. Chenoweth, M.B. and McCarthy, L.P. (1963) On the mechanism of the pharmacophoric effect of halogenation. *Pharmacol. Rev.* **15**: 673–707.

104. Goldman, P. (1969) The carbon-florine bond in compounds of biological interest. *Science* **164**: 1123–1130.

105. Peters, R.A. (1963) *Biochemical Lesions and Lethal Synthesis.* Pergamon Press, Oxford.

106. Chapman, N.B., James, J.W., Graham, J.D.P. and Lewis, G.P. (1952) Chemical reactivity and pharmacological activity among 2-haloethylamine derivatives with a naphtylmethyl group. *Chem. Ind (London)* 805–807.

107. Vaughan, J.R.J., Anderson, G.W., Clapp, R.C., Clark, J.H., English, J.P., et al. (1949) Antihistamine agents. IV. Halogenated N,N-dimethyl-N'-benzyl-N-(2-pyridyl)-ethylenediamines. *J. Org. Chem.* **14**: 228–234.

108. Schatz, V.B. (1963) Isosterism and bioisosterism. In Burger, A. (ed.). *Medicinal Chemistry*, pp. 72–88. Interscience Publishers, New York.

109. Fessenden, R.J. and Fessenden, J.S. (1967) The biological properties of silicon compounds. In Harper, N.J. and Simmonds, A.B. (eds). *Advances in Drug Research*, pp. 95–132. Academic Press, London.

110. Tacke, R. and Zilch, H. (1986) Sila-substitution—a useful strategy for drug design? *Endeavour, New Series* **10**: 191–197.

111. Tacke, R. and Zilch, H. (1986) Drug-design by sila-substitution and microbial transformations of organosilicon compounds: some recent results. *L'Actualité Chimique* 75–82.

112. Ricci, A., Seconi, G. and Taddei, M. (1989) Bioorganosilicon chemistry: trends and perspectives. *Chimica Oggi-Chemistry Today* **7**: 15–21.

113. Metcalf, R.L. and Fukuto, T.R. (1965) Silicon-containing carbamate insecticides. *J. Econ. Entomol.* **58**: 1151.

114. Anonymous (1994) Zifrosilone. *Drugs Fut.* **19**: 854–855.

115. Hornsperger, J.M., Collard, J.N., Heydt, J.G., Giacobini, E., Funes, S., et al. (1994) Trimethylsilylated trifluoromethyl ketones, a novel class of acetylcholinesterase inhibitors: biochemical and pharmacological profile of MDL 73,745. *Biochem. Soc. Trans.* **22**: 758–763.

116. Fessenden, R.J. and Coon, M.D. (1965) Silicon-substituted medicinal agents. Silacarbamates related to meprobamate. *J. Med. Chem.* **8**: 604–608.

117. Tacke, R. (1980) Sila-pharmaka XIX. Sila-pridinol und pridinol: darstellung und eigenschaften sowie strukturen im kristallinen und gelösten Zustand. *Chem. Ber.* **113**: 1962–1980.

118. Moberg, W. K (1985) Fungicidal 1,2,3-triazole derivatives. *US Patent* 4 510 136 (Apr. 9, 1985 to E.I. Du Pont).

119. Damour, D.M.B., Dutruc-Rosset, G., Doble, A., Piot, O., et al. (1994) 1,1-Diphenyl-3-dialkylamino-1-silacyclopentane derivatives: a new

class of potent and selective 5-HT$_{2A}$ antagonists. *Bioorg. Med. Chem. Lett.* **4**: 415–420.

120. Gotteland, J.-P., Brunel, I., Gendre, F., Désiré, J., Delhon, A., *et al.* (1995) (Aryloxy)methylsilane derivatives as new cholesterol biosynthesis inhibitors: synthesis and hypocholesterolemic activity of a new class of squalene epoxidase inhibitors. *J. Med. Chem.* **38**: 3207–3216.

121. Gotteland, J.-P., Delhon, A., Junquéro, D., Oms, P. and Halazy, S. (1996) Design and synthesis of new hypocholesteremic organosilanes with antioxidant properties. *Bioorg. Med. Chem. Lett.* **6**: 533–538.

122. Heinonen, P., Sipilä, H., Neuvonen, K., Lönnberg, H., Cockroft, V.B., *et al.* (1996) Synthesis and pharmacological properties of 4(5)-(2-ethyl-2,3-dihydro-2-silainden-2-yl)imidazole, a silicon analogue of atipamezole. *Eur. J. Med. Chem.* **31**: 725–729.

123. Bom, D., Curran, D.P., Kruszewski, S., Zimmer, S.G., Strode, J.T., *et al.* (2000) The novel silatecan 7-*tert*-butyldimethylsilyl-10-hydroxycamptothecin displays high lipophilicity, improved human blood stability, and potent anticancer activity. *J. Med. Chem.* **43**: 3970–3980.

124. Sommer, L.H., Bennet, O.F., Campbell, P.G. and Weyenberg, D.R. (1957) Stereochemistry of hydride ion displacement from silicon. Enhanced rates at Bridgehead and 4-ring silicon atoms. *J. Am. Chem. Soc.* **79**: 3295–3296.

125. Fessenden, R.J. and Ahlfors, C. (1967) The metabolic fate of some silicon-containing carbamates. *J. Med. Chem.* **10**: 810–812.

126. Landauer, W. (1954) On the chemical production of developmental abnormalities and of phenocopies in chicken embryos. *J. Cell. Comp. Physiol.* **43**: 261–305.

127. Landauer, W. and Clark, E.M. (1964) On the role of riboflavin in the teratogenic activity of boric acid. *J. Exp. Zool.* **156**: 307–312.

128. Browning, E. (1969) *Toxicity of Industrial Metals*, second edn., pp. 90–97. Appleton-Century-Crofts, New York.

129. Caujolle, P.H.C. (1968) Structure chimique et activité spasmolytique des organoboriques. *Arch. Int. Pharmacodyn. Ther.* **172**: 467–474.

130. Mubarak, S.I.M., Stanford, J.B. and Sugden, J.K. (1984) Some aspects of the antimicrobial and chemical properties of phenyl boronate esters of chloramphenicol. *Drug Dev. Ind. Pharm.* **10**: 1131–1160.

131. Dünitz, J.D., Hawley, D.M., Miklos, D., White, D.N.J., Berlin, Y., *et al.* (1971) Structure of boromycin. *Helv. Chim. Acta* **54**: 1709–1713.

132. Kinder, D.H. and Katzenellenbogen, J.A. (1985) Acylamino boronic acids and difluoroborane analogues of amino acids: potent inhibitors of chymotrypsin and elastase. *J. Med. Chem.* **28**: 1917–1925.

133. Alam, F., Soloway, A.H., Bapat, B.V., Barth, R.F. and Adams, D.M. (1989) Boron compounds for neutron capture therapy. *Basic Life Sci.* **50**: 107–111.

134. Kahl, S.B., Joel, D.D., Finkel, G.C., Micca, P.L., Nawrocky, M.M., *et al.* (1989) A carboranyl porphyrin for boron neutron capture therapy of brain tumors. *Basic Life Sci.* **50**: 193–203.

135. Gabel, D. (1989) Tumor-seeking for boron neutron capture therapy: synthesis and biodistribution. *Basic Life Sci.* **50**: 233–241.

136. Klayman, D.L. and Günther, W.H.H. (1973) *Organic Selenium Compounds: Their Chemistry and Biology.* Wiley-Interscience, New York.

137. Kang, S.I. and Spears, C.P. (1987) Linear free energy relationships and cytotoxicities of para-substituted 2-haloethyl aryl selenides and bis(-chloroethyl) selenides. *J. Med. Chem.* **30**: 597–602.

138. Fischer, H., Terlinden, R., Löhr, J.P. and Römer, A. (1988) A novel biologically active selenoorganic compound. VIII. Biotransformation of ebselen. *Xenobiotica* **18**: 1347–1359.

139. Parnham, M.J. and Graf, E. (1987) Seleno-organic compounds and the therapy of hydroperoxide-linked pathological conditions. *Biochem. Pharmacol.* **36**: 3095–3102.

14

RING TRANSFORMATIONS

Camille G. Wermuth

| Loschmidt 1861 | Kekulé 1865 | Pauling, Corey, Koltun 1950 |

The Loschmidt formula for the toluene ring system[1]

I INTRODUCTION

When active molecules contain cyclic systems, these can be opened, expanded, contracted, modified in many other ways, or even abolished. Conversely, noncyclic molecules can be cyclized, attached to, or included in, ring systems.

In the daily practice of medicinal chemistry three kinds of approaches are currently used. The first approach does not affect the global complexity of the cyclic system and yields generally close analogues (or 'me-too' compounds) of the original active principle. For this we propose the term *analogical approach*. It consists of ring–chain transformations, ring contractions or expansions and various other ring transformations. The second strategy, called the *disjunctive*

approach[2] aims at the progressive simplification of the original active principle (which is often a natural compound). The objective is to extract information about the minimal structure that is required for activity. Finally, the *conjunctive* approach[2] is based on the creation or addition of supplementary rings. The objective is to constrain an originally flexible compound and to impose precise conformations and configurations. The preparation of such molecules is of prime importance in the exploration of ligand–receptor interaction and for molecular modelling studies.

II ANALOGICAL APPROACHES

A Analogy by ring opening: open-chain analogues

Open ring analogues of cyclic active principles (open drugs, open-chain analogues) can be designed and synthesized. The usefulness of such compounds is rather questionable, and it appears that most of them were prepared for me-too purposes. Actually two possibilities can be foreseen:

(a) The open analogue is again cyclized after oxidation or dehydration by a metabolic enzyme. We deal here with potential rings, which represent nothing more than

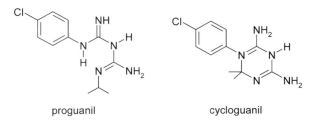

Fig. 14.1 Cycloguanil is the active metabolite of proguanil.[4]

metabolic precursors of the active species.

(b) The open analogue does not cyclize *in vivo* but can present some conformational analogy with the ring-containing active principle. These kinds of analogues are known as *pseudocycles*.

Potential rings: in vivo *return to the cyclic derivative*

Compounds generating the active form after *in vivo* cyclization are in fact prodrugs and will be discussed in a more detailed manner in Chapter 33.

Proguanyl. A historical example is the antimalarial drug proguanil[3] (Fig. 14.1). It was observed that this compound is inactive *in vitro* cultures of *Plasmodium gallinaceum*, but that the serum of animals treated with proguanil is active in these cultures. It was concluded that the actual active principle was a metabolite,[5] which was subsequently identified as cycloguanil.[4] In tropical medicine, proguanil is preferred to cycloguanil, the latter compound being too rapidly eliminated by the kidney.

Potassium canrenoate. This is a water-soluble prodrug and can be administered parenterally (Fig. 14.2). It has no intrinsic activity, but it can exert its diuretic activity (as an aldosterone antagonist) because of its interconversion with

canrenone. Canrenone is itself the major metabolite of spironolactone.[6]

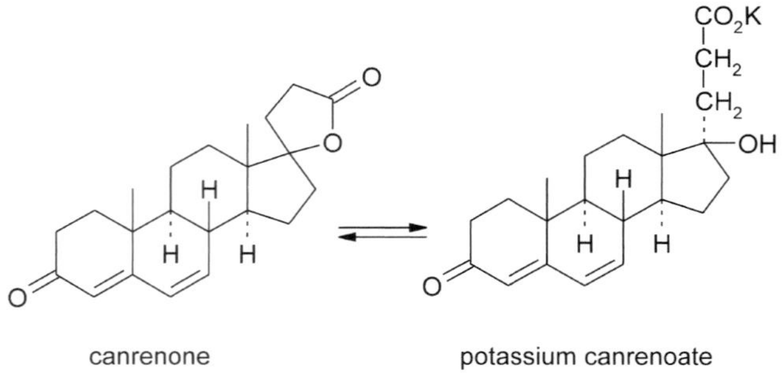

Irreversible compounds: pseudocycles

The open analogue is assumed to present a similar conformation to that of the cyclic one. Before adopting the term pseudocycles it must be ascertained by NMR or X-ray crystallography that they really mimic the ring-closed analogue.

Diethylstilbestrol. It is currently accepted that the theory of pseudocycles was elaborated to account for the estrogenic activity of compounds such as bisdehydrodoisynolic acid, allenoestrol and diethylstilbestrol (Fig. 14.3). The similarity with the natural hormone estradiol is striking. Nevertheless, it is highly probable that for receptor binding, the general shape of the molecules, and the distances separating the functional groups are more important than their degree of cyclization.[7]

Clonidine. Open analogues of the centrally acting hypotensive agent clonidine (Fig. 14.4) have a similar activity profile but with a 30- to 100-fold loss in potency.[8] Surprisingly, seco-clonidine, which is the closest analogue of clonidine, was found to be less active than the corresponding monomethyl derivative.

Cromakalim. A more recent example is the open-chain analogue of cromakalim, which was prepared as a more flexible pyrrolidone replacement (Fig. 14.5). It retains about a third of the potency of cromakalim.[9]

Tetrahydrofolic acid. Open-chain analogues of 5,6,7,8-tetrahydrofolic acid were prepared by researchers from Burroughs Wellcome.[10] None of the ring-opened analogues was as potent as the ring-closed lead structure in inhibiting tumour cell growth.

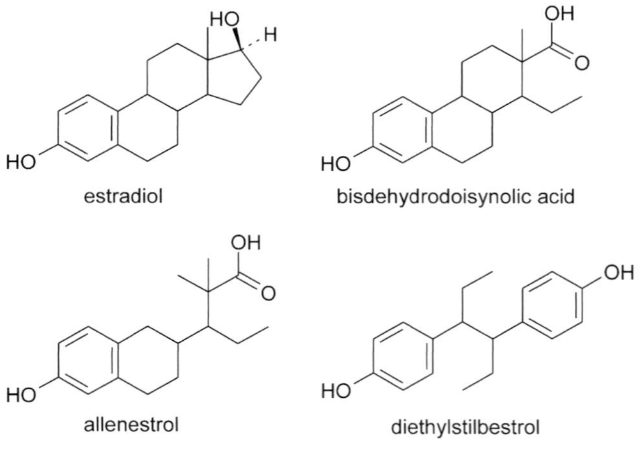

Fig. 14.2 *In vivo* the inactive potassium canrenoate cyclizes to canrenone.[6]

Fig. 14.3 Open analogues of estradiol.

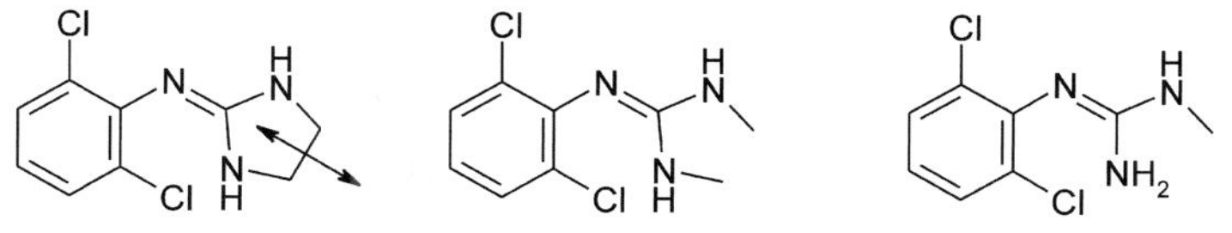

clonidine	open analogue 'secoclonidine'	monomethyl analogue
(-19% at 0.3 μmole/Kg)	(-7% at 10 μmole/Kg)	(-23% at 10 μmole/Kg)

Fig. 14.4 Open analogues of clonidine (the hypotensive activity is expressed as variation of the arterial blood pressure 30 minutes after i.v. injection in pentobarbital anaesthetized rats).[8]

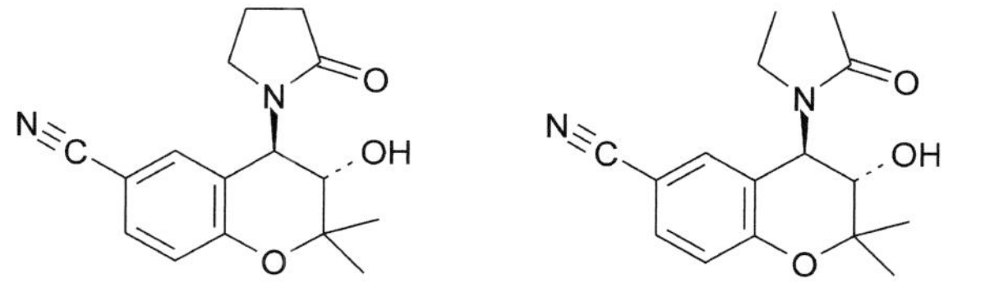

	CROMAKALIM	**OPEN CHAIN ANALOGUE**
Dose mg/Kg, p.o.	**0.3**	**1.0**
Maximal decrease in blood pressure (%)	**39 ± 4**	**22 ± 5**

Fig. 14.5 Cromakalim and its open-chain analogue.[9]

B Analogy by ring closure

Cyclizing open structures or creating an additional ring system in a given structure represents one of the useful methods in the search for biologically active conformations. The end result is a more constrained molecule, with an imposed conformation. The inconvenience is that additional isomeric centres may be introduced and that the selected cyclization mode might not lead to the active conformation adopted by the open-chain drug. A particularly convincing example is given by the ring-closed analogue of the thrombin inhibitor NAPAP which is 100 times more potent than the corresponding open-chain drug (Fig. 14.6).[11]

Inhibition of Gastric H^+/K^+ ATPase by substituted Imidazo[1,2-α]pyridines. Apparently, the substitution in the para position of the pyridine ring is detrimental (R = H → R = Me, Fig. 14.7). However, the ring closure achieving a conformational restriction yields a highly potent compound.[12]

Mevinolin and compactin. An example of reversible ring closure is found with mevinolin and compactin, which are both potent inhibitors of hydroxymethylglutaryl-coenzyme A reductase (HMG-CoA reductase), the rate-determining enzyme in the *de novo* biosynthesis of cholesterol. *In vivo* these ring-closed derivatives (Fig. 14.8) are hydrolysed to

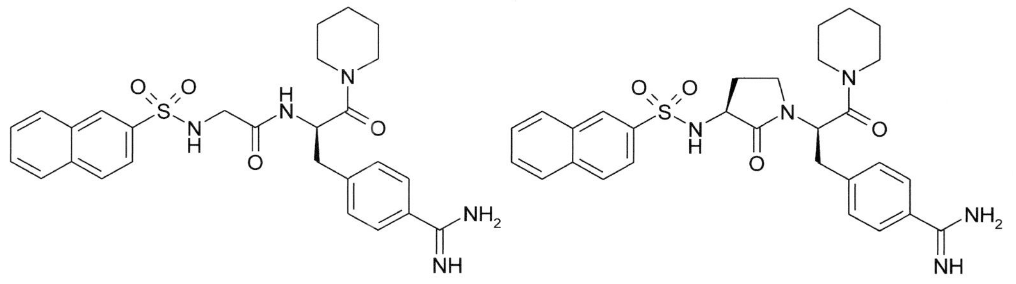

Fig. 14.6 The ring-closed analogue of the thrombin inhibitor NAPAP is 100 times more potent than the corresponding open-chain drug. (Kubinyi. (1998) *Pharmazie in unserer Zeit*, p. 167. Fig. 16.)

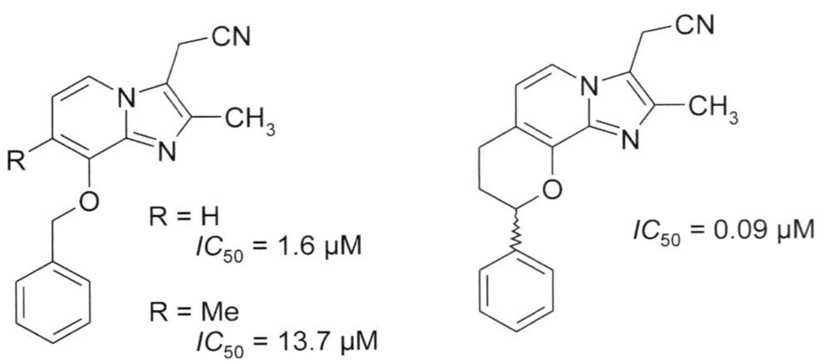

Fig. 14.7 Ring closure achieves a conformational restriction and yields a highly potent compound.

the open chain 3,5-dihydroxyvaleric acid form that mimics the structure of the proposed intermediate in the reduction of HMG-CoA by HMG-CoA reductase.[13]

Arylpropionic analgesic and anti-inflammatory drugs. The potent analgesic benzoylindane carboxylic acid, TAI-901,[14] is the cyclized analogue of the well-known anti-inflammatory analgesic agent ketoprofene (Fig. 14.9). The corresponding heterocyclic analogues were also prepared.[15] The compounds show potent analgesic activities with low gastric irritation.

The sulpiride side-chain. Among the numerous benzamide drugs developed by the Delagrange scientists, some have diethylaminoethyl side-chains (e.g. tiapride) whereas others, such as sulpiride, have *N*-ethylpyrrolidinylmethyl side-chains that can be considered as the corresponding ring-closed analogues (Fig. 14.10). Both compounds are dopaminergic antagonists with neurotropic and antiemetic activity. Note that the ring closure creates an asymmetric

centre; the commercial form is the racemate, the slightly more active isomer being the *S*-(−)-sulpiride.[16] An additional constraining factor results from the establishment of a hydrogen bond between the amidic N—H hydrogen and the methoxy oxygen.[17] A conformationally restricted remoxipride analogue in which the intramolecular hydrogen bond is replaced by a covalent bond (Fig. 14.10) is equipotent in D$_2$-receptor preparations.[18]

Cyclized dopamine: the ADTNs. The 2-amino-5,6-dihydroxy- and the 2-amino-6,7-dihydroxy-1,2,3,4-tetrahydronaphtalenes are cyclized analogues of dopamine, corresponding to the α- and the β-rotamer, respectively (Fig. 14.11). As the cyclization generates a chiral centre, four different ADTNs are possible, showing differential affinities for the dopamine receptors.[19] An extensive study of the aminotetralins and analogues containing additional rings (octahydrobenz[g]quinolines) and compounds resulting from ring enlargements was published by Seiler *et al.*[20]

GABA-ergic agonists. The transition from GABA to *trans*-4-aminocrotonic acid, followed by cyclization into isoguvacine, and finally into THIP (Fig. 14.12), achieves simultaneously the rigidification of the flexible GABA molecule and the production of THIP, a metabolically stable and still potent GABA agonist.[21]

Ring-closed analogue of nicotine. The following example illustrates a very intriguing result: with natural (S)-(−)-nicotine.[22] The ring-closure strategy yielded a compound which possesses the same pharmacological profile as nicotine but which totally fails to compete for [³H]-nicotine

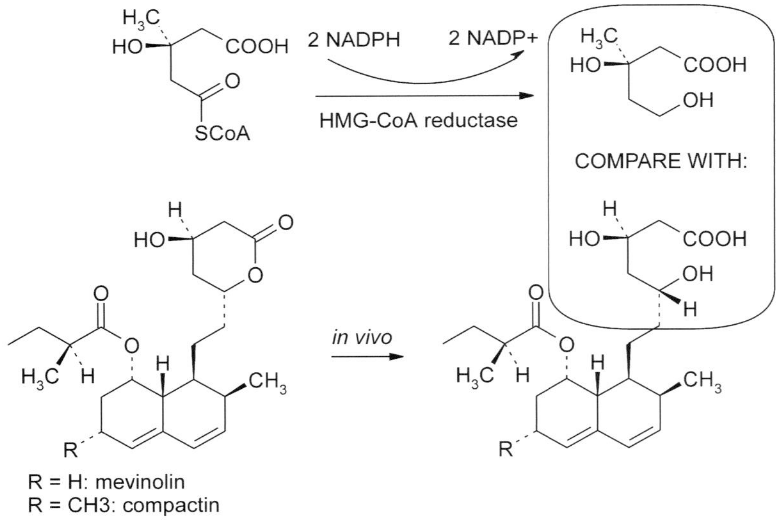

Fig. 14.8 The chemically stable lactones mevinolin and compactin represent the ring-closed forms of the *in vivo* active parent substituted 3,5-dihydroxyvaleric acid.

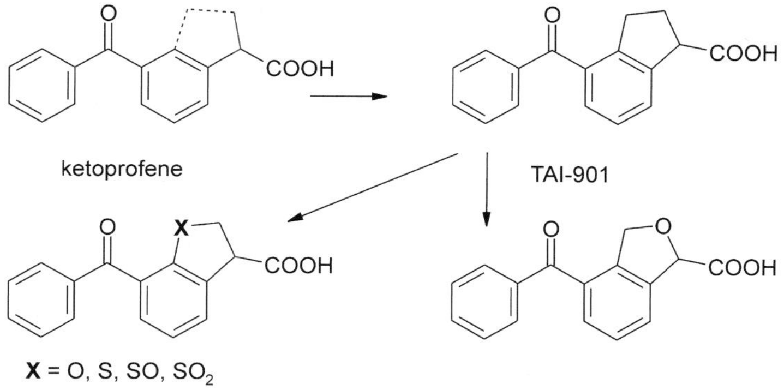

Fig. 14.9 Ring-closed me-toos of ketoprofene.

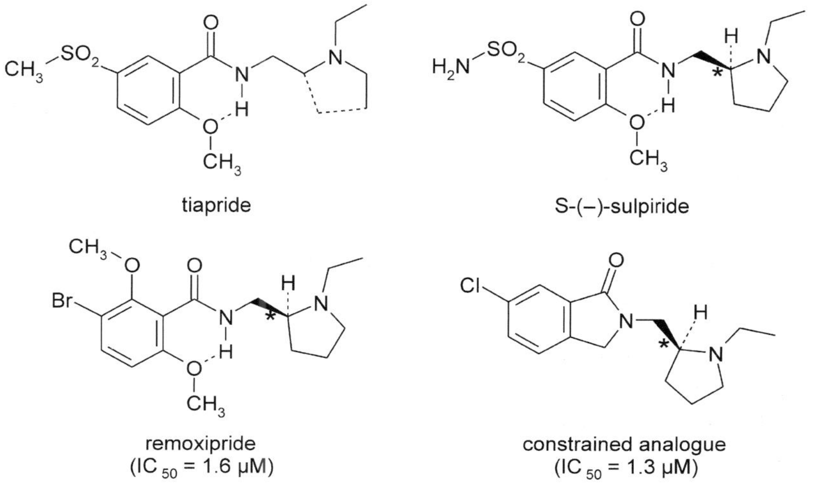

Fig. 14.10 The typical sulpiride side-chain results formally from a ring closure of the diethylaminoethyl side-chain of earlier prepared derivatives such as tiapride. Observe the intramolecular hydrogen bond that creates a pseudocycle[14] and which can be mimicked by a covalent and constraint analogue.[15]

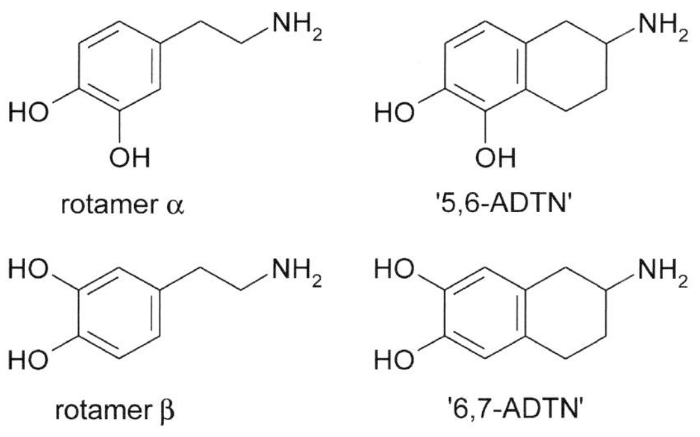

Fig. 14.11 Ring-closed analogues of the two rotamers of dopamine.

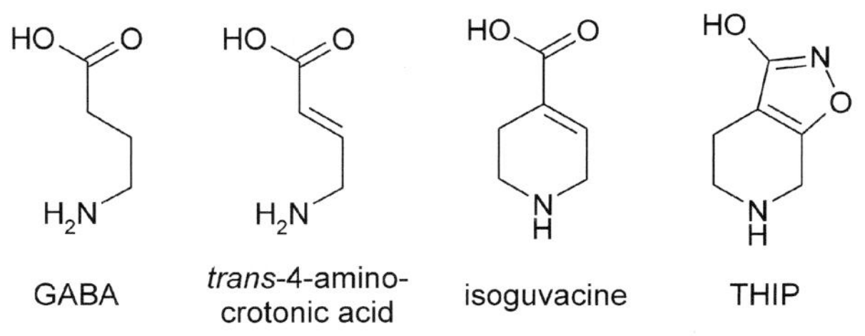

Fig. 14.12 GABA-ergic agonists.[16]

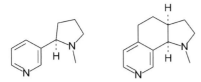

Fig. 14.13 Natural (S)-(−)-nicotine and its bridged analogue.[2]

binding (Fig. 14.13). The conformationally restricted (+)-*cis*-2,3,3a,4,5,9b-hexahydro-1-methyl-1*H*-pyrrolo-[3,2-*h*]isoquinoline has the 3aR,9bS configuration as determined by X-ray crystallography and corresponds thus to (S)-(−)-nicotine. *In vivo* this compound demonstrates a pharmacological profile similar to that of (S)-(−)-nicotine with an ED$_{50}$ of 7.13 μM kg^{-1} for the inhibition of spontaneous activity and 7.45 μM kg^{-1} for antinociception (tail flick test) compared with 4.44 and 4.81 μM kg^{-1} respectively for (S)-(−)-nicotine. However, the failure of mecamylamine (a well-established nicotinic antagonist) to antagonize the effects of the bridged analogue and the absence of competition for [^{3}H]-nicotine binding suggest that either it binds to an as-yet-unidentified nicotinic receptor or that it represents a novel class of non-nicotinic analgesics.

Cyclic analogues of β-blockers. Conventional β-blockers possess a number of pharmacological properties, e.g. β-blocking, quinidine-like, local anaesthetic and hypotensive. With the hope of achieving some specificity, Basil *et al.*[23] considered the possibility of synthesizing ring-closed analogues (Fig. 14.14). One of the compounds prepared, 3,4-dihydro-3-hydroxy-6-methyl-1,5-benzoxazocine was a potent β-blocker. This activity is unlikely to be due to hydrolysis to the open-chain derivative since the corresponding primary amine, formed by hydrolysis of

the benzoxazocine ring has less than 0.25 the activity of the latter. On the other hand, it is difficult to reconcile the benzoxazocine configuration with the structural requirements associated with the occupation of β-receptors.

Later studies by Evans *et al.* also envisaged the synthesis of cyclized analogues of the phenylpropanolamine type of β-blockers.[24,25] The authors hoped that by restricting the conformation (by cyclizing the carbon atom bearing the terminal amino group to the aromatic ring, see Fig. 14.14), β-blocking activity would be lost but antihypertensive activity might be retained. This turned out to be true in animal tests and in double-blind clinical studies, so the potassium channel activator cromakalim was developed.[25]

Cyclized diphenhydramine. Nefopam (Fig. 14.15) is the representative of a new class of centrally acting skeletal muscle relaxants, also possessing a benzoxazocine structure.[26] Formally, nefopam is a cyclized analogue of orphenadrine and diphenhydramine. In contrast to the parent molecules, nefopam has no antihistaminic activity, retaining only the muscle relaxant effects. Clinically, nefopam is used as a muscle relaxant, but also, and this was originally not anticipated, it is an antidepressant and an analgesic. These clinical indications can be explained by the interference of nefopam with the serotonergic transmission. More precisely, studies of the separated stereoisomers of nefopam explain its serotonin uptake properties and suggest that descending serotonergic pathways are involved in its antinociceptive activity.[27]

Ring variations around phenylbutazone. The anti-inflammatory drug phenylbutazone (**1**, Fig. 14.16) led to many me-too copies, such as the ring-opened analogue bumadizon (**2**) (Ca^{2+} salt = Eumotol$^{®28}$) or the ring-closed analogue apazone (**3**) (Prolixan$^®$).[29] The cinnoline derivative (**4**)[30] results again from a ring closure and served as a model for

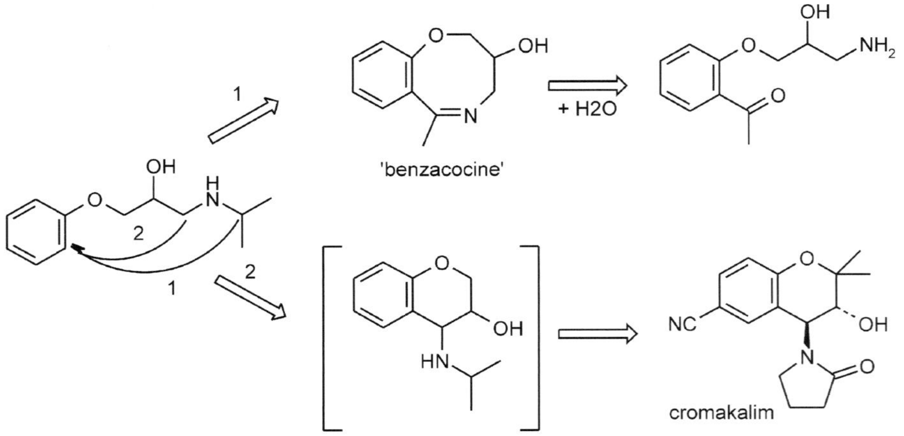

'benzacocine'

+ H$_2$O

cromakalim

Fig. 14.14 Cyclized analogues of β-blocking phenylpropanolamines.[21,23]

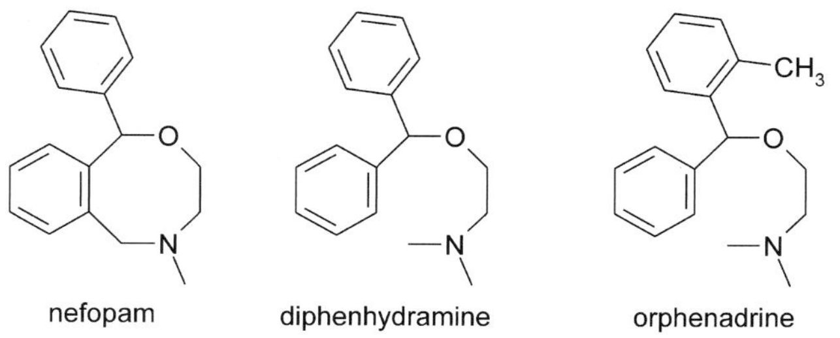

Fig. 14.15   Nefopam is a cyclized analogue of orphenadrine and diphenhydramine.[24]

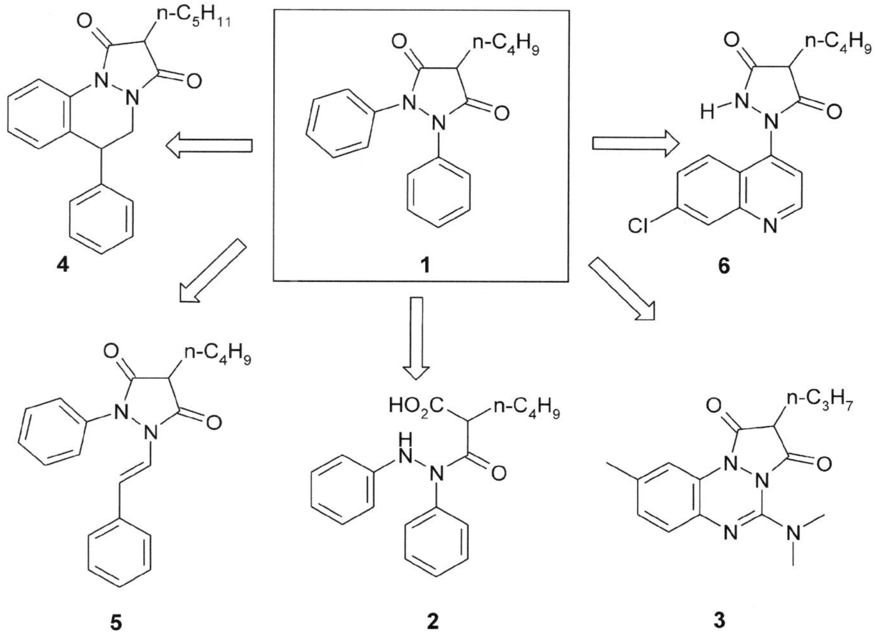

Fig. 14.16   Successive ring openings and closures in phenylbutazone-derived anti-inflammatory drugs.

the design of its open counterpart, the styrylbutazone (**5**).[31] The quinolinyl-3,5-dioxopyrazolidine (**6**)[32] represents another interesting ring variation with a 7-chloroquinoline moiety in the butazone portion.

C   Other analogies

Applied to ring systems, the following molecular modifications seem to be conducted mainly with the objective of

bypass patent protections and to allow the synthesis of me-too products.

Ring enlargement and ring contraction

Ring enlargement and ring contraction can be considered as homologous variations in the cyclic series and have already been mentioned in Chapter 12.

In Fig. 14.17 two additional examples taken from the barbituric and from the opiate series, respectively, are shown. In the case of the change of the barbiturics to

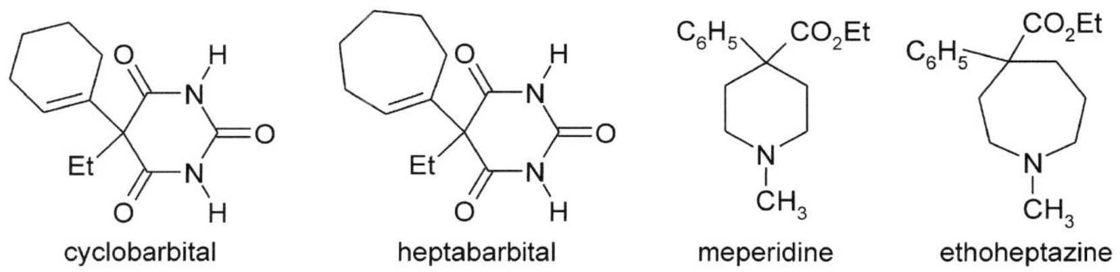

Fig. 14.17   Six-membered rings exchanged for seven-membered rings.

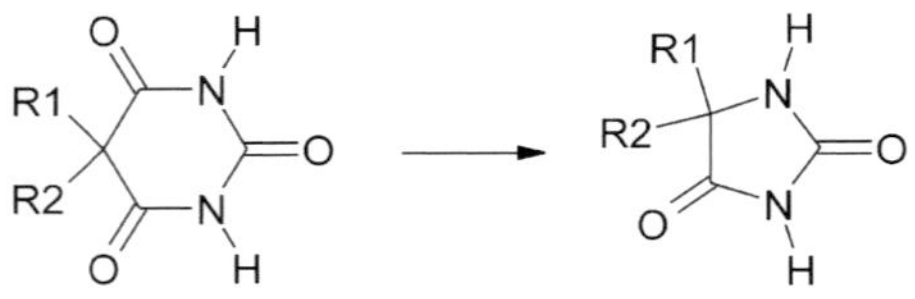

Fig. 14.18 Barbiturates (left) and hydantoins (right).

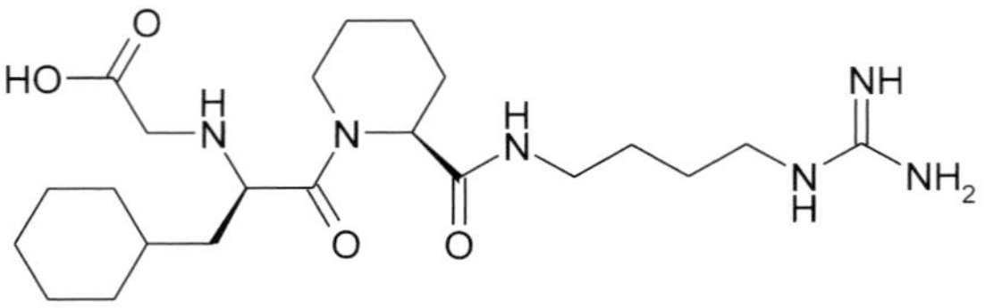

inogatran

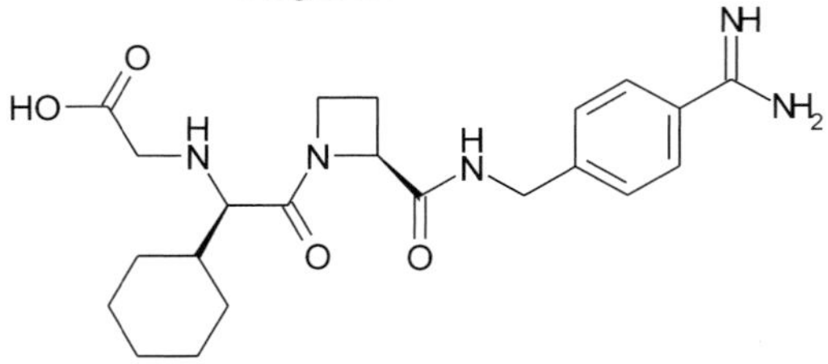

melagatran

Fig. 14.19 Structures of the thrombin inhibitors inogatran and melagatran.[3,4]

hydantoins the contraction is accompanied by the loss of a carbonyl group (Fig. 14.18). Nevertheless, potent antiepileptics are found in both series.

The classical motif of thrombin inhibitors is the D-Phe-Pro-Arg sequence mimicking thrombin's natural substrate, fibrinogen. In development candidates such as inogatran[33] and melagatran[34] (Fig. 14.19), the proline unit was replaced by its ring-expanded and its ring-contracted equivalent, respectively.

Several 2,3-benzodiazepin-4-ones such as GYKI 52 4966 and GYKI 53 655 (Fig. 14.20) are AMPA receptor antagonists and possess noteworthy anticonvulsant and neuroprotective properties. The corresponding ring-contracted analogues, 6,7-methylene-dioxydihydrophtalazines and 6,7-methylene-dioxyphtalazin-1(2*H*)-ones, possess a similar activity profile.[35,36]

Oxotremorine and its ring-opened analogue oxo-2 (Fig. 14.21) are partial muscarinic agonists producing large guanine nucleotide shifts in the heart (32 and 23, respectively), suggesting strong M_2 agonist-like effects.[37]

The corresponding piperidinic analogue oxo-pip[38] having a predicted antagonist [3H]QNB/[3H] CD ratio of 2.2, produced only a weak shift (5.0) in the concentration–response curve with the addition of the stable guanine nucleotide analogue.[37] Thus the change from a pyrrolidine to a piperidine ring is able to change a partial agonist into an antagonist.

The replacement of the core cyclopentane ring of the prostaglandin FP agonist cloprostenol with a cyclohexane ring (Fig. 14.22) yielded a clearly less active agent (EC_{50} = 319 nM instead of 1 nM).[39]

Reorganization of ring systems

The four molecular variations described below represent some more 'exotic' approaches to the design and manipulation of the original ring systems. They may provide useful

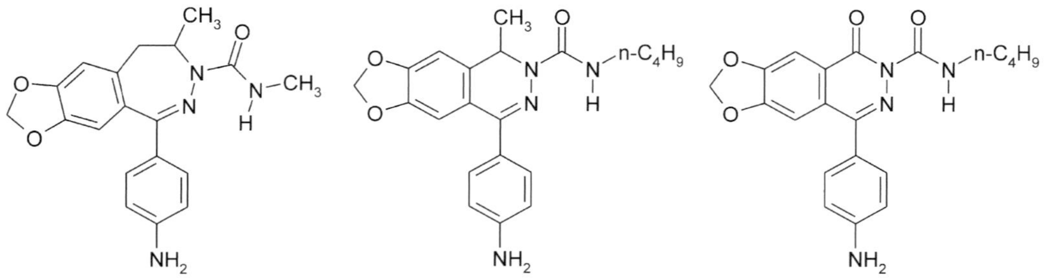

GYKI 53 655 SYM 2207 'Compound 21'

Fig. 14.20 The heptacyclic AMPA antagonist GYKI 53 655 and ring-contracted analogues.[5,6]

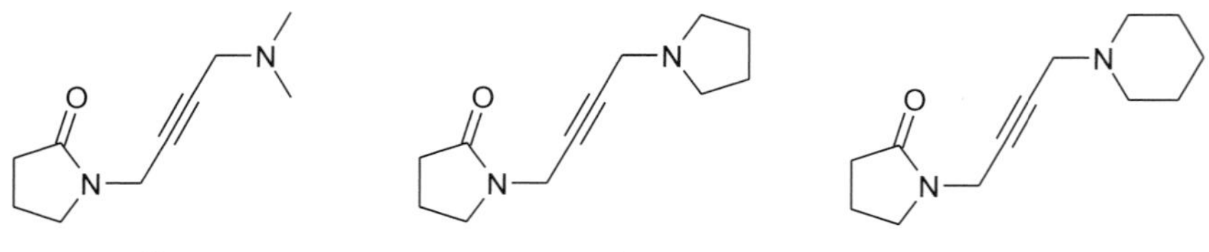

oxo-2 oxotremorine oxo-pip

Fig. 14.21 Oxotremorine yields ring-opened and ring-extended analogues.[31]

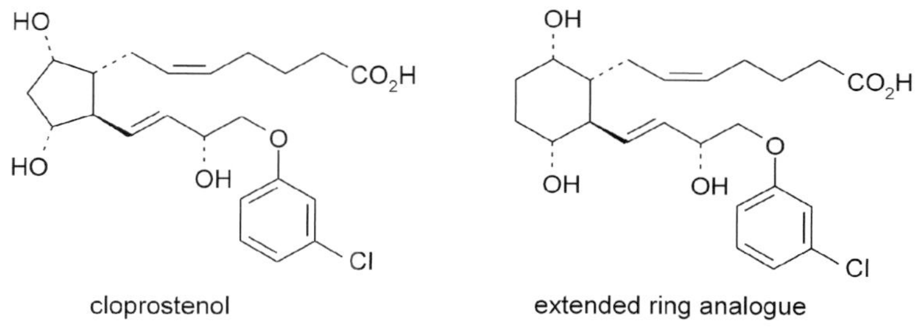

Fig. 14.22 Cloprostenol and ring-extended analogue.[7]

alternatives allowing escape from overcrowded avenues of research.

Transforming simple rings into spiro derivatives or into bi- or tricyclic systems. A first example is in the guanethidine analogues.[40] As the original guanethidine patents covered ring sizes varying from five-membered to ten-membered rings a possible way to get round them was the design of isolipophilic *spiro* systems (Dausse compounds a and b,[41,42] Fig. 14.23). Another possibility, originating from Takeda scientists, involves the use of an azetidine surrogate for the ethylene-diamine chain.[43] Finally, polycyclic systems can replace the octahydroazocine ring, as illustrated by the bicyclic compounds c and d from Dausse[44] or by the tricyclic compound from Lumière Laboratories.[45,46] Many other imaginative solutions have been proposed; they are well reviewed by Mull and Maxwell.[40] More recently, similar

variations were applied to the design of an impressive number of analogues of the anticonvulsant drug gabapentin[47,48] (Fig. 14.24; see also Chapter 15, Fig. 15.8.)

Splitting benzo compounds ('Benzo Cracking'). Dissociation of a fused ring system (Fig. 14.25), particularly by splitting a benzo compound, can sometimes improve its solubility but only slightly alter its pharmacokinetic profile and its long-term toxicity.

Restructuring ring systems. Among the above-mentioned molecular variations on ring systems, some can be used *simultaneously*. Thus the splitting of the benzimidazole heterocycle in the anthelmintic thiabendazole and the concomitant association of the two five-membered rings yields tetramizole (Fig. 14.26). One of the two enantiomers of tetramizole, the L-(−)-form, or levamizole, is also a potent anthelmintic.

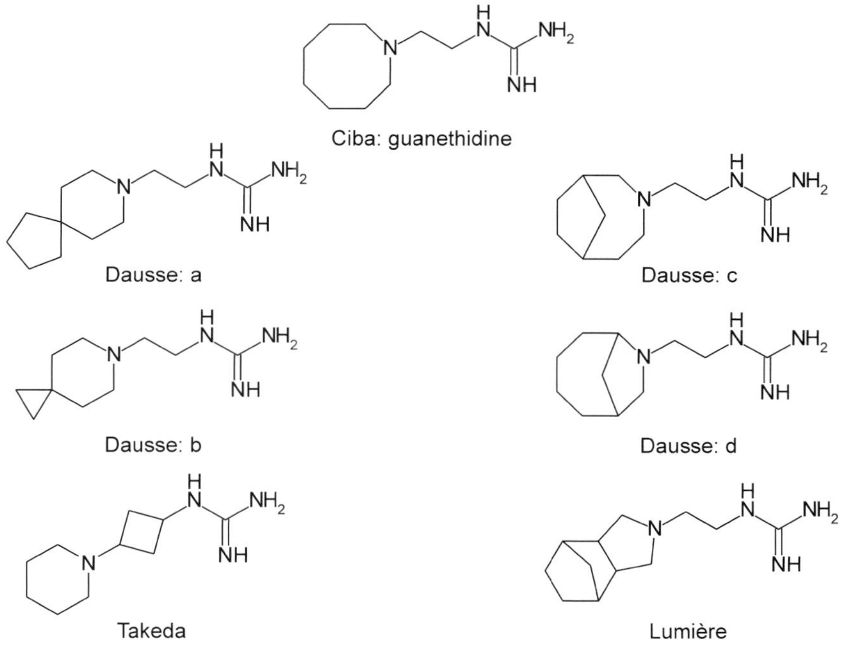

Fig. 14.23 Alternative possibilities in the design of guanethidine analogues.

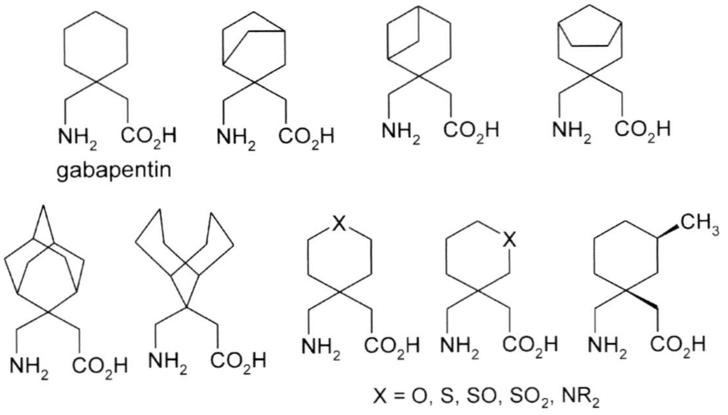

Fig. 14.24 Alternative possibilities in the design of gabapentin analogues.

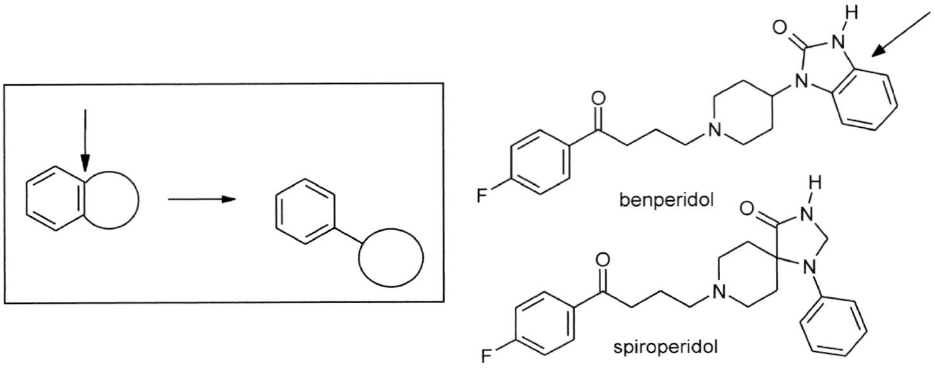

Fig. 14.25 Splitting of fused rings often yields drugs with similar activity, sometimes with improved solubility and/or less toxicity.

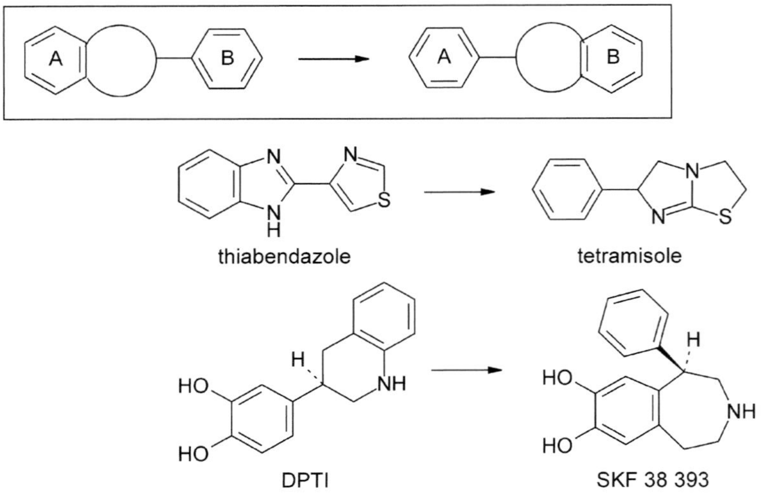

Fig. 14.26 Restructured ring systems.

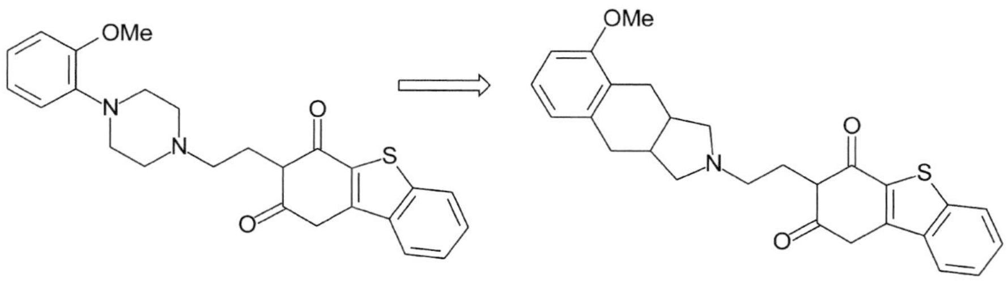

Fig. 14.27 Hexahydroisoindole as a surrogate of the very frequently used orthomethoxy-*N*-arylpiperazine.

The change from the D_1-selective dopaminergic agonist DPTI (3-(3,4-dihydroxyphenyl)-1,2,3,4-tetrahydroisoquinoline) to the equally D_1-selective compound SKF 38 393 is a combination of benzo cracking, a new benzo fusion and ring enlargement (Fig. 14.26). As a result, the compounds still resemble each other and are both recognized by the dopamine D_1-receptor.[20]

An interesting example of a restructured ring system was designed by Meyer *et al.*[49] for the design of 2-methoxybenz-(e)-hexahydroisoindole as a surrogate of the very frequently used orthomethoxy-*N*-arylpiperazine. It led both to an enhancement of affinity and improved selectivity for α_{1A} receptor antagonism (Fig. 14.27).

Ring dissociation. The natural compound khellin generated two families of cardioactive drugs: on one side the benzopyrones, illustrated by the 3-methyl-chromone[50] and chromonar (carbochromen); on the other side the benzofur-

ans, illustrated by amiodarone (Fig. 14.28). Both families possess antiarrhythmic and antianginal properties.

III DISJUNCTIVE APPROACHES

Starting from a polycyclic structure (which is often from natural origin), the chemist proceeds to progressive simplifications of the molecule ('molecular strip tease'). Sometimes very simple reasoning guides the medicinal chemist and the final compound has only a remote resemblance to the model compound. Such an exercise led to the transformation of the natural compound asperlicin to a totally synthetic simplified benzodiazepine derivative (see Fig. 14.32).

Cocaine-derived local anaesthetics. Fig. 14.29 illustrates how simplified synthetic copies of cocaine were designed. The change from cocaine to procaine retains the local anaesthetic effects without the narcotic properties.

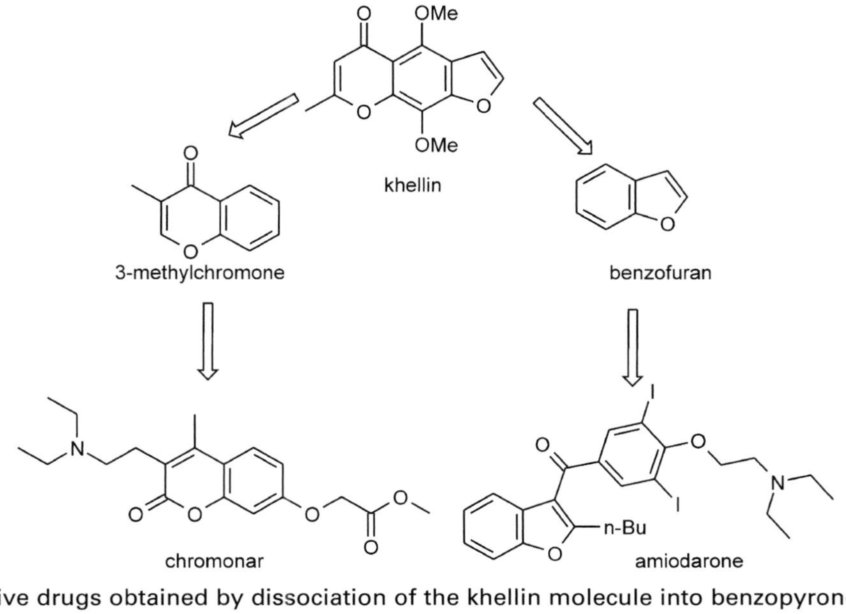

Fig. 14.28 Cardioactive drugs obtained by dissociation of the khellin molecule into benzopyrones and benzofurans.

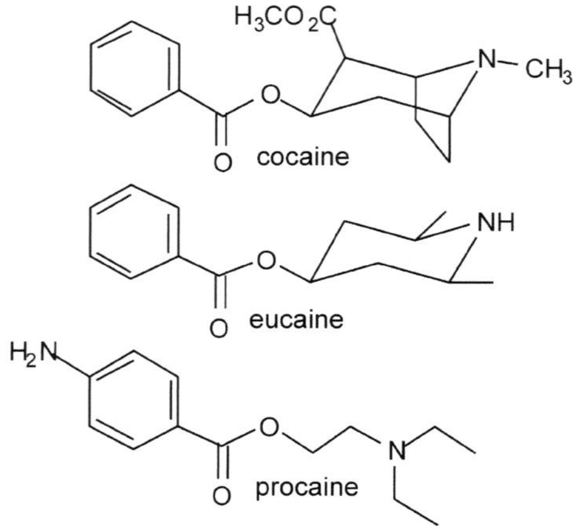

Fig. 14.29 Progressive simplification of the cocaine molecule.

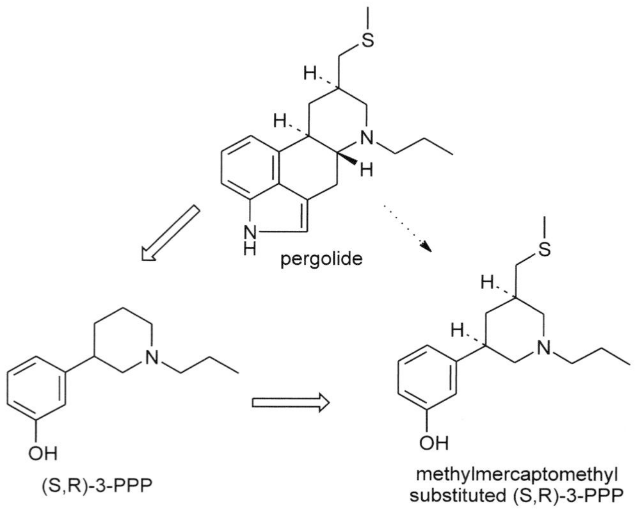

Morphinic analgesics. Probably more than a thousand more or less simplified analogues of the alkaloid morphine have been investigated.[51] Many of them were inactive, but it was soon recognized that the phenylpiperidine unit was crucial for the central analgesic properties (Fig. 14.30). In contrast to what was observed for cocaine, no clear discrimination between the analgesic and the narcotic properties could be achieved.

Dopamine autoreceptor agonists. The discovery of 3-(3-hydroxyphenyl)-*N*-*n*-propyl piperidine ((±)-3-PPP), a centrally acting dopamine receptor agonist with selectivity for dopaminergic autoreceptors,[52] offers a potential alternative to neuroleptics in the treatment of schizophrenia. The structure of (±)-3-PPP (Fig. 14.31) can be considered as resulting from a disjunctive approach applied to pergolide.[54,55] Surprisingly, an increase in the

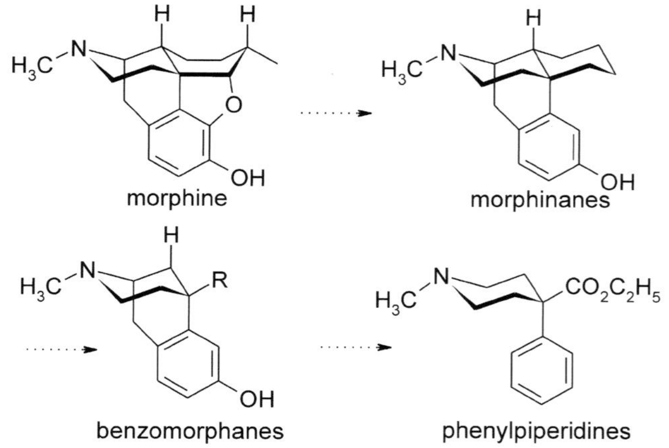

Fig. 14.30 Progressive simplification of the morphine molecule.

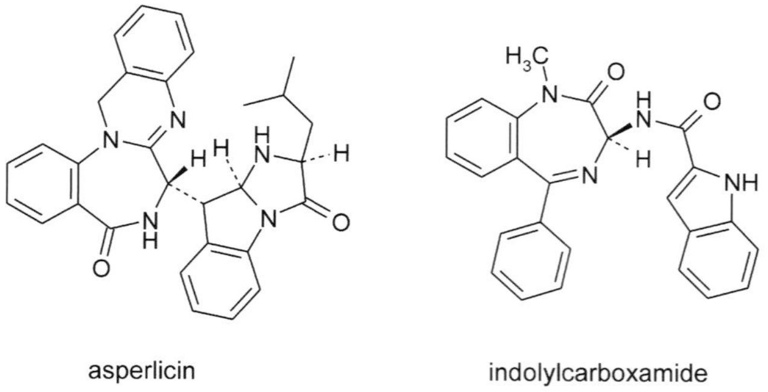

Fig. 14.31 3-PPP is a result of the disjunctive approach applied to pergolide.[52,53]

pergolide-like character of 3-PPP, through incorporation of a methylmercaptomethyl group, did not improve the potency.[53]

CCK antagonists. After the discovery of the potent CCK antagonistic activity of the natural compound asperlicin,[56] scientists from the Merck group first prepared some simple semisynthetic derivatives.[57] Then, recognizing in asperlicin the elements of a benzodiazepinone and a tetrahydroindole, they followed the hunch that these elements alone may confer some CCK antagonistic activity.[58] This reasoning proved to be valid (Fig. 14.32).

IV CONJUNCTIVE APPROACHES

As already mentioned at the beginning of this chapter, the purpose of the conjunctive method lies in the design

Fig. 14.32 Productive disjunction of the asperlicin molecule.[58]

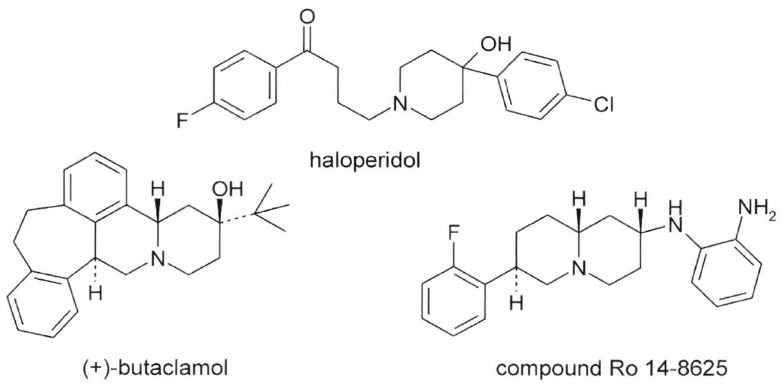

Fig. 14.33 The conjunctive method applied to the design of haloperidol-derived dopaminergic antagonists.

of compounds structurally more complex than the lead compound. In practice, this is generally achieved in creating and/or adding supplementary ring systems to constrain the molecule and to impose specific conformations.

Dopaminergic antagonists. Starting from the flexible haloperidol molecule, Humber and colleagues designed the rigid (+)-butaclamol that contains three clearly defined stereocentres (Fig. 14.33).[59] The same stereochemical requirements as in (+)-butaclamol are found in compound Ro 14–8625 prepared by Imhof *et al.*[60]

Glutamate NMDA and AMPA receptor antagonists. The progressive change of glutamic acid to D-AP5,[61] to the piperidine analogue CGS 19755[62]and finally to the

tetrahydroisoquinoline PD 134705,[63] led to NMDA receptor antagonists. Similarly, rigidification into the perhydroquinolines 7, 8, and 9[64] (Fig. 14.34) illustrates another application of the conjunctive approach that led to potent AMPA antagonists. In addition to the elements enhancing structural rigidity, these latter compounds contain three new chiral centres.

Norfloxacin analogues. Since the development in 1980 of norfloxacin[65] as a useful antibacterial agent, a large number of analogues has been synthesized. Among them, the conjunctive approach led to highly potent tetracyclic analogues[66] (Fig. 14.35). In a comparable way, a number of annelated analogues of the 5-HT₃ antagonist ondansetron were investigated (Fig. 14.36). Among them, cilansetron

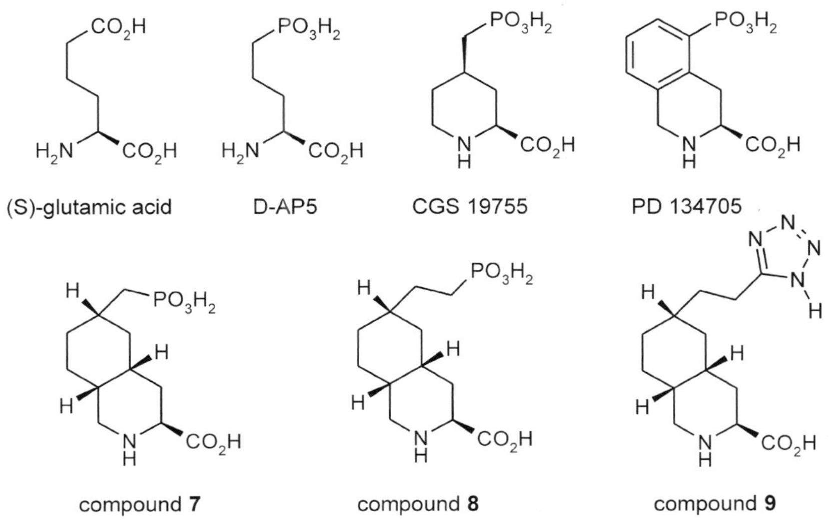

Fig. 14.34 The conjunctive method applied to the design of NMDA and AMPA receptor antagonists.

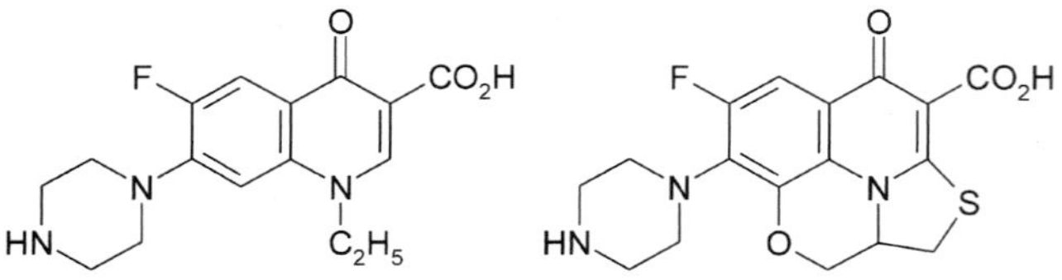

norfloxacin   tetracyclic analogue

Fig. 14.35 Norfloxacin and its tetracyclic analogue.[66]

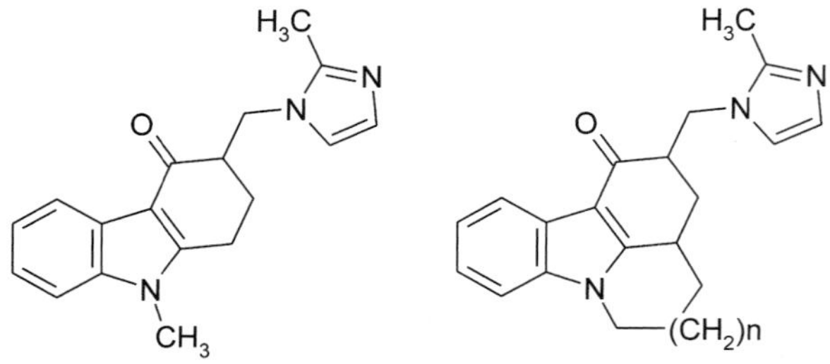

ondansetron   (n = 1): cilansetron

Fig. 14.36 Cilansetron, an annelated analogue of ondansetron.[67]

($n = 1$) was found to be about ten times more potent without loss in selectivity.[67]

Melatonin analogues. Tetracyclic analogues in which the melatonin indole ring is fused to a dihydroindole, a tetrahydroisoquinoline and a benzazepine system, respectively (**7c, 21c** and **25c**; Fig. 14.37), represent ligands for the melatonin MT2 receptor as potent as melatonin but selective with regard to the mt1 receptor. Interestingly, the passage from the six-membered (**21c**) to the seven-membered fused ring system (**25c**) produced a switch from agonist to antagonist activity.[68]

The well-known angiotensin-converting enzyme (ACE) inhibitor, captopril, has a relatively simple structure and therefore represents excellent starting material for conjunctive approaches (Fig. 14.38).

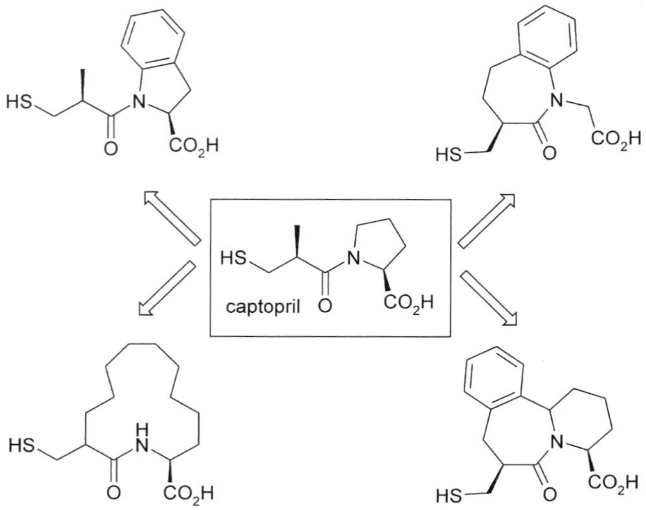

Fig. 14.38 Angiotensin-converting enzyme inhibitors derived from captopril.

Modelling studies based on a template structure constructed from the superposition of the energy-minimized benzo-fused ACE inhibitors shown in Fig. 14.38 suggested the synthesis of the 13-membered heterocyclic lactam analogue.[69]

V CONCLUSION

Molecular variations involving the study of homologous series or the application of the vinylogy concept, induce relatively minor changes of the pharmacological profile and rather result in optimizing the potency. Modifying ring systems — ring-chain transformation, ring contractions and expansions, reorganization of cyclic systems — represents a highly efficient approach in the exploration of the requirements governing drug–receptor interactions and sometimes provides even novel drug candidates.

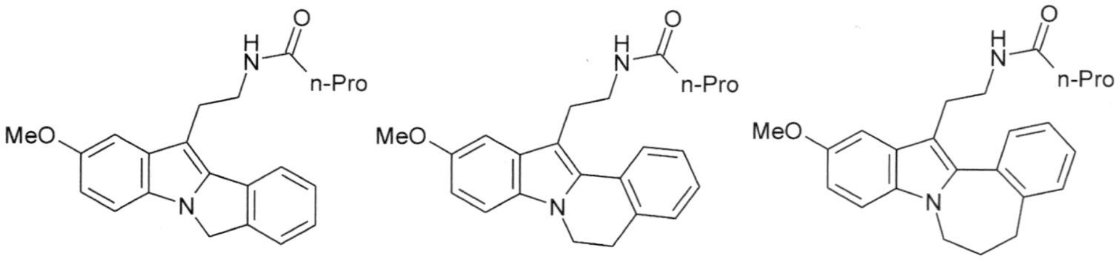

compound '7c'   compound '21c'   compound '25c'

Fig. 14.37 Ring-fused melatonin analogues.[68]

REFERENCES

1. Loschmidt, J. (1861) *Chemische Studien. I. A. Constitutions-Formeln der Organischen Chemie in geographischer Darstellung. B. Das Mariotte'sche Gesetz.* Carl Gerold's Sohn. [Reprinted in 1989 by Aldrich Chemical Company, Inc., Milwaukee, Aldrich Catalogue Number Z-18, 576-0, Wien].

2. Schueller, F.W. (1960) *Chemobiodynamics and Drug Design.* McGraw-Hill, New York.

3. Curd, F.H.S., Davey, D.G. and Rose, F.L. (1945) Studies on synthetic antimalarial drugs-X. Some biguanide derivatives as new types of antimalarial substances with both therapeutic and causal prophylactic activity. *Ann. Trop. Med.* **39**: 208–214.

4. Crowther, A.F. and Levi, A.A. (1953) Proguanyl—the isolation of a metabolite with high antimalarial activity. *Br. J. Pharmacol.* **8**: 93–101.

5. Hawking, F. and Perry, W.L.M. (1948) Activation of Paludrine. *Br. J. Pharmacol.* **3**: 320–331.

6. Weiner, I.M. (1990) Drugs affecting renal function and electrolyte metabolism. In Goodman-Gilman, A., Rall, T.W., Nies, A.S. and Taylor, P. (eds). *Goodman and Gilman's The Pharmacological Basis of Therapeutics*, 8th edn., pp. 708–731. Pergamon Press, New York.

7. Buzetta, B. and Hospital, M. (1978) Relations structure–activité. *Pharmacologie Moléculaire*, pp. 27–36. Masson & Cie, Paris.

8. Rouot, B., Leclerc, G., Wermuth, C.G., Miesch, F. and Schwartz, J. (1978) Synthèse et essais pharmacologiques d'arylguanidines, analogues ouverts de la clonidine. *Eur. J. Med. Chem.* **13**: 337–342.

9. Ashwood, V.A., Cassidy, F., Coldwell, M.C., Evans, J.M., Hamilton, T.C., Howlett, D.R., Smith, D.M. and Stemp, G. (1990) Synthesis and antihypertensive activity of 4-(substituted-carbonylamino)-2H-1-benzopyrans. *J. Med. Chem.* **33**: 2667–2672.

10. Bigham, E.C., Hodson, S.T., Mallory, W.R., Wilson, D., Duch, D.S., Smith, G.K. and Foreign, R. (1992) Synthesis and biological activity of open-chain analogues of 5,6,7,8-tetrahydrofolic acid-potential antitumor agents. *J. Med. Chem.* **35**: 1399–1410.

11. Mack, H., Pfeifffer, T., Hornberger, W., Böhm, H.J. and Höffken, H.W. (1995) Design, synthesis and biological activity of novel rigid amidinophenylalanine derivatives as inhibitors of thrombin. *J. Enzyme Inhib.* **9**: 73–86.

12. Kaminski, J.J., Wallmark, B., Briving, C. and Andersson, B.-M. (1991) Antiulcer agents. 5. Inhibition of gastric H+/k+-Atpase by substituted imidazo[1,2-a]pyridines and related analogues and its implication in modeling the high affinity potassium ion binding site of the gastric proton pump enzyme. *J. Med. Chem.* **34**: 533–541.

13. Nakamura, C.E. and Abeles, R.H. (1985) Mode of interaction of β-hydroxy-β-methylglutaryl coenzyme A reductase with strong binding inhibitors: compactin and related compounds. *Biochemistry* **24**: 1364–1376.

14. Kawai, K., Tamura, S., Morimoto, S., Ishii, H. and Kuzuna, S. (1982) Pharmacology of 4-benzoyl-1-indancarboxylic acid (TAI-901) and 4-(4-methylbenzoyl)-1-indancarboxylic acid (TAI-908). *Arzneim.-Forsch.* **32**: 113–117.

15. Boyle, E.A., Mangan, F.R., Markwell, R.E., Smith, S.A., Thomson, M.J., Ward, R.W. and Wyman, P.A. (1986) 7-Aroyl-2,3-dihydrobenzo[b]furan-3-carboxylic acids and 7-benzoyl-2,3-dihydrobenzo[b]thiophene-3-carboxylic acids as analgesic agents. *J. Med. Chem.* **29**: 894–898.

16. Jenner, P., Clow, A., Reavill, C., Theodoru, A. and Marsden, C.D. (1980) Stereoselective actions of substituted benzamide drugs on cerebral dopamine mechanisms. *J. Pharm. Pharmacol.* **32**: 39–44.

17. Waterbeemd van de, H. and Testa, B. (1983) Theoretical conformational studies of some dopamine antagonistic benzamide drugs: 3-pyrrolidyl- and 4-piperidyl derivatives. *J. Med. Chem.* **26**: 203–207.

18. Norman, M.H., Kelley, J.L. and Hollingsworth, E.B. (1993) Conformationally restricted analogues of remoxipride as potential antipsychotic agents. *J. Med. Chem.* **36**: 3417–3423.

19. Cannon, J.G., Lee, T., Goldman, H.D., Costall, B. and Naylor, R.J. (1977) Central dopamine agonist properties of some 2-aminotetralin derivatives after peripheral and intracerebral administration. *J. Med. Chem.* **20**: 1111–1116.

20. Seiler, M.P., Bölsterli, J.J., Floersheim, P., Hagenbach, A., Markstein, R., Pfaffli, P., Widmer, A. and Wüthrich, H. (1993) Recognition at dopamine receptor subtypes. In Testa, B., Kyburz, E., Fuhrer W. and Giger, G. (eds). *Perspectives in Medicinal Chemistry*, pp. 221–237. Verlag Helvetica Chemica Acta and VCH, Basel and Weinheim.

21. Krogsgaard-Larsen, P., Hjeds, H., Falch, E., Jørgensen, F.S. and Nielsen, L. (1988) Recent advances in GABA agonists, antagonists and uptake inhibitors: structure–activity relationships and therapeutic potential. In Testa, B. (ed.). *Advances in Drug Research*, pp. 381–456. Academic Press, London.

22. Glassco, W., Suchocki, J., George, C., Martin, B.R. and May, E.L. (1993) Synthesis, optical resolution, absolute configuration, and preliminary pharmacology of (+)- and (−)-cis-2,3,3a,4,5,9b-hexahydro-1-methyl-1H-pyrrolo-[3,2-h]isoquinoline, a structural analog of nicotine. *J. Med. Chem.* **36**: 3381–3385.

23. Basil, B., Coffee, E.C.J., Gell, D.L., Maxwell, D.R., Sheffield, D.J. and Woolridge, K.R.H. (1970) A new class of sympathic β-receptor blocking agents. 3,4-dihydro-3-hydroxy-1,5-benzoxazocines. *J. Med. Chem.* **13**: 403–406.

24. Evans, J.M., Fake, C.S., Hamilton, T.C., Poyser, R.H. and Watts, E.A. (1983) Synthesis and antihypertensive activity of substituted trans-4-amino-3,4-dihydro-2,2-dimethyl-2H-1-benzopyran-3-ols. *J. Med. Chem.* **26**: 1582–1589.

25. Stemp, G. and Evans, J.M. (1993) Discovery and development of cromokalim and related potassium channel activators. In Ganellin, C.R. and Rakerts, S.M. (eds). *Medicinal Chemistry*, pp. 141–162. Academic Press, London.

26. Heel, R.C., Brogden, R.N., Pakes, G.E., Speight, T.M. and Avery, G.S. (1980) Nefopam: a review of its pharmacological properties and therapeutic efficacy. *Drugs* **19**: 249–267.

27. Glaser, R. and Donnel, D. (1989) Stereoisomer differentiation for the analgesic drug nefopam hydrochloride using modeling studies of serotonin uptake area. *J. Pharm. Sci.* **78**: 87–90.

28. Pfister, R., Sallmann, A. and Hammerschmidt, W. (1969) Substituted malonic acid hydrazides. US patent 3 455 999 (July 15, 1969 to Geigy Chemical Corporation).

29. Mixich, G. (1968) Zum chemischen Verhalten des Antiphlogisticums 'Azapropazon' (Mi 85) = 3-dimethylamino-7-methyl-1,2-(n-propylmalonyl)-1,2-dihydro-1,2,4-benzotriazin. *Helv. Chim. Acta* **51**: 532–538.

30. Jahn, U. and Wagner-Jauregg, T. (1986) Vergleich von Zwei neuen Klassen Antiphlogisticher Substanzen im Collier-Test. *Arzneim.-Forsch.* **18**: 120–121.

31. Yamamoto, H. and Kaneko, S.-I. (1970) Synthesis of 1-phenyl-2-styryl-3,5-dioxopyrazolidines as anti-inflammatory agents. *J. Med. Chem.* **13**: 292–295.

32. Wermuth, C.G. and Choay, J (1973) Nouveaux dérivés de la pyrazolidine, leur procédé de fabrication et médicaments contenant ces nouveaux dérivés. French patent 2 244 513 (July 12, 1973 to Choay S.A.).

33. Teger-Nilsson, A.C., Bylund, R., Gustafsson, D., Gysander, E. and Eriksson, U. (1997) *In vitro* effects of inogatran, a selective low molecular weight thrombine inhibitor. *Thromb. Res.* **85**: 133.

34. Gustafsson, D., Antonsson, T., Bylund, R., Eriksson, U., Gyander, E., *et al.* (1998) Effects of melagatran, a low-molecular-weight thrombin inhibitor, on thrombin and fibrinolytic enzymes. *Thromb. Haemostasis* **79**: 110.

35. Pelletier, J.C., Hesson, D., Jones, K.A. and Costa, A.M. (1996) Substituted 1,2-dihydrophtalazines: potent, selective, and noncompetitive inhibitors of the AMPA receptor. *J. Med. Chem.* **39**: 343–346.

36. Grasso, S., De Sarro, G., De Sarro, A., Micale, N., Zappala, M., *et al.* (2000) Synthesis and anticonvulsant activity of novel and potent 6,7-methylenedioxyphtalazin-1 (2*H*)-ones. *J. Med. Chem.* **43**: 2851–2859.

37. Trybulski, E.J., Zhang, J., Kramss, R.H. and Mangano, R.M. (1993) The synthesis and biochemical pharmacology of enantiomerically pure methylated oxotremorine derivatives. *J. Med. Chem.* **36**: 3533–3541.

38. Ringdahl, B. and Jenden, D.J. (1983) Pharmacological properties of oxotremorine and its analogues. *Life Sci.* **32**: 2401–2413.

39. Klimko, P.G., Davis, T.L., Griffin, B.W. and Sharif, N.A. (2000) Synthesis and biological activity of a novel 11a-homo (cyclohexyl) prostaglandin. *J. Med. Chem.* **43**: 3400–3407.

40. Mull, R.P. and Maxwell, R.A. (1967) Guanethidine and related adrenergic neuronal blocking agents. In Schlittler, E. (ed.). *Antihypertensive Agents*, pp. 115–149. Academic Press, New York.

41. Giudicelli, R., Najer, H. and Lefèvre, F. (1965) Comparaison des durées d'action du N-β-guanidino-éthyl-aza-6 spiro[2,5] octane (LD 3598) et de la guanéthidine. *Compt. Rend. Acad. Sci.* **260**: 726–729.

42. Najer, H., Giudicelli, R. and Sette, J. (1964) Guanidines douées d'action antihypertensive, 4e mémoire: N-β-guanidinoéthyl azaspiro alcanes (1ère partie). *Bull. Soc. Chim. France* 2572–2581.

43. Toda, N., Usui, H. and Shimamoto, K. (1972) Modification by AZ-55, guanethidine and bretylium of responses of atria and aortic strips to transmural stimulation. *Jpn. J. Pharmacol.* **22**: 125–135.

44. Najer, H., Giudicelli, R. and Sette, J. (1962) Guanidines douées d'activité antihypertensive, 3ᵉ mémoire: N-β-guanidinoéthyl azabicyclo alcanes. *Bull. Soc. Chim. France* 1593–1597.

45. Anonymous (1964) New basic derivatives of 6,9-endomethylene-3-azabicyclo-[4.3.0]-nonane. British patent 972 088 (February 19, 1962 to Laboratoire Lumière, S.A.).

46. Anonymous (1964) New basic derivatives of 6,9-endoxo-3-azabicyclo-[4.3.0]-nonane. British patent 973 533 (February 21, 1962 to Laboratoire Lumière, S.A.).

47. Horwell, D. C., Bryans, J. S., Kneen, C. O., Morrell, A. I. and Ratcliffe, G. S. (1997) Preparation of novel bridged cyclic amino acids as pharmaceutical agents. WO 9733859 (Warner-Lambert Company, USA).

48. Bryans, J.S. and Morrell, A.I. (1999) Novel stereoselective processes for the preparation of gabapentin analogues. WO 9914184 (Warner-Lambert Company, USA).

49. Meyer, M.D., Altenbach, R.J., Basha, F.Z., Caroll, W.A., Condon, S., *et al.* (2000) Structure–activity studies for a novel series of tricyclic substituted hexahydrobenz[e]isoindole α1$_A$ adrenoceptor antagonists as potential agents for the symptomatic treatment of benign prostatic hyperplasia (BPH). *J. Med. Chem.* **43**: 1586–1603.

50. Jongebreur, G. (1952) Relation between the chemical constitution and the pharmacological action, especially on the coronary vessels of the heart, of some synthesized pyrones and khellin. *Arch. Int. Pharmacodynamie* **90**: 384–411.

51. Janssen, P.A.J. and Van der Eycken, C.A.M. (1968) The chemical anatomy of potent morphine-like analgesics. In Burger, A. (ed.). *Drugs Affecting the Central Nervous System*, pp. 25–60. Marcel Dekker, Inc., New York.

52. Hjorth, S., Carlsson, A., Wikström, H., Lindberg, P., Sanchez, D., Hacksell, U., Arvidsson, L.-E., Svensson, U. and Nilsson, J.L.G. (1981) 3-PPP, A new centrally acting DA-receptor agonist with selectivity for autoreceptors. *Life Sci.* **28**: 1225–1238.

53. Kelly, T.R., Howard, H.R., Koe, K. and Sarges, R. (1985) Synthesis and dopamine autoreceptor activity of a 5-(methylmercapto)methyl-substituted derivative of (±)-3-PPP (3-(3-hydroxyphenyl)-1-*n*-propylpiperidine). *J. Med. Chem.* **28**: 1368–1371.

54. Fuller, R.W., Clemens, J.A., Kornfeld, E.C., Snoddy, H.D., Smalstig, E.B. and Bach, N.J. (1979) Effects of (8β)-8-[(methylthio)methyl]-6-propylergoline on dopaminergic function and brain dopamine turnover in rats. *Life Sci.* **24**: 375–382.

55. Bach, N.J., Kornfeld, E.C., Jones, N.D., Chaney, M.O., Dorman, D.E., Paschal, J.W., Clemens, J.A. and Smalstig, E.B. (1980) Bicyclic and tricyclic ergoline partial structures. Rigid 3-(2-aminoethyl)pyrazoles as dopamine agonists. *J. Med. Chem.* **23**: 481–491.

56. Chang, R.S.L. and Lotti, V.Y. (1986) Biochemical and pharmacological characterization of an extremely potent and selective non-peptide CCK antagonist. *Proc. Natl Acad. Sci. USA* **83**: 4923–4926.

57. Bock, M.G., DiPardo, R.M., Rittle, K.E., Evans, B.E., Freidinger, R.M., Veker, D.F., Chang, R.S.L., Chen, T.-B., Keegan, M.E. and Lotti, V.J. (1986) Cholecystokinin antagonists. Synthesis of asperlicin analogues with improved potency and water solubility. *J. Med. Chem.* **29**: 1941–1945.

58. Evans, B.E., Rittle, K.E., Bock, M.G., DiPardo, R.M. and Freidinger, R.M. (1988) Methods for drug discovery: development of potent, selective, orally effective cholecystokin antagonists. *J. Med. Chem.* **31**: 2235–2246.

59. Bruderlein, F.T., Humber, L.G. and Voith, K. (1975) Neuroleptic agents of the benzocycloheptapyridoisoquinoline series. 1. Syntheses and stereochemical and structural requirements for activity of butaclamol and related compounds. *J. Med. Chem.* **18**: 185–191.

60. Imhof, R., Kyburz, E. and Daly, J.J. (1984) Design, synthesis, and X-ray data of novel potential antipsychotic agents. Substituted 7-phenylquinolizidines: stereospecific, neuroleptic, and antinociceptive properties. *J. Med. Chem.* **27**: 165–175.

61. Evans, R.H., Francis, A.A., Jones, A.W., Smith, D.A.S. and Watkins, J.C. (1982) The effects of a series of ω-phosphonic α-carboxylic amino acids on electrically evoked and excitant amino acid-induced responses in isolated spinal cord preparations. *Br. J. Pharmacol.* **75**: 65–75.

62. Hutchinson, A.J., Williams, M., Angst, C., de Jesus, R., Blanchard, L., *et al.* (1989) 4-(Phosphonoalkyl)- and 4-(phosphonoalkenyl)-2-piperidinecarboxylic acids: synthesis, activity at N-methyl-D-aspartic acid receptors, and anticonvulsant activity. *J. Med. Chem.* **32**: 2171–2178.

63. Humblet, C., Johnson, G., Malone, T. and Ortwine, D. F (1990) Design, synthesis and molecular modeling of phosphonoalkyl-substituted tetrahydroisoquinolines as competitive NMDA antagonists. In *Proceedings of the XIth International Symposium on Medicinal Chemistry*, Jerusalem, Israel.

64. Ornstein, P.L., Arnold, M.B., Augenstein, N.K., Lodge, D., Leander, J.D. and Schoepp, D.D. (1993) (3SR,4aRS,6RS,8aRS)-6-[2-(1H-Tetrazol-5-yl)ethyl]decahydroisoquinoline-3-carboxylic acid: a structurally novel, systematically active, competitive AMPA receptor antagonist. *J. Med. Chem.* **36**: 2046–2048.

65. Koga, H., Itoh, A., Murayama, S., Suzue, S. and Irikura, T. (1980) Structure–activity relationships of antibacterial 6,7- and 7,8-disubstituted 1-alkyl-1,4-dihydro-4-oxoquinoline-3-carboxylic acids. *J. Med. Chem.* **23**: 1358–1363.

66. Jinbo, Y., Taguchi, M., Inoue, Y., Kondo, H., Miyasaka, T., Tsujishita, H., Sakamoto, F. and Tsukamoto, G. (1993) Synthesis and antibacterial activity of a new series of tetracyclic pyridone carboxylic acids, 2. *J. Med. Chem.* **36**: 3148–3153.

67. Wijngaarde van, I., Hamminga, D., Hes, R.v., Standaar, P.J., Tipker, J., Tulp, M.T.M., Mol, F., Olivier, B. and Jongede, A. (1993) Development of high-affinity 5-HT$_3$ receptor antagonists. Structure–affinity relationships of novel 1,7-annelated indole derivatives. 1. *J. Med. Chem.* **36**: 3693–3699.

68. Faust, R., Garrat, P.J., Yeh, L.-K., Tsotinis, A., Panoussopoulou, M., *et al.* (2000) Mapping the melatonin receptor. 6. Melatonin agonists and antagonists derived from 6H-isoindolo[2,1-a]indoles, 5,6-dihydroindolo[2,1-a]isoquinolines, and 6,7-dihydro-5*H*-benzo[c]azepino[2,1-a]indoles. *J. Med. Chem.* **43**: 1050–1061.

69. Stanton, J.L., Sperbeck, D.M., Trapani, A.J., Cote, D., Sakane, Y., Berry, C.J. and Ghai, R.D. (1993) Heterocyclic lactam derivatives as dual angiotensin converting enzyme and neutral endopeptidase 24.11 inhibitors. *J. Med. Chem.* **36**: 3829–3833.

15

CONFORMATIONAL RESTRICTION AND/OR STERIC HINDRANCE IN MEDICINAL CHEMISTRY

André Mann

If a change occurs in one of the conditions of a system in equilibrium, the system will adjust itself so as to nullify,
as far as possible, the effect of that change. **Henri-Louis Le Chatelier (1850–1936)**

I   INTRODUCTION

Once a lead compound has been identified for a targeted biological receptor, its optimization towards potency and/or selectivity is usually the next step. Several strategies are at hand to achieve this goal. The choice will mainly depend on the chemical structure of the identified lead. This chapter will focus on a popular tactic in medicinal chemistry, namely the introduction of conformational restriction and/or steric hindrance into a given ligand. In the present topic mainly low molecular weight ligands will be considered.

Initially, some theoretical points will be addressed in order to establish the approach, and then several actual examples will be presented where the introduction of rigidity has produced positive or negative effects in structure-based drug design. This approach can tentatively be conceptualized as follows: it consists in the optimization of the *free energy* gained during the association of a ligand with a receptor by the relative spatial disposition of functional groups. This can be achieved in introducing *bulkiness, unsaturation* or *cyclization*. The expected benefits will be a rise in receptor selectivity, an increase in potency, progress in the pharmacophore identification, or increased metabolic stability. Sometimes a chemical *bonus* can be expected in producing original compounds with innovative chemistry. However conformational restriction also entails some risks, insofar that any structural change in an active compound may alter more than one property with unknown consequences. In practice this approach is valuable in cases where the lead compounds are exhibiting flexibility and where low energy conformations are not representative of a good fit at receptor level.

A   Theoretical points

In biology noncovalent interactions are of primary importance and many of the issues in medicinal chemistry hinge on understanding the bimolecular noncovalent interactions between a biomacromolecule (receptor) and a small ligand (drug). Under equilibrating conditions the association constant (K) is experimentally accessible by direct measure of the energy of the interaction, using the Gibbs-van't Hoff relation (equation 1):

$$\text{Log } K = -\Delta G/RT \quad with: \quad \Delta G = \Delta H - T\Delta S \qquad (1)$$

The values of two state functions H (enthalpy, a measure of the heat content of a substance) or S (entropy, a measure of the unusable energy, or disorder in a system) are essential to predict if a reaction or an equilibrium between two partners will take place.[1] As a general consideration if ΔG is negative, it means that a chemical process (a reaction, for the forming of covalent bonds, or an equilibrium, for the forming of an association) are favoured. If, for the formation of a covalent bond, the contribution to ΔG is usually dominated by ΔH (the bond is formed in irreversible processes), in the case of a noncovalent interaction or equilibrium the parts taken by ΔH and $T\Delta S$ in the energy balance are often comparable.

The preceding considerations are the result of the respective bond strengths from 25 to 200 kcal mol^{-1} for the covalent bond (the enthalpy term is largely prevailing: $\Delta H \gg T\Delta S$) and, from 1 to 10 kcal mol^{-1} for the irreversible binding (the enthalpy and entropy are of similar importance: $\Delta H \approx T\Delta S$).[2-9]

The following example, reporting the binding of a series of agonists and antagonists of the β-adrenergic receptor (Table 15.1) illustrates this observation in comparing the thermodynamic parameters of agonists and antagonists binding.[1] β-Adrenergic agonists and/or partial agonists are found to bind with large negative enthalpies ($\Delta H°$ from -18.8 to -4.1 kcal mol^{-1}), indicating strong electrostatic interactions between the bound conformation on the receptor, associated with a large loss of motional entropy (a tight complex is formed: $\Delta S°$ from -35.3 to $+6.6$ cal mol^{-1} degree^{-1}). Antagonists on the other hand do not fulfil the requirements of good complementarity as

shown by the criterion of small enthalpies of binding ($\Delta H°$ from -5.1 to $+3.9$ kcal mol^{-1}). The loss of motional entropy is therefore less than that of the agonists (a loose complex is formed $\Delta S°$ from $+13.0$ to $+40.5$ cal mol^{-1} degree^{-1}). However, the agonists and antagonists have similar affinity constants, supporting the enthalpy/entropy compensation.

B Constrained analogues

Let us consider the interaction of a biological receptor with a small ligand. Before the interaction, the two entities have translational and rotational flexibility contributive to their respective entropy. Once the association occurs, degrees of motion, as well as internal rotations about single bonds are lost. The consequence is an entropy cost of about -14 kcal mol^{-1} at room temperature for a small ligand (MW less than 1000 daltons). A general statement for the binding of a ligand to a receptor is that an entropy cost is associated with any bimolecular interaction. This is a consequence of the loss of degrees of motion when two molecules are rigidly constrained within a complex. The torsional entropy (the free rotation of a bond) is related to the rotatable bonds present in the ligand, and the freezing of one of them has a cost of 0.7 kcal mol^{-1} at room temperature.[5,8,9] Considering equation (1), the entropy penalty will render ΔS negative, thus reducing the binding energy. As a consequence for building an association constant of a reasonable value ($\Delta G < 0$), the free energy costs involved in bringing about conformational order, must be offset by favourable intermolecular interactions such as hydrogen bonds, van der Waals packing, hydrophobic and Coulombian interactions. One consequence of this qualitative thermodynamic analysis is that a flexible ligand has to compensate the entropic parameter by means of enthalpic parameters. Therefore, binding optimization can be accomplished either by rendering ΔH more negative, or making ΔS more positive, or an appropriate combination of both. Theoretically, the highest value for a given binding constant is accessible to a flexible ligand if the receptor recognizes the ligand in its low energy conformation and with an optimal orientation of the functional groups. This means that all information gained towards the knowledge of the active conformation of the ligand will serve the design of a better lead. There are some illustrative examples obtained by the incorporation of constrained amino acids into bioactive compounds.[10] However, as yet, the access to thermodynamic parameters of ligand/receptor interactions are still in their infancy, structure-based drug design rests mostly on semi-empirical strategies using structure–activity relationships for optimization steps. The restriction of flexibility is one of them. Search for rigid analogues, introduction of conformational constraints, creation of steric hindrance, or

Table 15.1 Thermodynamic parameters of ligand binding to the β-adrenergic receptor of turkey erythrocytes and the equilibrium constants K_D at 37°C (adapted from reference 1)

	$\Delta G°$ (kcal mol^{-1})	$\Delta H°$ (kcal mol^{-1})	$\Delta S°$ (cal mol^{-1} deg^{-1})	K_D mol at 37°C
Agonists				
(−)Isoproterenol	−9.3	−13.3	−12.9	$10^{-6.6}$
(−)Norephedrine	−7.9	−18.8	−35.3	$10^{-5.6}$
Partial agonists				
Soteronol	−8.2	−7.8	+1.2	$10^{-5.8}$
Fenoterol	−7.7	−6.0	+5.6	$10^{-5.4}$
Terbutaline	−6.1	−4.1	+6.6	$10^{-4.4}$
Antagonists				
(−)Propranolol	−12.5	−3.8	+27.9	$10^{-8.8}$
IPS-339	−12.3	+0.2	+40.5	$10^{-8.6}$
Pindolol	−11.8	−5.0	+21.8	$10^{-8.4}$
Atenolol	−7.4	−3.4	+13.0	$10^{-5.3}$
Practolol	−7.4	+3.9	+36.7	$10^{-5.2}$
Sotalol	−8.2	−2.1	+19.5	$10^{-5.8}$

preparation of locked conformations are the terms used in medicinal chemistry to account for this approach. Finally, the constraints introduced by chemical synthesis are expected to be compensated by the increase of the binding affinity.

C Conformational analysis

Many possible calculation methods to find accessible conformation for a flexible ligand are described. However, the available methods calculate the energy for *in vacuo* isolated ligands, and thus do not account for the real binding conditions.[11]

It is obvious that during the binding process, flexible ligands are deformed when binding to receptors. This deformation is a general phenomenon and the reason for it, is presumed to lie in the ligand's search for hydrogen bonds with the protein, in order to replace the solute-solvent hydrogen bonds which are lost as the molecule enters the binding site. A study has been performed on 33 ligands whose single X-ray structures, as well as their conformations obtained by co-crystallization with their receptor have been recorded.[12] From this study it appears that the degree of conformational change depends somewhat on the number of rotatable bonds in the ligand. For ligands with five or more rotatable bounds, the crystal structure seldom represents the same shape as that of the protein-bound conformation. Therefore the solid state structure of flexible ligand remains only as an existing conformation, but with limited use. Of course, in the more favourable cases when a crystal structure of the complex−ligand receptor can be obtained, the design of improved ligands is facilitated. In this case the main interactions can be quantified and then optimized. However, for the present, and for membrane located receptors (RCGPs), the available crystal complexes are still limited.

D Steric effects

Steric effects may arise in a number of ways.[13] Primary steric effects result from repulsions between valence electrons or non-bonded atoms. Such repulsions can only result from the energy increase of a group of atoms. The overall steric effect of a chemical reaction may be either favourable or unfavourable. For example, if steric effects in the reactant are larger than in the product (or transition state), then the reaction is favoured (steric augmentation); if the reverse is the case, the reaction is disfavoured (steric diminution). One can expect the same result in biological systems for the formation of receptor−ligand complex. In comparing the binding of a ligand to a biological receptor with or without a subsequent chemical reaction, say an enzyme or an hormonal receptor, there are some obvious

differences. The enzymatic reactions involve only structural groups in proximity to those atoms which are actually participating in bond making or breaking. Therefore the enzyme tolerates structurally different ligands provided that the fragment of the ligands, implied in the reaction is accessible. In some cases, even if the direct affinity of a ligand for its enzyme is low, the subsequent chemical transformation can take place. In the formation of a ligand−receptor complex any group of atoms which is in van der Waals contact with the receptor or the biopolymer can be (or is) involved in the binding event. If the receptor site is a pocket which can adjust any bioactive substance, no matter what its size or shape, then no steric effect will be observed. However, if the parent biopolymer has a limited conformational flexibility, and, as is likely, this flexibility is not the same in all directions, then steric effect will be observed. Furthermore, the steric effect will be conformationally dependent, and it is probable that the minimal steric interaction principle will be observed. This principle states that a substituent whose steric effect is conformationally dependent, will prefer that conformation which minimizes steric repulsions and which will give rise to the smallest steric effects. Finally there are some secondary steric effects on receptor binding that are produced by a substituent: (1) lowering the accessibility to an important group; (2) changing the concentration of a conformer; (3) shielding the active site from attack by a bulky reactant; (4) modifying the electronical resonance of a π-bonded substituent by out of plane repulsion.

E Rigid compounds and bioavailability

If the management of potency and selectivity of a set of drug candidates is often an easy task, the prediction of bioavailability has no real rationale. Therefore any effort in this direction is of great importance in the drug discovery process. An intriguing correlation between the bioavailability of a compound and the number of its rotatable bounds was found by Veber *et al.*[14] using an empirical approach based on a set of 1100 drug candidates (this study was conducted in the rat). Indeed a drug candidate possessing 10 or fewer rotatable bonds together with a polar surface of 140 $Å^2$ (or 12 or fewer H-bond donors and/or acceptors) irrespective of the molecular weight, has a high probability of presenting good oral bioavailability in the rat. This finding is of great importance because until now the limit of the molecular weight for a bioavailable drug candidate was set at 500 daltons by the Lipinski rules.[15] Veber *et al.* suggest that by freezing some of the rotatable bounds, the molecular weight (although the upper limit is not yet predictable) is no longer an essential parameter to be considered. Of course further data have to be accumulated in order to validate the above observations, however it seems that the introduction of conformational constraints in a drug candidate has to be considered when pharamacokinetics is an issue to be solved.

II SELECTED EXAMPLES

To illustrate the above theoretical considerations, concrete examples are presented in various topics of medicinal chemistry, such as enzyme inhibitors, neurotransmitter mimics or nucleotide analogues.

A Protease inhibitors

An important step in the rational design of enzyme inhibitors is the optimization of a known ligand, using structural information gained from the enzyme–ligand complex. The aim of such 'second-generation' design is to enhance the hydrophobic or electrostatic interactions between the inhibitor and the target, or to reduce the conformational flexibility of the ligand. If the predominant conformation of the free ligand is the same as in the binding site, the entropic disadvantage of the complex formation is reduced and its affinity is enhanced in comparison to a more conformationally mobile analogue. The following example illustrates this strategy. Several phosphorous-containing peptide analogues are potent inhibitors of the aspartic proteases. The X-ray crystal structure of the complex between penicillopepsin and the phosphonate peptide 1 (Fig. 15.1) was determined at 1.7 Å resolution: the phosphonate function in 1 is oriented towards the catalytic aspartate of the enzyme, and the backbone of the inhibitor is bound in the extended conformation which is typical for aspartic peptidase ligands. A salient feature of this conformation is the proximity of the side chains on alternating residues (e.g. P_3 and P_1, and P_2 and $P_{1'}$). These side chains share extended, hydrophobic binding pockets on opposite sides of the peptide, and there is room to connect them together. By chemical simplification of the original structure of 1, compounds 2, 3 and 4 were designed by introducing a napthyl ring between P_1 and P_2. Interestingly compounds 2 and 3 are almost equipotent, but two orders of magnitude less active than the parent structure (1). Compound 4 was far less active, reinforcing the conformational constraint strategy. Furthermore the X-ray structures of the enzyme complexed with 2, 3 and 4 have been determined, and it appears that compound 2 has a favourable conformation for binding, but the orientation of the naphtalene bridge reduces the advantage conveyed by the macrocyclic constraint. There are other possibilities for performing conformational restrictions for instance between the P_2 and $P_{1'}$ residues. A CAVEAT search for a unit that would link

the C—C bond of the P_2-valine side chain with the meta C—H bond of the $P_{1'}$ phenyl moiety (in the orientation they adopt in the bound complex of 4 with penicillopepsin) readily identified three and four atom chains as possible linkers. When the linkers **a–h** were checked for the best fit, it was found that the amide containing linkage (**a**) was the most attractive, with small conformational space and half of its low energy conformers matching the backbone of inhibitor 1. Additionally, the synthetic accessibility for introduction of link (**a**) in compound 1 was also part of its attractiveness.

Indeed the presence of the macrocyclic ring has a dramatic effect on the binding of the inhibitors in strongly enhancing the affinity of 5 over the acyclic analogues. Moreover 5 is a significantly better inhibitor than 1. The K_i differences between the macrocycle 5 and the acyclic analogues 6 and 7 represent a profit of 3.6 to 5.6 kcal mol^{-1} in free energy, only attributable to the entropic contribution. These values are the highest reported for a conformational constraint.[16–19] This splendid example confirms that macrocyclic conformational constraint shows particular promise for peptidomimetic drug design, since the affinity can be enhanced without substantially increasing molecular weight or hydrophobicity.

Several examples using a similar strategy were reported in the literature. The design of inhibitors of matrix metalloproteinases using macrocyclization between P1 and P2′, and the design, using the program GrowMol, of protease inhibitors using macrocyclization between P1 and P3 are the most significant.[20–23] Recently a new approach has been designed for the identification of the bioactive conformation of larger peptides: positional cyclization scanning. This strategy was applied to determine the secondary structures in the bioactive conformation of glucagon. A positional cyclization was performed at different parts of the 29 amino acid-containing peptide by means of disulphide or lactam bridges. The biological evaluation on the glucagon receptor revealed matched and mismatched conformations, and conclusions could be obtained about the bioactive conformations which could not be gained from earlier biophysical studies of glucagon.[24]

B Glutamate

Although glutamate (Glu) is the neurotransmitter of a vast majority of synapses in the central nervous system, there are presently no drugs available acting on specific glutamate receptors with a clear therapeutic application. Glu acts on ionotropic (ion channel: 17 subtypes) and metabotropic (protein G coupling: 8 subtypes) receptors. The chemical structure of Glu reveals several rotatable bonds, paving the way for a great deal of constraint analogues that can be designed for finding selective ligands for each subtype. Indeed Mother Nature did a great job in supplying medicinal chemists and pharmacologists with several rigid analogues of Glu (Fig. 15.2), among them kainic acid (**7**) allowed the identification of an ionotropic subtype.

The objective of various groups performing SAR studies on Glu was to incorporate rings or unsaturation into the amino acid backbone and to define, as usual, steric tolerances and spatial orientation for optimal receptor binding. These studies as well as the isosteric replacement of the γ-carboxylate by an γ-phosphonate were very fruitful for the harvesting of potent and selective ligands of both types of Glu receptors.

Applied to the synthesis of Glu analogues, the creativity of the medicinal chemists seems to be unlimited. Compounds **8–18** (Fig. 15.3) represent a small sample of all the compounds that were reported. Some of the ligands show binding constants in the nanomolar range, whereas Glu itself is only in the micromolar range. This example confirms that conformational rigidification is a powerful tool when applied to neurotransmitters, and also confirms that the creativity of the medicinal chemists is endless as long as chemical pathways are at hand.[25–29]

C Cocaine

The ability of cocaine to bind at the dopamine transporter (DA-T) and to inhibit the re-uptake of dopamine has been implicated in the reinforcing properties of this drug. Numbers of highly potent cocaine analogues, together with informations concerning their structure–activity relationships at the DA-T, have been reported. However, the precise details of the binding interactions between these analogues and the DA-T is still a matter of debate. It has been shown that the C-3 benzoate can be replaced by a phenyl group (Fig. 15.4). Similarly, the C-2 carboxylate can be replaced by alkyl or alkenyl groups. In both cases the obtained analogues **1**, **2** and **3** retain high binding affinity. Little is known about the spatial requirements of the nitrogen lone pair of these cocaine analogues. The directionality of the nitrogen lone pair is likely to be of some importance for binding affinity. Kozikowski *et al.* have reported the design and biological evaluation of a series of cocaine analogues in which the conformation of the nitrogen lone pair is fixed by means of a tether to either the three- or two-carbon bridge of the tropane moiety (compounds **3–11**).[30]

As compounds **1–9a–c** (the front-bridged or the back-bridged structures — either isomer at C-6 or C-7) were more potent than cocaine, it seems that the orientation of the nitrogen lone pair is not a crucial determinant for high affinity binding, but can be of importance for the transporter

selectivity between the 5-H-T, the NE-T or the DA-T. Interestingly compounds **10a/b** and **11**, devoid of DA-T binding, are new potent ligands and selective inhibitors for the 5-HT uptake protein.[30–34]

D Rigidified ribose in nucleotide construction

The sugar ring of nucleosides and nucleotides equilibrates in solution between two extreme forms oscillating from a 2'-*exo*/3'-*endo* (**A**) conformation to the opposite 2'-*endo*/3'-*exo* (**B**) conformation. Preference for any of these specific conformations in solution is determined by the interplay of important interactions resulting from anomeric and *gauche* effects. When a nucleoside or nucleotide binds to its target, in general one form is expected to be present, and its selection is additionally determined by packing forces. While the gap between the two conformations **A** and **B** is in the neighbourhood of 4 kcal mol^{-1}, such a disparity can explain a difference between micro- and nanomomolar binding affinities.

Recently oligonucleotides containing conformationally locked nucleosides have attracted considerable attention. Indeed it has been established that the sugar pucker of DNA in the natural DNA-**RNA** duplex tends to adopt a 3'-*endo* conformation (**A**) while the 2'-*endo* (**B**) sugar pucker is predominant in the natural DNA-**DNA** duplex. Therefore it is anticipated that conformationally restricted 3'-*endo* nucleosides would enhance hybridization of chemically modified oligonucleotides to the complementary RNA, while conformationally restrained 2'-*endo* nucleosides would enhance hybridization to the complementary DNA. This assumption is based on the concept that pre-organization of the oligonucleotide strand would produce an entropic advantage for duplex formation and stability.

Several groups have followed this concept and found valuable compounds such as **1**, a methano-carbacyclic thymidine, with an inversion barrier of 5–6 kcal mol^{-1} higher than the original thymidine **2**. Excellent antiviral activity was obtained with carbacyclic thymidine **1**. The other rigidified furanoses (**3–11**) were incorporated in oligomers and the melting temperatures (T_m) determined with complementary sequences of DNA or RNA. Depending on the number of modified nucleotides and their position in the polymer chain, an enhancement of near $\Delta T_m = +2°C$, per modified nucleotide was observed for the melting temperature in hybridation experiments with RNA. As locked nucleic acids, these compounds are potential candidates for antisense therapy and diagnostic probes[35–43] (Fig. 15.5).

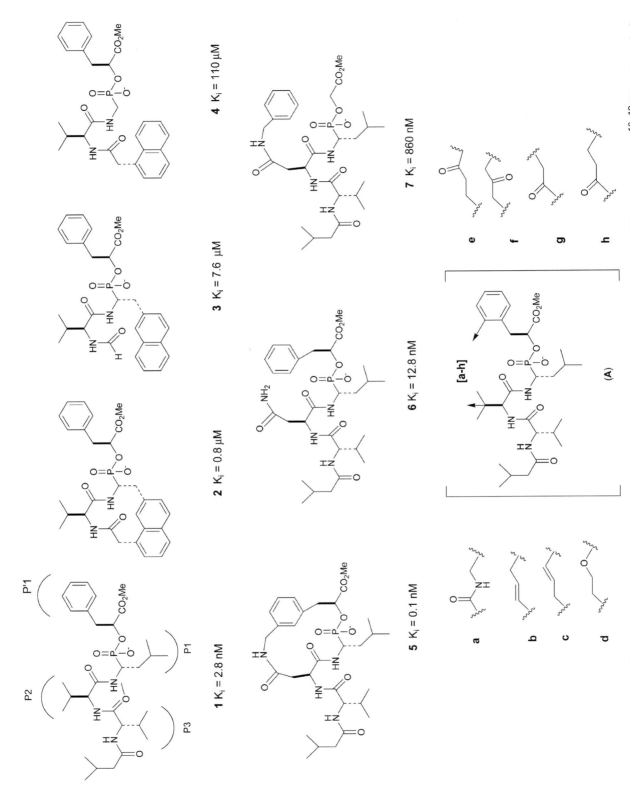

Fig. 15.1 Design of rigid ligands of the penicillopepsin protease and the values of the corresponding inhibition constants.[16–18] The short chains **a–h** introduced between the two arrows on structure (**A**) were designed using the CAVEAT program.

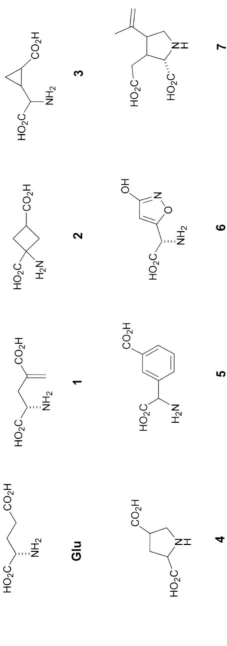

Fig. 15.2 Naturally occurring constrained Glu analogues. Mother Nature has designed and provided us with original examples that were adapted and largely extended.

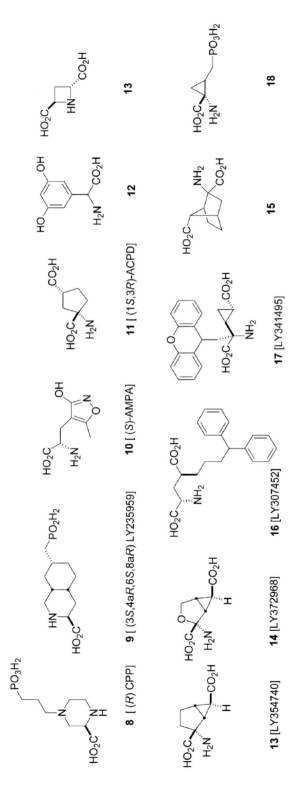

Fig. 15.3 Synthetic analogues of Glu. Several tactics were used for conformational restriction: the introduction of liphophilic groups for reducing the flexibility and of cycles for limiting the rotatable bounds.

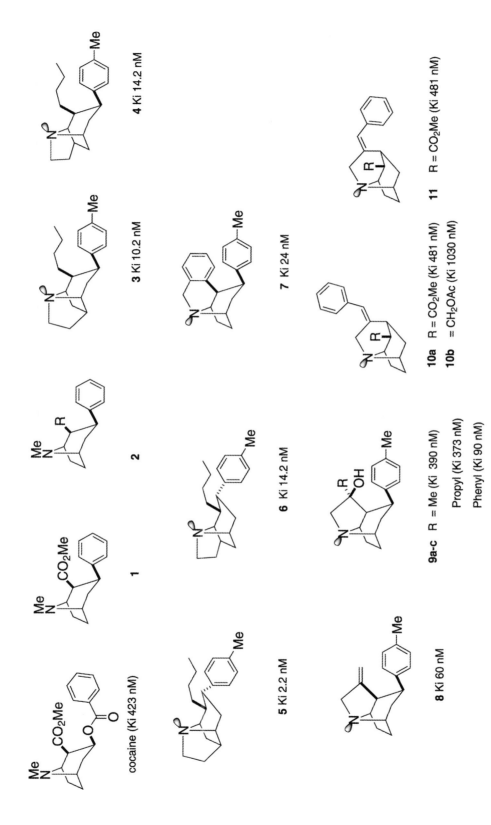

Fig. 15.4 Rigid analogues of cocaine and binding data versus [³H] Win 35,428. The compounds were designed for a defined orientation of the nitrogen lone pair.

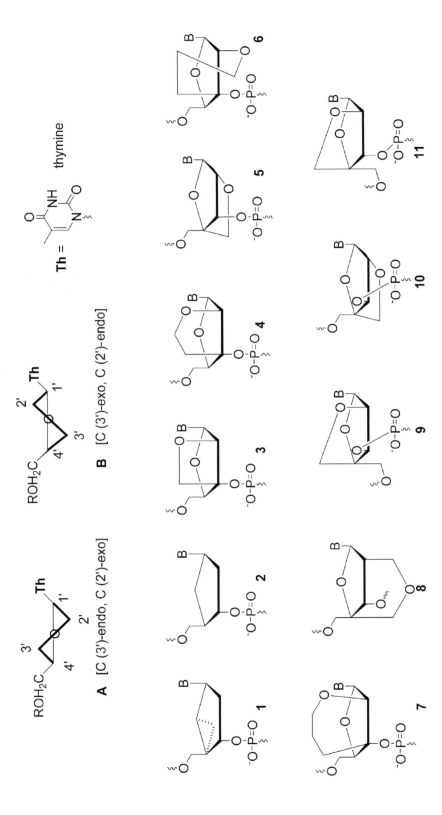

Fig. 15.5 Structure of rigid nucleotide furanose residues. The desoxy ribose part of thymidine was modified for freezing its accessible conformations.

E Conformationally restricted tetrahydro-cannabinols

Structure–activity studies of conventional cannabinoids have established that the alkyl chain at the C_3 position is an essential part of the pharmacophore for cannabimimetic activity. Naturally occurring cannabinoids such as Δ^9-THC possess an *n*-pentyl side chain at this position. However it is also known that a heptyl side is optimal for activity on the cannabinoid CB_1 or CB_2 receptors. Furthermore *n*-heptyl-Δ^8-THC (**1**, Fig. 15.6) was found to have high affinity for the CB receptors. Analogues of **1** were designed and synthesized: compounds **2** and **3** incorporate a cyclohexane ring fused to the C_2—C_3 position carrying either an *n*-pentyl or *n*-hexyl chain at a suitable position so as to produce analogues in which the *n*-heptyl chain is conformationally restricted around the $C_{1'}$—$C_{2'}$ and/or C_3—$C_{1'}$ bond; in a third analogue **4**, the six-membered ring is fused to the C_3—C_4 bond of the phenyl ring.[44,45]

When the binding affinities of **2**, **3** or **4** were compared with those of *n*-heptyl-Δ^8-THC (**1**), the prototypical ligand with a totally unrestrained seven-membered side chain, the data indicate that incorporation of the first one or two carbons of the side chain into a six-membered ring leads to a significant reduction in affinities for both CB_1 and CB_2. This may be attributed either to the conformational restriction imposed on the early part of the chain or, alternatively, to the presence of additional carbons of the six-membered ring which may interfere with the binding of the ligand at the active site. Analogue **2** exhibits an 18-fold and a three-fold higher affinity for the CB_1 and CB_2 receptors respectively, than **3**. These results suggests that the cannabinoid receptor affinities decreases significantly when the side chain is forced into a lateral orientation and further away from the phenol ring. The low binding of analogue **4** can be attributed to its undesirable side-way chain orientation and to steric constraints of the C_3—C_4 ring with the receptor binding site. If these results are compared with other analogues such as **5**, **6**, or **7** in which more subtle modifications have been made, such as the introduction of an alkyne (**5**), an *exo*-methylene (**6**) or a dithiane (**7**) units at the $C_{1'}$ position, the following conclusions can be taken: the good binding for analogue **5**, possessing a triple bond at $C_{1'}$ restricts rotation around the $C_{1'}$—$C_{2'}$ highlights the crucial importance of the benzylic position in the orientation of the side (the side chain needs some flexibility near the $C_{1'}$ position); compound **6** confirms this observation and indicates the tolerance for a hydrophobic group. The introduction of the moderate bulky dithiane group at $C_{1'}$ (**7**) completely supports that assumption.

Additionally it has to be emphasized that compound **7** is a cannabinomimetic with one of the highest affinities reported to date.[46] This is a convincing example of how

constraints and group modulation can be combined to get important information for the structure-based drug design.

F Baclofen

Baclofen (**1**, Fig. 15.7) is a selective ligand for the $GABA_B$ receptor subtype. Its structure–activity profile is rather 'flat' in the sense that almost all changes lead to inactive compounds. Even modest structural modifications such as the positional change of the chlorine atom on the aromatic ring, or its replacement by another halogen, produce a complete loss of biological activity. In order to understand more of baclofene's activity, it was accepted that the solid state conformation of **1** could represent a possible bio-active conformation. Indeed, solid-state conformations reduce van der Waals interactions by constraints in the crystalline network and are potential data sources for the design of rigid analogues. The examination of the X-ray data for (*R*)-baclofen, the eutomer, revealed that the phenyl ring is almost perpendicular to the GABA backbone. This conformation lowers the steric interaction between the phenyl ring and the two ortho hydrogens ('ortho effect'), and also corresponds to a local minimum as shown from molecular modelling calculations. In the hope of trapping the baclofen active conformation in this way, compounds **2**, **3** and **4** were designed and prepared. In these compounds the ortho effect is greatly enhanced either by the presence of chlorine atoms (compounds **3** and **4**), or by insertion of the benzylic carbon in a five-membered indane ring (compound **2**). The indane ring was selected because of comparable values for the length (1.51 Å) of the benzylic bond in **2** and **1**. In comparison, the benzocyclobutene ring presents a longer bond length (1.56 Å) than the tetrahydronapthalene ring which has a similar length (1.51 Å), but a too large extra volume with respect to **1**. In binding experiments, only compound **3** reveals marginal affinity whereas the two others did not bind to the receptor preparation.

Given this disappointing result, a single crystal X-ray analysis was performed on **2**. When the two solid-state structures of **1** and **2** were superimposed, a perfect fit was observed (RMS = 0.08).

From this study it was concluded that baclofen needs flexibility for activating its receptor. It could be that the binding kinetics are governed by a multi-step process demanding ligand flexibility.[47] Recently the $GABA_B$ receptor has been cloned and the proposed fixation of the ligand was shown to be driven by a dimeric protein.[48,49] The above negative results in an attempt to rigidify **1**, illustrate a possible limitation of the methodology when the biological receptor is of complex nature.

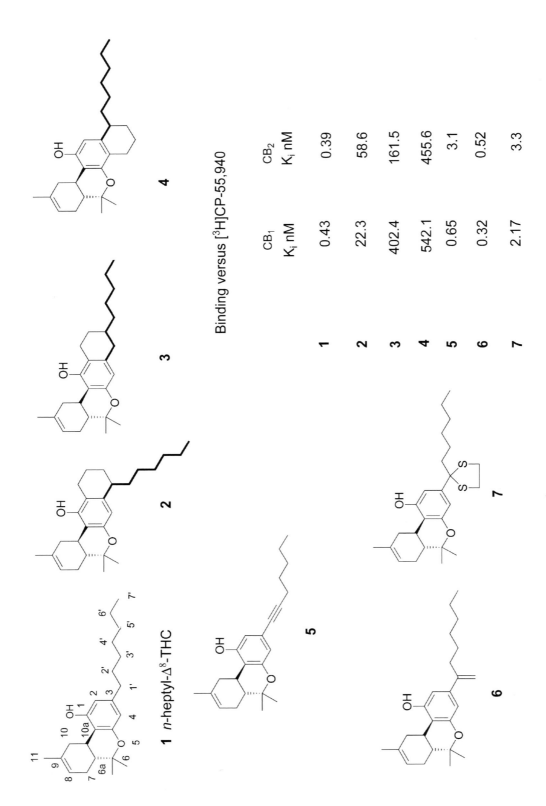

Binding versus [³H]CP-55,940

	CB₁ Kᵢ nM	CB₂ Kᵢ nM
1	0.43	0.39
2	22.3	58.6
3	402.4	161.5
4	542.1	455.6
5	0.65	3.1
6	0.32	0.52
7	2.17	3.3

Fig. 15.6 THC analogues with binding data towards [³H] CP-55,940. Starting from the reference compound **1**, the heptyl side chain was incorporated in various surrogates in order to determine its orientation during receptor activation.

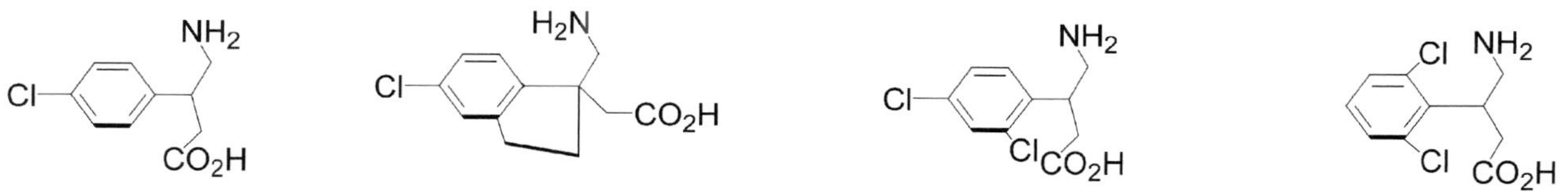

Fig. 15.7 Rigid analogues of baclofen. The solid state of baclofen (X-ray structure) has the conformation depicted in **1**. Compounds **2**, **3** and **4** were designed according to this naturally occurring conformation.

G Gabapentin

Gabapentin (**1**, Fig. 15.8), originally designed as a lipophilic GABA analogue, is an anticonvulsant with as yet an uncertain mechanism of action. Despite its structural analogy with GABA, gabapentin does not interact with any of the GABA receptors, nor is it an inhibitor of GABA uptake or degradation. Recently a high affinity binding site for gabapentin, located on a subunit of a calcium channel has been reported, allowing rapid screening for analogues, and suggesting possible pharmacological actions. In order to identify the active conformation of gabapentin, two questions have to be answered: (1) which residue attached at the quaternary carbon of the cyclohexane – the aminomethyl or the acetyl – is axial (**1a** versus **1b**)? (2) what is the relative orientation of the amino and carboxylate groups (extended, folded or in between)?

To answer these questions, a series of analogues were prepared having methyl group(s) appended to the cyclohexane ring with the aim of locking the ring, and orienting the aminomethyl moiety either axially or equatorially. In another series, the GABA part was subjected to conformational constraints in order to investigate the preferred binding conformation of the GABA moiety of gabapentin. The couple of diastereomeric molecules **2a/2b**, **3a/3b** and **4a/4b**, the spiro adducts **5**, **6a/b**, **7** and **8** and the fused bicycles **8** and **9** were chosen as targets. As can be seen from the binding results the analogues (**2a**, **4a'**, **4a''**) possessing the aminomethyl residue in equatorial orientation and the acetic residue in axial orientation (these conformations were obtained by extensive NMR analysis) are far better ligands than the analogues with the reverse situation. Enantiomerically pure **4a'** is the most potent compound. Among the spiro analogues the enantiomerically pure compound **6a**, had a similar binding affinity as gabapentin, whereas compounds **5**, **6b**, **7**, **8** and **9** all show significantly lower affinities. These results suggested that the restriction of the nitrogen atom and carboxylic group in the pyrrolidine ring of **6a** could mimic the biologically active binding conformation of the GABA portion in gabapentin. Thus the above results were obtained simply by introducing small substituents in selected positions of the cyclohexane ring, and by ring closing strategies with

dedicated chemistry. Finally the use of NMR analysis on the compounds for proper identification of their stereochemistry, assisted by computer modelling allowed the proposal of a rather convincing pharmacophore for gabapentin.[50–53]

H Serotonin

Serotonin (5-HT, **1**, Fig. 15.9) is a widely distributed neurotransmitter in the brain and peripheral tissues. There are 14 distinct receptors identified by their primary sequence and their pharmacological profile (5-HT1-7, with subtypes). In the past, a multitude of non-selective ligands has been identified, but recently the focus of many research groups was devoted to identifying potent subtype-selective 5-HT receptor ligands.

The sequence similarity of the different 5-HT receptor subtypes is around 60% and their identity is around 30%. This significant difference supports the expectation that selective ligands could be discovered. As the aminoethyl side chain of serotonin is highly flexible, several approaches based on the synthesis of constraint or rigid analogues were developed. This was especially the case in the search for agonists for the 5-HT$_1$ subtype, which is involved in the anti-migraine and anti-depressive therapies. Lysergic acid (**2**, LSD), a 5-HT agonist without selectivity, was used by the Lilly group to design LY228729 (**3**) a potent and selective 5-HT$_{1a}$ ligand. This example shows how natural products can stimulate the design of compounds embedding conformational rigidity.

In the search for ligands selective for the 5-HT$_{1D}$ receptor subtype, many examples demonstrate the ability of medicinal chemists to design powerful ligands. In some cases the structures suggested by computer modelling are sophisticated, and demand great efforts towards their synthesis. In Fig. 15.9 analogues of 5-HT, illustrating tactics for the introduction of restrictions of the rotatable bonds, are reported. Interestingly the introduction of various substituents at C-5 (**4**, **6**, **12** and **14**) and rigidification are responsible for selectivity for the 5-HT$_{1D}$ subtype. Conformational restriction of the aminoethyl function by means of a 2,7-diazabicyclooctane ring system has led to compounds (**14**) with high selectivity towards 5-HT$_{1D}$ receptors, high

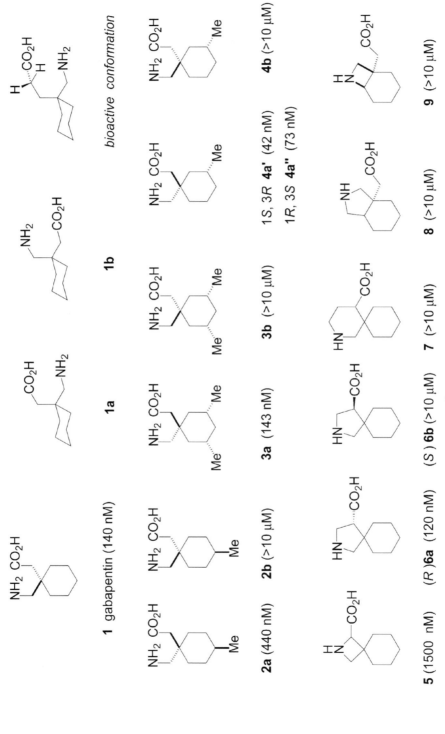

Fig. 15.8 Analogues of gabapentin with binding data against [^{3}H] gabapentin. In compounds **2a**, **3a**, **4a′** and **4a″** the methyl substituent(s) and the aminomethyl side chain are in 1,3 equatorial orientation (for drawing simplification purposes this appears as a *cis* relationship). The assumed bioactive conformation is representative for those conformations.

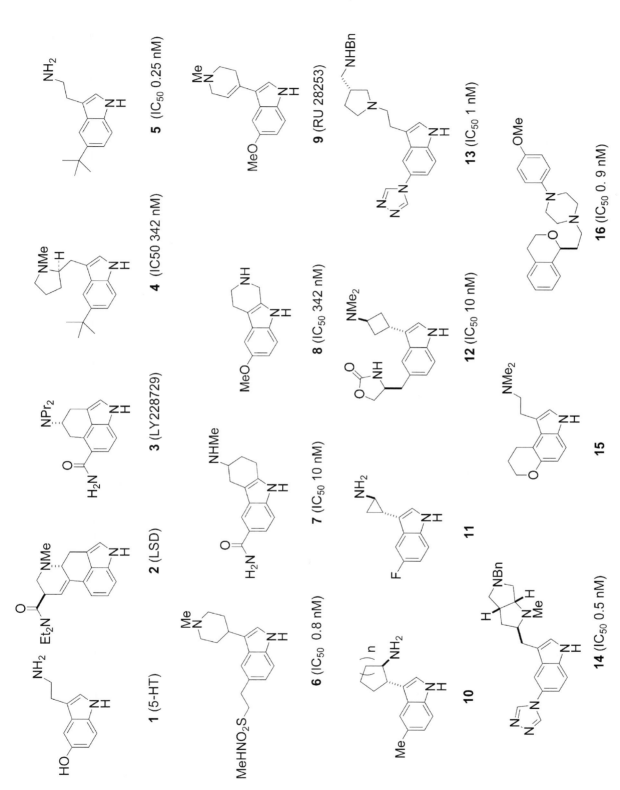

Fig. 15.9 Constrained 5-HT analogues (IC$_{50}$ values for the 5-HT$_{1D}$ receptor). The compounds were designed starting from the indole nucleus except for compound **16**.

metabolic stability and excellent bioavailability. Finally, the structure of the newly reported compound (16) confirms, if necessary, that other drug design approaches are also valuable, such as side-effect optimization,[54-57] for detecting potent ligands.[58-68]

III OUTLOOK

As illustrated by the above nonexhaustive examples, conformational restriction is one of the powerful weapons in the arsenal of medicinal chemists when potency and selectivity are the goals for a given target. An additional advantage of ring formation is that rigid compounds tend to be more orally bioavailable.[14] From a thermodynamic point of view, the designed, constrained ligand must be pre-organized for optimum recognition. Obvious limitations are access to the parameters controlling the binding process such as diffusion, solvation or water clustering which are included in the entropy factors, but there are some techniques amenable to determine precisely the thermo-dynamics of ligand–receptor interactions.[69,70] Therefore, before finding the right template for a lead compound several rigid analogues have to be synthesized, mainly by nontrivial chemical processes. It is expected that in the near future when more crystal structures of proteins complexed with ligands become accessible (even RCPGs), the efficiency of the design will be increased.

REFERENCES

1. Miklavc, A., Kocjan, D., Mavri, J., Koller, J. and Hadzi, D. (1990) On the fundamental difference in the thermodynamics of agonist and antagonist interactions with α-adrenergic receptors and the mechanism of entropy driven binding. *Biochem. Pharmacol.* **40**: 663–669.
2. Jencks, W. (1981) On the attribution and additivity of binding energies. *Proc. Natl Acad. Sci. USA* **78**: 4046–4050.
3. Raffa, R.B. and Porreca, F. (1989) Thermodynamic analysis of the drug–receptor interaction. *Life Sci.* **44**: 245–258.
4. Andrews, P.R., Craik, D.J. and Martin, J.L. (1984) Functional group contributions to drug-receptor interactions. *J. Med. Chem.* **27**: 1657–1684.
5. Page, M.I. (1977) Entropie, Bindungsenergie und enzymatische Katalyse. *Angew. Chem.* **89**: 456–467.
6. Bardi, S.J., Luque, I. and Freire, E. (1997) Structure-based thermo-dynamic analysis of HIV-1 protease inhibitors. *Biochemistry* **36**: 6588–6596.
7. Velazquez-Campoy, A., Todd, M.J. and Freire, E. (2000) HIV-1 protease inhibitors: enthalpic versus entropic optimization of the binding affinity. *Biochemistry* **39**: 2201–2207.
8. Searle, M.S. and Williams, D.H. (1992) The cost of conformational order: entropy changes in molecular associations. *J. Am. Chem. Soc.* **114**: 10690–10697.
9. Williams, D.H. and Westwell, M.S. (1998) Aspects of weak interactions. *Chem. Soc. Rev.* **27**: 57–63.
10. Nicklaus, M.C., Wang, S., Driscoll, J.S. and Milne, G.W.A. (1995) Conformational changes of small molecules binding to proteins. *Bioorg. Med. Chem.* **3**: 411–428.
11. Leach, A.R. and Lewis, R.A. (1994) A racing-bracing approach to computer-assisted ligand design. *J. Comp. Chem.* **15**: 233–240.
12. Hruby, V.J., Li, G., Hacksell-Luevano, C. and Shenderovich, M. (1997) Design of peptides, proteins and peptidomimetics in chi space. *Biopolymers* **43**: 219–266, and references cited therein.
13. Charton, M. and Motoc, I. (1983) Steric effects in drug design. *Top. Curr. Chem.* **114**: 1–6.
14. Veber, D.F., Johnson, S.R., Cheng, H.-Y., Smith, B.R., Ward, K.W. et al. (2002) Molecular properties that influence the oral bioavailability of drug candidates. *J. Med. Chem.* **45**: 2615–2623.
15. Lipinski, C.A., Lombardo, F., Dominy, B.W. and Feeney, P.J. (1997) Experimental and computational approaches to estimate solubility and permeability in drug discovery and development settings. *Adv. Drug Del. Rev.* **23**: 4–25.
16. Meyer, J.H. and Bartlett, P.A. (1998) Macrocyclic inhibitors of penicillopepsin. 1. Design, synthesis, and evaluation of an inhibitor bridged between P1 and P3. *J. Am. Chem. Soc.* **120**: 4600–4609.
17. Ding, J.D., Fraser, M.E., Meyer, J.H., Bartlett, P.A. and James, M.N.G. (1998) Macrocyclic inhibitors of penicillopepsin. 2. X-Ray cristallo-graphic analyses of penicillopepsin complexed with a P3-P1 macrocyclic peptidyl inhibitor and its two acyclic analogues. *J. Am. Chem. Soc.* **120**: 4610–4621.
18. Smith, W.W. and Bartlett, P.A. (1998) Macrocyclic inhibitors of penicillopepsin. 3. Design, synthesis, and evaluation of an inhibitor bridged between P2 and P1′. *J. Am. Chem. Soc.* **120**: 4622–4628.
19. Khan, A.R., Parrish, J.C., Fraser, M.E., Smith, W.W., Bartlett, P.A. et al. (1998) Lowering the entropic barrier for binding conformation-ally flexible inhibitors to enzymes. *Biochemistry* **37**: 16839–16845.
20. Cherney, R.J., Wang, L., Meyer, D.T., Xue, C.B., Wassermann, Z.R. et al. (1998) Macrocyclic amino carboxylates as selective MMP-8 inhibitors. *J. Med. Chem.* **41**: 1749–1751.
21. Xue, C.B., He, X., Roderick, J., Degrado, W.F., Cherney, R.J. et al. (1998) Design and synthesis of cyclic inhibitors of matrix metallopro-teinases and TNF-α production. *J. Med. Chem.* **41**: 1745–1748.
22. Ripka, A.S., Satyshur, K.A., Bohacek, R.S. and Rich, D.H. (2001) Aspartic protease inhibitors designed from computer-generated templates bind as predicted. *Org. Lett.* **3**: 2309–2312.
23. Duan, J.J.-W., Chen, L., Xue, C.-B., Wasserman, Z.R., Hardman, K.D. et al. (1999) P1,P2′-Linked macrocyclic amine derivatives as matrix metalloproteinase inhibitors. *Bioorg. Med. Chem. Lett.* **9**: 1453–1458.
24. Anh, J.-M., Gitu, P.M., Medeiros, M., Swift, J.R., Trivedi, D. et al. (2001) A new approach to search for bioactive conformation of glucagon: positional cyclization scanning. *J. Med. Chem.* **44**: 3109–3116.
25. Räuner-Osborne, H., Egebjerg, J., Nielsen, E.O., Madsen, U. and Krogsgaard-Larsen, P. (2000) Ligands for glutamate receptors: design and therapeutic prospects. *J. Med. Chem.* **43**: 2609–2645.
26. Costantino, G., Macchiarulo, A. and Pelliciari, R. (1999) Pharmaco-phore models of group I and group II metabotropic glutamate recep-tor agonists. Analysis of conformational, steric, and topological parameters affecting potency and selectivity. *J. Med. Chem.* **42**: 2816–2827.
27. Ornstein, P.L. and Klimbowski, V.J. (1992) Competitive NMDA receptor antagonists. In *Excitatory Amino Acid Receptors, Design of Agonists and Antagonists*, pp. 183–200. Ellis Horwood, New York.
28. Monn, J.A., Valli, M., Massey, S.M., Hansen, M.M., Kress, T.J. et al. (1999) Synthesis and pharmacological characterization and molecular modeling of heterobicyclic amino acids related to (LY354740): identification of LY379268 and LY389795: two new potent, selective

and systematically active agonists for group 2 metabotropic glutamate receptors. *J. Med. Chem.* **42**: 1027–1040, and references cited therein.

29. Kingston, A.E., Ornstein, P.L., Wright, R.A., Johnson, B.G., Mayne, N.G. *et al.* (1996) LY34195 is a nanomolar potent an selective antagonist of group II metabotropic glutamate receptors. *Neuropharmacology* **37**: 1–12.

30. Smith, P.M., Johnson, K.M., Zhang, M., Flippen-Anderson, J.L. and Kozikowski, A.P. (1998) Tuning the selectivity of monoamine transporter inhibitors by stereochemistry of the nitrogen lone pair. *J. Am. Chem. Soc.* **120**: 9072–9073.

31. Kozikowski, A.P., Araldi, G.L., Boja, J., Meil, W.M., Johnson, K.M. *et al.* (1998) Chemistry and pharmacology of the piperidine-based analogues of cocaine. Identification of potent DA-T inhibitors lacking the tropane skeleton. *J. Med. Chem.* **41**: 1962–1969.

32. Smith, M.P., George, C. and Kozikowski, A.P. (1998) The synthesis of tricyclic cocaine analogs via the 1,3 dipolar cycloaddition of oxidopyridinium. *Tetrahedron Lett.* **39**: 197–200.

33. Tamiz, A.P., Smith, M.P. and Kozikowski, A.P. (2000) Design, synthesis and biological evaluation of 7-azatricyclodecanes: analogue of cocaine. *Bioorg. Med. Chem. Lett.* **10**: 297–300.

34. Hoepping, A., Johnson, K.M., George, C., Flippen-Anderson, J. and Kozikowski, A.P. (2000) Novel conformationally constrained tropane analogues by 6-endo-trig radical cyclization and Stille coupling-switch of activity toward the serotonin and/or norepinephrine transporter. *J. Med. Chem.* **43**: 2064–2071.

35. Singh, S.K., Nielsen, P., Koshkin, A.A. and Wengel, J. (1998) LNA (locked nucleic acids): synthesis and high-affinity nucleic acid recognition. *Chem. Commun.* 455–456.

36. Nielsen, P., Pfundheller, H.M. and Wengel, J. (1997) A novel class of conformationally restricted oligonucleotide analogues: synthesis of 2', 3'-bridged monomers and RNA-selective hybridation. *Chem. Commun.* 825–826.

37. Wang, G., Gunic, E., Girardet, J.L. and Stoisavljevic, V. (1999) Conformationally locked nucleosides. Synthesis and hybridization properties of oligodeoxynucleotides containing 2',4'-C-bridged 2'-deoxynucleotides. *Bioorg. Med. Chem. Lett.* **9**: 1147–1150.

38. Nielsen, P., Pfundheller, H.M., Olsen, C.E. and Wengel, J. (1997) Synthesis of 2'-O,3'-C-linked bicyclic nucleosides and bicyclic oligonucleotides. *J. Chem. Soc. Perkin Trans.* **1**: 3423–3433.

39. Koshkin, A.A., Singh, S.K., Nielsen, P., Rajwanshi, V.K., Kumar, R. *et al.* (1998) LNA (locked nucleic acids): synthesis of adenine, cytosine, guanine, 5-methylcytosine, thymine and uracil bicyclo-nucleoside monomers, oligomerisation, and unprecedented nucleic acid recognition. *Tetrahedron Lett.* **54**: 3607–3630.

40. Marquez, V.E., Siddiqui, M.A., Ezzitouni, A., Russ, P., Wang, J. *et al.* (1996) Nucleosides with a twist. Can fixed forms of sugar ring pucker influence biological activity in nucleosides and oligonucleotides? *J. Med. Chem.* **39**: 3739–3747.

41. Obika, S., Nanbu, D., Hari, Y., Andoch, J., Morio, K. *et al.* (1998) Stability and structural feature of the duplexes containing nucleoside analogues with a fixed N-type conformation, 2'-O,4'-C-methylene-ribonucleosides. *Tetrahedron Lett.* **39**: 5401–5404.

42. Christensen, N.K., Petersen, M., Nielsen, P., Jacobsen, J.P., Olsen, C.E. *et al.* (1998) A novel class of oligonucleotide analogues containing 2'-O, 3'-C-linked [3.2.0] bicycloarabinonucleoside monomers: synthesis, thermal affinity studies, and molecular modeling. *J. Am. Chem. Soc.* **120**: 5458–5463.

43. Rajwanshi, V.K., Hakansson, A.E., Sorensen, M.D., Pitsch, S., Singh, S.K. *et al.* (2000) The eight stereoisomers of LNA (locked nucleic acid): a remarquable family of strong RNA binding molecules. *Angew. Chem. Int. Ed.* **39**: 1656–1659, and references cited therein.

44. Khanolkar, A.D., Lu, D., Fan, P., Tian, X. and Makriyannis, A. (1999) Novel conformationally restricted analogs of Δ^8-tetrahydrocannabinol. *Biorg. Med. Chem. Lett.* **9**: 2119–2124.

45. Huffman, J.W. and Yu, S. (1999) Synthesis of a tetracyclic conformationally constrained analog of Δ^8-THC. *Biorg. Med. Chem. Lett.* **6**: 2281–2288.

46. Papahatjis, D.P., Kourouli, T., Abadji, V., Goutopoulos, A. and Makriyannis, A. (1998) Pharmacophoric requirements for cannabinoid side chains: multiple bond and C1'-substituted Δ^8-tetrahydrocannabinols. *J. Med. Chem.* **41**: 1195–1200.

47. Mann, A., Boulanger, T., Brandau, B., Durant, F., Evrard, G. *et al.* (1991) Synthesis and biochemical evaluation of baclofen analogues locked in the baclofen solid state conformation. *J. Med. Chem.* **34**: 1307–1313.

48. Galvez, T., Parmentier, M.-L., Joly, C., Malitschek, B., Kaupmann, K. *et al.* (1999) Mutagenesis and modeling of the GABA$_B$ receptor extracellular domain support a venus flytrap mechanism for ligand binding. *J. Biol. Chem.* **274**: 13362–13369.

49. Kaupmann, K., Huggel, K., Heid, J., Flo, P.J., Bischoff, S. *et al.* (1997) Expression cloning of GABA$_B$ receptors uncovers similarity to metabotropic glutamates receptors. *Nature* **386**: 239–246.

50. Receveur, J.M., Bryans, J.S., Field, M., Singh, L. and Horwell, D.C. (1999) Synthesis and biological evaluation of conformationally restricted gabapentin analogues. *Bioorg. Med. Chem. Lett.* **9**: 2329–2339.

51. Bryans, J.S., Davies, N., Gee, N.S., Dissnayake, V.U.S., Ratcliffe, G.S. *et al.* (1998) Identification of novel ligands for the gabapentin binding site on the $\alpha_2\delta$ subunit of a calcium channel and their evaluation as anticonvulsant agents. *J. Med. Chem.* **41**: 1838–1845.

52. Bryans, J.S., Horwell, D.C., Ratcliffe, G.S., Receveur, J.M. and Rubin, J.R. (1999) An *in vitro* investigation into conformational aspects of gabapentin. *Bioorg. Med. Chem.* **7**: 715–721.

53. Bryans, J.S., Davies, N., Gee, N.S., Horwell, D.C., Kneen, C.O. *et al.* (1997) Investigation into the preferred conformation of gabapentin for interaction with its binding site on the $\alpha_2\delta$ subunit of a calcium channel. *Bioorg. Med. Chem. Lett.* **7**: 2481–2484.

54. Wermuth, C.G. (1998) Search for new lead compounds: the example of the chemical and pharmacological dissection af Aminopyridazines. *J. Heterocyclic Chem.* **35**: 1091–1100.

55. Wermuth, C.G. and Clarence-Smith, K. (2000) 'Drug-like' leads: bigger is not always better. *Pharmaceut News* **7**: 53–57.

56. Poroikov, V., Akimov, D., Shabelnikova, E. and Filimonov, D. (2001) Top 200 medicines: can new actions be discovered through computer-aided prediction? *SAR QSAR Environ. Res.* **12**: 327–344.

57. Wermuth, C.G. (2001) The 'SOSA' approach: an alternative to high-throughput screening. *Med. Chem. Res.* **10**: 431–439.

58. Ghosh, A., Wang, W., Freeman, J.P., Althaus, J.S., von Voigtlander, P.F. *et al.* (1991) Stereocontrolled syntheses of some conformationally restricted analogs of serotonin. *Tetrahedron Lett.* **47**: 8653–8662.

59. Vangveravong, S., Kanthasamy, A., Lucateis, V., Nelson, D.L. and Nichols, D.E. (1998) Synthesis and serotonin affinities of a series of trans-2-(indol-3-yl)cyclopropylamine derivatives. *J. Med. Chem.* **41**: 4995–5001.

60. Xu, Y.C., Schaus, J.M., Walker, C., Krushinski, J., Adham, N. *et al.* (1999) N-Methyl-5-tert-butyltryptamine: a novel, highly potent 5-HT$_{1D}$ receptor agonist. *J. Med. Chem.* **42**: 526–531.

61. Glennon, R.A., Hong, S.S., Bondarev, M., Law, H., Dukat, M. *et al.* (1996) Binding of O-alkyl derivatives of serotonin at human 5-HT$_{1D}$ receptors. *J. Med. Chem.* **36**: 314–322.

62. Castro, J.L., Street, L.J., Guiblin, A.R., Jelley, R.A., Russell, M. *et al.* (1997) 3-[2-5(Pyrrolidin-1-yl)ethyl]indoles and 3-[3-(piperidin-yl) propyl]indoles: agonists for the h5-HT$_{1D}$ receptor with high selectivity over the h5-HT$_{1B}$ subtype. *J. Med. Chem.* **40**: 3497–3500.

63. Castro, J.L., Collins, I., Russell, M.G.N., Watt, A.P., Sohal, B. *et al.* (1998) Enhancement of oral absorption in selective 5-HT$_{1D}$ receptors agonists: fluorinated 3-[3-(piperidin-1-yl)propyl]indoles. *J. Med. Chem.* **41**: 2667–2670.

64. Ennis, M.D., Ghazal, N.B., Hoffman, R.L., Smith, M.W., Schlachter, S.K. *et al.* (1998) Isochroman-6-carboxamides as highly selective 5-HT$_{1D}$ agonists: potential new treatment for migraine without cardiovascular side effects. *J. Med. Chem.* **41**: 2180–2183.

65. Jandu, K.S., Barrett, V., Brockwell, M., Cambridge, D., Farrant, D.R. *et al.* (2001) Discovery of 4-[3-(trans-3-dimethylaminocyclobutyl)-1*H*-indol-5-ylmethyl]-(4S)-oxazolidin-2-one (4991W93), a 5-HT$_{1B/1D}$ partial agonist and a potent inhibitor of electrically induced plasma extravasation. *J. Med. Chem.* **44**: 681–693.

66. Ngo, J., Mealy, N. and Castaner, J. (1997) Zolmitriptan. *Drugs in the Future* **22**: 260–269.

67. Russell, M.G.N., Beer, M.S., Stanton, J.A., Sohal, B., Mortishire-Smith, R.J. *et al.* (1999) 2,7-diazabicyclo[3.3.0]octanes as novel h5-HT$_{1D}$ receptor agonists. *Bioorg. Med. Chem. Lett.* **9**: 2491–2496.

68. Bourrain, S., Neduvelil, J.G., Beer, M.S., Stanton, J.A., Showell, G.A. *et al.* (1999) 4-Hydroxy-1-[3-(5-(1,2,4-triazol-yl)-1H-indol-3-yl) propyl]piperidine: selective h5-HT$_{1D}$ agonists for the treatment of migraine. *Bioorg. Med. Chem. Lett.* **9**: 3369–3374.

69. Parker, M.H., Lunney, E.A., Ortwine, D.F., Pavlowsky, A.G., Humblet, C. *et al.* (1999) Analysis of the binding of hydroxamic acid and carboxylic acid inhibitors to the stromlysin-1 (matrix metalloproteinase-3) catalytic domain by isothermal titration calorimetry. *Biochemistry* **38**: 13592–13601.

70. Parker, M.H., Ortwine, D.F., O'Brien, P.M., Lunney, E.A., Banotai, C.A. *et al.* (2000) Stereoselective binding of an enantiomeric pair of stromelysine-1 inhibitors caused by conformational entropy factors. *Bioorg. Med. Chem. Lett.* **10**: 2427–2430.

16

IDENTICAL AND NON-IDENTICAL TWIN DRUGS

Jean-Marie Contreras and Jean-Jacques Bourguignon

There is in living organisms, human beings and societies a balance between these main two forces, between creative asymmetry, imagination, or revolution and cooperative symmetry logic or order; between Dionysios and Apollon

Jean-Pierre Changeux

I INTRODUCTION

Drugs containing two pharmacophoric groups covalently bounded in a single molecule are called twin drugs. The combination of two identical pharmacophoric entities will lead to an identical twin drug, whereas the association of two different drug entities will generate a non-identical twin drug (Fig. 16.1). The first strategy consists of molecular variations based on duplication, whilst the second one results from associative synthesis.

Identical and non-identical twin drugs may be combined by a linker, a no linker or in overlap mode (Fig. 16.2). The spacer group can be a single bond, a polymeric chain (usually a methylenic chain) or in somes cases an aromatic or nonaromatic cycle.

Two examples of twin drugs are given in Fig. 16.3. Duplication of aspirin led to the identical twin drug diaspirin where the connection mode is a no linker mode. Association of salicyclic acid and paracetamol with an overlap mode gave the non-identical twin drug acetaminosalol.

Identical twin drugs may have different modes of connection of the two drugs entities. Referring to polymer chemistry nomenclature, each molecule can be formally represented with a head and a tail. Thus, a head-to-head, a tail-to-tail or a head-to-tail connection is possible (Fig. 16.4). The first and the second modes generate symmetrical compounds (i.e. glycol salicylate, dibozane) which represent the major part of the identical twin drugs described in the literature. However, nonsymmetrical drugs such as amentoflavone are not uncommon.

Non-identical twin drugs are also named dual acting drugs or hybrids because of the different pharmacological responses targeted by the two pharmacophoric moieties. The design of dual acting drugs, called the symbiotic approach,[1] can be accomplished according two strategies (Fig. 16.5). The first one combines two non-identical pharmacophoric moieties into a hybrid molecule as illustrated by the sulfonamidic derivative. The associative synthesis of a chlorobenzenesulfonamide with an indole derivative through a methylenic linker generates a β-blocking/diuretic agent.[2] The second strategy starts with a lead compound found to already exhibit both activities. A rational optimization will lead to an intrinsically dual acting drug such as the benzocycloheptapyridinylene piperidine.[3] In this case, it is hard to distinguish the area of the molecule responsible of a single biological activity.

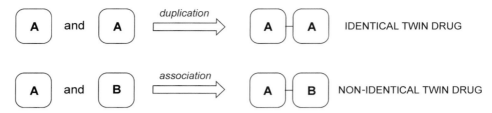

Fig. 16.1 Identical and non-identical twin drugs.

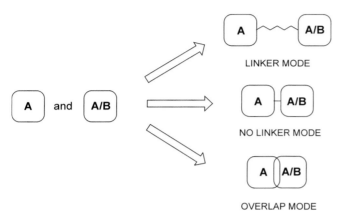

Fig. 16.2 Modes of combining two moities in twin drugs.

The administration of twin drugs can be favourable compared with the two separated drugs. The new entity will have its own pharmacokinetic property (absorption, distribution and metabolism) and thus possible improved efficacy *in vivo*. This aspect represents the main advantage of designing dual acting drugs in addition to the beneficial therapeutic combination of the two active principles. The twin drug must express both activities in

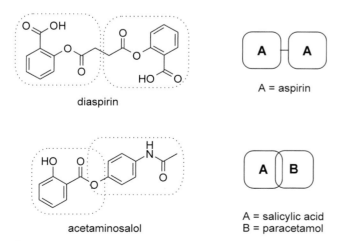

Fig. 16.3 Examples of identical and non-identical twin drugs.

an appropriate balance: a stochiometric association of diazepam (2–20 mg per day) with aspirin (200–2000 mg per day) would be nonsense.

It has also to be considered that the twin drug may produce its constituents after its administration (metabolism). In this case, the twin drug is actually a prodrug of the pharmacophoric entities. If the twin drug is not split *in vivo*, then the mechanism of action can be as followed: (1) for identical twin drugs, the interaction of the pharmacophoric moities with a symmetric macromolecule is conceivable especially when the targeted protein exists as a dimer; an additional interaction with another binding site (an allosteric site for example) on the targeted macromolecule is also possible (Fig. 16.6) and will trigger the biological response; (2) in the case of non-identical twin drugs or dual acting drugs, the simultaneous interaction with both targeted macromolecules is of course excluded; the main advantage resides in the new pharmacological profile of the hybrid derivative.

II IDENTICAL TWIN DRUGS

A Symmetry in nature

Nature is efficient in producing compounds with a high degree of symmetry which allows reduction of information and complexity levels.[4] Natural symmetry is observed for the assembly of macromolecules (oligomers)[5] as for HIV protease, hemoglobin and insulin. The aggregation of insulin monomers to hexamers in presence of zinc affords a macromolecular complex with a high degree of symmetry[6] (C_3 symmetry). DNA, by means of its symmetrical double-stranded structure, determines the cell's morphology and function. These well-organized macromolecular systems constitute binding sites for smaller molecules, including water and ions.

Symmetrical natural compounds generally present a C_2 symmetry axis (Fig. 16.7), like the alkaloids lobelanine (treatment for drug addicts), sparteine (Grave's disease treatment) or isochondrodendrine, and the anticoagulant dicumarol and antispermatogenic gossypol.[7,8]

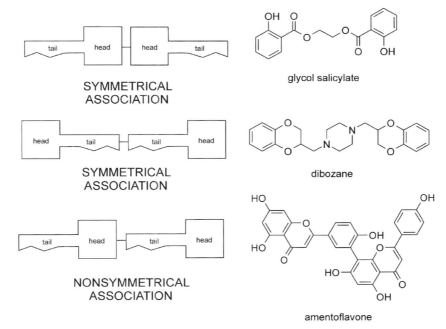

glycol salicylate

dibozane

amentoflavone

SYMMETRICAL
ASSOCIATION

SYMMETRICAL
ASSOCIATION

NONSYMMETRICAL
ASSOCIATION

Fig. 16.4 Modes of association for identical twin drugs.

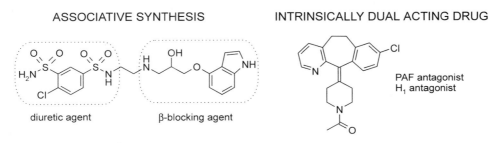

ASSOCIATIVE SYNTHESIS

INTRINSICALLY DUAL ACTING DRUG

PAF antagonist
H₁ antagonist

diuretic agent

β-blocking agent

Fig. 16.5 The symbiotic approach.[1]

Some examples of C_3 symmetrical compounds are known. Valinomycin,[9] a cyclic peptide lactone antibiotic, is a highly selective K^+-carrier. It consists of a cyclic trimer containing L-valine, D-α-hydroxyisovaleric acid (Hyi), D-valine and L-lactate. The C_3 symmetrical agent enterobactin is a cyclic lactonic *N*-acylated serine trimer.

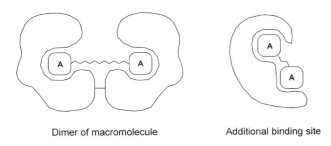

Dimer of macromolecule

Additional binding site

Fig. 16.6 Possible mechanisms of action of twin drugs.

B Twin drugs acting on receptors

Identical twin drugs have shown increasing potencies and/or modified selectivity profiles as receptor ligands, when compared with their corresponding single drug.

Ligands of biogenic amine receptors

Several receptors of biogenic amines (catecholamines), quaternary ammoniums (acetylcholine, PAF-acether), and peptides (angiotensin, endothelin) belong to the well-known class of G-protein coupled receptors (GPCR). They present one subunit with seven trans-membrane spanning domains, three extracellular and three intracellular loops, which are coupled to G-proteins. Duplication of drugs within this series of ligands has been efficient in several cases.

Adrenergic receptors. The development of polyamine disulfides, such as benextramine, an adrenergic blocking agent, allowed the hypothesis of symmetrical properties

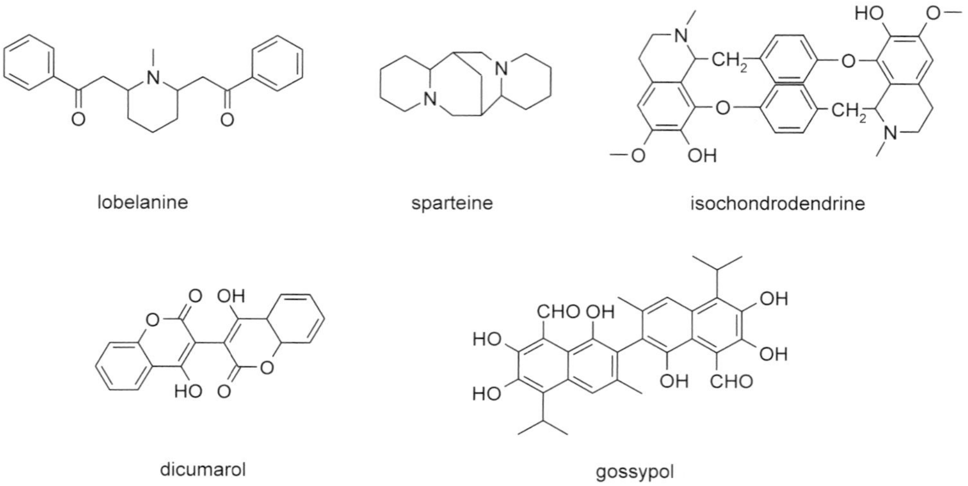

Fig. 16.7   Natural compounds presenting a C_2 symmetry axis.

of α-adrenergic receptors[10] (Fig. 16.8). Duplication of the α1-adrenergic antagonist piperoxan led to dibozane.[7] Several β1-selective adrenoreceptor antagonists have been designed by duplication of the well-known propanolamine structure, oxprenolol.[11]

Acetylcholine receptors. Methoctramine, a selective M2 antagonist,[12] is a useful probe for characterizing acetylcholine muscarinic receptor (mAChR) subtypes. The replacement of the 2-methoxybenzyl group in methoctramine by the hydrophobic moiety of pirenzepine leads to a very potent M2 antagonist, whereas pirenzepine is known as a selective M1 antagonist (Fig. 16.9). The duplication of pirenzepine leads to a potent antagonist with a reversed M2/M1 selectivity profile. Recently,

studies[13,14] have shown that the length of the polymethylene chain of methoctramine polyamines derivatives could be important to convert muscarinic antagonist into selective nicotinic antagonists. Thus, starting from a common pattern, symmetrical twin drugs acting as antagonists at different receptors, have been designed.

PAF-acether receptors. Veraguensin (Fig. 16.10), a symmetrical natural product, was found to inhibit the binding of the platelet-activating factor (PAF) to its specific receptor site (IC50 of 1.1 μM). Symmetrical potent compounds were designed, such as L 652,731 2 (IC50 of 19 nM).[15] The replacement of the tetrahydropyran core by a piperazine led to potent PAF antagonists.[16]

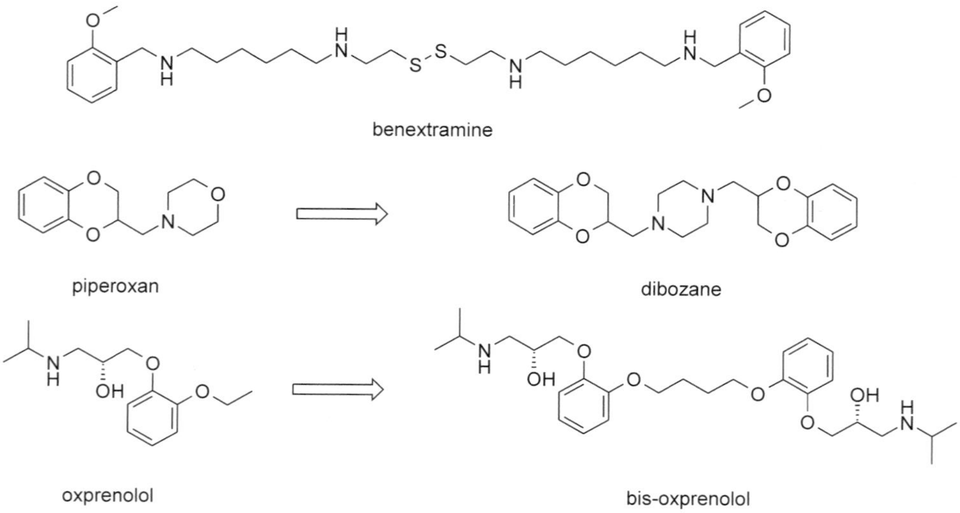

Fig. 16.8   Duplication of adrenergic ligands.

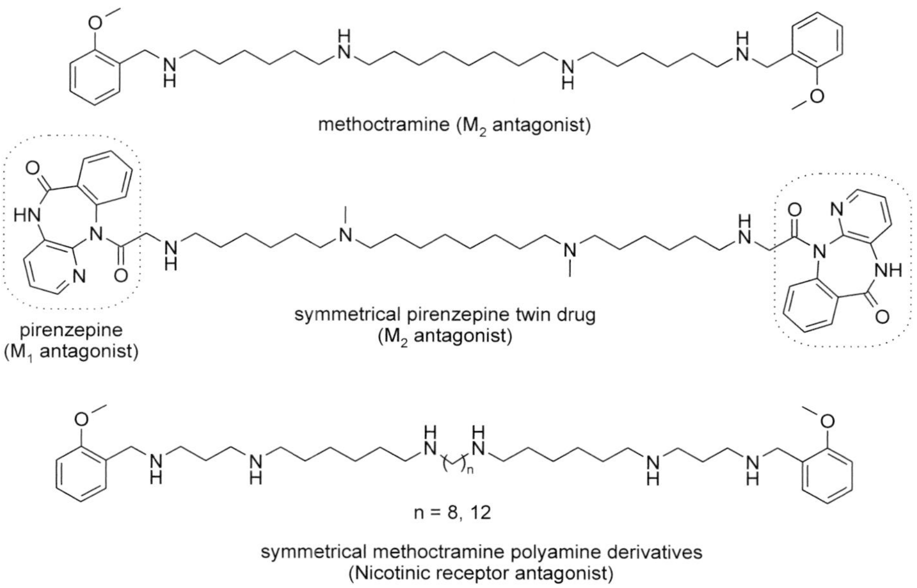

Fig. 16.9 Ligands of acetylcholine receptor.

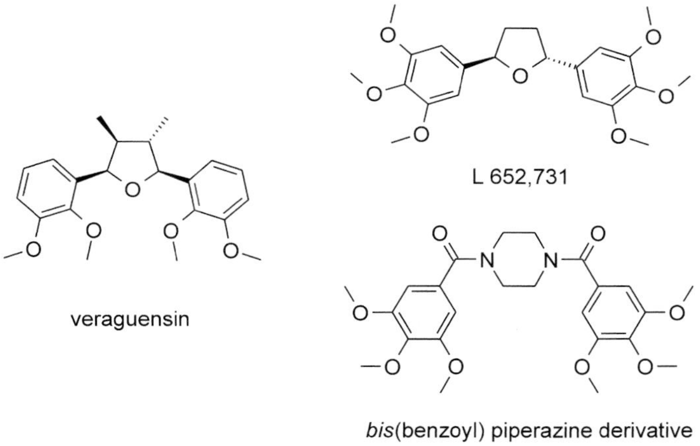

Fig. 16.10 PAF antagonists.

Ligands of peptide receptors

Several studies have previously reported that dimerization of peptidergic receptor ligands can result in an increase in affinity, potency, and/or metabolic resistance.[17–19] The design of non-peptidergic antagonists of bradykinin B$_2$ receptor, potential therapeutic agents in the treatment of inflammation and pain, led to a symmetrical *bis*-phosphonium salt[20] (Fig. 16.11). The 10 Å-distance between the two positive charges is in good agreement with that found between guanidinium cations of Arg (1) and Arg (9) of bradykinin.

In the case of enkephalines, dimerization has shown better analgesic properties when compared with their monomeric counterparts (Fig. 16.12). The increase in

potency and the selectivity profile depend on the length of the methylenic chain.[21]

The design of a bridged opioid dimer (Fig. 16.12) led to a κ-selective opioid receptor antagonist such as norbinaltorphimine.[22,23] In this case, the two identical morphonic units were linked by the mean of a pyrrole ring.

Other examples of identical twin drugs as receptor ligands

The symmetrical *bis*-1,4-dihydropyridine BDHP (Fig. 16.13) is about ten times as potent as the corresponding entity nitrendipine, a highly potent calcium channel antagonist (IC$_{50}$ of 0.2 nM).[24]

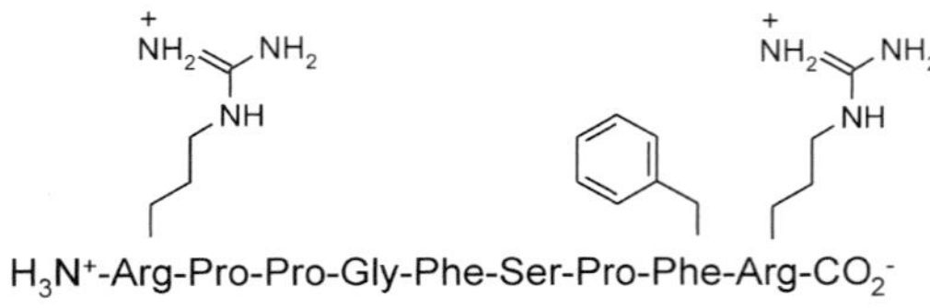

H_3N^+-Arg-Pro-Pro-Gly-Phe-Ser-Pro-Phe-Arg-CO_2^-

bradykinin

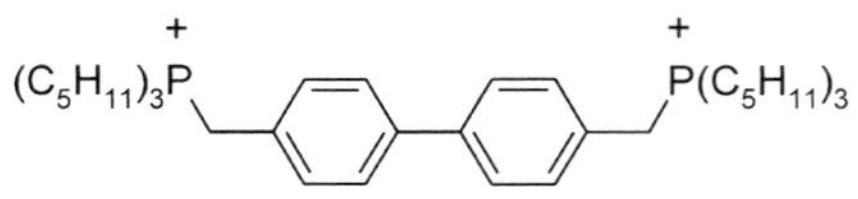

$(C_5H_{11})_3P^+$ ... $P^+(C_5H_{11})_3$

symmetrical *bis*-phosphonium salt

Fig. 16.11 Bradykinin antagonists.

Recently, *bis*-azomacrocycles such as the bicyclam AMD3100 (Fig. 16.13) were reported to exhibit potent and selective inhibition of the HIV (human immunodeficiency virus) replication. These derivatives bind to a receptor called CXCR4 necessary for the entry of the virus into the cell.[25]

C Twin drugs as enzyme inhibitors

Within the symmetrical arrangement of enzymes into homodimers or tetramers lies the active site of the enzyme in a highly symmetrical fashion. Thus, symmetrical inhibitors will correspond generally to the binding site of the enzyme.

HIV-enzyme inhibitors

HIV reverse transcriptase (HIV RT) and protease (HIV PR) are essential for the maturation and production of infectious viral particles. A combined inhibition of HIV

RT and HIV PR is capable of reducing the viral load in blood patients.[26] These enzymes, which exist respectively as a heterodimer and a homodimer for HIV RT and HIV PR, are well-characterized: more than 170 structures of HIV PR and its complexes with various inhibitors have been solved by protein crystallography techniques.[27,28] Thus, a dipalmitoylated derivative of 2,7-naphthalene disulfonic acid demonstrated micromolar activity for both HIV-1 and HIV-2 RT[29] (Fig. 16.14). Symmetrical nature of HIV PR was used in the search for novel anti-HIV drugs that would embody the predicted characteristic of the active site. The design of inhibitors of HIV PR has led to symmetrical compounds, which can be divided into two groups: (1) pseudosymmetrical compounds, like derivatives A 74704[30] and L 700,417 which contain asymmetric atoms in close proximity to the inhibitor two-fold axis; (2) fully C_2-symmetrical inhibitors like the cyclic urea[31] and the diol derivatives[32] (Fig. 16.14).

H-Tyr-D-Ala-Gly-Phe-NH_2 $\Longrightarrow$

H-Tyr-D-Ala-Gly-Phe-NH
|
$(CH_2)_n$
|
H-Tyr-D-Ala-Gly-Phe-NH

enkephaline twin drugs (n = 2, 12)

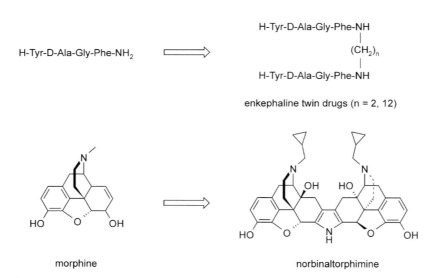

morphine

norbinaltorphimine

Fig. 16.12 Twin drugs for opoid receptors.

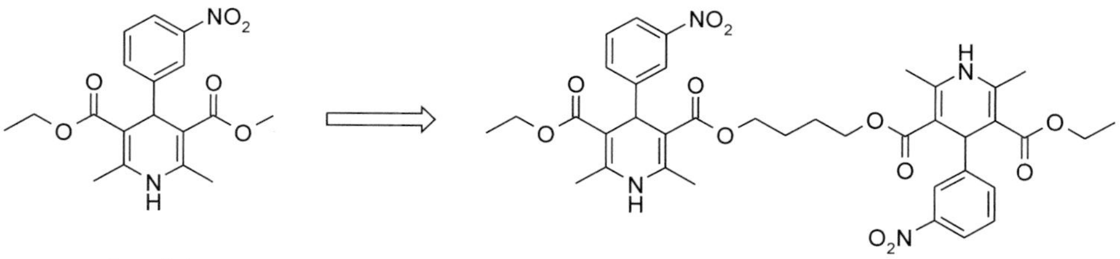

nitrendipine

BDHP

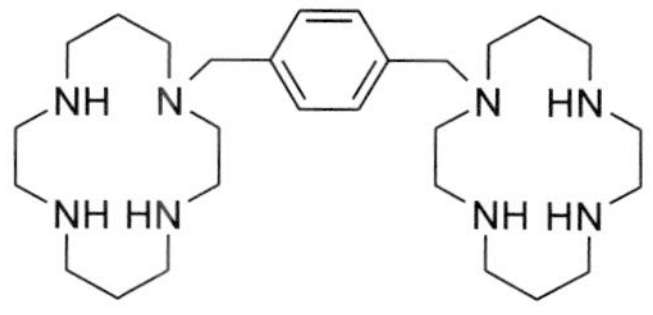

AMD3100

Fig. 16.13 Other symmetrical receptor ligands.

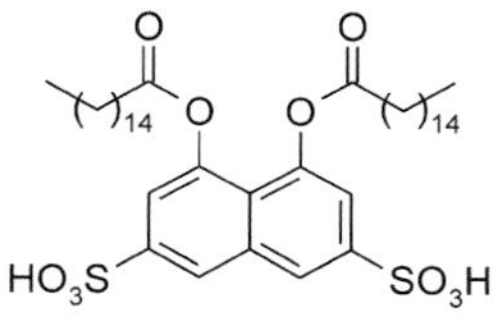

symmetrical disulfonate

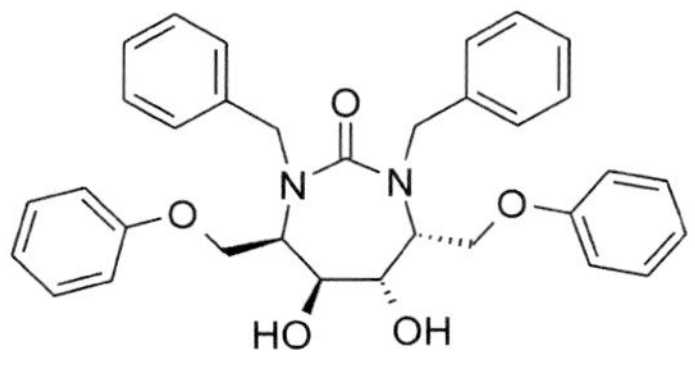

cyclic urea derivative

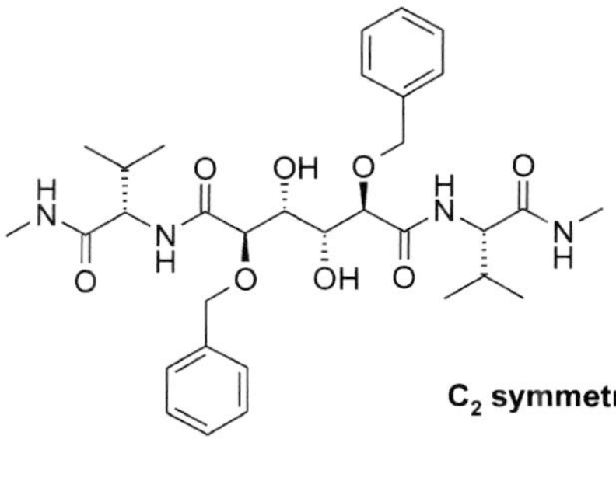

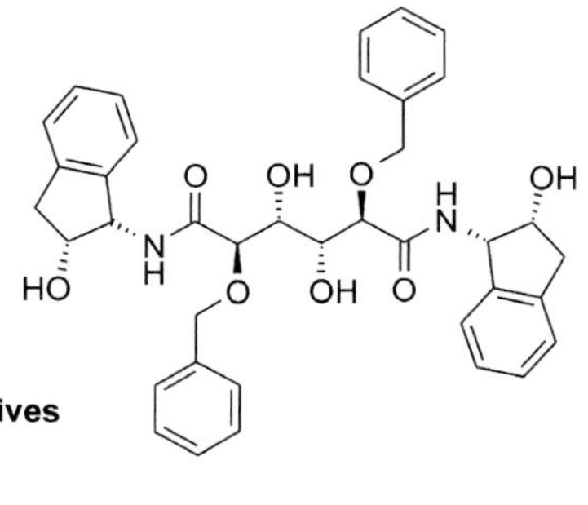

C₂ **symmetric diol derivatives**

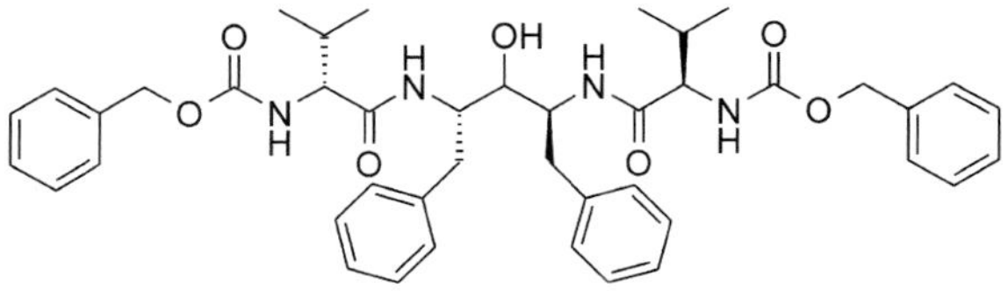

A 74704

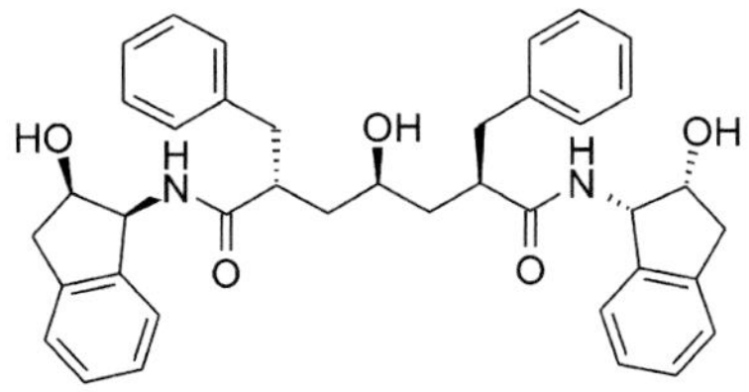

L 700,417

Pseudo-symmetrical inhibitors

Fig. 16.14 HIV reverse transcriptase and protease inhibitors.

Protein kinase inhibitors

The protein kinase (PK) enzyme family plays a pivotal role in the signal transduction pathways of hormones, neurotransmitters and other endogenous substances. Thus PKC, a serine/threonine kinase, is an interested target for new chemotherapeutic agents because of its function in tumour formation and metastasis. Starting from the aglycone part of staurosporine, a nonselective PKC inhibitor, symmetric diindolylmaleimide[33–35] and dianilinophthalimide[36] derivatives were designed and found to be potent and selective protein kinase inhibitors (Fig. 16.15). Dequalinium (DECA) analogues with longer and saturated linkers exhibited enhanced potency for inhibition of PKCα. The presence of a two-point contact on the enzyme by DECA analogues may explain the potency and the selectivity of such compounds.[37]

Piceatannol, an antileukemic principle known to inhibit proteine-tyrosine kinase (PTK)[38] was chosen as a lead compound for SAR analysis. The fully symmetrical 5-hydroxy isomer was found to be four times more potent than piceatannol. When compared with the reference monomer A46, the *bis*-tyrphostin was found to be 150 times more potent in protein tyrosine kinase assays.[39]

Other identical twin drugs as enzyme inhibitors

The symmetrical bipyridine dicarboxylic acid was reported as a potent inhibitor of prolyl hydrolase, an enzyme involved in collagen biosynthesis[40] (Fig. 16.16). The beneficial effect of drug duplication observed here may result of a single binding mode in each catalytic subunit.

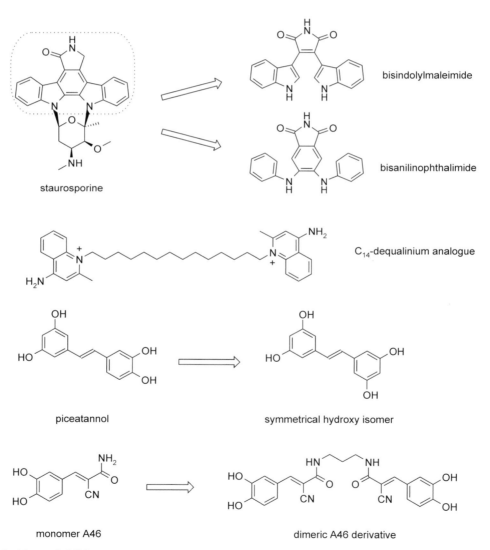

staurosporine

bisindolylmaleimide

bisanilinophthalimide

C₁₄-dequalinium analogue

piceatannol

symmetrical hydroxy isomer

monomer A46

dimeric A46 derivative

Fig. 16.15 Protein kinase inhibitors.

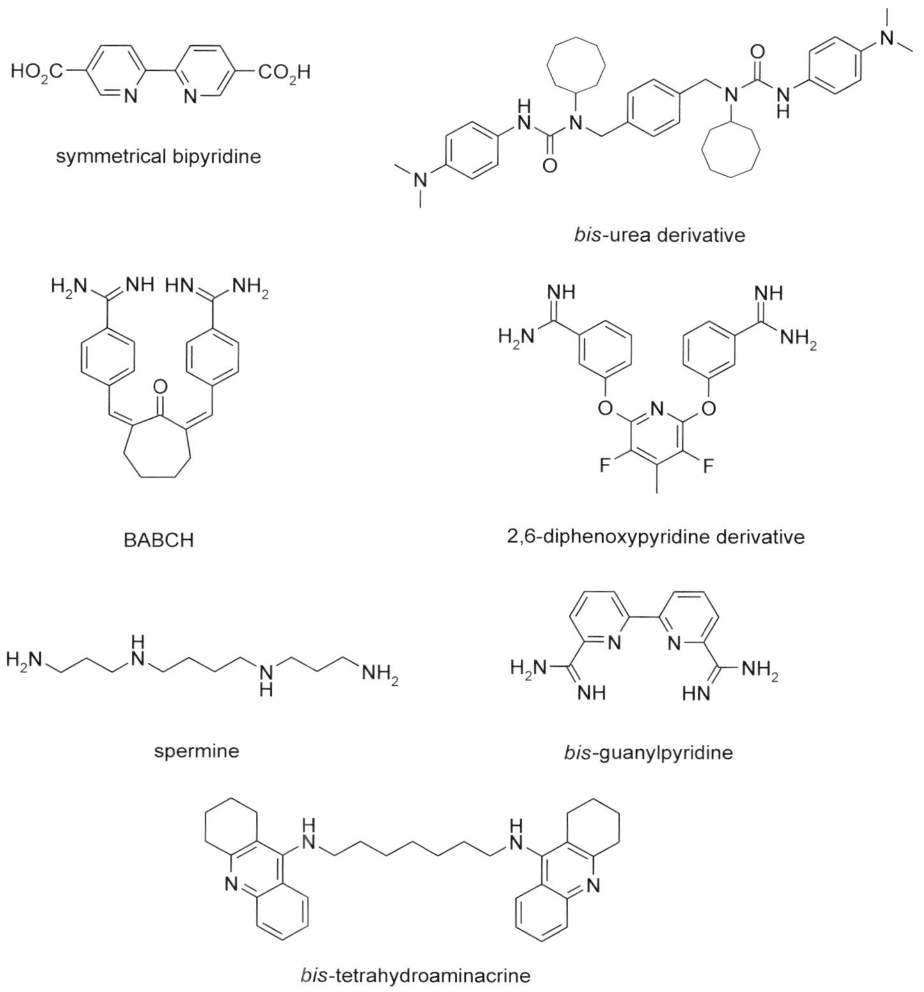

Fig. 16.16 Various enzyme inhibitors.

The *bis*-urea derivative YM 17E has been described as an inhibitor of acylcoenzyme A cholesterol transferase (ACAT), a potential anti-cholesterolemic drug.[41]

Active inhibitors of factor Xa (FXa), a serine protease, would allow control of the coagulation process and minimize bleeding problems.[42] A recent study[43] shown that replacement of the cycloheptanone central scaffold of the BABCH derivative by a pyridine heterocycle led to more potent and selective FXa inhibitors such as the diphenoxypyridine compound. Inhibitors of enzymes involved in the polyamine metabolic pathway have been designed as potential antitumour and antiparasitic agents. The polyamines spermidine and spermine exhibit promising antitumour activity against cultured human lung cancer cells by inhibiting spermidine-spermine-N_1-acetyltransferase (SSAT) an enzyme in polyamine catabolism.[44] The symmetrical *bis*-guanylpyridine analogue was found to be a potent and selective S-adenosylmethionine decarboxylase

(SAMDC) inhibitor.[45] SAMDC is another rate-limiting enzyme of polyamine biosynthesis.

The design of bifunctional acetylcholinesterase inhibitors was achieved in order to obtain potent and selective derivatives. *Bis*-tetrahydroaminacrine showed a simultaneous interaction with the active site and the peripheral site[46] (allosteric site) of the enzyme resulting in an improvement of potency and selectivity.[47]

D   Twin drugs acting as DNA ligands

The DNA molecule is the primary target of many antitumour agents. Small molecules, which bind to DNA by intercalation, require polycyclic systems in their structure for efficient binding. Because of the symmetrical arrangement of the helical double strand, symmetry is found in the structure of DNA ligands. Polycyclic systems bearing symmetrical polyamine side chains, such as mitoxantrone[48]

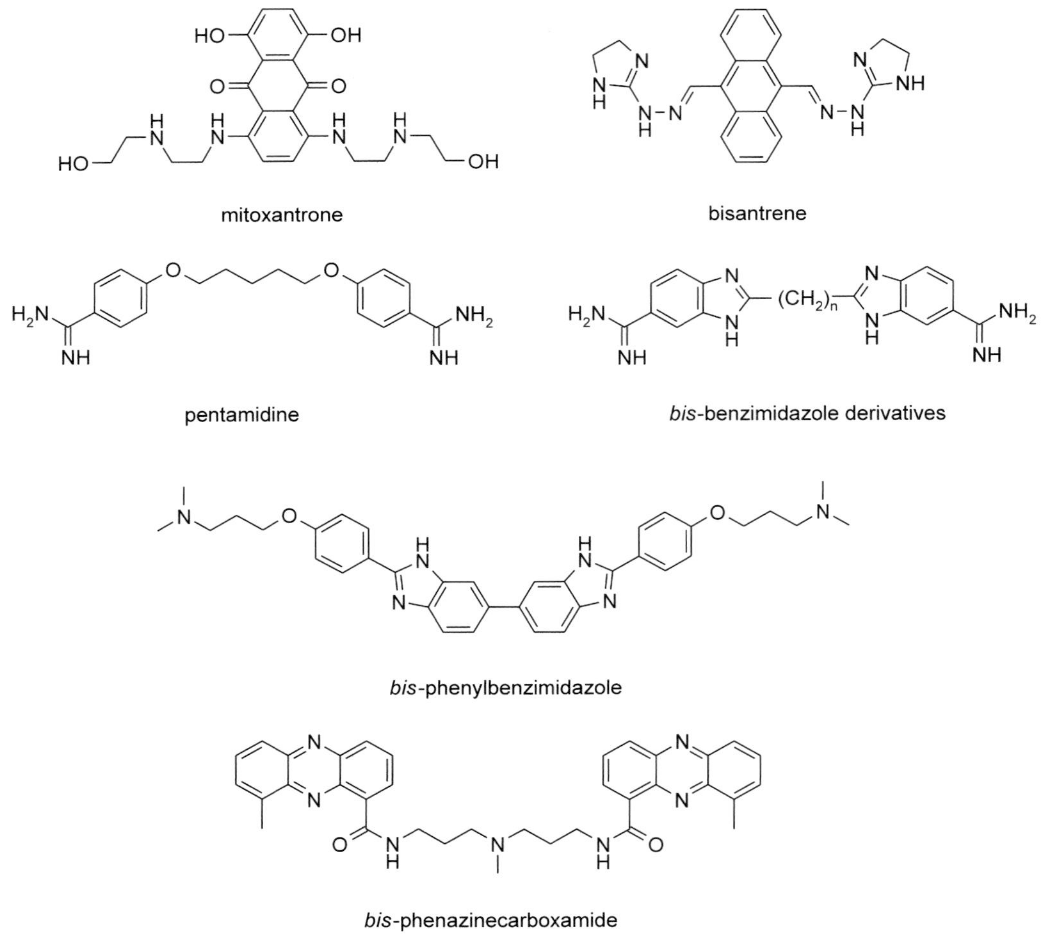

mitoxantrone

bisantrene

pentamidine

bis-benzimidazole derivatives

bis-phenylbenzimidazole

bis-phenazinecarboxamide

Fig. 16.17 DNA ligands.

and bisantrene,[49] interact as polycations with the phosphate groups of DNA (Fig. 16.17). Pentamidine and the *bis*-amidinobenzimidazoles bind to the minor groove of DNA and show higher affinity for AT-base pairs rich regions.[50] Compounds with an even number of methylenes connecting benzimidazole rings have a higher affinity for DNA than those with an odd number of methylenes. Recently, a symmetric head-to-head arrangement of two phenylbenzimidazole groups led to a potent antitumour agent.[51] *Bis*(phenazine-1-carboxamide) derivatives showed a large increase in potency over their monomeric counterparts.[52] Such compounds were found to have an action on both topoisomerase I and II.

E Other identical twin drugs of pharmacological interest

Chlorhexidine is a very effective antiseptic drug used in disinfectant soaps. Hexachlorophene, a twin drug of 2,4,5-trichlorophenol, is used as a bactericide and veterinary flukicide (Fig. 16.18). It is more potent and less toxic than the corresponding monomer.[7] Well-known symmetrical compounds such as dioxin (TCDD: tetra-chlorodibenzodioxin, a contaminant produced during the manufacture of 2,4,5-trichlorophenol) or the herbicide paraquat, are highly toxic. Disulfides sulbutiamine and pyritinol are used respectively as twin drugs of vitamin B_1 and B_6.[53]

Several bisquinolines, such as piperaquine[54] have notable activity and longer duration of action against chloroquine-resistant malaria (Fig. 16.19). In a recent study, bis(aminoacridine) derivatives with different linker were evaluated for their antiparasitic activity.[55] The activity profile of these compounds was strongly dependent upon the nature and the length of the connecting linker between the heterocyclic rings.

Meprobamate, a dicarbamate derivative, has been developed for its anxiolytic and antiepileptic properties (Fig. 16.20). Disulfiram is used to treat alcoholism.

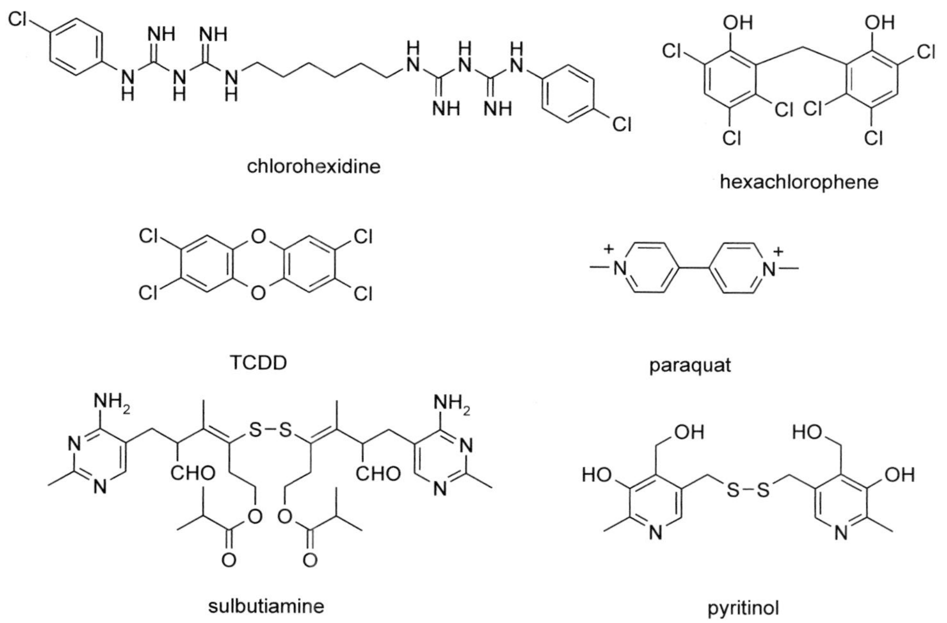

Fig. 16.18 Other identical twin drugs.

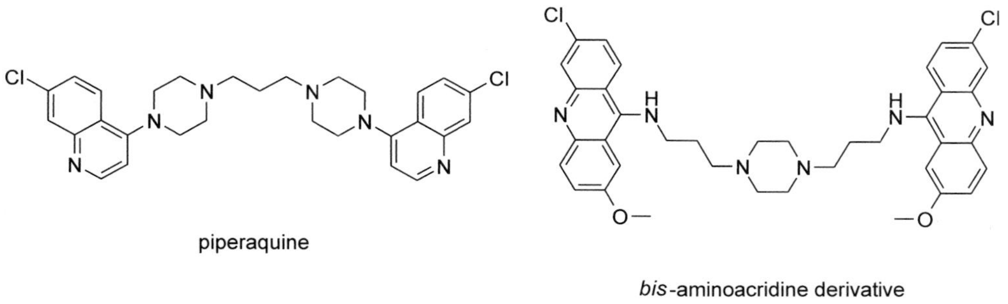

chlorohexidine

hexachlorophene

TCDD

paraquat

sulbutiamine

pyritinol

Fig. 16.19 Bis-quinoline and aminoacridine twin drugs.

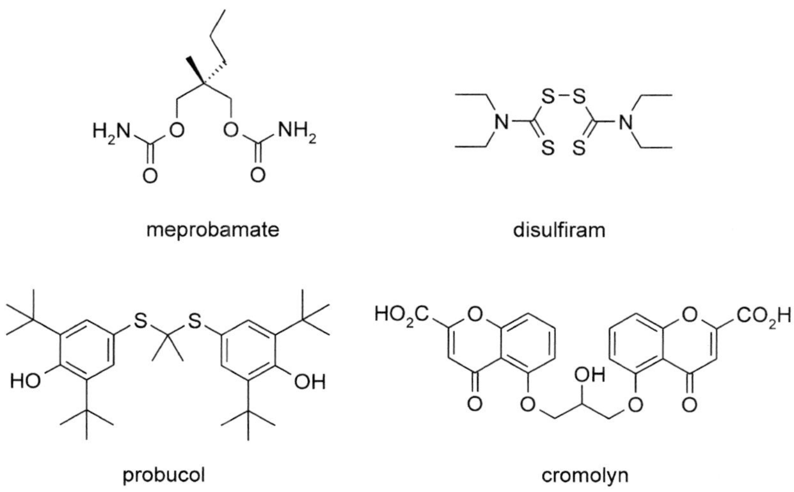

piperaquine

bis-aminoacridine derivative

Fig. 16.20 Various identical twin drugs.

meprobamate

disulfiram

probucol

cromolyn

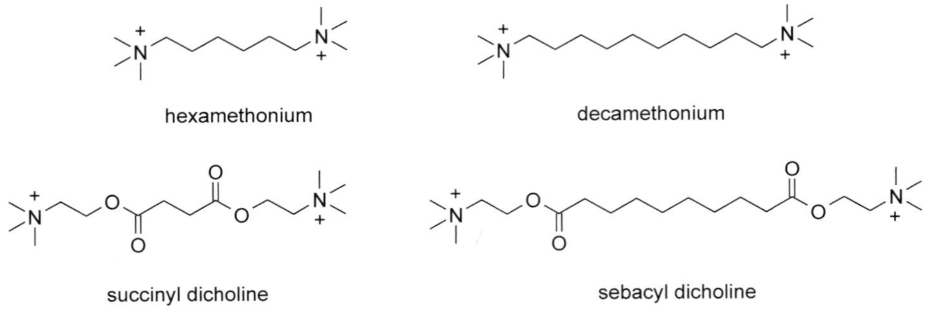

Fig. 16.21 Cholinergic twin drugs.

Diethyl dithiocarbamate, the major metabolite of disulfiram has recently been shown to block dopamine β-hydroxylase, and may be the reason for the hypotensive effects characteristic of the disulfiram-mediated modulation of ethanol metabolism.[56] The antioxidant probucol lowers the cholesterol level in blood. Cromolyn, a chromone heterocycle, is useful in the inhalation treatment of bronchial asthma.

The search of cationic cholinergic agents has led to numerous twin drugs (Fig. 16.21). The *bis*-quaternary ammonium salts hexamethonium and decamethonium are potent blockers in ganglia and in neuromuscular junction, respectively. Other neuromuscular blocking agents such as succinyl and sebacyl dicholines can be regarded as pure acetylcholine twin drugs.

III NON-IDENTICAL TWIN DRUGS: DUAL-ACTING DRUGS

Dual-acting drugs exert their dual action on two different targets. These targets could be receptors (GPCR, receptor subtypes in the same family or different receptor families), enzymes or a combination of both. The association of two physiological effects is aimed at obtaining a synergic response in the treatment of a disease or a disorder. As we see before, hybrid molecules could result from the association of two distinct active principles (associative synthesis) or from a compound with an intrinsic dual acting profile. In the first case, the two pharmacophoric entities are linked and recognizable, whereas in the second case it is often difficult to identify the chemical component of the molecule responsible for biological activity.

A Hybrid molecules as ligands of two different receptors

GPCRs possess physical, biochemical and structural similarities. Thus, selectivity toward biogenic amines such as noradrenaline (NA), serotonin (5-HT), dopamine (D) and histamine (H) depends on a typical Asp interaction (amino acid localized in the third transmembranar helix) and additional binding interactions. Because pharmacophores of all these ligands are similar, the control of their selectivity constitutes an important challenge for medicinal chemists. Thus, it may be of interest to synthesize hybrid drugs that bind potently to different GPCRs as an agonist, antagonist or mixed agonist/antagonist. Labetalol lowers blood pressure by blocking α- as well as β-adrenergic receptors,[57] whereas D2343 acts as a combined β-agonist and α-antagonist[58] (Fig. 16.22).

It has been proposed that a combined administration of a 5-HT$_2$ antagonist (for example, ritanserine) and a D$_2$ antagonist could be efficient in the treatment of

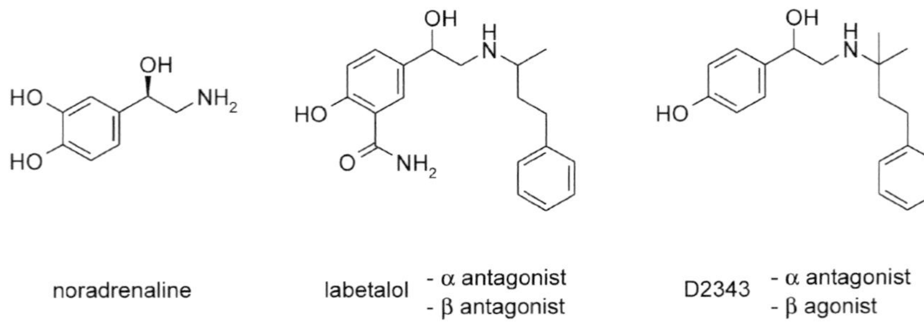

Fig. 16.22 Adrenergic dual acting drugs.

schizophrenic patients. SAR study on bridged γ-carbo-lines, reported for its potent affinity for 5HT$_2$ and moderate affinity for D$_2$ receptors, was achieved and led to the design of a compound with equipotent and nanomolar affinity for both receptors[59] (Fig. 16.23).

Positive inotropic drugs, such as arpromidine, have been developed by combining in a same molecule histamine H$_1$ antagonistic and H$_2$ agonistic properties[60] (Fig. 16.24). The H$_1$ antihistaminic drug loratadine presents weak PAF antagonistic property. Taking into account the physiological importance of PAF in asthma, it was of therapeutic potential interest to antagonize the action of both mediators in a single molecule.[3] Thromboxane A$_2$ (TXA$_2$) is also implicated in the pathophysiological conditions of asthma. Therefore efforts have been made to design TXA$_2$ receptor

antagonists. The symbiotic approach concept led to the recent discovery of dibenzoxepin derivatives, such as KF 15766, with both histamine H$_1$ and TXA$_2$ dual antagonizing activity.[61]

The octapeptide hormone angiotensin II (AT) is involved in vascular smooth muscle contraction and release of other endogenous substances. Since AT$_1$ and AT$_2$ receptors are present in varying proportions in many tissues and organs, dual antagonists may constitute efficient pharmacological tools. Starting from losartan (Fig. 16.25), a selective AT$_1$ antagonist, and PD 123317, a selective AT$_2$ antagonist, potent and dual AT$_1$ and AT$_2$ antagonists were designed.[62]

Dual-acting drugs can also result from the combination of ligands belonging to completely distinct pharmacophores.

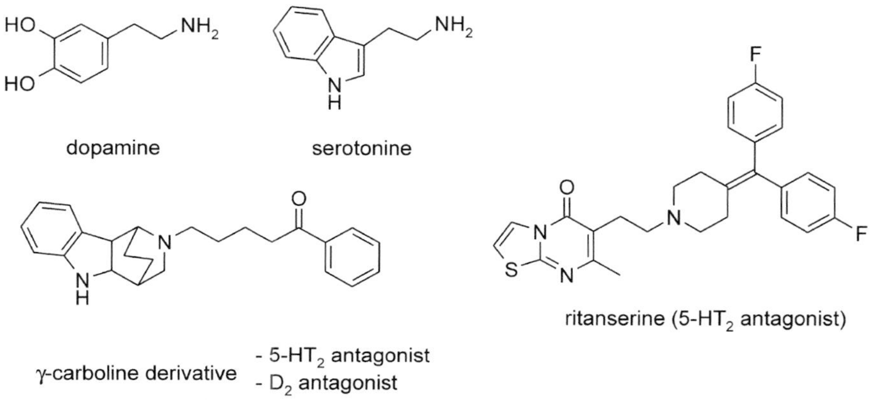

Fig. 16.23 Dopaminergic and serotonergic hybrid drugs.

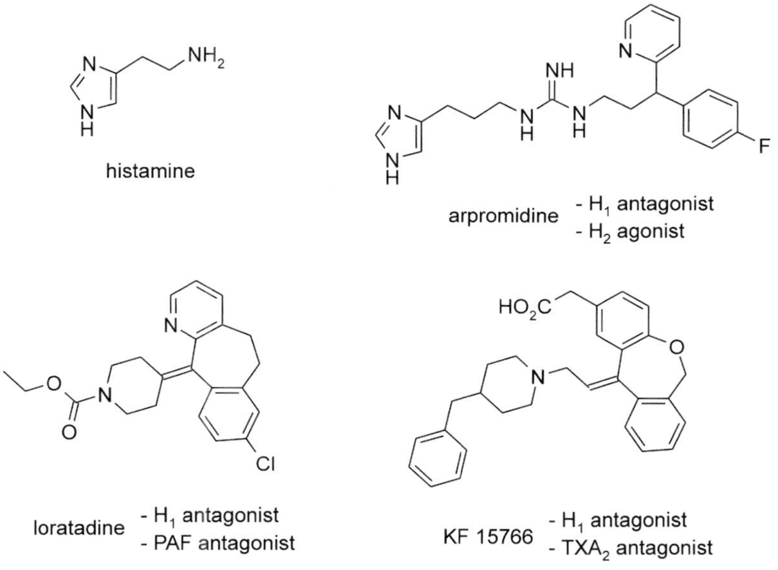

Fig. 16.24 Histaminergic hybrid drugs.

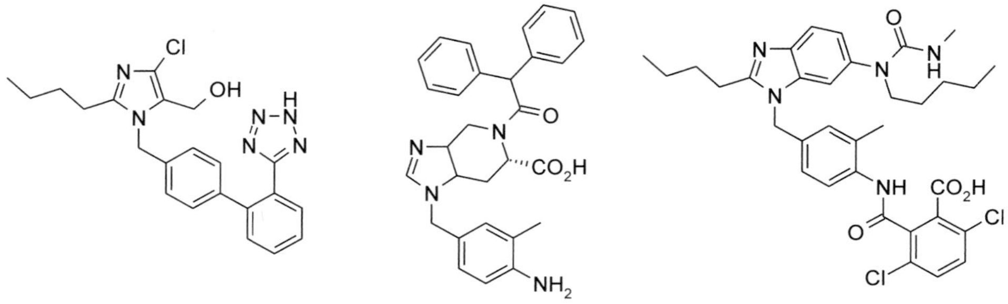

Fig. 16.25 Angiotensine receptor ligands.

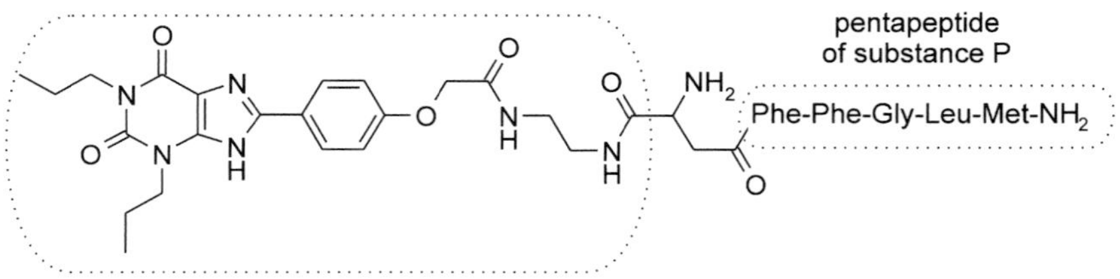

losartan (AT$_1$ antagonist)   PD 123317 (AT$_2$ antagonist)   BIBS 222 $-$ AT$_1$ antagonist / $-$ AT$_2$ antagonist

Fig. 16.26 Substance P and adenosine hybrid ligand.

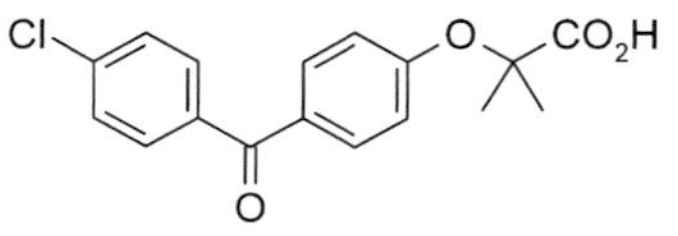

xanthine derivative

pentapeptide of substance P

Phe-Phe-Gly-Leu-Met-NH$_2$

fenofibric acid (PPARα agonist) 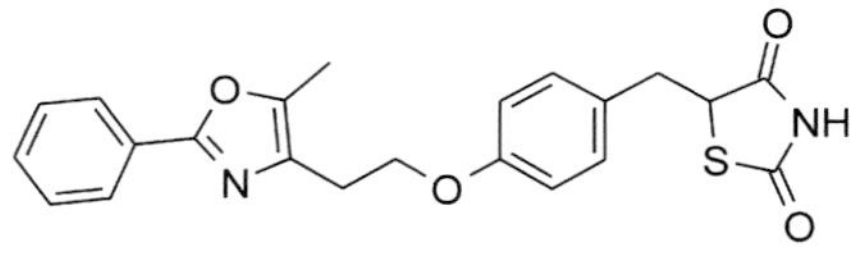   thiazolidinedione derivative (PPARγ agonist)

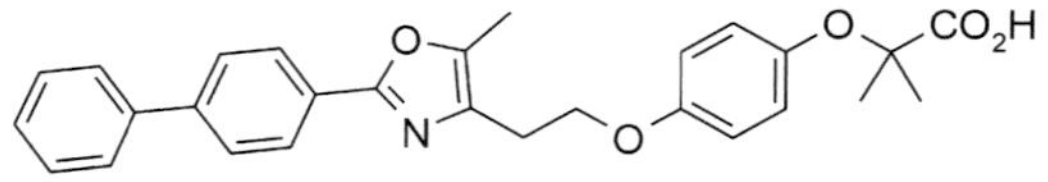

aryloxazole derivative $-$ PPARα agonist / $-$ PPARγ agonist

Fig. 16.27 Dual PPAR agonist.

As activation of substance P and adenosine receptors produce the same effect (e.g., hypotension and analgesia), it was of therapeutic interest to combine these properties in a single compound. A xanthine derivative, an A$_1$ adenosine receptor antagonist, was linked with the pentapeptide terminal part of substance P to give a twin drug (Fig. 16.26) with similar magnitude for both receptors.[63]

Peroxisome proliferator activated receptors[64] (PPARs) are members of the superfamily of nuclear receptor that includes receptors for steroid, retinoid and thyroid hor-

mones. The demonstration that PPARα and PPARγ were the receptors through which, respectively, the fibrate (lipid lowering activity) and the glitazone (insulin sensitization) drugs mediate their biological activity has led to the design of a new generation of dual acting drugs. Dual PPARα/γ agonists appear well suited for treatment of hyperglycemia with prevention of cardiovascular disease in type 2 diabetes. The aryloxazole derivative (Fig. 16.27) is a typical example of the combination of the fenofibric acid and the thiazolidinedione derivative.[65]

B Hybrids as enzyme inhibitors

As observed with receptors, enzymatic systems can be subdivided into enzyme families, and each enzymatic type presents several isoforms. Thus, for pharmacological and therapeutic purposes, it may be of interest to combine structural characteristics for inhibition of two different isoenzymes in a same molecule, two enzymes belonging to the same family, or two enzymes for which inhibitors are showing pharmacophore similarities.

Dual inhibitors of both cyclooxygenase (CO) and 5-lipoxygenase (5-LO), enzymes involved in the biosynthesis of prostaglandins and leukotrienes, are being studied as anti-inflammatory agents with an improved safety profile in comparison with nonsteroidal anti-inflammatory drugs (Fig. 16.28). The thiazolone CI-1004 for example was identified as a nonulcerogenic water-soluble and orally active anti-inflammatory agent.[66] Simultaneous inhibition of P450 aromatase (P450 arom) and thromboxane A_2 synthase (P450 TxA$_2$) represents a novel strategy for the treatment of mammary tumours and the prophylaxis of metastases. Imidazolyl dihydroquinoline derivative[67] is a potent and dual inhibitor of both P450 arom and P450 TxA$_2$. The inactivation of the endogenous opioid peptide *enkephaline* is one of the physiological roles of neutral endopeptidase (NEP). It has been suggested that simultaneous inhibition of the related metallopeptidases angiotensin I-converting enzyme (ACE) and NEP might be advantageous in the treatment of congestive heart failure or hypertension. Thiorphan, the well-known inhibitor of NEP, has dual NEP-ACE inhibiting properties, but it is a hundred times less potent as an ACE inhibitor than as a NEP inhibitor. A rigid benzazepinone was designed as a PheLeu mimetic and was shown to be a potent dual NEP/ACE inhibitor.[68]

Nucleoside reverse transcriptase inhibitors (NRTIs), such as zidovudine (AZT) or zalcitadine (ddC), have proven to have high efficacy when they are used in combination with HIV PR inhibitors. Because of increasing adverse side-effects, research has been focused on nonnucleoside reverse transcriptase inhibitors (NNRTIs), inhibitors of a noncompetitive binding site. NRTI-NNRTI combination therapy will exhibit a synergic activity and have greater efficacy. Thus, conjugates containing both nucleoside analogue components and nonnucleoside type inhibitors (Fig. 16.29) were designed and showed a different inhibition mode.[69]

Administration of acetylcholinesterase (AChE) inhibitors again represent a promising approach for treating Alzheimer's disease, since the correlation between this enzyme and amyloid formation has been demonstrated. New potent and selective AChE inhibitors such as the tacrine-huperzine hybrid[70] derivative or the *bis*-interacting galanthamine ligand[71] has been design and synthesized (Fig. 16.30). The first analogue is a combination in a molecule of two well-known AChE inhibitors, whereas the second represents

CI-1004
- CO inhibitor
- 5-LO inhibitor

imidazolyl dihydroquinoline
- P450 TxA$_2$ inhibitor
- P450 arom inhibitor

benzazepinone
- NEP inhibitor
- ACE inhibitor

Fig. 16.28 Dual acting inhibitors.

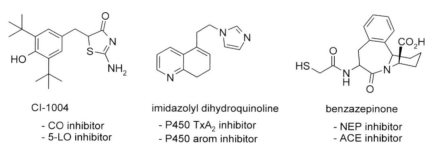

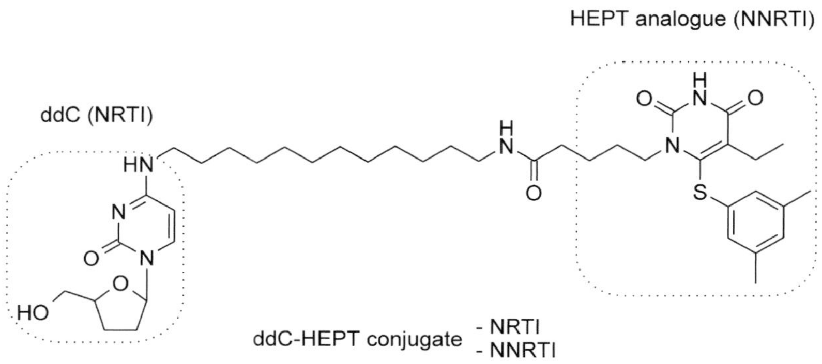

HEPT analogue (NNRTI)

ddC (NRTI)

ddC-HEPT conjugate — NRTI — NNRTI

Fig. 16.29 Mixed HIV protease inhibitor.

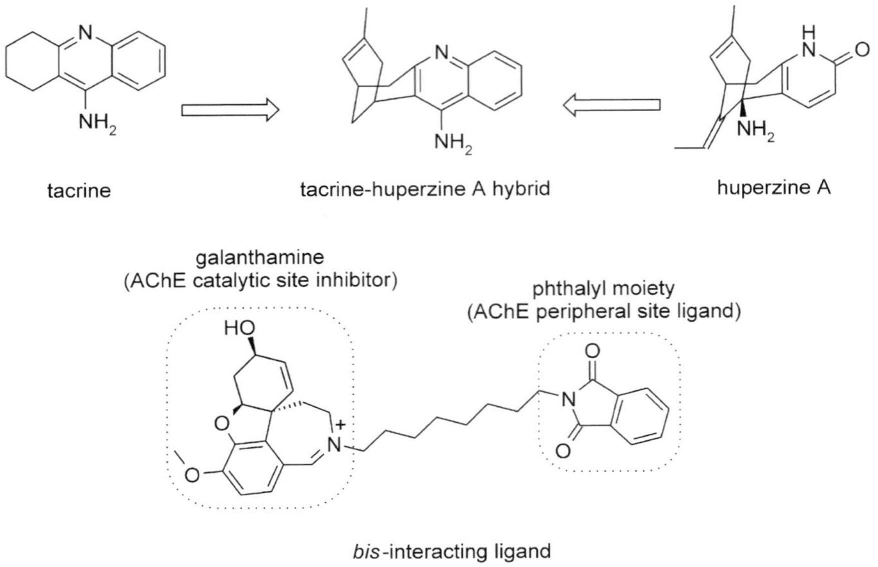

tacrine   tacrine-huperzine A hybrid   huperzine A

galanthamine
(AChE catalytic site inhibitor)

phthalyl moiety
(AChE peripheral site ligand)

bis-interacting ligand

Fig. 16.30 AChE hybrid inhibitors.

an association of two pharmacophoric entities able to interact with the active and the peripheral site of the enzyme.

C Hybrids acting at one receptor and one enzyme

Hybrid molecules acting simultaneously on a receptor and on an enzyme may produce potent synergistic effects. An illustration is given by the example of derivatives interfering with thromboxane A_2 (TXA$_2$), a powerful inducer of platelet aggregation and vascular smooth muscle

contraction. The inhibition of TXA$_2$ synthase (TxS) and the selective blockade of TXA$_2$ receptors (TxR) have been pursued as alternative therapeutic strategies to prevent the thrombotic action of TXA$_2$. Thus, TxS-inhibiting and TxR antagonistic properties have been combined in a single molecule, such as samixogrel[72] starting from isbogrel, a synthase inhibitor, and daltroban, a TXA$_2$ receptor antagonist (Fig. 16.31). This hybrid compound was then optimized into the guanidine derivative terbogrel.[73]

Depending on the physiological hypothesis, both targets may belong to different systems (Fig. 16.32). Thus TxS

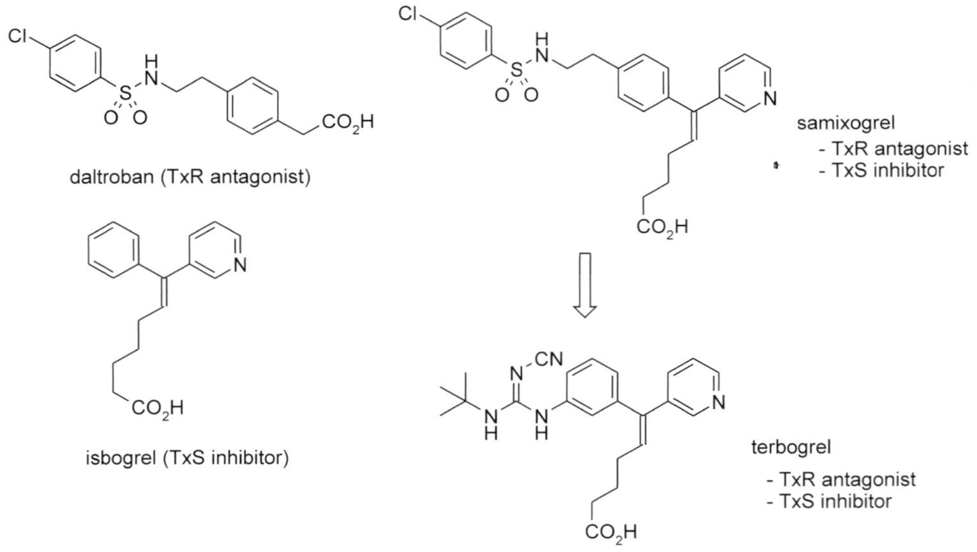

daltroban (TxR antagonist)

samixogrel
- TxR antagonist
- TxS inhibitor

isbogrel (TxS inhibitor)

terbogrel
- TxR antagonist
- TxS inhibitor

Fig. 16.31 TXA$_2$ hybrid drugs.

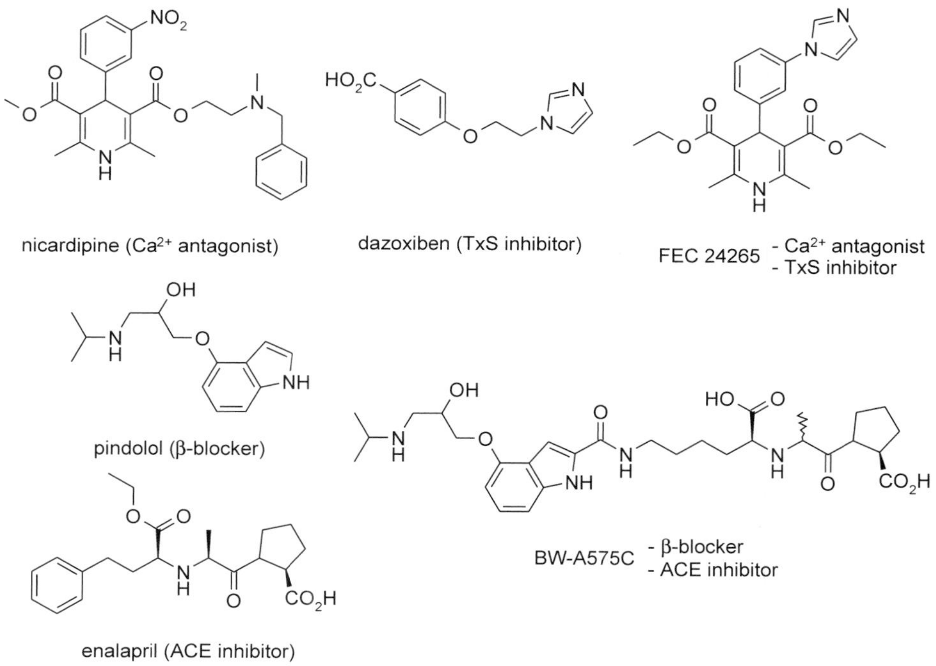

Fig. 16.32 Hybrid drugs with synergic effects.

inhibition and dihydropyridine calcium antagonistic properties led to novel compounds, such as FEC 24265, with relatively more favourable *in vivo* pharmacological profiles, particularly in pathologies where both enhanced TXA_2 synthesis and cellular calcium overload are involved.[74] Hybridization of the β-blocker pindolol and the angiotensin-converting enzyme (ACE) inhibitor enelapril led to the derivative BW-A575C which expresses both activities.

Minaprine, an atypical antidepressant drug that enhances both serotonergic and dopaminergic transmissions[75]

presents cholinomimetic properties.[76] Thus, this derivative is a partial muscarinic M_1 receptor agonist with a weak anticholinesterase activity. Minaprine represents an original lead for the search of dual-acting drugs for the treatment of Alzheimer's disease. A first SAR study (Fig. 16.33) allowed the design of potent and selective M_1 agonists such as the derivative SR 42559-A.[77,78] Recently, a new exploration achieved from, again, minaprine, led to the preparation of selective mixed-type acetylcholinesterase inhibitors[79,80] with micromolar affinity for muscarinic M_1 and M_2 receptors.

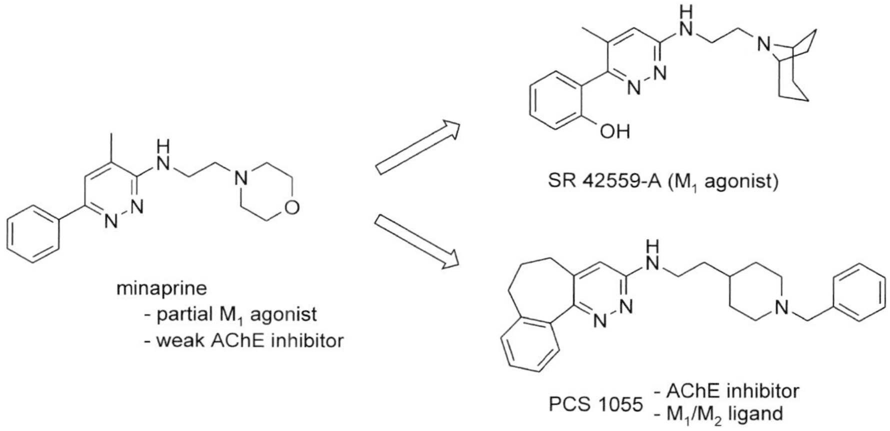

Fig. 16.33 Intrinsically dual acting drug.

D Other examples of dual-acting drugs

Combined treatment is necessary in the long-term treatment of essential hypertension. β-Blocker and diuretic properties in a same molecule would present a great interest for hypertension management. Few attempts to synthesize hybrid molecules by combining the structures of a β-adrenoreceptor antagonist and a diuretic were described (Fig. 16.34). A hybrid sulfonamide was achieved by linking the β-blocker propranolol derivative with the 2-chlorobenzene sulfonamide moiety of mefruside.[2] In prizidilol a typical hydrazino pyridazine core, known for its vasodilator property, is combined with propanolol, leading to a dual vasodilator/β-blocker.[58]

Dual-acting antibacterial drugs linking quinolones (Fig. 16.35), such as ciprofloxacin, to cephalosporin demonstrated potent activity against a broad spectrum of Gram-positive and G-negative bacteria including β-lactam-resistant strains.[81]

In the search for new and efficient antidepressants, dual acting drugs with selective serotonin (5-HT) re-uptake (SSR) inhibition properties and 5-HT$_{1A}$ receptor antagonism were designed (Fig. 16.36). A combination of the SSR inhibitor duloxetine and an arylpiperazine derivative, a class of derivatives known to have high affinity for 5-HT$_{1A}$ receptor (for example, NAN-190), led to the hybrid benzothiophene piperazine, a potential class of antidepressant with a dual mechanism of action.[82] In the same manner, blockade of terminal 5-HT$_{1B/1D}$ receptors by selective antagonists would in theory prevent the initial decrease. Thus, coadministration of SSR inhibitors and 5-HT$_{1B/1D}$ antagonists would lead to a large increase of 5-HT

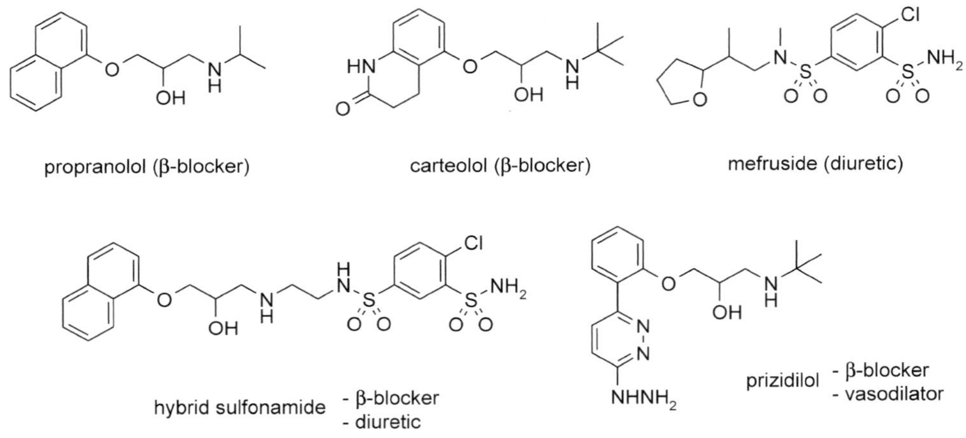

propranolol (β-blocker)   carteolol (β-blocker)   mefruside (diuretic)

hybrid sulfonamide - β-blocker / - diuretic

prizidilol - β-blocker / - vasodilator

Fig. 16.34 Hypotensive hybrid drugs.

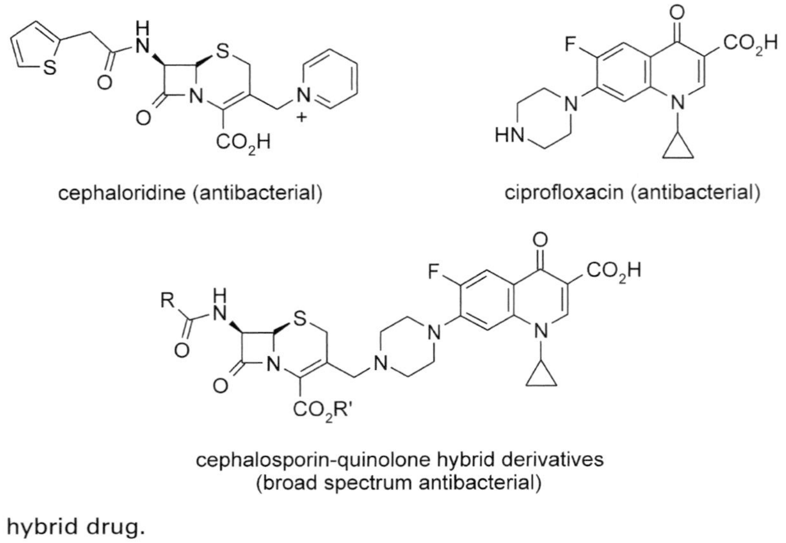

cephaloridine (antibacterial)   ciprofloxacin (antibacterial)

cephalosporin-quinolone hybrid derivatives
(broad spectrum antibacterial)

Fig. 16.35 Antibacterial hybrid drug.

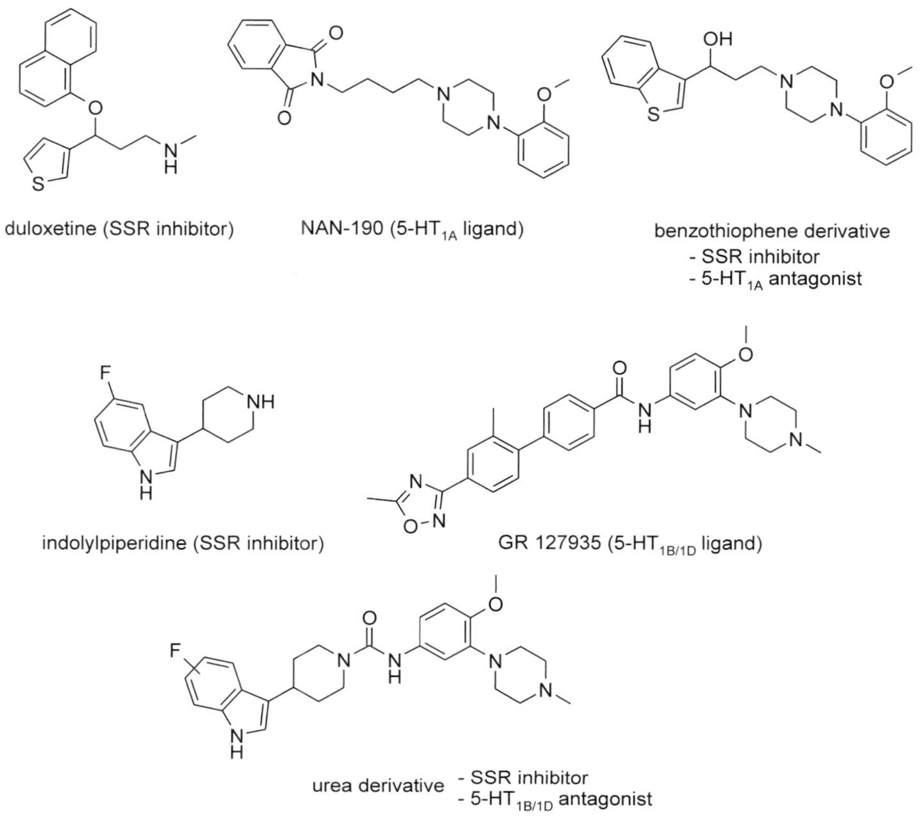

Fig. 16.36 Antidepressant hybrid drugs.

extracellular concentration and would be efficient in the treatment of depressive disorders. Coupling of a selective inhibitor of 5-HT re-uptake (for example the indolylpiperidine derivative) with GR 127935, the first selective 5-HT$_{1B/1D}$ antagonist reported, allowed a urea derivative to be obtained, showing both 5-HT re-uptake inhibition and 5-HT$_{1B/1D}$ antagonism *in vitro*.[83]

IV CONCLUSION

The principle of combining two pharmacophoric groups in a single molecule can be extended to other classes of drugs such as pro-drugs, molecular chameleons,[84] ligand-bearing metal complexes.[85] Peptidic hormones can also be considered as dual acting drugs, with specific amino acid residues needed for molecular recognition by the receptor ('message') and others responsible of the transport and access to the receptor ('address'[86]).

Twin drugs combining two structural components in a single molecule have been described in numerous domains of medicinal chemistry. Historically they resulted from empirical structural modifications, but today rational design of dual binding ligands may involve the knowledge of the structure of the protein which contains these binding sites. Protein X-ray crystallography has revealed in several cases a high degree of symmetry resulting from the existence of dimeric (C-2), trimeric (C-3) or tetrameric protein assemblies. It is associated with the increasing potency of identical twin drugs, that can simultaneously fit to these protein complex symmetrical binding sites. Recent works dealing with the design of HIV proteases strongly supported this hypothesis. However, combining in one molecule two non-identical pharmacophores leads to a new compound, which may not bind simultaneously to each relevant binding site. Recent findings in molecular pharmacology, molecular biology, enzymology and physiology will help to select pertinent pairs of targets involved in different pathologies. For therapeutic purposes the earlier search of selective drugs is replaced today by the design of nonselective drugs, but with control of their selectivity profile, using the 'dual acting drug' concept. The design of dual acting drugs is promising, but is much more difficult than the conventional design of a compound with a single activity, as follows:[1,58] (1) Combining two pharmacophore components in a single molecule may lead to an inactive compound. As a typical example, the combination of the vasodilator nifedipine and a β-blocker led to a hybrid devoid of the expected pharmacological activities. This example emphasizes the importance of a good knowledge of SAR data within each pharmacophore (types of interaction, steric hindrance-sensitive regions, local hydrophilic and hydrophobic areas), and the choice of the linker (nature, position of the linkage). However, in some cases, the two respective SARs may prove mutually exclusive. (2) The hybrid showed the awaited pharmacological profile, but the attempt failed because of unpredicted toxicological problems. (3) The balanced potency of the dual acting drugs has to be carefully evaluated. Design of agonist/antagonist hybrids has to take into consideration that drugs with antagonistic activity on receptors usually have to be given in concentrations significantly less than those needed for agonists (affinities in the nM and μM range, respectively). In a similar manner, design of hybrids combining a receptor ligand and an enzyme inhibitor should take into account both efficacies, and particularly kinetic properties of the considered enzyme. However, the approach is workable and successful attempts have been obtained in cardiovascular research, particularly for the treatment of hypertension. The gastrointestinal system represents another interesting field of application for dual-acting drugs. Concerning the central nervous system, only a few hybrids are at present clinically used. However, the approach should be applied for the treatment of diseases that need restoration of dopaminergic or cholinergic balance. In spite of this, numerous recent works dealing with the design of twin drugs acting in various systems have been reported in this chapter, and account for the increasing interest of this approach in drug design.

REFERENCES

1. Baldwin, J.J., Lumma, W.C., Lundell, G.F., Ponticello, G.S., Raab, A.W., Engelhardt, E.L., Hirschmann, R., Sweet, C.S. and Scriabine, A. (1979) Symbiotic approach to drug design: antihypertensive β-adrenergic blocking agents. *J. Med. Chem.* 22: 1284–1290.
2. Cecchetti, V., Fravolini, A., Schiaffella, F., Tabarrini, O., Bruni, G. and Segre, G. (1993) *o*-Chlorobenzenesulfonamidic derivatives of (aryloxy) propanolamines as β-blocking/diuretic agents. *J. Med. Chem.* 36: 157–161.
3. Piwinski, J.J., Wong, J.K., Green, M.J., Ganguly, A.K., Billah, M.M., West, R.E. Jr, and Kreutner, W. (1991) Dual antagonists of platelet activating factor and histamine. Identification of structural requirements for dual activity of *N*-acyl-4-(5,6-dihydro-11*H*-benzo[5,6]cyclohepta-[1,2-*b*]pyridin-11-ylene)piperidines. *J. Med. Chem.* 34: 457–461.
4. Changeux, J.P. (1969) Remarks on the symmetry and cooperative properties of biological membranes. In Engström, A. and Strandberg, B.

(eds). *Symmetry and Function of Biological Systems at the Macromolecular Level*, pp. 235–256. Almqvist and Wiksell, Stockholm.
5. Blundell, T., Sewell, T. and Turnell, B. (1981) Symmetry in the structure and organization of proteins. In Dodson, G., Glusker, J.P. and Sayre, D. (eds). *Struct. Stud. Mol. Biol. Interest*, pp. 390–403. Oxford University Press, London.
6. Sidddle, K. (1992) The insulin receptor. In Burgen, A. and Barnard, E.A. (eds). *Receptor Subunits and Complexes*, pp. 261–351. Cambridge University Press, Cambridge.
7. Ariëns, E.J. (1971) Drug design — a general introduction. In Ariëns, E.J. (ed.). *Drug Design*, pp. 1–270. Academic Press, New York.
8. Bell, M.R., Batzold, F.H. and Winneker, R.C. (1986) Chemical control of fertility. In Bailey, D.M. (ed.). *Annual Reports in Medicinal Chemistry*, pp. 169–177. Academic Press, San Diego.
9. Neupert-Laves, K. and Dobber, M. (1975) The crystal structure of a K + complex of valinomycin. *Helv. Chim. Acta* 58: 432–442.
10. Melchiorre, C. (1981) Tetramine disulfides: a new tool in α-adrenergic pharmacology. *Tr. Pharmacol. Sci.* 2: 209–211.
11. Kierstead, R.W., Faraone, A., Mennona, F., Mullin, J., Guthrie, R.W. and Crowley, H. (1983) Beta-1-selective adrenoceptor antagonists. Blocking activity of a series of binary (aryloxy) propanolamines. *J. Med. Chem.* 26: 1561–1569.
12. Melchiorre, C., Bolognesi, M.L., Chiarini, A., Minarini, A. and Spampinato, S. (1993) Synthesis and biological activity of some methoctramine-related tetraamines bearing a 11-acetyl-5,11-dihydro-6*H*-pyrido[2,3-*b*][1,4]-benzodiazepin-6-one moiety as antimuscarinics: A second generation of highly selective M₂ muscarinic receptor antagonists. *J. Med. Chem.* 36: 3734–3737.
13. Rosini, M., Budriesi, R., Bixel, M.G., Bolognesi, M.L., Chiarini, A., Hucho, F., Krogsgaard-Larsen, P., Mellor, I.R., Minarini, A., Tumiatti, V., Usherwood, P.N.R. and Melchiorre, C. (1999) Design, synthesis, and biological evaluation of symmetrically and unsymmetrically substituted methoctramine-related polyamines as muscular nicotinic receptor noncompetitive antagonists. *J. Med. Chem.* 42: 5212–5223.
14. Rosini, M., Bixel, M.G., Marucci, G., Budriesi, R., Krauss, M., Bolognesi, M.L., Minarini, A., Tumiatti, V., Hucho, F. and Melchiorre, C. (2002) Structure–activity relationships of methoctramine-related polyamines as muscular nicotinic receptor noncompetitive antagonists. 2. Role of polymethylene chain lengths separating amine functions and of substituents on the terminal nitrogen atoms. *J. Med. Chem.* 45: 1860–1878.
15. Biftu, T., Chabala, J.C., Acton, J., Beattle, T., Brooker, D. and Bugianesi, R. (1988) Synthesis and structure–activity relationship of 2, 5-diaryltetrahydrofurans as PAF antagonists. Prostaglandins 35: 846.
16. Lamouri, A., Heymans, F., Tavet, F., Dive, G., Batt, J.P., Blavet, N., Braquet, P. and Godfroid, J.-P. (1993) Design and modeling of new platelet-activating factor antagonists. 1. Synthesis and biological activity of 1,4-bis(3′,4′,5′-trimethoxybenzoyl)-2-[[(substituted carbonyl and carbamoyl)oxy]methyl]piperazines. *J. Med. Chem.* 36: 990–1000.
17. Roth, R.A., Cassell, D.J., Morgan, D.O., Tatnell, M.A., Jones, R.H., Schüttler, A. and Brandenburg, D. (1984) Effects of covalently linked insulin dimers on receptor kinase activity and receptor down regulation. *FEBS Lett.* 170: 360–364.
18. Fauchère, J.C., Rossier, M., Capponi, A. and Vallotton, M.B. (1985) Potentiation of the antagonistic effect of ACTH₁₁₋₂₄ on steroidogenesis by synthesis of covalent dimeric conjugates. *FEBS Lett.* 183: 283–286.
19. Chino, N., Yoshizawa-Kumagaye, K., Noda, Y., Watanabe, T.X., Kimura, T. and Sakakibara, S. (1986) Synthesis and biological properties of antiparrallel and parallel dimers of human α-human atrial natriuretic peptide. *Biochem. Biophys. Res. Comm.* 141: 665–672.
20. Salvino, J.M., Seoane, P.R., Douty, B.D., Awad, M.A., Dolle, R.E., Houck, W.T., Faunce, D.M. and Sawutz, D.G. (1993) Design of potent

non-peptide competitive antagonists of the human bradykinin B$_2$ receptor. *J. Med. Chem.* **36**: 2583–2584.

21. Costa, T., Wüster, M., Herz, A., Shimohigashi, Y., Chen, H.-C. and Rodbard, D. (1985) Receptor binding and biological activity of bivalent enkephalins. *Biochem. Pharmacol.* **34**: 25–30.

22. Lin, C.-E., Takemori, A.E. and Portoghese, P.S. (1993) Synthesis and κ-opioid antagonist selectivity of a norbinaltorphimine congener. Identification of the address moiety required for κ-antagonist activity. *J. Med. Chem.* **36**: 2412–2415.

23. Portoghese, P.S. (1989) Bivalent ligands and the message–address concept in the design of selective opioid receptor antagonists. *Trends Pharmacol. Sci.* **10**: 230–235.

24. Joslyn, A.F., Luchowski, E. and Triggle, D.J. (1988) Dimeric 1,4-dihydropyridines as calcium channel antagonists. *J. Med. Chem.* **31**: 1489–1492.

25. Bridger, G.J., Skerlj, R.T., Padmanabhan, S., Martellucci, S.A., Henson, G.W., Struyf, S., Witvrouw, M., Schols, D. and De Clerq, E. (1999) Synthesis and structure–activity relationships of phenylenebis(methylene)-linked bis-azamacrocycles that inhibit HIV-1 and HIV-2 replication by antagonism of the chemokine receptor CXCR4. *J. Med. Chem.* **42**: 3971–3981.

26. De Clerq, E. (1995) Toward improved anti-HIV chemotherapy: therapeutic strategies for intervention with HIV infections. *J. Med. Chem.* **38**: 2491–2517.

27. Appelt, K. (1993) Crystal structures of HIV-1 protease-inhibitor complexes. *Persp. Drug Discov. Des.* **1**: 23–48.

28. Bone, R., Vacca, J.P., Anderson, P.S. and Holloway, M.K. (1991) X-ray Crystal Structure of the HIV protease complex with L-700,417, an inhibitor with pseudo C$_2$ symmetry. *J. Am. Chem. Soc.* **113**: 9382–9384.

29. Tan, G.T., Wickramasinghe, A., Verma, S., Singh, R., Hughes, S.H., Pezzuto, J.M., Baba, M. and Mohan, P. (1992) Potential anti-AIDS naphtalenesulfonic acid derivatives. Synthesis and inhibition of HIV-1 induced cytopathogenesis and HIV-1 and HIV-2 reverse transcriptase activity. *J. Med. Chem.* **35**: 4846–4853.

30. Erickson, J.W., Neidhardt, D.J., VanDrie, J., Kempf, D.J., Wang, X.C., Norbeck, D.W., Plattner, J.J., Rittenhouse, J.W., Turon, M., Wideburg, N., Kohlbrenner, W.E., Simmer, R., Helfrich, R., Paul, D.A. and Knigge, M. (1990) Design, activity, and 2.8 angstrom crystal structure of a C$_2$ symmetric inhibitor complexed to HIV-1 protease. *Science* **249**: 527–533.

31. Hulten, J., Bonham, N.M., Nillroth, U., Hansson, T., Zuccarello, G., Bouzide, A., Aqvist, J., Classon, B., Danielson, U.H., Karlen, A., Kvarnstrom, I., Samuelsson, B. and Hallberg, A. (1997) Cyclic HIV-1 protease inhibitors derived from mannitol: synthesis, inhibitory potencies, and computational predictions of binding affinities. *J. Med. Chem.* **40**: 885–897.

32. Alterman, M., Bjorsne, M., Muhlman, A., Classon, B., Kvarnstrom, I., Danielsson, H., Markgren, P.-O., Nillroth, U., Unge, T., Hallberg, A. and Samuelsson, B. (1998) Design and synthesis of new potent C$_2$-symmetric HIV-1 protease inhibitors. Use of the L-mannaric acid as a peptidomimetic scaffold. *J. Med. Chem.* **41**: 3782–3792.

33. Davis, P.D., Hill, C.H., Lawton, G., Nixon, J.S., Wilkinson, S.E., Hurst, S.A., Keech, E. and Turner, S.E. (1992) Inhibitors of protein kinase C. 1. 2,3-bisarylmaleimides. *J. Med. Chem.* **35**: 177–184.

34. Davis, P.D., Elliott, L., Harris, W., Hill, C.H., Hurst, S.A., Keech, E., Kumar, H., Lawton, G., Nixon, J.S. and Wilkinson, S.E. (1992) Inhibitors of protein kinase C. 2. Substituted bisindolylmaleimides with improved potency and selectivity. *J. Med. Chem.* **35**: 994–1001.

35. Bit, R.A., Davis, P.D., Elliott, L., Harris, W., Hill, C.H., Keech, E., Kumar, H., Lawton, G., Maw, A., Nixon, J.S., Vesey, D.R., Wadsworth, J. and Wilkinson, S.E. (1993) Inhibitors of protein kinase C. 3. Potent and highly selective bisindolylmaleimides by conformational restriction. *J. Med. Chem.* **36**: 21–29.

36. Trinks, U., Buchdunger, E., Furet, P., Kump, W., Mett, H., Meyer, T., Müller, M., Regenass, U., Rihs, G., Lydon, N. and Traxler, P. (1994) Dianilinophthalimides: Potent and selective, ATP-competitive inhibitors of the EGF-receptor protein tyrosine kinase. *J. Med. Chem.* **37**: 1015–1027.

37. Qin, D., Sullivan, R., Berkowitz, W.F., Bittman, R. and Rotenberg, S.A. (2000) Inhibition of protein kinase Cα by dequalinium analogues: Dependence on the linker length and geometry. *J. Med. Chem.* **43**: 1413–1417.

38. Thakkar, K., Geahlen, R.L. and Cushman, M. (1993) Synthesis and protein-tyrosine kinase inhibitory activity of polyhydroxylated stilbene analogues of piceatannol. *J. Med. Chem.* **36**: 2950–2955.

39. Levitzki, A. and Gilon, C. (1991) Tyrphostins as molecular tools and potential antiproliferative drugs. *Trends Pharmacol. Sci.* **12**: 171–174.

40. Hales, N.J. and Beattie, J.F. (1993) Novel inhibitors of prolyl 4-hydroxylase. 5. The intriguing structure–activity relationships seen with 2,2′-bipyridine and its 5,5′-dicarboxylic acid derivatives. *J. Med. Chem.* **36**: 3853–3858.

41. Ito, N., Yasunaga, T., Iizumi, Y. and Araki, T (1990) Bis(ureidoalkyl)benzenes for inhibition of acylcoenzyme A cholesterol transferase (ACAT). European Patent Application EP 325 397 (26 July 1989, to Yamanouchi Pharmaceutical Co., Ltd) (*Chem. Abstr.* **112**: 55271a).

42. Mao, S.-S. (1993) Factor Xa inhibitors. *Persp. Drug Discov. Des.* **1**: 423–430.

43. Phillips, G., Davey, D.D., Eagen, K.A., Koovakkat, S.K., Liang, A., Ng, H.P., Pinkerton, M., Trinh, L., Whitlow, M., Beatty, A.M. and Morrissey, M.M. (1999) Design, synthesis, and activity of 2,6-diphenoxypyridine-derived factor Xa inhibitors. *J. Med. Chem.* **42**: 1749–1756.

44. Saab, N.H., West, E.E., Bieszk, N.C., Preuss, C.V., Mank, A.R., Casero, R.A. Jr, and Woster, P.M. (1993) Synthesis and evaluation of unsymmetrically substituted polyamine analogues as modulators of human spermidine/spermine-N^1-acetyltransferase (SSAT) and as potential antitumor agents. *J. Med. Chem.* **36**: 2998–3004.

45. Stanek, J., Caravatti, G., Capraro, H.-G., Furet, P., Mett, H., Schneider, P. and Regenass, U. (1993) S-adenosylmethionine decarboxylase inhibitors: New aryl and heteroaryl analogues of methylglyoxal bis(guanylhydrazone). *J. Med. Chem.* **36**: 46–54.

46. Taylor, P. and Lappi, S. (1975) Interaction of fluorescence probes with acetylcholinesterase. The site and specificity of propidium binding. *Biochemistry* **14**: 1989–1997.

47. Pang, Y.-P., Quiram, P., Jelacic, T., Hong, F. and Brimijoin, S. (1996) Highly potent, selective, and low cost bis-tetrahydroaminacrine inhibitors of acetylcholinesterase. *J. Biol. Chem.* **271**: 23646–23649.

48. Stefanska, B., Dzieduszycka, M., Martelli, S., Tarasiuk, J., Bontemps-Gracz, M. and Borowski, E. (1993) 6-[(Aminoalkyl)amino]-substituted 7H-benzo[e]perimidin-7-ones as novel antineoplastic agents. Synthesis and biological evaluation. *J. Med. Chem.* **36**: 38–41.

49. Wunz, T.P., Dorr, R.T., Alberts, D.S., Tunget, C.L., Einspahr, J., Milton, S. and Remers, W. (1987) New antitumor agents containing the anthracene nucleus. *J. Med. Chem.* **30**: 1313–1321.

50. Fairley, T.A., Tidwell, R.R., Donkor, I., Naiman, N.A., Ohemeng, K.A., Lombardy, R.J., Bentley, J.A. and Cory, M. (1993) Structure, DNA minor groove binding, and base pair specificity of alkyl- and aryl-linked bis(amidinobenzimidazoles) and bis(amidinoindoles). *J. Med. Chem.* **36**: 1746–1753.

51. Mann, J., Baron, A., Opoku-Boahen, Y., Johansson, E., Parkinson, G., Kelland, L.R. and Neidle, S. (2001) A new class of symmetric bisbenzimidazole-based DNA minor groove-binding agents showing antitumor activity. *J. Med. Chem.* **44**: 138–144.

52. Spicer, J.A., Gamage, S.A., Rewcastle, G.W., Finlay, G.J., Bridewell, D.J.A., Baguley, B.C. and Denny, W.A. (2000) Bis(phenazine-1-carboxamide): Structure–activity relationships for a new class of dual

topoisomerase I/II-directed anticancer drugs. *J. Med. Chem.* **43**: 1350–1358.

53. Fröstl, W. and Maître, L. (1989) The families of cognition enhancers. *Pharmacopsychiatry* **22**(Suppl.): 54–100.

54. Vennerström, J.L., Ellis, W.Y., Ager, A.L. Jr, Anderson, S.L., Gerena, L. and Milhous, W.K. (1992) Bisquinolines. 1. N,N-bis(7-chloroquinolin-4-yl) alkanediamines with potential against chloroquine-resistant malaria. *J. Med. Chem.* **35**: 2129–2134.

55. Girault, S., Grellier, P., Berecibar, A., Maes, L., Mouray, E., Lemière, P., Debreu, M.-A., Davioud-Charvet, E. and Sergheraert, C. (2000) Antimalarial, antitrypanosomal, and antileishmanial activities and cytotoxiciy of bis(9-amino-6-chloro-2-methoxyacridines): Influence of the linker. *J. Med. Chem.* **43**: 2646–2654.

56. Rall, T.W. (1990) Hypnotics and sedatives; ethanol. In Goodman-Gilman, A., Rall, T.W., Nies, A.S. and Taylor, P. (eds). *Goodman and Gilman's The Pharmacological Basis of Therapeutics*, 8th edn., pp. 345–382. Pergamon Press, New York.

57. Brittain, R.T., Drew, G.M. and Levy, G.P. (1980) The α- and β-adrenoceptor blocking potencies of labetalol and its individual stereoisomers. *Br. J. Pharmacol.* **69**: 282p–283p.

58. Nicolaus, B.J.R. (1983) Symbiotic approach to drug design. In Gross, F. (ed.). *Decision Making in Drug Research*, pp. 173–186. Raven Press, New York.

59. Mewshaw, R.E., Silverman, L.S., Mathew, R.M., Kaiser, C., Sherrill, R.G., Cheng, M., Tiffany, C.W., Karbon, E.W., Bailey, M.A., Borosky, S.A., Ferkany, J.W. and Abreu, M.E. (1993) Bridged γ-carbolines and derivatives possessing selective and combined affinity for 5-HT$_2$ and D$_2$ receptors. *J. Med. Chem.* **36**: 1488–1495.

60. Buschauer, A. (1989) Synthesis and in vitro pharmacology of arpromidine and related phenyl(pyridylalkyl)guanidines, a potential new class of positive inotropic drugs. *J. Med. Chem.* **32**: 1963–1970.

61. Ohshima, E., Sato, H., Obase, H., Miki, I., Ishii, A., Kawakage, M., Shirakura, S., Karasawa, A. and Kubo, K. (1993) Dibenzoxepin derivatives: Thromboxane A$_2$ synthase inhibition and thromboxane A$_2$ receptor antagonism combined in one molecule. *J. Med. Chem.* **36**: 1613–1618.

62. Zhang, J.C., Entzeroth, M. and Wienen, W. (1992) Characterization of BIBS 39 and BIBS 222: two new non peptide angiotensine II receptor antagonists. *Eur. J. Pharmacol.* **218**: 35–41.

63. Jacobson, K.A., Lipkowski, A.W., Moody, T.W., Padgett, W., Pijl, E., Kirk, K.L. and Daly, J.W. (1987) Binary drugs: Conjugates of purines and a peptide that bind to both adenosine and substance P. *J. Med. Chem.* **30**: 1529–1532.

64. Willson, T.M., Brown, P.J., Sternbach, D.D. and Henke, B.R. (2000) The PPARs: from orphan receptors to drug discovery. *J. Med. Chem.* **43**: 527–550.

65. Brooks, D.A., Etgen, G.J., Rito, C.J., Shuker, A.J., Dominianni, S.J., Warshawsky, A.M., Ardecky, R., Paterniti, J.R., Typhonas, J., Karanewsky, D.S., Kauffman, R.F., Broderick, C.L., Oldham, B.A., Montrose-Rafizadeh, C., Winneroski, L.L., Faul, M.M. and McCarthy, J.R. (2001) Design and synthesis of 2-methyl-2-(4-[2-(5-methyl-2-aryloxazol-4-yl)ethoxy]phenoxy)propionic acids: A new class of dual PPARα/γ agonists. *J. Med. Chem.* **44**: 2061–2064.

66. Unangst, P.C., Connor, D.T., Cetenko, W.A., Sorenson, R.J., Kostlan, C.R., Sircar, J.C., Wright, C.D., Schrier, D.J. and Dyer, R.D. (1994) Synthesis and biological evaluation 5-[[3,5-Bis(1,1-dimethylethyl)-4-hydroxyphenyl]methylene]oxazoles, -thiazoles, and imidazoles: Novel dual 5-lipoxygenase and cyclooxygenase inhibitors with anti-inflammatory activity. *J. Med. Chem.* **37**: 322–328.

67. Jacobs, C., Frotscher, M., Dannhardt, G. and Hartmann, R.W. (2000) 1-Imidazolyl(alkyl)-substituted di- and tetrahydroquinolines and analogues: Syntheses and evaluation of dual inhibitors of thromboxane A$_2$ synthase and aromatase. *J. Med. Chem.* **43**: 1841–1851.

68. Flynn, G.A., Beight, D.W., Mehdi, S., Koehl, J.R., Giroux, E.L., French, J.F., Hake, P.W. and Dage, R.C. (1993) Application of a conformationally restricted Phe-Leu dipeptide mimetic to the design of a combined inhibitor of angiotensin I-converting enzyme and neutral endopeptidase 24.11. *J. Med. Chem.* **36**: 2420–2423.

69. Pontikis, R., Dollé, V., Guillaumel, J., Dechaux, E., Note, R., Nguyen, C.H., Legraverend, M., Bisagni, E., Aubertin, A.-M., Grierson, D.S. and Monneret, C. (2000) Synthesis and Evaluation of 'AZT-HEPT', 'AZT-Pyridinone', and 'ddC-HEPT' conjugates as inhibitors of HIV reverse transcriptase. *J. Med. Chem.* **43**: 1927–1939.

70. Camps, P., El Achab, R., Gorbig, D.M., Morral, J., Munoz-Torrero, D., Badia, A., Banos, J.E., Vivas, N.M., Barril, X., Orozco, M. and Luque, F.J. (1999) Synthesis, *in vitro* pharmacology, and molecular modeling of very potent tacrine-huperzine A hybrids as acetylcholinesterase inhibitors of potential interest for the treatment of Alzheimer's disease. *J. Med. Chem.* **42**: 3227–3242.

71. Mary, A., Renko, D.Z., Guillou, C. and Thal, C. (1998) Potent acetylcholinesterase inhibitors: Design, synthesis, and structure–activity relationships of *bis*-interacting ligands in the galanthamine series. *Bioorg. Med. Chem.* **6**: 1835–1850.

72. Soyka, R., Heckel, A., Nickl, J., Eisert, W., Muller, T.H. and Weisenberger, H. (1994) 6,6-Disubstituted hex-5-enoic acid derivatives as combined thromboxane A$_2$ receptor antagonists and synthetase inhibitors. *J. Med. Chem.* **37**: 26–39.

73. Soyka, R., Guth, B.D., Weisenberger, H.M., Luger, P. and Muller, T.H. (1999) Guanidine derivatives as combined thromboxane A$_2$ receptor antagonists and synthase inhibitors. *J. Med. Chem.* **42**: 1235–1249.

74. Cozzi, P., Carganico, G., Fusar, D., Grossoni, M., Menichincheri, M., Pinciroli, V., Tonani, R., Vaghi, F. and Salvati, P. (1993) Imidazol-1-yl and pyridin-3-yl derivatives of 4-phenyl-1,4-dihydropyridines combining Ca^{++} antagonism and thromboxane A$_2$ synthase inhibition. *J. Med. Chem.* **36**: 2964–2972.

75. Wermuth, C.G., Schlewer, G., Bourguignon, J.-J., Maghioros, G., Bouchet, M.-J., Moire, C., Kan, J.-P., Worms, P. and Bizière, K. (1989) 3-Aminopyridazine derivatives with atypical antidepressant, serotonergic, and dopaminergic activities. *J. Med. Chem.* **32**: 528–537.

76. Worms, P., Kan, J.P., Steinberg, R., Terranova, J.-P., Pério, A. and Bizière, K. (1989) Cholinomimetic properties of minaprine. *Naunyn-Schmiedeberg's Arch. Pharmacol.* **340**: 411–418.

77. Wermuth, C.G. (1993) Aminopyridazines — An alternative route to potent muscarinic agonist with no cholinergic syndrome. *Il Farmaco* **48**: 253–274.

78. Wermuth, C.G., Bourguignon, J.J., Hoffmann, R., Boigegrain, R., Brodin, R., Kan, J.P. and Soubrié, P. (1992) SR 46559 A and related aminopyridazines are potent muscarinic agonists with no cholinergic syndrome. *Bioorg. Med. Chem. Lett.* **2**: 833–838.

79. Contreras, J.-M., Parrot, I., Sippl, W., Rival, Y.M. and Wermuth, C.G. (2001) Design, synthesis, and structure–activity relationships of a series of 3-[2-(1-benzylpiperidin-4-yl)ethylamino]pyridazine derivatives as acetylcholinesterase inhibitors. *J. Med. Chem.* **44**: 2707–2718.

80. Contreras, J.-M., Rival, Y.M., Chayer, S., Bourguignon, J.-J. and Wermuth, C.G. (1999) Aminopyridazines as acetylcholinesterase inhibitors. *J. Med. Chem.* **42**: 730–741.

81. Albrecht, H.A., Beskid, G., Christenson, J.G., Deitcher, K.H., Georgopapadakou, N.H., Keith, D.D., Konzelmann, F.M., Pruess, D.L. and Wei, C.C. (1994) Dual-action cephalosporins incorporating a 3$^{\prime}$-tertiary-amine-linked quinolone. *J. Med. Chem.* **37**: 400–407.

82. Martinez-Esparza, J., Oficialdegui, A.-M., Perez-Silanes, S., Heras, B., Orus, L., Palop, J.-A., Lasheras, B., Roca, J., Mourelle, M., Bosch, A., Del Castillo, J.-C., Tordera, R., Del Rio, J. and Monge, A. (2001) New 1-aryl-3-(4-arylpiperazin-1-yl)propane derivatives, with dual action at

5-HT$_{1A}$ serotonin receptors and serotonin transporter, as a new class of antidepressants. *J. Med. Chem.* **44**: 418–428.

83. Matzen, L., van Amsterdam, C., Rautenberg, W., Greiner, H.E., Harting, J., Seyfriend, C.A. and Bottcher, H. (2000) 5-HT reuptake inhibitors with 5-HT1B/1D antagonistic activity: A new approach toward efficient antidepressants. *J. Med. Chem.* **43**: 1149–1157.

84. Carrupt, P.-A., Testa, B. and Bechalany, A., El Tayar, N., Descas, P. and Perrissoud, P. (1991) Morphine-6-glucuronide and morphine-3-glucuronide as molecular chameleons with unexpected lipophilicity. In Silipo, C. and Vittoria, A. (eds). *QSAR: Rational Approaches to the Design of Bioactive Compounds*, pp. 541–544. Elsevier Science, Amsterdam.

85. Chi, D.Y., O'Neil, J.P., Anderson, C.J., Welch, M.J. and Katzenellenbogen, J.A. (1994) Homodimeric and heterodimeric bis(amino thiol) oxometal complexes with rhenium(V) and technetium(V). Control of heterodimeric complex formation and an approach to metal complexes that mimic steroid hormones. *J. Med. Chem.* **37**: 928–937.

86. Portoghese, P.S., Sultana, M. and Takemori, A.E. (1989) Design of peptidomimetic δ opioid receptor antagonists using the message-adress concept. *J. Med. Chem.* **33**: 1714–1720.

17

OPTICAL ISOMERISM IN DRUGS

Camille G. Wermuth

Most natural organic compounds, the essential products of life, are asymmetric and possess such asymmetry that they are not superposable on their images... This establishes perhaps the only well marked line of demarcation that at present can be drawn between the chemistry of dead matter and the chemistry of living matter

(Pasteur, Van t'Hoff, Le Bel and Wislicenus, Memoirs, 1901)

I INTRODUCTION

This chapter is concerned with bioactive compounds bearing one or more asymmetric carbon atom(s) on their skeleton. For such compounds the term *configuration* defines the implantation mode of the four covalent linkages on the central, asymmetric, carbon atom. The terms *optical isomers*, *optical antipodes*, *enantiomorphs* or *enantiomers*, are synonyms and relate to molecules which are mirror images of one another and are not, therefore, superimposable. Owing to their non-identical 3-D structure, enantiomers may elicit differentiated biological responses and thus provide useful information on drug–receptor interactions and on receptor characteristics. A great number of books deal with chirality and drug design.[1–7]

II EXPERIMENTAL FACTS AND THEIR INTERPRETATION

A Stereoselectivity in biologically active compounds

Towards a biological target, the potency of two enantiomers can sometimes differ considerably and sometimes be very similar (Table 17.1). Often the activity is concentrated in only one enantiomer. When such a high stereoselectivity arises, it is admitted that the mechanism of action at the molecular level involves a highly specific interaction between the ligand — a chiral molecule — and the recognition site — a chiral environment. It is to be expected that the most active isomer, in terms of affinity, achieves a better steric complementarity to the receptor than the less active one.

When considering *in vivo* activities, the difference in activity observed for the two enantiomers is neither always nor exclusively the result of the quality of

Table 17.1 *Differences in activity or in affinity (eudismic index) between couples of enantiomers*

Substance		Eudismic index
R-thiorphan	K_i values for inhibition of neutral endopeptidase 24.11 (former enkephalinase)	$S/R = 1,2$[8]
R(−)-noradrenaline [R(−)-norepinephrine] natural enantiomer	Adrenergic activity at rat aorta α_1 sites	$R(-)/S(+) = 33$[9]
S(−)-nicotine (natural nicotine)	Affinity constants for rat brain thalamus sites labelled by [^{3}H]-(−)-nicotine	$S(-)/R(+) = 35$[10]
S(+)-chlorpheniramine (Polaramine)	K_i values for human brain frontal lobe sites labelled by [^{3}H]-mepyramine	$S(+)/R(-) = 83$[11]
S(+)-oxaprotiline	IC$_{50}$ values for noradrenaline uptake into rat brain synaptosomes	$S(+)/R(-) = 1000$[12]
S(+)-dexetimide	[^{3}H]-dexetimide binding to brain muscarinic receptors	$S(+)/R(-) = 2000$[13]
(+)-lysergide (LSD)	(+)-[^{3}H] lysergic acid diethylamide binding on rat forebrain suspension	$R(+)/S(-) = 24000$[14]

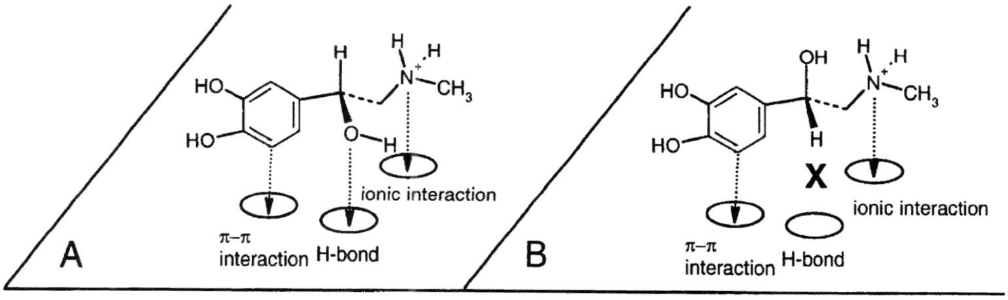

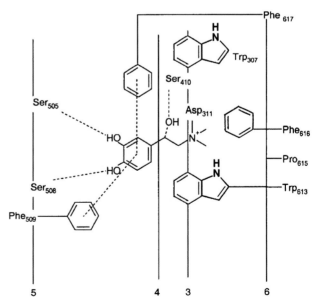

Fig. 17.1 Interaction capacities of the natural R(+)-epinephrine and its S(−)-antipode. In simply assuming that the natural R(+)-epinephrine establishes a three point interaction with its receptor (A) the combination of the donor–acceptor interaction, the hydrogen bond and the ionic interaction will be able to generate energies in the order 12 to 17 kcal mole^{-1}, that corresponds to binding constants of 10^{-9} to 10^{-12} M.[17] The less active isomer, S(+)-epinephrine, may establish only a two point contact (B). The loss of the hydrogen bond interaction equals to approximately 3 kcal mole^{-1}, this isomer should therefore possess an approximately 100-fold lesser affinity. The experience confirms this estimate. If we consider less abstract models it becomes apparent that the less potent enantiomer is also able to develop three intermolecular bonds to the receptor, provided that it approaches the receptor in a different manner. However, the probability of this alternative binding mode to trigger the same biological response is close to null.

the ligand–receptor fit. It must be kept in mind that *in vivo* the pharmacokinetic processes (ADME) may account for the observed difference in activity. The interpretation of pharmacological data obtained from *in vivo* assays should thus be questioned and does not allow anticipation of the quality of the ligand–receptor interaction.

B   The three-point contact model

When, in a compound exhibiting stereoselectivity, only one asymmetric centre is present in the molecule, it is thought that the substituents on the chiral carbon atom make a three-point contact with the receptor. Such a fit ensures a very specific molecular orientation which can only be obtained for one of the two isomers. A three-point fit of this type was first suggested by Easson and Stedman,[15] and the corresponding model proposed by Beckett[16] in the case of R(−)-adrenaline (R(−)-epinephrine). The more active natural R(−)-adrenaline establishes contacts with its receptor through the following three interactions (Fig. 17.1):

(1)   Acceptor–donor or hydrophobic interaction between the aromatic ring of adrenaline and an aromatic ring of the receptor protein
(2)   A hydrogen bond at the alcoholic hydroxyl
(3)   An ionic bond between the protonated amino group and an aspartic or glutamic carboxylic group of the receptor.

The combination of these interactions can generate binding energies in the order of 12–17 kcal M^{-1}, corresponding to binding constants in the order of 10^{-9} to 10^{-12} M.[17] The biologically weak optical isomer, S(+)-adrenaline, can

make contact through only two groups. According to this hypothesis, it would be anticipated that deoxyadrenaline (epinine) has much the same activity as S(+)-adrenaline. This has been found to be basically true.[15,18,19]

Computer-generated receptor models for protein-G-linked receptors are now available,[20,21] and Fig. 17.2 illustrates the fit of R(−)-adrenaline into the active site of the β^2-adrenergic receptor. It appears clearly that the docking involves more interactions than only the above-mentioned three points:

Fig. 17.2 Interaction capacities of the natural R(+)-epinephrine with a model of its receptor (after References 20 and 21).

(1) The two phenolic hydroxyl groups exchange hydrogen bonds with Ser[505] and Ser[508], respectively

(2) The aromatic ring of adrenaline is stabilized by means of $\pi-\pi$ interactions with Phe[509] and Phe[617]

(3) The cationic head exerts a coulombic interaction with the Asp311 carboxylate and is located in a hydrophobic pocket made of Trp[307], Phe[616], and Trp[613]

(4) Finally, the secondary benzylic hydroxyl exchanges a hydrogen bond with Ser[410].

Even in taking into account these newer findings, it can be speculated that the the Easson–Stedman hypothesis still holds. The non-natural S(+)-adrenaline, having the wrong orientation of its benzylic hydroxyl, is unable to exchange a hydrogen bond with Ser[410], and therefore achieves a weaker interaction with the receptor.

C Diastereoisomers

When more than one asymmetric centre is involved, the complexity of the problem increases rapidly. For the four isomers of ephedrine, which represent a set of diastereoisomers, the R configuration of the β-carbon (as found for adrenaline, noradrenaline, nordefrin, phenylephrin and octopamine, see Patil[22] is not automatically associated with the highest alpha-agonistic activity. Both (−)-ephedrine and (−)-pseudo-ephedrine possess the β-(R) configuration, yet only R(−)-ephedrine acts as an agonist. This anomaly can be explained if one takes into account the preferred conformations of these two compounds, calculated by using the ETH (Extended Hückel Theory).[23,24] In the (−)-ephedrine molecule, the methyl group attached to the carbon in the alpha-position to the amino function is projected above the plane of the phenyl-ethyl-amino group, whereas in (−)-pseudo-ephedrine the methyl group is oriented below the plane and thus prevents an efficient interaction of the drug with the receptor (Fig. 17.3).

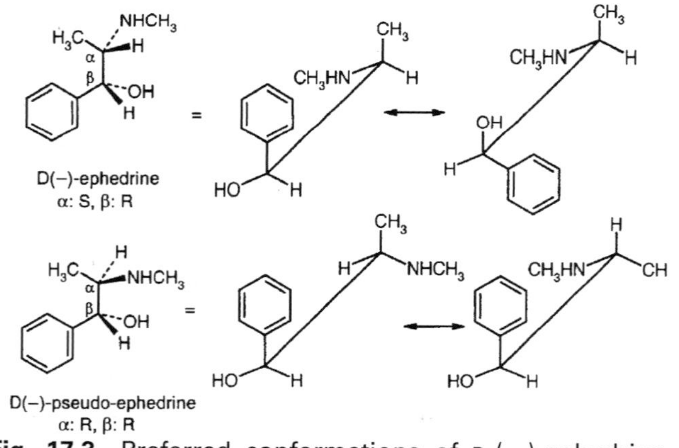

Fig. 17.3 Preferred conformations of D-(−)-ephedrine and of D-(−)-pseudo-ephedrine.

The antibacterial activity of chloramphenicol isomers represents a similar example. Significant activity is only found for the (−)-threo-chloramphenicol.[25] The clinical formulation of the adrenergic receptor-blocking agent labetalol consists of a mixture of equal proportions of the four optical isomers (RR, SS, RS and SR). Each possesses different pharmacological properties. The most active RR enantiomer was developed some years ago as Dilevalol,[26] but had to be withdrawn after some months due to a slightly higher than average degree of hepatic toxicity.[27]

The antihistaminic drug clemastine (Tavegyl), despite the fact that it contains two chiral centres, provides one of the few examples of chiral antihistamines employed clinically in the form of a single isomer. Data on the antihistaminic activity of clemastine and its isomers[28] are summarized in Table 17.2.

The stereoisomers of some oxotremorine analogues containing two chiral centres and acting as oxotremorine antagonists show *in vivo* (tremorolytic activity) stereoselectivity ratios as high as 1 to 200.[29]

D Stereoselectivity ratios

Stereoselectivity was defined by Rauws[30] as follows: 'Stereoselectivity is the extent to which an enzyme or other macromolecule, or macromolecular structure (antibody or receptor) exhibits affinity towards one molecule of a pair of isomers in comparison with and in contrast to the other isomer.' Lehmann[31,32] has expressed this in a mathematical form: the ratio of activity of the better fitting enantiomer (*eutomer*; Greek, 'eu' = good), to that of the less fitting enantiomer (*distomer*; Greek, 'dys' = bad) is defined *eudismic ratio*. From this a *eudismic affinity quotient* can be derived (Table 17.3).

Table 17.2 *Antihistamine activities of clemastine and its isomers*

R,R-clemastine

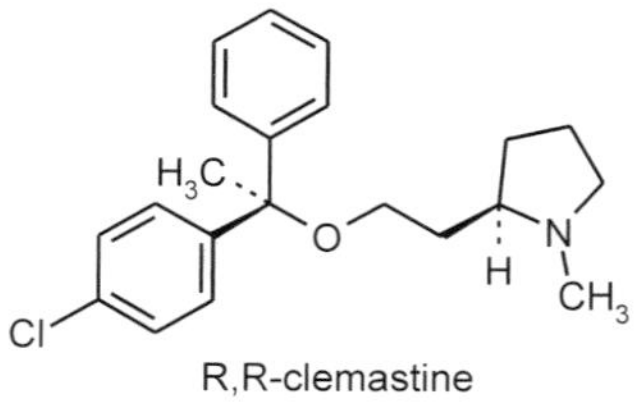

Isomer	Prevention of histamine toxicity ED$_{50}$ (mg/kg sc)	Prevention of histamine spasm	pA$_2$
RR (clemastine)	0.04	~ +7	9.45
SS	5.1	~ −1.5	7.99
SR	11.0	~ −6	8.57
RS	0.28	~ +5	9.40

Table 17.3 *Nomenclature and definitions in drug stereoselectivity*

Eutomer (Eu): enantiomer presenting the *highest* affinity
  (or activity)
Distomer (Dis): enantiomer presenting the *lowest* affinity
  (or activity)
Eudismic quotient: Affin·*Eu*/Affin·*Dis*
Eudismic index (EI): Log Affin.*Eu*-Log Affin.*Dis*

In a series of agonists or antagonists one can write Equation (1).

$$EI = a + b \text{ Log Affin} \cdot Eu \qquad (1)$$

in which: a is a constant and b is the quotient of eudismic affinity (QEA) which precisely accounts for the stereoselectivity.

When the activity of the eutomer 'Eu' is compared with that of the racemic mixture 'Rac', four possibilities can arise:[33,34]

(1) The activity ratio is equal to 2: Eu/Rac = 2/1. In this case, the activity is only concentrated in the eutomer and the distomer does not contribute significantly to the observed activity. The chiral compound shows stereoselectivity.

(2) The activity ratio is higher than 2: Eu/Rac > 2 (for example Eu/Rac = 2/0.3). This means that the distomer represents a competitive antagonist of the eutomer. In practise, such a situation is rarely encountered (see below III. Practical considerations, B. The distomer counteracts the eutomer).

(3) If the activity ratio is lower than 2: Eu/Rac < 2 (for example Eu/Rac = 2/1.6), we are in the presence of two active isomers. The distomer reinforces the activity of the eutomer. Such a situation indicates a decrease of the receptor selectivity.

(4) The activity ratio is Eu/Rac = 1; in this case both isomers are equipotent and no stereoselectivity is observed. This can be explained by the assumption that (1) the compounds act through a non-specific mechanism; (2) the active compound and the receptor make only a two-point contact with the chiral centre; (3) the chiral centre is not involved in the contact (is located in a 'silent region').

E   Pfeiffer's rule

One usually admits that the discriminative effect between the two enantiomers increases with the proximity of the chiral centre to the site of interaction with the receptor. An empirical rule published by Pfeiffer in 1956[35] states that the isomeric activity ratio (eudismic ratio) of a highly active couple of isomers is always superior to that of a less active couple. In other words: 'The greater the difference between the pharmacological activity of the R and the S isomers, the greater is the potency of the active isomer'. However, there are some exceptions to Pfeiffer's rule. Some of the reasons are conformational flexibility of the ligands,[36] others reside in an improper selection of 'homologous' sets of compounds as illustrated with muscarinic agonists and antagonists.[37] Quantitative analyses of the correlations between biological activity and the structure of stereoisomeric compounds are difficult.[38,39]

III   OPTICAL ISOMERISM AND PHARMACODYNAMIC ASPECTS

The biological response induced by a pair of enantiomers can differ in potency (quantitative difference) or in nature (qualitative difference). In the latter case it is assumed that one enantiomer acts at one receptor site, whereas its antipode is recognized by other sites and possesses a different activity and toxicity profile.

A   Differences in potency and antagonism between two enantiomers

Two optical isomers are never antagonists, at least at comparable dosages. This comes from the space relationship required for the interaction with the receptor site, which is only slightly altered by passing from S to R forms, or vice versa. If one of the enantiomers achieves the optimal fit to the receptor site in exchanging the highest number of noncovalent linkages, its antipode can only give rise to a weaker interaction, even in the most favourable conditions (Fig. 17.1). From a practical point of view, this absence of *stoichiometric* antagonism entails two consequences:

(1) If a racemic mixture does not show any activity, it is useless to carry out the separation of the two antipodes.

(2) A racemic mixture usually has the average potency of both constituents, thus, the maximal benefit one can achieve in resolving racemic mixtures is to increase the activity of one of the antipodes to twice that of the racemate.

B   Differences in the pharmacological profile of two enantiomers

Besides the difference in potency, it often happens that two enantiomers show differences in their pharmacological profile. In such a case, resolving the racemic mixture can generate two pharmacologically different and useful compounds and also separate the more active compound

Table 17.4 *Differences in pharmacological profile of couples of enantiomers*

Racemate	Levorotatory enantiomer	Dextrorotatory enantiomer	References
Quinine/quinidine (racemate not in use)	quinine: antipyretic, antimalarial	quinidine: antiarrythmic antimalarial	White *et al.*[42] White *et al.*[43] Alexander *et al.*[44]
Sotalol	(−)-sotalol β-adrenoceptor blocker	(+)-sotalol antiarrythmic agent	Drayer[26]
Racemorphane	(−)-*N*-methyl-3-methoxy-morphinane antitussive	(+)-*N*-methyl-3-methoxy-morphinane analgesic	Benson *et al.*[41]
Indacrinone	R(−)-indacrinone diuretic	S(+)-indacrinone uricosuric	Drayer[26]
Propoxyfene	α-levopropoxyfene (Novrad) antitussive	α-dextropropoxyfene (Darvon) analgesic	Drayer[26]
Tetramisole	S(−)-levamisole: nematocidal, immunostimulant	R(+)-dexamisole: antidepressant	Bullock *et al.*[45] Schnieden [46]
3-Amino-1-hydroxypyrrolid-2-one(HA-966)	3R-(+)-HA-966 partial agonist at the glycine site of the NMDA receptor	3S-(−)-HA-966 γ-butyrolactone-like sedative	Singh *et al.*[47]
3-Methoxy-cyproheptadine	(−)-3-Methoxy-cyproheptadine anticholinergic activity	(+)-3-Methoxy-cyproheptadine antiserotonin activity	Remy *et al.*[48]

from its less well-tolerated or more toxic isomer. In the quinine–quinidine couple (Table 17.4), both isomers share antimalarial, antipyretic, oxytoxic, as well as skeletal and cardiac muscle depressant activities. However, whereas antipyresis and treatment of malaria represent the main use of quinine, quinidine is more effective on the cardiac muscle and is used in the therapy of atrial fibrillation and in certain other arrhythmias.[40] In the *N*-methyl-3-methoxy morphinane racemate (racemorphane), most of the analgesic and addictive properties are concentrated in the (+)-isomer. The corresponding (−)-isomer is nonaddictive and retains only antitussive properties.[41] The same kind of discrimination is found for the antitussive levopropoxyphene and its well-known analgesic enantiomer dextropropoxyphene.[26]

The substituted imidazo-thiazole, dexamisole, has anti-depressant properties and its isomer, levamisole, possesses anthelmintic and immunostimulant properties.[45,46] Enantiomers of HA-966 (3-amino-1-hydroxypyrrolid-2-one) exhibit distinct central nervous system effects: (+)-HA-966 is a selective glycine/*N*-methyl-D-aspartate receptor antagonist, but (−)-HA-966 is a potent γ-butyrolactone-like sedative.[47] A comparison of (+) and (−)-3-methoxycyproheptadine shows that all of the anticholinergic activity of the (±)-3-methoxycyproheptadine resides solely in the dextrorotatory enantiomer, while the antiserotonin activity resides in the levorotatory enantiomer.[48]

Table 17.5 shows some experimental data for the active isomers of the fluoro analogues of the tricyclic neuroleptic

Table 17.5 *Comparison of the racemate and the two enantiomers of a fluoro analogues of the tricyclic neuroleptic clotepin*[50]

Test	Measurements	RS(±) 1	S(+) 2	R(−) 3
Increase in brain homovanillic acid	At 100 mg kg^{-1}	256%	316%	128%
Adenylate-cyclase inhibition	$c = 10^{-6}$ M	48%	72%	27%
Inhibition of conditioned flight reflexes in rats	ED$_{50}$ (mg kg^{-1}) per os	14	10	>100
Inhibition of apomorphine-induced emesis in dogs	ED$_{50}$ (mg kg^{-1}) per os	20	12	>30
Acute toxicity in mice	LD$_{50}$ (mg kg^{-1}) per os	200	515	68

clotepin as compared with the corresponding racemate.[49] In this example it appears clearly that the neuroleptic activity is concentrated in the dextro-rotary compound (+) **2** whereas the toxicity resides in the (−) **3** antipode.

In the present case, the 1:10 therapeutic index of the racemate, unsatisfactory for a clinical outlook, was risen to the much more acceptable 1:50 ratio for the isolated S(+) antipode.[49]

IV OPTICAL ISOMERISM AND PHARMACOKINETIC EFFECTS

After administration and before it arrives in the vicinity of its receptor site, a drug is subjected to a variety of physiological processes: absorption, distribution metabolism, uptake at storage sites and excretion. Many of these processes are stereoselective (for reviews see Jamali *et al.*[50] and Kroemer *et al.*[51]).

A Isomer effects on absorption and distribution

The higher narcotic potency *in vivo* of the S(+)-isomer of hexobarbital was shown to be related to higher CNS levels than for the R(−) form, this seems to be due to an improved crossing of the blood–brain barrier (BBB).[52] In a distribution study of [^{14}C] (+)- and (−)-alpha-methyl-DOPA in the rat after intravenous injection,[53] the (−)-isomer attained higher concentrations than the (+)-form in most organs, in accordance with the fact that of the two isomers, only the (−) isomer has hypotensive activity.[54]

B Isomer effects on metabolism

Since all enzymes are chiral in nature, and therefore probably possess some degree of asymmetry at the reactive centre, it is not surprising that most metabolic reactions of isomers lead to qualitative and quantitative differences in the metabolites formed (for review articles see Testa,[1,2] Vermeulen,[55] and Kroemer *et al.*[51]).

Differential metabolism of two antipodes

The levo isomers of 3-hydroxy-*N*-methyl-morphinan and of methadone are demethylated by rat liver two to three times more rapidly than the corresponding dextro antipodes.[56,57] The S(+)-enantiomer of hexobarbital (Fig. 17.4) is metabolized almost twice as rapidly as the R(−)-enantiomer by allylic hydroxylation[58] and, in the dog, the dextrorotatory isomer of 5-ethyl-5-phenyl-hydantoin affords ten times more *p*-hydroxy-metabolite than the levorotatory isomer.[59] Hydroxylation takes place alpha to a carbonyl in the dextrorotary enantiomer of glutathimide, whereas the

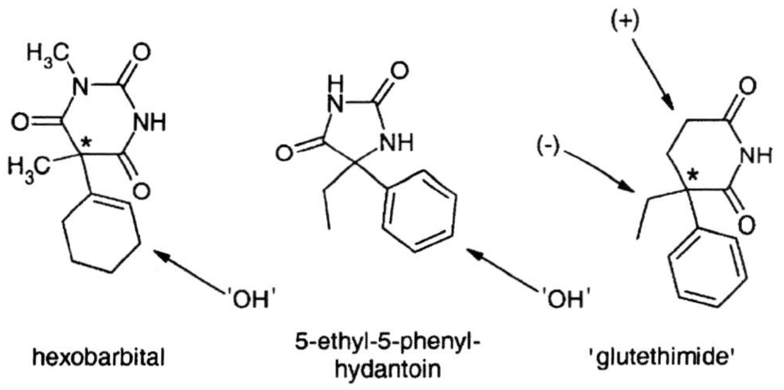

Fig. 17.4 The structures of hexobarbital, 5-ethyl-5-phenyl-hydantoin, and glutethimide.

levorotamer is hydroxylated on the methylene group of the ethyl side chain.[60] Numerous other examples are found in the literature.[61,62]

Enzymatic inversion

The energy requirements necessary for the conversion of a given sp^3 configuration into its optical antipode imply the formation of an intermediary carbenium ion, carbanion, or free radical and are unlikely to arise in biological systems. Thus racemization or epimerization involving nonoxygenated sp^3 carbon atoms are generally not encountered in mammals. They are usually restricted to microorganisms (e.g. alanine-racemase). One case of this unusual phenomenon is described in mammals for arylpropionic acids. More precisely, for the nonsteroidal anti-inflammatory agent ibuprofen (R,S-*p*-isobutyl-hydratropic acid), it has been demonstrated that only the S(+)-isomer is active *in vitro* as inhibitor of the prostaglandin-synthesizing enzyme cyclo-oxygenase. Surprisingly no significant differences could be observed *in vivo* between the S(+) or the R(−)-enantiomers and the racemate (ibuprofen).[63] It was therefore concluded that *in vivo* there must be an almost complete inversion of the poorly active R(−) form to the much more active S(+)-isomer. In man, the main metabolites isolated after administration of (racemic) ibuprofen were dextrorotatory[64] and also the pure R(−)-enantiomer is converted to the S(+)-isomer.[65] A biochemical investigation, using deuterium-labelled R(−)-isomer, led to the hypothesis of the existence of an R-arylpropionic acid isomerase ('R-APAI') enzyme system proceeding via the enzymes of lipid catabolism and anabolism as outlined in Fig. 17.5.

It is assumed that the coenzyme A ester of the R(−)-enantiomer acts as a substrate for the fatty acid deshydrogenase, thus eliminating the chiral centre. The next step may, or may not take place, depending whether or not the CoA-ester must be transferred to an acyl-carrier protein or another site in the fatty acid synthetase system, so that a stereoselective reduction by an enoylreductase can take

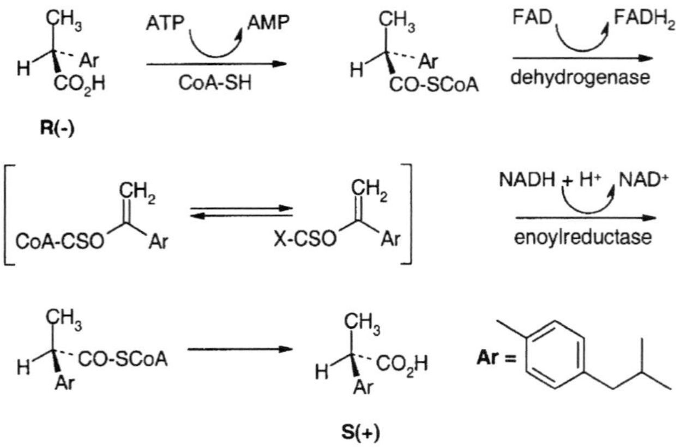

Fig. 17.5 Mechanism of the enzymatic inversion of R(−)-ibuprofen.[66]

place. Thus the nature of X is unknown.[66] Similar epimerization reactions were also described for some other arylpropionic acids, such as benoxaprofen,[67] carprofen,[68] and isopropyl-indanyl-propionic acid.[69] It was demonstrated that the configural inversion does not take place in the liver, and that the responsible enzyme, R-(−)-arylpropionic acid isomerase, is located in the gut wall.[66,70]

C Isomer effects on uptake

As drugs are usually absorbed by passive diffusion and since enantiomers do not differ in their aqueous and lipid solubilities, absorption is not usually considered to be a stereoselective process. However, stereoselectivity has been described for drugs that are transported by a carrier-mediated process. Typical uptake selectivity is observed for neurotransmitter reuptake inhibitors such as nipecotic acid, oxaprotiline, fluoxetine and venlafaxine. Uptake of drugs by various organs can also be enantioselective, for example the liver/plasma concentration ratios of S(−) and R(+)-phenprocoumon in the rat were found to be different (6.9 and 5.2, respectively), indicating a preferential uptake of the more potent isomer.[71]

D Isomer effects on excretion

The kinetics of excretion are a direct consequence of the kinetics of metabolic transformations. The faster a drug is metabolized, the faster its elimination can be expected. In accordance with this assertion, rats given R,S(±), S(+), and R(−)-amphetamine, were found to excrete less (+)-p-hydroxy-amphetamine than its (−)-isomer; this may be the basic explanation of the more pronounced pharmacological properties of the dextro-, compared with the levo-amphetamine.[72] For the hypnotic agent hexobarbital, the elimination half-life in man is about three times longer for the

more active (+)-isomer than for the less active (−)-isomer. This was attributed to a difference in hepatic metabolic clearance and not in volumes of distribution or plasma binding between the enantiomers.[73]

V PRACTICAL CONSIDERATIONS

A Racemates or enantiomers?

Many drugs having a centre of asymmetry are still used in clinical practice as racemates, and racemic mixtures were estimated to represent 10–15% of all the marketed drugs.[3] For certain types of therapeutics, such as the β-adrenergic agents, β-adrenergic blockers, anti-epileptics and oral anticoagulants, up to 90% of the compounds are, according to Ariëns,[3] in fact racemic mixtures. For antihistaminics and local anaesthetics this holds true for about 50% of the drugs currently used.[3] Often racemic drugs were introduced in clinical practice because the animal and the clinical pharmacology, the toxicology, and the teratology were performed with the racemates. The reasons for that is that at the time of the discovery of the drug, the resolution (or the chiral synthesis) appeared to be too difficult or too costly, or even impossible.

The question now arises to decide when and why to use either racemic mixtures or pure enantiomers. Although it seems good sense to use pure eutomers and to consider the distomer as an unwanted xenobiotic (a kind of pollution, or even an impurity), there are however instances where it is recommended to use racemates rather then eutomers. Thus racemates may be more stable, more active or less toxic or present a favourable combination of the properties of each separate isomer (see below). Finally one can ask if it would not be wise to design effective drugs without centres of asymmetry.

B The distomer counteracts the eutomer

Contrary to a well-established belief, there are no examples of inactive racemates in which the distomer antagonizes, in a stoichiometric manner, the activity of the eutomer.

Thus, a dihydropyridine-derived calcium inhibitor, the R(−) enantiomer of compound Sandoz 202-791 inhibits the uptake of [$^{45}Ca^{++}$] with an IC$_{50}$ of 4.3×10^{-8}, whereas its S(+) enantiomer increases the uptake with an IC$_{50}$ of approximately 10^{-6} to 10^{-7} M.[74] The corresponding racemic mixture inhibits the uptake with an IC$_{50}$ of 1.7×10^{-7} M. Some other examples are reported in Table 17.6.

As shown in Table 17.6, a more or less important residual activity is always present in the racemate but resolution would generally be beneficial. Picenadol (LY150,720) seems to be an exception to the rule that pure eutomers

Table 17.6 *Antagonism in couples of enantiomers*

Compound	Eutomer	Distomer	Racemate	Reference
N-Isopropyl-norepinephrine	(−) α-adrenergic agonist	(+) inactive competitive antagonist	(±) partial agonist	Page 15 of ref 3
5-Ethyl-5(1,3-dimethylbutyl) barbituric acid	(+) convulsant	(−) depressant	(±) convulsant	Ho and Harris[75]
Ozolinone (metabolite of etazoline)	(−) diuretic	(+) inhibits low doses of (−) or of furoxemide	(±) diuretic	Greven *et al.*[76]
Picenadol	(+) morphinomimetic	(−) narcotic antagonist	(±) partial agonist	Zimmerman and Gesellchen[77]
Alpha-(2,4,5)-trichloro-phenoxy-propionic acid	(+) auxin-like plant growth regulator	(−) decreases activity of (+)	(±) auxin-like plant growth regulator	Smith *et al.*[78]
6-Ethyl-9-oxaergoline (EOE)	(−) dopamine agonist	(+) dopamine antagonist	(±) dopamine agonist	Lotti and Taylor[79]

should be used when the distomer shows antagonistic properties. For clinical trials as narcotic analgesics, the racemate was the preferred preparation owing to its partial agonist profile.

C   Racemic switches

Presently a general trend in the pharmaceutical industry is to switch from racemates to single enantiomers. Examples are given by (R)-(−)-verapamil, (S)-fluoxetin, (S)-ketoprofen, (R)-albuterol, levofloxacin, omeprazole, cisapride and many others.[80,81] In addition to the quality improvement of the drug, this switch also represents a means to prolong its life insofar as the isolated eutomer is legally considered as a new drug entity. As a consequence, drug companies are increasingly adopting racemic switches as a management strategy. The company first develops a chiral drug as a racemate, and later on patents and develops the single isomer.[82] This strategy does not always work successfully. This is illustrated by the S(−)-eutomer of propranolol. This compound shows reduced β-blocking activity when administered as single isomer compared with its bioavailability when administered as a racemate, suggesting that the presence of R(+)-propranolol had a beneficial effect on the availability of S(−)-propranolol.[82] The same phenomenon happened when the racemate of fluoxetin was compared with its eutomer. The consequence was that the management at Eli Lilly decided not to practise the racemic switch for this compound.

D   The distomer is metabolized to unwanted or toxic products

Racemic deprenyl, a monoamine-oxidase inhibitor used in the treatment of depression, is metabolized to (+)- and (−)-metamphetamine,[83] the latter being much more active than its (−)-isomer as a central stimulant leading to drug abuse (Table 17.7).

On the other hand, the (−)-isomer of deprenyl is a much more potent MAO-B inhibitor than the (+)-isomer. For these reasons racemic deprenyl has been replaced by (−)-deprenyl in clinical practice. In the racemic local anaesthetic prilocaïne (Fig. 17.6) only the R-(−)-isomer is metabolized to an aniline derivative (ortho-toluidine) and to the corresponding para- and ortho-aminophenols that are highly toxic and responsible for met hemoglobinemia.[84]

The S-(+) enantiomer is not a substrate for the metabolizing enzyme and would probably be chemically safe.

Many of the side-effects (e.g. granulocytopenia) encountered with racemic DOPA were not seen with levo-DOPA and therefore can be attributed to the (+)-enantiomer.[85] For this reason the racemate is no longer given. Post-anaesthesia

Table 17.7 *Activities of deprenyl enantiomers and their metabolites*

Formula	Activity	Ratio
deprenyl	MAO-B inhibition	(−) ≫ (+)
metamphetamine	amphetamine effects	(−) < (±)
amphetamine	amphetamine effects	(−) ≪ (+)

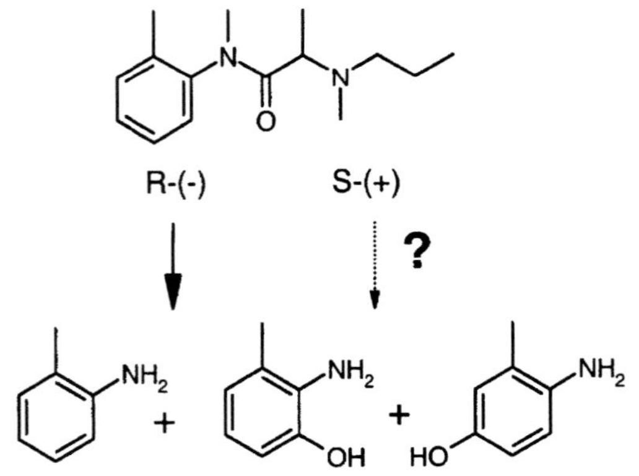

Fig. 17.6 Stereoselective metabolic attack yielding toxic metabolites.[84]

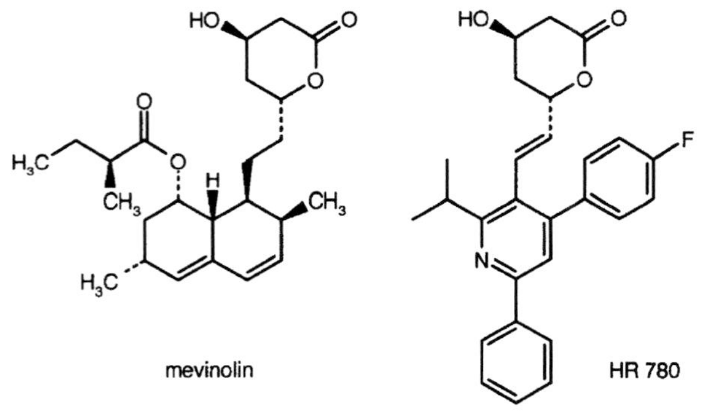

Fig. 17.7 Deletion of five out of seven chiral centres still yield highly potent mevinolin analogues.[91]

reactions to the anaesthetic and analgesic agent ketamine are overwhelmingly associated with the R(−) antipode.[86]

In vitro studies suggest that the beneficial antiarrhythmic properties of disopyramide are concentrated in the S(+)-isomer, whereas the negative inotropic effect predominates in the R(−)-isomer.[87] In addition, the pharmacokinetics (clearance and protein binding) differ.[88,89] For these reasons, selection of the S(+) isomer may have led to the development of a very effective drug with significantly fewer therapeutic problems.

E   Deletion of the chiral centre

Nowadays it is well accepted that racemates and both enantiomers are usually three different pharmacological entities and that it requires extensive pharmacological, toxicological and clinical pharmacological research before it can be decided whether it is advantageous to use racemates or enantiomers in clinical practice. According to Soudijn,[90] these research efforts could be reduced to about one third when drugs without centres or planes of asymmetry could be developed with the same or higher affinity.

Effectively asymmetry is far from being an absolute requisite for activity! The alkaloid morphine possesses five chiral centres, on the other hand its synthetic derivative fentanyl is devoid of any asymmetry centre and nonetheless belongs to the most potent analgesics known. In some instances the chiral centres can at least partially be eliminated. This is the case for the synthetic analogues of the HMG-CoA reductase inhibitor mevinolin. Mevinolin itself (Fig. 17.7) has eight asymmetric centres, but SAR studies rapidly revealed that the six chiral centres contained in the hexahydronaphtalene unit are unnecessary for HMG-CoA inhibition. The second generation of mevinolin analogues, illustrated in Fig. 17.7 by the compound HR 780, retains only two of the initial seven chiral centres.[91]

Usually chiral centres are eliminated in creating symmetry. Thus, in a series of muscarinic agonists derived from 3-aminopyridazines, one of the most favourable side chains, was the racemic 2-*N*-ethylpyrrolidinylmethyl chain, that is the side chain of sulpiride (Fig. 17.8). The 5-methyl-6-phenylpyridazine bearing this basic chain at its 3-amino function presented a 0.26 micromolar affinity for M_1 muscarinic receptor preparations.[91]

After resolution of the racemate, the corresponding enantiomers show only a six-fold difference in M_1 affinity. It was therefore decided to eliminate the chiral centre by introducing symmetry either by ring opening, or by ring closure or even by replacing the 2-*N*-ethylpyrrolidinyl-methyl unit by the non-chiral tropane ring. The modified structures show affinities similar to that of the corresponding chiral molecule.[92]

F   Usefulness of racemic mixtures

In practice, if both optical isomers are of similar potency and have similar pharmacokinetic profiles, it may be useless to proceed to the resolution of the racemic mixture. Such situations are infrequent but may occur. An example is given by the antithrombotic acids 21-X and 21-Y (Fig. 17.9).[93] The corresponding pure enantiomers were first compared with the corresponding racemates for their *in vitro* activities.

In both series almost equipotent activities were observed for thromboxane receptor antagonism and thromboxane synthase inhibition ($IC_{50} = 2-30$ nM). Upon oral administration to guinea pigs, the enantiomers inhibited the *ex vivo* U-46619-induced platelet aggregation with potencies similar to that of the corresponding racemates. This indicates that the enantiomers have pharmacologic profile and bioavailability similar to that of the corresponding racemic compound.

The racemates can even be more potent than either of the enantiomers used separately, this is observed with

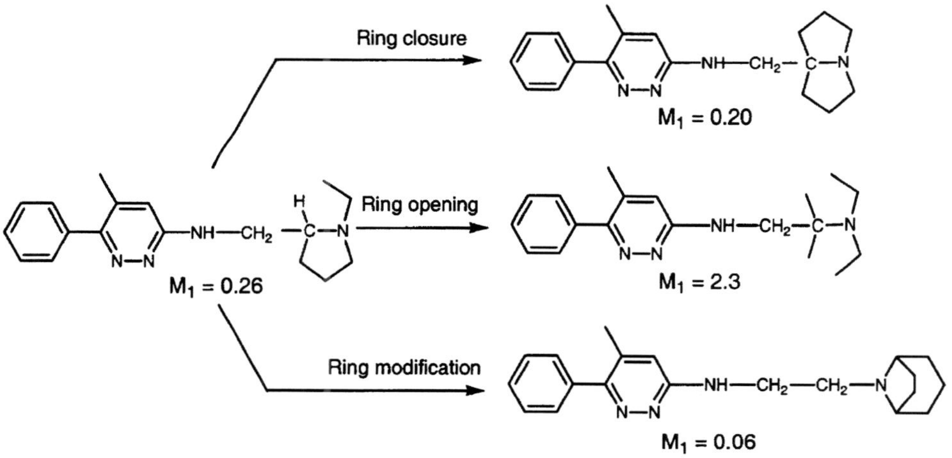

Fig. 17.8 Introducing symmetry and thus abolishing a chiral centre (affinity values for M_1 receptor preparations expressed as micromoles).

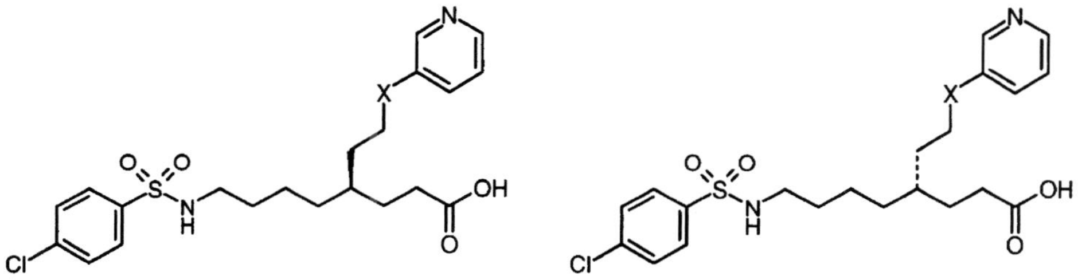

21-X: X = CH_2; 21-Y: X = O

Fig. 17.9 Isoactive antithrombotic enantiomers.[93]

the antihistaminic drug, isothipendyl.[94] In other cases it may be of interest to racemize a natural optically active molecule. Thus, to warrant a constant pharmacological activity of ergotamine, which is racemized in solution, producing inactive ergotaminine, the commercial solution is produced as an equilibrium mixture of the two antipodes.[95] Another example of the utility of a racemic mixture is given by the lysine salts of aspirin. The acetylsalicylate prepared from (R,S)-lysine is a stable, crystalline white powder which is freely soluble in water giving a tasteless, odourless and colourless solution, suitable for parenteral injection. Surprisingly, the corresponding salts of pure (R)-lysine or pure (S)-lysine do not crystallize (Baetz, J., personal communication). Finally, when the distomer is converted to the eutomer *in vivo*, as seen above for ibuprofen and its analogues, it becomes also preferable to commercialize the racemate.

The recommendations of the European Community Working Party on drug quality, safety and efficacy, take into account two situations.[96] For already well-established racemates, clinical use can continue as such, no specific study of the isolated enantiomers is required. For newly introduced chiral drugs, both enantiomers have to be prepared and studied separately with regard to their activity as well as their disposition *in vivo*. However, the final decision to introduce the drug on to the market as enantiomer or as racemate belongs to the producer.

REFERENCES

1. Testa, B. (1986) Chiral aspects of drug metabolism. *Trends Pharmacol. Sci.* **7**: 60–64.
2. Testa, B. and Mayer, J.M. (1988) Stereoselective drug metabolism and its significance in drug research. In Jucker, E. (ed.). *Progress in Drug Research*, pp. 249–303. Birkhäuser, Basel.
3. Ariëns, E.J. (1983) Stereoselectivity of bioactive agents: general aspects. In Ariëns, E.J., Soudjin, W. and Timmermans, P.B.M.W.M. (eds). *Stereochemistry and Biological Activity of Drugs*, pp. 11–32. Blackwell Scientific Publications, Oxford.
4. Ariëns, E.J. (1988) Stereospecificity of bioactive agents. In Ariëns, E.J., Van Rensen, J.J.S. and Welling, W. (eds). *Stereoselectivity of Pesticides*. Elsevier, Amsterdam.

5. Simonyi, M. (1990) *Problems and Wonders of Chiral Molecules*, pp. 400. Akadémia Kiado, Budapest.

6. Brown, C. (1990) *Chirality in Drug Design and Synthesis*, pp. 243. Academic Press, London.

7. Casy, A.F. and Dewar, G.H. (1993) *The Steric Factor in Medicinal Chemistry Dissymmetric Probes of Pharmacological Receptors.* Plenum Press, New York.

8. Fournié-Zaluski, M.C., Lucas-Soroca, E., Devin, J. and Roques, B.P. (1986) ^{1}H NMR Configural correlation for retro-inverso dipeptides: application to the determination of the absolute configuration of 'Enkephalinase' inhibitors. *J. Med. Chem.* **29**: 751–757.

9. Jordan, R., Midgeley, J.M., Thonoor, C.M. and Williams, C.M. (1987) Beta-adrenergic activities of octopamine and synephrine stereoisomers on guinea-pig atria and trachea. *J. Pharm. Pharmacol.* **39**: 752–754.

10. Martino-Barrows, A.M. and Kellar, K.J. (1987) [3H]Acetylcholine and [3H](−)nicotine label the same region in brain. *Mol. Pharmacol.* **31**: 169–174.

11. Chang, R.S.L., Tran, V.T. and Snyder, S.H. (1979) Heterogenicity of histamine H1-receptors: species variations in [3H]mepyramine binding of brain membranes. *J. Neurochem.* **32**: 1653–1663.

12. Waldmeier, P.C., Baumann, P.A., Hauser, K., Maitre, L. and Storni, A. (1982) Oxaprotiline, a noradrenaline uptake inhibitor with an active and an inactive enantiomer. *Biochem. Pharmacol.* **31**: 2169–2176.

13. Laduron, P.M., Vervimp, M. and Leysen, J.E. (1979) Stereospecific *in vitro* binding of [^{3}H]-dexetimide to brain muscarinic receptors. *J. Neurochem.* **32**: 421–427.

14. Lovell, R.A. and Freedman, D.X. (1976) Stereospecific receptor sites for *d*-lysergic acid diethylamide in rat brain: effects of neurotransmitters, amine antagonists, and other psychotopic drugs. *Mol. Parmacol.* **12**: 620–630.

15. Easson, L.H. and Stedman, E. (1933) Studies on the relationship between chemical constitution and physiological action. V. Molecular dissymmetry and physiological activity. *Biochem. J.* **27**: 1257–1266.

16. Beckett, A.H. (1959) Stereochemical factors in biological activity. In Jucker, E. (ed.). *Fortschritte der Arzneimittel Forschung*, pp. 455–530. Birkhäuser Verlag, Basel.

17. Farmer, P.S. and Ariëns, E.J. (1982) Speculations on the design of nonpeptide peptidomimetics. *Trends Pharmacol. Sci.* **3**: 362–365.

18. Patil, P.N., LaPidus, J.B., Campbell, D. and Tye, A. (1967) Steric aspects of adrenergic drugs. II. Effects of DL isomers and deoxy derivatives on the reserpine-pretreated vas deferens. *J. Pharmacol. Exp. Ther.* **155**: 13–23.

19. Ruffolo, R.R., Jr. (1938) Stereoselectivity in adrenergic agonists and adrenergic blocking agents. In Ariëns, E.J., Soudjin, W. and Timmermans, P.B.M.W.M. (eds). *Stereochemistry and Biological Activity of Drugs*, pp. 103–125. Blackwell Scientific Publications, Oxford.

20. Hibert, M.F., Trumpp-Kallmeyer, S., Bruinvels, A. and Hoflack, J. (1992) Three-dimensional models of neurotransmitter G protein coupled receptors. *Mol. Pharmacol.* **40**: 8–15.

21. Trumpp-Kallmeyer, S., Hoflack, J., Bruinvels, A. and Hibert, M. (1992) Modelling of G protein-coupled receptors. Application to dopamine, adrenaline, serotonin, acetylcholine and mammalian opsin receptors. *J. Med. Chem.* **35**: 3448–3462.

22. Patil, P.N. (1968) Steric aspects of adrenergic drugs. 8. Optical isomers of beta adrenergic receptor antagonists. *J. Pharmacol. Exp. Ther.* **160**: 308–314.

23. Kier, L.B. (1968) The preferred conformations of ephedrine isomers and the nature of the alpha adrenergic receptor. *J. Pharmacol. Exp. Ther.* **164**: 75–81.

24. Portoghese, P.S. (1967) Stereochemical studies on medicinal agents. IV Conformational analysis of ephedrine isomers and related compounds. *J. Med. Chem.* **10**: 1057–1063.

25. Maxwell, R.E. and Nickel, V.S. (1954) The antibacterial activity of the isomers of chloramphenicol. *Antibiot. Chemother.* **4**: 289–295.

26. Drayer, D.E. (1986) Pharmacodynamic and pharmacokinetic differences between drug enantiomers in humans: an overview. *Clin. Pharmacol. Ther.* **40**: 125–133.

27. Fell, A.F. (1998) Current perspectives on chiral drug development. *EUFEBS News Lett.* **7**: 1–2.

28. Ebnöther, A. and Weber, H.-P. (1976) Synthesis and absolute configuration of clemastine and its isomers. *Helv. Chim. Acta* **59**: 2462–2468.

29. Ringdahl, B., Resul, B. and Dahlbom, R. (1979) Stereoselectivity of some oxotremorine antagonists containing two chiral centers. *J. Pharm. Pharmacol.* **31**: 837–839.

30. Rauws, A.G. (1983) Origin and basis of stereoselectivity in biology. In Ariëns, E.J., Soudjin, W. and Timmermans, P.B.M.W.M. (eds). *Stereochemistry and Biological Activity of Drugs*, pp. 1–10. Blackwell Scientific Publications, Oxford.

31. Lehman, F.P.A., Ariëns, E.J. and Rodrigues de Miranda, J.F. (1976) Stereoselectivity and affinity in molecular pharmacology. In Jucker, E. (ed.). *Progress in Drug Research*, pp. 101–142. Birkhäuser Verlag, Basel.

32. Lehman, P.A. (1986) Stereoisomerism and drug action. *Trends Pharmacol. Sci.* **7**: 281–285.

33. Casy, A.F. (1970) Stereochemistry and biological activity. In Burger, A. (ed.). *Medicinal Chemistry*, pp. 81–107. Wiley-Interscience, New York.

34. Schröder, E., Rufer, C. and Schmiechen, R. (1976) *Arzneimittelchemie*, pp. 48. Georg Thieme Verlag, Stuttgart.

35. Pfeiffer, C.C. (1956) Optical isomerism and pharmacological action, a generalization. *Science* **124**: 29–31.

36. Barlow, R.B. (1990) Enantiomers: how valid is Pfeiffer's Rule? *Trends Pharmacol. Sci.* **11**: 148–150.

37. Gualtieri, F. (1990) Pfeiffer's Rule OK? *Trends Pharmacol. Sci.* **11**: 315–316.

38. Lien, E.J., Rodrigues de Miranda, J.F. and Ariëns, E.J. (1976) Quantitative structure–activity correlation of optical isomers: a molecular basis for Pfeiffer's Rule. *Mol. Pharmacol.* **12**: 598–604.

39. Portoghese, P.S. and Williams, D.A. (1970) Stereochemical studies on medicinal agents. VIII. Absolute stereochemistries of isomethadol isomers. *J. Med. Chem.* **13**: 626–630.

40. Roden, D.M. (1995) Antiarrhythmic drugs. In Hasdman, J.G., Limbird, L.E., Molinoff, P.B., Ruddon, R.W. and Goodman Gilman, A. (eds). *The Pharmacological Basis of Therapeutics*, 9th edn., p. 1905. McGraw-Hill, New York.

41. Benson, W.M., Stefko, P.L. and Randall, L.O. (1953) Comparative pharmacology of levorphan, racemorphan and dextrorphan and related methyl ethers. *J. Pharmacol. Exp. Ther.* **109**: 189–200.

42. White, N.J., Looareeswan, S. and Warrel, D.A. (1981) Quinidine in falciparum malaria. *Lancet* **2**: 1069–1071.

43. White, N.J., Looareeswan, S. and Warrel, D.A. (1983) Quinine and quinidine: a comparison of EKG effects during the treatment of malaria. *J. Cardiovasc. Pharmacol.* **5**: 173–175.

44. Alexander, F., Gold, H. and Katz, L.N. (1947) The relative value of synthetic quinidine, dihydroquinidine, commercial quinidine and quinine in the control of cardiac arrhythmias. *J. Pharmacol. Exp. Ther.* **90**: 191–201.

45. Bullock, M.W., Hand, J.J. and Waletzky, E. (1968) Resolution and racemization of dl-tetramisole, dl-6-phenyl-2,3,5,6-tetrahydroimidazo-[2,1-b]thiazole. *J. Med. Chem.* **11**: 169–171.

46. Schnieden, H. (1981) Levamisole — a general pharmacological perspective. *Int. J. Immunopharmacol.* **3**: 9–13.

47. Singh, L., Donald, A.E. and Foster, A.C. (1990) Enantiomers of HA-966 (3-amino-1-hydroxypyrrolid-2-one) exhibit distinct central nervous system effects: (+)-HA-966 is a selective glycine/*N*-

methyl-D-aspartate receptor antagonist, but $(-)$-HA-966 is a potent γ-butyrolactone-like sedative. *Proc. Natl Acad. Sci. USA* **87**: 347–351.

48. Remy, D.C., Rittle, K.E., Hunt, C.A., Anderson, P.S., Engelhardt, E.L., *et al.* (1977) $(+)$ and $(-)$-3-methoxycyproheptadine: A comparative evaluation of the antiserotonin, antihistaminic, anticholinergic, and orexigenic properties with cyproheptadine. *J. Med. Chem.* **20**: 1681–1684.

49. Aschwanden, W., Kyburz, E. and Schönholzer, P. (1976) Stereospezifizität der neuroleptischen Wirkung und Chiralität von $(+)$-3-{2-[Fluor-2-methyl-10,11-dihydrodibenzo[*b,f*]thiepin-10-yl)-1-piperazinyl]-äthyl}-2-oxazolidinon (16). *Helv. Chim. Acta* **59**: 1245–1252.

50. Jamali, F., Mehvar, R. and Pasutto, F.M. (1989) Enantioselective aspects of drug action and disposition: therapeutic pitfalls. *J. Pharm. Sci.* **78**: 695–715.

51. Kroemer, H.K., Gross, A.S. and Eichelbaum, M. (1994) Enantioselectivity in drug action and drug metabolism: influence on dynamics. In Cutler, N.R., Sramek, J.J. and Narang, P.K. (eds). *Pharmacodynamics and Drug Development*, pp. 103–114. John Wiley & Sons, Chichester.

52. Buch, H., Rummel, W. and Brandenburger, V. (1967) Versuche zur Aufklärung der Ursachen der unterschiedlichen narkotischen Wirksamkeit von $(+)$- und $(-)$-Evipan. *Arch. Pharmakol. Exp. Pathol.* **257**: 270–271.

53. Duhm, B., Maul, W., Medenwald, H., Platzschke, K. and Wegner, L.A. (1967) Experimental animal studies with alpha-methyldopa-14C, with special attention to the optical isomers. II. Organ distribution. *Z. Naturforsch.* **22b**: 70–84.

54. Sjoerdsma, J. and Udenfriend, S. (1961) Pharmacology and biochemistry of α-methyl-dopa in man and experimental animals. *Biochem. Pharmacol.* **8**: 164.

55. Vermeulen, N.P.E. (1986) Stereoselective biotransformation: its role in drug disposition and drug action. In Harms, A.F. (ed.). *Innovative Approaches in Drug Research*, pp. 393–416. Elsevier Science Publishers, Amsterdam.

56. Axelrod, J. (1956) The enzymatic N-demethylation of narcotic drugs. *J. Pharmacol. Exp. Ther.* **117**: 322–330.

57. Elison, C., Elliott, H.W., Look, M. and Rapoport, H. (1963) Some aspects of the fate and relationship of the *N*-methyl group of morphine to its pharmacological activity. *J. Med. Chem.* **6**: 237–246.

58. Degwitz, E., Ullrich, V., Staudinger, H. and Rummel, W. (1969) Metabolism and cytochrome P-450 binding spectra of $(+)$ and $(-)$ hexobarbital in rat liver microsomes. *Hoppe-Seylers Z. Physiol. Chem.* **350**: 547–553.

59. Kupfer, A., Bircher, J. and Preisig, R. (1977) Stereoselective metabolism, pharmacokinetics and biliary elimination of phenylethylhydantoin (Nirvanol) in the dog. *J. Pharmacol. Exp. Ther.* **203**: 493–499.

60. Keberle, H., Riess, W. and Hoffman, K. (1963) The stereospecific metabolism of the optical antipodes of α-phenyl-α-ethylglutarimide (Doriden). *Arch. Int. Pharmaco.* **142**: 117–124.

61. Jenner, P. and Testa, B. (1973) The influence of stereochemical factors on drug metabolism. *Drug Metab. Rev.* **2**: 117–184.

62. Test, B. and Jenner, P. (1980) *Concepts in Drug Metabolism. Part A.* Marcel Dekker, New York.

63. Adams, S.S., Bresloff, P. and Mason, C.G. (1976) Pharmacological differences between the optical isomers of ibuprofen: evidence for metabolic inversion of the $(-)$-isomer. *J. Pharm. Pharmacol.* **28**: 256.

64. Adams, S.S., Cliffe, E.E., Lessel, B. and Nicholson, J.S. (1967) Some biological properties of 2-(4-isobutylphenyl)-propionic acid. *J. Pharm. Sci.* **56**: 1686.

65. Vane, J.R. (1971) Inhibition of prostaglandin synthesis as a mechanism of action for aspirin-like drugs. *Nature* **231**: 232–235.

66. Wechter, W.J., Loughhead, D.G., Reisher, R.J., van Geissen, G.J. and Kaiser, D.G. (1974) Enzymatic inversion at saturated carbon: nature and mechanism of the inversion of R$(-)$*p-iso*-butyl hydratropic acid. *Biochem. Biophys. Res. Commun.* **61**: 833–837.

67. Bopp, R.J., Nash, J.F., Ridolfo, A.S. and Shepard, E.R. (1979) Stereoselective inversion of (R)-$(-)$-benoxaprofen to the (S)-$(+)$-enantiomer in humans. *Drug Metab. Disp.* **7**: 356–359.

68. Kemmerer, J.M., Rubio, F.A., McClain, R.M. and Koechlin, B.A. (1979) Stereospecific assay and stereospecific disposition of carprofen in rats. *J. Pharm. Sci.* **68**: 11274–11280.

69. Tanaka, Y. and Hayashi, R. (1980) Stereospecific inversion of configuration of 2-(2-isopropylindan-5-yl)-propionic acid in rats. *Chem. Pharm. Bull.* **28**: 2542–2545.

70. Simmonds, R.G., Woodage, T.J., Duff, S.M. and Green, J.N. (1980) Stereospecific inversion of (R)-$(-)$-benoxafen in rat and in man. *Eur. J. Drug Metab. Pharmaco.* **5**: 169–172.

71. Schmidt, W. and Jahnchen, E. (1977) Stereoselective drug distribution and anticoagulant potency of phenprocoumon in rats. *J. Pharm. Pharmacol.* **29**: 266–271.

72. Gunne, L.M. and Galland, L. (1967) Stereoselective metabolism of amphetamine. *Biochem. Pharmacol.* **16**: 1374–1377.

73. Vermeulen, N.P.E. and Breimer, D.D. (1983) Stereoselective drug and xenobiotic metabolism. *Stereochemistry and Biological Activity of Drugs.* Blackwell Scientific Publications, Oxford.

74. Hof, R.P., Rüegg, U.T., Hof, A. and Vogel, A. (1985) Stereoselectivity at the calcium channel: opposite action of the enantiomers of a 1,4-dihydropyridine. *J. Cardiovasc. Pharmacol.* **7**: 689–693.

75. Ho, I.K. and Harris, R.A. (1981) Mechanism of action of barbiturates. *Ann. Rev. Pharmacol. Toxicol.* **21**: 93–111.

76. Greven, J., Defrain, W., Glaser, G., Meywald, K. and Heidenreich, O. (1980) Studies with the optically active isomers of the new diuretic drug ozolinone. *Pflüger's Arch.* **384**: 57–60.

77. Zimmerman, D.M. and Gesellchen, P.D. (1982) Analgesics (peripheral and central), endogenous opioids and their receptors. *Ann. Rep. Med. Chem.* **17**: 21–30.

78. Smith, S.M., Wain, R.L. and Wightman, F. (1952) Studies on plant growth-regulating substances. V. Steric factors in relation to mode of action of certain aryloxyalkylcarboxylic acids. *Ann. Appl. Biol.* **39**: 295–307.

79. Lotti, V.J. and Taylor, D.A. (1982) α2-Adrenergic agonist and antagonist activity of the respective $(-)$ and $(+)$-enantiomers of 6-ethyl-9-oxaergoline. *Eur. J. Pharmacol.* **85**: 211–215.

80. Stinson, S.C. (1995) Chiral drugs. *Chem. Eng. News* 44–74.

81. Stinson, S.C. (1999) Chiral drug interactions. *Chem. Eng. News* 101–120.

82. Lindner, W., Rath, M., Stochitzky, K. and Semmelrock, H.J. (1989) Pharmacokinetic data of propranolol enantiomers in a comparative human study with (S)- and (R,S)-propranolol. *Chirality* **1**: 10–13.

83. Reynolds, G.P., Elsworth, J.D., Blau, K., Sandler, M., Lees, A.J., *et al.* (1978) Deprenyl is metabolized to metamphetamine and amphetamine in man. *Br. J. Clin. Pharmacol.* **6**: 542–544.

84. Akerman, B., Astrom, A., Ross, S. and Telc, A. (1966) Studies on the absorption, distribution and metabolism of labeled prilocaine and lidocaine in some animal species. *Acta Pharmacol. Toxicol.* **24**: 389–403.

85. Cotzias, G.C., Papavasiliow, P.S. and Gellene, R. (1969) Modification of parkinsonism: chronic treatment with L-dopa. *New Engl. J. Med.* **280**: 337–345.

86. White, P.F., Ham, J., Way, W.L. and Trevor, A.J. (1980) Pharmacology of ketamine isomers in surgical patients. *Anesthesiology* **52**: 231–239.

87. Kidwell, G.A., Lima, J.J., Schaal, S.F. and Muir, W.M. (1989) Hemodynamic and electrophysiologic effects of disopyramide enantiomers in a canine blood superfusion model. *J. Cardiovasc. Pharmacol.* **13**: 644–655.

88. Lima, J.J., Boudoulas, H. and Shields, B.J. (1985) Stereoselective phamacokinetics of disopyramide enantiomers in man. *Drug Metabol. Disp.* **13**: 572–577.

89. Giacomini, K.M., Nelson, W.L., Pershe, R.A., Valdevieso, L., Turner-Tamayasu, K., *et al.* (1986) *In vivo* interactions of the enantiomers of disopyramide in human subjects. *J. Pharmacokin. Biopharmacol.* **14**: 335–356.

90. Soudijn, W. (1983) Advantages and disadvantages in the application of bioactive racemates or specific isomers as drugs. In Ariëns, E.J., Soudijn, W. and Timmermans, P.B.M.W.M. (eds). *Stereochemistry and Biological Activity of Drugs*, pp. 89–102. Blackwell Scientific Publications, Oxford.

91. Baader, E., Bartmann, W., Beck, G., Bergmann, A., *et al.* (1990) Rational approaches to enzyme inhibitors: new HMG-CoA reductase inhibitors. In Claassen, V. (ed.). *Trends in Drug Research*, pp. 49–71. Elsevier, Amsterdam.

92. Wermuth, C.G. (1993) Aminopyridazines — an alternative route to potent muscarinic agonists with no cholinergic syndrome. *Il Farmaco* **48**: 253–274.

93. Bhagwat, S.S., Gude, C., Cohen, D.S., Dotson, R., Mathis, J., *et al.* (1993) Thromboxane receptor antagonism combined with thromboxane synthase inhibition. 5. Synthesis and evaluation of enantiomers of 8-{[(4-chlorophenyl)sulfonyl]amino}-4-(3-pyridinylalkyl)octanoic acid. *J. Med. Chem.* **36**: 205–210.

94. Yamamura, S., Oda, K., Mizoguchi, T., Saito, S., Iwasawa, Y., *et al.* (1979) Study on the structure–activity relationships of adrenergic, beta-mimetic benzylamine derivatives. V. 9-aryl-1*H*-2,3,7,8,9,10-hexahydro-benzo[d,e]quinolines. *Chem. Pharm. Bull.* **27**: 858–869.

95. Stoll, A. (1945) Über Ergotamin. *Helv. Chim. Acta* **28**: 1283–1308.

96. Knabe, J. (1995) Synthetische Enantiomere als Arzneistoffe. *Pharmazie in unserer Zeit* **24**: 324–330.

18

APPLICATION STRATEGIES FOR PRIMARY STRUCTURE–ACTIVITY RELATIONSHIP EXPLORATION

Camille G. Wermuth

Le bon sens est la chose du monde la mieux partagée: car chacun pense en être si bien pourvu, que ceux même qui sont les plus difficiles à contenter en tout autre chose, n'ont point coutume d'en désirer plus qu'ils n'en ont. Common sense is the worldly thing which is the best shared: as each of us thinks to be so well provided with, that even those who are the most difficult to satisfy in any other thing, don't want to desire more of it than they already have.

René Descartes (1596–1650)[1]

When confronted with a new lead structure or when, for patent reasons, it is necessary to enlarge the protection perimeter around newly discovered structures, the medicinal chemist may be daunted by the immensity of his task. Effectively the possibilities of molecular variations around the lead structure are immense and a priori the synthesis of several thousand potential analogues can be envisaged. The aim of the present chapter is to provide some guidelines and strategies rendering easier and more efficacious the decision on which compounds to prepare and which ones to reject. The proposed guidelines derive essentially from common-sense reasoning, a feature that may explain why they are often forgotten. In addition to his personal experience, the author was inspired by the articles of Messer,[2] Cavalla,[3] Craig,[4] and Austel.[5]

I PRELIMINARY CONSIDERATIONS

Before considering the different possibilities for molecular variation presented in the previous chapters (homology, isostery, conformational restriction, optical isomerism, ring system modifications and synthesis of twin drugs, etc.), one has to decide what kind of general strategy should be applied. Depending on the lead structure's size and on its degree of complexity, the strategy may involve a simplification (disjunctive approach), conservation of the same level of complexity (analogical approach) or enlargement through additional elements (conjunctive approach).

Simplification of the original lead compound is especially appropriate for natural substances. This approach, known as the *disjunctive approach*[6] (see Chapter 14) consists of a molecular dissection that deletes functions, structural elements or cycles. Classical examples of disjunctive approaches are found in the pruning of the acetylcholinesterase inhibitor physostigmine to yield neostigmine and, later on, rivastigmine (Fig. 18.1) or the change from somatostatin to a simplified hexapeptide (see Chapter 23).[7] Other examples of disjunctive approaches are collected in Table 18.1. The main result of the methodology is the identification of the portions of the molecule that are essential for the expected biological activity and of those that are not.

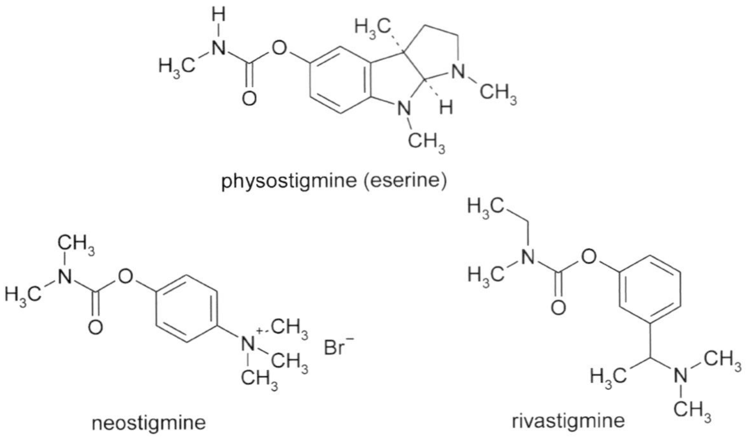

physostigmine (eserine)

neostigmine

rivastigmine

Fig. 18.1   Neostigmine and rivastigmine are the result of disjunctive approach applied to physostigmine.

Conservation of the lead compound's degree of complexity usually proceeds through isosteric exchanges or functional inversions and can be considered as being the *analogical approach*[8] (Table 18.2).

Finally, when additional moieties are grafted on to the molecule, one speaks of *conjunctive approaches*.[6] These can also consist in the attachment of additional structural elements, as in the association of two separate drugs (associative synthesis, nonsymmetrical twin drugs, the symbiotic approach[9]) or in the duplication of the parent drug (symmetrical twin drugs).

The change of the GABA$_B$-receptor agonist CGP 27 492 to the GABA$_B$-receptor antagonist CGP 54 062 (Fig. 18.2) represents a typical example of conjunctive approach resulting from the attachment of additional structural elements.[10] A similar case is provided by the design of the H$_2$-receptor agonist impromidine.[11]

Table 18.1   *Drugs resulting from disjunctive manipulations*

Lead	Derivative
Cocaine	Procaine
Tubocurarine	Decamethonium
Morphine	Morphinanes
	Benzomorphanes
	Phenylpiperidines
Atebrine	Chloroquine
Asperlicin	Benzodiazepine analogue
Phylloquinone	Menadione
(vitamine K$_1$)	(vitamine K$_3$)
Triampterene	Amiloride
Cimetidine	Roxatidine
Somatostatine	Simplified peptide
Bothrops jaraca	Teprotide
Venin	Captopril

Table 18.2   *Drug analogues possessing a similar size to the model compound*

Initial drug	Analogue
Chlorpromazine	Thioridazine
Imipramine	Amitryptyline
Propranolol	Pindolol
Furosemide	Bumetamide
Enalapril	Perindopril
Cimetidine	Ranitidine
Pravastatine	Fluindostatine

Examples of drugs resulting from *associative synthesis* (nonsymmetrical twin drugs) and of *duplication* of the parent drug (symmetrical twin drugs) are listed in Tables 18.3 and 18.4. A more detailed study is presented in Chapters 16 and 28.

II HIT OPTIMIZATION STRATEGIES

The strategy of hit optimization depends heavily on the amount of information available at the start of the study. Particularly, if some knowledge of the three-dimensional structure of the target is available, the synthesis programme can immediately take into account the corresponding information. This will also be the case if some earlier SAR studies are available. However, in the most frequent cases, the target is new and original. In such a situation all the initial SAR exploration rests in the hands of the medicinal chemist. The purpose of this chapter is to let him benefit from more than 30 years experience and to help him in the choice of the most appropriate strategy.

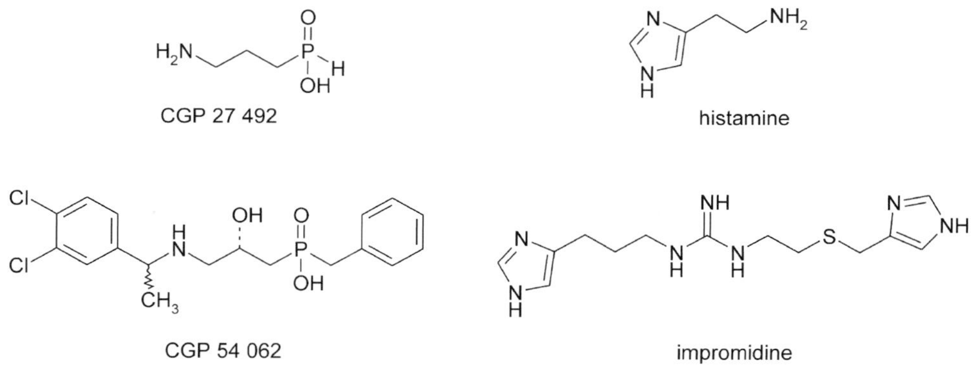

Fig. 18.2 Conjunctive approach in drug design. The attachment of two benzylic groups and a hydroxyl in *S*-configuration changes a GABA$_B$ receptor agonist into a GABA$_B$ antagonist,[10] similarly the H$_2$-histaminergic agonist impromidine is the result of a conjunctive approach applied to histamine.[11]

A   Some information about the target is available

The first point to consider is to ascertain if the hit (or the lead) that has to be optimized is relevant to a computer-aided design. If the X-ray three-dimensional structure of the target protein is already described, it becomes possible to match the different candidate molecules with the target structure and eliminate those which evidently are too bulky or which possess an inadequate geometry. A similar situation occurs if a nonexperimental three-dimensional model of the target is available (see Chapters 24, 26 and 27).

In the absence of any target structure, some information can be gathered in comparing the hit structure with the endogenous ligand (if known) or that other structures showing affinities for the same target (if known). If several compounds exist that are recognized by the target it becomes possible to practice the active analogue approach (Chapter 24).

B   No information about the target is available

When one has to deal with a real new target, the only way to progress is to perform enough syntheses to identify the

molecular features which are favourable and those which are detrimental to the activity. Such molecular variation programmes can be undertaken in several ways. Below, two are presented that are both characterized by their systematic aspects: the first one is a topological exploration, the second one is a more open and global approach.

The topological exploration

One of the most fruitful strategies revealing the features associated with high activity consists of what we call the *topological exploration of the lead compound*. In this approach, the possible modifications of the molecule are considered from the four cardinal points: the south, the north, the west and the east, and from the centre of the molecule (Fig. 18.3). In order to illustrate the point, let us assume that the starting lead compound is the muscarinic antagonist pirenzepine. We can then consider successively the different modification sites.

East-side modifications. In the present example, the east-side modifications mainly concern changes at the level of the pyridine ring, and the questions that should be answered are as follows (Fig. 18.4):

Table 18.3 *Nonsymmetrical twin drugs*

Drug 1	Drug 2	Twin drug
Caffeine	Amphetamine	Fenethylline
Aspirin	Paracetamol	Benorylate
Clofibric acid	Nicotinic acid	Etofibrate
Hydrazinopyridazine	β-blocker	Prizidilol
Pindolol	Captopril	BW-B385C[12]

Table 18.4 *Symmetrical twin drugs*

Drug	Activity
Bialamicol	Antiamoebic
Ethambucol	Tuberculostatic
Probucol	Antihyperlipoproteinaemic
Thiamine disulphide	Vitamin
Dicumarol	Anticoagulant
Netropsin	DNA binding agent
Succinylcholine	Skeletal muscle relaxant

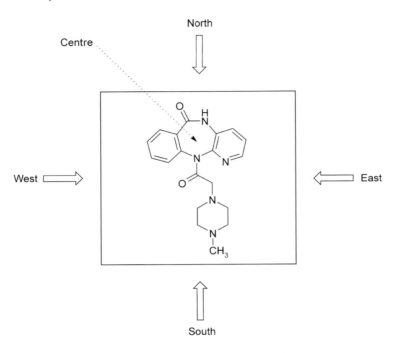

Fig. 18.3 Scheme of the topological exploration of the pirenzepine molecule.

- Is the pyridine nitrogen necessary? Can the pyridine ring be replaced by a phenyl ring (a)?
- Can the pyridine nitrogen be displaced to other positions (b, c, d)?
- What is the influence of substituents on the pyridine ring (e)? Can it be substituted by various functions (associated with typical electronic, steric or lipophilic changes)?
- Can the pyridine ring be changed to other aromatic heterocycles: pyridazine (f), pyrimidine (g), pyrazine (h), triazines (i), thiazoles (j) etc?

North-side modifications. On the north side of pirenzepine one can consider the NH and the $C=O$ groups separately or together, as an amide function (Fig. 18.5).

- Can the NH group be substituted (a) (possible role as hydrogen bond donor)?
- Can the NH group be replaced by a CH_2 group (b) (or any other possible bioisosteric group)?
- Is the carbonyl group necessary? Can it be changed to a CH_2 group (c)?
- Can the amide be replaced by any bioisostere such as described in Chapter 13 (d, e, f)?

West-side modifications. Owing to the almost symmetrical structure of pirenzepine, essentially the same kind of modifications can be applied as those which were suggested for the east side.

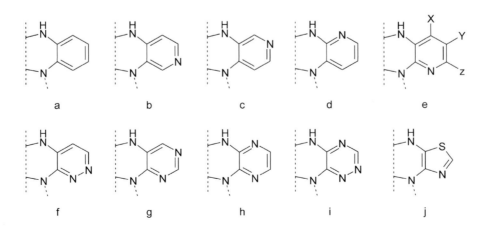

Fig. 18.4 East-side modifications on the pirenzepine molecule.

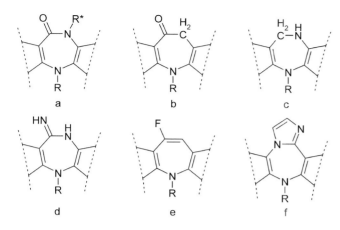

Fig. 18.5 North-side modifications on the pirenzepine molecule.

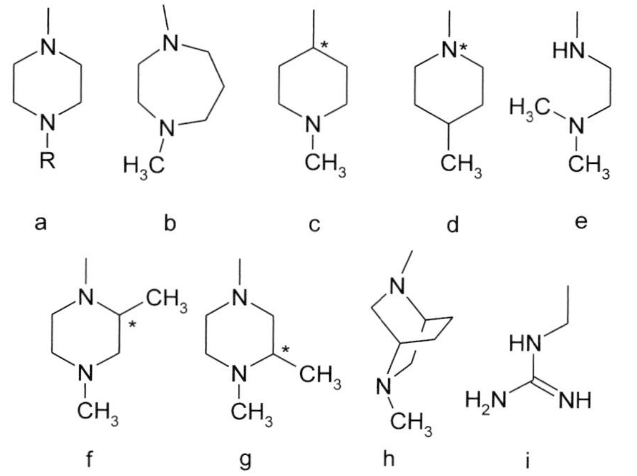

Fig. 18.7 Variations on the *N*-methylpiperazine ring.

- For both east and west sides, 'benzo-splitting' (see Chapter 14) can be considered (Fig. 18.6).

South-side modifications. They concern the changes made on the basic side-chain, with a huge number of possibilities existing at different levels:

- On the piperazine ring (Fig. 18.7):
  Can the *N*-methyl group be replaced by higher alkyl, aralkyl or aryl groups? (a)

Can it be replaced by homopiperazine (b), by piperidine (c and d), by ring-opened diamines (e)?
Can it be substituted (g and h) or bridged (i)?
Can it be replaced by some vague bioisosteric equivalent such as a guanidino group (*N*-methyl piperazine was used as a guanidine substitute in the design of thrombin inhibitors)?

- On the carboxamido function (Fig. 18.8):
  Can the carbonyl group be reduced to a CH_2 group (a)?
  Can the carboxamido function be replaced by a carbon–carbon double bond (b)?

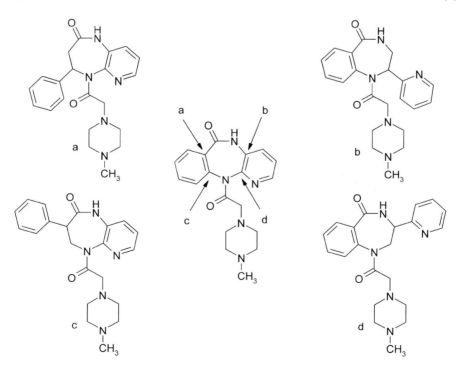

Fig. 18.6 'Benzo-splitting' applied to the east and the west side of pirenzepine.

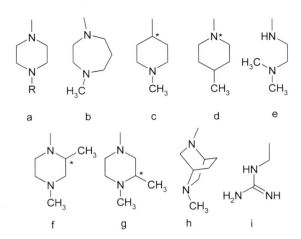

Fig. 18.8 Variations on the side-chain carboxamido group of pirenzepine.

Can it be included in a bioisosteric and constraint ring system (c)?

Centre modifications. The literature available for the tricyclic psychotropic drugs proposes numerous possibilities of variations potentially applicable to the present central diazepinone ring (Fig. 18.9). They include ring contraction (a), ring extension (b), ring bridging (c) and changes in the nature and number of the heteroatoms.

The global approach

This approach consists in applying, in a systematic manner, the molecular variation methodologies presented as the primary exploration of structure–activity relationships in Chapters 12 to 17. It starts with 'paper chemistry', drawing all the conceivable structural analogues of the lead. For this purpose, all the possibilities offered by the molecular variation methodology should be freely used: the reader is encouraged to apply successively isomeric, homologous, vinylogous and, of course, isosteric replacements; and also to try functional exchanges, ring modifications, stereochemical changes, conformational restriction, symmetrical

and nonsymmetrical twin drugs, and so on. This strategy allows one, starting from a given lead compound, to envisage an extremely high number of structural variations in a purely formal manner. Applied systematically, it proves to be very useful because it stimulates the imagination and generates new ideas.

The richness of this purely formal procedure is illustrated in Fig. 18.10, where it is applied to the well-known anxiolytic molecule diazepam. In combining the different individual modifications, a bewildering number of structures can be imagined! Even when applied to a known structure such as a benzodiazepine, the *systematic* application of the stratagem allows the identification of unknown analogues and generates interesting new ideas. However, a disadvantage of this approach is that others can also follow the same reasoning and end up with the same ideas. It is therefore recommended to use this approach preferentially with leads that are your own.

III APPLICATION RULES

It would certainly be tiresome and beyond the possibilities of a medicinal chemistry team to be forced to prepare all the compounds imagined by means of 'paper' chemistry. The following rules aim to codify precisely the use of all the strategies and to increase their efficacy by establishing priorities and selection rules.

A Rule 1: The minor modifications rule

This rule[8] can be defined as giving priority to the design of analogues that are close to the lead structure and that result from only minor changes. Minor changes are achieved by very simple organic reactions such as hydrogenations, hydroxylations, methylations, acetylations, racemate resolutions, changes in substituents and isosteric replacements. The modification can produce either an increase in potency or an increase in selectivity or even sometimes the suppression of unwanted toxic or side-effects (Table 18.5).

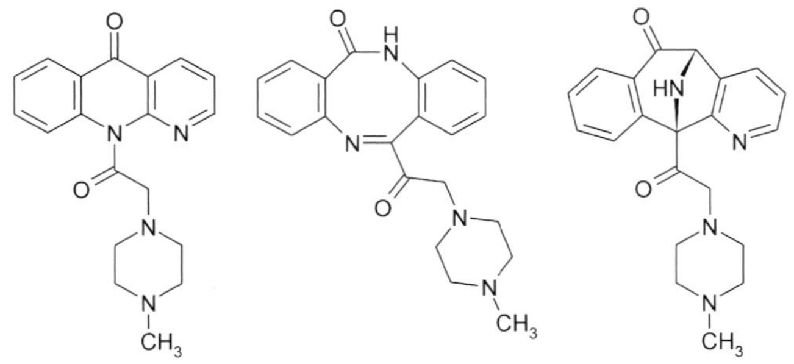

Fig. 18.9 Modifications of the central ring of pirenzepine.

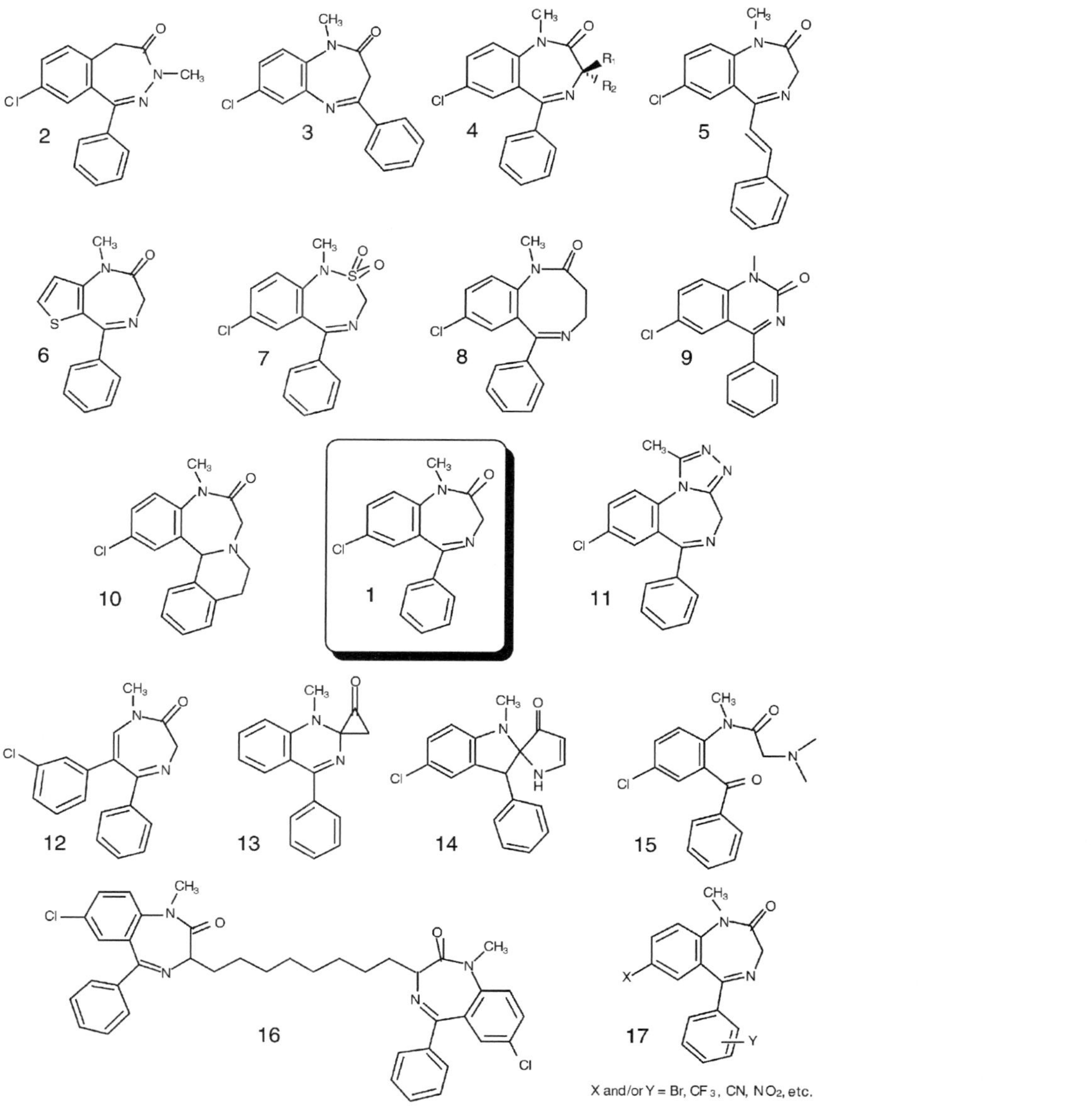

Fig. 18.10 Molecular variations applied to diazepam (**1**). Compounds (**2**) and (**3**) are positional isomers; optical isomerism is introduced in (**4**). Compound (**5**) is a vinylogue (vinylogy is also used to render amidic the carbonyl of (**14**). Compounds (**6**) and (**7**) are isosteres. Compounds (**8**) to (**15**) result from various ring modifications: enlargement (**8**), contraction (**9**), introduction of additional rings (**10** and **11**), benzo splitting (**12**), use of spiro systems (**13** and **14**), ring opening (**15**). Compound (**16**) is a symmetrical twin drug and compound (**17**) underlines that many substituent variations can be used. The presented structures do not strictly correspond to existing molecules.

A simple change in the aliphatic chain length can abolish *mutagenic* properties. Thus in a series of muscarinic M_1 partial agonists (Fig. 18.11), the compound with a dimethylene side-chain is potent (IC_{50} [^{3}H]-pirenzepine = 3 nM), but mutagenic. The higher trimethylenic homologue is less potent (IC_{50} [^{3}H]-pirenzepine = 15 nM), but safe in terms of mutagenicity.[14]

In some instances very slight changes such as hydrogenation or de-hydrogenation can induce dramatic changes in the activity profile of drug molecules. Examples are found in the imidazoline I_3-receptor ligands which act on insulin secretion. The imidazolinic compound efaroxan acts as an agonist,[15] and the corresponding imidazole acts as an antagonist[16] (Fig. 18.12). A similar transition is found in that of the agonistic benzofuranic compound 2-BFI[17] to its antagonistic dihydro derivative.[18]

The minor modifications rule, even supported by prestigious results, is largely unrecognized. Making use of

Table 18.5 *Minor modifications*

Original compound	Modified compound	Result[a]
Ergotamine	Dihydroergotamine	Increase in potency as α-adrenergic antagonist Decrease in toxicity
Chlorothiazide	Hydrochlorothiazide	20-Fold increase in potency
Chloroquine	Hydroxychloroquine	Decrease in toxicity
Morphine	Codeine	Change in activity profile (analgesic $\rightarrow$ antitussive)
Carbachol	Bethanechol	Increase in selectivity (exclusively muscarinic)
Imipramine	Desmethylimipramine	Change in activity profile (noradrenergic $\rightarrow$ serotonergic)
Tolbutamide	Chlorpropamide	Longer duration of action (5–7 h $\rightarrow$ 24–48 h)
Racemic amphetamine	Dexamphetamine	Fewer cardiovascular side-effects

[a] Taken from Reference 13.

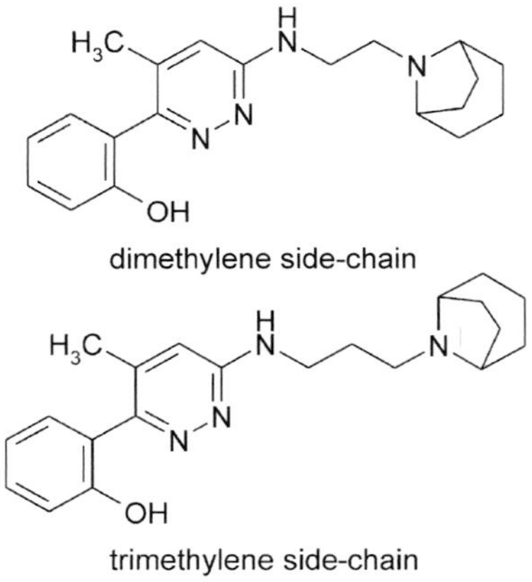

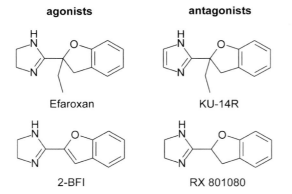

Fig. 18.11 The mutagenicity of the original dimethylenic muscarinic M₁ partial agonist could be abolished in changing the dimethylene to a trimethylene side-chain.

Fig. 18.12 Hydrogenations and dehydrogenations can induce switches from agonist to antagonist profiles.

ordinary chemistry, it is not always accepted with enthusiasm by organic chemists. The very simple reactions that are involved do not add much to their fame and they are more fascinated by the challenge of a total synthesis, especially of a natural substance bearing many chiral centres. Seen from a practical point of view, priority has, nevertheless, to be given to this principle. Its simplicity of implementation, and especially the spectacular results that it brings, militate in its favour.

B Rule 2: The biological logic rule

The second rule of application rests on the earliest possible utilization of biochemical data. Indeed, even when a medicinal chemist ignores all of the biochemistry of the substance that he studies, he learns to consider the former from the biochemical angle which will provide him with a good number of subjects to think about and may sometimes allow anticipation of the behaviour of the molecule. In particular, biological activity may be rationalized if it stems from the chemical or physicochemical properties of the series.

Very general properties can be foreseen in so far as functions or moieties present in the structure can suggest interference with a biological system; for example, hydrazines or hydroxylamines and pyridoxal-containing coenzymes, complexing agents and metallic coenzymes, electron donors or acceptors and oxido-reduction coenzymes, tensioactive amphiphilic substances and production of hemolysis in erythrocytes. In a more precise manner, the alkylating properties of compounds such as the nitrogen mustards, the nitrosoureas and the mitomycins, or the crosslinking properties of *cis*-platinum derivatives relate to their anticancer activity. The activity of anthracyclines, ellipticine, and the anthracene-diones has been shown to be due to intercalation in the double helix of DNA. The anti-coccidial polyether ionophores are potent complexing agents for mono- and divalent cations.

The biological action of a compound is also readily explicable if it mimics a natural substrate or mediator. This is the case for enzymes with inhibitors (angiotensin I and captopril), suicide substrates (GABA and vigabatrin), antimetabolites (*p*-aminobenzoic acid and *p*-aminobenzenesuphonamide) and for receptors with agonists (acetylcholine and muscarine), antagonists (GABA and gabazine) and uptake inhibitors (GABA and nipecotic acid). The analogy with the endogenous substance can sometimes be very vague, as seen with quaternary ammonium compounds which all present more or less affinity for the cholinergic receptors, or with purines, which are often recognized by the phosphodiesterases.

The pathways of drug metabolism follow some general rules (see Chapter 31) and the metabolites of a given substance can, at least qualitatively, be imagined in advance. As a consequence, various measures can be taken to favour or, conversely, to slow down the biodegradation. Some chemical groupings are more prone than others to yield unwanted toxic metabolites (see Chapter 32). Among the best known are the aromatic nitro, nitroso, azo and amino compounds, the bromoarenes, the hydrazines and the hydroxylamines, and the polyhalogenated aliphatic or aromatic compounds. A proposed explanation for arylamine toxicities (the so-called para effect) is their facile oxidation to an electrophilic quinonic system followed by addition of thiol nucleophiles; a process that models well-known hapten formation reactions.[19] Finally, if the active principle is an acid or a base, the choice of the salifying counterion also has to follow some selection criteria (see Chapter 35). Oxalates and nitrates, for example, are not very popular whereas hydrochlorides represent a satisfactory compromise.

C Rule 3: The structural logic rule

This rule implies that, as soon as some structural data are available (intercharge distances, *E*- or *Z*-conformations, axial or equatorial substituent orientations, misoriented substituents, etc.), they have to be fed back into drug design. When dealing with enzymes or receptors of unknown structure, one route to such information consists of comparing already known active compounds, recognized by the same molecular target, and to deduce the important stereoelectronic features associated with potency and selectivity. This approach is referred to as pharmacophore identification or receptor mapping (see Chapters 22 and 24). Initially presented by Marshall *et al.,*[20] it has some predictive merit[21,22] and, at least avoids unnecessary syntheses of *a priori* inactive compounds. In practice, the most efficient methods consist of steady comings and goings between synthetic and computer chemistry in order to achieve the ideal interplay between intuition and computer assistance.

A structural guide is also available for drugs designed to bind to the neurotransmitters of the central nervous system. On the basis of a comparison of the crystal structure of recognized representative compounds from each of eight major CNS active drug classes, Andrews and Lloyd[23,24] identified a common structural basis, essentially characterized by an aromatic plane distant by about 5 Å from a nitrogen moiety. The explanation for this finding resides in the biochemical origin of the neurotransmitters.[22] Most of them are of the arylethylamine type, as a result of the decarboxylation of aromatic amino acids such as DOPA or histidine.

Another example is given by acetylcholine which can be considered as bioisosteric with GABA (Fig. 18.13); this property explains the observation that a compound such as the GABA$_A$ receptor antagonist bicuculline is recognized by both the GABA$_A$ and the nicotinic receptors.[22]

D Rule 4: The right substituent choice

Half of all existing drugs contain easily substituted aromatic rings. The replacement, in such rings, of a hydrogen by a substituent (alkyl, halogen, hydroxyl, nitro, cyano, alkoxy, amino, carboxylate, etc.) can dramatically modify the intensity, the duration, and perhaps even the nature of the pharmacological effect (see Chapters 19, 20 and 22). It becomes therefore of prime importance to proceed to the optimal choice of substituents so as to explore with the smallest set possible the three-dimensional space formed by lipophilic, electronic and steric parameter coordinates (Fig. 18.14).

The right substituent choice minimizes the number of test compounds that have to be synthesized to ensure a significant space volume. Fig. 18.13 represents a three-dimensional extension of the Craig plot, discussed by Craig[4] and by Austel.[5] In this context, the decision tree proposed by Topliss[25] allows a rapid identification of the substituents associated with the highest potency. Application examples of the Topliss scheme are discussed by Martin and Dunn[26] and in Chapter 22.

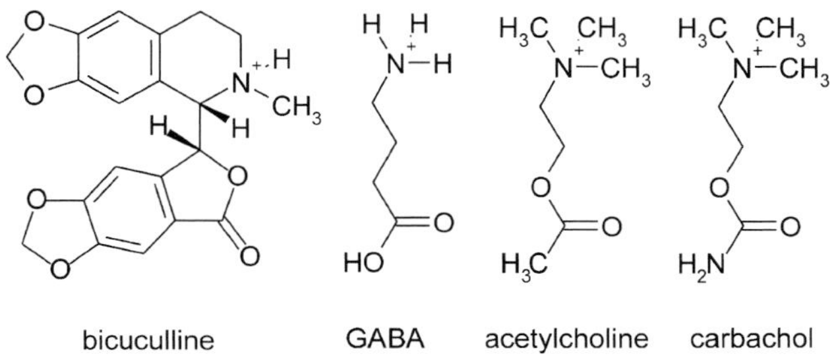

bicuculline   GABA   acetylcholine   carbachol

Fig. 18.13 Similar intercharge distances between the protonated nitrogen and the carbonyl dipole exist in bicuculline, GABA, acetylcholine and carbachol.

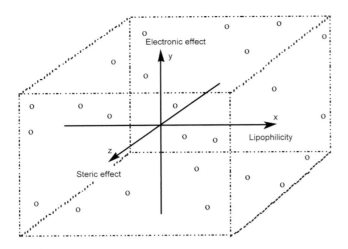

Fig. 18.14 Three-dimensional space formed by lipophilic, electronic and steric coordinates.

E Rule 5: The easy organic synthesis (EOS) rule

Synthesis of new compounds is a costly and lengthy process, therefore any measure able to render it more efficacious is welcome. Thus, for example, when the decision is taken to prepare a given set of compounds, why not first prepare those whose synthesis is the easiest? Along the same line of thought why not prepare first compounds for which intermediates are commercially available?

A particular recommendation is to synthesize heterocycles. In statistics established in 1982 on 1522 drug molecules, Kleemann and Engel[27] highlighted the fact that, among the synthetic drugs, 62% contained at least one heterocyclic ring, the percentage within natural compounds being even higher (77%). Indeed, heterocycles present many advantages: (1) they allow the insertion of elements capable of giving interactions that the carbocycles do not give; (2) they allow a greater number of combinations. It therefore becomes easier to be original; (3) they represent rigid analogues of endogenous substances that themselves are often nitrogenous metabolites of amino acids; and (4) their facile synthesis often permits the preparation of large series.[28] One of the major problems when dealing with isosteric or bioisosteric replacements in heterocyclic systems is the selection of the a priori most promising candidate among several dozens of possible rings. A simple clue, which reflects the dipolar moment, can be given by knowledge and comparison of the boiling points of the basic heterocycles (see Chapter 13).

F Rule 6: Eliminate the chiral centres

Although optical isomerism has already been discussed in Chapter 17, some practical considerations on chiral molecules are appropriate here. Nowadays it is accepted that racemates and both enantiomers are usually three different pharmacological entities and that it requires extensive pharmacological, toxicological and clinical pharmacological research before it can be decided whether it is advantageous to use racemates or enantiomers in clinical practice. According to Soudijn,[29] these research efforts could be reduced to about one-third when drugs without centres or planes of asymmetry could be developed with the same or higher affinity. Effectively asymmetry is far from being an absolute requisite for activity! The alkaloid morphine possesses five chiral centres, however, its synthetic derivative fentanyl is devoid of any asymmetric centre but nonetheless belongs to the most potent analgesics known. Usually chiral centres are eliminated in creating symmetry (see Chapter 17). A typical example of this process is the design of nonchiral immunosuppressive 2-aminopropane-1,3-diols starting from the natural compound myriocin[30] (Fig. 18.15).

In some instances chiral centres can be, at least, partially eliminated. This is the case for the synthetic analogues of the HMG-CoA reductase inhibitor mevinolin. Mevinolin itself (see Chapter 17) has seven asymmetric centres but the five chiral centres contained in the hexahydronaphtalene ring system are unnecessary for HMG-CoA inhibition. The second generation of mevinolin analogues retains only two of the initial seven chiral centres.[31]

When one nevertheless has to deal with chiral centres, why not first prepare the racemic compound and start with an enantioselective synthesis only if an interesting activity is found? This latter point is tricky, because many people believe that two enantiomers might happen to antagonize

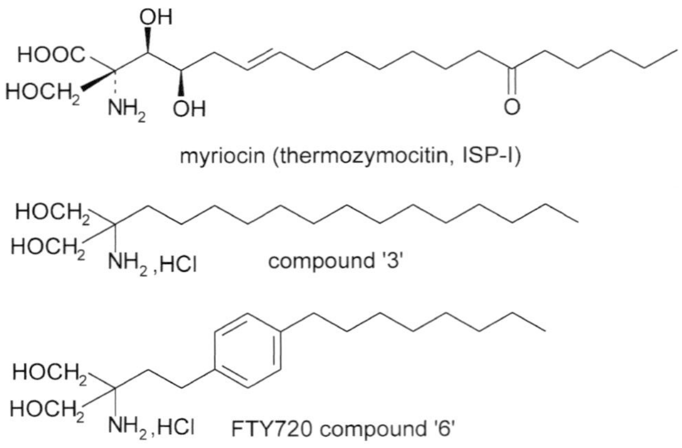

myriocin (thermozymocitin, ISP-I)

compound '3'

FTY720 compound '6'

Fig. 18.15 Suppression of chiral centres through introduction of symmetry.[30]

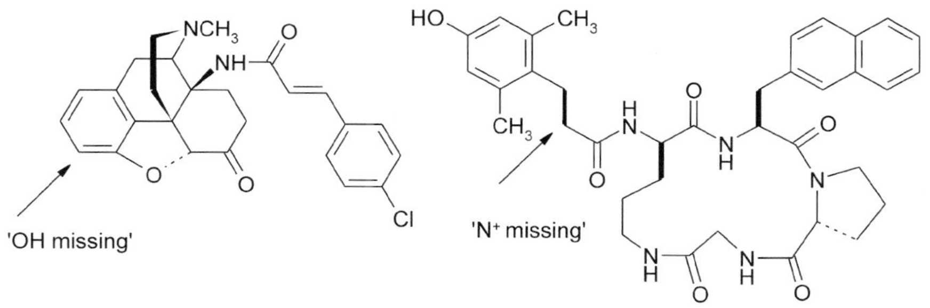

Fig. 18.16 Nonconventional, OH missing (left) or N^+ missing (right), opiates.

each other. They refer to the numerous examples published in the literature.[29,32–34] In reality, two optical isomers *are never antagonists at comparable dosages.* This comes from the space-relationship required for the interaction with the receptor site, which is only slightly altered by passing from *S*- to *R*- forms, or vice versa. If one of the enantiomers achieves the optimal fit to the receptor site in exchanging the highest number of non-covalent linkages, its antipode only gives rise to weaker interactions, even under the most favourable conditions (see Chapter 17).

From a practical point of view, this absence of *stoichiometric* antagonism entails two consequences: (a) if a racemic mixture does not show any activity, it is useless to carry out the separation of the two antipodes; (b) a racemic mixture usually has the average potency of both constituents, thus, the maximal benefit one can achieve in resolving racemic mixtures is an increase of the activity to twice that of the racemate.

G Rule 7: The pharmacological logic rule

In Chapter 2, we insisted that a correctly performed pharmacological study must satisfy certain criteria (relationship between dose and effect, presentation of the confidence limits, comparison with a reference compound, determination of the time of the peak action). On the chemical side it is also extremely important to provide pharmacologists with reference compounds published by the competitor laboratories. Even if it is felt tedious and time-consuming to resynthesize an already described compound, the operation is always worthwhile and sometimes surprising. How often a good-looking published molecule, for which attractive activities are claimed, loses much of its charm once it is reinvestigated by one's own team!

An open attitude, without prejudice, towards active molecules is a highly appreciated quality for all medicinal chemists. This means that they should not always take for granted apparently well-established dogmas. For example,

during one century it was accepted that the critical factors for morphinic compounds were the presence of a phenolic hydroxyl group and of a protonated nitrogen. Two articles, published in 2000, described novel compounds lacking either a phenolic hydroxyl group,[35] or a positively loaded nitrogen,[36] and which nevertheless present submicromolar affinities for the μ receptor (Fig. 18.16).

Another point that may contribute to increasing the credibility of your work is the so-called *a contrario* probe. In other words, when your own structure–activity relationship studies allow identification of the molecular features associated with high activity, proceed, of course, to the synthesis of the most interesting representatives, but also prepare at least one compound that, according to your results, should be inactive.

REFERENCES

1. Descartes, R. (1972) Discours de la méthode, 1ére partie 1637. In Robinet, A. (ed.). *Les Nouveaux Classiques Larousse*, p. 27. Librairie Larousse, Paris.
2. Messer, M. (1984) Traditional or pragmatic research. In Jolles, G. and Woodridge, C.R.M. (eds). *Drug Design: Fact or Fantasy?* pp. 217–224. Academic Press, London.
3. Cavalla, J.F. (1983) Drug design valuable for refining an active drug. In Gross, F. (ed.). *Decision Making in Drug Research*, pp. 165–172. Raven Press, New York.
4. Craig, P.N. (1980) Guidelines for drug and analog design. In Wolffe, M.E. (ed.). *The Basis of Medicinal Chemistry*, pp. 331–348. Wiley-Interscience, New York.
5. Austel, V. (1984) Features and problems in practical drug design. In Charton, M. and Moto, I. (eds). *Steric Effects in Drug Design*, pp. 8–19. Lange and Springer, Berlin.
6. Schueller, F.W. (1960) *Chemobiodynamics and Drug Design.* McGraw-Hill, New York.
7. Freidinger, R.M. and Veber, D.F. (1984) Design of novel cyclic hexapeptide somatostatin analogs from a model of the bioactive conformation. In Vida, J.A. and Gordon, M. (eds). *Conformationally Directed Drug Design*, pp. 169–187. American Chemical Society, Washington DC.
8. Wermuth, C.G. (1966) Modifications chimiques des médicaments en vue de l'amélioration de leur action. *Agressologie* **7**: 213–219.

9. Nicolaus, B.J.R. (1983) Symbiotic approach to drug design. In Gross, F. (ed.). *Decision Making in Drug Research*, pp. 173–186. Raven Press, New York.

10. Froestl, W., Furet, P., Hall, R.G., Mickel, S.J., Strub, D., *et al.* (1993) GABA$_B$ antagonists: novel CNS-active compounds. In Testa, B., Kyburz, E., Fuhrer, W. and Giger, R. (eds). *Perspectives in Medicinal Chemistry*, pp. 259–272. VHC, Weinheim.

11. Durant, G.J., Duncan, W.A.M., Ganellin, C.R., Parsons, M.E., Blakemore, R.C. and Rasmussen, A.C. (1978) Impromidine (SK&F 92 676) is a very potent and specific agonist for histamine H$_2$ receptors. *Nature* **276**: 403–405.

12. Hardy, G.W. and Allan, G. (1988) BW-B3895C. *Drugs Fut.* **13**: 204–206.

13. Goodman-Gilman, A., Rall, T.W., Nies, A.S. and Taylor, P. (eds) (1990) *Goodman and Gilman's The Pharmacological Basis of Therapeutics*, 8th edn. Pergamon Press, New York.

14. Kan, J. P. (1997) Sanofi-Synthelabo Recherche, Montpellier, France, personal communication, 17 January, 1997.

15. Chan, S.L., Pallett, A.L. and Morgan, N.G. (1997) Clotrimazole and efaroxan stimulate insulin secretion by different mechanisms in rat pancreatic islets. *Naunyn Schmiedebergs Arch. Pharmacol.* **356**: 763–768.

16. Chan, S.L., Atlas, D., James, R.F. and Morgan, N.G. (1997) The effect of the putative endogenous imidazoline receptor ligand, clonidine-displacing substance, on insulin secretion from rat and human islets of Langerhans. *Br. J. Pharmacol.* **120**: 926–932.

17. Morgan, N.G., Chan, S.L., Mourtada, M., Monks, L.K. and Ramsden, C.A. (1999) Imidazolines and pancreatic hormone secretion. *Ann. NY Acad. Sci.* **881**: 217–228.

18. Brown, C.A., Chan, S.L., Stillings, M.R., Smith, S.A. and Morgan, N.G. (1993) Antagonism of the stimulatory effects of efaroxan and glibenclamide in rat pancreatic islets by the imidazoline, RX801080. *Br. J. Pharmacol.* **110**: 1017–1022.

19. Sanner, M.A. and Higgins, T.J. (1991) Chemical basis for immune mediated idiosyncratic drug hypersensitivity. In Bristol, J.A. (ed.). *Annual Reports in Medicinal Chemistry*, Vol. 26. Academic Press, San Diego.

20. Marshall, G.R., Barry, C.D., Bosshard, H.E., Dammkoehler, R.A. and Dunn D.A. The conformational parameter in drug design: the active analog approach. In Olson, E.C. and Christofferson, R.E. (eds). *Computer-assisted Drug Design*, pp. 205–226. American Chemical Society, Washington DC.

21. Marshall, G.R. and Cramer III, D.R. (1988) Three-dimensional structure–activity relationships. *Trends Pharmacol. Sci.* **9**: 285–289.

22. Wermuth, C.G. and Langer, T. (1993) Pharmacophore identification. In Kubinyi, H. (ed.). *3D QSAR in Drug Design—Theory, Methods, and Applications*, pp. 117–136. ESCOM, Leiden.

23. Andrews, P.R. and Lloyd, E.J. (1983) A common structural basis for CNS drug action. *J. Pharm. Pharmacol.* **35**: 516–518.

24. Lloyd, E.J. and Andrews, P.R. (1986) A common structural model for central nervous system drugs and their receptors. *J. Med. Chem.* **29**: 453–462.

25. Topliss, J.G. (1972) Utilization of operational schemes for analog synthesis in drug design. *J. Med. Chem.* **15**: 1006–1011.

26. Martin, Y.C. and Dunn, W.J. (1973) Examination of the utility of the Topliss schemes for analog synthesis. *J. Med. Chem.* **16**: 578–579.

27. Kleemann, A. and Engel, J. (1982) *Pharmazeutische Wirkstoffe-Synthese, Patente, Anwendungen* (Preface). Georg Thieme Verlag, Stuttgart.

28. Messer, S. (1982) Personal communication, Rhône–Poulenc Recherches, Centre de Recherches Nicolas Grillet.

29. Soudijn, W. (1983) Advantages and disadvantages in the application of bioactive racemates or specific isomers as drugs. In Ariëns, E.J., Soudijn, W. and Timmermans, P.B.M.W.M. (eds). *Stereochemistry and Biological Activity of Drugs*, pp. 89–102. Blackwell Scientific Publications, Oxford.

30. Kiuchi, M., Adachi, K., Kohara, T., Minoguchi, M., Hanano, T., Aoki, Y., Mishina, T., Arita, M., Nakao, N., Ohtsuki, M., Hoshino, Y., Teshima, K., Chiba, K., Sasaki, S. and Fujita, T. (2000) Synthesis and immunosuppressive activity of 2-substituted 2-aminopropane-1,3-diols and 2-aminoethanols. *J. Med. Chem.*, **43**: 2946–2961.

31. Baader, E., Bartmann, W., Beck, G., Bergmann, A., Granzer, E. *et al.* (1990) Rational approaches to enzyme inhibitors: new HMG–CoA reductase inhibitors. In Claasen, V. (ed.). *Trends in Drug Research*, pp. 49–71. Elsevier, Amsterdam.

32. Towart, R., Wehninger, E. and Meyer, H. (1981) Effects of unsymmetrical ester substituted 1,4-dyhydropyridine derivatives and their optical isomers on contraction of smooth muscle. *Naunyn-Schmiedeberg's Arch. Pharmacol.* **317**: 183–185.

33. Lotti, V.J. and Taylor, D.A. (1982) α_2-Adrenergic agonist and antagonist activity of the respective (−)- and (+)-enantiomers of 6-ethyl-9-oxaergoline. *Eur. J. Pharmacol.* **85**: 211–215.

34. Hof, R.P., Rüegg, U.T., Hof, A. and Vogel, A. (1985) Stereoselectivity at the calcium channel: opposite action of the enantiomers of a 1,4-dihydropyridine. *J. Cardiovasc. Pharmacol.* **7**: 689–693.

35. Derrick, I., Neilan, C.L., Andes, J., Husbands, S.M. and Woods, J.H. (2000) 3-Deoxyclocinnamox: the first high-affinity, nonpeptide μ-opioid antagonist lacking a phenolic hydroxyl group. *J. Med. Chem.* **43**: 3348–3350.

36. Schiller, P.W., Berezowska, I., Nguyen, T.M.-D., Schmidt, R., Lemieux, C., Chung, N.N., Falcone-Hindlay, M.L., Yao, W., Liu, J., Iwama, S., Smith, A.B. III, and Hirschmann, R. (2000) Novel ligands lacking a positive charge for the δ- and μ-opioid receptors. *J. Med. Chem.* **43**: 551–559.

PART IV

SUBSTITUENTS AND FUNCTIONS: QUALITATIVE AND QUANTITATIVE ASPECTS OF STRUCTURE–ACTIVITY RELATIONSHIPS

19

SPECIFIC SUBSTITUENT GROUPS

Camille G. Wermuth

Fifty percent of the currently used drugs contain at least one aromatic ring that can be a matter of substitution
John Taylor[1]

I INTRODUCTION

The replacement, in an active molecule, of a hydrogen atom by a substituent (alkyl, halogen, hydroxyl, nitro, cyano, alkoxy, amino, carboxylate, etc.) can deeply modify the potency, the duration, perhaps even the nature of the pharmacological effect. Structure–activity relationship studies implying substituent modifications therefore represent a common practice in medicinal chemistry, all the more since half of the existing drugs contain easy-to-substitute aromatic rings. The perturbations brought by the substituent can affect various parameters of a drug molecule, such as its *partition coefficient, electronic density, steric environment, bioavailability, pharmacokinetics*, and, finally its capacity to establish *direct interactions* between the substituent and the receptor or the enzyme (Fig. 19.1).

Fig. 19.1 The replacement in an active molecule of a hydrogen atom by another atom or by a functional group can affect various parameters of a drug molecule such as its partition coefficient, its electronic density, its steric environment, its bioavailability and pharmacokinetics and finally, its capacity to establish direct interactions between the substituent and the receptor or the enzyme.

In reality it is impossible to modify only one alone of these five parameters. Thus for example, the replacement of a hydrogen atom by a methyl group is going to play simultaneously on the five parameters listed above. Nevertheless, through a careful selection of the adequate substituent, it is possible to vary one of the considered parameters in a dominant manner.

To illustrate the repercussions on biological activity resulting from substituent effects, we will study the effects of methyl groups, of unsaturated groups, and of halogen substitution successively. Hydroxy groups, thiols and acidic or basic functions will be discussed more briefly. The final section deals with the attachment of large lipophilic additional binding moieties.

II METHYL GROUPS

In this section we show how a methyl group, so often considered as chemically inert, is able to alter deeply the pharmacological properties of a molecule. We will envisage successively effects on the solubility, conformational effects, electronic effects, and effects on the bioavailability and the pharmacokinetics. In the final paragraph we will present some replacement possibilities of the methyl group by related groups and extend the study to some larger alkyl groups.

A Effects on solubility

As a rule, the grafting of one or several methyl groups on an active molecule renders the former more lipophilic and therefore less soluble in water. However, in some particular cases, grafting one or several methyl groups to a molecule results in an increase of the water solubility by mechanisms such as the increase of hydrophobic bonding possibilities or diminution of the crystal lattice energy.

Increase in lipophilicity

Normally one expects that methyl groups increase the lipophilicity. Indeed, the partition coefficient P between *n*-octanol and water is P = 490 for toluene, compared with P = 135 for benzene.[2] Similarly one will find (for the system olive oil–water) P = 83 and P = 360 for acetamide and propionamide, or P = 15 and P = 44 for urea and *N*-methyl-urea, respectively.[3] More generally, the passage of (M)—H to (M)—CH$_3$ gives place to a positive increment of 0.52 in Hansch constants calculations (see Chapter 22).

Hydrophobic interactions

There are exceptions to this rule: grafting one or several methyl groups can render the molecule more compact (more 'globular'). A good example is provided by aliphatic alcohols.[4] As expected, one observes that the increase in lipophilicity when passing from *n*-butanol to *n*-pentanol is accompanied by a decrease of the water solubility. However, 2-pentanol, and even more neopentanol, although possessing one methyl more than *n*-butanol, are *less* lipophilic, which means *more* soluble in water (Table 19.1).

How do we explain this anomaly? It has simply to be attributed to an *entropic effect*.[5] In aqueous solution the particle is imprisoned in a tridimensional network (a cluster) of *structured* water molecules. On the other hand, a smaller amount of structured water molecules is needed to create a

Table 19.1 *Solubility in water at 20°C of* n-*butanol (1), isobutanol (2), tert-butanol (3),* n-*pentanol (4), 2-pentanol (5) and neopentanol (6)*[4]

Compound	Solubility
1	8.2 g/100 g H$_2$O
2	~ 5 g/100 g H$_2$O
3	100% (miscible)
4	2.4 g/100 g H$_2$O
5	4.9 g/100 g H$_2$O
6	12.2 g/100 g H$_2$O

cluster around a compact molecule than around an extended one (Fig. 19.2). This new structural arrangement is energetically favourable.

For the same reason some basic side chains linked to an aromatic ring are more soluble than anticipated. In these derivatives, the chain folds in such a manner (Fig. 19.3) that the cationic head becomes placed under the aromatic ring and can establish a typical donor–acceptor interaction with the π cloud of the aromatic ring ('folding effect').[6]

Crystal lattice cohesion

A greater water solubility can also result from a decrease of the crystal lattice energy, the methyl groups hindering the various intermolecular interactions (hydrogen bonds, dipole–dipole bonds, etc.). In the antibacterial sulfonamide series, the substitution of the pyrimidine ring of sulfadiazine by one, then two, methyl groups causes an increase in

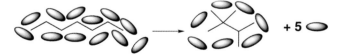

Fig. 19.2 Fewer structured water molecules are needed to wrap a compact molecule (2,2,3-trimethylbutane) than to wrap an extended one (*n*-heptane).

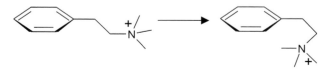

Fig. 19.3 Aralkylamine salts are more water-soluble than expected because they can adopt a folded, more compact, conformation that favours solvatation.[6]

Table 19.2 *Increased water solubility caused by insertion of methyl groups*[7]

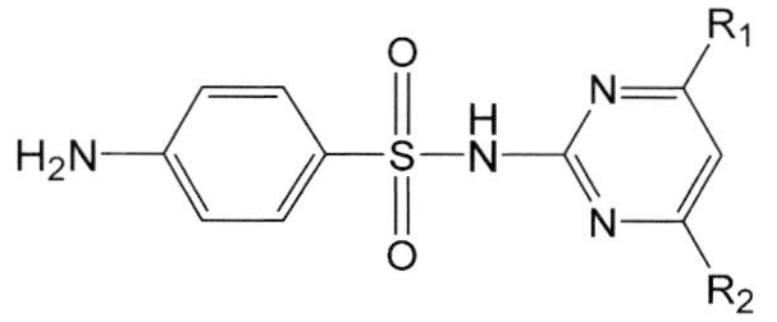

R_1	R_2	Drug	pK (acidic)	Percent Ionized at pH 5.2	Solubility pH 5.2, 37°C (M)
H	H	Sulfadiazine	6.5	3.9	0.0005
CH_3	H	Sulfamerazine	7.1	1.4	0.0013
CH_3	CH_3	Sulfamidine	7.4	0.7	0.0024

solubility (Table 19.2).[7] A priori, one would expect that the methyl substituted derivatives are less soluble, for the dual reason that they show increased lipophilicity and that they are less dissociated than the parent molecule. Indeed the inductive character of the methyl groups disfavours ionization and the nonionized form of a molecule is always less soluble than the corresponding ionized form. Despite this unfavourable electronic effect, sulfamidine is approximately five times more soluble than sulfadiazine. Similarly, the grafting of only one methyl group to the herbicide *simazine* provides *atrazine* that is 14 times more soluble in water (Fig. 19.4).[8]

B Conformational effects

The steric hindrance generated by a methyl group can create constraints and impose particular conformations that may be favourable or unfavourable for ligand–receptor interactions.

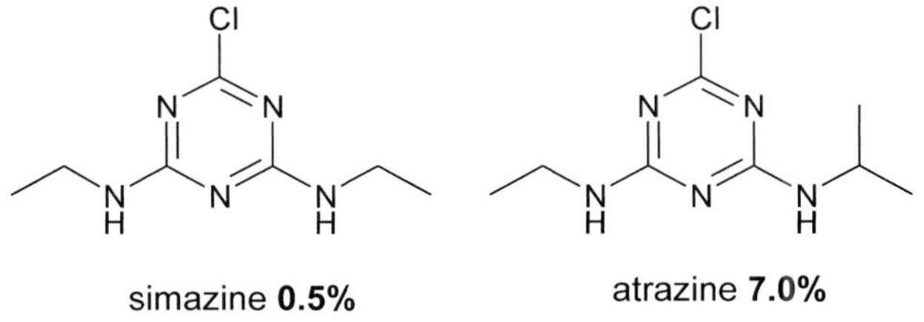

simazine **0.5%**   atrazine **7.0%**

Fig. 19.4 Comparison of the water solubility at 25°C of simazine and atrazine.[8]

Harms and Nauta[9] have studied the effects of methyl substitution on the aromatic ring of the spasmolytic *diphenhydramine*. The presence of a methyl in *para* position corresponds to a 3.7-fold increase in antihistaminic activity compared with the non-substituted derivative (Fig. 19.5). Conversely, the presence of a methyl in *ortho* position inactivates the molecule (one fifth of the activity of the non-substituted derivative) prevents the side chain adopting the favourable coplanar conformation as found in phenindamine.[9]

The explanation proposed by authors is as follows: the methyl group in *ortho* position would prevent the side chain from adopting the usual 'antihistaminic' conformation such as found for phenindamine, for example. Curiously the *ortho-ortho'*-disubstituted analogue of diphenhydramine shows local anaesthetic properties (40 times those of diphenhydramine).

In steroids, the two angular methyl groups in position 18 and 19 stand on the surface and form a screen above the β face (Fig. 19.6). This entails selective attacks on the rear face (α face) of the molecule.[10] In addition, the presence of the methyl in position 18 imposes a preferential conformation to the methylketone chain placed in position 17.[10]

The antihypertensive imidazoline *clonidine* (Fig. 19.7; $R_1 = R_2 = Cl$) and its analogues activate norepinephrine, as well as specific receptors of the central nervous system. The maximal activity in this series is always observed when both

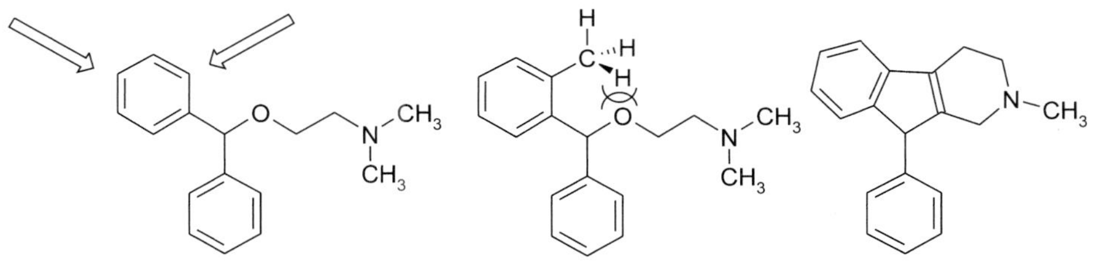

diphenhydramine   ortho-Me diphenhydramine   phenindamine

Fig. 19.5 The presence on diphenhydramine of an *ortho*-methyl group.

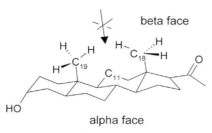

Fig. 19.6 The two angular methyl groups in position 18 and 19 of the steroidal skeleton protect the carbon-11 from β face attacks.[10]

the *ortho* positions are substituted ($R_1 = R_2$ = methyl, chlorine or ethyl, etc.). This situation implies a restrained rotation of the atropisomery type and the impossibility for the two cycles to lie in a coplanar situation. Correspondingly the geometry of the molecule becomes close to that of the norepinephrine.[11]

Leuprotide, deslorelin, and *nafarelin* are synthetic analogues of luteinizing hormone-releasing hormone (LHRH; pGlu-His-Trp-Ser-Tyr-Gly-Leu-Arg-Pro-GlyNH₂). They possess agonist properties and are currently used in the treatment of prostate cancer, endometriosis, precocious puberty, and other indications that are testosterone- or estrogen-dependent. In substituting separately each peptide bond of these analogues with N-methyl groups various interesting results were observed:[12]

(1) The introduction of an *N*-methyl group in the peptide's backbone at position 7 favours the bioactive conformation, i.e. a β-turn extending from residues 5 to 8 of the peptide. Thus, the N-Me-Leu[7] analogue of nafarelin (pGlu-His-Trp-Ser-Tyr-D2Nal-Leu-Arg-Pro-Gly-GlyNH₂; D2Nal = D-3-(2-naphtyl)-alanine), is *20 times more potent* than nafarelin.

(2) The other *N*-methyl analogues are generally less active than the nonmethylated compounds.

(3) For some of the compounds, *N*-methylation resulted in a conversion of LHRH agonists to antagonists.

(4) *N*-methylation also improved the pharmacokinetics, mainly in increasing the stability against enzymatic degradation (see below, Section II. D).

C Electronic effects

The methyl group and, more generally all alkyl groups, are the only substituents acting by an inductive electron-donating effect. All the other groups are electron donors by mesomeric effects. This means that the methyl and the alkyls are electron donors in any environment, while a basic group, dimethylaminoethyl for example, will be a mesomeric donor in basic or neutral medium, but will become strongly electron attracting by protonation in gastric medium (pH ≈ 2). Table 19.3, taken from Chu[13] presents some numerical values for substituents commonly met in medicinal chemistry. Hansch's π constant accounts for the contribution of lipophilicity, Hammet's σ constants reflect the electronic effects and the molecular refraction, MR, is related to the volume of the substituent. The table illustrates clearly the dramatic change in Hammet's σ parameter when passing from a free amino group ($\sigma = -0.66$) to a protonated one ($\sigma = +0.60$).

Table 19.3 *Some common substituent constants (taken from Chu[13])*

Group	π	σ	MR
H	0.00	0.00	1.03
CH₃	0.56	−0.17	5.65
CF₃	0.88	0.54	5.02
Cl	0.71	0.23	6.03
OH	−0.67	−0.37	2.85
OCH₃	−0.02	−0.27	7.87
NH₂	−1.23	−0.66	5.42
NH₃⁺	—	0.60	—
NO₂	−0.28	0.78	7.36
CN	−0.57	0.66	6.33
CO₂H	−0.32	0.45	6.93
COCH₃	−0.55	0.50	11.18

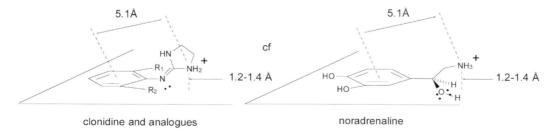

clonidine and analogues   noradrenaline

Fig. 19.7 The restricted rotation resulting from *o*- and *o'*-substitution imposes a quasi-perpendicular orientation of the imidazolinic ring towards the phenyl ring.[11]

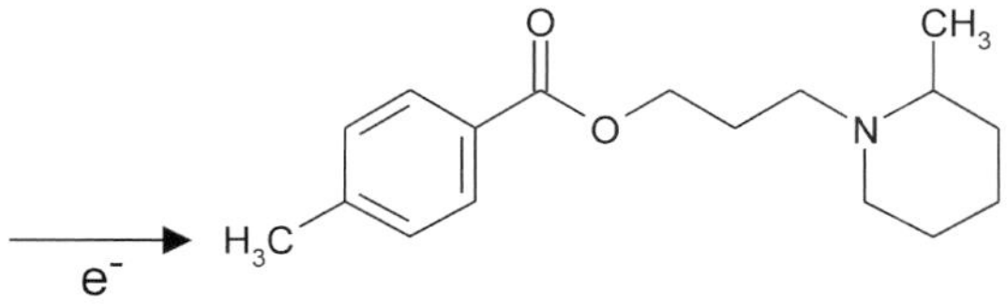

Fig. 19.8 Stabilization of the ester function thanks to an inductive effect of the *p*-methyl group.[14]

A practical consequence is that it is always judicious to include a methyl (or an alkyl) group in a SAR study. Thus in any series of R-substituted molecules, when one wants to vary R, the methyl group is generally chosen as a representative of an electron-donating group, the second substituent being chosen among the electron attractors (Cl, CN, NO$_2$, CF$_3$, etc.).

Compared with methycaine (Fig. 19.8), the 4-methyl derivative is more active *in vivo*, thanks to the inductive electron-donating effect of its methyl group. This latter exerts two synergic effects.[14] The first one is a lesser reactivity of the ester carbonyl, retarding the ester group hydrolysis. The second one is an increased capacity of the carbonyl group to form hydrogen bonds, and therefore to achieve better interaction with the receptor.

The pharmacological profile of histamine is profoundly altered by methylation. Thus 2-methylhistamine is active in stimulating the guinea-pig ileum (H$_1$ site) with 17% of the histamine potency (Fig. 19.9), but is only slightly active as stimulant of gastric secretion in the rat (H$_2$ site). The reversed type of selectivity is shown, to a remarkable degree, by 4-methylhistamine that has only 0.2% of the potency of histamine on the ileum, yet having nearly half the potency of histamine as a stimulant of gastric secretion. Thus a 200-fold discrimination towards the H$_2$ receptor is achieved.[15]

The electron-donating effect of methyl groups is also reflected by the pK_a values of burimamide compared with those of methylburimamide (Fig. 19.9).[16]

Phenanthrene itself is not carcinogenic, whereas *5,6-dimethyl-phenanthrene* is highly carcinogenic. Methyl groups in position 5 and 6 increase the electronic density of the carbon atoms 9 and 10 ('K region'). This activation allows the formation of an epoxide under biological conditions (Fig. 19.10). Such epoxides, like other carcinogenics, are alkylating agents.

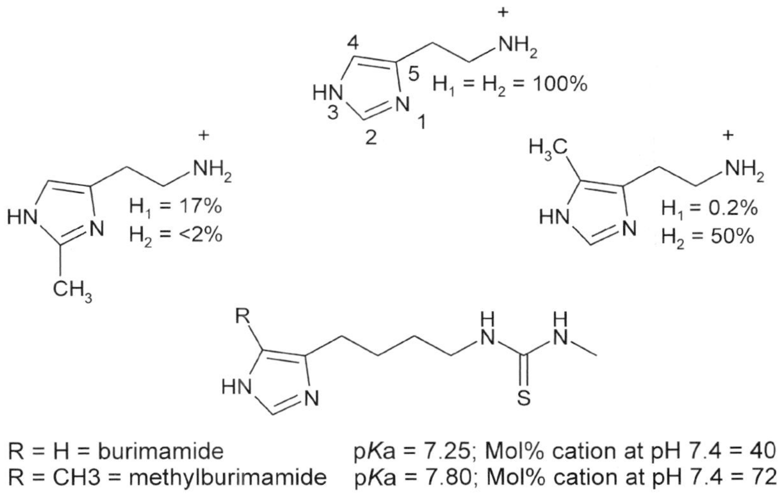

R = H = burimamide   pK_a = 7.25; Mol% cation at pH 7.4 = 40
R = CH3 = methylburimamide   pK_a = 7.80; Mol% cation at pH 7.4 = 72

Fig. 19.9 Methyl effects in histamine derivatives.

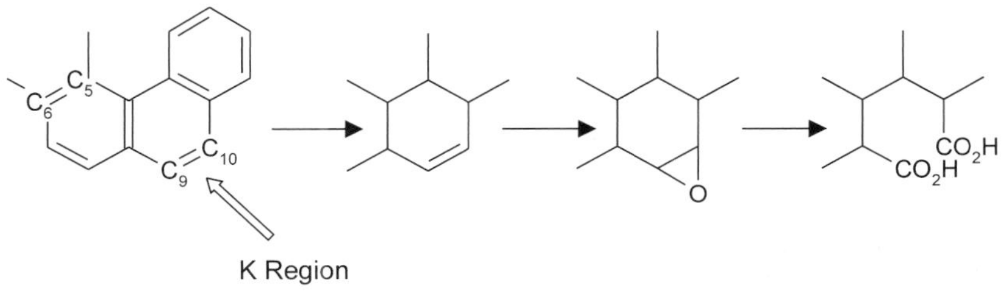

K Region

Fig. 19.10 Activation of the K region in 5,6-dimethyl-phenanthrene by an inductive effect of the methyl groups in position 5 and 6.[17]

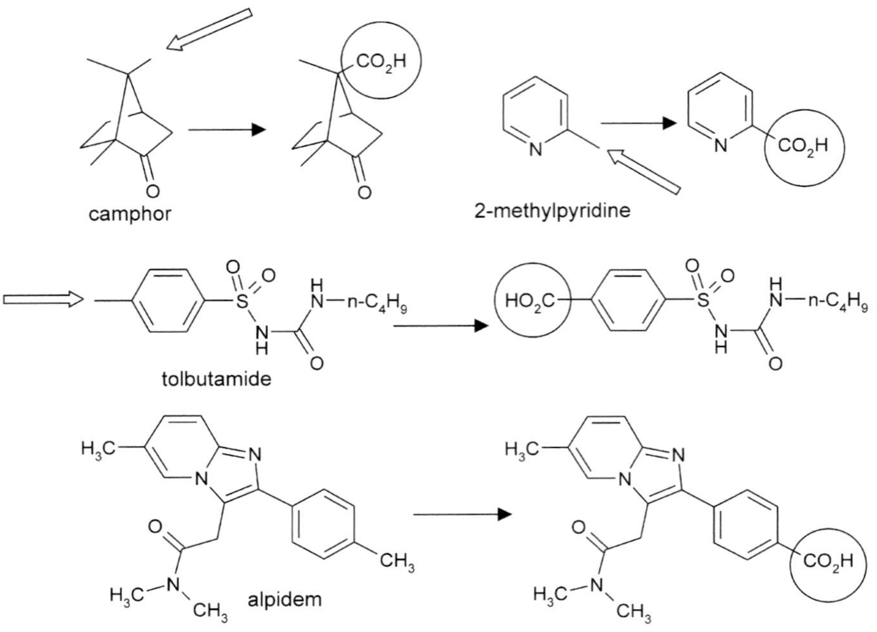

Fig. 19.11 Examples of oxidation of methyl into carboxyl groups.

The activation of the double bond also explains the formation of ring-opened compounds found in animals, bound to the plasmatic proteins, after administration of polycyclic hydrocarbons.[17]

D Effects on metabolism

Seen from the metabolic point of view, the methyl group plays a particularly important role. There are three possibilities: (1) the methyl group is oxidized; (2) the methyl group is transported; and (3) the methyl group is not (or only slightly) attacked and can then serve as blocking group.

Oxidation of the methyl group

The oxidation of the methyl group continues usually up to the carboxyl step (Fig. 19.11). This is observed for camphor, for 2-methyl-pyridine and for the drugs tolbutamide and alpidem, explaining the relatively short half-life of these latter compounds.

The grafting of a methyl group, especially on aromatic rings, often represents a good mean of detoxification. It is rapidly oxidized to an inactive and easy-to-eliminate carboxylic group. When the grafted chains are longer than methyl, the attack takes place preferentially at the benzylic position or at position $\omega - 1$ (Fig. 19.12).

Angular methyl groups of steroids are usually resistant to metabolic oxidation, probably in relation to local steric hindrance.

The methyl group is transported

A methyl group, when grafted on a nitrogen or sulfur atom, can transform this latter in an 'onium', able to act as methyl donor. In living organisms the usual suppliers of methyl rests are choline and methionine. Methionine is first activated *in vivo* by combination with adenosine to yield *S*-adenosyl-methionine (SAM; Fig. 19.13).

More generally, any *S*- or *N*-methylated drug can a priori constitute a methyl donor. On the other hand, when the methyl (or alkyl) rest is linked to a good leaving group, as found for alkyl sulfates or sulfonates, such as methyl sulfate or busulfan, alkylating reagents are produced and there exists a huge risk of carcinogenicity.

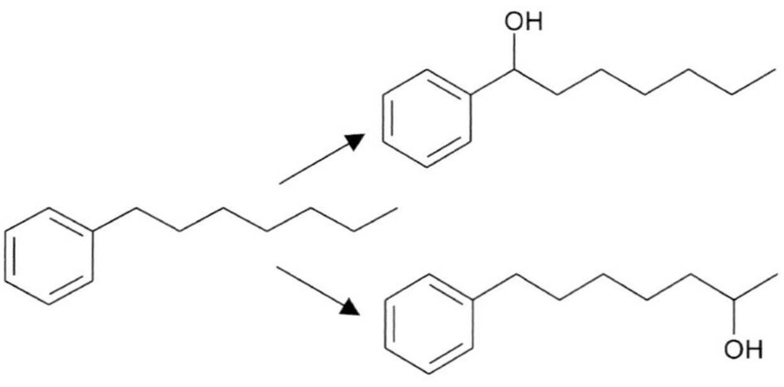

Fig. 19.12 Privileged oxidative attacks of long chains.

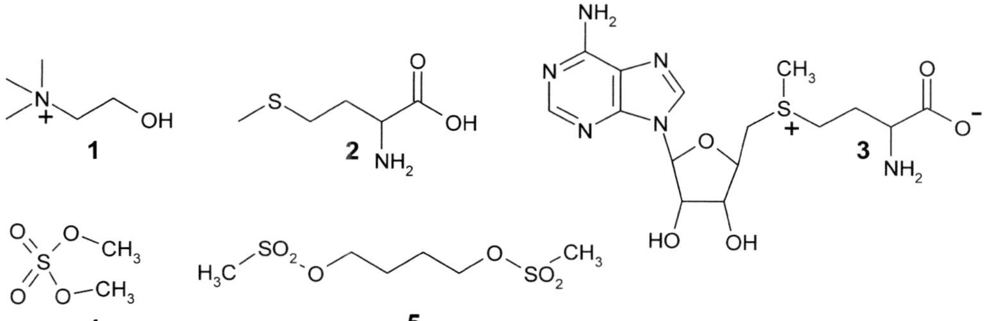

Fig. 19.13 Methyl donors and alkylating compounds choline (**1**), methionine (**2**), *S*-adenosyl-methionine (**3**), dimethyl sulfate (**4**) and butane-1,4-diol bis-methane sulfonate (**5**) (busulfan).

The methyl serves to block a reactive function

A reactive function, such as an active hydrogen belonging to a hydroxyl, thiol or amino, can be masked by methylation. Methyl groups can serve to protect even carbon atoms from metabolic hydroxylation.

The ene–diol function is essential to the antioxidant properties of vitamin C, it is therefore not surprising that its methylation leads to an inactive compound (Fig. 19.14).

Ethylene-bis-dithiocarbamic acid (nabam) is a fungicide. Through bioactivation it gives birth to an alkylating di-isothiocyanate, able to block the reactive thiol functions of the parasitic fungus. *N*-Methylation of nabam inactivates the compound because it prevents its transformation into di-isothiocyanate[18] (Fig. 19.15).

As such, the endogenous peptides, methionine- and leucine enkephalin are inactive by the oral route. Starting

from Met-enkephalin, Roemer *et al.*[19] prepared a less vulnerable analogue of methionine enkephalin (Tyr-D-Ala-Gly-*N*-Me-Phe-Met(*O*)-ol), with prolonged parenteral and oral analgesic activity. As depicted in Fig. 19.16, several modifications were needed: replacement of glycine by the unnatural D-alanine, *N*-methylation of the Gly-Phe amide bond, oxidation of methionine to the sulfoxide, and reduction of the C-terminus to the alcohol.

For other examples of drug development starting from peptide leads, see the excellent reviews of Plattner and Norbeck[20] and of Fauchère.[21]

Tetramethylbenzidine is a safe substitute for benzidine. For many years, benzidine has been used as a sensitive and specific reagent for the detection of blood. However, its extreme carcinogenicity, due to the metabolite 3,3'-dihydroxybenzidine (Fig. 19.17), has curtailed its use. In the search of a safe substitute tetramethylbenzidine (3,3',5,5'-tetramethylbenzidine) was selected. With this compound *ortho*-hydroxylation is impossible, and it does not produce tumours. On the other hand, it proved to be as sensitive as benzidine in routine clinical tests for detection of occult blood in urine or faeces.[22] Comparable reasoning guided the synthesis of mono- or di-substituted analogues of paracetamol. Methyl or higher alkyl groups in *ortho*

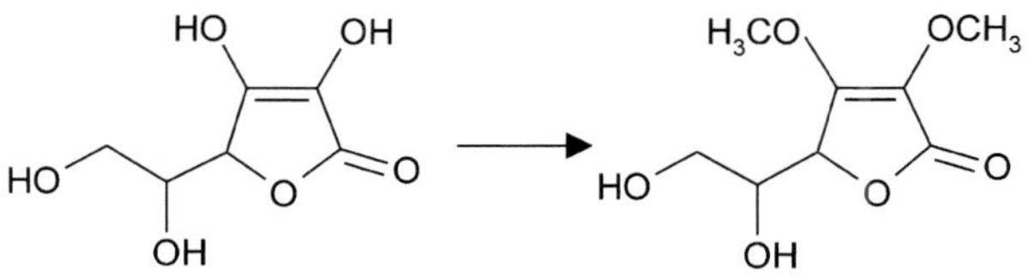

Fig. 19.14 The methylation of the ene–diol function of ascorbic acid leads to a chemically stable, but pharmacologically inactive compound.

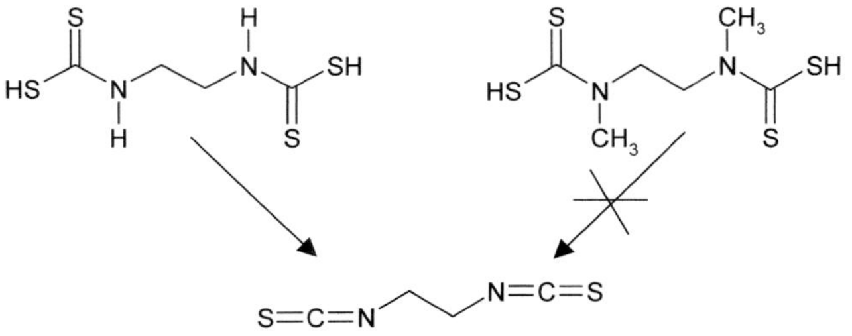

Fig. 19.15 *N*-Methylation of ethylene-bis-dithiocarbamic acid.

Met-enkephalin = Tyr-Gly-Gly-Phe-Met

Leu-enkephalin = Tyr-Gly-Gly-Phe-Leu

Tyr-D-Ala-Gly-*N*-MePhe-Met(*O*)-ol

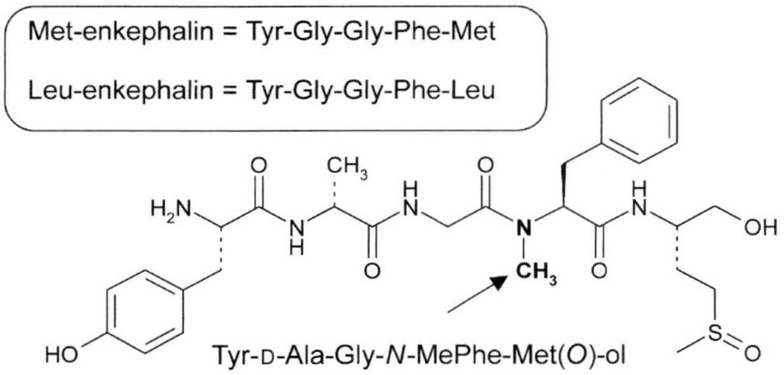

Fig. 19.16 Less vulnerable analogue of Met-enkephalin.[19]

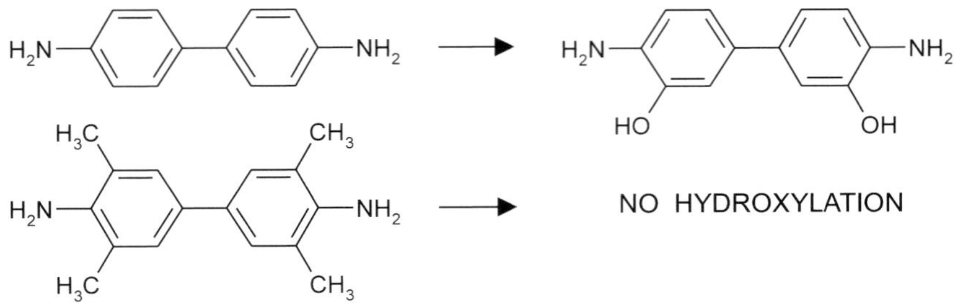

Fig. 19.17 Protection by methyl groups of aromatic carbon hydroxylation.[22]

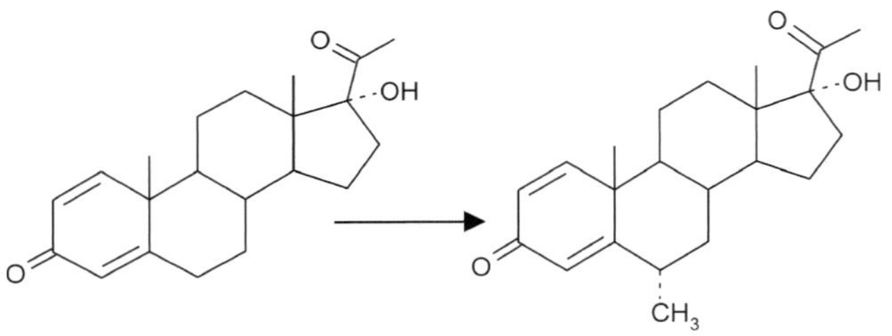

Fig. 19.18 Protection of prednisolone against metabolic hydroxylations.

Table 19.4 *Lipophilic, steric, and electronic descriptors for some current aliphatic groups*[24]

Group	π	π	E
Methyl	0.50	− 0.17	0.00
Isopropyl	1.30	− 0.19	− 1.08
Cyclopropyl	1.20[a]	− 0.30[b]	—
Cyclobutyl	1.80	− 0.20	− 0.67
Tertiobutyl	1.98	− 0.30	− 2.46
Cyclopentyl	2.14	− 0.20	− 1.12
Cyclohexyl	2.51	− 0.15	− 1.40

[a] Taken from Hansch and Anderson.[27]
[b] Taken from Martin.[28]

position (with regard to the phenolic hydroxyl) yielded safer analgesics showing less hepatotoxicity.[23]

In steroids the 6α position (e.g. prednisolone, Fig. 19.18) is a position that is normally hydroxylated. Grafting a methyl in this position prevents this hydroxylation. Halogens (particularly fluorine) serve even better because they are not sensitive at all to oxidative attacks.

E Extensions — cognate groups

The methyl group is the prototype of a saturated aliphatic substituent with lipophilic and electron-donor inductive effect. In some instances it can advantageously be replaced by related groups bringing either symmetry, or more lipophilicity or an increased inductive effect. We cite some possibilities below.

Numerical values

The values reported in Table 19.4, taken from Tute,[24] allow comparison of some characteristic alkyl groups. Note the comparable bulkiness (E values) of the isopropyl and cyclopentyl groups, while the tertiobutyl group is far more voluminous. Furthermore it is remarkable to observe that the electron-donor effect of the cyclopentyl group, is superior to that of the cyclohexyl group.

Gem-dimethyl and spiro-cyclopropyl

Gem-dimethyl and spiro-cyclopropyl are useful for rendering a carbon atom quaternary and therefore resistant to metabolic attack. They can also constitute solutions to

introduce symmetry into a chiral centre. The spiro-cyclopropyl moiety has been used to synthesize a more lipophilic analogue of glycine (Fig. 19.19).[25]

Isopropyl and cyclopropyl

The cyclopropyl group is less bulky than the isopropyl group for a maximal electron-donor effect. For a review on cyclopropane derivatives in medicinal chemistry, see Cussac *et al.*[26]

The cyclopentyl group

The cyclopentyl group creates the maximal inductive effect for a relatively reasonable bulkiness. It is often a good filling of a hydrophobic pocket as illustrated for the cAMP phosphodiesterase inhibitor rolipram (Fig. 19.20). The inhibitory activity towards type IV cAMP-phosphodiesterase is increased ten times when the *meta*-methoxy group is replaced by a *meta*-cyclopentyl group (rolipram).[29] Presumably the cyclopentyl group fills, in an optimal manner, a hydrophobic pocket of the active site of the enzyme.

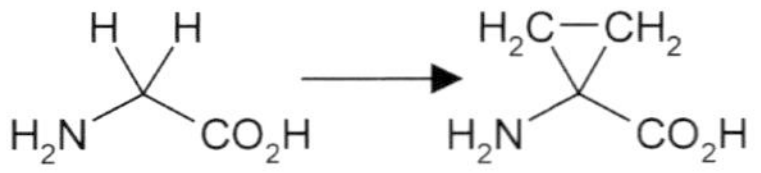

Fig. 19.19 *spiro*-Cyclopropyl glycine derivative with agonist properties to the glycine receptor.[25]

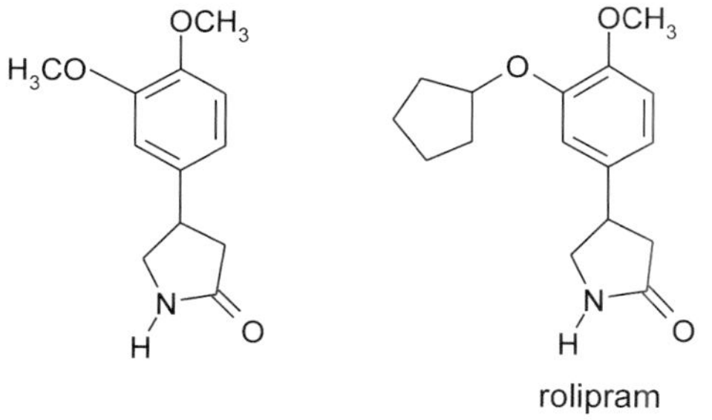

Fig. 19.20 Structures of rolipram and of its dimethoxy analogue.[29]

The cyclopentyl group has also proven advantageous in replacing a *gem*-dimethyl in a series of inhibitors of acyl-CoA-cholesterol acyltransferase, which is an enzyme involved in the absorption of alimentary cholesterol (Fig. 19.21).[30]

III EFFECTS OF UNSATURATED GROUPS

The introduction of an unsaturated rest (vinyl, ethynyl, allyl, etc.) in a drug molecule generally entails one or several of the following consequences.[31,32]

(1) Increase of the narcotic power and toxicity in comparison with the corresponding saturated compound. Ethylene, acetylene, trichlorethylene, divinyl oxide and, by extension, cyclopropane are examples of unsaturated narcotics.

(2) Possibility of the existence of a geometrical isomery.

(3) Existence of electronic effects: the unsaturated groups behave as electron attractors through inductive effects. Furthermore direct interactions of donor–acceptor type are possible because of the π electron cloud surrounding multiple bonds.

(4) Possibility of activation through conjugation: the association of several unsaturated functions in conjugated position (dienes, enynes, enones, enolides, polyunsaturated derivatives) renders the corresponding molecules very reactive. It especially facilitates the addition of biological nucleophiles and notably of thiols.

(5) Facilitation of the metabolism. The unsaturated element often constitutes the vulnerable site of the molecule, that will be attacked first (for example by formation of an epoxide that evolves into a diol which, in its turn, can undergo oxidative cleavage), but this is not always the case.

With regard to their classification, we will distinguish four series of unsaturated derivatives: the *vinyl* series, the *allyl* series, the *acetylenic* series, and the *ring unsaturated derivatives* that are bioisosteric to aromatic rings.

A Vinyl series

Besides active substances containing actual vinyl groups, this series comprises molecules containing substituted vinyl groups as well as cyclopropyl groups.

Vinyl groups

Vinyl groups are not excessively used in medicinal chemistry. Divinyl oxide is an excellent general anaesthetic but it polymerizes easily and forms peroxides. Stabilization of the compound is usually achieved by addition of 0.01% of *N*-phenyl α-naphtylamine. On the other hand, compounds such as kainic acid, vinylbital, quinine, 17α-vinyl-testosterone, compound SKF 100047, and vigabatrin (Fig. 19.22) are perfectly stable vinyl derivatives.

Synthetic or natural vinylic epoxides (kapurimycin A₃) produce radical-induced DNA cleavage, even in presence of the protecting glutathionyl radicals. This property may constitute a starting point for the treatment of radiotherapy-resistant tumours for which the resistance is due to a local increase of the glutathione level.[33]

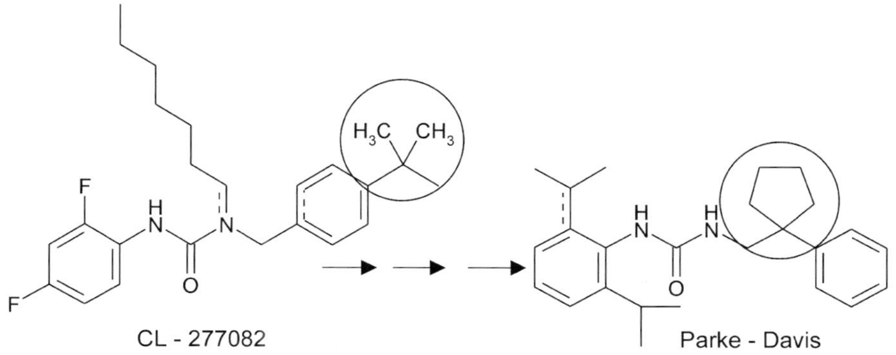

Fig. 19.21 Replacement of a *gem*-dimethyl by a cyclopentyl group.[30]

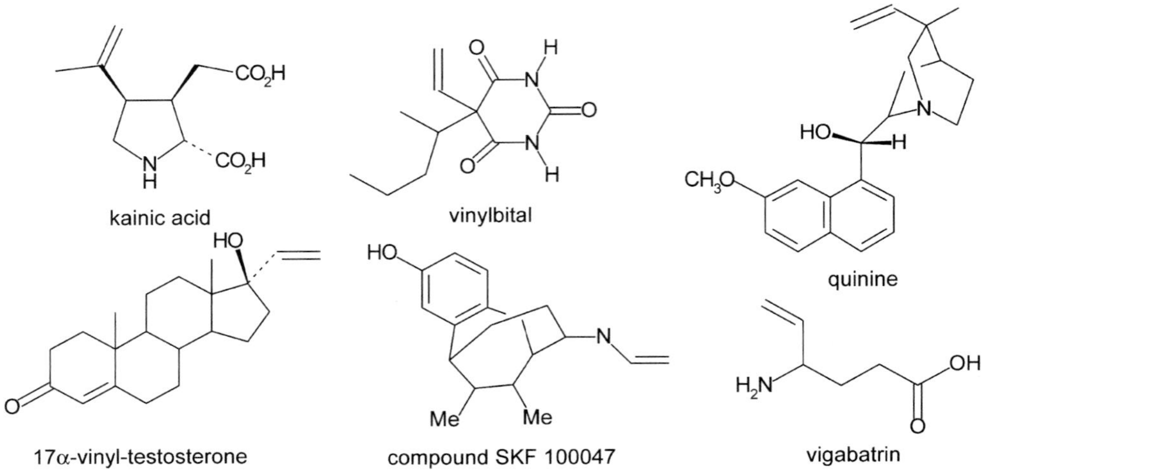

Fig. 19.22 Drugs containing a vinyl group.

Substituted vinyl derivatives comprise unsaturated barbiturates such as vinbarbital, the acetylcholinesterase inhibitor huperzine A2, various antifungal imidazoles bearing *gem-* and *vic-*dialkylvinyl groups[34] (Fig. 19.23). Some chlorovinyl derivatives were proposed as morphinic antagonists (see Fig. 19.26). For the α-styryl carbinol derived antifungal agents DuP 860 and DuP 991 oxidation into vinyl epoxides has been observed in solubilized systems (water and propylene-glycol, or glycerol, or polyethylene-glycol 400).[35]

Cyclopropyl groups

Cyclopropyl rings can constitute interesting substitutes for vinyl groups when they are too fragile or when they give place to unwanted isomeries or tautomeries. Thus tranylcypromine, an antidepressant acting by inhibition of the

monoamine-oxidases, is a stable compound, while its ethylenic analogue no longer is (Fig. 19.24). A supplementary advantage in the use of cyclopropanic analogues comes from their fixed stereochemistry: there is no spontaneous conversion from *cis* to *trans* isomer as frequently observed with ethylenic derivatives.

B Allylic series

All allylic derivatives are relatively hepatotoxic and irritant. Allylic alcohol itself serves to create experimental hepatic lesions that allow testing of hepatoprotecting drugs. We consider three categories of allylic derivatives: *C*-allyl derivatives, *N*-allyl derivatives, and *O*- and *S*-allyl derivatives which often possess alkylating poroperties.

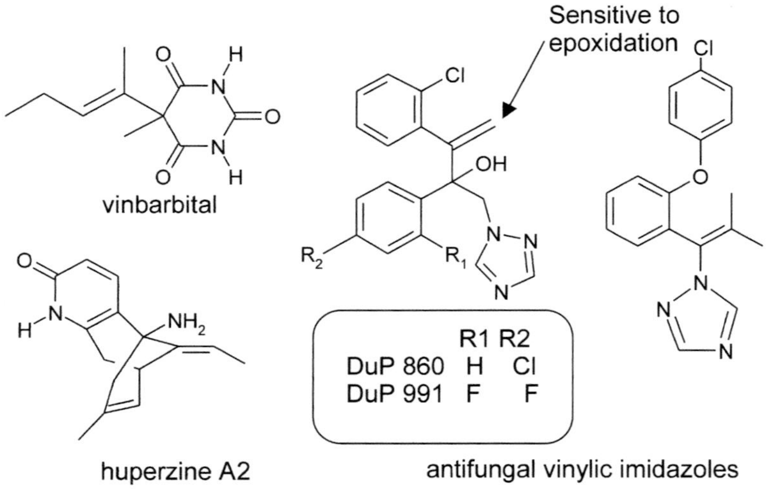

Fig. 19.23 Substituted vinyl derivatives are less sensitive to epoxidation than unsubstituted ones, such as the compounds DuP 860 and DuP 991.

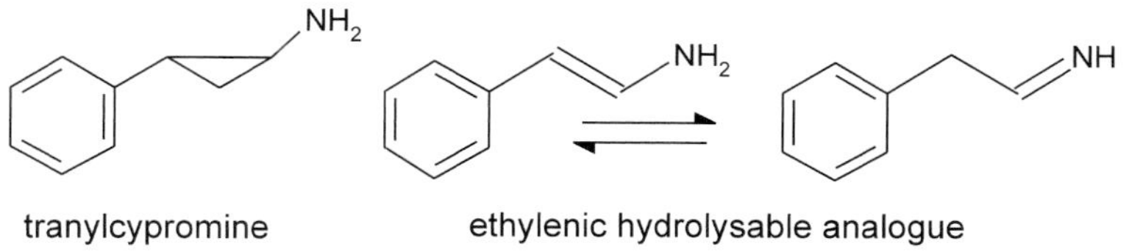

tranylcypromine   ethylenic hydrolysable analogue

Fig. 19.24 Tranylcypromine represents a stable substitute of the enamine aminostyrene.

C-*Allyl derivatives*

These present the double advantage of being lipophilic (rapid onset) and giving fast biodegradation (short duration of action). However, they often conserve the intrinsic hepatotoxicity of the allyl group.[36] Allobarbital is a sedative-hypnotic that is no longer used; allylestrenol acts as a pure progestative hormone and alprenolol is a β-blocker (Fig. 19.25).

Acetamidoeugenol is an intravenous anaesthetic of ultra-short duration of action, it has been withdrawn because it provokes irritations and lesions of the vascular wall. Acetamidoeugenol is oxidized very rapidly *in vivo* into the corresponding aryl-acetic acid. This observation was the basis of the synthesis of another intravenous short-acting anaesthetic: propanilide (Fig. 19.25).[37]

N-*Allyl derivatives*

The replacement in morphine, and in some of its simplified analogues, of the *N*-methyl group by a *N*-allyl group (and subsequently by some related groups)[38] has constituted a decisive step in the study of opiate analgesics. Indeed this modification achieved for the first time the passage

of morphinic receptor agonists to the corresponding antagonists (Fig. 19.26).

Aloxidone and albutoine are anticonvulsivant *N*-allyl derivatives of hydantoin and thiohydantoin (Fig. 19.27). The dibenzazepine azapetine is an α-adrenergic blocking agent used as a peripheral vasodilator.

O- *and* S-*Allyl derivatives*

Several of these compounds appear in the Merck Index. The β-blocking oxprenolol, the arylacetic analgesic–anti-inflammatory drug aclofenac and the fungicide enilconazole are *O*-allyl derivatives. Penicillin O and penicillin S are both *S*-allyl derivatives.

Alkylating allyl derivatives. When the allyl group bears a good leaving group, it easily generates the allylic cation. This cation is stabilized by mesomery, and is an excellent electrophile. Many natural compounds can release allylic alcohols. A first example is found in allicine, the antibacterial principle of garlic, that results from the action of alliinase on alliine (Fig. 19.28). Several varieties of senecio (*Senecio vulgaris* L., Compositae, *Senecio platyphyllus*,

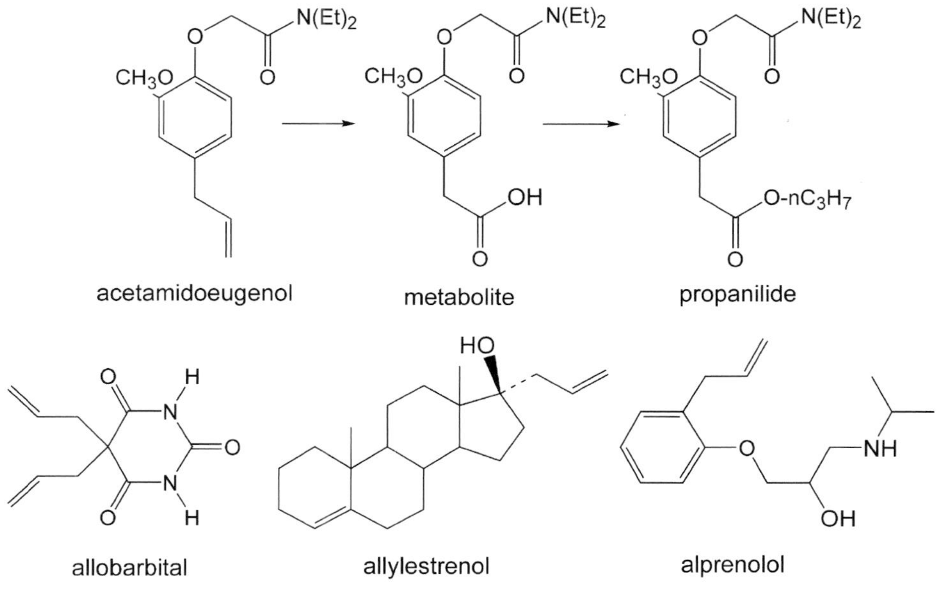

acetamidoeugenol   metabolite   propanilide

allobarbital   allylestrenol   alprenolol

Fig. 19.25 *C*-Allyl derivatives.

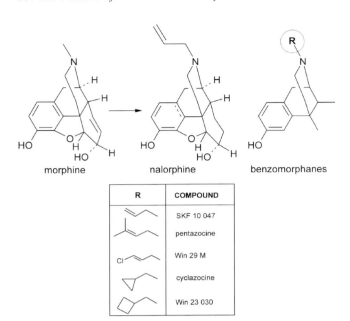

Fig. 19.26 Nalorphine and cognate derivatives.[38]

R	COMPOUND
	SKF 10 047
	pentazocine
Cl	Win 29 M
	cyclazocine
	Win 23 030

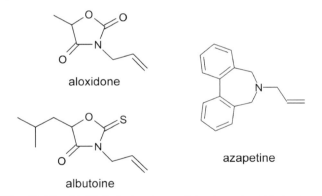

aloxidone

albutoine

azapetine

Fig. 19.27 *N*-Allyl derivatives of hydantoin, thiohydantoin and dibenzazepine.

etc.) contain alkaloids derived from pyrrolizidine: senecionine, seneciphylline, etc. These substances provoke hepatic cancers, notably in cattle. Here too, the alkylating properties are ascribed to the allylic structure (Fig. 19.28).[39] It is reasonable also to implicate the formation of allylic carbamates in mitomycins *A*, *B* and *C*, which are antimitotic drugs used in cancerology. Presumably the allyl function is created by an elimination reaction involving the departure of an acetalic hydroxy or methoxy group (Fig. 19.28).

C Acetylenic series

Acetylenic groups are used for their electronic effects, as equivalents of aromatic rings and to impose structural constraints.

Electronic effects

The acetylene function exerts an electron-attracting effect. This effect can be reinforced by substitution of the acetylenic hydrogen. Ethynyl compounds are essentially found among the light sedative-hypnotic drugs (CNS-depressing effect of the unsaturated derivatives) and in steroid series, where their fixing in position 17a provides orally active steroids (Fig. 19.29).

In the sedative-hypnotic series, most of the acetylenic alcohols are used as carbamic esters: meparfynol, ethinamate, etc. The bromoethynyl moiety confers an acidity comparable to that induced by a trichloromethyl group (compare for example chlorobutanol with 3-methyl-pentyne-3-ol). Bromoethynylcyclohexanol is sufficiently acidic to form salts; its bismuth salt was used for a while as an antisyphilitic drug under the name of biarsamide. In the steroid series the metabolism of the ethynyl group can lead (by hydration of the triple bond) to 17 α-methylketones.

Aromatic ring equivalents

Thanks to their π electron clouds and to their small volume, ethynyl groups can sometimes function as bioisosteres of

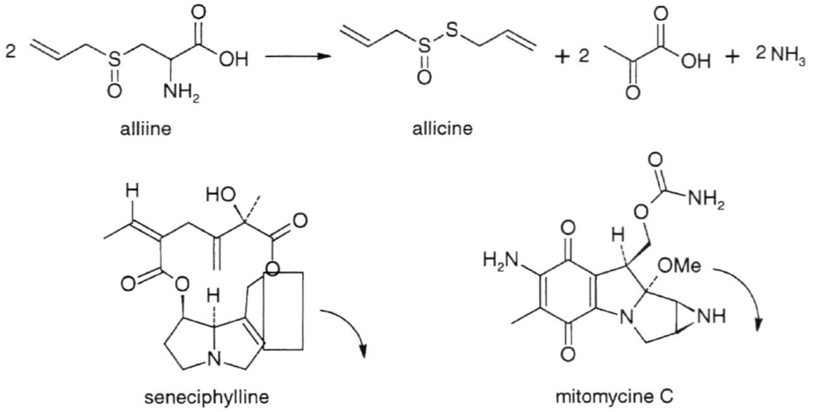

alliine   allicine

seneciphylline   mitomycine C

Fig. 19.28 Alkylating allylic derivatives.

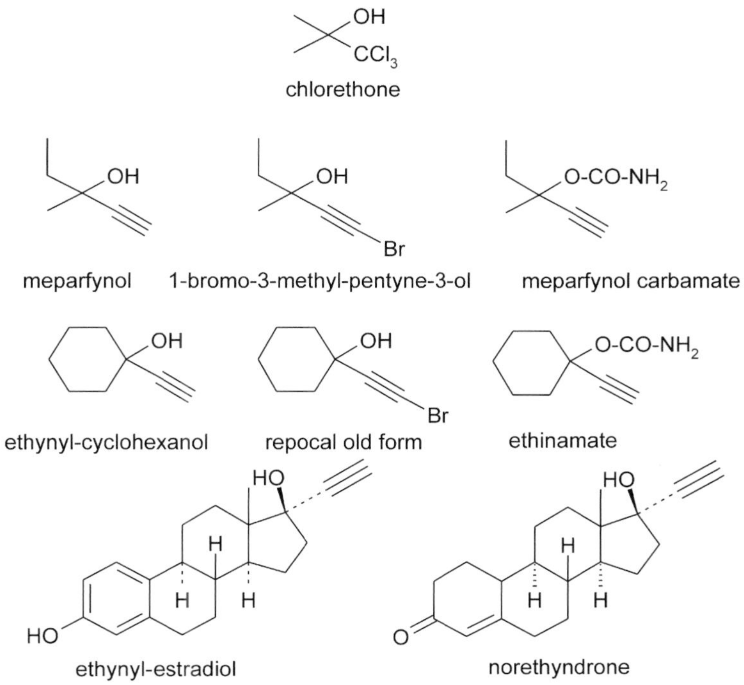

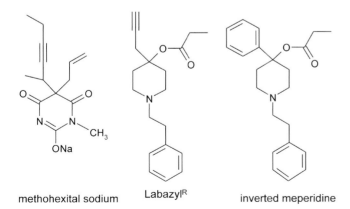

Fig. 19.29 Ethynylated drugs.

aromatic rings and give similar donor–acceptor interactions. However, such acetylenic analogues are more rapidly metabolized. Sodium methohexital, for example (Fig. 19.30) is used as an injectable barbituric for very short anaesthesia. The obsolete antitussive drug Labazyl® is an analogue of the reversed pethidines in which the phenyl ring has been replaced by a propargyl group.

Structural constraints

In inserting an acetylenic function between two carbon atoms, one achieves a structure with four 'on-line' atoms

representing a rigid entity with a distance of 4.2 Å between the two extreme atoms (Fig. 19.31). This kind of arrangement is found in the cholinergic agonist oxotremorine and in the GABA analogue 4-amino-tetrolic acid. This compound is recognized, like GABA itself, by the enzyme GABA-transaminase for which it acts as an inhibitor[40] (Fig. 19.31).

D   Cyclenic equivalents of the phenyl ring

The cyclohexenyl ring and, to a lesser extent the cyclopentenyl and cycloheptenyl rings can possibly replace a phenyl ring. This is the case for the barbiturics cyclobarbital

Fig. 19.30 Acetylenic groups as aromatic ring equivalents.

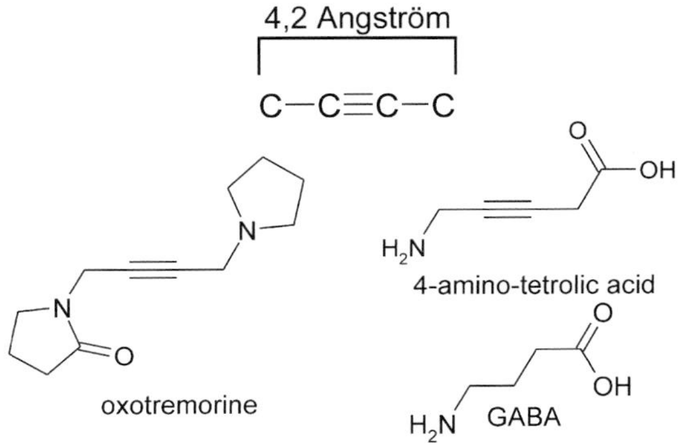

Fig. 19.31 Rigidity and extension imposed by a triple bond.

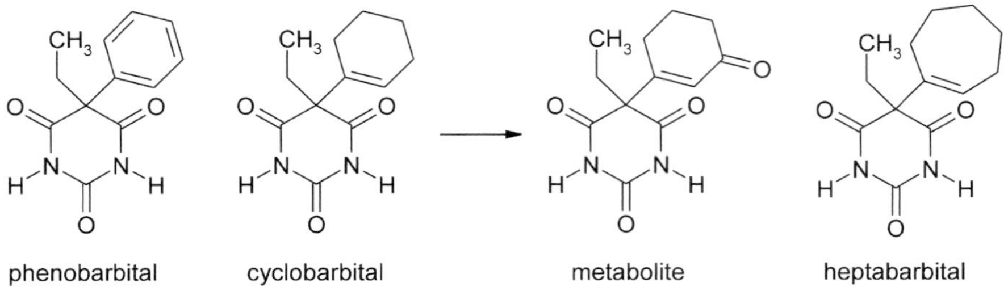

Fig. 19.32 Cyclenic equivalents of the phenyl ring.

and heptabarbital which are entirely comparable to phenobarbital. From the metabolic point of view, the cyclohexenyl ring is oxidized in position α to the double bond to produce the corresponding cyclohexenone (Fig. 19.32).

Another example comes from the benzodiazepine series where tetrazepam can be compared with diazepam. However, in this case a slight difference in the activity profile exists: tetrazepam is less sedative, hypnotic and anticonvulsant than diazepam; on the other hand, it has more muscle relaxant and analgesic effects, which indicates its use in visceral and articular pain. From the chemical point of view, the cyclohexenic double bond is introduced in a rather unexpected manner by means of a radicalar rearrangement of an *N*-chloroamide[41] (Fig. 19.33).

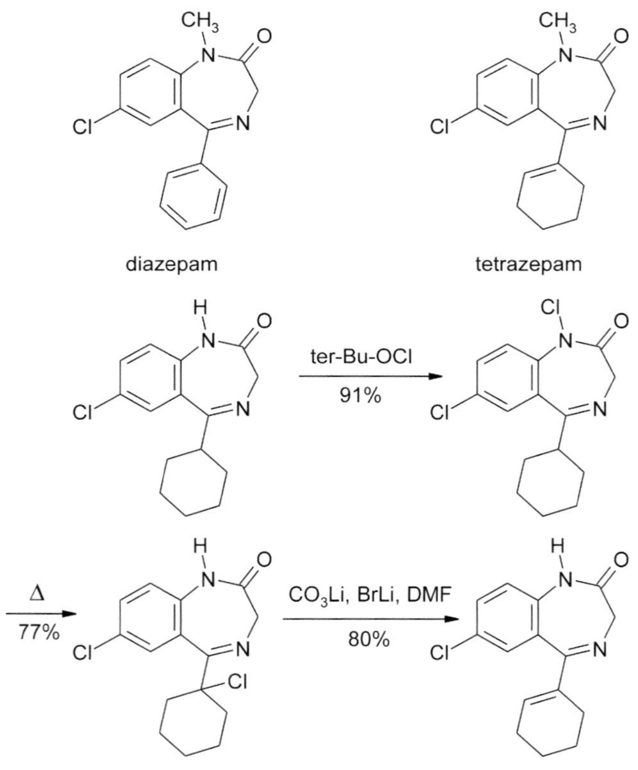

Fig. 19.33 Cyclenic equivalents in the benzodiazepine series.

IV EFFECTS OF HALOGENATION

Currently, one drug out of three is a halogenated derivative and halogens are found in drugs belonging to practically all therapeutic classes. That has not always been the case; indeed, in the past medicines were mostly of natural origin and natural-substance chemistry is relatively scarce in halogenated substances. Halogen-containing drugs have entered into usage only since 1820. The first organic halogenated drugs have mainly been used for their depressive action on the central nervous system: production of general anaesthesia with chloroform, sedation or hypnosis with chloral and bromural. From the twentieth century on a regular increase in the number of halogenated drugs is observed, and it became explosive after the end of the Second World War. Even halogenated drugs from natural origin became available, such as chlortetracycline or chloramphenicol, and also substances from marine origin or from fermentation broths.

A The importance of the halogens in the exploration of structure–activity relationships

Steric effects

The obstruction of a molecule by means of halogen substitution can impose certain conformations or mask certain functions. In the case of clonidine the bulky halogen atoms prevent free rotation and maintain the planes of the aromatic rings in a perpendicular position to each other (Fig. 19.34A).[11]

Also in a series of benzodiazepine receptor ligands derived from CGS-9896, strong steric effects are described by Fryer *et al.* (Fig. 19.34B).[42] Indeed, the *ortho* and *para* isomers can be considered as having the same lipophilicity and very similar electronic effects. Thus the reduction of binding affinity of the *ortho*–chloro compound is attributed to the steric effect.

Electronic effects

The electronic effects of the halogens are ascribed to their inductive electron attracting properties. These latter are

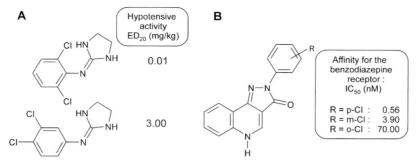

Fig. 19.34 (A) The *ortho–ortho'* substitution in clonidine maintains the planes of the aromatic rings in a perpendicular position to each other. (B) The *ortho*-chloro isomer of the benzodiazepine ligand CGS-9896 has a 125-fold lower affinity than the parent molecule.

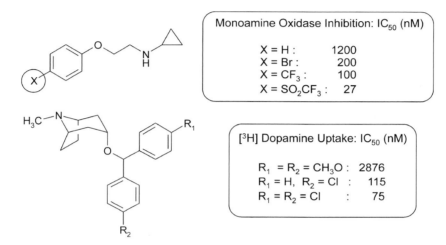

Fig. 19.35 Influence of halogenated substituents on monoamine-oxidase inhibition potency *in vitro*.

maximal for chlorine and bromine, less marked for iodine, and very weak for fluorine. The mesomeric donor effect of the halogen atoms is usually not involved in biological media. The influence of halogens on the potency of monoamine oxidase inhibition and of dopamine uptake blockade *in vitro* are shown in Fig. 19.35. The choice of the optimal substituent allows noticeable gains in potency, compared with the parent molecule.[43,44]

Progressive mono and di-substitution of diazepam-related benzodiazepinones enhances the affinity for the mitochondrial benzodiazepine receptor (MBR) by a factor of 233 (Fig. 19.36).[45]

N-Acylation of L-tryptophan benzyl esters, followed by replacement of the 3,5-dimethyl groups by their trifluoromethyl analogues, achieved an almost thousand-fold increase in potency in a series of substance P receptor antagonists (Fig. 19.37).[45]

Hydrophobic effects

The predominantly lipophilic influence of halogen substitution is seen in the classical cases of the halocarbon

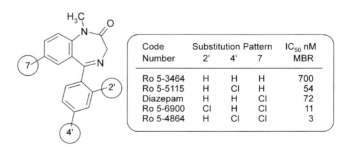

Fig. 19.36 Chlorine effects in the benzodiazepine series.[45]

Code Number	Substitution Pattern			IC$_{50}$ nM
	2'	4'	7	MBR
Ro 5-3464	H	H	H	700
Ro 5-5115	H	Cl	H	54
Diazepam	H	H	Cl	72
Ro 5-6900	Cl	H	Cl	11
Ro 5-4864	H	Cl	Cl	3

anaesthetics, the halogenophenol antiseptic and the halogenated insecticides (Fig. 19.38). For these compounds there is direct correlation between biological activity and certain physicochemical parameters such as partition coefficient, surface tension or vapour pressure. The accumulation of halogen atoms favours the passage of the biomembranes and access to the CNS.

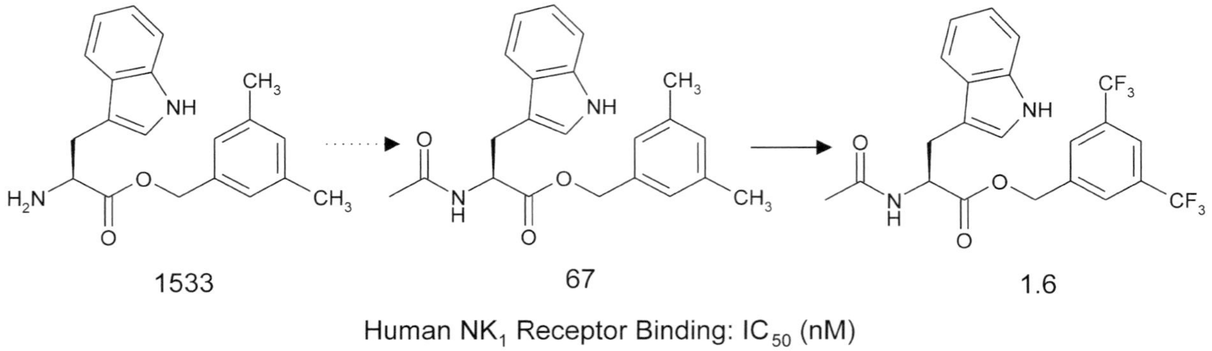

1533 → 67 → 1.6

Human NK$_1$ Receptor Binding: IC$_{50}$ (nM)

Fig. 19.37 Successive *N*-acetylation and CH$_3$ → CF$_3$ replacement achieve a 958-fold increase in affinity.[45]

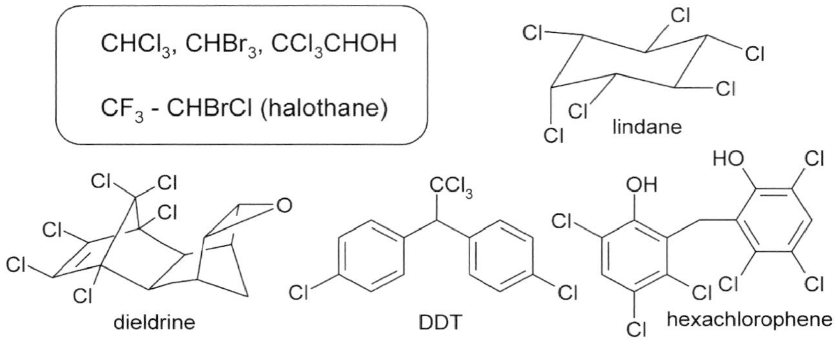

CHCl$_3$, CHBr$_3$, CCl$_3$CHOH

CF$_3$ - CHBrCl (halothane)

lindane

dieldrine   DDT   hexachlorophene

Fig. 19.38 Compounds in which the halogens play essentially a lipophilic role.

Reactivity of the halogens

In terms of bond strength, all C-halogen bonds, except C—F are weaker than C—H (Table 19.5).

Usefulness of the halogens and of cognate functions

Depending upon their physical properties and their reactivity, the derivatives of fluorine, chlorine, bromine and iodine present various degrees of usefulness (Table 19.6).

(1) The most utilized halogens in medicinal chemistry are chlorine and fluorine attached to a nonactivated carbon atom. Fluorine presents the advantage of its small

bulkiness (van der Waals radius comparable to that of hydrogen). It will be used essentially to block metabolically sensitive positions of a molecule.

The CF$_3$ group is comparable in size to chlorine and can advantageously replace it when it is placed in an activated position (e.g. R—CO—Cl → R—CO—CF$_3$). A chlorine substituent simultaneously produces an increase in lipophilicity, an electron-attracting effect and a metabolic obstruction.

Table 19.5 *Atomic radii and characteristics of carbon-halogen bonds (taken from Buu-Hoï[46])*

Atomic radius (Å)	Bond[a]	Interatomic distance (Å)	Bond strength (kcal mol^{-1})
H: 0.29	C—H	1.14	93
F: 0.64	C—F	1.45	114
Cl: 0.99	C—Cl	1.74	72
Br: 1.14	C—Br	1.90	59
I: 1.33	C—I	2.12	45

[a] In aliphatic series.

Table 19.6 *Some substituent constants for halogens and equivalent functions (taken from Chu[13])*

Group	π	σ	MR	F	R
H	0.00	0.00	1.03	0.00	0.00
F	0.14	0.06	0.92	0.43	−0.34
Cl	0.71	0.23	6.03	0.41	−0.15
Br	0.86	0.23	8.88	0.44	−0.17
I	1.12	0.18	13.94	0.40	0.19
CF$_3$	0.88	0.54	5.02	0.38	0.19
CH$_3$	0.56	−0.17	5.65	−0.04	−0.13
CN	−0.57	0.66	6.33	0.51	0.19
SO$_2$CF$_3$	0.55	0.93	12.86	0.73	0.26
SCF$_3$	1.44	0.50	13.81	0.35	0.18
SCN	0.41	0.52	13.40	0.36	0.19

(2) In certain active molecules the role of the fluorine or chlorine atoms is not apparent at first glance. Thus for example two compounds, chemically as different as *m*-trifluoromethylphenylethylamine and 5-hydroxy-tryptamine, show many pharmacological analogies. In this case the explanation lies in the similitude of the electrostatic potential maps (Fig. 19.39). Conversely two closely related pyrazoloquinolines, compounds CGS 8216 and its *para*-chloro analogue CGS 9896 present totally opposed activity profiles on the same benzodiazepine receptor.[47] A dramatic effect resulting from chlorine substitution is also found in the change from β-phenyl-GABA to β-(*p*-chlorophenyl)-GABA.[48]

(3) Bromine is the less used halogen, and is usually incorporated as a bromoaryl. The disadvantage of

using bromine is that it generates alkylating reactive intermediates more easily than chlorine or fluorine (see Chapter 32) and therefore it can confer, during long-term treatment, toxic potentialities to the molecule that bears it. This was the case for the anti-inflammatory-analgesic drug bromfenac sodium withdrawn from the US market due to reports of hepatotoxicity.[49]

(4) Although even less tolerated than bromine, iodine is indispensable in the treatment of certain thyroidal deficiencies. Administered by the internal route, iodine and iodine derivatives trigger either acute hypersensitivity reactions (larynx oedema, cutaneous haemorrhages, fever, arthralgies, etc.) or chronic reactions ('iodism'). In addition to its use in certain dysfunctions of the thyroid gland, iodine presents to specific uses: covalent iodine derivatives serve as radiological

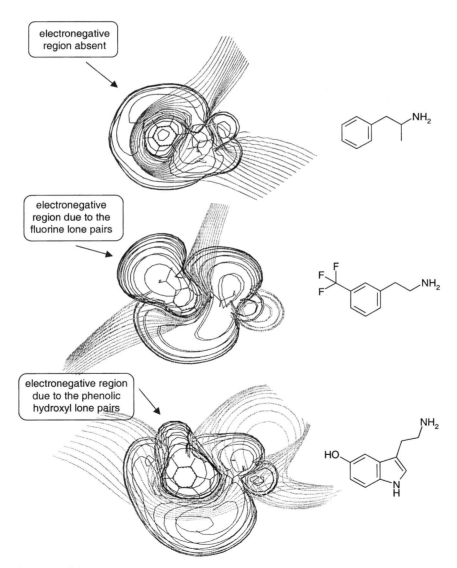

Fig. 19.39 Electrostatic potential maps.

contrast substances and [131]iodine (half-life: 8 days) is used as radioactive tracing agent.

(5) Extensions—cognate groups: Chlorine, trifluoromethyl, cyano or azido groups are more or less bioisosteres. Other possible candidates are: SCN, SCF$_3$, SO$_2$CF$_3$ and CH=CF$_2$ (see Chapter 13).

V   EFFECTS OF HYDROXYLATION

The substitution of OH for H affects biological activity profoundly, as in the conversion of ethane to ethanol or benzene to phenol. Simple alcohols have narcotic, and simple phenols bacteriostatic properties. Polyfunctional compounds can act as chelating or complexing agents.

A   Effects on solubility

The introduction of an alcoholic or a phenolic hydroxy group into an active molecule changes the partition coefficient towards more hydrophilicity and renders the molecule more water-soluble. Thus in changing benzene to phenol, or benzamide to *p*-hydroxybenzamide, the partition coefficient drops from log P = 2.13 to log P = 1.46 and from log P = 0.64 to log P = 0.33, respectively.[2] The value of the Hansch π constant for a hydroxy group is -0.67. This means that to compensate the loss in lipophilicity due to the monohydroxylation of an active compound, it is necessary to attach at an appropriate site on the molecule a chlorine atom ($\pi = 0.71$), for example.

B   Effects on the ligand–receptor interaction

For some hydroxylated drugs like morphine, dopamine, haloperidol, γ-hydroxybutyrate, serotonine or most of the steroids the hydroxy group is an essential element for hydrogen bonding with the receptor. For others the attachment of a hydroxy group can result in potency changes. Examples are found in hycanthone, which is 10 times more active against schistosomes than lucanthone,[50] or in hydroxylated minaprine analogues which show a 10 times better affinity for M$_1$ muscarinic receptors than the parent drugs.[51]

C   Hydroxylation and metabolism

As a rule, metabolic hydroxylation of an active compound represents a detoxication mechanism. It results generally from a first pass effect and can be followed or not by a conjugation reaction (see Chapters 30 and 31). Classical examples of drugs detoxified through hydroxylation are paracetamol, oxyphenbutazone and hydroxychloroquine. Other important reactions of hydroxy compounds, whether alcoholic or phenolic, are based on their capacity to accept activated groups through the action of group-transferring enzymes (methylation, sulfation, phosphorylation, glycosylation, etc.).

VI   EFFECTS OF THIOLS AND OTHER SULFUR-CONTAINING GROUPS

Thiol and disulfide groups occur widely in natural products. They are found in small molecules such as lipoic acid, glutathione and thiamine as well as in cysteine-containing peptides and proteins (hormones, enzymes, antibiotics). In all these substances, thiol and disulfide groups are clearly associated either with high chemical reactivity or with consolidation of peptide and protein architecture. Being too reactive, the thiol and the disulfide groups are normally not used in medicinal chemistry as substituents in QSAR studies. Occasionally methylthio substitution on aromatic rings is practised, but even then, the thioethers obtained are very reactive. They are easily converted to sulfoxides and vice versa (see sulindac, Chapter 32).

Drugs containing thiol groups are mainly used for the strong affinity that the thiolate anion presents towards heavy metals. This is the case for thiol-containing angiotensin-converting enzyme inhibitors which bind to a zinc-containing enzyme (see Chapter 6 and Ganellin and Roberts[15]).

The heavy metal chelating properties of thiols were taken advantage of in the design of dimercaprol ('British Anti-Lewisite', BAL) as counterpoison of the arsenical war gas lewisite (Fig. 19.40). Today dimercaprol is used to treat poisoning by compounds of gold, mercury, antimony and arsenic. The toxic nature of the heavy metals is masked and chelate is stable enough to be excreted as such in the urine.

Penicillamine (D-β,β-dimethylcysteine) is an effective chelator of copper, mercury, zinc and lead which promotes the excretion of these metals in the urine. It is clinically used in patients with Wilson's disease, with rheumatoid arthritis and with heavy-metal intoxications.[52] Ziram and ferbam are

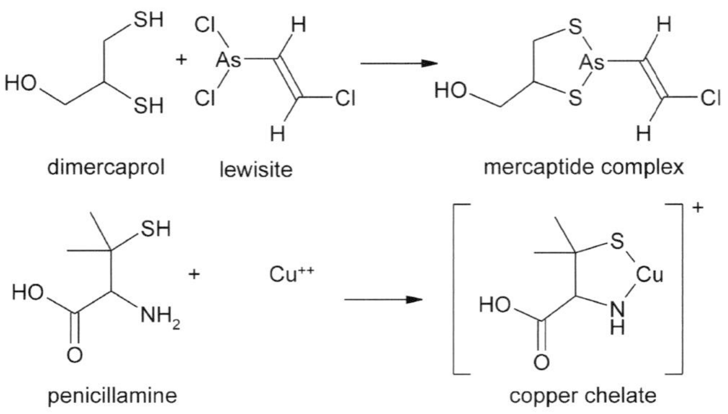

Fig. 19.40   Chelating properties of thiol derivatives.

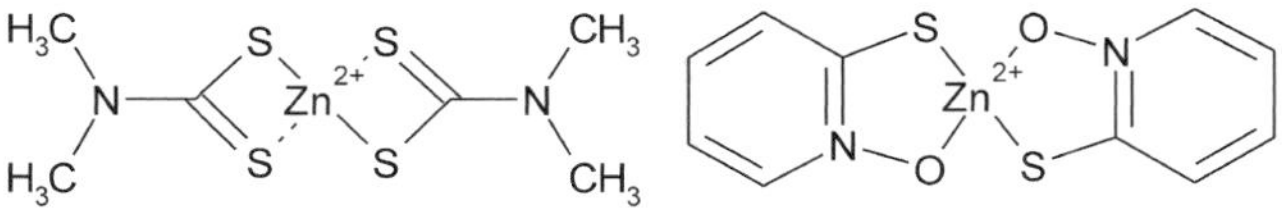

Fig. 19.41 Ziram and pyrithione-zinc are chelated salts.

the zinc and the iron salts, respectively, of dimethyldithio-carbamic acid (Fig. 19.41). They are widely used as selective fungicides in agriculture. Pyrithione [1-hydroxy-2(1*H*)-pyridinethione], as its zinc salt (Fig. 19.41) is used in dermatology as an antiseborrheic.

The sulfoxide and the sulfonamide functions are very polar and usually confer mediocre CNS bioavailability. However, sulfonamides presenting moderate brain penetration ($\sim 10\%$) are known. An example is found with the 5-HT$_6$ receptor antagonist SB-271046[53] (Fig. 19.42).

Covalently bound thiocyanates are rather unusual. Recent examples are the corticotropin-releasing factor (CRF) antagonist NPC 22009[54] and 4-phenoxy-phenoxyethyl

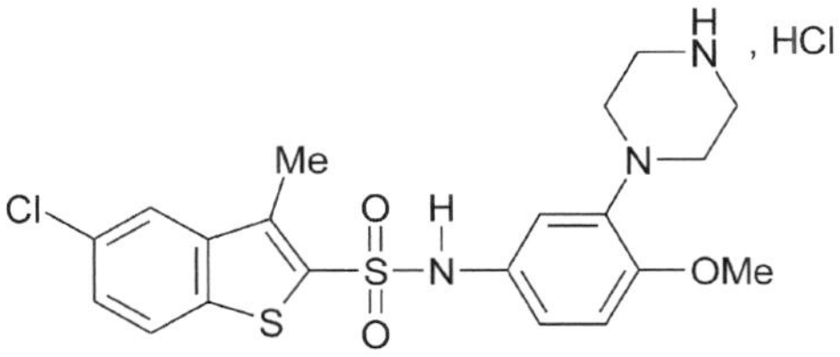

Fig. 19.42 A sulfonamide presenting moderate brain penetration.

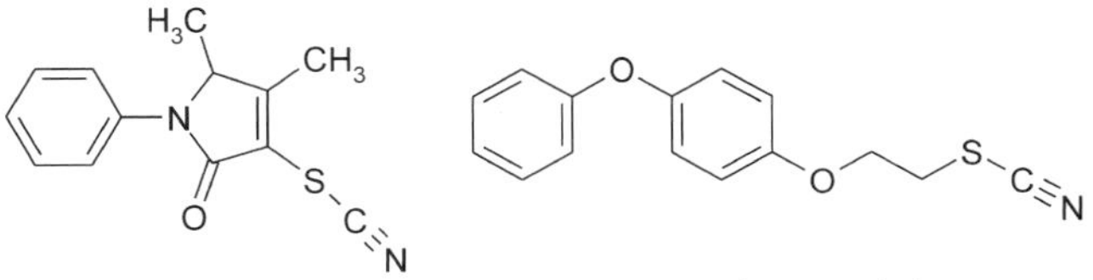

Fig. 19.43 Covalently bound thiocyanates.[54,55]

thiocyanate (Fig. 19.43) which act as inhibitors of the sterol biosynthesis in *Trypanosoma cruzi*,[55] the protozoon agent responsible for Chagas disease. Sometimes thiocyanates are used as counteranions for the preparation of pharmaceutically acceptable salts.[56]

Heterocyclic thioureas such as 6-propylthiouracil, methimidazole and carbimidazole (Fig. 19.44) are used as antithyroid drugs. They inhibit the formation of thyroid hormones; one of the presumed mechanisms is the inhibition of the iodine incorporation into the tyrosyl residues of thyroglobulin. It was proposed that the iodine atom is bound to a protein as a sulfenyl iodide. The thioureas may act in establishing covalent —S—S— bonds and in displacing iodine as HI.[57]

More generally, thiourea and its simpler aliphatic derivatives, as well as thioamides, produce goiter and have to be avoided in drug design. A natural compound, L-5-vinyl-2-thiooxazolidone (goitrin) is responsible for the goiter of cattle eating turnips or cruciferous plants.[58] Besides their affinity for metallic ions, thiol groups have other characteristics such as the ability to interconvert to disulfides through redox reactions, to add to conjugated double bonds and to form complexes with the pyridine nucleotides by nucleophilic attack at the 4-position of the pyridine ring.

VII ACIDIC FUNCTIONS

The prototypical representatives of the group are the carboxylic acids. However, a large number of bioisosteres such as sulfonic or phosphonic acids, tetrazoles or 3-hydroxyisoxazoles are available (see Chapter 13). In addition functions like esters, amides, peptides, aldehydes, primary alcohols and related functions can work as prodrugs or bioprecursors (see Chapter 33).

The introduction of an acidic group into a biologically active compound which does not already contain such a group has essentially a solubilizing effect. This effect can even be enhanced through salt formation (see Chapter 35). Carboxylic acids are often highly ionized at the physiological pH values and this is even more the case for sulfonic acids. As a rule, strong and highly ionized acids cannot cross the biological membranes which are permeable only to non-dissociated molecules; they are then subject to a rapid

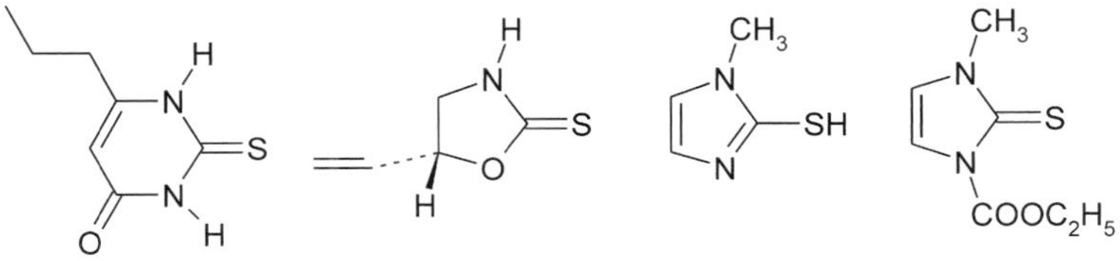

Fig. 19.44 Heterocyclic thioureas with antithyroid effects.

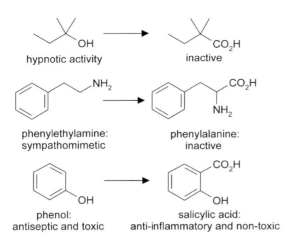

Fig. 19.45 Introduction of a carboxylic group in a small molecule can destroy the activity originally present or, conversely, create the conditions necessary for activity.

clearance from the animal body. However, once absorbed, they can establish strong ionic interactions with the basic amino acids, especially with lysine, contained in the blood serum albumin, or the enzyme and receptor proteins.

Changes in biological activity distinguish the sulfonic from the carboxylic acids. Broadly speaking, the sulfonic acids as a class are generally not biologically active. Exceptions are certain complex dyes or trypanocides (Trypan blue, suramin, etc.) and sulfonic amino acids such as taurine and hypotaurine for which an active transport mechanism exists. For carboxylic acids the situation depends on whether the carboxylic function is introduced

in small or large molecules. In small molecules, the introduction of a carboxylic group fundamentally changes the biological activity. Very often the initial biological activity is destroyed and the toxicity of the parent compound is reduced. In a series of cyproheptadine analogues the replacement of a chlorine substituent on a benzo ring by a carboxylic group resulted in a 4000-fold loss in affinity for the spiperone labelled dopamine receptor.[59] Conversely, the presence of the carboxylic group can sometimes create the conditions necessary for activity (Fig. 19.45).

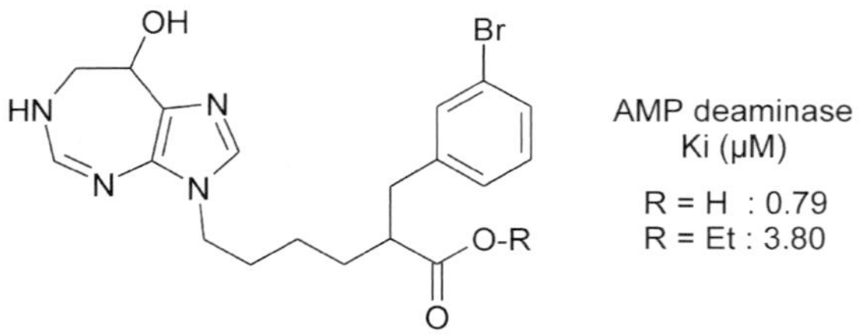

In large molecules, high pharmacological activity is maintained despite the presence of the carboxylic group. Examples are the anti-inflammatory arylacetic acids, the prostaglandins, cromolin and the related anti-asthmatics, and finally the β-lactam antibiotics. In these drugs the relative weight of the carboxyl is notably smaller. An illustrative example is found in a series of acids and esters derived from coformycin and acting as AMP deaminase inhibitors (Fig. 19.46). In this series the free acid is clearly more potent than the corresponding ethyl ester.[60]

Antitumour sulfonamides targeting G1 phase provide an *a contrario* example. Despite the relatively large size of the molecule, the presence of a carboxylic function completely abolishes the *in vitro* antiproliferative activity,[61] whereas the corresponding carboxamide and sulfonamide are highly active (Fig. 19.47).

With carboxyl-derived functions such as esters and amides, the initial activity of the drug, lost in introducing the carboxyl group, is often regained. Amides, ureides, hydantoins and barbiturates share CNS-depressing properties and are frequently indispensable elements of sedative, tranquillizing and anticonvulsant drugs. Nitriles as substituents are often comparable to chlorine atoms, but sometimes more toxic.

VIII BASIC GROUPS

The basic groups met in medicinal chemistry are the amines, the amidines, the guanidines and practically all nitrogen-containing heterocycles. Basic groups are polar and one would expect that highly ionized bases (especially quaternary ammonium salts) would resemble the sulfonic acids and

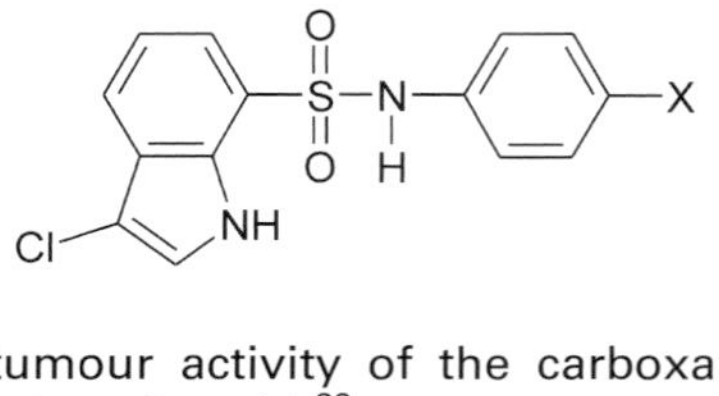

AMP deaminase
Ki (μM)

R = H : 0.79
R = Et : 3.80

Fig. 19.46 The free carboxylic acid is more active than the corresponding ethyl ester.

	antiproliferative activity IC$_{50}$ (μg/mL)	
	colon 38 murine adenocarcinoma	P388 murine leukemia
X = CO$_2$H :	>100	>100
X = CONH2 :	0.19	0.87
X = SO$_2$NH$_2$:	0.10	0.45

Fig. 19.47 The antitumour activity of the carboxamide and of the sulfonamide is completely abolished for the corresponding free carboxylic acid.[60]

show limited activity due to their mediocre membrane permeability. In practice, bases with pK_a values superior to 10 have a very limited chance of reaching the CNS.

As seen for the acidic groups, the introduction of a basic group into a biologically active compound that does not already contain such a group has essentially a solubilizing effect. This effect can also be enhanced through salt formation (see Chapter 35). In drug–protein interactions, the classical counteranions of organic bases are the aspartic and the glutamic carboxylates.

The biological activity of amines and basic heterocycles is immense and justifies the adage: 'no biological activity without nitrogen'. Steroids, prostaglandins, nonsteroidal anti-inflammatory drugs are, of course, exceptions. Primary amines often demonstrate less specific effects than secondary or tertiary amines. Acylation deactivates the amines strongly, as does the introduction in some other site of the molecule of a carboxylic or sulfonic group (formation of zwitterions: bipolar ions). Diamines and polyamines are usually more active than monoamines. Aromatic amines are always more hazardous than aliphatic amines and form toxic metabolites, examples are 2-naphtylamine, benzidine, aniline (see Chapter 32). They are easy to detoxicate by introducing a carboxyl group, as shown by the change from aniline to the nontoxic *p*-aminobenzoic acid.

IX ATTACHMENT OF ADDITIONAL BINDING SITES

Many polar active molecules, selected through *in vitro* screening tests are unable to cross the biomembranes and their bioavailability is particularly low. Attaching a very lipophilic moiety can sometimes help to overcome this drawback. A typical example is given by the development of the anticonvulsant drug thiagabine[62] starting from nipecotic acid (Fig. 19.48). Cyclic amino acids such as

nipecotic acid and guvacine have been shown to inhibit GABA uptake. However, these small amino acids do not readily cross the blood–brain barrier thus limiting their potential clinical usefulness. A considerable improvement has been the discovery by Yunger *et al.*[63,64] of compound SKF 89976A, a *N*-(4,4-diphenyl-3-butenyl) substituted nipecotic acid. *In vitro* this compound is approximately 20 times more potent than the parent amino acid as inhibitor of [³H] GABA uptake. Moreover, it is active orally as an anticonvulsant in mice and rats with ED_{50} around 8 mg kg^{-1}.

Further developments demonstrated that the 4,4-diphenyl-3-butenyl moiety could be replaced by ether-type analogues[65,66] and that the attachment of a lipophilic 3,3-diphenylpropyl side chain can take place at the carbon atom at the 6-position of the amino acid.[67] However, in this latter case, despite a reasonable *in vitro* activity ($IC_{50} = 100$ nM), no *in vivo* activity is observed. Final optimization of SKF 89976A led to the *bis*-thiophene tiagabine.[62]

In a similar way, ω, ω-diphenyl-alkyl chains ('butterflies') were also attached to L-glutamic acid to yield Glu m3 glutamate metabotropic m3 receptor-selective agonists[68] and to histamine, to yield the potent histamine H_1 receptor agonists histaprodifen and methylhistaprodifen[69] (Fig. 19.49).

Instead of ensuring high lipophilicity, aralkyl groups sometimes serve to achieve additional interactions with the target macromolecule. This is typically the case for the

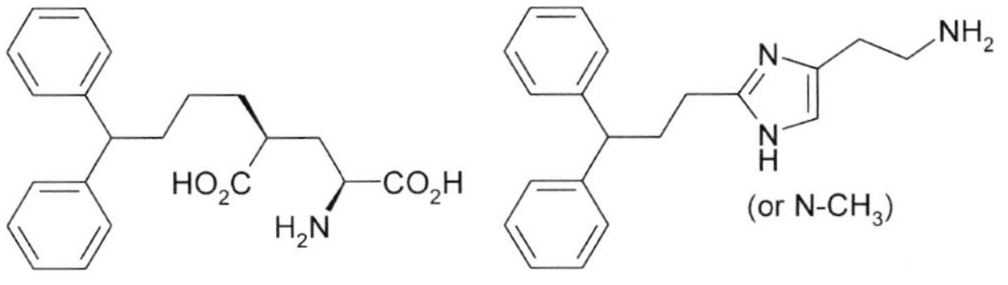

Fig. 19.49 The lipophilic diphenylalkyl moiety attached to glutamic acid and to histamine, respectively.

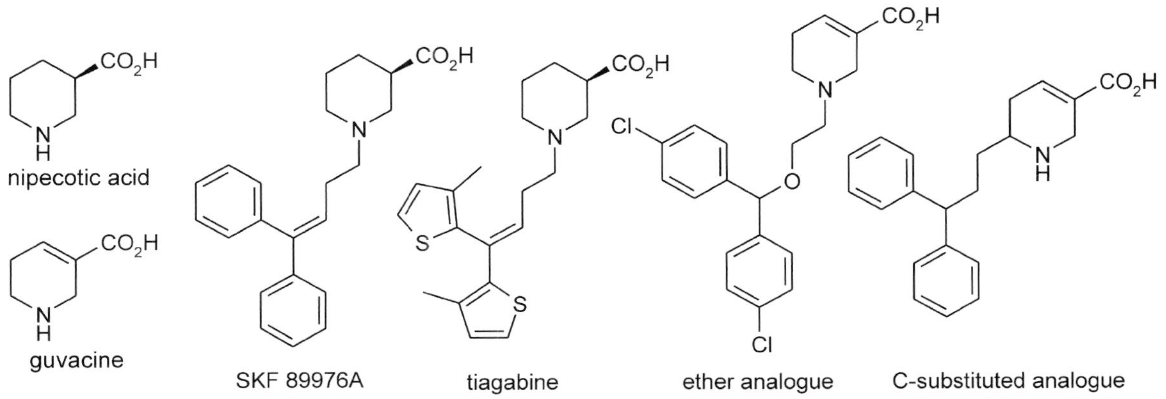

Fig. 19.48 Lipophilic derivatives of nipecotic acid and of guvacine.

angiotensin-converting enzyme inhibitor enalaprilat. The exchange in captopril of the thiol function for a carboxylic group as ligand for the enzyme zinc atom entails an important decrease in activity (Chapter 5). This decrease could be compensated for by the attachment of a phenethyl moiety.

REFERENCES

1. Taylor, J (1984) The Topliss Approach—Opinion of an Industrial Scientist on QSAR Methods; Lecture given at the Louis Pasteur University, Stasbourg, March 8, 1984.
2. Rekker, R.F. and Mannhold, R. (1992) *Calculation of Drug Lipophilicity*. Verlag VCH, Weinheim, Germany.
3. Albert, A. (1979) *Selective Toxicity*, p. 60. Chapman & Hall, London.
4. Ginnings, P. and Baum, R. (1937) Aqueous solubilities of the isomeric pentanols. *J. Am. Chem. Soc.* **59**: 1111–1113.
5. Nemethy, G. (1967) Hydrophobic interactions. *Angew. Chem. Int. Ed. Eng.* **6**: 195–206.
6. Hansch, C. and Anderson, S.M. (1967) The effect of intramolecular hydrophobic bonding on partition coefficients. *J. Org. Chem.* **32**: 2583–2586.
7. Gilligan, D. and Plummer, N. (1943) Comparative solubilities of sulfadiazine, sulfamerazine and sulfamethazine and their N4-acetyl derivatives at varying pH levels. *Proc. Soc. Exp. Biol. Med.* **53**: 142–145.
8. Albert, A. (1979) *Selective Toxicity*, p, 41. Chapman & Hall, London.
9. Harms, A.F. and Nauta, W.T. (1960) The effects of alkyl substitution in drugs. I. Substituted dimethylaminoethyl benzhydryl ethers. *J. Med. Pharm. Chem.* **2**: 57–77.
10. Velluz, L. (1961) La structure nor-stéroïde. In Cheymol, J. and Hazard, R. (eds). *Actualités Pharmacologiques*, pp. 221–243. Masson, Paris.
11. Wermuth, C.G., Schwartz, J., Leclerc, G., Garnier, J.P. and Rouot, B. (1973) Communication préliminaire: Conformation de la clonidine et hypothèses sur son interaction avec un récepteur α-adrénergique. *Eur. J. Med. Chem.* **8**: 115–116.
12. Haviv, F., Fitzpatrick, T.D., Swenson, R.E., Nichols, C.J., Mort, N.A., *et al.* (1993) Effect of *N*-methyl substitution of the peptide bonds in luteinizing hormone-releasing hormone antagonists. *J. Med. Chem.* **36**: 363–369.
13. Chu, K.C. (1980) The quantitative analysis of structure–activity relationships. In Wolf, M.E. (ed.). *The Basis of Medicinal Chemistry/Burger's Medicinal Chemistry*, pp. 393–418. John Wiley, New York.
14. McElvain, S.M. and Carney, T.P. (1946) Piperidine derivatives. XVII. Local anesthetics derived from substituted piperidinoalcohols. *J. Am. Chem. Soc.* **68**: 2592–2600.
15. Ganellin, C.R. and Roberts, S.M. (1993) *Medicinal Chemistry, The Role of Organic Chemistry in Drug Research*, 2 edn. Academic Press, London.
16. Black, J.W., Durant, G.J., Emmet, J.C. and Ganellin, C.R.C. (1974) Sulphur-methylene isosterism in the development of metiamide, a new histamine H2-receptor antagonist. *Nature* **248**: 65–67.
17. Arcos, J.C. and Arcos, M. (1962) Molecular geometry and mechanisms of action of chemical carcinogens. In Jucker, E. (ed.). *Progress in Drug Research*, pp. 407–581. Birkhäuser Verlag, Basel.
18. Albert, A. (1979) *Selective Toxicity*, p. 48. Chapman & Hall, London.
19. Roemer, D., Beuscher, H.H., Hill, R.C., Pless, J., Bauer, W., *et al.* (1977) A synthetic enkephalin analogue with prolonged parenteral and oral activity. *Nature* **268**: 547–549.

20. Plattner, J.J. and Norbeck, D.W. (1990) Obstacles to drug development from peptide leads. In Clark, C.R. and Moos, W.H. (eds). *Drug Discovery Technologies*, pp. 92–126. Ellis Horwood, Chichester.
21. Fauchère, J.L. (1986) Elements for the rational design of peptide drugs. In Testa, B. (ed.). *Advances in Drug Research*, pp. 29–69. Academic Press, London.
22. Holland, V.R., Saunders, B.C., Rose, F.L. and Walpole, A.L. (1974) A safer substitute for benzidine in the detection of blood. *Tetrahedron* **30**: 3299–3302.
23. Van de Straat, R., de Vries, J., Groot, E.J., Zijl, R. and Vermeulen, N. (1987) Paracetamol. 3-Monoalkyl- and 3,5-dialkyl derivatives: comparison of their hepatotoxicity in mice. *Toxicol. Appl. Pharmacol.* **89**: 183–189.
24. Tute, M.S. (1971) Principles and practice of Hansch analysis: a guide to structure–activity correlation for the medicinal chemist. In Harper, N.J. and Simmonds, A.B. (eds). *Advances in Drug Research*, pp. 1–77. Academic Press, London.
25. Nadler, V., Kloog, Y. and Sokolovsky, M. (1988) 1-Aminocyclo-propane-1-carboxylic acid (ACC) mimics the effect of glycine on the NMDA receptor ion channel. *Eur. J. Pharmacol.* **157**: 115–116.
26. Cussac, M., Pierre, J.L., Boucherle, A. and Favier, F. (1975) Intérêt des dérivés du cyclopropane en chimie thérapeutique. *Ann. Pharm. Franç.* **33**: 513–529.
27. Hansch, C. and Anderson, S.M. (1967) The structure–activity relationship in barbiturates and its similarity to that in other narcotics. *J. Med. Chem.* **10**: 745–753.
28. Martin, Y.C. (1978) *Quantitative Drug Design, A Critical Introduction*. Marcel Dekker, New York.
29. Marivet, M.C., Bourguignon, J.J., Lugnier, C., Mann, A., Stoclet, J.C. and Wermuth, C.G. (1989) Inhibition of cyclic adenosine-3′,5′-monophosphate phosphodiesterase from vascular smooth muscle by rolipram analogues. *J. Med. Chem.* **32**: 1450–1457.
30. Trivedi, B.K., Holmes, A., Stoeber, T.L., Blankley, C.J., Roark, H.W., *et al.* Inhibitors of acyl-CoA: cholesterol acyltransferase. 4. A novel series of urea ACAT inhibitors as potential hypocholesterolemic agents. *J. Med. Chem.* **36**: 3300–3307.
31. Sexton, W.A. (1963) *Chemical Constitution and Biological Activity*, 3rd edn, p. 103. E. & F. N. Spon, London.
32. Craig, P.N. (1980) Guidelines for drug and analog design. In Wolf, M.E. (ed.). *The Basis of Medicinal Chemistry—Burge's Medicinal Chemistry*, pp. 331–348. John Wiley, New York.
33. Breen, A.P. and Murphy, J.A. (1993) Radical-induced DNA cleavage mediated by a vinyl epoxide. *J. Chem. Soc. Chem. Commun.* 193–194.
34. Ogata, M., Matsumoto, H., Shimizu, S., Kida, S., Shiro, M. and Tawara, K. (1987) Synthesis and antifungal activity of new 1-vinylimidazoles. *J. Med. Chem.* **30**: 1348–1354.
35. Maurin, M.B., Addicks, W.J., Rowe, S.M. and Hogan, R. (1993) Physical chemical properties of alpha styryl carbinol antifungal agents. *Pharm. Res.* **10**: 309–312.
36. Browning, E. (1965) *Toxicity and Metabolism of Industrial Solvents*, pp. 377–381. Elsevier, New York.
37. Scholtan, W. and Sy, L. (1966) Kolloid-chemische Eigenschaften eines neuen Kurznarkotikums. Teilgewicht der Mizelle und Verteilungs-gleichgewicht des Wirkstoffes (Colloidal chemical properties of a new short-acting anesthetic. Particle weight of the micelle and distribution equilibrium of the active compound). *Arzneimit.-Forsch.* **16**: 679–691.
38. Milne, G.M., Jr. and Johnson, M.R. (1967) Narcotic antagonists and analgesics, *Annual Reports in Medicinal Chemistry*, pp. 23–32. Academic Press, New York.
39. Culvenor, C.C.J., Dann, A.T. and Dick, A.T. (1962) Alkylation as the mechanism by which the hepatotoxic pyrrolizidine alkaloids act on cell nuclei. *Nature* **195**: 570–573.

40. Beart, P.M., Uhr, M.L. and Johnston, G.A.R. (1972) Inhibition of GABA transaminase activity by 4-aminotetrolic acid. *J. Neurochem.* **19**: 1855–1861.

41. Schmitt, J. (1967) Sur un nouveau myorelaxant de la classe des benzodiazépines: le tétrazepam. *Eur. J. Med. Chem.* **2**: 254–259.

42. Fryer, R.I., Zhang, P., Rios, R., Gu, Z.-Q., Basile, A.S., *et al.* (1993) Structure–activity relationship studies at the benzodiazepine receptor (BZR): a comparison of the substituent effects of pyrazoloquinolinone analogs. *J. Med. Chem.* **36**: 1669–1673.

43. Taylor, J.B. and Kennewell, P.D. (1981) *Introductory Medicinal Chemistry*, p. 89. Ellis Horwood, Chichester.

44. Newman, A.H., Allen, A.C., Izenwasser, S. and Katz, J. (1994) Novel 3α-(diphenylmethoxy)tropane analogs: potent dopamine uptake inhibitors without cocaine-like behavioral profiles. *J. Med. Chem.* **37**: 2258–2261.

45. MacLeod, A.M., Merchant, K.J., Cascieri, M.A., Sadowski, S., Ber, E., Swain, C.J. and Bakes, R. (1994) *N*-Acyl-L-tryptophan benzyl esters: potent substance P receptor antagonists. *J. Med. Chem.* **36**: 2044–2045.

46. Buu-Hoï, N.P. (1961) Les dérivés organiques du fluor d'intérêt pharmacologique. In Jucker, E. (ed.). *Progress in Drug Research*, pp. 9–74. Birkhäuser, Basel.

47. Gee, K.W. and Yamamura, H.I. (1982) A novel pyrazoloquinoline that interacts with brain benzodiazepine receptors: characterization of some *in vitro* and *in vivo* properties of CGS 9896. *Life Sci.* **30**: 2245–2252.

48. Bowery, N.G., Bittiger, H., Olpe, H.-R. (eds) (1990) *GABA_B Receptors in Mammalian Function*. John Wiley, Chichester.

49. Anonymous. (1998) Bromfenac sodium, Duract^R. *Drugs Fut.* **23**: 1234–1235.

50. Rosi, D., Peruzzotti, G., Dennis, E.W., Berberian, D.A., Freele, H. and Archer, S. (1965) A new, active metabolite of 'Miracil D'. *Nature* **208**: 1005–1006.

51. Wermuth, C.G. Aminopyridazines—an alternative route to potent muscarinic agonists with no cholinergic syndrome. *Il Farmaco* **48**: 253–274.

52. Klaassen, C.D. (1990) Heavy metals and heavy-metal antagonists. In Goodman-Gilmah, A., Rall, T.W., Nies, A.S. and Taylor, P. (eds). *Goodman and Gilman's The Pharmacological Basis of Therapeutics*, pp. 1592–1614. Pergamon Press, New York.

53. Bromidge, S.M., Brown, A.M., Clarke, S.E., Dogson, K., Gager, T., *et al.* (1999) 5-Chloro-*N*-(4-methoxy-3-piperazin-1-yl-phenyl)-3-methyl-2-benzothiophenesulfonamide (SB-271046): A potent, selective, and orally bioavailable 5-HT_6 receptor antagonist. *J. Med. Chem.* **42**: 202–205.

54. Abreu, M.E., Rzeszotarski, W., Kyle, D.J. and Hiner, R. (1991) Preparation of oxopyrazolinyl thiocyanates and bis(oxopyrazolinyl) disulfides as corticotropin-releasing factor antagonists. US Patent 5063245 A to Nova Pharmaceutical Corp., USA.

55. Szajnman, S.H., Yan, W., Bailey, B.N., Docampo, R., Elhalem, E., *et al.* (2000) Design and synthesis of aryloxyethyl thiocyanate derivatives as potent inhibitors of *Trypanosoma cruzi* proliferation. *J. Med. Chem.* **43**: 1826–1840.

56. Stahl, P.H. and Wermuth, C.G. (2002) *Pharmaceutical Salts*, Helvetica Chimica Acta, John Wiley, Zürich.

57. Jirousek, L. and Pritchard, E. (1971) On the chemical iodination of tyrosine with protein sulfenyl iodide and sulfenyl periodide derivatives. *Biochim. Biophys. Acta* **243**: 230–238.

58. Haynes, R.C., Jr. (1990) Thyroid and antithyroid drugs. In Goodman-Gilman, A., Rall, T.W., Nies, A.S. and Taylor, P. (eds). *Goodman and Gilman's The Pharmacological Basis of Therapeutics*, pp. 1361–1383. Pergamon Press, New York.

59. Remy, D.C., Britcher, S.F., King, S.W., Anderson, P.S., Hunt, C.A., *et al.* (1983) Synthesis and receptor binding studies relevant to the neuroleptic activities of some 1-methyl-4-piperidylene-9-substituted-pyrrolo[2,1-*b*][3]benzazepine derivatives. *J. Med. Chem.* **26**: 974–980.

60. Bookser, B.C., Kasibhatla, S.R., Appleman, J.R. and Erion, M.D. (2000) AMP deaminase inhibitors. 2. Initial discovery of a non-nucleotide transition state inhibitor. *J. Med. Chem.* **43**: 1495–1507.

61. Owa, T., Yoshino, H., Okauchi, T., Yoshimatsu, K., Ozawa, Y., *et al.* (1999) Discovery of novel antitumor sulfonamides targeting G1 phase of the cell cycle. *J. Med. Chem.* **42**: 3789–3799.

62. Andersen, K.E., Braestrup, C., Groenwald, F.C., Joergensen, A.S., Nielsen, E.B., *et al.* The synthesis of novel GABA uptake inhibitors. 1. Elucidation of the structure-activity studies leading to the choice of (*R*)-1.[4,4-Bis(3-methyl-2-thienyl)-3-butenyl]-3-piperidinecarboxylic acid (Tiagabine) as an anticonvulsant drug candidate. *J. Med. Chem.* **36**: 1716–1725.

63. Yunger, L.M., Fowler, P.J., Zarevics, P. and Setler, P.E. (1984) Novel inhibitors of γ-aminobutyric acid (GABA) uptake: anticonvulsant actions in rats and mice. *J. Pharmacol. Exp. Ther.* **228**: 109–115.

64. Ali, F.E., Bondinell, W.E., Dandridge, P.A., Frazee, J.S., Garvey, E., *et al.* (1985) Orally active and potent inhibitors of γ-aminobutyric acid uptake. *J. Med. Chem.* **28**: 653–660.

65. Falch, E. and Krogsgaard-Larsen, P. (1991) GABA uptake inhibitors. Syntheses and structure-activity studies on GABA analogs containing diarylbutenyl and diarylmethoxyalkyl *N*-substituents. *Eur. J. Med. Chem.* **26**: 69–78.

66. Pavia, M.R., Lobbestael, S.J., Nugiel, D., Mayhugh, D.R., Gregor, V.E., *et al.* (1992) Structure–activity studies on benzhydrol-containing nipecotic acid and guvacine derivatives as potent, orally-active inhibitors of GABA uptake. *J. Med. Chem.* **35**: 4238–4248.

67. N'Goka, V., Schlewer, G., Linget, J.-M., Chambon, J.-P. and Wermuth, C.G. (1991) GABA-uptake inhibitors: construction of a general pharmacophore model and successful prediction of a new representative. *J. Med. Chem.* **34**: 2547–2557.

68. Wermuth, C.G., Mann, A., Schoenfelder, A., Wright, R.A., Johnson, B.G., *et al.* (1996) (2S,4S)-2-Amino-4-(4,4-diphenylbut-1-yl)-pentane-1,5-dioic acid: a potent and selective antagonist for metabotropic glutamate receptors negatively linked to adenylate cyclase. *J. Med. Chem.* **39**: 814–816.

69. Elz, S., Kramer, K., Pertz, H.H., Detert, H., ter Laak, A.M., *et al.* (2000) Histadiprofens: synthesis, pharmacological *in vitro* evaluation, and molecular modeling of a new class of highly active and selective histamine H_1 receptor agonists. *J. Med. Chem.* **43**: 1071–1084.

20

THE ROLE OF FUNCTIONAL GROUPS IN DRUG–RECEPTOR INTERACTIONS

Peter Andrews

Alice remained looking thoughtfully at the mushroom for a minute, trying to make out which were the two sides of it; and, as it was perfectly round, she found this a very difficult question.

Lewis Carroll, *Alice's Adventures in Wonderland*

I INTRODUCTION

The strength of the interaction between a drug molecule and its receptor can be determined directly from the equilibrium constant for the interaction:

$$K_d = \frac{[\text{drug}][\text{receptor}]}{[\text{complex}]} \qquad (1)$$

by expressing it in terms of the corresponding free energy change:

$$\Delta G = -2.303 \, RT \log K_d \qquad (2)$$

Under physiological conditions ($T = 310$ K) this is approximated (in kJ mol^{-1}) by

$$\Delta G = 5.85 \log K_d \qquad (3)$$

The observed equilibrium constant thus provides a direct measurement of ΔG. For example, drug binding with a K_d of 10^{-10} M requires $(-5.85) \times (-10) = 58.5$ kJ mol^{-1} to dissociate from the receptor. The purpose of this chapter is to provide an understanding of the physical and chemical factors which contribute most significantly to the strength of drug–receptor interactions. The first part consists of a physical description of the influence of electrostatic and steric match on the various types of nonbonded drug–receptor interactions. The second part provides a more chemical interpretation, concentrating on the intrinsic strengths of individual functional group contributions to the affinity of drugs for their receptors. Finally, the relevance of these numbers to the practising medicinal chemist is discussed using inhibitors of the viral enzyme HIV-protease as typical examples.

II THE IMPORTANCE OF THE ELECTROSTATIC AND STERIC MATCH BETWEEN DRUG AND RECEPTOR

What determines K_d? Basically, it depends on two factors. One is the electrostatic match between drug and receptor, which is primarily a function of electron density. The other is the steric match between drug and receptor, which is primarily dependent on conformation. How do these two factors contribute to the strengths of the various bonds that make up drug–receptor interactions? In fact, most non-covalent interactions depend to some degree on both electrostatic and steric properties, but for convenience in the following discussion we will divide them into those which are primarily electrostatic and those which are primarily steric.

A Electrostatic interactions

Electrostatic interactions are the net result of the attractive forces between the positively charged nuclei and the negatively charged electrons of the two molecules. The attractive force between these opposite charges leads to three main bond types: charge–charge, charge–dipole and dipole–dipole interactions.

Ionic bonds

The strength of any electrostatic interaction can be calculated from equation (4), where q_i and q_j are two charges separated by a distance r_{ij} in a medium of dielectric constant D. This equation applies equally to ionic interactions, where the charges q_i and q_j are integer values, or to polar interactions, in which the total energy is summed over the contributions calculated from the partial charges on all the individual atoms.

$$E = \frac{q_i q_j}{D r_{ij}} \qquad (4)$$

It follows from equation (4) that the strengths of ionic interactions are crucially dependent on the dielectric constant D of the surrounding medium. In hydrophobic environments, like the interior of a protein molecule, the dielectric constant may be as low as four, whereas in bulk-phase water the corresponding value is 80. In other environments, intermediate values are appropriate, e.g. for interactions occurring near the surface of a protein, a D value of 28 is commonly used. It also follows from equation (4) that the strengths of ionic bonds are inversely proportional to the distance separating the two charges. However, since the strengths of other noncovalent bonds are even more sharply dependent on distance than that of ionic bonds, ionic attraction frequently dominates the initial long range interactions between drugs and receptors.

Charge–dipole and dipole–dipole interactions

Although charge–dipole and dipole–dipole interactions are weaker than ionic bonds, they are nevertheless key contributors to the overall strengths of drug–receptor interactions, since they occur in any molecule in which electronegativity differences between atoms result in significant bond, group or molecular dipole moments.

The key differences between ionic and dipolar interactions relate to their dependence on distance and orientation. For charge–dipole interactions, the strength of the interaction depends inversely on the square of the distance, while for dipole–dipole interactions it reduces with the cube of the distance separating the dipoles.

Similarly, while steric effects are of little importance in ionic interactions, stricter geometric requirements apply to dipolar interactions, which may be either attractive or repulsive, depending on the orientation of the dipole moments.

Inductive interactions

The formation of a drug–receptor complex is often accompanied by intramolecular and/or intermolecular redistributions of charge. In the intramolecular case, this redistribution is referred to as an induced polarization, whereas a redistribution of charge between two molecules is described as a charge transfer interaction. In either case, the resulting interactions are always attractive and strongly dependent on the distance separating the two molecules. An interesting example of the importance of inductive inter-actions is the recent calculation by Bajorath et al.[1] on the binding of folate and dihydrofolate to dihydrofolate reductase. This revealed a shift in net charge equivalent to half an electron from the pteridine ring to the glutamate moiety on binding to the enzyme, with the major change in density being focused on the bonds that are catalytically reduced.

Hydrogen bonds

The most important noncovalent interactions in biological systems, hydrogen bonds, are also best described as electrostatic interactions. The approximate strength of individual hydrogen bonds can therefore be calculated from the partial charges of the atoms in the interacting groups using equation (4), which makes hydrogen bonds, like ionic bonds, important long-range recognition factors between drugs and receptors. Unlike ionic bonds, however, hydrogen bonds are dependent to some extent on steric orientation. Thus, statistical studies[2–4] of hydrogen bonds in small molecule crystal structures from the Cambridge Crystallographic Database show clear directional prefer-ences, reflecting conventional hybridization concepts. On the other hand, the atomic surroundings of uncharged hydrogen bonding groups in ligand–protein structures recorded in the Brookhaven Protein Databank showed no strong directional preferences,[5] suggesting that the energy differential favouring conventional hybridization states is small relative to other components of the interaction.

Stronger directional preferences are observed for hydrogen bond reinforced ionic interactions, for which analysis of protein–ligand interactions in the Brookhaven

Protein Databank showed that ligand carboxyl groups participate in two distinct types of binding: a close chelate-type interaction with the guanidino group of arginine, and a lateral interaction between one of the carboxyl oxygen atoms and a nitrogen atom from a variety of positively charged amino acid residues. This is consistent with the fact that the strongest hydrogen bonds are formed between groups with the greatest electrostatic character.[6,7]

B Steric interactions

While electrostatic interactions are the dominant interactions involving polar molecules, there are also strong interactions between nonpolar molecules, particularly at short intermolecular distances.

Dispersion forces

Dispersion or London forces are the universal forces responsible for attractive interactions between nonpolar molecules. Their occurrence is due to the fact that any atom will, at any given instant, be likely to possess a finite dipole moment as a result of the movement of electrons around the nuclei. Such fluctuating dipoles tend to induce opposite dipoles in adjacent molecules, thus resulting in a net attractive force. Although the individual interactions between pairs of atoms are relatively weak, the total contribution to binding from dispersion forces can be very significant if there is a close fit between drug and receptor. The quality of the steric match is thus the dominant factor in nonpolar interactions.

Short-range repulsive forces

The short-range repulsive forces resulting from the overlap of the electron clouds of any two molecules increase exponentially with decreasing internuclear separation. The balance between these repulsive interactions and the dispersion forces thus determines both the minimum and the most favourable non-bonded separation between any pair of atoms. The equilibrium distance can be determined from crystal data, and is equivalent to the sum of the van der Waals radii of the two interacting atoms. For nonpolar molecules this balance between the attractive dispersion forces and the short-range repulsive forces is generally defined in terms of the Buckingham (6–exp) potential given in equation (5) or the alternative Lennard–Jones 6–12 potential, given in equation (6).

$$E = \frac{Ae^{-Br}}{r^d} - \frac{C}{r^6} \qquad (5)$$

$$E = \frac{A^r}{r^{12}} - \frac{C}{r^6} \qquad (6)$$

Conformational energy

While intramolecular interactions within the drug molecule are the primary factors in determining the lowest energy conformation of the unbound drug, intermolecular interactions with the receptor also have a significant effect on conformation. If the bound conformation of a flexible molecule is also its lowest energy conformation, there is no conformational energy cost involved in binding. If, on the other hand, the optimal interaction between drug and receptor requires a higher energy conformation, this energy difference will reduce the apparent strength of the interaction between the two molecules.

C The role of entropy

Hydrophobic interactions

When a nonpolar molecule is surrounded by water, stronger than normal water–water interactions are formed around the solute molecule to compensate for the weaker interactions between solute and water. This results in an increasingly ordered arrangement of water molecules around the solute and thus a negative entropy of dissolution. The decrease in entropy is roughly proportional to the nonpolar surface area of the molecule. The association of two such nonpolar molecules in water reduces the total nonpolar surface area exposed to the solvent, thus reducing the amount of structured water, and therefore providing a favourable entropy of association.

As for van der Waals forces, hydrophobic interactions are individually weak (0.1 to 0.2 kJ mol^{-1} for every square angstrom of solvent-accessible hydrocarbon surface[8]), but the total contribution of hydrophobic bonds to drug–receptor interactions is substantial. Similarly, the overall strength of the hydrophobic interaction between two molecules is very dependent on the quality of the steric match between the two molecules. If this is not sufficiently close to squeeze all of the solvent from the interface, a substantial entropy penalty must be paid for each of the trapped water molecules.

Rotational and translational entropy

There are also substantial entropic penalties associated with the loss of three rotational and three translational degrees of freedom of the drug molecule on binding to the receptor. Since these are replaced by six vibrational degrees of freedom in the complex, the resulting entropy change is dependent on the relative tightness of the complex. For a typical ligand–protein interaction, the estimated change in free energy resulting from the loss of entropy on binding (at 310 K) ranges from 12 kJ mol^{-1} for a very weak interaction to 60 kJ mol^{-1} for a tightly bound complex.[9]

Conformational entropy

In the case of flexible drug molecules, there is a further entropy loss due to the conformational restriction which accompanies binding. Based on the observed entropy changes accompanying cyclization reactions, the extent of this entropy loss is estimated[10] at 5–6 kJ mol^{-1} per internal rotation, although the actual figure again depends on the overall strength of the interaction between the drug and the receptor. In the case of rigid analogues, there is no such loss of conformational entropy on binding. Provided that they offer a good steric and electrostatic match to the receptor, rigid analogues should therefore have a free energy advantage relative to more flexible drugs.

III THE STRENGTHS OF FUNCTIONAL GROUP CONTRIBUTIONS TO DRUG–RECEPTOR INTERACTIONS

The total free energy of interaction between a drug and its receptor provides a measure of the strength of the association between the two molecules, but tells us little or nothing about the overall quality of their match. Does the observed binding reflect a composite of interactions between every part of the drug and its receptor, or is it a case of one or two strong interactions contributing sufficient energy to disguise an otherwise mediocre fit? Is the observed increase in interaction energy resulting from the addition of a new functional group consistent with what might have been anticipated? To answer these questions we need some means of estimating the individual functional group contributions to drug–receptor interactions.

A Measuring functional group contributions

The free energy of binding, ΔG, can be defined in terms of the binding energies for the individual functional groups which make up the drug molecule according to equation (7)

$$\Delta G = T\Delta S_{rt} + n_r E_r = \sum n_x E_x \qquad (7)$$

where $T\Delta S_{rt}$ is the loss of overall rotational and translational entropy associated with binding of the drug molecule, n_r is the number of internal degrees of conformational freedom lost on binding the drug molecule, and E_r is the energy equivalent of the entropy loss associated with the loss of each degree of conformational freedom on receptor binding.

Intrinsic binding energy

The final term in equation 7 is the sum of the binding energies E_x associated with each functional group X, of which there are n_x present in the drug. In the ideal case, when the specified functional group is aligned optimally and without strain with the corresponding functional group in the receptor, E_x is referred to as the *intrinsic binding energy*.[11] In other cases, the term *apparent binding energy* is used.

It should be noted that each binding energy E_x is actually a combination of the various enthalpic and entropic interactions outlined above. These include the enthalpy of interaction between the functional group and its corresponding binding site on the receptor, the enthalpy changes associated with the removal of water of hydration from the functional group and its target site and the subsequent formation of bonds between the displaced water molecules, and the corresponding entropy terms associated with the displacement and subsequent bonding of water molecules.

It is apparent that these intrinsic binding energies may be regarded, at least approximately, as properties of the functional group that should be relatively independent of the groups to which the particular functional group is attached. Such intrinsic binding potentials might thus reasonably be used in an additive manner to provide an overall estimate of the drug–receptor interaction.

Anchor principle

It follows from equation (7) that the binding energy, E_x, due to the interaction between the receptor and a specific functional group, X, can be estimated by comparing the binding energies for pairs of compounds which differ only in the presence or absence of the specific functional group. This approach was first applied by Page[9] who referred to it as the 'anchor principle'. It is based on the premise that the difference in binding of a drug molecule with or without the particular functional group is due to factors associated solely with that group, i.e. the binding energy E_x plus any degrees of conformational freedom lost specifically as a result of binding of group X. Other degrees of conformational freedom lost on binding and the loss of overall rotational and translational entropy associated with the remainder of the drug molecule (the anchor) are assumed to be unaffected by the presence or absence of X.

Similarly, the impact of a single amino acid substitution in the active site of an enzyme on transition state stabilization, as determined by the change in either catalytic efficiency or inhibitor binding, provides a measure of the relative binding energy of the two side chains.

Clearly, the magnitude of the binding energies obtained using the anchor principle will vary widely with the quality of the interaction. If the functional groups are not properly aligned, as might reasonably be expected in many mutant proteins, a small or even repulsive interaction may result. Alternatively, the strength of the additional bond may be offset by a reduction in the strengths of the existing bonds. Under these circumstances the anchor principle will lead to an underestimate of the true bond strength.

Average binding energy

An alternative to the pair-by-pair approach inherent in the anchor principle was developed by Andrews *et al.*,[12] who sought to average the contributions of individual functional groups to the observed binding energies of 200 ligand–protein interactions in aqueous solution. For this purpose, the average loss of overall rotational and translational entropy accompanying drug-receptor binding, $T\Delta S_{rt}$ in equation (7), was estimated at 58.5 kJ mol^{-1} (14 kcal mol^{-1}) at 310 K. Regression analysis against n_r (obtained by counting the number of degrees of conformational freedom in each of the 200 ligand structures) and n_x (the number of occurrences of each functional group, X, in each of the 200 ligand structures) as the independent variables was then used to obtain average values of the binding energies associated with each functional group and for the loss of entropy associated with each degree of conformational freedom.

The results of this analysis showed that the loss of entropy associated with each internal rotation on receptor binding is equivalent to a reduction in the free energy of binding by an average of 3 kJ mol^{-1}, which may be compared with the estimated[10] value for the total loss of conformational freedom around a single bond of $5–6 \text{ kJ mol}^{-1}$. The smaller number obtained empirically implies that conformational freedom is not fully lost for all the bonds in an average drug–receptor interaction, and is consistent with experimental estimates of $1.6–3.6 \text{ kJ mol}^{-1}$ for the entropic cost of restricting rotations in hydrocarbon chains.[13]

The corresponding binding energies obtained by the averaging process were C (sp^2 or sp^3) 3 kJ mol^{-1}; O, S, N, or halogen, 5 kJ mol^{-1}; OH and C=O, 10 and 14 kJ mol^{-1} respectively; and CO_2^-, OPO_3^{2-}, and N^+, 34, 42 and 48 kJ mol^{-1}, respectively. Once again, it should be stressed that these are not intrinsic binding energies in the sense defined above. This would be the case only if each functional group in each drug in the series was optimally aligned with a corresponding functional group in the receptor. In fact, since every functional group of every drug was included in the analysis, the calculated values are averages of apparent binding energies, including those for some groups which may not interact with the receptor at all.[14] The calculated averages are thus almost certainly smaller than the corresponding intrinsic binding energies, although they follow expected trends in that charged groups lead to stronger interactions than polar groups, which in turn are stronger than nonpolar groups such as sp^2 or sp^3 carbons.

B The methyl group and other nonpolar substituents

The initial application of the anchor principle described by Page[9] related to data on the selectivity of amino

Fig. 20.1 Isoleucine **1** and desmethyl-isoleucine **2**.

acid–tRNA synthetases, from which he estimated intrinsic binding energies for the methylene group in the range $12–14 \text{ kJ mol}^{-1}$. For example, the calculated binding energies (equation 3) for isoleucine **1** (Fig. 20.1) and its desmethyl analogue **2** to isoleucyl–tRNA synthetase are 29.7 and 15.9 kJ mol^{-1}, respectively, indicating that the methyl group contributes a total of 13.8 kJ mol^{-1} to the overall interaction.

This estimate, having been derived from observations on a highly selective enzyme–substrate interaction, is probably also approaching the intrinsic limit for the binding contribution of a methyl group. There are unfortunately relatively few active site mutagenesis data available for mutations involving a single methyl group. The mutation of glycine to alanine in subtilisin BPN results in free energy differences of $1–3 \text{ kJ mol}^{-1}$, although larger changes have been observed in other enzymes.[15]

For longer hydrocarbon side chains, the positive contribution due to dispersion forces and hydrophobic interactions tends to be offset by the loss of conformational entropy on binding. Thus, the 'average' binding energy of 3 kJ mol^{-1} obtained by Andrews *et al.*[12] for sp^2 and sp^3 carbon groups is identical to the 'average' reduction in free energy of binding estimated for the loss of conformational freedom around a single bond. Clearly, this effect will be greater in saturated hydrocarbon chains than their more conformationally constrained unsaturated or cyclic analogues.

C The hydroxyl group and other hydrogen bond-forming substituents

The most extensive studies of hydroxyl group contributions to drug–receptor interactions are those of Wolfenden and Kati,[16] on the contribution of hydrogen bonds formed by hydroxyl groups in transition state analogues. In a series of 13 examples of paired ligands with and without hydroxyl groups, they used the anchor principle to determine apparent binding energies for single hydroxyl groups ranging from $20–42 \text{ kJ mol}^{-1}$.

Thus, in comparing the binding of 1,6-dihydropurine ribonucleoside **3** (Fig. 20.2) and its 6-hydroxy derivative **4** to adenosine deaminase, they observed[17] K_i values for these two inhibitors of 5.4×10^{-10} and 3×10^{-13} M respectively, reflecting a difference in binding energy of 41 kJ mol^{-1}.

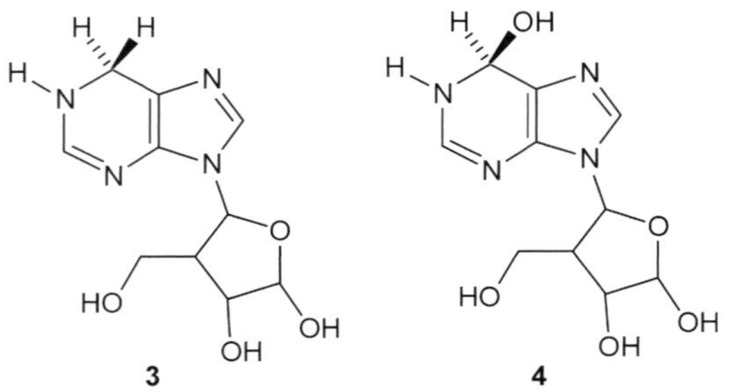

Fig. 20.2 1,6-Dihydropurine ribonucleoside **3** and its 6-hydroxy analogue **4**.

As noted by Kati and Wolfenden,[17] this remarkable affinity appears to suggest that the 6-hydroxyl group, which has very limited freedom of movement, is likely to be in almost ideal alignment with the active site, and that at least one charged active site residue is also likely to be involved in its hydrogen bonding interaction. This conjecture has since been verified by the determination of the crystal structure[18] of the inhibitory complex between adenosine deaminase and 6-hydroxy-1,6-dihydropurine ribonucleoside, which showed that the 6-hydroxyl group interacts with a zinc atom, with a protonated histidyl residue, and with an aspartic acid residue at the enzyme's active site.

Once again, the data from active-site mutagenesis studies are less striking, but nevertheless reveal some very substantial hydrogen-bonding interactions. In Fersht's studies[19] on tyrosyl tRNA synthetase, for example, hydrogen bonds between this enzyme and uncharged substrate groups contributed between 2 and 6 kJ mol^{-1} towards specificity, while hydrogen bonds to charged groups contributed between 15 and 19 kJ mol^{-1} corresponding to a factor of 1000 in specificity. These numbers are comparable to the 'average' functional group contributions determined by Andrews *et al.* which ranged from 5 kJ mol^{-1} for uncharged H-bond acceptors (O, N, S) through 10 kJ mol^{-1} for a hydroxyl group to 14 kJ mol^{-1} for a carbonyl.

D   Acidic and basic substituents

Application of the anchor principle to data on the selectivity of amino acid–tRNA synthetases gives estimated intrinsic binding energies for the carboxyl and amino groups of 18 and >28 kJ mol^{-1}, respectively. However, since the sidechains, rather than the ionic groups, are the primary determinants of amino acid/tRNA synthetase specificity, these energies are likely to be underestimates.

An indication of this likelihood may be obtained using simple observations on the interactions of individual charged groups with appropriate enzymes. The phosphate ion, for example, binds alkaline phosphatase[20] with a dissociation constant of 2.3×10^{-6} M, equivalent to a ΔG value of approximately 33 kJ mol^{-1}. Taking the most conservative estimate for the loss of rotational and translational entropy associated with this interaction, 12 kJ mol^{-1} for a loosely bound complex, equation (7) then gives a lower estimate for binding of the phosphate ion of 45 mol. If the same value of $T\Delta S_{rt}$ is applied to the binding of oxalate ion to transcarboxylase,[21] for which the dissociation constant is 1.8×10^{-10} M (33 kJ mol^{-1}), equation (7) gives an apparent binding energy of 24 kJ mol^{-1} per carboxylate group after allowance for a minimal conformational entropy loss of 3 kJ mol^{-1}.

Similar results were obtained by Fersht *et al.*[19] from observed k_{cat}/K_m values in active-site mutagenesis studies. Charge–charge interactions in the tyrosyl–tRNA synthetase system ranged from 12 to 25 kJ mol^{-1} for groups interacting with the substrate pyrophosphate moiety, and up to 33 kJ mol^{-1} for the interaction between Asp78 and the substrate amino group, although in the latter case it was recognized that removal of the aspartate residue probably results in some structural reorganization in the active site.[22]

Once again, these figures are broadly consistent with the 'average' values of Andrews *et al.*,[12] which were in the range 34–48 kJ mol^{-1} for charged phosphate, amine and carboxyl groups.

IV   PRACTICAL APPLICATIONS FOR THE MEDICINAL CHEMIST

The apparent contributions of different functional groups and/or bond types to overall binding energies derived from the various studies reviewed above, are summarized in Table 20.1. Also included are corresponding values used or suggested for the overall loss of rotational and translational entropy, $T\Delta S_{rt}$, and the loss of conformational entropy resulting from restriction of free rotation, E_r.

The variations in these estimates demonstrate that they are still far from definitive, but as a rule of thumb we may state that the higher values derived using the anchor principle are the best estimates of the intrinsic binding energies of groups that are optimally aligned with matching groups in the receptor. For 'goodness of fit' calculations, on the other hand, the optimal binding contributions determined from highly specific applications of the anchor principle are not appropriate, since the absence of detailed structural data means that the summation in equation 7 is necessarily done over all the functional groups in the drug molecule, regardless of whether or not they are directly involved in binding to the receptor. The 'average' values derived previously by Andrews *et al.*[12] are thus a better starting point for 'goodness of fit' calculations.

Table 20.1 *Functional group contributions to drug–receptor interactions (kJ mol^{-1})*

Functional group type	Technique employed to determine interaction energy		
	Anchor principle	Site-directed mutagenesis	Average energy
Nonpolar (per carbon atom)	12–14	1–3	3–6
H-bonding (uncharged)	16	2–6	5–14
H-bonding (charge-assisted)	20–42	15–19	
Charged (carboxyl, amine)	18–28 +	12–25	34–48
$T\Delta S_{rt}$	12–60		58.5
E_r (internal rotation)	5–6		3

A   Assessing a lead compound

Summation of the average contributions of individual binding groups, including allowance for conformational, rotational and translational entropy terms as shown in equation (7), provides a simple back-of-the-envelope calculation of the strength of binding which might be expected for a drug forming a typical interaction with a receptor. This figure, when compared with the observed affinity of the drug for the target receptor, then gives a direct indication of the actual quality of the electrostatic and steric match between the drug and the receptor.

Binding is tighter than expected

If the observed binding of a drug to its receptor turns out to be substantially stronger than that calculated from equation (7), it is reasonable to expect that the drug structure offers a good fit to the receptor in a reasonably low energy conformation. The structure should therefore provide an excellent starting point for the development of even more bioactive compounds. A good example of this is biotin **5** (Fig. 20.3), which was the most extreme case of a positive deviation from the calculated 'average' in the original

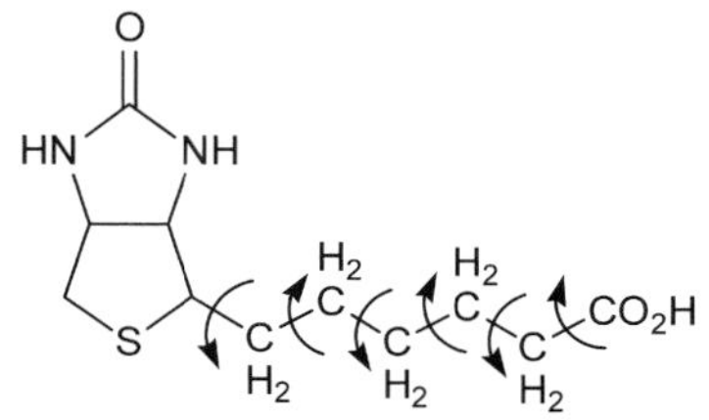

5

Fig. 20.3   Structure of biotin.

$$\Delta G_{av} = T\Delta S_{rt} + 5E_r + 8E_{Csp3} + 2E_N + E_{C=O} + 2E_{COOH}$$

$$= -58.5 + 5(-3) + 8(3) + 2(5 + 5 + 14 + 34)48$$

$$= 13.5 \text{ kJ mol}^{-1}$$

$$\Delta G_{obs} = -5.85 \log K_d = -5.85(-15) = 87.75 \text{ kJ mol}^{-1}.$$

set of 200 ligand–protein interactions studied by Andrews *et al.*[12]

Application of equation (7) to biotin (see above) gives an 'average' binding energy of 13.5 kJ mol^{-1}, whereas substitution into equation (3) of the experimentally observed binding constant to the protein avidin (10^{-15} mol l^{-1}) gives a binding energy of 87.75 kJ mol^{-1}. The difference of almost 75 kJ mol^{-1} implies an exceptionally good fit between biotin and the structure of the protein. It has since been established that this is indeed the case, with polarization of the biotin molecule by the protein actually leading to an ionic interaction where a neutral hydrogen bonding interaction had been assumed.

Binding is looser than expected

If the observed binding is significantly weaker than anticipated on the basis of an 'average' energy calculation, the fit between the drug and the receptor is less than perfect. In some cases this will be because the match between drug and receptor is less a matter of hand and glove than of 'square peg and round hole', and the only realistic option for the drug designer is to start again.

In other cases, simpler remedies may be followed:

(1) The fit may be unsatisfactory because only part of the drug is interacting with the receptor. This situation applies particularly to large drug molecules (e.g. peptide hormones), for which selective pruning of unused parts of the structure may produce simpler compounds without loss of affinity.

(2) The drug may be binding to the receptor in a comparatively high energy conformation. In this case, the design of more rigid structures which are already fixed in the desired conformation will give an increase in binding energy equivalent to the conformational energy cost of binding the more flexible analogue. Once again, an extreme case from the original set of 200 ligand–protein interactions studied by Andrews *et al.*[12] offers a simple example. Application of equation (7) to methotrexate **6** gives the methotrexate (Fig. 20.4).

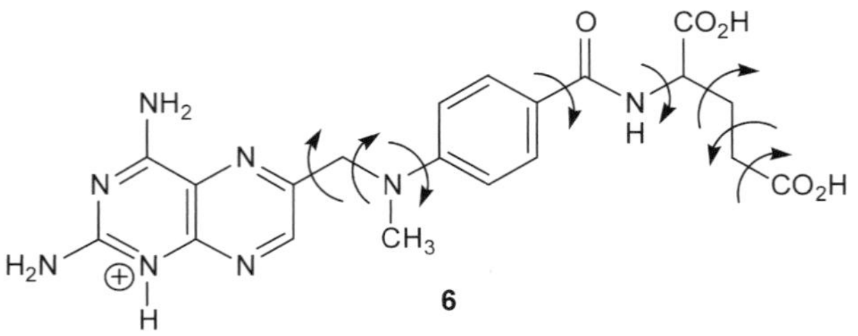

Fig. 20.4 Structure of methotrexate.

$$\Delta G_{av} = T\Delta S_{rt} + 9E_r + 5E_{Csp3} + E_{Csp2} + 7E_N + E_{C=O}$$

$$+ 2E_{COOH} + E_{N+}$$

$$= -58.5 + 9(-3) + 5(3) + 12(3) + 7(5) + 14$$

$$+ 2(34) + 48 = 130.5 \text{ kJ mol}^{-1}$$

$$\Delta G_{obs} = -5.85 \log K_d = -5.85(-11) = 64.35 \text{ kJ mol}^{-1}.$$

The fact that methotrexate binds to dihydrofolate reductase some 66 kJ mol^{-1} less tightly than anticipated suggests that despite its exceptional affinity for the enzyme ($K_d = 10^{-11}$ mol l^{-1}) the drug does not offer a good overall fit to the active site of the enzyme. Again, the direct evidence of the crystal structure verifies this suggestion, with substantial parts of the structure, including one of the carboxylic acid groups, being exposed to solvent rather than utilized in binding to the enzyme.

B   Assessing the effectiveness of substituents

Equally simple back-of-the-envelope calculations based on equation (7) can be used to predict the increase in binding energy which might be expected upon the addition of a functional group which is optimally aligned with a corresponding group in the receptor. This figure, when compared with the observed increase in affinity, gives direct feedback on whether or not the new group is actually performing the function anticipated in the design strategy.

An interesting example of how this approach can be used to assess the validity of a drug design hypothesis is provided by the receptor-based design of sialidase inhibitors as potential anti-influenza drugs.

Starting from the knowledge[23] of the structurally invariant active site of influenza A and sialidases, von Itzstein *et al.*[24] postulated that substitution of the 4-hydroxyl group of the nonselective sialidase inhibitor 2-deoxy-2,3-didehydro-D-acetylneuraminic acid **7** (Fig. 20.5) with a positively charged substituent would fill an occupied pocket lined with anionic residues. Synthesis and testing of the 4-guanidino analogue **8** revealed a reduction in K_i from 10^{-6} to 10^{-10} mol l^{-1}, equivalent (from equation 20.3) to an additional binding energy of 23 kJ mol^{-1}. Although not at the upper limit of the increments in binding energy

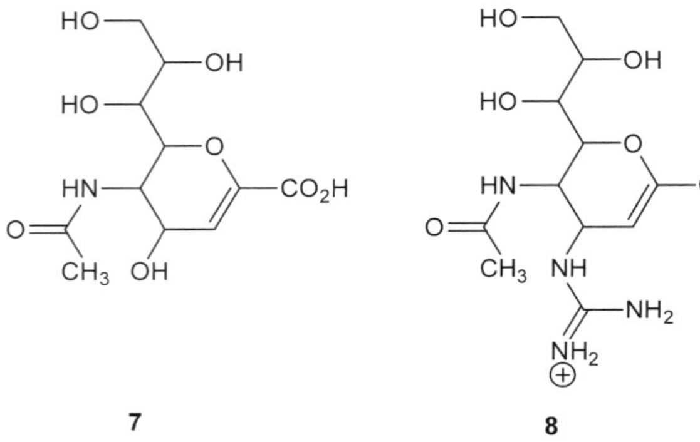

Fig. 20.5 Sialidase inhibitors: 2-Deoxy-2,3-didehydro-D-*N*-acetylneuraminic acid **7** and its 4-guanidino analogue **8**.

anticipated for well-aligned ionic interactions on the basis of the data in Table 20.1, this figure is certainly consistent with the design hypothesis, as is borne out by the crystal structure of the complex.[24] This shows that the guanidino lies between two target carboxyl groups in the active site of the enzyme, although only one appears to be optimally placed for a strong interaction.

C   HIV-protease inhibitors: a practical example

The aspartic protease HIV-protease (HIV-PR) is an essential enzyme for the replication of human immunodeficiency virus (HIV), and is widely regarded as one of the most promising targets for the design of new drugs for the treatment of AIDS. As a result, many hundreds of inhibitor classes have already been identified, and crystal structures have been determined for at least 160 complexes of inhibitors with the enzyme.[25] HIV-PR inhibitors are thus excellent test cases for the validity or otherwise of simple 'average' binding energy calculations based on equation (7).

HIV-PR is a 99 amino acid homodimer which uses a pair of aspartic acid carboxyl residues to cleave key peptide bonds in its polyprotein substrate. The likely nature of its interaction with a typical substrate, based on known enzyme-inhibitor structures, is summarized in Fig. 20.6.

Peptidomimetic inhibitors

Most HIV-PR inhibitor designs have focused on transition state analogues, i.e. analogues based on the known sequences of peptide substrates, but with the scissile bond replaced by a peptide bond isostere. Among the most effective peptide bond isosteres introduced into HIV-PR inhibitors have been the hydroxyethylene and hydroxyethyl-amine moieties (Fig. 20.7), in both of which the isosteric hydroxyl group is thought to mimic the tetrahedral

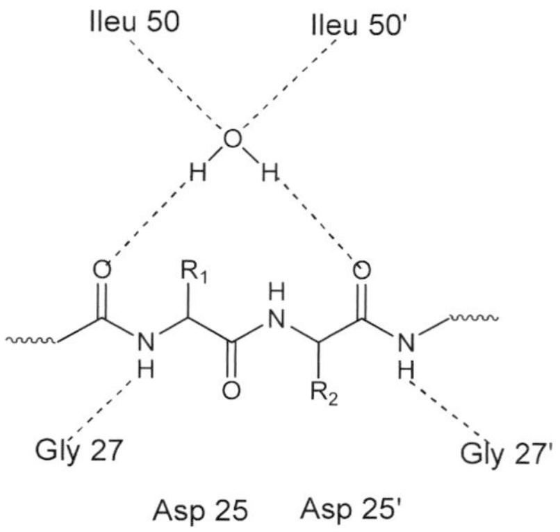

Fig. 20.6 The interaction between HIV-protease and its polyprotein substrate involves a pair of hydrogen bonds (via a buried water molecule) from substrate carbonyls of Gly 27 and Gly 27'; a pair of hydrogen bonds (via a buried water molecule) from substrate C=O groups to main-chain groups of Ileu 50 and Ileu 50'; and nonplar interactions with hydrophobic residues lining the pockets that accommodate side chains R and R'. In the transition state the carboxyl of the central (scissile) bond forms a tetrahedral hydroxy intermediate which interacts with the carboxyl groups of Asp 25 and Asp 25'.

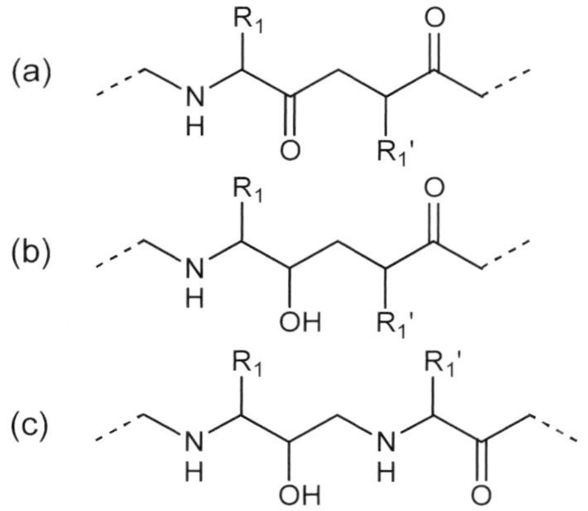

Fig. 20.7 The peptide bond (a) and its hydroxyethylene (b) and hydroxyethylamine (c) isosteres.

arrangement of the amide carbonyl in the transition state by binding to the catalytically active aspartic acid carboxyl groups of HIV-PR.

A good example of this strategy is the hydroxyethyl-amine isostere JG-365 **9** (Fig. 20.8). IC_{50} values[26] for inhibition of HIV-PR by JG-365 and its diastereomer are given in Table 20.2. Also listed are the corresponding binding energies calculated from equation (3) (using the IC_{50} values as very approximate estimates of K_d) and equation (7). Following the general principles outlined above, these data lead directly to the following conclusions:

(1) The role of the hydroxyl group is confirmed by the 20-fold difference (8 kJ mol^{-1}) between diastereomers, consistent with the 'average' contribution for a hydroxyl group of 11 kJ mol^{-1}.

(2) Inhibition by JG-365 is substantially less than would be expected if all of its functional groups made 'average' interactions with the enzyme.

(3) It should be possible to remove a significant part of the structure of JCG-365 without loss of activity.

The validity of the last conclusion is evident in the somewhat more optimized hydroxyethylamine structure Ro-31-8959 **10** (Fig. 20.9), in which inhibitory activity is improved[27] and the importance of the correct orientation of the hydroxyl group is reflected in a difference in binding energy of 14 kJ mol^{-1} between the two isomers (Table 20.2). Nevertheless, the fact that the experimentally observed binding energy of the more active isomer (55 kJ mol^{-1}) is still 50 kJ mol^{-1} less than 'average' (105 kJ mol^{-1}, equation 7) suggests that further pruning of this structure should be possible without significant loss of activity.

Nonpeptidic inhibitors

Unfortunately, although many peptidomimetic analogues, such as JG-365 and Ro-31-8959, are potent and selective inhibitors of HIV-PR, they have particularly poor oral bioavailability due to rapid breakdown in the gut and bloodstream. Many laboratories have therefore employed alternative strategies in an effort to identify simpler nonpeptidic structures that might lead to better starting

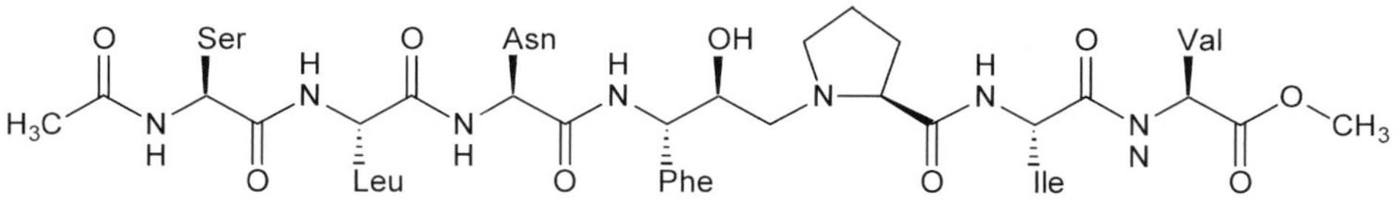

9

Fig. 20.8 Structure of the HIV-protease inhibitor JG-365 **9**.

Table 20.2 *Observed and calculated ('average') binding energies for JG-365 and RO-31-8959 diastereomers*

	Structure	IC$_{50}$ (nM)	Binding energy (kJ mol^{-1})	
			Observed (equation 3)	Calculated (equation 7)
JG-365	Ac-Ser-Leu-Asn-Phe-HEA (*S*)-Pro-Ile-Val-OMe	3.4	50	144
	Ac-Ser-Leu-Asn-Phe-HEA (*R*)-Pro-Ile-Val-OMe	65	42	144
RO-31-8959	Qua-Asn-Phe-HEA(*R*)-Diq-NHtBu	<0.4	55	105
	Qua-Asn-Phe-HEA(*S*)-Diq-NHtBu	>100	41	105

Qua, quinoline-2-carboxylic acid; Diq, [(4aS,8aS)-decahydroisoquinoline-3(S)-yl]carbonyloxy.

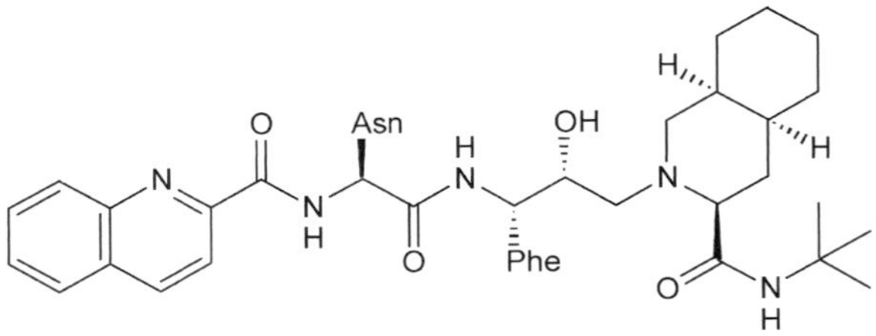

10

Fig. 20.9 Structure of the HIV-protease inhibitor Ro-31-8959.

points for design. Some of these strategies and their outcomes are outlined below.

Receptor-based lead discovery using shape-matching criteria. DesJarlais *et al.*[28] used the program DOCK to search structural databases for compounds which were complementary in shape to the active site of HIV-PR. This process led to a range of possible inhibitors, of which one of the best was bromperidol, a close relative of the antipsychotic drug haloperidol **11** (Fig. 20.10). Subsequent testing of haloperidol showed weak inhibition ($K_i = 100 \, \mu$M) of the enzyme. Does this mean that haloperidol should be a good lead? Unfortunately, it does not. In fact, the observed binding energy of haloperidol (23 kJ mol^{-1}, equation 3) is considerably less than would be expected for an 'average' interaction (67 kJ mol^{-1}, equation 7), suggesting that the fit of the drug to HIV-PR is far from optimal. This conclusion is borne out by the fact that three crystallographically observed binding modes of a closely related haloperidol derivative are all quite distinct from that initially deduced from the DOCK study.[29]

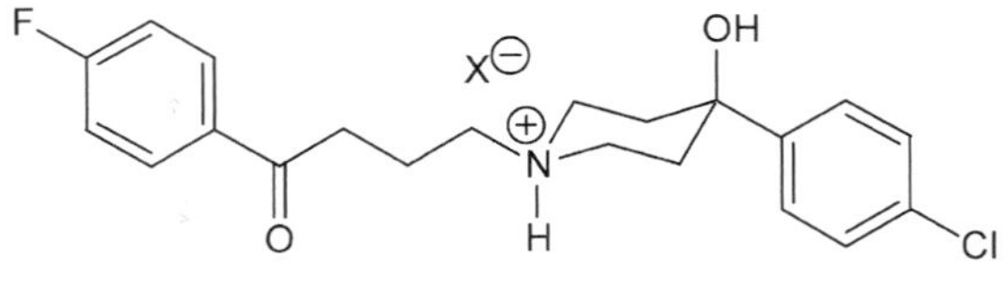

Receptor-based lead discovery using functional group matching criteria. X-ray crystallographic data for various peptidomimetic inhibitors binding to HIV-PR have revealed a common binding mode incorporating the following features:

(1) A central hydroxyl or diol group which binds to the catalytic aspartic acid carboxyls.
(2) Hydrogen bond donors (generally amide NH groups) on either side of the central hydroxyl responsible for binding to main-chain carbonyls in the active site of the enzyme.
(3) Hydrophobic groups which fill the P1 and P1' pockets of the active site of the enzyme.
(4) A buried water molecule that forms hydrogen bonds to two main-chain amide NH groups in the enzyme and two carbonyl oxygens in the inhibitor.

These constant features of the peptidomimetic inhibitors, all of which are consistent with the substrate binding mode illustrated in Fig. 20.6, clearly provide a most valuable guide to the design of nonpeptidic analogues, and numerous pharmacophore-based studies have used these data as the starting point. Of particular interest has been the prospect of incorporating the buried water molecule directly into the inhibitor structure, thus maintaining the contribution of the two hydrogen bonds mediated by the water molecule, but eliminating the entropic cost of immobilizing it in the active site of the enzyme.

This strategy was specifically adopted by Bures *et al.*[30] who developed a pharmacophore comprising the central hydroxyl group, the adjacent hydrogen bond donors, the buried water molecule and an optional hydrophobic moiety. A systematic search of structural databases using this pharmacophore led to the discovery of a series of dibenzophenones (Table 20.3), which inhibited HIV-PR at concentrations between 10 and 100 μM.

11

Fig. 20.10 Structure of the antipsychotic drug haloperidol **11**.

Table 20.3 *Dibenzophenone inhibitors based on a pharmacophore model derived from the binding modes of peptidomimetic inhibitors (adapted from reference 25)*

R_1'	R_2	R_3	IC$_{50}$ (μM)	Binding energy (kJ mol^{-1})		
				Observed (equation 3)	Calculated (NH$_3^+$)	Calculated NH$_2$
OCH$_2$COOEt	Cl	CH$_2$NH$_2$	11	29	117	31
OCH$_2$COOH	Cl	CH$_2$NH$_2$	85	24	135	49
OCH$_2$COOEt	H	Cl	15	28	68	25

Comparison of experimental and 'average' binding energies for these compounds leads to an interesting question on how best to apply 'average' energy calculations. If we were to assume that the two positively charged amines formed ionic interactions with appropriate anionic groups in the active site, then the observed binding would be between 40 and 111 kJ mol^{-1} less than anticipated (Table 20.3), suggesting a less than optimal match to the active site. However, since the design hypothesis is based on the premise that only a main-chain hydrogen bonding interaction is available to the amines, the numbers should be recalculated for neutral amine groups (5 kJ mol^{-1} rather than 48 kJ mol^{-1}). These numbers (Table 20.3) suggest that the dibenzophenone nucleus is a reasonable starting point for further optimization, although incorporation of a charged carboxyl group in the substituent is clearly not the way to go!

A similar pharmacophore model, but based on a diol as the catalytic aspartate binding group, led Lam *et al.*[31] to a series of cyclic urea inhibitors (Table 20.4) in which the urea carbonyl takes the place of the buried water molecule. In this case, the binding of the lead diallyl compound (49 kJ mol^{-1}) is already better than might be anticipated for an 'average' interaction (38 kJ mol^{-1}), suggesting that the core structure is indeed a good match to the active site of HIV-PR. The quality of this match is confirmed by the crystal structure, which shows that the diol oxygens are positioned to interact with the aspartyl carboxylates, while the urea carbonyl forms two hydrogen bonds to main-chain amides.

Further elaboration of the lead by replacement of the allyl groups shows that while the quality of the match is retained in the dicyclopropylmethyl analogue, extension

of these side chains to aromatic substituents does not enhance binding to the extent anticipated, although the data suggest that there is certainly space in the active site for further optimization of these substituents.

Bioassay-based lead discovery using high throughput screening of known chemicals. Screening of large numbers of compounds in the Parke-Davis collection against HIV-PR led[32] to the discovery of the series of pyran-2-ones shown in Table 20.5. The lead compound in this series has an experimental binding energy of 32 kJ mol^{-1} (compared with a calculated 'average' value of only 14 kJ mol^{-1}),

Table 20.4 *Simple nonpeptide HIV-PR inhibitors based on a pharmacophore derived from peptidomimetic inhibitors*[31]

R	K_i (nM)	Binding energy (kJ mol^{-1})	
		Observed (equation 3)	Calculated (equation 7)
Allyl	4.7	49	38
Cyclopropylmethyl	2.1	51	42
β-Naphtylmethyl	0.31	56	62
p-Hydroxymethylbenzyl	0.27	56	58

Table 20.5 *Nonpeptide HIV-PR inhibitors discovered by screening a known compound library* [32]

R	R′	IC$_{50}$ (μM)	Binding energy (kJ mol^{-1})	
			Observed (equation 3)	Calculated (equation 7)
C$_6$H$_5$	C$_6$H$_5$	3.0	32	14
C$_6$H$_5$	CH$_2$C$_6$H$_5$	1.67	34	14
C$_6$H$_5$	CH$_2$CH$_2$C$_6$H$_5$	1.26	35	15
C$_6$H$_5$	CH$_2$CH$_2$CH$_2$C$_6$H$_5$	1.41	34	15
C$_6$H$_4$(4-OCH$_2$COOH)	CH$_2$CH$_2$C$_6$H$_5$$_5$	0.16	40	48

making it the best structural match to the HIV-PR active site of the various leads considered here. This is confirmed by the crystal structure of the closely related benzyl derivative, which shows that the enolic hydroxyl group binds to both aspartic carboxyls, while the lactone moiety of the inhibitor takes the place of the buried water molecule, forming multiple interactions with the amide NH groups of Ile$_{50}$ and Ile$_{150}$.

Further elaboration of this lead by homologous extension of R gives only a very slight increase in the observed binding, and the same trend is evident in the calculated ('average') binding energies. This is because the van der Waals interaction associated with each additional methylene is balanced by a corresponding loss of conformational entropy. In contrast to this observation, extension at the R position by introducing a tethered carboxyl group, although giving an order of magnitude improvement in binding, does not provide anything like the increase that would be anticipated if the carboxyl group were to interact as postulated with an arginine residue in the active site.

REFERENCES

1. Bajorath, J., Kraut, J., Li, Z., Kitson, D.H. and Hagler, A.T. (1991) Theoretical studies on the dihydrofolate reductase mechanism: Electronic polarization of bound substrates. *Proc. Natl Acad. Sci. USA* **88**: 6423–6426.
2. Taylor, R., Kennard, O. and Versichel, W. (1983) Geometry of the N—H···O=C hydrogen bond 1. Lone-pair directionality. *J. Am. Chem. Soc.* **105**: 5761–5766.
3. Taylor, R., Kennard, O. and Versichel, W. (1984) Geometry of the N—H···O=C hydrogen bond. 2. Three-center ('bifurcated') and four-center ('trifurcated') bonds. *J. Am. Chem. Soc.* **106**: 244–248.
4. Murray-Rust, P. and Glusker, J.P. (1984) Directional hydrogen bonding to sp^2- and sp^3-hybridized oxygen atoms and its relevance to ligand-macromolecule interactions. *J. Am. Chem. Soc.* **106**: 1018–1025.
5. Tintelnot, M. and Andrews, P.R. (1989) Geometries of functional group interactions in enzyme–ligand complexes: Guides for receptor modelling. *J. Comput. Aided Mol. Design* **3**: 67–84.
6. Taylor, R., Kennard, O. and Versichel, W. (1984) The geometry of the N–H···O=C hydrogen bond. 3. Hydrogen-bond distances and angles. *Acta Crystallogr.* **B40**: 280–288.
7. Gorbitz, C.H. (1989) Hydrogen-bond distances and angles in the structures of amino acids and peptides. *Acta Crystalogr.* **B45**: 390–395.
8. Sharp, K.A., Nicholls, A., Friedman, R. and Honig, B. (1991) Extracting hydrophobic free energies from experimental data: Relationship to protein folding and theoretical models. *Biochemistry* **30**: 9686–9697.
9. Page, M.I. (1977) Entropy, binding energy, and enzymic catalysis. *Angew. Chem. Int. Ed. Engl.* **16**: 449–459.
10. Page, M.I. and Jencks, W.P. (1971) Entropic contributions to rate accelerations in enzymic and intramolecular reactions and the chelate effect. *Proc. Natl Acad. Sci USA* **68**: 1678–1683.
11. Jencks, W.P. (1981) On the attribution and additivity of binding energies. *Proc. Natl Acad. Sci. USA* **78**: 4046–4050.
12. Andrews, P.R., Craik, D.J. and Martin, J.L. (1984) Functional group contributions to drug–receptor interactions. *J. Med. Chem.* **27**: 1648–1657.
13. Searle, M.S. and Williams, D.H. (1992) The cost of conformational order: Entropy changes in molecular associations. *J. Am. Chem. Soc.* **114**: 10690–10697.
14. Andrews, P.R. (1993) Drug–receptor interactions. In Kubinyi, H. (ed.). *3D QSAR in Drug Design: Theory, Methods and Applications*, pp. 13–40. ESCOM Science, Leiden.
15. Wells, J.A. (1990) Additivity of mutational effects in proteins. *Biochemistry* **29**: 8509–8517.
16. Wolfenden, R. and Kati, W.M. (1991) Testing the limits of protein-ligand binding discrimination with transition-state analogue inhibitors. *Acc. Chem. Res.* **24**: 209–215.
17. Kati, W.M. and Wolfenden, R. (1989) Contribution of a single hydroxyl group to transition-state discrimination by adenosine deaminase: Evidence for an 'entropy trap' mechanism. *Biochemistry* **28**: 7919–7927.
18. Wilson, D.K., Rudolph, F.B. and Quiocho, F.A. (1991) Atomic structure of adenosine deaminase complexed with a transition-state analog: Understanding catalysis and mutations. *Science* **252**: 1278–1284.
19. Fersht, A.R., Shi, J.P., Knill-Jones, J., Lowe, D.M., Wilkinson, A.J., *et al.* (1985) Hydrogen bonding and specificity analysed by protein engineering. *Nature* **134**: 235–238.

20. Levine, D., Reid, T.W. and Wilson, I.B. (1969) The free energy of hydrolysis of the phosphoryl–enzyme intermediate in alkaline phosphatase catalyzed reactions. *Biochem.* **8**: 2374–2380.

21. Northrop, D.B. and Wood, H.G. (1969) Transcarboxylase VII. Exchange reactions and kinetics of oxalate inhibition. *J. Biol. Chem.* **244**: 5820–5827.

22. Ward, W.H.J., Timms, D. and Fersht, A.R. (1990) Protein engineering and the study of structure-function relationships in receptors. *Trends Pharmacol. Sci.* **11**: 280–284.

23. Colman, P.M. (1989) Neuraminidase: enzyme and antigen. In Krug, R.M. (ed.). *The Influenza Virus*, pp. 175–218. Plenum Press, New York.

24. Von Itzstein, M., Wu, W.-Y., Kok, G.B., Pegg, M.S., Dyason, J.C., *et al.* (1993) Rational design of potent sialidase-based inhibitors of influenza virus replication. *Nature* **363**: 418–423.

25. Wlodawer, A. and Erickson, J.W. (1993) Structure-based inhibitors of HIV-1 protease. *Annu. Rev. Biochem.* **62**: 543–585.

26. Rich, D.H., Sun, C.-Q., Vara Prasad, J.V.N., Pathiasseril, A., Toth, M.V., *et al.* (1991) Effect of hydroxyl group configuration in hydroxyethylamine dipeptide isosteres on HIV protease inhibition. Evidence for multiple binding modes. *J. Med. Chem.* **34**: 1222–1225.

27. Krohn, A., Redshaw, S., Ritchie, J.C., Graves, B.J. and Hatada, M.H. (1991) Novel binding mode of highly potent HIV-proteinase inhibitors incorporating the (R)-hydroxyethylamine isostere. *J. Med. Chem.* **34**: 3340–3342.

28. DesJarlais, R.L., Seibel, G.L., Kuntz, I.D., Furth, P.S., Alvarez, J.C., Ortiz de Montellano, P.R., DeCamp, D.L., Bake, L.M. and Craik, C.S. (1990) Structure-based design of nonpeptide inhibitors specific for the human immunodeficiency virus 1 protease. *Proc. Natl Acad. Sci. USA* **87**: 6644–6648.

29. Rutenber, E., Fauman, E.B., Keenan, R.J., Fong, S., Furth, P.S., *et al.* (1993) Structure of a non-peptide inhibitor complexed with HIV-1 protease. *J. Biol. Chem.* **268**: 15343–15346.

30. Bures, M.G., Hutchins, C.W., Maus, M., Kohlbrenner, W., Kadam, S. and Erickson, J.W. (1990) *Tetrahedron Comp. Methodol.* **3**: 673–680.

31. Lam, P.Y.S., Jabhav, P.K., Eyermann, C.J., Hodge, C.N., Lee, Y.R., *et al.* (1994) Rational design of potent, bioavailable, nonpeptide cyclic ureas as HIV protease inhibitors. *Science* **263**: 380–384.

32. Vara Prasad, J.V.N., Para, K.S., Lunney, E.A., Ortwine, D.F., Dunbar, Jr, J.B., *et al.* (1994) Novel series of achiral, low molecular weight, and potent HIV-1 protease inhibitors. *J. Am. Chem. Soc.* **116**: 6989–6990.

21

COMPOUND PROPERTIES AND DRUG QUALITY

Christopher A. Lipinski

Our achievements speak for themselves. What we have to keep track of are our failures, discouragements, and doubts. We tend to forget the past difficulties, the many false starts, and the painful groping. We see our past achievements as the end result of a clean forward thrust, and our present difficulties as signs of decline and decay.
Eric Hoffer

I INTRODUCTION

Compound properties required for a successful orally active drug have not changed in the last decade. A compound now, as in the past, must still have adequate solubility and permeability relative to its clinical potency to succeed as an oral drug. However, what has changed in the last decade is the practice of organic synthesis as applied to medicinal chemistry. As a result of synthetic chemistry practice changes in the last decade, compounds now typically tend to be larger and more lipophilic or larger with more hydrogen-bonding functionality than in the previous decade. These changes respectively translate into poorer aqueous solubility and poorer intestinal permeability than in the previous era. As a result, there is a greater divergence in compound properties between the early discovery stage and the discovery–development interface than in the era covered by the previous edition of *The Practice of Medicinal Chemistry*. At the risk of repetition, the vast majority of recent chemistry changes is in the early discovery stage and very few are at the discovery–development interface stage.

Accordingly, the discussions in the previous edition of this book on the properties of crystalline, well-characterized compounds intended for oral dosing which are entering the development phase, still remain valid. What is new is the phenomena of early stage discovery compounds with properties frequently far removed from those in the discovery–development interface stage. This chapter focuses on the recent changes in synthetic chemistry practices, shows how these changes impact on compound properties, and provides guidance on how the pattern of early compound properties can be improved towards those described in the previous edition of *The Practice of Medicinal Chemistry*. Changes in medicinal chemistry synthetic practice can be broadly characterized into two main areas: changes related to synthesis of one compound at a time and moderate output parallel synthesis and the more radical changes related to combinatorial chemistry. This chapter primarily focuses on issues related to the design of combinatorial libraries. A highly technical discussion in book chapter format tends to rapidly become out of date. Thus, the main focus of both chapter sections will be on general principles, which tend not to be well covered in primary journal technical publications.

II COMBINATORIAL LIBRARIES

The design of combinatorial libraries with drug-like properties is a trade-off between efficient chemistry and therefore better numerical compound output on one side and increasingly difficult chemistry but better physicochemical properties on the other. This trade-off is not solely a technical issue but is strongly influenced by people attitudes and organizational timing issues. The author believes that there is a hierarchy of properties that need to be controlled in a quality library. Simply put, those properties that are under poorest chemistry control and that are the most difficult to fix in chemistry optimization should be given highest priority. In this context, poor chemistry control equates to chemistry SAR which is difficult to control. Avoiding chemistry functionality known to be associated with toxicity is an example of chemistry SAR that is under good control. Figure 21.1 shows a chemical structure that we call 'the molecule from hell'. It was created by my computational colleague Dr. Beryl Dominy and contains chemistry functionality that in a single molecule triggers 27 different filters for chemistry functionality known to be associated with mutagenic activity. A molecule like this is used to test that a computational filter will detect the undesirable chemistry functionality. An example of chemistry SAR that is under poor control is the SAR relating to aqueous solubility and gastrointestinal tract permeability. Physicochemical profiling of solubility and permeability has recently been reviewed.[1] These properties are difficult but not impossible to control by medicinal chemistry. In a combinatorial chemistry setting, poor aqueous solubility is almost universally a problem[2] and must be explicitly addressed by both computational and experimental intervention strategies.

A Library design

Library design viewed as a computational process and the experimental implementation of a library design can result in quite different outcomes. In a library design one might consider a core template which can be modified by a variety of substituent groups. The physicochemical profile of the library is determined by the chemical structure of the core and the range of physicochemical properties of the substituent groups. For example, there might only be one core in a design but many substituent groups. As a result

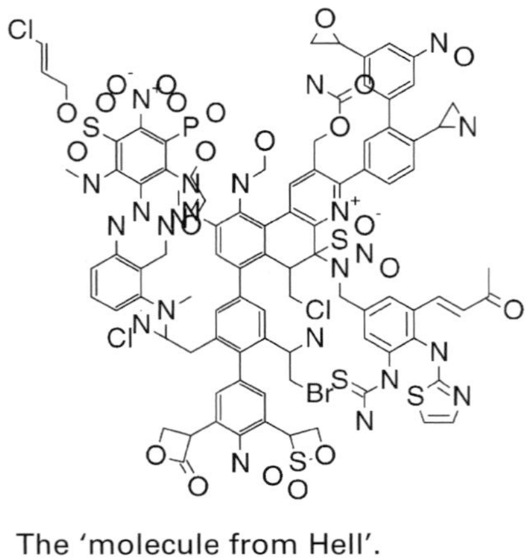

the physicochemical envelope of the library is very much influenced both by the availability of the chemical pieces that will become the substituents and by the physicochemical properties of those substituents. If the experimental implementation (synthesis) of a design is 100% successful then the actual experimental physicochemical envelope profile is the same as in the library design. In actuality, the chemical synthesis is seldom 100% successful. Thus what matters is whether there is any bias in the experimental synthesis success rate such that the experimental library differs significantly from the design library. A priori, the chemical synthesis success rate is usually not known so that, in general, libraries are over designed. More compounds are designed than will actually ever be synthesized.

Experimental synthesis success rate

The experimental synthesis success rate almost always biases the experimental library so that the physicochemical profile relative to aqueous solubility is significantly worse than that in the design library. The fewest chemistry problems are found in lipophilic substituent moieties lacking polar functionality. In almost all cases polar functionality is electron withdrawing so that reactions of a substituent moiety like reactive amination, acylation or nucleophilic substitution proceed more poorly. Blocking and deblocking of a polar group adds to the complexity and length of a synthesis. As a result polar reagents which require blocking and deblocking are experimentally selected against. Robotic pipettors perform poorly or not at all on slurries of precipitates so any factor that increases the insolubility of a reagent in an organic solvent will bias the library outcome.

Poor solubility and library design

Poor solubility in an organic solvent arises from two quite different factors: solvation energy and disruption of intermolecular crystal packing forces in the crystalline reagent. Solvation of a lipophilic reagent in an organic solvent is typically not a problem. However, disruption of intermolecular crystal packing forces is very much a problem in an organic solvent, especially if the reagent has a high melting point. This type of problem is most likely to be present in a reagent with polar hydrogen bond acceptor/donor functionality. Thus the reagent insolubility problem tends to bias a library towards a more lipophilic and hence more aqueous insoluble profile. To accommodate diversity considerations a range of substituent moieties is selected. A large structural range translates into a broad molecular weight distribution. The combination of reagent solubility and diversity considerations results in an experimental library that is biased towards higher lipophilicity and higher molecular weight relative to the design library. The bias occurs because high lipophilicity and high

Fig. 21.1 The 'molecule from Hell'.

molecular weight are the worst combination of 'Rule-of-5' parameters in terms of leading to poor aqueous solubility.

Importance of the synthesis rate-determining step

Effectiveness of ADME (absorption, distribution, metabolism, excretion) design implementation depends on whether chemistry protocol development or chemistry production is rate determining. If chemistry production is rate determining there will be excess validated protocols relative to library production. This means that protocols can be prioritized as to their ADME attractiveness and the least attractive protocols from an ADME perspective may never be translated into actual library production. However, protocol development and not library production is often the rate-determining step. This eventuality is unfortunate because there is an understandable reluctance to discontinue chemistry synthetic efforts because of a poor ADME experimental profile if considerable chemistry effort has already been expended. Consider the following situation. The effort towards library production is 70% complete. The experimental ADME profile is poor. Would you discontinue completion of library synthesis because of poor ADME if 70% of the chemistry effort had already been completed? Thus a key issue becomes how much chemistry experimental effort takes place before exemplars are experimentally profiled in ADME screens?

If protocol development is rate determining

If protocol development is rate determining, the effectiveness of ADME experimental assays depends very much on how the early exemplars are synthesized. In theory, the most effective method would be to obtain a well-spaced subset of the library in an experimental design sense. A traditional, noncombinatorial, synthesis would accomplish this but would not fit in well with a combinatorial optimization process. A possible means around this problem is to institute some type of early automated clean-up of combinatorial exemplars from partially optimized reaction schemes. This is not a tidy solution because the most efficient process would be an automated clean-up on the entire library after the optimization process was complete.

The least effective method of providing samples for experimental ADME profiling is a late-stage selection from the optimized combinatorial libraries. It is least effective, not because of chemistry efficiency considerations. A late-stage selection from the optimized combinatorial libraries is actually very chemistry efficient. However, the inefficiency comes from the people aspect. The data come too late to prevent poor ADME quality compounds from being made. The timing problem in obtaining combinatorial exemplars is one of the driving forces that makes computational ADME profiling so attractive.

Poor ADME properties — business aspects

There is a business aspect here for a company selling combinatorial libraries if the combinatorial libraries contain compounds experimentally verified to have poor ADME properties. What do you do with the poor ADME compounds? Do you sell them to an unsuspecting customer or do you swallow the cost by not selling these compounds? At the present time there is still a way out. Not all customers want all compounds to have good ADME properties. There is still a sizeable market (perhaps 30% but dwindling) for compounds perceived to be possibly active in an *in vitro* assay irrespective of ADME properties. The argument is that one cannot proceed anywhere without that first 'lead'. So try to get the 'lead' first and then worry about ADME later. I am personally not a proponent of this viewpoint, but I can see how people who are involved in screening against very difficult targets might adopt this position.

If library production is rate determining

If library production is rate determining then the effectiveness of experimental ADME assays becomes much simpler. Operationally, it is fairly straightforward to de-prioritize a library that profiles poorly in ADME assays if there are more protocols than can be translated into production. One simply executes the best protocols in an ADME sense and drops from production those that profile the worst. There is a clear message here if one really cares about ADME properties. The manning and effort should be greater at the protocol validation stage as opposed to the library production stage. This ADME-derived message is contrary to that which one receives at the vendor trade shows. The vendor trade shows tend to emphasize the production side. This is hardly surprising if there is much more money to be made in selling production hardware and technology as opposed to selling tools for protocol design and validation.

Relative importance of ADME assays

I believe that ADME assays are not all equal in terms of contributing to drug quality. ADME design is most important for those properties under the poorest chemistry control. Good chemistry control equates with chemists' ability to control SAR. Good control of SAR means that a chemist can make a small molecular change with a resultant large change in the measured property. Poor control of SAR means a loss of relationship between molecular structure and the measured property. I believe it is inadvisable to totally filter out compounds with poor properties if they can easily be fixed by chemistry. The goal in drug research is to discover inherently active compounds with appropriate ADME properties. There is little value in being so restrictive that few inherently active compounds will be discovered.

III CHEMISTRY CONTROL OF INTESTINAL PERMEABILITY

Chemistry control of intestinal permeability is poor. The good news is that, in general, poor permeability is not a problem in combinatorial libraries. One really has to go out of one's way to introduce enough polar functionality in a combinatorial compound to make a compound impermeable via passive *trans*-cellular processes (the most common absorptive pathway). The bad news is that if a compound has poor intestinal permeability there is virtually nothing to fix the problem in terms of pharmaceutical formulation technology. Chemistry control of intestinal permeability is poor because, except for a few very specific exceptions, chemistry SAR is blunt.

Improving permeability

The best general guide for improving permeability is to reduce the polar surface area or to reduce the sum of the hydrogen-bond accepting and donating moieties in the compound. Some authors[3] recommend a polar surface area (PSA) cutoff of about 120 $Å^2$. In my experience a compound with PSA of less than about $140-150 Å^2$ should have reasonable intestinal permeability unless something else is wrong. The 'something else is wrong' includes excessive basicity (above pK_a 11.5) or excessive acidity (below pK_a 3) or log D at pH 6.5 (or 7) below about 0.0. However, excessive acidity may be compatible with acceptable permeability if the anion exhibits extensive charge delocalization, e.g. a vinylogous enolic acid-like system. In cases like this acceptable permeability can occur via a charge delocalized ion pair. The blunt SAR feature comes from the phenomenon that permeability improves only gradually as a physicochemical property is moved in a desired direction. One does not see a sudden improvement in permeability from a properly positioned methyl group as one might see with respect to *in vitro* SAR. Biological phenomena might be responsible for apparent poor permeability. For example, the compound is extensively metabolized by cytochrome P-450 3A4 in the gut intestinal wall or the compound is actively effluxed by *p*-glycoprotein in the gut wall. Unfortunately, both these biological systems tend to show broad substrate specificity so the chemistry SAR can be blunt.

Hydrogen bonding and permeability

Permeability is intensely affected by the presence of an intramolecular hydrogen bond.[4] In this area tight SAR can occur. Any structural change that causes an internal hydrogen bond to form (or that prevents one from forming) can have a dramatic effect on permeability. The effect of a single intramolecular hydrogen bond on increasing permeability is great and can easily be a factor of 10 or more. This type of structural effect is currently not predictable computationally. Using bond order and structure one can predict the possibility that an intramolecular hydrogen bond is possible. However, to my knowledge, no existing commercially available, reasonably fast batch mode programme can calculate whether an intramolecular hydrogen bond is likely to form in reality, i.e. that the hydrogen bond formation is energetically likely. In theory, some type of rule-based calculation might be able to predict intramolecular hydrogen bonding, but this would require a sizeable database of experimental data.

Intramolecular hydrogen bonds

Experimental assays can correctly predict the permeability enhancing effect of an intramolecular hydrogen bond. These assays can be cell based as in Caco-2 cell culture assays or MDCK cell assays; or they can be nonbiological as in a traditional log P or log D, a parallel artificial membrane permeation assay (PAMPA) assay[5] or even a surface tension measurement.[6] Experimental assays work because they are sensitive to the actual hydrogen-bonding properties of a compound. There is considerable variation in the extent to which biological and nonbiological assays have been adopted in the pharmaceutical industry as a predictor for intestinal permeability. In my opinion, this may partly reflect differences in chemistry across organizations. An organization with many conformationally flexible compounds bearing hydrogen bond donor and acceptor groups might be particularly likely to use a nonbiological type of experimental assay because of the need to identify the more permeable compounds due to intramolecular hydrogen bonding. Conversely, an organization with many heterocyclic compounds having limited possibilities for intramolecular hydrogen bonding might not see the need for this type of assay.

Permeability testing

There is no uniformity as to how permeability testing is carried out in the pharmaceutical industry. This suggests that there may not be a great deal of difference in the effectiveness of the various experimental and computational permeability prediction methods. The people factor can easily be as (or even more) important than purely technical factors. The goal in a permeability assay or calculation is to influence chemistry behaviour, i.e. to direct chemistry synthesis towards more permeable compounds. Thus the best assay or calculation may be the one that chemists (for whatever reason) believe and act on. This means that issues such as capacity, ease of use, ease of interpretation and internal credibility can be the deciding factor for effectiveness.

IV CHEMISTRY CONTROL OF AQUEOUS SOLUBILITY

Chemistry control of aqueous solubility is poor. The good news is that if a compound has poor aqueous solubility, methods do exist to fix the problem in terms of pharmaceutical formulation technology. However, these are always expensive in time and manning, and depending on the degree of the solubility problem, may have limited or no precedent in terms of existing approved products. By far the preferred solution to poor solubility is to fix it in chemistry. Formulation fixes are a last resort. The bad news is that, in general, poor aqueous solubility is by far the most common ADME problem in combinatorial libraries. It takes no effort at all to introduce poor aqueous solubility into a combinatorial compound library. The combination of high molecular weight and high lipophilicity outside the 'Rule-of-5' limits[7] is an almost certain guarantee of poor aqueous solubility. Based on our experimental screening, lipophilicity above the 'Rule-of-5' log P limit of 5 by itself carries with it a 75% chance of poor aqueous solubility.

The definition of poor solubility

The definition of 'poor solubility' has reduced in the combichem/HTS era. A classic pharmaceutical science textbook might have defined poor solubility as anything below a solubility of 1 g mL^{-1} at pH 6.5 (or pH 7). Currently, most drug researchers would be very excited by a solubility as high as 1 g mL^{-1}. In general, with average permeability and a projected clinical potency of 1 mg kg^{-1}, a drug needs a minimum aqueous solubility of 50–100 μg mL^{-1} to avoid the use of nonstandard solubility fixing formulation technology. We find the guidelines published by Pfizer's Curatolo on maximum absorbable dose to be an excellent guide for the combination of permeability, solubility and potency required in an orally active drug.[8]

Figure 21.2 is a bar graph illustrating the minimum acceptable solubility as a function of compounds projected clinical potency and permeability in medicinal chemistry. The middle set of bars show that a compound has to have a minimum thermodynamic solubility of 52 μg mL^{-1} when the permeability is average (avg K_a) and the projected clinical potency is 1 mg kg^{-1}.

Aqueous solubility and blunt SAR

Chemistry control of aqueous solubility is poor because, except for a few very specific exceptions, chemistry SAR is blunt. In this respect, control of solubility like that of permeability is poor. Solubility due to excessive lipophilicity improves only gradually as the lipophilicity is moved in the desired downward direction. Trying to decrease lipophilicity by incorporating polar functionality may or may not work. The potential solubility improvement attendant on introdu-

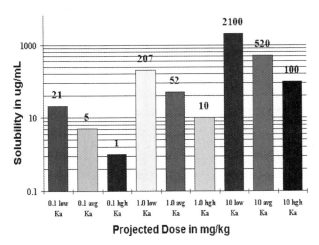

Fig. 21.2 Minimum acceptable drug solubility as a function of projected clinical potency (0.1 to 10 mg kg^{-1}) and intestinal permeability (low–avg–high K_a).

cing polar hydrophilic functionality can easily be more than counterbalanced by a decrease in solubility due to increased intermolecular crystal packing forces arising from the new polar functionality. In our experimental solubility screening, about 60% of poor solubility is unrelated to excessive lipophilicity. This 60% of poor solubility arises from high crystal packing forces due to intramolecular interactions in the solid crystal state that make disruption of the crystal energetically costly. Thus the blunt SAR feature in solubility comes from the phenomenon that solubility improves only gradually (or not at all) as a physicochemical property (lipophilicity) is moved in a desired downward direction.

Changing the pK$_a$

Changing the pK_a of an acidic or basic group in a molecule so that more of the compound exists in the ionized form at physiological pH lowers log D (at about pH 7) and, in general, should improve aqueous solubility. The improvement in solubility is limited, however, if the solubility of the neutral form of the compound (the inherent solubility) is very low. The situation is worsened if the starting pK_a is far from 7. We find this to be a particular problem with weak bases. Weakly basic pyridines, quinolines, quinazolines and thiazoles seem to be frequent members of combinatorial libraries.

The extent of poor aqueous solubility may be experimentally underestimated in a combinatorial library. No combinatorial library is purified by traditional crystallization. The vast majority of compounds purified by an automated process will be isolated in amorphous form. Compounds in an amorphous solid form exist in a much higher energy state than a true crystalline solid, and aqueous solubilities of amorphous solids are always higher than those of crystalline solids. This phenomenon may only be recognized if there is a high degree of interest in a combinatorial compound. The combinatorial compound is scaled-up and purified by crystallization. The newly crystallized compound can easily be an order of magnitude more insoluble, and hence more poorly absorbed than the original sample.

Improving aqueous solubility

Aqueous solubility can be improved by medicinal chemistry despite the blunt SAR feature and our internal record has been quite successful in this respect. However, to improve solubility requires commitment to a combination of computational and experimental interventions and a real effort on the part of chemists to incorporate solubility information into synthesis design. The importance of rapid experimental feedback is particularly important given the current inability to predict computationally poor solubility arising from crystal packing interactions. It is critical not to miss a serendipitous improvement in solubility attendant on a molecular change. Owing to the blunt SAR feature, the easiest way to improve solubility with respect to library design is to try to design the best solubility profile right at the start.

V   *IN VITRO* POTENCY AND CHEMISTRY CONTROL

In vitro potency has always been under excellent chemistry control; the hallmark of good control being tight chemistry SAR. With respect to combinatorial chemistry several exceptions should be noted. Compounds are often encountered as leads in HTS screening that can be characterized as 'phony HTS leads'. These types of compounds should be avoided at all cost as templates in combinatorial chemistry. A common attribute of these leads is that the chemistry SAR is flat and fuzzy if they are subjected to lead optimization. Large chemistry changes can be made with only very small changes in activity. Often these types of problems can be avoided by similarity searches on the initial apparent lead. A loss or gain of activity related to a small structural change over the initial lead is a good sign. A change in activity of a factor of 10 between two compounds differing by only a single methyl group is a classic example of good SAR. A flat SAR among analogues is not a good sign. Often these 'leads' are not very active, perhaps in the low micromolar range. Sometimes the exact same compound appears active in multiple HTS screens. Alternatively, as we have observed, members of a well-defined structural series appear as apparent HTS actives but are not necessarily the exact same compound across different HTS screens. Some phony HTS 'leads' are removable by compound quality filters. The mechanisms for the 'phony HTS lead' phenomenon are largely unknown. Some of our experience

suggests that nonspecific lipid membrane perturbation of intramembrane protein receptor targets could be a factor. Experienced medicinal chemistry 'intuition' works quite well in avoiding 'phony HTS leads' but it would clearly be advantageous to use computational filters for this problem. Software, even if it worked no better than chemists' intuition, would be advantageous from the people viewpoint. Biologists generally do not understand or appreciate chemists' exquisitely tuned sense of what constitutes a 'good' chemical structure. Thus many hard feelings and miscommunication between chemists and biologists could be avoided by a computation that merely mimics chemists' structural intuition.

Lead complexity

The other major limitation on good *in vitro* potency control relates to the complexity of the apparent lead. Some apparent leads simply lack a critical structural feature and cannot be easily optimized by traditional medicinal chemistry means. These are most likely to be detected among very weak actives in traditional HTS screens and form the majority of actives in techniques such as SAR by NMR. The increasing probability of a missing critical piece as MWT decreases probably sets the lower size range for lead discovery libraries. Several recent publications provide an excellent perspective on lead generation libraries.[9,10] The general theme is that the properties of a lead must allow for the almost inevitable increase in molecular weight and lipophilicity that accompany *in vitro* activity optimization. Andrews' binding energy[11] can be used as a rough indicator of functional group complexity, i.e. does the apparent lead have enough 'stuff' on it to interact with a receptor target. We find no difference in the overall sum of Andrews' binding energy between phase II drugs and combinatorial libraries. However, we do find large differences in the density of functionalization. Phase II drugs are much more compact and densely functionalized than combinatorial compounds. This is easily seen by simply plotting the ratio of Andrews' binding energy to molecular weight for the members of the two types of libraries. The ratio is much larger for the phase II compounds. The same functionality is placed on a more compact smaller structure; hence the ratio is larger.

VI   METABOLIC STABILITY

Metabolic stability is generally on all lists of ADME filters.[12] The chemistry control is highly situational in the sense that in some chemical series control is excellent, i.e. the chemistry SAR is very tight. A specific example might be the blocking of hydroxylation by a fluorine atom. In other cases the control is poor, for example, the biology target SAR dictates the presence of a metabolically unstable moiety. A specific example is the hydroxamic acid moiety, found in many early lipoxygenase inhibitors, which was readily metabolically converted to the parent carboxylic acid. Where the chemistry control is highly dependent on the chemistry context, I believe it is dangerous to implement exclusionary filters for highly probable metabolic events. For example, perhaps as many as 40 to 50% of compounds might be substrates to some extent for cytochrome P-450 3A4-mediated oxidation. As this event is very probable, I think it is unwise to implement a blanket combinatorial exclusionary filter for cytochrome P-450 3A4 substrates. What might make more sense would be to factor in the probability that the problem could be fixed with chemistry. Thus in a combinatorial library one might want to allow compounds with a single cytochrome P-450 3A4 potentially metabolically unstable site but filter out those compounds with two or more sites of metabolic instability.

One could envision a quite different use of the same information depending on the research stage. In lead optimization one would want to use all available metabolic stability information because the immediate goal is to improve the compound's drug-like properties. However, in a biology lead-seeking step such as the preparation of a combinatorial library for an HTS screen one is much more interested in the lead generation process. Here, one would want to apply metabolic stability criteria in a looser sense taking into account the probability of a subsequent medicinal chemistry fix.

ADME space is of low dimensionality

ADME space is of low dimensionality. For example, if one computes the types of physicochemical properties likely to be important to oral absorption one seldom finds more than six or perhaps seven significant independent properties (as for example in a principal components analysis). Typically, these are properties related to size, lipophilicity, polarity, hydrogen-bonding properties and charge status. The low dimensionality of ADME space explains the effectiveness of simple filtering algorithms like the 'Rule-of-5'. The low dimensionality of ADME space contrasts with the very high dimensionality of chemistry space. For example, description of a large diverse chemical library by electrotopological parameters does not result in low dimensionality. A principal components analysis on a large diverse chemical library might reveal that eight independent properties (components) derived from electrotopological properties describe less than 50% of the variance in chemistry space.

ADME computational models

Computational models for ADME properties work best when the models are based on single mechanism experimental assays. Scientists approaching an ADME compu-

tational model are often influenced by their familiarity with therapeutic target computational models. A typical biological HTS screen consists of a single mechanism assay. For example, a compound is screened to determine whether it is an agonist or antagonist for a single receptor subtype. For the single mechanism screen more experimental data usually mean a better computational model, so with this history it is very easy to fall into the trap of believing that more data in any ADME screen will result in a better computational model. In a therapeutic target assay one does not deliberately mix targets so there is no history of what to expect if the experimental endpoint is due to multiple mechanisms. Suppose one were to deliberately mix half a dozen biological targets in a single HTS screen such that a hit on any of the targets gave a common analytical endpoint. Could one develop a computational model for the endpoint if the experimental response was based on half a dozen unrelated structure activity patterns? I think this would be very difficult, especially as the number of experimental data points were increased. However, this is exactly what occurs if one tries to build a computational model based on experimental data in a multimechanism ADME assay.

Limitations of Caco-2 cell culture

Our experience with trying to build computational models based on experimental permeability screening in Caco-2 cell culture illustrates the problem introduced by multiple mechanisms.[13] We found that deviation from a single mechanism could arise either in the assay *per se* or could arise from the compounds that were screened in the assay. One aspect of the multiple mechanism problem is the presence of active multiple biological transport mechanisms for both enhancing and reducing absorption in cell culture assays. This issue is well documented and is outside the scope of this chapter.

Poor aqueous solubility and permeability assay noise

Poor aqueous solubility, a compound-related factor rather than an assay-related factor, has a major effect by introducing 'noise' into absorption screening and thus has an effect on making computational model building very difficult. It must be stressed that the compound solubility factor virtually never appears as an explicit consideration in the published permeability literature. Compound sets are published that are used to validate *in vitro* cell-based absorption assays. Validation usually means obtaining an acceptable correlation between human fraction absorbed data and *in vitro* permeability data. The absorption data always include the experimentally well-controlled but compound number-limited human fraction-absorbed data that are used to define absorption ranges in the FDA bioavailability waiver guidelines. This limited compound set is then supplemented with additional compounds chosen from published human absorption literature. In our own work we have been able to accumulate literature human fraction-absorbed data on a total of about 330 compounds. Larger datasets of up to about 1000 compounds exist, which are based on published reference texts[14] or intensive literature searches supplemented by detective work to differentiate the absorption and metabolism components in oral bioavailability.[15] The hallmark of compounds with human absorption data is that they are well-behaved compounds from a 'drug-like' viewpoint. The fraction absorbed is heavily biased to the high percentage absorbed range and the compounds are almost universally soluble in aqueous media. This simply reflects the compound quality-filtering process that must be passed for a compound to enter the types of studies likely to generate human fraction-absorbed data. In short, literature compound permeability validation sets are completely appropriate and say a good deal about assay issues in a permeability screen, but they have almost no relevance to assay reproducibility issues related to poor compound solubility.

Physiologically relevant screening concentration

Table 21.1 sets the stage for the types of solubility among currently synthesized compounds that are likely to be submitted to a permeability screen such as a Caco-2 assay. In this type of assay a variety of biological transporters are present that mediate both absorption and efflux. The movement of a compound through the Caco-2-polarized cell layer through the action of these transporters can be saturated if the drug concentration is high enough. Thus it is important to screen at a physiologically relevant concentration. If the dose

is too low the permeability estimate will be too low because the importance of efflux transporters will be overestimated.

VII ACCEPTABLE SOLUBILITY GUIDELINES FOR PERMEABILITY SCREENS

Table 21.1 maps the acceptable solubility ranges as defined by pharmaceutical science to the molar concentration range of biological screening. For an average potency compound of about 1 mg kg^{-1}, the screening dose in a Caco-2 screen should be somewhere in the range of 100 μM. This concentration is the minimum required for adequate absorption. However, pharmaceutical industry Caco-2 screening doses are typically 10–25 μM. This dose range is chosen for practical reasons. If the assays were run at 100 μM a high incidence of insoluble or erratically soluble compounds would be encountered. In Caco-2 screening in our Groton, USA laboratories, one-third of compounds screened at 10 μM are insoluble in an aqueous medium. When one-third of compounds screened in an assay are insoluble in aqueous media, assay reproducibility becomes a major issue. I think it is entirely reasonable to question the value of permeability screening of combinatorial libraries given their general tendency towards poor solubility.

Batch-mode solubility prediction

The reader quickly realizes whether poor solubility might be a confounding factor for permeability screening of a combinatorial library. In my experience, the existing batch-mode solubility calculation programmes generate similar

Table 21.1 *Solubility ranges among currently synthesized compounds*

μg mL^{-1}	μM (MWT 300)	μM (MWT 400)	μM (MWT 500)	μM (MWT 600)	Compounds
1	3.33	2.50	2.00	1.67	30% of Groton compounds are in this solubility range
3	10.00	7.50	6.00	5.00	
5	**16.67**	**12.50**	**10.00**	**8.33**	
10	33.33	25.00	20.00	16.67	10% of Groton compounds are in this solubility range
20	66.67	50.00	40.00	33.33	
30	100.00	75.00	60.00	50.00	
40	133.33	100.00	80.00	66.67	
50	**166.67**	**125.00**	**100.00**	**83.33**	Solubility acceptable for 1 mg kg^{-1} potency
60	200.00	150.00	120.00	100.00	
70	233.33	175.00	140.00	116.67	
80	266.67	200.00	160.00	133.33	60% of Groton compounds are in this solubility range
90	300.00	225.00	180.00	150.00	
100	333.33	250.00	200.00	166.67	
200	666.67	500.00	400.00	333.33	
300	1000.00	750.00	600.00	500.00	FDA 1 mg kg^{-1} solubility in 250 mL water
500	1666.67	1250.00	1000.00	833.33	
1000	3333.33	2500.00	2000.00	1666.67	

and quite reasonable solubility histogram profiles when run on thousands of compounds (although I would not trust numerical prediction results for small numbers of compounds). Experimental permeability screening (especially if it is manning intensive) might not be worthwhile because of the solubility noise factor if a significant fraction of the library is predicted to be insoluble at 10 μM (the low end of the typical screening concentration range).

In summary, poor aqueous solubility is the single physicochemical property that is most likely to be problematical in a combinatorial library. It can be avoided in part by intelligent use of batch-mode solubility calculations. The solubility problem is not simply a technical issue in library design. It is exacerbated by chemistry synthesis considerations and by the timing of the availability of combinatorial exemplars. Formulation fixes are available unless the solubility is extremely poor, but these should be avoided as much as possible. Poor permeability is seldom a problem in combinatorial libraries, but is disastrous if present since effectively formulation fixes do not currently exist.

REFERENCES

1. Avdeef, A. (2001) Physicochemical profiling (solubility, permeability and charge state). *Curr. Top. Med. Chem (Hilversum, Neth.)* **1**: 277–351.
2. Lipinski, C.A. (2001) Avoiding investment in doomed drugs. Is poor solubility an industry wide problem? *Curr. Drug Disc.* 17–19.
3. Kelder, J., Grootenhuis, P.D., Bayada, D.M., Delbressine, L.P. and Ploemen, J.P. (1999) Polar molecular surface as a dominating determinant for oral absorption and brain penetration of drugs. *Pharm. Res.* **16**: 1514–1519.
4. Gudmundsson, O.S., Jois, S. D. S., Vander Velde, D.G., Siahaan, T.J., Wang, B. and Borchardt, R.T. (1999) The effect of conformation on the membrane permeation of coumarinic acid and phenylpropionic acid-based cyclic prodrugs of opioid peptides. *J. Pept. Res.* **53**: 383–392.
5. Kansy, M., Senner, F. and Gubernator, K. (1998) Physicochemical high throughput screening: parallel artificial membrane permeation assay in the description of passive absorption processes. *J. Med. Chem.* **41**: 1007–1010.
6. Kibron micro-tensiometer, http://www.kibron.com/Applications/Research/index.html.
7. Lipinski, C.A., Lombardo, F., Dominy, B.W. and Feeney, P.J. (1997) Experimental and computational approaches to estimate solubility and permeability in drug discovery and development settings. *Adv. Drug Delivery Rev.* **23**: 3–25.
8. Curatolo, W. (1998) Physical chemical properties of oral drug candidates in the discovery and exploratory development settings. *Pharm. Sci. Technol. Today* **1**: 387–393.
9. Hann, M.M., Leach, A.R. and Harper, G. (2001) Molecular complexity and its impact on the probability of finding leads for drug discovery. *J. Chem. Inf. Comput. Sci.* **41**: 856–864.
10. Teague, S.J., Davis, A.M., Leeson, P.D. and Oprea, T. (1999) The design of leadlike combinatorial libraries. *Angew. Chem. Int. Ed.* **38**: 3743–3748.
11. Andrews, P.R., Craik, D.J. and Martin, J.L. (1984) Functional group contributions to drug–receptor interactions. *J. Med. Chem.* **27**: 1648–1657.
12. Thompson, T.N. (2001) Optimization of metabolic stability as a goal of modern drug design. *Med Res. Rev.* **21**: 412–449.
13. Lipinski, C.A. (2001) Drug-like properties and the causes of poor solubility and poor permeability. *J. Pharmacol. Toxicol. Methods* **44**: 235–249.
14. Advanced Algorithm Builder: http://www.pion-inc.com/products.htm.
15. Oraspotter: http://www.zyxbio.com.

22

QUANTITATIVE APPROACHES TO STRUCTURE–ACTIVITY RELATIONSHIPS

Han van de Waterbeemd and Sally Rose

Life is the act of drawing sufficient conclusions from insufficient premises

Samuel Butler (1835–1902) *Notebooks*

I INTRODUCTION

The rational design of new chemical entities intended for use as drugs can be based on several methods. For the optimization of binding to the molecular target, structure-based design has been very successful. However, a good drug has not only high and selective affinity for its target, it should also have appropriate pharmacokinetic[1,2] and biopharmaceutic properties.[3]

Qualitative structure–activity relationships (SAR) for binding affinity or ADME (absorption, distribution, metabolism, excretion) parameters can be developed within a particular compound series. Even more powerful though to guide the chemist in the design process are quantitative structure–activity relationships (QSAR), using various statistical and mathematical tools.

First attempts to express quantitatively relationships between chemical structure and bioactivity go back to the beginning of the previous century.[4] But only when computers and relevant mathematical methods became available were these approaches widely applied. The credit goes to Corwin Hansch and Toshio Fujita for introducing these quantitative methods into medicinal chemistry in the 1960s.[5–7] Initially in their pioneering work, Hansch and coworkers focused their attention on the role of octanol/water partition coefficients ($\log P$) in drug transport processes which were thought to contribute crucially to the measured activities. As we know now, $\log P$ is the most predominant descriptor in many structure–activity correlation studies. However, as we will see below, many other ways can be used to describe chemical structure in a quantitative way.

The Practice of Medicinal Chemistry
ISBN 0-12-744481-5

First we need some more reflection on the terminology. Studies aimed at broadening the quantitative understanding of correlations between intrinsic, physical and chemical or biological molecular properties, are called QSAR studies.[8,9] Nevertheless, in some cases no 'activity' is involved and it would then be better to speak of quantitative structure–property relationships (QSPR). In another context 'P' may stand for pharmacokinetic or permeability, and 'A' for absorption.

A number of reviews have documented the history, strategy and successes of quantitative drug design, i.e. design using QSAR methods.[4–13] The impact of QSAR methods on drug discovery and lead optimization may be manifold (see Fig. 22.1).[12] Quantitative models may be derived, assisting the medicinal chemist in potency or ADME optimization and in the generation of new ideas.

Early QSAR studies were mainly focused on analysing the effect of aromatic substituents on activity using substituent constants to describe the steric, electronic and lipophilic characteristics of the substituents. However, with the advent of combinatorial chemistry and the associated increased interest in molecular similarity (or dissimilarity), the diversity of compounds in a library has generally increased. More recently attention also focused on 'drug-likeness' of libraries. This has brought about a change in the types and increase in the numbers of descriptors which are used in QSAR studies as discussed in Section II.

Progress in chemometrics has made available a number of new statistical techniques, which are increasingly being used.[14] This concerns both new supervised and unsupervised (or pattern recognition) techniques. Chemometrics was defined about twenty years ago as the chemical discipline which uses mathematical, statistical and related techniques to design optimal measurement procedures and experiments, and to extract maximum relevant information from chemical data. The science of chemometrics has been developed to promote applications of statistics in analytical, organic and medicinal chemistry.

In this chapter only QSAR methods which use physico-chemical or structural features of molecules will be discussed, while in Chapter 25 3D-QSAR approaches will be presented. These so-called 3D-QSAR techniques, e.g. CoMFA, use the basic statistical principles, such as partial least squares (PLS), of QSAR methods, but in addition use the three-dimensional characteristics of a molecule specifically related to electronic, steric and lipophilic field effects. In these methods the molecular superposition believed relevant to binding to the target is crucial.

In Fig. 22.2 the elements of QSAR or QSPR studies are depicted. On one side high quality and relevant biological data are required, while on the other relevant chemical descriptors should be defined. A further critical element is the proper choice of a model to investigate relationships between these data. If the right prerequisites are met, relevant information may be extracted from the data, which can be used to get better understanding of the molecular structures and possibly the mode of action at the molecular level. This information may then be used to predict the properties and activities of new compounds.

The design of new compounds may be based on either a lead compound or structural information about the target, e.g. the crystal structure of an enzyme, or a combination of both. In the case where no structural information about the target is available, new compounds could be designed to include variation of substituents at a particular site or the variation of larger building blocks to give greater diversity. The choice of appropriate substituents and building blocks depends on synthetic feasibility and costs, but should also be based on an understanding of the physicochemical properties of the substituents, as well as the predicted properties of the targeted compound. Therefore we discuss below how these choices may be made as rational as possible by considering experimental design techniques, preceded by an overview of datatypes of physicochemical and biological descriptors.

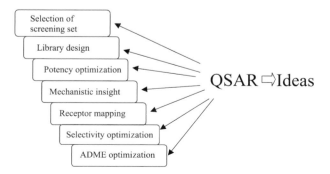

Fig. 22.1 The impact of QSAR studies on drug discovery.

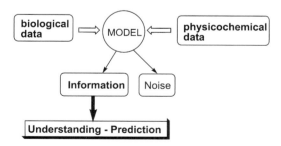

Fig. 22.2 Structure–activity and structure–property correlations using different statistical data modelling techniques may provide the basis for understanding and prediction of biological activity and physicochemical features.

Next we will present a section on methods used to find quantitative relationships or correlations between biological and physicochemical data. These methods are generally termed 'supervised' methods as they use the activity data in the generation of the model. Besides two classical methods, namely the Hansch and Free-Wilson approach, more modern approaches using partial least squares (PLS) regression, neural networks and decision trees are discussed. If the compound set is sufficiently large, then the starting point of an analysis should ideally be the selection of a training or calibration set and a test or validation set. The idea is to keep some of the compounds separate to test the quantitative model derived with the rest of the set. Often compounds in a training set are chosen based on their diversity in molecular structure. Statistical experimental design (see Section III) may be used to make more rational and well-balanced choices. An alternative approach is to validate the predictive power of a model using cross-validation (see Section IV).

It is often of interest to investigate data sets by so-called 'pattern recognition' or 'unsupervised' methods in order to detect clustering within a set of compounds or activities and these are discussed in Section V.

The biological and physicochemical data relevant to a certain project may be represented as two tables and may be analysed in various ways (see Fig. 22.3). Taking biological or physicochemical data either separately or combined, pattern recognition or classification studies may be useful to detect redundancy in the test systems or classify the compounds in a particular way which may be related to their specific mechanism of action. Clustering and classification of compounds based on their properties is central to molecular similarity studies. Regression or correlation studies between the biological and chemical data are of course useful to rationalize structure–activity relationships. Both kinds of studies, regression or pattern recognition, are called multivariate statistical data analysis, or QSAR studies.

II DESCRIPTORS

A Biological and physicochemical descriptors

The targets of drug action may be quite diverse, including for example, membrane-bound receptors, ion channels, enzymes and DNA. Biologists developed a broad variety of biological and pharmacological test systems producing different kinds of data. Some are quite simple and accurate, e.g. IC_{50} values as a measure of ligand affinity, while others are more complex with large errors e.g. *in vivo* data. Activity may be expressed as a continuous measure, e.g. IC_{50} or % inhibition at a specified concentration or as categorical data such as active versus inactive, agonist versus antagonist, or as strong, medium and weak. Both continuous as well as categorical data can be used in QSAR studies. The proper choice of a mathematical model to relate biological to chemical data depends on the quality and kind of data to be analysed. Therefore, for example, the classical Hansch approach using multiple linear regression (see below) is not suited for all purposes.

Molecular structures may be considered at different levels, each containing certain types of information.[15,16] The simplest representation is the empirical chemical formula, while a molecular electrostatic potential (MEP) representation on the Van der Waals surface includes both steric and electronic information. Molecular properties can be divided into various categories (see Fig. 22.4). There are experimental and calculated properties. Intrinsic properties are directly related to the structure without considering any interaction, such as molecular weight. Some properties are related to a substituent or fragment. When a compound interacts with a chemical or biological environment, we may define physicochemical properties, e.g. lipophilicity or ionization constants, biochemical properties, such as binding constants, and biological properties, such as activity or toxicity.

Chemical descriptors may contain structural, also called global, information, or local information for substructural parts of the molecule. A large set of chemical descriptors of molecular structures and fragments has been reported in the literature.[6,17] Parameterization of chemical structures or substructures is not only of great interest to QSAR studies, but has much current interest in definitions of molecular similarity and diversity. This information may be used in molecular modelling studies or in combinatorial chemistry

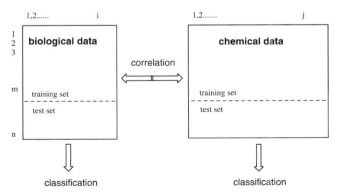

Fig. 22.3 The chemometric analysis of multivariate data tables. Two major types of studies can be defined: (1) correlation between biological and (physico)chemical data using regression techniques; and (2) classification of compounds or descriptors using pattern recognition methods.

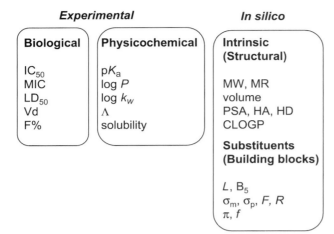

Fig. 22.4 Molecular properties can be divided into experimental (subdivided into biological and physico-chemical) and *in silico* (subdivided into structural and substructural) properties, physicochemical and biological properties. Examples of experimental data are IC_{50} (binding affinity), MIC (antibacterial minimum inhibition concentration), LD_{50} (lethal dose), Vd (volume of distribution), F% (bioavailability), pK_a (ionization constant), $\log P$ (partition coefficient from shake flask determination), $\log k_w$ (lipophilicity from HPLC measurement), Λ (hydrogen bond capability), solubility. Examples of calculated properties, either for whole molecule or for substituents or buildings blocks, are MW (molecular weight), MR (molar refractivity), molecular volume, PSA (polar surface area), HA (number of H-bond acceptors), HD (number of H-bond donors), CLOGP (calculated $\log P$ values), L (substituent length), B5 (substituent width), σ (Hammett constant), F, R (field and resonance parameters), π (Hansch constant), f (hydrophobic fragmental constant).

projects aimed at generating large molecular diversity in order to improve lead finding chances. Descriptors may be experimental or calculated, as well as pure or composed. The traditional way to subdivide substituent or whole molecule properties is in terms of lipophilic, steric, electronic/electrostatic and H-bonding effects. These properties are usually considered in the systematic variation of a selected substitution site. More advanced methods using derived properties, so-called principal properties, are discussed below in Section III.

The *lipophilicity* of a compound is often considered as an important design factor since it is related to processes such as absorption, brain uptake, volume of distribution and protein binding. This property is often expressed as the partition coefficient ($\log P$) or distribution coefficient ($\log D$, typically measured at pH 7.4). By several lines of evidence it was shown that $\log P$ values should be

considered as composed of two factors (see equation 1), namely a steric and polar contribution.[16,18]

$$\log P = aV - \Lambda \qquad (1)$$

The *size descriptor* molar volume V can be calculated. Thus the polarity factor Λ is obtained indirectly from $\log P$ measurements, where $\Lambda = 0$ for nonpolar compounds. Λ appears to reflect the *H-bonding capacity* of a compound.

The 1-octanol/water system is often taken as the reference or standard for partition coefficients. However, other partitioning systems may give useful information too. It has been recommended by Leahy and colleagues[19] to use a 'critical quartet' of solvent systems for lipophilicity measurements. They suggested that any biological membrane can be modelled by one of the four solvents: alkane (inert), octanol (amphiprotic), chloroform (proton donor) and PGDP (propylene glycol dipelargonate) (proton acceptor). It has also been found that the differences between $\log P$ values measured in two different solvent systems ($\Delta \log P$) may contain relevant information related to the H-bonding capacity of a compound.[20] However, tedious measurements of partitioning data are not required on a larger scale, since it has been demonstrated that the calculation of the polar surface area of a compound is an adequate substitute reflecting the hydrogen bonding capacity of a mole.[21] In practice, the lipophilicities of series of compounds are often measured by RP-HPLC[22] or centrifugal partition chromatography (CPC),[23] and more recently by high throughput shake plate approaches. Within a series of closely related compounds $\log P$ values are correlated to $\log k_w$ values from RP-HPLC. For more diverse molecules this is often not the case. However, one should consider each lipophilicity scale as unique and reflecting, as seen above, a combination of the steric and H-bonding properties of a compound.

It is also good to realize that properties like fragment lipophilicity contributions are additive properties, but may be very much dependent on the structural environment (Fig. 22.5). Some substitutions may have a more dramatic effect than expected. Radioactive labelling with ^{125}I is quite common for biological studies. One should be aware, however, that aromatic iodination increases the $\log P$ of the compound by approximately 1 $\log P$ unit, and thus a different tissue distribution may result. An aromatic fluoro substituent has very little effect on the lipophilicity, but mainly serves in drugs to avoid oxidative biotransformation.

Inversely related to lipophilicity is the aqueous solubility of a compound. High throughput methods are now available to measure solubility.[24,25] Less successful are approaches to estimate solubility from computational models. These are not sufficiently accurate to predict poor solubility. Moreover, solubility is influenced by many factors such as ionic strength and type of buffer.

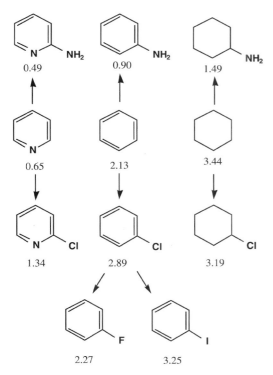

Fig. 22.5 Impact of substitution on the lipophilicity of a compound depends on its structural environment.

B Selection of descriptors

The choice of descriptors is dependent on several factors:

- What property data can be readily measured
- What can be readily calculated
- What is particularly relevant to the therapeutic target
- What variation is relevant to the compound series.

Ideally, a QSAR model is preferred which uses properties that can provide information on mechanism of action. Thus properties which are readily understood, such as log P and pK_a, are ideal, while connectivity indices for example can be harder to interpret. Measured properties are generally more accurate than calculated properties, but have the limitation that they are not available to predict the activity of novel analogues not yet synthesized.

It is commonly the case that a wide variety of properties can be included in a QSAR analysis and a decision must be made on whether to include all possibilities or limit the number of properties. This decision depends on the size of the data set and the correlation matrix between the properties. Large sets of property data contain much redundancy of information. For example, molecular weight, surface area and molar refraction are always highly correlated, therefore a decision only to use molecular

weight could be made. Some multivariate statistical analysis methods are tolerant of data sets which contain more properties than compounds, e.g. partial least squares, while others are not, e.g. linear discriminant analysis. Ideally, a set of uncorrelated properties is desirable as this is most likely to give a robust, interpretable model.

The choice of properties has a major influence on pattern recognition methods and different property sets will result in different patterns of similarity between compounds.[26]

Several methods are available to make selections of subsets of uncorrelated properties which can be used for QSAR studies. These include:

- Forward or backward elimination (as used in multiple linear regression)
- Principal component analysis (PCA)
- Cluster analysis (CA)
- Genetic algorithms[27]
- Simulated annealing
- *k*-Nearest-neighbours[28]
- Neural networks
- Evolutionary approaches.

Some of these methods are discussed in this chapter. An example using principal component analysis for variable selection is given in Section V. The application of principal component analysis to identify and combine individual correlated properties into new generalized descriptors, termed principal properties, is described in Section III.

III EXPERIMENTAL DESIGN

A Overview

Optimization covers many aspects in medicinal chemistry. The optimization of the affinity for the biological target and the pharmacokinetic properties of a lead compound is the primary goal of most preclinical research projects. Secondly, optimization strategies may be applied to synthesis procedures to minimize the cost of goods.[29] In both cases a number of variables have to be taken into account simultaneously. Strategies changing only one variable at a time take much time and many experiments are needed. In contrast to the sequential approach, by a proper selection of a limited number of experiments the full variable space can be covered by far fewer experiments. Experimental design schemes are therefore of great help to focus on the most informative experiments. These techniques have been applied in two types of synthetic programs, namely peptide design and substituent variation. Below we will describe how physicochemical descriptors for aromatic and aliphatic substituents may be used for substituent selection. In a similar way, relevant descriptors for amino acids may be used. It has been pointed out that a series of compounds based on some experimental design plan are most likely to produce significant QSAR equations. Therefore, careful selection of appropriate substituents and variables is an important matter. More recently design strategies have been applied to the creation of virtual and combinatorial libraries and selection of sets of compounds for purchase or screening. Specifically, attention has been focused on the relevance of such compounds for drug discovery purposes.

B Topliss tree and Craig plot

Various strategies have been advocated in order to cover the physicochemical parameter space of a series of new compounds as well as possible. Familiar strategies go back to proposals of Topliss[30] and Craig.[31] Both are schemes used for substituent variation at a selected site.

The Topliss substitution scheme can be used to optimize aromatic and aliphatic substituents using a fixed set of substituents. It starts with the assumption that the lead compound has an unsubstituted phenyl ring. In the first step a *p*-chloro derivative is made and its activity measured. Depending on the activity, less, equal or better than the parent, the next step is made. This consists of replacing the *p*-chloro substituent by either a methoxy or methyl group, or by adding an additional chlorine substituent. As before the activity of this compound is compared with the previous one and this defines which compound should be made in the following step. In this manner, the medicinal chemist follows a particular path through a predefined tree diagram (dendrogram). This scheme applies manually the basic features of a good design plan, without statistical considerations, making it appealing to most medicinal chemists.

A Craig plot is a two-dimensional plot of selected descriptors, e.g. Hammett σ and Hansch π values (see Fig. 22.6). From this plot substituents can be selected from each quadrant such that they vary widely in their properties, e.g. lipophilic and hydrophilic, electron–donor and electron–acceptor.

A further extension would be to consider a three-dimensional Craig plot using three descriptors, e.g. reflecting steric, lipophilic and electronic properties of the substituents. In that case substituents may be chosen from the eight octants. If one wants to consider even more descriptors, this approach becomes impractical. In that case more advanced experimental design techniques may be applied.

Hansch and Leo have used cluster analysis to define sets of aliphatic and aromatic substituents useful in the design of compounds, such that various aspects of the substituents are taken into account in a balanced way.[32]

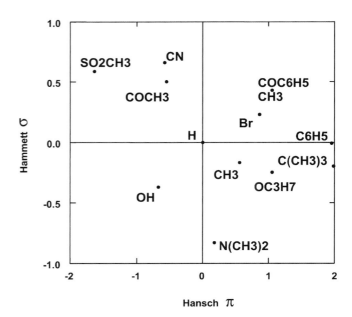

Fig. 22.6 Two-dimensional Craig plot. Electronic properties (Hammett σ_p constants) are plotted against lipophilic (Hansch–Fujita π values) substituent properties.

C Factorial, central composite and D-optimal designs

More complex statistical designs are necessary when more than three properties are to be systematically varied. In order to limit the number of combinations, each variable may be considered at two levels, e.g. lipophilic versus hydrophilic. A two-level factorial design (FD) with k variables requires 2^k experiments. A Craig plot is an example of a 2^2 FD. In other words, the minimum number of compounds to synthesize using two descriptors is four. As stated above, with many variables this number rapidly becomes impractical and fractional factorial designs (FFD) should be preferred. Using a reduction factor r, the number of experiments then becomes 2^{k-r}. This reduction factor is in practice chosen rather pragmatically, such that one has to consider 8 or 16 compounds. Further design schemes are known as central composite and D-optimal design. In this latter method the determinant of the variance–covariance matrix of considered properties is calculated. This determinant has a maximum value for those combinations of substituents which have a maximum variance and minimum covariance in their physicochemical descriptors. The variance–covariance matrix is an important mathematical cornerstone in matrix operations used in multiple linear regression and principal component analysis, and to obtain a correlation matrix among a set of selected variables.

D Principal properties of substituents

In the literature, a large number of substituent descriptors have been reported.[16,32] In order to use this information for substituent selection, appropriate statistical methods may be used. Pattern recognition or data reduction techniques, such as principal component analysis (PCA) or cluster analysis (CA) are good choices. As explained in Section V in more detail, PCA consists of condensing the information in a data table into a few new descriptors made of linear combinations of the original ones. These new descriptors are called principal components or latent variables. This technique has been applied to define new descriptors for amino acids, as well as for aromatic or aliphatic substituents, which are called principal properties (PPs). The principal properties can be used in factorial design methods or as variables in QSAR analysis.

In the case of amino acids, the principal properties are known as z-scales, which can be used for the design of peptides.[33]

An analogous approach using PPs has been used for aliphatic and aromatic substituents, using databases with 59 substituents and 121 descriptors[34] or a set of 100 substituents and nine descriptors.[35] An attempt has been made to combine both former studies to derive one unique set.[36] In this example, the data matrix was initially subdivided into four groups based on the descriptor type: lipophilic, steric, electronic and H-bonding properties. PCA was then applied to each group in turn. The first two principal components of each set were extracted and combined in a new data matrix of eight disjoint principal properties (DPPs). This approach has the advantage that the principal properties are easier to interpret as they are not a mix of, for example, steric and electronic data, however, the descriptors are not orthogonal.

An illustration of the use of principal properties, is a PLS (partial least squares) analysis of dopamine antagonistic clebopride analogues.[37] The paper describes the use of experimental design techniques to select a training set from a series of 20 synthesized and tested compounds for establishing a quantitative structure–activity relationship with good predictability. The substituents and substituent sites are given in Fig. 22.7. Five significant PPs were derived from nine substituent descriptors, namely σ_m and σ_p Hammett constants, F and R Swain and Lupton field and resonance parameters, the Hansch π lipophilicity constant, MR molar refractivity, and Verloop steric parameters L, B_1 and B_5. These PPs were used in a fractional factorial design to select the training series in a PLS analysis. It was shown that with eight well-selected compounds the predictivity of the model is the same as with a model derived for all 20 compounds. This example shows that PPs are good design variables and that experimental design may limit the

H,Et,Cl,Br,OMe

H,OH

OCH$_3$

H,Et,Cl,Br,OMe

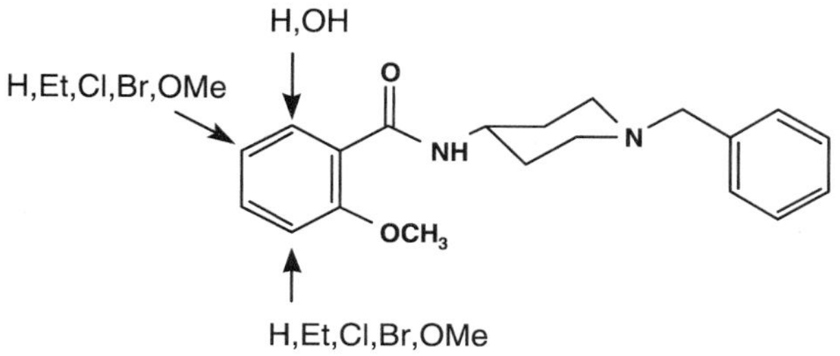

Fig. 22.7 Substituted benzamides of the clebopride type.

number of compounds that need to be synthesized in a series to a strict minimum.

E Drug-like properties

Various analyses of historical data on existing drugs have led to a better understanding of what constitutes drug-like properties.[24,38] Several groups have trained neural networks to distinguish between 'drug-like' and 'non-drug-like' compounds.[38,39] The training sets for the network consist of one large database of drug molecules (e.g. from Derwent's World Drug Index, WDI) and one database on non-drugs (e.g. MDL's Available Chemicals Directory, ACD). Both databases are pre-processed to remove unwanted substances. The trained network can then be used to predict the drug-likeness of combinatorial arrays and screening sets. More information on supervised neural networks is given in Section IV.

An analysis of compounds in the World Drug Index has resulted in the rule-of-five,[24] which states that poor oral absorption is expected for compounds with a molecular weight above 500, a calculated octanol/water partition coefficient >5, the number of hydrogen bond donors >5, and the number of hydrogen bond acceptors >10. Exceptions to this rule may be compounds with an active uptake mechanism. Such simple rules are now widely used in the design of combinatorial libraries. More detailed knowledge on relevant properties can be used to optimize oral absorption and pharmacokinetics,[1] an approach which has been called property-based design.[40]

F Combinatorial libraries

An ideally designed set of compounds covering a wide range of the available property space should consist of a variety of structural shapes and molecular properties, while avoiding redundancies.[41] The above principles have been extended for use in the design of combinatorial libraries, as well as for the selection of compounds for high throughput screening (HTS). Basically for this purpose a large set of

1D, 2D and 3D molecular representations can be used together with appropriate statistical tools as discussed in this chapter.[41] There has been much debate as to whether combinatorial library design should be based on reagent (monomer) diversity or on the diversity of the final product.[42] Some relative merits of each approach are outlined below:

Advantages of using final product designs:

- Can calculate the physicochemical property profile of the final array
- Suitable for pharmacophore-based designs
- More descriptors are relevant to final products than reagents
- 'Greater diversity' can be obtained using a final product design.[42]

Advantages of using reagent designs:

- Easy substitution of reagent if it is found to be poorly reactive
- Reagent diversity is quicker to calculate than fully enumerated final product libraries
- Clustering studies on reagent sets can be re-used for new arrays
- More sympathetic to parallel synthesis logistics.

An interesting hybrid approach, termed reagent-biased product design, selects reagent sets based on final product diversity. This approach was initially proposed by Good and Lewis[43] and their method, HARPick, is summarized in Fig. 22.8. The method uses a genetic algorithm (see Section IV) to optimize the selection of sets of monomers in an iterative fashion to maximize diversity in the final all-combinations array.

Reagent-biased product design is now quite widely used and commonly the method of choice.[2,44-46] The methods most commonly applied to measure diversity in large compounds sets (combinatorial libraries or screening collections) include 2D fingerprints (for substructural

diversity) and 3D three-point pharmacophores[43] (to measure the range of pharmacophoric groups that are present within the compound set).

G Virtual screens

Virtual screens are applied to assist in library (array) design, decisions on commercial library acquisition and the selection of screening sets of compounds. They are used to measure and focus both diversity and relevance to drug discovery as part of the overall experimental design process. One approach is to test for 'drug-likeness' properties as described in Section III. Other virtual screens or filters may include, for example, screening compounds against pharmacophores derived for a specific target,[45] high throughput docking studies,[47] screening for unwanted reactive groups, screening for suitable physico-chemical properties,[44,46] screening for diversity of 3D pharmacophores, etc.

IV DETERMINING RELATIONSHIPS BETWEEN CHEMICAL AND BIOLOGICAL DATA

A Overview

This section looks at methods for analysing biological activity in terms of physicochemical properties and/or structural features. QSAR studies were initially focused on understanding activity in terms of variation in substituents using multiple linear regression and typified by the Hansch and Free-Wilson approaches. Both Hansch and Free-Wilson analysis are now considered as traditional or classical QSAR methods. Recently, chemometrics has had a significant impact on QSAR analysis, and the methodology has been even further augmented by advances in artificial neural networks, genetic algorithms and artificial intelligence. The various methods described below have different data requirements and different degrees of ease of interpretability. It is essential to select the appropriate analysis methods for the data set (see Table. 22.1). For example, multiple linear regression is not suitable for analysing data sets containing large numbers of properties or for properties which are inter-correlated. Both these data attributes are acceptable to partial least squares. These two approaches are the most widely used, so have been described in more depth than the other supervised methods.

B The Hansch approach

In the 1960s, Hansch and Fujita proposed a method to describe quantitatively relationships between biological

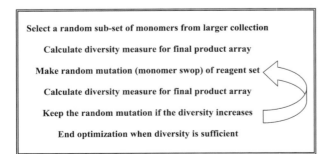

Fig. 22.8 The HARPick approach to library design.

Table 22.1 *Characteristics of the supervised QSAR methods described in Section IV*

Method	Suggested minimum ratio of Compounds:variables	Biological Data	Interpretability
MLR (Hansch)	5	Continuous (or classified)	High
Free-Wilson	5	Continuous	High
PLS	No limit	Continuous (or classified)	Good
LDA	3 × smallest class	Classified	Good
SIMCA	No limit	Classified	Quite good
Back propagation	Guidelines are unclear	Either	Difficult due to non-linearity
Decision trees	Guidelines are unclear	Classified	Difficult due to multiple recurrences of properties

activity and chemical descriptors.[48] This can be expressed as follows:

biological activity

= function (molecular or fragmental descriptors)   (2)

The Hansch-Fujita approach is also called linear free-energy relationship (LFER) or extrathermodynamic approach, since most of the descriptors are derived from rate or equilibrium constants.

The simplest means to obtain such a quantitative relationship is to use multiple linear regression (MLR) available in any statistical software package. In order to avoid statistically insignificant relationships or chance correlations, one should always apply the following rules of thumb: (1) the ratio of compounds to descriptors should be >5; (2) the descriptors should not be intercorrelated (inter-descriptor correlation coefficient should be less than $r^2 < 0.5$).

Also nonlinear regression, i.e. using quadratic terms such as $(\log P)^2$, may be used. However, as described in detail,[48] there are a number of pitfalls to this method. A statistically more robust method which could be used instead of MLR is the partial least squares (PLS) regression method (see below).

There are numerous examples of traditional Hansch QSAR studies in the literature. Some include large sets of descriptors, while others explore just a few. If descriptor values are not readily available, indicator or dummy variables, denoting presence or absence of a certain structural feature, may be of help. This is the basis of the Free-Wilson method (see below).

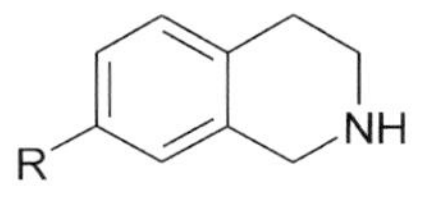

Fig. 22.9 Optimization of activity and specificity of 7-substituted tetrahydroquinolines by MLR.

An example of a Hansch analysis using MLR is a study on substituted tetrahydroisoquinolines with affinity for both phenylethanolamine *N*-methyltransferase (PNMT) and the α2-adreno-receptor[49] (see Fig. 22.9). The multiple regression equations obtained were:

$$pK_i(\text{PNMT}) = 0.599(\pm 0.167)\pi - 0.0725(\pm 0.0268)\text{MR}$$
$$+ 1.55(\pm 0.917)\sigma_m + 5.80(\pm 0.48) \quad (3)$$

$$n = 27; \ r = 0.885; \ s = 0.573; \ F = 27.61$$

$$pK_i(\alpha 2) = 0.599(\pm 0.129)\pi - 0.054(\pm 0.019)\text{MR}$$
$$- 0.951(\pm 0.623)\sigma_m + 6.45(\pm 0.34) \quad (4)$$

$$n = 27; \ r = 0.917; \ s = 0.397; \ F = 40.53$$

From such equations two types of information are obtained, namely about statistical quality and relevance,[48] and chemical implications. The standard deviation of each coefficient is given in parentheses; n is the number of compounds in the study; r is the correlation coefficient, where 1.0 indicates perfect correlation and 0 is totally uncorrelated and a good relationship should ideally be >0.8 and where r^2 ($\times 100\%$) is the variance explained by the equation; s is the standard deviation of the regression, which should have a value near to the experimental error in the biological dependent variable (here pK_i); finally the F value is a measure for the statistical significance of the regression model and is calculated as the ratio between regression and residual variances. This value should be higher than a value which can be found in a Fisher F statistics table, and is a function of the degrees of freedom and the significance level.[48] In more recent papers, the cross-validated correlation coefficients q^2 is often added as explained under Section IV. The error in the regression coefficients should not be larger than the coefficient itself. It is preferable to report 95% confidence intervals, rather than standard deviations which are about a factor of two smaller and may give a too optimistic figure. In practice r and s are the most informative statistical parameters. In QSAR studies, it is common to transform biological activities to their negative logarithmic form, e.g. $\log 1/IC_{50} = -\log IC_{50} = pIC_{50}$ and similarly for binding affinities, pK_i is used. Thus, the most active compounds have the largest values.

Among all the descriptors evaluated in this study, only those which are relevant appear in the final equation. π is the lipophilic constant for the substituent, MR is its molar refraction and σ_m is the electron-donating effect. This is an interesting example as the equations show that the hydrophobicity (π) and bulk (MR) effects are similar, but that the sign for the coefficient in σ_m is opposite for the two targets. Thus selectivity can be achieved by exploiting differences in the electron-donating properties of the substituents.

A second example is the Hansch–Fujita analysis of cyclooxygenase inhibition.[50] The goal of the study was to understand the physicochemical background of the effect of substituents R_1, R_2 and R_3 (see Fig. 22.10) for a rational choice in the selection of compounds for further development. This detailed understanding was obtained by developing correlations for each varied position. For example, for R_1 substituted compounds, in which the thiazole ring is unsubstituted, equation (5) was obtained.

$$pIC_{50} = 1.08(\pm0.50)\pi(R1) + 1.18(\pm0.97)\sigma_I(R1)$$
$$+ 5.64(\pm0.38) \tag{5}$$

$$n = 11; \ r = 0.89; \ s = 0.28; \ F = 15.81$$

In this equation σ_I is Charton's electronic parameter for inductive effects of the substituents. The positive coefficient indicates that electron-attracting substituents are favourable. It is demonstrated in the paper that π values are not position independent. Besides an intrinsic hydrophobic factor, intramolecular steric and hydrogen-bonding components are also included. The final equation including all compounds is equation (6).

$$pIC_{50} = 1.03(\pm0.42)\pi(R1) - 4.48(\pm1.64)\sigma_R(R2)$$
$$- 0.86(\pm0.13)\Delta L(R2) + 0.44(\pm0.27)\pi(R2,3)$$
$$- 0.40(\pm0.26)\Delta L(R3) - 1.48(\pm0.52)I_{iso}$$
$$+ 6.11(\pm0.21) \tag{6}$$

$$n = 45; \ r = 0.95; \ s = 0.38; \ F = 54$$

This equation is statistically relevant and informative. Although six variables are used, this can be accepted since the number of compounds is sufficiently large. Furthermore, the intercorrelation between the independent variables is nonsignificant. Optimal substituents need to be lipophilic, but small, since the ΔL terms have a negative coefficient. ΔL is the length of the substituent compared with a hydrogen atom. An indicator variable (I_{iso}) was assigned for compounds with the i-Pr group in the R_2 position, which appears to be detrimental to affinity for cyclooxygenase. One of the best compounds is given in Fig. 22.10. Although this compound was synthesized at an early stage of the project, the analysis clarified the physicochemical background of the substituents. This can be useful to fill gaps in patent coverage or in the decision to stop a project.

C Free-Wilson analysis and related methods

The Free-Wilson (FW) model was proposed in 1964 at the same time as the Hansch model, but is far less widely used.[48,51] It uses indicator values, having a value of unity for the presence of a substructural feature, e.g. a *p*-chloro substituent and zero for its absence, as sole parameters in a Hansch model-like regression equation. FW has also been named the additivity model. Several closely related approaches, such as the Fujita–Ban variant, have been proposed which offer mathematical advantages. Details will not be discussed here. FW analysis has the limitation that it does not allow extrapolation to substructural fragments which are not contained in the training set of compounds. The greatest interest of this method lies in its mixed use with Hansch analysis, as illustrated above. Thus, the combination of physicochemical properties with substructure indicators is often the best way to proceed. A good example of a comparative study using MLR, FW and PLS is given in reference 52.

D Partial least squares (PLS)

Traditional Hansch analysis using multiple linear regression suffers from several shortcomings. One of the problems is that often one has more variables than compounds. Furthermore there is often a need to consider correlations between chemical descriptors and several biological tests

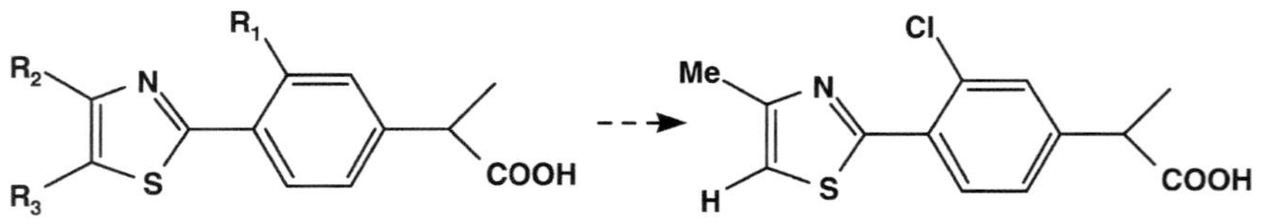

Fig. 22.10 Optimization of cyclooxygenase inhibitors by MLR.

simultaneously. PLS (partial least squares projections to latent structures) is a generalization of regression which is appropriate to treat these problems.[53,54] Further alternatives, such as continuum regression, are being evaluated,[55] but will not be discussed here. PLS can handle numerous and even collinear (inter-correlated) variables, and allows for a certain amount of missing data. An important part of a PLS data modelling study is the cross-validation of the results to determine the robustness of the model (see later). PLS considers all independent descriptors together and calculates their modelling power, i.e. their contribution to the regression. PLS is particularly useful when many descriptors are taken into consideration. Above we have seen that the alternative is MLR or PCA (see Section V) combined with MLR. However, experience has shown that PLS gives the most relevant and statistically significant results and should be the preferred default technique in SAR correlation studies.

Modern validation techniques used in PLS and other multivariate statistical methods are called bootstrapping or cross-validation. Cross-validation (CV) evaluates a model not by how well it fits the data, but by how well it predicts data. The data set consisting of *n* compounds is divided into groups. Leaving out one group by a fixed or random pattern, for the reduced data set the MLR or PLS model is recalculated and the missing values are predicted. This is repeated until every compound is left out once and only once. When each time only one compound is left out, this is referred to as the leave-one-out (LOO) method. Many authors use the LOO method, although it has been shown that the leave-several-out (LSO) approach is preferable.[56] A recommendation is to divide the data set into seven groups. Using the predicted values the PRESS (predictive residual sum of squares) and SD values are obtained as

$$PRESS = \sum(activity_{observed} - activity_{predicted})^2 \quad (7)$$

$$SD = \sum(activity_{observed} - activity_{mean})^2 \quad (8)$$

and the cross-validated correlation coefficient is calculated as

$$q^2 = r^2_{cv} = (SD - PRESS)/SD \quad (9)$$

Note: PRESS and q^2 may relate to any property that is being modelled and not just 'activity'. q^2 will always be smaller than r^2. When $q^2 > 0.3$, a model is considered significant. Although cross-validation may seem a robust validation technique, some difficulties should not be overlooked. Variables that do not contribute to prediction, i.e. cause noise in the model, may have detrimental effects on CV. This may particularly play a role when many variables have to be considered, such as in a 3D-QSAR CoMFA analysis (see Chapter 25). A procedure for variable selection in the case of many variables has been developed and is named GOLPE (generating optimal linear PLS estimations).[57]

When compounds are strongly grouped, CV may not work well. Recent examples have shown that CV is misleading when it is applied after variable selection in stepwise MLR.[57] Thus although cross-validation is considered as the state-of-the-art statistical validation technique, its results are only relevant when correctly applied.

An illustration of the PLS approach is given in Fig. 22.11.

A PLS study has been performed on a series of 99 1,5-substituted-3,4-dihydroxybenzenes as catechol *O*-methyltransferase (COMT, EC 2.1.1.6) inhibitors[58] (see Fig. 22.12). A set of 19 variables to characterize the inhibitors was considered. This is a typical situation where MLR would pick out a few variables which together are best correlated to the biological activity. However, there may be several good combinations of descriptors with similar statistical relevance. PLS gives a better feeling of the contribution of each of the variables considered. In this case, the best PLS model showed that inhibition activity is nonlinearly related to the size of the R_5 substituent and greatly depends on the electronic nature of both R_1 and R_5 substituents. Electron-withdrawing substituents enhance activity.

A second example concerns a PLS analysis of 83 thiolcarbamates with fungicidal and herbicidal activities[59] (see Fig. 22.12). With PLS both activities, fungicidal and herbicidal, can be correlated together with physicochemical properties. Remember that with the Hansch method, using MLR, only one biological variable at a time is studied. Thus, PLS can be used to optimize an activity profile. For each of the three substitution sites, a steric and a lipophilic descriptor were used. Three significant components, called t_1, t_2 and t_3, have been obtained, explaining together 75.6% of the activity variance. Activity can now be expressed for the individual activities in terms of these components or

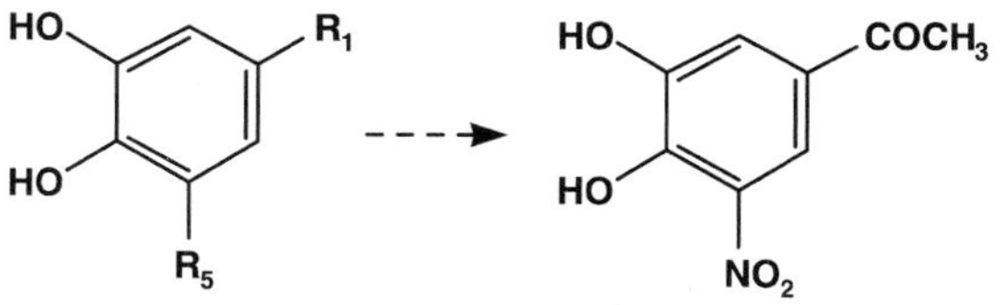

Fig. 22.11 Optimal substituent selection of soluble COMT inhibitors by PLS.

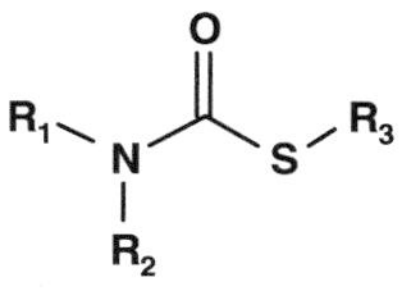

Fig. 22.12 Profile optimization using PLS for fungicidal and herbicidal thiolcarbamates.

latent variables, i.e.

fungicidal activity

$$= -0.482t_1 + 0.278t_2 - 0.184t_3 + 1.723 \quad (10)$$

herbicidal activity

$$= 0.329t_1 + 0.498t_2 + 0.467t_3 + 2.133 \quad (11)$$

Using these equations, the activity of 81% and 76% of the compounds in the data set is correctly predicted for the fungicidal and herbicidal activities, respectively. In order to understand which structural parameters should be modified to improve activity, the t-values can be translated back in terms of original variables, and then substituted in equations (10) and (11). This finally gives a MLR-like model relating biological activity to physicochemical descriptors. The opposite signs of the coefficients in t_1 and t_3 means selectivity can be built into the series. Insight in the data structure of the descriptors can be obtained from PLS projections plotting, for example, t_1 against t_2 (see Fig. 22.13). This information cannot be obtained using principal component analysis (PCA, see below) of the same independent variables.

E Linear discriminant analysis (LDA)

In many biological experiments only discrete (or classification, or categorical) data are obtained, such as inactive/active, agonist/antagonist. In some data sets a clear separation may be found between such classes in multidimensional space. An appropriate method to describe separation between classes is linear discriminant analysis (LDA). For example, using LDA it was possible to distinguish 24 calmodulin inhibitors in three different activity groups (with different associated binding modes) using descriptors which described the positive potential surface area on the side chain, as well as the total and neutral surface areas on the ring in the inhibitor molecules.[60] This group assignment information was used to classify 29 additional inhibitors. A related method used to analyse categorical data is adaptive least squares (ALS).

F SIMCA and related methods

It often occurs that active compounds cannot be well separated from inactive ones using linear models such as PLS or LDA. This may be because the active compounds cluster together in an area of property space and they are surrounded by inactive compounds. Such data are called embedded or asymmetric data. Several methods have been developed to treat such data sets, the best known is the SIMCA algorithm. The SIMCA (soft independent modelling of class analogy) method is a tool for pattern

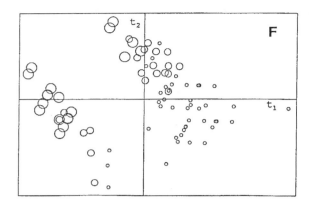

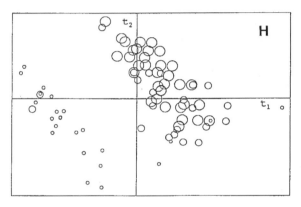

Fig. 22.13 PLS projections of thiolcarbamates data of the first latent variable t_1 against the second latent variable t_2. The size of the circles denote the fungicidal (F) or herbicidal (H) activity.

recognition in a data set.[54] The basic idea is to build several local class models using disjoint PCA (see Section V) from a training data set. For new test compounds predictions can be made to which activity class the new compound belongs. A related method for analysing embedded data is single-class discrimination.[61]

G Back propagation neural networks

A number of techniques related to artificial intelligence and natural computing have been investigated in quantitative structure–activity relationships. The methods use so-called natural algorithms, which are based upon principles in nature, such as natural selection in evolution. Examples are artificial neural network,[62] learning machines,[63] rule-induction[64] and genetic algorithms (see below). Different types of neural networks have been conceived, but for QSAR studies back-propagation (BP) appears to be the most suitable approach.[62] A BP network generally consists of three layers: input nodes (one for each property), hidden nodes and output nodes (the activity), linked by weights. The network is trained, i.e. the values of the weights are optimized, using the property values of the compounds to accurately reproduce the activity of the compounds in the training set. An example of the architecture of a 5-2-1 back propagation network is given in Fig. 22.14.

The advantage of neural networks is that few statistical assumptions are made a priori. Among the disadvantages are that no real statistical validation method has been developed. There is a danger of over-fitting of the data, resulting in poor predictions outside the training set. The current best practice is to divide the data set into three: one training set, one test set and one validation set. The network is trained using the training set and training is terminated once the predictions for the test set start to deteriorate due to over-fitting of the training set. The ability of the network to make accurate predictions is then tested using the validation set. The results are not always easy to interpret in terms of chemistry due to the complexity and non-linearity of the neural network model. However, the ability to deal with non-linear relationships is an attractive feature of neural networks.

H Decision trees

The term decision trees[64] is used here to cover a variety of related methods including rule induction and recursive partitioning. Decision trees classify compounds into local activity groups using sets of rules. The data set is sequentially divided using rules which maximize the separation of active and inactive compounds at each step. For example, a simple rule which states $\log P > 2$ may serve to distinguish the majority of active compounds from more hydrophilic inactive ones. The data set is sequentially split in this manner, resulting in a tree diagram where the final nodes contain, ideally, a single activity class. One active group of compounds may be defined by a set of rules which state, for example:

$$\log P > 2, \text{ mol.wt.} < 450, \text{ dipole moment} < 4$$

while another group of active compounds may be defined by a very different set of rules, for example:

$$\log P < 2, \text{ p}K_a > 7.4, \text{ mol.wt.} < 300$$

Because of this, decision trees can sometimes result in models which are hard to interpret. One useful approach is to limit the number of descriptors to those felt to be relevant to the problem at hand and in this way less complex trees may be derived which provide interpretable information.

I Genetic algorithms

In multiple linear regression automatically a 'best' set of descriptors is selected, although within a larger set of descriptors there may be several 'best' solutions. This problem occurs in particular when large descriptor sets are being considered. The selection of the most appropriate descriptor set solutions in a QSAR study can be done by several techniques, including genetic algorithms. These algorithms are inspired by population genetics and are believed to produce higher-quality predictive models.[27] Genetic algorithms are being widely applied in drug discovery software for QSAR and molecular modelling for optimization applications.[65]

V PATTERN RECOGNITION IN DATA SETS

A Overview

Pattern recognition or unsupervised multivariate methods are used in QSAR to visualize similarity and clustering in a data set. The methods can be used to look for potential clustering of variables or compounds, by considering chemical or biological data separately or together. When compounds cluster together in a multivariate parameter space, this means that they are 'similar' with respect to the variables considered. This forms the basis of the science of molecular diversity studies. Alternatively, if we perform a pattern recognition using biological data, clustered compounds have a similar activity profile. This section describes

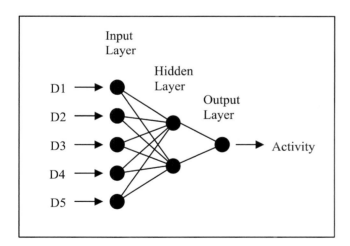

Fig. 22.14 Schematic representation of a 5-2-1 back propagation network. The network consists of an input layer containing five nodes into which values for descriptors D1 to D5 are input. This is connected by weights to a hidden layer of two nodes, which is connected by weights to the final output layer which has one node which is the activity value. The network is trained in an iterative fashion by adjusting the weights until the predicted activity values best match the measured activity values.

the methods of principal component analysis (PCA), cluster analysis (CA), non-linear mapping (NLM) and Kohonen mapping (KM). PCA is the most widely used method in QSAR and has therefore been described in most detail in this section. A brief overview of the other methods has been provided. A comparison of PCA, cluster analysis, non-linear mapping and Kohonen mapping is given in reference 66.

B Principal component analysis (PCA)

Large data tables may hide information which is not easily detected by simple inspection of the various columns. Principal component analysis and some closely related techniques such as factor analysis (FA), correspondence factor analysis (CFA) and non-linear mapping (NLM), reduce a data matrix to new supervariables retaining a maximum of information or variance from the original data matrix. These new variables are called latent variables or principal components, and are orthogonal vectors composed of linear combinations of the original variables. This concept is shown schematically in Fig. 22.15.

The typical use of PCA is illustrated by an example from antibacterial research. In studies of the antibacterial effects

of sulfones and sulfonamides in whole-cell and cell-free systems, missing data (19%) have been estimated by an iterative process using PCA.[67] However, estimation of missing values should be done with care and preferably avoided.

Using the minimal inhibition concentration (MIC) data for nine different strains, two significant principal components could be obtained, accounting for 77.1% and 16.1%, respectively, of the data variance. The loading plot, i.e. a plot of the calculated principal components with respect to the descriptors, shows that the first component is mainly related to the seven cell-free test systems, while the second one represents the two whole-cell test results. In other words, much redundant information was obtained by measuring in nine test systems; two would have been sufficient. This separation means that the potency in both test systems is governed by different physicochemical properties.

The principal components derived from the activities can be correlated to physicochemical properties using MLR. Thus it was found that component one appears to be dominated by electronic factors (equation 12), while in component two transport (lipophilicity) properties (equation 13) play a role. The following parameters are used: Δppm(NH$_2$) is the NMR chemical shift of the amino protons relative to the unsubstituted congener, f_i is the fraction ionized at pH 7.4, and log k' is the lipophilicity measured by HPLC. Equation (13) shows an example of a nonsignificant constant term, since the standard deviation is larger than the term itself. In such cases the equation should be forced through the origin.

$$PC1 = -7.02(\pm 1.25)\Delta ppm(NH_2) + 1.81(\pm 0.42)f_i$$
$$- 0.93(\pm 0.19) \qquad (12)$$

$$PC2 = 1.40(\pm 0.52)\log k' - 3.49(\pm 1.32)\log(0.098k'$$
$$+ 1) + 0.51(\pm 0.73) \qquad (13)$$

Thus PCA may be used to filter out the most relevant information in a data set. Applications of variations of PCA, such as correspondence factorial analysis and non-linear mapping may have small advantages with particular data sets, but require expert support.

As mentioned in Section II, PCA can also be used to select a subset of uncorrelated variables from a large starting set. This reduced set of descriptors can then be used in a MLR or PLS analysis to find a correlation between biological and chemical data. An illustration of this procedure is the optimization of aqueous solubility of xanthine antagonists (Fig. 22.16) by QSAR methods.[68]

Preprocessing of 28 parameters by PCA produced a selection of 11 significant descriptors. The final QSAR using only these 11 descriptors obtained by MLR was

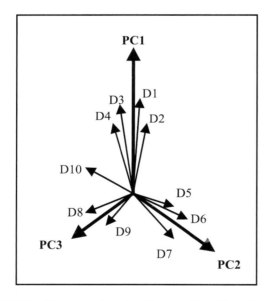

Fig. 22.15 Diagram showing the relationship between a set of ten descriptors (D1 to D10) and their first three principal components (PC1 to PC3). PC1 explains the greatest % of variation (depicted below by a longer vector), PC2 is orthogonal to PC1 and smaller, PC3 is the smallest PC and is orthogonal to PC1 and PC2. PC1 is highly correlated with descriptors D1 to D4, PC2 is correlated with descriptors D5 to D7, PC3 is correlated with descriptors D8 and D9. Descriptor 10 is not well represented by any of the PCs.

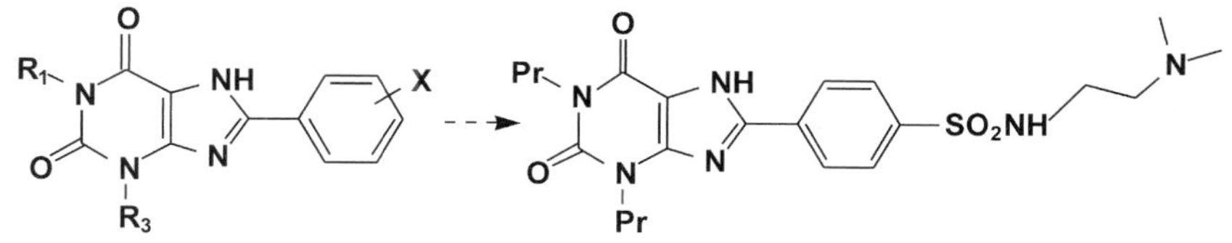

Fig. 22.16 Design of water soluble xanthine antagonists by PCA/MLR.

equation (14), in which only six descriptors remained to build a robust model.

$$PTNCY_{cow} = -0.99(\pm 0.13)HACCEPT_m$$

$$+ 0.81(\pm 0.18)\pi_{R3} - 1.16(\pm 0.18)MR_0$$

$$- 0.88(\pm 0.20)\sigma_0 - 1.57(\pm 0.24)ACID$$

$$- 1.17(\pm 0.24)HBOND + 2.22 \qquad (14)$$

$$n = 56; \; r^2 = 0.83; \; s = 0.40; \; F = 40.0$$

The quality of this equation is not excellent, but explains nevertheless 83% of the variance. The s value should be close to the experimental error in the biological assay, here approximately 0.15. $PTNCY_{cow}$ reflects the potency of the compounds measured in a bovine membrane assay. The interpretation of this equation is as follows. A H-bond accepting group in the meta position of the phenyl ring (e.g. 3-NH_2) is unfavourable. Lipophilic groups in position R_3 are favourable. In the ortho position on the phenyl ring large groups and H-bond acceptors (OH, and OMe) are detrimental to activity, while electron-donating groups are favourable. The phenyl ring should not be substituted with carboxylic acid function. HACCEPT, ACID and HBOND are so-called indicator values, denoting absence (0) or presence (1) of a specific feature.

Since no descriptor for the para position in the phenyl ring appears in the equation, it was successfully attempted to improve the aqueous solubility of the compounds by introducing basic sulfonamide groups in that position, while conserving the excellent biological activity.

C Cluster analysis

Among the mathematical tools to investigate patterns and clustering behaviour in data sets, two techniques are widely established, namely principal component analysis and cluster analysis. Both can be used to reduce the dimensionality of a problem. Or in other words, cluster analysis can be used for variable or descriptor selection from a larger set. On the other hand cluster analysis may be used to investigate similarity among compounds. Cluster analysis is often used complementary to PCA.

Similarity and dissimilarity among points in multidimensional space can be quantified by calculating their Euclidean distance. A number of different hierarchical clustering algorithms are available, e.g. single linkage or complete linkage. The difference lies in the definition of the spatial distance among pairs of data points. The results are presented as a dendrogram, i.e. a tree-like figure, where very similar compounds or descriptors are close together. Good results are often obtained using Ward's method. Clustering may still be partially due to chance and unrelated to the underlying chemical or biological meaning.

Cluster significance analysis (CSA) is a related, supervised method that can be used to determine subsets of properties that cause active compounds to cluster together.[69,70]

D Non-linear mapping

Like PCA, non-linear mapping (NLM), or multidimensional scaling, is a method for visualizing relationships between objects, which in the medicinal chemistry context often are compounds, but could equally be a number of measured activities.[71] It is an iterative minimization procedure which attempts to preserve interpoint distances in multidimensional space in a 2D or 3D representation. Unlike PCA however, the axes are not orthogonal and are not clearly interpretable with respect to the original variables. Non-linear mapping has been used to cluster aromatic and aliphatic substituents.[72,73]

E Kohonen mapping

A certain advantage over other pattern recognition techniques, such as PCA, may exist in the data reduction capability of neural networks. The Kohonen neural network, also known as a self-organizing map, is a technique to map multidimensional properties such as surface properties on a 2D plot.[74] This is an unsupervised learning process which can be used to detect patterns in large data sets, and to predict whether a compound belongs to a particular activity class. A Kohonen map (KM) consists of a 2D grid of nodes and compounds with similar properties cluster in the same or adjacent nodes following training of the map. Other more

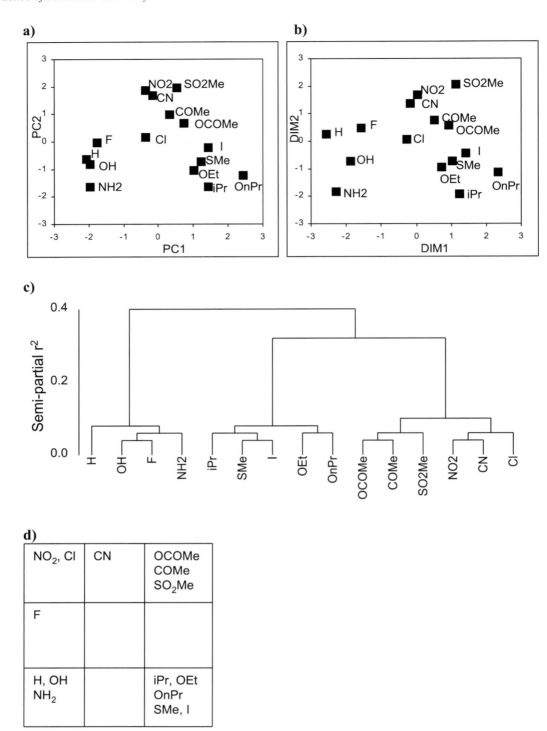

Fig. 22.17 Comparison of (a) principal components analysis, (b) non-linear mapping, (c) Ward's cluster analysis and (d) Kohonen mapping to display similarity of 15 substituents characterized by five para substituent constants (π, *F*, *R*, MR and *L*).

complex representations of KMs have been reported, such as 3D torus (doughnut-like) maps.

A comparison of the results obtained by applying PCA, cluster analysis, NLM and KM to a data set of 15 substituents characterized by five substituent constants (π, F, R, MR and Verloop's L) is given in Fig. 22.17.

It can be seen that the PCA and NLM plots are quite similar. This is to be expected when the first two principal components explain a high percentage of the variation in the data (73% in this case). The main difference between the plots is on the left-hand side where the H, F, OH and NH_2 are less closely clustered in the NLM than in the PCA plot. Similar clustering is apparent in the cluster analysis and KM. In fact all four approaches give similar, but not identical, representations of similarity in this example.

VI PRACTICAL HINTS AND PERSPECTIVES

A wide variety of chemometric statistical tools may be used to investigate quantitative structure−activity relationships or more general structure−property correlations. Some of these techniques require expert support. However, the bench chemist may successfully use a number of techniques, when some basic guidelines as discussed in this chapter are followed. The most important methods are:

Correlation studies:

- Hansch analysis using MLR
- Partial least squares (PLS) regression
- Back-propagation neural networks.

Pattern recognition studies:

- Principal component analysis (PCA)
- Non-linear mapping (NLM)
- Cluster analysis
- Kohonen neural networks.

Which physicochemical or structural descriptors should be used? The answer may be rather pragmatic: all those available. First of all, any experimental physicochemical property can be used, such as lipophilicity data,[75] aqueous solubility, membrane transport properties and ionization constants. A number of descriptors can be easily calculated, such as the molecular weight, octanol/water partition coefficients, molar refractivity, molecular volume, surface area, polar surface area, number of H-bond donating and accepting groups. Using quantum chemical methods, a number of electronic properties may be calculated, such as partial atomic charges. As discussed the number of descriptors may be too large to generate confident models

and some prior work needs to be done for which a number of variable selection tools are available.

Since drugs and their targets are three-dimensional objects, it is of course appropriate to consider 3D molecular properties. This is the objective of several approaches, combining statistical and modelling techniques, referred to as 3D-QSAR. The comparative molecular field analysis (CoMFA) method has been widely used and will be discussed elsewhere in this book (see Chapter 25).

Although molecular modelling is more appealing to most medicinal chemists, QSAR studies are of considerable interest to many projects. In many cases the two approaches are complementary. Nevertheless, one should be aware that in QSAR as well as in modelling studies, most models are being developed and used under the assumption of a single binding mode. However, X-ray studies have shown very elegantly that different binding modes may occur, even within a series of closely related structures, such as thrombin inhibitors.

Genomics, bio- and chemoinformatics, combinatorial chemistry, and high-throughput screening are key contributors to modern medicinal chemistry, particularly for finding new targets and lead compounds. Using parallel and traditional synthesis methods, it is attempted to optimize these leads to drug candidates with high target affinity and selectivity, but also with optimal pharmacokinetic and physicochemical properties, which taken together produces the desired therapeutic profile.[2] Quantitative approaches to structure−activity, structure−permeability, structure−pharmacokinetic and structure−metabolism relationships will continue to play a role in this optimization process. The wide recognition of the importance of the latter three types of relationships to molecular structure, as well as library design and HTS has renewed a keen interest in various QSAR approaches.[76,77]

REFERENCES

1. Smith, D.A., Jones, B.C. and Walker, D.K. (1996) Design of drugs involving the concepts and theories of drug metabolism and pharmacokinetics. *Med. Res. Rev.* **16**: 243−266.
2. In Testa, B., Van de Waterbeemd, H., Folkers, G., Guy, R. (eds), (2001) *Pharmacokinetic Optimization in Drug Research: Biological, Physicochemical and Computational Strategies.* Wiley−VCH, Weinheim, Zurich.
3. Chan, O.H. and Stewart, B.H. (1996) Physicochemical and drug-delivery considerations for oral drug bioavailability. *Drug Disc. Today* **1**: 461−473.
4. Tute, M.S. (1990) History and objectives of quantitative drug design. In Hansch, C., Sammes, P.G. and Taylor, J.B. (eds). *Comprehensive Medicinal Chemistry*, vol. 4, Quantitative Drug Design, pp. 1−31. Pergamon Press, New York.
5. Hansch, C. (1981) The physicochemical approach to drug design and discovery (QSAR). *Drug Dev. Res.* **1**: 267−309.
6. Hansch, C. (1984) On the state of QSAR. *Drug Inf. J.* **18**: 115−122.

7. Craig, P.N. (1984) QSAR — origins and present status: a historical perspective. *Drug Inf. J.* **18**: 123–130.

8. Hansch, C. (1993) Quantitative structure–activity relationships and the unnamed science. *Acc. Chem. Res.* **26**: 147–153.

9. Hyde, R.M. and Livingstone, D.J. (1988) Perspectives in QSAR: computer chemistry and pattern recognition. *J. Comput.-Aid. Mol. Des.* **2**: 145–155.

10. Van de Waterbeemd, H. (1992) The history of drug research: From Hansch to the present. *Quant. Struct.-Act. Relat.* **11**: 200–204.

11. Martin, Y.C. (1981) A practioner's perspective of the role of quantitative structure–activity analysis in medicinal chemistry. *J. Med. Chem.* **24**: 229–237.

12. Topliss, J.G. (1993) Some observations on classical QSAR. *Perspect. Drug Disc. Des.* **1**: 253–268.

13. Van de Waterbeemd, H. (1993) Recent progress in QSAR-technology. *Drug Des. Disc.* **9**: 277–285.

14. In Van de Waterbeemd, H. (ed.), (1995) *Chemometric Methods in Molecular Design.* VCH, Weinheim.

15. Testa, B. and Kier, L.B. (1991) The concept of molecular structure in structure-activity relationship studies and drug design. *Med. Res. Rev.* **11**: 35–48.

16. Van de Waterbeemd, H. and Testa, B. (1987) The parametrization of lipophilicity and other structural properties in drug design. *Adv. Drug Res.* **16**: 85–225.

17. Todeschini, R. and Consonni, V. (2000) *Methods and Principles in Medicinal Chemistry*, vol. 11, Handbook of Molecular Descriptors. Wiley–VCH, Weinheim.

18. El Tayar, N., Testa, B. and Carrupt, P.A. (1992) Polar intermolecular interactions encoded in partition coefficients: an indirect estimation of hydrogen-bond parameters of polyfunctional solutes. *J. Phys. Chem.* **96**: 1455–1459.

19. Leahy, D.E., Morris, J.J., Taylor, P.J. and Wait, A.R. Membranes and their models: Towards a rational choice of partitioning system. In Silipo, C. and Vittoria, A. (eds). *QSAR: Rational Approaches to the Design of Bioactive Compounds*, pp. 75–82. Elsevier, Amsterdam.

20. Young, R.C., Mitchell, R.C., Brown, T.H., Ganellin, C.R., Griffiths, R., Jones, M., Rana, K.K., Saunders, D., Smith, I.R., Sore, N.E. and Wilks, T.J. (1988) Development of a new physicochemical model for brain penetration and its application to the design of centrally acting H_2 receptor histamine antagonists. *J. Med. Chem.* **31**: 656–671.

21. Van de Waterbeemd, H. and Kansy, M. (1992) Hydrogen-bonding capacity and brain penetration. *Chimia* **46**: 299–303.

22. Lombardo, F., Shalaeva, M.Y., Tupper, K.A., Gao, F. and Abraham, M.H. (2000) ElogPoct: A tool for lipophilicity determination in drug discovery. *J. Med. Chem.* **43**: 2922–2928.

23. El Tayar, N., Marston, A., Bechalany, A., Hostettmann, K. and Testa, B. (1989) Use of centrifugal partition chromatography for assessing partition coefficients in various solvent systems. *J. Chromatogr.* **469**: 91–99.

24. Lipinski, C.A., Lombardo, F., Dominy, B.W. and Feeney, P.J. (1997) Experimental and computational approaches to estimate solubility and permeability in drug discovery and development settings. *Adv. Drug. Deliv. Rev.* **23**: 3–25.

25. Bevan, C.D. and Lloyd, R.S. (2000) A high-throughput screening method for the determination of aqueous drug solubility using laser nephelometry in microtiter plates. *Anal. Chem.* **72**: 1781–1787.

26. Rose, V.S., Rahr, E. and Hudson, B.D. (1994) The use of Procrustes analysis to compare different property sets for the characterization of a diverse set of compounds. *Quant. Struct.-Act. Relat.* **13**: 152–158.

27. Rogers, D. and Hopfinger, A.J. (1994) Application of genetic function approximation to quantitative structure–activity relationships and quantitative structure–property relationships. *J. Chem. Inf. Comp. Sci.* **34**: 854–866.

28. Zheng, W. and Tropsha, A. (2000) Novel variable selection quantitative structure-property relationship approach based on the k-nearest-neighbour principle. *J. Chem. Inf. Comput. Sci.* **40**: 185–194.

29. Carlson, R. and Nordahl, A. (1993) Exploring organic synthetic experimental procedures. *Top. Curr. Chem.* **166**: 1–64.

30. Topliss, J.G. (1972) Utilization of operational schemes for analog synthesis in drug design. *J. Med. Chem.* **15**: 1006–1011.

31. Craig, P.N. (1971) Interdependence between physical parameters and selection of substituent groups for correlation studies. *J. Med. Chem.* **14**: 680–684.

32. Hansch, C. and Leo, A. (1979) *Substituent Constants for Correlation Analysis in Chemistry and Biology.* Wiley, New York.

33. Hellberg, S., Sjöström, M., Skagerberg, B. and Wold, S. (1987) Peptide quantitative structure–activity relationships, a multivariate approach. *J. Med. Chem.* **30**: 1127–1135.

34. Van de Waterbeemd, H., El Tayar, N., Carrupt, P.A. and Testa, B. (1989) Pattern recognition study of QSAR substituent descriptors. *Comput.-Aid. Mol. Des.* **3**: 111–132.

35. Skagerberg, B., Bonelli, D., Clementi, S., Cruciani, G. and Ebert, C. (1989) Principal properties for aromatic substituents. A multivariate approach for design in QSAR. *Quant. Struct.-Act. Relat.* **8**: 32–38.

36. van de Waterbeemd, H., Costantino, G., Clementi, S., Cruciani, G. and Valigi, R. (1995). In Van de Waterbeemd, H. (ed.). *Methods and Principles in Medicinal Chemistry*, vol. 2, Chemometric Methods in Molecular Design, pp. 103–112. VCH: Weinheim.

37. Norinder, U. and Högberg, Th. (1992) PLS-based quantitative structure–activity relationship for substituted benzamides of clebopride type. Application of experimental design in drug design. *Acta Chem. Scand.* **46**: 363–366.

38. Sadowski, J. and Kubinyi, H. (1998) A scoring scheme for discriminating between drugs and nondrugs. *J. Med. Chem.* **41**: 3325–3329.

39. Ajay, A., Walter, W.P. and Murko, M.A. (1998) Can we learn to distinguish between 'drug-like' and 'nondrug-like' molecules? *J. Med. Chem.* **41**: 3314–3324.

40. Van de Waterbeemd, H., Smith, D.A., Beaumont, K. and Walker, D.K. (2001) Property-based design: Optimisation of drug absorption and pharmacokinetics. *J. Med. Chem.* **44**: 1313–1333.

41. Gorse, D. and Lahana, R. (2000) Functional diversity of compound libraries. *Curr. Opin. Chem. Biol.* **4**: 287–294.

42. Gillet, V.J. and Nicolotti, O. (2000) Evaluation of reactant-based and product-based approaches to the design of combinatorial libraries. *Perspect. Drug Disc. Des.* **20**: 265–287.

43. Good, A.C. and Lewis, R.A. (1997) A new methodology for profiling combinatorial libraries and screening sets: Cleaning up the design process with HARPick. *J. Med. Chem.* **40**: 3926–3936.

44. Martin, E. and Wong, A. (2000) Sensitivity analysis and other improvements to tailored combinatorial library design. *J. Chem. Inf. Comput. Sci.* **40**: 215–220.

45. Sheridan, R.P., SanFeliciano, S.G. and Kearsley, S.K. (2000) Designing targeted libraries with genetic algorithms. *J. Mol. Graph. Model.* **18**: 320–334.

46. Brown, R.D., Hassan, M. and Waldman, M. (2000) Combinatorial library design for diversity, cost efficiency and drug-like character. *J. Mol. Graph. Model.* **18**: 427–437.

47. Baxter, C.A., Murray, C.W., Waszkowycz, B., Li, J., Sykes, R.A., Bone, R. G. A., Perkins, T. D. J. and Wylie, W. (2000) New approach to molecular docking and its application to virtual screening of chemical databases. *J. Chem. Inf. Comput. Sci.* **40**: 254–262.

48. Kubinyi, H. (1993) *Methods and Principles in Medicinal Chemistry*, vol. 1, QSAR: Hansch Analysis and Related Approaches, VCH, Weinheim.

49. Grunewald, G.L., Dahanukar, V.H., Jalluri, R.K. and Criscione, K.R. (1999) Synthesis, biochemical evaluation and classical

and three-dimensional quantitative structure–activity relationship studies of 7-substituted-1,2,3,4-tetrahydroisoquinolines and their relative affinities toward phenylethanolamine *N*-methyltransferase and the α2-adrenoreceptor. *J. Med. Chem.* **42**: 118–134.

50. Naito, Y., Yamaura, Y., Inoue, Y., Fukaya, C., Yokoyama, K., Nakagawa, Y. and Fujita, T. (1992) Quantitative structure–activity relationships of 2-[4-(thiazol-2-yl)phenyl]propionic acid derivatives inhibiting cyclooxygenase. *Eur. J. Med. Chem.* **27**: 645–654.

51. Kubinyi, H. (1990) The Free-Wilson method and its relationship to the extrathermodynamic approach. In Hansch, C., Sammes, P.G. and Taylor, J.B. (eds). *Comprehensive Medicinal Chemistry*, vol. 4, Quantitative Drug Design, pp. 589–643. Pergamon Press, New York.

52. Fleischer, R., Frohberg, P., Büge, A., Nuhn, P. and Wiese, M. (2000) QSAR analysis of substituted 2-phenylhydrazonoacetamides acting as inhibitors of 15-lipoxygenase. *Quant. Struct.-Act. Relat.* **19**: 162–172.

53. Cramer, R.D. (1993) Partial least squares (PLS): its strength and limitations. *Perspect. Drug Disc. Des.* **1**: 269–278.

54. Eriksson, L. and Johansson, E. (1996) Multivariate design and modeling in QSAR. *Chemom. Int. Lab. Syst.* **34**: 1–19.

55. Malpass, J.A., Salt, D.W., Ford, M.G., Watcyn, E.W. and Livingstone, D.J. (1995) Continuum regression. In van de Waterbeemd, H. (ed.). *Methods and Principles in Medicinal Chemistry*, vol. 3, Advanced Computer-Assisted Techniques in Drug Discovery, pp. 163–189. VCH, Weinheim.

56. Wold, S. and Eriksson, L. (1995) Validation tools. In van de Waterbeemd, H. (ed.). *Methods and Principles in Medicinal Chemistry*, vol. 2, Chemometric Methods in Molecular Design, pp. 309–318. VCH, Weinheim.

57. Baroni, M., Costantino, G., Cruciani, G., Riganelli, D., Valigi, R. and Clementi, S. (1993) Generating optimal linear PLS estimations (GOLPE): an advanced chemometric tool for handling 3D-QSAR problems. *Quant. Struct.-Act. Relat.* **12**: 9–20.

58. Lotta, T., Taskinen, J., Bäckström, R. and Nissinen, E. (1992) PLS modelling of structure–activity relationships of catechol *O*-methyltransferase inhibitors. *J. Comput.-Aided Mol. Des.* **6**: 253–272.

59. Miyashita, Y., Ohsako, H., Takayama, C. and Sasaki, S. (1992) Multivariate structure–activity relationships analysis of fungicidal and herbicidal thiolcarbamates using partial least squares method. *Quant. Struct.-Act. Relat.* **11**: 17–22.

60. Liu, Q., Hirono, S. and Moriguchi, I. (1990) Quantitative structure–activity relationships for calmodulin inhibitors. *Chem. Pharm. Bull.* **38**: 2184–2189.

61. Rose, V.S., Wood, J. and MacFie, H. J. H. (1991) Single class discrimination using principal component analysis (SCD-PCA). *Quant. Struct.-Act. Relat.* **10**: 359–368.

62. Manallack, D.T. and Livingstone, D.J. (1999) Neural networks in drug discovery: have they lived up to their promise? *Eur. J. Med. Chem.* **34**: 195–208.

63. King, R.D., Hirst, J.D. and Sternberg, M. J. E. (1993) New approaches to QSAR: neural networks and machine learning. *Perspect. Drug Disc. Des.* **1**: 279–290.

64. A-Razzak, M. and Glen, R.C. (1995) Rule induction applied to the derivation of quantitative structure–activity relationships. In van de Waterbeemd, H. (ed.). *Methods and Principles in Medicinal Chemistry*, vol. 3, Advanced Computer-Assisted Techniques in Drug Discovery, pp. 319–331. VCH, Weinheim.

65. Clarke, D.E. (2000) *Evolutionary Algorithms in Molecular Design*. Wiley–VCH, Weinheim.

66. Rose, V.S., Croall, I.F. and MacFie, H. J. H. (1991) An application of unsupervised neural network methodology (Kohonen topology-preserving mapping) to QSAR analysis. *Quant. Struct.-Act. Relat.* **10**: 6–15.

67. Coats, E.A., Cordes, H.-P., Kulkarni, V.M., Richter, M., Schaper, K.J., Wiese, M. and Seydel, J.K. (1985) Multiple regression and principal component analysis of antibacterial activities of sulfones and sulfonamides in whole cell and cell-free systems of various DDS sensitive and resistant bacterial strains. *Quant. Struct.-Act. Relat.* **3**: 99–109.

68. Hamilton, H.W., Ortwine, D.F., Worth, D.F., Badger, E.W., Bristol, J.A., Bruns, R.F., Haleen, S.J. and Steffen, R.P. (1985) Synthesis of xanthines as adenosine antagonists, a practical quantitative structure–activity relationship application. *J. Med. Chem.* **28**: 1071–1079.

69. McFarland, J.W. and Gans, D.J. (1986) On the significance of clusters in the graphical display of structure–activity data. *J. Med. Chem.* **29**: 505–514.

70. Rose, V.S. and Wood, J. (1998) Generalized cluster significance analysis and stepwise cluster significance analysis with conditional probabilities. *Quant. Struct.-Act. Relat.* **17**: 348–356.

71. Kowalski, B.R. and Bender, C.F. (1973) Pattern recognition. II Linear and non-linear methods for displaying chemical data. *J. Am. Chem. Soc.* **95**: 686–693.

72. Domine, D., Devillers, J. and Chastrette, M. (1994) A nonlinear map of substituent constants for selecting test series and deriving structure–activity relationships. I. Aromatic series. *J. Med. Chem.* **37**: 973–980.

73. Domine, D., Devillers, J. and Chastrette, M. (1994) A nonlinear map of substituent constants for selecting test series and deriving structure–activity relationships. II. Aliphatic series. *J. Med. Chem.* **37**: 981–987.

74. Anzali, S., Barnickel, G., Krug, M., Sadowski, J., Wagener, M., Gasteiger, J. and Polanski, J. (1996) The comparison of geometric and electronic properties of molecular surfaces by neural networks: Application to the analysis of corticosteroid-binding globulin activity of steroids. *J. Comput.-Aid. Mol. Des.* **10**: 521–534.

75. In Pliska, V., Testa, B. and Van de Waterbeemd, H. (eds). (1996) *Lipophilicity in Drug Action and Toxicology*. VCH, Weinheim.

76. Livingstone, D. (1995) *Data Analysis for Chemists*. Oxford University Press, Oxford.

77. Hansch, C. and Leo, A. (1995) *Fundamentals and Applications in Chemistry and Biology*. ACS, Washington.

PART V

SPATIAL ORGANIZATION, RECEPTOR MAPPING AND MOLECULAR MODELING

23

STEREOCHEMICAL ASPECTS OF DRUG ACTION I: CONFORMATIONAL RESTRICTION, STERIC HINDRANCE AND HYDROPHOBIC COLLAPSE

Daniel H. Rich, M. Angels Estiarte and Philip A. Hart

Man's mind once stretched by a new idea, never regains its original dimension   **Oliver Wendell Homes**

I INTRODUCTION

Conformational restriction, steric hindrance and hydrophobic collapse represent important concepts used by medicinal chemists to optimize biological activity for a target receptor or enzyme or to gain receptor selectivity. Although these concepts were formulated independently, they are interdependent, since the lack of biological activity exhibited by synthetic analogs can, in principle, be rationalized in terms of each concept. We begin with a brief look at the origin of conformational restriction and end by illustrating how these ideas are used today.

II ORIGIN OF CONFORMATIONAL RESTRICTION

The importance of multiple functional groups and their spatial arrangement to effective receptor binding has been recognized for many years. The hypothesis that a fixed and specific two-dimensional distance between two oxygen functions in estrogen analogs was needed to elicit estrogenic activity[1] marks the very early thinking about the correlation of the spatial disposition of important functional groups with biological activity. It was not until the early 1950s, when our present concepts of conformational analysis were being developed,[2] that conformation was suspected to be a determinant of the spatial arrangement of functional groups. It was suggested soon thereafter that enzymes or drug receptors might prefer specific ligand conformations or distributions of conformations. Schueler hypothesized that conformational flexibility might be the important determinant for the muscarinic activity of acetylcholine, which was consistent with the attenuated activity of rigid piperidinium analogs of acetylcholine.[2-5] To rationalize how flexible ligands might bind to the muscarinic receptor more effectively than rigid counterparts, Schueler proposed that the receptor must also be flexible so as to permit better structural correlation between ligand and receptor. In contrast, Archer et al.[6] adopted the point of view that the nicotinic and muscarinic actions of acetylcholine might be due to different conformations of the flexible ligand, which could be differentiated by conformational restriction (Archer described it as "configurationally frozen"). He synthesized various isomers of 2-tropanyl acetate and concluded that a transoid conformation is favored at the muscarinic receptor and that a cisoid form is favored at the nicotinic receptor. These two studies, one emphasizing the dynamics of ligand–receptor interactions and the other, the time-independent complementarity of ligand and receptor have evolved into the present. Today, the concept of rigid complementarity has led to attempts to develop "rigid", or conformationally restricted (conformationally constrained) analogs of inherently conformationally flexible substances in order to delineate the preferred conformation that a ligand would adopt upon binding to a given receptor, which we will call the "bioactive conformation". In addition, conformational restriction has been applied with the expectation that very active analogs can be obtained, and as a requisite first step to simplify lead structures in the search for new drug candidates. Finally, conformational restriction is used to discover specificity among the members of multireceptor families. Throughout this chapter, the reader is cautioned to recognize that even highly conformationally restricted molecules can adopt multiple, closely related conformations so that the terms *bioactive conformation* or *constrained analog* more accurately describe an average of a limited population of closely related conformers.

III USE WITH SMALL LIGANDS

The principle of conformational restriction was first applied to characterize the bioactive conformation of acetyl choline acting at the muscarinic and nicotinic receptors. Conformational restriction has been applied to many other small ligands (e.g. see reviews by Martin-Smith *et al.*,[7] Portoghese,[8] Mutschler and Lambrecht,[9] and Casy *et al.*[10]), but the work on acetylcholine analogs exemplifies the necessary ideas and techniques required to understand the general approach, as well as its strengths and limitations, when applied to small molecules.

Acetylcholine has four bonds labelled χ_1, χ_2, χ_3, and χ_4 (see (**23.1**), Fig. 23.1) about which conformational change can take place. If one ignores the methyl group rotations and corrects for the identical conformations formed by rotations about the carbon-nitrogen bond (χ_4), a total of nine distinct acetylcholine conformations are possible. In principle, each of these conformations could have a measurably different ligand–receptor binding constant, which could be tested by synthesizing the appropriate restricted analog. X-ray analysis of various acetylcholine crystals reveals conformational variations in the crystal depending upon the nature of the counterion.[8] Conformational analysis of acetylcholine in solution, suggests that the gauche or near gauche conformation about χ_3 is the most probable one in aqueous medium, but the method is unable to specify the remaining conformational parameters for the other torsion angles.[8,11]

Early work on the synthesis and bioassay of conformationally restricted ACh analogs (Fig. 23.1) suggested that *trans* rigid analogs are preferred by the muscarinic receptor.[7,8] Results pertinent to the nicotinic receptor were not conclusive. Many of the compounds (e.g. (**23.2**)–(**23.4**)) are analogs derived from muscarine. The only rigid analog of acetylcholine to have muscarinic activity comparable to that of acetylcholine is the *trans* cyclopropyl analog (**23.5**)[12,13] and the corresponding *cis* cyclopropyl analog is

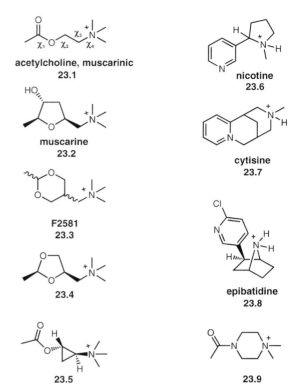

Fig. 23.1 Conformationally restricted analog of acetylcholine.

much less active. How does one interpret such data? Here is the first example where the interdependency of steric hindrance and conformational restriction complicate the interpretation of structure–activity data. The observed loss in biological activity for an analog could be caused by steric hindrance between the ligand and receptor due to the added constraining atoms, or it could be caused by the inability of the ligand to attain the proper conformation for binding to the receptor. Since the added restricting atoms could cause either effect, one cannot differentiate between these two possibilities when the derivative has no biological activity. In contrast, when the synthetic analog retains agonist activity, one can conclude that the proper conformation was realized in the ligand–receptor complex in spite of the restriction atoms.

These results illustrate the strengths and weaknesses of the conformationally restricted analog approach. In practice, many conformationally restricted analog must be prepared and tested to explore conformational space and to corroborate 3D models for the bioactive conformation of a small compound. Some of the active nicotinic and muscarinic agonists and antagonists studied up to 1970 are shown in Fig. 23.1. The muscarine analog provide information about the torsion angles, χ_1, χ_2, χ_3, in the bioactive conformation, whereas the cyclopropyl analog (**23.5**) provides information about χ_4. Bioassay of these and related compounds shows that the nicotinic and muscarinic receptors have different structural specificities. Models for the conformations that activate each receptor were deduced by using analog of acetylcholine in which the distances between the positive nitrogen and a hydrogen bond acceptor were constrained by the molecular structure to within certain specified distances.[14] These more or less "rigid" structures exemplify some of the possible conformations that acetylcholine can adopt. Muscarinic activity was correlated with a conformation in which the quaternary ammonium group (or with an equivalent group such as alkyl sulfonium, $-S^+(CH_3)_2$, and an unshared pair of electrons were separated by a distance of approximately 4.4 Å. Nicotinic activity was correlated with a different conformation in which the quaternary ammonium group and the ester carbonyl group are separated by about 5.9 Å. Other authors have established similar though not identical correlations.[15,16] Work with epibatidine (**23.8**), the pure nicotinic agonist and central analgesic, shows that its semi-rigid structure is well correlated with the classical nicotinic agonist, nicotine.[17–19]

Another type of conformational restriction reported by McGroddy *et al.*,[20,21] is derived from amide analogs of ACh, e.g. **23.9**. These are selective nicotinic agonists that manifest slow *cis–trans* amide rotation that influences biological activity. The nicotinic activity of these and more constrained analogs was measured using *Torpedo* electroplaque or BC$_3$H-1 cell-membrane or intact cells. The authors were able to differentiate the rate of initial binding (probably a function of the slow amide conformational exchange), relative antagonism (a measure of the rate of channel opening) and desensitization (a measure of the rate of ligand–receptor structural transition to a nonfunctional state). Related effects of *cis–trans* isomerization are known for tentoxin,[22,23] certain ACE inhibitors, and cyclosporin A (*vide infra*).

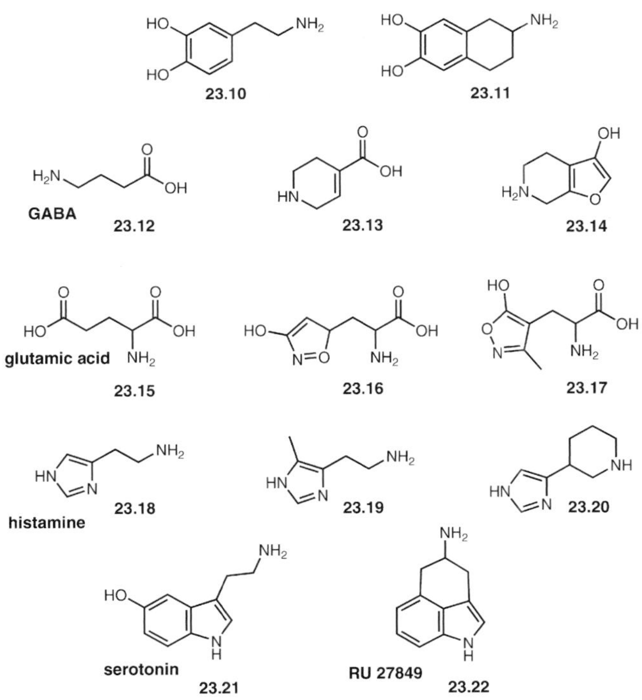

Conformationally restricted analogs of other low molecular mass agonists have been synthesized or discovered in attempts to characterize the bioactive conformation of the ligand. This field is vast and selected examples of constrained agonist analogs of dopamine (**23.10**) versus (**23.11**),[24] GABA (**23.12**) versus (**23.13**)–(**23.14**),[25] glutamic acid (**23.15**) versus (**23.16**) and (**23.17**),[26] histamine (**23.18**) versus (**23.19**) and (**23.20**),[27] serotonin (**23.21**) versus (**23.22**)[28] that have been discovered by application of this approach are shown in Fig. 23.2.

IV CONFORMATIONALLY RESTRICTED RECEPTOR ANTAGONISTS

The relationship between the structures of agonists and antagonists has always fascinated medicinal chemists, and has led to a number of suggestions that some inhibitors are conformationally restricted analogs of the agonist, except for the additional atoms that prevent receptor activation.[29] New methods to test whether peptide agonists and nonpeptide antagonists bind to the same atoms in a receptor have been developed by applying site-directed mutagenesis to G-protein-coupled receptors and comparing the effects of receptor mutation on agonist and antagonist binding. It appears that some peptide agonists and antagonists bind to different subsites on the protein receptor.[30,31] These initial results suggest that correlations of the structures of tachykinin agonists with the structures of the nonpeptide antagonists may not be valid. If these results prove to be general, they will discourage attempts to superimpose agonist and antagonist structures in other receptor systems.[32]

V CONFORMATIONAL RESTRICTION OF PEPTIDES

The use of conformational restriction to probe the bioactive conformation of a molecule has been much more productive with peptides than with small molecules. For the most part, this results from the fact that small peptides have so many flexible torsion angles that enormous numbers of conformations are possible in solution. For example, a tripeptide such as thyrotropin releasing hormone (TRH; (**23.23**), Fig. 23.3) with six flexible bonds could have over 65 000

Fig. 23.2 Conformationally restricted receptor antagonists.

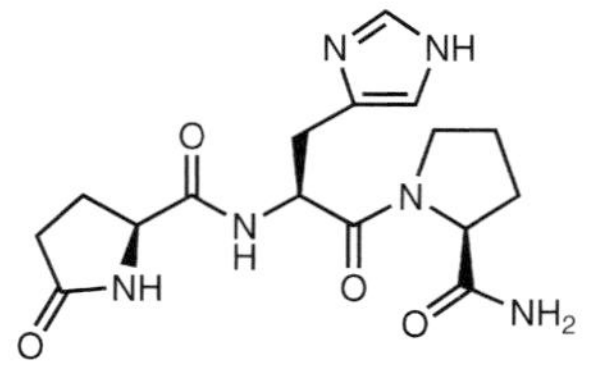

TRH
23.23

Fig. 23.3 Structure of TRH tripeptide.

possible conformations. The number of potential conformers for larger peptides is enormous and some method is needed to select the more plausible. Modern biophysical methods, e.g. X-ray crystallography or isotope-edited NMR,[33] have been used to characterize enzyme-bound conformations of peptides bound to small proteins, but most biophysical methods cannot determine the conformation of ligands bound to membrane-bound receptors of the G-protein coupling type.[34,35] Therefore, conformational restriction remains an important and powerful method for characterizing the bioactive conformation of peptides when biophysical characterization of the conformation is not possible.

Cyclization is one of the earliest techniques applied to restrict the conformational mobility of a peptide. Cyclic peptides are more stable to amide bond hydrolysis and allow less conformational flexibility. Consequently the resulting analogs are anticipated to be more selective and less toxic.

The first successful application of conformational restriction to peptide chemistry was carried out by Veber and his associates at Merck,[36] who were trying to simplify the structure of somatostatin (**23.24**) to produce an orally active derivative. Their approach was to introduce conformational restraints into the macrocyclic peptide ring system in order to reduce the number of conformations available to the analog. Not all substitutions were expected to produce biologically active products, but those that retained activity were assumed to be able to adopt conformations close to the normal bioactive conformation. This work began from the earlier discovery by Rivier[37] that replacement of L-tryptophan in the 8-position of somatostatin by D-tryptophan produced an analog which retained biological activity. This unusual biological result is produced when a D,L sequence (DTrp-Lys) replaces an L,L sequence (Trp-Lys) in a peptide at a type II′ β-turn because the topography of the amino acid side chains at these positions is essentially identical in these turns.[38] These results led Veber and associates to postulate that the amino acid sequence Phe-Trp-Lys-Thr might be part of a type II′ β-turn, and that this tetrapeptide sequence might comprise the active pharmacophore. Although this hypothesis was highly speculative for its time, it was shown to be essentially correct by applying the principle of conformational restriction (Fig. 23.4). Deletion of the *N*-terminal dipeptide, followed by insertion of the D-Trp at position-8, and replacement of the disulfide sulfurs with carbons produced analog (**23.25**).

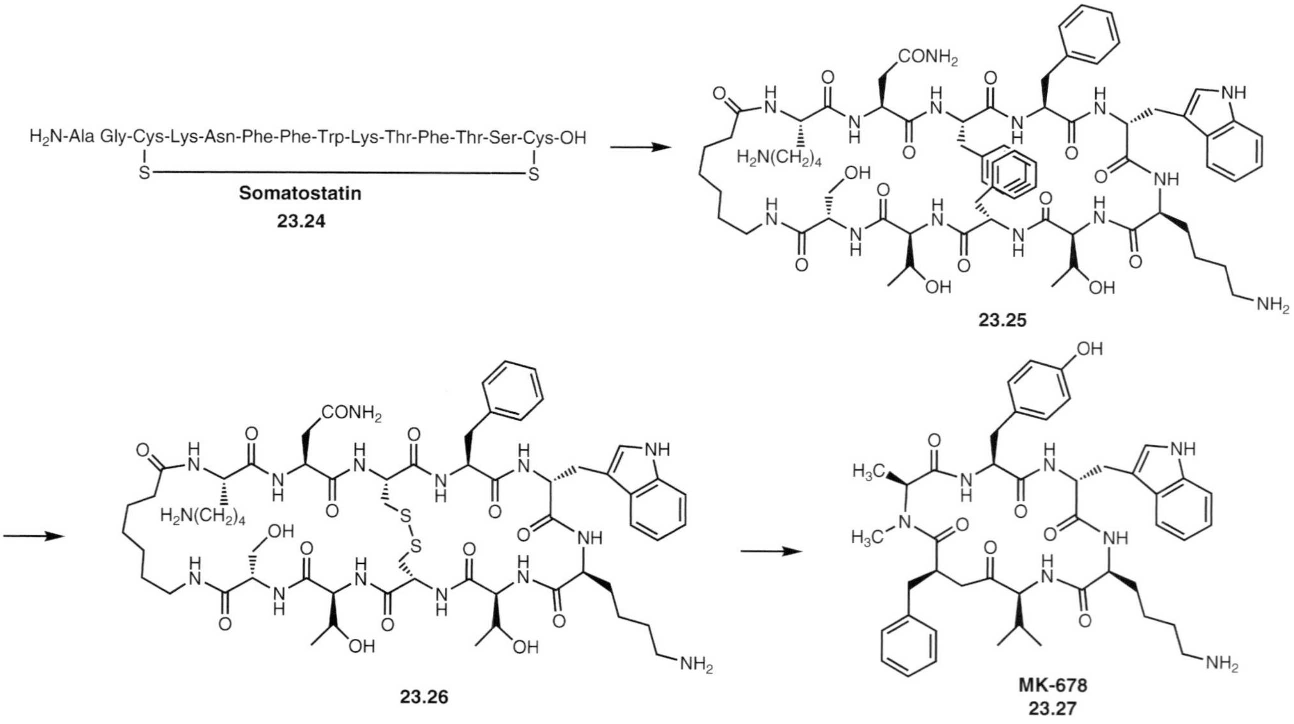

H₂N-Ala Gly-Cys-Lys-Asn-Phe-Phe-Trp-Lys-Thr-Phe-Thr-Ser-Cys-OH

Somatostatin
23.24

23.25

23.26

MK-678
23.27

Fig. 23.4 Conformationally restricted somatostatin analogs.

NMR and other data suggested that the two phenylalanine side chains were clustered and might be replaceable by other bridging groups. This led to the analog (**23.26**) in which a transannular disulfide bond limited the available conformations. When analog (**23.26**) was found to retain biological activity, the disulfide units were replaced with the corresponding stable carbon derivatives, more constraints were introduced, and the process was repeated. Some analogs were designed specifically to be inactive according to the pending hypothesis, in order to provide controls. After several iterations, a biologically active cyclic hexapeptide, (**23.27**) was discovered in which only six of the original 14 amino acids in somatostatin were needed to produce a fully active derivative. Veber also realized that the accessible surface area of the cyclic peptide was approximately the same as that of traditional drugs, e.g. benzodiazepines, and suggested that any biological activity that can be elicited by a cyclic hexapeptide or a smaller peptide could be mimicked by a nonpeptide, heterocyclic system.[39]

The work of Veber and coworkers established that valuable information about the bioactive conformation of a flexible peptide could be obtained by applying the principles of conformational restriction, and several additional examples were soon developed by following this strategy. An unusually active analog of α-melanotropin, e.g. (**23.28**) (Fig. 23.5), was formed by cyclizing the more flexible precursor.[40] Conformationally restricted enkephalin analogs e.g. (**23.29**), have been formed by cyclizing between positions 2 and 5 of enkephalin, and small cyclic analogs of endothelin, e.g. (**23.30**) have been discovered by applying these methods.[41]

Conformational restriction can be introduced into flexible peptides by a variety of methods. For example, Marshall *et al.* introduced α-methyl amino acid substituents into peptides as a way to decrease the conformational space available to the resulting peptide.[42] Freidinger *et al.* developed a cyclic lactam moiety (**23.31**) that stabilized β- and γ-turn structures and applied this to LH-RH (e.g. (**23.32**)) to show that a β-turn about residues 6−7 was compatible with activity.[43] Conformational restriction has been applied to determine the bioactive conformation of enzyme-inhibitor systems for which no X-ray crystal structure is available. Thorsett[44] synthesized conformationally restricted bicyclic lactam derivatives of the angiotensin-converting enzyme (ACE) inhibitors enalapril (**23.33**) and enalaprilat (**23.34**) (Fig. 23.6) in order to characterize torsion angles in the bioactive conformation. Analog (**23.35**) was used to constrain

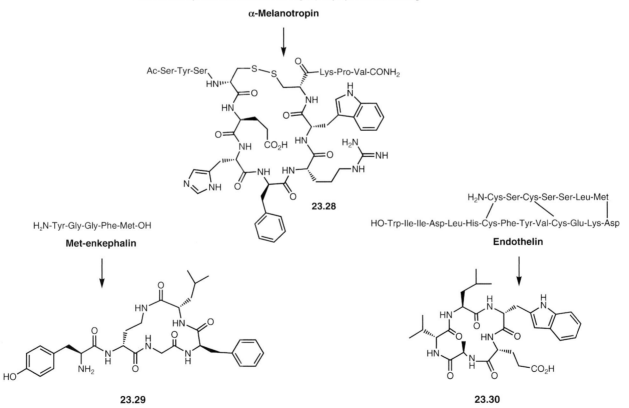

Fig. 23.5 Cyclic hormone peptide analogs.

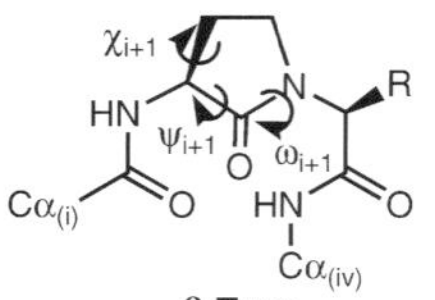

β-Turn
23.31

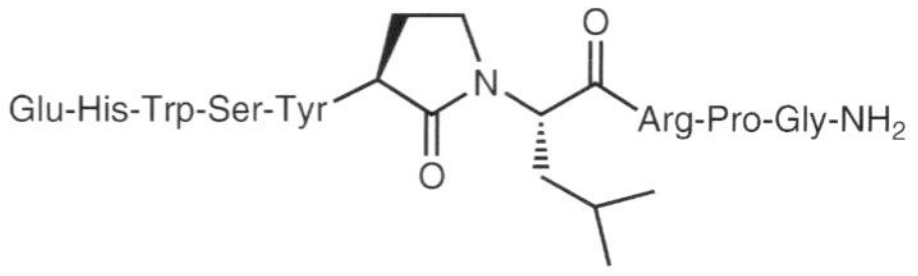

LH-RH β-turn mimetic
23.32

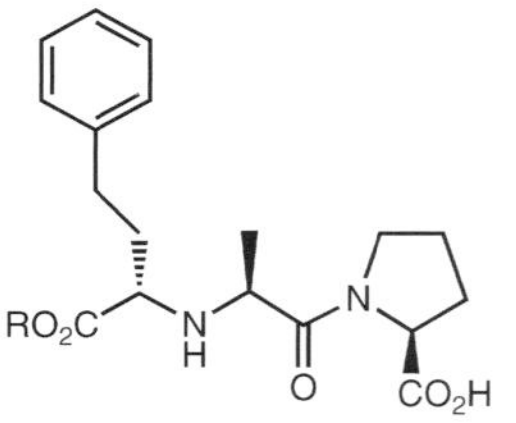

Enalapril, R = Et, **23.33**
Enalaprilat, R = H, **23.34**

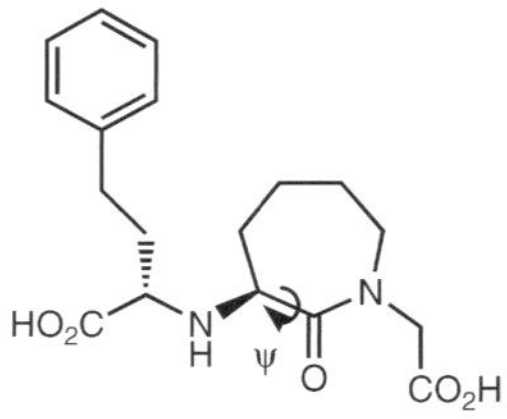

23.35

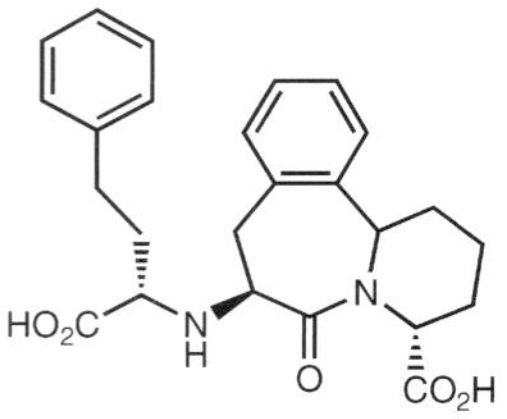

23.36

Fig. 23.6 Conformationally restricted inhibitors of LH-RH and ACE.

the torsion angle ψ. Flynn *et al.*[45] extended this principle to prepare the very tight-binding tricyclic ACE inhibitor (**23.36**). Numerous additional conformational constraints have been developed and the reader is encouraged to consult these reviews for additional examples.[46–53]

VI WHEN IS THE BIOACTIVE CONFORMATION FORMED?

The emergence of NMR as a method for characterizing the solution conformation of peptides was accompanied by the realization that the solution conformation might not be the same as the bioactive conformation. In the early years, the limitations in NMR technology precluded studies in highly aqueous media and only recently has it become possible to determine the enzyme-bound conformation of ligands bound to small proteins.[54] Scepticism about the relevance of solution conformation to bioactive conformation has increased in recent years, but this is probably because many solution studies have been carried out in organic solvents, especially chloroform, because of the insolubility of organic compounds in water at the concentrations needed for NMR analysis. Except for highly conformationally restricted compounds, there is no particular reason to expect that the chloroform conformation of an organic molecule, especially one with hydrophobic groups attached via flexible tethers, should be predictive of the aqueous or bioactive conformation of that molecule. Developments provide evidence that in some cases the bioactive conformation of a flexible peptide may exist to an appreciable extent in *aqueous* media prior to binding of the ligand to the protein. These observations, if general, merit careful examination because they illustrate how conformational studies should be carried out to determine the bioactive conformation of a peptide, and offer encouragement that these efforts will be productive.

Conformational restriction played an important role in the discovery that water induces the bioactive conformation of CsA. Cyclosporin A (Sandimmune®, CsA, (**23.37**), Fig. 23.7), is a major drug for preventing rejection of transplanted human organs and has been the subject of many synthetic, conformational and mechanism of action studies.[55–57] To produce immunosuppression, CsA first binds to cyclophilin A (CyP A),[58] a peptidyl prolyl *cis–trans* isomerase (PPIase),[59,60] to form the CsA–CyP complex, which then binds to and inhibits calcineurin (CaN), a calmodulin-dependent serine/threonine protein phosphatase,[61,62] thereby inhibiting interleukin-2 (IL-2) synthesis.[63,64]

The conformations of CsA in chloroform and when bound to cyclophilin differ dramatically. CsA adopts closely related conformations in three different crystal forms and in two different solvent systems,[65] characterized by a *cis*

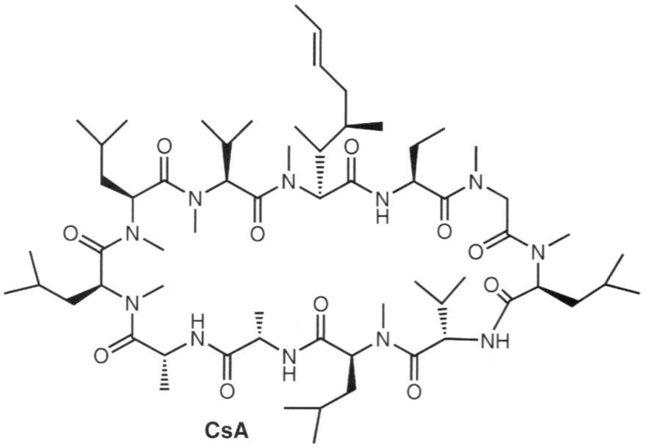

CsA
23.37

Fig. 23.7 Structure of cyclosporin A.

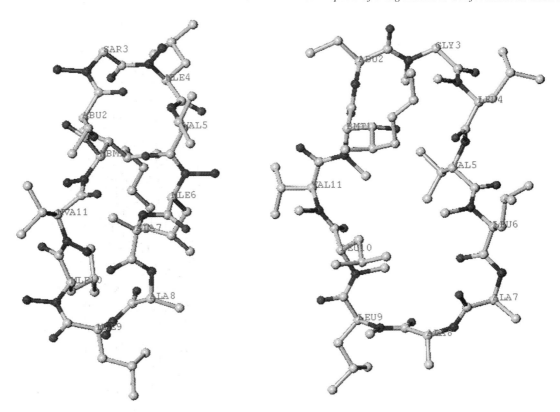

Plate 4 Crossed stereo views of chloroform (left, 23.37c) and enzyme-bound (right, 23.37t) conformations of CsA.

peptide bond between MeLeu residues 9 and 10. However, numerous attempts to prepare conformationally restricted CsA derivatives based on modifying the chloroform conformation of CsA (Goodfellow and Rich, unpublished data, and reference 66), according to the strategies developed by Veber (*vide supra*) were unsuccessful. These negative results prompted a reassessment of the bioactive conformation of CsA by isotope edited NMR methods, which led to the discovery that in CsA bound to CyP the amide bond between the 9,10 residues is *trans*.[67] Complete structures of CsA bound to Cyp were subsequently reported.[68,69] Plate 4 (see Plate section) shows stereo representations of the chloroform (**23.37c**) and the enzyme-bound (**23.37t**) conformers of CsA, and illustrates the remarkable difference in overall shape of the molecule produced by the *cis* to *trans* isomerization.

VII USE OF TIME-RESOLVED CONFORMATIONAL RESTRICTION

A The bioactive conformation of CsA is formed in water

The dissociation constants of slowly interconverting populations of conformations can be determined when the biological response rate is faster than the rate of conformational interconversion, and in favorable cases this process can provide information about the bioactive conformation of a molecule. This principle was first demonstrated with analogs of tentoxin,[22,23] a phytotoxic cyclic tetrapeptide. Because certain conformers of D-MeAla-tentoxin interconvert slowly, it was possible to isolate and bioassay different conformational populations of the molecule, and show selective inhibition of chloroplast coupling factor-1.

In a similar fashion, it has been possible to isolate different conformations of the immunosuppressive drug, cyclosporin A. By preparing anhydrous THF solutions of CsA, with and without 0.4 M LiCl, it was possible to restrict the peptide ring system conformation in CsA to *trans* and *cis* conformers. Addition of these conformers separately to the assay buffer enabled the dissociation constants of both the *cis* and *trans* amide bond conformers in CsA for inhibition of PPIase to be determined.[70] The *trans* conformer (**23.37t**) is very active; the *cis* conformer (**23.37c**) is not, and these plus other data were used to show that CsA adopts a conformation in water that is very close to the enzyme-bound CsA conformation.[51] Subsequently, Wenger and coworkers at Sandoz (now Novartis) showed by NMR experiments that a 3-substituted CsA derivative, e.g. (**23.38**) (Fig. 23.8) adopts a conformation in water that is essentially identical

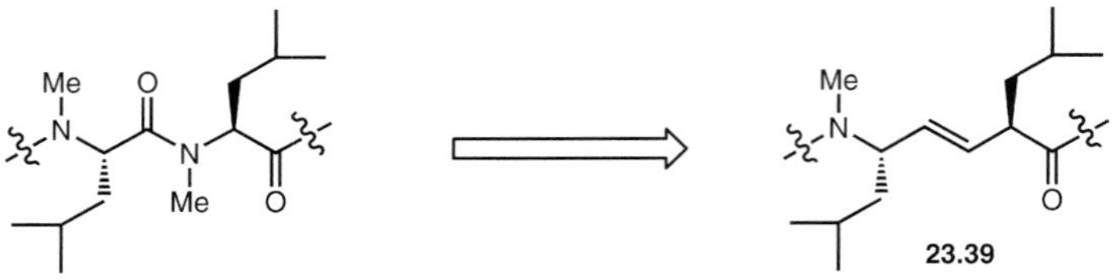

23.39

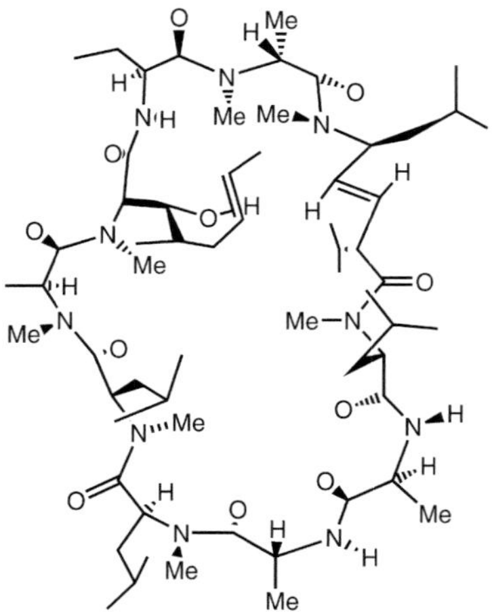

[D-MeAla]^{3}CsA

23.38

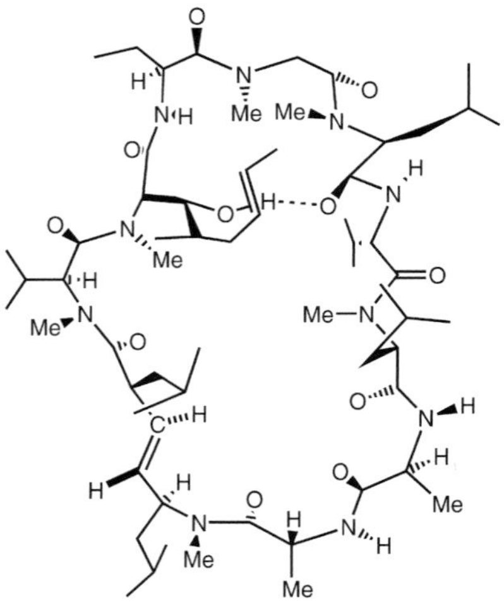

Olefin Analog of Cyp-bound CsA

Ki > 10,000 x10^{-9}

23.40

Fig. 23.8 D-MeAla3 and olefin cyclosporin analogs.

with the enzyme-bound CsA conformation.[71,72] Thus, formation of the bioactive conformations of CsA or (**23.38**) is driven by dissolution in water; the enzyme binds the preformed, *trans* amide conformer and does not catalyze its formation. Although multiple conformations of CsA exist in hydrogen bonding solvents, an appreciable amount of the correct conformation comparable to that of D-Ala3-CsA (**23.38**) must exist in solution prior to binding to the enzyme. This prediction, based on the enzyme kinetic data, is consistent with the modest boost in activity reported for another conformationally constrained CsA derivative.[73] Interestingly, the D-MeAla-CsA derivative (**23.38**) corresponds to a CsA analog in which one amino acid has added a single methyl group to stabilize conformational interconversions of the peptide ring system. This is an example of the α-substitution strategy introduced by Marshall.

The process of constructing conformationally restricted analogs may fail for reasons other than steric hindrance or incorrect conformation. An enlightening example was encountered when the olefin isostere[74,75] was used in place of the *trans* amide bond in the enzyme-bound conformation of CsA. As noted previously, the amide bond between positions 9,10 in CsA switches from *cis* in organic solution to *trans* in water and in the enzyme-bound conformation. When the amide bond between MeLeu-MeLeu was replaced with the *trans* olefin isostere (**23.39**), a remarkably inactive CsA analog (**23.40**) was formed (Fig. 23.8) (Bohnstedt, Flentke and Rich, unpublished data). Although amide replacement by olefins is successful in other systems, notably in enkephalin analogs,[73] the CsA derivative had lost over four orders of magnitude in potency against cyclophilin. Subsequently, the X-ray crystal structure of CsA bound to Cyp showed that the carbonyl group in the MeLeu–MeLeu unit of CsA is hydrogen bonded to the indole NH in tryptophan-121 in cyclophilin.[76] Presumably this missing hydrogen bond is a significant factor in the loss of potency of (**23.40**), but the major point here is that conformational restriction can fail because essential groups have been deleted in the course of designing the restricted analog.

VIII HYDROPHOBIC COLLAPSE

The discovery that the bioactive conformation of CsA is induced by dissolution in water led to the idea that 'hydrophobic collapse' might be a determinant of bioactive conformation of flexible, hydrophobic peptides and peptidomimetics in water. The medicinal chemical use of the term *hydrophobic collapse* is defined here to mean a significant conformational change in a molecule produced by dissolving the molecule in water, relative to the conformation observed for this same molecule in organic solution or *in vacuo*. With very hydrophobic molecules such as cyclosporin, it is assumed that hydrophobic clustering of side chains plus hydrophilic interactions between the amide bonds and water help stabilize the bioactive conformation. Whether hydrophobic clustering in these conformations is purely a hydrophobic effect (see Gellman *et al.* for examples of where it appears not to be)[77,78] or results from a combination of multiple interactions (e.g. $\pi-\pi$, H-bonding; hydrophobic interactions) is not known.[79] The fact that the DMSO conformation of CsA is similar to the conformation of CsA in water suggests that hydrogen bonding to the secondary amides stabilizes the observed conformation; hydrophobic collapse would not be expected to be strong under these experimental conditions. It is safe to say that the effects of water on both pre-and post-binding conformations are critical to the pre-binding conformation of many classes of molecules and should be considered when attempting to design and interpret the biological properties of conformationally restricted analogs or peptidomimetics.

Since this concept was first proposed, several flexible molecules have been found that adopt aqueous conformations in which hydrophobic side chains are clustered.[80,81] In the cases of thrombin inhibitors and taxol derivatives, hydrophobic clustering may help stabilize the bioactive conformations.[82,83] However, many factors over and above pure hydrophobic interactions can stabilize the aqueous conformation. In contrast, many receptor antagonists that bind to a variety of receptor systems are constructed from templates that should resist hydrophobic collapse. The ubiquitous diphenylmethyl pharmacophoric group (**23.41**) shown in Fig. 23.9 along with some close variants, provides a hydrophobic surface that cannot internally associate due to conformational constraints.

IX CONFORMATIONALLY CONSTRAINED RGD PEPTIDES AND DERIVATIVES

Some outstanding examples of the use of conformational restriction to characterize bioactive conformations of Arg-Gly-Asp (RGD) antagonists illustrate the present state-of-the-art. Members of the integrin family of receptors[84] recognize and bind the peptide sequence, Arg-Gly-Asp-, as an important step in platelet aggregation and other physiological processes. Competitive antagonists for this process have been recognized as potential drug candidates and much effort has been directed toward identifying small ligands that might mimic the RGD peptide sequence. This drug design concept was supported by the fact that protein antagonists of integrin receptors are known that contain the RGD sequence[85] and that small peptide sequences containing the RGD moiety weakly antagonize the endogenous ligand.[86] Consequently, several groups synthesized conformationally restricted derivatives of small peptides as starting points for developing metabolically stable peptides or peptidomimetics. Ali *et al.*[87] synthesized a series of disulfide derivatives of the RGD sequence, which were designed by analogy with the somatostatin work (*vide supra*). Excellent antagonists related to (**23.42**) were obtained. Further constraint of the peptide system by use of the *o*-thiol benzene derivatives shown led to the novel antagonist SKF 107260 (**23.43**; Fig. 23.10), a good inhibitor of both platelet aggregation and binding to GPII$_B$III$_A$. Burnier's group followed a similar strategy but utilized cyclic sulfides as the conformationally restricting element.[88] These derivatives had the advantage of being rapidly synthesized by solid phase methods. Systematic structure−activity studies with respect to the amino acid preceding the RGD sequence and the chirality of sulfoxide derivatives led to the discovery of G-4120 (**23.44**), a potent, biologically active derivative. The conformation of (**23.44**) in water was found to be highly constrained and a single predominant conformation could be characterized in waqueous solution by use of NMR methods and computational chemistry.[89] The bioactive conformation derived from **23.44** in water, which defined the topographical placement of the arginine quanidine group and the aspartic carboxyl group, was superimposed on to a conformationally restricted template of a class of compounds with generally suitable pharmacodynamic properties. In this case, the benzodiazepine ring system was used and the strategy

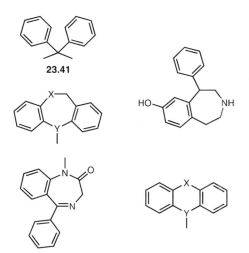

Fig. 23.9 Variations of the diphenylmethyl moiety found in drugs.

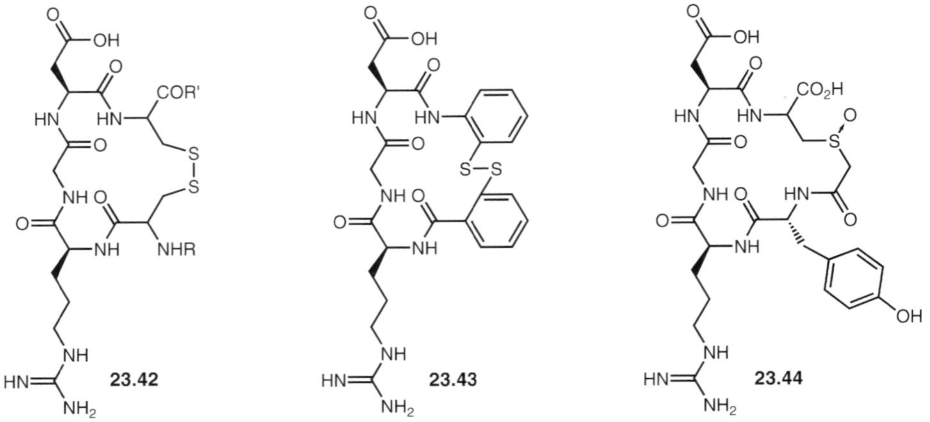

cyclo(Cys-Arg-Gly-Asp-Cys) cyclo(Man-Arg-Gly-Asp-Map) cyclo(Ac-DTyr-Arg-Gly-Asp-Cys(O))

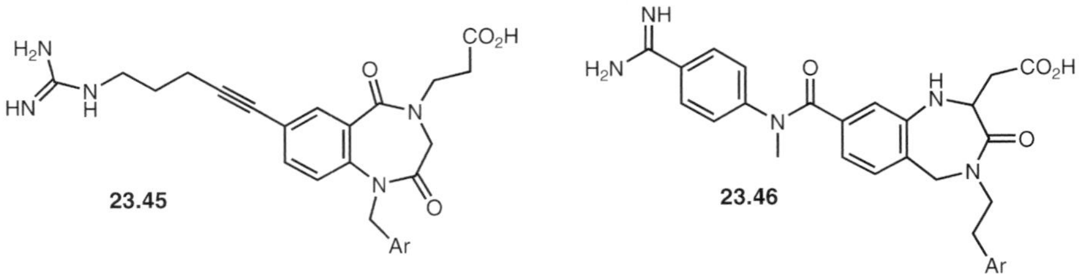

Fig. 23.10 Conformationally restricted analogs of RGD peptides.

generated the low molecular nonpeptide RGD receptor antagonist **23.45**,[90] which contains at least two conformational restrictions, the bicyclic heterocycle and the acetylene linker. The compounds shown in Fig. 23.10 represent what can be achieved by applying the principles of conformational restriction to peptides when no X-ray or NMR structural information is available for the complex between ligand and receptor. Benzodiazepine **23.45** represents the first peptidomimetic (see Chapter 29) designed *de novo* by systematically modifying a natural receptor-binding peptide.

X EXAMPLES OF CONFORMATIONAL RESTRICTION IN IMPORTANT BIOLOGICAL TARGETS

During the last few years a large variety of highly sophisticated conformational constraints have been successfully applied to the design of peptidase inhibitors. Selected examples from the recent literature are summarized in Fig. 23.11. For instance, Veber and coworkers have recently described the design and synthesis of

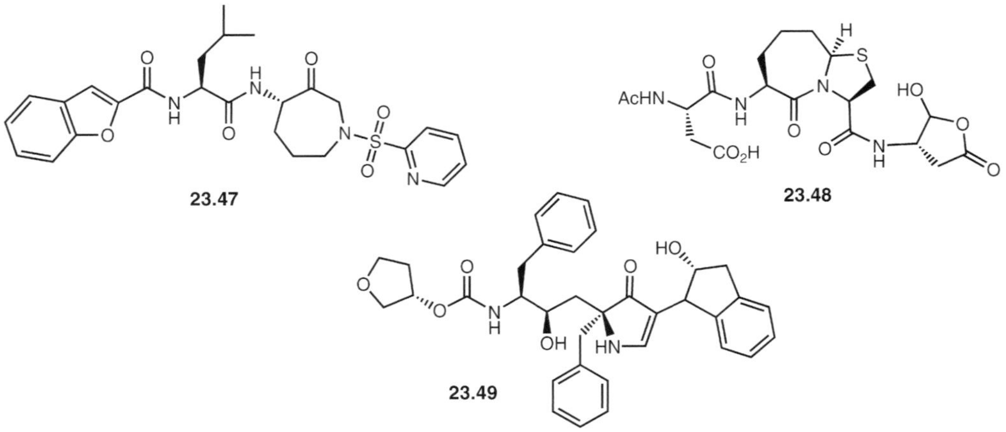

Fig. 23.11 Selected conformationally restricted peptidase inhibitors.

conformationally constrained cyclic ketones as highly potent and selective Cathepsin K inhibitors. The azepanone derivative **23.47** (Fig. 23.11) constrains the diaminopropanone moiety into the bioactive conformation for selective inhibition of this cysteine protease (K_i = 4.8 pM).[91] This strategy has also been applied to prepare peptidomimetic derivatives of the tetrapeptide caspase-1 (ICE) substrate sequence (X-Y-Z-Asp-NHR) which gave the conformationally constrained caspase-1 inhibitor **23.48**.[92] Hirschmann *et al.* have prepared different pyrrolinone analogs, designed to adopt the typical enzyme-bound extended β-conformation and reduce amide bond character.[93,94] Compound (**23.49**) is a highly modified, conformationally restricted HIV protease inhibitor.

HTS for nonpeptide inhibitors of farnesyl transferase led to the potent inhibitor SCH 66701 (**23.50**, Fig. 23.12) which was crystallized within the enzyme active site.[95] This completely nonpeptidic inhibitor is reminiscent of the classical tricyclic CNS agents and contains one of the classical pharmacophores resistant to hydrophobic collapse; it contains only two readily rotatable bonds. This inhibitor shows that potency can be achieved even without the "essential" cysteine or sulfhydryl mimic. More recently, a series of macrocyclic 3-aminopyrrolidinone farnesyltransferase inhibitors, (e.g. **23.51**), has shown that cyclization provides enhanced plasma half-life, attenuates metabolic

clearance and could increase potency in comparison with related acyclic pyrrolidone-based inhibitors.[96]

Conformational restriction was also used to develop potent antagonists of Endothelin-1 receptors. One successful approach assumed that the phenyl groups of a screening lead might mimic two of the aromatic side chains, (Tyr,[13] Phe[14] or Trp[21]) of ET-1.[97] Knowing that the carboxylic acid was also necessary for good activity, researchers at SmithKline overlaid their inhibitor with the aromatic groups Tyr,[13] Phe[14] and Asp[18] in ET-1. Use of NMR-derived conformations in solution and iterative SAR, eventually produced (**23.52**) a potent antagonist of both the ET_A and ET_B receptors with K_i = 0.43 nM and 14.7 nM, respectively. Analogs based on a pyrrolidine scaffold are also effective (e.g. **23.53**, Fig. 23.13).[98]

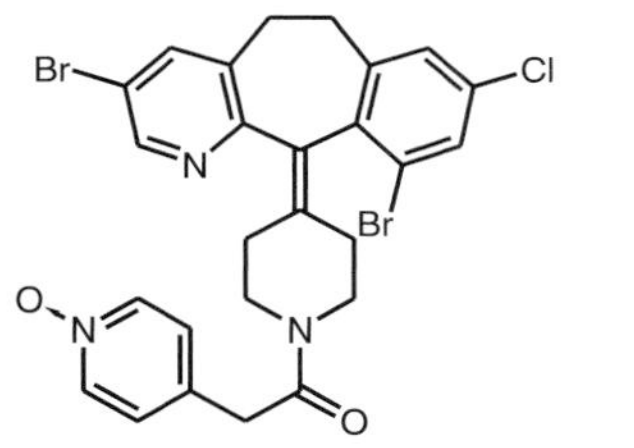

REFERENCES

1. Schueler, F.W. (1946) Sex-hormonal action and chemical constitution. *Science* **103**: 221–223.
2. Barton, D.H.R. (1950) The conformation of the steroid nucleus. *Experientia* **6**: 316–329.
3. Schueler, F.W. (1953) The interaction of statistical and coulombic factors in the characterization of pharmacophoric moieties. *Arch. Intern. Pharmacodyn.* **95**: 376–397.
4. Schueler, F.W. (1953) The statistical nature of the intramolecular distance factor of the muscarinic moiety. *Arch. Intern. Pharmacodyn.* **93**: 417–426.
5. Schueler, F.W. (1956) Two cyclic analogs of acetylcholine. *J. Am. Pharm. Assoc.* **45**: 197–199.
6. Archer, S., Lands, A.M. and Lewis, T.R. (1962) Isomeric 2-acetoxytropine methiodides. *J. Med. Pharm. Chem.* **5**: 423–430.
7. Martin-Smith, M., Smail, G.A. and Stenlake, J.B. (1967) Conformational isomerism in the biological actions of acetylcholine. *J. Pharm. Pharmacol.* **19**: 561–589.
8. Portoghese, P.S. (1970) Relations between stereostructure and pharmacological activities. *Annu. Rev. Pharmacol.* **10**: 51–76.
9. Mutschler, E. and Lambrecht, G. (1983) Stereoselectivity and conformation: flexible and rigid compounds. *Stereochem. Biol. Act. Drugs* 63–79.
10. Casy, A.F., Hassan, M.M.A. and Wu, E.C. (1971) Conformation of some acetylcholine analogs as solutes in deuterium oxide and other solvents. *J. Pharm. Sci.* **60**: 67–73.
11. Frydenvang, K. and Jensen, B. (1996) Conformational analysis of acetylcholine and related choline esters. *Acta Crystallogr., Sect. B: Struct. Sci.* **B52**: 184–193.
12. Armstrong, P.D., Cannon, J.G. and Long, J.P. (1968) Conformationally rigid analogs of acetylcholine. *Nature* **220**: 65–66.
13. Cannon, J.G., Rege, A.B., Gruen, T.L. and Long, J.P. (1972) 1,2-Disubstituted cyclopropane and cyclobutane derivatives related to acetylcholine. *J. Med. Chem.* **15**: 71–75.
14. Beers, W.H. and Reich, E. (1970) Structure and activity of acetylcholine. *Nature* **228**: 917–922.
15. Chothia, C. (1970) Interaction of acetylcholine with different cholinergic nerve receptors. *Nature* **225**: 36–38.
16. Schulman, J.M., Sabio, M.L. and Disch, R.L. (1983) Recognition of cholinergic agonists by the muscarinic receptor. 1. Acetylcholine and other agonists with the NCCOCC backbone. *J. Med. Chem.* **26**: 817–823.

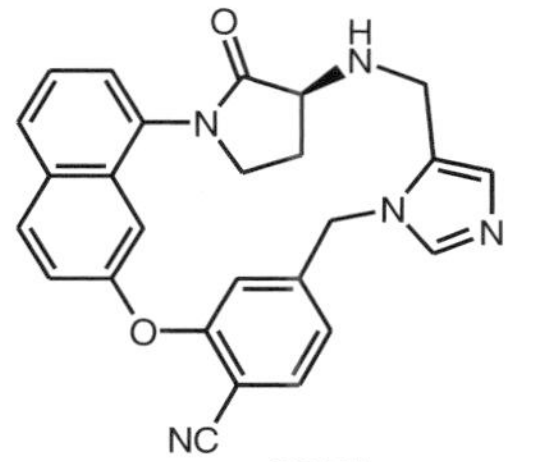

Fig. 23.12 Conformationally restricted FTase inhibitors.

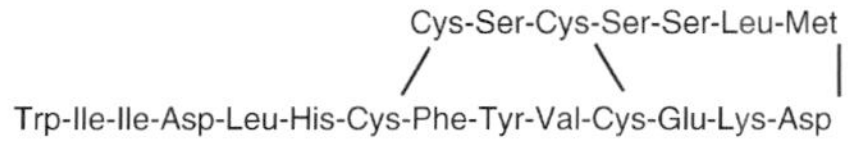

Cys-Ser-Cys-Ser-Ser-Leu-Met

Trp-Ile-Ile-Asp-Leu-His-Cys-Phe-Tyr-Val-Cys-Glu-Lys-Asp

Endothelin-1

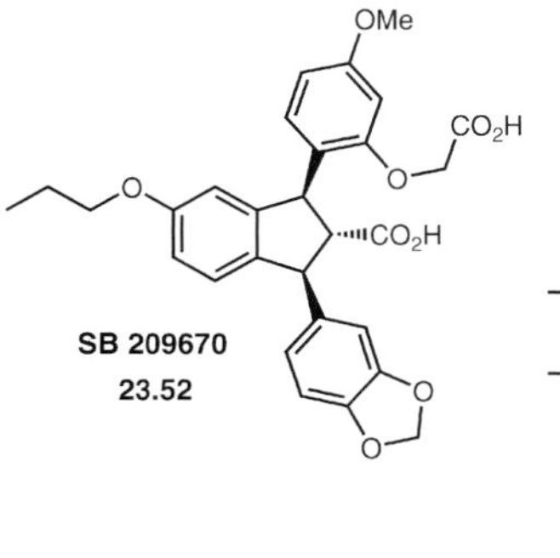

SB 209670
23.52

A-306552
23.53

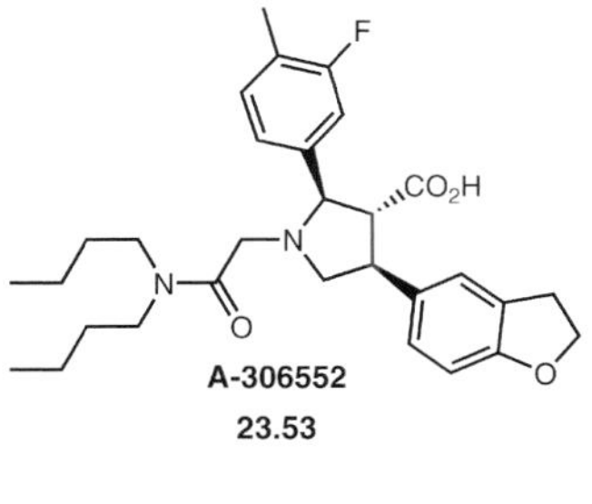

Fig. 23.13 Conformationally restricted ET inhibitors.

17. Spande, T.F., Garraffo, H.M., Edwards, M.W., Yeh, H.J.C., Pannell, L. and Daly, J.W. (1992) Epibatidine: a novel (chloropyridyl)azabicyclo-heptane with potent analgesic activity from an Ecuadoran poison frog. *J. Am. Chem. Soc.* **114**: 3475–3478.

18. Qian, C., Li, T., Shen, T.Y., Libertine-Garahan, L., Eckman, J., Biftu, T. and Ip, S. (1993) Epibatidine is a nicotinic analgesic. *Eur. J. Pharmacol.* **250**: R13–R14.

19. Dukat, M., Damaj, M.I., Glassco, W., Dumas, D., May, E.L., Martin, B.R. and Glennon, R.A. (1994) Epibatidine: A very high affinity nicotine-receptor ligand. *Med. Chem. Res.* **4**: 131–139.

20. McGroddy, K.A. and Oswald, R.E. (1993) Solution structure and dynamics of cyclic and acyclic cholinergic agonists. *Biophys. J.* **64**: 314–324.

21. McGroddy, K., Carter, A.A., Tubbert, M.M. and Oswald, R.E. (1993) Analysis of cyclic and acyclic nicotinic cholinergic agonists using radioligand binding, single channel recording, and nuclear magnetic resonance spectroscopy. *Biophys. J.* **64**: 325–338.

22. Rich, D.H., Bhatnagar, P.K., Jasensky, R.D., Steele, J.A., Uchytil, T.F. and Durbin, R.D. (1978) Two-conformations of the cyclic tetrapeptide, [D-MeAla1]-tentoxin have different biological activities. *Bioorg. Chem.* **7**: 207–214.

23. Rich, D.H. and Bhatnagar, P.K. (1978) Conformational studies of tentoxin by nuclear magnetic resonance spectroscopy. Evidence for a new conformation for a cyclic tetrapeptide. *J. Am. Chem. Soc.* **100**: 2212–2218.

24. Horn, A.S. and Rodgers, J.R. (1980) 2-Amino-6,7-dihydroxytetrahy-dronaphthalene and the receptor-site preferred conformation of dopamine. A commentary. *J. Pharm. Pharmacol.* **32**: 521–524.

25. Krogsgaard-Larsen, P. (1981) Gamma-aminobutyric acid agonists, antagonists, and uptake inhibitors. Design and therapeutic aspects. *J. Med. Chem.* **24**: 1377–1383.

26. Tamura, N., Iwama, T. and Itoh, K. (1992) Synthesis and glutamate-agonistic activity of (S)-2-amino-3-(2,5-dihydro-5-oxo-3-isoxazolyl)-propanoic acid derivatives. *Chem. Pharm. Bull.* **40**: 381–386.

27. Schunack, W. (1973) Structure–action relationship of histamine analogs 1. Histamine-like compounds with cyclized side chain. *Arch. Pharm.* **306**: 934–942.

28. Friedman, E., Meller, E. and Hallock, M. (1981) Effects of conformationally restrained analogs of serotonin on its uptake and binding in rat brain. *J. Neurochem.* **36**: 931–937.

29. Horn, A.S. and Synder, S.H. (1971) Chlorpromazine and dopamine. Conformational similarities that correlate with the antischizophrenic activity of phenothiazine drugs. *Proc. Natl Acad. Sci. USA* **68**: 2325–2328.

30. Fong, T.M., Cascieri, M.A., Yu, H., Bansal, A., Swain, C. and Strader, C.D. (1993) Amino-aromatic interaction between histidine 197 of the neurokinin-1 receptor and CP 96345. *Nature* **362**: 350–353.

31. Gether, U., Yokota, Y., Emonds-Alt, X., Breliere, J.C., Lowe, J.A., III, Snider, R.M., Nakanishi, S. and Schwartz, T.W. (1993) Two nonpeptide tachykinin antagonists act through epitopes on correspond-ing segments of the NK1 and NK2 receptors. *Proc. Natl Acad. Sci. USA* **90**: 6194–6198.

32. Gether, U. (2000) Uncovering molecular mechanisms involved in activation of G protein-coupled receptors. *Endocrine Rev.* **21**: 90–113.

33. Erickson, J.W. and Fesik, S.W. (1992) Macromolecular X-ray crystallography and NMR as tools for structure-based drug design. *Annu. Rep. Med. Chem.* **27**: 271–289.

34. Mierke, D.F. and Giragossian, C. (2001) Peptide hormone binding to G-protein-coupled receptors: Structural characterization via NMR techniques. *Med. Res. Rev.* **21**: 450–471.

35. Muller, G. (2000) Towards 3D structures of G protein-coupled receptors: A multidisciplinary approach. *Curr. Med. Chem.* **7**: 861–888.

36. Veber, D. (1979) Conformational considerations in the design of somatostatin analogs showing increased metabolic stability. *Pept. Struct. Biol. Funct.* 409–419.

37. Rivier, J., Brown, M. and Vale, W. (1975) D-Trp8-somatostatin, analog of somatostatin more potent than the native molecule. *Biochem. Biophys. Res. Commun.* **65**: 746–751.

38. Rose, G.D., Gierasch, L.M. and Smith, J.A. (1985) Turns in peptides and proteins. *Adv. Protein Chem.* **37**: 1–109.

39. Veber, D.F. (1992) Design and discovery in the development of peptide analogs. *Pept. Chem. Biol.* 3–14.

40. Sawyer, T.K., Hruby, V.J., Darman, P.S. and Hadley, M.E. (1982) [half-Cys4,half-Cys10]-α-Melanocyte-stimulating hormone: A cyclic α-melanotropin exhibiting superagonist biological activity. *Proc. Natl Acad. Sci. USA* **79**: 1751–1755.

41. Schiller, P.W. (1984) *The Peptides: Analysis, Synthesis and Biology.* Academic Press, San Diego.

42. Marshall, G.R., Gorin, F.A. and Moore, M.L. (1978) Peptide conformation and biological activity. *Annu. Rep. Med. Chem.* **13**: 227–238.

43. Freidinger, R.M., Veber, D.F., Perlow, D.S., Brooks, J.R. and Saperstein, R. (1980) Bioactive conformation of luteinizing hor-mone-releasing hormone: evidence from a conformationally con-strained analog. *Science* **210**: 656–658.

44. Thorsett, E.D. (1986) Conformationally restricted inhibitors of angiotensin converting enzyme. *Actual. Chim. Ther.* **13**: 257–268.

45. Flynn, G.A., Giroux, E.L. and Dage, R.C. (1987) An acyl-iminium ion cyclization route to a novel conformationally restricted dipeptide mimic: applications to angiotensin-converting enzyme inhibition. *J. Am. Chem. Soc.* **109**: 7914–7915.

46. Estiarte, A.M. and Rich, D.H. (2002) Peptidomimetics for drug design. In Abraham, D. (ed.). *Burger's Medicinal Chemistry and Drug Discovery*, vol. 4: *Drug Discovery Technologies*, Wiley-Interscience, New York (in press).

47. Hölzemann, G. (1991) Peptide conformation mimetics (Part 1). *Kontakte* **1**: 3–12.

48. Hölzemann, G. (1991) Peptide conformation mimetics (Part 2). *Kontakte* **2**: 55–63.

49. Hruby, V.J. and Balse, P.M. (2000) Conformational and topographical considerations in designing agonist peptidomimetics from peptide leads. *Curr. Med. Chem.* **7**: 945–970.

50. Kahn, M. (1993) Peptide secondary structure mimetics: recent advances and future challenges. *Synlett.* 821–826.

51. Rich, D.H. (1993) Effect of hydrophobic collapse on enzyme-inhibitor interactions. Implications for the design of peptidomimetics. *Perspect. Med. Chem.* 15–25.

52. Rich, D.H. (1989) *Peptidase Inhibitors*. Pergamon Press, Oxford.

53. Tyndall, J.D.A. and Fairlie, D.P. (2001) Macrocycles mimic the extended peptide conformation recognized by aspartic, serine, cysteine and metallo proteases. *Curr. Med. Chem.* **8**: 893–907.

54. Hicks, R.P. (2001) Recent advances in NMR: expanding its role in rational drug design. *Curr. Med. Chem.* **8**: 627–650.

55. Borel, J.F. (1989) Pharmacology of cyclosporine (Sandimmune) IV. Pharmacological properties *in vivo*. *Pharmacol. Rev.* **41**: 259–371.

56. Georgiev, V.S. (1991) Immunomodulating peptides of natural and synthetic origin. *Med. Res. Rev.* **11**: 81–90.

57. Sigal, N.H. and Dumont, F.J. (1993) *Immunosupression*. Raven Press, New York.

58. Handschumacher, R.E., Harding, M.W., Rice, J., Drugge, R.J. and Speicher, D.W. (1984) Cyclophilin: a specific cytosolic binding protein for cyclosporin A. *Science* **226**: 544–547.

59. Takahashi, N., Hayano, T. and Suzuki, M. (1989) Peptidyl-prolyl cis-trans isomerase is the cyclosporin A-binding protein cyclophilin. *Nature* **337**: 473–475.

60. Fischer, G., Wittmann-Liebold, B., Lang, K., Kiefhaber, T. and Schmid, F.X. (1989) Cyclophilin and peptidyl-prolyl cis-trans isomerase are probably identical proteins. *Nature* **337**: 476–478.

61. Fruman, D.A., Klee, C.B., Bierer, B.E. and Burakoff, S.J. (1992) Calcineurin phosphatase activity in T lymphocytes is inhibited by FK 506 and cyclosporin A. *Proc. Natl Acad. Sci. USA* **89**: 3686–3690.

62. Friedman, J. and Weissman, I. (1991) Two cytoplasmic candidates for immunophilin action are revealed by affinity for a new cyclophilin: one in the presence and one in the absence of CsA. *Cell* **66**: 799–806.

63. Schreiber, S.L. and Crabtree, G.R. (1992) The mechanism of action of cyclosporin A and FK506. *Immunol. Today* **13**: 136–142.

64. O'Keefe, S.J., Tamura, J., Kincaid, R.L., Tocci, M.J. and O'Neill, E.A. (1992) FK-506- and CsA-sensitive activation of the interleukin-2 promoter by calcineurin. *Nature* **357**: 692–694.

65. Loosli, H.R., Kessler, H., Oschkinat, H., Weber, H.P., Petcher, T.J. and Widmer, A. (1985) Peptide conformations. Part 31. The conformation of cyclosporin A in the crystal and in solution. *Helv. Chim. Acta* **68**: 682–704.

66. Lee, J.P., Dunlap, B. and Rich, D.H. (1990) Synthesis and immunosuppressive activities of conformationally restricted cyclosporin lactam analogs. *Int. J. Pept. Protein Res.* **35**: 481–494.

67. Fesik, S.W., Gampe, R.T., Jr, Holzman, T.F., Egan, D.A., Edalji, R., Luly, J.R., Simmer, R., Helfrich, R., Kishore, V. and Rich, D.H. (1990) Isotope-edited NMR of cyclosporin A bound to cyclophilin: evidence for a trans 9,10 amide bond. *Science* **250**: 1406–1409.

68. Fesik, S.W., Gampe, R.T., Jr, Eaton, H.L., Gemmecker, G., Olejniczak, E.T., Neri, P., Holzman, T.F., Egan, D.A., Edalji, R., *et al.* (1991) NMR studies of [U-13C] cyclosporin A bound to cyclophilin: bound conformation and portions of cyclosporin involved in binding. *Biochemistry* **30**: 6574–6583.

69. Weber, C., Wider, G., Von Freyberg, B., Traber, R., Braun, W., Widmer, H. and Wuethrich, K. (1991) NMR structure of cyclosporin A bound to cyclophilin in aqueous solution. *Biochemistry* **30**: 6563–6574.

70. Kofron, J.L., Kuzmic, P., Kishore, V., Gemmecker, G., Fesik, S.W. and Rich, D.H. (1992) Lithium chloride perturbation of cis–trans peptide bond equilibria: effect on conformational equilibria in cyclosporin A and on time-dependent inhibition of cyclophilin. *J. Am. Chem. Soc.* **114**: 2670–2675.

71. Wenger, R.M., France, J., Bovermann, G., Walliser, L., Widmer, A. and Widmer, H. (1994) The 3D structure of a cyclosporin analog in water is nearly identical to the cyclophilin-bound cyclosporin conformation. *FEBS Lett.* **340**: 255–259.

72. Altschuh, D., Vix, O., Rees, B. and Thierry, J.C. (1992) A conformation of cyclosporin A in aqueous environment revealed by the X-ray structure of a cyclosporin–Fab complex. *Science* **256**: 92–94.

73. Alberg, D.G. and Schreiber, S.L. (1993) Structure-based design of a cyclophilin-calcineurin bridging ligand. *Science* **262**: 248–250.

74. Bohnstedt, A.C., Prasad, J.V.N.V. and Rich, D.H. (1993) Synthesis of E- and Z-alkene dipeptide isosteres. *Tetrahedron Lett.* **34**: 5217–5220.

75. Spatola, A.F. (1983) Peptide backbone modifications: A structure–activity analysis of peptides containing amide bond surrogates. Conformational restrains and related backbone replacements. In Weinstein, B. (ed.). *Chemistry and Biochemistry of Amino Acids, Peptides and Proteins*, vol. VII, pp. 267–357. Marcel Dekker, New York.

76. Mikol, V., Kallen, J., Pflugl, G. and Walkinshaw, M.D. (1993) X-ray structure of a monomeric cyclophilin A-cyclosporin A crystal complex at 2.1 A resolution. *J. Mol. Biol.* **234**: 1119–1130.

77. McKay, S.L., Haptonstall, B. and Gellman, S.H. (2001) Beyond the hydrophobic effect: Attractions involving heteroaromatic rings in aqueous solution. *J. Am. Chem. Soc.* **123**: 1244–1245.

78. Newcomb, L.F. and Gellman, S.H. (1994) Aromatic stacking interactions in aqueous solution: Evidence that neither classical hydrophobic effects nor dispersion forces are important. *J. Am. Chem. Soc.* **116**: 4993–4994.

79. Tsang, K.Y., Diaz, H., Graciani, N. and Kelly, J.W. (1994) Hydrophobic cluster formation is necessary for dibenzofuran-based amino acids to function as β-sheet nucleators. *J. Am. Chem. Soc.* **116**: 3988–4005.

80. Bogusky, M.J., Brady, S.F., Sisko, J.T., Nutt, R.F. and Smith, G.M. (1993) Synthesis and solution conformation of c(D-Trp-D-Cys(SO₃-Na⁺)-Pro-D-Val-Leu), a potent endothelin-A receptor antagonist. *Int. J. Pept. Protein Res.* **42**: 194–203.

81. Kemmink, J., Van Mierlo, C.P.M., Scheek, R.M. and Creighton, T.E. (1993) Local structure due to an aromatic–amide interaction observed by proton nuclear magnetic resonance spectroscopy in peptides related to the N terminus of bovine pancreatic trypsin inhibitor. *J. Mol. Biol.* **230**: 312–322.

82. Lim, M.S.L., Johnston, E.R. and Kettner, C.A. (1993) The solution conformation of (D)Phe-Pro-containing peptides: implications on the activity of Ac-(D)Phe-Pro-boroArg-OH, a potent thrombin inhibitor. *J. Med. Chem.* **36**: 1831–1838.

83. Vander Velde, D.G., Georg, G.I., Grunewald, G.L., Gunn, C.W. and Mitscher, L.A. (1993) 'Hydrophobic collapse' of taxol and Taxotere solution conformations in mixtures of water and organic solvent. *J. Am. Chem. Soc.* **115**: 11650–11651.

84. Ruoslahti, E. and Pierschbacher, M.D. (1987) New perspectives in cell adhesion: RGD and integrins. *Science* **238**: 491–497.

85. Dennis, M.S., Henzel, W.J., Pitti, R.M., Lipari, M.T., Napier, M.A., Deisher, T.A., Bunting, S. and Lazarus, R.A. (1990) Platelet glycoprotein IIb-IIIa protein antagonists from snake venoms: evidence for a family of platelet-aggregation inhibitors. *Proc. Natl Acad. Sci. USA* **87**: 2471–2475.

86. Haverstick, D.M., Cowan, J.F., Yamada, K.M. and Santoro, S.A. (1985) Inhibition of platelet adhesion to fibronectin, fibrinogen, and von Willebrand factor substrates by a synthetic tetrapeptide derived from the cell-binding domain of fibronectin. *Blood* **66**: 946–952.

87. Ali, F.E., Bennett, D.B., Calvo, R.R., Elliott, J.D., Hwang, S.-M., Ku, T.W., Lago, M.A., Nichols, A.J., Romoff, T.T., *et al.* (1994) Conformationally constrained peptides and semipeptides derived from RGD as potent inhibitors of the platelet fibrinogen receptor and platelet aggregation. *J. Med. Chem.* **37**: 769–780.

88. Barker, P.L., Bullens, S., Bunting, S., Burdick, D.J., Chan, K.S., Deisher, T., Eigenbrot, C., Gadek, T.R., Gantzos, R., *et al.* (1992) Cyclic RGD peptide analogs as antiplatelet antithrombotics. *J. Med. Chem.* **35**: 2040–2048.

89. McDowell, R.S. and Gadek, T.R. (1992) Structural studies of potent constrained RGD peptides. *J. Am. Chem. Soc.* **114**: 9245–9253.

90. McDowell, R.S., Blackburn, B.K., Gadek, T.R., McGee, L.R., Rawson, T., Reynolds, M.E., Robarge, K.D., Somers, T.C., Thorsett, E.D., *et al.* (1994) From peptide to non-peptide. 2. The *de novo* design of potent, non-peptidal inhibitors of platelet aggregation based on a benzodiazepinedione scaffold. *J. Am. Chem. Soc.* **116**: 5077–5083.

91. Marquis, R.W., Ru, Y., LoCastro, S.M., Zeng, J., Yamashita, D.S., *et al.*, (2001) Azepanone-based inhibitors of human and rat cathepsin K. *J. Med. Chem.* **44**: 1380–1395.

92. Karanewsky, D.S., Bai, X., Linton, S.D., Krebs, J.F., Wu, J., Pham, B. and Tomaselli, K.J. (1998) Conformationally constrained inhibitors of caspase-1 (interleukin-1β converting enzyme) and of the human CED-3 homolog caspase-3 (CPP32, apopain). *Bioorg. Med. Chem. Lett.* **8**: 2757–2762.

93. Smith, A.B. III, Hirschmann, R., Pasternak, A., Yao, W., Sprengeler, P.A., Holloway, M.K., Kuo, L.C., Chen, Z., Darke, P.L. and Schleif, W.A. (1997) An orally bioavailable pyrrolinone inhibitor of HIV-1 protease: Computational analysis and X-ray crystal structure of the enzyme complex. *J. Med. Chem.* **40**: 2440–2444.

94. Tyndall, J.D.A., Reid, R.C., Tyssen, D.P., Jardine, D.K., Todd, B., *et al.* (2000) Synthesis, stability, antiviral activity, and protease-bound structures of substrate-mimicking constrained macrocyclic inhibitors of HIV-1 protease. *J. Med. Chem.* **43**: 3495–3504.

95. Strickland, C.L., Weber, P.C., Windsor, W.T., Wu, Z., Le, H.V., *et al.* (1999) Tricyclic farnesyl protein transferase inhibitors: Crystallographic and calorimetric studies of structure–activity relationships. *J. Med. Chem.* **42**: 2125–2135.

96. Bell, I.M., Gallicchio, S.N., Abrams, M., Beese, L.S., Beshore, D.C., *et al.* (2002) 3-Aminopyrrolidinone farnesyltransferase inhibitors: Design of macrocyclic compounds with improved pharmacokinetics and excellent cell potency. *J. Med. Chem.* **45**: 2388–2409.

97. Elliott, J.D., Lago, M.A., Cousins, R.D., Gao, A.M., Leber, J.D., *et al.* (1994) 1,3-Diarylindan-2-carboxylic acids, potent and selective non-peptide endothelin receptor antagonists. *J. Med. Chem.* **37**: 1553–1557.

98. Jae, H.-S., Winn, M., von Geldern, T.W., Sorensen, B.K., Chiou, W.J., Nguyen, B., Marsh, K.C. and Opgenorth, T.J. (2001) Pyrrolidine-3-carboxylic acids as endothelin antagonists. 5. Highly selective, potent, and orally active ETA antagonists. *J. Med. Chem.* **44**: 3978–3984.

24

PHARMACOPHORE IDENTIFICATION AND RECEPTOR MAPPING

Hans-Dieter Höltje

One cannot guess how a model functions. One has to look at its use and learn from that.

L. Wittgenstein (1950)

I INTRODUCTION

It is well known that the large majority of drugs exert their action via specific binding to biomacromolecules. Above all, proteins like enzymes and receptors, but also nucleic acids, serve as physiological binding partners. In all cases a specific and unique three-dimensional structure of the drug molecule is a prerequisite for a certain pharmacological activity. The initial step in the formation of drug–receptor interaction complexes is a recognition event.[1,2] The receptor has to recognize whether an approaching molecule possesses the properties necessary for binding. This characteristic three-dimensional set of structural elements is called the pharmacophore.[3] Because experimental knowledge about the particular three-dimensional structure of receptors for most of the drugs in therapeutic use is still unavailable, corresponding hypothetical pharmacophore models are an important source for understanding drug–receptor interactions at the molecular level. In order to describe the pharmacophoric pattern correctly, the steric and the electronic features of the bioactive conformations of drug molecules have to be

determined. The availability of a set of compounds from a distinct class of candidates showing a large variety in chemical structure and which interact via the same binding mechanism with the same receptor is an ideal starting point for the identification of a pharmacophore.

In the course of the pharmacophore identification process, clearly two different steps have to be taken in succession. First, a conformational analysis has to be carried out. After this initial step, a common three-dimensional arrangement of functional groups is determined through a superpositioning procedure. During the second step this preliminary sterical pharmacophore model has to be checked and consolidated by electron density calculations and establishment of the corresponding molecular electrostatic potentials (MEPs).

Using programs such as GRIN/GRID or HINT, the electrostatic fields can be translated and extended to molecular interaction fields which mimic the interaction potential of the pharmacophore. Based on this three-dimensional description of the physicochemical properties, a map of suitable receptor binding sites can be constructed. This hypothetical receptor model is then used for the calculation of interaction energies between all members of the series and the receptor. The calculated binding energies should correctly reflect equivalent biological affinities.

At this point, experimental structure–activity relationship data must be used to optimize and refine the pharmacophore. It is essential that as well as active substances, inactive analogues or at least congeners with low activity can also be described by the theoretical interaction energies. If this is achieved, then the receptor model can be used to predict chemical structures and biological activities of new and hitherto unknown compounds.

The synthesis and subsequent confirmation of the predicted structures by pharmacological testing is the final and decisive proof for the correctness or, more precisely, the usefulness of the deduced receptor map of the pharmacophore. The derived pharmacophore model then allows conclusions regarding the structure and properties of the unknown receptor binding site.

II METHODS

A Conformational analysis

Some introductory remarks are necessary on ensuring that the developed pharmacophore possesses sufficient reliability. In the first place, it is important to collect a training set of at least 10 structures (around 15 or more would be better) which, as already mentioned, should represent the greatest possible structural variety in the series. In addition, the pharmacological potencies of these congeners should span a relative scale of at least 1 to 1000. These features increase the confidence level for the prediction of new structures. Another important factor which has to be carefully considered is the conformational flexibility of the drug molecules. It is extremely desirable that a semirigid, or better a rigid, highly potent congener belongs to the series to be studied. This compound may be used as a template for the more flexible ones. If potent conformationally restricted molecules are not available, the determination of the pharmacophoric conformation requires much greater effort and the use of statistical methods such as principal component analysis (PCA) or factor analysis is imperative in order to resolve this multidimensional problem[4-6] (see Chapter 22 for further discussion).

As a first step, all molecular structures have to be generated using the computer. If possible, experimental datas from X-ray crystallography, available through the Cambridge Crystallographic Data Bank,[7] are used. If structures are not included in this source, they may be constructed using fragment libraries and/or building routines offered by commercial molecular modelling software packages.[8,9] In any case, the generated geometries (bond lengths, valence, angles, etc.) have to be optimized using molecular mechanics routines before a systematic conformational search can be executed. During this search operation, rotations around all single bonds are performed in 30° or, better, 10° steps. The rotamers produced are checked using Van der Waals energy as the criterion for strong steric hindrance, and all conformers which do not pass this test are discarded.

This systematic procedure is not feasible for large systems because the amount of output data can become extremely large. In these cases the conformational space should preferably be analysed using a molecular dynamics operation.[10] Using a random conformation, a molecular dynamics process is initiated to produce an initial dynamic structure. Then a short (e.g. 5 ps) molecular dynamics run is performed at high temperature (e.g. 900 K); subsequently, the system is cooled to 300 K and molecular dynamics is continued for 10 ps. The final structure is extracted and the energy is minimized. The total cycle is repeated at least 20 times, so that 20 energy-minimized structures are produced.

Independently of the method used for conformational analysis, the conformations generated are analysed for conformational similarity and classified according to energy content, and only dissimilar conformers within a defined energy window are retained for further investigation. Depending on the molecular mechanics force field used, the window may extend from 5 to 10 kcal mol^{-1} above the absolute energy minimum.

The next task is to search for a unique common conformation of all congeners, where most if not all pharmacophoric elements of the molecules are presented superimposed. Thus we first have to define the pharmacophoric elements. This can be done best on the basis of known structure−activity relationship data. The information thus obtained greatly facilitates the subsequent superpositioning procedure, because only conformations with common positions of the important pharmacophoric elements need to be considered.

Several different superpositioning procedures are available. They comprise manual or automatic fitting by rigid-body rotation, or flexible fitting procedures where both r.m.s. (root mean square) deviation between the fitted atom pairs and conformational energies are minimized. We always use the FITIT[11] method, which has been developed by our group. This program fits each energetically accessible conformation of one molecule with each allowed conformation of a second one. The resulting fit pairs are sorted according to r.m.s. values and only fit pairs with low r.m.s. values are saved. This procedure is repeated for the complete list of molecules, and in general, finally yields only a small number of different pharmacophore models (Fig. 24.1).

Other methodologies for pharmacophore identification which must be mentioned are the distance geometry approach developed by Crippen[12-17] and Marshall's active analogue approach.[18,19]

B Molecular electrostatic potentials

Since the initial step in the formation of a drug−receptor interaction complex is a recognition event, which is highly dependent on polar electrostatic interactions, one very easy and efficient way to test the so far purely steric pharmacophore model for significance is calculation of the molecular electrostatic potentials (MEPs)[20] for all

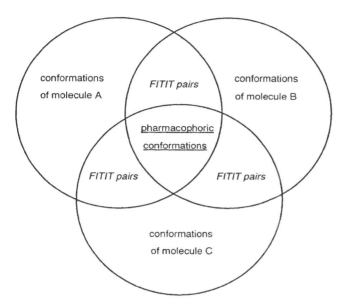

Fig. 24.1 Schematic drawing representing the FITIT procedure. The conformational space is reduced by pairwise superpositioning of energetically allowed conformations of the different chemical classes. A second superpositioning step for all selected FITIT pairs finally yields the common pharmacophoric conformations.

compounds in the training set. This must, of course, be carried out for the individual pharmacophoric conformations determined.

It is well known that MEPs provide an informative means for assessing the electronic structure of molecules; however, this is extremely dependent on the method employed for the calculation of the molecular charge distribution. Two conceptually different procedural treatments of this problem are possible. Charge distributions can be determined using a topological or a quantum-chemical method. Topological methods, for example that developed by Gasteiger and Marsili,[21] which is included in SYBYL, are very fast. Molecular electron densities here are calculated solely on the basis of electronegativity differences along a bond and include experimental NMR data. Only directly bonded atoms are considered and conformational aspects are completely neglected. However, if mesomeric effects are not important for the molecule investigated, the results may compare favourably with dipole moments from experimental sources. Nevertheless, it is very dangerous to use this kind of procedure without a careful check of the applicability to each individual problem.

Quantum-chemical methods, on the other hand, are divided into semiempirical and *ab initio* procedures. Semiempirical methods such as the various procedures collected in the MOPAC package (MNDO,[22a] AM1,[22b] PM3[22c]) are generally used because they are much less time

consuming and may also be applied to rather large molecules. However, it should be noted that these methods might lead to erroneous results for hetero atoms.[23,24] Considering the increasing computational power of workstations and the availability of mathematically optimized *ab initio* procedures (i.e. SPARTAN,[25] D-mol[26]), STO-3G, 4-31G or even 6-31G* basis set calculations for structures with the normal size of drug molecules can now be performed as standard procedure. The subsequent calculation of the MEPs should then be done on the basis of the wavefunctions[27] and not, as usually found in the commercial molecular modelling packages, using the monopole charges. The latter procedures may lead to incorrect results in regions of space close to and inside Van der Waals surfaces of molecules. Further outside, the monopole charge-derived MEPs, in general, give rather reliable pictures of the electrostatic behaviour of the molecules and can be used with caution as a fast and easily accessible tool.

Regardless of the method used, it is essential to demonstrate the correctness of calculated charge densities against experimental dipole moments. As dipole moments are conformation dependent, this test should only be performed for rigid molecules or small and conformationally undemanding model compounds.

C Molecular interaction fields

So far we have taken into account only the steric and electrostatic fit of congeneric drug molecules that bind at the same receptor site. This is clearly not sufficient if one aims towards a more detailed description of the structural properties of a receptor. Besides the already mentioned characteristics, hydrophobic areas, regions of charge transfer or several types of different polar interactions can also be distinguished. A receptor map which contains three-dimensional information of this kind is very helpful in the interpretation and understanding of known experimental results but also, which is even more desirable, for the prediction of new, hitherto unknown structures.

The generation of molecular interaction fields, which are the basis for the construction of a receptor map, can be performed using a variety of programs. The most widely used are GRIN/GRID,[28] CoMFA[29] and HINT.[30,31] CoMFA is implemented in a commercial 3D QSAR program package which allows the automated treatment of large numbers of compounds under constant conditions. A prerequisite for this procedure is, of course, some sort of alignment for the molecules to be studied. This can be attained by determination of a pharmacophore as described earlier. However, according to the authors, the CoMFA method can itself be used for this purpose. Therefore, a typical CoMFA study starts with only a very rough alignment of the compounds, or possibly none at all. After calculation of interaction energies at gridded space points between the molecule of interest and a probe atom simulating, for example, hydrophobic or hydrogen bond donor properties, interaction fields can be defined. The relative three-dimensional position of these fields in the space surrounding the molecules is found with the help of statistical and chemometric methods (e.g. PLS[32]). (For detailed information about CoMFA, see Chapter 25.) After calculation of different fields, the superpositional fit for the training set can subsequently be optimized. This means that compounds with different structures which have been identified to interact with the same receptor, but cannot be superimposed using an atom-by-atom fit procedure, might nevertheless be superimposable on the basis of their corresponding molecular interaction fields.

Molecular interaction fields can also be generated using GRIN/GRID. This method is especially reliable because it is based on a very careful parametrization of the interaction terms. The parameters are founded from experimental crystallographic data; that is, the direction, type and typical strength of a particular interaction are classified according to actual crystals.

Numerous different probes are available for a graded description of the molecular properties. As in CoMFA, molecules are located in a cubic grid and interaction energies between the molecule and the probe are calculated for each grid point outside the molecular Van der Waals volume. The resulting fields may be analysed by calculation of isoenergy contours at any given energy level. Comparison of the contours for all pharmacophoric conformations and all different types of probes allows the definition of a common receptor map for a drug family.[33-35] Typical results for this kind of study are shown in Fig. 24.2 for the 5-HT$_{2a}$ antagonists[36] using an aromatic CH-probe (hydrophobic interaction) and an aliphatic OH-probe (hydrogen bond donor and acceptor interaction), respectively.

The isoenergy level contours are -1.5 kcal mol^{-1} for the hydrophobic and -4.0 kcal mol^{-1} for the polar contacts.

D Receptor mapping

The next step is to translate this interaction field into a model of the receptor which is composed from single isolated amino acids with chemical properties that satisfy the different types of binding present in the pharmacophore. The relative three-dimensional positions of the amino acid binding sites are defined by the corresponding GRIN/GRID results. The resulting amino acid receptor model is sometimes called a pseudoreceptor.[37-39]

If experimental knowledge of the amino acid sequence of the receptor protein is absent, it may be that several different models can be constructed. The choice of one or another hypothetical receptor map is possible on the basis of calculated interaction energies and their subsequent correlation with the known binding affinities. The model producing the most significant agreement is selected for prediction purposes. Of course, the selection procedure and, coupled to this, the quality of the model are superior if structural information from molecular biochemistry can be used. This is true, for example, for the G-protein-coupled receptors like the serotoninergic 5-HT$_{2a}$ receptor.[40,41] Homology searching and sequence alignment operations have been performed intensively, so there are some ideas about selected amino acids in binding positions in the active site of the receptor.[42,43] (For further information, see Chapter 26.)

The next step is calculation of interaction energies and comparison with experimentally determined binding affinities. This can be efficiently carried out using force field methods. For example, the DOCKING procedure and the MAXIMIN module of the SYBYL software package can be employed for optimization of interaction geometries and energy calculation. Other programs can be used equally as well. As long as only relative energy differences are of interest, the results are quite reliable. However, one must be aware that the regular force field methods only describe two different types of binding forces adequately—the dispersion and the electrostatic terms.[44] The latter dramatically depends on the dielectric constant employed and it is extremely important to choose the one appropriate to the situation considered. Inside a protein environment, for example in the core of the G-protein-coupled receptor channel, the prevailing dielectric constant is ensured between 3 and 5. Binding sites at protein surfaces are better treated with a value around 10. Only in special cases should the constant for vacuum conditions be used. This would be reasonable, for example, if one assumes hydrogen bonds to be of crucial importance for the binding. Since force fields naturally can only simulate the electrostatic part of hydrogen bonds and neglect the covalent part, this drawback can be roughly compensated for through an overestimation of the electrostatic interaction. We will return to the subject of energy terms which are not included in force field interaction energies in the discussion of the serotoninergic receptor model. Interaction energies are determined according to

$$IE = E_{RL} - (E_R + E_L)$$

where IE = interaction energy; E_{RL} = energy of the receptor–ligand complex; E_R = energy of the isolated receptor protein; and E_L = energy of the isolated ligand.

In order to obtain comparable energy data, the interaction geometries of the complexes are generated for all the ligands in an absolutely corresponding manner. All ligands are kept in the pharmacophoric conformation and location. Hydrophobic and polar amino acids mimicking equivalent

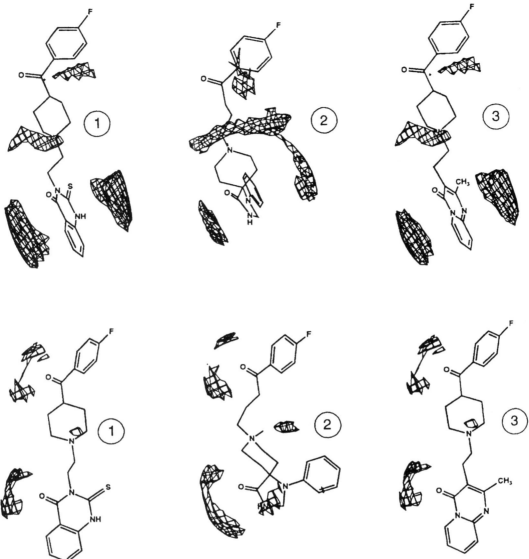

Fig. 24.2 Interaction fields calculated using GRIN/GRID: (1) altanserin; (2) pirenperone; (3) spiperone. The upper row presents hydrophobic fields calculated using an aromatic CH-probe, the contour level is -1.55 kcal mol^{-1}. The lower row shows the same molecules interacting with an OH-probe; the contour level is -4.0 kcal mol^{-1}.

receptor binding sites are positioned according to the GRIN/GRID contours. Each individual receptor model–ligand complex is then geometry-minimized in the MAX-IMIN force field. No constraints are employed. The procedure therefore simulates an induced fit between ligand and receptor which can be assumed to occur in reality. An energy cut-off of 0.01 kcal mol^{-1} should be used.

One word of caution is necessary with respect to the experimentally derived biological activities. These should constitute pure receptor binding affinities and must stem from a single laboratory. Since the computer models simulate molecular interaction events in a highly simplified

manner, the experimental data which are combined with them in a correlation equation must be as close to the molecular level as possible. It is therefore impermissible and virtual nonsense to correlate calculated interaction energies with pharmacological *in vivo* (whole animal) data, because the receptor interaction can be blurred or even completely hidden by pharmacokinetics and biotransformation of the drug molecules. Sometimes even the use of functional *in vitro* data is dangerous if a reaction cascade separates the measured event from the receptor-binding interaction. If receptor map and interaction complex have been generated carefully, a rough but nevertheless correct

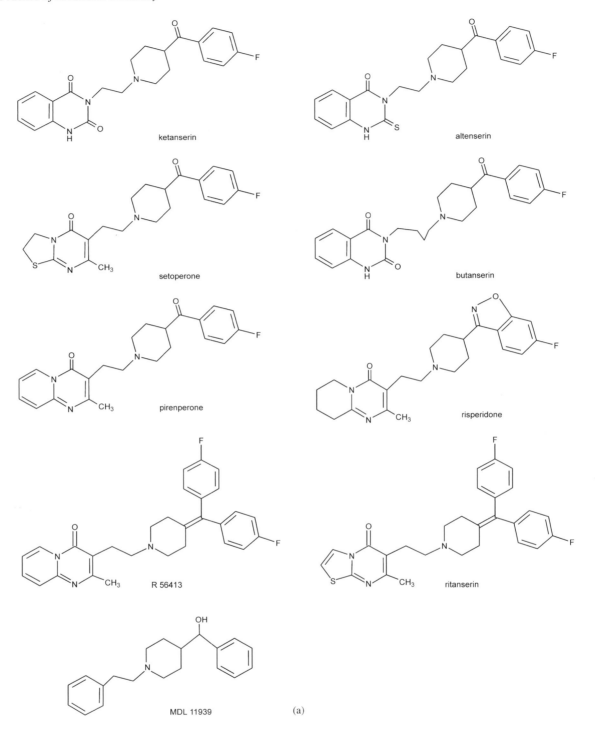

Fig. 24.3 Structural formulae of 5-HT$_{2a}$ antagonists used for pharmacophore identification: (a) ketanserin analogues; (b) spiperone analogues; (c) cyproheptadiene analogues; (d) irindalone analogues.

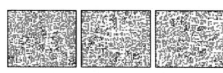

256-dim.   512-dim.   1024-dim.

Plate 1 Kohonen maps of the entire dataset of 5513 hydantoins, represented by Daylight 2D fingerprints of three different lengths. Magenta indicates neurons that contain hits; red neurons that have only received non-hits.

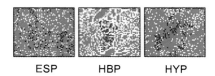

ESP   HBP   HYP

Plate 2 Kohonen maps of the entire dataset of 5513 hydantoins, represented by autocorrelation of the electrostatic (ESP), the hydrogen bonding (HBP), and the hydrophobicity potential (HYP) on the molecular surface. Magenta indicates hits, red non-hits.

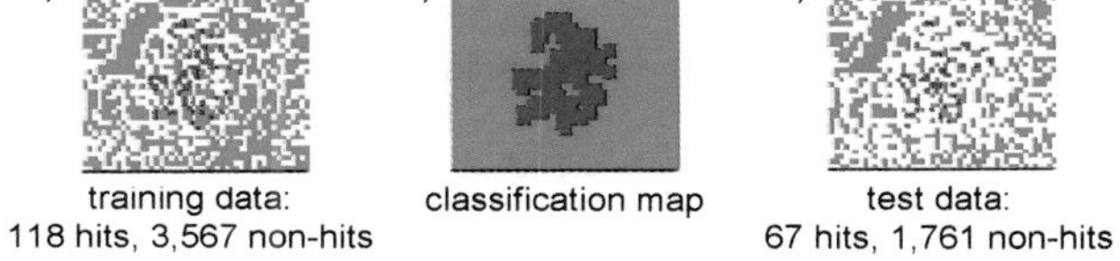

a)   b)   c)

training data: classification map   test data:
118 hits, 3,567 non-hits   67 hits, 1,761 non-hits

Plate 3 Kohonen maps obtained by representing molecules by autocorrelation of the hydrogen bonding potential. (a) Training set; (b) filter obtained by recolouring (see text); (c) test set.

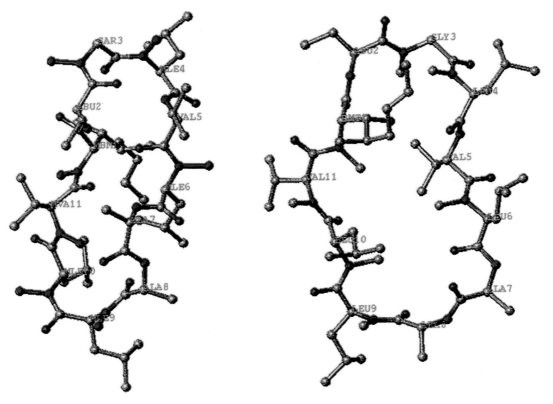

Plate 4 Crossed stereo views of chloroform (left, 23.37c) and enzyme-bound (right, 23.37t) conformations of CsA.

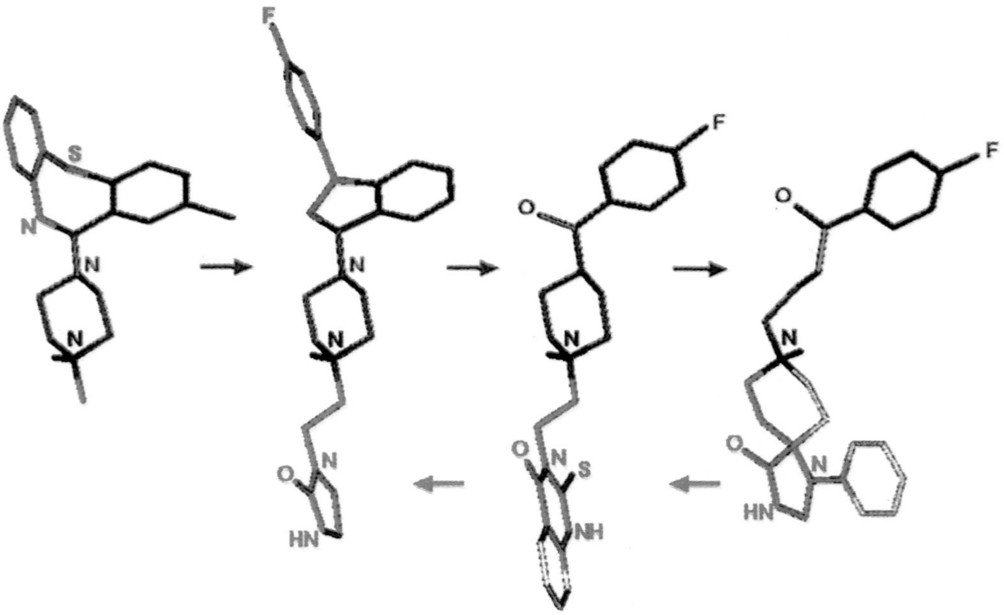

Plate 5

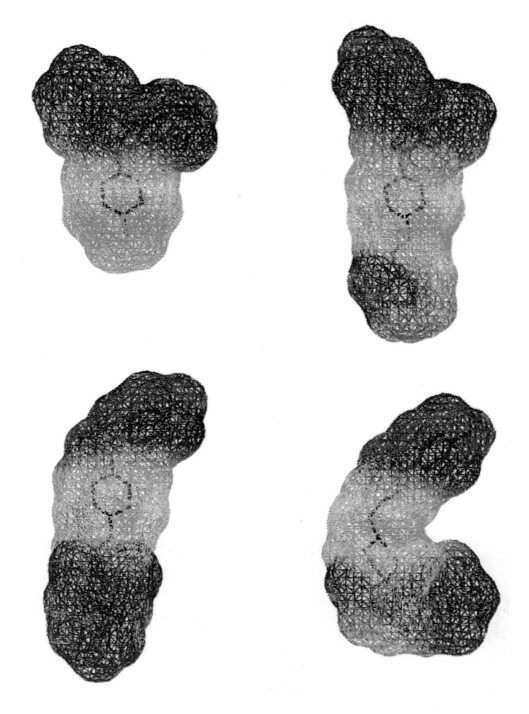

Plate 6

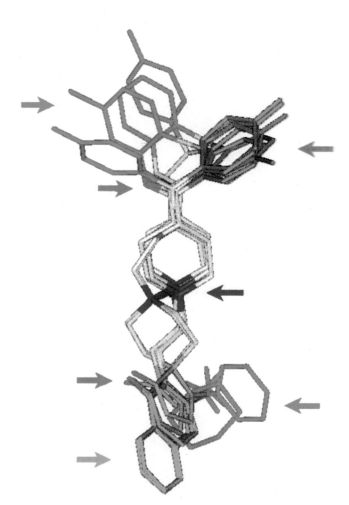

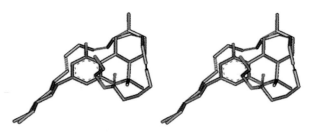

Plate 9 Schematic alignment of Δ^9-THC compound with anandamide molecule.

Plate 7

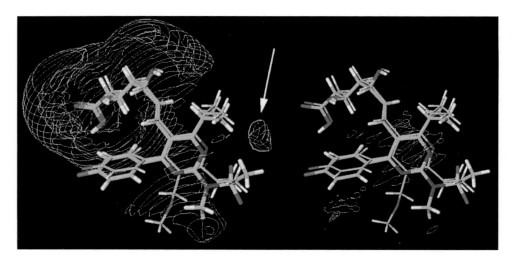

Plate 8 Molecular interaction fields (MIF) produced using the X-ray structures of Cerivastatin and Rosuvastatin drugs superimposed on the active site of the HMG-CoA re-educates protein. Green refers to hydrophobic attractive interaction energies between the HMG-CoA protein and the two drug molecules. Blue refers to bond donor attractive energies. The small blue region indicated with the arrow is due to the interaction between the Ser565 of the protein and the sulfide moiety of the Rosuvastatin, and is responsible for the increased binding affinity of Rosuvastatin for HMG-CoA protein.

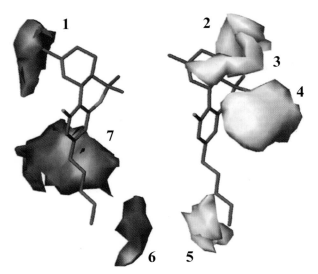

Plate 10 Grid plot of PLS partial weights for the CB_1 receptor model (see text for discussion).

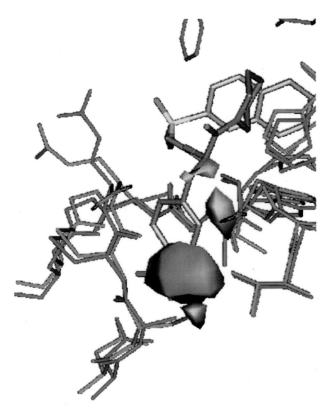

Plate 11 Differential pseudofield plot showing MIF differences between thrombin (blue) and trypsin (red) for the GRID carbonyl (O) probe within the S1 pocket. Negative interaction energies are shown in blue, while energetically unfavourable fields are shown in a yellow contour. Reference inhibitor is NAPAP.[30]

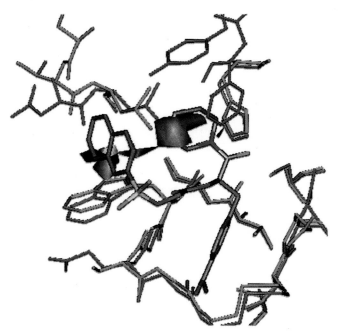

Plate 12 Pseudofield plot showing MIF differences between thrombin (blue) and trypsin (red) for the GRID DRY probe within the P pocket. Energetically unfavourable fields are shown in a yellow contour. Reference inhibitor is NAPAP.[30]

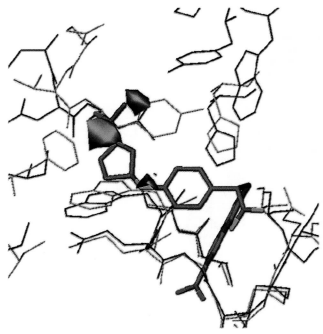

Plate 13 Pseudofield plot showing MIF differences between thrombin (blue) and factor Xa (green) for the GRID trimethyl ammonium (NM3) probe within the D pocket. Favourable interaction energies are shown in blue. Reference inhibitor is DX 9065 A.[30]

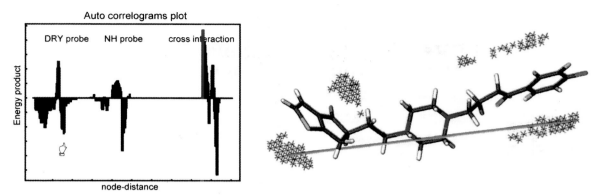

Plate 14 Left, correlogram profile obtained for a butyrophenone derivative [52] molecule, using a DRY (hydrophobic) probe and a N1 (amide N) probe. From left to right the plot contains the auto-correlogram DRY–DRY, the auto-correlogram N1–N1 and the cross-correlogram DRY–N1. Right, the isocontour plot of a GRID MIF computed with a DRY probe around a molecule. The contour encloses points with negative values under -1.0 kcal mol^{-1}. The hand points to the variable represented in the figure with a line. These variables show an unfavourable interaction (for the pharmacological activity) between hydrophobic regions present in long compounds like the one represented in the figure (for details see reference 52).

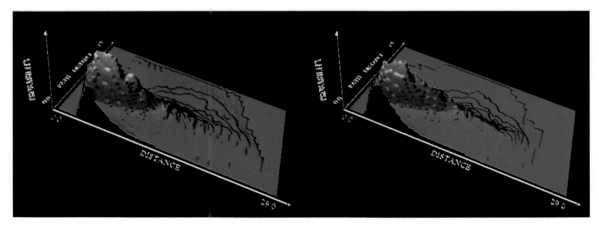

Plate 15 MIF transformation obtained with PathFinder program. Left, the transformation for the NAD site of the protein L-aspartate oxidase. Right, the MIF transformation for the NAD ligand molecule alone. The two maps are very similar, although the NAD molecule was roto-translated from the initial true position. The map comparison obtained from PathFinder can substitute virtual docking procedures.

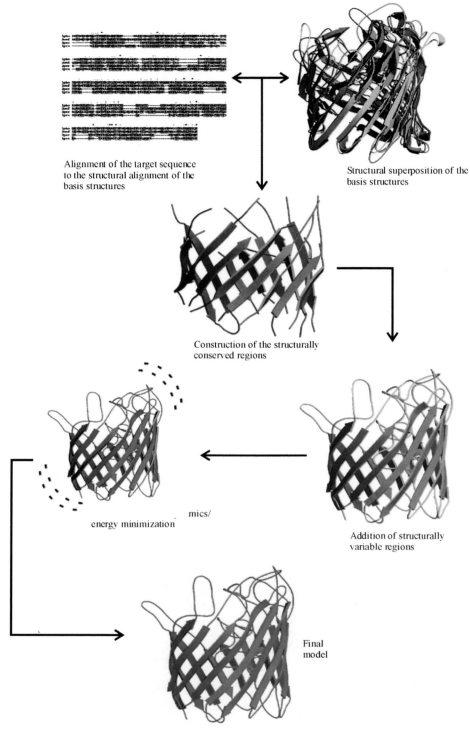

Alignment of the target sequence
to the structural alignment of the
basis structures

Structural superposition of the
basis structures

Construction of the structurally
conserved regions

Addition of structurally
variable regions

mics/
energy minimization

Final
model

Plate 16 A summary of the fragment-based modelling approach.

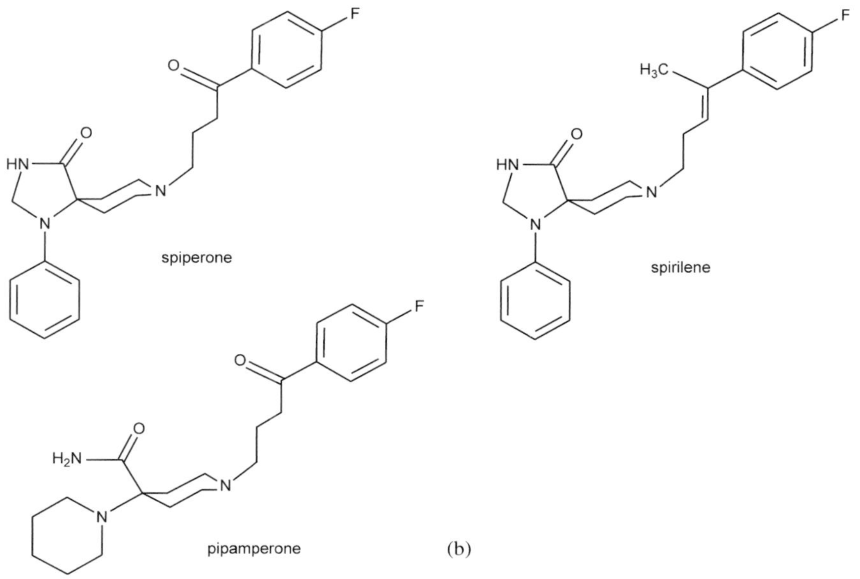

spiperone

spirilene

pipamperone

(b)

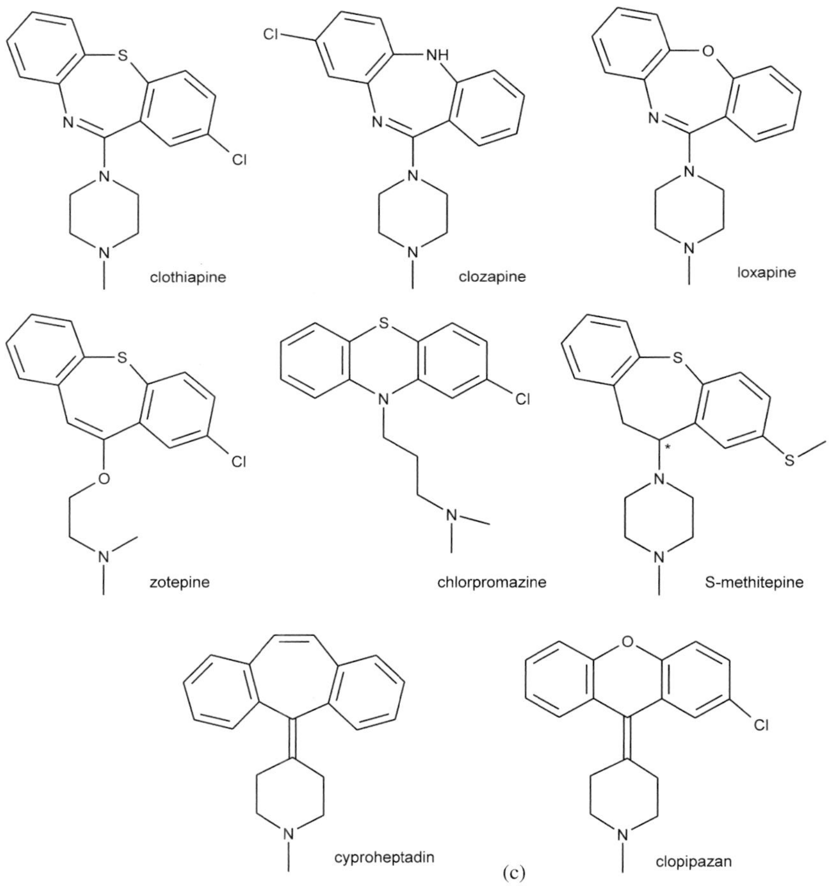

clothiapine

clozapine

loxapine

zotepine

chlorpromazine

S-methitepine

cyproheptadin

clopipazan

(c)

Fig. 24.3 (*continued*)

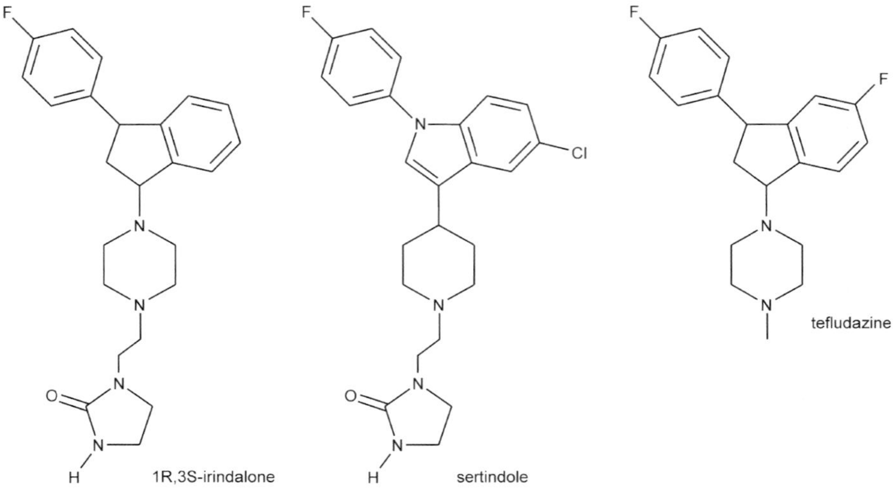

Fig. 24.3 (*continued*)

picture of the reality may be given. However, as long as the real receptor is still unknown the efficiency and meaning of the model cannot be assessed except by prediction of new substances. This should always be the ultimate test of usefulness for each hypothetically derived receptor map.

III CASE STUDY: THE 5-HT$_{2a}$ SEROTONINERGIC RECEPTOR MAP

To portray the routes to be followed for pharmacophore identification and receptor mapping in practice, the complete suite of procedures will be demonstrated using an example from the author's laboratory.

The example to be described and discussed in detail deals with antagonists of the serotoninergic 5-HT$_{2a}$ receptor. Compounds that interact with this receptor subtype have been known for many years.[45] They can be divided into four different groups in terms of chemical structure (Fig. 24.3):

(1) butyrophenone derivatives: spiperone analogues;
(2) 4-(phenylketo)piperidines: ketanserin analogues;
(3) tricyclic compounds: cyproheptadiene analogues;
(4) irindalone analogues.

Unfortunately, this set of altogether 20 compounds does not contain any rigid molecules, but some of them are, at least in parts, conformationally restricted. This is true, for example, for clothiapine and irindalone as well as spiperone, while the members of the ketanserin subfamily, which contain five major rotatable bonds, show a high degree of conformational freedom.

Experimental structure activity data for the 5-HT$_{2a}$ antagonist can be summarized as follows. The pharmacological results suggest that two planar aromatic or heterocyclic ring systems separated by a certain distance and connected by an aliphatic or alicyclic chain, which contains a basic protonatable nitrogen, seem to constitute a potent 5-HT$_{2a}$ ligand.[46,47] Additional hydrophobic substituents or a carbonyl group in the geterocyclic ring enhance the antagonistic potency[48] (Fig. 24.4).

This knowledge was taken into account for the conformational analysis. In the case considered here, an additional advantage could be drawn from the fact that various partial structural elements of the different conformationally constrained molecules can be matched with diverse regions of the highly flexible congeners. In Plate 5 (see Plate section), comparable structural elements of the four main groups of 5-HT$_{2a}$ antagonists are colour coded and the stepwise superpositioning approach which was developed is indicated.

Next, the electronic properties for all derived pharmacophoric conformations have to be investigated. Using AMI-derived charges, we have determined and compared MEPs of all compounds. Plate 6 (see Plate section) presents typical examples for the different groups. The high degree of similarity is evident. However, the superpositioning operation using the FITIT routine resulted in two slightly different pharmacophores. However, both of them have to be treated as equally meaningful because r.m.s. values and total agreement of the MEPs within the two sets are comparably good.

In general, in a situation like this a decision on one or the other model can only be taken by considering the three-dimensional structure of the receptor-binding site. If such information is missing, no decision is possible. However, a

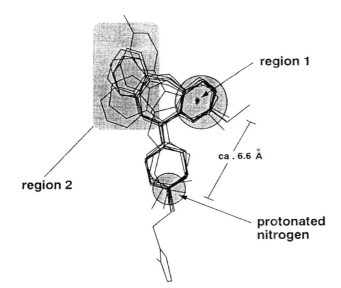

Fig. 24.4 Structure–activity relationship studies can be summarized as follows: for 5-HT$_{2a}$ affinity an aromatic centre (region 1) must be present at a distance of 6.6 Å from a protonatable nitrogen. A second hydrophobic area (region 2) close to the aromatic centre enhances receptor binding.

closer inspection of the two pharmacophores for 5-HT$_{2a}$ antagonists did bring to light one slight but significant structural divergence. In one of the two models all the protons at the pharmacophorically important cationic tertiary nitrogens are pointing in the same direction. In the other model this is not the case. Assuming that the cationic

protonated nitrogen is involved in a hydrogen-bond-enforced ionic interaction with an anionic receptor-binding site, the first pharmacophore model would clearly be favoured. Therefore, only this pharmacophore will be considered in the further investigation.

As described previously, the evaluation of molecular interaction fields was performed using GRIN/GRID. The results have already been presented in Fig. 24.2. Using a variety of different probes, a detailed picture of the molecular interaction potential for 5-HT$_{2a}$ antagonists can be derived (Plate 7) (see Plate section). Careful inspection of Plate 7 leads almost automatically to the selection of suitable binding partners needed for the construction of the 5-HT$_{2a}$ receptor map. Hydrophobic amino acids like phenylalanine, tryptophan, valine, leucine or isoleucine should be positioned on both 'sides' of the planar cyclic systems. Opposite to the protonated nitrogen, an acidic amino acid (e.g. aspartic acid) should be used to fill the location marked by the interaction field created with a hydroxylic probe. The other regions of this field close to the two carbonyl groups found in most of the ligands should be filled with serine, threonine or tyrosine. At this point, of course, we do not know whether all the interaction possibilities discovered are in fact realized at the receptor level, and we are unable to know this with certainty until the three-dimensional structure of the receptor protein has been elucidated. Nevertheless, structure–activity relationship (SAR) studies are very helpful in deciding on the existence or absence of binding sites.

In the case under study, SAR data tell us that the carbonyl group of the fluorobenzoyl partial structural element is not

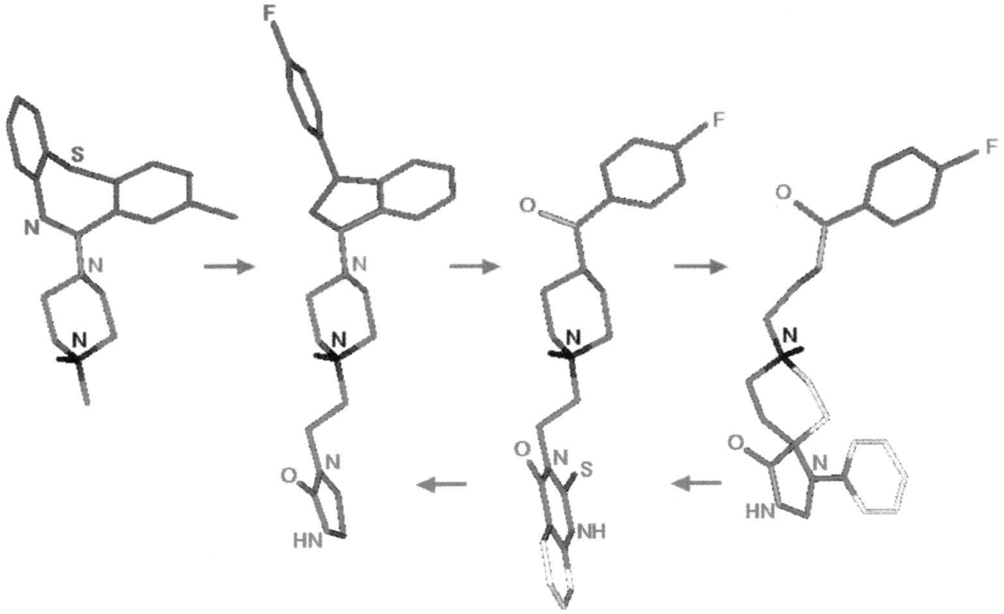

Plate 5

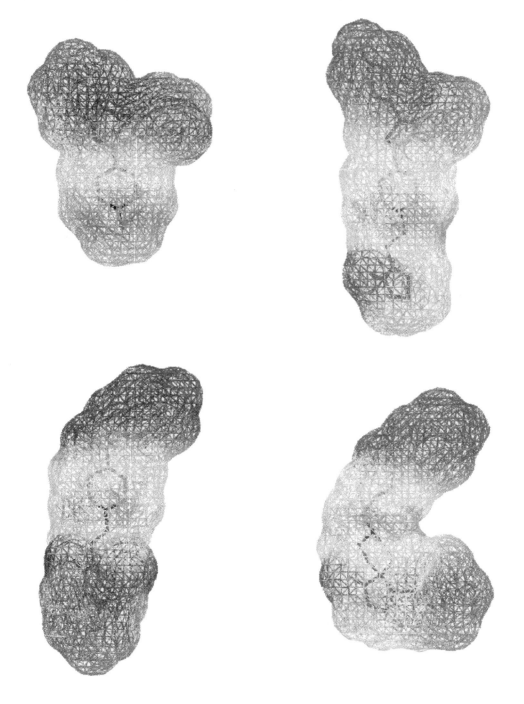

Plate 6

essential and may be omitted without detrimental effect on the binding strength.[49] It is therefore concluded that a corresponding hydrogen-bond-donating binding site will probably not be present in the receptor. The same is true for the carbonyl element involved in the heterocyclic system. This can be deduced from the fact that ketanserin derivatives with undiminished affinity are known which instead present a thiocarbonyl[50] group or even possess a

naphthyl system in place of the heterocycle (Elz, unpublished results). In conclusion, from the three interaction sites for hydrogen-bond contacts between the ligands and the receptor protein, only the hydrogen-bond-enforced ionic interaction exerted by the protonated nitrogen will be present in the amino acid model.

One additional correction of the interaction field-derived receptor map is necessary. The serotoninergic 5-HT$_{2a}$

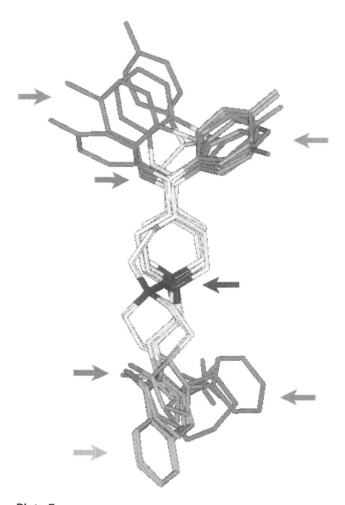

Plate 7

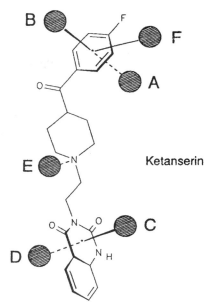

Ketanserin

Fig. 24.5 The receptor map contains six positions for receptor contact. The map has been constructed on the basis of interaction field calculations as well as experimental structure–activity relationship data. Positions A, B, C, D and F depict hydrophobic contacts, position E is an ionic interaction.

pharmacophore tells us that the aromatic part of the fluorobenzoyl system may be extensively substituted with hydrophobic elements and that this type of substitution leads to increasing affinity. So far this fact has not been accounted for in the receptor map, and so we have to add a third hydrophobic amino acid binding site to this region. The final receptor map then looks as shown in Fig. 24.5.

As mentioned before, the sites A to F of the map now can be occupied by different amino acids presenting the necessary chemical properties. The available biochemical information, such as amino acid sequence, bacteriorhodopsin homology, alignment studies, etc., has led us to construct the amino acid model of the 5-HT$_{2a}$ receptor presented in Fig. 24.6. Interaction energies for the complexes formed between ligands and the receptor model were calculated as described in Section II. D.

The biological data for the 5-HT$_{2a}$ ligands were taken from Elz;[51] most of the substances were also synthesized by that author. The correlation for the 15 ligands used is

shown in Table 24.1 together with calculated interaction energies as well as experimental and theoretical binding affinities.

Theoretical binding affinities can be derived from the correlation equation, which is shown graphically in Fig. 24.7. The correlation seems to be quite significant.

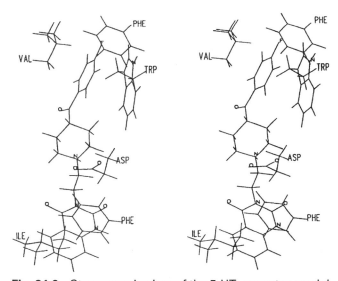

Fig. 24.6 Stereoscopic view of the 5-HT$_2$ receptor model constructed from single isolated amino acids on the basis of the receptor map (see Fig. 24.5).

Eighty-nine per cent of the variation in the biological data can be explained with the receptor model. On the other hand, a systematic deviation can be noticed. Some compounds (denoted by bold type in Table 24.1) are described by the model as being too weak by a constant factor of 1.0 or 1.5 orders of magnitude when compared with the experimental affinities. The inspection of the molecular formulae identifies those showing the larger deviation as fluorobenzoyl derivatives and those with the smaller deviation as benzoyl derivatives. In all cases, substances which are not optimally described by the model, possess an electron-deficient aromatic system. This leads to the hypothesis that for binding at the real receptor a type of interaction may be important which is

not accounted for in the force field energies. According to biochemical data, the model is constructed with a tryptophan molecule as one of the binding partners of the aromatic ring system. This electron-rich amino acid could very well be involved in a charge–transfer interaction to the electron-deficient phenyl system. This type of interaction energy is not included in the force field energy. To check whether charge–transfer interactions can in fact explain the missing binding energies for the compounds mentioned, HOMO and LUMO energies were calculated for some test complexes between a truncated tryptophan (3-ethylindole) and models of the electron-deficient structures of the ligand molecules. The results are shown in Fig. 24.8. It can clearly be seen that an electron transfer

Table 24.1 *Fifteen 5-HT$_{2a}$-antagonistic substances and corresponding interaction energies calculated using the 5-HT$_{2a}$ receptor model^a*

Compound	Interaction energy (kcal mol^{-1})	pK_B calculated	pK_B experiment
Ketenserin	23.44	7.94	9.55
EZS 15	22.38	7.60	8.60
EZS 21	22.54	7.65	7.60
EZS 32	21.62	7.35	7.40
Fluorbenzoylpiperidine (FTB)	13.62	4.73	6.25
Benzoylpiperidine (BP)	12.56	4.39	5.30
EZS 8	10.04	3.57	3.50
EZS 9	10.57	3.74	3.60
EZS 11	11.98	4.20	4.15
EZS 12	12.69	4.43	4.50
EZS 40	21.49	7.31	7.00
EZS 22	22.44	7.62	7.20
EZS 13	20.94	7.13	7.32
EZS 34	24.60	8.32	8.40
EZS 39	23.38	7.92	8.40

^{a} Interaction energies are transformed into binding affinities (pK_B calculated) and compared to experimentally derived affinities (pK_B experiment).

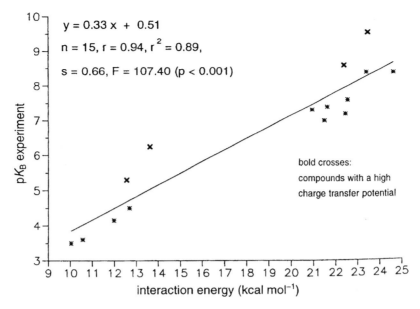

$y = 0.33\,x + 0.51$

$n = 15$, $r = 0.94$, $r^2 = 0.89$,

$s = 0.66$, $F = 107.40$ ($p < 0.001$)

bold crosses:
compounds with a high
charge transfer potential

Fig. 24.7 Correlation between experimentally determined binding affinities and interaction energies calculated from complexes constructed using the receptor model. Some compounds (bold crosses) do not fit the correlation line satisfactorily.

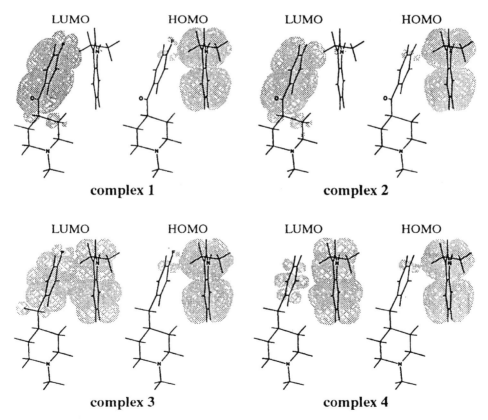

Fig. 24.8 HOMOs and LUMOs calculated using AM1 for complexes formed between ethylindole representing tryptophan and four truncated drug molecules. Charge–transfer interactions may occur in complexes 1, 2 and 3; they are impossible in complex 4.

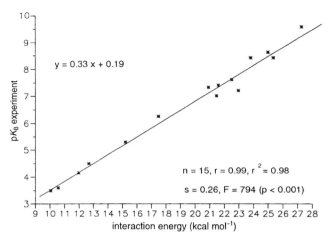

y = 0.33 x + 0.19

n = 15, r = 0.99, r^2 = 0.98

s = 0.26, F = 794 (p < 0.001)

Fig. 24.9 Correlation between binding affinities and interaction energies including a charge–transfer correction term.

from the HOMO of electron-rich tryptophan to the LUMO of the electron-deficient partner only takes place for the fluorobenzoyl- (complex 1) and the benzoyl-*N*-methylpiperidine (complex 2) model. In only these complexes are HOMOs and LUMOs neatly separated between the interacting molecules. In the other two cases, fluorobenzyl- (complex 3) and benzyl-*N*-methylpiperidine (complex 4), the indole system participates not only in the HOMO but also in the LUMO, so that electron transfer may be drastically diminished (complex 3) or absent (complex 4). For this reason the force field interaction energies have to be corrected with a special charge–transfer correction factor for fluorobenzoyl and benzoyl derivatives. Direct combination of calculated energy values stemming from different methods is impossible, because the absolute numbers will almost certainly not match. Therefore, an adaption procedure has to be found to allow correct inclusion of the charge–transfer part into the total interaction energy. In the case reported here this could be accomplished by correlating differences between LUMO and HOMO energies of the charge–transfer complexes with experimentally derived binding affinities of the ligands. Consideration of this charge–transfer correction factor leads to a significantly improved correlation equation (see Fig. 24.9).

The *F*-test value is extremely high and now 98% of the variation of the biological activities for the 15 compounds in the series can be explained on the basis of the receptor

Table 24.2 *Thirteen substances including 10 newly predicted structures and corresponding interaction energies calculated on the basis of the receptor model*[a]

Altanserin   Pirenperone   Risperidone   EZS 57   EZS 155 (R=F) / EZS 156 (R=H)   4-Benzylpiperidine

EZS 56   EZS 302   EZS 300   EZS 301   EZS 151 (R=F) / EZS 63 (R=H)

Table 24.2 (*continued*)

Compound	Interaction energy (kcal mol^{-1})	pK_B predicted	pK_B experiment
Altanserin	27.62	9.31	9.80
Pirenperone	27.16	9.15	9.30
Risperidone	28.69	9.66	9.70
EZS-57	22.27	7.54	7.20
EZS-155	23.91	8.08	8.50
EZS-156	21.61	7.32	8.20
EZS-302	22.09	7.48	7.70
EZS-300	23.23	7.86	7.65
EZS-301	19.37	6.58	6.25
4-Benzylpiperidine	12.64	4.36	5.00
EZS-56	25.53	8.62	8.60
EZS-151	23.64	7.99	7.40
EZS-63	21.41	7.26	7.20

model. As mentioned earlier, as long as the real receptor is unknown the efficiency and meaning of the model cannot be assessed except by prediction of new substances. Of course, the compounds must subsequently be synthesized and tested pharmacologically in order to prove or disprove the hypothesis. Based on the receptor map described, altogether 10 new structures (see Table 24.2) were predicted and the respective interaction energies were calculated (including charge–transfer correction).

These energies are presented together with the predicted binding affinities collected from the correlation curve and the experimental binding affinities reported by Elz (unpublished results). The graphical representation of the relation between predicted and experimental receptor affinities (Fig. 24.10) is remarkable and shows that the receptor

model developed seems to be consistent with the 'real' serotoninergic 5-HT$_{2a}$ receptor.

The example reported demonstrates that theoretical methods such as pharmacophore identification and receptor mapping can be successfully employed as predictive tools in medicinal chemistry.

REFERENCES

1. Kier, L.B. and Höltje, H.-D. (1975) A stochastic model of the remote recognition of preferred conformation in a drug–receptor interaction. *J. Theor. Riol.* **49**: 401–416.
2. Höltje, H.-D. (1992) Pharmacophore identification based on molecular electrostatic potentials. In Wermuth, C.G., Koga, N., Koenig, H. and Metcalf, B. (eds). *Medicinal Chemistry for the 21st Century*, pp. 181–189. Blackwell Scientific, Oxford.
3. Humblet, C. and Marshall, G.R. (1980) Pharmacophore identification and receptor mapping. *Annu. Rep. Med. Chem.* **15**: 267–276.
4. Cosentino, V., Moro, G., Pitea, D., *et al.* (1992) Pharmacophore identification by molecular modeling and chemometrics: the case of HMG-CoA reductase inhibitors. *J. Computer-Aided Mol. Design* **6**: 47–60.
5. Belvisi, L., Brossa, S., Salimbeni, A., Scolastico, C. and Todeschini, R. (1991) Structure–activity relationship of Ca^{2+} channel blockers: a study using conformational analysis and chemometric methods. *J. Computer-Aided Mol. Design* **5**: 571–584.
6. Cosentino, V., Moro, G., Pitea, D., Todeschini, R., Brossa, S., Gualandi, F., Scolastico, C. and Gienessi, F. (1990) Pharmacophore identification in amnesia-reversal compounds using conformational analysis and chemometric methods. *Quant. Struct.-Act. Relat.* **9**: 195–201.
7. Allen, F.H. and Kennard, O. (1993) 3D Search and researching using the Cambridge Structural Database. *Chem. Des.-Automat. News*, **8**: 31–37; Cambridge Structural Database (CSD), http://www.ccdc.cam.ac.uk
8. SYBYL 6.7.1 TRIPOS Inc., St Louis, MO, USA; http://www.tripos.com
9. INSIGHT II, Accelrys Inc.; http://www.accelrys.com

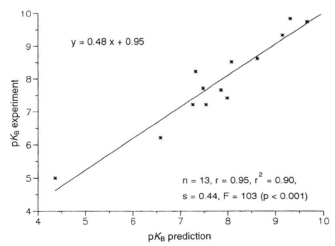

$y = 0.48 \, x + 0.95$

$n = 13$, $r = 0.95$, $r^2 = 0.90$, $s = 0.44$, $F = 103$ ($p < 0.001$)

Fig. 24.10 Correlation between predicted and experimentally determined binding affinities at the 5-HT$_{2a}$ receptor for 13 antagonists including 10 newly predicted structures.

10. Auffinger, P. and Wipff, G. (1990) High temperature annealed molecular dynamics simulation as a tool for conformational sampling. *J. Comput. Chem.* **11**: 19–31.

11. Hoeltje, H.-D. and Jendretzky, U.K. (1995) Construction of a detailed serotoninergic 5-HT$_{2a}$ receptor model. *Arch. Pharm.* **328**: 577–584.

12. Crippen, G.M. (1977) A novel approach to calculation of conformation: distance geometry. *J. Comput. Phys.* **24**: 96–107.

13. Kuntz, I.D., Crippen, G.M. and Kollman, P.A. (1979) Applications of distance geometry to protein tertiary structure calculations. *Biopolymers* **18**: 939–957.

14. Donné-Op den Kelder, G.M. (1987) Distance geometry analysis of ligand binding to drug receptor sites. *J. Computer-Aided Mol. Des.* **1**: 257–264.

15. Ghose, A.K. and Crippen, G.M. (1985) Geometrically feasible binding modes of a flexible ligand molecule at the receptor site. *J. Comput. Chem.* **6**: 350–359.

16. Sheridan, R.P., Nilakantan, R., Dixon, J.S. and Venkataraghavan, R.J. (1986) The ensemble approach to distance geometry: application to the nicotinic pharmacophore. *J. Med. Chem.* **29**: 899–906.

17. Crippen, G.M. (1982) Prediction of new leads from a distance geometry binding site model. *Quant. Struct.-Act. Relat.* **2**: 95–100.

18. Marshall, G.R., Barry, C.D., Bosshard, H.E., Dammkoehler, R.A. and Dunn, D.A. (1979) The conformational parameter in drug design: the active analog approach. In Olson, E.C. and Christoffersen, R.E. (eds) *Computer-Assisted Drug Design*, vol. 112, ACS Symp. Series, pp. 205–226. American Chemical Society, Washington DC.

19. Marshall, G.R. (1987) Computer-aided drug design. *Annu. Rev. Pharmacol Toxicol.* **27**: 193–213.

20. Scrocco, E. and Tomasi, J. (1978) Electronic molecular structure, reactivity and intermolecular forces: a heuristic interpretation by means of electrostatic molecular potentials. In Lödwin, P. (ed.). *Advances in Quantum Chemistry*, vol. II, pp. 115–193. Academic Press, New York.

21. Gasteiger, J. and Marsili, M. (1980) Iterative partial equalization of orbital electronegativity—a rapid access to atomic charges. *Tetrahedron* **36**: 3219–3228.

22. (a) Dewar, M. J. S. and Thiel, W. (1977) Ground states of molecules. 38. The MNDO method. Approximations and parameters. *J. Am. Chem. Soc.* **99**: 4899–4907; (1977) Ground states of molecules. 39. MNDO results for molecules containing hydrogen, carbon, nitrogen and oxygen. *J. Am. Chem. Soc.* **99**: 4907–4917. (b) Dewar, M.J.S., Zoebisch, E.G., Healy, E.F. and Stewart, J. J. P. (1985) A new general purpose quantum mechanical molecular model. *J. Am. Chem. Soc.* **107**: 3902–3909. (c) Stewart, J. J. P. (1989) Optimization of parameters for semiempirical methods. I. Method. *J. Comput. Chem* **10**: 209–220. (d) Stewart, J.J.P. (1989) Optimization of parameters for semiempirical methods. II. Applications. *J. Comput. Chem.* **10**: 221–264.

23. Van de Waterbeemd, H., Carrupt, P.A. and Testa, B. (1986) Molecular electrostatic potential of orthopramides: implications for their interaction with the D-2 dopamine receptor. *J. Med. Chem.* **29**: 600–606.

24. Kocjan, D., Hodoscek, M. and Hadzi, D. (1986) Dopaminergic pharmacophore of ergoline and its analogs. A molecular electrostatic potential study. *J. Med Chem.* **29**: 1418–1423.

25. SPARTAN, Wavefunction Inc., Irvine, CA, USA; http://www.wavefun.com

26. DMol (Modul in INSIGHT II), Accelrys Inc.; http:// accelrys.com

27. Alemán, C., Luque, F.J. and Orozco, M. (1993) A new scaling procedure to cortect semiempirical MEP and MEP-derived properties. *J. Computer-Aided Mol. Des.* **7**: 721–742.

28. Goodford, P.J. (1985) A computational procedure for determining energetically favorable binding sites on biologically important macromolecules. *J. Med. Chem.* **28**: 849–857.

29. Cramer, R.D., III, Patterson, D.E. and Bunce, J.D. (1988) Comparative molecular field analysis (CoMFA). 1. Effect of shape on binding of steroids to carrier proteins. *J. Am. Chem. Soc.* **110**: 5959–5967.

30. Kellogg, G.E., Semus, S.F. and Abraham, D.J. (1991) HINT — A new method of empirical field calculation for CoMFA. *J. Computer-Aided Mol. Des.* **5**: 545–552.

31. Kellogg, G.E., Joshi, G.S. and Abraham, D.J. (1992) New tools for modeling and understanding hydrophobicity and hydrophobic interactions. *Med. Chem. Res.* **1**: 444–453.

32. Wold, S., Johansson, E. and Cocchi, M. (1993) PLS — partial least-squares projections to latent structures. In Kubinyi (ed.) *3D QSAR in Drug Design—Theory Methods and Applications*, pp. 523–550. ESCOM, Leiden.

33. Höltje, H.-D. and Anzali, S. (1992) Molecular modelling studies on the digitalis binding site of the Na$^+$/K$^+$-ATPase. *Die Pharmazie* **47**: 691–697.

34. Höltje, H.-D. and Dall, N. (1993) A molecular modelling study on the hormone binding site of the estrogen receptor. *Die Pharmazie* **48**: 243–249.

35. Höltje, M. and Höltje, H.-D. (1991) Molecular modelling study on the negative inotropic potencies of 1,4-dihydropyridines. *Pharm. Pharmacol. Lett.* **1**: 19–22.

36. Höltje, H.-D. and Jendretzki, U. (1992) Conformational analysis of 5-HT$_2$ receptor antagonists. *Pharm. Pharmacol Lett.* **1**: 89–92.

37. Snyder, J.P., Rao, S.N., Koehler, K.F. and Pellicciari, R. (1992) Drug modeling at cell membrane receptors: the concept of pseudoreceptors. In Angeli, P., Gulini, U. and Quaglia, W. (eds). *Trends in Receptor Research*, pp. 367–403. Elsevier, Amsterdam.

38. Snyder, J.P. and Rao, S.N. (1989) Pseudoreceptors: a bridge between receptor fitting and receptor mapping in drug design. *CDA News* **4/10**: 1/13–15.

39. Rao, S.N. and Snyder, J.P. (1990) Pseudoreceptor modeling: an experiment in large scale computing. *Cray Channels* **11**: 4–12.

40. Guan, X.-M., Peroutka, S.J. and Kobilka, B.K. (1992) Identification of a single amino acid residue responsible for the binding of a class of β-adrenergic receptor antagonists to 5-hydroxytryptamine$_{1A}$ receptors. *Mol. Pharmacol.* **41**: 695–698.

41. Kao, H.-T., Adham, N., Olsen, M.A., Weinshank, R.L., Branchek, T.A. and Hartig, P.R. (1992) Site-directed mutagenesis of a single residue changes the binding properties of the serotonin 5-HT$_2$ receptor from a human to a rat pharmacology. *FEBS Lett.* **307**: 324–328.

42. Humblet, C. and Mirzadegan, T. (1992) Models of G protein-coupled receptors. *Annu. Rep. Med. Chem.* **27**: 291–300.

43. Trumpp-Kallmeyer, S., Hoflack, J., Bruinvels, A. and Hibert, M. (1992) Modeling of G-protein-coupled receptors: application to dopamine, adrenaline, serotonin, acetylcholine, and mammalian opsin receptors. *J. Med. Chem.* **35**: 3448–3462.

44. Andrews, P.R. (1993) Drug-receptor interactions. In Kubinyi, H. (ed.) *3D QSAR in Drug Design — Theory Methods and Applications*, pp. 13–40. ESCOM, Leiden.

45. Hibert, M.F., Mir, A.K. and Fozard, J.R. (1990) Serotonin (5-HT) receptors. In Emmert, J.C. (ed.). *Comprehensive Medicinal Chemistry*, Vol. 3, pp. 567–601. Pergamon Press, New York.

46. Watanabe, Y., Usui, H., Shibano, T., Tanaka, T. and Kanao, M. (1990) Syntheses of monocyclic and bicyclic 2,4(*1H,3H*)-pyrimidinediones and their serotonin 2 antagonist activities. *Chem. Pharm. Bull.* **38**: 2726–2732.

47. Ketanserin patent, Janssen Pharmaceutica N.V., European Patent Office. Kennis, L.E.J., Van der Aa, M.J.M., Van Heertum, A.H.M. and Jones, A.J. (1980) Nr. 0013612, Appl. Nr. 803000595.

48. Glennon, R.A. (1991) Serotonin receptor subtypes: basic and clinical aspects. In Peroutka, S.J., Venter, J.C. and Harrison, L.C. (eds).

Receptor Biochemistry and Methodology, vol. 15, pp. 19–64. Wiley-Liss, New York.

49. Herndon, J.L., Ismaiel, A., Ingher, S.P., Teitler, M. and Glennon, R.A. (1992) Ketanserin analogues: structure–affinity relationships for 5-HT$_2$ and 5-HT$_{1C}$ serotonin receptor binding. *J. Med. Chem.* **35**: 4903–4910.

50. Bogeso, K.P., Arnt, J., Boeck, V., Christensen, A.V., Hyttel, J. and Jensen, K.G. (1988) Antihypertensive activity in a series of I-piperazino-3-phenylindans with potent 5-HT$_2$-antagonistic activity. *J. Med. Chem.* **31**: 2247–2256.

51. Elz, S (1992) Synthesis and *in vitro* pharmacology of antiserotoninergic$_2$ ketanserin analogues and N-imidazolylalkyl substituted 4-(4-fluorobenzoyl)piperidine-1-carboxamidines. *XIIth International Symposium on Medicinal Chemistry*, Basel, Switzerland, poster-128.C, abstract book, p. 349.

25

THREE-DIMENSIONAL QUANTITATIVE STRUCTURE–PROPERTY RELATIONSHIPS

Gabriele Cruciani, Emanuele Carosati and Sergio Clementi

By asking for the impossible, we obtain the best possible **Italian Proverb**

I INTRODUCTION

In medicinal chemistry, simple mathematical models provide means for identifying which structural features may be important in determining biochemical activity over a group of compounds in a complex biochemical system. This is achieved by forming relationships between those variables that describe the structural variation within the group of chemicals and those that describe the bio-activities of the compounds. The relationship is denoted as a quantitative structure–activity relationship (QSAR).

Traditional QSAR, also known as 2D QSAR, accurately forecasts the potency of additional compounds and has led to the development of several commercial drugs and pesticides.[1] Statistical analysis weighs the molecular descriptors and the effects of substituents on biological activity. This strategy identifies which structural variations are the dominant ones influencing the change in biological properties.

However, the traditional approach may be considered to have some limitations. Besides the obvious requirements of the additional thermodynamic relationship, where only series of compounds with a common skeleton framework should be considered, conformational equilibria are not taken into account, and, in general, information on the 3D structure is not employed at all.

On the other hand, molecular modelling techniques have become extremely popular, especially because of increasing computation power. These methods are aimed at calculating the energy of a number of conformations for each molecule at different levels of approximation, and then to study the possible interactions between the molecule and its binding site. It became possible from this approach to describe each molecule/conformation by a series of theoretically computed parameters, some of which are intrinsically 3D in nature.

A 3D QSAR is, therefore strictly speaking, a QSAR relationship in which the structural descriptors have 3D nature. Several compounds are studied at the same time within the framework of a regression model, with the objective of ascertaining which structural features significantly affect the biological response. Notably, these 3D descriptors are usually derived from the different modelling techniques.

A particularly interesting example of such a 3D QSAR is comparative molecular field analysis (CoMFA),[2] or its variant called GRID-Golpe,[3] in which reactants with common structural backbones varying by residue substitutions may be aligned with one another, and physical characteristics such as electrostatic potential, hydrogen bond potential and steric energies may be measured for each reactant on a common grid with thousands of points. This is particularly interesting because it explicitly uses self-computed descriptors that depend only on the 3D character of a molecule rather than any information on its topology.

The critical factor for biological activity is the 3D spatial arrangement of these chemical and physical properties. The result of a 3D QSAR analysis is typically supplied as a 3D graphics image superimposed on a molecule of the data set. The images represent the coefficients of a pseudo-regression equation with the original variables, here as locations in the 3D space.

Thus, another key difference between traditional and 3D QSAR is the form of the output. This visualization of the results increases the fidelity of the communication between the QSAR modeller and the synthetic chemists.[4] Moreover, specialized graphics tools help the users to model, interpret and understand at best one particular problem.[5]

Some of the aspects of 3D QSAR methods remain to be addressed, such as the superposition of the compounds, the flexibility and the choice of bioconformation.[6] Work is in progress and this paper highlights some of the recent advances in this field.

II CALCULATION OF 3D MOLECULAR DESCRIPTORS

To design and optimize a drug, the medicinal chemists must know at least what building blocks to use (organic substituents or blocks of reagents) and how to connect them. In 3D QSAR, those building blocks are described only by means of self-computed descriptors, called molecular interaction fields, that depend only on the 3D character of a molecule rather than on its topology.

When computed on biomolecules (e.g. protein active sites), the molecular interaction fields identify regions where certain chemical groups can interact favourably, suggesting positions where a ligand should place similar chemical groups. MIF can also be computed starting from the ligands themselves. In this case, the regions showing favourable energy of interaction represent positions where groups of a potential receptor would interact favourably with the ligand. Using different probes, one can obtain for a certain ligand a set of such positions which defines a 'virtual receptor site' (VRS). This abstract entity defines an ideal complementary site for a certain chemical compound and represents its potential ability to bind a biomolecule. If a compound is known to bind a certain receptor, some of the regions defined in its VRS should actually overlap groups of the real receptor site and, therefore, at least a subset of the VRS regions would be relevant for representing the binding properties of the ligand. For the last statement to be true, the VRS must have been obtained from the bioactive conformation of the ligand

and the probes used to compute it should represent chemical groups present in the binding site.

MIF are computed by moving a chemical group, called a probe, around in a rectangular box of grid points and through a target molecule, to produce a three-dimensional box of interaction energy fields. Depending on the computational procedure used, these fields may represent total interaction energies (GRID),[7] steric or electrostatic fields (CoMFA),[2] molecular lipophilic potential (CoMPA, MLP),[8,9] hydrophobic interactions (HINT),[10] electron densities, etc. These fields may be used as point descriptors of the 3D molecular structure and physicochemical behaviour of the target molecule. Most properties related to molecular interactions can be represented in a 3D molecular field. Moreover, graphical analysis allows simple interpretation of the fields, such as the visualization of the regions where the probe interacts most strongly with the target either by attraction or repulsion (Plate 8) (See Plate section).

The interaction of molecules with biological receptors is mediated by surface properties such as shape, electrostatic forces, H-bonds and hydrophobicity. Above all, the importance of hydrogen bonding structure and functions provide convincing evidence today. Therefore, the GRID force field, which uses well-tested hydrogen bond potential, is largely used to characterize putative polar and hydrophobic interaction sites around target molecules.

Another convincing reason for using GRID in this field is the presence of more than a hundred different precomputed chemical probes the user can select, together with the ability to project and build any new chemical probe of interest. Among them, the hydrophilic probe group is always used to simulate solvation–desolvation processes and hydrogen-bonding interaction with the receptors, when known. Hydrophobic probes are used to simulate drug–receptor hydrophobic interactions. Hydrophobic energy[11] is very important to simulate binding processes. The hydrophobic energy component is computed at each grid point as $E_{\text{entropy}} + E_{\text{LJ}} - E_{\text{HB}}$, where E_{entropy} is the ideal entropic component of the hydrophobic effect in an aqueous environment, E_{LJ} measures the induction and dispersion interactions occurring between any pair of molecules, and E_{HB} measures the H-bonding interactions between water molecules and polar groups on the target surface.

Once calculated, the molecular interaction field maps can be visually compared and studied. However, problems arise when a number of molecules are studied at the same time. In this case, a simple graphics analysis and visualization is not sufficient to provide the necessary information to understand the observed trend in the biological properties of a series of compounds. In such a case, appropriate chemometric tools may be extremely useful in order to condense and extract hidden information.

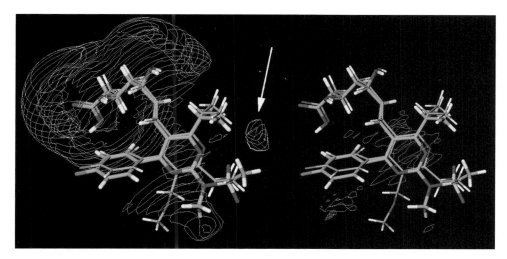

Plate 8 Molecular interaction fields (MIF) produced using the X-ray structures of Cerivastatin and Rosuvastatin drugs superimposed on the active site of the HMG-CoA re-educates protein. Green refers to hydrophobic attractive interaction energies between the HMG-CoA protein and the two drug molecules. Blue refers to bond donor attractive energies. The small blue region indicated with the arrow is due to the interaction between the Ser565 of the protein and the sulfide moiety of the Rosuvastatin, and is responsible for the increased binding affinity of Rosuvastatin for HMG-CoA protein.

III 3D QSAR, MOLECULAR CONFORMATION AND SUPERPOSITION

The ideal choice of conformers of the structures being modelled for 3D QSAR would be the biologically relevant ones. When experimental structural data are available revealing ligands bound to protein (i.e. X-ray data from co-crystallized ligand–target complexes) and when the bioactive conformation of the ligands are available, these should be used to derive the alignment rule.[12]

For the vast majority of cases, however, the bioactive conformation of a compound is not known. This is especially problematic for membrane bound enzymes and integral membrane proteins such as transporters for which there is no crystallized structure. Furthermore, the relevant conformation of a compound that traverses the lipid membrane can only be inferred or speculated upon.[13]

In order to deduce bioactive conformations based on ligand structures in the absence of structural information (e.g. for a receptor), it is normally considered the receptor like a rigid moiety, and it is assumed that the ligands, regardless of chemical composition, bind in conformations which present similar steric and electrostatic potential patterns to the target. This is the basis of the *pharmacophore concept*. The pharmacophore is defined as the critical three-dimensional arrangement of functional groups in the ligand that is responsible for creating biological response. Thus the structural features of a pharmacophore are complementary to the binding features in the target site(s), however this is

dependent on the tolerance applied and a single receptor may recognize multiple pharmacophores.[14] For example, CYP active sites allow some degree of conformational flexibility[15] and therefore may allow multiple pharmacophores to co-exist.

In making the connection between ligand and target structures, a problem arises in the inherent conformational flexibility of most molecules. This typically results in a multitude of possible patterns that the ligand can potentially present to the target. The solution to this problem requires at least two steps. First, one decides by means of chemical modification the relative importance of each functional group in the ligand. Then one should consider correspondence between different functional groups across the series of compounds. An assortment of computational tools exists to force molecules to assume a conformation as similar as possible to that of a rigid template molecule in a presumably bioactive conformation. In cases where pharmacophoric groups are present, these groups can be tethered during molecular mechanics minimization routines forcing similar groups to be maximally superimposed while allowing the rest of the molecule to relax to a low energy state. Alternatively, algorithms which force potential energy fields of molecules, not atoms, to be as similar as possible, can be used. Field fit minimization[16] and the SEAL procedure[17] are two examples of automated alignment techniques in CoMFA studies.[18,19] A recent review presents a detailed description of many alternative computational methods for molecular alignment.[20]

An important draw-back of all 3D molecular descriptors is that they are sensitive to the position and orientation of the molecular structures in the space. Accordingly, the initial alignment of the compounds in the series is widely recognized as one of the most difficult and time-consuming steps. Moreover, when the ligands are intrinsically flexible, superposition becomes impracticable if not completely subjective. The problem is that no statistical tools can inform the users of a bad superposition procedure, or on the other hand, there is no statistical procedure able to 'validate' the model independently from the superposition performed by the user. The only strong validation method, in our opinion, is the model interpretation linked with a careful analysis of experimental tests. In the last part of this chapter, the new concept of alignment-free descriptors will be better explained and addressed.

IV UNFOLDING THREE-DIMENSIONAL ENERGY MAPS

Ordinary statistical methods require a two-way table of objects and variables. An object is often a physically distinguishable entity, such as target molecule or a probe, while a variable represents the results of an observation, or a measurement, or a computation undertaken with this object. However, 3D QSAR methods produce three-way matrices of molecular descriptors. In order to transform three-way matrices into a two-way table, an unfolding procedure and simple reorganization of the data are required. The unfolding procedure is a method for transforming a multiway matrix into a one-dimensional vector of numbers, while data reorganization is a procedure for organizing one-dimensional vectors into a two-way data table.

The interaction energies between a target molecule and a probe produced by traditional 3D QSAR methods may be viewed as descriptors of the probe, or as descriptors of the target molecule behaviour. The two-fold interpretation of descriptors leads to different methods for producing the two-way data reorganization. Usually a 3D descriptor matrix for a single molecule is organized as a three-way table where the rows, the columns and the sheets are variables; the table itself represents the object and this three-way table can be easily rearranged as a one-dimensional vector.

In the presence of several molecules the procedure should be repeated for all the molecules and the vectors of variables assembled together in a two-way table in order to obtain a target matrix. Thus, the target matrix will contain the interaction energies between all the molecules and one specific interacting group.

With only one target molecule different computations may result by varying the probe, and in this case, a probe matrix is obtained.[21] The probe matrix contains information about the interactions of different chemical groups with the same target molecule. With such a problem, multivariate statistics may be used to select the most suitable probes in order to design selective target ligands.

Target matrices can be combined, thus, obtaining only one larger matrix. In the CoMFA procedure, two probes are employed as blocks of descriptors and the resulting two target matrices are combined to form a unique matrix containing the same number of objects and twice the number of variables. Similarly, by using GRID, several probe matrices can be combined by keeping the number of variables constant and increasing the number of objects. Clearly, the choice of using either the target or the probe or combined matrices for individual studies depends on what is to be deduced from the data, and is closely related with the problem in question.[22]

V STATISTICAL TOOLS

Principal component analysis[23] and partial least squares analysis[24] are chemometric tools for extracting and rationalizing the information from any multivariate description of a biological system. Complexity reduction and data simplification are two of the most important features of such tools. PCA and PLS condense the overall information into two smaller matrices, namely the score plot (which shows the pattern of compounds) and the loading plot (which shows the pattern of descriptors). Because the chemical interpretation of score and loading plots is simple and straightforward, PCA and PLS are usually preferred to other nonlinear methods, especially when the noise is relatively high.[25]

Score and loading plots are interconnected so that any descriptor change in the loading plot is reflected by changes in the position of compounds in the score plot. Pairwise comparison can be made directly with interactive plots[3] as developed in the Golpe program, and the relative contributions to the property under study is shown in the related descriptors space.

PCA is a least square method and for this reason its results depend on data scaling. The initial variance of a column variable partly determines its importance in the model. In order to avoid this problem, column variables were scaled to unit variance before analysis. The column average was then substracted from each variable. From a statistical point of view, this corresponds to moving the multivariate system to the centre of the data, which becomes the starting point of the mathematical analysis. The same autoscaling and centring procedures were applied to the PLS discriminant analysis.

Once the PCA model is developed, PCA predictions for new compounds or external test compounds are made by projecting the compound descriptors into the PCA model.

This is made by calculating the score vector T of descriptors X and average $\bar{x}$ for the new compounds, using the loading P of the PCA model, according to equation (1):

$$T = (X - \bar{x}) \cdot P' \cdot (P \cdot P')^{-1} \qquad (1)$$

For the PLS discrimination, external predictions were made using the following equation (2):

$$Y = \bar{y} - \bar{x} \cdot P' \cdot (P \cdot P') \cdot B \cdot Q$$
$$+ X \cdot P' \cdot (P \cdot P')^{-1} \cdot B \cdot Q \qquad (2)$$

where $\bar{y}$ is the Y column average and Q is the loading vector for the y space and B the coefficient between the X and Y spaces.

VI   APPLICATIONS

A   3D QSAR study on the structural requirements for binding to CB1 cannabinoid receptor

In this example the GRID-Golpe procedure[3] has been used in order to predict the structural requirements for binding to CB1 cannabinoid receptor. Since the three-dimensional structure of cannabinoid receptors is still unknown and little knowledge is available on the nature of the ligand–receptor interaction, this work was aimed at studying a set of molecules belonging to structurally different series by a 3D QSAR, including noncyclic compounds such as anandamide and its derivatives.[26] To obtain general structure information about the CB1 receptor and to derive a model able to predict the potency of compounds not included in the data set, receptor affinity was chosen as the dependent variable. The modelled molecules were selected from literature data reporting the binding affinities (K_i) with respect to CB1 for three series of significantly different structures such as THC and derivatives, anandamide and derivatives and indole or azole derivatives. The selected set of molecules appeared to be suitable for the investigation due to the homogeneity of the biological response and due to the wide variation of both structure and activities.

The structure of Δ^9-THC was built from the structure of Δ^9-tetrahydrocannabinolic acid B as determined by X-ray crystallography. THC analogues were modelled by modifying the basic structure. Aminoalkylindoles and azoles were subject to conformational analysis in order to select initial low energy conformations for all molecules.

Δ^9-THC was chosen as the structural template for the alignment process. THC analogues possessing the aromatic ring and the aliphatic side chain were aligned to Δ^9-THC by superimposing the above common groups, while for indole

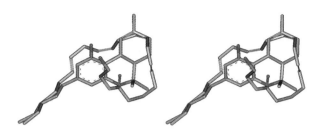

Plate 9   Schematic alignment of Δ^9-THC compound with anandamide molecule.

derivatives different alignments were considered (see Plate 9) (see Plate section).

The GRID program was used to describe the previously superposed molecular structures. A CB1 pseudoreceptor model reported in the literature suggests that the aspartic acid and histidine residues are involved in the interaction with cannabinoids. Therefore, in the present work, the carboxy anion and the N-sp$_2$ probe typical of hystidine moiety were chosen to mimic the receptor. The energy calculations were performed using 1.0 Å spacing between the grid points, and the three-dimensional matrix, arranged as a one-dimensional vector, was used as input for the statistical analysis.

CB₁ receptor model

The structure–activity correlation obtained by using PLS in GOLPE procedure identified the significant GRID variables corresponding to the regions of the molecules involved in the binding to the CB$_1$ receptor. The PLS model derived on the 1295 variables selected from the initial 17 550, was optimal with only two PLS components accounting for 95% of variance in the CB$_1$ receptor binding affinity and being highly predictive ($Q^2 = 0.85$). Figure 25.1 reports a graphical representation of the model fitting ability. The model is therefore successful in explaining a set of 20 molecules belonging to completely different structural classes, with good prediction ability. Hence it provided safe grounds for a unique pharmacophoric interpretation of the regions responsible for the CB$_1$ receptor affinity of all compounds.

The GRID plot of the partial weights (Plate 10) (see Plate section) identifies areas in space that contribute most to the CB$_1$ receptor binding affinity model. Plate 10 highlights seven areas representing the regions of interaction between the target molecules and the receptor and it includes, as an example, the structure of Δ^9-THC. The pale regions highlight areas where a polar group (e.g. an H bond donor in the ligand structure) causes an increase in the binding affinity, while the dark regions highlight areas where interactions result in an overall binding affinity decrease. Conversely, an apolar group (e.g. a methyl in the ligand structure) interacting with light or dark regions causes,

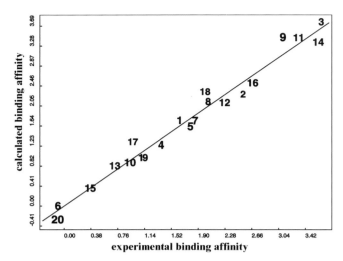

Fig. 25.1 Experimental versus calculated inhibition constant (log K_i) from the CB_1 receptor model.

respectively, a decrease or an increase in the binding affinity.[26]

The methyl group attached to C_9 of Δ^9-THC interacts with region 1 with a positive binding contribution. A bulky substituent at C_9 would protrude more deeply into the above interaction region with an expected activity increase.

The methyl groups at C_6 of Δ^9-THC interact, respectively, with region 4 and with region 2, providing a negative contribution for the binding affinity. Therefore removal of these methyls would favour binding to the CB_1 receptor while their substitution with polar groups (NH_2, OH), giving a positive binding contribution, would result in a further affinity increase.

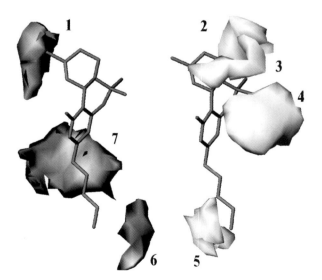

Plate 10 Grid plot of PLS partial weights for the CB_1 receptor model (see text for discussion).

The well-known high activity of 9-nor-9β-OH-hexa-hydrocannabinol (HHC), where the C_9 hydroxy substituent, if interacting with region 1, would produce a negative contribution for binding, could be due to the interaction of OH with region 3. The above interaction would, in that particular case, be possible due to the absence of the C_9–C_{10} double bond which makes possible an arrangement of the OH group closer to region 3.

Interaction of the carbamido nitrogen of anandamide and derivatives with region 4 provides a relevant modulation of the binding affinities. The lack of the above interaction for other derivatives (see Chapter 26), the conformation of which is such that the amido nitrogen lies away from region 4, results in great affinity decrease. The presence of region 7, where alkyl groups would improve the binding with the receptor, confirms literature data which point out that branching at the first atom of the alkyl chain increases binding affinity. Consistently, the highly active Δ^9-THC derivatives exhibit both branching methyls (attached at C_1 of the linear chain) interacting with region 7 and a chain extension (seven carbon atoms) which is optimal for interaction with region 6.

B 3D QSAR as a tool to design selective ligands

Selectivity towards a single biological target is an essential requirement for potential drugs. In the design of antibiotics or anticancer drugs, the selective activity is implicit in their pharmacological profile, but in other fields, selectivity is also critically important to minimize potential side-effects. Therefore, it is desirable to involve selectivity considerations as early as possible in the drug design efforts. In the past, the design of selective drugs relied on trial and error or on the close inspection of structural data. This task was especially difficult when targeting a single protein within a highly homologous family.

With the same purpose, QSAR or 3D QSAR approaches were often used in the past. However, this requires a series of compounds that have already been synthesized and tested and relies critically on the alignment and superposition of the compounds. Moreover, this approach is difficult if the structural variation within the test compounds is too high.

The availability of three-dimensional structures of the target proteins considerably simplifies the search for selective ligands. Then it is possible to compare the binding sites of different targets, looking for differences in sequence or structure that can be exploited for selective target–ligand interactions. However, this is by no means trivial, as competing contributions may have to be balanced, and there is always the danger of overlooking structural differences that might have a strong impact on selectivity.

To avoid these problems, a new 3D QSAR approach based on GRID/PCA algorithms was developed and successfully applied to investigate the selectivity between pairs of biomolecular targets.[27–29] The new method starts from one or more 3D structures of several target proteins, uses a multivariate description of the binding sites performed using the program GRID and analyses the molecular interaction fields using consensus principal component analysis (CPCA). As a result one obtains contour plots highlighting both the regions and the type of interactions in these regions that can be used to introduce selectivity into a potential ligand of these protein targets.[30]

Selectivity in serine proteases

Thrombin and factor Xa are prominent players in the blood clotting cascade.[31] They are, therefore, important targets for the development of new anticoagulant/antithrombotic drugs.[32–37]Trypsin is an enzyme excreted by the pancreas for helping in digestion, and has classically been used as a model enzyme for the whole serine protease family. Thus, in order to minimize side-effects of thrombin/factorXa inhibitors, and to enhance their bioavailability, potential drugs should exhibit selectivity towards thrombin/factorXa with respect to trypsin.[38]

Crystallographic structures for all three enzymes show a similar structure of their active sites.[39] The main determinant for the specificity of ligands towards proteins of the chymotrypsin family is the deep hydrophobic S1 pocket. Asp189 (the numbering scheme of the amino acid residues follows the chymotrypsin scheme[18]) is located at the bottom of the S1 pocket, where salt bridges can be formed with basic residues of the substrate peptides. This is the most conserved region of the three studied serine proteases, all residues are conserved but for a A190S mutation in trypsin which makes its pocket slightly more hydrophilic. Two other pockets on the unprimed side of the catalytic triad are important for substrate and inhibitor binding. Both the P and D pocket are mostly hydrophobic, and the differences in their amino acid residues have been used in the past to enhance selectivity in potential protein ligands.

The GRID/PCA analysis and subsequent extensions[27–30] was one of the first methods proposed to rationally design selective ligands. In this method, two or more target proteins can be studied and characterized using GRID with n different probes representing different chemical groups. The problem is formulated, from a chemometric point of view, as a collection of objects equal to the number of proteins $\times$ the number of probes, each one representing a different target–probe interaction. In order to extract relevant information from this X matrix (which contains all target–probe interaction energies collected at the grid points), the PCA or the most advanced CPCA are used to decompose the matrix into a product of two smaller matrices

T (score matrix) and P' (loading matrix) that explain at best the overall variance of the original X matrix. From a practical point of view, the PCA analysis of the X matrix allows on one hand a simplified view of the system: in the scores plot some clusters, corresponding to the target proteins, can be recognized. On the other hand, the loading plot can be used to identify the variables with highest participation in the PC that discriminates these two clusters. When these variables are represented in the space around the targets using appropriate isocontour plots, they identify selectivity-relevant regions.

Contour plots. The PCA or CPCA loading can be translated into contour plots describing the interaction fields between a GRID probe and the target protein structure. For a selectivity study, the loading discriminating different target proteins is of interest. Unfortunately, often more than one PC contributes to separate these objects in the scores plot, and therefore, any single loadings plot can only partially explain the structural features which were found as important by the model. In this respect, GOLPE offers the possibility of using active plots.[40] Here, one draws a vector linking pairs of objects in a 2D scores plot which is then translated into isocontour plots highlighting those variables that contribute most to differentiate the selected objects. In order to obtain such isocontours, GOLPE calculates the difference between the two points for the first and second principal component and projects these differences back into the original space (a pseudo-field) using the PCA loading. The result is a grid plot of the differences in the pseudo fields that highlight the object differences for the corresponding probe. Using these plots, one is able to answer: (1) where are the regions that can produce selective interaction with respect to the starting and endpoint of the drawn vector (which, for example, could connect a pair of protein targets); and (2) which interaction (i.e. probe) is responsible for this difference. It should be emphasized that these plots are different for each probe, while in the original GRID/PCA method, only a single loadings plot was obtained.

Computational details and results

The three-dimensional structures of thrombin, trypsin and factor Xa used in this study were taken from the Research Collaboratory for Structural Bioinformatics (RCSB) protein database.[41]Crystallographic water molecules, bound ligands and counterions were removed. Then the protein structures were aligned according to their Cα traces using the INSIGHTII[42] modelling software. The GRID calculations were performed with version 17 of the GRID software.[43] The proteins were considered to be rigid. The GRID box dimensions were chosen to encompass all relevant residues within the respective actives sites of the proteins.

Table 25.1 *List of the GRID probe types used in the selectivity study*

Name	Chemical group	Some additional properties
DRY	Hydrophobic probe	Empirical term for entropy
C3	Methyl group	
N1	Neutral flat NH, e.g. amide	H-bond donor probe
N:	Sp^3 N with lone pair bonded to three heavy atoms	Lone pair donor probe
NH=	Sp^2 NH with lone pair	
N1+	Sp^3 amine NH cation	Charge +1
NM3	Trimethyl-ammonium cation	Charge +1
O	Sp^2 carbonyl oxygen	
OH	Phenol or carboxy OH	Aromatic hydroxyl group
OS	Oxygen of sulfone/sulfoxide	
COO−	Carboxylic acid anion	Multi-atom probe, charge − 1
O::	Sp^2 carboxy oxygen atom	

The GRID probes were chosen to represent all relevant interactions (hydrophobic, charge–charge and hydrogen bond donor/acceptor) and to cover the most common chemical groups used in known thrombin and factor Xa inhibitors. Table 25.1 lists the GRID probes used and gives a short description, including which functional groups they represent. For each of the three main binding pockets, we first try to establish which interaction type could infer selectivity, then the exact region of this interaction is identified. The first task is accomplished by examining the distribution of the x-variable values within the binding pocket, the exact region is then displayed as a CPCA differential pseudofield contour.

S1 pocket. The S1 pockets are very similar in all three proteins and in fact from the distribution of the calculated interaction energies for 10 different GRID probes, it can be concluded that none of the interactions represented by the GRID probes is particularly favourable in one of the proteins. Plate 11 (see Plate section) shows the pseudofield differences for the GRID carbonyl oxygen probe between thrombin and trypsin. The large cyan contour indicates a region where interaction between the O probe and the enzyme is more favourable in trypsin than in thrombin. Therefore, making a potential ligand more hydrophobic, should shift selectivity towards thrombin. Indeed, there is experimental evidence for this behaviour.[44,45] Therefore, even though a hydrophilic group will be electrostatically favoured by trypsin, the steric restraints will eventually shift selectivity away from trypsin towards thrombin or factor Xa. This observation has also been reported in the literature.[46,47]

P pocket. In contrast to the results for the S1 pocket, several probes have particularly high interaction energies in one of

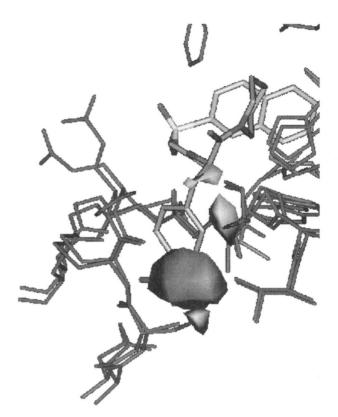

Plate 11 Differential pseudofield plot showing MIF differences between thrombin (blue) and trypsin (red) for the GRID carbonyl (O) probe within the S1 pocket. Negative interaction energies are shown in blue, while energetically unfavourable fields are shown in a yellow contour. Reference inhibitor is NAPAP.[30]

the proteins. For both the hydrophobic DRY and C3 probes, the highest interaction energies are found in the thrombin P pocket. The cationic probes N1+ and NM3 are also particularly favourable for thrombin. Therefore, a ligand with either hydrophobic or positively charged functional groups in the region of the P pocket should improve selectivity towards thrombin.

Plate 12 (see Plate section) shows the CPCA differential plot between thrombin and trypsin for the DRY probe. Immediately below the thrombin insertion loop residues Tyr60A and Trp60D a large yellow contour indicates that the introduction of hydrophobic groups in a potential ligand at that position would increase its selectivity for thrombin. As can be seen in Plate 12, the hydrophobic piperidine group of NAPAP, for example, occupies the same region as the yellow contour, thus contributing to NAPAP's selectivity for thrombin.

The results of the interaction energy distributions suggest that a positive interaction (N1+ and NM3 probes) could also be exploited to increase selectivity towards thrombin. However, inspection of the CPCA differential plots for both

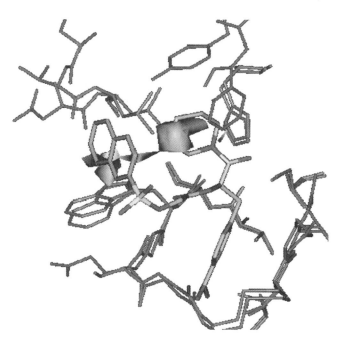

Plate 12 Pseudofield plot showing MIF differences between thrombin (blue) and trypsin (red) for the GRID DRY probe within the P pocket. Energetically unfavourable fields are shown in a yellow contour. Reference inhibitor is NAPAP.[30]

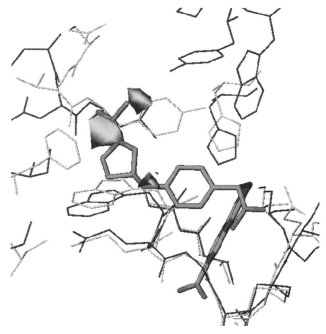

Plate 13 Pseudofield plot showing MIF differences between thrombin (blue) and factor Xa (green) for the GRID trimethyl ammonium (NM3) probe within the D pocket. Favourable interaction energies are shown in blue. Reference inhibitor is DX 9065 A.[30]

the N1+ and the NM3 probe show that besides a favourable region for thrombin in the P pocket, nearby regions favour interactions both with factor Xa and trypsin. Therefore, it would probably be difficult to alter the specifity of ligands with this type of interaction. This is in agreement with experimental findings that a positive charge does not improve selectivity towards thrombin.[48]

D pocket. The DRY, C3 and NM3 probes show the highest interaction energies for factor Xa, and therefore, these interactions are expected to be the most important ones for selectivity towards this enzyme. In addition, for trypsin the O and OS probes are particularly favourable.

In factor Xa, the D pocket is lined by aromatic residues in addition to the negatively charged Glu97, therefore, the finding that hydrophobic or cation–π interactions are important again underlines the strength of our x-variable weighting procedure.

Plate 13 (see Plate section) shows the CPCA pseudo-difference field plot for the cationic NM3 probe. One large favourable cyan contour for the NM3 probe is located in the hydrophobic box, pointing to the possible cation–π interaction in factor Xa. Therefore, the introduction of positively charged or polarized groups at either position will increase selectivity for factor Xa over thrombin or trypsin. Indeed, a number of highly active and specific factor Xa

inhibitors have positively charged groups directed towards the contours in the D pocket.[49–51]

The GRID/CPCA analysis allows a detailed analysis of structural differences important for selectivity within a given family of target proteins. Based on the three-dimensional structures of the proteins, this method analyses selectivity differences from the view of the receptor, and is therefore independent of the availability of appropriate ligands for a ligand-based QSAR analysis. The new procedure overcomes several of the disadvantages identified in selectivity analyses by the GRID/PCA approach which has been used in the past. Graphical representation of the pertinent differences responsible for selectivity can be directly translated into suggestions of how an existing ligand can be modified (or which functional groups a novel ligand should have) to enhance selectivity towards a given target.[30]

VII THE FUTURE OF 3D QSAR

3D QSAR methods are based on the mechanistic underlying assumptions that the modelled compounds should bind in similar mode and with similar bioactive conformation. Moreover, the underlying assumptions on molecular description are the congruency of the descriptor matrix

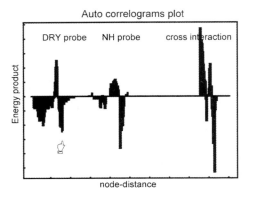
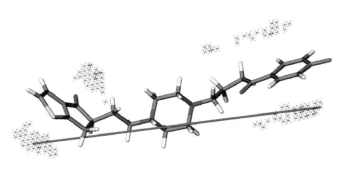

Auto correlograms plot

DRY probe   NH probe   cross interaction

Energy product

node-distance

Plate 14 Left, correlogram profile obtained for a butyrophenone derivative [52] molecule, using a DRY (hydrophobic) probe and a N1 (amide N) probe. From left to right the plot contains the auto-correlogram DRY–DRY, the auto-correlogram N1–N1 and the cross-correlogram DRY–N1. Right, the isocontour plot of a GRID MIF computed with a DRY probe around a molecule. The contour encloses points with negative values under -1.0 kcal mol^{-1}. The hand points to the variable represented in the figure with a line. These variables show an unfavourable interaction (for the pharmacological activity) between hydrophobic regions present in long compounds like the one represented in the figure (for details see reference 52).

and the presence of continuity constraints between the fields computed at neighbouring grid nodes.

Accordingly traditional 3D-QSAR relies upon a time-consuming alignment step that can also introduce a certain degree of user bias into the results. There are several methods which try to overcome this problem, but in general the necessary transformations prevent a simple interpretation of the resultant models in the original descriptor space (i.e. 3D molecular coordinates). We have recently presented a novel class of molecular descriptors which we have termed grid independent descriptors (GRIND).[52] They are derived in such a way as to be highly relevant for describing biological properties of compounds whilst being alignment independent, chemically interpretable and easy to compute. GRIND are obtained starting from a set of molecular interaction fields, computed by the program GRID or by other programs. The procedure for computing the descriptors involves a first step, in which the fields are simplified, and a second step, in which the results are encoded into alignment-independent variables using a particular type of autocorrelation transform (see Plate 14) (see Plate section). The molecular descriptors thus obtained can be used to obtain graphical diagrams called 'correlograms' and can be used in different chemometric analyses, such as principal component analysis or partial least squares.

An important feature of GRIND is that, with the use of appropriate software, the original descriptors (molecular interaction fields) can be regenerated from the autocorrelation transform and, thus, the results of the analysis are represented graphically, together with the original molecular structures in three-dimensional plots. The use of the methodology was recently illustrated in examples from the

field of 3D QSAR. Highly predictive and interpretable models are obtained showing the promising potential of the novel descriptors in drug design.[52,53]

The increased speed of the procedure may also help the medicinal chemist to discover compounds with high affinity towards predefined biological targets. Modern high-throughput techniques have rendered this strategy immensely successful, but still the road that leads from a high-affinity ligand to a pharmacokinetically and toxicologically well-behaved drug candidate is very long.

In order to decrease the costly and time-consuming development of active compounds ultimately doomed by hidden pharmacokinetic and toxicological defects, medicinal chemists must integrate metabolic considerations into drug design and lead optimization strategies. For that purpose, a novel family of 3D molecular descriptors based on path-distances of molecular interaction fields has been produced (Cruciani, G, *et al.*, in preparation). The new descriptors are characterized for being alignment-independent, highly relevant for describing the pharmacological properties of the compounds, well suited for describing the macromolecules, thus defining a method to compare macrostructures with their potential ligands without the need for virtual docking, and for being relatively fast to compute and easy to interpret (see Plate 15). Path-distance descriptors are well suited for 3D quantitative structure-metabolism relationships in which substrates can be structurally so different as to make impossible the relative superposition using standard techniques, and/or the mechanism of action can be different for different substrates.

In our experience, this new generation of descriptors can be used either for obtaining high quality 3D QSMR models

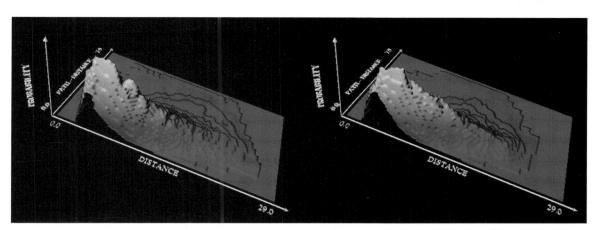

Plate 15 MIF transformation obtained with PathFinder program. Left, the transformation for the NAD site of the protein L-aspartate oxidase. Right, the MIF transformation for the NAD ligand molecule alone. The two maps are very similar, although the NAD molecule was roto-translated from the initial true position. The map comparison obtained from PathFinder can substitute virtual docking procedures.

or to propose pharmacophore hypotheses of the metabolic sites, without any superimposition and in a short time. Work is in progress to validate and test these preliminary results.

ACKNOWLEDGEMENTS

We would like to thank the M. Hann group (GSK, Stevenage, UK) for granting funds to the research projects in the PathFinder program, our team of coworkers in Perugia and Molecular Discovery (London) for support in implementing the programs.

REFERENCES

1. (a) Hansch, C., Leo, A. and Khoekman, D. (eds). (1995) *Exploring QSAR: Hydrophobic, Electronic, and Steric Constants*, p. 348. American Chemical Society, Washington, DC; (b) Fujita, T. (1984) The role of QSAR in drug design. In Jolles, G. and Wolldridge, K. R. H. (eds). *Drug Design: Fact or Fantasy?* pp. 19–33. Academic Press, London; (c) Boyd, D. B. (1990) Successes of computer-assisted molecular design. In Lipkowitz, K. B. and Boyd, D. B. (eds). *Reviews in Computational Chemistry*, pp. 355–371. VCH, New York; (d) Hansch, C., Fujita, T. (eds). (1995) Classical and three-dimensional QSAR in agrochemistry, pp. 342. American Chemical Society, Washington, DC.

2. Cramer, R.D., III, Patterson, D.E. and Bunce, J.D. (1988) Comparative molecular field analysis (CoMFA): 1. Effect of shape on binding of steroids to carrier proteins. *J. Am. Chem. Soc.* **110**: 5959–5967.

3. Pastor, M., Cruciani, G. and Clementi, S. (1997) Smart region definition: a new way to improve the predictive ability and interpretability of three-dimensional quantitative structure–activity relationships. *J. Med. Chem.* **40**: 1455–1464.

4. Martin, Y. C. (1998) 3D QSAR: current state, scope and limitations. In Kubinyi, H. (ed.) *3D QSAR in Drug Design*, pp. 3–23. Kluwer, Dordrecht.

5. Cruciani, G., Clementi, S., Baroni, M. and Pastor, M. (1998) Recent developments in 3D QSAR methodologies. In Liljefors, T., Jorgensen, F.S. and Krogsgaard-Larsen, P. (eds). *Rational Molecular Design in Drug Research*, Alfred Benzon Symposium **42**, pp. 87–97. Munksgaard, Copenhagen.

6. Clementi, S., Cruciani, G., Riganelli, D. and Valigi, R. (1995) GOLPE: merits and drawbacks in 3D-QSAR. In Sanz, F. (ed.). *QSAR and Molecular Modeling: Concepts, Computational Tools and Biological Applications*, pp. 408–414. Prous Sci., Barcelona.

7. (a) Goodford, P. J. (1985) A computational procedure for determining energetically favourable binding sites on biologically important macromolecules. *J. Med. Chem.* **28**: 849–857; (b) Boobbyer, D. N. A., Goodford, P. J., Mcwhinnie, P. M. and Wade, R. C. (1989) New hydrogen-bond potentials for use in determining energetically favourable binding sites on molecules of known structure. *J. Med. Chem.* **32**: 1083–1094; (c) Wade, R., Clerk, K. J. and Goodford, P. J. (1993) Further development of hydrogen bond function for use in determining energetically favourable binding sites on molecules of known structure. Ligand probe groups with the ability to form two hydrogen bonds. *J. Med. Chem.* **36**: 140–147.

8. Floershein, P., Nozulak, J. and Weber, H.P. (1993) Experience with comparative molecular field analysis. In Wermuth, C.G. (ed.). *Trends in QSAR and Molecular Modeling '92*, pp. 227–232. ESCOM, Leiden.

9. Carrupt, P.A., Gaillard, P., Billois, F., Weber, P., Testa, B., Meyer, C. and Perez, S. (1995) The Molecular lipophilicity potential (MLP): a new tool for logP calculation and docking, and in comparative molecular field analysis (CoMFA). In Pliska, V., Testa, B. and van de Waterbeemd, H. (eds). *Lipophilicity in Drug Action and Toxicology*, pp. 195–215. VCH, Weinheim.

10. Kellogg, G.E. and Abraham, D.J. (1992) KEY, LOCK and LOCKSMITH: complementary hydropathic map predictions of drug structure from a known receptor–receptor structure from known drugs. *J. Mol. Graph.* **10**: 212–217.

11. GRID v.20, manual; www.moldiscovery.com.

12. Cruciani, G. and Watson, K.A. (1994) Comparative molecular field analysis using GRID force-field and GOLPE variable selection methods in a study of inhibitors of glycogen phosphorylase b. *J. Med. Chem.* **37**: 2589–2601.

13. Elkins, S., Waller, C.L., Swaan, P.W., Cruciani, G., Wrighton, S.A. and Wikel, J.H. (2000) Progress in predicting human ADME parameters *in silico. J. Pharmacol. Toxicol. Meth.* **44**: 251–272.

14. Anzenbacherova, E., Bec, N., Anzenbacher, P., Hudecek, J., Soucek, P., Jung, C., Munro, A.W. and Lange, R. (2000) Flexibility and stability of the structure of cytochromes P450 3A4 and BM-3. *Eur. J. Biochem.* **267**: 2916–2920.

15. Arnold, G.E. and Ornstein, R.L. (1997) Molecular dynamics study of time-correlated protein domain motions and molecular flexibility: cytochrome P450BM-3. *Biophys. J.* **73**: 1147–1159.

16. Clark, M., Cramer, R., Jones, D., Patterson, D. and Simeroth, P. (1990) Comparative molecular field analysis (CoMFA): 2. Toward its use with 3D-structural databases. *Tetrahedron Comp. Meth.* **3**: 47.

17. Kearsley, S. and Smith, G. (1990) An alternative method for the alignment of molecular structures: maximizing electrostatic and steric overlap. *Tetrahedron Comp. Meth.* **3**: 615.

18. Waller, C.L., Oprea, T.I., Chae, K., Rhee-Park, H.-K., Korach, K.S. *et al.* (1996) Ligand-based identification of environmental estrogens. *Chem. Res. Toxicol.* **9**: 1240.

19. Klebe, G., Mietzner, T. and Weber, F. (1994) Different approaches toward an automatic structural alignment of drug molecules: applications to sterol mimics, thrombin and thermolysin inhibitors. *J. Comput.-Aided Mol. Design.* **8**: 751.

20. Lemmen, C. and Lengauer, T. (2000) Computational methods for the structural alignment of molecules. *J. Comput.-Aided Mol. Design* **14**: 215–232.

21. Cruciani, G. and Clementi, S. (1994) GOLPE philosophy and applications in 3D-QSAR. In Van de Waterbeemd, H. (ed.). *Advanced Computer-Assisted Tecniques in Drug Discovery*, pp. 305–330. VCH, Weinheim.

22. Cruciani, G. and Goodford, P.J. (1994) A search for specificity in DNA–drug interactions. *J. Mol. Graph.* **12**: 116–129.

23. Wold, S., Esbensen, K. and Geladi, P. (1987) Principal component analysis. *Chemometrics Intellig. Lab. Syst.* **2**: 37–52.

24. Westerhuis, J.A., Kourti, T. and Macgregor, J.F. (1998) Analysis of multiblock and hierarchical PCA and PLS models. *J. Chemometrics* **12**: 301–321.

25. Crivori, P., Cruciani, G., Carrupt, P.A. and Testa, B. (2000) Predicting blood–brain barrier permeation from the three-dimensional molecular structure. *J. Med. Chem.* **43**: 2204–2216.

26. Fichera, M., Cruciani, G., Bianchi, A. and Musumarra, G. (2000) A 3D-QSAR study on the structural requirements for binding to CB_1 and CB_2 cannabinoid receptors. *J. Med. Chem.* **43**: 2300–2309.

27. Cruciani, G. and Goodford, P.J. (1994) A search for specificity in DNA–drug interactions. *J. Mol. Graph.* **12**: 116–129.

28. Pastor, M. and Cruciani, G. (1995) A novel strategy for improving ligand selectivity in receptor-based drug design. *J. Med. Chem.* **38**: 4637–4647.

29. Matter, H. and Schwab, W. (1999) Affinity and selectivity of matrix metalloproteinase inhibitors: a chemometrical study from the perspective of ligands and proteins. *J. Med. Chem.* **42**: 4506–4523.

30. Kastenholz, M.K., Pastor, M., Cruciani, G., Haaksma, E.E.J. and Fox, T. (2000) GRID/CPCA: A new computational tool to design selective ligands. *J. Med. Chem.* **43**: 3033–3044.

31. Davie, E.W., Fujikawa, K. and Kisiel, W. (1991) The coagulation cascade: initiation, maintenance, and regulation (review). *Biochemistry* **30**: 10363–10370.

32. Kaiser, B. (1998) Thrombin and factor Xa inhibitors. *Drugs Fut.* **23**: 423–436.

33. Wiley, M.R. and Fisher, M.J. (1997) Small-molecule direct thrombin inhibitors. *Exp. Opin. Ther. Patents* **7**: 1265–1282.

34. Tapparelli, C., Metternich, R., Ehrhardt, C. and Cook, N.S. (1993) Synthetic low-molecular weight thrombin inhibitors: molecular design and pharmacological profile (review). *Trends Pharmacol. Sci.* **14**: 366–376.

35. Hauptmann, J. and Stürzebecher, J. (1999) Synthetic inhibitors of thrombin and factor Xa: from bench to bedside. *Thromb. Res.* **93**: 203–241.

36. Babine, R.E. and Bender, S.L. (1997) Molecular recognition of protein–ligand complexes: applications to drug design. *Chem. Rev.* **97**: 1359–1472.

37. Claeson, G., Elgendy, S., Cheng, L., Chino, N., Goodwin, C.A., Scully, M.F. and Deadman, J. (1993) Design of novel types of thrombin inhibitors based on modified D-Phe-Pro-Arg sequences. *Adv. Exp. Med. Biol.* **340**: 83–89.

38. Krishnan, R., Zhang, E., Hakansson, K., Arni, R.K., Tulinsky, A. *et al.* (1998) Highly selective mechanism-based thrombin inhibitors: structures of thrombin and trypsin inhibited with rigid peptidyl aldehydes. *Biochem.* **37**: 12094–12103.

39. Stubbs, M.T. and Bode, W. (1993) A player of many parts: The spotlight falls on thrombin's structure. *Thromb. Res.* **69**: 1–58.

40. Cruciani, G., Pastor, M. and Clementi, S. (1999) Handling information from 3D grid maps for QSAR studies. In Gundertofte, K. and Jorgensen, F.S. (eds). *Molecular Modeling and Prediction of Bioactivity*. Kluwer, New York.

41. RCSB, Protein Data Bank, operated by the Research Collaboratory for Structural Bioinformatics. http://www.rcsb.org/pdb/index.html.

42. InsightII 98.0: Molecular Simulations, Inc., San Diego.

43. Goodford, P. (1998) *GRID, version 17*.

44. Feng, D.-M., Gardell, S.-J., Lewis, S.D., Bock, M.G., Chen, Z. *et al.* (1997) Discovery of a novel, selective, and orally bioavailable class of thrombin inhibitors incorporating aminopyridyl moieties at the P1 position. *J. Med. Chem.* **40**: 3726–3733.

45. Sanderson, P.E.J., Cutrona, K.J., Dorsey, B.D. and Dyer, D.L. (1998) L-374,087, an efficacious, orally bioavailable, pyridinone acetamide thrombin inhibitor. *Bioorg. Med. Chem. Lett.* **8**: 817–822.

46. Malikayil, J.A., Burkhart, J.P., Schreuder, H.A., Broersma, R.J., Jr, Tardif, C., *et al.* (1997) Molecular design and characterization of an Alpha-thrombin inhibitor containing a novel PI moiety. *Biochemistry* **36**: 1034–1040.

47. Rewinkel, J.B.M., Lucas, H., van Galen, P.J.M., Noarch, A.B.J., van Dinther, T.G. *et al.* (1999) 1-aminoisoquinoline as benzamidine isostere in the design and synthesis of orally active thrombin inhibitors. *Bioorg. Med. Chem. Lett.* **9**: 685–690.

48. Stürzebecher, J., Prasa, D., Hauptmann, J., Vieweg, H. and Wikström, P. (1997) Synthesis and structure–activity relationships of potent thrombin inhibitors: piperazides of 3-amidinophenylalanine. *J. Med. Chem.* **40**: 3091–3099.

49. Al-Obeidi, F. and Ostrem, J.A. (1999) Factor Xa inhibitors. *Exp. Opin. Thes. Pat.* **9**: 931–953.

50. Phillips, G., Davey, D.D., Eagen, K.A., Koovakkat, S.K., Liang, A., *et al.* (1999) Design, synthesis, and activity of 2,6-diphenoxypyridine-derived factor Xa inhibitors. *J. Med. Chem.* **42**: 1749–1756.

51. Gabriel, B., Stubbs, M.T., Bergner, A., Hauptmann, J., Bode, W., *et al.* (1998) Design of benzamidine-type inhibitors of factor Xa. *J. Med. Chem.* **41**: 4240–4250.

52. Pastor, M., Cruciani, G., McLay, I., Pickett, S. and Clementi, S. (2000) GRid-INdependent descriptors (GRIND): a novel class of alignment-independent three-dimensional molecular descriptors. *J. Med. Chem.* **43**: 3233–3243.

53. Benedetti, P., Mannhold, R., Cruciani, G. and Pastor, M. (2002) GBR compounds and mepyramines as cocaine abuse therapeutics: chemometric studies on selectivity using grid independent descriptors (GRIND). *J. Med. Chem.* **44**: 2486–2489.

26

PROTEIN CRYSTALLOGRAPHY AND DRUG DISCOVERY

Jean-Michel Rondeau and Herman Schreuder

If you can look into the seeds of time
And say which grain will grow and which will not
Speak then to me ...

Shakespeare, Macbeth[1]

I INTRODUCTION

A Why is protein crystallography important?

Recent advances in drug discovery technologies, such as high-throughput screening and combinatorial chemistry, have substantially shortened the time needed for the identification and optimization of lead structures. However, the discovery of a suitable molecule fulfilling all the criteria required to become a real drug remains a daunting challenge. A potential new drug not only needs to be safe and effective, but has to be better than drugs already on the market. Therefore, it is very important for a pharmaceutical company to be first on the market with a new drug and to reduce the time needed for research and development as much as possible.

These challenges have prompted the pharmaceutical industry to harness the tremendous potential of computational approaches to drug discovery such as structure-based drug design, database mining and virtual screening.[2–12] These approaches rely heavily on a thorough understanding of the drug target at the molecular level. Protein crystallography is able to provide just that. Most drug targets are proteins. Examples include receptors for hormones and neurotransmitters, specific proteases like those involved in apoptosis, neurodegenerative diseases, or the blood coagulation cascade, protein kinases regulating cell cycle division or signal transduction pathways, bacterial enzymes involved in cell wall formation, and so on. Many of these proteins have already been crystallized and, with the advent of structural genomics,[13-16] efforts to crystallize many more are under way. As will be explained in this chapter, crystallographers are able to calculate electron-density maps which are, in fact, images of the molecules which make up the crystals, enlarged about a hundred million times. These electron-density maps are examined using computer graphics and an atomic model is fitted (Fig. 26.1), which, after refinement, allows the investigator to examine the three-dimensional structure of the protein in great detail. Most of what is currently known about how proteins are folded has been derived from crystal structures.

X-ray crystallography is a powerful technique. There are no theoretical limits to the size of the molecules or complexes to be studied, although a practical limit is imposed by the necessity of obtaining crystals of sufficient quality. The structure determination of ribosomal particles is undoubtedly one of the most stunning achievements of protein crystallography, to date, and is an invaluable contribution to anti-infective research.[17-21]

B What kind of information do we get from protein crystallography?

X-ray crystallographic studies can be broadly divided into the following four categories.

(1) *The determination of hitherto unknown structures.* This is still the mainstay of protein crystallography and serves as the basis for the other types of studies mentioned below. The sequencing of the genome from many organisms, including humans, has left us with the problem of understanding the biological role of the many predicted genes with an unknown function. Recent examples indicate that in some, but not all, cases the determination of the three-dimensional structure of a novel gene product may provide a clue to its biological function.[22-24]

(2) *Analysis of protein–ligand and protein–protein interactions.* Small-molecule ligands like substrates, inhibitors and co-factors are often accommodated by the crystalline protein without major changes of the crystal packing. Crystals of complexes of proteins with other macromolecules (protein, DNA, etc.) can be obtained by co-crystallization. Crystal structures of these complexes show in great detail how the protein interacts with its ligand(s) and, in contrast to methods studying the ligand alone, show the biologically relevant bound conformation of the ligand, information which is very important for drug design.

(3) *Studies of enzymatic mechanisms.* Enzymes can often be crystallized in the presence of substrates, substrate analogues or reaction products, which allows the identification of reactive groups in the active site (acidic and basic groups, nucleophiles, metal ions, co-factors, etc.) in close proximity to the substrate. In addition, it is often possible to obtain structural information on intermediate steps of the reaction by co-crystallization with transition-state analogues,[25] or by arresting the enzymatic reaction halfway by the use of low temperatures, inactive mutants or pH values which prevent completion

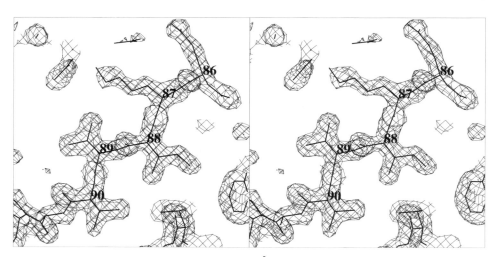

Fig. 26.1 Stereoview showing a small portion of an 1.8 Å electron-density map of porcine pancreatic elastase contoured at 1σ (1 standard deviation) above mean. The electron-density map, drawn in thin lines, is the direct experimental result while the atomic model, drawn in thick lines, has been built into the electron-density map by the crystallographers. (To learn how to view stereoimages, see Further Reading).

of the catalytic reaction (e.g. see references 26 and 27). This information is used, in combination with biochemical and spectroscopic data, to postulate a reaction mechanism which can be verified, for example, by mutating putative catalytic residues or by synthesizing and testing putative transition-state analogues.

(4) *Analysis of mobility and conformational changes.* This may seem impossible at first glance, since crystallography produces basically static pictures. A crystal structure represents the average of the $\approx 10^{16}$ protein molecules present in a typical crystal (space-average). It also represents the average conformation of these molecules during the 1–5 days it normally takes to collect a full diffraction pattern (time-average). However, movement of molecules in the crystal, either in the form of slightly different conformations for neighbouring molecules (static disorder) or movements with time of individual molecules (dynamic disorder), leads to a smearing of the electron density. In extreme, but not rare, cases certain flexible loops are totally invisible in the electron-density maps. Crystallographers not only refine the position of each atom in space (*xyz* coordinates), but also determine a factor which indicates how much each atom moves around in the crystal. This factor is called the temperature-factor or *B*-factor. Regions that are flexible in the crystal are likely to be flexible in solution and are often important for biological function. Conformational changes that are induced by ligand binding or by pH changes are studied by determining crystal structures in the absence and presence of the ligand, or at different pH values. Similarly, the determination of the structure of an enzyme in different activation states provides insights into the regulation of its activity and can have important implications for drug design.[28,29]

C How can protein crystallography contribute to drug discovery?

All the information mentioned above is valuable for drug design. The determination of a new protein structure often provides new insights into its biological function and the structural basis thereof. Sophisticated molecular modelling tools[30,31] can be used to display and analyse the structural and physicochemical features of the target receptor site. Detailed knowledge about the catalytic mechanism of an enzyme facilitates the classical approaches in rational drug design, such as the design of mechanism-based inhibitors,[32] multisubstrate analogue inhibitors[33] and transition-state analogues.[34] In addition, crystal structures of the target protein and of its complexes with known ligands can be exploited by chemists for the design of focused combinatorial chemistry libraries,[35–39] and by molecular modellers for lead optimization, database mining using substructure searches or 3D pharmacophores, *de novo* drug design and virtual screening using automated, high-throughput docking algorithms.[10] It is usually important to have detailed information about conformational changes occurring upon ligand binding, as this may profoundly influence the results.

In other cases, the protein itself is viewed as a potential drug. A variety of monoclonal antibodies, cytokines, growth factors, growth hormones and other peptidic hormones fall into this category. In general, the ultimate goal of structural analysis in this case is to facilitate the discovery of simpler, nonpeptidic molecular surrogates for the naturally occurring molecules, or to provide supporting information for patenting purposes. In practice, the contribution of protein crystallography to drug discovery programmes may be classified into the following categories.

(1) *Elucidation of the mechanisms of drug action at the molecular level.* In some cases, protein crystallography has provided the molecular basis for the observed activity of compounds discovered either by serendipity or by (natural) product screening. For example, it showed how a series of compounds with antiviral activity against picornaviruses, notably the human common cold virus, were able to interfere with the viral disassembly step of the infection cycle.[40] More recently, X-ray analyses of ribosomal particles in complex with several important classes of antibiotics have for the first time provided insights into the molecular mechanisms leading to the blockade of the protein synthesis machinery.[18,19] In the pharmaceutical industry, protein crystallography is routinely used to unveil the mode of action of high-throughput screening hits, in particular when these are not related in structure to previously known substrates or ligands. Sometimes this analysis reveals new binding sites which can be exploited for drug design.[41]

(2) *Analysis of structure–activity relationships.* X-ray structure analysis clearly facilitates the interpretation of the structure–activity data, which is sometimes far from straightforward. In particular, structure–activity relationships may be obscured by the fact that the bound conformation of a ligand is drastically different from that determined for the free molecule in solution, as happened with cyclosporin[42–44] and another immuno-suppressant, FK-506.[45] Also, structural variations on a chemical lead sometimes result in different binding modes. Such changes are hard to predict. It is important in such a situation to realize quickly what is happening on and to characterize fully the binding modes of what should be considered as separate series of compounds. Many such examples have been reported, including the human rhinovirus inhibitors[46] mentioned above, HIV-1 proteinase,[47,48] human thrombin,[49–51] collagenase,[52] acetylcholinesterase,[53] neuraminidase[54] and cdk2 inhibitors.[55]

(3) *Structure-assisted lead optimization.* Protein crystallography has proved extremely valuable in guiding and accelerating further chemical elaboration of an existing lead structure.[56-70] It often allows one to identify and to discard doomed drug design strategies and to focus the synthetic efforts on the most promising analogues, or, alternatively, to use this information in the design of focused combinatorial chemistry libraries.[35-38] Unfavourable interactions, and putative additional favourable ones, can be identified. In this way, compounds with improved potency may be generated quickly. Furthermore, the availability of crystallographic data allows detailed analysis of the bound conformation of a compound. Unfavourable high-energy conformations, when present, can be identified and ways to avoid them can be sought.[71,72] In addition, one may try to design conformationally restricted analogues, which in general show improved potency for entropic reasons.[73-79] A further advantage of enhancing the binding of an inhibitor through the introduction of conformational restraints is the potential for reducing its size.[74] When compound selectivity is an issue, protein crystallography provides a powerful means to pinpoint structural differences between structurally related enzymes or receptors which can be exploited to improve selectivity.[38,39,49,58-60,80-84] This approach has proved particularly useful for protein kinases. Structural variations within the highly conserved ATP-binding site can easily be modelled from the many known protein kinase structures, and can then be exploited to generate selective inhibitors.[80]

In general, the information gained from crystal structure analysis is not directly relevant to drug development issues such as patentability, toxicity, stability, solubility, bioavailability, formulation, and so on, but when used in combination with the appropriate expertise it allows one to predict what kind of structural modifications can be tried to improve the pharmacological properties of the compound without jeopardizing potency.[73,85] A particular example here relates to the inhibition of proteases, many of which are of pharmaceutical importance. Using substrate-based, rational drug design and/or combinatorial peptide libraries, it is usually possible to obtain within a short time-frame highly potent, highly selective peptidomimetics with K_i values in the nanomolar to subnanomolar range. Unfortunately, these peptide-based molecules usually exhibit very poor pharmacological properties. Several recent examples illustrate beautifully how protein crystallography can be used to overcome these problems by directing the design of orally active, nonpeptidic inhibitors.[63,74,86,87]

(4) *Structure-based drug design.* Structure-based drug design refers to an iterative procedure which uses the crystal structure of a target protein to design a novel lead structure, which is further elaborated during subsequent cycles of X-ray analysis followed by design, synthesis and biological testing (Fig. 26.2). The pioneering work on thymidylate synthase,[88-90] purine nucleoside phosphorylase,[67-70] HIV-1 proteinase,[65,74, 91] influenza neuraminidase,[66] phospholipase A2,[84] and the more recent results obtained with renin,[63] inosine 5'-monophosphate dehydrogenase,[92] Grb2-SH2,[93] β-lactamase,[94] and other targets illustrate the tremendous potential of this approach. While chemical expertise and intuition are still key in *de novo* structure-based drug design, more and more powerful computational approaches are being developed to guide this process.[5, 95] Compound database mining using 3D pharmacophores or substructure searches and virtual screening have become an integral part of this approach.

(5) *X-ray crystallographic screening.* An emerging application of protein crystallography in drug discovery is the so-called 'X-ray crystallographic screening' approach.[96] Protein crystals of the unliganded target protein are soaked in mixtures of 10–100 compounds, and ligand binding is detected by monitoring changes in the electron-density maps. The compound mixtures are prepared in such a way that bound ligands can be unambiguously identified by visual inspection of the electron density. This approach is restricted to protein crystals for which the active site of the target macromolecule is unhindered and accessible to small

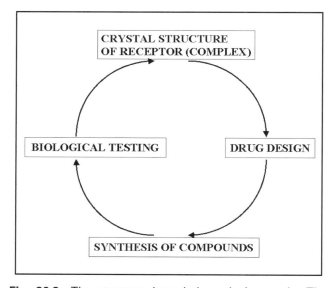

Fig. 26.2 The structure-based drug design cycle. The cycle can start either with a chemical lead, which is co-crystallized with the target protein, or with the structure of the target protein alone for *de novo* design. A number of cycles is usually necessary before one arrives at a suitable clinical candidate.

molecule ligands through the solvent channels in the crystal. It has been successfully used in the discovery of a new and orally bioavailable urokinase inhibitor[96] (see selected example).

D How to gain access to crystallographic information

The results of crystallographic studies can be retrieved from the Protein Data Bank, at www.rcsb.org. Links to a wealth of other databases offering clear and detailed analyses of all known protein–ligand complexes are available from the RCSB site.

II PROTEIN CRYSTALS AND DATA COLLECTION

A What are protein crystals?

Protein crystals, like any crystal of organic or inorganic compounds, are regular three-dimensional arrays of identical molecules or molecular complexes (Fig. 26.3). Depending on the space group, all molecules in a crystal have a limited number of unique orientations with respect to the crystal lattice. This means that the diffraction of all individual molecules adds up to yield intensities which are sufficiently strong to be measured.

B How do we get crystals?[97,98]

Obtaining X-ray quality crystals is often the most difficult step of a structure determination. Strong and dedicated support in molecular biology and biochemistry is an absolute must for the production of sufficient quantities of 'crystallization grade' protein. When the crystallization conditions are not known, one often needs as much as 100 mg of protein to find the right crystallization conditions. In most cases, several protein constructs or protein variants need to be generated until one is found which is amenable to crystallization. When crystallization conditions are already known and one is lucky, 1 mg of protein may be enough to produce a series of crystals with different inhibitors. One should bear in mind, however, that published crystallization protocols are often difficult to reproduce! In such cases, particular attention should be paid to small differences in the protein construct or in the purification protocol, which can have a dramatic influence on crystallization behaviour.

Crystals are produced by slowly precipitating the protein from solution. Under the right conditions, the protein will not form an amorphous precipitate but will instead settle in a well-ordered crystalline array. Methods for precipitating proteins involve dialysing away the salt, if salt is necessary

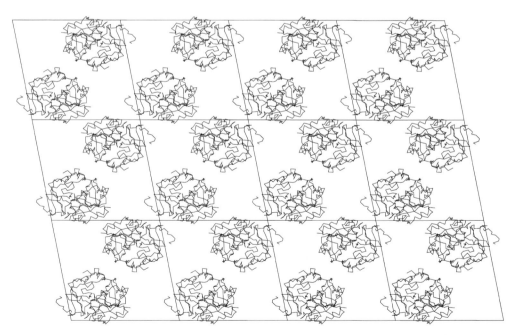

Fig. 26.3 Crystal packing of a human thrombin complex. Twelve unit cells with one layer of molecules are shown. By looking carefully, one can see that the two molecules in each unit cell are rotated 180° with respect to each other. Protein crystals used for X-ray diffraction extend into three dimensions and consist of many layers of molecules. The next layer of thrombin molecules fits into the holes present in the layer shown.

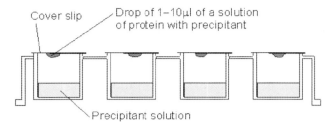

Fig. 26.4 The hanging drop crystallization setup. A drop of a solution of protein with precipitant (≈ 5–$10 \ \mu l$) is suspended above a reservoir containing a much larger amount (≈ 0.5–1.0 ml) of a more concentrated precipitant solution and the drop is allowed to equilibrate with the reservoir. Water moves from the less concentrated protein drop to the more concentrated reservoir solution via the vapour phase, causing the drop to shrink. The concentrations of protein and precipitant in the drop increase until the saturation point is reached and the protein starts to precipitate slowly. The precipitate is usually amorphous, but crystals will form in successful experiments.

to solubilize the protein, concentrating a nearly saturated protein solution by evaporation (usually in a hanging drop setup, see below) and the addition of precipitants such as poly(ethylene glycol) or ammonium sulfate. Other possibilities, which are used less often, are temperature and pH gradients. The method most often used for screening for crystallization conditions is the hanging drop method (Fig. 26.4).

C Preparation of crystals of protein–ligand complexes

An important part of protein crystallography which is especially relevant to drug design is the determination of protein–ligand complexes. If the ligand is a relatively small molecule, it is often possible to obtain crystals of the complex by soaking crystals of the native protein in a mother liquor containing the inhibitor. Protein crystals usually contain solvent channels which are large enough to allow the inhibitor to diffuse into the interior of the crystals. A soaking experiment requires little material (1 μmol compound is usually enough), but solubility is often a problem. The high protein content of the crystallization drop usually requires ligand concentration in the range of 0.1– 1.0 mM. The problem is often overcome by dissolving the (hydrophobic) ligand in a suitable organic solvent such as dimethyl sulfoxide or acetonitrile and adding this solution to the crystallization drop to a final concentration of solvent of up to 10%. Purity may not be a problem if none of the contaminants binds to the protein. The chemical structure of the compound under study has to be known, and when chiral centres are present it is preferable to know their absolute

configuration beforehand. At 2.5 Å or better resolution, it is often possible to deduce the absolute configuration from the electron-density maps. When epimerization occurs, it is in general faster than the time required to prepare the crystal and collect the data.

Soaking has some practical advantages: it is relatively fast (the soaking time ranges from a few hours to a few days); it requires minimal amounts of the ligand; and, as the crystallization conditions for the native protein are often well established, it is relatively easy to obtain large, well-diffracting crystals. However, the soaking method also has several disadvantages. Possible conformational changes induced by ligand binding may be hindered by the crystal packing and may therefore remain unobserved. The crystal lattice may be incompatible with ligand binding, causing the crystals to crack and/or dissolve upon soaking, or not to bind the ligand at all.

For this reason, if enough material is available, it is preferable to try first to obtain protein–ligand complexes via the second method: co-crystallization. For large ligands, this is the only possible method of obtaining crystals from the complex. With co-crystallization, the complex is formed in solution by adding the ligand to the protein solution, and the complex is subsequently crystallized. The advantages of this method are that, since the complex is formed in solution, conformational changes induced by the ligand are unhindered and will show up in the crystal structure. Also, the diffraction quality of the crystals is not compromised by the soaking procedure. A disadvantage is that the crystallization conditions for a complex are often different from the crystallization conditions for the unliganded protein. As a consequence, a new crystallization screening is required for each and every compound. Another disadvantage is that it often takes several weeks to several months for a protein to crystallize and not all ligands are stable for such a lengthy period of time. Also, in the context of a structure-based drug design programme, the crystallographic results should be obtained before the next round of synthesis and biological testing of compounds, and not afterwards.

D Data collection[99,100]

Protein crystals contain on average 50% solvent and, if exposed to air, they will dry out and completely disintegrate. Protein crystals are therefore either mounted in sealed X-ray capillaries which contain a little mother liquor to maintain the same humidity as during crystallization or, what has become today's standard, the crystal is picked up in a loop and is flash-frozen to liquid nitrogen temperatures (Fig. 26.5). The crystal is placed on a goniometer, a device which makes it possible to rotate the crystal in all directions, and the crystal is exposed to an intense X-ray beam (Fig. 26.6). Protein crystals are normally not larger than

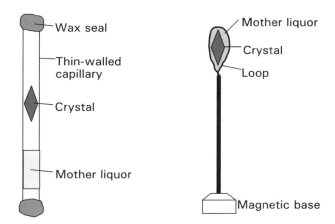

Fig. 26.5 Prior to a diffraction experiment, protein crystals are mounted either in an X-ray capillary or in a nylon loop, for data collection at room temperature or 100 K, respectively.

0.3–0.5 mm and the diameter of the beam is chosen to match the size of the crystal so as to minimize background scattering. The crystal is slowly rotated to bring all reflections into diffracting condition.

The diffraction spots are usually recorded on CCD detectors or on detectors based on imaging plates. These detectors are able routinely to collect high-quality X-ray data sets from single crystals in 1 to 5 days. The diffraction images of these detectors are fed directly into a computer, which produces a list of reflection intensities. Ten thousand to several hundred thousand reflections are recorded per

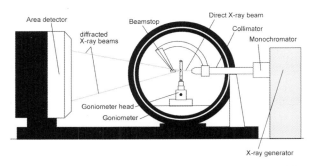

Fig. 26.6 The setup for an X-ray experiment. The X-ray generator produces a powerful beam. The monochromator selects X-rays of a single wavelength (1.54 Å for copper targets) and the collimator limits the diameter of the beam to 0.3–0.5 mm. This beam hits the crystal and some of the X-rays are diffracted by the crystal. Most X-rays pass straight through and are stopped by a small piece of lead, the beam stop. The diffracted X-rays are detected by an area detector, an imaging plate, or by other detection systems. The goniometer, shown here as a big black circle, has four rotation axes and allows the crystal to be positioned in any orientation with respect to the X-ray beam.

crystal, depending on the quality of the crystal and the size of the unit cell.

III   FROM DIFFRACTION INTENSITIES TO A MOLECULAR STRUCTURE[99,100]

A   Light microscopy and X-ray crystallography share the same basic principle

A light microscope allows us to study in great detail small objects such as insects or cell slices, but it is physically impossible to resolve any details which are smaller than half the wavelength of the light used. For blue light this limit is about 200 nm. To resolve atomic details, which are on the order of 1–5 Å (0.1–0.5 nm), electromagnetic radiation with a much shorter wavelength than light is required: X-rays. A light microscope and an X-ray setup share the same basic principle, although the practical implementation is quite different, owing to the different properties of X-rays and visible light.

In a microscope, light from a light source shines on the sample and is scattered in all directions. A set of lenses is used to reconstruct from this scattered light an enlarged image of the original sample. In an X-ray experiment, X-rays from an X-ray source hit the crystal and are scattered in all directions, just as with the light microscope. However, to date, no lenses exist which are able to bring the scattered X-rays into focus to reconstruct an enlarged image of the sample. All the crystallographer can do is to record directly the scattered X-rays (the diffraction pattern) and to use computers to reconstruct the image of the sample.

B   X-Rays are scattered by electrons

Although X-rays interact only weakly with matter, they are occasionally absorbed by electrons, which start to oscillate. These oscillating electrons serve as X-ray sources which can send an X-ray photon in any direction. X-ray photons, scattered from different parts of the crystal have to add up constructively in order to produce a measurable intensity. The condition under which the scattered X-rays add up constructively are laid down in Bragg's law, which treats crystals in terms of sets of parallel planes (Fig. 26.7).

Sets of planes from the different unit cells in the crystal have to scatter in phase as well. This is true only for a limited subset of planes and results in a characteristic pattern of diffraction spots (Fig. 26.8).

C   The diffraction pattern corresponds to the Fourier transform of the crystal structure

Each diffraction spot is caused by reflection of X-rays by a particular set of planes in the crystal. If the crystal contains

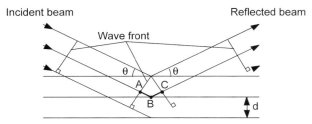

Fig. 26.7 Bragg's law. X-rays are reflected by sets of parallel planes, separated by a distance *d*, which run through the crystal. Diffraction is only observed under the following conditions. (i) The angle of the incident beam with the planes is identical to the angle of the reflected beam with the planes. This angle is indicated by θ. (ii) The path difference for beams reflected from subsequent planes (path ABC in the figure), should equal an integer times the wavelength, such that the different beams remain in phase.

layers of atoms with the same spacing and orientation as a particular set of planes which would satisfy Bragg's law (if the set of planes is physically present), the corresponding diffraction spot will be strong. On the other hand, if nothing is physically present in a crystal which corresponds to a particular set of planes, the corresponding reflection will be weak. The complicated structure present in the crystal is transformed by the diffraction process into a set of diffraction spots which correspond to sets of planes (more precisely, sinusoidal density waves), just as our ear converts a complicated sound signal into a series of (sinusoidal) tones when we listen to music. This conversion of a complicated function into a series of simple sine- and cosine functions is called a Fourier transformation.

D The phase problem

The original function, in our case the electron-density distribution in the crystal, can be reconstructed with computers by performing the inverse Fourier transformation, i.e. by adding together the corresponding density waves for all reflections (see Fig. 26.9). However, in order to make this addition, we need to know not only the amplitude of the density wave but also its relative position with respect to all other density waves (the phase). The amplitude can be measured, because it is calculated from the intensity of the corresponding diffraction spot, but there is currently no way to measure the phases directly. This so-called 'phase problem' can be solved by one of the following techniques.

(1) *Direct methods.* These methods only work for small molecules and use phase relationships which exist between certain sets of reflections.[101]
(2) *Multiple isomorphous replacement (MIR).* Crystals are soaked in solutions with 'heavy' atoms (Hg, Pt, Au salts, etc.) under such conditions that a few of these heavy atoms attach themselves to well-defined spots on the protein molecule. The heavy atom positions are found by

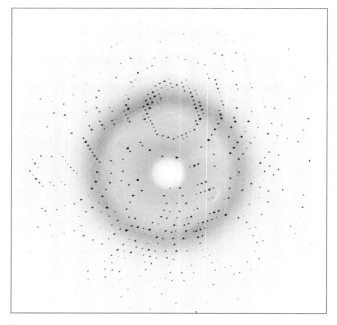

Fig. 26.8 X-ray diffraction pattern as recorded in the laboratory of the authors. The characteristic pattern of diffraction spots is caused by the fact the X-rays diffracted from different unit cells in the crystal have to scatter in phase.

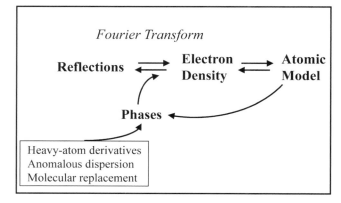

Fig. 26.9 The phase problem. The experimental data obtained in an X-ray experiment are the intensities of the reflections. By using an inverse Fourier transform, it is possible to calculate electron-density maps from these intensities. However, it is essential for this calculation to know the phase associated with each reflection. Approximate 'starting' phases can be obtained from heavy-atom derivatives, anomalous dispersion or molecular replacement (see text). More accurate phases can be derived from the refined model, once it has been obtained.

analysing the differences between the native diffraction pattern and the diffraction patterns of crystals treated with heavy atom reagents. When two or more suitable heavy atom derivatives are found, it is possible to calculate phases and an electron density map.

(3) *Anomalous scattering (AS)*. This method makes use of the fact that some inner electrons of the heavier elements have absorption edges in the range of X-ray wavelengths. The method is used to supplement the phase information of a single heavy atom derivative, but also to obtain full phase information from proteins which are labelled with, for example, selenomethionine, a selenium-containing amino acid. Recently, crystals were soaked for a short time in solutions containing 0.5–1.0 M bromine or iodine and the crystal structure was solved using the anomalous signal of bound halide ions.[102] The AS method has become the preferred method to solve *de novo* protein structures.

(4) *Molecular replacement*. When a suitable model of the unknown crystal structure is available, it can be used to solve the phase problem. Examples are the use of the structure of human thrombin to solve the structure of bovine thrombin, the use of a known antibody fragment to solve the structure of an unknown antibody, or the use of the structure of an enzyme to solve the structure of an inhibitor complex in a different crystal form. The model is oriented and positioned in the unit cell of the unknown crystal with the use of rotation and translation functions, and the oriented model is subsequently used to calculate phases and an electron density map.

Method (1) can not (yet?) be used for proteins. Method (2) can be used to solve any protein structure *de novo*, but this method requires testing and measuring dozens of heavy atom derivatives, and is hardly used any more. Method (3) allows the fast determination of *de novo* protein structures, provided they can be labelled with seleno methionine (expression in *E. coli*). Extremely promising new methods use the anomalous signal of soaked-in halide ions, or the signal of native sulfurs. When these methods deliver what they promise, they allow any new structure to be solved in a matter of weeks. Method (4) is very fast; it usually takes less than a week, but it requires that the structure of a similar protein be available, which is not always the case. This method is extremely useful in solving the structures of a protein complexed with a series of inhibitors or other ligands.

E Model building and refinement

Once an electron-density map is obtained, it is interpreted by the crystallographer. In the case of a MIR(AS) map, a complete model of the protein has to be fitted to the electron density. The Cα atoms are placed first (chain tracing), and subsequently the complete main-chain and the side-chains are built, a process which has become more and more automated in recent years. In the case of molecular replacement, the model used to solve the phase problem has to be rebuilt to reflect the molecule present in the crystal. The model is usually of a similar protein and the changes involve substitution of amino acids, making insertions and deletions, changing the orientation of loops, and so on.

After (re)building, the model is refined; that is, the differences between the observed diffraction amplitudes (F_o), and the diffraction amplitudes calculated from the model (F_c) are minimized, while the geometry of the model is optimized. The ratio between observations and parameters in protein crystallography is usually quite low, which makes it impossible to do a free atom refinement. It is always necessary to restrain the bond lengths, angles, etc. towards ideal values. The refinement is done by computer programs, either with least squares methods (often referred to as energy minimization), with molecular dynamics or, more recently, with maximum likelihood methods.[103] Phases calculated from the refined model allow the calculation of improved electron-density maps, which are analysed by the crystallographer to again rebuild (improve) the model. Cycles of refinement and rebuilding are repeated until convergence is reached and a set of coordinates is obtained which is ready for deposition with the Protein Data Bank.[104]

F Most-used types of electron-density maps

Since the direct experimental result of a crystallographic analysis is an electron-density map, and since the model is based on a (subjective) interpretation of this map, we will discuss here the types of electron-density maps most often used.

(a) $F_o − F_c$ or difference maps. These maps are obtained after subtracting the calculated structure factors (F_c) from the observed structure factors (F_o), an operation which is in a first approximation equivalent to subtracting the calculated electron density from the observed electron density. Features that are present in the 'observed' density, but not in the calculated density will give peaks, while atoms present in the model (in the F_c), but not in the 'observed' electron density will result in holes (Fig. 26.10). These maps are frequently used to detect errors in the model and can also be used to obtain an unbiased electron density of a bound inhibitor, for example, by removing the inhibitor completely from the model. In this case, the resulting electron density for the inhibitor is entirely caused by the experimental data, and not by any model bias present in the phases. These maps are often referred to as 'omit maps'.

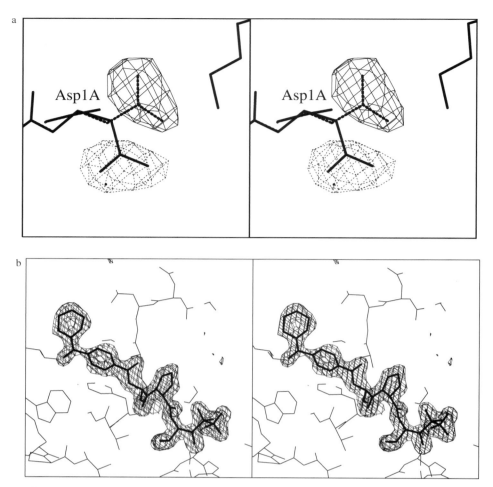

Fig. 26.10 $F_o - F_c$ difference maps. (a) Detection of errors in the model. The side-chain of Asp1A in human thrombin was deliberately moved to a wrong position and an $F_o - F_c$ difference map was calculated. The model used to calculate F_c is indicated in thick solid lines, the correct position of the side-chain is indicated in thick broken lines. Negative contours (4σ below mean) are drawn in thin broken lines and positive contours (4σ above mean) are drawn in thin solid lines. Strong negative difference density is present around the wrongly placed side-chain while strong positive density is present at the position where the side-chain should be according to the experimental data. These maps are extremely useful to spot errors in the model. (b) $F_o - F_c$ 'omit map' (see text) contoured at 3.5σ of an inhibitor bound to porcine pancreatic elastase. Protein atoms, used for calculation of F_c are indicated in thin lines, the inhibitor which has been removed from the model is drawn in thick lines. This $F_o - F_c$ density map has been calculated with a model which contains no information whatsoever about the inhibitor. The difference density shown is therefore entirely due to the experimental data and can be used to verify the correctness of the placement of the inhibitor.

(b) $2F_o - F_c$ maps. These are the standard electron-density maps. Because of model bias, maps calculated with F_o and model phases tend to show only electron density associated with the model. As discussed above, $F_o - F_c$ maps show everything which is in F_o but not in the model. By combining an F_o map with an $F_o - F_c$ map, a $2F_o - F_c$ electron-density map is obtained, which shows both electron density for the model and electron density for features which are not yet accounted for in the model, such as bound water molecules, carbohydrates (Fig. 26.11) and other molecules associated with the protein.

Several weighting schemes exist to remove model bias. Examples are figure-of-merit, σ_A and maximum likelihood weighting.

IV USING CRYSTAL STRUCTURES: QUALITY CRITERIA

A Errors in crystal structures

The theory of the diffraction of X-rays by crystals and of the Fourier transformation is rigorous and exact; it involves no

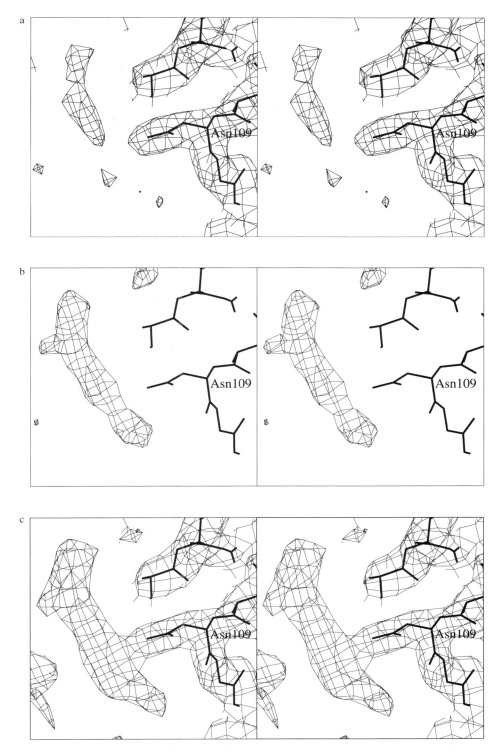

Fig. 26.11 The rationale behind the $2F_o - F_c$ maps. Shown is a carbohydrate attachment site at Asn109 in human leukocyte elastase. The carbohydrate chain has not been added to the model. The F_o map in (a) mainly shows the electron density of the model, owing to the model bias. Some electron density is present for the carbohydrate, but it is very weak and not connected to the Asn residue. The $F_o - F_c$ map in (b) shows only features which have not been accounted for in the model, in this case the carbohydrate moiety. The $2F_o - F_c$ map in (c) is a combination of the two and shows both density belonging to the model and strong connected density for the carbohydrate. From these figures, it is clear why crystallographers prefer a $2F_o - F_c$ map over an F_o map.

approximations. Consequently, crystal structures have proved to be very reliable and errors are rare. However, when using a protein crystal structure, one should keep in mind that they do have certain limitations.

The direct experimental result of a crystallographic analysis is an electron-density map, and not the atomic model everybody looks at! If errors occur in crystal structures, they most often occur at the level of the (subjective) interpretation of the electron-density maps by the crystallographer. A severe problem, especially at low resolution (lower than 3.0 Å), is the so-called model bias. To calculate an electron-density map, one needs amplitudes and phases. The amplitudes are determined experimentally, but the phases cannot be measured directly. In later stages of refinement, they are calculated from the model, which means that if the model contains errors, the phases will contain the same errors. Since phases make up at least 50% of the information which is used to calculate the electron-density maps, wrong features may still have reasonable electron density because of these phase errors.

Errors in crystal structures can be divided into three categories:

(1) *Complete garbage.* Completely wrong structures may arise from wrong space group assignments, errors in the sign of the electron-density maps and interpretation of uninterpretable maps. This has happened a few times in the past but nowadays powerful software exists to validate protein structures and this should no longer happen.

(2) *Localized errors (loops and connectivity).* Loops are often flexible and in these cases poorly defined in the electron-density maps. Crystallographers, trying to build these loops, sometimes connect wrong parts of the protein and place loops into density which do not belong to those loops. These kinds of errors occur most often with 'hot structures', when people rush to solve the structure in the shortest possible time in order to stay ahead of the competition.

(3) *Errors in main-chain dihedral angles and side-chain conformations.* When comparing structures from the same protein, independently solved by different groups, one often observes discrepancies in main-chain dihedral angles and side-chain conformations. There may be several reasons for this. Electron-density maps, especially at lower resolution (≈ 3.0 Å or less) do not show individual atoms, but rather the overall shape of the main-chain and side-chains. From these maps, it is not always clear how to orient a peptide plane, or which side-chain conformation to choose. In addition, particular side-chains often have multiple conformations in the crystal while only a single conformation has been fitted by the crystallographer. Also, it is not possible with protein crystals to distinguish on the basis of the electron density alone between C, O and N, which means that the side-chain orientations of His, Asn, Thr and Gln are never uniquely defined by the electron density.

(4) *Other errors and limitations.* These include the assignment of bound solvent molecules and metal ions, the protonation or oxidation state of certain prosthetic groups and the choice of incorrect parameters for the refinement of nonstandard groups. One should bear in mind, however, that these latter errors are usually below the accuracy of 0.3–0.5 Å of the protein structure. Protons and other light elements (Li^+, Be^{2+}) cannot be seen by protein X-ray crystallography. It is, therefore, not possible to determine the protonation state of active site residues and bound ligands with X-ray studies.

While errors of the first two categories should not happen, one cannot blame the crystallographer for errors of the third type because (i) these errors are within the error limits of the structure determination and (ii) side-chain conformations are often not wrong but merely represent one out of a number of conformations which occur in the crystal and in solution.

B Quality criteria

When using crystal structures, it is important to know whether a structure is very accurate, for example, allowing conclusions about the strengths of hydrogen bonds from the observed hydrogen bond length, or less accurate, which would make such conclusions quite hazardous. Below we discuss a number of parameters which crystallographers normally publish along with the description of a crystal structure, and which should allow others to judge the quality of a crystal structure.

Quality of the experimental data

The quality of a crystal structure cannot be better than the quality of the experimental data on which it is based. The experimental data can be judged by the following criteria.

(1) *Resolution.* This corresponds to the shortest spacing of planes (d) whose reflections have been used in map calculation and refinement (see Fig. 26.7). The smaller this spacing, the sharper and more detailed the electron-density maps will be. The resolution is probably the single most important criterion determining the quality of a crystal structure. At high resolution (better than 2.0 Å) the protein and bound water molecules are well defined and it is very unlikely that the structure will contain any serious errors. At low resolution (2.8–3.5 Å), it is usually not possible to assign bound water molecules with certainty and the crystallographer needs

to be careful not to make any errors. For drug design, a resolution of 2.5 Å or better is highly desirable. However, successful structure-based drug design has been done on purine nucleoside phosphorylase at 3.2 Å.[67–70]

(2) *Completeness of the data.* One can calculate the total number of reflections to a certain resolution and, ideally, one would like to measure them all. However, for various reasons it is in practice often not possible to measure all reflections. If only a small fraction of the reflections is missing ($\approx 10\%$), and the missing reflections are weak, the electron-density maps will hardly be affected. However, if a significant fraction of the reflections is missing, this may lead to artefacts in the electron-density maps and also the problem of model bias will become more severe.

(3) *R-sym.* This is the error between multiple measurements of the same reflection. The lower the R-sym, the better. R-syms up to 10% are tolerable.

Quality of the model

Not only should the experimental data be of good quality, but the model also should be fitted correctly to the electron-density map. Below we list some of the most widely used criteria for judging the quality of the model.

(1) *R-factor.* This is the error between the observed amplitudes (F_o), and the amplitudes calculated from the model (F_c). Well refined structures have *R*-factors below 20%.

(2) *Free R-factor.* Modern refinement programs are very powerful and capable of producing reasonable refinement statistics with wrong models. In order to be sure that refinement is progressing correctly and one is not merely reducing the *R*-factor of a wrong model, Brünger proposed to exclude a subset of reflections from refinement and to use these reflections only for the calculation of 'free' *R*-factors.[105] If refinement is progressing correctly, the free *R*-factor will drop as well, but if the model contains serious errors it will remain essentially random ($\approx 55\%$). For correct structures, the free *R*-factor is generally below 30%.

(3) *Deviations from ideality of bond lengths and bond angles.* A correctly fitted model is generally not strained. Significant deviations from ideal values for bond lengths and bond angles point to problems with the structure. Root-mean-square (r.m.s.) deviations from ideality should not be much larger than 0.02 Å for bond lengths, and 3° for bond angles. The bond lengths and angles are biased towards the target values which are used during refinement. *For protein crystals, it is not possible to obtain accurate unbiased values for these parameters.*

(4) *φ, ψ plot.* Because of steric hindrance, only certain combinations of the main-chain dihedral angles φ and ψ are 'allowed' (Fig. 26.12). The protein folding may force some residues to assume unallowed φ, ψ values, and this may have functional significance for active site residues.[106,107] However, if more than a few per cent of all the residues have φ, ψ values completely outside allowed regions, one should suspect errors.

C Flexibility and temperature factors

Most proteins are flexible and many crystalline proteins contain some very flexible regions. The mobility of atoms in a crystal is expressed in terms of temperature-factors, or *B*-factors, which are optimized during refinement. The relationship between mean total displacement and *B*-factors is given in Fig. 26.13. The mean displacement of atoms with *B*-factors in excess of 60 Å² is larger than 1.5 Å, which is the length of a carbon—carbon bond. These atoms are generally poorly defined in the electron-density maps (Fig. 26.14), although in the case of static disorder these atoms may still have well-defined, albeit weak, electron density.

From the discussion above, one should not get the impression that all X-ray structures are inaccurate. On the contrary! Most X-ray structures, especially at a resolution of 2.5 Å or better, are very accurate (estimated mean coordinate errors of less than 0.5 Å) and allow researchers to look at enzymes and other proteins with unprecedented detail and clarity (see, for example, Figs 26.1 and 26.10b).

V SOME SELECTED EXAMPLES OF STRUCTURE-BASED DRUG DESIGN

Very early on, protein crystallography had a profound impact on drug design. The crystal structure of haemoglobin, the second protein structure to be solved, provided insights into the molecular basis of sickle cell anaemia and allowed the study of the mechanisms of action of drugs with antisickling properties.[108] Some general rules governing the interactions between drugs and proteins have emerged from these studies.[108] In another area, X-ray analyses of insulin led to the design of a monomeric form of the hormone which showed an improved rate of absorption.[109] However, the major contribution of protein crystallography to date relates to the design of enzyme inhibitors. Since the early days, numerous enzyme–inhibitor complexes have been studied and many more are under way. These studies have contributed a great deal to the general understanding of the basic principles underlying enzyme inhibition and they have unravelled, for many different functional classes of enzymes, the critical interactions

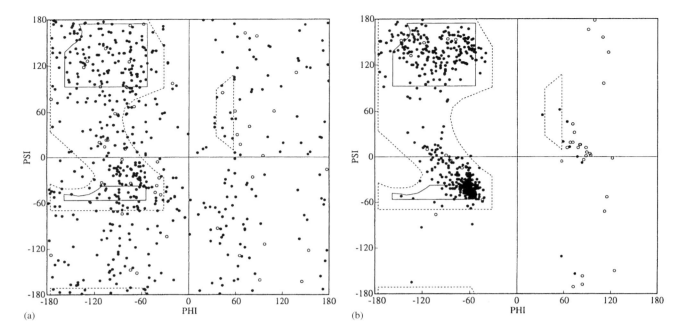

Fig. 26.12 φ, ψ plot. Open circles indicate glycine residues, filled circles indicate nonglycine residues. Residues with φ, ψ angles outside the allowed regions (indicated by broken lines) are strained. Much larger regions are allowed for glycine residues because they do not have a side-chain. When more than a few per cent of the residues have φ, ψ outside the allowed regions, one should be suspicious of errors in the structure. Panel (a) shows the φ, ψ plot of plant RuBisCo, which has been incorrectly fitted to the electron-density map.[146] Panel (b) shows the φ, ψ plot of the same protein after it has been properly fitted and refined.[147]

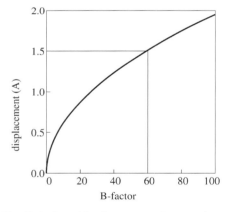

Fig. 26.13 Relationship between the total root mean square displacement and the temperature factor B. At temperature factors of 60 Å^2 and higher, the displacement becomes larger than 1.5 Å and the electron density of those atoms becomes very poor (see Fig. 26.14). The formula used in the figure is derived from the relationship $B = 8\pi^2 \langle u^2 \rangle$ where $\langle u^2 \rangle$ represents the displacement perpendicular to the diffracting planes. The total mean square displacement $\langle u_{\text{tot}}^2 \rangle = 3 \langle u^2 \rangle$. Hence, $\langle u_{\text{tot}}^2 \rangle^{1/2} = (3 B/8\pi^2)^{1/2}$.

involved in enzyme–inhibitor complexes. The know-how gained from the study of one particular enzyme also proved

useful for the design of inhibitors targeting a related enzyme of unknown three-dimensional structure.[57,80,110] For instance, the discovery of captopril[111] and cilazapril,[112,113] two antihypertensive drugs acting on the angiotensin-converting enzyme (ACE), was inspired in part by crystallographic analyses of carboxypeptidase A and thermolysin. Similarly, the search for renin inhibitors has long been based on crystallographic information gleaned from fungal aspartic proteinases[114–117] until direct structure-based drug design finally became possible.[63] Other prominent examples of the use of protein crystallography for enzyme inhibitor design include dihydrofolate reductase,[83] α-thrombin,[37,49–51] elastase,[56,87,118] carbonic anhydrase,[71] purine nucleoside phosphorylase,[67–70] thymidylate synthase,[88–90] phospholipase A2,[64,84] influenza neuraminidase,[54,66,119–122] trypanosomal glyceraldehyde-3-phosphate dehydrogenase,[39,59–60] HIV-1 proteinase[65,74,123–125] and HIV-1 reverse transcriptase,[62] inosine 5'-monophosphate dehydrogenase,[92] β-lactamase[94] and urokinase.[61,96] The majority of current structure-based drug design programs still deals with enzymes,[38,52,53,58,81,82,126–130] with a strong emphasis on protein kinases.[80]

We will proceed by giving some examples, selected from the literature, which illustrate the multiple facets of the use of crystallographic data in pharmaceutical research.

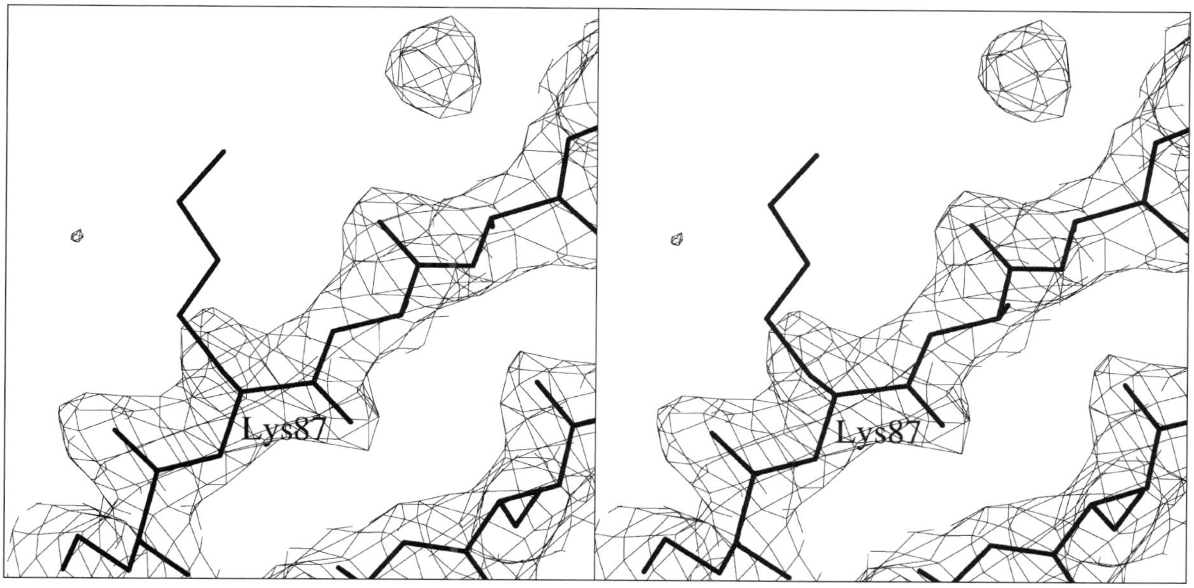

Fig. 26.14 Long and flexible side-chains (such as Arg, Lys, Glu and Gln) which are exposed to the solvent often move freely around. As a result, these side-chains have very high temperature factors and are very poorly or not at all defined in the electron-density map. Lys87, located at the surface of human thrombin, is shown as an example. One should bear in mind that many exposed surface residues do not have a well-defined orientation if one uses protein structures for drug design.

A Purine nucleoside phosphorylase

Purine nucleoside phosphorylase (PNP) catalyses the phosphorolysis of purine ribonucleosides and 2′-deoxyribonucleosides to the purine base and ribose- or 2-deoxyribose-α-1-phosphate. PNP is a key enzyme in the T-cell-mediated immune response. As such, it is an attractive target in a number of therapeutic areas, such as organ transplantation, T-cell-mediated autoimmune disorders and T-cell proliferative diseases. PNP has been the subject of a thorough and successful structure-based drug design study,[67-70] of which the following gives only a flavour.

Why combining two improvements may sometimes be a failure

Both 8-aminoguanine and 9-deazaguanine analogues are superior inhibitors of PNP compared with the corresponding guanine analogues (Fig. 26.15). Compounds combining these two features, namely 8-amino-9-deazaguanine derivatives, were expected to be even better, but surprisingly they were not. The structural reason for this medicinal chemistry riddle was elucidated through X-ray analysis (Fig. 26.15). The N-7 position of the guanine derivatives is a hydrogen-bond acceptor, while it is a hydrogen-bond donor in the 9-deazaguanine series.

Optimal hydrogen-bonding interactions between the enzyme and the 9-deazaguanine analogues are achieved through a movement of the side-chain of Asn243, which also triggers a concomitant shift in the position of Thr242. These conformational changes no longer allow the formation of a hydrogen-bond between the 8-amino substituent and the side-chain hydroxyl group of Thr242. Furthermore, an unfavourable contact with the side-chain methyl group of Thr242 is, instead, generated (Fig. 26.15).

A chlorinated benzene ring as a replacement for a ribose moiety!

The X-ray analysis of PNP also revealed that the binding site of the ribose moiety of the substrate is essentially hydrophobic, thus unravelling the hitherto unexplained potency of 9-(arylmethyl)guanine analogues. Indeed, a 'herringbone' packing interaction is observed between the benzyl group of these inhibitors and two phenylalanine side-chains located in the ribose binding site of PNP. The search for appropriate substituents and positions on the benzyl ring was greatly facilitated by X-ray analysis. It led to the discovery of a series of 9-(arylmethyl)-9-deazapurines that are potent, membrane-permeable inhibitors of PNP. The most potent compound in this series,

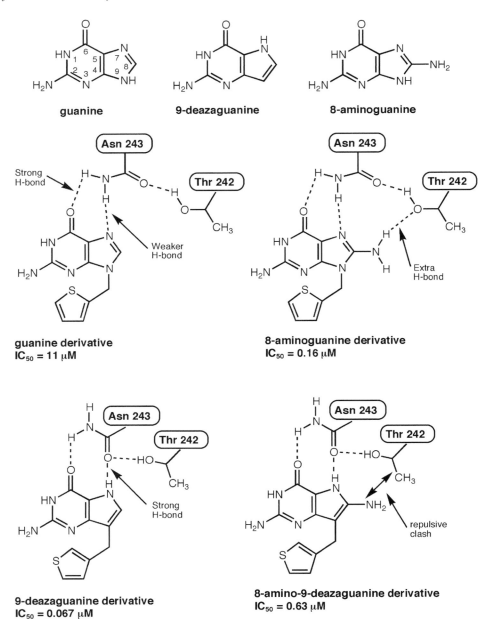

Fig. 26.15 Structure-based design of purine nucleoside phosphorylase inhibitors. For explanation, see text. (Adapted, with permission, from Montgomery, J.A. *et al.* (1993) *J. Med. Chem.* **36**: 55–69. Copyright © 1993 American Chemical Society.)

(*S*)-9-[1-(3-chlorophenyl)-2-carboxyethyl]-9-deazaguanine has an IC$_{50}$ of 6 nM.[69] One of the compounds generated during this programme, BCX-34, is in clinical trials.[131]

B Human leukocyte elastase

How to convert a peptide-based lead into an orally active, nonpeptidic drug

Human leukocyte elastase (HLE) is a serine proteinase involved in inflammation and tissue degradation. It is

believed that suitable HLE inhibitors would be useful for the treatment of a number of disease states, such as emphysema and cystic fibrosis. ICI 200 880 is a mechanism-based inhibitor of HLE which forms a stable hemiketal adduct to the active site serine Ser195 (Fig. 26.16). This compound had shown encouraging effects *in vivo* in animal models following aerosol administration, but was not orally active, presumably because of its peptidic nature. Orally active, nonpeptidic inhibitors of HLE were designed[87] by taking advantage of

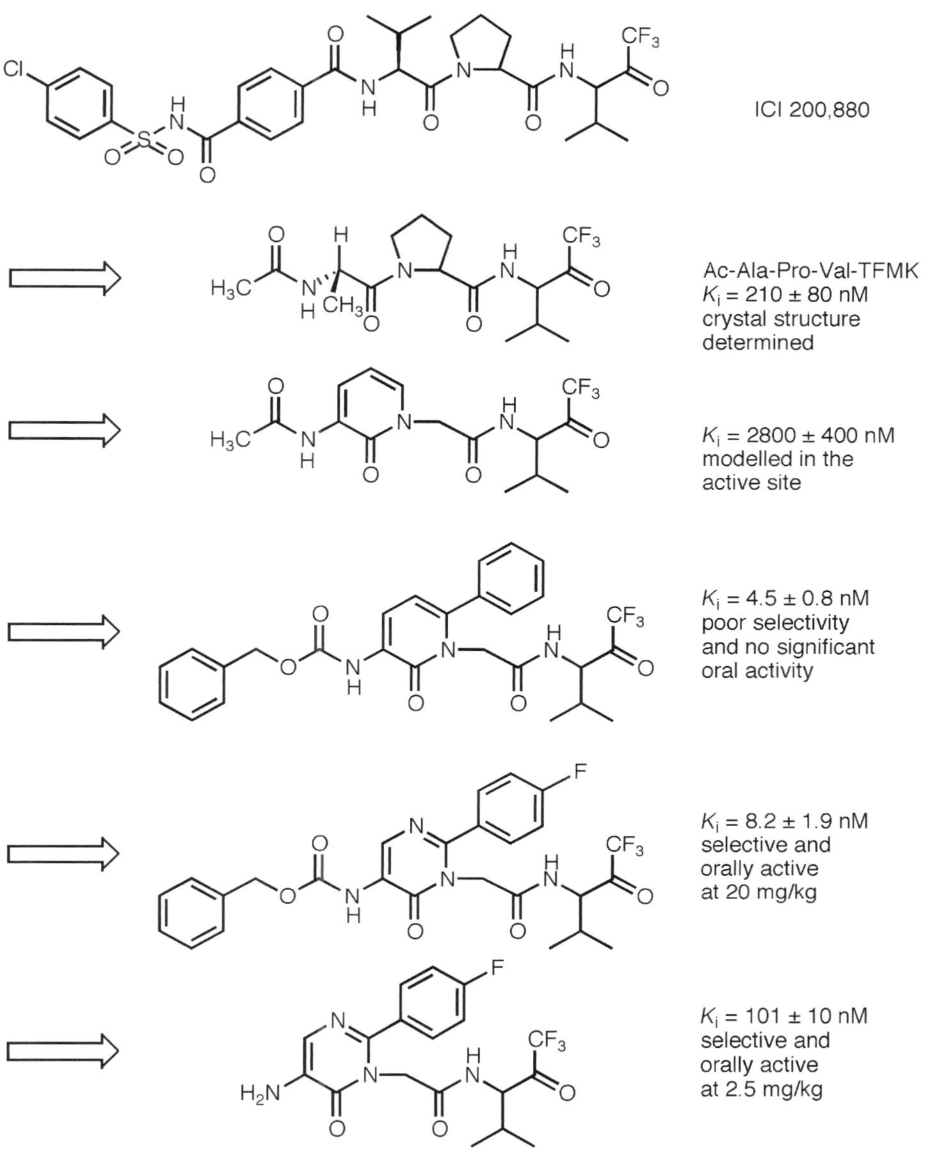

Fig. 26.16 Structure-based design of a human leukocyte elastase inhibitor. For explanation, see text.

the structural information provided by the X-ray analysis of porcine pancreatic elastase in complex with Ac-Ala-Pro-Val-trifluoromethyl ketone.[118] Careful examination of this crystal structure led to the conclusion that a substituted pyridone, 3-amino-2-oxo-1,2-dihydro-1-pyridylacetic acid, would be a suitable surrogate for the Ala-Pro portion of the parent compound (Fig. 26.16). Indeed, the designed molecule retained considerable activity against HLE ($K_i = 2.8\ \mu M$).[87] Incorporation of a phenyl substituent on the pyridone ring was suggested by modelling the pyridone in the active site of the enzyme. This modification, together with the replacement of the *N*-acetyl group by an N-CBZ group, resulted in nanomolar potency ($K_i = 4$ nM). However, further structure–activity studies were necessary to

achieve oral activity. First, a search for heterocycles which would retain the key features of the pyridone led to the identification of the corresponding pyrimidone derivative, which showed oral activity at 20 mg kg^{-1}. Then, it was found that the introduction of a *p*-fluoro substituent on the phenyl ring increased selectivity. Finally, a number of *N*-substituents were tried in an attempt to further improve oral activity. This region was targeted for chemical modification on synthetic grounds but also because modelling studies had indicated the *N*-terminal protecting group to be rather mobile at the enzyme surface. The unprotected 3-aminopyrimidone, despite a somewhat lower potency ($K_i = 101$ nM) *in vitro*, was found to exhibit sustained oral activity (ED$_{50} = 7.5$ mg kg^{-1}) for more than 4 hours following oral administration, with excellent bioavailability and selectivity.[87] Thus, by combining structure-based drug design and classical structure–activity studies, a novel class of orally active, nonpeptidic HLE inhibitors was discovered.

C Thymidylate synthase

Thymidylate synthase (TS) catalyses the methylation of deoxyuridylate to thymidylate using 5,10-methylenetetrahydrofolate (Fig. 26.17a) as a coenzyme. TS inhibitors have potential as chemotherapeutic agents for the treatment of cancer as TS provides the sole biosynthetic source of thymidylate, a precursor in DNA biosynthesis. The design of TS inhibitors is a superb example of structure-based drug design. The reader is urged to read the original papers, which describe the work in great detail.[88–90]

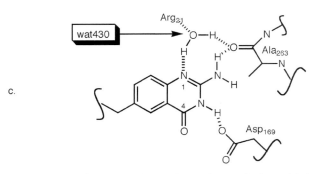

Fig. 26.17 Structure-based design of thymidylate synthase inhibitors. (a) The coenzyme 5,10-methylenetetrahydrofolate. (b) The classical antifolate CB3717. (c) Interactions of the pteridine moiety of CB3717 with active site residues Asp169 and Ala263 and with wat430, a bound water molecule. For explanation, see text. (Adapted, with permission, from Reich, S.H. and Webber, S.E. (1993) *Perspect. Drug Dis. Des.* **1**: 371–390. © 1993 Escom Science Publishers B.V.)

When a water molecule makes the difference

Certain water molecules, especially those which interact with active site residues or participate in ligand binding, may be key players in the drug design strategy. Water-mediated hydrogen bonds and desolvation effects have a profound influence on binding, and structural mimics for active site water molecules may be included in the design of novel ligands (see the example on HIV proteinase). However, the identification of bound water molecules requires medium- to high-resolution crystallographic data (say 2.5 Å or better). The thymidylate synthase story provides a nice illustration of this point. An initial analysis at 2.8 Å resolution of TS with a bound inhibitor, the antifolate CB3717 (Fig. 26.17b), did not reveal a water molecule hydrogen-bonded to N-1 of CB3717, to the carbonyl oxygen of Ala263 and the side-chain guanidinium of Arg21 (Fig. 26.17c). In the meantime, analogues of CB3717 bearing a—CH in position 1 were prepared and found to be less potent than the parent compound.[90] The reason for this became apparent at a later stage, when 2.3 Å resolution data became available and this key water molecule (wat430) was identified.[132]

How to circumvent drug toxicity and mechanisms of drug resistance

The glutamate side-chain of CB3717 (Fig. 26.17b) and related compounds causes hepatic toxicity through intracellular accumulation of polyglutamylated products. In addition, glutamate-containing inhibitors require active transport to penetrate cells, which can lead to the development of drug resistance. Crystallographic analysis showed that the binding site of the benzoyl glutamate moiety was fairly hydrophobic and that this part of the inhibitor could be redesigned. The design was based on the 2-methyl-2-desamino derivative of CB3717, because of its greater solubility (Fig. 26.18a). This compound inhibits human TS with a K_i of 8.5 nM. Removal of the CO-L-glutamate moiety resulted in a dramatic loss of binding affinity ($K_i = 2.2 \mu M$) (Fig. 26.18b). Through iterative structure-based drug design, novel lipophilic TS inhibitors were discovered, including a series of diphenylsulfone and N-sulfonylindole derivatives[90] (Fig. 26.18c and d). One of the compounds designed, AG85, was selected for clinical evaluation (Fig. 26.18d).

De novo structure-based design of novel lead structures

The discovery of structurally novel naphthostyril-based TS inhibitors[88] represents a beautiful example of *de novo* structure-based design (Fig. 26.19). The design was based on a crystal structure of a ternary complex of TS, from which the bound inhibitor CB3717 (Fig. 26.17b) was subsequently removed for careful examination of its binding cavity. In this way, conformational changes in the binding site region triggered by substrate or cofactor binding are

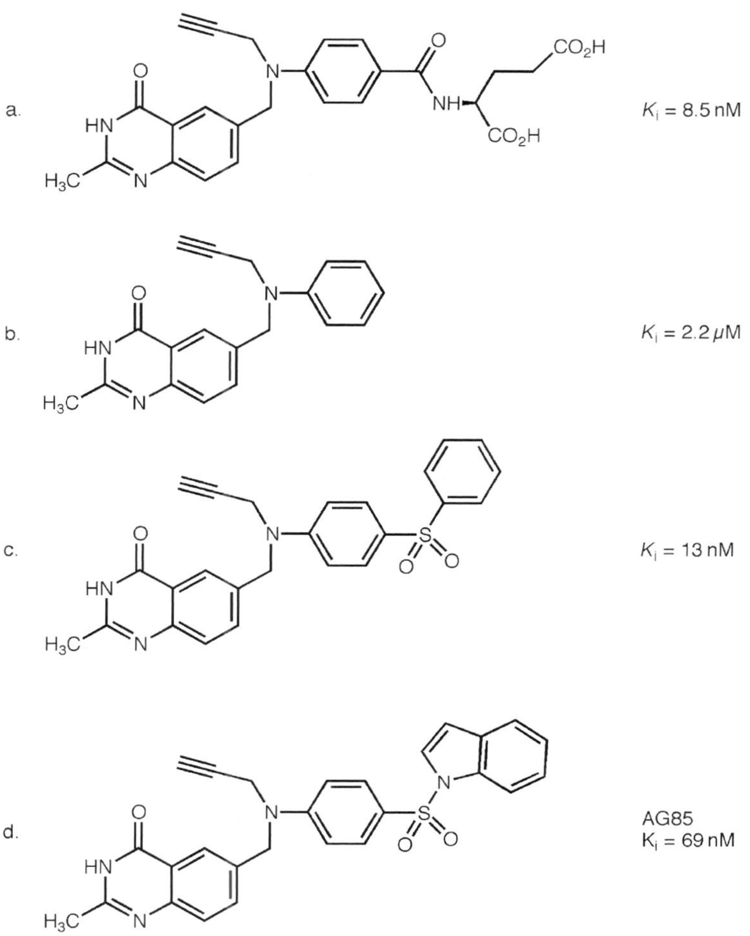

Fig. 26.18 Structure-based design of thymidylate synthase inhibitors. (a) Drug lead 2-methyl-2-desamino-N^{10}-propargyl-5,8-dideazafolic acid. (b) Derivative lacking the CO-L-glutamate moiety. (c) Designed diphenylsulfone derivative. (d) AG85, a designed *N*-sulfonylindole derivative selected for clinical evaluation.

taken into account. In a first step, Appelt *et al.*[133] made use of the program GRID to locate, within the enzyme active site, regions interacting favourably with an aromatic CH functional probe. Naphthalene was identified as a simple chemical structure which seemed to fill the binding pocket of the pteridine moiety of CB3717. Next, hydrogen-bonding groups were added in order to retain two key interactions which had been identified in the structure with CB3717, one with wat430, a bound water molecule already mentioned above, and the other with the side-chain of Asp169 (Fig. 26.17c). A carbonyl group was therefore introduced in position 1 to accept a hydrogen bond from wat430, and an NH was placed in position 8 to donate a hydrogen bond to Asp169. This design strategy led to the naphthostyryl scaffold, a promising novel substructure[88,133] (Fig. 26.19). A substituent was then introduced in position 5 in order to fill up empty space in the enzyme active site. A dialkylated amine was chosen to avoid the introduction of a chiral centre and also because of ease of synthesis and further

chemical modifications. Modelling of this fragment in the active site suggested the choice of a benzyl group as one of the alkyl substituents. Furthermore, it showed that position 4 of this benzene ring was suitable for the introduction of a solubilizing group. A (phenylsulfonyl)piperazine moiety was chosen because it was known from previous structure–activity studies to improve both binding and solubility.[133] The final compound was found to inhibit human TS with a K_i of 1.6 μM (Fig. 26.19). Its binding mode was analysed by X-ray crystallography and, surprisingly, this analysis revealed that the designed hydrogen-bonded interactions were not achieved in the complex. The compound was slightly shifted with respect to its modelled position, resulting in an unfavourable interaction with the main-chain carbonyl oxygen of Ala263 and the loss of wat430. The carbonyl group of the lactam was replaced by an amino group, in an attempt to restore wat430 binding and favourable interactions with the carbonyl oxygen of Ala263, and also to strengthen the interaction with Asp169

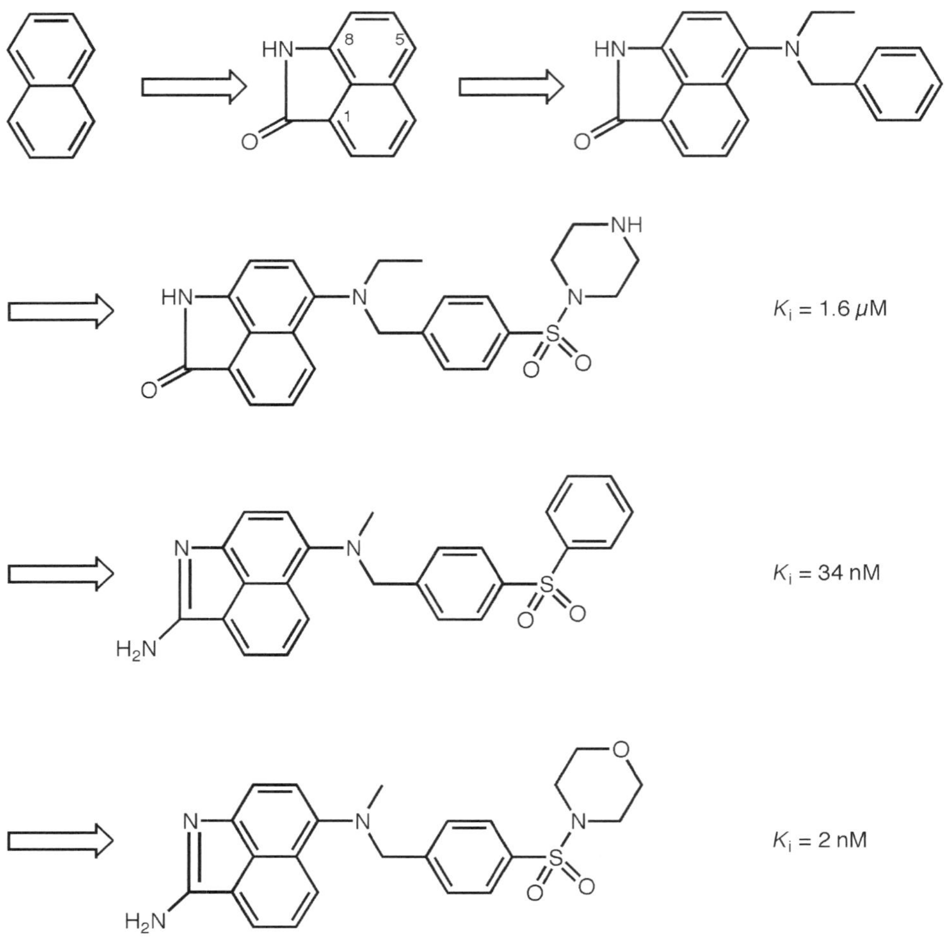

Fig. 26.19 Structure-based design of thymidylate synthase inhibitors: *de novo* design of a novel lead structure. See text for explanation.

(the amidine moiety was expected to be protonated). The resulting compound was found to inhibit human TS with a K_i of 34 nM (Fig. 26.19). This time, the X-ray analysis of the complex showed the predictions to be fully correct: wat430 had returned to its original location and the modelled hydrogen-bonding pattern was indeed observed. Replacement of the solubilizing piperazine ring by a morpholine group led to the final compound, a 2 nM inhibitor of human TS selected for clinical trials as an antitumour drug (Fig. 26.19).

D Human immunodeficiency virus protease

The human immunodeficiency virus protease (HIV-PR), an aspartic acid protease, is involved in the processing of viral polyproteins and is therefore essential for the production of new infectious virions. This enzyme is one of the best-characterized macromolecules from the vantage of drug design, with several hundred crystal structures determined to date.[65] Protein crystallography has contributed in many

ways to drug discovery programmes in this field. It has revealed the three-dimensional structure of this small homodimeric enzyme (2×99 residues) which features a single active site with perfect two-fold molecular symmetry.[134,135] X-ray analysis of HIV-PR/inhibitor complexes has provided the structural basis for the puzzling substrate specificity of this enzyme.[136,137] It has shed light on intriguing structure–activity data, notably those concerning stereochemical preferences for hydroxyethylamine-based peptidomimetics.[47,48,72,136] In several instances, protein crystallography proved useful for improving existing lead molecules.[125,137–141] Moreover, the structural information was used to design novel lead compounds, notably C_2-symmetric or pseudo-C_2-symmetric inhibitors[91,142] and cyclic urea derivatives.[74]

Taking advantage of a strategically located water molecule

A remarkable feature of all HIV-PR/inhibitor complexes initially determined was the presence of a tetrahedrally coordinated solvent molecule which mediated hydrogen-bonding interactions between the inhibitor and the flaps of the enzyme. This key water molecule had no counterpart in the related mammalian aspartic proteinases. Early on, it was realized that incorporation into an inhibitor of a structural mimic for this water molecule would be entropically favourable and should also result in improved selectivity. In a landmark structure-based drug design study by Lam and coworkers,[74] nonpeptide cyclic ureas were discovered which, for the first time, fulfilled this requirement. In addition, these molecules showed high potency combined with high oral bioavailability and high selectivity.

The design strategy was based on previous structure–activity studies on linear C_2-symmetric diols.[143] These potent inhibitors could not be developed into useful drugs owing to their poor oral bioavailability.[143] The goal of Lam *et al.*[74] was to find novel, small molecular weight HIV-PR inhibitors that would not retain any peptide character. They hypothesized that a conformationally restricted, cyclic structure which would incorporate a diol as the transition-state mimic as well as a mimic for the structural water could meet these requirements. The design strategy involved the following steps[74] (Fig. 26.20). (i) A model of a C_2-symmetric diol bound to the HIV-PR was built using the crystal structure

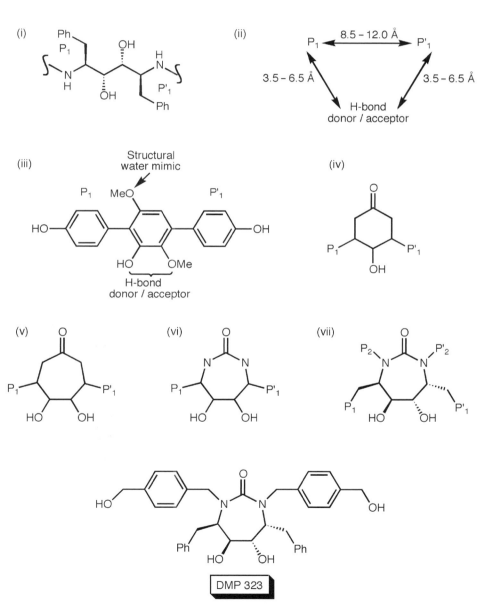

Fig. 26.20 Structure-based design of nonpeptide inhibitors of the HIV proteinase. For explanation, see text. (Adapted, with permission, from Lam, P.Y.S. *et al.* (1994) *Science* **263**: 380–384. Copyright © 1994 by the AAAS.)

of a HIV-PR complex with a hydroxyethylamine inhibitor.[144] (ii) A pharmacophore model was generated, which featured two symmetry-related hydrophobic groups and a hydrogen-bond donor/acceptor group. (iii) A 3D database search was carried out using the pharmacophore model. (iv) One hit from the 3D search suggested that a cyclohexanone ring could represent a promising new substructure. (v) Since diol-containing C_2-symmetric peptidomimetics were known to be superior inhibitors of the HIV-PR in comparison to the mono-ol derivatives,[143] the cyclohexanone was modified to a cycloheptanone to incorporate the diol functionality. (vi) The target structure was further modified to a cyclic urea, in order to strengthen the hydrogen-bonded interactions with the flaps. (vii) The optimal stereochemistry and conformation for these cyclic ureas was correctly inferred from molecular modelling analyses. This is a remarkable result, since there are four chiral centres in these molecules and also because the preferred configuration for the P1/P1′ substituents counterintuitively corresponds to the non-natural D-phenylalanine. (viii) Several substituted cyclic ureas derivatives were synthesized and tested. One of them, DMP323 (Fig. 26.20), showed good potency against the HIV-PR and the HIV virus *in vitro,* combined with significant oral bioavailability.[74] This compound was thus selected for clinical studies.[74]

E   Urokinase

A novel class of orally bioavailable inhibitors by X-ray crystallographic screening

Human urokinase (urokinase-type plasminogen activator) is a trypsin-like serine proteinase involved in tissue remodelling, a biological process playing an important role in cancer, arthritis, atherosclerosis and other disease states. Urokinase inhibitors have been shown to slow tumour growth and metastasis, and these results have been corroborated using urokinase knockout mice. A new drug discovery approach, called 'X-ray crystallographic screening' was recently introduced[96] and applied very successfully to the identification of new urokinase inhibitors showing a better pharmacokinetic profile than previously known leads.[96] In a first step, human urokinase was re-engineered to provide a crystal form amenable to ligand soaking experiments.[145] Then, diffraction data were collected from several crystals exposed to mixtures of compounds. Three criteria were used in the design of the mixtures: (i) the shapes of the compounds had to be different enough to allow unambiguous identification of any bound ligand by visual inspection of the electron density maps; (ii) all selected compounds had a positively charged functionality to interact with Asp 189 in the primary binding pocket of urokinase; and (iii) only weakly basic compounds were included, to increase the probability of being absorbed by oral route. The X-ray crystallographic screening led to the identification of five novel ligands, belonging to three compound classes (Fig. 26.21a). The *in vitro* activity of these ligands was measured and the more potent one, 8-hydroxy-2-aminoquinoline, was selected for lead optimization. The optimization step was greatly aided by previous crystallographic analyses of urokinase complexes with small molecules inhibitors, which had revealed the presence of a subsite adjacent to the primary binding pocket[61] (Fig. 26.21b). In particular, it was known that this subsite could be utilized by 2-naphthamidine derivatives with suitable substituents on the 8-position.[61] Notably, a substantial (150-fold) increase in binding potency had been obtained with an amino-pyrimidyl group linked to the 8-position of 2-naphthamidine. By combining these previous observations with the crystallographic screening data, it became immediately apparent that 8-aminopyrimidyl-2-aminoquinoline had the potential to be a very effective urokinase inhibitor. Indeed, this derivative showed a 100-fold increase in inhibitory potency as compared with the crystallographic screening hit. More importantly, this compound showed good oral bioavailability, in sharp contrast to the naphthamidine analogue.

VI   FUTURE PROSPECTS

Protein crystallography is a mature technique, which now routinely contributes to many drug discovery projects. Once the crystallization and soaking conditions have been established, it is usually possible to get three-dimensional structures of inhibitor complexes in a matter of weeks. In the aftermath of the human genome project, many large-scale academic and industrial initiatives have been established with the goal of solving as many different crystal structures in as short a time as possible.[13–16] These efforts are focused on improving the speed and throughput of all aspects of protein crystallography.[2] For protein expression, methods have been developed to detect the presence of (mis)folded proteins, either by expressing green fluorescent fusion proteins, or by the activation of special reporter genes. For protein purification, automated methods using tags are being developed. For crystallization, robotics using nano-liter technology and automatic imaging techniques are being developed, which will allow the automatic screening of a large number of crystallization conditions using very little protein. For data collection, robots are being installed at synchrotrons that will allow automated data collection from a large number of crystals in a short time. Finally, software to process data and solve crystal structures is constantly being improved and further automated to make the determination of new protein structures faster.

The impact of these structural genomics initiatives will be two-fold: (i) a large database of crystal structures will be available for immediate use in structure-based drug design

A. Lead finding by crystallographic screening

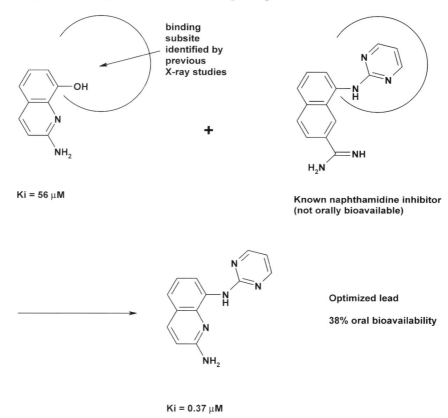

Fig. 26.21 Structure-based discovery of novel urokinase inhibitors. For explanation, see text.

projects; and (ii) most of the technology that is being developed will become available to all protein crystallography departments. With the increase in automation and speed of data processing, structure determination and refinement, structural biology is poised to make a major contribution to drug discovery in the new millennium.

FURTHER READING

Green, N.M. (1994) Stereoimages. A practical approach. *Structure* **2**: 85–87.

Ducruix, A. and Giege, R. (eds) (1999) *Crystallization of Nucleic Acids and Proteins: A Practical Approach,* 2nd edn. Oxford University Press, Oxford.

McRee, D.E. (1999) *Practical Protein Crystallography,* 2nd edn. Academic Press, San Diego.

REFERENCES

1. Shakespeare, W. (1997) *Macbeth*. Editions Aubier Montaigne, Paris.
2. Goodwill, K.E., Tennant, M.G. and Stevens, R.C. (2001) High-throughput X-ray crystallography for structure-based drug design. *Drug Discovery Today* **6**(Suppl.): S113–S118.

3. Antel, J. (1999) Integration of combinatorial chemistry and structure-based drug design. *Curr. Opin. Drug Discovery Dev.* **2**: 224–233.

4. Amzel, L.M. (1998) Structure-based drug design. *Curr. Opin. Biotechnol.* **9**: 366–369.

5. Gane, P.J. and Dean, P.M. (2000) Recent advances in structure-based rational drug design. *Curr. Opin. Struct. Biol.* **10**: 401–404.

6. Klebe, G. (2000) Recent developments in structure-based drug design. *J. Mol. Med.* **78**: 269–281.

7. Murcko, M.A., Caron, P.R. and Charifson, P.S. (1999) Structure-based drug design. *Annu. Rep. Med. Chem.* **34**: 297–306.

8. Lunney, E.A. (1998) Structure-based drug design begins a new era. *Med. Chem. Res.* **8**: 352–361.

9. DeCamp, D., Ogden, R., Kuntz, I. and Craik, C.S. (1996) Site-directed drug design. *Protein Eng.* 467–505.

10. Gschwend, D.A., Good, A.C. and Kuntz, I.D. (1996) Molecular docking towards drug discovery. *J. Mol. Recognit.* **9**: 175–186.

11. Marrone, T.J., Briggs, J.M. and McCammon, J.A. (1997) Structure-based drug design: computational advances. *Annu. Rev. Pharmacol. Toxicol.* **37**: 71–90.

12. Bohacek, R.S., McMartin, C. and Guida, W.C. (1996) The art and practice of structure-based drug design: a molecular modeling perspective. *Med. Res. Rev.* **16**: 3–50.

13. Kim, S.H. (1998) Shining a light on structural genomics. *Nature Struct. Biol.* **5**: 643–645.

14. Sali, A. (1998) 100 000 Protein structures for the biologist. *Nature Struct. Biol.* **5**: 1029–1032.

15. Shapiro, L. and Lima, C.D. (1998) The Argonne structural genomics workshop: Lamaze class for the birth of a new science. *Structure* **6**: 265–267.

16. Montelione, G.T. and Anderson, S. (1999) Structural genomics: keystone for a human proteome project. *Nature Struct. Biol.* **6**: 11–12.

17. Ramakrishnan, V. and Moore, P.B. (2000) Atomic structures at last: the ribosome in 2000. *Curr. Opin. Struct. Biol.* **11**: 144–154.

18. Brodersen, D.E., Clemons, W.M., Jr, Carter, A.P., Morgan-Warren, R.J., *et al.* (2000) The structural basis for the action of the antibiotics tetracycline, pactamycin, and hygromycin B on the 30S ribosomal subunit. *Cell* **103**: 1143–1154.

19. Pioletti, M., Schlunzen, F., Harms, J., Zarivach, R., Gluhmann, M., *et al.* (2001) Crystal structures of complexes of the small ribosomal subunit with tetracycline, edeine and IF3. *Embo J.* **20**: 1829–1839.

20. Ban, N., Nissen, P., Hansen, J., Moore, P.B. and Steitz, T.A. (2000) The complete atomic structure of the large ribosomal subunit at 2.4 A resolution. *Science* **289**: 905–920.

21. Nissen, P., Hansen, J., Ban, N., Moore, P.B. and Steitz, T.A. (2000) The structural basis of ribosome activity in peptide bond synthesis. *Science* **289**: 920–930.

22. Moult, J. and Melamud, E. (2000) From fold to function. *Curr. Opin. Struct. Biol.* **10**: 384–389.

23. Hwang, K.Y., Chung, J.H., Kim, S.-H., Han, Y.S. and Cho, Y. (1999) Structure-based identification of a novel NTPase from *Methanococcus jannaschii. Nature Struct. Biol.* **6**: 691–696.

24. Yang, F., Gustafson, K.R., Boyd, M.R. and Wlodawer, A. (1998) Crystal structure of *Escherichia coli* HdeA. *Nature Struct. Biol.* **5**: 763–764.

25. Lolis, E. and Petsko, G.A. Transition-state analogues in protein crystallography: probes of the structural source of enzyme catalysis. *Ann. Rev. Biochem.* **59**: 597–630.

26. Verschueren, K.H.G., Seljée, F., Rozeboom, H.J., Kalk, K.H. and Dijkstra, B.W. (1993) Crystallographic analysis of the catalytic mechanism of haloalkane dehalogenase. *Nature* **363**: 693–698.

27. Strynadka, N.C.J., Adachi, H., Jensen, S.E., Johns, K., Sielecki, A., *et al.* (1992) Molecular structure of the acyl-enzyme intermediate in β-lactam hydrolysis at 1.7 Å resolution. *Nature* **359**: 700–705.

28. Davies, T.G., Tunnah, P., Meijer, L., Marko, D., Eisenbrand, G., *et al.* (2001) Inhibitor binding to active and inactive CDK2. The

crystal structure of CDK2-cyclin A/indirubin-5-sulphonate. *Structure* **9**: 389–397.

29. Schindler, T., Bornmann, W., Pellicena, P., Miller, W.T., *et al.* (2000) Structural mechanism for STI-571 inhibition of Abelson tyrosine kinase. *Science* **289**: 1938–1942.

30. Boyd, D.B. (1995) Compendium of software for molecular modeling. *Rev. Comput. Chem.* **6**: 383–437.

31. Martin, Y.C. (1995) Accomplishments and challenges in integrating software for computer-aided ligand design in drug discovery. *Perspect. Drug Discovery Des.* **3**: 139–150.

32. Walsh, C.T. (1984) Suicide substrates, mechanism-based enzyme inactivators: recent developments. *Ann. Rev. Biochem.* **53**: 493–535.

33. Radzicka, A. and Wolfenden, R. (1995) Transition state and multi-substrate analog inhibitors. *Methods Enzymol.* **249**: 284–312.

34. Schramm, V.L. (1998) Enzymic transition states and transition state analog design. *Annu. Rev. Biochem.* **67**: 693–720.

35. Kirkpatrick, D.L., Watson, S. and Ulhaq, S. (1999) Structure-based drug design: combinatorial chemistry and molecular modeling. *Comb. Chem. High Throughput Screening* **2**: 211–221.

36. Ghose, A.K., Viswanadhan, V.N. and Wendoloski, J.J. (1999) Adapting structure-based drug design in the paradigm of combinatorial chemistry and high-throughput screening: an overview and new examples with important caveats for newcomers to combinatorial library design using pharmacophore models or multiple copy simultaneous search fragments. *ACS Symp. Ser.* **719**: 226–238.

37. Illig, C., Eisennagel, S., Bone, R., Radzicka, A., Murphy, L., *et al.* (1998) Expanding the envelope of structure-based drug design using chemical libraries: application to small-molecule inhibitors of thrombin. *Med. Chem. Res.* **8**: 244–260.

38. Aronov, A.M. Munagala, N.R. Ortiz de Montellano, P.R., Kuntz, I.D. and Wang, C.C. (2000) Rational design of selective submicromolar inhibitors of tritrichomonas foetus hypoxanthine-guanine-xanthine phosphoribosyltransferase. *Biochemistry* **39**: 4684–4691.

39. Bressi, J.C., Verlinde, C.L.M.J., Aronov, A.M., Shaw, M.L., Shin, S.S., *et al.* (2001) Adenosine analogues as selective inhibitors of glyceraldehyde-3-phosphate dehydrogenase of trypanosomatidae via structure-based drug design. *J. Med. Chem.* **44**: 2080–2093.

40. Smith, T.J., Kremer, M.J., Luo, M., Vriend, G., Arnold, E., *et al.* (1986) The site of attachment in human rhinovirus 14 for antiviral agents that inhibit uncoating. *Science* **233**: 1286–1293.

41. Kallen, J., Welzenbach, K., Ramage, P., Geyl, D., Kriwacki, R., *et al.* (1999) Structural basis for LFA-1 inhibition upon lovastatin binding to the CD11a I-domain. *J. Mol. Biol.* **292**: 1–9.

42. Weber, C., Wider, G., von Freyberg, B., Traber, R., Braun, W., *et al.* (1991) The NMR structure of cyclosporin A bound to cyclophilin in aqueous solution. *Biochemistry* **30**: 6563–6574.

43. Fesik, S.W., Gampe, R.T., Jr, Eaton, H.L., Gemmecker, G., Olejniczak, E.T., *et al.* (1991) NMR studies of [U-13C]cyclosporin A bound to cyclophilin: bound conformation and portions of cyclosporin involved in binding. *Biochemistry* **30**: 6574–6583.

44. Pflügl, G., Kallen, J., Schirmer, T., Jansonius, J.N., *et al.* (1993) X-ray structure of a decameric cyclophilin-cyclosporin crystal complex. *Nature* **361**: 91–94.

45. Van Duyne, G.D., Standaert, R.F., Karplus, P.A., Schreiber, S.L. and Clardy, J. (1991) Atomic structure of FKBP-FK506, an immunophilin–immunosuppressant complex. *Science* **252**: 839–842.

46. Badger, J., Minor, I., Kremer, M.J., Oliveira, M.A., Smith, T.J., *et al.* (1988) Structural analysis of a series of antiviral agents complexed with human rhinovirus 14. *Proc. Natl. Acad. Sci. USA* **85**: 3304–3308.

47. Rich, D.H., Sun, C.Q., Vara Prasad, J.V.N., Pathiasseril, A., Toth, M.V., *et al.* (1991) Effect of hydroxyl group configuration in hydroxyethylamine dipeptide isosteres on HIV protease inhibition. Evidence for multiple binding modes. *J. Med. Chem.* **34**: 1222–1225.

48. Krohn, A., Redshaw, S., Ritchie, J.C., Graves, B.J. and Hatada, M.H. (1991) Novel binding mode of highly potent HIV-proteinase inhibitors uncorporating the (R)-hydroxyethylamine isostere. *J. Med. Chem.* **34**: 3340–3342.

49. Banner, D.W. and Hadváry, P. (1991) Crystallographic analysis at 3.0 Å resolution of the binding to human thrombin of four active site-directed inhibitors. *J. Biol. Chem.* **266**: 20085–20093.

50. Stubbs, M.T. and Bode, W. (1993) Crystal structures of thrombin and thrombin complexes as a framework for antithrombotic drug design. *Perspect. Drug Dis. Design* **1**: 431–452.

51. Chirgadze, N.Y., Sall, D.J., Briggs, S.L., Clawson, D.K., Zhang, M., *et al.* (2000) The crystal structures of human α-thrombin complexed with active site-directed diamino benzo[b]thiophene derivatives: a binding mode for a structurally novel class of inhibitors. *Protein Sci.* **9**: 29–36.

52. Brandstetter, H., Engh, R.A., Von Roedern, E.G., Moroder, L., Huber, R., *et al.* (1998) Structure of malonic acid-based inhibitors bound to human neutrophil collagenase. A new binding mode explains apparently anomalous data. *Protein Sci.* **7**: 1303–1309.

53. Bartolucci, C., Perola, E., Pilger, C., Fels, G. and Lamba, D. (2001) Three-dimensional structure of a complex of galanthamine (nivalin) with acetylcholinesterase from *Torpedo californica*: implications for the design of new anti-Alzheimer drugs. *Proteins. Struct. Funct. Genet.* **42**: 182–191.

54. Chand, P., Babu, Y.S., Bantia, S., Chu, N., Cole, L.B., *et al.* (1997) Design and synthesis of benzoic acid derivatives as influenza neuraminidase inhibitors using structure-based drug design. *J. Med. Chem.* **40**: 4030–4052.

55. Schulze-Gahmen, U., Brandsen, J., Jones, H.D., Morgan, D.O., Meijer, L., *et al.* (1995) Multiple modes of ligand recognition: crystal structures of cyclin-dependent protein kinase 2 in complex with ATP and two inhibitors, olomoucine and isopentenyladenine. *Proteins* **22**: 378–391.

56. Powers, J.C., Oleksyszyn, J., Narasimhan, S.L., Kam, C.M., Radhakrishnan, R., *et al.* (1990) Reaction of porcine pancreatic elastase with 7-substituted 3-alkoxy-4-chloroisocoumarins: design of potent inhibitors using the crystal structure of the complex formed with 4-chloro-3-ethoxy-7-guanidinoisocoumarin. *Biochemistry* **29**: 3108–3118.

57. Singh, J., Dobrusin, E.M., Fry, D.W., Haske, T., Whitty, A. and McNamara, D.J. (1997) Structure-based design of a potent, selective, and irreversible inhibitor of the catalytic domain of the erbB receptor subfamily of protein tyrosine kinases. *J. Med. Chem.* **40**: 1130–1135.

58. Maignan, S. and Mikol, V. (2001) The use of 3D structural data in the design of specific Factor Xa inhibitors. *Curr. Top. Med. Chem.* **1**: 161–174.

59. Aronov, A.M., Suresh, S., Buckner, F.S., Van Voorhis, W.C., Verlinde, C.L.M.J., *et al.* (1999) Structure-based design of submicromolar, biologically active inhibitors of trypanosomatid glyceraldehyde-3-phosphate dehydrogenase. *Proc. Natl Acad. Sci. USA* **96**: 4273–4278.

60. Aronov, A.M., Verlinde, C.L.M.J., Hol, W.G.J. and Gelb, M.H. (1998) Selective tight binding inhibitors of trypanosomal glyceraldehyde-3-phosphate dehydrogenase via structure-based drug design. *J. Med. Chem.* **41**: 4790–4799.

61. Nienaber, V.L., Davidson, D., Edalji, R., Giranda, V.L., Klinghofer, V., *et al.* (2000) Structure-directed discovery of potent non-peptidic inhibitors of human urokinase that access a novel binding subsite. *Structure* **8**: 553–563.

62. Mao, C., Sudbeck, E.A., Venkatachalam, T.K. and Uckun, F.M. (2000) Structure-based drug design of non-nucleoside inhibitors for wild-type and drug-resistant HIV reverse transcriptase. *Biochem. Pharmacol.* **60**: 1251–1265.

63. Rahuel, J., Rasetti, V., Maibaum, J., Rueger, H., Goschke, R., *et al.* (2000) Structure-based drug design: the discovery of novel nonpeptide orally active inhibitors of human renin. *Chem. Biol.* **7**: 493–504.

64. Mihelich, E.D. and Schevitz, R.W. (1999) Structure-based design of a new class of anti-inflammatory drugs: secretory phospholipase A2 inhibitors, SPI. *Biochim. Biophys. Acta* **1441**: 223–228.

65. Wlodawer, A. and Vondrasek, J. (1998) Inhibitors of HIV-1 protease: a major success of structure-assisted drug design. *Annu. Rev. Biophys. Biomol. Struct.* **27**: 249–284.

66. Wade, R.C. (1997) 'Flu' and structure-based drug design. *Structure* **5**: 1139–1145.

67. Montgomery, J.A. Niwas, S. Rose, J.D. Secrist, J.A., III, Babu, Y.S. *et al.* (1993) Structure-based design of inhibitors of purine nucleoside phosphorylase. Part 1. 9-(Arylmethyl) derivatives of 9-deazaguanine. *J. Med. Chem.* **36**: 55–69.

68. Secrist, J.A., III, Niwas, S., Rose, J.D., Babu, Y.S., Bugg, C.E., *et al.* (1993) Structure-based design of inhibitors of purine nucleoside phosphorylase. Part 2. 9-Alicyclic and 9-heteroalicyclic derivatives of 9-deazaguanine. *J. Med. Chem.* **36**: 1847–1854.

69. Erion, M.D., Niwas, S., Rose, J.D., Ananthan, S., Allen, M., *et al.* (1993) Structure-based design of inhibitors of purine nucleoside phosphorylase. Part 3. 9-Arylmethyl derivatives of 9-deazaguanine substituted on the methylene group. *J. Med. Chem.* **36**: 3771–3783.

70. Guida, W.C., Elliott, R.D., Thomas, H.J., Secrist, J.A., III, Babu, Y.S., *et al.* (1994) Structure-based design of inhibitors of purine nucleoside phosphorylase. Part 4. A study of phosphate mimics. *J. Med. Chem.* **37**: 1109–1114.

71. Baldwin, J.J., Ponticello, G.S., Anderson, P.S., Christy, M.E., Murcko, M.A., *et al.* (1989) Thienothiopyran-2-sulfonamides: novel topically active carbonic anhydrase inhibitors for the treatment of glaucoma. *J. Med. Chem.* **32**: 2510–2513.

72. Appelt, K. (1993) Crystal structures of HIV-1 protease-inhibitor complexes. *Perspect. Drug Dis. Design* **1**: 23–48.

73. Vacca, J.P., Fitzgerald, P.M.D., Holloway, M.K., Hungate, R.W., Starbuck, K.E., Chen, L.J., Darke, P.L., Anderson, P.S. and Huff, J.R. (1994) Conformationally constrained HIV-1 protease inhibitors. *Bioorg. Med. Chem. Lett.* **4**: 499–504.

74. Lam, P.Y.S., Jadhav, P.K., Eyermann, C.J., Hodge, C.N., Ru, Y., *et al.* (1994) Rational design of potent, bioavailable, nonpeptide cyclic ureas as HIV protease inhibitors. *Science* **263**: 380–384.

75. Morgan, B.P., Holland, D.R., Matthews, B.W. and Bartlett, P.A. (1994) Structure-based design of an inhibitor of the zinc peptidase thermolysin. *J. Am. Chem. Soc.* **116**: 3251–3260.

76. Thaisrivongs, S., Blinn, J.R., Pals, D.T. and Turner, S.R. (1991) Conformationally constrained renin inhibitory peptides: cyclic (3-1)-1-(carboxymethyl)-L-prolyl-L-phenylalanyl-L-histidinamide as a conformational restriction at the P$_2$–P$_4$ tripeptide portion of the angiotensinogen template. *J. Med. Chem.* **34**: 1276–1282.

77. Weber, A. E., Halgren, T. A., Doyle, J. J., Lynch, R. J., Siegl, P. K. S., *et al.* Design and synthesis of P2–P1' -linked macrocyclic human renin inhibitors. *J. Med. Chem.* 34: 2692–2701.

78. Alberg, D.G. and Schreiber, S.L. (1993) Structure-based design of a cyclophilin-calcineurin bridging ligand. *Science* **262**: 248–250.

79. Furet, P., Garcia-Echeverria, C., Gay, B., Schoepfer, J., *et al.* (1999) Structure-based design, synthesis, and X-ray crystallography of a high-affinity antagonist of the Grb2-SH2 domain containing an asparagine mimetic. *J. Med. Chem.* **42**: 2358–2363.

80. Toledo, L.M., Lydon, N.B. and Elbaum, D. (1999) The structure-based design of ATP-site directed protein kinase inhibitors. *Curr. Med. Chem.* **6**: 775–805.

81. Borkakoti, N. (1998) Matrix metalloproteases: variations on a theme. *Prog. Biophys. Mol. Biol.* **70**: 73–94.

82. Somoza, J.R., Skillman, A.G., Jr, Munagala, N.R., Oshiro, C.M., Knegtel, R.M.A., *et al.* (1998) Rational design of novel antimicrobials: blocking purine salvage in a parasitic protozoan. *Biochemistry* **37**: 5344–5348.

83. Gschwend, D.A., Sirawaraporn, W., Santi, D.V. and Kuntz, I.D. (1997) Specificity in structure-based drug design: identification of a novel, selective inhibitor of *Pneumocystis carinii* dihydrofolate reductase. *Proteins Struct. Funct. Genet.* **29**: 59–67.

84. Schevitz, R.W., Bach, N.J., Carlson, D.G., Chirgadze, N.Y., Clawson, D.K., *et al.* (1995) Structure-based design of the first potent and selective inhibitor of human non-pancreatic secretory phospholipase A2. *Nat. Struct. Biol.* **2**: 458–465.

85. Kim, E.E., Baker, C.T., Dwyer, M.D., Murcko, M.A., Rao, B.G., *et al.* (1995) Crystal structure of HIV-1 protease in complex with VX-478, a potent and orally bioavailable inhibitor of the enzyme. *J. Am. Chem. Soc.* **117**: 1181–1182.

86. MacPherson, L.J., Bayburt, E.K., Capparelli, M.P., Bohacek, R.S., Clarke, F.H., *et al.* (1993) Design and synthesis of an orally active macrocyclic neutral endopeptidase 24.11 inhibitor. *J. Med. Chem.* **36**: 3821–3828.

87. Brown, F.J., Andisik, D.W., Bernstein, P.R., Bryant, C.B., Ceccarelli, C., *et al.* (1994) Design of orally active, non-peptidic inhibitors of human leukocyte elastase. *J. Med. Chem.* **37**: 1259–1261.

88. Varney, M.D., Marzoni, G.P., Palmer, C.L., Deal, J.G., Webber, S., *et al.* (1992) Crystal-structure-based design and synthesis of benz[cd]indole-containing inhibitors of thymidylate synthase. *J. Med. Chem.* **35**: 663–676.

89. Reich, S.H., Fuhry, M.A.M., Nguyen, D., Pino, M.J., Welsh, K.M., *et al.* (1992) Design and synthesis of novel 6,7-imidazotetrahydro-quinoline inhibitors of thymidylate synthase using iterative protein crystal structure analysis. *J. Med. Chem.* **35**: 847–858.

90. Reich, S.H. and Webber, S.E. (1993) Structure-based drug design (SBDD): every structure tells a story. *Perspect. Drug Dis. Design* **1**: 371–390.

91. Erickson, J.W. (1993) Design and structure of symmetry-based inhibitors of HIV-1 protease. *Perspect. Drug Dis. Design* **1**: 109–128.

92. Sintchak, M.D. and Nimmesgern, E. (2000) The structure of inosine 5′-monophosphate dehydrogenase and the design of novel inhibitors. *Immunopharmacology* **47**: 163–184.

93. Fretz, H., Furet, P., Garcia-Echeverria, C., Rahuel, J. and Schoepfer, J. (2000) Structure-based design of compounds inhibiting Grb2-SH2 mediated protein–protein interactions in signal transduction pathways. *Curr. Pharm. Des.* **6**: 1777–1796.

94. Gubernator, K., Heinze-Krauss, I., Angehrn, P., Charnas, R.L., Hubschwerlen, C., *et al.* Structure-based design of potent β-lactamase inhibitors. *Methods Princ. Med. Chem.* **6**: 89–103.

95. Joseph-McCarthy, D. (1999) Computational approaches to structure-based ligand design. *Pharmacol. Ther.* **84**: 179–191.

96. Nienaber, V.L., Richardson, P.L., Klighofer, V., Bouska, J.J., *et al.* (2000) Discovering novel ligands for macromolecules using X-ray crystallographic screening. *Nat. Biotechnol.* **18**: 1105–1108.

97. Ducruix, A. and Giege, R. (1999) *Crystallization of Nucleic Acids and Proteins: A Practical Approach*, 2nd edn. Oxford University Press, Oxford.

98. McPherson, A. (1999) *Crystallization of Biological Macromolecules.* Cold Spring Harbor Laboratory Press, Cold Spring Harbor, NY.

99. McRee, D.E. (1999) *Practical Protein Crystallography*, 2nd edn. Academic Press, San Diego.

100. Drenth, J. (1999) *Principles of Protein X-ray Crystallography*, 2nd edn. Springer Verlag, New York.

101. Viterbo, D. (1992) Solution and refinement of crystal structures. In Giacovazzo, C. (ed.). *Fundamentals of Crystallography*, pp. 319–401. Oxford University Press, Oxford.

102. Dauter, Z. and Dauter, M. (2001) Entering a new phase: using solvent halide ions in protein structure determination. *Structure* **9**: R21–R26.

103. Bricogne, G. (1993) Direct phase determination by entropy maximization and likelihood ranking: status report and perspectives. *Acta Crystallogr. Sect. D* **D49**: 37–60.

104. Berman, H.M., Westbrook, J., Feng, Z., Gilliland, G., Bhat, T.N., *et al.* (2000) The protein data bank. *Nucleic Acids Res.* **28**: 235–242.

105. Brünger, A.T. (1992) Free *R* value: a novel statistical quantity for assessing the accuracy of crystal structures. *Nature* **355**: 472–475.

106. Jia, Z., Vandonselaar, M., Quail, J.W. and Delbaere, L. T. J. (1993) Active-centre torsion-angle strain revealed in 1.6 Å-resolution structure of histidine-containing phosphocarrier protein. *Nature* **361**: 94–97.

107. Chevrier, B., Schalk, C., D'Orchymont, H., Rondeau, J.-M., *et al.* (1994) Crystal structure of Aeromonas proteolytica aminopeptidase: a prototypical member of the co-catalytic zinc enzyme family. *Structure* **2**: 283–291.

108. Perutz, M. (1992) *Protein Structure. New Approaches to Disease and Therapy.* W. H. Freeman and Company, New York.

109. Brange, J., Ribel, U., Hansen, J.F., Dodson, G., Hansen, M.T., *et al.* (1988) Monomeric insulins obtained by protein engineering and their medical implications. *Nature* **333**: 679–682.

110. Ring, C.S. Sun, E. McKerrow, J.H. Lee, G.K. Rosenthal, P.J., *et al.* (1993) Structure-based inhibitor design by using protein models for the development of antiparasitic agents. *Proc. Natl. Acad. Sci. USA*, **90**: 3583–3587.

111. Cushman, D.W., Cheung, H.S., Sabo, E.F. and Ondetti, M.A. (1977) Design of potent competitive inhibitors of angiotensin-converting enzyme. Carboxyalkanoyl and mercaptalkanoyl amino acids. *Biochemistry* **16**: 5484–5491.

112. Hassall, C.H., Kröhn, A., Moody, C.H. and Thomas, W.A. (1982) The design of a new group of angiotensin-converting enzyme inhibitors. *FEBS Lett.* **147**: 175–179.

113. Attwood, M.R., Hassall, C.H., Kröhn, A., Lawton, G. and Redshaw, S. (1986) The design and synthesis of the angiotensin-converting enzyme inhibitor cilazapril and related bicyclic compounds. *J. Chem. Soc. Perkins Trans.* **1**: 1011–1019.

114. Blundell, T.L., Cooper, J., Foundling, S.I., Jones, D.M., *et al.* (1987) On the rational design of renin inhibitors: X-ray studies of aspartic proteinases complexed with transition-state analogues. *Biochemistry* **26**: 5585–5590.

115. Greenlee, W.J. (1990) Renin inhibitors. *Med. Res. Rev.* **10**: 173–236.

116. Lunney, E.A., Hamilton, H.W., Hodges, J.C., Kaltenbronn, J.S., Repine, J.T., *et al.* (1993) Analyses of ligand binding in five endothiapepsin crystal complexes and their use in the design and evaluation of novel renin inhibitors. *J. Med. Chem.* **36**: 3809–3820.

117. Iizuka, K., Kamijo, T., Harada, H., Akahane, K., Kubota, T., *et al.* (1990) Orally potent human renin inhibitors derived from angiotensinogen transition state: design, synthesis, and mode of interaction. *J. Med. Chem.* **33**: 2707–2714.

118. Takahashi, L.H., Radhakrishnan, R., Rosenfield, R.E., Jr., Meyer, E.F., Jr., *et al.* (1988) X-ray diffraction analysis of the inhibition of porcine oancreatic elastase by a peptidyl trifluoromethylketone. *J. Mol. Biol.* **201**: 423–428.

119. Luo, M., Jedrzejas, M.J., Singh, S., White, C.L., Brouillette, W.J., *et al.* (1995) Benzoic acid inhibitors of influenza virus neuraminidase. *Acta Crystallogr. Sect. D* **D51**: 504–510.

120. Bethell, R.C. and Smith, P.W. (1997) Sialidase as a target for inhibitors of influenza virus replication. *Expert Opin. Invest. Drugs* **6**: 1501–1509.

121. Luo, M., Air, G.M. and Brouillette, W.J. (1997) Design of aromatic inhibitors of influenza virus neuraminidase. *J. Infect. Dis.* **176**(Suppl. 1): S62–S65.

122. Babu, Y.S., Chand, P., Bantia, S., Kotian, P., Dehghani, A., *et al.* (2000) Bcx-1812 (rwj-270201): discovery of a novel, highly potent, orally active, and selective influenza neuraminidase inhibitor through structure-based drug design. *J. Med. Chem.* **43**: 3482–3486.

123. Watenpaugh, K.D. (1996) Structure-based drug design and informatics: the development of potent HIV protease inhibitors as clinical drug candidates. *Folding Des.* **1**(Suppl. 1): S67.

124. Aristoff, P.A. (1988) Dihydropyrone sulfonamides as a promising new class of HIV protease inhibitors. *Drugs Future* **23**: 995–999.

125. Thaisrivongs, S. and Strohbach, J.W. (1999) Structure-based discovery of Tipranavir Disodium (PNU-140690E): a potent, orally bioavailable, nonpeptidic HIV protease inhibitor. *Biopolymers* **51**: 51–58.

126. Rastelli, G., Vianello, P., Barlocco, D., Costantino, L., *et al.* (1997) Structure-based design of an inhibitor modeled at the substrate active site of aldose reductase. *Bioorg. Med. Chem. Lett.* **7**: 1897–1902.

127. Iwata, Y., Arisawa, M., Hamada, R., Kita, Y., Mizutani, M.Y., *et al.* (2001) Discovery of novel aldose reductase inhibitors using a protein structure-based approach: 3D-database search followed by design and Synthesis. *J. Med. Chem.* **44**: 1718–1728.

128. Du, X., Hansell, E., Engel, J.C., Caffrey, C.R., *et al.* (2000) Aryl ureas represent a new class of anti-trypanosomal agents. *Chem. Biol.* **7**: 733–742.

129. Matthews, D.A., Dragovich, P.S., Webber, S.E., Fuhrman, S.A., Patick, A.K., *et al.* (1999) Structure-assisted design of mechanism-based irreversible inhibitors of human rhinovirus 3C protease with potent antiviral activity against multiple rhinovirus serotypes. *Proc. Natl Acad. Sci. USA* **96**: 11000–11007.

130. Frecer, V., Maliar, T. and Miertus, S. (2000) Protease inhibitors as anticancer drugs: role of molecular modelling and combinatorial chemistry in drug design. *Int. J. Med. Biol. Environ.* **28**: 161–173.

131. Morris, P.E., Jr. and Omura, G.A. (2000) Inhibitors of the enzyme purine nucleoside phosphorylase as potential therapy for psoriasis. *Curr. Pharm. Des.* **6**: 943–959.

132. Matthews, D.A., Appelt, K., Oatley, S.J. and Xuong, Ng. H. (1990) Crystal structure of *Escherichia coli* thymidylate synthase containing bound 5-fluoro-2′-deoxyuridylate and 10-propargyl-5,8-dideazafolate. *J. Mol. Biol.* **214**: 923–936.

133. Appelt, K., Bacquet, R.J., Bartlett, C.A., *et al.* (1991) Design of enzyme inhibitors using iterative protein crystallographic analysis. *J. Med. Chem.* **34**: 1925–1934.

134. Wlodawer, A., Miller, M., Jaskolski, M., Sathyanarayana, B.K., Baldwin, E., *et al.* (1989) Conserved folding in retroviral proteases: crystal structure of a synthetic HIV-1 protease. *Science* **245**: 616–621.

135. Lapatto, R., Blundell, T., Hemmings, A., Overington, J., Wilderspin, A., *et al.* (1989) X-ray analysis of HIV-1 proteinase at 2.7 Å resolution confirms structural homology among retroviral enzymes. *Nature* **342**: 299–302.

136. Wlodawer, A. and Erickson, J.W. (1993) Structure-based inhibitors of HIV-1 protease. *Annu. Rev. Biochem.* **62**: 543–585.

137. Clare, M. (1993) HIV protease: structure-based design. *Perspect. Drug Dis. Des.* **1**: 49–68.

138. Kalish, V.J., Tatlock, J.H., Davies, J.F., Kaldor, S.W., Dressman, B.A., *et al.* (1995) Structure-based drug design of nonpeptidic P2 substituents for HIV-1 protease inhibitors. *Bioorg. Med. Chem. Lett.* **5**: 727–732.

139. Romero, D.L., Tommasi, R.A., Janakiraman, M.N., Strohbach, J.W., Turner, S.R., *et al.* (1996) Structure-based design of HIV protease inhibitors: 5,6-dihydro-4-hydroxy-2-pyrones as effective, nonpeptidic inhibitors. *J. Med. Chem.* **39**: 4630–4642.

140. Skulnick, H.I., Turner, S.R., Strohbach, J.W., Tommasi, R.A., Johnson, P.D., *et al.* (1996) Structure-based design of HIV protease inhibitors: sulfonamide-containing 5,6-dihydro-4-hydroxy-2-pyrones as non-peptidic inhibitors. *J. Med. Chem.* **39**: 4349–4353.

141. Skulnick, H.I., Johnson, P.D., Aristoff, P.A., Morris, J.K., Lovasz, K.D., *et al.* (1997) Structure-based design of nonpeptidic HIV protease inhibitors: the sulfonamide-substituted cyclooctylpyranones. *J. Med. Chem.* **40**: 1149–1164.

142. Babine, R.E., Zhang, N., Jurgens, A.R., Schow, S.R., Desai, P.R., *et al.* (1992) The use of HIV-1 protease structure in inhibitor design. *Bioorg. Med. Chem. Lett.* **2**: 541–546.

143. Kempf, D.J., Codacovi, L., Wang, X.C., Kohlbrenner, W.E., Wideburg, N.E., *et al.* (1993) Symmetry-based inhibitors of HIV protease structure–activity studies of acylated 2,4-diamino-1,5-diphenyl-3-hydroxypentane and 2,5-diamino-1,6-diphenylhexane-3,4-diol. *J. Med. Chem.* **36**: 320–330.

144. Swain, A.L., Miller, M.M., Green, J., Rich, D.H., Schneider, J., *et al.* (1990) X-ray crystallographic structure of a complex between a synthetic protease of human immunodeficiency virus 1 and a substrate-based hydroxyethylamine inhibitor. *Proc. Natl Acad. Sci. USA* **87**: 8805–8809.

145. Nienaber, V., Wang, J., Davidson, D. and Henkin, J. (2000) Re-engineering of human urokinase provides a system for structure-based drug design at high resolution and reveals a novel structural subsite. *J. Biol. Chem.* **275**: 7239–7248.

146. Knight, S., Andersson, I. and Brändén, C.-I. (1989) Reexamination of the three-dimensional structure of the small subunit of RuBisCO from higher plants. *Science* **244**: 702–705.

147. Curmi, P.M.G., Cascio, D., Sweet, R.M., Eisenberg, D. and Schreuder, H. (1992) Crystal structure of the unactivated form of ribulose-1,5-bisphosphate carboxylase/oxygenase from tobacco refined at 2.0 Å resolution. *J. Biol. Chem.* **267**: 16980–16989.

27

PROTEIN COMPARATIVE MODELLING AND DRUG DISCOVERY

Charlotte M. Deane and Tom L. Blundell

There are some enterprises in which a careful disorderliness is the true method   **Hermann Melville**

I INTRODUCTION

For the rational design of drugs, structural information concerning the target protein and ligand binding is vital. Although the database of protein crystal structures is growing exponentially due to important methodological improvements in structure-determination techniques,[1] there are still no structural data for the majority of the relevant targets. The large number of membrane proteins which are potential targets makes the challenge greater as these have proved particularly difficult to crystallize. However, in cases where structure elucidation is not possible, comparative models can be built for the target protein by exploiting sequence similarity to a homologous protein or other proteins of known structure.

Comparative modelling predicts a protein's structure using its sequence together with the three-dimensional (3D) structures of homologues. Sequence data are far more available than structural data and sequencing itself is rapid. According to the Protein International Resource (PIR) — International Sequence Database Release June 2000,[2] 160 729 nonredundant protein sequences have been deposited versus 23 291 protein structures. Furthermore, although there are 18 577 sets of protein domain coordinates in the protein data bank (PDB)[1] release used for CATH version 1.6,[3] they represent only 1028 homologous superfamilies and 672 folds.[3] Calculated slightly differently, but highlighting the same discrepancy, Moult[4] states that ~ 1% of the 200 000 or so known unique amino-acid sequences of proteins have a corresponding experimentally determined structure. The conflict between the demand for structural information of known sequences and the relatively slow rate of accumulating 3D coordinates of proteins from NMR and X-ray analysis highlights the need for rapid reliable modelling techniques.

Although comparative modelling is less accurate than high-resolution experimental methods, it can be helpful in proposing and testing hypotheses in molecular biology. Identifying putative binding sites by modelling is far faster and cheaper than by experiment. Thus, even though comparative modelling techniques still need significant improvements, they can and already are being used to address many practical problems. In the coming years an increasing number of structurally undefined proteins, with sequence similarity to examples of known 3D structures, will be targets for drug design, and so comparative modelling is likely to be central to drug discovery for some time to come.

The practice of comparative modelling exploits knowledge of the general principles of protein 3D structure as well as our understanding of the evolution of homologous protein families. We, therefore, begin by outlining these principles briefly before discussing what is known about the structural relationships between members of divergently evolved

The Practice of Medicinal Chemistry
ISBN 0-12-744481-5

protein families and superfamilies. We then go on to describe the most generally used approaches to comparative modelling, discussing the accuracy of the models constructed. When sequence similarities are low, for example below 20% identity, much of this depends on the identification of homologues. Furthermore, correct alignment of the sequence of interest with that of the homologue of known structure remains a challenge even when sequence identities are as high as 40%. We briefly discuss these important aspects of comparative modelling. Finally, we review some examples of comparative modelling that have been useful in drug discovery.

II PRINCIPLES OF PROTEIN STRUCTURE

Many of the important features of protein structure arise from the regular repeating nature of the polypeptide main chain. All naturally occurring amino acids, which constitute a protein, are the same enantiomer, L (Fig. 27.1). The peptide bond (between two amino acids in a polypeptide chain) is planar, due to the delocalization of the p

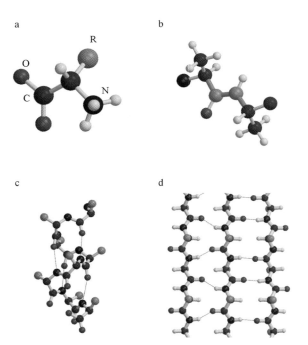

Fig. 27.1 (a) L-amino acid. All amino-acids are the same enantiomer. If the amino acid is viewed along the hydrogen bond to Cα bond — with the hydrogen towards the observer, the groups read clockwise CORN; (b) Ala — Ala *trans* peptide; (c) The right-handed α helix found in proteins, only the Cα, CO and NH atoms of the backbone and the Cβ atom of the side-chain are shown. The dotted lines indicate the hydrogen bonds; (d) β-sheet with antiparallel and parallel β-strands.

electrons over the N—C—O of the peptide and almost always *trans* with respect to the polypeptide backbone (Fig. 27.1); the *trans* arrangement is normally favoured over *cis* by 10^3-fold.[5]

A Regular secondary structure

Secondary structure refers to regions of proteins where amino-acid residues adopt similar conformations and regular hydrogen-bonding patterns. The two most important types of secondary structure are the right-handed α-helix (formally known as the 3.6_{13} helix) and the extended β-strand (Fig. 27.1). In the right-handed α-helix all the CO groups point towards the *C*-terminus, the NH groups towards the *N*-terminus of the helix and there are 3.6 residues per helical turn (Fig. 27.1). The α-helix is stabilized by hydrogen bonds between the CO of residue i and the NH of residue i + 4 making a ring of 13 atoms. These helices are found widely in globular proteins and in fibrous proteins such as keratin. Tighter helices with three residues per turn (3_{10}-helix) or looser with 4.4 residues per turn (4.4_{16} helix) are also possible, but have either less optimized hydrogen bonds or less well packed cores, and are consequently less stable. Only short helices of these types are found in globular proteins, and then often at the termini of the more stable α-helices. The helix with the reverse twist, i.e. the left-handed α-helix, is found only in short stretches of repeated Gly, since the residue side-chains (absent in Gly) are brought far closer to and may clash with the backbone CO groups. The extended (β)-strand is technically also a helix, with slightly more than two (2.2) residues per turn. Here, the NH and CO groups of the backbone are involved in inter-chain H-bonds to other β-strands to form β-sheets (Fig. 27.1). This can contribute to two kinds of slightly twisted sheet, one in which the strands run antiparallel and the other in which they run parallel; the side-chains project alternately above and below the plane of the sheet in both.

B Loops

It is estimated that 30% of globular protein structure comprises nonrepeating irregular regions connecting regular secondary structure. These are known as loops. They tend to be the least well defined and therefore the most error-prone regions of protein structure in both X-ray and NMR analyses; they are also the most difficult regions to model using knowledge-based procedures. Loop regions have been clustered and classified. Many contain turns that reverse chain direction; these include β-turns of four residues in length that create a 180° chain reversal between two β-strands,[6] and γ-turns of three residues that also give a chain reversal.[7] Some loop conformations frequently recur in protein structure; these include β-hairpins, which link sequential pairs of antiparallel β-strands,[8] and many other defined loop families.[9-13] Irregular loops can also be classified according to their linearity, planarity and location of the *N*- and *C*-termini.[14]

C Side-chain conformers

Side-chain conformations tend to exist in a limited number of canonical combinations or rotamers, which are dependent on the local environmental features like secondary structure, solvent accessibility and hydrogen bonding.[15-18] Thus, side-chain conformations are often defined by libraries of rotamers calculating the mean rotamer angle for a residue with a particular backbone φ/ψ combination.[19-22] Such libraries have been widely used both in modelling and in verification of new protein structures. However, Lovell *et al.*[23] have shown that rotamers included in many libraries give rise to substantial van der Waals clashes, and their use in modelling into electron density by protein crystallographers has tended to lead to bias in the experimental coordinates deposited in the PDB. A more detailed discussion of protein structure can be found in reference 24.

III PROTEIN FOLDING AND *AB INITIO* MODELLING OF PROTEINS

An understanding of the mechanism by which a polypeptide chain folds from the denatured coil state to the native protein structure remains an elusive goal of structural biology. Despite considerable effort, both theoretical and experimental, we are still not able to give a detailed mechanistic description of any such folding process.[4,25] Within a cell, myriads of amino-acid sequences fold effortlessly into precise 3D structures, yet at present no algorithm exists which can identify the lowest free energy state for a protein under the conditions of the cell, due to various complications.

According to the Levinthal paradox,[26] all conformations of the polypeptide chain, except for the native state, are equally probable, so that the native state can only be found by an unbiased random search. This proposes a protein potential energy surface like a golf course.[27] For such a surface, the time to find the native state of a protein is given by the number of possible configurations of the polypeptide chain (say 10^{70} for a 100-residue protein), multiplied by the time required to find one configuration (say, $10-11$ seconds), which leads to a very long folding time (about 10^{59} seconds, or approximately 10^{52} years for the example).[28] Proteins, however, generally fold within milliseconds to seconds

(barring special slowing factors such as Pro isomerization[29]): there is indeed a paradox.

The exact nature of this problem has been rethought, and a random search is now not considered to be a good description of folding. Harrison and Durbin[30] suggested that there are many ways of reaching the native state in reasonable time, so that a specific pathway (from unfolded to folded state) does not have to be postulated. Recently, this kind of model has been further developed using lattice modelling studies and statistical mechanical models.

Perhaps the most serious problem for simulation of a protein fold from physical principles is the fact that folded and unfolded states of proteins differ very little in overall energy (only about $20-65 \text{ kJ mol}^{-1}$ difference), whereas the structures have total energies of the order of $40 \text{ million kJ mol}^{-1}$;[31] any successful energy function therefore needs to be enormously accurate (and becomes prohibitively computationally expensive).

The protein-folding problem has been split in two: the prediction of the 3D structure of a protein from its sequence; and the understanding of the kinetics and thermodynamics of the folding process. Most methods for modelling the 3D structure of proteins attempt only to answer the first of these problems.

The central premise upon which the modelling of proteins rests is that the 3D structure of a protein is determined by only its sequence and its environment, without the obligatory role of external factors. Reality is known to be more complex than this because, although many proteins have been refolded into their active forms,[32] chaperones and disulphide interchange enzymes have been identified as assisting the folding process. However, it is generally considered that such molecules merely aid rather than dictate the final fold.[31]

IV PROTEIN EVOLUTION: FAMILIES AND SUPERFAMILIES

The number of conformations accessible to a polypeptide chain — the protein structure space — is restrained not only by physical factors, such as amino acid and peptide bond geometry, but also by the evolutionary history of the organism. There is good evidence that many new genes are produced in evolution by gene duplication:[33] new genes are modified from extra copies of existing genes.[34] Functional genes may be duplicated accidentally, and both copies of the gene may then be subject to selection. As a consequence, many proteins can easily be classed into families, which are likely to have a common evolutionary ancestor and to have diverged evolutionarily, e.g. the serine proteinases (Fig. 27.2).[35] Indeed, if two sequences that have evolved naturally are greater than 25% identical and there are few insertions and deletions in the alignment between them, we

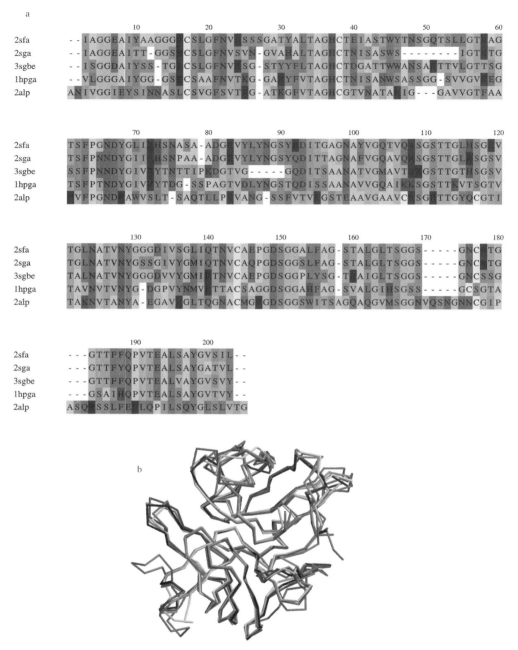

Fig. 27.2 The serine proteases are a typical homologous family demonstrating the clear family relationship in sequence (a) and structure (b).

can be fairly sure that they will have very similar 3D structures.[36]

Selective pressure in evolution is mainly on the functions of the proteins as manifested in the whole organism. This is dependent on conserving the identity of a tiny percentage of amino acids, such as those that constitute catalytic groups or substrate binding sites, and those that maintain a stable 3D structure. The overall fold can be maintained with conservation of very little sequence identity, sometimes only 3–4% of all residues. Thus, as first suggested by Rossmann and Argos,[37] there may well be very little constraint on amino-acid sequences over long periods of time, and proteins much less similar than the serine proteinases may be related by divergent evolution, particularly if pseudogenes, where evolutionary pressures are relaxed, are considered.[38] Such distantly related families of proteins are usually known as protein superfamilies.

Superfamily members will often have sequences that are less than 25% identical — they are in the so-called 'twilight zone'.[39] It is often impossible to be sure that proteins are divergently, rather than convergently, evolved on the basis of sequence alone.[40] Thus, in practice, we define members of superfamilies as those that have similar structures and structurally equivalent functional groups.

V THE PROCESS OF COMPARATIVE MODELLING

A The principle

Knowledge of the general principles of protein structure and of the general features of protein families and superfamilies provides a useful approach to modelling proteins[41] which is generally characterized as comparative modelling.[42-50]

Comparative modelling generates a protein structure from its sequence using information from (an)other homologous structure(s).[51] By extension of the evolutionary principles discussed above, a model for a sequence of unknown structure can be built by extrapolating structural features from the known (basis) homologous structures for sequence stretches that are considered to be similar; regions considered dissimilar must be modelled using other techniques. The process can be elaborated if multiple structures are available as a basis, since different pieces from different basis structures may be locally more similar: combining the information from these different sources may generate a better overall model. Automated and rule-based modelling procedures have been developed in order to minimize subjective manual decisions. In general, comparative modelling techniques are divided into several distinct procedures: finding basis structures; aligning the target sequence to the basis structures; and then finally, either assembling fragments or imposing restraints based on this alignment to generate a 3D model of the target. Some of the stages are discussed only briefly here; more detail can be found in Sternberg[31] and Johnson *et al.*[51]

B Historical perspective on comparative modelling

The use of homologous or other proteins with a common fold as a basis for modelling studies began in the late 1960s and early 1970s. Models were built with wire and plastic.[52] Computer graphics, firstly interactively[53] and then automatically,[54-56] have now taken over. Significant advances came from the realization that information could be usefully derived from more than one homologous structure, particularly in variable regions[57] but also in relatively conserved regions.[54]

The first models were built as exact copies of the backbone coordinates of a single homologous structure, altering only those side-chains which were not identical in the protein to be modelled. However, as sequences become more dissimilar (i.e. $<30\%$ sequence identity), deletions and insertions between the target sequence and basis structure caused significant problems. Browne *et al.*[52] were the first to publish a report of comparative modelling. The bovine α-lactalbumin sequence was modelled on the 3D structure of hen egg white lysozyme. The sequence identity of 39% without insertions and a conserved pattern of disulphides allowed the modelling to proceed. This was repeated by Warme *et al.*[58] Once the X-ray structure was available, Acharya *et al.*[59] compared both models with it and reported that they were generally correct except for the carboxy terminal regions where they differed both from one another and the true structure. After this auspicious start, many modelling studies were carried out such as the early studies on α-lytic protease,[59] insulin-related structures,[60-62] serine proteases[57] and renin.[63-67] Since the mid-1980s a large number of other protein models have been constructed and reported in the literature; with the ever-increasing automation of the software the number of modelled protein structures and their uses have escalated rapidly.

C Identifying structural templates

The first step in comparative modelling is the identification of the known 3D structure(s) which can act as the basis structure(s) for the target sequence; this can be done from structural knowledge, similarity of function, gene grouping, sequence similarity and evolutionary relationship.

This may be trivial if sequence identity between the target sequence and a known 3D structure is high (say $>30\%$), as then simple pair-wise sequence comparison methods (FASTA,[68] SSEARCH[68]) will easily identify the relationship.[69,70] Where sequence identity is lower and a superfamily or even fold relationship must be identified, recognition of the structural similarity between two sequences may be very difficult. Sequence-only methods, such as PSI-BLAST,[71] hidden Markov models[72,73] and intermediate sequence search,[74] use information from multiple sequence alignments to represent the characteristics shared by related sequences (sequence profiles), and use this to search for structural homologues. These profiles can then be augmented by secondary structure prediction.[75,76]

On the other hand, there are methods that make use of 3D information instead to find structural templates for sequences. There are many ways to use 3D information to aid prediction. 'Threading' methods[77] exploit the difference in the distribution of inter-residue distances in protein structures for different pairs of amino acids; this is central to several successful fold recognition techniques.[78-80] Other methods do not rely on such potentials but on profiles generated from multiple sequence alignments, scored using propensities or substitution tables.[51,81-85] It should be noted that whereas threading can genuinely be termed 'fold recognition', evolutionary-relationship techniques by definition operate at the 'family' or 'superfamily' rather than 'fold' level. Proteins in the same family or superfamily are homologous, i.e. derived from a common ancestor.

D Aligning target to basis structures

Once the basis set has been identified, the next task is alignment of the target sequence to these known structures or structure; this lies at the heart of comparative modelling and is still the most frequent and serious source of errors.[42,43,49] While the correct sequence alignment is trivial if the percentage identity between compared sequences is high (say 35%), it becomes extremely difficult when sequence identity is low (25% or below).

All algorithms assess sequence alignments using some scheme that scores the pairing of all aligned residues. In general, these schemes contain scores for the 210 possible pairs of amino acids, stored in a 20 by 20 matrix of similarities, where the alignment of identical residues (e.g. Ile versus Ile) and those of a similar character (e.g. Ile versus Leu) is given a higher score than dissimilar pairings (e.g. Ile versus Asp). Many different scoring schemes have been devised including identity scoring,[86] observed substitution scoring,[86] genetic code[87] and chemical similarity scoring.[88] The general consensus is[89] that matrices derived from observed substitutions (e.g. Dayhoff[86] or BLOSSUM[90]) are superior to identity, genetic code or physical property matrices. However, as yet the reliability of sequence alignments is not very high in cases of lower sequence identity and depends strongly upon parameters that influence the alignment score.[91] Shi *et al.*[84] have shown that the use of environment-specific substitution tables is an effective way of incorporating knowledge of the 3D structure of the basis structures into the process of alignment of their sequences with the sequence to be modelled.

The problem of generating the optimal alignment (i.e. which has the best score) may be divided into methods for two sequences, methods for multiple sequences, and methods that incorporate additional nonsequence information (for example from the tertiary structure of the protein). Dynamic programming algorithms, one of the most common alignment methods, were first developed by Needleman and Wunsch.[92] They operate on pair-wise alignments and identify the optimal alignment for two sequences under a given scoring scheme, including penalties for gaps and extensions of gaps between the sequences; the method can also be optimized for local rather than global alignment.[93] Multiple sequence alignment is a more difficult problem and can be overcome by extensions of pair-wise dynamic programming algorithms (the first of these as a practical approach was by Sankoff *et al.*[94]); hierarchical extensions of pair-wise methods (e.g. Sander and Schneider[36]); segment methods (e.g. Bacon and Anderson[95]); or consensus or 'regions' methods. This alignment must then be used as a basis for implying the structural conformation of the target from the basis structures.

E Assembly of rigid fragments

The strategy of modelling by rigid fragment assembly is based on the demonstration by Jones and Thirup,[96] that most protein structures can be built from a combination of segments from other proteins, which was also supported by subsequent studies.[17,97,98] Thus, the backbone is pieced together from fragments or weighted average fragments of the basis structures, or from structure fragments of other proteins where necessary. Side-chains are then added to this backbone.

In terms of the alignment to the basis structures, the backbone fragments can be divided into two types: structurally conserved regions (SCRs) and structurally variable regions (SVRs),[57] illustrated in Plate 16 (see Plate section).

The stages of fragment assembly are (Plate 16):

- *Structural alignment of the basis structures:* There are many algorithms for performing structural superposition (e.g. Sali and Blundell[47]). This structure-based alignment is vital to elucidate conserved structural and sequence features.
- *Alignment of the target sequence on to the basis structures:* The creation of a correct alignment is still considered the most challenging stage of comparative modelling. As described above, there are many different scoring schemes, the most successful of which use sequence and structure information.[84] However, at low sequence identity the optimal alignment can often only be achieved with manual intervention.
- *Building those regions of the backbone thought to be similar to one or more of the basis structures:* There are three methods for building the 'core' of the model. The conserved secondary structure elements in the basis set;[99] fully aligned equivalent residues (these may or may not be in secondary structure) in the basis set;[55] the selection of sections from the basis structures based on an estimation of the likelihood of similarity of local environments for individual amino acids.[100]

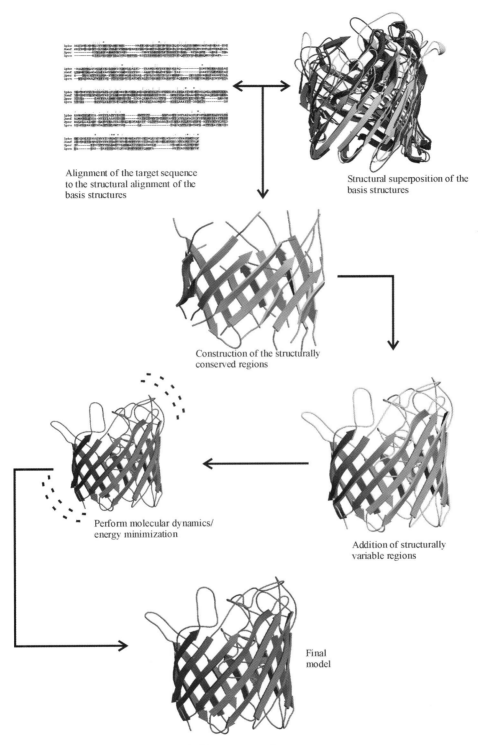

Alignment of the target sequence
to the structural alignment of the
basis structures

Structural superposition of the
basis structures

Construction of the structurally
conserved regions

Addition of structurally
variable regions

Perform molecular dynamics/
energy minimization

Final
model

Plate 16 A summary of the fragment-based modelling approach.

- *Building those regions of the backbone not similar to any parent structure:* Those regions not taken from the basis structures are the most inaccurate parts of any comparative model.[42,43,49] There have been two main types of

methods for the prediction of these regions of proteins, those that adopt a knowledge-based approach (use of fragments from protein structures) (e.g. Jones and Thirup,[96] Rufino *et al.*,[101] Deane and Blundell[102]) and those that use an *ab initio* or conformation searching methods (e.g. Deane and Blundell,[103] Bruccoleri and Karplus,[104] Zhang *et al.*[105]). These methods can be combined to give a consensus prediction.[102]

- *Correctly orientating side-chains:* Side-chain orientation is strongly dependent upon backbone conformation and local environmental features. Thus, the accuracy of side-chain placement in the model will depend upon the accuracy of the backbone. In general, in comparative modelling the most accurate placement of side-chains is achieved with the use of rotamer libraries (e.g. Lovell *et al.*,[23] Dunbrack and Karplus[106]).

- *Refining the model:* To remove steric clashes or to remove minor inconsistencies in the model, energy minimization with programs such as CHARMM[107] is often carried out. It is not clear that this improves the model in terms of similarity to the true structure.

F Use of spatial restraints

The basic problem with the fragment assembly method is the use of the least-squares superposition, which means that the proteins are being treated as rigid bodies. This may result in only a small number of equivalent positions being used to pinpoint a large part of the model, and information other than the Cα-positions in known structures is often neglected. Thus, another technique was developed to allow a more flexible representation of protein structure, both for comparison and modelling purposes.

Analysis of protein structure has led to the discovery of many rules and filters, all of which should be included in a protein structure prediction method, but are largely ignored in fragment assembly methods. Spatial restraint methods[56,108–113] allow a more flexible approach to protein structure prediction and naturally include much of the known information that can be used to restrain a model.

In these methods, the most probable structure for a target sequence is once again found by using its alignment with related structures, but the 3D model is obtained by optimally satisfying spatial restraints derived from/implied by the alignment. All restraint-based approaches to comparative modelling extract distance and/or dihedral restraints for the target sequence from the alignment with related structures; other restraints can then be implied by the covalent topology (stereochemical restraints). The model is then calculated by minimizing the violations of all restraints (Fig. 27.3). This introduces an additional advantage: the ease with which constraints or restraints can be added from other sources, both theoretical and experimental. Theoretical restraints include rules of secondary structure packing,[114,115] hydrophobicity,[116]

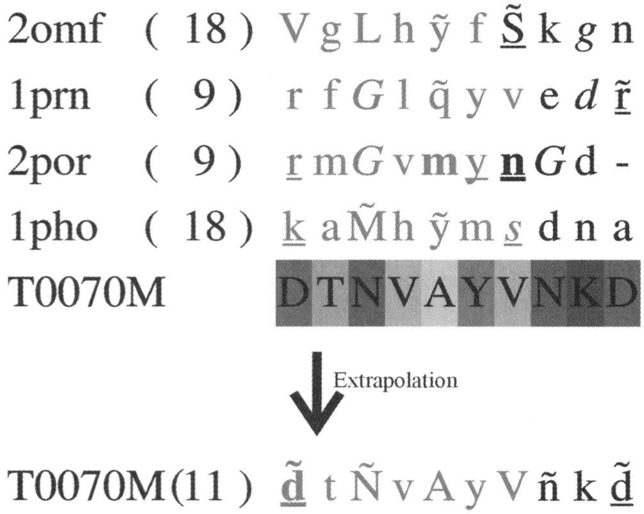

2omf	(18)	V g L h ỹ f S̲̃ k *g* n
1prn	(9)	r f G l q̃ y v e *d r̲̃*
2por	(9)	r̲ m *G* v m y **n** *G* d -
1pho	(18)	k̲ a M̃ h ỹ m s̲ d n a
T0070M		DTNVAYVNKD

Extrapolation

T0070M(11) d̲̃ t Ñ v A y V ñ k d̲̃

Key to JOY

solvent inaccessible	UPPER CASE	X
solvent accessible	lower case	x
α-helix	red	x
β-strand	blue	x
3₁₀-helix	maroon	x
hydrogen bond to main chain amide	**bold**	**x**
hydrogen bond to main chain carbonyl	underline	x̲
hydrogen bond to other sidechain	tilde	x̃
disulphide bond	cedilla	ç
positive φ	*italic*	*x*
cis-peptide	breve	x̆

Fig. 27.3 Comparative modelling by satisfaction of spatial restraints. The starting point is the alignment of basis structure (2mf, 1prn, 2por and 1pho) with the target sequence (T0070M). Various spatial features of the basis structures such as hydrogen bonds, main chain side-chain dihedral angles are indicated by different font types as show in the key. Spatial features are transferred from the basis structures to the target. Thus, a number of restraints on its structure are obtained, for example, the β-strand character of the first residue in all the structures in this subsection of the alignment. This provides a spatial restraint on the target sequence. The 3D model is obtained by satisfying as well as possible all the restraints.

correlated mutations[117] and empirical potentials of mean force;[77] while experimental restraints may come from NMR,[118] crosslinking experiments, fluorescence spectroscopy, electron microscopy, or site-directed mutagenesis.[119] This means that comparative modelling could be improved consistent with experimental data and/or knowledge of protein structure.[47]

G Validation and accuracy

Model building allows the examination of the proposed 3D structure, to determine whether or not it explains the available data. Models should be treated as hypotheses which may be accepted, rejected or modified if they are inconsistent with the sequence or other experimental data. This leads to an iterative approach to modelling such that any model should, once built, be examined in the light of all available experimental data and model validation techniques.

Models can be validated using software such as PROCHECK,[120] VERIFY3D,[121] PROSA II,[122] visual inspection using 3D graphics software. The structural realignment of the model to its basis structures and the examination of conservation of both residues and structural environments using software such as JOY[123] can also be used to authenticate the model.

It is important to have objective tests of comparative modelling in general. The blind tests CASP (critical assessment of methods of protein structure prediction) (http://predictioncenter.llnl.gov/)[124–126] and CAFASP (critical assessment of fully automated methods of protein structure prediction) (http://www.cs.bgu.ac.il/ ~ dfischer/ CAFASP2/)[127] are very useful in this respect. These are community-wide experiments to assess methods of protein structure prediction. The first assesses the ability of the protein modelling software with expert manual intervention to predict structure. The second concentrates on the performance of automatic prediction methods available through web servers. All the CASP experiments have highlighted that the major problems for comparative modelling are alignment and the bridging of insertions and deletions.[42,43]

VI EXAMPLES

We will briefly illustrate the modelling process following the general strategic scheme detailed above: (i) choice of the target; (ii) sequence analysis; (iii) 3D modelling; and (iv) experimental validation. Readers are referred to the original publications for more details. Many drug targets have been modelled using the comparative methods described here. We illustrate the methodology by describing two related aspartic acid systems: renin, a target for antihypertensives, provides an example of comparative modelling based on several closely related homologues; and human immunodeficiency virus (HIV) proteinase, a target for AIDS antivirals, is an example of the more challenging task of constructing a model of a superfamily member with little sequence similarity.

A Renin, a target for antihypertensives

The target

In the early 1980s interest in human renin as a target for antihypertensives resulted from the success of captopril and enalopril, inhibitors of angiotensin-converting enzyme (ACE). Renin cleaves angiotensinogen to give angiotensin I from which two carboxy-terminal amino acids are then cleaved by ACE to give angiotensin II, responsible for vasoconstriction and increase of water and salt leading to high blood pressure. Inhibition of the production of angiotensin II is effective in the control of blood pressure. Renin is more specific than ACE, and this was thought originally to be a potential advantage of renin inhibitors as a useful antihypertensive (see Whittle and Blundell[128] for review).

Sequence analysis

In the late 1970s renin was recognized to be a member of the aspartic proteinase family on the basis of its sensitivity to pepstatin-type inhibitors. Aspartic proteinases include mammalian pepsin and chymosin, as well as a number of exocellular fungal enzymes which had been extensively studied because of their interest as renins for cheese making. The sequencing of mouse and then human renin confirmed the classification as an aspartic proteinase. These enzymes have around 320 amino acids divided into two evolutionarily related lobes, each containing the amino acid motif Asp-Thr-Gly, which provide the two catalytic aspartates at the active site. Although pepsin would have provided the closest basis structure for modelling, the first X-ray analysis did not provide reliable coordinates and so the first models were constructed using the structure of the fungal homologue, endothipepsin.[63,129] When reliable high-resolution models became available for pepsin and chymosin, the structures were remodelled.[130]

Modelling

The first models of rennin[63,129] were constructed from endothiapepsin (Fig. 27.4) using the interactive graphics program FRODO[131] implemented on an Evans and Sutherland work station. As the sequence identity of fungal enzymes with renin was around 30%, few of the loops could be taken from homologues, as recommended by Greer.[57] Most of the modelling was therefore achieved by using the principles of protein structure and keeping conformation and space occupation close to that of the basis structures. In later models, for example see reference 130, fragment assembly methods were exploited with the program COMPOSER.[41,55] Candidate fragments for variable regions were selected on geometric criteria from a database derived from real structures and the fragments were ranked using propensity tables.[132] Substrate analogues and inhibitors were modelled into the renin structure using experimental

a

b

Fig. 27.4 Structures of the aspartic proteinases super-family: (a) HIV proteinases; (b) rennin.

data for the 3D structure of other aspartic proteinases, for example see reference 133.

Experimental validation

Models of renin were used in drug discovery during the 1980s until the 3D structures of uncomplexed[134] and complexed[135,136] were experimentally defined using X-ray analysis. This allowed a restrospective analysis of their accuracy. In general, and as expected, models constructed with mammalian enzymes were more accurate than those with fungal enzymes, reflecting their sequence similarity. Significant hinge-bending movements of the *C*-domain[137] led to unexpected differences in structure, especially for pepsin inhibitor complexes, which were not observed in renin inhibitor complexes. However, these had relatively small effects on the active site region and the specificity subsites that were rather well modelled, even in those derived from endothiapepsin.[130]

B HIV protease

The target

In 1985 Toh *et al.*[138] recognized a sequence in the Rous sarcoma virus (RSV) genome that resembled that of the Asp-Thr-Gly motif of aspartic proteinases, and a similar sequence was subsequently identified in HIV.[139] This led to the suggestion that retroviruses contained a dimer which resembled the putative dimeric ancestor of the pepsin-like aspartic proteinases.[140] The analogy, supported by observations that an HIV proteinase was sensistive to pepstatin, led to interest in the development of HIV proteinase inhibitors.[141] It was later noted that HIV requires a specific protease for the maturation of its components and therefore inhibition of this protease could represent a therapeutic approach to the treatment of AIDs.

Sequence analysis

Although the sequence similarity between the HIV protease and each of the two symmetrical domains of several cellular aspartic proteinases was low, several residues, in addition to the catalytic triad of Asp-Thr/Ser-Gly, critical to the structure and conserved in aspartic proteinase domains and retroviral proteinases, were identified.[142–144] There were many uncertainties in the alignment arising from the fact that HIV proteinase had a sequence of only 99 amino acid residues compared to about 160 for a lobe of the pepsin-like aspartic proteinases, but the majority of the deletions could be accommodated in surface regions.

3D modelling

The first 3D models of the HIV protease by[142–144] Pearl and Taylor used one of the two symmetrical domains of endothiapepsin[145] (Fig. 27.4) as their basis structure. The structure of the aspartic protease from the RSV was then solved.[146] This provided a nearer structural homologue. This protease was a dimer with an active site very similar to monomeric cellular proteases. A second comparative model of the HIV protease was built with this new basis structure,[146] enabling the active site to be described in detail and decreasing the uncertainty concerning the modelling of the long insertions and deletions which were made in constructing the models on the basis of pepsin-like aspartic proteinases.

Experimental validation

The proposed models were consistent with site-directed mutagenesis results and later verified by the crystal structure[147,148] (Fig. 27.4). In particular, the location of the active site and substrate binding residues were accurately predicted.[147] This story demonstrates that modelling of unknown structures based on related proteins can give accurate results. In this case the two original models provided sufficient insight into the structure and function of HIV protease to be immediately useful for the rational design of potential inhibitors of this enzyme. Useful molecules are now exploited as cocktails in the treatment of AIDS with encouraging results, although it is evident that mutation in HIV allows the virus to escape quickly if challenged with a single antiviral agent.

VII GENERAL CONCLUSIONS

Structure-based approaches will undoubtedly be important in the design of new proteins, drugs and vaccines. Structural determination of proteins remains difficult so comparative modelling represents a valuable alternative for the generation of 3D structure. However, scientists must remain aware of the limitations of these methods and the models that they generate if they are to be of significance.

REFERENCES

1. Berman, H.M., Westbrook, J., Feng, Z., Gilliland, G., Bhat, T.N., *et al.* (2000) The protein data bank. *Nucleic Acids Res.* **28**: 235–242.
2. Barker, W.C., Garavelli, J.S., Huang, H., McGarvey, P.B., Orcutt, B.C., *et al.* (2000) The protein information resource (PIR). *Nucleic Acids Res.* **28**: 41–44.
3. Pearl, F.M., Lee, D., Bray, J.E., Sillitoe, I., Todd, A.E., *et al.* (2000) Assigning genomic sequences to CATH. *Nucleic Acids Res.* **28**: 277–282.
4. Moult, J. (1999) Predicting protein three-dimensional structure. *Curr. Opin. Biotechnol.* **10**: 583–588.
5. Pal, D. and Chakrabarti, P. (1999) *Cis* peptide bonds in proteins: residues involved, their conformations, interactions and locations. *J. Mol. Biol.* **294**: 271–288.
6. Wilmot, C.M. and Thornton, J.M. (1990) Beta-turns and their distortions: a proposed new nomenclature. *Protein Eng.* **3**: 479–493.
7. Milner-White, E.J. (1990) Situations of gamma-turns in proteins. Their relation to alpha-helices, beta-sheets and ligand binding sites. *J. Mol. Biol.* **216**: 386–397.
8. Sibanda, B.L. and Thornton, J.M. (1985) Beta-hairpin families in globular proteins. *Nature* **316**: 170–174.
9. Li, W.Z., Liang, S., Wang, R.X., Lai, L.H. and Han, Y.Z. (1999) Exploring the conformational diversity of loops on conserved frameworks. *Protein Eng.* **12**: 1075–1086.
10. Kwasigroch, J.-M., Chomilier, J. and Mornon, J.-P. (1996) A global taxonomy of loops in globular proteins. *J. Mol. Biol.* **259**: 855–872.
11. Donate, L.E., Rufino, S.D., Canard, L.H.J. and Blundell, T.L. (1996) Conformational analysis and clustering of short and medium size loops connecting regular secondary structures: a database for modeling and prediction. *Protein Sci.* **5**: 2600–2616.
12. Oliva, B., Bates, P.A., Querol, E., Aviles, F.X. and Sternberg, M.J. (1997) An automated classification of the structure of protein loops. *J. Mol. Biol.* **266**: 814–830.
13. Burke, D.F., Deane, C.M. and Blundell, T.L. (2000) Browsing the SLoop database of structurally classified loops connecting elements of protein secondary structure. *Bioinfomatics* **16**: 513–519.
14. Ring, C.S., Kneller, D.G., Langridge, R. and Cohen, F.E. (1992) Taxonomy and conformational analysis of loops in proteins [published erratum appears in *J. Mol. Biol.* 1992 Oct 5; 227(3):977]. *J. Mol. Biol.* **224**: 685–699.
15. Tuffery, P., Etchebest, C. and Hazout, S. (1997) Prediction of protein side chain conformations: a study on the influence of backbone accuracy on conformation stability in the rotamer space. *Protein Eng.* **10**: 361–372.
16. Janin, J. and Wodak, S. (1978) Conformation of amino acid side-chains in proteins. *J. Mol. Biol.* **125**: 357–386.
17. Levitt, M. (1992) Accurate modeling of protein conformation by automatic segment matching. *J. Mol. Biol.* **226**: 507–533.
18. Moult, J. and James, M.N.G. (1986) An algorithm for determining the conformation of polypeptide segments in proteins by systematic search. *Proteins* **1**: 146–163.
19. Bower, M.J., Cohen, F.E. and Dunbrack, R.L., Jr (1997) Prediction of protein side-chain rotamers from a backbone-dependent rotamer library: a new homology modeling tool. *J. Mol. Biol.* **267**: 1268–1282.
20. De Maeyer, M., Desmet, J. and Lasters, I. (1997) All in one: a highly detailed rotamer library improves both accuracy and speed in the modelling of sidechains by dead-end elimination. *Fold Des.* **2**: 53–66.
21. Ponder, J.W. and Richards, F.M. (1987) Tertiary templates for proteins. Use of packing criteria in the enumeration of allowed sequences for different structural classes. *J. Mol. Biol.* **193**: 775–791.
22. Schrauber, H., Eisenhaber, F. and Argos, P. (1993) Rotamers: to be or not to be? An analysis of amino acid side-chain conformations in globular proteins. *J. Mol. Biol.* **230**: 592–612.
23. Lovell, S.C., Word, J.M., Richardson, J.S. and Richardson, D.C. (2000) The penultimate rotamer library. *Proteins* **40**: 389–408.
24. Branden, C. and Tooze, J. (1991) *Introduction to Protein Structure*, 2nd edn. Garland Publishing, New York.
25. Karplus, M. and Shakhnovich, E. (1992) Protein folding: theoretical studies of thermodynamics and dynamics. In Creighton, T. (ed.). *Protein Folding*, pp. 127–195. W.H. Freeman & Sons, New York.
26. Levinthal, C. (1969) How to fold graciously. In Debrunner, P., Tsibris, J.C.M. and Munck, E. (eds. *Mossbauer Spectroscopy in Biological Systems, Proceedings of a meeting held at Allerton House, Monticello, Illinois*, pp. 22. University of Illinois Press, Urbana.
27. Bryngelson, J.D. and Wolynes, P.G. (1989) Intermediates and barrier crossing in a random energy-model (with applications to protein folding). *J. Phys. Chem.* **93**: 6902–6915.
28. Karplus, M. (1997) The Levinthal paradox: yesterday and today. *Fold Des.* **2**: S69–S75.
29. Schmid, F.X. (1992) Kinetics of folding and refolding of single domain proteins. In Creighton, T. (ed.). *Protein Folding*, pp. 197–241. W.H. Freeman & Sons, New York.
30. Harrison, S.C. and Durbin, R. (1985) Is there a single pathway for the folding of a polypeptide chain? *Proc. Natl Acad. Sci. USA* **82**: 4028–4030.
31. Sternberg, M.J.E. (1996) *Protein Structure Prediction*. Oxford University Press, Oxford.
32. Seckler, R. and Jaenicke, R. (1992) Protein folding and protein refolding. *FASEB J.* **6**: 2545–2552.
33. Ohno, S. (1970) *Evolution by Gene Duplication*. Springer-Verlag, New York.
34. Teichmann, S.A., Park, J. and Chothia, C. (1998) Structural assignments to the Mycoplasma genitalium proteins show extensive gene duplications and domain rearrangements. *Proc. Natl Acad. Sci. USA* **95**: 14658–14663.
35. Chothia, C. and Lesk, A.M. (1986) The relation between the divergence of sequence and structure in proteins. *EMBO J.* **5**: 823–826.
36. Sander, C. and Schneider, R. (1991) Database of homology-derived protein structures and the structural meaning of sequence alignment. *Proteins* **9**: 56–68.
37. Rossmann, M.G. and Argos, P. (1975) A comparison of the heme binding pocket in globins and cytochrome b5. *J. Biol. Chem.* **250**: 7525–7532.

38. Gibson, L.J. (1994) Pseudogenes and Origins. *Origins* **21**: 91–108.

39. Doolittle, R.F. (1981) Similar amino acid sequences: chance or common ancestry? *Science* **214**: 149–159.

40. Bajaj, M. and Blundell, T. (1984) Evolution and the tertiary structure of proteins. *Ann. Rev. Biophys. Bioeng.* **13**: 453–492.

41. Blundell, T.L., Sibanda, B.L., Sternberg, M.J. and Thornton, J.M. (1987) Knowledge-based prediction of protein structures and the design of novel molecules. *Nature* **326**: 347–352.

42. Jones, T.A. and Kleywegt, G.J. (1999) CASP3 comparative modeling evaluation. *Proteins* **37**:(Suppl. 3): 30–46.

43. Martin, A.C.R., MacArthur, M.W. and Thornton, J.M. (1997) Assessment of comparative modeling in CASP2. *Proteins* (Suppl. 1): 14–28.

44. Mosimann, S., Meleshko, R. and James, M.N. (1995) A critical assessment of comparative molecular modeling of tertiary structures of proteins. *Proteins* **23**: 301–317.

45. Rost, B., Schneider, R. and Sander, C. (1993) Progress in protein structure prediction? *Trends Biochem. Sci.* **18**: 120–123.

46. Sali, A., Potterton, L., Yuan, F., Vanvlijmen, H. and Karplus, M. (1995) Evaluation of comparative protein modeling by modeler. *Proteins* **23**: 318–326.

47. Sali, A. and Blundell, T.L. (1990) Definition of general topological equivalence in protein structures. A procedure involving comparison of properties and relationships through simulated annealing and dynamic programming. *J. Mol. Biol.* **212**: 403–428.

48. Sanchez, R. and Sali, A. (1997) Advances in comparative protein-structure modelling. *Curr. Opin. Struct. Biol.* **7**: 206–214.

49. Srinivasan, N. and Blundell, T.L. (1993) An evaluation of the performance of an automated procedure for comparative modelling of protein tertiary structure. *Protein Eng.* **6**: 501–512.

50. Sternberg, M.J., Bates, P.A., Kelley, L.A. and MacCallum, R.M. (1999) Progress in protein structure prediction: assessment of CASP3. *Curr. Opin. Struct. Biol.* **9**: 368–373.

51. Johnson, M.S., Srinivasan, N., Sowdhamini, R. and Blundell, T.L. (1994) Knowledge-based protein modeling. *Crit. Rev. Biochem. Mol. Biol.* **29**: 1–68.

52. Browne, W.J., North, A.C.T., Phillips, D.C., Brew, K., Vanaman, T.C., *et al.* (1996) A possible three dimensional structure of alpha-lactalbumin based on that of Hens egg-white lysozyme. *J. Mol. Biol.* **42**: 65–86.

53. Isaacs, N., James, R., Niall, H., Bryant-Greenwood, G., Dodson, G., *et al.* (1978) Relaxin and its structural relationship to insulin. *Nature* **271**: 278–281.

54. Sutcliffe, M.J., Hayes, F.R.F. and Blundell, T.L. (1987) Knowledge based modelling of homologous proteins. Part II: rules for the conformations of substituted sidechains. *Protein Eng.* **1**: 385–392.

55. Sutcliffe, M.J., Haneef, I., Carney, D. and Blundell, T.L. (1987) Knowledge based modelling of homologous proteins. Part I: three-dimensional frameworks derived from the simultaneous superposition of multiple structures. *Protein Eng.* **1**: 377–384.

56. Sali, A. and Blundell, T.L. (1993) Comparative protein modelling by satisfaction of spatial restraints. *J. Mol. Biol.* **234**: 779–815.

57. Greer, J. (1980) Model for haptoglobin heavy chain based upon structural homology. *Proc. Natl Acad. Sci. USA* **77**: 3393–3397.

58. Warme, P.K., Momany, F.A., Rumball, S.V., Tuttle, R.W. and Scheraga, H.A. (1974) Computation of structures of homologous proteins. Alpha-lactalbumin from lysozyme. *Biochemistry* **13**: 768–782.

59. Acharya, K.R., Stuart, D.I., Walker, N.P., Lewis, M. and Phillips, D.C. (1989) Refined structure of baboon alpha-lactalbumin at 1.7Å resolution. Comparison with C-type lysozyme. *J. Mol. Biol.* **208**: 99–127.

60. Blundell, T.L., Bedarkar, S., Rinderknecht, E. and Humbel, R.E. (1978) Insulin-like growth factor: a model for tertiary structure accounting for immunoreactivity and receptor binding. *Proc. Natl Acad. Sci. USA* **75**: 180–184.

61. Bedarkar, S., Turnell, W.G., Blundell, T.L. and Schwabe, C. (1977) Relaxin has conformational homology with insulin. *Nature* **270**: 449–451.

62. Eigenbrot, C., Randal, M., Quan, C., Burnier, J., O'Connell, L., *et al.* (1991) X-Ray structure of human relaxin at 1.5Å. Comparison to insulin and implications for receptor binding determinants. *J. Mol. Biol.* **221**: 15–21.

63. Blundell, T., Sibanda, B.L. and Pearl, L. (1983) Three-dimensional structure, specificity and catalytic mechanism of renin. *Nature* **304**: 273–275.

64. Sibanda, B.L., Blundell, T., Hobart, P.M., Fogliano, M., Bindra, J.S., *et al.* (1984) Computer graphics modelling of human renin. Specificity, catalytic activity and intron–exon junctions. *FEBS Lett.* **174**: 102–111.

65. Carlson, W., Karplus, M. and Haber, E. (1985) Construction of a model for the three-dimensional structure of human renal renin. *Hypertension* **7**: 13–26.

66. Akahane, K., Umeyama, H., Nakagawa, S., Moriguchi, I., Hirose, S., *et al.* (1985) Three-dimensional structure of human renin. *Hypertension* **7**: 3–12.

67. Hutchins, C. and Greer, J. (1991) Comparative modeling of proteins in the design of novel renin inhibitors. *Crit. Rev. Biochem. Mol. Biol.* **26**: 77–127.

68. Pearson, W.R. and Lipman, D.J. (1988) Improved tools for biological sequence comparison. *Proc. Natl Acad. Sci. USA* **85**: 2444–2448.

69. Park, J., Karplus, K., Barrett, C., Hughey, R., Haussler, D. *et al.* (1998) Sequence comparisons using multiple sequences detect three times as many remote homologues as pairwise methods. *J. Mol. Biol.* **284**: 1201–1210.

70. Brenner, S.E., Chothia, C. and Hubbard, T.J. (1998) Assessing sequence comparison methods with reliable structurally identified distant evolutionary relationships. *Proc. Natl Acad. Sci. USA* **95**: 6073–6078.

71. Altschul, S.F., Madden, T.L., Schaffer, A.A., Zhang, J.H., Zhang, Z., *et al.* (1997) Gapped BLAST and PSI-BLAST: a new generation of protein database search programs. *Nucleic Acids Res.* **25**: 3389–3402.

72. Karplus, K., Sjolander, K., Barrett, C., Cline, M., Haussler, D., *et al.* (1997) Predicting protein structure using hidden Markov models. *Proteins* (Suppl): 134–139.

73. Eddy, S.R. (1998) Profile hidden Markov models. *Bioinformatics* **14**: 755–763.

74. Park, J., Teichmann, S.A., Hubbard, T. and Chothia, C. (1997) Intermediate sequences increase the detection of homology between sequences. *J. Mol. Biol.* **273**: 349–354.

75. Jones, D.T. (1999) GenTHREADER: an efficient and reliable protein fold recognition method for genomic sequences. *J. Mol. Biol.* **287**: 797–815.

76. Koretke, K.K., Russell, R.B., Copley, R.R. and Lupas, A.N. (1999) Fold recognition using sequence and secondary structure information. *Proteins* **37**: 141–148.

77. Sippl, M.J. (1990) Calculation of conformational ensembles from potentials of mean force. An approach to the knowledge-based prediction of local structures in globular proteins. *J. Mol. Biol.* **213**: 859–883.

78. Panchenko, A., Marchler-Bauer, A. and Bryant, S.H. (1999) Threading with explicit models for evolutionary conservation of structure and sequence. *Proteins* **37**: 133–140.

79. Domingues, F.S., Koppensteiner, W.A., Jaritz, M., Prlic, A., Weichenberger, C. *et al.* (1999) Sustained performance of knowledge-based potentials in fold recognition. *Proteins* **37**: 112–120.

80. Jones, D.T., Taylor, W.R. and Thornton, J.M. (1992) A new approach to protein fold recognition. *Nature* **358**: 86–89.

81. Bowie, J.U., Luthy, R. and Eisenberg, D. (1991) A method to identify protein sequences that fold into a known three-dimensional structure. *Science* **253**: 164–170.

82. Fischer, D. and Eisenberg, D. (1996) Protein fold recognition using sequence-derived predictions. *Protein Sci.* **5**: 947–955.

83. Fischer, D., Rice, D., Bowie, J.U. and Eisenberg, D. (1996) Assigning amino acid sequences to 3-dimensional protein folds. *FASEB J.* **10**: 126–136.

84. Shi, J., Blundell, T.L. and Mizuguchi, K. (2001) FUGUE: sequence-structure homology recognition using environment-specific substitution tables and structure-dependent gap penalties. *J. Mol. Biol.* **310**: 243–257.

85. Kelley, L.A., MacCallum, R.M. and Sternberg, M.J. (2000) Enhanced genome annotation using structural profiles in the program 3D-PSSM. *J. Mol. Biol.* **299**: 499–520.

86. Dayhoff, M.O., Schwartz, R.M. and Orcutt, B.C. (1978) *Atlas of Protein sequence and Structure*, pp. 345–358. National Biomedical Research Foundation, Washington DC.

87. Fitch, W.M. (1966) An improved method of testing for evolutionary homology. *J. Mol. Biol.* **16**: 9–16.

88. Macarthur, M.W., Laskowski, R.A. and Thornton, J.M. (1994) Knowledge-based validation of protein-structure coordinates derived by X-ray crystallography and NMR-spectroscopy. *Curr. Opin. Struct. Biol.* **4**: 731–737.

89. Barton, G.J. (1996) Protein sequence alignment and database scanning. In Sternberg, M.J. (ed.). *Protein Structure Prediction*, pp. 31–63. Oxford University Press, Oxford.

90. Henikoff, S. and Henikoff, J.G. (1992) Amino-acid substitution matrices from protein blocks. *Proc. Natl Acad. Sci. USA* **89**: 10915–10919.

91. Vingron, M. and Waterman, M.S. (1994) Sequence alignment and penalty choice. Review of concepts, case studies and implications. *J. Mol. Biol.* **235**: 1–12.

92. Needleman, S.B. and Wunsch, C.D. (1970) A general method applicable to the search for similarities in the amino-acid sequence of two proteins. *J. Mol. Biol.* **48**: 443–453.

93. Smith, T.F. and Waterman, M.S. (1981) Identification of common molecular subsequences. *J. Mol. Biol.* **147**: 195–197.

94. Sankoff, D., Cedergren, R.J. and Lapalme, G. (1976) Frequency of insertion–deletion, transversion, and transition in the evolution of 5S ribosomal RNA. *J. Mol. Evol.* **7**: 133–149.

95. Bacon, D.J. and Anderson, W.F. (1986) Multiple sequence alignment. *J. Mol. Biol.* **191**: 153–161.

96. Jones, T.A. and Thirup, S. (1986) Using known substructures in protein model building and crystallography. *EMBO J.* **5**: 819–822.

97. Unger, R., Harel, D., Wherland, S. and Sussman, J.L. (1989) A 3D building blocks approach to analyzing and predicting structure of proteins. *Proteins* **5**: 355–373.

98. Claessens, M., Van Cutsem, E., Lasters, I. and Wodak, S. (1989) Modelling the polypeptide backbone with 'spare parts' from known protein structures. *Protein Eng.* **2**: 335–345.

99. Bates, P.A. and Sternberg, M.J. (1999) Model building by comparison at CASP3: using expert knowledge and computer automation. *Proteins* **37**: 47–54.

100. Deane, C.M., Kaas, Q. and Blundell, T.L. (2001) SCORE: predicting the core of protein models. *Bioinformatics* **17**: 541–550.

101. Rufino, S.D., Donate, L.E., Canard, L.H.J. and Blundell, T.L. (1997) Predicting the conformational class of short and medium size loops connecting regular secondary structures: application to comparative modelling. *J. Mol. Biol.* **267**: 352–367.

102. Deane, C.M. and Blundell, T.L. (2001) CODA: a combined algorithm for predicting the structurally variable regions of protein models. *Protein Sci.* **10**: 599–612.

103. Deane, C.M. and Blundell, T.L. (2000) A novel exhaustive search algorithm for predicting the conformation of polypeptide segments in proteins. *Proteins* **40**: 135–144.

104. Bruccoleri, R.E. and Karplus, M. (1987) Prediction of the folding of short polypeptide segments by uniform conformational sampling. *Biopolymers* **26**: 137–168.

105. Zhang, H., Lai, L., Wang, L., Han, Y. and Tang, Y. (1997) A fast and efficient program for modeling protein loops. *Biopolymers* **41**: 61–72.

106. Dunbrack, R.L., Jr and Karplus, M. (1993) Backbone-dependent rotamer library for proteins. Application to side-chain prediction. *J. Mol. Biol.* **230**: 543–574.

107. Brooks, B.R., Bruccoleri, R.E., Olafson, B.D., States, D.J., Swaminathan, S., *et al.* (1983) A program for macromolecular energy minimisation and dynamics calculations. *J. Comp. Chem.* **4**: 187–217.

108. Snow, M.E. (1993) A novel parameterization scheme for energy equations and its use to calculate the structure of protein molecules. *Proteins* **15**: 183–190.

109. Brocklehurst, S.M. and Perham, R.N. (1993) Prediction of the three-dimensional structures of the biotinylated domain from yeast pyruvate carboxylase and of the lipoylated H-protein from the pea leaf glycine cleavage system: a new automated method for the prediction of protein tertiary structure. *Protein Sci.* **2**: 626–639.

110. Havel, T.F. (1993) Predicting the structure of the flavodoxin from *escherichia-coli* by homology modeling, distance geometry and molecular-dynamics. *Mol. Sim.* **10**: 175–210.

111. Abagyan, R., Totrov, M. and Kuznetsov, D. (1994) ICM — a new method for protein modeling and design — applications to docking and structure prediction from the distorted native conformation. *J. Comput. Chem.* **15**: 488–506.

112. Havel, T.F. and Snow, M.E. (1991) A new method for building protein conformations from sequence alignments with homologues of known structure. *J. Mol. Biol.* **217**: 1–7.

113. Srinivasan, S., March, C.J. and Sudarsanam, S. (1993) An automated method for modeling proteins on known templates using distance geometry. *Protein Sci.* **2**: 277–289.

114. Nagarajaram, H.A., Reddy, B.V.B. and Blundell, T.L. (1999) Analysis and prediction of inter-strand packing distances between beta-sheets of globular proteins. *Protein Eng.* **12**: 1055–1062.

115. Reddy, B.V.B., Nagarajaram, H.A. and Blundell, T.L. (1999) Analysis of interactive packing of secondary structural elements in alpha/beta units in proteins. *Protein Sci.* **8**: 573–586.

116. Aszodi, A. and Taylor, W.R. (1994) Secondary structure formation in model polypeptide chains. *Protein Eng.* **7**: 633–644.

117. Taylor, W.R. and Hatrick, K. (1994) Compensating changes in protein multiple sequence alignments. *Protein Eng.* **7**: 341–348.

118. Sutcliffe, M.J., Dobson, C.M. and Oswald, R.E. (1992) Solution structure of neuronal bungarotoxin determined by two-dimensional NMR spectroscopy: calculation of tertiary structure using systematic homologous model building, dynamical simulated annealing, and restrained molecular dynamics. *Biochemistry* **31**: 2962–2970.

119. Boissel, J.P., Lee, W.R., Presnell, S.R., Cohen, F.E. and Bunn, H.F. (1993) Erythropoietin structure–function relationships. Mutant proteins that test a model of tertiary structure. *J. Biol. Chem.* **268**: 15983–15993.

120. Laskowski, R.A., Macarthur, M.W., Moss, D.S. and Thornton, J.M. (1993) PROCHECK — a program to check the stereochemical quality of protein structures. *J. Appl. Crystallog.* **26**: 283–291.

121. Luthy, R., Bowie, J.U. and Eisenberg, D. (1992) Assessment of protein models with 3-dimensional profiles. *Nature* **356**: 83–85.

122. Sippl, M.J. (1993) Recognition of errors in three-dimensional structures of proteins. *Proteins* **17**: 355–362.

123. Mizuguchi, K., Deane, C.M., Blundell, T.L., Johnson, M.S. and Overingon, J.P. (1998) JOY: protein sequence-structure representation and analysis. *Bioinformatics* **14**: 617–623.

124. Moult, J., Pedersen, J.T., Judson, R. and Fidelis, K. (1995) A large-scale experiment to assess protein-structure prediction methods. *Proteins* **23**: R2–R4.

125. Moult, J., Hubbard, T., Fidelis, K. and Pedersen, J.T. (1999) Critical assessment of methods of protein structure prediction (CASP): Round III. *Proteins: Structure, Function and Genetics* 2–6.

126. Moult, J., Hubbard, T., Bryant, S.H., Fidelis, K. and Pedersen, J.T. (1997) Critical assessment of methods of protein structure prediction (CASP): round II. *Proteins* (Suppl): 2–6.

127. Fischer, D., Barret, C., Bryson, K., Elofsson, A., Godzik, A. *et al.* (1999) CAFASP-1: Critical assessment of fully automated structure prediction methods. *Proteins* **37** (Suppl. 3): 209–217.

128. Whittle, P.J. and Blundell, T.L. (1994) Protein structure-based drug design. *Annu. Rev. Biophys. Biomol. Struct.* **23**: 349–375.

129. Sibanda, B.L., Hemmings, A.M. and Blundell, T.L. (1985) Computer graphics modelling and the subsite specifies of human and mouse renins. In Kostka, V. (ed.). *Aspartic Proteinases and their Inhibitors*, pp. 339–349. Walter de Gruyter & Co, Berlin/New York.

130. Frazao, C., Topham, C., Dhanaraj, V. and Blundell, T.L. (1994) Comparative modelling of human renin: a retrospective evaluation of the model with respect to the X-ray crystal structure. *Pure Appl. Chem.* **66**: 43–50.

131. Jones, T.A. (1978) A graphics model building and refinement system for macromolecules. *J. Appl. Cryst.* **11**: 268–272.

132. Topham, C.M., McLeod, A., Eisenmenger, F., Overington, J.P., Johnson, M.S., *et al.* (1993) Fragment ranking in modelling of protein structure. Conformationally constrained environmental amino acid substitution tables. *J. Mol. Biol.* **229**: 194–220.

133. Foundling, S.I., Cooper, J., Watson, F.E., Cleasby, A., Pearl, L.H., *et al.* (1987) High resolution X-ray analyses of renin inhibitor-aspartic proteinase complexes. *Nature* **327**: 349–352.

134. Sielecki, A.R., Hayakawa, K., Fujinaga, M., Murphy, M.E., Fraser, M., *et al.* (1989) Structure of recombinant human renin, a target for cardiovascular-active drugs, at 2.5Å resolution. *Science* **243**: 1346–1351.

135. Rahuel, J., Priestle, J.P. and Grutter, M.G. (1991) The crystal structures of recombinant glycosylated human renin alone and in complex with a transition state analog inhibitor. *J. Struct. Biol.* **107**: 227–236.

136. Dhanaraj, V., Dealwis, C.G., Frazao, C., Badasso, M., Sibanda, B.L., *et al.* (1992) X-ray analyses of peptide–inhibitor complexes define the structural basis of specificity for human and mouse renins. *Nature* **357**: 466–472.

137. Sali, A., Veerapandian, B., Cooper, J.B., Moss, D.S., Hofmann, T., *et al.* (1992) Domain flexibility in aspartic proteinases. *Proteins* **12**: 158–170.

138. Toh, H., Kikuno, R., Hayashida, H., Miyata, T., Kugimiya, W., *et al.* (1985) Close structural resemblance between putative polymerase of a Drosophila transposable genetic element 17.6 and pol gene product of Moloney murine leukaemia virus. *EMBO J.* **4**: 1267–1272.

139. Ratner, L., Haseltine, W., Patarca, R., Livak, K.J., Starcich, B., *et al.* (1985) Complete nucleotide sequence of the AIDS virus, HTLV-III. *Nature* **313**: 277–284.

140. Tang, J. and Wong, R.N. (1987) Evolution in the structure and function of aspartic proteases. *J. Cell Biochem.* **33**: 53–63.

141. Wlodawer, A. and Erickson, J.W. (1993) Structure-based inhibitors of HIV-1 protease. *Annu. Rev. Biochem.* **62**: 543–585.

142. Pearl, L.H. and Taylor, W.R. (1987) A structural model for the retroviral proteases. *Nature* **329**: 351–354.

143. Tang, J., James, M.N., Hsu, I.N., Jenkins, J.A. and Blundell, T.L. (1978) Structural evidence for gene duplication in the evolution of the acid proteases. *Nature* **271**: 618–621.

144. Blundell, T., Carney, D., Gardner, S., Hayes, F., Howlin, B., *et al.* (1988) 18th Sir Hans Krebs lecture. Knowledge-based protein modelling and design. *Eur. J. Biochem.* **172**: 513–520.

145. Pearl, L. and Blundell, T. (1984) The active site of aspartic proteinases. *FEBS Lett.* **174**: 96–101.

146. Weber, I.T., Miller, M., Jaskolski, M., Leis, J., Skalka, A.M., *et al.* (1989) Molecular modeling of the HIV-1 protease and its substrate binding site. *Science* **243**: 928–931.

147. Wlodawer, A., Miller, M., Jaskolski, M., Sathyanarayana, B.K., Baldwin, E., *et al.* (1989) Conserved folding in retroviral proteases: crystal structure of a synthetic HIV-1 protease. *Science* **245**: 616–621.

148. Lapatto, R., Blundell, T., Hemmings, A., Overington, J., Wilderspin, A., *et al.* (1989) X-ray analysis of HIV-1 proteinase at 2.7 A resolution confirms structural homology among retroviral enzymes. *Nature* **342**: 299–302.

28

THE TRANSITION FROM AGONIST TO ANTAGONIST ACTIVITY: SYMMETRY AND OTHER CONSIDERATIONS

David J. Triggle

One side will make you grow taller, and the other side will make you grow shorter. One side of what?, the other side of what?, thought Alice.

Lewis Carroll (1832–1898)[*]

I INTRODUCTION

Ours is an asymmetric world and life is not even-handed. This asymmetry plays out in our understanding and definition of biologically important interactions: these interactions are also asymmetric. Receptors are chiral entities and the interactions of ligands, physiological and pharmacological, at these specific entities exhibit chirality of interaction. Indeed, so common is this phenomenon that its occurrence is frequently used as one component of the definition of receptors, of specific ligand–receptor interactions and in their quantitation. Chirality of ligand–receptor interaction is the rule rather than the exception.[1–5]

---

[*] *Alice's Adventures in Wonderland.*

Louis Pasteur[6,7] recognized quite explicitly in 1860 the critical issue of biological stereochemistry when he demonstrated that the levorotatory and dextrorotatory forms of tartaric acid, separable by differential crystallization, were destroyed at different rates by molds and yeasts: 'There cannot be the slightest doubt that the only and exclusive cause of this difference in the fermentation of the two tartaric acids is caused by the opposite molecular arrangements of the tartaric acids. In this way, the idea of the influence of the molecular asymmetry of natural organic molecules is introduced into physiological studies, this important characteristic being perhaps the only distinct line of demarcation which we can draw today beyond dead and living matter. I have in fact set up a theory of molecular asymmetry, one of the most important and wholly surprising chapters of the science, which opens up a new, distant but definite horizon for physiology'.

Somewhat paradoxically, however, this underlying molecular asymmetry translates to a considerable symmetry of internally and externally expressed biological anatomy. Our visual consideration of this external anatomy translates to an apparent preference for the symmetrical over the asymmetrical as, for example, in the definition and recognition of beauty, possibly because the presentation of such symmetry is an index of biological fitness or quality.[8–10]

The recognition that enantiomeric and diastereomeric drugs frequently differ both quantitatively and qualitatively in their therapeutic and toxicological effects has generated substantial interest at the regulatory level. Surveys of the chiral character of natural/semisynthetic and totally synthetic drugs reveal that, as anticipated, the majority of the former are chiral and available as single enantiomers or isomers.[4] For synthetic drugs, the majority are still available as racemates; however, both the extent of availability and the conversion of racemates to single enantiomers are increasing. With new regulatory demands it is clear that the introduction of single-enantiomer species as new chemical entities will continue to increase. Accordingly, issues of the stereochemistry of drug–receptor interactions and of quantitative and qualitative differences between drug isomers will continue to be of major scientific, clinical and regulatory importance.

The concept of specificity in drug–receptor interactions was quite explicit in early considerations of these interactions. Crum-Brown and Fraser[11] in their pioneering structure–activity studies on alkaloids implicitly defined stereoselectivity as a component of the relationship between biological activity and chemical constitution. Similarly, Paul Ehrlich[12] and John Newton Langley[13] in their original work on receptor definition recognized the importance of specific receptor-mediated interactions defined by, 'a law for which both their (drugs) relative mass and chemical affinity for the substance (receptor) are factors',[13] and the significance of the receptor as a transducer which 'receives the stimulus and, by transmitting it, causes contraction (response)'.[14]

The stereochemical basis of drug–receptor interactions was early investigated by Arthur Cushny[15] at the beginning of the twentieth century. He was able to demonstrate the stereochemistry of action of atropine and related compounds, including the hyoscyines and hyoscyamines. Although the enantiomeric pairs behaved in qualitatively similar fashion, they were quantitatively distinct. Similar differentiation was made between the enantiomers of the catecholamines. This quantitative distinction between the enantiomers of a drug is a common, although not inevitable, accompaniment to stereoselective drug action. A number of reviews of this area are available.[1–5,16–19] Additionally, Cushny revealed a clinical enantiomeric distinction whereby the (−)-enantiomer, but not the (+)-enantiomer, of scopolamine was effective in producing 'twilight sleep'.

Specificity, stereoselectivity and the factoring of ligands and drugs as agonists and antagonists with qualitative and quantitative differences in activity have thus long been recognized as fundamental properties of drug–receptor interactions. Increasingly, it is recognized that ligands need to be considered in terms of a continuum of agonist–partial agonist–antagonist properties and that these properties are, furthermore, not invariant but can be modified, both qualitatively and quantitatively, according to receptor density, receptor expression and the coupling factors involved. Furthermore, a certain symmetry is imposed on ligand–receptor interactions with the existence of agonists, antagonists and neutral antagonists. This concept is well illustrated with the benzodiazepine receptor, where agonist, antagonist and inverse agonists serve as anti-

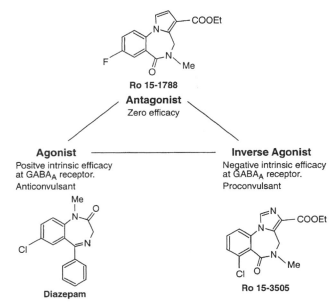

Fig. 28.1 Agonists, antagonists and inverse agonists active at the benzodiazepine receptor.

convulsant, neutral and proconvulsant species, respectively (Fig. 28.1).[20,21] It is likely, in fact, that the majority of 'simple' competitive antagonists are inverse agonists.[22,23]

II RECEPTORS AS CHIRAL ENTITIES

The defined asymmetry of ligand–receptor interactions derives from the chirality of the molecular building blocks of receptors — the L-amino acids — sometimes decorated with D-sugars. The origin of this fundamental chirality remains to be determined although there is no shortage of hypotheses.[24–26] Of particular interest is the proposal by Martin Rees that the very existence of our cosmos depends on the 'fine tuning' of a small number of physical constants, including gravity, the rates of expansion and contraction of the universe, the number of extended dimensions and the strong nuclear force.[27] In particular, a change in the latter would not have permitted our carbon-based biology and a two-dimensional world would have been extraordinarily limiting and a four- or more dimensional world too complex. The limitations of a two-dimensional world have been explicitly defined by Edwin Abbott in *Flatland: A Romance of Many Dimensions*.[28,29] Thus, our carbon-based three-dimensional world stems from a very particular set of physical constants.

Of particular interest is the question of whether this terrestrial chirality is universe-wide or whether alternative 'mirror worlds' exist where receptors are composed of D-amino acids and L-sugars. However, given that the underlying chirality of ligand–receptor interactions derives from the chirality of the protein substrate, it would be

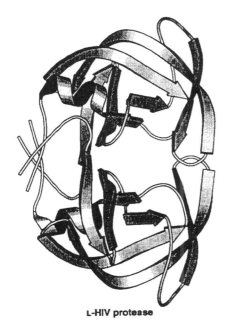
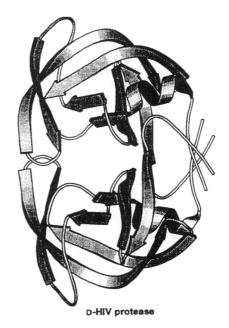

L-HIV protease   **D-HIV protease**

Fig. 28.2 Ribbon representations of the polypeptide backbone of the homodimeric HIV protease. (Reproduced with permission from reference 31.)

expected that 'alternative' proteins composed of D(R)-amino acids should have the same folding properties but will exhibit the opposite chirality of interaction to their naturally occurring terrestrial counterparts.[30] Exactly this has been achieved with HIV-1 protease (Fig. 28.2) where the D- and L-forms exhibit reciprocal chiral specificity with their substrates.[31] This reciprocal interaction is doubtless of importance to the appropriate regulatory authority in any 'mirror world', but the role of D-amino acids in modifying the biological properties of L-peptides and L-proteins is also of importance on earth. In fact, D-amino acid containing peptides are relatively common in nature being found in antibiotics,[32,33] neuroactive peptides[34] and spider toxins.[35,36] This relatively widespread occurrence suggests that there exists a family of peptidyl-aminoacyl-L, D-isomerases with broad substrate selectivity.[37]

The D-peptides and proteins are resistant to naturally occurring proteases, a property that can confer enhanced stability and prolonged duration of action. Such D-species cannot interact with the mirror-image surface of the natural L-protein. However, retro-enantiomers in which the D-amino acids run in the opposite direction to that of the natural L-peptide or protein will have the same side-chain orientation and a similar topochemical surface. Several studies have shown that such retro-enantiomers can have similar biological activities to the parent L-material.[38]

In some systems chirality of interaction may be apparent when two molecules interact, but upon dissociation of the complex it is lost. In appropriately designed systems there exists not only chirality of recognition, but also chiral memory. The substituted non-planar porphyrin of Fig. 28.3 complexes with two molecules of mandelic acid in a single diastereomeric complex. Following dissociation of the complex, however, the porphyrin retains its chiral preference, a process that can be modified by light. The existence of chiral memory, albeit in a simple model of a receptor, may be of substantial importance to the study of drug–receptor interactions at pharmacological receptors as a chiral form of 'molecular imprinting'.[39]

Fig. 28.3 A D2 symmetric octaalkyl substituted porphyrin as chirality sensor and memory. (Reproduced with permission from Chemical and Engineering News, 1997.)

III CONSEQUENCES OF CHIRAL INTERACTIONS AT RECEPTORS

There is ample documentation of the generality of chirality of drug interactions at receptors.[1,2,4,5,16,17,19] In principle, isomers may differ in biological activity in four principal ways:

(1) Both (all) isomers are of equal activity and there is no observed stereochemistry of interaction.
(2) The isomers differ quantitatively in their activities, one enantiomer being more active then the other but facilitating the same overall response. In the extreme situation one enantiomer may exhibit zero biological activity. The majority of chiral ligands and drugs fall into this category.
(3) The isomers differ qualitatively in their activities and express different overall biological activities.
(4) The behaviour of the isomers in the racemate is different than indicated by the properties of the single isomers alone.

Examples of all four situations are known and the same molecule (isomer) may under certain conditions exhibit different biological activities.[5,40] Of particular interest are those instances where the isomers may exhibit agonist, partial agonist and antagonist properties. This is clearly seen with, for example, the catecholamine series active at a variety of α- and β-adrenoceptors. Thus with isoproterenol the $(-)$ and $(+)$ enantiomers are agonist and antagonist respectively at α-adrenoceptors.[1] In this same series of catecholamines derived from norepinephrine there is with increasing *N*-alkyl substitution a progressive loss of agonist activity and a corresponding gain in antagonist activity at α-adrenoceptors with an inversion in stereoselectivity (Table 28.1). A further example is provided by the α,β-blocker labetalol with two asymmetric centres and with different chiral demands at α- and β-receptors (Table 28.2). A number of other examples of stereoselectivity are depicted in Fig. 28.4. The isomers of promethazine are virtually identical in their activities at H_1 receptors and in their toxicity:[41] propranolol is some 40 times more active as

Table 28.1 *Stereoselectivity indices of catecholamines at α- and β-adrenoceptors*

HO—⟨⟩—CHOHCH₂NHR
  HO

R	Ratio $(-)/(+)$	
	α	β
H	4	20
Me	8	50
Pr^{I}	$(-)$ agonist:	>500
	$(+)$ antagonist	
—CHMeCH₂—C₆H₄—OH-4	0.1	>1000

Data from Ariens *et al.*[1]

Table 28.2 *Stereoselectivity of labetalol enantiomers at adrenoceptors*

H₂NOC—⟨⟩—ĊHOH—CH₂NHĊHMeCH₂CH₂—⟨⟩
  HO

Stereoisomer (*,*)	α₁ (Rabbit aorta)	β₁ (Guinea pig atrium)	β₂ (Guinea pig trachea)
	pA₂	pA₂	pA₂
RR	5.87	8.26	8.52
SS	5.98	6.43	<6.0
RS	5.5	6.97	6.33
SR	7.18	6.37	<6.0

Data from Ariens *et al.*[1]

the S-enantiomer at β-adrenoceptors, but the enantiomers are equal in their local anaesthetic properties:[42] the enantiomers of the barbiturate *N*-methyl-5-propylbarbituric acid are convulsant and anti-convulsant respectively;[43] the $(+)$- and $(-)$-enantiomers of propoxyphene are a morphine-like analgesic and anti-tussive agent respectively;[44] and, the enantiomers of the 1,4-dihydropyridine Bay K 8644 show Ca^{2+} channel activation and antagonism.[45] From the clinical perspective, major issues of stereoselectivity of action arise when one enantiomer possesses the desired therapeutic effect and the other exerts a contrary and deleterious action. One example is the β₂-agonist (RS)-albuterol used in the relief of asthma where the R-enantiomer is a smooth muscle relaxant (anti-asthmatic), and the S-enantiomer has been claimed to have pro-contractile or pro-inflammatory properties.[46,47]

Examples are known where a ligand may exhibit different stereoselectivities according to the system. Thus, 17-β-estradiol is the active estrogenic agent whose actions are mediated through nuclear receptors. However, 17-β-estradiol and 17-α-estradiol are equally active as relaxants of arterial smooth muscle contraction, an action presumably mediated through non-specific membrane interactions.[48] Pheromones are widely employed across the animal kingdom. 5-(1-decenyl)oxacyclopentan-2-one is a pheromone employed by scarab beetles: the species *Anomala osakana* and *Popillia japonica* differentially employs the enantiomers as attractants.[49] The female *Anomala* employs the (S,Z)-$(+)$ enantiomer as an attractant and the antipode as an antagonist. In, contrast, *Popillia* shows the opposite stereoselectivity for male attraction — an interesting example of chiral sexual discrimination.

Examples are also known where a non-physiological enantiomer exhibits higher efficacy than its physiological counterpart. Pregnenolone sulfate is a memory enhancing neurosteroid, but the synthetic $(-)$-enantiomer is more potent, actions apparently exerted actions different mechanisms since the effects of the physiological enantiomer are

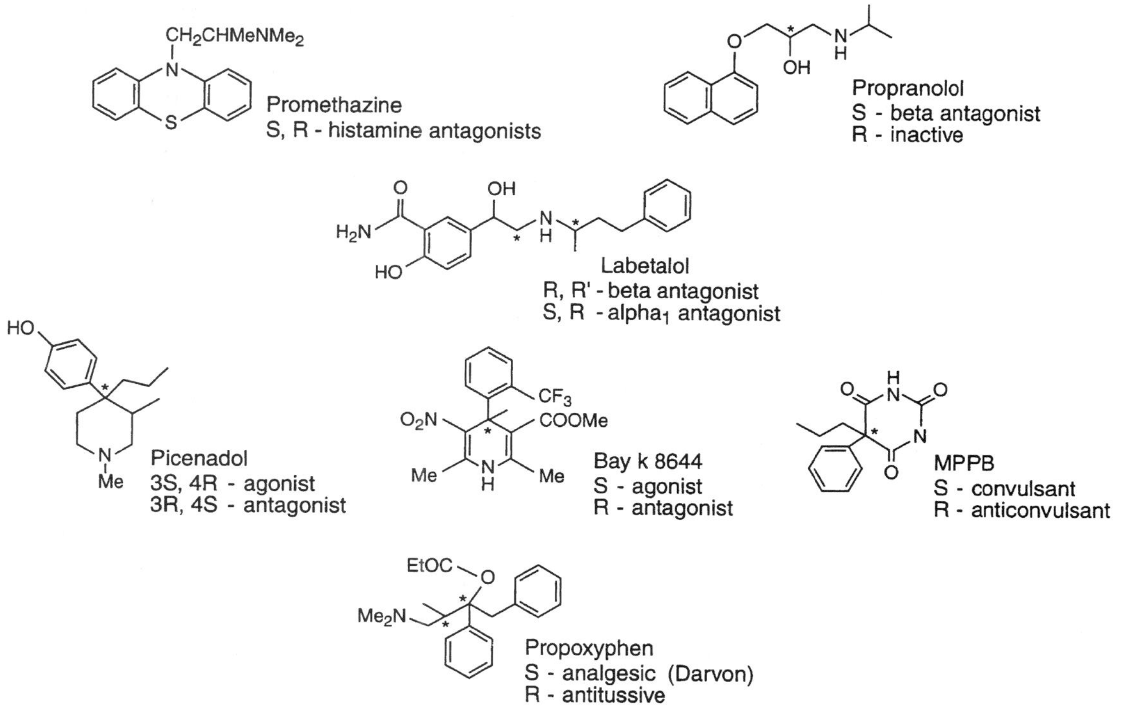

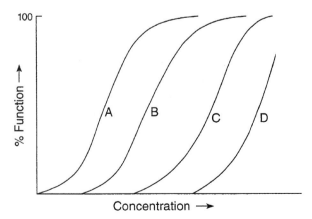

Fig. 28.4 Chiral drugs that show differences in pharmacological activity.

blocked by an NMDA antagonist, but those of the non-natural steroid are insensitive to NMDA receptor blockade.[50]

IV PROBLEMS OF DEFINITION

The linkage of biological response to the initial drug–receptor interaction and the definition of quantitative relationships presents several major issues. Response is the end stage of a multistep series of events that may also represent several different pathways with a common starting point and with multiple end-points:

multiple steps

$$A + Receptor \rightarrow [A\text{-}receptor] \rightarrow Response(s)$$

multiple paths

Accordingly, there are likely to be several different dose–response curves for a single agonist according to both pathway and response in question (Fig. 28.5). Many specific examples have been analyzed.[51,52] The dose–response curves may depend also upon the cellular concentrations of receptor(s) and coupling proteins, factors that are increasingly important in bio-engineered cells and in pathological cellular states, and a drug that appears as an agonist in one cell or tissue may appear as an antagonist or partial agonist in another. This is particularly well illustrated with G-protein-coupled receptors.[51-53] Figure 28.6 depicts muscarinic receptor-mediated increases in phosphatidylinositol turnover and inhibition of adenylate cyclase before and after reduction in receptor number by an irreversible muscarinic receptor antagonist.[54] The full agonist carbachol has been converted to a partial agonist by this process. Similarly, in a cell line expressing only the cloned m2 muscarinic receptor, quite separate dose–response curves for carbachol-mediated

Fig. 28.5 Dose–response curves for agonist–receptor interaction depicting binding, several intermediate steps and the final measured response: A represents a tissue response, B and C intermediate events and D agonist binding.

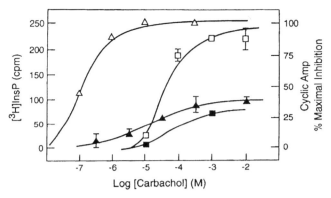

Fig. 28.6 Dose–response curves for carbachol-mediated inhibition of cAMP formation (△,▲) and stimulation of phosphatidylinositol (InsP) hydrolysis (□,■) before (open symbols) and after (closed symbols) receptor inactivation in chick heart cells. (Reproduced with permission from reference 54.)

inhibition of adenylate cyclase and stimulation of phosphatidylinositol turnover occur (Fig. 28.7).[55] This reflects the different efficiencies of coupling of a single receptor subtype to different G-protein effectors. This is usefully considered in terms of 'ensemble' theory according to which ligand–receptor complexes are capable of interacting with multiple response-defining effectors.[56] Thus, an agonist ligand may have different abilities to produce a physiological response or responses, to induce desensitization and down-regulation. Examples are known where these differences are profound. Thus at the CCR5 chemokine receptor the chemokine peptide

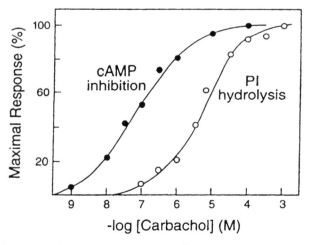

Fig. 28.7 Dose–response curve for carbachol-mediated inhibition of cAMP formation and stimulation of phosphatidylinositol hydrolysis (PI) in cells expressing cloned m2-muscarinic receptors. (Reproduced with permission from reference 55.)

RANTES both activates the receptor and promotes internalization, whereas the analog aminoxypentane-RANTES promotes only internalization.[57]

A particularly illuminating example of this issue is provided by the opiate receptor system, a member of the G-protein-coupled receptor family. At the μ-subclass of opiate receptor, opiate drugs produce analgesia and also tolerance and addiction. However, these properties are clearly dissociable since endogenous opiates produce analgesia, but not addiction. The underlying processes appear to be that all opiate drugs interact with the receptor to produce analgesia, but that subsequent development of tolerance or addiction depends upon the ability of the opiate–receptor complex to undergo desensitization and endocytosis to produce down-regulation. The latter are 'adaptive' responses to continued opiate receptor signaling, whereby the receptor becomes uncoupled from the G-protein, is phosphorylated and internalized via a clathrin-coated pit process. Opiates differ in their ability to produce acute and adaptive responses and those that produce tolerance and addiction, including morphine, are those that generate prolonged signaling and do not readily induce desensitization and down-regulation of the receptors. In turn the prolonged receptor signaling enables the cell to generate the compensatory responses that contribute to tolerance and addiction.[58,59] The ability to examine the several structure–activity relationships, including affinity, efficacy and stereoselectivity, that govern these several events involved in analgesia is thus critically important.

That competitive agonists and antagonists share common binding sites has long been accepted and validated in a number of receptor systems, notably G-protein-coupled receptors with small neurotransmitters such as acetylcholine, histamine and norepinephrine.[60] Clearly, occupancy cannot be identical since agonists promote responses involving protein conformational changes and antagonists do not. However, occupancy of a common site is certainly not an obligatory prerequisite for agonist–antagonist interactions. Thus, for the substance P receptor (neurokinin-1), mutants have been identified that bind non-peptide antagonists such as CP 96345, but not substance P itself (Fig. 28.8).[61] Similarly, the binding of the benzodiazepine receptor antagonists L 365360 and L 364718 (Fig. 28.9) to the brain cholecystokinin B/gastrin receptors in humans and rodents differ according to the residue at equivalent positions 319 (human) and 355 (rodent) — valine and leucine, respectively. However, agonist binding is insensitive to this difference.[62] These observations indicate both the importance of key critical residues in the distinctions between agonist– and antagonist–receptor interactions and that these critical residues can control major differences in interspecies pharmacology. The rat and human histamine H₃ receptors differ by only five residues in the transmembrane domains, and two key residues — alanine (r) and threonine

	Wild type NK-1	CR NK1 (NK3-TM7)	CR NK1 (NK3-TM6-7)	CR NK1 (NK3-TM5-7)	Wild type NK-3
Substance P	0.14±0.03	0.49±0.04	1.1±0.2	0.9±0.1	300±80
Eledoisin	16±5	4.6±1.3	12±3.5	4.5±0.6	4.7±1.1
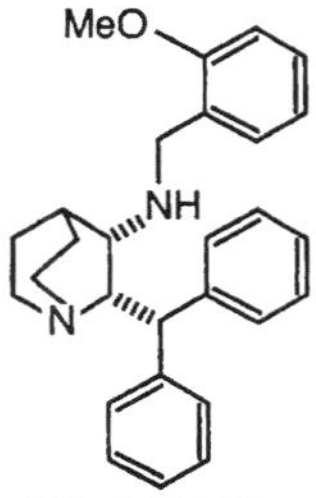 (CP 96345)	14±6	5.4±0.9	330±60	>>10,000	>>10,000

Fig. 28.8 Substance P and CP 96345 interaction (K_D, nM) at chimeric NK-1/NK-3 receptors. (Data from reference 61.)

(h) at position 119 and valine (r) and alanine (h) at position 122, yet these differences control significant differences in ligand affinities.[63]

Other factors that determine the potency and the quality of a drug–receptor interaction include receptor clustering, whereby ligand-induced receptor association determines response,[64] membrane potential whereby agonist–antagonist transitions occur according to the level of the membrane potential of the cell set by physiological or pathological factors,[65] and mutations in receptors and channels so that receptors become constitutively active.[66]

V QUANTITATIVE APPROACHES TO AGONISM AND ANTAGONISM

To measure agonism and antagonism it is necessary to define both the affinity of the ligand for its receptor and the ability of the ligand to induce the measured response. This is no simple matter for there are significant difficulties in defining the quantitative formalism necessary to define agonism, inverse agonism and antagonism and it is clear that a ligand–receptor interaction may trigger several effector pathways, each of which may have its own structure–activity relationships, stereoselectivity, affinity and efficacy. Nonetheless, these quantitative approaches are necessary.[60,67]

In the original formalism of Clark it was assumed that a dose–response curve represented a receptor occupancy relationship and that biological response was directly proportional to this occupancy:[68]

$$R_A/R_{max} = [R.A]/[R_{tot}]$$

Clearly, this simple relationship does not accommodate many commonly observed phenomena such as the graded transitions in an homologous series with a transition from agonism to antagonism (Fig. 28.10). The term 'intrinsic activity' introduced by Ariens[69] provided a phenomenological descriptor for the relative ability of a drug to produce response relative to some standard on a scale of 0 to 1:

$$R_A/R_{max} = i.a \, [R.A]/[R_{tot}]$$

The experimental observation that a 'receptor reserve' existed, whereby only a fraction of the available receptors needed to be occupied to produce a response was accommodated by Stephenson in his parameter stimulus,

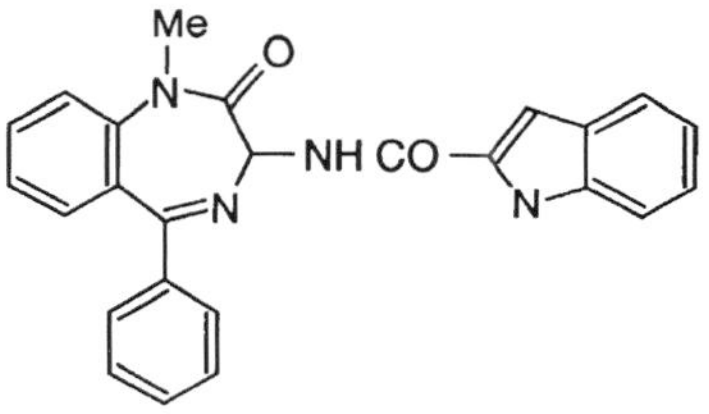

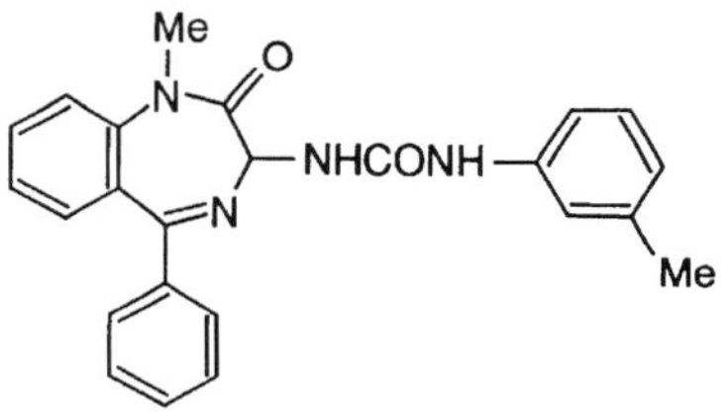

Fig. 28.9 Structural formulae of benzodiazepines active at human and rodent cholecystokinin/gastrin B receptors.

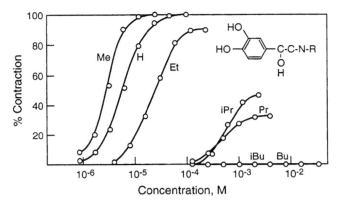

Fig. 28.10 Dose–response curves for a homologous series of catecholamines depicting the gradual transition in properties from agonist, through partial agonist to inactive with increasing molecular substitution. Quite frequently the molecules in series that are inactive as agonists function as competitive antagonists.

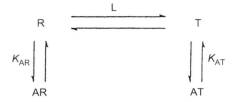

Fig. 28.11 The allosteric model of drug–receptor interaction where the receptor exists (minimally) in two states R and T. K_{AR} and K_{AT} are the equilibrium constants for drug binding to states R and T. L is the equilibrium constant for the R to T transition.

S, defined as:[70]

$$S = \varepsilon[R{\cdot}A]/[R_{tot}]$$

where ε is a dimensionless parameter, efficacy, that denotes the ability of a drug to produce response:

$$R_A/R_{max} = f[S] = f\varepsilon[A]/[A] + K_A$$

By dissociating receptor stimulus and tissue response as directly proportional quantities, the existence of a receptor reserve was accommodated. An immediate corollary of this formalism is that response can be generated from drugs with low efficacy and high occupancy and from drugs with high efficacy and correspondingly low occupancy. Subsequently, Furchgott introduced the term '*intrinsic efficacy*' where:[71]

$$E = \varepsilon/[R_{tot}]$$

All of these treatments are, however, phenomenological: they permit analysis, but do not provide any clear definition of the underlying molecular mechanisms involved in agonist and antagonist behaviour.

A better definition of agonist and antagonist properties can be gained by considering the receptor protein as conformationally mobile around a ground state. Ligands interact with the receptor to bias its conformation and these biased conformations are then capable of interacting with other effectors as in the well-established G-protein-coupled systems, or permeating ions where the protein is an ion channel.[52,56] Thus, a receptor existing in the interconvertible R and T states with an equilibrium constant L and microscopic affinities for drug A of K_{AT} and K_{AR} will exist in the equilibrium shown in Fig. 28.11. If R and T represent active and inactive conformations, respectively, of the

receptor then selective and nonselective interactions of drugs with these states will be associated with varying degrees of agonism and antagonism.

VI OPERATIONAL DEFINITIONS AND EXAMPLES

A G-protein-coupled receptors

The G-proteins are a major family of heterotrimers made up of α-, β- and γ-subunits which cycle between GDP- and GTP-ligated states and which have intrinsic GTPase activity.[72] This enzymatic activity is critical to the cascade of events initiated by receptor–ligand interaction and to the termination of such events (Fig. 28.12). The interaction of a ligand-bound conformation of the receptor with the G_α subunit initiates a cycle of events in which bound GDP is replaced by GTP to produce the activated $G\alpha$-GTP subunit and a β,γ subunit that can both interact with discrete effectors. These systems provide additional definitions of both efficacy and of agonist–antagonist transitions in activity. Thus, for β-adrenergic and opiate receptors coupled through $G_\alpha(s)$ and $G_\alpha(i)$, respectively, intrinsic activity at the level of the cyclase correlates with intrinsic activity at the level of GTPase (Fig. 28.13).[73,74]

The function of the agonist in these G-protein-coupled receptors is to promote coupling of the receptor to G_α in ternary complexes of ligand, receptor and G-protein. This activated coupling of the receptor is achievable in the absence of agonist ligand in constitutively active receptors. Such constitutively active receptors derive from the equilibrium shown in Fig. 28.11, where there is always a fraction of the active receptor state. In mutated receptors this fraction can increase and form the basis of a number of disease states.[75,76] However, the over-expression of wild-type receptors can also generate constitutive activity if the receptors are present at a sufficiently high density. Thus, overexpression of β_1-adrenoceptors in mice leads to a phenotype of dilated cardiomyopathy and heart failure,

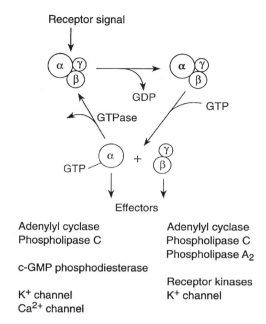

Receptor signal

GDP

GTPase

GTP

GTP — α + γ/β

Effectors

Adenylyl cyclase	Adenylyl cyclase
Phospholipase C	Phospholipase C
	Phospholipase A$_2$
c-GMP phosphodiesterase	
	Receptor kinases
K$^+$ channel	K$^+$ channel
Ca^{2+} channel	

Fig. 28.12 The receptor—G-protein complex. An activated receptor interacts with the trimeric GDP-ligated complex to produce an interchange of GDP with GTP and dissociation into the activated G$_\alpha$GTP and G$_{\beta\gamma}$ subunits: these subunits then interact with a variety of effectors. The activated receptor thus acts as a switch for the G-protein complex. Constitutively active receptors do not require a ligand to interact with the G-protein.

parallel to the pathology produced by over-administration of catecholamines. The enhanced activity of the receptor is, however, independent of catecholamines and is insensitive to neutral antagonists such as propranolol.[77] Receptors active without ligand pose an interesting challenge to the conventional definitions of agonism and antagonism.

The ternary complex model of drug–receptor interactions also permits a symmetry-based view of agonists and antagonists in which agonists promote receptor–effector coupling, neutral antagonists do not affect receptor–effector coupling but prevent the actions of agonists, and negative antagonists that reduce spontaneous receptor–effector coupling.[78–80] In a constitutively active β$_2$-adrenoceptor mutant, some β-antagonists, including betaxolol and ICI 118551, are negative antagonists that depress basal (spontaneous) adenylate cyclase activity, a process that is competitively blocked by the neutral antagonist propranolol (Fig. 28.14).[81] Similarly, negative competitive antagonists have also been found for the G-coupled delta-opiate receptor.[82]

B Mono- and bivalent interactions

For an increasing number of receptor systems dimerization and aggregation appear to be critical components of the activation and transduction pathways. Indeed, in a number of cases the dimerization or aggregation process is itself a sufficient signal for receptor activation which may thus occur in the absence of an agonist ligand.[83] This has been well explored for the tyrosine kinase class of receptors to which insulin and related growth factors bind.[64,84]

The general assumption that receptor dimerization and oligomerization were important processes only for large polypeptide hormones and growth factors has been tempered with the realization that G-protein-coupled receptors may also function physiologically and pathologically as homo- and hetero-dimers.[85–87]

Gonadotropin-releasing hormone (GnRH; Fig. 28.15) stimulates pituitary luteinizing hormone release by a process that involves receptor microassociation.[88,89] The hexapeptide D-pyroGlu-D-His-D-Trp-D-Ser-D-Tyr-D-Lys

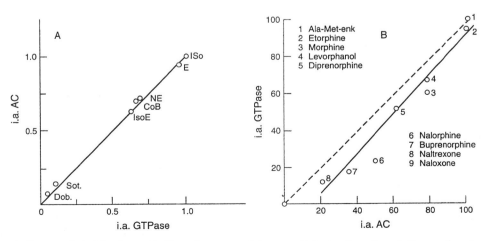

Fig. 28.13 Correlations of intrinsic activities for agonists mediated via stimulation of adenylyl cyclase by catecholamines (A) or by opiates (B) with GTPase activity. (Reproduced with permission from references 73 and 74.)

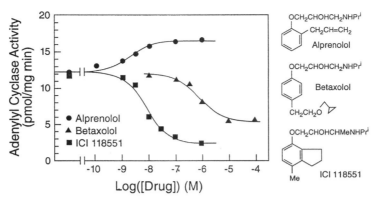

Fig. 28.14 The actions of β-adrenoceptor antagonists on the activity of adenylyl cyclase at constitutively active β2-adrenoceptors. (Reproduced with permission from reference 81.)

is a GnRH antagonist with no detectable agonist properties. The dimeric hexapeptide is also an antagonist; however, the product of this dimeric antagonist hexapeptide and an antibody to GnRH yields an agonist complex that serves to release LH from pituitary cells (Fig. 28.15).[90,91] An antagonist–agonist transition has thus been achieved through a symmetrical arrangement of monovalent antagonist species at a peptide receptor. Subsequent to these early observations a considerable weight of evidence now indicates the generality of receptor dimerization for G-protein-coupled receptors.[85–87] These interactions are of importance both to ligand recognition specificity and efficacy and to the transduction process itself. The GABA$_B$ receptor consists, in the native form, of a heterodimer between the GABA$_{B(!a)}$ and GABA$_{B(2)}$ units and it is this form only that binds GABA agonists with high affinity.[92,93] Of particular interest are heterodimers between non-similar G-protein-coupled receptors. Thus, heterodimers between the polypeptide angiotensin II and bradykinin receptors represent the opposing physiological functions of vasoconstriction, and vasodilatation, respectively, and have an enhanced response to angiotensin II.[94] Of particular interest the expression of this heterodimer contributes to the hypertension in preeclampsia of pregnancy.[95]

C Receptor–ligand entities

Thrombin is a protease and a potent physiological activator of platelet aggregation, a process of key significance to both hemostasis and thrombosis. Thrombin acts on protease-activated receptors.[96,97] At least four protease-activated receptors are known — PAR$_{1-4}$ — all members of the G-protein-coupled receptor family.[98] Thrombin and other proteases activate these receptors by cleaving the *N*-terminal region at a proteolytic site some 41 residues from the terminus. Cleavage reveals a masked 'endogenous' or 'tethered' ligand that now constitutes the new *N*-terminal region.[98–100] and that serves to activate the receptor (Fig. 28.16). The PARs are actually self-contained ligand–receptor entities. The structure–function relationships for

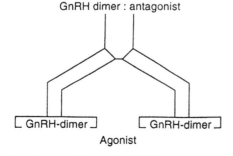

Fig. 28.15 The GnRH and GnRH receptor interaction. A dimer of the GnRH antagonist also serves as an antagonist, but when expressed linked to a divalent antibody it functions as an agonist. (Based on the work of Conn, see references 90 and 91.)

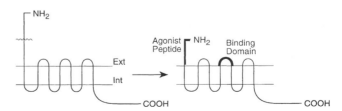

Fig. 28.16 The thrombin receptor, a member of the PAR family of receptors, expresses an endogenous ligand — a tethered ligand — at the extracellular *N*-terminus subsequent to proteolytic cleavage.

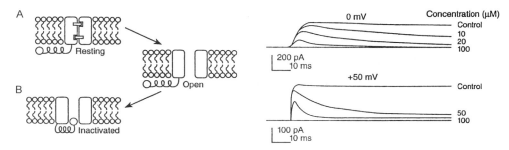

Fig. 28.17 Schematic representation of the 'ball and chain' model of K$^+$ channel inactivation. (A) The *N*-terminal sequence serves as an inactivation particle. (B) The effects of an *N*-terminal peptide administered intracellularly on a mutant K$^+$ channel lacking this terminal peptide. (Reproduced with permission from reference 108.)

agonist activity of the proteolytically exposed *N*-terminal peptide have been established. Thus, at the PAR$_1$ receptor the pentapeptide Ser-Phe-Leu-Leu-Arg-NH$_2$ is an agonist and the 14-residue Ser-Phe-Leu-Leu-Arg-Asn-Pro-Asn-Asp-Lys-Tyr-Glu-Pro-Phe-NH$_2$ is a full agonist. Similar structure–activity relationships are available for the other members of the PAR family.[98,101–103] The critical agonist binding domains are extracellular and associated with the loop linking transmembrane domains 4 and 5.[104]

The existence of a receptor that contains a permanently tethered agonist ligand and that is activated by a necessarily all-or-none proteolytic process raises questions of fundamental importance to the issues of graded responses and response termination.[105] How does a cell distinguish between low and high thrombin concentrations, since all receptors will ultimately be cleaved to generate the agonist peptide? How is the response turned off given the permanent presence of the ligand? How does one generate competitive antagonists? A correlation has been observed between receptor cleavage and the accumulation of messenger IP$_3$, indicating that each 'quantum' of phosphatidylinositol hydrolysis is accompanied by receptor 'turn-off' despite the 'permanent' presence of the tethered agonist ligand.[105] Thus, cells generate dose–response curves through different rates of receptor cleavage, rather than through conventional fractional receptor occupancy.

Similar tethered ligand systems exist in ion channels: in these systems, however, the tethered ligand is an antagonist or channel blocker. Early work by Armstrong and Bezanilla on the inactivation of K$^+$ channels led to a physical 'ball and chain' model in which this cytoplasmic component of the channel would physically occlude the channel gate.[106] Subsequent structural work has confirmed this model. Thus, the *N*-terminal 1–19 sequence of the K$_V$ channel is characterized by a sequence of 11 consecutive hydrophobic or uncharged residues and a subsequent sequence of eight hydrophilic and charged residues. This sequence may be mimicked by the organic ligands, including quaternary ammonium species that produce an open channel block.

Indeed, an *N*-terminal peptide alone can restore inactivation to a channel rendered noninactivating by mutation (Fig. 28.17).[107,108] A similar tethered antagonist exists for the voltage-gated Na$^+$ channel where the cytoplasmic loop linking domains III and IV is critical to the inactivation process. A peptide derived from this sequence is capable of producing Na$^+$ channel inactivation.[109–111]

D The immune receptor

Immune protection in mammalian species is provided by the T and B lymphocytes. Antigen receptors on these cells have a remarkably diverse and virtually unlimited ligand recognition capacity. T cell recognition is unique since it involves recognition of processed or degraded antigen by the T cell receptor (TCR) in association with molecules encoded by the major histocompatibility complex (MHC) and further accessory species. The MHC molecules are extremely polymorphic and serve as an effective guidance system for the T cells.[112] An obviously critical issue given this extraordinary recognition capacity is the ability to recognize 'self' from 'non-self' and to avoid autoimmune responses. The T cell recognition process was long considered to be simply a binary event: either the presented peptide was of sufficient affinity to be recognized or it was not. It is now recognized that the process is in fact quite parallel to that of other receptor systems, that agonist, partial agonist and antagonist peptides exist, and that their presentation has extremely important and different biological consequences for the immune response.[113–115]

Antagonist ligands have potential in the control of autoimmune diseases since they may prevent the T cell activation response (Fig. 28.18). Of particular importance is the role of peptide affinity in T cell education and selection as it matures in the thymus gland. Positive selection results in the maturation of the cell and subsequent differentiation into functional lymphocytes, whereas negative selection generates apoptotic cell death.[116] Peptide ligands that interact but weakly favor positive selection and those ligands that

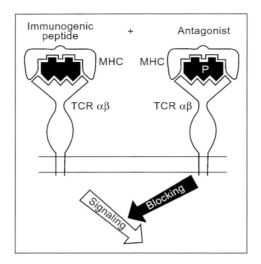

Fig. 28.18 Representation of T cell agonism and antagonism. (Reproduced with permission from reference 113.)

interact strongly promote cell death, thus favoring a system that is directed against self-recognition.[117–119]

E Ion channels

Ion channels constitute two families of pharmacological receptors — voltage-gated and ligand-gated. Members of both channel families are sensitive to ligands, the principal difference being that the primary stimulus for the voltage-gated and ligand-gated families are membrane potential and chemical potential respectively.[40,120,121]

Ion channels may, in fact, be best regarded as multiple receptors since they typically contain binding sites for multiple structurally distinct ligands as depicted for the voltage-gated L-type Ca^{2+} channel (Fig. 28.19).[122] Since ion channels exist in multiple states or families of states — resting, open and inactivated — according to membrane potential and ligation state, the interpretation of structure–activity relationships is complex. According to the modulated receptor treatment of structure–activity interactions at ion channels, drugs may bind to or access preferentially one or other channel state, with the following consequences:[120,122–124]

(1) Different channel states have different affinities for ligands
(2) Ligands may exhibit structure–activity relationships, including stereoselectivity, that exhibit qualitative and quantitative differences according to channel state.

These factors have been discussed in detail elsewhere.[40, 125,126] Two examples are illustrative of the issues. The local anesthetic RAC 109 shows a stereoselectivity of blockade of voltage-gated Na^+ channels that differs according to stimulus mode (Table 28.3).[127] This may be interpreted as an absence of stereoselectivity when RAC 109 interacts with the resting channel state and an enhanced stereoselectivity when the ligand interacts with the inactivated channel state favored by depolarization. A similar change of stereoselectivity can be observed in derivatives of tocainide (Fig. 28.20) as a function of stimulus frequency (Table 28.4).[128]

The local anesthetic agents interacting at Na^+ channels provide an example of quantitative changes in structure–activity relationships as a function of channel state. The 1,4-dihydropyridines interacting at the L-type of voltage-gated Ca^{2+} channel provide an example where both

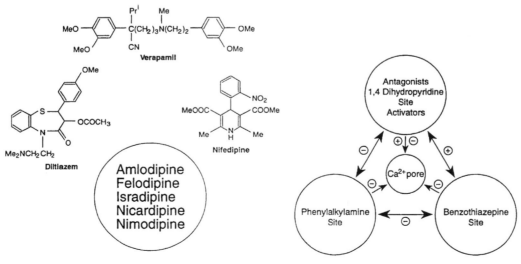

Fig. 28.19 Multiple drug binding sites on the L-type voltage-gated calcium channel.

Table 28.3 *Stereoselectivity of local anesthetic (RAC-109 and RAC-421) action*

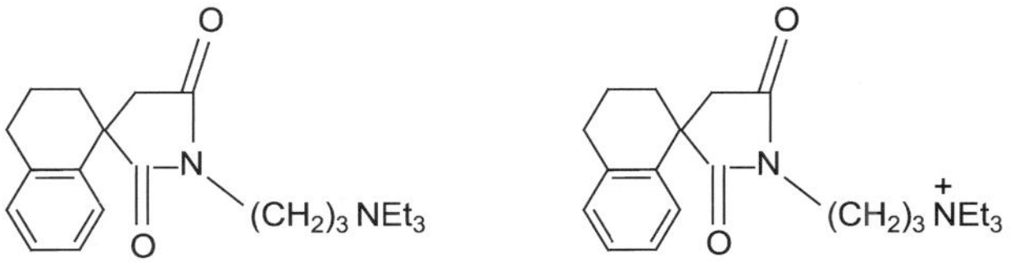

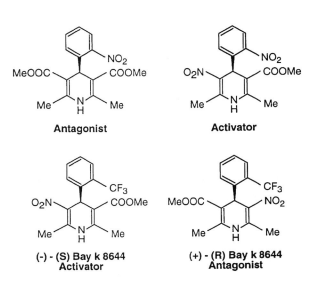

	RAC – 109		RAC – 421

Enantiomer	IC$_{50}$, mM		
	Rest	0 mV	+80 mV
(−) 109	0.85	0.14	0.034
(+) 109	1.30	0.79	0.49
(−) 421	1.48	0.10	0.042
(+) 421	2.09	0.98	0.42

Data from Yeh *et al.* [127]

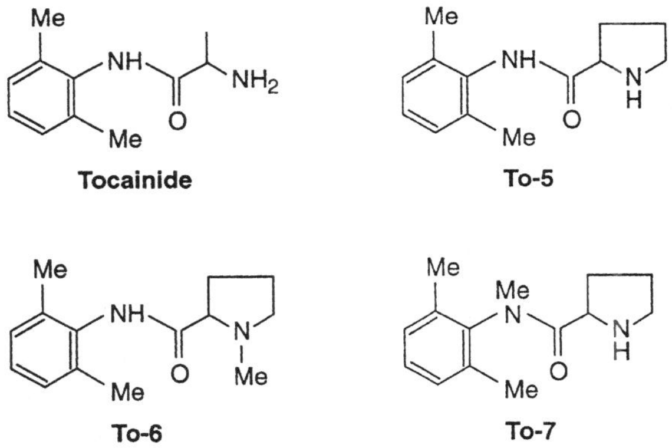

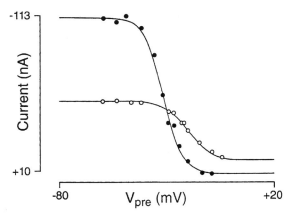

Fig. 28.20 The structure of tocainide and some derivatives.

Table 28.4 *Stereoselectivity of tocainide local anesthetics (Fig. 28.20)*

Compound	IC$_{50}$, mM		
	Rest	Stimulus frequency	
		2 Hz	10 Hz
R-Tocainide	0.58	0.38	0.27
S-Tocainide	0.50	0.30	0.22
R-To5	0.12	0.03	0.01
S-To5	0.23	0.08	0.05
R-To6	0.25	0.18	0.07
S-To6	0.17	0.07	0.03
R-To7	0.34	0.27	0.22
S-To7	0.10	0.10	0.07

Data from Talon *et al.* [128]

quantitative and qualitative changes in structure–activity relationship occur.[40,129–133] Very small structural changes distinguish activator from antagonist properties in the 1,4-dihydropyridines (Fig. 28.19) and in some series the enantiomeric forms have opposing biological activities (Fig. 28.21).[129–132] Of even greater interest the property of activation or antagonism is voltage-dependent switching according to the level of membrane potential (Fig. 28.22).[133] The 1,4-dihydropyridines are thus chemical chameleons

Fig. 28.21 Activator/antagonist pairs of 1,4-dihydropyridines.

Fig. 28.22 The availability of Ca^{2+} current in cardiac cells as modified by membrane potential and the ligand Bay K 8644. Channel current is expressed as a function of holding (pre) potential to a test pulse to 0 mV in the presence (●) or absence (○) of Bay K 8644. (Reproduced with permission from reference 133.)

switching their properties according to the background potential.

VII SUMMARY AND CONCLUSIONS

The issue of agonist-to-antagonist transition is a central theme of medicinal chemistry and is intimately linked to those factors that determine the symmetry of drug–receptor interactions. It is quite clear that agonist/antagonist properties are not simple functions of the ligand alone, but rather are also properties of the receptor and its associated coupled effectors. The nature and stoichiometry of the receptor–effector components are critical determinants of the characteristics of the response. Additionally, constitutively active receptors initiate response in the absence of ligand and receptors may also contain ligand functionality as an integral component of the receptor structure itself. Receptors and ligands enjoy a complementarity and symmetry of interaction that befits their mutual dependence. The fundamental question of whether the ligand or the receptor evolved first remains, however, to be determined.

REFERENCES

1. Ariens, E.J., Soudjin, W. and Timmermans, P.B.M.W.M. (eds). (1983) *Stereochemistry and Biological Activity of Drugs*. Blackwell Scientific, Oxford.
2. Crossley, R. (1993) The relevance of chirality to the study of biological activity. *Tetrahedron* 48: 8155–8178.
3. Eliel, E.L. and Wilen, S.H. (1994) *Stereochemistry of Organic Compounds*. Wiley, New York.
4. Stinson, S.C. (2001) Chiral pharmaceuticals. *Chem. Eng. News* 79–98.
5. Triggle, D.J. (2002) Chirality in drug design and development. In Lough, J. and Wainer, I. (eds) *Chirality in Natural and Applied Science*, pp. 109–138. Blackwell, Oxford.
6. Pasteur, L. (1901) On the asymmetry of naturally occurring organic compounds (two lectures delivered before the Chemical Society of Paris, 20 January and 3 February, 1860. In Richardson, G.M. (ed.). *Memories on Stereochemistry*, pp. 1–33. American Book Company, London.
7. Holmstedt, B. (1990) The use of enantiomers in biological studies: a historical review. In Holmstedt, B., Frank, H. and Testa, B. (eds). *Chirality and Biological Activity*, pp. 1–14. Alan R Liss, New York.
8. Enquist, M. and Arak, A. (1994) Symmetry, beauty and evolution. *Nature* 372: 169–172.
9. Johnstone, R.A. (1994) Female preference for symmetrical males as a by-product of selection for mate recognition. *Nature* 372: 172–175.
10. Rikowski, A. and Grammer, K. (1999) Human body odour, symmetry and attractiveness. *Proc. Roy. Soc. Lond. Ser. B* 266: 869–874.
11. Crum-Brown, A. and Fraser, T.R. (1865) *On the Connection between Chemical Constitution and Biological Activities*. Neill and Company, Edinburgh.
12. Ehrlich, P. (1910) Croonian lecture. On immunity with special reference to cell life. *Proc. Roy. Soc. Lond. Ser. B* 66: 424–448.
13. Langley, J.N. (1878) On the physiology of the salivary secretion. *J. Physiol.* 1: 339–367.
14. Langley, J.N. (1906) Croonian lecture. On nerve endings and on special excitable substances in cells. *Proc. Roy. Soc. Lond. Ser. B* 78: 170–194.
15. Cushny, A.R. (1926) *Biological Relation of Optically Isomeric Substances*. Ballière, Tindall and Cox, London.
16. Smith, D.F. (ed.) (1984) *Handbook of Stereoisomers: Drugs in Psychopharmacology*, CRC Press, Boca Raton.
17. Smith (ed.) D.F. (1989) *Handbook of Stereoisomers: Therapeutic Drugs*. CRC Press, Boca Raton.
18. Casy, A.F. (1993) The steric factor in medicinal chemistry, *Dissymmetric Probes of Pharmacological Receptors*. Plenum Press, New York.
19. Lough, J. and Wainer, I. (eds). (2002) *Chirality: The Legacy of Louis Pasteur*. Blackwell, Oxford.
20. Fryer, R.I. (1990) Ligand interactions at the benzodiazepine receptor. In Emmet, J.C. (ed.). *Comprehensive Medicinal Chemistry*, vol. 3, *Membranes and Receptors*, pp. 539–566. Pergamon Press, Oxford.
21. Kenakin, T. (2002) Efficacy at G-protein-coupled receptors. *Nat. Rev. Drug Disc.* 1: 103–109.
22. Milligan, G. and Bond, R.A. (1997) Inverse agonism and regulation of receptor number. *Trends Pharmacol. Sci.* 18: 468–474.
23. de Light, R.A.F., Kourounakis, A.P. and Ijzerman, A.P. (2000) Inverse agonism at G protein-coupled receptors: (patho)physiological relevance and implications for drug discovery. *Br. J. Pharmacol.* 130: 1–12.
24. Bonner, W. (1996) Homochirality and life. In Cline, D.B. (ed.). *Physical Origin of Homochirality*, pp. 266–282. American Institute of Physics, New York.
25. Palyi, G., Zucchi, C. and Caglioti, L. (eds) (1999) *Advances in Biochirality*. Elsevier, Amsterdam.
26. Bailey, J.M. (1998) RNA-directed amino acid homochirality. *FASEB J.* 12: 503–507.
27. Rees, M. (1999) *Just Six Numbers*. Basic Books, New York.
28. Abbott, E.A. (1884) *Flatland: A Romance in Many Dimensions*. Seeley, London.
29. Stewart, I. (2001) *The Annotated Flatland*. Perseus Publishing, New York.
30. Petsko, G.A. (1992) On the other hand… *Science* 256: 1403–1404.
31. Milton, R.C., Milton, S.C.F. and Kent, S.B.H. (1992) Total chemical synthesis of a D-enzyme: the enantiomers of HIV-1 protease show demonstration of reciprocal chiral substrate specificity. *Science* 28: 343–350.
32. Jung, G. (1992) Proteins from the D-chiral world. *Angewande Chemie* 31: 1457–1459.
33. Kreil, G. (1997) D-Amino acids in animal peptides. *Ann. Rev. Biochem.* 66: 337–345.
34. Broccardo, M., Erspamer, V., Falconieri-Erspamer, G., Improta, G., Linari, G. *et al.* (1981) Pharmacological data on dermorphins, a new class of potent opiod peptides. *Br. J. Pharmacol.* 73: 625–631.
35. Kuwuda, M., Teramoto, T., Kumagaye, K.Y., Nakajima, K., Watanabe, T., *et al.* (1994) w-Agatoxin-TK containing D-serine at position 46, but not synthetic w-[ser⁴⁶]agatoxin, exerts blockade of P-type calcium channels in cerebellar Purkinje neurons. *Mol. Pharmacol.* 46: 587–593.
36. Kreil, G. (1994) Conversion of L- to D-amino acids: a post-translational modification. *Science* 266: 1065–1068.
37. Kreil, G. (2001) D-Amino acids in secretory peptides of animals. *Pharmaceutical News* 8: 32–35.
38. Guichard, G., Benkirane, N., Zeder-Lutz, G., van Regenmortel, M.H., Briand, J.P., *et al.* (1994) Antigenic mimicry of natural L-peptides with retro-inverse peptidomimetics. *Proc. Natl Acad. Sci. USA* 91: 9765–9769.
39. Furusho, Y., Kimurat, T., Mizuno, Y. and Ada, T. (1997) Chirality-memory molecule: a D2-symmetric fully substituted porphyrin as a conceptually new chirality sensor. *J. Am. Chem. Soc.* 119: 5267–5268.

40. Kwon, Y.W. and Triggle, D.J. (1991) Chiral aspects of drug action at ion channels. *Chirality* **3**: 393–404.

41. Toldy, I., Wargha, L., Tolh, L. and Borsy, J. (1959) Uber untersuchungen von promethazin. *I. Acta Chem. Acad. Sci. Hung.* **19**: 273–277.

42. Ruffolo, R.R. (1991) Chirality in alpha- and beta-adrenoceptor agonists and antagonists. *Tetrahedron* **47**: 4953–4980.

43. Buch, H.P., Schneider-Affeld, F. and Rummel, A. (1973) Stereochemical dependence of pharmacological activity in a series of optically active *N*-methylated barbiturates. *Naunyn-Schmied. Arch. Pharmacol.* **277**: 191–198.

44. Hyneck, M., Dent, J. and Hook, J.B. (1990) Introduction to homogeneous and asymmetric catalysis. In Brown, C. (ed.). *Chirality in Drug Design and Synthesis*, pp. 1–27. Academic Press, New York.

45. Goldman, S. and Stoltefuss, J. (1991) 1,4-Dihydropyridines: effects of chirality and conformation on the calcium antagonist and calcium agonist activities. *Angewandte Chemie* **30**: 1555–1578.

46. Handley, D. (2001) The therapeutic advantages achieved through single-isomer drugs. *Pharmaceutical News* **8**: 18–24.

47. Waldeck, B. (2001) Chiral aspects of β-adrenergic agonists. *Pharmaceutical News* **8**: 25–31.

48. Naderli, E.K., Walker, A.B., Doyle, P. and Williams, G. (1999) Comparable vasorelaxant effects of 17α- and 17β-estradiol on rat mesenteric resistance arteries: an action independent of the oestrogen receptor. *Clin. Sci.* **97**: 649–655.

49. Leal, W.S. (1996) Chemical communication in scarab beetles: reciprocal behavioral agonist–antagonist activities of chiral pheromones. *Proc. Natl Acad. Sci. USA* **93**: 12112–12115.

50. Akwa, Y., Ladurelle, N., Covey, D.F. and Baulieu, E.-E. (2001) The synthetic enantiomer of pregnenolone sulfate is very active on memory in rats and mice, even more so than its physiological neurosteroid counterpart. Distinct mechanisms. *Proc. Natl Acad. Sci. USA* **98**: 14033–14037.

51. Kenakin, T. (1993) *Pharmacological Analysis of Drug–Receptor Interactions*, 2nd edn. Raven Press, New York.

52. Kenakin, T. (2001) Inverse, protean, and ligand-selective agonism: matters of receptor conformation. *FASEB J.* **15**: 598–611.

53. Hoyer, D. and Boddeke, H.W.G.M. (1993) Partial agonists, full agonists, antagonists: dilemmas of definition. *Trends Pharmacol. Sci.* **14**: 270–273.

54. Brown, J.H. and Goldstein, D. (1986) Differences in muscarinic receptor reserve for inhibition of adenylate cyslcase and stimulation of phosphoinositide hydrolysis in heart cells. *Mol. Pharmacol.* **30**: 566–570.

55. Ashkenazi, A., Winslow, J.W., Perlata, E.G., Peterson, G.L., Schimerlik, M.I., Capon, D.J. and Ramachandran, J. (1987) An M2 muscarinic receptor subtype coupled to both adenylate cyclase and phosphoinositide turnover. *Science* **238**: 672–674.

56. Kenakin, T. (2002) Efficacy at G-protein-coupled receptors. *Nat. Rev. Drug Disc.* **1**: 103–110.

57. Rodriguez-Frades, J.M., Vila-Coro, A.J., Martin, A., Nieto, M., Sanchez-Madrid, F., *et al.* (1999) Similarities and differences in RANTES- and AOP-RANTES-triggered signals: implications for chemotaxis. *J. Cell Biol.* **144**: 755–765.

58. Whistler, J.L., Chuang, H.-H., Chu, P., Jan, L.Y. and von Zastrow, M. (1999) Functional dissociation of μ opiod receptor signaling and endocytosis: implications for the biology of opiate tolerance and addiction. *Neuron* **23**: 737–746.

59. Finn, A.K. and Whistler, J.L. (2001) Endocytosis of the Mu opiod receptor reduces tolerance and a cellular hallmark of opiate withdrawal. *Neuron* **32**: 829–839.

60. Triggle, D.J. (2002) Drug Receptors: a perspective. In Williams, D. and Lemke, T. (eds). *Foye's Principles of Medicinal Chemistry*, 5th edn. Lippincott and Williams and Wilkins, Baltimore.

61. Getler, V., Johansen, T.E., Snider, R.M., Lowe, J.A., III, Nakanishi, S., *et al.* (1993) Different binding epitopes on the NK1 receptor for substance P and a nonpeptide antagonist. *Nature* **362**: 345–348.

62. Beinhrorn, M., Lee, Y.-M., McBride, E.W., Quinn, S.M. and Kopin, A.S. (1993) A single amino acid of the cholecystokinin-B1 gastrin receptor determines specificity for non-peptide antagonists. *Nature* **362**: 348–351.

63. Ligneau, X., Morisset, S., Tardivel-Lacombe, T., Ghabou, F., Ganellin, C.R., *et al.* (2000) Distinct pharmacology of rat and human histamine H3 receptors: role of two amino acids in the third transmembrane domain. *Br. J. Pharmacol.* **131**: 1247–1250.

64. Weiss, A. and Schlessinger, J. (1998) Switching signals on or off by receptor dimerization. *Cell* **94**: 277–280.

65. Triggle, D.J. (2002) Ion Channels: structure, function and pharmacology. In Krogsgaard-Larsen, P. and Madsen, U. (eds). *Textbook of Drug Design*, 3rd edn. Taylor and Francis, London.

66. Farzel, Z., Bourne, H.R. and Iri, T. (1999) Mechanisms of disease: the expanding spectrum of G protein diseases. *New Engl. J. Med.* **340**: 1012–1020.

67. Kenakin, T. (2002) Drug efficacy at G protein-coupled receptors. *Ann. Rev. Pharmacol.* **42**: 349–379.

68. Clark, A.J. (1937) General pharmology. In *Heffners Handbuch der Experimental Pharmakologie*, vol. 4. Springer, Berlin.

69. Ariens, E.J. (1954) Affinity and intrinsic activity in the theory of competitive antagonism. *Arch. Int. Pharmacodyn.* **99**: 32–49.

70. Stephenson, R.P. (1956) A modification of receptor theory. *Br. J. Pharmacol.* **11**: 379–393.

71. Furchgott, R.F. (1966) The use of 2-haloalkylamines in the differentiation of dissociation constants of receptor agonist complexes. *Adv. Drug Res.* **3**: 21–55.

72. Sprang, S.R. (1997) G protein mechanisms. *Ann. Rev. Biochem.* **66**: 639–678.

73. Pike, L.J. and Lefkowitz, R.J. (1981) Correlation of beta-adrenergic receptor-stimulated [3H]GDP release and adenylate cyclase activation. *J. Biol. Chem.* **256**: 2207–2212.

74. Koski, G., Streaty, R.A. and Klee, W.A. (1981) Modulation of sodium-sensitive GTPase by partial opiate agonists. *J. Biol. Chem.* **257**: 14035–14040.

75. Spiegel, A.M. (1996) Defects in G protein-coupled signal transduction in human disease. *Ann. Rev. Physiol.* **58**: 143–170.

76. Speiegel, A.M. (2000) G protein defects in signal transduction. *Hormone Res.* **53** (Suppl. 3): 17–22.

77. Koch, W.J., Lefkowitz, R.J. and Rockman, H.A. (2000) Functional consequences of altering myocardial adrenergic receptor signaling. *Ann. Rev. Physiol.* **62**: 237–260.

78. Karlin, A. (1967) On the application of a 'plausible' model of allosteric proteins to the receptor for acetylcholine. *J. Theor. Biol.* **16**: 306–320.

79. Costa, T., Ogino, Y., Munson, P.J., Onaran, H.O. and Rodbard, D. (1992) Drug efficacy of guanine nucleotide-binding regulatory protein-linked receptors: thermodynamic interpretation of negative antagonism and of receptor activity in the absence of ligand. *Mol. Pharmacol.* **41**: 549–560.

80. Schutz, W. and Friessmuth, M. (1992) Reverse intrinsic activity of antagonists on G protein-coupled receptors. *Trends Pharmacol. Sci.* **13**: 376–380.

81. Samama, P., Pei, G., Costa, T., Cotecchia, S. and Lefkowitz, R.J. (1994) Negative antagonists promote an inactive conformation of the β2-adrenergic receptor. *Mol. Pharmacol.* **45**: 390–394.

82. Costa, T. and Herz, A. (1989) Antagonists with negative intrinsic activity at opiod receptor coupled to GTP binding proteins. *Proc. Natl Acad. Sci. USA* **86**: 7321–7325.

83. Lauffenburger, D.A. and Lindermann, J.J. (1993) *Models for Binding, Trafficking and Signaling*. Oxford University Press, Oxford.

84. Chan, F.K.-M., Chun, H.J., Zheng, L., Siegel, R.M., Bui, K.L. and Lenardo, M.J. (2000) A domain in TNF receptors that mediates ligand-independent receptor assembly and signaling. *Science* **288**: 2351–2358.

85. Gomes, I., Jordan, B.A., Gupta, A., Rios, C., Trapaidze, N. and Devi, L.A. (2001) G protein coupled receptor dimerization: implications in modulating receptor function. *J. Mol. Med.* **79**: 226–242.

86. Devi, L.A. (2001) Heterodimerization of G-protein-coupled receptors: pharmacology, signaling and trafficking. *Trends Pharmacol. Sci.* **22**: 532–537.

87. Angers, S., Salahpour, A. and Bouvier, M. (2002) Dimerization: an emerging concept for G-protein-coupled receptor ontogeny and function. *Ann. Rev. Pharmacol.* **42**: 409–436.

88. Hopkins, C.R., Semoff, S. and Gregory, H. (1981) Regulation of gonadotropin secretion in the anterior pituitary. *Phil. Trans. Roy. Soc. Lond. Ser B.* **296**: 73–81.

89. Gregory, H., Taylor, C.L. and Hopkins, C.R. (1982) Luteinizing hormone release from dissociated pituitary by dimerization of occupied LHRH receptors. *Nature* **300**: 269–271.

90. Conn, P.M., Rogers, D.C., Stewart, J.M., Niedel, J. and Sheffield, T. (1982) Conversion of a gonadotropin-releasing hormone antagonist to an agonist. *Nature* **296**: 653–655.

91. Blum, J.J. and Conn, P.M. (1982) Gonadotropin-releasing hormone stimulation of luteinizing hormone release: a ligand-receptor-effector model. *Proc. Natl Acad. Sci. USA* **79**: 7307–7311.

92. White, J.H., Wise, A., Main, M.J., Green, A., Fraser, N.J., Disney, G.H., Barnes, A.A., Emerson, P., Foord, S.M. and Marshall, F.H. (1998) Heterodimerization is required for the formation of a functional GABA$_B$ receptor. *Nature* **396**: 679–682.

93. Kuner, R., Kohr, G., Grunewald, S., Eisenhardt, G., Bach, A. and Kornau, H.C. (1999) Role of heteromer formation in GABA$_B$ receptor function. *Science* **283**: 74–77.

94. AbdAllah, S., Lother, H. and Quitterer, U. (2000) AT$_1$-receptor heterodimers show enhanced G-protein activation and altered receptor sequestration. *Nature* **407**: 94–98.

95. AbdAlla, S., Lother, H., Massiery, A.E. and Quitterer, U. (2001) Increased AT$_1$ receptor heterodimers in preeclampsia mediate enhanced angiotensin II responsiveness. *Nat. Med.* **7**: 1003–1009.

96. Cocks, T.M. and Moffatt, J.D. (2000) Protease-activated receptors: sentries for inflammation? *Trends Pharmacol. Sci.* **21**: 103–108.

97. Dery, O., Corvera, C.U., Steinhoff, M. and Bunnett, N.W. (1998) Proteinase-activated receptors: novel mechanisms of signaling by serine proteases. *Am. J. Physiol.* **274**: C1429–C1452.

98. Vergnolle, N., Wallace, J.L., Bunnett, N.W. and Hollenberg, M.D. (2001) Protease-activated receptors in inflammation, neuronal signaling and pain. *Trends Pharmacol. Sci.* **22**: 146–152.

99. Scarborough, R.M., Naughton, M.A., Teng, W., Hing, D.T., Rose, J., Vu, T.-K.H., Wheaton, V.I., Turck, C.W. and Coughlin, S.R. (1992) Tethered ligand agonist peptides. Structural requirements for thrombin receptor activation reveal mechanism of proteolytic unmasking of agonist function. *J. Biol. Chem.* **267**: 13146–13149.

100. Coughlin, S.R. (1999) How the protease thrombin talks to cells. *Proc. Natl Acad. Sci. USA* **96**: 11023–11027.

101. Hollenberg, M.D., Saifeddine, M., Al-Ani, B. and Kawabata, A. (1997) Proteinase-activated receptors: structural requirements for activity, receptor cross-reactivity, and receptor selectivity of receptor-activating peptides. *Can. J. Physiol. Pharmacol.* **75**: 832–841.

102. Kawabata, A., Saifeddine, M., Al-Ani, B., Leblond, L. and Hollenberg, M.D. (1999) Evaluation of proteinase-activated receptor-1 (PAR1) agonists and antagonists using cultured cell receptor assay: activation of PAR2 by PAR1 ligand. *J. Pharmacol. Exp. Ther.* **288**: 358–370.

103. Vergnolle, N. (2000) Review article: proteinase-activated receptors — novel signals for gastrointestinal pathophysiology. *Alim. Pharmacol. Ther.* **14**: 257–266.

104. Gerszten, R.E., Chen, J., Ishii, M., Ishii, K., Wang, L., Nanevicz, T., Turck, C.W., Vu, T.-K. and Coughlin, S.R. (1994) Specificity of the thrombin receptor for agonist peptide is defined by its extracellular surface. *Nature* **368**: 648–651.

105. Ishii, K., Hein, L., Kobilka, B. and Coughlin, S.R. (1993) Kinetics of thrombin receptor cleavage in intact cells. Relation to signaling. *J. Biol. Chem.* **268**: 9780–9786.

106. Armstrong, C.M. and Bezanilla, F. (1977) Inactivation of the sodium channel. II. Gating current experiments. *J. Gen. Physiol.* **70**: 567–590.

107. Hoshi, T., Zagota, W.N. and Aldrich, R.W. (1990) Biophysical and molecular mechanisms of Shaker potassium channel inactivation. *Science* **250**: 533–538.

108. Zagota, W.N., Hoshi, T. and Aldrich, R.W. (1990) Restoration of inactivation in mutants of Shaker potassium channels by a peptide derived from ShB. *Science* **250**: 568–571.

109. Eaholtz, G., Scheuer, T. and Catterall, W.A. (1994) Restoration of inactivation and block of open sodium channels by an inactivation gate peptide. *Neuron* **12**: 1041–1048.

110. Eaholz, G., Zagotta, W.N. and Catterall, W.A. (1998) Kinetic analysis of block of open sodium channels by a peptide containing the isoleucine, phenyalanine, and methionine (IFM) motif from the inactivation gate. *J. Gen. Physiol.* **111**: 75–82.

111. Zhou, M., Morals-Cabral, J.H., Mann, S. and MacKinnon, R. (2001) Potassium channel receptor site for the inactivation gate and quaternary amine inhibitors. *Nature* **411**: 657–662.

112. Nelson, D.L. and Cox, M.M. (eds). (2000) Protein function. In *Lehninger: Principles of Biochemistry*, 3rd edn. Worth Publishers, New York.

113. Evavold, B.D., Sloan-Lancaster, J. and Allen, P.M. (1993) Tickling the TCR: selective T-cell functions stimulated by altered peptide ligands. *Immunol. Today* **14**: 602–609.

114. Madrenas, J., Wange, R.L., Wang, J.L., Isakov, N., Samelson, L.E. and Germain, R.N. (1995) Phosphorylation without ZAP-70 activation induced by TCR antagonists or partial agonists. *Science* **267**: 515–518.

115. Marrack, P. and Parker, D.C. (1994) A little of what you fancy. *Nature* **368**: 397–398.

116. Rothenberg, E.V. (1992) The development of functionally responsive T cells. *Adv. Immunol.* **51**: 85–214.

117. Hogquist, K.A., Jameson, S.C., Heath, W.R., Howard, J.L., Bevan, M.J., *et al.* (1994) T cell receptor antagonist peptides induce positive selection. *Cell* **76**: 17–27.

118. Ashton-Rickard, P.G., Bandiera, A., Delaney, J.R., Van Kaer, L., Pircher, H.-P., *et al.* (1994) Evidence for a differential avidity model of T cell selection in the thymus. *Cell* **76**: 651–663.

119. Alam, S.M., Travers, P.J., Wung, J.L., Nasholds, W., Redpath, S., *et al.* (1996) T cell receptor affinity and lymphocyte positive selection. *Nature* **381**: 616–620.

120. Hille, B. (2002) *Ion Channels*, 3rd edn. Sinauer Press, Sunderland, MA.

121. Striessnig, J., Grabner, M., Mitterdorfer, J., Hering, S., Sinneger, M.J., *et al.* (1998) Structural basis of drug binding at L-type Ca^{2+} channels. *Trends Pharmacol. Sci.* **19**: 108–114.

122. Hille, B. (1977) Local anesthetics: hydrophilic and hydrophobic pathways for the drug–receptor reaction. *J. Gen. Physiol.* **69**: 497–515.

123. Hondeghem, L.M. and Katzung, B.G. (1977) Time- and voltage-dependent interactions of antiarrhythmic drugs with sodium channels. *Biochim. Biophys. Acta* **472**: 373–398.

124. Hondeghem, L.M. and Katzung, B.G. (1984) Antiarrhythmic agents: the modulated receptor mechanism of action of sodium and calcium channel-blocking drugs. *Annu. Rev. Pharmacol.* **24**: 387–423.

125. Triggle, D.J. (1989) Structure–function correlations of 1,4-dihydropyridine calcium channel antagonists and activators. In Hondeghem,

L.M. (ed.). *Molecular and Cellular Mechanisms of Antiarrhythmic Agents*, pp. 269–292. Futura Publishing, Mt Kiscoe, NY.

126. Triggle, D.J. (1994) On the other hand: the stereoselectivity of drug action at ion channels. *Chirality* **6**: 58–62.

127. Yeh, J.Z. Blockade of sodium channels by stereoisomers of local anesthetics. In Fink, B.R (ed.) *Molecular Mechanisms of Anesthesia*, pp. 35–44. Raven Press: New York.

128. Talon, S., De Luca, A., De Bellis, M., Desaphy, J.-F., Lentini, G., *et al.* (2001) Increased rigidity of the chiral centre of tocainide favours stereoselectivity and use-dependent block of skeletal muscle Na^+ channels enhancing the antimyotonic activity *in vivo*. *Br. J. Pharmacol.* **134**: 1523–1531.

129. Triggle, D.J., Langs, D.A. and Janis, R.A. (1989) Ca^{2+} channel ligands. Structure–function relationships of the 1,4-dihydropyridines. *Med. Res. Rev.* **9**: 123–180.

130. Goldmann, S. and Stoltefuss, J. (1991) 1,4-Dihydropyridines: effects of chirality and conformation on the calcium antagonist and calcium agonist activities. *Angew. Chem. Int. Ed.* **30**: 1559–1578.

131. Frankowiak, G., Bechem, M., Schramm, M. and Thomas, G. (1985) The optical isomers of the 1,4-dihydropyridine Bay K 8644 show opposite effects on Ca channels. *Eur. J. Pharmacol.* **114**: 223–226.

132. Wei, X.-Y., Luchowski, E.M., Rutledge, A., Su, C.M. and Triggle, D.J. (1986) Pharmacologic and radioligand binding analysis of the actions of 1,4-dihydropyridine activator-antagonist pairs in smooth muscle. *J. Pharmacol. Exp. Ther.* **239**: 144–153.

133. Sanguinetti, M.C., Krafte, D.S. and Kass, R.S. (1986) Voltage-dependent modulation of Ca channel current in heart cells by Bay K 8644. *J. Gen. Physiol.* **88**: 369–392.

29

DESIGN OF PEPTIDOMIMETICS

Hiroshi Nakanishi and Michaël Kahn

It's not just what we inherit from our mothers and fathers that haunts us. It's all kinds of old defunct theories, all sorts of old defunct beliefs, and things like that.
Henrik Ibsen, *Ghosts* (1881)

I INTRODUCTION

Peptides and proteins control all biological processes at some level (transcriptional, translational or posttranslational). Yet, at the molecular level, our understanding of the relationship between structure and function remains rudimentary. Moreover, the problems involved in clarifying these issues are somewhat different for peptides and proteins. The dissection of multidomain proteins into small conformationally restricted components is an important step in the design of low molecular weight nonpeptides that mimic the activity of the native protein. Mimetics of critical functional domains might possess beneficial properties in comparison to the intact proteinaceous species with regard to specificity and therapeutic potential, and are valuable probes for the study of molecular recognition events.[1] On the other hand, peptides are characteristically highly flexible molecules whose structure is strongly influenced by their environment.[2] Their random conformations in solution complicate their use in determining their receptor-bound or bioactive structures.[3] Conformational constraints can significantly aid this determination.[4] Peptidomimetics are powerful tools for the study of molecular recognition and provide a unique opportunity to dissect and investigate structure–function relationships in both peptides and complex proteins.[5-7]

II NONPEPTIDE MIMETICS

The isolation and identification in 1975 of the endogenous opioid pentapeptides, methionine and leucine enkephalin,[8] represents the intellectual groundbreaking in the field of peptidomimetics. The work demonstrated that, despite their highly disparate structures, these linear pentapeptides and the condensed heterocyclic species morphine elicit their biological response, analgesia, by binding to the opiate receptor. However, despite intensive research efforts to understand this relationship, it is fair to say that more than a quarter century later it remains far from clear.[9]

III SCREENING

To date, unquestionably the greatest success in developing nonpeptide leads for peptide ligands has come from a screening approach. A wide array of ligands, often bearing little, if any, structural resemblance to the endogenous peptide ligand which they mimic, have been uncovered through receptor-based screening programs.[10] What lessons can be learned from these screening leads? Foremost, they have validated many of the critical concepts which underlie the rational design of peptidomimetics, in that they prove

The Practice of Medicinal Chemistry
ISBN 0-12-744481-5

that compounds lacking amide bonds, obvious pharmacophore similarity and flexibility, can be potent and selective ligands for peptide receptors. However, a sobering note is that these relationships provide us with very limited information with regard to developing generic solutions for rationally traversing the pathway from peptides to mimetics.[10]

IV   DESIGN OF NONPEPTIDE MIMETICS

There has been an increasing effort to rationally design and synthesize highly active analogues of biologically significant peptides and proteins.[11] It is anticipated that these drugs of the future will possess greater selectivity, and fewer side-effects than their present-day counterparts.[12] The complex problems associated with the rational design of mimetics are constantly being reduced owing to advances in molecular biology, spectroscopy and computational chemistry. The determination of the receptor-bound conformation of a peptide or protein ligand is invaluable for the rational design of mimetics. However, with few exceptions this information is not readily available. Conformational constraints constitute one of the most promising avenues for the solution of this problem, particularly if the constraint is such that only one conformation of the ligand is significantly populated. Rigid analogues 'prepay' an entropy cost before binding to their receptor and therefore should bind more avidly, assuming appositive placement of pharmacophoric residues.[13] Selectivity can be enhanced by the preclusion of conformers that produce undesired bioactivity.[14] An additional advantage is that proteolytic enzymes generally prefer conformationally adaptable substrates; therefore, constrained analogues are generally endowed with increased proteolytic stability.

V   THE SECONDARY STRUCTURE APPROACH

One approach to the design of peptidomimetics has been guided by the simple elegance which nature has employed in the molecular architecture of proteinaceous species.[15] Three basic structural building blocks (α-helices, β-sheets and reverse turns) are utilized for the construction of all proteins. The design and synthesis of peptidomimetic prosthetic units to replace these three architectural motifs affords the opportunity to dissect and investigate complex structure–function relationships in proteins through the use of small synthetic conformationally restricted components. This is a critical step towards the rational design of low molecular weight nonpeptidic pharmaceutical agents which are devoid of the shortcomings of conventional peptides.[16]

The progress in this field has been reviewed in several comprehensive articles.[11,13,17–21]

A   Reverse turn (β-turns and γ-turns)

The surface localization of turns in proteins, and the predominance of residues containing potentially critical pharmacophoric information, has led to the hypothesis that turns play critical roles in a myriad of recognition events.[22] Reverse turns are classified into γ-turns, consisting of three residues (sometimes referred to as a C7 conformation), and the more common β-turns (C10 conformation), formed by a tetrapeptide which are further divided into a variety of subtypes.[23] An excellent review by Ball *et al.* in 1990[24] summarized the progress to that point in reverse-turn mimetics. More recently, Baca *et al.*[25] made use of Nagai and Sato's thiazolidine lactam type II' β-turn mimetic structure[26] to replace the type I' β-turn at Gly16 and Gly17 in HIV-1 protease. The β-turn mimetic-containing enzyme dimerized similarly to the native enzyme. It was fully active, possessed the same substrate specificity as the native enzyme, and showed enhanced thermal stability.[25]

Hirschmann *et al.* utilized 3-deoxy-β-D-glucose (**1.1**) in Fig. 29.1 as a scaffold to synthesize the first nonpeptidic analogue of the receptor-recognizing β-turn (Phe7, Trp8, Lys9, Thr10) of somatostatin and its cyclic analogue (**1.2**).[27–29] Mimetic (**1.1**) binds to the pituitary gland somatostatin receptor with an IC_{50} value of 1.3 μM. Interestingly, in a functional assay this mimetic displayed agonist activity at 3 μM (less frequently observed with peptidomimetics).

Since then, numerous nonpeptidic agonists have been reported for peptide hormone and neuropeptide GPCR receptors,[10,20] including angiotensin-II,[30] cholecystokinin-A (CCK-A),[31] bradykinin,[32] melanocortin,[33,34] C5a,[35] growth hormone secretagogue[36] and vasopressin receptors.[37–41] It should be noted that the majority of these peptide hormones and neuropeptides are postulated to form a reverse turn as their bioactive conformation, as determined either from NMR studies or by using cyclic peptide analogues. Some examples of the involvement of β-turns as the bioactive conformation for a peptide hormone and/or neuropeptide ligand for GPCRs are

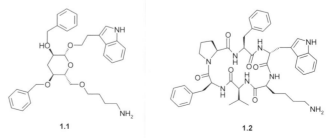

Fig. 29.1 Somatostatin receptor agonists.

listed in Table 29.1.[14,42−69] Although many of these agonists have been discovered through high-throughput screenings of large compound libraries with no predetermined secondary structure motifs, many of them possess pharmacophores which are superimposable with their corresponding peptide agonists in the β-turn conformation, thus consequentially mimicking a reverse-turn motif. Therefore, the screening of libraries based on a β-turn template should provide many interesting hits against the GPCR superfamily.

Interestingly, cyclic pentapeptides with amino-acid configurations of DLDDL were proposed as scaffolds developing antagonists for GPCRs,[70] based on the naturally occurring cyclic pentapeptide, BE-18257 and its synthetic analogue BQ-123 (Table 29.1). This cyclic peptide is known to form a rigid structure with a type II′ β-turn with two consecutive D-amino acids and an inverse γ-turn at the other end. Thus it serves as an ideal template system to present functional groups in either β-turn or γ-turn conformations. As an example, the cyclic peptide cyclo[D-Trp-Pro-D-Lys-D-Trp-Phe] was synthesized based on the known structure−activity relationships of NK1 receptor antagonists.[58] Functional assessment verified, even though weak, its micromolar antagonistic activity at the NK1 receptor.[70]

Piperidines and piperazines are recurring motifs among nonpeptidic GPCR agonists and antagonists.[71] Researchers at Merck extensively utilized substituted piperidine derivatives, in particular spiropiperidines, as their 'privileged scaffold'[72] for targeting GPCRs (Fig. 29.2). Compounds (2.1) and (2.2) were selective agonists for the human somatostatin-2 receptor with K_i values of 1.0 and 0.01 nM, respectively.[73] The structure of (2.1) was energy minimized, incorporating available NMR data. The three pharmacophores presented on the spiropiperidine urea scaffold, i.e. phenyl, indole and amine moieties of (2.1) can be superimposed very well with the FWK motif of the cyclic peptide (1.2), thus mimicking the β-turn conformation. Interestingly, a minor modification at the 4-piperidine position converted agonists (2.1) and (2.2) to an antagonist (2.3) of sst-2.[74]

Slightly different variations of the piperidine scaffold were found to bind to growth hormone secretagogue[75] with a modest agonist activity (EC_{50} = 300 nM). Since tryptophan contributes significantly to the activity of the potent GH-releasing hexapeptide, GHRP-6 (His-D-Trp-Ala-Trp-D-Phe-Lys-NH$_2$; EC_{50} = 10 nM),[76,77] the initial lead was optimized by incorporation of an indole-mimicking spiroindanyl group, providing a more potent analogue (2.4) with an EC_{50} of 1.3 nM. The new agonist (2.4) released growth hormone from rat pituitary cells in culture and its pharmacological profile was identical to that of GHRP-6, but clearly distinguishable from the natural growth hormone-releasing hormone.[78] Since the discovery of this orally active clinical candidate (2.4), several other potent GHS agonists, e.g. (2.5) with a 3,3-disubstituted piperidine scaffold and (2.6) with a fused pyrazolidinone scaffold, have been reported.[79−81] Other variations of this piperidine scaffold (2.7) and (2.8) were recently published in the patent literature as agonists of the melanocortin subtype-4 receptor. Particularly, (2.8) was reported to be a very potent agonist, selective for MC4R over all other subtypes of MC receptors, with an EC_{50} of 2.1 nM.[33,34] Surprisingly, small molecule compounds found to date exhibiting any binding affinity towards the chemotaxin C5a receptor, including compound (2.9), are all agonists,[35] even though a hexapeptide NMe-Phe-Lys-Pro-D-Cha-Trp-D-Arg-OH has been identified as a full antagonist.[82]

Table 29.1 *Examples of β-turn motif in ligands for neuropeptide and peptide hormone GPCRs*

GPCR	Studied sequence	β-turn motif	Reference
Angiotensin	Y-IHPF-OH	IHPF	42
Bombesin	DNal-cyclo[CYwKV]-Nal-NH$_2$	YwKV	43
Bradykinin	R-PPGF-TPFR	TPFR	44
C5aR	F-cyclo[KPChaWR]	PChaWR	45
Calcitonin/CGRP	FVPTDVGPFAF-NH$_2$	VPTD, VGPF	46
Cholecystokinin	DYMG-WMDF	WMDF	47, 48
Endothelin	cyclo[vLweA], BQ123 [wdPvL]	vLwe(d)	49
Gastrin	Y(SO$_3$)GYGWMDF	GYGW, WMDF	50, 51
Ghrelin/GHS	HwAWfK-NH2	AWfK	52
Gonadotropin/LHRH	EHWS-YGLR-PG-NH$_2$	YGLR	53, 54
Melanocortin	Ac-Nle-cyclo[DHfRWK]-NH$_2$	HfRW	55, 56
Neurokinin/Tachykinin	cyclo[MDWFDapL]	FFGL	57, 58
Opioid	enkephalin and endomorphin	YGGFL, YPWF	59−61
Oxytocin	cyclo[CYIQNC]PLG	YIQN	62, 63
Somatostatin	cyclo-[FPYWKT], [PFwKTF]	PYWK	14, 64
Urotensin II	AGTAD[CFWKYC]	FWKY	65
Vasopressin	cyclo[CYFQNC]PRG	YFQN	66−69

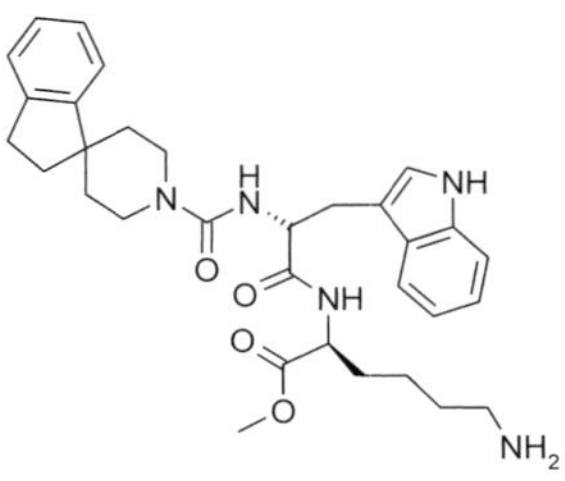

2.1 sst2 agonist

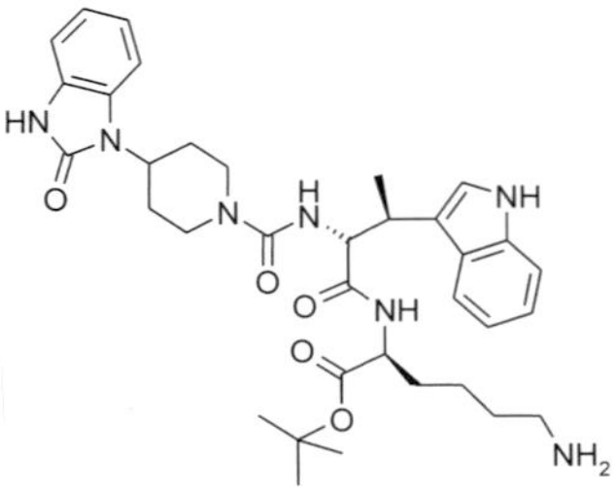

2.2 sst2 agonist

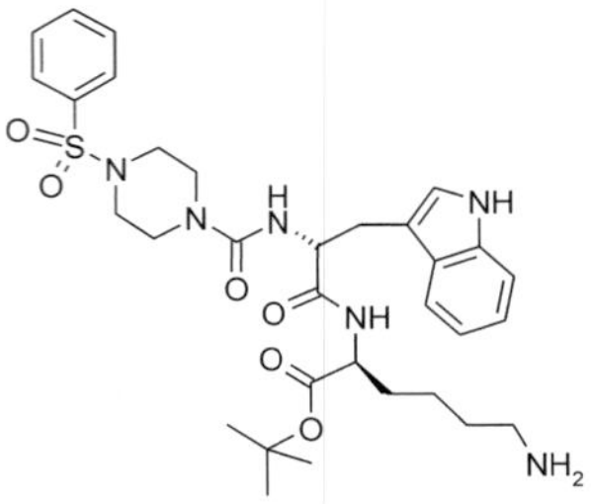

2.3 sst2 antagonist

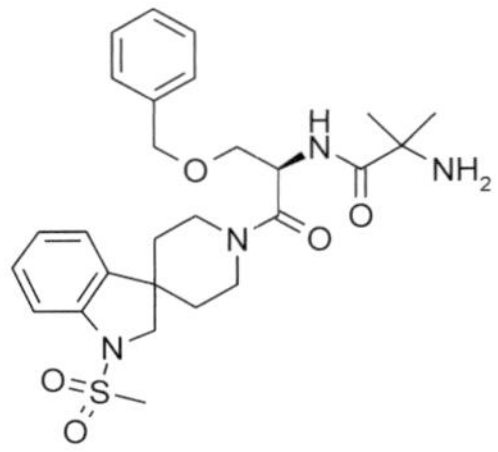

2.4 MK-0677 GHS agonist

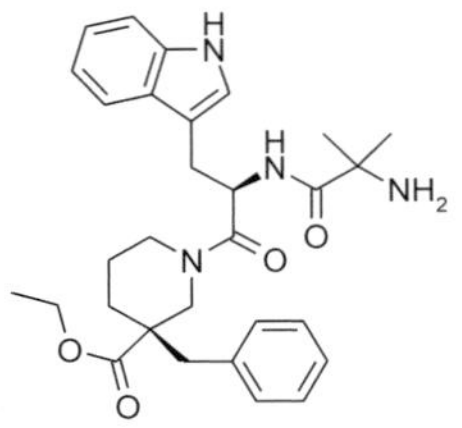

2.5 L-163,540 GHS agonist

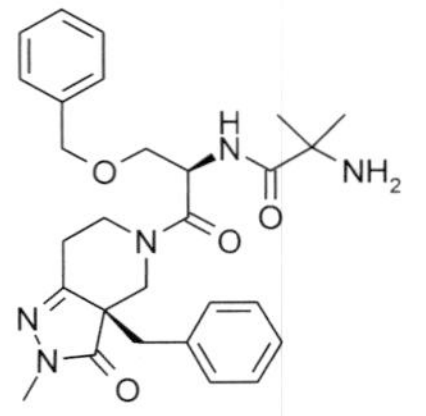

2.6 CP-424,391 GHS agonist

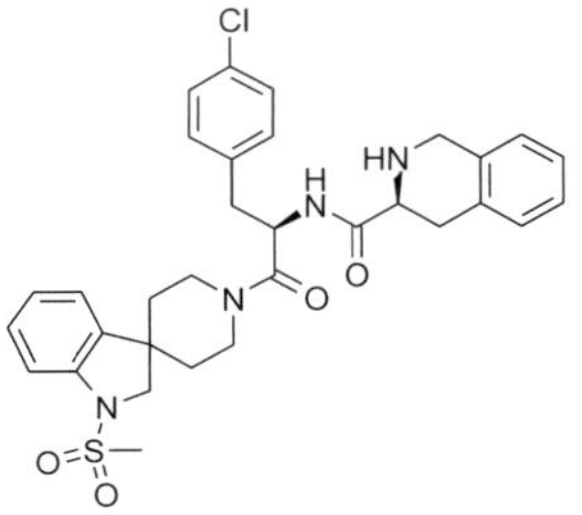

2.7 MC4 agonist

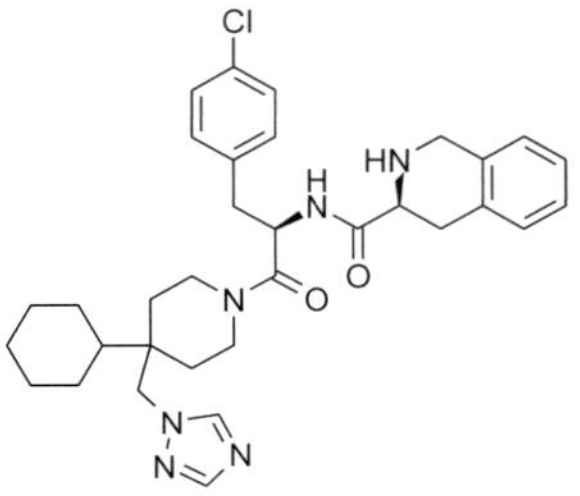

2.8 MC4 agonist

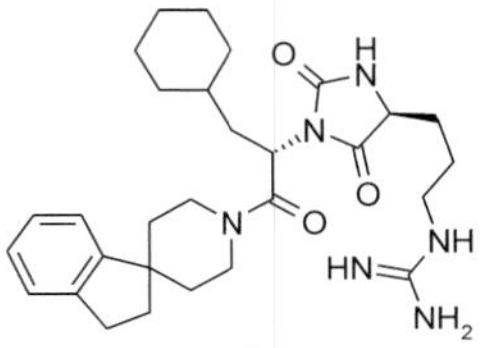

2.9 C5a agonist

Fig. 29.2 Piperidine derivatives as GPCR agonists.

Benzoazepines and benzodiazepines are another recurring 'privileged scaffold' in agonists and antagonists of GPCRs, notably agonists for the CCK-A, opioid[83] and vasopressin receptors.[38,39] Although the gastrointestinal hormone CCK shares the same *C*-terminal peptide sequence, i.e. Gly[29]-Trp-Met-Asp-Phe-NH$_2$, with another related hormone, gastrin, a β-turn motif formed by the *C*-terminal tetrapeptide has been incorporated into the design of CCK-A and CCK-B specific agonists/antagonists (Fig. 29.3).[84] The benzodiazepine system is extensively used to present the critical aromatic and hydrophobic pharmacophores of Trp30, Met31 and Phe33 from CCK.[10,31,85] A series of 1,5-benzodiazepine CCK-A agonists (**3.4–3.6**) were identified based on the earlier discovery of a CCK-A antagonist (**3.1**). Interestingly, a replacement of the R1 methyl moiety with an isopropyl group converted the full antagonist (**3.3**) of CCK-A to a full agonist (**3.4**) with an ED$_{50}$ of 1.6 µM.[86] Compound (**3.6**) was a full agonist of the human CCK-A

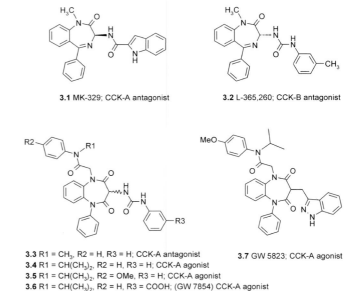

3.1 MK-329; CCK-A antagonist

3.2 L-365,260; CCK-B antagonist

3.3 R1 = CH$_3$, R2 = H, R3 = H; CCK-A antagonist
3.4 R1 = CH(CH$_3$)$_2$, R2 = H, R3 = H; CCK-A agonist
3.5 R1 = CH(CH$_3$)$_2$, R2 = OMe, R3 = H; CCK-A agonist
3.6 R1 = CH(CH$_3$)$_2$, R2 = H, R3 = COOH; (GW 7854) CCK-A agonist

3.7 GW 5823; CCK-A agonist

Fig. 29.3 Benzodiazepine CCK receptor agonists.

receptor expressed on the CHO cells ($EC_{50} = 60$ nM), and a potent antagonist of CCK-B receptor ($IC_{50} = 0.8$ nM).[87] Further modification by incorporating 3-indazolyl as an indole mimetic and an *N*-isopropyl-*N*-phenylacetamide agonist 'trigger' moiety into the 1,5-benzodiazepine scaffold led to a potent agonist (**3.7**) of CCK-A with an IC_{50} of 23 nM in a binding assay and an ED_{50} of 109 nM. It showed 100% of the intrinsic activity level and demonstrated oral efficacy by the suppression of food intake in a rat feeding model.[86,88,89] Another fascinating finding which emerged from a recent study was that the CCK-A and CCK-B receptors selectively recognized enantiomeric dispositions of the Trp30 indole,[84] explaining the subtype preferences observed earlier in optical isomers of a number of nonpeptide ligands such as (**3.1**) and (**3.2**).

An interesting recent application involved the design of the benzodiazepine scaffold to inhibit the enzyme *Ras* farnesyl transferase (**4.1**) (Fig. 29.4). The benzodiazepine system is intended to mimic the proposed β-turn in the CaaX motif (where C is a Cys with an essential thiol group, a is any aliphatic residue and X is usually Met or Ser) of the enzyme substrate[90] and has an IC_{50} less than 1 nM. Another example of a nonpeptidic *Ras*-farnesyl-transferase inhibitor based on the benzodiazepine was reported by BMS. The essential sulfhydryl group was replaced by imidazole[91] and the optimized compound (**4.2**) with an IC_{50} value of 1.4 nM advanced into human clinical trials.[92,93] An initial screening hit from the Schering-Plough compound library, with low micromolar activity, had a tricyclic amides structure with little resemblance to the benzodiazepines. This was optimized to give a high potency with $IC_{50} = 1.9$ nM without the critical cysteine moiety (**4.3**) (Fig. 29.5).[94,95]

Interestingly, the structurally related to (**4.1**), a lipophilic conformationally restricted benzoazopine mimic (**5.1**) of the tripeptide (Z)-Phe-His-Leu is a potent inhibitor of angiotensin-converting enzyme (ACE) and neutral endopeptidase.[96] Presumably in this instance, by analogy with other metalloprotease substrate interactions,[97] the enzyme-bound substrate adopts an extended conformation. A 2-oxo-1,5-benzodiazepine was recently utilized as a conformationally constrained β-strand peptidomimetic with three conserved key hydrogen-bonding capabilities to bind to the active site

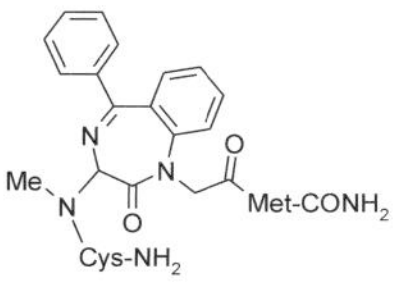

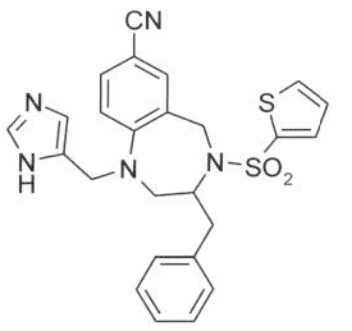

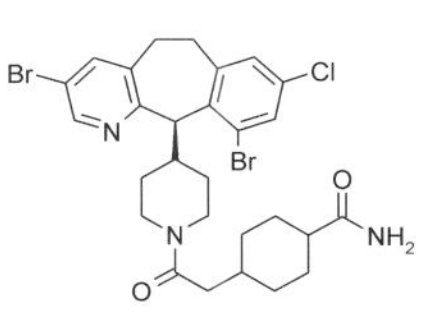

Fig. 29.4 Examples of benzodiazepine as a β-turn mimicking scaffold.

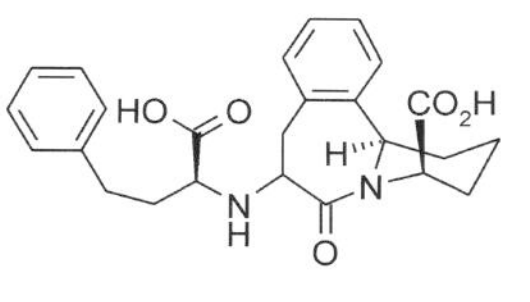

Fig. 29.5 Examples of benzoazepine as a β-strand mimicking scaffold.

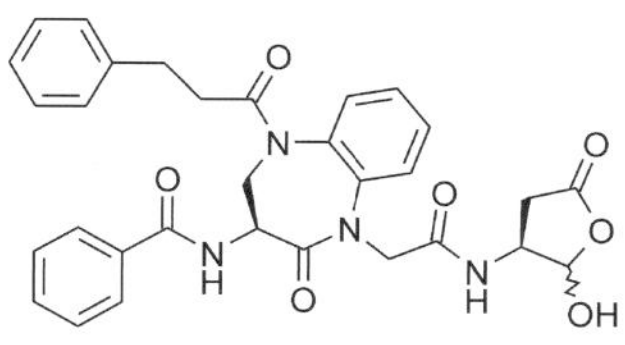

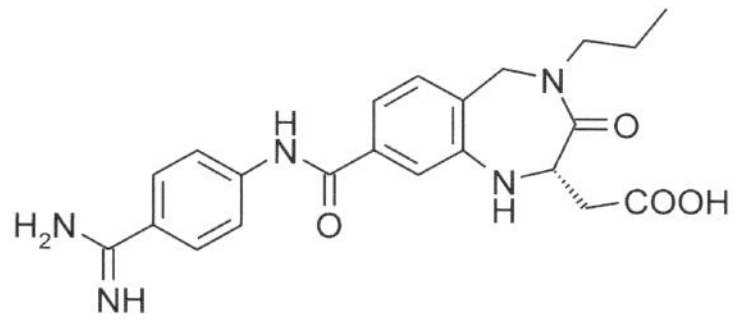

of a cysteine protease, caspase-1 (ICE).[98] Compound (**5.2**) was found to be a potent inhibitor of ICE with a K_i value of 90 nM. Benzodiazepine scaffolds have also been employed in many other applications, for example, to display key acidic and basic groups in an RGD motif as a fibrinogen receptor antagonist (**5.3**).[99]

B   Recent advances

In the design of reverse-turn mimetic systems, there are a number of concerns and criteria which need to be addressed. β-Turns comprise a rather diverse group of structures. β-Turns are classified according to the φ and ψ torsional angles of the middle $i + 1$ and $i + 2$ residues. In addition to a number of turn types (I, I', II, II', III, III', IV, V, Va, VIa, VIb and VIII) the C_i^α to C_{i+3}^α distance varies from 4 to 7 Å.[23] From this cursory discussion, it should be readily apparent that no one structure can accurately mimic this diversity in β-turns. The interaction of the amino acid side-chains with their complementary receptor binding sites is the critical determinant of biological specificity. A successful peptidomimetic must correctly position the appropriate functional groups on a relatively rigid framework. Therefore, an idealized mimetic design should incorporate the ability to accurately display critical pharmacophoric information in the same manner in which it is presented in native reverse turns. This is a far from trivial synthetic problem, in that it requires the stereo- and enantiocontrolled introduction of a minimum of four noncontiguous asymmetric centres. Furthermore, the nonpeptidic character of these molecules best promises to overcome the inherent problems of peptides. Synthetic expediency is a major concern that should not be lightly regarded, particularly at an early stage when the delineation of structure–activity relationships is critical and requires the synthesis and evaluation of a series of related structures.

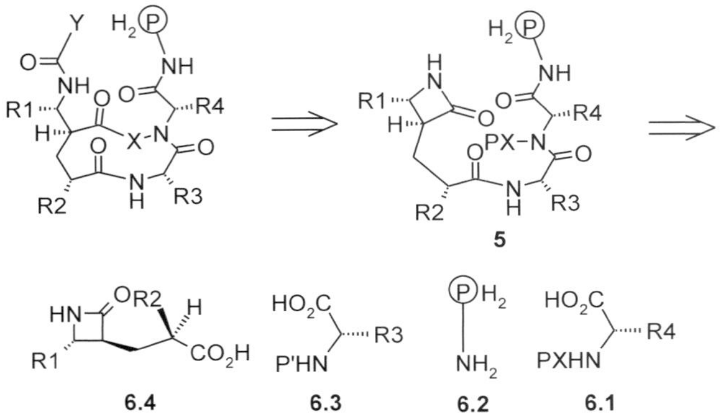

Fig. 29.6 Retro-solid-phase-synthetic scheme of cyclic β-turn mimetics.

The major breakthrough in peptide synthesis was the development of a solid-phase methodology by Merrifield, for which he was awarded the Nobel Prize in 1985.[100] We wished to capture the modular component nature and automation potential of SPPS (solid-phase peptide synthesis) in our approach to peptidomimetics. An additional advantage is that libraries of conformationally constrained peptidomimetics are readily generated.[101,102] With this mandate, we developed a retrosynthetic strategy to accomplish this goal (Fig. 29.6). The synthesis of the reverse-turn mimetics can be performed in solution; however, it is also fully compatible with SPPS protocols. In essence, it involves the coupling of the first modular component piece (6.1) to the amino terminus of a growing peptide chain (6.2). Coupling of the second component (6.3), removal of the protecting group P′, and subsequent coupling of the third modular component (6.4) provides the nascent β-turn (6.5). The critical step in this sequence involves the use of an azetidinone as an activated ester to effect the macrocyclization reaction.[103] Upon nucleophilic opening of the azetidinone by the X moiety, a new amino terminus is generated for continuation of the synthesis. An important feature of this scheme is the ability to alter the X-group linker, with regard to both the length and degree of rigidity/flexibility. Solution structures of this template with X = $CH_2C(CH_3)_2$ were determined by 2D NMR and found to mimic a β-turns well.[104] The requisite stereogenic centres

are readily derived, principally from the 'chiral pool'. The synthesis readily allows for the introduction of natural or non-natural amino acid side-chain functionality in either L or D configuration. Additionally, deletion of the second modular component (6.3) provides access to γ-turn mimetics[105,106] (Fig. 29.7).

We have used this mimetic system to explore SAR among molecules of the immunoglobulin gene superfamily.[107] Immunoglobulins are constructed from a series of antiparallel β-pleated sheets connected by loops.[108] The specificity of these molecules is determined by the sequence and size of the canonical hypervariable complementarity-determining regions (CDRs).[109]

The monoclonal antibody 87.92.6 (mAb 87.92.6) is an anti-idiotype antibody which binds to the cellular receptor of the type 3 reovirus. Sequence analysis revealed an intriguing sequence homology between the two proteinaceous ligands which bind the receptor.[110] In particular, a region within the CDR2 of the light chain of mAb 87.92.6 and the haemaggluttinin of the type 3 reovirus exhibited strong primary sequence homology. The V_L CDR2 canonically exists in a reverse-turn conformation.[111] On the basis of this analysis, we designed and synthesized a reverse-turn mimetic (Fig. 29.8), which incorporated the sequence Tyr-Ser-Gly-Ser-Ser that is conserved in both ligands. Importantly, the mimetic displayed similar binding properties to the cellular reovirus receptor and to mAb 9BG5, and had the same inhibitory effect on cell proliferation as did the antibody 87.92.6.

We have also used this mimetic system to design a mimetic of the CDR2-like region of human CD4 (Fig. 29.9).[112] CD4 is a 55-kDa glycoprotein, primarily found on the cell surface of the helper class of T cells. It binds the human immunodeficiency virus glycoprotein (HIV gp120) with high affinity (K_d = 1–4 nM), and is an important route of cellular entry for the virus. Extensive mutagenesis[113,114] and peptide mapping[115] experiments have shown that the region of amino acids 40–45 within the CDR2-like domain of CD4 is critical for gp120 binding. X-ray crystallographic analysis showed that residues Gln40 to Phe43 reside on a highly surface-exposed β-turn connecting the C′ and C″ β-strands. A first-generation mimetic of this region (9.1) was designed and synthesized. Importantly, this small-molecule mimic (9.1) (molecular weight = 635) abrogates the binding of

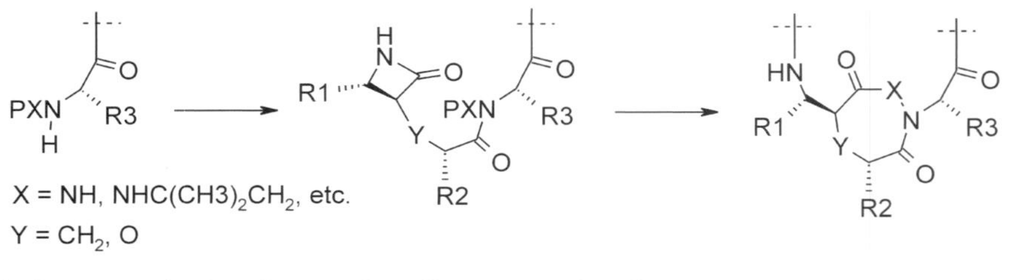

Fig. 29.7 Retro-solid-phase-synthetic scheme of cyclic γ-turn mimetics.

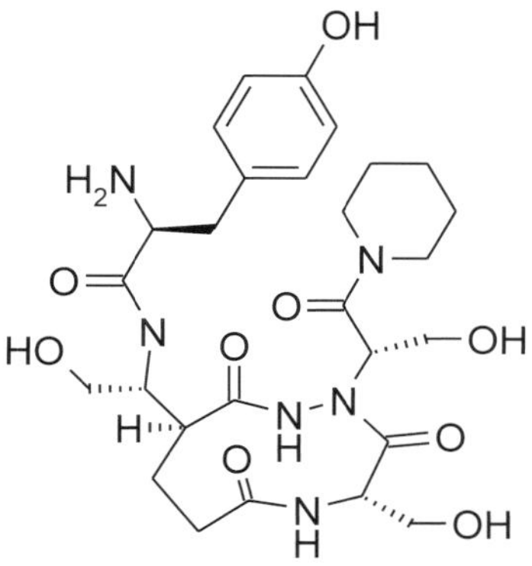

Fig. 29.8 β-Turn mimetic used for presenting a conserved epitope in the CDR2 domain of the antibody.

HIV-1 (IIIB) gp120 to CD4$^+$ cells at low micromolar levels and reduces syncytium formation 50% at 250 μg/ml.[1] A different CD4 mimetic (**9.2**) was also designed to incorporate another key functionality of Arg59 at the beginning of the DE loop by using homo-Arg residue, in addition to the Gln40–Phe43 of the highly surface-exposed β-turn connecting the C′ and C″ strands.[116] The aqueous solution structure of the 10-membered ring system in (**9.2**) was determined by distance geometry calculations using 36 observed rOe (rotational frame nuclear Overhauser effect) distance constraints and 138 antidistance constraints from 2D NMR experiment.[117] The best 12 NMR structures are shown in Fig. 29.10 overlaid with an idealized type II′ β-turn in bold.

While this monocyclic reverse-turn template clearly mimics the backbone conformation and side-chain presentations of β- and γ-turns quite well, one concern with such a large macrocycle is that the relatively flexible peptide backbone conformations are strongly influenced by the side-chain–side-chain interactions, as investigated by NMR solution structure (Fig. 29.11).[104] We therefore needed a new type of turn mimetic bearing a constrained and rigid skeleton which could potentially enhance binding and/or improve bioavailability. Based on the conformational analyses of several heterobicyclic systems, we envisioned

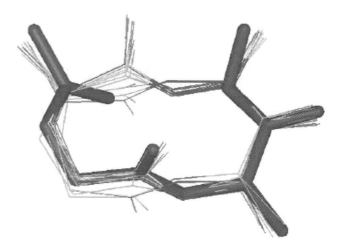

Fig. 29.10 Twelve NMR aqueous solution structures of a 10-membered β-turn template overlaid with idealized type II′ β-turn shown in bold.

that (**11.1**) and (**11.2**) could be good candidates to fulfil our requirements.[102,118,119] For the solid-phase preparation of (**11.1**), two types of linker L, i.e. acetal type (A) or olefin type (B), were employed as masked aldehydes (Fig. 29.12). The key transformation involves cleavage from the resin accompanied by acid-catalysed tandem cyclization of the corresponding acetal or aldehyde via an acyliminium intermediate (**12.2**). The resulting product is a single diastereomer (**12.1**) in which the hydrogen at the ring junction is *trans* to the α-hydrogen at the i + 2 position, as confirmed by both NMR and X-ray crystallography.[102] With this synthetic scheme, four functional side-chains are readily introduced from four commercially available components and simple synthetic procedures.[120] This template with a sulfonylamide group at the i position was crystallized for X-ray structural analysis. The X-ray structure (Fig. 29.13) was compared with the idealized β-turn conformations at seven atom positions, i.e. S, N1, C3, C6, C11, N8 and C12 against the corresponding $C_\alpha(i)$, C(i), $C_\alpha(i + 1)$, $C_\alpha(i + 2)$, $C_\beta(i + 2)$, N(i + 3) and $C_\alpha(i + 3)$ atoms of ideal β-turns. It appears that this bicyclic β-turn template closely mimics the most common type I β-turn

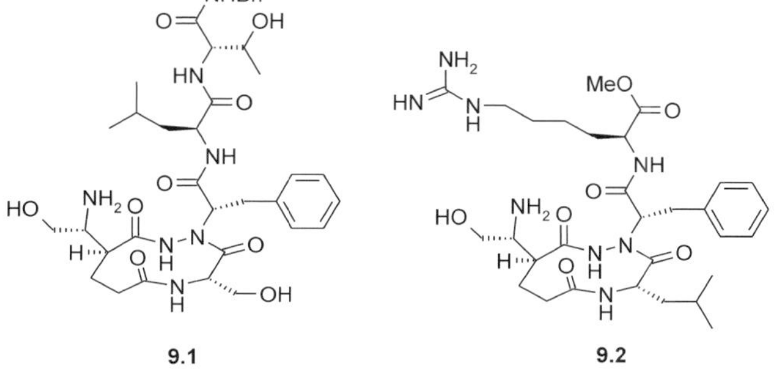

Fig. 29.9 β-Turn mimetics of the CDR2-like domain of CD4 that inhibit HIV-1 gp120 binding to CD4$^+$ cells.

9.1

9.2

X = SO₂, COO or CONH

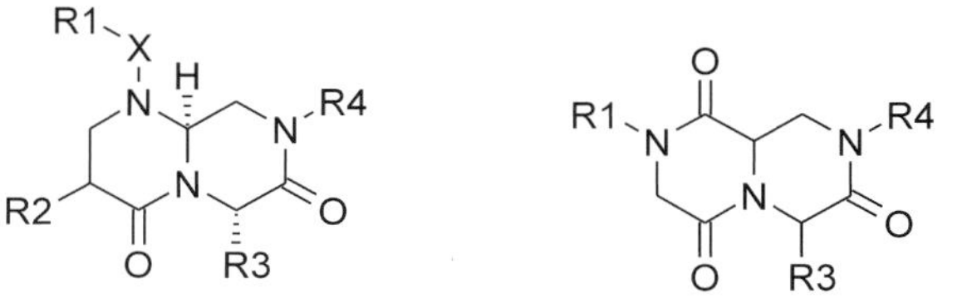

11.1

11.2

Fig. 29.11 Bicyclic β-turn mimicking scaffold with three or four points of functional diversity.

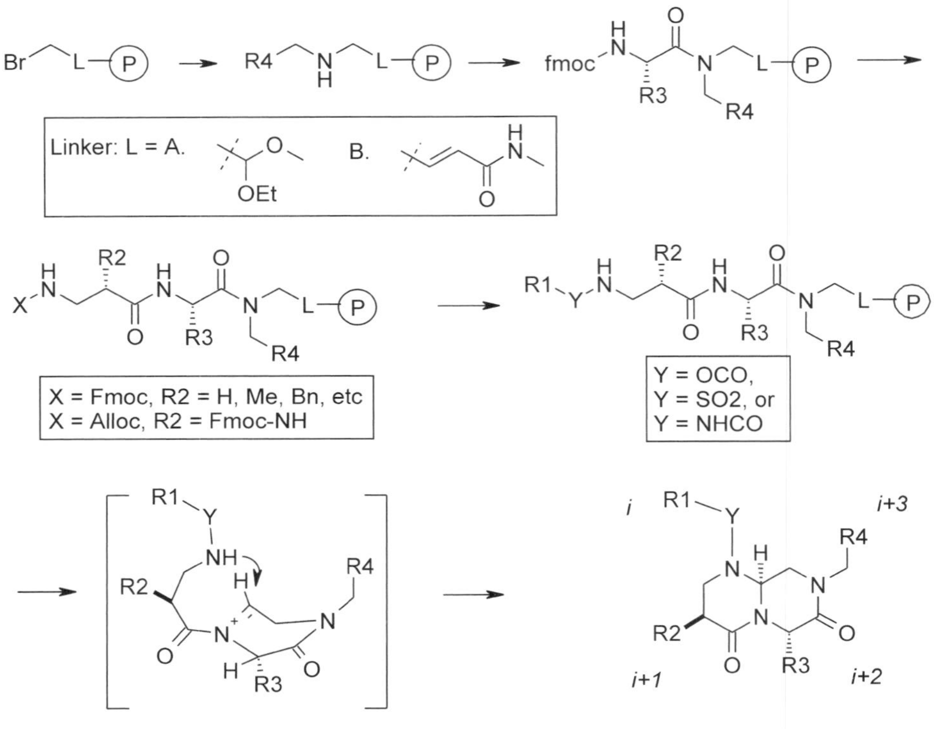

Fig. 29.12 Solid-phase synthetic scheme for the bicylic β-turn scaffold with four points of diversity.

(rmsd = 0.55 Å). Utilizing this chemistry and the IRORI minikan™ system, we have produced more than 40 000 reverse-turn mimetics based on this privileged chemotype. All compounds are examined for identity and purity by LC-MS, with a success rate of more than 75% at a purity level of greater than 80%.[121]

Recently, we have addressed a problem previously described concerning the relationship between morphine and enkephalin. The inherent mobility[122,123] of the enkephalin linear peptide framework, its rapid degradation *in vivo*,[124] and the existence of multiple receptor subtypes,[125] have hampered the assessment of its bioactive conformations. Conformationally constrained peptides or peptidomimetics should facilitate this task. Several turn conformations have been proposed based on computational models,[126–130] X-ray crystallography[131,132] and spectroscopic studies.[122,123,133,134]

In 1976, Bradbury *et al.*[135] proposed a β-turn model stabilized by an intramolecular hydrogen bond between the amide nitrogen of Phe4 and the carbonyl oxygen of Tyr1 (4 → 1 β-turn), which in turn produces a spatial disposition between the Phe4 aromatic ring and the tyramine segment of Tyr1, analogous to that existing between the corresponding moieties in the potent morphine analogue PEO (7-(1-hydroxy-1-methyl-3-phenylpropyl)-6,14-*endo*-ethenotetra-

hydrooripavine).[136] Further support for the relevance of this conformation was provided by energy calculations on the potent [D-Ala2, Met5] enkephalin analogue, the lowest energy conformer[137,138] of which contained a folded structure with a reverse-turn centred on residues D-Ala/Gly2 and Gly3. However, conformation analysis of 13- and 14-membered cyclic analogues by DiMaio

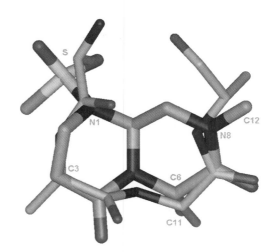

Fig. 29.13 X-ray structure of β-turn scaffold in thick tube overlaid on the type I β-turn in thin tube.

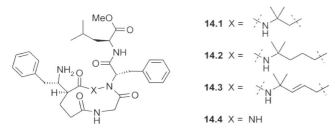

Fig. 29.14 Enkephalin mimetics with a β-turn at Gly2 and Gly3.

and Schiller showed the presence of a type II′ 5 → 2 β-turn centred at Gly3 and Phe4 instead of the 4 → 1 β-turn centred at Gly2 and Gly3.[139,140]

To examine further this hypothetical bioactive conformation, we have synthesized a family of 4 → 1 β-turn mimetics (**14.1–14.4**) (Fig. 29.14). The lowest energy conformer of the 10-membered ring system is an excellent mimic of an idealized type I′ β-turn (6 atom rmsd = 0.22 Å) and displays excellent overlap with the critical Phe4 aromatic ring and tyramine moieties of the morphine analogue PET (7-(1-phenyl-3-hdroxybutyl-3-)*endo*ethenotetrahydrothebaine), yet it is essentially devoid of biological activity.[101,141] Only the 14-membered ring analogue (**14.3**), which has a rather expanded loop structure, demonstrates any, albeit minimal, binding activity at the μ opioid receptor. The results of this investigation can be interpreted as casting significant doubt on the biological relevance of a 4 → 1 β-turn conformation for enkephalin.

The more rigid bicyclic β-turn template (**11.1**) was recently utilized to elucidate the bioactive conformation of enkephalin and newly isolated opioid peptides, endomorphin-1 (YPWF) and endomorphin-2 (YPFF).[142,143] By employing our privileged β-turn scaffold, we can investigate the following hypothetical bioactive β-turn conformation models: enkephalin 4 → 1 β-turn model (A), enkephalin 5 → 2 β-turn model (B) and endomorphin 4 → 1 β-turn model (C) (Fig. 29.15). To examine these models, we incorporated the designed diversity elements into the scaffold at the *i* to *i + 3* positions. A focused library of opioid mimetics was constructed from 12 primary amines at the *i + 3* position (R4), 13 α-amino acids at the *i + 2* position (R3), one β-amino acid (β-alanine) at the *i + 1*

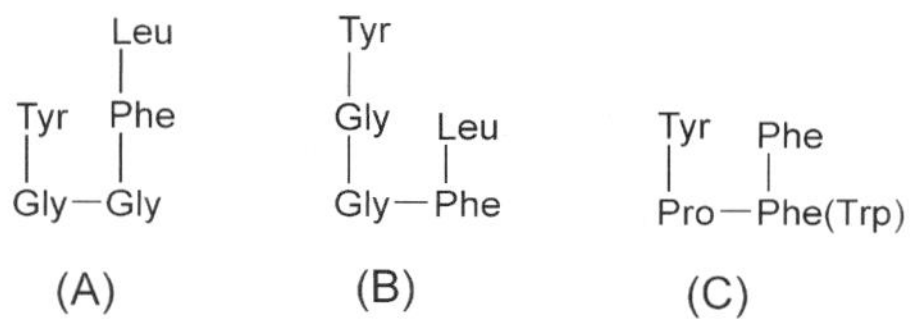

```
        Leu            Tyr
         |              |
  Tyr   Phe      Gly         Tyr   Phe
   |     |        |    Leu     |     |
  Gly — Gly      Gly — Phe   Pro — Phe(Trp)

   (A)            (B)            (C)
```

Fig. 29.15 Schematic bioactive conformational models of enkephalin and endomorphin.

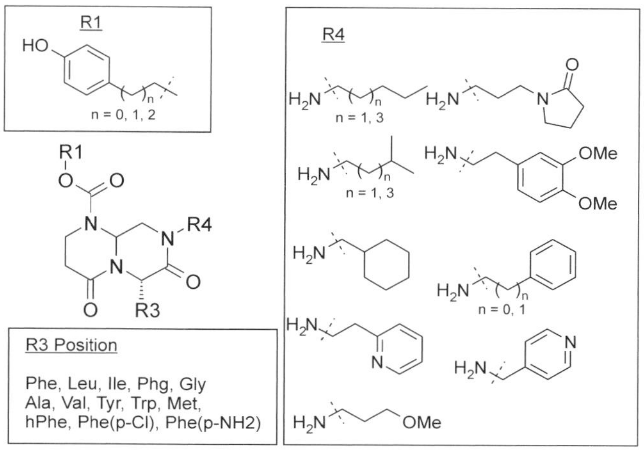

Fig. 29.16 Diversity components for enkephalin and endomorphin β-turn mimetics.

position (R2), and a phenol component with three different lengths of spacer at the *i* position, as shown in Fig. 29.16.[144]

Based on the preliminary SAR result from the focused library, four compounds (**1–4**) along with the linear analogue (**5**) in Table 29.2 were resynthesized, and IC$_{50}$ values were determined for the inhibition of specific radioligand (^{3}H-naloxone) binding to the rat cerebral cortex relatively nonselective opioid receptors. Compounds (**3**) and (**4**) were further studied using human μ-, δ-, and κ-receptors, and found to be very μ-selective. These compounds represent an enkephalin (B) model with one methylene length shorter at the *i* position and an endomorphin (C) model with one methylene length longer at the *i* position. Therefore it is not surprising that the compounds (**3**) and (**4**) exhibit strong μ-selectivity. For comparison, a linear peptide (**5**) corresponding to (**3**) displayed no binding activity at 1 μM level. This confirms that the correct spatial orientations and conformational rigidity of these functional groups are critical for high receptor affinity. Compound (**3**) was selected for an *in vivo* analgesic efficacy test using the mouse hot plate model. A 10 mg kg^{-1} i.v. dose produced 300% analgesic effect equipotent to morphine at the initial time-point, while the *in vivo* half-life (∼ 25 minutes) was almost two-fold less than that of morphine. It is interesting to note that the compound (**4**) achieved very high potency (IC$_{50}$ = 9 nM) without incorporating the presumed critical protonated nitrogen atom in the molecule.

It is hoped that this type of a systematic approach to the synthesis of constrained reverse-turn analogues in conjunction with multiple peptide synthetic strategies will clarify the situation regarding the biological significance of these and other proposed receptor-bound reverse-turn conformations.

Table 29.2 *Nonselective opioid receptor binding assay of enkephalin and endomorphin β-turn mimetics*

	Compounds				Inhibition (%)		IC$_{50}$ (nM)
	R1	R2	R3	R4	0.1 μM	1 μM	
1	p-OH-Phenethyl	H	Bn	nBu		84	150
2	p-OH-Phenethyl	H	Bn	Bn		96	80
3	p-OH-Phenethyl	H	Phenethyl	nBu	79	99	27
4	p-OH-Phenethyl	H	Phenethyl	Bn	91		9
5	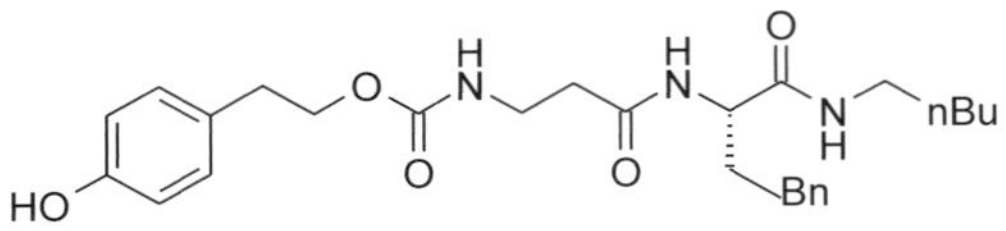					< 10	> 1000

C   α-Helix

Helices are the most common secondary structural element found in globular proteins, accounting for just over one-third of all residues.[145] Extensive effort has been directed towards an understanding of helix formation, its stability and amino acid propensities.[146,147] Yet, as noted by Kemp and Curran in 1988,[148] little attention has been given to the design of helical templates compared with the other secondary structure templates, i.e. β-strand and reverse-turn templates. The situation has not changed dramatically since. This is due in large part to the inherently more difficult task of mimicking the approximately 12 amino acids (i.e. three turns of an α-helix) required to form a stabilized, isolated helical peptide. The formation of an α-helix involves two steps: initiation and propagation. To date, most of the effort in the design and synthesis of α-helix mimetics has centred on *N*-terminal initiation motifs.

Arrhenius and Satterthwait[149,150] and recently Cabezas and Satterthwait[151] used a hydrazone-ethylene bridge to replace the 5 → 1 backbone hydrogen bond in one turn of an α-helix to afford the cyclic peptide shown in Fig. 29.17. Conformational analysis of both the methyl ester and amide were performed in CDCl$_3$ and DMSO-d$_6$ by NMR spec-

troscopy. Based upon 1D nuclear Overhauser effect (nOe) measurements, both compounds seem to prefer the *cis-N*-methyl peptide bond conformation in DMSO and an equilibrium mixture in CDCl$_3$. Based on this analysis, the formation of an α-helix inducing conformation in these macrocyclic compounds is inconclusive. A peptapeptide Ala-(Glu-γ-ethylester)$_4$-ethyl ester was subsequently added to the carboxy terminus of the macrocyclic template (Fig. 29.17). The conformations of the pentapeptide both with and without the attached macrocyclic template were monitored by NMR utilizing 3J coupling constants, sequential nOe, and H/D exchange rates of the amide protons in deuterated trifluoroethanol. The observed 3.8 Hz 3J coupling constant[152] observed for the alanine residue, together with reduced amide proton exchange rates, tends to indicate the formation of an α-helical conformation. Another hexapeptide, Ala-Glu-Ala-Ala-Lys-Ala-NH$_2$, was constructed by solid-phase synthesis,[151] attached to the template, and found to form a full-length α-helix in water as indicated by NMR spectroscopy. Substitution of Ala with Pro within the cyclic template enhanced α-helix nucleation and shifted the equilibrium further toward a full-length α-helix, as evidenced by the strong d$_{\alpha\beta}$(*i, i + 3*) nOe signals.

The proper alignment of two or more hydrogen-bond acceptors is crucial for the design of a successful helix nucleation template (Fig. 29.18). Kemp and Curran restrained two proline rings with a thiamethylene bridge

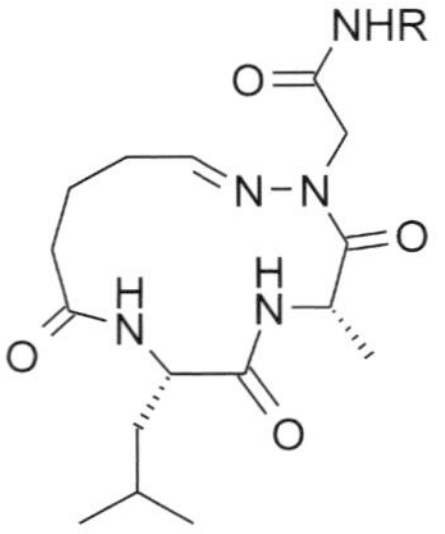

R = Ala-(Glu-γ-ethyl ester)$_4$-ethyl ester
R = Ala-Glu-Ala-Ala-Lys-Ala-NH$_2$

Fig. 29.17 Macrocyclic α-helix initiator.

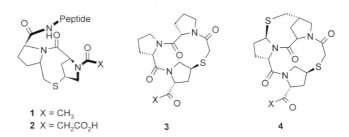

1 X = CH$_3$
2 X = CH$_2$CO$_2$H

Fig. 29.18 Cyclic polyproline triacid α-helix initiators.

to form a tricyclic template Ac-Hel (**18.1**),[148] with the intention of orienting the three amide carbonyls at the proper pitch and spacing for a right-handed α-helix. Observation of an nOe from the acetyl methyl of (**18.1**) to the following proline α-proton in CDCl$_3$[153] leads one to believe that the carbonyl oxygen of the acetyl group at the *N*-terminus is improperly positioned (i.e. oriented *cis* instead of the desired *trans*), presumably to avoid the dipole–dipole repulsion of the aligned carbonyls. However, NMR investigation of (poly-alanine) ($n = 1-6$) conjugated to this Ac-Hel template in CDCl$_3$ indicated the existence of a conformational equilibrium between an intramolecularly hydrogen-bonded structure with a *trans N*-terminal acetamide bond, and a nonhydrogen bonded structure derived from the *cis* conformer.[154] The *trans*-to-*cis* ratio (which corresponds with the observed helix-to-random coil ratio) was found to be length and solvent dependent for poly(alanine) oligomers.[155] Use of an anionic N-cap at X[156] (**18.2**) was explored and found to be superior to *N*-acetyl at inducing α-helix structure, as determined by monitoring the CD spectra.[157]

In a subsequent design, the number of hydrogen-bond donor sites was increased to four by using cyclic triproline helix templates (**18.3**, **18.4**).[158,159] X-ray and NMR studies determined that the templates (**18.3**) and (**18.4**) exist largely in a nonhelical conformation. Appending a poly(alanine) ($n = 1-3$) sequence to the template generated a peptide that largely adopted a 3$_{10}$ helical conformation. A similar template was synthesized in an attempt to align the carboxamide dipoles for α-helix initiation, but without success.[160,161] A template that affords rigid alignment of three carbonyl oxygens was devised by Müller *et al.*[162] They utilized the cage compound (**19.1**) which was readily accessible via a series of electrocyclic addition reactions (Fig. 29.19). A nonapeptide (mixture of Ala and Aib) coupled to this cage compound showed significantly increased α-helicity compared with the linear nonapeptide in a 1:1 water–TFE solution, as judged by CD spectra. On the other hand, the same nonapeptide coupled to the enantiomeric template (**19.2**), where the three carbonyl groups are aligned as in a left-handed helix, exhibited decreased helicity under the same conditions. Estimated

α-helicity based on the CD ellipticities at 222 nm are 40%, 70% and 25% for the *N*-Boc protected linear peptide, right-handed conjugate (**19.1**) and left-handed conjugate (**19.2**), respectively.[162]

Bicyclic diacid template (Fig. 29.20) has also been designed and synthesized to stabilize the hydrogen-bonding pattern of a conjugated peptide in an α-helix conformation.[163] A hexapeptide EALAKA-NH$_2$ was attached to either the bicyclic amide or ester derivatives, and the influence of the template on the peptide conformation was evaluated in aqueous solution by using both CD and NMR. The ester-linked template enhanced the helicity of the peptide to 49% at 0°C compared with the reference hexapeptide value of 8%, while the amide-linked template was only slightly effective.

We have designed and synthesized dipeptide α-helix initiators such as (**21.1**) and (**21.2**) in Fig. 29.21.[164–166] The conformationally constrained dipeptide mimetic (**21.1**) was inserted into the 'AA' position of C-terminal analogues of NPY (Neuropeptide Y) to examine the correlation between peptide helicity and biological activity proposed by Jung *et al.*[167] CD studies showed that the linear peptides with Ala-Ala or Ala-Aib in the highlighted sequence Ac-^{25}RAAANLITRQRY-NH$_2$ had similar helicities of 3% in water, while the analogous peptide containing the α-helix mimetic (**21.1**) displayed enhanced helicity of 13%.[164]

D β-Strand

A wide array of potential pharmaceutical applications for β-strand mimetics (e.g. enzyme inhibitors,[19] antigen

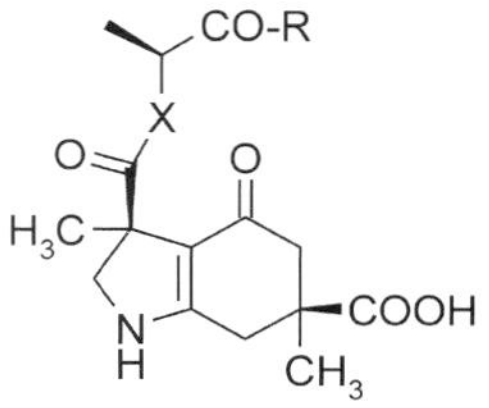

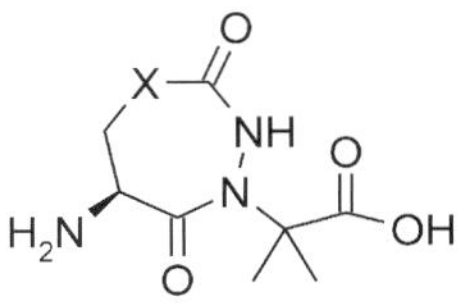

X = NH or O
R = Glu-Ala-Leu-Ala-Lys-Ala-NH$_2$

Fig. 29.20 Bicyclic diacid α-helix initiators.

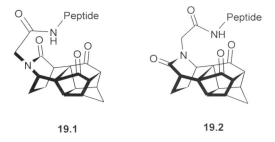

19.1   **19.2**

Fig. 29.19 Right- and left-handed α-helix initiators.

21.1 X = CH$_2$
21.2 X = NH

Fig. 29.21 Monocyclic α-helix initiators.

presentation,[168] disruption of dimerization[169,170] and second messenger signalling[171]) has encouraged a recent flurry of activity in the development of various dipeptide β-strand mimetics. The use of peptidomimetics as enzyme inhibitors has been recently reviewed by Rich's group.[13,19] In proteins and enzyme-substrate interfaces, β-sheet formation is the result of both an extensive hydrogen-bond network, and side-chain interactions. Providing hydrogen-bond donors, acceptors and side-chain functional groups in the proper arrangement presents a significant synthetic challenge in designing β-strand templates.

Kemp *et al.* have prepared β-sheet mimetics based on a diacylaminoepindolidione template. This template was linked to a Pro-D-Ala dipeptide, which is presumed to adopt a β-turn conformation. Subsequent residues are attached via a urea linkage, which inverts the directionality of the peptide chain, permitting the formation of an antiparallel β-sheet (**22.1**) (Fig. 29.22).[172–174] Attachment of residues via amide instead of urea linkage allows for the formation of a parallel β-sheet (**22.2**).[172] The existence of a β-sheet conformation in (**22.1**) was confirmed by the observation of nOes between the α-protons of glycine and the H-1 and H-7 of the epindolidione, and between the protons of the *N*-methyls and H-4 and H-10 of the epindolidione in DMSO. Additional supporting evidence was provided by the temperature dependence of amide proton chemical shifts, and geminal coupling constants. The β-turn-forming sequence Pro-D-Ala was replaced by Sar-Gly (Sar = *N*-methylglycine) to examine the effect of the β-turn on the formation of the antiparallel β-sheet. The CD spectrum for this compound, in solvents ranging in polarity from THF to DMSO, exhibited no bands in the range 300–500 nm, indicating no significant interaction between the two amino acid asymmetric centres and the epindolidione, consistent with there being no significant secondary structure.[175] A similar approach has been recently taken by Nowick's group to construct a stable artificial β-sheet with the use of 5-amino-2-methoxybenzamide (**23.1**) and 5-amino-2-methoxybenzoic hydrazide (**23.2**) β-strand scaffolds (Fig. 29.23).[176–180]

Although the epindolidiones and amino-methoxybenzoic hydrazide β-strand mimetics may provide valuable information on the nucleation and stability of parallel and antiparallel β-sheets, a significant shortcoming is the difficulty of incorporating side-chain functionality into this template, which would be required for many biological applications. The use of 3,5-linked pyrrolin-4-ones[181,182] overcomes this problem at the expense of a displaced NH group (Fig. 29.24). Initial modelling indicated that a β-strand is the favoured conformation and that the side-chain orientations closely mimic a natural β-strand found in angiotensin. It was also determined from examination of the unit cell that the mimetic dimer adopts an antiparallel β-pleated sheet.

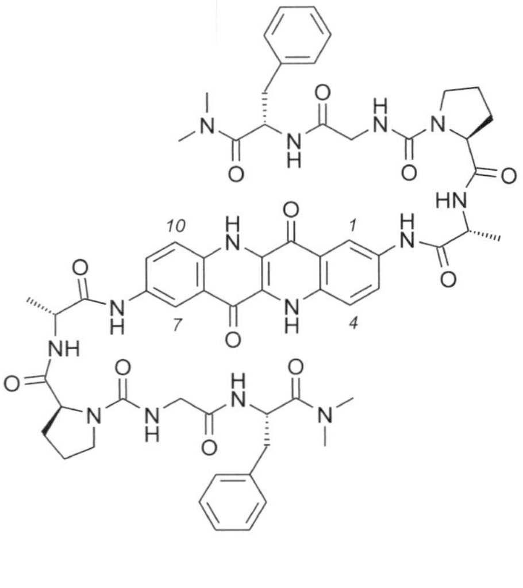

22.1 antiparallel β-sheets

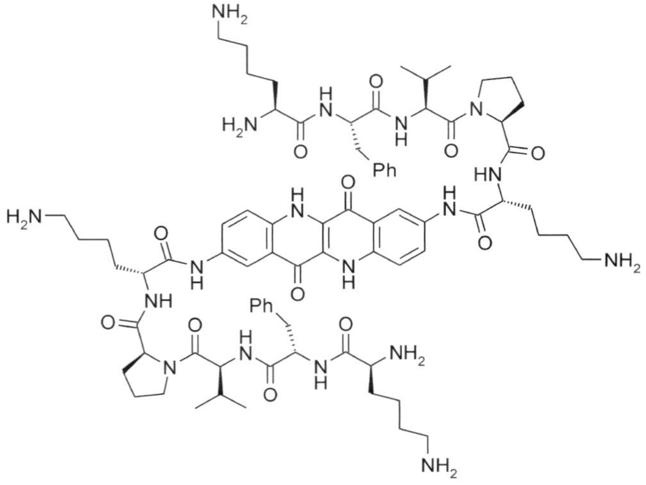

22.2 parallel β-sheets

Fig. 29.22 β-Sheet nucleator.

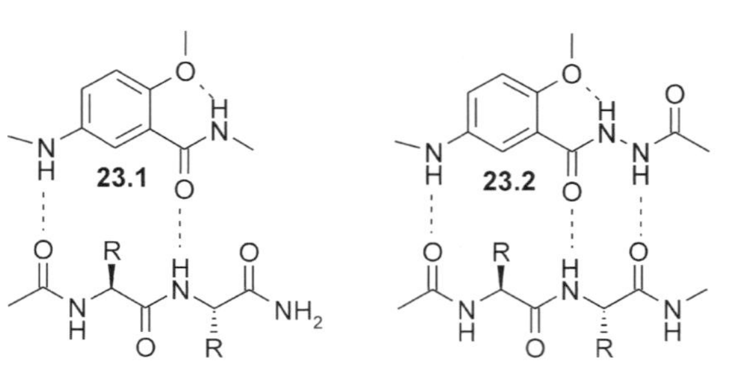

Fig. 29.23 Use of amino-methoxy-benzamide as a β-strand scaffold.

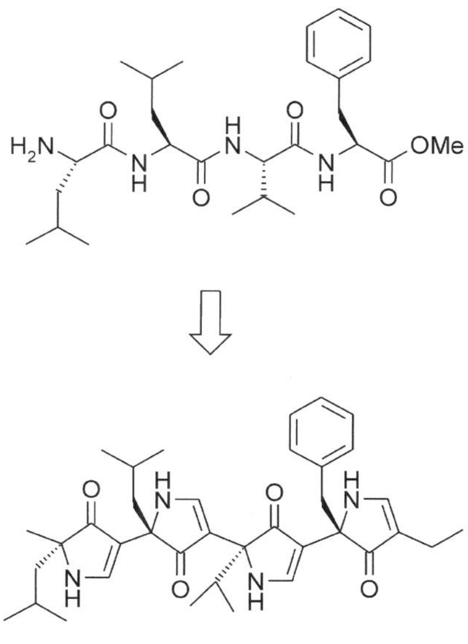

Fig. 29.24 Angiotensin inhibitor with pyrrolinone β-strand scaffold.

Aspartic proteinase inhibitors (for renin and HIV-1 protease) were constructed using the pyrrolinone template based on previously reported peptidic inhibitors (Fig. 29.25). The first generation inhibitor (**25.1**) of HIV-1 protease based on the potent peptidic inhibitor L-682,679 (**25.3**) ($IC_{50} = 0.6$ nM, $CIC_{95} = 6$ μM) was less effective than (**3**) in an *in vitro* enzyme assay ($IC_{50} = 10$ nM), but displayed higher cellular antiviral activity ($CIC_{95} = 1.5$ μM) presumably due to the higher cell permeability of (**25.1**).[183] The

most active compound in this series (**25.2**) ($IC_{50} = 1.3$ nM, $CIC_{95} = 0.8$ μM) was, however, not orally bioavailable in dogs. The second-generation inhibitor (**25.4**) was thus designed to improve the bioavailability by reducing the molecular weight from 735 to 583 employing only one pyrrolinone moiety.[184] The structure of (**25.4**) was based on another potent peptidic inhibitor L-697,807(**25.5**). In contrast with the bispyrrolinone inhibitors, the mono-pyrrolinone inhibitor (**25.4**) was found to be less active ($IC_{50} = 2.0$ nM, $CIC_{95} = 100$ nM) in both *in vitro* and cell-based assays than the corresponding peptide analogue (**25.5**) ($IC_{50} = 0.03$ nM, $CIC_{95} = 3$ nM). However, (**25.4**) was orally bioavailable in dogs (F = 13%) with a half-life of 35 min and a clearance rate of 20 mL min^{-1} kg^{-1}. X-ray analysis of (**25.4**) bound to the protease revealed an unexpected hydrogen bond between catalytic Asp25 and the pyrrolinone NH, in addition to a unique water molecule bridging the indanol hydroxyl and the NH of Asp29 compensating for the loss of hydrogen bonding from Gly27. This entropically unfavourable inclusion of a water molecule in the active site may be one of the reasons for the lower *in vitro* activity observed for (**25.4**). Recently, utility of this template was further expanded to successful construction of a competitive peptidomimetic ligand for the rheumatoid arthritis-associated class II major histocompatibility complex protein HLA-DR1.[185] The exercises presented here demonstrated that useful *in vivo* activity and bioavailability can be attainable by replacing the peptide backbone by rigid peptidomimetics such as the pyrrolinone scaffold.

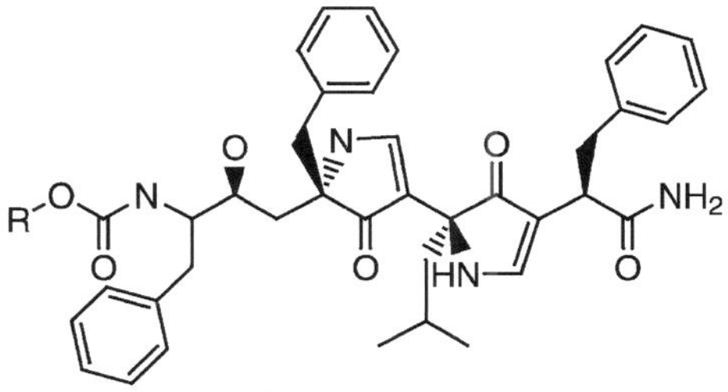

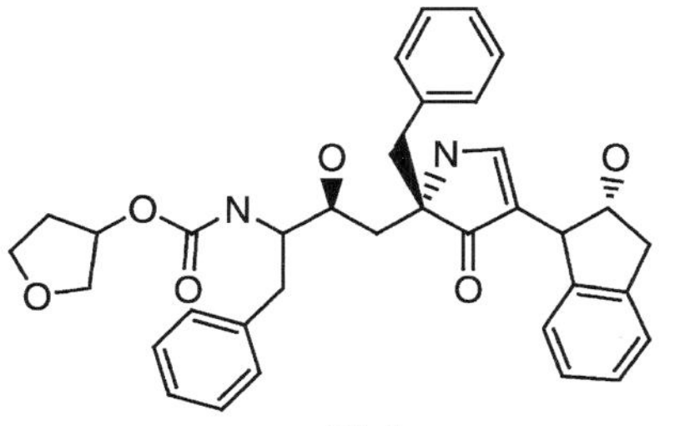

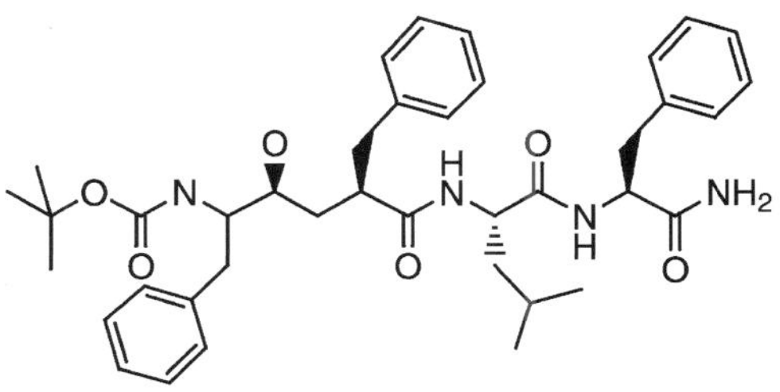

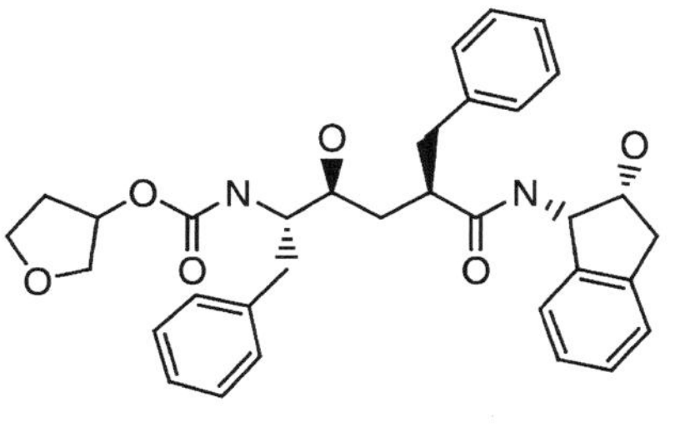

Fig. 29.25 HIV protease inhibitors with pyrrolinone β-strand scaffold.

E   Recent advances

Since our last review of this field,[186] there has been a plethora of reports on peptidomimetic inhibitors targeted towards the numerous enzymes and receptors which interact with peptide substrates and ligands that are in either extended or β-strand conformation. Several excellent reviews[13,19,187,188] are devoted to the design of peptidomimetics for inhibiting cysteine proteases, aspartyl proteases, especially HIV-1 protease,[189] metallo-proteases and serine proteases. Here, we will limit our discussions to the use of β-strand scaffolds for the design of thrombin inhibitors.[190] Thrombin is a serine protease that belongs to the chymotrypsin family. It plays a critical role in thrombosis and haemostasis as the terminal enzyme in the coagulation cascade. An X-ray co-crystal structure of the irreversible inhibitor PPAC (D-Phe-Pro-Arg chloromethylketone) with human α-thrombin[191] revealed that the tripeptide binds in a β-strand conformation with three key hydrogen-bond interactions, hydrophobic/aromatic interactions at the P2 and P3/D pockets and a salt-bridge interaction at the P1 pocket as schematically shown in Fig. 29.26 This binding motif has served as an excellent starting point for the design of many of the peptidomimetic inhibitors reported to date.

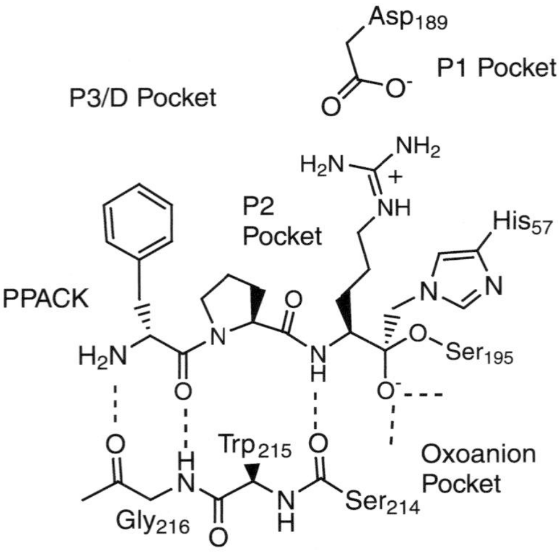

Monte Carlo conformational analysis suggested that the bicyclic structures (**27.1**) and (**27.2**) would rigidify the enzyme-bound conformation of the PPACK backbone yet preserve the key hydrogen-bond donors and acceptor, and orient the encompassed functional groups to approximate an idealized peptide in a β-strand conformation (Fig. 29.27). Boltzmann weighted average rmsd values at the seven superimposable heavy atom positions were 0.4 and 0.5 Å for (**27.1**) and 0.5 and 0.5 Å for (**27.2**) against parallel and antiparallel β-strands, respectively. More importantly, the templates (**27.1**) and (**27.2**) superimposed against PPACK and against BPTI (bovine pancreatic trypsin inhibitor, which represents a canonical loop conformation as a proteolysis substrate mimic), at the P2–P3 sites with a Boltzmann-averaged rmsd values of 0.4 and 0.2 Å, respectively. Encouraged by the molecular modelling study, a direct

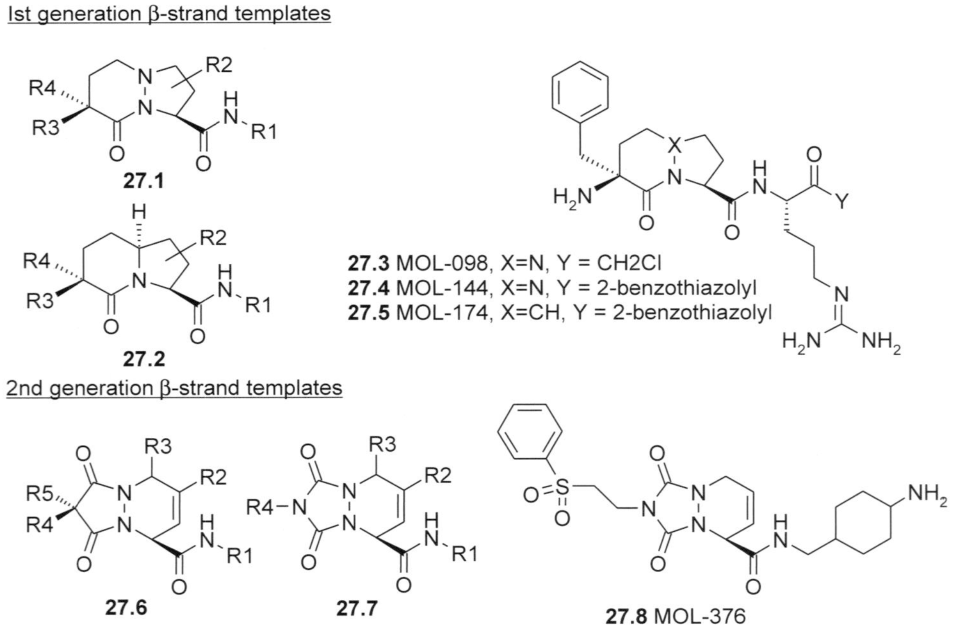

Fig. 29.26 Binding motif of PPACK at the thrombin catalytic site.

Fig. 29.27 Evolution of bicyclic β-strand scaffolds and application to inhibit the thrombin active site.

analogue of PPACK, MOL-098 (**27.3**), was synthesized with the bicyclic β-strand template (**27.1**).[192,193] MOL-098 was found to be equipotent to PPACK against thrombin (IC$_{50}$ = 1.2 nM versus 1.5 nM for PPACK), and showed slightly improved selectivity against other coagulation and anticoagulation enzymes (Table 29.3). As expected, an X-ray co-crystal structure of (**27.3**) with human α-thrombin revealed exactly the same binding mode as PPACK[191] with a rmsd value of 0.55 Å at all superimposable heavy atoms (30 atoms). The structure also showed that the β-strand template (**27.1**) formed three key hydrogen bonds to the thrombin, exactly as intended.

Although the thrombin inhibitory activity of (**27.3**) was good, the possibility of toxicity due to nonspecific alkylation of the highly reactive chloromethylketone led us to pursue alternative inhibitory strategies. Replacement of the chloromethylketone with an α-ketobenzothiazole P1′ moiety gave reversible inhibitors MOL-144 (**27.4**) with the β-strand template (**1**) and MOL-174 (**27.5**) with the β-strand template (**27.2**). Inhibitory activities of (**27.4**) and (**27.5**) against thrombin and selectivities against other coagulation and anticoagulation enzymes are listed in Table 29.3. Improved activity of the derivatives can be explained by a favourable stacking interaction between the aromatic heterocycle and Trp60D, and by a hydrogen bond to His57 of thrombin from the benzothiazole amine, as demonstrated by X-ray structural analysis of similar compounds containing an α-ketobenzothiazole group and by molecular modelling studies. Comparison of the *in vitro* data for PPACK with those of the constrained β-strand indicates that the template imparts greater selectivity for thrombin than the more flexible peptide. The selectivity of the α-keto-heterocycle inhibitors is even greater with (**27.5**) showing over 1000-fold selectivity for thrombin over the anticoagulation enzymes (Table 29.3).

Compounds (**27.4**) and (**27.5**) were evaluated in an *in vivo* arterio-venous shunt thrombosis model in baboons,[194,195] and both compounds at 1 μg min^{-1} were found to be effective in blocking platelet deposition, as monitored in real time by accumulation of [111]Indium oxine-labelled baboon platelets on a dacron graft. Encouraged by these results, compounds (**27.4**) and (**27.5**) were further evaluated

for their Caco-2 cell permeability (P$_{app}$ = 0.30 and 0.25 × 10^{-6} cm s^{-1} respectively). Bioavailability was evaluated in rats and monkeys. Although (**27.4**) and (**27.5**) displayed similar characteristics in the *in vitro* assays, the bioavailability of (**27.4**) was estimated to be 25% in both rats and monkeys, while that of (**27.5**) was 2% in both species.[192]

The first-generation β-strand templates (**27.1**) and (**27.2**) served very well to demonstrate the concept of utilizing a bicyclic structure to mimic a β-strand motif and were successfully incorporated into a variety of potent protease inhibitors. However, they lacked the means to readily introduce diverse functionality around the P2 site. Alternative templates (**27.6**) and (**27.7**) were chosen to take full advantage of the benefits of solid-phase organic synthesis while using a modular synthetic approach, and were constructed using the robust hetero-Diels-Alder (D-A) cycloaddition reaction as the key step.[196,197] Conformational analysis of (**27.6**) with R5 = NHCOCH$_3$, R4 = CH$_3$, and (**7**) with R4 = CH$_3$ (R1 = CH$_3$, R2 = R3 = H) indicates that the best conformers to mimic a β-strand were within 1.2 kJ mol^{-1} of the global energy minimum. The rmsd values of the best conformers overlaid with an ideal antiparallel β-strand were quite good, with 0.37 and 0.51 Å at 7 atom positions for templates (**27.6**) and (**27.7**), respectively. The dienophiles required to produce (**27.7**), *1,2,4-triazole-3,5-diones*, are much more reactive than the pyrazolidinyl diones used to synthesize (**27.6**), allowing the D-A reaction to proceed at room temperature or even below it. In addition, the resulting template becomes stereochemically simpler, although at the expense of a potentially important hydrogen-bond donor at the R4 position. To allow for a high degree of structural diversity in the synthesized template, as well as high efficiency and reproducibility to incorporate D-A partners, dienes can be prepared on the solid-support.[198]

The attractive features of this approach are the ability to rapidly introduce several different diversity elements within the template as well as the ability to link the template to a variety of components. Additionally, any of the groups attached to the template may include a chemically reactive functional group that can be further elaborated.

Table 29.3 *Thrombin inhibitors based on the β-strand scaffolds and their selectivity against coagulation enzymes*

		K_i (nM) (or IC$_{50}$ (nM) for PPACK and MOL-098)							
		Thrombin	Trypsin	Factor Xa	Factor VIIa	Protein C	Plasmin	Urokinase	t-PA
	PPACK	1.50	ND	170	200	280	700	508	106
27.3	MOL-098	1.20	ND	390	140	530	980	927	635
27.4	MOL-144	0.65	0.64	270	270	3300	420	600	495
27.5	MOL-174	0.085	0.85	19	200	1250	250	340	90
27.8	MOL-376	0.78	3400	> 40 000	> 40 000	> 40 000	> 40 000	> 40 000	> 40 000

A 100-member library utilizing template (**27.6**) and a 5000-member library with template (**27.7**) were constructed. From these libraries of compounds, several potent and selective inhibitors of a wide range of serine and cysteine proteases have been discovered.[21] As has been seen with a series of thrombin inhibitors derived from Argatroban[199] and NAPAP,[200] thrombin can well accommodate a nonelectrophilic moiety at the P1. Further modification of lead compounds by incorporating nonelectrophilic P1 moieties led to MOL-376 (**27.8**), which was found to be a highly potent and selective inhibitor of thrombin, with subnanomolar K_i, greater than 4000-fold selectivity against trypsin, and even higher selectivity against all coagulation and anticoagulation enzymes (Table 29.3). The P1 cyclohexyl group linker used in (**27.8**) to present a basic group to Asp189 at the bottom of the P1 pocket helps increase selectivity against trypsin, since the thrombin P1 pocket prefers more bulky hydrophobic groups while the trypsin P1 pocket is more accommodating of aromatic groups.[201] The phenylsulphonyl group at R4 in (**27.8**) was chosen to optimize thrombin activity while minimizing undesirable nonspecific plasma protein binding. The crystal structure of (**27.8**) bound to thrombin was determined along with the structures of other nonelectrophilic active-site inhibitors.[202] As predicted from the modelling study, the urazole β-strand template forms two hydrogen-bond interactions with thrombin at Gly216 and Ser214. Judging from the binding affinity of (**27.8**), the loss of one hydrogen-bond donor was well compensated for by the more rigid template and other binding interactions. The P1 cyclohexyl amine formed direct hydrogen-bond interactions with the O_δ of Asp189 and the carbonyl oxygen of Gly219, in contrast with the interaction observed in other inhibitors containing a P1 lysine residue, where the hydrogen bond between the lysine amine and Asp189 was mediated through a water molecule.[203] The sulphonyl group of (**27.8**) was located on the surface and had no interactions with thrombin. However, its solubility and acute geometrical angle probably helped to place the phenyl group in the thrombin D pocket in the proper manner. ADME properties of (**27.8**) will be reported in due course along with results from *in vivo* studies.

The synthetic scheme developed for the β-strand template (**27.7**) has proved to be robust and efficient. Utilizing this chemistry and the IRORI minikan™ system, we have produced well over 70 000 β-strand mimetics based on this privileged chemotype (**27.7**). All compounds are examined for identity and purity by LC-MS, with a success rate of over 85% at a purity level of greater than 80%.[121]

Numerous variations of β-strand templates have been reported as thrombin active-site inhibitors capable of presenting P1 and P3 groups in proper orientations while maintaining proper hydrogen-bonding capabilities.[190] Several examples of monocyclic β-strand templates (**28.1**),[204,205] (**28.2**),[206,207]

(**28.3**),[208,209] (**28.4**),[210] (**28.5**)[211] and (**28.6**),[212] and bicyclic β-strand templates (**28.7**),[213] (**28.8**),[214] (**28.9**),[215,216] (**28.10**),[217,218] (**28.11**)[219,220] and (**28.12**)[221] are listed in Fig. 29.28.

VI   TERTIARY STRUCTURE MIMETICS

The interacting surfaces of proteins or 'functional epitopes' often consist of a few residues that are spatially close due to the folding of the protein yet discontinuous and distant in primary sequence or even belonging to different protein subunits. This concept has been elegantly demonstrated by protein shaving experiments of the growth hormone receptor with its ligand.[222,223] Mimicking the functional epitopes can therefore be reduced to finding an appropriate molecular architecture to present the necessary functional groups in the proper spatial orientations. The problem then becomes to identify suitable molecular scaffolds capable of not only presenting the necessary functional groups, but also reproducing the C_α positions and the C_α–C_β vectors of the epitope functional groups. The case-by-case variation in the spatial arrangement of epitope makes generalizing the unique strategy for translating a functional epitope into a small molecule more challenging.[224,225] Contributions from high-throughput screening, computational pattern matching and screening/SAR by NMR[226–228] greatly help to accelerate this process. Gadek *et al.* recently employed a similar strategy to generate a small molecule inhibitor of the LFA-1/ICAM interaction by transferring an ICAM-1 functional epitope to a small molecule.[229] ICAM-1 residues of E34, K39, M64, Y66 and Q73 in the first domain had previously been identified as essential for its binding to LFA-1. By examining the spatial arrangement, the authors took a two-step approach to mimic at least two of these functional groups (Fig. 29.29). They first constructed and optimized cyclic peptides, then incorporated the SAR of the optimized cyclic-peptide (**29.1**) ($IC_{50} = 1.6$ μM) into a small molecule compound (**29.2**) ($IC_{50} = 1.4$ μM) that had been identified from a selectivity counterscreen in a related program. A small focused library was built based on the ortho-bromobenzoyl tryptophan moiety of (**29.2**). The best compound (**29.3**) exhibited an IC_{50} of 1.4 nM in an LFA-1 ELISA assay. More importantly, it showed very potent activity ($IC_{50} = 3$ nM, seven times more potent than the Fab fragment of a humanized anti-CD11, and 20 times more potent than cyclosporin A) in the MLR assay that is considered more reflective of realistic conditions. Furthermore, superimposition of (**29.3**) on to the crystal structure of the first domain of ICAM-1[230] demonstrated that compound (**29.3**) displayed the ICAM-1 epitope of E34, M64, Y66, N68 and Q73 with amazing efficiency,

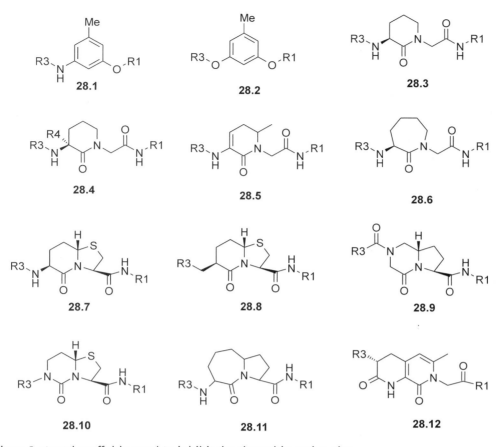

Fig. 29.28 Various β-strand scaffolds used to inhibit the thrombin active site.

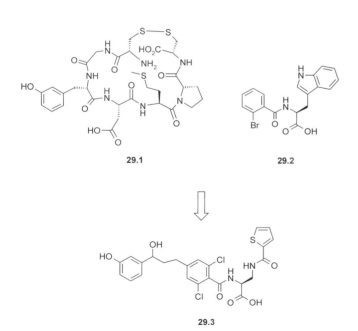

Fig. 29.29 A tertiary structure mimetic of ICAM-1.

showing the dichlorophenyl, thiophene, and carboxyl moieties overlaid perfectly on to the functional groups of Y66, M64 and E34, respectively, and the *m*-phenol group placed in between Q73 and N68.[229]

It is hoped that concerted efforts with computational chemistry, focused combinatorial chemistry, high-throughput screening, and SAR by NMR and X-ray will accelerate the rational translation of functional epitopes displayed by protein tertiary structures on to a small molecular scaffold and make this approach more feasible in the near future.[231]

VII CONCLUSION

The advent of genomics, molecular biology (in particular cDNA cloning and monoclonal antibodies) and ever accelerating structural biology with high-throughput X-ray crystallography and high-field NMR has provided enormous opportunities for structural as well as functional analysis of a wide array of proteinaceous species. The critical roles that proteins play at all levels of biological regulation have opened virtually limitless potential for therapeutic intervention with recombinant proteins. However, with some notable exceptions (EPO, tPA, soluble TNFR, etc.), the therapeutic applications of proteinaceous species have been severely restricted. Proteins are subject to poor bioavailability and rapid proteolytic degradation and clearance, and are potentially antigenic. One approach to overcome these liabilities is to develop small molecule mimetics.

Recently, there has been an increasing effort to rationally design and synthesize biologically active nonpeptide analogues of the peptides and proteins newly identified by genomics and proteomics studies. It is anticipated that these drugs of tomorrow will possess greater selectivity, and hence fewer side-effects, than their present-day counterparts. A number of approaches to accomplish this task has been outlined in this chapter. Perhaps the most fascinating feature to emerge in the field of peptidomimetics is the enormous structural diversity and creativity that is being utilized to mimic the ingenious simplicity of nature.

REFERENCES

1. Chen, S., Chrusciel, R.A., Nakanishi, H., Raktabutr, A., Johnson, M.E., *et al.* (1992) Design and synthesis of a CD4 β-turn mimetic that inhibits human immunodeficiency virus envelope glycoprotein gp120 binding and infection of human lymphocytes. *Proc. Natl Acad. Sci. USA* **89**: 5872–5876.

2. Marshall, G.R., Barry, C.D., Bosshard, H.E., Dammkoehler, R.A. and Dunn, D.A. (1979) The conformational parameter in drug design: the active analog approach. *Computer-Assisted Drug Design*, pp. 205–226. American Chemical Society.

3. Hruby, V.J. (1987) Implications of the X-ray structure of deaminooxytocin to agonist/antagonist-receptor interactions. *Trends Pharmacol. Sci.* **8**: 336–339.

4. Hruby, V.J., Al-Obeidi, F. and Kazmierski, W. (1990) Emerging approaches in the molecular design of receptor-selective peptide ligands: conformational, topographical and dynamic considerations. *Biochem. J.* **268**: 249–262.

5. Giannis, A. and Kolter, T. (1993) Peptidomimetics for receptor ligands-discovery, development, and medical perspectives. *Ang. Chem. Int. Ed. Engl.* **32**: 1244–1267.

6. Gante, J. (1994) Peptidomimetics — tailored enzyme inhibitors. *Angew. Chem. Int. Ed. Engl.* **33**: 1699–1720.

7. Hruby, V.J., Li, G.G., HaskellLuevano, C. and Shenderovich, M. (1997) Design of peptides, proteins, and peptidomimetics in chi space. *Biopolym.* **43**: 219–266.

8. Hughes, J., Smith, T.W., Kosterlitz, H.W., Fothergill, L.A., Morgan, B.A. and Morris, H.R. (1975) Identification of two related pentapeptides from the brain with potent opiate agonist activity. *Nature* **258**: 577–579.

9. Schiller, P.W. (1993) Development of receptor-selective opioid peptide analogs as pharmacologic tools and as potential drugs. In Herz, A. (ed.). *Handbook of Experimental Pharmacology*, vol. 104/I Opioids I, pp. 681–710. Springer-Verlag, Berlin.

10. Sugg, E.E. (1997) Nonpeptide agonists for peptide receptors: Lessons from ligands. In Bristol, J.A. (ed.). *Annual Reports in Medicinal Chemistry*, vol. 32, pp. 277–283. Academic Press, London.

11. Damewood, J.R., Jr. (1996) Peptide mimetic-design with the aid of computational chemistry. In Lipkowitz, K.B. and Boyd, D.B. (eds). *Reviews in Computational Chemistry*, vol. 9, pp. 1–80. VCH Publishers, New York.

12. Fauchère, J.L. (1986) Elements for the rational design of peptide drugs. *Adv. Drug Res.* **15**: 29–69.

13. Bursavich, M.G. and Rich, D.H. (2002) Designing non-peptide peptidomimetics in the 21st century: inhibitors targeting conformational ensembles. *J. Med. Chem.* **45**: 541–558.

14. Veber, D.F., Holly, F.W., Nutt, R.F., Bergstrand, S.J., Brady, S.F., *et al.* (1979) Highly active cyclic and bicyclic somatostatin analogues of reduced ring size. *Nature* **280**: 512–514.

15. Kaiser, E.T. and Kezdy, F.J. (1983) Secondary structures of proteins and peptides in amphilic environments (a review). *Proc. Natl Acad. Sci. USA* **80**: 1137–1143.

16. Kahn, M. (ed.) (1993) *Peptide Secondary Structure Mimetics*, Tetrahedron Symposia-in-Print. Number 50. Pergamon Press, Oxford.

17. Olson, G.L., Bolin, D.R., Bonner, M.P., Bös, M., Cook, C.M., *et al.* (1993) Concepts and progress in the development of peptide mimetics. *J. Med. Chem.* **36**: 3039–3049.

18. Fairlie, D.P., West, M.L. and Wong, A.K. (1998) Towards protein surface mimetics. *Curr. Med. Chem.* **5**: 29–62.

19. Ripka, A.S. and Rich, D.H. (1998) Peptidomimetic design. *Curr. Opin. Chem. Biol.* **2**: 441–452.

20. Hruby, V.J. and Balse, P.M. (2000) Conformational and topographical considerations in designing agonist peptidomimetics from peptide leads. *Curr. Med. Chem.* **7**: 945–970.

21. Kim, H.-O. and Kahn, M. (2000) A merger of rational drug design and combinatorial chemistry: development and application of peptide secondary structure mimetics. *Comb. Chem. High Throughput Screen* **3**: 167–183.

22. Rose, G.D., Gierasch, L.M. and Smith, J.A. (1985) Turns in peptides and proteins. *Ad. Pro. Chem.* **37**: 1–109.

23. Wilmot, C.M. and Thornton, J.M. (1988) Analysis and prediction of the different types of β-turn in proteins. *J. Mol. Biol.* **203**: 221–232.

24. Ball, J.B., Andrews, P.R., Alewood, P.F. and Hughes, R.A. (1990) A one-variable topographical descriptor for the β-turns of peptides and proteins. *FEBS Lett.* **273**: 15–18.

25. Baca, M., Alewood, P.F. and Kent, S.B.H. (1993) Structural engineering of the HIV-1 protease molecule with a β-turn mimic of fixed geometry. *Pro. Sci.* **2**: 1085–1091.

26. Nagai, U. and Sato, K. (1985) Synthesis of a bicyclic dipeptide with the shape of β-turn central part. *Tet. Lett.* **26**: 647–650.

27. Nicolaou, K. C., Salvino, J. M., Raynor, K., Pietranico, S., Reisine, T., *et al.* (1990) Design and synthesis of a peptidomimetic employing β-D-glucose for scaffolding. *Peptide Chem. Struct. Biol. Proc. 11th Am. Peptide Symp.* pp. 881–884.

28. Hirschmann, R., Nicolaou, K.C., Pietranico, S., Salvino, J., Leahy, E.M., *et al.* (1992) Nonpeptidal peptidomimetics with a β-D-glucose scaffolding. A partial somatostatin agonist bearing a close structural relationship to a potent, selective substance P antagonist. *J. Am. Chem. Soc.* **114**: 9217–9218.

29. Hirschmann, R., Nicolaou, K.C., Pietranico, S., Leahy, E.M., Salvino, J., *et al.* (1993) De novo design and synthesis of somatostatin non-peptide peptidomimetics utilizing β-D-glucose as a novel scaffolding. *J. Am. Chem. Soc.* **115**: 12550–12568.

30. Perlman, S., Costa-Neto, C.M., Miyakawa, A.A., Schambye, H.T., Hjorth, S.A., *et al.* (1997) Dual agonistic and antagonistic property of nonpeptide angiotensin AT1 ligands: susceptibility to receptor mutations. *Mol. Pharmacol.* **51**: 301–311.

31. de Tullio, P., Delarge, J. and Pirotte, B. (1999) Recent advances in the chemistry of cholecystokinin receptor ligands (agonists and antagonists). *Curr. Med. Chem.* **6**: 433–455.

32. Asano, M., Hatori, C., Sawai, H., Johki, S., Inamura, N., *et al.* (1998) Pharmacological characterization of a nonpeptide bradykinin B2 receptor antagonist FR165649, and agonist, FR190997. *Br. J. Pharmacol.* **124**: 441–446.

33. Nargund, R.P., Ye, Z., Palucki, B., Bakshi, R., Patchett, A.A. and Van der Ploeg, L.H. (1999) *WO 99/64002*. Novel spiropiperidine derivatives with melanocortin-4-receptor agonist activity. Assigned to Merck, Sharp and Dohme Ltd.

34. Bakshi, R.K., Barakat, K.J., Nargund, R.P., Palucki, B.L., Patchett, A.A., Sebhat, I., Ye, Z. and Van der Ploeg, L.H. (2000) *WO 00/74679*. Substituted piperidines useful as melanocortin-4-receptor agonists. Merck, Sharp and Dohme Ltd.

35. de Laszlo, S.E., Allen, E.E., Li, B., Ondeyka, D., Rivero, R., *et al.* (1997) A nonpeptidic agonist ligand of the human C5a receptor: synthesis, binding affinity optimization and functional characterization. *Bioorg. Med. Chem. Lett.* **7**: 213–218.

36. DeVita, R.J., Bochis, R., Frontier, A.J., Kotliar, A., Fisher, M.H., *et al.* (1998) A potent, orally bioavailable benzazepinone growth hormone secretagogue. *J. Med. Chem.* **41**: 1716–1728.

37. Ogawa, H., Kondo, K., Shinohara, T., Kan, K., Tanada, Y., Kurimura, M., Morita, S. and Uchida, M. (1997) *WO-09722591*. Benzazepine derivatives with vasopressin agonistic activity. Otsuka Pharmaceutical Co Ltd.

38. Failli, A.A., Shumsky, J.S. and Steffan, R.J. (1998) *WO-09906403*. Novel tricyclic vasopressin agonists. Wyeth.

39. Failli, A.A., Shumsky, J.S. and Trybulski, E.J. (2000) *WO-00046224*. Tricyclic pyridine *N*-oxides vasopressin agonists. Wyeth.

40. Steffan, R.J. and Failli, A.A (2000) *WO-00046228*. Pyrrobbenzodiazepine carboxyamide vasopressin agonists. Wyeth.

41. Roux, R. and Serradeil-Le, G.C. (2001) *WO200198295-A1*. Novel 1,3-dihydro-2H-indol-one derivatives, method for preparing same and pharmaceutical compositions containing them. Sanofi-Synthelabo.

42. Johannesson, P., Lindeberg, G., Tong, W.M., Gogoll, A., Karlen, A. and Hallberg, A. (1999) Bicyclic tripeptide mimetics with reverse turn inducing properties. *J. Med. Chem.* **42**: 601–608.

43. Cristau, M., Devin, C., Oiry, C., Chaloin, O., Amblard, M., *et al.* (2000) Synthesis and biological evaluation of bombesin constrained analogues. *J. Med. Chem.* **43**: 2356–2361.

44. Sawutz, D.G., Salvino, J.M., Dolle, R.E., Casiano, F., Ward, S.J., *et al.* (1994) The nonpeptide WIN64338 is a bradykinin B2 receptor antagonist. *Proc. Natl Acad. Sci. USA* **91**: 4693–4697.

45. Wong, A.K., Finch, A.M., Pierens, G.K., Craik, D.J., Taylor, S.M. and Fairlie, D.P. (1998) Small molecular probes for G-protein-coupled C5a receptors: conformationally constrained antagonists derived from the C terminus of the human plasma protein C5a. *J. Med. Chem.* **41**: 3417–3425.

46. Carpenter, K.A., Schmidt, R., von Mentzer, B., Haglund, U., Roberts, E. and Walpole, C. (2001) Turn structures in CGRP C-terminal analogues promote stable arrangements of key residue side chains. *Biochem.* **40**: 8317–8325.

47. Goudreau, N., Weng, J.H. and Roques, B.P. (1994) Conformational analysis of CCK-B agonists using ^{1}H-NMR and restrained molecular dynamics: comparison of biologically active Boc-Trp-(N-Me)Nle-Asp-Phe-NH$_2$ and inactive Boc-Trp-(N-Me)Phe-Asp-Phe-NH$_2$. *Biopolym.* **34**: 155–169.

48. de la Figuera, N., Martin-Martinez, M., Herranz, R., Garcia-Lopez, M.T., Latorre, M. *et al.* (1999) Highly constrained dipeptoid analogues containing a type II' beta-turn mimic as novel and selective CCK-A receptor ligands. *Bioorg. Med. Chem. Lett.* **9**: 43–48.

49. Doi, M., Asano, A., Ishida, T., Katsuya, Y., Mezaki, Y., *et al.* (2001) Caged and clustered structures of endothelin inhibitor BQ123, cyclo(-D-Trp-D-Asp-Pro-D-Val-Leu-)-Na$^+$, forming five and six coordination bonds between sodium ions and peptides. *Acta Crystallogr. D Biol. Crystallogr.* **57**: 628–634.

50. Mammi, S., Mammi, N.J. and Peggion, E. (1988) Conformational studies of human Des-Trp1,Nle12-minigastrin in water-trifluoroethanol mixtures by ^{1}H NMR and circular dichroism. *Biochem.* **27**: 1374–1379.

51. Nikiforovich, G.V., Liepinia, I.T., Tseitin, V.M., Shenderovich, M.D. and Galaktionov, S.G. (1991) Conformational–functional relationships of tetragastrin analogs. *Bioorg. Khim.* **17**: 626–636.

52. Silva Elipe, M.V., Bednarek, M.A. and Gao, Y.D. (2001) ^{1}H NMR structural analysis of human ghrelin and its six truncated analogs. *Biopolym.* **59**: 489–501.

53. Cho, N.B., Harada, M., Imaeda, T., Imada, T., Matsumoto, H., *et al.* (1998) Discovery of a novel, potent, and orally active nonpeptide antagonist of the human luteinizing hormone-releasing hormone (LHRH) receptor. *J. Med. Chem.* **41**: 4190–4195.

54. Karten, M.J. and Rivier, J.E. (1986) Gonadotropin-releasing hormone analog design. Structure–function studies toward the development of agonists and antagonists: rationale and perspective. *Endocr. Rev.* **7**: 44–66.

55. Fan, W., Boston, B.A., Kesterson, R.A., Hruby, V.J. and Cone, R.D. (1997) Role of melanocortinergic neurons in feeding and the agouti obesity syndrome. *Nature* **385**: 165–168.

56. Haskell-Luevano, C., Shenderovich, M.D., Sharma, S.D., Nikiforovich, G.V., Hadley, M.E. and Hruby, V.J. (1995) Design, synthesis, biology, and conformations of bicyclic alpha-melanotropin analogues. *J. Med. Chem.* **38**: 1736–1750.

57. Pavone, V., Lombardi, A., Pedone, C., Quartara, L. and Maggi, C.A. (1994) A new potent and highly selective, long lasting, peptide based neurokinin A antagonist: Rational design of MEN 10627. In Hodges, R.S. and Smith, J.A. (eds). *Peptides: Chemistry, Structure, and Biology – Proc. 13th Am. Peptide Symp*, pp. 487–489. ESCOM, Leiden.

58. Hirschmann, R., Yao, W.Q., Cascieri, M.A., Strader, C.D., Maechler, L., Cichyknight, M.A., Hynes, J., Vanrijn, R.D., Sprengeler, P.A. and Smith, A.B. (1996) Synthesis of potent cyclic hexapeptide NK-1 antagonists. Use of a minilibrary in transforming a peptidal somatostatin receptor ligand into an NK-1 receptor ligand via a polyvalent peptidomimetic. *J. Med. Chem.* **39**: 2441–2448.

59. Mosberg, H.I., Hurst, R., Hruby, V.J., Gee, K., Yamamura, H.I., Galligan, J.J. and Burks, T.F. (1983) Bis-penicillamine enkephalins possess highly improved specificity toward delta opioid receptors. *Proc. Natl Acad. Sci. USA* **80**: 5871–5874.

60. Mosberg, H.I., Hurst, R., Hruby, V.J., Galligan, J.J., Burks, T.F., Gee, K. and Yamamura, H.I. (1983) Conformationally constrained cyclic enkephalin analogs with pronounced delta opioid receptor agonist selectivity. *Life Sci.* **32**: 2565–2569.

61. Podlogar, B.L., Paterlini, M.G., Ferguson, D.M., Leo, G.C., Demeter, D.A., Brown, F.K. and Reitz, A.B. (1998) Conformational analysis of the endogenous µ-opioid agonist endomorphin-1 using NMR spectroscopy and molecular modeling. *FEBS Lett.* **439**: 13–20.

62. Shenderovich, M.D., Kover, K.E., Wilke, S., Collins, N. and Hruby, V.J. (1997) Solution conformations of potent bicyclic antagonists of oxytocin by nuclear magnetic resonance spectroscopy and molecular dynamics simulations. *J. Am. Chem. Soc.* **119**: 5833–5846.

63. Dettin, M., De Rossi, A., Autiero, M., Guardiola, J., Chieco-Bianchi, L. and Di Bello, C. (1993) Structural studies on synthetic peptides from the principal neutralizing domain of HIV-1 gp120 that bind to CD4 and enhance HIV-1 infection. *Biochem. Biophys. Res. Comm.* **191**: 364–370.

64. Veber, D.F., Freidlinger, R.M., Perlow, D.S., Paleveda, W.J., Jr., Holly, F.W., *et al.* (1981) A potent cyclic hexapeptide analogue of somatostatin. *Nature* **292**: 55–58.

65. Bern, H.A., Pearson, D., Larson, B.A. and Nishioka, R.S. (1985) Neurohormones from fish tails: the caudal neurosecretory system. I. 'Urophysiology' and the caudal neurosecretory system of fishes. *Recent. Prog. Horm. Res.* **41**: 533–552.

66. Jard, S., Lombard, C., Seyer, R., Aumelas, A., Manning, M. and Sawyer, W.H. (1987) Iodination of vasopressin analogues with agonistic and antagonistic properties: effects on biological properties and affinity for vascular and renal vasopressin receptors. *Mol. Pharmacol.* **32**: 369–375.

67. Manning, M., Chan, W.Y. and Sawyer, W.H. (1993) Design of cyclic and linear peptide antagonists of vasopressin and oxytocin: current status and future directions. *Regul. Pept.* **45**: 279–283.

68. Manning, M. and Sawyer, W.H. (1993) Design, synthesis and some uses of receptor-specific agonists and antagonists of vasopressin and oxytocin. *J. Recept. Res.* **13**: 195–214.

69. Iwadate, M., Nagao, E., Williamson, M.P., Ueki, M. and Asakura, T. (2000) Structure determination of [Arg8]vasopressin methylene-dithioether in dimethylsulfoxide using NMR. *Eur. J. Biochem.* **267**: 4504–4510.

70. Porcelli, M., Casu, M., Lai, A., Saba, G., Pinori, M., Cappelletti, S. and Mascagni, P. (1999) Cyclic pentapeptides of chiral sequence DLDDL as scaffold for antagonism of G-protein coupled receptors: synthesis, activity and conformational analysis by NMR and molecular dynamics of ITF 1565 a substance P inhibitor. *Biopolym.* **50**: 211–219.

71. Patchett, A.A and Nargund, R.P. (2000) Chapter 26. Privileged structures — an update. In Doherty, A.M. (ed.). *Annual Reports in Medicinal Chemistry*, vol. 35, pp. 289–298. Academic Press, San Diego.

72. Evans, B.E., Rittle, K.E., Bock, M.G., DiPardo, R.M., Freidinger, R.M., *et al.* (1988) Methods for drug discovery: development of potent, selective, orally effective cholecystokinin antagonists. *J. Med. Chem.* **31**: 2235–2246.

73. Yang, L.H., Berk, S.C., Rohrer, S.P., Mosley, R.T., Guo, L.Q., *et al.* (1998) Synthesis and biological activities of potent peptidomimetics selective for somatostatin receptor subtype 2. *Proc. Natl Acad. Sci. USA* **95**: 10836–10841.

74. Hay, B.A., Cole, B.M., DiCapua, F.M., Kirk, G.W., Murray, M.C., *et al.* (2001) Small molecule somatostatin receptor subtype-2 antagonists. *Bioorg. Med. Chem. Lett.* **11**: 2731–2734.

75. Nargund, R.P., Patchett, A.A., Bach, M.A., Murphy, M.G. and Smith, R.G. (1998) Peptidomimetic growth hormone secretagogues. Design considerations and therapeutic potential. *J. Med. Chem.* **41**: 3103–3127.

76. Bowers, C.Y., Momany, F.A., Reynolds, G.A. and Hong, A. (1984) On the *in vitro* and *in vivo* activity of a new synthetic hexapeptide that acts on the pituitary to specifically release growth hormone. *Endocrinology* **114**: 1537–1545.

77. Momany, F.A., Bowers, C.Y., Reynolds, G.A., Hong, A. and Newlander, K. (1984) Conformational energy studies and *in vitro* and *in vivo* activity data on growth hormone-releasing peptides. *Endocrinology* **114**: 1531–1536.

78. Patchett, A.A., Nargund, R.P., Tata, J.R., Chen, M.-H., Barakat, K.J., Johnston, D.B.R., *et al.* (1995) Design and biological activities of L-163,191 (MK-0677): A potent, orally active growth hormone secretagogue. *Proc. Natl Acad. Sci. USA* **92**: 7001–7005.

79. Pan, L.C., Carpino, P.A., Lefker, B.A., Ragan, J.A., Toler, S.M., *et al.* (2001) Preclinical pharmacology of CP-424,391, an orally active pyrazolinone-piperidine growth hormone secretagogue. *Endocrine* **14**: 121–132.

80. Palucki, B.L., Feighner, S.D., Pong, S., McKee, K.K., Hreniuk, D.L., *et al.* (2001) Spiro(indoline-3,4′-piperidine) growth hormone secretagogues as ghrelin mimetics. *Bioorg. Med. Chem. Lett.* **11**: 1955–1957.

81. Tokunaga, T., Hume, W.E., Umezome, T., Okazaki, K., Ueki, Y., *et al.* (2001) Oxindole derivatives as orally active potent growth hormone secretagogues. *J. Med. Chem.* **44**: 4641–4649.

82. Konteatis, Z.D., Siciliano, S.J., Van Riper, G., Molineaux, C.J., Pandya, S., *et al.* (1994) Development of C5a receptor antagonists. Differential loss of functional responses. *J. Immunol.* **153**: 4200–4205.

83. Nicholls, I.A., Craik, D.J. and Alewood, P.F. (1994) NMR and molecular modelling based conformational analysis of some N-alkyl 1-and 2-benzazepinones: useful central nervous system agent design motifs. *Biochem. Biophys. Res. Comm.* **205**: 98–104.

84. Low, C.M., Black, J.W., Broughton, H.B., Buck, I.M., Davies, J.M., *et al.* (2000) Development of peptide 3D structure mimetics: rational design of novel peptoid cholecystokinin receptor antagonists. *J. Med. Chem.* **43**: 3505–3517.

85. de Tullio, P., Delarge, J. and Pirotte, B. (2000) Therapeutic and chemical developments of cholecystokinin receptor ligands. *Expert Opin. Investig. Drugs* **9**: 129–146.

86. Willson, T.M., Henke, B.R., Momtahen, T.M., Myers, P.L., Sugg, E.E., *et al.* (1996) 3-[2-(*N*-phenylacetamide)]-1,5-benzodiazepines: orally active, binding selective CCK-A agonists. *J. Med. Chem.* **39**: 3030–3034.

87. Hirst, G.C., Aquino, C., Birkemo, L., Croom, D.K., Dezube, M., *et al.* (1996) Discovery of 1,5-benzodiazepines with peripheral cholecysto-kinin (CCK-A) receptor agonist activity (II): Optimization of the C3 amino substituent. *J. Med. Chem.* **39**: 5236–5245.

88. Henke, B.R., Willson, T.M., Sugg, E.E., Croom, D.K., Dougherty, R.W., *et al.* (1996) 3-(1H-Indazol-3-ylmethyl)-1,5-benzodiazepines: CCK-A agonists that demonstrate oral activity as satiety agents. *J. Med. Chem.* **39**: 2655–2658.

89. Henke, B.R., Aquino, C.J., Birkemo, L.S., Croom, D.K., Dougherty, R.W., Jr., *et al.* (1997) Optimization of 3-(1H-indazol-3-ylmethyl)-1,5-benzodiazepines as potent, orally active CCK-A agonists. *J. Med. Chem.* **40**: 2706–2725.

90. James, G.L., Goldstein, J.L., Brown, M.S., Rawson, T.E., Somers, T.C., *et al.* (1993) Benzodiazepine peptidomimetics: potent inhibitors of *ras* farnesylation in animal cells. *Science* **260**: 1937–1942.

91. Ding, C.Z., Batorsky, R., Bhide, R., Chao, H.J., Cho, Y., *et al.* (1999) Discovery and structure–activity relationships of imidazole-containing tetrahydrobenzodiazepine inhibitors of farnesyltransferase. *J. Med. Chem.* **42**: 5241–5253.

92. Hunt, J.T., Ding, C.Z., Batorsky, R., Bednarz, M., Bhide, R., *et al.* (2000) Discovery of (R)-7-cyano-2,3,4,5-tetrahydro-1-(1H-imidazol-4-ylmethyl)-3-(phenylmethyl)-4-(2-thienylsulfonyl)-1H-1,4-benzo-diazepine (BMS-214662), a farnesyltransferase inhibitor with potent preclinical antitumor activity. *J. Med. Chem.* **43**: 3587–3595.

93. Rose, W.C., Lee, F.Y., Fairchild, C.R., Lynch, M., Monticello, T., *et al.* (2001) Preclinical antitumor activity of BMS-214662, a highly apoptotic and novel farnesyltransferase inhibitor. *Cancer Res.* **61**: 7507–7517.

94. Njoroge, F.G., Taveras, A.G., Kelly, J., Remiszewski, S., Mallams, A.K., *et al.* (1998) (+)-4-[2-[4-(8-Chloro-3,10-dibromo-6,11-dihydro-5H-benzo[5,6]cyclohepta[1,2-b]-pyridin-11(R)-yl]-1-piperidinyl]-2-oxo-ethyl]-1-piperidinecarboxamide (SCH-66336): a very potent farnesyl protein transferase inhibitor as a novel antitumor agent. *J. Med. Chem.* **41**: 4890–4902.

95. Liu, M., Bryant, M.S., Chen, J., Lee, S., Yaremko, B., *et al.* (1998) Antitumor activity of SCH 66336, an orally bioavailable tricyclic inhibitor of farnesyl protein transferase, in human tumor xenograft models and wap-ras transgenic mice. *Cancer Res.* **58**: 4947–4956.

96. Flynn, G.A., Beight, D.W., Mehdi, S., Koehl, J.R., Giroux, E.L., *et al.* (1993) Application of a conformationally restricted Phe-Leu dipeptide mimetic to the design of a combined inhibitor of angiotensin I-converting enzyme and neutral endopeptidase 24.11. *J. Med. Chem.* **36**: 2420–2423.

97. Borkakoti, N., Winkler, F.K., Williams, D.H., D'Arcy, A., Broad-hurst, M.J., *et al.* (1994) Structure of the catalytic domain of human fibroblast collagenase complexed with an inhibitor. *Structr. Biol.* **1**: 106–110.

98. Lauffer, D.J. and Mullican, M.D. (2002) A practical synthesis of (S) 3-*tert*-butoxycarbonylamino-2-oxo-2,3,4,5-tetrahydro-1,5-benzodiazepine-1-acetic acid methyl ester as a conformationally restricted dipeptide-mimetic for Caspase-1 (ICE) inhibitors. *Bioorg. Med. Chem. Lett.* **12**: 1225–1227.

99. Keenan, R.M., Callahan, J.F., Samanen, J.M., Bondinell, W.E., Calvo, R.R., *et al.* (1999) Conformational preferences in a benzodiazepine series of potent nonpeptide fibrinogen receptor antagonists. *J. Med. Chem.* **42**: 545–559.

100. Merrifield, R.B. (1985) Solid phase synthesis (Nobel Lecture). *Angew. Chem. Int. Ed. Engl.* **24**: 799–810.

101. Gardner, B., Nakanishi, H. and Kahn, M. (1993) Conformationally constrained nonpeptide β-turn mimetics of enkephalin. *Tetrahedron* **49**: 3433–3448.

102. Eguchi, M., Lee, M.S., Nakanishi, H., Stasiak, M., Lovell, S. and Kahn, M. (1999) Solid-phase synthesis and structural analysis of bicyclic β-turn mimetics incorporating functionality at the *i* to *i* + *3* positions. *J. Amer. Chem. Soc.* **121**: 12204–12205.

103. Wasserman, H.H. (1982) Transamidation reactions using β-lactams. The synthesis of homaline. *Tet. Lett.* **23**: 465–486.

104. Lee, M.S., Gardner, B., Kahn, M. and Nakanishi, H. (1995) The three-dimensional solution structure of a constrained peptidomimetic in water and in chloroform—observation of solvent induced hydrophobic cluster. *FEBS Lett.* **359**: 113–118.

105. Sato, M., Lee, J.Y.H., Nakanishi, H., Johnson, M.E., Chrusciel, R.A. and Kahn, M. (1992) Design, synthesis and conformational analysis of γ-turn peptide mimetics of bradykinin. *Biochem. Biophys. Res. Commun.* **187**: 999–1006.

106. Ferguson, M.D., Meara, J.P., Nakanishi, H., Lee, M.S. and Kahn, M. (1997) The development of hydrazide γ-turn mimetics. *Tet. Lett.* **38**: 6961–6964.

107. Saragovi, H.U., Fitzpatrick, D., Raktabutr, A., Nakanishi, H., Kahn, M. and Greene, M.I. (1991) Design and synthesis of a mimetic from an antibody complementarity-determining region. *Science* **253**: 792–795.

108. Amzel, L.M. and Poljak, R.J. (1979) Three-dimensional structure of immunoglobulins. *Annu. Rev. Biochem.* **48**: 961–997.

109. Chothia, C., Lesk, A.M., Tramontano, A., Levitt, M., Smith-Gill, S.J., *et al.* (1989) Conformations of immunoglobulin hypervariable regions. *Nature* **342**: 877–883.

110. Bruck, C., Co, M.S., Slaoui, M., Gaulton, G.N., Smith, T., *et al.* (1986) Nucleic acid sequence of an internal image-bearing monoclonal anti-idiotype and its comparison to the sequence of the external antigen. *Proc. Natl Acad. Sci. USA* **83**: 6578–6582.

111. Chothia, C. and Lesk, A.M. (1987) Canonical structures for the hypervariable regions of immunoglobulins. *J. Mol. Biol.* **196**: 901–917.

112. Chen, S., Chrusciel, R.A., Nakanishi, H., Raktabutr, A., Johnson, M.E., *et al.* (1992) Design and synthesis of a CD4 β-turn mimetic that inhibits human immunodeficiency virus envelope glycoprotein gp120 binding and infection of human lymphocytes. *Proc. Natl Acad. Sci. USA* **89**: 5872–5876.

113. Landau, N.R., Warton, M. and Littman, D.R. (1988) The envelope glycoprotein of the human immunodeficiency virus binds to the immunoglobulin-like domain of CD4. *Nature* **334**: 159–162.

114. Ashkenazi, A., Presta, L.G., Marsters, S.A., Camerato, T.R., Rosenthal, K.A., Fendly, B.M. and Capon, D.J. (1990) Mapping the CD4 binding site for human immunodeficiency virus by alanine-scanning mutagenesis. *Proc. Natl Acad. Sci. USA* **87**: 7150–7154.

115. Jameson, B.A., Rao, P.E., Kong, L.I., Hahn, B.H., Shaw, G.M., *et al.* (1988) Location and chemical synthesis of a binding site for HIV-1 on the CD4 protein. *Science* **240**: 1335–1339.

116. Ramurthy, S., Lee, M.S., Nakanishi, H., Shen, R. and Kahn, M. (1994) Peptidomimetic antagonists designed to inhibit the binding on CD4 to HIV gp120. *Bioorg. Med. Chem.* **2**: 1007–1013.

117. Nakanishi, H., Ramurthy, S., Kahn, M. and Lee, M.S. (1996) Solution structure and conformational analysis of a β-turn CD4 peptidomimetic, unpublished result.

118. Kim, H.O., Nakanishi, H., Lee, M.S. and Kahn, M. (2000) Design and synthesis of novel conformationally restricted peptide secondary structure mimetics. *Org. Lett.* **2**: 301–302.

119. Kahn, M., Eguchi, M. and Kim, H.-O. (2000) US Patent No. 6 013 458.

120. Nagula, G., Huber, V.J., Lum, C. and Goodman, B.A. (2000) Synthesis of alpha-substituted β-amino acids using pseudoephedrine as a chiral auxiliary. *Org. Lett.* **2**: 3527–3529.

121. Goodman, B.A. (1999) Managing the workflow of a high-throughput organic synthesis laboratory: a marriage of automation and information management technologies. *J. Assoc. Lab. Automat.* **4**: 48–52.

122. Garbay-Jaureguiberry, C., Roques, B.P. and Oberlin, R. (1977) ^{1}H and ^{13}C NMR studies of conformational behaviour of Leu-enkephalin. *FEBS Lett.* **76**: 93–98.

123. Roques, B.P., Garbay-Jaureguiberry, C., Oberlin, R., Anteunis, M. and Lala, A.K. (1976) Conformation of Met5-enkephalin determined by high field PMR spectroscopy. *Nature* **262**: 778–779.

124. Patel, A., Smith, H.J. and Sewell, R.D. (1993) Inhibitors of enkephalin-degrading enzymes as potential therapeutic agents. *Prog. Med. Chem.* **30**: 327–378.

125. Satoh, M. and Minami, M. (1995) Molecular pharmacology of the opioid receptors. *Pharmacol. Ther.* **68**: 343–364.

126. Hassan, M. and Goodman, M. (1986) Computer simulations of cyclic enkephalin analogues. *Biochem.* **25**: 7596–7606.

127. Smith, P.E. and Pettitt, B.M. (1992) Amino acid side-chain populations in aqueous and saline solution: bis-penicillamine enkephalin. *Biopolym.* **32**: 1623–1629.

128. Malicka, J., Groth, M., Czaplewski, C., Wiczk, W. and Liwo, A. (2002) Conformational studies of cyclic enkephalin analogues with L-or D-proline in position 3. *Biopolym.* **63**: 217–231.

129. Shenderovich, M.D., Liao, S., Qian, X. and Hruby, V.J. (2000) A three-dimensional model of the delta-opioid pharmacophore: comparative molecular modeling of peptide and nonpeptide ligands. *Biopolym.* **53**: 565–580.

130. Carlacci, L. (1998) Conformational analysis of [Met(5)]-enkephalin: solvation and ionization considerations. *J. Comput. Aid. Molec. Design* **12**: 195–213.

131. Marraud, M. and Aubry, A. (1996) Crystal structures of peptides and modified peptides. *Biopolym.* **40**: 45–83.

132. Griffin, J.F. and Smith, G.D. (1988) X-ray diffraction studies of enkephalins and opiates. *NIDA Res. Monogr.* **87**: 41–59.

133. Hruby, V.J., Kao, L.F., Pettitt, B.M. and Karplus, M. (1988) The conformational properties of the delta opioid peptide [D-Pen[2], D-Pen[5]]enkephalin in aqueous solution determined by NMR and energy minimization calculations. *J. Am. Chem. Soc.* **110**: 3351–3359.

134. Mosberg, H.I., Sobczyk-Kojiro, K., Subramanian, P., Crippen, G.M., Ramalingam, K. and Woodard, R.W. (1990) Combined use of stereospecific deuteration, NMR, distance geometry, and energy minimization for the conformational analysis of the highly δ opioid receptor selective peptide [D-Pen[2],D-Pen[5]]enkephalin. *J. Am. Chem. Soc.* **112**: 822–829.

135. Bradbury, A.F., Smyth, D.G. and Snell, C.R. (1976) Biosynthetic origin and receptor conformation of methionine enkephalin. *Nature* **260**: 165–166.

136. Loew, G.H. and Burt, S.K. (1978) Energy conformation study of Met-enkephalin and its D-Ala2 analogue and their resemblance to rigid opiates. *Proc. Natl Acad. Sci. USA* **75**: 7–11.

137. De Coen, J.L., Humblet, C. and Koch, M.H. (1977) Theoretical conformational analysis of Met-enkephalin. *FEBS Lett.* **73**: 38–42.

138. Manavalan, P. and Momany, F.A. (1981) Conformational energy calculations on enkephalins and enkephalin analogs. Classification of

conformations to different configurational types. *Int. J. Pept. Protein Res.* **18**: 256–275.

139. DiMaio, J. and Schiller, P.W. (1980) A cyclic enkephalin analog with high *in vitro* opiate activity. *Proc. Natl Acad. Sci. USA* **77**: 7162–7166.

140. Schiller, P.W. and DiMaio, J. (1982) Opiate receptor subclasses differ in their conformational requirements. *Nature* **297**: 74–76.

141. Kahn, M., Lee, M.S., Nakanishi, H., Urban, J. and Gardner, B. (1993) Peptide secondary structure mimetics: Recent advances and future challenges. In Hodges, R.S. and Smith, J.A. (eds). *Peptides: Chemistry, Structure, and Biology — Proc. 13th Am. Peptide Symp.*, pp. 271–274. ESCOM, Leiden.

142. Zadina, J.E., Hackler, L., Ge, L.J. and Kastin, A.J. (1997) A potent and selective endogenous agonist for the μ-opiate receptor. *Nature* **386**: 499–502.

143. Zadina, J.E., Martin-Schild, S., Gerall, A.A., Kastin, A.J., Hackler, L., *et al.* (1999) Endomorphins: novel endogenous μ-opiate receptor agonists in regions of high mu-opiate receptor density. *Ann. N. Y. Acad. Sci.* **897**: 136–344.

144. Eguchi, M., Shen, R.Y., Shea, J.P., Lee, M.S. and Kahn, M. (2002) Design, synthesis, and evaluation of opioid analogues with non-peptidic β-turn scaffold: Enkephalin and endomorphin mimetics. *J. Med. Chem.* **45**: 1395–1398.

145. Barlow, D.J. and Thornton, J.M. (1988) Helix geometry in proteins. *J. Mol. Biol.* **201**: 601–619.

146. Rohl, C.A. and Baldwin, R.L. (1998) Deciphering rules of helix stability in peptides. In Ackers, G.K. and Johnson, M.L. (eds). *Energetics of Biological Macromolecules; Methods in Enzymology*, vol. 295, pp. 1–26. Academic Press, San Diego.

147. Padmanabhan, S., Marqusee, S., Ridgeway, T., Laue, T.M. and Boldwin, R.L. (1990) Relative helix-forming tendencies of nonpolar amino acids. *Nature* **344**: 268–270.

148. Kemp, D.S. and Curran, T.P. (1988) (2S,5S,8S,11S)-1-acetyl-1,4-diaza-3-keto-5-carboxy-10-thia-tricyclo-[2.8.0^{4,8}]-tridecane, *1*, synthesis of prolyl-proline-derived, peptide-functionalized templates for α-helix formation. *Tet. Lett.* **29**: 4931–4934.

149. Arrhenius, T., Lerner, R.A. and Satterthwait, A.C. (1987) The chemical synthesis of structured peptides using covalent hydrogen-bond mimics. In Oxender, D.L. (ed.). *Protein Structure, Folding, and Design*, vol. 2, pp. 453–465. Alan R. Liss, Inc.

150. Arrhenius, T. and Satterthwait, A.C. (1989) The substitution of an amide–amide backbone hydrogen bond in an α-helix peptide with a covalent hydrogen bond mimic. In *Peptide Chem. Struct. Biol. Proc. 11th Am. Peptide Symp.* pp. 453–465.

151. Cabezas, E. and Satterthwait, A.C. (1999) The hydrogen bond mimic approach: Solid-phase synthesis of a peptide stabilized as an α-helix with a hydrazone link. *J. Amer. Chem. Soc.* **121**: 3862–3875.

152. Pardi, A., Billeter, M. and Wüthrich, K. (1984) Calibration of the angular dependence of the amide proton-C_α proton coupling constants, $3J_{HN\alpha}$, in a globular protein. *J. Mol. Biol.* **180**: 741–751.

153. Kemp, D.S. and Curran, T.P. (1988) The preferred conformation of *1* (*1* = αTemp-OH) and its peptide conjugates αTemp-L-(Ala)n-OR (n = 1 to 4) and αTemp-L-Ala-L-Phe-L-Lys(εBoc)-L-Lys (εBoc)-NHMe: studies of templates for α-helix formation. *Tet. Lett.* **29**: 4935–4938.

154. Kemp, D.S., Allen, T.J. and Oslick, S.L. (1991) Development of a 3-state equilibrium model for the helix-nucleation template Ac-Hel$_1$-OH. In Smith, J.A. and Rivier, J.E. (eds). *Proceedings of the 12th American Peptide Symposium*, pp. 352–355. ESCOM, Leiden.

155. Kemp, D.S., Curran, T.P., Boyd, J.G. and Allen, T.J. (1991) Studies of N-terminal templates for a-helix formation. Synthesis and conformational analysis of (2S, 5S, 8S, 11S)-1-Acetyl-1,4-diaza-3-keto-5-carboxy-10-thiatricyclo [2.8.1.0^{4,8}]-tridecane (Ac-Hel$_1$-OH). *J. Org. Chem.* **56**: 6672–6682.

156. McClure, K.F., Renold, P. and Kemp, D.S. (1995) An improved synthesis of a template for α-helix formation. *J. Org. Chem.* **60**: 454–457.

157. Maison, W., Arce, E., Renold, P., Kennedy, R.J. and Kemp, D.S. (2001) Optimal N-caps for N-terminal helical templates: effects of changes in H-bonding efficiency and charge. *J. Am. Chem. Soc.* **123**: 10245–10254.

158. Kemp, D.S. and Rothman, J.H. (1995) Synthesis and analysis of a macrocyclic triproline-derived template containing a local conformational constraint. *Tet. Lett.* **36**: 4019–4022.

159. Kemp, D.S. and Rothman, J.H. (1995) Efficient helix nucleation by a macrocyclic triproline-derived template. *Tet. Lett.* **36**: 4023–4026.

160. Lewis, A., Wilkie, J., Rutherford, T.J. and Gani, D. (1998) Design, construction and properties of peptide N-terminal cap templates devised to initiate α-helices. Part 2. Caps derived from N-[(2S)-2-chloropropionyl]-(2S)-Pro-(2S)-Pro-(2S,4S)-4-thioPro-OMe. *J. Chem. Soc. Perkin Trans.* **1**: 3777–3793.

161. Lewis, A., Rutherford, T.J., Wilkie, J., Jenn, T. and Gani, D. (1998) Design, construction and properties of peptide N-terminal cap templates devised to initiate α-helices. Part 3. Caps derived from N-[(2S)-2-chloropropionyl]-(2S)-Pro-(2R)-Ala-(2S,4S)-4-thioPro-OMe. *J. Chem. Soc. Perkin Trans.* **1**: 3795–3806.

162. Müller, K., Obrecht, D., Knierzinger, A., Stankovic, C., Spiegler, C., *et al.* (1993) Building blocks for the induction or fixation of peptide conformations. In Testa, B., Kyburz, E., Fuhrer, W. and Giger, R. (eds). *Perspectives in Medicinal Chemistry*, pp. 513–531. Verlag, Basel.

163. Austin, R.E., Maplestone, R.A., Sefler, A.M., Liu, K., Hruzewicz, W.N., *et al.* (1997) Template for stabilization of a peptide α-helix: Synthesis and evaluation of conformational effects by circular dichroism and NMR. *J. Am. Chem. Soc.* **119**: 6461–6472.

164. Meara, J.P., Cao, B., Urban, J., Nakanishi, H., Yeung, E. and Kahn, M. (1994) Design and synthesis of small molecule initiators of α-helix structure in short peptides in water. In Maia, L.S. (ed.). *Proceedings of the 23rd European Peptide Symposium*, pp. 692–693. ESCOM, Leiden.

165. Kahn, M. (1995) US Patent No. 5 446 128.

166. Kahn, M. (1998) US Patent No. 5 710 245.

167. Jung, G., Beck-Sickinger, A.G., Durr, H., Gaida, W. and Schnorrengerg, G. (1991) α-Helical small molecular size analogues of Neuropeptide Y: structure–activity relationships. *Biopolym.* **31**: 613–619.

168. Bjorkman, P.J., Saper, M.A., Samraoui, B., Bennett, W.S., Strominger, J.L. and Wiley, D.C. (1987) Structure of the human class I histocompatibility antigen, HLA-A2. *Nature* **329**: 506–512.

169. Schramm, H.J., Nakashima, H., Schramm, W., Wakayama, H. and Yamamoto, N. (1991) HIV-1 reproduction is inhibited by peptides derived from the N- and C-termini of HIV-1 protease. *Biochem. Biophys. Res. Commun.* **179**: 847–851.

170. Zutshi, R. and Chmielewski, J. (2000) Targeting the dimerization interface for irreversible inhibition of HIV-1 protease. *Bioorg. Med. Chem. Lett.* **10**: 1901–1903.

171. Waksman, G., Kominos, D., Robertson, S.C., Pant, N., Baltimore, D., *et al.* (1992) Crystal structure of the phosphotyrosine recognition domain SH2 of *v-src* complexed with tyrosine-phosphorylated peptides. *Nature* **358**: 646–653.

172. Kemp, D.S., Blanchard, D.E. and Muendel, C.C. (1991) Studies on the nucleation in DMSO and water of β-sheet structures with peptide-epindolidione conjugates. In Smith, J.A. and Rivier, J.E. (eds). *Proceedings of the 12th American Peptide Symposium*, pp. 319–322. ESCOM, Leiden.

173. Kemp, D.S. and Bowen, B.R. (1988) Synthesis of peptide-functionalized diacylaminoepindolidiones as templates for β-sheet formation. *Tet. Lett.* **29**: 5077–5080.

174. Kemp, D.S. and Bowen, B.R. (1988) Conformational analysis of peptide-functionalized diacylaminoepindolidiones: 1H NMR evidence for β-sheet formation. *Tet. Lett.* **29**: 5081–5082.

175. Kemp, D.S. and Bowen, B.R. (1990) Diacylaminoepindolidiones as templates for β-sheets. In Gierasch, L.M. and King, J. (eds). *Protein Folding: Deciphering the Second Half of the Genetic Code*, pp. 293–303. AAAS, Washington DC.

176. Smith, E.M., Holmes, D.L., Shaka, A.J. and Nowick, J.S. (1997) An artificial antiparallel β-sheet containing a new peptidomimetic template. *J. Org. Chem.* **62**: 7906–7907.

177. Holmes, D.L., Smith, E.M. and Nowick, J.S. (1997) Solid-phase synthesis of artificial β-sheets. *J. Am. Chem. Soc.* **119**: 7665–7669.

178. Tsai, J.H., Waldman, A.S. and Nowick, J.S. (1999) Two new β-strand mimics. *Bioorgan. Med. Chem.* **7**: 29–38.

179. Nowick, J.S., Cary, J.M. and Tsai, J.H. (2001) A triply templated artificial β-sheet. *J. Am. Chem. Soc.* **123**: 5176–5180.

180. Nowick, J. S. and Maitra, S. (2001) *WO200114412*. Compounds useful to mimic peptide beta-strands. University of California.

181. Smith, A.B., III, Keenan, T.P., Holcomb, R.C., Sprengeler, P.A., Guzman, M.C., *et al.* (1992) Design, synthesis, and crystal structure of a pyrrolinone-based peptidomimetic possessing the conformation of a β-strand: potential application to the design of novel inhibitors of proteolytic enzymes. *J. Am. Chem. Soc.* **114**: 10672–10674.

182. Smith, A.B., III, Holcomb, R.C., Guzman, M.C., Keenan, T.P., Sprengeler, P.A. and Hirschmann, R. (1993) An effective synthesis of scalemic 3,5,5-trisubstituted pyrrolin-4-ones. *Tet. Lett.* **34**: 63–66.

183. Smith, A.B., Hirschmann, R., Pasternak, A., Guzman, M.C., Yokoyama, A., *et al.* (1995) Pyrrolinone-based HIV protease inhibitors. Design, synthesis, and antiviral activity: evidence for improved transport. *J. Am. Chem. Soc.* **117**: 11113–11123.

184. Smith, A.B., Hirschmann, R., Pasternak, A., Yao, W.Q., Sprengeler, P.A., *et al.* (1997) An orally bioavailable pyrrolinone inhibitor of HIV-1 protease: Computational analysis and x-ray crystal structure of the enzyme complex. *J. Med. Chem.* **40**: 2440–2444.

185. Smith, A.B., Benowitz, A.B., Guzman, M.C., Sprengeler, P.A., Hirschmann, R., *et al.* (1998) Design, synthesis, and evaluation of a pyrrolinone-peptide hybrid ligand for the class II MHC protein HLA-DR1. *J. Am. Chem. Soc.* **120**: 12704–12705.

186. Nakanishi, H. and Kahn, M. (1996) Design of Peptidomimetics. In Wermuth, C.G. (ed.). *The Practice of Medicinal Chemistry*, pp. 571–590. Academic Press, London.

187. Wlodawer, A. and Vondrasek, J. (1998) Inhibitors of HIV-1 protease: a major success of structure-assisted drug design. *Annu. Rev. Biophys. Biomol. Struct.* **27**: 249–284.

188. Thompson, L.A. and Tebben, A.J. (2001) Pharmacokinetics and design of aspartyl protease inhibitors. In Trainor, G.L. (ed.). *Annual Reports in Medicinal Chemistry*, vol. 36, pp. 247–256. Academic Press, San Diego.

189. Lebon, F. and Ledecq, M. (2000) Approaches to the design of effective HIV-1 protease inhibitors. *Curr. Med. Chem.* **7**: 455–477.

190. Wiley, M.R. and Fisher, M.J. (1997) Small-molecule direct thrombin inhibitors. *Expert Op. Ther. Patents* **7**: 1265–1282.

191. Bode, W., Mayr, I., Baumann, U., Huber, R., Stone, S.R. and Hofsteenge, J. (1989) The refined 1.9 Å crystal structure of human a-thrombin: interaction with D-Phe-Pro-Arg chloromethylketone and significance of the Tyr-Pro-Pro-Trp insertion segment. *Embo J.* **8**: 3467–3475.

192. Boatman, P.D., Ogbu, C.O., Eguchi, M., Kim, H.O., Nakanishi, H., *et al.* (1999) Secondary structure peptide mimetics: Design, synthesis, and evaluation of β-strand mimetic thrombin inhibitors. *J. Med. Chem.* **42**: 1367–1375.

193. Kahn, M. (2000) US Patent No. 6 020 331.

194. Scott, N.A., Nunes, G.L., King, S.B., Harker, L.A. and Hanson, S.R. (1994) Local delivery of an antithrombin inhibits platelet-dependent thrombosis. *Circulation* **90**: 1951–1955.

195. Harker, L.A., Kelly, A.B. and Hanson, S.R. (1991) Experimental arterial thrombosis in nonhuman primates. *Circulation* **83**: IV41–IV55.

196. Ogbu, C.O., Qabar, M.N., Boatman, P.D., Urban, J., Meara, J.P., *et al.* (1998) Highly efficient and versatile synthesis of libraries of constrained β-strand mimetics. *Bioorg. Med. Chem. Lett.* **8**: 2321–2326.

197. Qabar, M. N., McMillan, M.K., Kahn, M.S., Tulinsky, J.E., Ogbu, C.O. and Mathew, J. US Patent No. 6 117 896.

198. Blaskovich, M.A. and Kahn, M. (1998) Solid-phase preparation of dienes. *J. Org. Chem.* **63**: 1119–1125.

199. Okamoto, S. and Hijikata, A. (1981) Potent inhibition of thrombin by the newly synthesized arginine derivative No. 805. The importance of stereostructure of its hydrophobic carboxamide portion. *Biochem. Biophys. Res. Commun.* **101**: 440–446.

200. Sturzebecher, J., Markwardt, F., Voigt, B., Wagner, G. and Walsmann, P. (1983) Cyclic amides of N alpha-arylsulfonylamino-acylated 4-amidinophenylalanine — tight binding inhibitors of thrombin. *Thromb. Res.* **29**: 635–642.

201. Hilpert, K., Ackermann, J., Banner, D.W., Gast, A., Gubernator, K., *et al.* (1994) Design and synthesis of potent and highly selective thrombin inhibitors. *J. Med. Chem.* **37**: 3889–3901.

202. Krishnan, R., Mochalkin, I., Arni, R. and Tulinsky, A. (2000) Structure of thrombin complexed with selective non-electrophilic inhibitors having cyclohexyl moieties at P1. *Acta Crystallogr. D Biol. Crystallogr.* **56**: 294–303.

203. Weber, P.C., Lee, S.-L., Lewandowski, F.A., Schadt, M.C., Chang, C.-H. and Kettner, C.A. (1995) Kinetic and crystallographic studies of thrombin with Ac-(D)Phe-Pro-boroArg-OH and its lysine, amidine, homolysine, and ornithine analogs. *Biochem.* **34**: 3750–3757.

204. von der Saal, W., Kucznierz, R., Leinert, H. and Engh, R.A. (1997) Derivatives of 4-amino-pyridine as selective thrombin inhibitors. *Bioorg. Med. Chem. Lett.* **7**: 1283–1288.

205. Weber, I.R., Neidlein, R., von der Saal, W., Grams, F., Leinert, H., Strein, K., Engh, R.A. and Kucznierz, R. (1998) Diarylsulfonamides as selective, non-peptidic thrombin inhibitors. *Bioorg. Med. Chem. Lett.* **8**: 1613–1618.

206. Soll, R.M., Lu, T., Tomczuk, B., Illig, C.R., Fedde, C., *et al.* (2000) Amidinohydrazones as guanidine bioisosteres: application to a new class of potent, selective and orally bioavailable, non-amide-based small-molecule thrombin inhibitors. *Bioorg. Med. Chem. Lett.* **10**: 1–4.

207. Lu, T.B., Tomczuk, B., Bone, R., Murphy, L., Salemme, F.R. and Soll, R.M. (2000) Non-peptidic phenyl-based thrombin inhibitors: exploring structural requirements of the S1 specificity pocket with amidines. *Bioorg. Med. Chem. Lett.* **10**: 83–85.

208. Tamura, S.Y., Shamblin, B.M., Brunck, T.K. and Ripka, W.C. (1997) Rational design, synthesis, and serine protease inhibitory activity of novel P-1-argininoyl heterocycles. *Bioorg. Med. Chem. Lett.* **7**: 1359–1364.

209. Tamura, S.Y., Semple, J.E., Reiner, J.E., Goldman, E.A., Brunck, T.K., *et al.* (1997) Design and synthesis of a novel class of thrombin inhibitors incorporating heterocyclic dipeptide surrogates. *Bioorg. Med. Chem. Lett.* **7**: 1543–1548.

210. Semple, J.E. (1998) Design and construction of novel thrombin inhibitors featuring P-3-P-4 quaternary lactam dipeptide surrogates. *Bioorg. Med. Chem. Lett.* **8**: 2501–2506.

211. Sanderson, P.E.J., Lyle, T.A., Cutrona, K.J., Dyer, D.L., Dorsey, B.D., *et al.* (1998) Efficacious, orally bioavailable thrombin inhibitors based on 3-aminopyridinone or 3-aminopyrazinone acetamide peptidomimetic templates. *J. Med. Chem.* **41**: 4466–4474.

212. Semple, J.E., Rowley, D.C., Owens, T.D., Minami, N.K., Uong, T.H. and Brunck, T.K. (1998) Potent and selective thrombin inhibitors featuring hydrophobic, basic P-3-P-4-aminoalkyllactam moieties. *Bioorg. Med. Chem. Lett.* **8**: 3525–3530.

213. Wagner, J., Kallen, J., Ehrhardt, C., Evenou, J.P. and Wagner, D. (1998) Rational design, synthesis, and X-ray structure of selective noncovalent thrombin inhibitors. *J. Med. Chem.* **41**: 3664–3674.

214. Bachand, B., DiMaio, J. and Siddiqui, M.A. (1999) Synthesis and structure–activity relationship of potent bicyclic lactam thrombin inhibitors. *Bioorg. Med. Chem. Lett.* **9**: 913–918.

215. Plummer, J.S., Berryman, K.A., Cai, C.M., Cody, W.L., DiMaio, J., *et al.* (1999) Potent and selective bicyclic lactam inhibitors of thrombin: Part 3: P1' modifications. *Bioorg. Med. Chem. Lett.* **9**: 835–840.

216. St Denis, Y., Augelli-Szafran, C.E., Bachand, B., Berryman, K.A., DiMaio, J., *et al.* (1998) Potent bicyclic lactam inhibitors of thrombin: Part I: P3 modifications. *Bioorg. Med. Chem. Lett.* **8**: 3193–3198.

217. DiMaio, J., Siddiqui, M.A., Gillard, J.W., St-Denis, Y., Tarazi, M., *et al.* (1996) WO09619483. Low molecular weight bicyclic thrombin inhibitors. Shire BioChem Inc.

218. Bachand, B., Doherty, A.M., Siddiqui, M.A. and Edmunds, J.J. (1998) WO09828326. Bicyclic thrombin inhibitors. Shire BioChem Inc.

219. Salimbeni, A., Paleari, F., Scolastico, C. and Criscuoli, M. (1997) WO09705160. Bicyclic lactam derivatives as thrombin inhibitors. A Menarini, Ind Pharm Riunite Srb.

220. Salimbeni, A., Paleari, F., Canevotti, R., Criscuoli, M., Lippi, A., *et al.* (1997) Design and synthesis of conformationally constrained arginal thrombin inhibitors. *Bioorg. Med. Chem. Lett.* **7**: 2205–2210.

221. Coburn, C.A., Rush, D.M., Williams, P.D., Homnick, C., Lyle, E.A., *et al.* (2000) Bicyclic pyridones as potent, efficacious and orally bioavailable thrombin inhibitors. *Bioorg. Med. Chem. Lett.* **10**: 1069–1072.

222. Clackson, T. and Wells, J.A. (1995) A hot spot of binding energy in a hormone–receptor interface. *Science* **267**: 383–386.

223. Lin, L. and Wells, J.A. (1994) Dissecting the energetics of an antibody–antigen interface by alanine shaving and molecular grafting. *Prot. Sci.* **3**: 2351–2357.

224. Cochran, A.G. (2000) Antagonists of protein–protein interactions. *Chem. Biol.* **7**: R85–R94.

225. Cochran, A.G. (2001) Protein–protein interfaces: mimics and inhibitors. *Curr. Opin. Chem. Biol.* **5**: 654–659.

226. Shuker, S.B., Hajduk, P.J., Meadows, R.P. and Fesik, S.W. (1996) Discovering high-affinity ligands for proteins: SAR by NMR. *Science* **274**: 1531–1534.

227. Peng, J.W., Lepre, C.A., Fejzo, J., Abdul-Manan, N. and Moore, J.M. (2001) Nuclear magnetic resonance-based approaches for lead generation in drug discovery. *Methods. Enzymol.* **338**: 202–230.

228. Erlanson, D.A., Braisted, A.C., Raphael, D.R., Randal, M., Stroud, R.M., *et al.* (2000) Site-directed ligand discovery. *Proc. Natl Acad. Sci. USA* **97**: 9367–9372.

229. Gadek, T.R., Burdick, D.J., McDowell, R.S., Stanley, M.S., Marsters, J.C. Jr. *et al.* (2002) Generation of an LFA-1 antagonist by the transfer of the ICAM-1 immunoregulatory epitope to a small molecule. *Science* **295**: 1086–1089.

230. Bella, J., Kolatkar, P.R., Marlor, C.W., Greve, J.M. and Rossmann, M.G. (1998) The structure of the two amino-terminal domains of human ICAM-1 suggests how it functions as a rhinovirus receptor and as an LFA-1 integrin ligand. *Proc. Natl Acad. Sci. USA* **95**: 4140–4145.

231. Toogood, P.L. (2002) Inhibition of protein–protein association by small molecules: approaches and progress. *J. Med. Chem.* **45**: 1543–1558.

30

THE FATE OF XENOBIOTICS IN LIVING ORGANISMS

Franz M. Belpaire and Mark G. Bogaert

To explain all nature is too difficult a task for any one man or even for any one age. 'T is much better to do a little with certainty, and leave the rest for others that come after you, than to explain all things.

Isaac Newton

Most drugs exert their effect through reversible binding to receptors, to inhibition of an enzyme or to an effect on an ion channel. Onset, duration and intensity of the effect will depend on the concentration of the drug in the fluid surrounding the effect site, i.e. the biophase. Only exceptionally, drugs are applied directly to the effect site; in most cases drugs have to be transferred from the site of administration to the biophase. After absorption, the drug is distributed via the circulation to the different parts of the organism, including the organ(s) in which the biophase for the drug is localized. The drug is also distributed to organs such as the kidneys which excrete it unchanged into the urine or the liver, which excrete it unchanged into the bile or metabolize it. The metabolites formed are either further metabolized or excreted as such. As a consequence of elimination (excretion and biotransformation) the concentration of the drug in the organism, i.e. in the biophase, decreases, resulting usually in a decrease of the effect. The processes involved in drug disposition, the description of drug absorption, distribution and elimination, are shown schematically in Fig. 30.1.

The concentrations of a drug in the biophase and elsewhere in the organism, depend upon the dose administered, and upon the rate and extent of absorption, distribution and elimination. Pharmacokinetics is the study of the drug concentration as a function of time in the different parts of the organism.

Drug disposition involves passage through biological membranes. In this overview, first the mechanisms of passage of drugs through membranes will be discussed, and then the different processes of drug disposition will be described.

I PASSAGE OF DRUGS THROUGH MEMBRANES

A drug often encounters several cell membranes before it reaches the biophase and other parts of the organism, e.g. the elimination sites. Cellular membranes are composed of an lipoidal matrix covered on both sides by proteins. Cell membranes contain small pores, and between cells narrow, water-filled channels exist. Drugs cross membranes by passive or, in some cases, carrier-mediated mechanisms.

A Passive processes

Filtration

If hydrostatic pressure is applied or if an osmotic gradient is imposed across a cell membrane, water, together with small solutes, will pass through the cell membrane pores. There is considerable variation in the estimated pore size of various membranes and, correspondingly, in the size of the molecules that can pass through these pores. In most cell membranes the pores have a diameter of about 7 Å

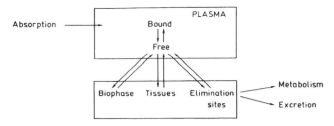

Fig. 30.1 Schematic representation of drug absorption, distribution and elimination.

and substances with a molecular weight below 100 dalton, such as water and urea, can pass through them. However, most drugs have a molecular weight above 100 dalton; hence passage by filtration through this type of pores is limited.

The channels between cells are usually rather narrow, but between capillary endothelial cells they measure about 40 Å. They allow passage of molecules with molecular weight below 60 kDa. Most drugs have a molecular weight lower than 1 kDa but most proteins have a molecular weight greater than 60 kDa; they cannot pass through these channels, and this is also true for proteins to which drugs are bound. Albumin (68 kDa) passes to a limited extent from the plasma to the extracellular fluid. The capillaries of the glomeruli of the kidney are composed of a particularly thin endothelium, which is very rich in intercellular channels, so that this membrane is far more permeable to solutes than are the capillaries elsewhere. The endothelial cells of the brain capillaries are surrounded by a layer of glial cells, which have tight intercellular junctions. This constitutes the so-called blood–brain barrier and filtration is not possible. To move from blood to brain and vice versa, materials must pass through the cells rather than between them.

Passive diffusion

Passive diffusion is the process by which molecules diffuse from a region of higher concentration to a region of lower concentration. It is the most important mechanism for passage of drugs through membranes. Lipid soluble drugs penetrate lipid membranes with ease. Polar molecules and all ionized compounds partition poorly into lipids, and are not able to pass through membranes, or do so at a much lower rate. Transmembrane diffusion is driven by the transmembranar concentration gradient of the drug. The rate of diffusion depends, apart from the lipid/water partition coefficient of the drug (P) and the concentration gradient ($C_{out} - C_{in}$), on membrane properties such as the membrane area (A) and thickness (h), and the diffusion coefficient (D) of the drug in the membrane, according to Fick's law:

$$\text{Rate of diffusion} = \frac{D.A.P.(C_{out} - C_{in})}{h}$$

Many drugs are acidic or basic compounds which are ionized to a certain degree in aqueous medium, depending on their dissociation constant (pK_a) and the pH of the solution, according to Henderson–Hasselbalch's equation.

For acidic drugs

$$pH = pKa + \log \frac{\text{Ionized concentration}}{\text{Unionized concentration}}$$

For basic drugs

$$pH = pKa + \log \frac{\text{Unionized concentration}}{\text{Ionized concentration}}$$

Very weak acids with pK_a values higher than 7.5, are essentially unionized at the pH values encountered in the organism. For these drugs, transport is rapid and independent of pH, provided the unionized form is lipid soluble. For acidic drugs with a pK_a value between 3.0 and 7.5, the fraction of unionized drug varies with the changes in pH encountered in the organism, and for these drugs a change in the rate of transport with pH is expected. For acidic drugs with a pK_a lower than 2.5, the fraction of unionized drug is low so that diffusion across membranes is very slow. A similar analysis can be done for bases.

At equilibrium, the concentrations of unionized molecules on both sides of a membrane are equal. If the pH on both sides of the membrane is equal, the concentration of ionized molecules, and thus the total concentration of the molecules will also be the same on both sides of the membrane. If there is a difference in pH, as for example between plasma (pH 7.4) and stomach contents (pH 1 to 3), the concentration of the ionized molecules at equilibrium and therefore the total concentration will be much higher at one side of the membrane than on the other. This phenomenon is called ion-trapping.

B Carrier-mediated processes

Many cell membranes possess specialized transport mechanisms that regulate entry and exit of physiologically important molecules and of drugs. Such transport systems involve a carrier molecule, i.e. a transmembrane protein, which binds one or more molecules and releases them on the other side of the membrane. Such systems may operate passively, without any energy source and operate along a concentration gradient; this is called 'facilitated diffusion'. Facilitated diffusion, however, seems to play only a minor role in drug transport. An example is the transport of vitamin B$_{12}$ across the gastrointestinal membrane. Alternatively, the system may require energy and the molecules can move against a concentration gradient; this mechanism is called 'active transport'. Active transport can explain renal and biliary excretion of many drugs, e.g. the renal tubular secretion of penicillins. At high drug concentrations the carrier sites become saturated and the rate of transport does not increase further. Furthermore, competitive inhibition of transport can occur if another drug for this carrier is present.

An important carrier protein mediating active transport of endogenous but also exogenous molecules, is P-glycoprotein (P-gp). This ATP-dependent transporter transports a wide variety of drugs out of the cells (efflux pump). It has first been characterized as a transporter responsible for efflux of chemotherapeutic agents from resistant cancer cells. It is also expressed in normal tissues, including the intestinal tract epithelia, the liver, the kidneys and the brain. The presence of P-gp in such tissues results for example in reduced absorption and enhanced elimination into the bile and urine, and in the prevention of entry into the brain. P-gp has been shown to transport a wide range of structurally unrelated drugs such as digoxin, quinidine, cyclosporine and HIV-1 protease inhibitors. The inhibition and induction of P-gp can lead to clinically significant drug interactions.

II ABSORPTION

Absorption can be defined as the passage of a drug from its site of administration into the circulation. If a drug is administered directly into the vascular system, e.g. by intravenous administration, absorption is not needed. Drugs can be administered by enteral and parenteral routes. Enteral administration occurs through contact of the drug with the buccal mucosa (sublingual), through swallowing (oral) or by rectal administration. With parenteral administration, the gastrointestinal tract is bypassed; examples are the intravenous and intramuscular routes. Drugs can also be absorbed through the skin or through the mucosa of various organs (bronchi, nose, vagina, etc.). In some cases a drug is applied for a local effect, and no absorption is intended.

In this chapter we will describe drug administration by the oral route in detail. Some characteristics of other common routes of drug administration are listed in Table 30.1. For more details on these various routes, the reader is referred to the literature.

The major components of the gastrointestinal tract are the stomach, the small intestine, and the large intestine or colon (Fig. 30.2). The small intestine includes the duodenum, jejunum and ileum. The major segments of the gastrointestinal tract differ from one another both anatomically and morphologically, as well as with respect to secretions and pH.

The rate and extent of absorption after oral administration are determined first of all by desintegration and dissolution. Indeed, many drugs are taken as tablets or capsules, and these have first to desintegrate. Therefore, liquid dosage forms are in general more rapidly absorbed than solid forms. Before a drug is absorbed from the gastrointestinal tract, it has to dissolve in the aqueous medium of the stomach and the intestine. Dissolution of the drug will depend on water solubility, particle size, chemical form and crystalline characteristics of the drug, and pH of the surrounding medium.

Disintegration can be controlled and in recent years various drug products have been modified to alter the timing of the release of the active drug from the drug product. The term 'controlled release' is used for various types of oral extended release rate dosage forms (such as sustained release, prolonged release) and delayed release rate dosage forms (e.g. enteric coated). One of the systems for controlled release is the osmotic pump: drug delivery is driven by an osmotically controlled device that pumps a constant amount of water through the system, dissolving and releasing a constant amount of drug per time unit.

Once the drug is dissolved, it can be absorbed, usually by passive diffusion, and pass into the capillaries of the gastrointestinal wall and so into the portal venous system. For some drugs with high lipid solubility, absorption via the lymphatic system is also possible. As drug molecules move through the gastrointestinal tract, they encounter environments which vary in the pH, enzymes and fluidity of contents, as well as in the area available for absorption. For acids and bases only the nonionized molecules can be absorbed. At all physiological pH values weak acids and bases exist mostly in the unionized form and can be absorbed as well from the stomach as from the intestine. Strong bases such as the quaternary ammonium compounds are to a large extent ionized at all physiological pHs, and are hardly absorbed at all.

Absorption is also influenced by the gastric emptying time. The residence time of drugs in the stomach varies from a few minutes to several hours and is dependent on the volume, viscosity and composition of the stomach content. A drug taken with food will stay longer in the stomach. Under normal conditions, gastric emptying is rapid and the stomach's role in drug absorption is modest. The small intestine is a much more important site for drug absorption. The intestinal mucosa is covered by numerous villi and microvili, providing a surface area of approximately 250 m^2 available for passive diffusion. In theory, weakly acidic drugs are a better substrate for passive diffusion at the pH of the stomach, than at that of the intestine. However, the limited residence time of the drug in the stomach and the relatively small surface area of the stomach more than balance the influence of pH in determining the optimal site of absorption. Factors that promote gastric emptying will increase the absorption rate of most drugs, but not necessarily the total amount of drug eventually absorbed. The motility of the intestine can also influence

Table 30.1 *Common routes of drug administration*

Route	Bioavailability	Advantages	Disadvantages
Parenteral routes			
Intravenous bolus (i.v.)	Complete (100%) systemic drug absorption. Rate of bioavailability considered instantaneous.	Drug is given for immediate effect.	Increased risk of adverse reaction. Possible anaphylaxis.
Intravenous infusion (i.v. inf)	Complete (100%) systemic drug absorption. Rate of drug absorption controlled by infusion pump.	Plasma drug levels more precisely controlled. May inject large fluid volumes. May use drugs with poor lipid solubility and/or irritating drugs.	Requires skill in insertion of infusion set. Tissue damage at site of injection (infiltration, necrosis, sterile abscess).
Intramuscular injection (i.m.)	Rapid from aqueous solution. Slow absorption from nonaqueous (oil) solutions.	Easier to inject than intravenous injection. Larger volumes may be used compared with subcutaneous solutions. Used, e.g. for insulin injection.	Irritating drugs may be very painful. Different rates or absorption depending upon muscle group injected and blood flow.
Subcutaneous injection (s.c.). Subcutaneous infusion	Prompt from aqueous solution. Slow absorption from repository formulations.	Used, e.g. for morphine.	Rate of drug absorption depends upon blood flow and injection volume.
Enteral routes			
Buccal or sublingual (s.l.)	Rapid absorption for lipid-soluble drugs.	No 'first-pass' effects.	Part of the drug may be swallowed. Not for most drugs or drugs with high doses.
Oral (p.o.)	Absorption may vary. Generally, slower absorption rate compared with i.v. bolus or i.m. injection.	Safest and easiest route of drug administration. May use immediate-release and modified-release drug products.	Some drugs may have erratic absorption, are unstable in the gastrointestinal tract, or metabolized by liver prior to systemic absorption.
Rectal (p.r.)	Absorption from suppository may vary. More reliable absorption from enema (solution).	Useful when patient cannot swallow medication. Used for local and systemic effects.	Absorption may be erratic. Suppository may migrate to different position. Some patient discomfort.
Other routes			
Transdermal	Slow absorption, rate may vary. Increased absorption with occlusive dressing.	Transdermal delivery system (patch) is easy to use. Used for lipid-soluble drugs with low dose and low MW.	Irritation by patch or drug possible. Permeability of skin variable with condition, anatomic site, age, and gender. Type of cream or ointment base affects drug release and absorption.
Inhalation	Rapid absorption. Total dose absorbed is variable.	May be used for local or systemic effects.	Particle size of drug determines site of placement in respiratory tract. May stimulate cough reflex. Part of the drug may be swallowed.

Reproduced with permission from reference 7.

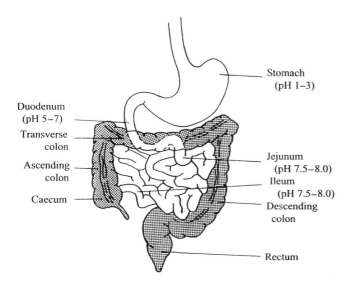

Fig. 30.2 Schematic representation of the gastrointestinal tract.

the absorption: when peristaltism increases, disintegration and dissolution of the drug are often accelerated.

To express the extent to which and at what rate a drug reaches the systemic circulation, the term bioavailability is used. Bioavailability of a drug is (1) the fraction of the administered dose which reaches the systemic circulation; and (2) the rate at which this occurs. After intravenous administration bioavailability is by definition 100%; for all other routes of administration, the bioavailability can vary between 0 and 100%.

Bioavailability is influenced by several factors, such as drug characteristics (solubility, particle size, crystalline

form, pK_a, lipid/water partition coefficient), formulation (pharmaceutical dosage form), changes at the site of administration (pH, blood flow), presence of other drugs or food, and the route of administration.

During absorption from the gastrointestinal tract, a drug has to pass from the gut lumen to the capillaries through the gut wall; the drug is then transported via the mesenteric vessels to the hepatic portal vein and then to the liver prior to reaching the systemic circulation. Biotransformation can occur before absorption, i.e. in the lumen, during absorption in the gut wall, and/or after absorption in the liver, but before reaching the systemic circulation (Fig. 30.3). This 'presystemic' metabolism, termed 'first pass effect', can result in an appreciable fraction of the dose not reaching the systemic circulation.

This first pass phenomenon also exists after intraperitoneal and, partially, after rectal administration. It does, of course, not exist for other routes of administration. Due to a hepatic first passage phenomenon, some drugs which are well absorbed through the gastrointestinal membrane still have a low bioavailability.

Recent data indicate that the P-glycoprotein in the gut wall limits bioavailability of some drugs. Since it is located on the epithelium of intestinal cells, it can act as a counter-transport pump that transports drugs back into the intestinal lumen as they begin to be absorbed across the gut wall.

III DRUG DISTRIBUTION

After absorption, drugs are distributed from plasma to the various organs. The rate and extent of distribution depend on

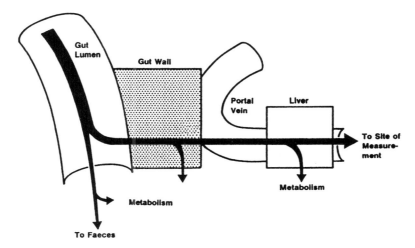

Fig. 30.3 After oral administration, a drug must pass from the gut lumen, through the gut wall and then through the liver before reaching the systemic circulation. Biotransformation may occur in the lumen before absorption, in the gut wall during absorption, and/or in the liver after absorption but before the systemic circulation is reached. (Reproduced with permission from reference 6.)

blood flow to different organs, tissue size, binding of drugs to plasma proteins and tissue components, and permeability of tissue membranes. The latter factor is related to the physicochemical properties of the drug. For lipid-soluble drugs the tissue membranes present no barrier, and distribution depends essentially on the perfusion rate of the tissue. For these drugs rapid equilibration occurs between blood and tissues such as lungs, kidney, liver, heart and brain, i.e. organs with a high blood flow. This is called 'perfusion rate-limited distribution'. Less rapid equilibration is found for skeletal muscle, bone and adipose tissue, which receive a considerably smaller volume of blood per unit mass. Tissue uptake of a drug continues until equilibrium is reached between the diffusible form of the drug in the tissue and the blood, i.e. until the free concentrations in plasma water and tissue water are equal. Drugs can be present in tissues in higher concentrations than in plasma as a consequence of pH gradients, but mainly as a consequence of binding to tissue constituents or of dissolution in fat.

A Volume of distribution

The volume of distribution is a proportionality constant, relating the total amount of drug present in the organism to its plasma concentration at the same moment. It is the volume in which the drug apparently distributes in a concentration equal to its concentration in plasma. This calculated value does not correspond to an anatomical or physiological part of the organism and can be much larger than the volume of total body water. It is therefore called 'apparent' volume of distribution. Drugs can have apparent volumes of distribution from $0.04 \, l \, kg^{-1}$ to $20 \, l \, kg^{-1}$, i.e. 2.8 to 1400 litres for a 70 kg person.

Total body water (42 litres in a normal 70 kg man) consists of plasma (3 litres), interstitial fluid (11 litres) and intracellular fluid (28 litres). If a drug which is not bound in plasma or tissues, and distributes over total body water, the apparent volume of distribution will be 42 litres per 70 kg; this is the case, for example, for antipyrine. If a drug is likewise not bound in plasma and tissues, but does not penetrate cells, the distribution will be limited to the extracellular space, equalling 14 litres. The apparent volumes of distribution of such drugs approximate their true volume of distribution. However, most substances bind to plasma and tissue proteins. For a drug which is preferentially bound to plasma proteins, for equal free concentrations the total concentrations will be higher in plasma than in the extracellular space. In this case the apparent volume of distribution will be smaller than 42 litres. If, however, a drug binds preferentially to tissue proteins, the total drug concentration will be lower in plasma than in tissues and the apparent volume of distribution will be larger

Table 30.2 *Apparent volume of distribution of some drugs in litres kg^{-1}*

Warfarin	0.11	Cimetidine	2.1
Ibuprofen	0.14	Propranolol	3.9
Salicylic acid	0.17	Digoxin	8.0
Gentamicin	0.25	Imipramine	30.0
Digitoxin	0.51	Chloroquine	235
Atenolol	0.70		

than 42 litres. A typical example is digoxin, which is highly bound in muscle and has an apparent volume of distribution of 600 litres.

The following equation describes the relationship between apparent volume of distribution, drug binding and anatomical volumes:

$$V_{\mathrm{d}} = V_{\mathrm{B}} + V_{\mathrm{T}} \frac{f_{\mathrm{B}}}{f_{\mathrm{T}}}$$

where V_{d} is the apparent volume of distribution, V_{B} is the blood volume, V_{T} is the extravascular volume and f_{B} and f_{T} are the free fractions of drug in blood and extravascular space, respectively. The apparent volume of distribution increases with increases in anatomical volumes or tissue binding, and decreases with increases in plasma or blood binding.

Many acidic drugs, e.g. salicylates, sulfonamides, penicillins and anticoagulants, are highly bound to plasma proteins or are not lipophilic enough to distribute intracellularly, and have therefore small volumes of distribution (<20 litres). Basic drugs on the other hand, are often highly distributed in tissues: their concentration in plasma is low and their apparent volume of distribution is large. Table 30.2 shows the apparent volume of distribution for some drugs.

B Plasma protein binding

Many drugs are bound to some extent to plasma proteins. The plasma protein binding is expressed as 'fraction bound', i.e. the ratio of bound concentration over total (bound plus free) concentration, or as 'percentage bound' if this value is multiplied by 100. Free fraction equals one minus bound fraction.

Many acidic drugs bind to albumin. Basic drugs bind also to α_1-acid glycoprotein and to lipoproteins. α_1-Acid glycoprotein is an acute phase reactant and its concentration in plasma rises in inflammation. For most drugs the binding of drugs to plasma proteins is a reversible process with extremely rapid rates of association and dissociation, and can be described by the law of mass action. The degree of binding is determined by affinity (expressed as the association constant), capacity (the number of binding sites per molecule protein), protein concentration and drug concentration. The number of binding sites is limited. At therapeutic drug concentrations, usually only a small fraction of the available binding sites is occupied; for a given protein concentration the free fraction is then rather constant and independent of drug concentration. In some instances the drug concentrations are so high, that most binding sites are occupied, and the free fraction becomes concentration-dependent. Concentration-dependent changes in drug binding are most likely to occur with drugs which have a high affinity for the proteins and which are given in large doses, e.g. acetylsalicylic acid, phenylbutazone, some penicillins and cephalosporins.

The plasma binding of drugs is altered in some physiological and pathological conditions. This often results from a change in plasma protein concentration. In various disease states (such as renal failure, liver disease, inflammation), in pregnancy and in the neonatal period, hypoalbuminemia is observed; α_1-acid glycoprotein concentrations rise in inflammatory diseases, stress and malignancy, and fall in liver disease. A change in affinity can lead to a change in binding. This can occur through competition between endogenous compounds and drug molecules for common binding sites. Free fatty acids, for example, bind strongly to albumin; when their concentration in plasma increases due to fasting, exercise or infection, the drug can be displaced from its binding sites. Accumulation of endogenous compounds also occurs in disease states, such as renal failure or liver diseases; hyperbilirubinemia, for example, can decrease the binding of other drugs. Binding can also be changed by competition between drugs. Such an interaction is to be expected when the molar concentration of the 'displacer' is about the same as that of the protein binding sites, resulting in a decrease of the binding sites available for the 'displaced' drug.

How do changes in drug binding, i.e. in free fraction, influence the free drug concentration in plasma? The changes of free concentration (calculated as total concentration times free fraction) will always be less than the changes of free fraction because of redistribution of the displaced drug to the tissues, and often, because of its more rapid elimination. The change of free drug concentration as a consequence of a change in plasma binding will be largest for drugs with a small initial distribution volume. Drug interactions due to competition for protein binding sites are more likely to be clinically significant when the displaced drug is highly plasma bound ($>90\%$), has a relatively small volume of distribution and a narrow therapeutic-toxic index. These conditions are not often met, however. Examples are the displacement of coumarin anticoagulant drugs by phenylbutazone or other nonsteroidal anti-inflammatory drugs.

IV DRUG ELIMINATION

Elimination covers both excretion (i.e. disappearance of unchanged drug from the body) and biotransformation (metabolism). Some drugs are excreted in the bile and may be eliminated with the faeces, others are excreted in saliva. General anaesthestics are often excreted by the lungs. However, renal excretion and hepatic biotransformation are the major routes of drug elimination and these processes will be discussed, together with biliary excretion.

A Excretion

Renal excretion

The basic anatomic unit of renal function is the nephron (Fig. 30.4A). Basic components are glomerulus, proximal tubule, loop of Henle, distal tubule, and collecting tubule. The renal excretion of drugs involves one or more of the processes of glomerular filtration, tubular reabsorption and active tubular secretion (Fig. 30.4B).

Glomerular filtration. Blood flow to the kidneys is about 1.2 to 1.5 l min^{-1}. About 10% of this volume is filtered through the glomeruli, which amounts to a filtrate of about 125 ml min^{-1} or 180 litres per 24 h. The pores of the glomerular capillaries are sufficiently large to permit passage of most drug molecules, but do not allow passage of blood cells and of large molecules (>60 kDa), such as plasma proteins. Drug molecules bound to plasma proteins are thus not filtered.

Tubular reabsorption. More than 99% of the original 180 litres of protein-free filtrate is reabsorbed via the tubular cells; only about 1.5 litres is excreted as urine. Solutes and drugs dissolved in this filtrate can also be reabsorbed. For different drugs, tubular reabsorption varies from almost absent to almost complete. For most drugs, reabsorption is a passive process (passive diffusion). The drugs diffuse from tubular fluid to plasma in accordance with their concentration gradient, lipid/water partition coefficient, degree of ionization and molecular weight. The pH of the urine varies between 4.5 and 7.0, and pH changes can influence passive reabsorption and thus the excretion of the drug (see the Henderson–Hasselbalch equation in the introduction). Acidifying the urine favours the reabsorption of weak acids such as salicylates and retards their excretion, whereas the reverse is true for weak bases. Alkalinization of the urine increases the excretion of weak acids. On the other hand, the urinary excretion of weak bases is low in alkaline urine. It is possible, for example, to shorten the half-life of phenobarbital, a weak acid, by administration of sodium bicarbonate to a patient in the case of overdose.

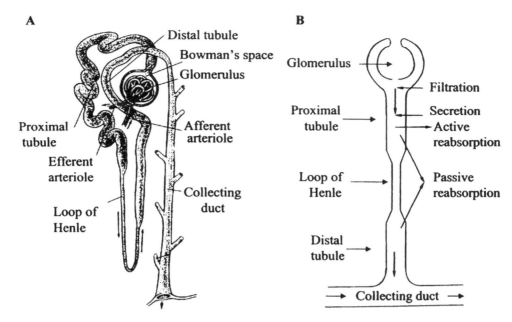

Fig. 30.4 Schematic representation of (A) a nephron and (B) the renal excretion of drugs.

Increasing the urinary volume (diuresis) also increases the renal excretion of drugs that are extensively reabsorbed; indeed, the concentration gradient between tubular fluid and plasma is decreased by the increased tubular water load.

Active tubular secretion. By this process drug is transported against a concentration gradient from the blood capillaries across the tubular membranes to the tubular fluid. In the proximal renal tubuli, two systems are primarily responsible for the active tubular secretion of drugs, one for organic anions and one for organic cations. The anionic system transports organic acids such as penicillins, indomethacin and glucuronides. The cationic system transports organic bases such as morphine, procaine and quaternary ammonium compounds. These transport systems are saturable at high drug concentrations and competition is possible between drugs which are transported by the same transport mechanism. This characteristic has been used to decrease the urinary excretion of penicillin by probenicid, a weak organic acid, thereby prolonging penicillin's effect.

P-glycoprotein is present in the brush border of the renal proximal tubules, and can play a role in the active tubular secretion of exogenous substances. This pump is involved in tubular secretion of, for example, digoxin, and can be inhibited by quinidine or verapamil, leading to an increase in digoxin serum concentrations.

Plasma protein binding does not affect the rate of active tubular secretion as the affinity of the drugs for this transport system is much higher than for the plasma proteins.

Renal clearance. A fraction of the drug presented to the kidneys along with the renal arterial blood, is removed by the above-mentioned processes. The efficiency of the renal excretion of a drug is expressed as 'renal clearance'. The renal clearance of a drug is the volume of plasma which is cleared of that substance per time unit. Substances such as inulin and creatinine are eliminated by glomerular filtration and are not subject to either tubular secretion or reabsorption, and are not bound to plasma proteins. Their renal clearance in adults with normal renal function will be around 125 ml min^{-1}, corresponding to the volume of plasma filtered. The clearance of inulin or creatinine can therefore be used as an index of the glomerular filtration rate. For substances which are filtered but also actively secreted, renal clearance is higher than 125 ml min^{-1} and can be as high as 650 ml min^{-1}, which is the total plasma flow through the kidneys. Such values are found for *p*-aminohippuric acid and penicillins, for example. For a drug which is filtered but reabsorbed, or if a drug is bound in plasma, clearance values lower than 120 ml min^{-1} are found. The renal clearance of a drug relative to the glomerular filtration rate therefore provides information on the mechanisms of renal excretion.

The renal clearance of a drug can be calculated by dividing the amount of drug excreted in the urine over a given time interval by the concentration of the drug in blood or plasma at the time corresponding to the midpoint of the urine collection interval.

Biliary excretion

Some drugs are actively secreted into the bile and pass as such into the intestine. In the rat, only compounds with a molecular weight greater than 350 Da are extensively

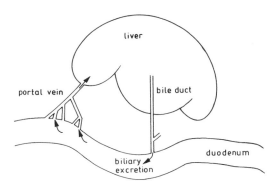

Fig. 30.5 Schematic representation of enterohepatic cycling of drugs.

secreted in the bile. In humans, the molecular weight threshold for appreciable biliary excretion is in the order of 400 to 500 Da. In order to be excreted into the bile, drugs usually also require a strongly polar group. Many drugs excreted into bile are metabolites, often glucuronide conjugates.

A drug (and/or its metabolites) entering the intestine from the bile may be excreted in the faeces. However, substances secreted with the bile can be reabsorbed from the intestine and thus undergo 'enterohepatic cycling' as shown in Fig. 30.5. Drug conjugates, e.g. glucuronides, can be hydrolysed in the gut by bacteria, with liberation and reabsorption of the parent drug. In particular this has been found for chloramphenicol and for steroids.

B Biotransformation

Many drugs are lipophilic and are only partially ionized at the pH values encountered in the organism. After glomerular filtration these compounds are largely reabsorbed from the renal tubules. The metabolites formed by biotransformation are in general more hydrophilic, therefore allowing their excretion by the kidneys. Biotransformation usually inactivates a drug, but in some cases active metabolites are formed. For some drugs, the activity may reside wholly in one or more metabolites: drugs that only become active after biotransformation are termed 'prodrugs'. Prodrugs are sometimes developed to improve absorption. They are often more lipid-soluble than the parent compound, and after absorption are rapidly converted to the parent compound in the gut wall or in the liver. An example is pivampicilline, an ester of ampicillin, which is rapidly and completely hydrolysed to ampicillin during absorption.

The main drug biotransformation site is the liver, but biotransformation can also take place in intestinal mucosa, lung, kidneys, skin, placenta and plasma. Most of biotransforming enzymes are found in the microsomes, a

cellular fraction derived from the endoplasmic reticulum. Liver or other organs are homogenized, and the homogenate is centrifuged at 9000 to 12 000 g for 30 min. The supernatant is centrifuged at 105 000 g for 1 h. The sediment collected is the microsomal fraction and the supernatant contains the cytosol.

Two phases can be distinguished in the pathways of biotransformation. Phase I involves addition of functionally reactive groups by oxidation, reduction or hydrolysis. Phase II consists of conjugation of reactive groups, present either in the parent molecule or after phase I transformation. Phenytoin, for example, is first hydroxylated by a phase I reaction and subsequently conjugated with glucuronic acid. The various phase I and phase II reactions are summarized in Tables 30.3 and 30.5.

Phase I reactions

Oxidations. Most oxidative processes take place in liver microsomes and are catalysed by mono-oxygenase enzymes known as mixed-function oxidases. These processes require reduced nicotinamide-adenine dinucleotide phosphate, molecular oxygen and a complex of enzymes in the endoplasmatic reticulum. The terminal oxidizing enzyme is cytochrome P450, a hemoprotein. The notation 'P450' refers to the ability of the reduced (ferrous) form of the hemoprotein to react with carbon monoxide, yielding a complex with absorption peak at 450 nm. For each molecule

Table 30.3 *Phase I reactions*

Oxidations via microsomal P450 system
  Aliphatic hydroxylation (pentobarbital)
  Aromatic hydroxylation (propranolol)
  Deamination (amphetamine)
  N-dealkylation (imipramine)
  O-dealkylation (phenacetin)
  S-dealkylation (thiopental)
  Dehalogenation (halothane)
  Desulfuration (parathion)
  Epoxidation (carbamazepine)
  N-hydroxylation (trimethylamine)
  Sulfoxidation (chlorpromazine)

Oxidations via non-microsomal P450 system
  Alcohol dehydrogenase (ethanol)
  Monoamine oxidase (tyramine)
  Purine oxidase (theophylline)

Reduction
  Azoreduction (sulfasalazine)
  Nitroreduction (chloramphenicol)

Hydrolysis
  Ester hydrolysis (procaine)
  Amide hydrolysis (procainamide)

of substrate oxidized, one molecule oxygen is consumed; one oxygen-atom is introduced into the substrate, and the other is reduced to form water. The P450s represent a superfamily of enzymes. Initially it was believed that there were only two forms, termed P450 and P448. Nowadays more than 30 different P450s have been identified in humans. To unify the nomenclature, P450s are grouped in families, designated by an arabic number, within which the amino acid sequence homology is higher than 40%. The majority of P450s involved in drug metabolism belong to three distinct families, CYP1, CYP2 and CYP3. Each P450 family is further divided into subfamilies, designated by capital letters, which in mammals contain proteins that share more than 55% amino acid sequence homology. In each subfamily, specific enzyme are denoted by an arabic number. Each isoenzyme has more or less distinct substrate specificity requirements. Only six of the numerous cytochromes P450 play a major role in the metabolism of drugs in common clinical use. Prominent among them in regard to the number of substrate drugs are CYP3A4 and CYP2D6, with smaller numbers of drugs metabolized by CYP2C9, CYP2C19, CYP1A2 and CYP2E1. Some selected substrates are listed in Table 30.4. Cytochrome CYP1A2 is particularly involved in the metabolism of environmental chemicals but also of drugs. CYP3A4 accounts for 30% of total P450 enzyme in the liver and is clinically the most important isoenzyme present in the liver. It is also substantially expressed in the mucosal epithelium of the intestine. Nearly 50% of all clinically used medications are metabolized by CYP3A4. This explains to a large extent why this enzyme is involved in many important drug interactions. Sometimes a single substrate is metabolized by

a single P450 enzyme, while other substrates can be oxidized to varying degrees by multiple P450 enzymes.

In addition to cytochrome P450s, hepatic microsomes contain another class of mono-oxygenases, the flavin-containing mono-oxygenases (FMO). These enzymes catalyse oxidation at nucleophilic nitrogen, sulfur and phosphorus atoms rather than oxidation at carbon atoms, e.g. for phenothiazines, ephedrine, norcocaine and the mono-ether and carbamate-containing pesticides.

Some oxidations are mediated by hepatic enzymes localized outside the microsomal system. Alcohol dehydrogenase and aldehyde dehydrogenase, which catalyse a variety of alcohols and aldehydes such as ethanol and acetaldehyde, are found in the soluble fraction of the liver. Xanthine oxidase, a cytosolic enzyme mainly found in the liver and in small intestine, but also present in kidneys, spleen and heart, oxidizes mercaptopurine to 6-thiouric acid. Monoamine oxidase, a mitochondrial enzyme found in liver, kidney, intestine and nervous tissue, oxidatively deaminates several naturally occurring amines (catecholamines, serotonin), as well as a number of drugs.

Reduction. Reduction, for example azo- and nitro-reduction, is a less common pathway of drug metabolism. Reductase activity is found in the microsomal fraction and in the cytosol of the hepatocyte. Anaerobic intestinal bacteria in the lower gastrointestinal tract are also rich in these reductive enzymes. A historical example concerns Prontosil[R], a sulfonamide prodrug. It is metabolized by azo-reduction to form the active metabolite, sulfanilamide. Sulfasalazine is also cleaved by azoreduction by intestinal bacteria to form aminosalicylate, the active component, and sulfapyridine. Chloramphenicol is metabolized by

Table 30.4 *Nonexhaustive list of substrates for, and inducers and inhibitors of some human liver cytochrome P450 isoenzymes*

Enzyme	Substrate	Inducer	Inhibitor
CYP1A2	caffeine, imipramine, paracetamol, theophylline	cigarette smoke, omeprazole	cimetidine, ciprofloxacin, enoxacin, fluvoxamine
CYP2C9	diclofenac, naproxen, piroxicam, tolbutamide, S-warfarin	phenobarbital, rifampicin	fluconazole, fluvoxamine
CYP2C19	diazepam, imipramine, S-mephenytoin, omeprazol, proguanil, propranolol	carbamazepine, rifampicin	cimetidine, felbamate
CYP2D6	codeine, debrisoquine, dextromethorphan, flecainide, paroxetine, perphenazine, thioridazine		fluoxetine, paroxetine, quinidine
CYP2E1	chlorzoxazone, ethanol, paracetamol	ethanol, isoniazid	diethyldithiocarbamate
CYP3A4	alprazolam, calcium channel blockers, cisapride, clarithromycin, cyclosporin A, erythromycin, HIV protease inhibitors, lidocaine, midazolam, simvastatin, terfenadine	carbamazepine, dexamethsone, phenobarbital, phenytoin, rifampicin, St John's wort	cimetidine, erythromycin, grapefruit juice, HIV protease inhibitors, itraconazole, ketoconazole

A more exhaustive list of substrates, inducers and inhibitors can be found in reference 3 and on the website www.drug-interactions.com.

nitro-reduction to an amine in bacteria and in a number of tissues.

Hydrolysis. Hydrolysis of esters and amides is a common pathway of drug metabolism. The liver microsomes contain non-specific esterases, as do other tissues and plasma. Hydrolysis of an ester results in the formation of an alcohol and an acid; hydrolysis of an amide results in the formation of an amine and an acid. The ester procaine, a local anaesthetic, is rapidly hydrolysed by plasma cholinesterases and, to a lesser extent, by hepatic microsomal esterase. An example of an amide which is hydrolysed, is the antiarrhythmic drug procainamide. Enalapril, a prodrug, is hydrolysed by esterases to the active metabolite enalaprilate, which inhibits the angiotensin-converting enzyme.

Phase II reactions

Compounds having polar constituents such as — OH, —NH_2 or —COOH, or acquiring them by a phase I reaction, may undergo a phase II or conjugation reaction. The major conjugation reactions are listed in Table 30.5.

The reactive group of the drug interacts with endogenous compounds such as glucuronic acid, sulfate, glycine, acetate or glutathione. Glucuronidation and sulfation are the most common conjugation process. Knowledge of the function, biochemistry, and molecular biology of the responsible enzymes, namely the UDP glycosyltransferases (UGTs) and sulfotransferases (STs), has increased extensively in recent years. UGTs are membrane-bound enzymes and are located in endoplasmatic reticulum, while STs are present in the cytosol. Both UGTs and STs comprise a superfamily of enzymes; in humans at least eight UGTs and two STs have been identified.

Conjugation reactions may involve an active, high-energy form of the conjugating agent, such as uridine diphosphoglucuronic acid (UDPGA) and acetyl CoA, which in the presence of the appropriate transferase enzyme combines with the drug to form the conjugate. For other conjugating reactions, the drug is activated to a high-energy compound that then reacts with the conjugating agent in the presence of a transferase enzyme. Glutathione, for example, reacts via the enzyme glutathione-*S*-transferase with reactive electrophilic oxygen intermediates of certain drugs, such as paracetamol.

Conjugates are usually pharmacologically inactive; they are more hydrophilic than the parent compounds and are easily excreted by the kidneys or the bile. Some conjugates, e.g. morphine-6-glucuronide and acetyl procainamide, are pharmacologically active.

Stereoselective metabolism

Synthesis of a drug with an asymmetrical or chiral centre usually results in two enantiomers, mirror images that cannot be superimposed. Such a 50:50 mixture is called a racemate. The enantiomers of a racemic drug often differ in their pharmacodynamic and/or pharmacokinetic properties as a consequence of stereoselective interaction with optically active biological macromolecules. Stereoselective metabolism of chiral xenobiotics is well recognized. Both phase I and phase II metabolic reactions are capable of discriminating between enantiomers. Stereoselective metabolism of chiral drugs implies the preferential enzymatic removal of one enantiomeric form over the other. Stereoselective drug metabolism may be divided in three groups: substrate stereoselectivity, product stereoselectivity and substrate-product stereoselectivity.

1. Substrate stereoselectivity is characterized by the preferential enzymatic metabolism of one enantiomer; metabolism can occur with retention or with loss of stereoisomerism. Most examples of stereoselective metabolism belong to this group. Several chiral non-steroidal anti-inflammatory agents of the 2-aryl propionic acid group undergo an unusual metabolic reaction whereby *R*-enantiomers are inverted to the active *S*-antipodes. The extent of inversion varies considerably depending on the drug, but is also species-dependent.
2. Product stereoselectivity is observed when a prochiral drug is preferentially metabolized to one or more chiral products. There are only a few examples of this type of stereoselectivity, e.g. the 5-hydroxylation of phenytoin and the 4-hydroxylation of debrisoquine, both with preferential formation of the *S*-enantiomer of the hydroxylated product.

Table 30.5 *Phase II reactions or conjugation reactions*

Conjugation	Localization	Conjugating agent	Functional groups
Glucuronidation	Microsomes	UDPGA	—OH, —COOH, —NH_2, SH
Sulfation	Cytosol	PAPS	—OH, —NH_2
Acetylation	Cytosol	Acetyl CoA	—COOH
Glutathion	Cytosol	Glutathion	Epoxides, arene oxides
Methylation	Cytosol	SAM	—OH, —NH_2
Amino-acid	Cytosol	Glycine	—COOH

UDPGA, uridine diphosphoglucuronic acid; PAPS, 3'-phospoadenosine-5'-phosphosulfate; SAM, *S*-adenosylmethionine.

3. Substrate-product stereoselectivity: the enantiomers of a drug which possess both asymmetrical and prochiral characteristics can undergo stereoselective metabolism, whereby a second chiral centre is introduced. Examples are the hydroxylation of perhexiline and the keto-reduction of warfarin.

Stereoselective metabolism is the most important process responsible for the stereoselectivity observed in pharmacokinetics. Verapamil has received considerable attention as an example of substrate stereoselective pharmacokinetics in humans. After oral administration, the drug undergoes an important stereoselective first pass metabolism, so that $(-)$-verapamil, the active enantiomer, has a two to three times lower bioavailability than its antipode. The $(-)/(+)$ plasma concentration ratio is therefore higher after intravenous than after oral administration.

Factors influencing biotransformation

Many factors can influence the metabolism of drugs, leading to large intra- and interindividual differences in the elimination of drugs. Some factors are genetically determined, others environmentally.

Species differences. There are large differences in drug metabolism between species. Rates of metabolism may differ; in general, smaller animals metabolize faster than larger animal species. Species may also show differences in individual metabolic pathways. In the rat, for example, amphetamine is mainly hydroxylated, whereas in humans and dogs it is largely deaminated. A specific pathway may be absent in a particular species; for example, the cat is unable to form glucuronic acid conjugates of some drugs; dogs are unable to *N*-acetylate aromatic amino groups and hydrazides.

The problems posed by species differences for the development and screening of new drugs are considerable. Knowledge of the mechanism of biotransformation of a new chemical agent in animals is, however, fundamental to its safety evaluation.

Genetic factors. It has been established by twin studies and other means that interindividual differences in drug metabolism are largely under genetic control. For drugs such as phenylbutazone, antipyrine and dicoumarol, fraternal twins show wide variations in metabolic rate, as generally seen in human populations, whereas each pair of identical twins shows rather similar metabolic rates. Family studies have suggested that metabolism is under polygenetic control. Frequency distribution plots of metabolic parameters usually yield continuous, unimodal curves similar to the normal distribution curve. The metabolism of some drugs, however, is under monogenetic control and shows a

bimodal distribution, with genetic polymorphism in the population.

Genetic polymorphism was first described for the *N*-acetylation of isoniazide and other drugs (e.g. procainamide, hydralazine, dapsone, sulfadimidine). Two phenotypes are found in the population: fast and slow acetylators. Later genetic polymorphism was also recognized for drug oxidations. The best-known example is the deficiency of the 4-hydroxylation of the antihypertensive agent debrisoquine by the CYP2D6 isoform. Two distinct modes in the frequency distribution of debrisoquine metabolic ratios are observed, which represent two distinct phenotypes: poor and extensive metabolizers (Fig. 30.6). The prevalence of poor metabolizers of debrisoquine in Caucasian populations varies between 5 and 10%; in other ethnic groups the frequencies range from 0% (amongst Japanese) to 30% (in a small population of Chinese living in Canada). Studies with subjects phenotyped for debrisoquine, have shown that the metabolism of more than 30 other drugs co-segregates with debrisoquine hydroxylation. These drugs include β-adrenergic blockers such as metoprolol and propranolol, the antitussive opioid dextromethorphan, tricyclic antidepressants, neuroleptics such as perphenazine and thioridazine and the antiarrhythmic agents propafenon and sparteine (Table 30.4).

Another intensively studied genetic polymorphism of drug oxidation concerns the antiepileptic drug mephenytoin. This deficiency, resulting from an inherited defect

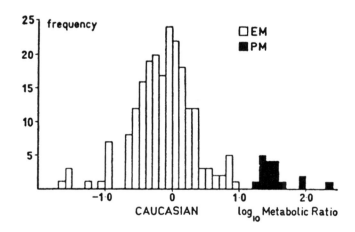

Fig. 30.6 Semilog frequency–distribution histogram of debrisoquin 4-hydroxylation (i.e. metabolic ratio) in 229 British Caucasians. Excretion as percentage of dose of debrisoquine and 4-hydroxydebrisoquine is measured after oral administration of 12.5 mg debrisoquine sulfate. The metabolic ratio was calculated by dividing the percentage excretion of debrisoquine by the percentage excretion of 4-hydroxydebrisoquine in urine collected during 8 h after administration of debrisoquine sulfate. (Reproduced with permission from reference 11.)

in the CYP2C19 isoenzyme, affects only the (*S*)-enantiomer of mephenytoin. (*S*)-mephenytoin undergoes a rapid and complete oxidation to a *p*-hydroxylated product, while the (*R*)-antipode is subject to a much slower *N*-demethylation pathway. The poor metabolizer phenotype occurs with a frequency of 2 to 5% in Caucasians, but with a frequency above 20% in Japanese people. The metabolism of drugs such as phenytoin, propranolol, diazepam and hexobarbital co-segregates with that of mephenytoin.

Two cytochrome P450 isoenzymes, CYP2D6 and CYP2C19, are involved in the metabolism of propranolol. The 4-hydroxylation of propranolol co-segregates with the debrisoquin polymorphism, but the side-chain oxidation to naphthoxylactic acid is catalysed in part by the mephenytoin isoenzyme.

The clinical consequences of polymorphism depend on the characteristics of the drug in extensive metabolizers (e.g. presence or absence of hepatic first pass), on the quantitative importance of the defective pathway(s) in the overall elimination of the drug, but also on the concentration–response relationship and the therapeutic index of the drug. The presence of the slow hydroxylator phenotype can result in accumulation of the parent drug, with stronger or prolonged effects and/or adverse reactions, or may lead to a decreased effect if the effect is related to an active metabolite. Codeine, for example, exerts its analgesic effect through the conversion of codeine to morphine by CYP2D6 and is thought to have no analgesic effect in poor metabolizers.

Gender. Male rats metabolize many drugs faster than females, and these differences occur for oxidative pathways and for glucuronide and glutathione conjugations of certain substrates. Studies in the rat have shown not only quantitative differences in the amount of cytochrome P450 isoenzymes between sexes, but also qualitative differences; there are male-specific and female-specific P450 isoenzymes. Sex hormones, especially androgens, play an important role in defining the differences in cytochrome P450 composition in male and female rats.

In humans too, sex differences in metabolism exist, but they are small and usually clinically not important. The clearance of benzodiazepines eliminated by metabolic conjugation (temazepam, oxazepam, lorazepam) is, for example, significantly smaller in women than in men.

Age. In most rodents the fetus is essentially devoid of hepatic cytochrome P450 linked drug-metabolizing enzymes. After birth, however, the system develops very rapidly and attains adult levels within a few days or weeks. Unlike the fetuses of common laboratory animals, the human fetus is able to oxidize drugs, although not all drugs and not at the rates achieved in adults. These activities begin to develop during the first trimester, reach a plateau at

parturition, and then rise gradually for several weeks until the adult level is reached. Sulfate conjugation seems to be as efficient in neonates as in adults, but most other conjugation pathways are poorly developed in the fetus. Conjugation with glucuronic acid reaches adult levels only at 3 years of age. This is responsible for the serious adverse reactions observed in neonates after administration of chloramphenicol, a drug that is ordinarily conjugated with glucuronic acid, leading to the 'grey syndrome'.

Certain drugs (e.g. antipyrine, diazoxide, carbamazepine, theophylline) are metabolized faster in older infants and children than in adults. In male rats oxidative metabolism of some drugs declines with ageing. In female rats the oxidative activities are well preserved during ageing. In humans, the age-related decrease in metabolism is usually minor in relation to the large interindividual variability in drug metabolism due to other reasons. The hepatic drug-metabolizing activity does not decrease significantly with increasing age. The observed age-related difference in drug clearance appears to be mainly accounted for by an age-related decrease in liver size or in liver blood flow.

Drug interactions. Numerous drugs or environmental chemicals can potentiate or inhibit the pharmacological and toxicological effects of other drugs by altering the activity of drug-metabolizing enzymes.

Enzyme induction. Chronic administration of a compound can stimulate its own biotransformation (auto-induction) or the biotransformation of other compounds (hetero-induction). Examples of inducers include drugs, steroids, industrial chemicals, pesticides, herbicides, polycyclic hydrocarbons and diet constituents. Phenobarbital and some polycyclic hydrocarbons such as 3-methylcholanthrene, benzpyrene, cigarette smoke have been extensively studied as inducers. Enzyme systems inducible by xenobiotics are cytochrome P450s, glutathione-*S*-transferase, glucuronyl transferase and epoxide hydrolase. The stimulation of the activity of enzymes involves new protein synthesis and, in most cases, an increase in the rate of gene transcription; it is true enzyme induction and not activation of latent enzyme. In the case of cytochrome P450, the effect of the inducer is to increase the microsomal concentration. However, not all P450 isoenzymes are inducible; for example, there are no known inducers of CYP2D6. Each inducer specifically increases the synthesis of certain P450 isoenzymes, as shown in Table 30.5. The former concept of 'phenobarbital-type inducers' and 'polycyclic hydrocarbon inducers' is no longer valid. At least three other distinct major inducer categories are now recognized. P450 induction by phenobarbital in the liver is accompanied by a substantial increase in the content of smooth endoplasmic reticulum within the liver cells and by an increase in liver weight. Such morphologic changes are, however, far less prominent with inducers from most other categories and during induction in non-hepatic tissues.

Clinically important enzyme inducers are carbamazepin, phenobarbital, phenytoin, rifampicin and St John's wort. A potent enzyme inducer such as rifampicin can markedly alter enzyme activity within 48 h after its administration, while for other inducers several days are necessary. Enzyme induction is dose-dependent.

Enzyme inhibition. The metabolism of drugs can be inhibited by exogenous and endogenous compounds. Some inhibitors of various CYPs are summarized in Table 30.5. Several mechanisms are involved. The most common mechanism is competitive inhibition. Any two drugs that are metabolized by the same P450 isoenzyme have the potential for competitive interaction. Moreover, some drugs act as competitive inhibitors of a particular P450 isoenzyme, although they are not metabolized by that P450 isoenzyme. This is the case in humans for quinidine which selectively inhibits CYP2D6, but is not metabolized by that isoenzyme. It is possible to predict such interactions by using an *in vitro* preparation of human hepatic microsomes or a recombinant preparation of the specific enzyme. The clinical significance of such a competitive interaction will depend on the drug's affinity for binding to the P450 isoenzyme, the concentrations of the drug in the endoplasmic reticulum, the dependence on the P450 isoenzyme for elimination, and the therapeutic-toxic index of the inhibited drug.

In addition to the reversible competitive inhibitors, other compounds can bind irreversibly to the enzyme via a reactive metabolic intermediate. This type of inhibition is designated as 'mechanism-based' or 'suicide' inhibition. Synthesis of new enzyme is the only means by which activity can be restored. Inhibition by macrolide antibiotics, such as erythromycin, is an example of this type of interaction. CYP3A4 demethylates and oxidizes the macrolide antibiotic into a nitrosoalkane that forms a stable, inactive complex with CYP3A4. Another mechanism of inhibition is by the binding of an imidazole or a hydrazine group to the haem portion of CYP450. In the case of cimetidine, the nitrogen in the imidazole ring binds to the haem portion of P450 causing nonselective inhibition of many P450 isoenzymes.

Inhibitors for nonmicrosomal enzymes are also known: disulfiram, for example, inhibits alcohol and aldehyde dehydrogenase, carbidopa inhibits dopa decarboxylase, allopurinol inhibits xanthine oxidase, phenelzine inhibits monoamine oxidase.

Hepatic clearance

The hepatic clearance (Cl_H) of a drug can be defined as the volume of blood perfusing the liver that is cleared of the drug per unit of time. The pharmacokinetic concept of hepatic clearance takes into consideration the anatomical and physiological fact that drug is transported to the liver by the portal vein and the hepatic artery, and leaves the organ by the hepatic vein. It diffuses from plasma water to reach the metabolic enzymes. There are therefore at least three major parameters to consider in quantifying drug elimination by the liver: blood flow through the organ (Q), which reflects transport to the liver; free fraction of drug in blood (f_u) which affects access of drug to the enzymes; and intrinsic ability of the hepatic enzymes to metabolize the drug, expressed as intrinsic clearance (Cl'_{int}). Intrinsic clearance is the ability of the liver to remove drug in the absence of flow limitations and blood binding. Taking in account these three parameters, the hepatic clearance can be expressed by

$$Cl_H = Q \frac{f_u . Cl'_{int}}{Q + f_u . Cl'_{int}}$$

It is obvious that the hepatic clearance cannot be larger than the total volume of blood reaching the liver per unit of time, i.e. the liver blood flow Q. The ratio of the hepatic clearance of a drug to the hepatic blood flow, is called the extraction ratio of the drug (E). The value of the extraction ratio can vary between 0 and 1. It is 0 when $f_u.Cl'_{int}$ is 0, i.e. when the drug is not metabolized in the liver. It is 1, when the hepatic clearance equals the hepatic blood flow (about $1.5 \, l \, min^{-1}$ in humans).

When $f_u.Cl'_{int}$ is very small in comparison to hepatic blood flow ($f_u.Cl'_{int} < Q$), the equation reduces to:

$$Cl_H = f_u . Cl'_{int}$$

In that case, clearance is not blood-flow-dependent but depends on enzymatic activity and blood binding. Binding to blood will limit the elimination. This is called 'restrictive elimination'. Drugs with a restrictive elimination have a low extraction ratio (< 0.3). Examples are antipyrine, phenytoin and warfarin. When $f_u Cl'_{int}$ is very large in comparison to hepatic blood flow ($f_u.Cl'_{int} > Q$), the equation reduces to

$$Cl_H = Q$$

In this case, clearance is dependent on hepatic blood flow and independent of Cl'_{int} and f_u. This is termed 'blood-flow-dependent' or 'nonrestrictive' elimination. Drugs with such an elimination have an high extraction ratio (> 0.7). Bound and free molecules are eliminated, as the affinity for the hepatic enzymes is larger than for the binding proteins in blood. The decrease in hepatic blood flow which occurs in hepatic disease and cardiac failure and after administration of β-blockers, leads to a decreased clearance. Examples of drugs with a high extraction ratio are nitroglycerin, propranolol and lidocaine. When a drug with a high extraction ratio is given orally, an important first pass elimination occurs. This is called 'presystemic elimination'. The higher the extraction ratio of a drug, the lower its bioavailability. For a drug which is completely absorbed by the gastrointestinal tract and metabolized in the liver, bioavailability equals $1 - E$.

REFERENCES

1. Gibaldi, M. (1991) *Biopharmaceutics and Clinical Pharmacokinetics.* Lea and Febiger, Philadelphia.
2. Gonzalez, F.J.G. and Idle, J.R. (1994) Pharmacogenetic phenotyping and genotyping. Present status and future potential. *Clin. Pharmacokinet.* **26**: 59–70.
3. McKinnon, R.A. and Evans, A.E. (2000) Cytochrome P450 3. Clinically significant drug interactions. *Austr. J. Hosp. Pharm.* **30**: 146–149.
4. Pelkonen, O. Mäenpää, J., Taavitsainen P., Rautio, A. and Raunio, H. (1998) Inhibition and induction of human cytochrome P450 (CYP) enzymes. *Xenobiotica* **28**: 1203–1253.
5. Pratt, W.B. and Taylor, P. (1990) *Principles of Drug Action.* Churchill Livingstone, New York.
6. Rowland, M. and Tozer, T.N. (1995) *Clinical Pharmacokinetics: Concepts and Application.* Lea and Febiger, Philadelphia.
7. Shargel, L. and Yu, A.B.C. (1999) *Applied Biopharmaceutics and Pharmacokinetics.* Appleton and Lange, Stamford.
8. Smith, D.A., Abel, S. M., Hyland, R. and Jones, B. C. (1998) Human cytochrome P450s: selectivity and measurement *in vivo. Xenobiotica* **28**: 1095–1128.
9. Tanigawara, Y. (2000) Role of P-glycoprotein in drug disposition. *Ther. Drug Monit.* **22**: 137–140.
10. Woolf, T. F. (1999) *Handbook of Drug Metabolism.* Marcel Dekker, New York.
11. Woodhome, N.M., *et al.* (1979) Debrisoquine hydroxylation polymorphism among Ghanaians and Caucasians. *Clin. Pharmacol. Ther.* **26**: 584–591.

31

BIOTRANSFORMATION REACTIONS

Jacques Magdalou, Sylvie Fournel-Gigleux, Bernard Testa and Mohamed Ouzzine

One will in one hand be able to establish laws allowing predictions on the fate of new compounds, and on the other hand gain increasing insight into the organism as a chemical agent. **M. Nencki, 1847–1901**

I INTRODUCTION

Drug biotransformation reactions are essential chemical processes, mainly mediated by enzymes, which lead to the formation of drug metabolites that are excreted from the body. Historically, this metabolism was known as a detoxication mechanism. However, these reactions which govern the pharmacological efficacy of the drugs, may in many cases, also be responsible for toxic unwanted effects. Due to the large variety of chemical structures, drug biotransformation involves numerous and sophisticated

reaction mechanisms and metabolic pathways that are catalysed by a large number of enzyme families. The aim of this chapter is to give medicinal chemists the essential information to understand, from a chemical point of view, the fate of drugs in human. This information is a prerequisite for predicting drug metabolism and reactivity, and therefore for designing safer and more efficient drugs. Besides drugs, xenobiotics present in our environment (pollutants, pesticides, food additives, etc.) also undergo similar biotransformation reactions.

Drugs are generally lipophilic substances. This property allows them to penetrate easily into the different cell compartments through the hydrophobic membrane barriers. In that context, hydrophilic substances (saccharine, carbohydrate-derived substances) that cannot enter are generally not subject to metabolic transformations. The general principle of drug metabolism is to convert the lipophilic drugs via several chemical steps, into hydrophilic, water-soluble derivatives which can easily be eliminated into urine. However, very highly lipophilic drugs, polyhalogenated xenobiotics, such as some insecticides that can concentrate in the membrane compartment are sterically shielded from metabolic attack.

The metabolism of xenobiotics involves two sequential steps known as phase I and phase II reactions (Fig. 31.1). During phase I, a *functionalization reaction* of the xenobiotic is achieved. New polar groups such as CO_2H, OH or NH_2 are introduced or unveiled from pre-existing functions through oxidative, reductive or hydrolytic reactions. The polar group created serves then as an anchor point for the second metabolic step. The phase II reactions, known as *conjugation reactions*, link an endogenous, generally hydrophilic moiety, either to the original drug (if polar functions are already present) or to

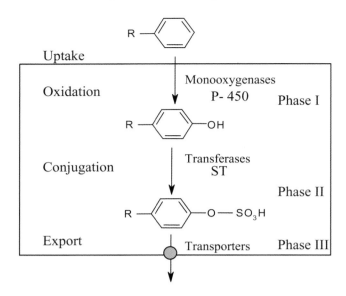

Fig. 31.1 Sequential steps of drug biotransformation. After uptake by the cell, a phenyl ring of a xenobiotic undergoes first a functionalization reaction (oxidation, phase I). The hydroxyl metabolite is then conjugated by addition of a sulfate group (phase II), before being exported from the cell by transporters (phase III) and excreted. P-450, cytochromes P-450; ST, sulfotransferases.

the phase I metabolite. Common endogenous groups are glucuronic acid, various amino acids, the tripeptide glutathione, or sulfate.

The water soluble conjugate is finally eliminated from the cell by transport proteins (organic anion transporters, multidrug resistance associated proteins), and finally excreted via the renal or the bile route. This transport step is considered as phase III of drug metabolism. It does not require further biotransformation, and will be not considered in this chapter.

The global result of phase I and II transformations should normally be the inactivation and detoxication of the xenobiotic. However, innumerable examples exist of metabolic activation. Phase I metabolite will possess its own activity which will be similar, higher or different from that of the parent drug. Phase 2 metabolites are generally less subject to activation. Metabolic precursors can even be designed to intentionally release the active species only *in vivo* upon transformation. Such compounds are called *prodrugs* (see Chapter 33). Other metabolites, such as electrophiles, may be highly reactive entities able to bind covalently to circulating or intracellular proteins (formation of adducts), to enzymes (mechanism-based, irreversible inactivation), or to DNA (mutagenic and carcinogenic

compounds). Metabolism-induced toxicity is discussed in Chapter 32.

In this context, it is important to predict, at an early stage of a drug's development, the metabolic pathway, the type of metabolites formed and their potential side/toxic effects. This challenge requires better knowledge, at a molecular level, of the enzymes that are implicated in drug biotransformation and the elucidation of the reaction mechanisms.

Drug biotransformation is catalysed by large families of enzymes (also known as phase I and II drug metabolizing enzymes). These proteins are also implicated in the metabolism of endogenous compounds. This situation, in which the drug or the xenobiotic and the natural substrate compete toward the same protein, may lead to cellular dysfunction and toxicity. Structural study of drug metabolizing enzymes is an increasing area of research. With the development of sophisticated technologies (genetic engineering: cloning, expression of cDNAs encoding these proteins, site-directed mutagenesis), protein chemistry, molecular modelling, X-ray crystallography, NMR etc., the organization of the active site can be elucidated and the amino acids that play a crucial role for catalysis and in the substrate recognition are identified. This chapter will illustrate the reaction mechanisms that have been established and which account for the biotransformation of drugs. Such information may help in predicting drug metabolism and provide a rational basis for the design of safer drugs.

II REACTIONS OF FUNCTIONALIZATION (PHASE I)

Phase I reactions, which often create anchor points in the xenobiotic molecule for subsequent conjugation, comprise oxidations (electron removal, dehydrogenation and hydroxylation), reductions (electron donation, hydrogenation and removal of oxygen), and hydrolytic reactions. Many metabolic reactions take place in the endoplasmic reticulum of the liver cells. Other organs, particularly kidneys and lungs, also participate in drug metabolism. In addition, a variety of other tissues (brain, skin, intestinal mucosa) have the capacity to metabolize xenobiotics.

A Oxidations catalysed by monooxygenases

The endoplasmic reticulum

The endoplasmic reticulum forms a dense network of membranes containing two structures called the smooth and the rough reticulum. The rough form contains small and dense organites, the ribosomes, which play an essential role in the biosynthesis of proteins. The smooth form is deprived of ribosomes, but contains oxidative enzymes.

When the hepatic tissue is homogenized, the endoplasmic reticulum is broken down into small membrane vesicles called microsomes. Microsomal preparations are obtained by centrifuging the tissue homogenate at 10 000 *g* for 10 min in order to sediment nuclei, mitochondria and debris, and then by centrifuging again at 100 000 *g* for 1 h. The microsomes are associated to the pellet and are separated from the supernatant (soluble) fraction (the cytosol).

The cytochromes P450 and the flavine monooxygenases

The major redox system present in the endoplasmic reticulum catalyses the reductive cleavage of molecular oxygen, transferring one atom of oxygen to the substrate and forming one molecule of H_2O with the other atom:

$$R—H + O_2 + NADPH + H^+ \rightarrow R—OH + NADP^+ + H_2O$$

This reaction is known as a monooxygenation and is supported by two main groups of monooxygenases: cytochromes P450 and flavine monooxygenases.

Cytochromes P450. Cytochromes P450 (CYP) are by far the most important xenobiotic- and endobiotic-metabolizing monooxygenases. They represent up to 25% of the total microsomal proteins and more than 50 cytochromes P450 monooxygenases are expressed by humans. They are involved in three main processes: (1) drug metabolism; (2) steroid metabolism (production of steroid hormones; (3) haem degradation (conversion of haem into biliverdin and bilirubin). Cytochromes P450 contain a molecule of haem, protoporphyrin IX, and a variable protein of MW

~ 50 kDa. Cytochromes P450 form a very large group of haemoproteins encoded by the CYP gene superfamily and classified in families and subfamilies. The major xenobiotic-metabolizing cytochromes P450 in humans are found in family 1 (CYP1A1 and CYP1A2), family 2 (CYP2B6, CYP2C8, CYP2C9, CYP2C18, CYP2D6 and CYP2E1), subfamily 3 (CYP3A), and family 4 (CYP4A9, CYP4A11 and CYP4B1). CYP3A4 is the most abundant and the most clinically important cytochrome in humans, as it metabolizes up to 50% of the available drugs. Cytochromes P450 belong to the haem-thiolate proteins in which the haem iron fifth ligand is a thiolate group, generally a cysteine residue. Such protein exhibits a Soret absorption band at 450 nm in the CO-difference spectrum of a dithionite-reduced form. The mechanism of the cytochrome P450 redox system is represented in Fig. 31.2. In microsomes, the two electrons necessary for monooxygenation are transferred by NADPH-cytochrome P450 reductase, the second electron in some cases coming from NADH-cytochrome b_5 reductase and cytochrome b_5. In mitochondria, which also contain cytochromes P450 devoted to the formation of steroid hormones from the hydroxylation of cholesterol or to the biosynthesis of bile acids and vitamin D, the electrons are supplied by the electron transfer chain composed of ferredoxin (adrenoxin) and ferredoxin reductase (adreno-xine reductase).

In the resting state, the central iron atom of proto-porphyrin IX is in a hexacoordinated, ferric form. The substrate R — H binds reversibly to the enzyme and the complex undergoes a reduction to the ferrous state. This allows molecular oxygen to bind as a third partner. Following the second reduction step, molecular oxygen is

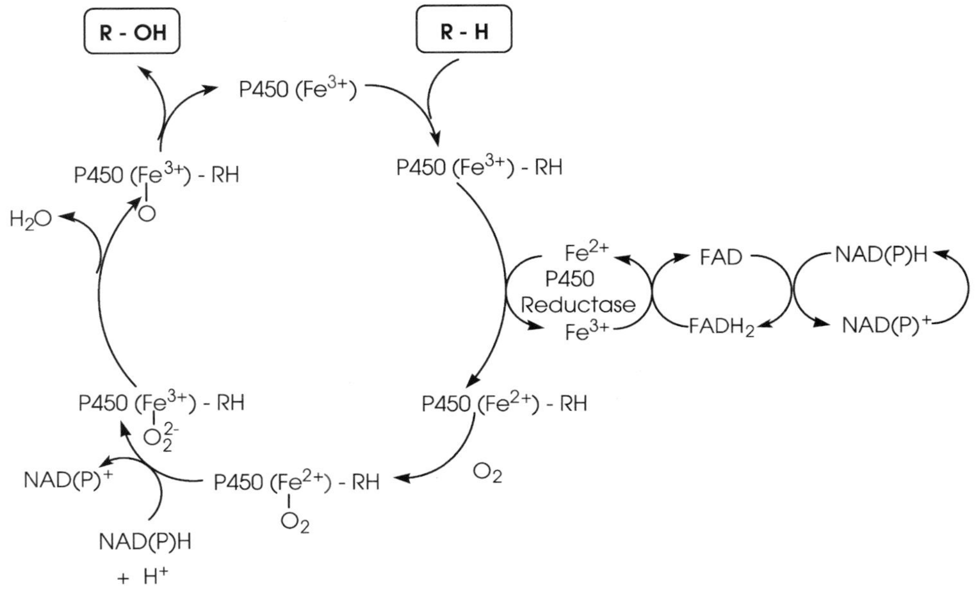

Fig. 31.2 The simplified cytochrome P450 redox cycle.

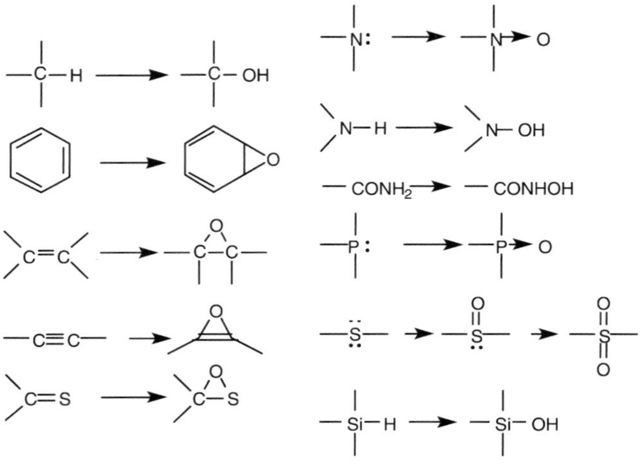

Fig. 31.3 Major reactions of oxygenation catalysed by cytochrome P450.

ultimately reduced to a hydroperoxide which is cleaved with liberation of H_2O and formation of a monooxygen known as oxene. The oxene, which is electrophilic and quite reactive, can act on the substrate in a manner which depends on the reactivity of the substrate itself. Thus, the oxene can either: (1) be transferred directly to the substrate (oxygen insertion or addition); (2) remove an electron, or more frequently; (3) pull a hydrogen radical away from the substrate and transfer back a formal HO° radical (a reaction known as oxygen rebound). The latter is the mechanism by which $RR'R''C$ — H is oxidized to $RR'R''C$ — OH. After release of the product, the regenerated cytochrome P450 is

ready for a new cycle. As illustrated below, the substrates to be oxidized are structurally unrelated, the oxidation involving C, Si, N, P, S, Se and other atoms (Fig. 31.3). The most common reaction catalysed by P450 is hydroxylation. However, it is also involved in a wide spectrum of reactions including epoxidation, *O*-, *N*-, and *S*-dealkylation, deamination, desulfuration, dehalogenation and peroxidation.

Flavine monooxygenases. Flavine monooxygenases (FMO) are a family of microsomal flavoproteins that catalyse the oxidation of numerous organic or inorganic compounds, including various structurally unrelated xenobiotics, in the presence of NADPH and oxygen. As for cytochrome P450, FMO are involved in detoxication and toxication reactions.

FMO are 58 kDa proteins that possess a NADPH and flavine binding sites. The catalytic cycle of the reaction involves four steps (Fig. 31.4): (1) rapid reduction of FAD by NADPH; (2) rapid introduction of molecular oxygen into FAD leading to the formation of the oxidating reagent, 4α-hydroxyperoxyflavine (E-FAD-OOH); (3) attack of the oxygen on the nucleophilic group of the xenobiotic (X) with the concomitant formation of 4α-hydroxyflavine; (4) generation of the starting enzyme (E-FAD) upon release of a water molecule and NADP$^+$.

Because the hydroperoxyflavine intermediate is not a strong oxidating reagent, FMO can oxidize molecules that contain a functional group rich in electrons and easily polarizable, such as amines or thiols (*N*-oxidation of tertiary amines, *S*-oxidation of thiols, thioamides, thiocarbamides, mercaptoimidazoles).

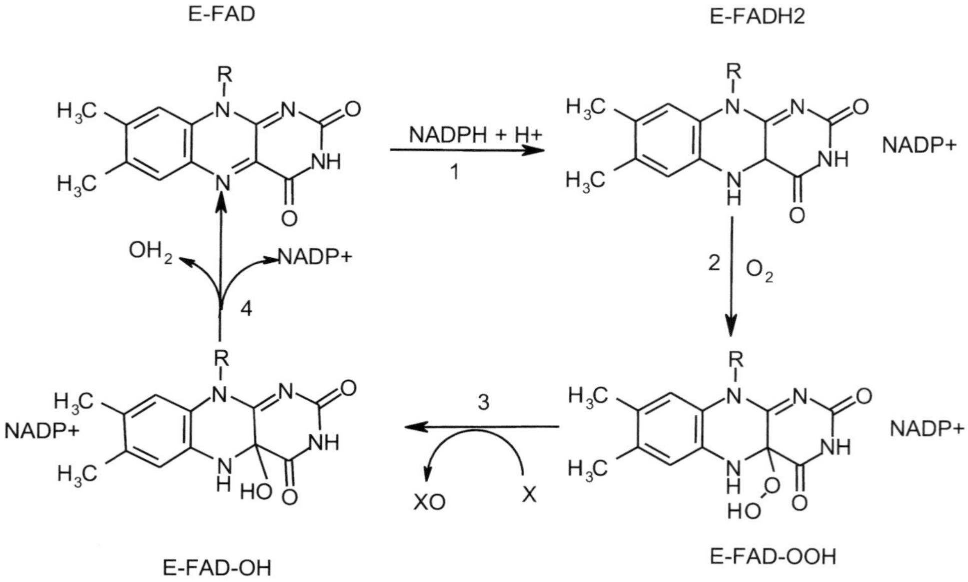

Fig. 31.4 Catalytic steps of flavine monooxygenase. The chemical structure of the isoalloxazine portion of FAD is shown.

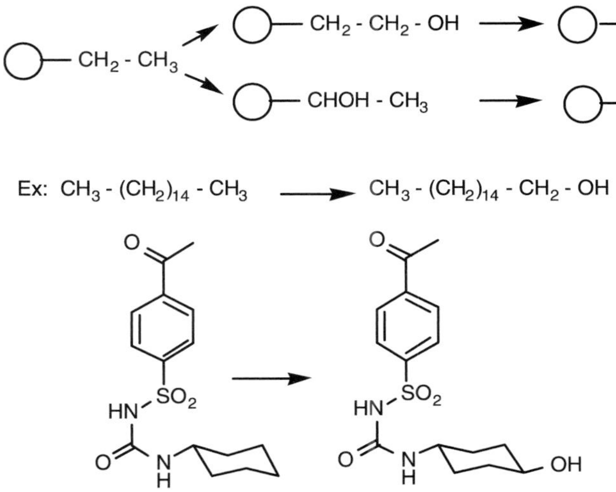

Fig. 31.5 Hydroxylations on saturated carbon atoms.

Although FMO and cytochrome P450 are involved in similar oxidation reactions, the reaction mechanism by FMO is radically different from that of cytochrome P450. Most importantly, the reactive hydroxyperoxyflavine is unusually stable with a rather long half-life and does not require the presence of the substrate to be oxidized. By contrast to cytochrome P450, the activated form of FMO is already present before introduction of the substrate.

Reactions of carbon oxidation

The reactions of C-oxidation represent the most common metabolic attacks on xenobiotics. From a mechanistic point of view, it is convenient to distinguish between saturated (sp³) and unsaturated (sp² and sp) carbon atoms.

Hydroxylation of saturated aliphatic carbon atoms. The saturated aliphatic carbons hydroxylated by cytochromes

P450 are found in complex molecules as well as in saturated hydrocarbons (alkanes and cycloalkanes). For example, the steroid progesterone is hydroxylated in positions 11β, 17α and 21 to yield hydrocortisone. In practice, a nonactivated alkyl group undergoes mainly ω- and ω − 1 oxidation. On the other hand, n-hexadecane is ω-hydroxylated in the liver to yield hexadecanol which is further oxidized to hexadecanoic acid. For shorter chains, both terminal and ω − 1 oxidations are observed (Fig. 31.5). Cyclic aliphatic systems are usually hydroxylated on the least hindered or most activated carbon atoms.

Hydroxylation at activated sp³ carbon atoms. For activated sp³ carbon atoms in allylic, propynylic or benzilic positions, the ω- and ω − 1 rule no longer holds and the activated atoms are hydroxylated preferentially (Fig. 31.6).

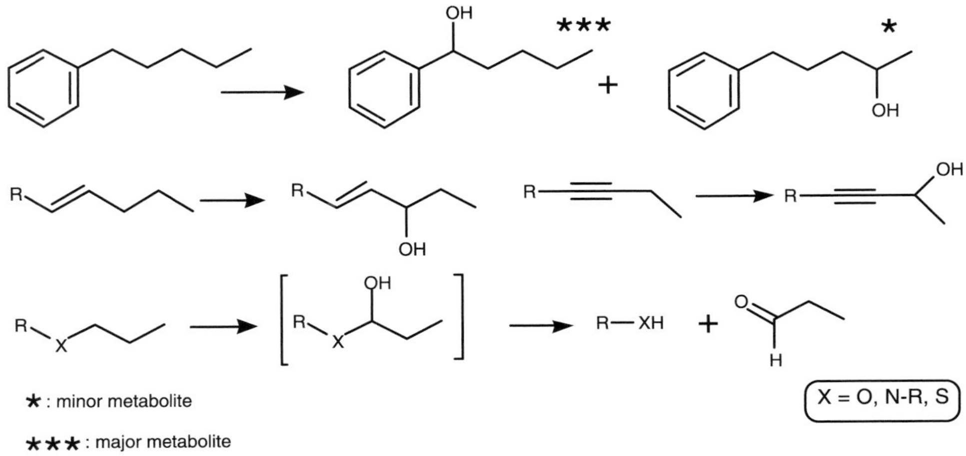

★ : minor metabolite

★★★ : major metabolite

Fig. 31.6 Regioselectivity of hydroxylation at activated sp³ carbon atoms.

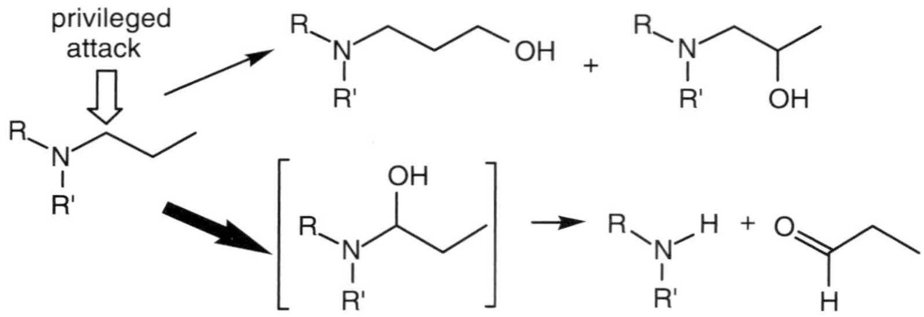

Fig. 31.7 For long-chain amines, α, ω- and $\omega - 1$ attacks coexist, but α attacks are privileged.

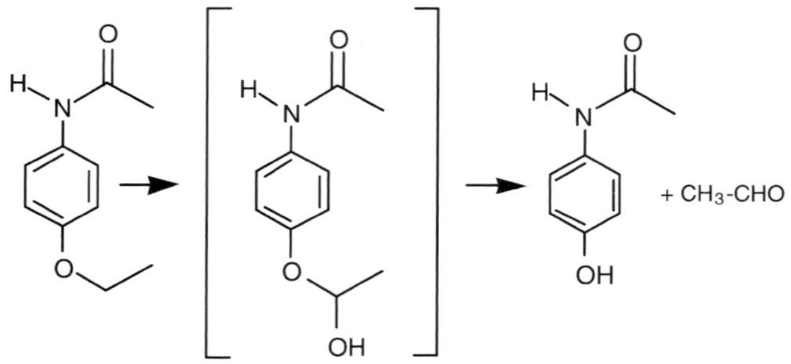

Fig. 31.8 The oxidative *O*-dealkylation of phenacetin yields paracetamol.

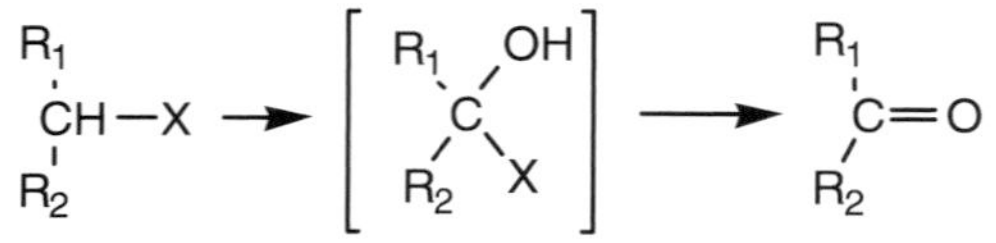

Fig. 31.9 Example of hydroxylation alpha to a halogen atom.

The same is true for carbons in a position alpha to a heteroatom such as N, O or S (Fig. 31.7). With amines, hydroxylation leads to a hydroxy-aminal which is immediately hydrolysed. The final result is *dealkylation* when a secondary or a tertiary amine loses an alkyl substituent, and *deamination* when the substrate loses an amino group. Here again, positions other than the alpha one can be hydroxylated, although to a lesser extent (Fig. 31.7).

Aromatic ethers undergo a similar α-hydroxylation, followed by hydrolysis of the hemiacetal to a phenol and an aldehyde (Fig. 31.8).

Note: Dealkylation reactions can also result from a *direct* oxidation of the heteroatom (N, S) as opposed to that of the α carbon (see below).

Chlorinated or brominated aliphatic derivatives can similarly be metabolized by successive hydroxylation and elimination of HCl or HBr. Dehalogenated carbonyl compounds are thus formed (Fig. 31.9).

Oxidative attack on unsaturated aliphatic systems. Carbon–carbon double bonds are oxidized by cytochromes P450 to reactive epoxides. Thus, vinyl chloride yields epoxychlorethane, an alkylating metabolite which can for example alkylate nucleic acids (Fig. 31.10).

The synthetic estrogen diethylstilbestrol undergoes a similar attack. Despite its relatively hindered character, the central double bond is converted to the corresponding epoxide which was once believed to be the ultimate carcinogenic metabolite (Fig. 31.11).

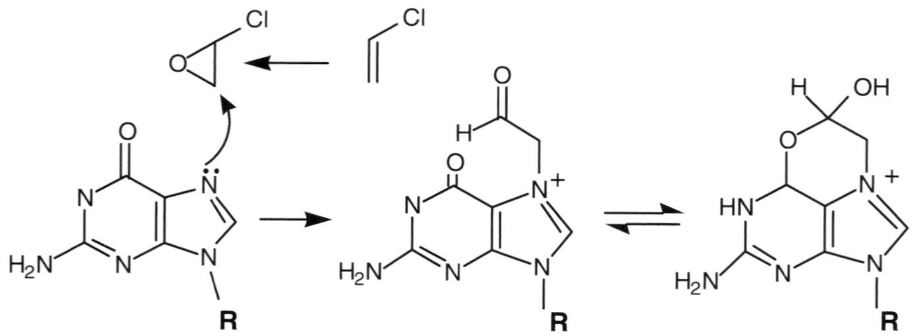

Fig. 31.10 Oxidative alkylation of a guanine moiety by vinyl chloride.

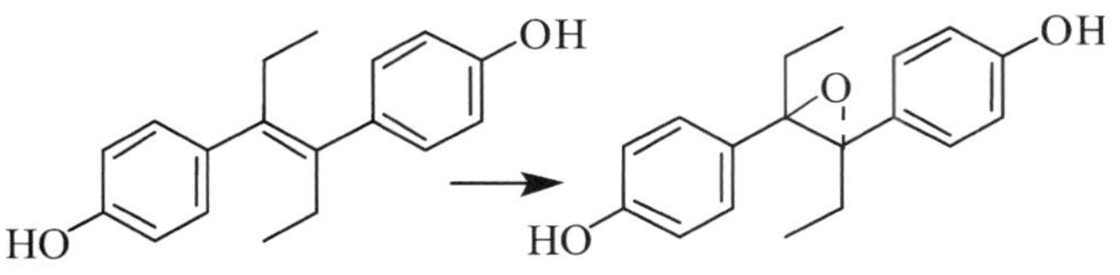

Fig. 31.11 Epoxidation of diethylstilbestrol.

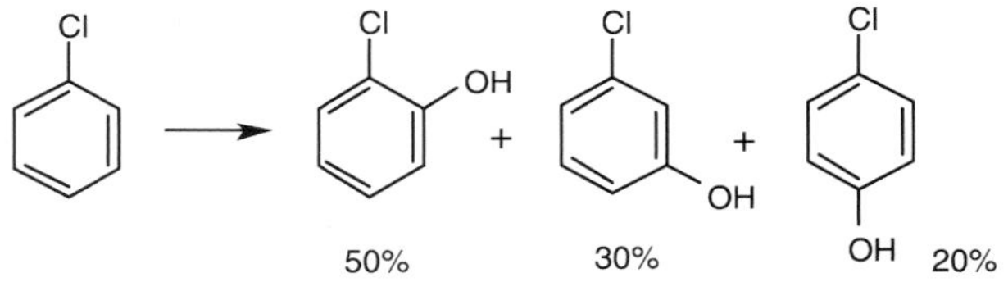

Fig. 31.13 Hydroxylation of chlorobenzene in the rat yields the three isomers, i.e. *ortho-*, *meta-* and *para-*chlorophenol.

With carbon–carbon triple bonds, oxygen insertion yields an oxirene which opens by heterolytic C — O bond cleavage to form a highly reactive intermediate which binds covalently to the enzyme. In the case of 17α-ethynyl steroids, the reaction can also result in an extension of ring D (Fig. 31.12).

Hydroxylation of aromatic rings. Aromatic rings are frequently oxidized into phenols followed by conjugation and excretion. The mechanism of the reaction is discussed later, and we shall first consider an example of phenol formation. In the hydroxylation of chlorobenzene, all three isomers are produced, i.e. *ortho-*, *meta-* and *para-*chlorophenol, but in different amounts (Fig. 31.13). As a rule, hydroxylation occurs on the less hindered site, usually the *para* position. Electronic factors are also operative. This is seen in the hydroxylation of many drugs, two of which are shown in Fig. 31.14.

The mechanism of the oxidative attack catalysed by cytochromes P450 involves the addition of an activated oxene species to the aromatic ring (Fig. 31.15). The resulting tetrahedral intermediate usually rearranges, forming an intermediate epoxide. Epoxides (also called arene oxides or oxiranes) are electrophiles which are detoxified by a number of routes, e.g. protonation and rearrangement to a phenol (often the most efficient pathway), hydration by epoxide hydrolase, and conjugation with glutathione. The toxicity of a number of aromatic compounds is due to the epoxide reacting with cellular constituents (cytotoxicity, mutagenesis), or to the further oxidation of phenols to diphenols and then to semiquinones and quinones.

The NIH-shift. The NIH-*shift* is a particular intramolecular rearrangement occurring during the hydroxylation of aromatic rings and resulting in the migration of a hydrogen or a halogen atom. In the case of a hydrogen atom, the migration is not apparent, but becomes observable with deuterium or tritium. This transposition was first observed in 1967 in the National Institutes of Health (Bethesda, MD, USA), when the enzymatic hydroxylation of 4-[^{3}H]-phenylalanine unexpectedly yielded 3-[^{3}H]-*para*-tyrosine with a retention of 90% of the radioactivity (Fig. 31.16).

The percentages of retention and migration depend on the nature of the label (hydrogen isotope or halogen atom) and of the substituent(s) carried by the aromatic ring. When the substituent R cannot provide a proton, good retention (40 to 60%) of the label is observed. This is the case for R = Cl, CN, OCH$_3$, NO$_2$, CONH$_2$ and C$_6$H$_5$, the predominant mechanism being that shown in Fig. 31.17.

When the R substituent is a proton donor (R = CO$_2$H, NH$_2$, SO$_2$NH$_2$, NH — CO — R$_1$, etc.), only little retention (0 to 30%) and migration is achieved. The mechanism shown in Fig. 31.18 accounts for the reaction. The acidic hydrogen competes with the deuterium atom, and two pathways are possible, namely elimination of H$^+$ (pathway 'a') or elimination of D$^+$ (pathway 'b').

The NIH-shift and obstructive halogenation. In order to decrease the metabolism of an aromatic ring, one can block the *para* position with a halogen atom. However, due to the NIH-*shift*, *para*-substituted rings can still be oxidized (Fig. 31.19). In drug design, obstructive halogenation is justified when the duration of action of an aromatic compound must be prolonged.

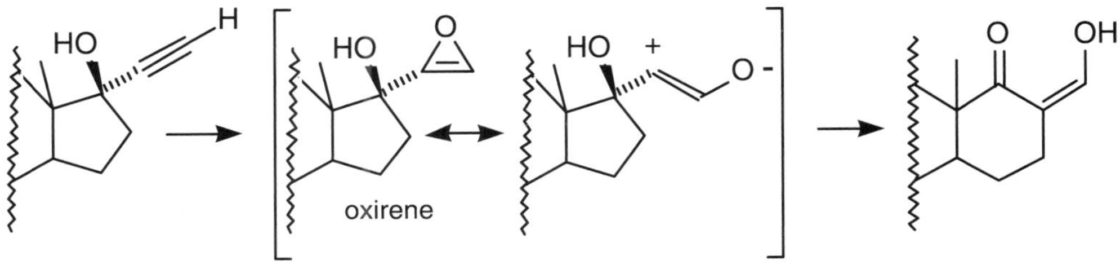

Fig. 31.12 The triple bond in a 17α-ethynyl steroid is first epoxidated to an oxirene which then undergoes a rearrangement and an extension of ring D.

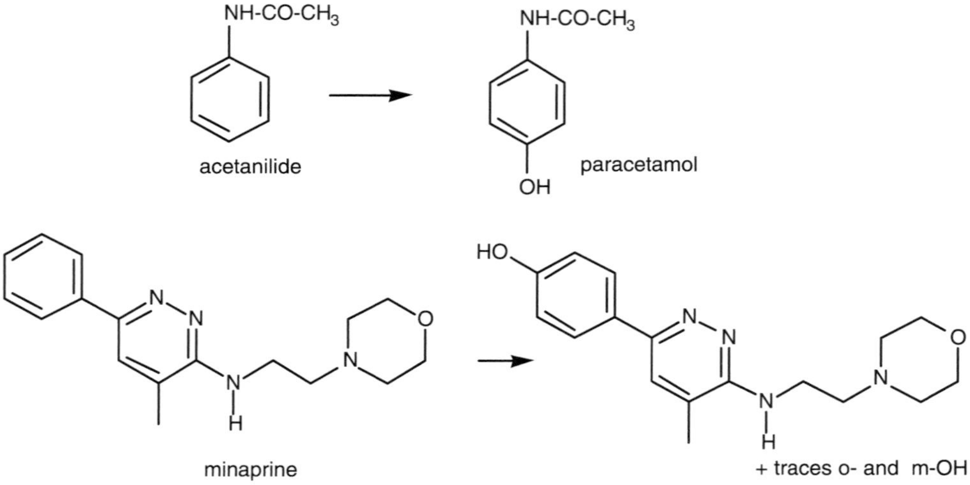

Fig. 31.14 Aromatic rings are predominantly hydroxylated in the *para* position.

Reactions of N-oxidation

Enzymatically, *N*-oxidation can be catalysed by cytochrome P450 and/or by the FAD-containing monooxygenases, depending on substrates and conditions. From a chemical point of view, the oxidation of nitrogen atoms in organic compounds can be summarized as follows:

Tertiary aliphatic amines are usually oxidized to the corresponding *N*-oxides, but the reaction is strongly affected by steric hindrance. Usual substrates are *N,N*-dimethylamino aliphatic and aromatic tertiary amines (e.g. *N,N*-dimethylaniline) (Fig. 31.20A), saturated *N*-methylazaheterocycles (e.g. *N*-methylpiperidines), and aromatic azaheterocycles (e.g. pyridine).

Secondary and primary amines are *N*-oxidated to hydroxylamines as shown in Fig. 31.20B. The intermediate is believed to be an *N*-oxide.

Note that a drug very seldom undergoes a single metabolic reaction, but is substrate of several competitive pathways. *N*-Benzylamphetamine offers an example in the present context (Fig. 31.21).[2]

Amides can by *N*-oxidated to hydroxylamides. This is true for primary and secondary amides. For example, urethane and *N*-acetylaminofluorene (Fig. 31.22) are converted into a highly pro-carcinogenic hydroxylamide. The toxicity of phenacetin has similarly been attributed to an *N*-hydroxylated metabolite.

Reactions of S-oxidation

Thiol compounds can be oxidized to disulfides, or to sulfenic, sulfinic and finally to sulfonic acids. Similarly sulfides are easily converted by monooxygenases to sulfoxides and then to sulfones. Thiocarbonyl derivatives are also substrates of monooxygenases, forming *S*-mon-oxides (sulfines) and then *S*-dioxides (sulfenes). The latter are highly reactive metabolites, specially towards nucleo-philic sites in biological macromolecules, and are believed

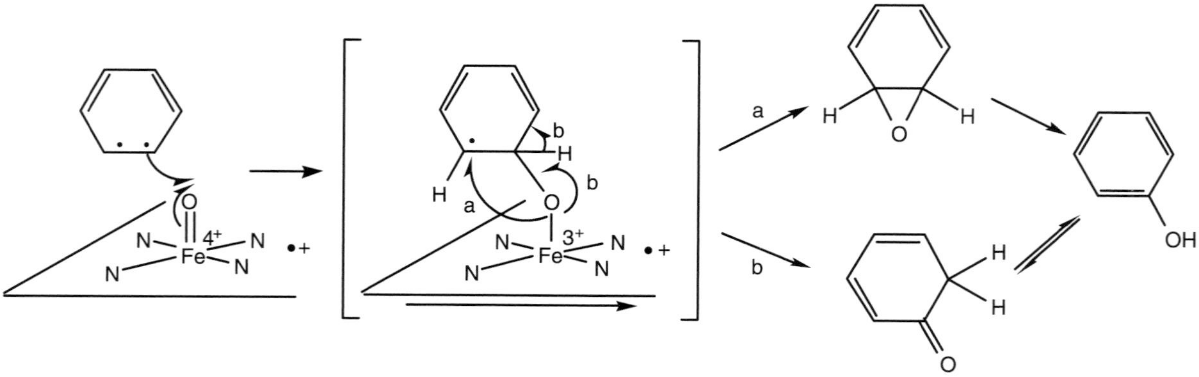

Fig. 31.15 Addition–rearrangement mechanism for arene oxide formation and proton-catalysed rearrangement of an arene oxide to a phenol (After Silverman[1]).

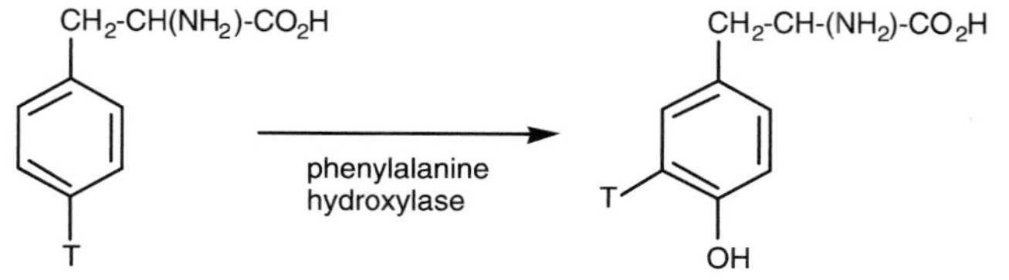

Fig. 31.16 NIH shift in the hydroxylation of tritiated phenylalanine to *p*-tyrosine.

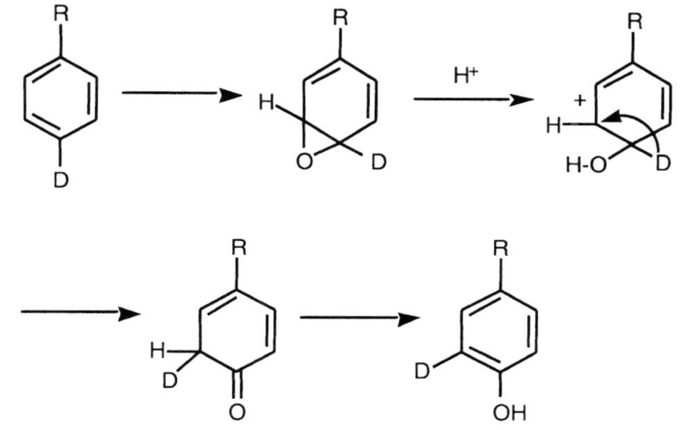

to account for the carcinogenicity of a number of thioamides. *S*-Monooxides can also rearrange to the corresponding carbonyl by expelling a sulfur atom (oxidative desulfuration of thioamides, thioureas and thiobarbiturates).

B Oxidations catalysed by other oxidoreductases

Enzymes

Besides the monooxygenases discussed above, a number of other oxidoreductases can oxidize xenobiotics. These enzymes are mostly but not exclusively nonmicrosomal, being present in the cytosol or mitochondria of the liver and extrahepatic tissues. The list includes: alcohol dehydrogenases, aldehyde dehydrogenases, dihydrodiol dehydrogenases, haemoglobin, monoamine oxidases, xanthine oxidase and aldehyde oxidase. Some of these enzyme systems are discussed below.

Fig. 31.17 Mechanism of the NIH shift when the substituent R is not a hydrogen donor.

Reactions of oxidation

Alcohol dehydrogenases catalyse the oxidation of primary and secondary alcohols to aldehydes and ketones, respectively. Typical primary alcohols acting as substrate are ethanol, benzylic alcohol, phenylethanol, geraniol and retinol. Methanol is a poor substrate and is only slowly oxidized at high concentrations. Secondary alcohols are more difficult to oxidize, while tertiary alcohols are resistant towards dehydrogenation. This explains in part the greater

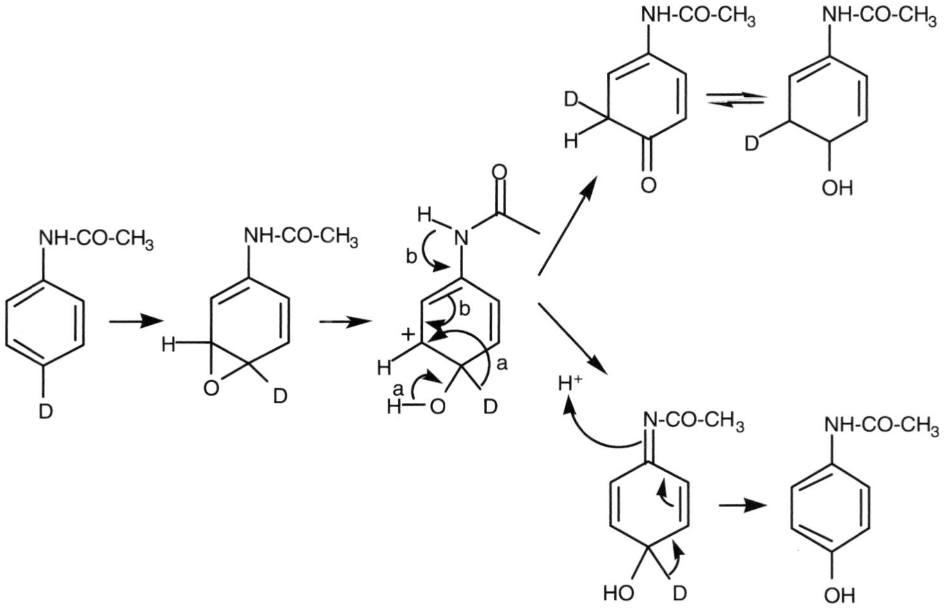

Fig. 31.18 Mechanism of the NIH shift when the substituent R is a proton donor.

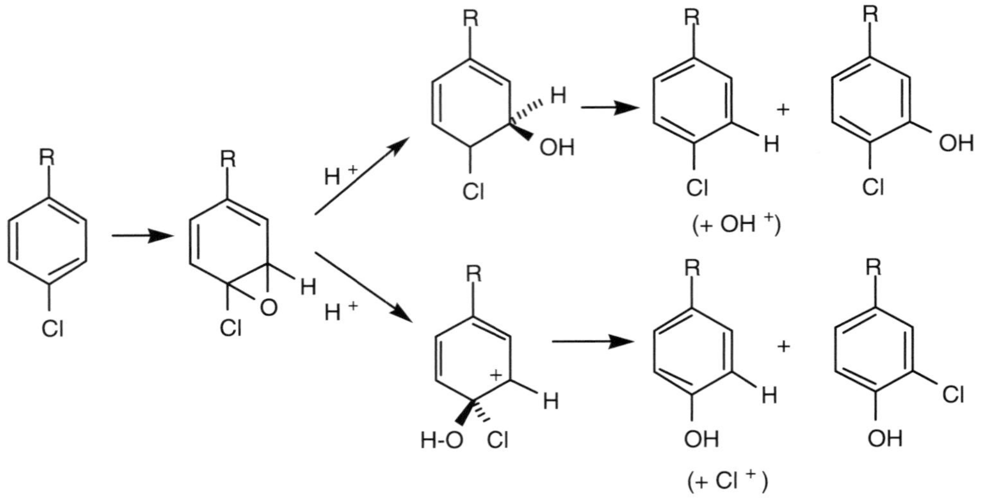

Fig. 31.19 Example of oxidation of a *para*-halogenated ring.

potency of tertiary alcohols as hypnotics (e.g. methylpentynol, amylene hydrate, etc.).

Alcohol dehydrogenases are zinc enzymes that use NAD^+ as coenzyme according to the reaction:

$$CH_3—CH_2—OH + NAD^+ \rightarrow CH_3—CHO + NADH + H^+$$

The reaction involves the transfer of a hydride ion to the nicotinamide part of NAD^+ and is stereospecific.

Aldehyde dehydrogenases transform aldehydes into carboxylic acids (e.g. succinaldehyde dehydrogenase). As with alcohol dehydrogenases, the key step of the reaction is the cleavage of the α-C — H bond, with a hydride transfer to NAD^+.

Monoamine oxidases (MAO) are mitochondrial enzymes existing in two forms, MAO-A and MAO-B. Their physiological function is to deaminate endogenous amines, in particular catecholamines, but their involvement in the oxidation of xenobiotics is exemplified with the toxication of MPTP (1-methyl-1,2,3,6-tetrahydropyridine)

and analogues to the corresponding pyridinium ions (MPP^+) (Fig. 31.23). Such species are known to induce Parkinson's disease by altering the function of dopamine nerves.

C Reductions

A number of reactions of reduction have been demonstrated in the metabolism of xenobiotics. From a quantitative viewpoint, they are less important than oxidations since our organism is mostly an aerobic one. From a qualitative point of view, however, reactions of reduction may be of great pharmacological or toxicological significance when they generate active metabolites or toxic metabolic intermediates.

Reductions at carbon atoms

The major reactions of reduction at carbon atoms can be found in Fig. 31.24. Thus, aldehydes and ketones are readily

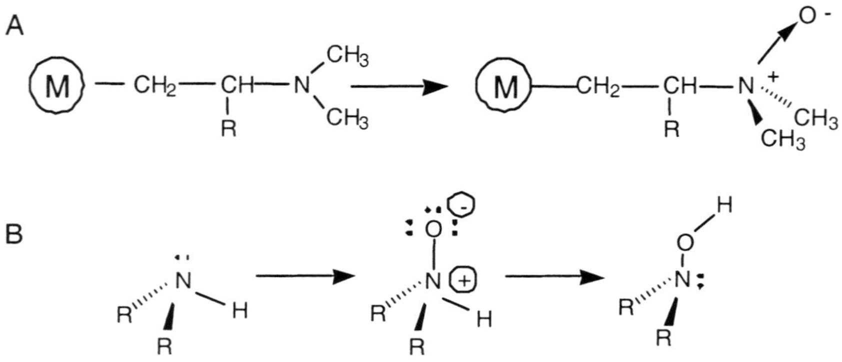

Fig. 31.20 *N*-oxidation of tertiary aliphatic amines (A) and secondary amines (B).

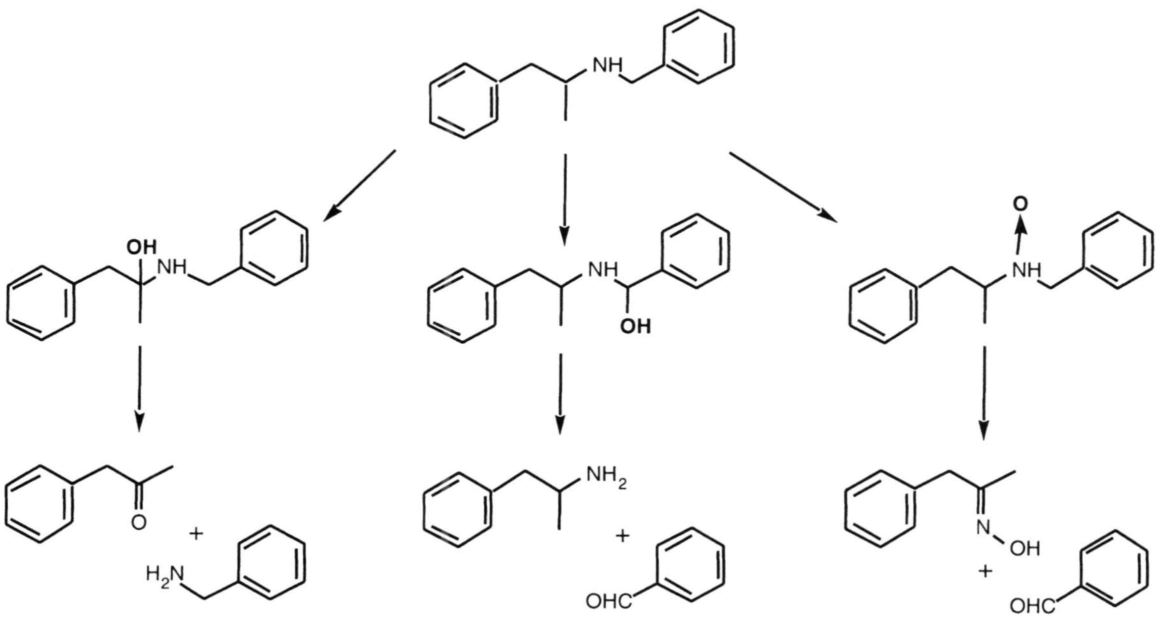

Fig. 31.21 The different metabolic pathways of *N*-benzylamphetamine (After Stenlake[3]).

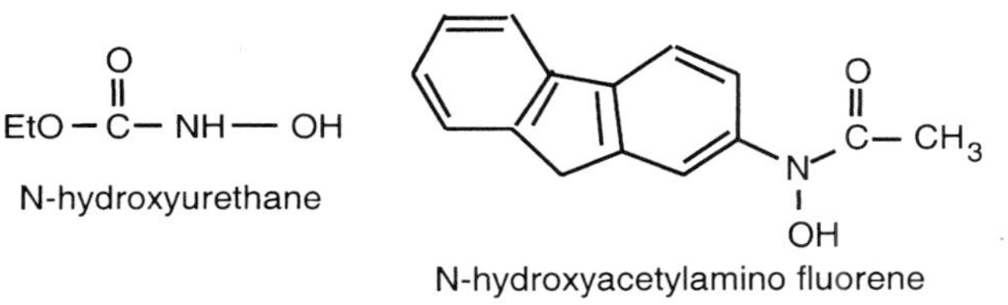

Fig. 31.22 Examples of hydroxylamides resulting from the *N*-oxidation of amides.

reduced to primary and secondary alcohols, respectively. Quinones can be reduced to dihydrodiols either by a two-electron mechanism (carbonyl reductase and quinone reductase) or by two single electron steps (cytochromes P450 and some flavoproteins). Reduction of olefinic groups, i.e. the reverse of desaturation, is documented for a few drugs bearing an α,β-ketoalkene function. Dehalogenation reactions can also proceed reductively. Reductive dehalogenations involve replacement of a halogen by a hydrogen, or *vic*-bisdehalogenation. Some radical species formed as intermediates may be of toxicological significance.

Reductions at other atoms
Various reactions of *N*-oxidation are reversible, cytochrome P450 and other reductases being able to deoxygenate

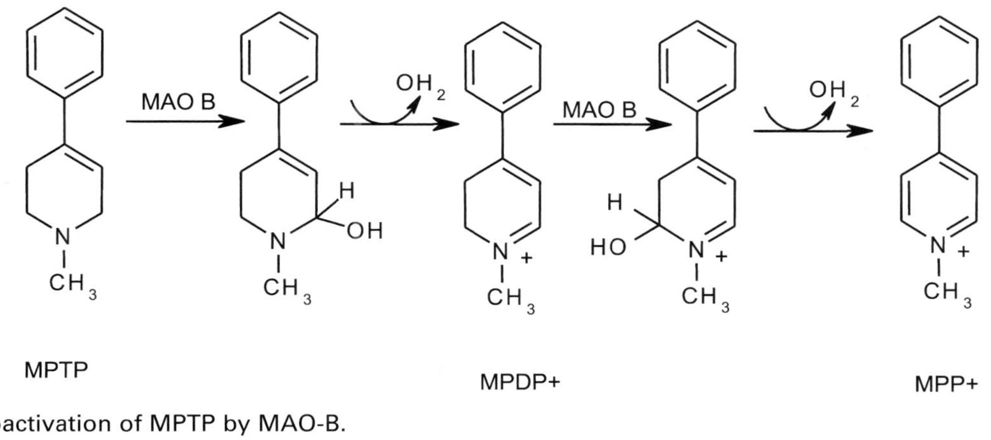

Fig. 31.23 Bioactivation of MPTP by MAO-B.

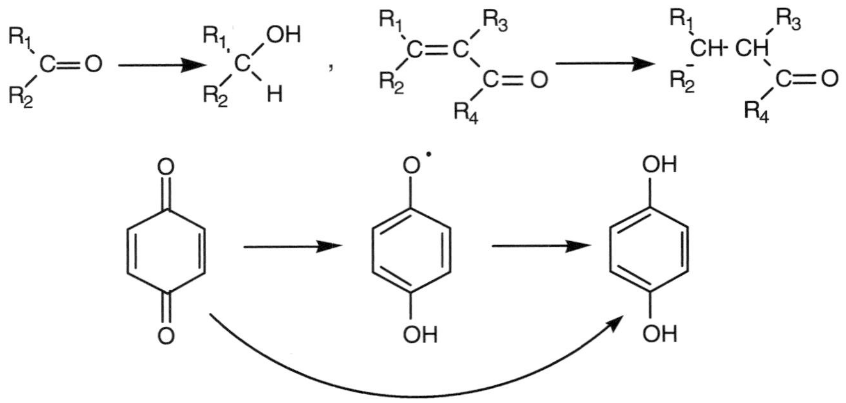

Fig. 31.24 Some reactions of reduction at carbon atoms.

N-oxides back to the amine (see Fig. 31.25). The same is true for aromatic nitro compounds, aromatic nitroso compounds and hydroxylamines, and imines and oximes, which can ultimately be reduced to primary amines. Azo and azoxy compounds can be reduced to hydrazines. An important pathway of hydrazines is their reductive cleavage to primary amines. A toxicologically significant pathway thus exists for the reduction of some aromatic azo compounds to potentially toxic primary aromatic amines.

Other reductions involve sulfur (Fig. 31.25) and a few other atoms. Thus, disulfides are reduced to thiols, and there are numerous examples of the reduction of sulfoxides. In contrast, the reduction of sulfones has never been found to occur.

D Hydrolytic reactions

Hydrolases

Hydrolases constitute a very complex ensemble of enzymes many of which are known or suspected to be involved in xenobiotic metabolism. Relevant enzymes among the serine

hydrolases include carboxylesterases, arylesterases, cholinesterases and a number of serine endopeptidases. Other hydrolases worth mentioning are arylsulfatases, aryldialkylphosphatases, β-glucuronidases, epoxide hydrolases, cysteine endopeptidases, aspartic endopeptidases and metallo-endopeptidases.

Arylsulfatases and β-glucuronidases expressed in various tissues and present in the bacterial flora of the intestine play a major role in the metabolism and disposition of drugs and xenobiotics. They can cleave the conjugates formed by the sulfo- and glucuronosyltransferases, leading back to the parent compound, thus extending the half-life of the drug. From a practical point of view, these enzymes are commonly used to detect and quantify drug conjugates in body fluid, upon hydrolysis.

The main reactions of hydrolytic cleavage (hydrolysis) are shown in Fig. 31.26. They are frequent for organic esters, inorganic esters such as nitrates, and amides. These reactions are catalysed by esterases, peptidases or other enzymes, but non-enzymatic hydrolysis is also known to occur for sufficiently labile compounds under biological conditions of pH and temperature. Such reactions are of

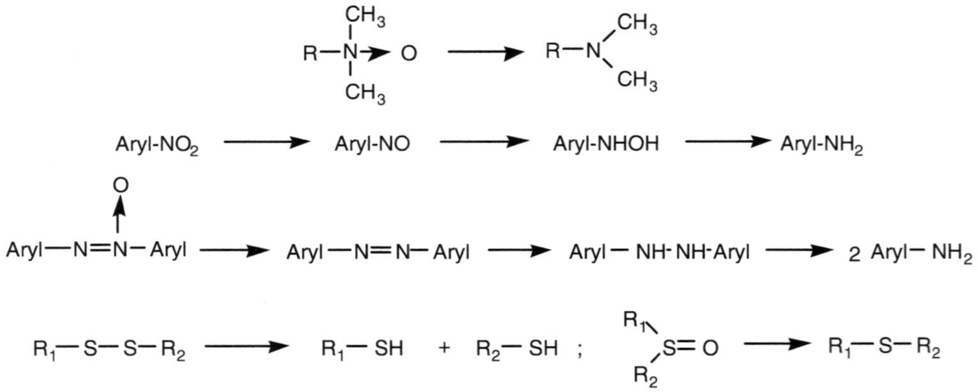

Fig. 31.25 Some reactions of reduction at nitrogen and sulfur atoms.

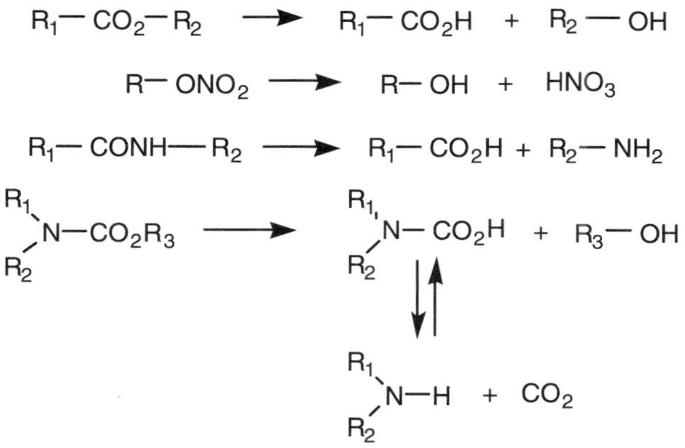

$$R_1-CO_2-R_2 \longrightarrow R_1-CO_2H + R_2-OH$$

$$R-ONO_2 \longrightarrow R-OH + HNO_3$$

$$R_1-CONH-R_2 \longrightarrow R_1-CO_2H + R_2-NH_2$$

$$\underset{R_2}{\overset{R_1}{>}}N-CO_2R_3 \longrightarrow \underset{R_2}{\overset{R_1}{>}}N-CO_2H + R_3-OH$$

$$\underset{R_2}{\overset{R_1}{>}}N-H + CO_2$$

Fig. 31.26 Some reactions of hydrolysis.

particular significance in the activation of ester prodrugs. As an example, the hypolipidemic agent clofibrate or the promising antitumour agent, irinotecan, contains an ester bond that is split by carboxylesterase to form a primary active carboxylic metabolite. Another reaction of interest is the hydrolysis of carbamate esters. Hydrolysis, which is often found to proceed readily, liberates the carbamic acid. The latter is unstable and breaks down to the amine and carbon dioxide.

Reactions of hydration and hydrolysis

Hydrolases catalyse the addition of a molecule of water to a variety of functional groups. Thus, epoxide hydrolases hydrate epoxides (three-membered oxygen compounds) to yield *trans*-dihydrodiols. This reaction is documented for many arene oxides, in particular metabolites of aromatic compounds, and epoxides of olefins. A molecule of water is added to the substrate without loss of a molecular fragment, hence the use of the term 'hydration' sometimes found in the literature. Two main groups of epoxide hydrolases are

located in the microsomal (mEH) and cytosolic (soluble, sEH) fractions. They present marked differences in substrate specificity (Fig. 31.27). mEH catalyses the hydrolysis of the procarcinogenic polycyclic aromatic compounds, epoxide derivatives of 1,3-butadiene and aflatoxin B1 or drug epoxide metabolites of the anticonvulsant drugs, phenytoin or carbamazepine. *trans*-Substituted epoxides and aliphatic epoxides from fatty acid metabolism are substrates of sEH.

The reaction mechanism catalysed by sEH has been recently elucidated from experiments using heavy isotopes, protein mass spectrometry, site-directed mutagenesis, and has been supported by the recent crystal structure determination at 2.8-Å resolution (Fig. 31.28). This two-step reaction mechanism involves a catalytic nucleophile (aspartic acid 333) which can attack the polarized epoxide ring by two tyrosyl residues (tyrosines 381 and 465) leading to the ring opening and the formation of an acyl–enzyme intermediate. The second step corresponds to hydrolysis of this intermediate by a water molecule activated by a histidine 523-aspartic acid 495 pair.[4]

III REACTIONS OF CONJUGATION (PHASE II)

A Introduction

As already defined above, conjugation reactions link an endogenous moiety either to the original drug (if polar functions are already present) or to the phase I metabolite. Conjugates are usually devoid of pharmacological activities, but there are notable exceptions. In addition, they are often more polar and water-soluble than the parent drug, and readily excreted via the renal route. However, certain conjugation reactions do not result in decreased lipophilicity, e.g. acetylations and

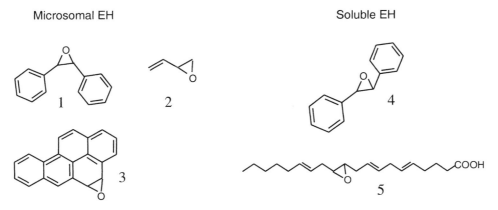

Fig. 31.27 Marker substrates of EH. (**1**) cis-stilbene oxide; (**2**) butadiene monoxide; (**3**) benzo[a]pyrene-4,5-oxide; (**4**) trans-stilbene oxide; (**5**) 11,12-epoxyeicosatrienoic acid.

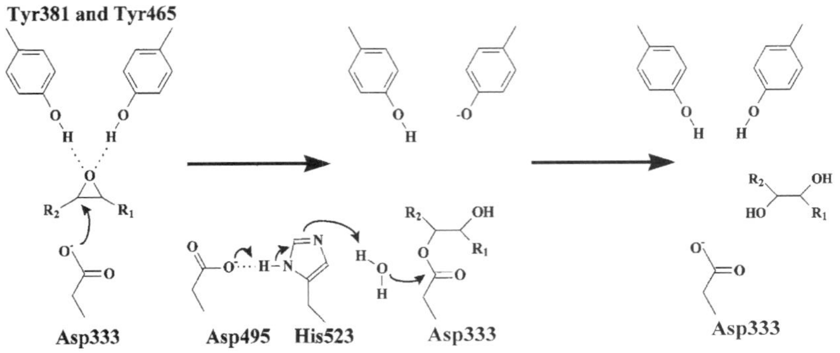

Tyr381 and Tyr465

Asp333   Asp495 His523   Asp333   Asp333

Fig. 31.28 Two step mechanism of soluble epoxide hydrolase (adapted from Yamada *et al.*,[4] with permission).

some reactions of methylation. Conjugation reactions also occur for endogenous compounds, such as the toxic bilirubin formed from the metabolism of haemoproteins or steroid hormones and retinoic acid which can compete with drugs toward the same binding site. By abolishing their physiological properties and modulating their concentration, conjugation reactions play a major role in the regulation of the biological activity of numerous compounds involved in cell growth and differentiation, and are believed to participate in the evolution of major diseases, such as cancer.[5]

Reactions of conjugation are characterized by the following criteria, none of which is a *sine qua non* condition:

(1) They are catalysed by families of enzymes known as *transferases.*
(2) They are two substrate enzymes: the drug and an endogenous molecule which bind on two vicinal binding sites. They transfer the endogenous molecule, or part of it, on the drug, leading to the formation of a conjugate.
(3) Except for methylation and acetylation, this endogenous molecule is highly polar and its size is comparable to that of the drug.

The transferases are present in numerous tissues, particularly in the liver. Extrahepatic conjugation reactions that are associated mainly to kidney, lung, skin, gastrointestinal tract and brain, also contribute to the final metabolism of drugs. The transferases are located in microsomes and in the cytosolic fraction (soluble forms). They can compete towards the same polar function of a drug, leading to the formation of several chemically distinct conjugates (Fig. 31.29). As an example, the structure of the antituberculosis drug, *p*-aminosalicylic acid contains a hydroxyl, carboxyl- and amine group that all can potentially undergo conjugation reactions. The hydroxyl group initiates the formation of an etherglucuronide or a sulfoconjugate by conjugation with a glucuronic acid or a sulfate group.

The carboxylic acid moiety can form an acylglucuronide or amino acid conjugate with a glucuronic acid or amino acid (glycine, glutamic acid) moiety. Finally three different conjugates (sulfo-, glucurono-, acetylconjugate) are expected from the primary amine group, by introduction of a sulfate, glucuronic acid or acetate group. The formation of such conjugates depends on several factors, among those are the concentration of the drugs, in relation to the kinetic properties of the corresponding transferases, and their lipophilicity (octanol/water coefficient), in relation to the localization of the enzymes within the cells (microsomal or cytosolic compartment). The majority of drugs and xenobiotics are excreted as monoconjugates. However, few examples of compounds forming very hydrophilic

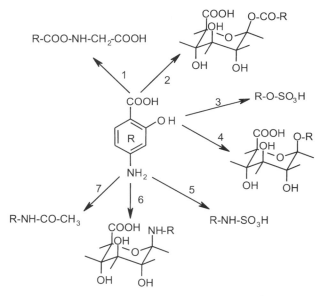

Fig. 31.29 Different metabolic pathways of *p*-aminosalicylic acid. Several conjugating enzymes are involved: glycine *N*-acyltransferases (reaction 1), UDP-glucuronosyltransferases (acyl-, ether-, *N*-glucuronides, reactions 2, 4 and 6, respectively), sulfotransferases (reactions 3 and 5), *N*-acetyltransferases (reaction 7).

diconjugates are known. The polycyclic aromatic hydrocarbon, benzo[a]pyrene-diphenol leads to formation of diglucuronide. The carboxylic nonsteroidal anti-inflammatory drug, desmethylnaproxen, is excreted, upon *O*-dealkylation to form 6-*O*-desmethylnaproxen as a mixed diconjugate (acylglucuronide-sulfate diconjugate).[6]

B Methylation

Biochemistry

Methylation is an important reaction in the biosynthesis of endogenous compounds such as adrenaline and melatonin, in the inactivation of biogenic amines such as the catecholamines, serotonine and histamine, and in modulating the activities of macromolecules, such as proteins and nucleic acids. The number of xenobiotics that are methylated is comparatively modest, yet this reaction is seldom devoid of pharmacodynamic consequences (toxication or detoxication). Reactions of methylation imply the transfer of a methyl group from the onium-type cofactor *S-adenosylmethionine* (SAM) to the substrate by means of a methyltransferase. The activated methyl group from SAM is transferred to the acceptor molecules R — XH or RX, as shown in Fig. 31.30.

A number of methyltransferases are able to methylate small molecules either on a phenolic, an amino or a thiol group. The major enzyme responsible for *O-methylations* is catechol *O*-methyltransferase (COMT), a cytosolic enzyme that also exists in membrane-bound form. This enzyme catalyses the *O*-methylation of catecholamines (dopamine, norepinephrine, epinephrine), L-dopa and related catechol drugs.

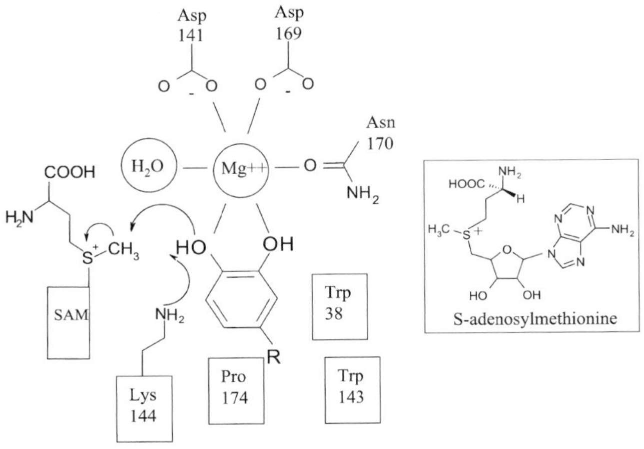

Fig. 31.30 Reaction mechanism of *O*-methylation of catechols catalysed by catechol *O*-methyltransferase. The structure of *S*-adenosylmethionine (SAM) is shown in the insert (from Männisto and Kaakkola, with permission).[8]

N-Methylations are catalysed by several enzymes such as nicotinamide *N*-methyltransferase, histamine methyltransferase, phenylethanolamine *N*-methyltransferase (noradrenaline *N*-methyltransferase) and a non-specific amine *N*-methyltransferase (arylamine *N*-methyltransferase, tryptamine *N*-methyltransferase). *S-Methylations* are catalysed by the membrane-bound thiol methyltransferase and the cytosolic thiopurine methyltransferase.

Methylation reactions

O-Methylation of xenobiotic catechols occurs preferentially at the *meta* position, L-dopa and isoproterenol being classical examples. Frequently *O*-methylation is a late event in the metabolism of aryl groups, after they have been oxidized to catechols. Thus the anti-inflammatory drug diclofenac yields in humans 3′-hydroxy-4′-methoxy-diclofenac as the major metabolite with a very long plasmatic half-life. Noncatechol diphenols are not subject to methylation, e.g. terbutaline. A few monophenols can also undergo methylation to a limited extent.

The catalytic machinery of COMT has been partly elucidated by means of determination of the crystal structure solved at 2.0 Å resolution, molecular dynamics simulations and site-directed mutagenesis.[7] The active site structure presented in Fig. 31.30 shows the crucial amino acid and other partners (a water molecule and a Mg^{++}) involved in the reaction or substrate recognition. Catechol-*O*-methylation proceeds via a nucleophilic reaction, whereby the lysine 144 residue is able to deprotonate the hydroxyl group of the catechol substrate with the help of the cation that renders this group more ionizable. The catecholate formed can attrack thereafter the methyl group of SAM, leading to the formation of the methylconjugate and *S*-adenosyl-homocysteine. Moreover Mg^{++} controls the orientation of the catechol moiety through establishment of an octahedral coordination with aspartic acids 141 and 169, asparagine 170, both catechol hydroxyls and with a water molecule. Finally the residues proline174, tryptophanes143 and 38 which form a hydrophobic pocket, are considered as gate-keepers of the active site and responsible for the selectivity of COMT toward the side chain R of the catechols *para* to the *O*-methylation site. On the basis of these results, very potent, highly selective and orally active COMT inhibitors have been recently developed. Nitrocatechol is the key structure of such molecules (Fig. 31.31).[8] They are used as adjuncts of levodopa for the treatment of Parkinson's disease characterized by a progressive loss in L-dopa biosynthesis. They act by reducing the methylation of dopamine in the brain.

The *N-methylation* of xenobiotics occurs with primary and secondary amines (e.g. amphetamine and tetrahydroisoquinolines, respectively), with pyrrol-type nitrogen

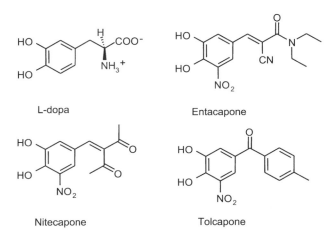

Fig. 31.31 Chemical structure of L-dopa and COMT nitrocatechol inhibitors.

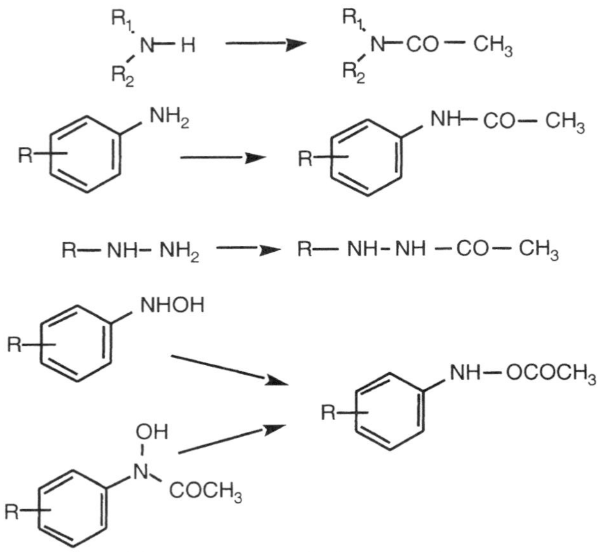

atoms (as exemplified by imidazole, histamine and thiobendazole), and with pyridine-type nitrogen atoms. The latter reaction of *N*-methylation is an effective route of detoxication. It leads to quaternary ammoniums which are stable to *N*-demethylation and more polar (having a permanent positive charge) than the parent compound. Known substrates are nicotinamide, pyridine, and a number of related heterocyclic compounds.

S-Methylation is known for aromatic sulfhydryl groups such as thiophenols, 6-mercaptopurine and propylthiouracil, as well as for aliphatic thiols like captopril. Once formed, such methylthio metabolites can be further processed to sulfoxides and sulfones before being excreted.

C Acetylation and acylation

All cases discussed in this section involve the reaction between an amine and an acyl group to yield an amide. The high-energy cofactor required is in most cases an acyl-coenzyme A derivative (acyl-S-CoA) where the acyl moiety is bound by a thioester linkage.

Acetylation reactions

The most common reactions of acylation are in fact acetylations of xenobiotics containing a primary amino group. The cofactor of acetylation is acetylcoenzyme A (acetyl-S-CoA), the reaction being catalysed by a variety of *N*-acetyltransferases. Arylamine *N*-acetyltransferases (NAT-1 and -2) are the most important enzyme, but aromatic-hydroxylamine *O*-acetyltransferase and *N*-hydroxyarylamine *O*-acetyltransferase are also involved in the acetylation of some aromatic amines and hydroxylamines.

The major reactions of *N*-acetylation are listed in Fig. 31.32. A large variety of primary aromatic amines are *N*-acetylated, often to a large extent; they include several

Fig. 31.32 Major reactions of *N*-acetylation.

drugs such as sulfonamides and *p*-aminosalicylic acid, not to mention various carcinogenic amines such as benzidine. Other substrates include a few aliphatic amines, cysteine conjugates, and mainly hydrazines and hydrazides. Medicinal examples of the latter include isoniazid and hydralazine. The metabolites resulting from *N*-acetylation are uncharged amides; when the parent xenobiotic was an amine of high basicity, acetylation may result in decreased water solubility.

A three-dimensional model of the human NAT-1 and -2 based on the crystal structure of the *Salmonella typhimurium* enzyme reveals that, interestingly, the proteins are members of the cysteine protease family (as papain).[9] They share a common mechanism involving a catalytic triad (cysteine-histidine-aspartic acid), whereby the cysteine sulfur atom activated by a charge-relay system via hydrogen bonds with histidine and aspartic acid residues can attack, in a nucleophilic way, the carboxyl group of the co-substrate.

Genetic differences do exist in some animal species and in humans, where one distinguishes between slow and fast acetylators. In humans, this common polymorphism associated with the expression of genetic variants of NAT-2 is known to be responsible for an increased susceptibility to drug toxicity. Several amino acid substitutions (point mutations) have been detected in the human population that lead to less efficient enzyme forms. Examples of drugs exhibiting acetylation polymorphism are the antibacterial drug sulfomethazine, the antituberculosis drug isoniazid, and the antileprosy agent dapsone.

Other reactions of acylation

Cholesteryl ester synthase and fatty acid synthase, but also sterol *O*-acyltransferase, perform conjugations of various xenobiotic alcohols (e.g. ethanol, tetrahydrocannabinol and codeine) with fatty acids such as palmitic, oleic, linoleic and linolenic acids. A limited number of *N*-formylation reactions of aromatic amines are catalysed by arylformamidase in the presence of *N*-formyl-L-kynurenine.

D   Acyl-coenzyme A thioesters as metabolic intermediates

Biochemistry

The reactions described in this section all have in common the fact that they involve xenobiotic carboxylic acids (R — COOH) coupling with coenzyme A (CoA-SH) to form an acyl-CoA metabolic intermediate (R — CO-S-CoA). The reaction requires ATP and is catalysed by various acyl-CoA ligases (or synthetases) of overlapping substrate specificity, e.g. acetate-CoA ligase, butyrate-CoA ligase, benzoate-CoA ligase and phenylacetate-CoA ligase. They are classified according to the length of the fatty acid substrates as short-chain (acetate, propionate, butyrate), medium-chain (C6-C8 fatty acids), long-chain (C12-C18 fatty acids) and very long-chain (C22 and > fatty acids) acyl-CoA ligases. Carboxylic acid-containing xenobiotics and drugs, such as the nonsteroidal anti-inflammatory drugs related to arylpropionic acid, or the hypolipidaemic drugs clofibric acid or ciprofibrate are also substrates of these enzymes. It should also be pointed out that these compounds can be metabolized through the glucuronidation reaction, leading to the formation of acylglucuronides (see below).

The acyl-CoA conjugates thus formed are seldom excreted, but undergo further transformation by a considerable variety of pathways, i.e.:

- Hydrolysis (by thioester hydrolases)
- Formation of amino acid conjugates
- Formation of hybrid triglycerides and phospholipids
- Formation of cholesteryl esters and bile acid esters
- Formation of acyl-carnitines
- Protein acylation
- Unidirectional chiral inversion of arylpropionic acids
- Dehydrogenation and β-oxidation
- 2-Carbon chain elongation.

Amino acid conjugation

This is a major route of metabolism for many xenobiotic acids, involving the formation of an amide bond between the xenobiotic acyl-CoA and the amino acid. The acyl-CoA ligases represent the rate-limiting step for conjugation of carboxylic acid xenobiotics to amino acids. Glycine is the amino acid most frequently used for conjugation, forming conjugates with the general structure R — CO — NHCH$_2$COOH. A few glutamine conjugates have also been characterized in humans. The enzymes catalysing these transfer reactions are various *N*-acyltransferases, for example glycine *N*-acyltransferase, glutamine *N*-phenylacetyltransferase, glutamine *N*-acyltransferase and glycine *N*-benzoyltransferase. The xenobiotic acids undergoing amino acid conjugation are mainly benzoic acids, such as benzoic acid itself and salicylic acid, which form hippuric acid and salicyluric acid, respectively. Phenylacetic acid derivatives can yield glycine and glutamine conjugates. In addition, other amino acids can be used for conjugation in various animal species, e.g. alanine and taurine, as well as some dipeptides.

Other reactions

Incorporation of xenobiotic acids into lipids via the formation of acyl-CoA intermediates has been characterized and forms highly lipophilic metabolites that accumulate for long periods of time in the adipose tissue. Triacylglycerol analogues (hybrid triglycerides) or cholesterol esters are formed and can be associated with membrane phospholipids. Such incorporation is susceptible to profoundly affect the properties and function of the membranes. In some cases, acyl-CoA conjugates formed from xenobiotic acids can also enter the physiological pathways of fatty acid catabolism or anabolism. Thus, intermediate metabolites of β-oxidation may be observed, as exemplified by valproic acid, whose structure is related to short-chain fatty acid. The enzymes catalysing this pathway are clearly those involved in the metabolism of fatty acids. In addition, xenobiotic alkanoic and arylalkanoic acids can undergo a two-carbon chain elongation, or chain shortening by two carbons or even four or six carbons.

Similarly, acyl-CoA conjugates can react with proteins (protein acylation). Fatty acids covalently bind to cysteine acylation sites. This process, which modifies the properties and the function of the protein, is believed to play a key role in the regulation of the cellular machinery. Drug acyl-CoA conjugates also interact covalently with plasma proteins, namely albumin.

Chiral inversion of carboxylic acid drugs containing an asymmetric centre is an important mechanism whereby the enantiomer (*R*) from drugs can be transformed into the (*S*)-antipode (Fig. 31.33). This unidirectional reaction (*S*-isomer is seldom inverted into *R*) depends on the animal species and on the compound considered, but requires the formation of an acyl-CoA intermediate. This situation stands for the drugs chemically related to 2-arylpropionic acid (propionates), such as the nonsteroidal anti-inflammatory drugs, ketoprofen or ibuprofen. As a consequence, the administration of these drugs as racemates (*RS*) may progressively lead to generation from the *R*-enantiomer of the *S*-antipode which is the active pharmacological species able to inhibit the cyclooxygenases. The reaction proceeds by three steps: acyl-CoA formation, epimerization and hydrolysis. The first

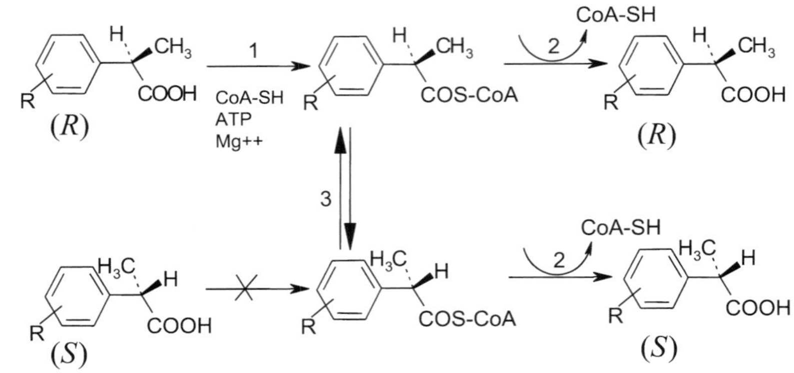

Fig. 31.33 Mechanism of the chiral inversion of (*R*)- or (*S*)-aryl-2-propionic acids. Reactions 1, 2 and 3 are catalysed by acyl-CoA ligases, acyl-CoA hydrolases and acyl-CoA epimerases, respectively.

step catalysed by acyl-CoA ligases is enantioselective, the *S*-isomer not being a substrate of the enzyme, whereas the other steps involving an acyl-CoA epimerase and hydrolase are not.

Reaction mechanism of formation of acyl-CoA intermediates by the acyl-CoA ligases

The transformation of the carboxylic acid substrate into acyl-CoA requires one ATP molecule and leads to the release of AMP and pyrophosphate. The reaction mechanism involves the formation of an acyl-AMP intermediate. Extensive and detailed kinetic studies provide details on the ordered addition of the substrate and ATP to the enzyme and the subsequent release of products (Fig. 31.34). The reaction proceeds via a Bi-Uni-Bi-Ping-Pong mechanism whereby ATP complexed to magnesium binds first, thus allowing the fixation of carboxylic acid to the enzyme. Acyl-AMP is thereafter formed and pyrophosphate released. Then

CoA-SH binds, leading to the formation and the release of the reaction products, acyl-CoA and AMP from the enzyme.

E Glucuronidation

Biochemistry

In glucuronidation (i.e. glucuronic acid conjugation), one molecule of glucuronic acid is transferred to the substrate from uridine-5′-diphospho-α-D-glucuronic acid (UDP-GlcA). This cofactor is generated by dehydrogenation of UDP-glucose, which is synthesized from glucose-1-phosphate and uridine triphosphate (Fig. 31.35).

The reaction is catalysed by UDP-glucuronosyltransferases (UGTs). These enzymes are endoplasmic reticulum, membrane-bound proteins,[11] which utilize UDP-GlcA as a sugar donor and transfer glucuronic acid to available substrates, a process that forms β-D-glucuronides. These metabolites can be formed from a wide variety of chemicals

A

$$\text{Carboxylic acid } + \text{ ATP } \xrightarrow{\text{Mg++}} \text{ Acyl-AMP } + \text{ PPi}$$

$$\text{Acyl-AMP } + \text{ CoA-SH } \longrightarrow \text{ Acyl-CoA } + \text{ AMP}$$

$$\text{Carboxylic acid } + \text{ ATP } + \text{CoA-SH } \longrightarrow \text{ Acyl-CoA } + \text{ AMP } + \text{ PPi}$$

B

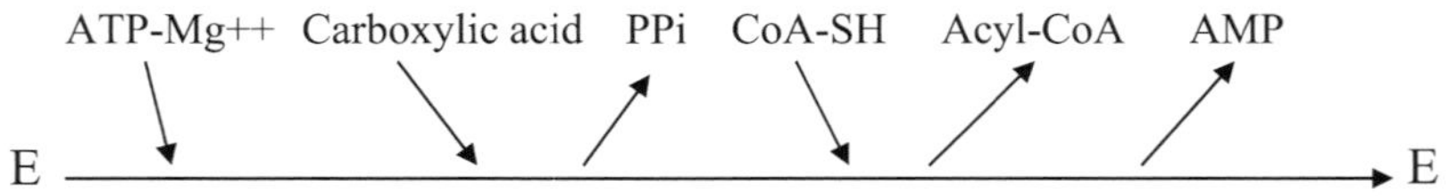

Fig. 31.34 Reaction scheme of the acyl-CoA formation catalysed by the acyl-CoA ligases. (A) Reaction equation; (B) suggested ordered mechanism. AMP, adenosine monophosphate; PPi, pyrophosphate (adapted from Vessey and Kelley[10]).

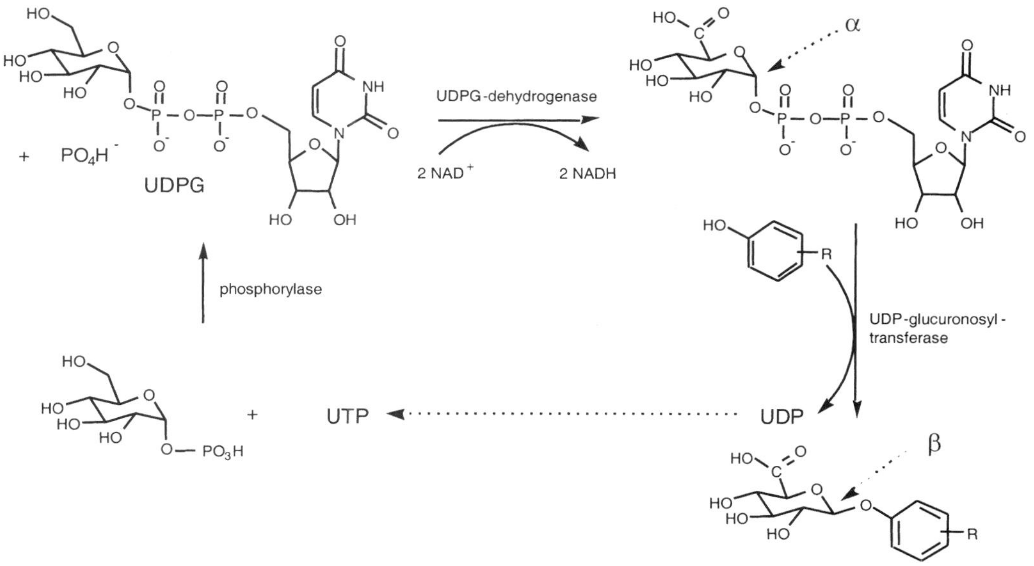

Fig. 31.35 Synthesis of uridine-5′-diphospho-α-D-glucuronic acid (UDP-GlcA) and glucuronyltransferase-catalysed glucuronidation of a phenol.

with the only common feature being the presence of an appropriate functional hydroxyl, carboxyl, amine, sulfhydryl or carbonyl group (Fig. 31.36). This structural diversity of the substrates allows glucuronidation of a large range of exogenous and endogenous compounds. The large substrate specificity of individual UGT isoforms facilitates the glucuronidation of structurally unrelated compounds.

A common characteristic of the functional groups, despite their great chemical variety, is their nucleophilic character. As a consequence of this diversity, the products of glucuronidation are classified as *O*-, *N*-, *S*- and *C*-glucuronides. The *O*-glucuronidation of phenolic xenobiotics such as 1-naphthol and endogenous compounds such as estrogens is often in competition with *O*-sulfation, with the latter reaction predominating at low doses and the former at high doses. Another major group of substrates undergoing *O*-glucuronidation is primary, secondary or tertiary alcohols. An interesting example is that of morphine, which is conjugated on its phenolic and secondary alcohol groups to form the 3-*O*-glucuronide (a weak opiate antagonist) and the 6-*O*-glucuronide (a strong opiate agonist), respectively.

Second in importance to *O*-glucuronides are the *N*-glucuronides formed from carboxamides, sulfonamides, and various amines. The reaction has special significance for antibacterial sulfanilamides, producing highly water-soluble metabolites which do not crystallize in the kidneys. *N*-Glucuronidation of aromatic and aliphatic amines and pyridine-type nitrogens has been observed in a few cases only. A reaction of greater significance in human drug metabolism is the *N*-glucuronidation of lipophilic, basic tertiary amines containing one or two methyl groups.

Some *S*-glucuronides are formed from aliphatic thiols, aromatic thiols and dithiocarboxylic acids. As for *C*-glucuronidation, this reaction is seen in humans for 1,3-dicarbonyl drugs, such as sulfinpyrazone.

Other important metabolites of *O*-glucuronidation are acylglucuronides. They are formed by esterification of carboxylic acids with glucuronic acid. Many therapeutic agents such as arylacetic acids (diclofenac, diflunisal), aliphatic acids (valproic acid) and arylpropionic acids (ketoprofen, naproxen) are metabolized as acylglucuronides (Fig. 31.37).

Due to the presence of the carbonyl group, acylglucuronides undergo spontaneous cleavage of the *O*-glycosidic bond concomitant to nucleophilic attack on potential nucleophilic groups of proteins such as NH_2 of lysines, SH of cysteines and OH of tyrosines (Fig. 31.38). In addition, intramolecular rearrangement has also been demonstrated in which the acyl group migrates from its initial C-1 position on glucuronic acid to position C-2 and subsequently, to C-3 and to C-4 resulting in the formation of position isomers that cannot be cleaved by β-glucuronidases, and in the opening of the pyranose ring. This process leads to covalent adducts on plasma and tissue proteins via imine formation (Fig. 31.38). Thus, acylglucuronide formation cannot be viewed solely as a reaction of inactivation and detoxication. A special class of acylglucuronides is that formed by carbamic acids which themselves are not stable enough to be characterized.

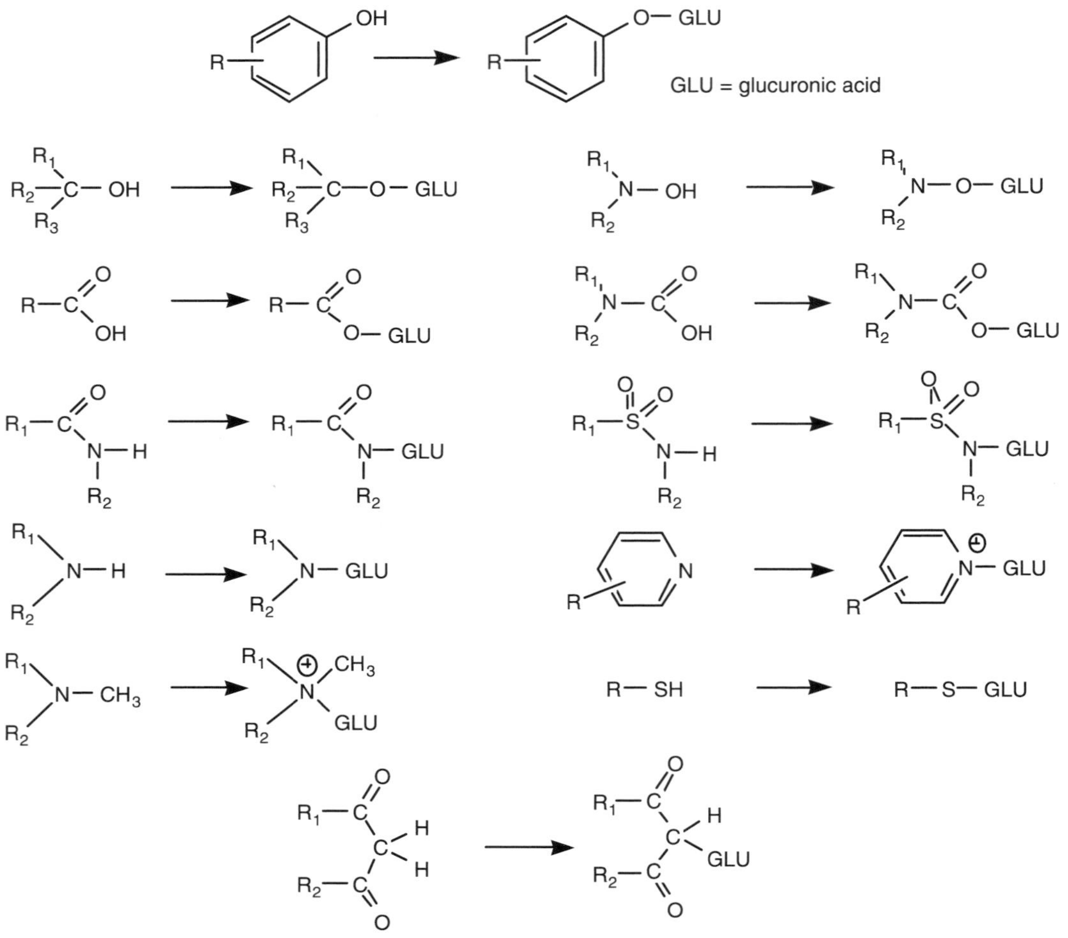

Fig. 31.36 Representative aglycones that undergo glucuronidation.

On the other hand, a few drugs and a number of xenobiotic aromatic amines are known to be *N*-hydroxylated and then *O*-glucuronidated. The reactivity of *N*-*O*-glucuronides, such as *N*-hydroxy-2-acetylaminofluorene glucuronide (Fig. 31.37) leads to heterolytic decomposition into a *N*-acetyl-*N*-arylnitrenium ion which then reacts with nucleic acids and proteins.[12]

UGTs have been classified into two main families, UGT1 and UGT2. The members of the human UGT1 family (UGT1A1, UGT1A3, UGT1A4, UGT1A5, UGT1A6, UGT1A7, UGT1A8, UGT1A9, UGT1A10) are encoded by a complex of 17 exons forming the UGT1A locus. These isoforms contain identical carboxyl terminal sequences encoded by the last four exons of the locus. In contrast to UGT1, human UGT2 members (UGT2B4, UGT2B7, UGT2B10, UGT2B11, UGT2B15, UGT2B17) are encoded by independent genes.

The known substrates for UGT1 enzymes include bilirubin, small planar phenols such as 1-naphthol and 4-nitrophenol, halogenated and bulky phenols, steroid hormones such as estrone and androsterone. The major endogenous substrates of the UGT2B isoforms are steroids and their metabolites such as estriol, hydroxyestradiol, androsterone and testosterone, and bile acids, such as hyodeoxycholic acid and lithocholic acid.

Reactions of glucuronidation

It has been proposed that the reaction mechanism accounting for glucuronidation is a SN_2 nucleophilic attack (with inversion of configuration of the substrate) on the C-1 of α-D-glucuronic acid, leading to β-D-glucuronide formation and the release of UDP. According to kinetic and chemical modification studies, UGTs may act as an acid/base catalyst. In the acceptor site, an amino acid residue acts as a base and abstracts a proton from the acceptor group. On the donor side, cleavage occurs at C1 of UDP-GlcA leading to the development of an additional negative charge on the leaving UDP, which would then be protonated by an acidic amino acid residue (Fig. 31.39).[14]

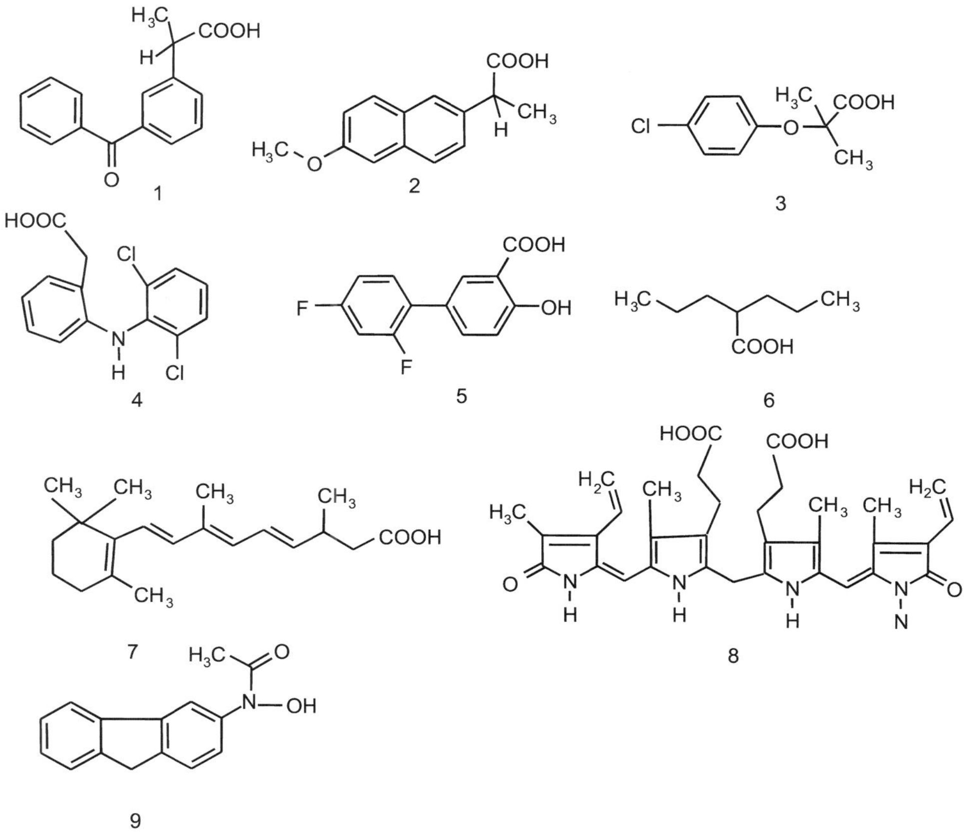

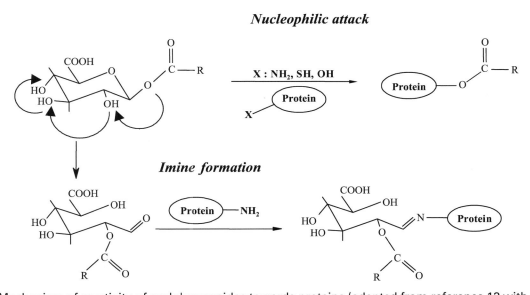

Fig. 31.37 Substrates that form reactive glucuronides: (**1**) ketoprofen; (**2**) naproxen; (**3**) clofibric acid; (**4**) diclofenac; (**5**) diflunisal; (**6**) valproic acid; (**7**) all-*trans* retinoic acid; (**8**) bilirubin; (**9**) *N*-hydroxy-2-acetylaminofluorene.

Fig. 31.38 Mechanism of reactivity of acylglucuronides towards proteins (adapted from reference 13 with permission).

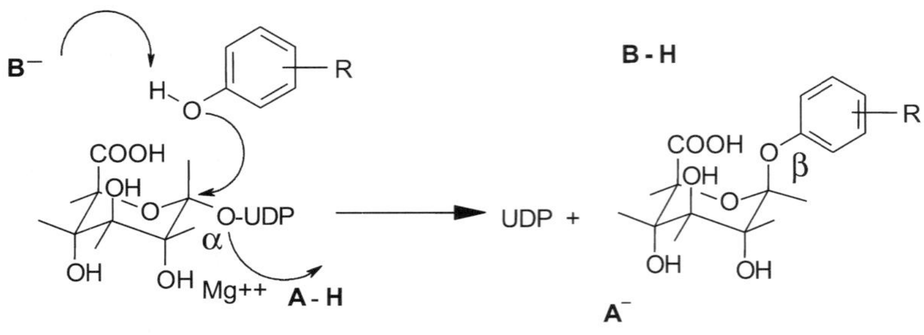

Fig. 31.39 Proposed mechanism for glucuronidation of phenols.

The development of selective inhibitors for different UGT isoforms has been useful in obtaining informations on the organization of the active site of these enzymes. 7,7,7-Triphenylheptanoic acid (Fig. 31.40) was developed as a probe for the acceptor binding site of the bilirubin isoform.[15] The presence of three aromatic rings, as well as that of a negatively charged group on the short aliphatic chain was a prerequisite for efficient inhibition.

In an attempt to develop effective UGT inhibitors, transition-state based inhibitors were synthesized taking advantage of the high affinity of UGT for the UDP moiety and the structural requirement for the acceptor substrate. ω,ω,ω-Triphenylalkyl-UDP derivatives (Fig. 31.40) have been shown to be powerful inhibitors of UGT bilirubin isoform. In the same manner, DMSU was an efficient inhibitor of phenol glucuronidation catalysed by UGT1A6.[16]

F Sulfate conjugation

Reaction of sulfoconjugation

Sulfuryl transfer, also referred to as sulfation or sulfonation is the transfer reaction of the sulfate group from the ubiquitous donor 3'-phosphoadenosine 5'-phosphosulfate (PAPS) to an oxygen atom of the acceptor substrate leading to a sulfuric acid ester ($R — O — SO_3^-$, $RR' — N — O — SO_3^-$) (Fig. 31.41).

The sulfate group can be transferred to other nucleophilic atoms, such as nitrogen or sulfur, resulting in the formation of sulfamates ($RR' — N — SO_3^-$) and thiosulfates ($R — S — SO_3^-$), respectively, rather than sulfates. The sulfotransferases (STs) that catalyse the sulfonation of drugs, hormones and neurotransmitters in animals, like those in plants that catalyse similar reactions with flavonols as

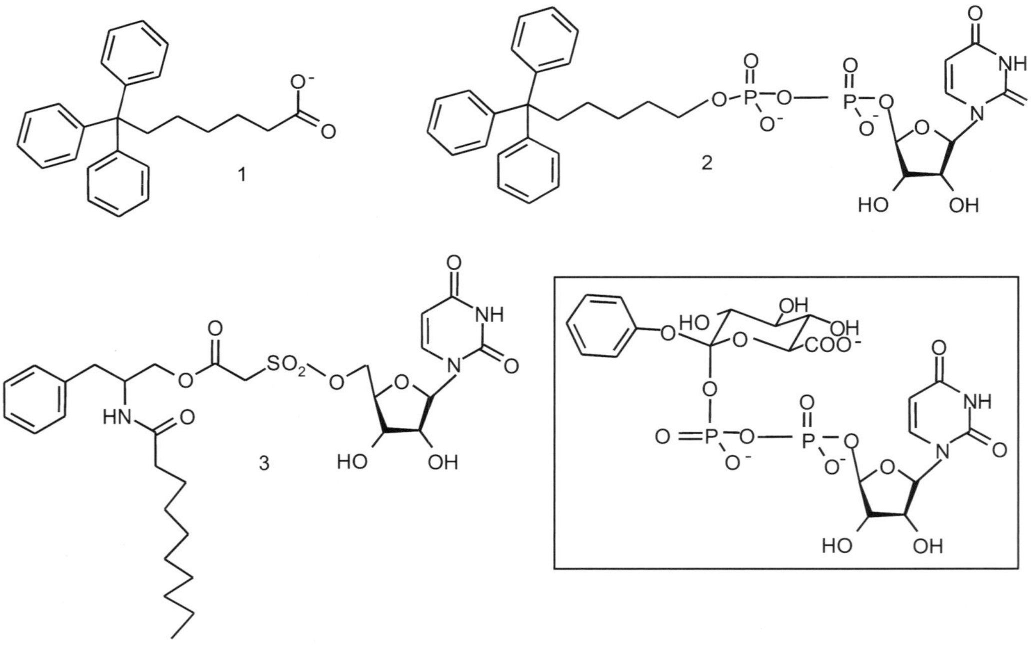

Fig. 31.40 Active site and transition-state analogues: (**1**) 7,7,7-triphenylheptanoic acid; (**2**) ω,ω,ω-triphenylalkyl-UDP; (**3**) 3-5'-*O*-[[(2-decanoyl-3-phenylpropyloxycarbonyl)methyl]sulfonyl]uridine (DMSU). In insert, structure of a putative transition state analogue for the phenol glucuronidation reaction catalysed by UGT.

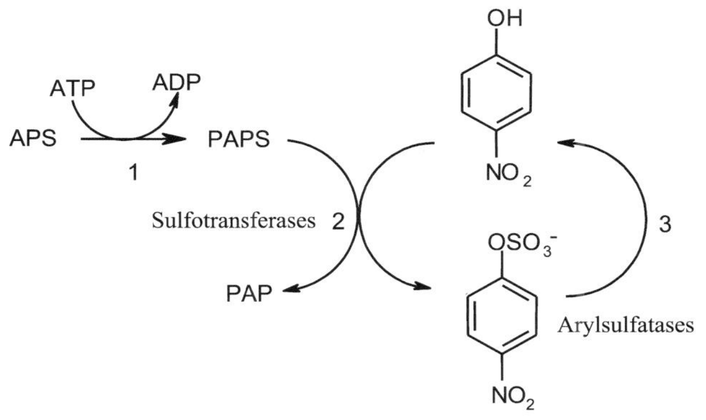

Fig. 31.41 The sulfoconjugation reaction: example of 4-nitrophenol as acceptor substrate. Step 1, PAPS formation from adenosine 5′-phosphosulfate (APS); step 2, sulfate conjugation with the subsequent release of 3′-phosphoadenosine 5′-phosphate (PAP); step 3, possible hydrolysis of the sulfoconjugate by arylsulfatases.

substrates, are cytosolic in subcellular localization. Although macromolecules substrates are sulfated by membrane-bound sulfotransferases, this review will primarily focus on cytosolic STs which are major contributors to the homeostasis and regulation of numerous biologically potent endogenous chemicals, such as catecholamines, steroids and iodothyronines as well as the detoxication of xenobiotics.[17] Included among the broad classes of compounds which undergo *O*-sulfation catalysed by soluble STs are 'simple phenols' such as 4-nitrophenol, phenolic catecholamine neurotransmitters such as dopamine, phenolic steroids such as the estrogens, non-phenolic hydroxysteroids such as dehydroepiandrosterone (DHEA) and in plants, flavonols such as quercetin. Cytosolic STs are also actively involved in the *N*-sulfation of alkyl- and arylamines, as well as alkyl- and arylamides leading in some cases to the formation of unstable metabolites.

Indeed, although the major physiological function of conjugation with an endogenous anionic moiety such as sulfate is the formation of inactive excretable products, numerous compounds are metabolized to chemically reactive metabolites via sulfation.[18] This can be rationalized by the fact that the sulfate group is electron-withdrawing and may be cleaved off heterolytically, thus leading to the formation of a strongly electrophilic cation. Such a mechanism has been shown to be involved in the genotoxicity mediated by arylamines such as 4-aminoazobenzene and arylamides such as 2-acetylaminofluorene. In contrast, significantly more stable products are obtained upon formation of sulfamates from primary and secondary alkylamines. Endogenous hydroxysteroids (i.e. cyclic secondary alcohols) also form relatively stable sulfates, while some secondary alcohol metabolites of xenobiotics (such as the alkenylbenzene derivatives safrole and estragole) form carbocations critically involved in DNA adduct formation. Primary alcohols, e.g. methanol and ethanol, can also form low amounts of sulfates whose alkylating capacity is well known. In contrast to some alcohols, phenols form stable sulfate esters.

The enzymes

Based on biochemical criteria, STs were classified into five classes which differ with regard to their substrate specificities, inhibitor sensitivities, thermal stability and regulation. Molecular cloning studies showed that cytosolic STs are members of a single superfamily now termed *SULT*, as judged from similarities in the nucleotide sequence of their genes/cDNAs. Two main families have been defined: the phenolsulfotransferases (SULT1) and the steroid sulfotransferases (SULT2). A classification into subfamilies 1A, 1B, 1C, 1E, 2A, 2B and 3A according to the degree of similarities of the deduced amino acid sequences is generally accepted. In humans, at least ten distinct SULT enzymes have been distinguished on the basis of their substrate specificity and/or amino acid sequence identity which ranges from 30 to 96%, most of them belong to the SULT1 family. A SULT catalysing the formation of a sulfamate has recently been isolated from rabbits that belongs to a third gene family. In addition, unique SULT cDNAs consisting of two new families (SULT4 and SULT5) are found in the DNA database. A brain SULT isoform has recently been suggested to belong to family 4.

STs enzymes are predominantly dimers, both homo- and heterodimers with MW values that vary from 30 to 36 kDa. The elucidation of the three-dimensional structure of several STs by X-ray crystallography revealed that they are globular proteins composed of a single α/β domain with a characteristic five stranded parallel β-sheet that constitutes the core of the PAPS-binding and catalytic sites.[19] The amino acid sequences of the 3′- and 5′-phosphate binding sites of PAPS are also strictly conserved and used as signatures for newly cloned STs sequences. Earlier kinetic studies showed that cytosolic STs catalyse sequential transfer reactions with the formation of a ternary complex between enzyme, PAPS and acceptor substrate.[20] Subsequently, crystallographic data suggested a common mechanism of the transfer reaction catalysed by membrane and cytosolic STs as illustrated in Fig. 31.42.

Conserved lysine, serine and histidine residues have been demonstrated to be determining factors for catalytic activity by site-directed mutagenesis. It is suggested that the side-chain nitrogen of the lysine residue (K47 in hEST) forms a hydrogen bond to an oxygen atom of the 5′-phosphate group of PAPS, whereas the hydroxyl group of the serine residue (S137) interacts with an oxygen atom of the 3′-phosphate group. Furthermore, a conserved histidine residue appears to function as the catalytic base. This histidine (H107 in hEST)

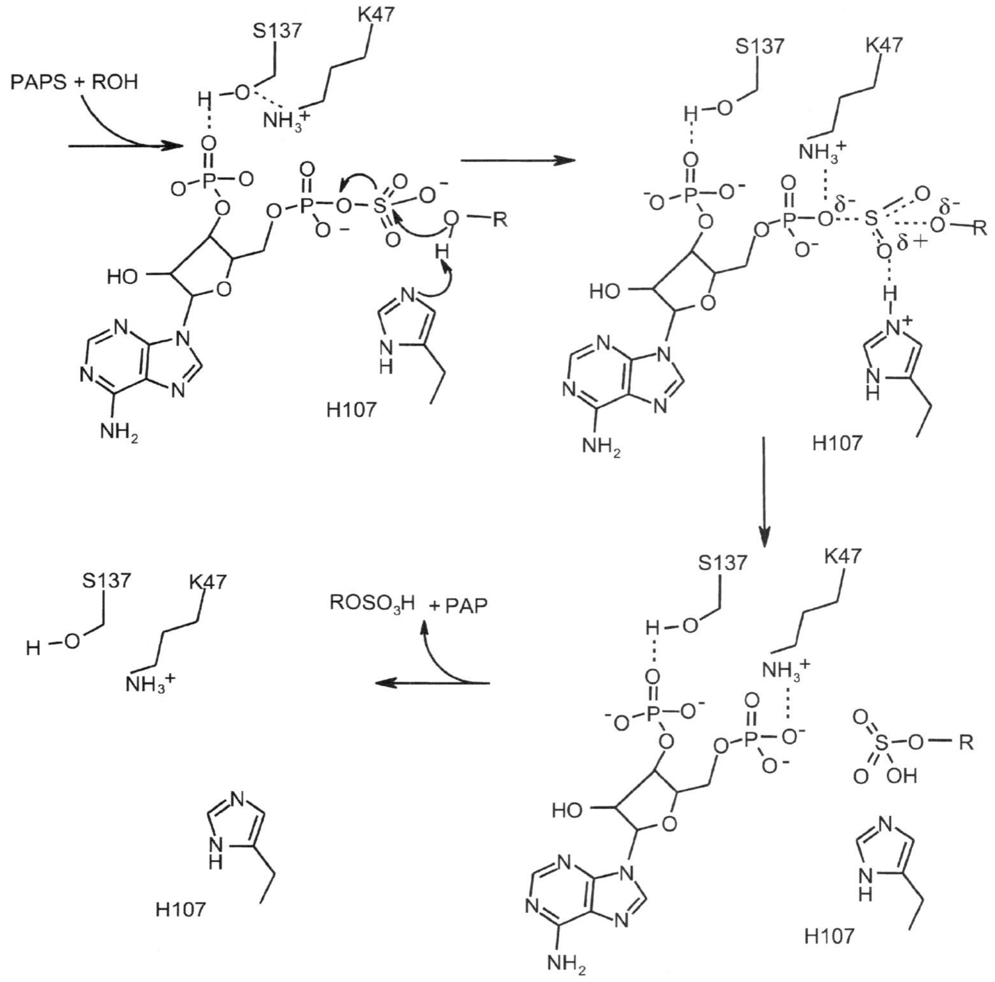

Fig. 31.42 Putative mechanism of sulfuryl transfer catalysed by STs. Residue numbers are taken from human estrogen ST (hEST). (From Negishi *et al.*[19] with permission).

is suggested to remove the proton from the acceptor group facilitating the nucleophilic attack on the sulfur atom of the PAPS molecule. The resulting developing negative charge on the bridging oxygen leads the side-chain nitrogen of the lysine residue to switch from the serine to the bridging oxygen promoting the dissociation of the sulfate group. Thus, the serine residue bound to the 3'-phosphate appears to prevent PAPS hydrolysis in the absence of substrate, favouring an ordered mechanism in which the donor substrate may bind first and the acceptor substrate follows.

The molecular properties that define the substrate recognition of individual STs are the subject of active investigation. A structural complementary of the acceptor substrate with a deep hydrophobic pocket common to cytosolic ST is likely to provide a principal determinant for the substrate specificity. Recently, critical amino acids governing the specificity of a mouse estrogen ST with respect to estradiol and DHEA as acceptor substrates, were

identified. Tyrosine 81 appears to form a structure-like gate with phenylalanine 142 providing steric interactions with the C-19 methyl group in DHEA. In addition to steric effects, the presence of charged groups such as that of the side-chain of glutamic acid 146 in SULT1A3 govern the selectivity of STs.[21] These authors suggest that glutamic acid 146 of SULT1A3 is involved not only in 'attracting' natural substrates such as dopamine, but also in 'repelling' many phenolic xenobiotics with hydrophobic or bulky substituents.

G Conjugation with glutathione

Biochemistry

Glutathione (GSH, γ-glutamyl-cysteinyl-glycine) is a thiol-containing tripeptide of capital significance in the detoxication and toxication of drugs and other xenobiotics. Glutathione reacts with endogenous and exogenous compounds in a variety of ways. First, the nucleophilic properties of the thiol group make it an effective conjugating agent. Second, glutathione can act as a reducing or oxidizing agent depending on its redox state (i.e. GSH or GSSG). Furthermore, the reactions of glutathione can be enzymatic (e.g. conjugations catalysed by glutathione-*S*-transferases, and peroxide reductions catalysed by glutathione peroxidases) or non-enzymatic (e.g. some conjugation and various redox reactions).

The glutathione transferase (GST) comprises multifunctional proteins coded by a multigene family. These enzymes are both cytosolic and microsomal and function as homodimers and heterodimers. They exist as four classes in mammals. The human enzymes comprise the following dimers: A1-1, A1-2, A2-2, A3-3 (alpha class), M1a-1a, M1a-1b, M1b-1b, M1a-2, M2-2, M3-3 (mu class), P1-1 (pi class), T1-1 (theta class), and three microsomal enzymes (MIC). The GST A1-1 and A1-2 are also known as ligandin when they act as binding or carrier proteins, a property also displayed by M1a-1a, M1b-1b and also by GSH (pi class).

GST are powerful detoxifying enzymes. The overexpression of GST that has been demonstrated in some human cancer cells, such as breast cancer cells, is associated with the multidrug resistance that impairs the efficacy of anticancer drugs. In this regard, GST inhibitors or substrate competitors, such as the diuretic drug ethacrynic acid, have been proposed as an adjuvant for cancer chemotherapy, thus enhancing the cytotoxicity of alkylating drugs in cancer cell lines.

The nucleophilic character of glutathione is due to its thiol or rather thiolate group. Indeed if the thiol group of GSH ($pK_a \sim 9$) is largely protonated at physiological pH, the binding to the enzyme is associated with the loss of the proton and to the electrophilic stabilization of the thiolate group. A hydroxyl group of a serine (GST theta class) or of a thyrosine residue (GST alpha, mu, pi classes) acts as hydrogen donor to the sulfur of GSH whereby lowering the pK_a of the thiol, leading to the presence of a predominantly ionized form at physiological pH. As a result, GSTs transfer glutathione to a very large variety of electrophilic groups (R — X, see Fig. 31.43) in nucleophilic reactions categorized as either substitutions or additions.

With compounds of sufficient reactivity, these reactions can also occur nonenzymatically. Once formed, glutathione conjugates (GS-R) are seldom excreted as such, but usually undergo further biotransformation. Cleavage of the glutamyl moiety by glutamyl transpeptidase and of the cysteinyl

moiety by cysteinylglycine dipeptidase or aminopeptidase M leaves a cysteine conjugate (Cys-S-R) which is further *N*-acetylated by cysteine-*S*-conjugate *N*-acetyltransferase to yield an *N*-acetylcysteine conjugate (CysAc-S-R). The latter type of conjugates are known as mercapturic acids. These may be either excreted or further transformed, since cysteine conjugates can be substrates of cysteine-*S*-conjugate β-lyase to yield thiols (R-SH). These in turn can rearrange or be *S*-methylated and then *S*-oxygenated to yield thiomethyl conjugates (R-S-Me), sulfoxides (R-SO-Me) and sulfones (R-SO$_2$-Me).

Reactions of conjugation

The major reactions of glutathione are shown in Fig. 31.44. Nucleophilic addition to epoxides yields nonaromatic conjugates which may undergo further transformation as described above. This reaction is well documented for the arene oxide metabolites of numerous drugs and xenobiotics containing an aromatic moiety. The same reaction can also occur readily for epoxides of olefins. An important pathway of substitution exists for — CH_2X moieties. Various electron-withdrawing leaving groups X may be involved, for example the chorine atom at the NCH_2CH_2Cl group of anticancer alkylating agents.

Addition at activated olefinic groups (e.g. β,γ-unsaturated carbonyls) are quite varied. A typical substrate is acrolein ($CH_2 = CH — CHO$). Quinones (*ortho* and *para*) and quinone imines react with glutathione by two distinct and competitive routes, namely nucleophilic addition to form a conjugate, and reduction to the hydroquinone or the aminophenol. The conjugates produced by addition may undergo reoxidation to *S*-glutathionylquinones or *S*-glutathionylquinone imines of considerable reactivity.

Haloalkenes may react with GSH either by substitution or by addition. Formation of mercapturic acids occurs as for other glutathione conjugates, but in this case S — C cleavage of the S-cysteinyl conjugates by the renal β-lyase yields thiols of significant toxicity since they rearrange to form highly reactive thioketenes ($XRC=C=S$) and/or thioacyl halides. With a good leaving group and adequate substituents, nucleophilic aromatic substitution reactions also occur at aromatic rings. A good example of the detoxication of acyl halides with glutathione is provided by phosgene ($O=CCl_2$), an extremely toxic metabolite of chloroform which is inactivated to the diglutathionyl conjugate $O=C(SG)_2$. The addition of glutathione to isocyanates and isothiocyanates is of significance due to its reversible character. Substrates of the reaction are xenobiotics such as the well-known toxin methyl isocyanate, whose glutathione conjugate behaves as a transport form able to carbamoylate various macromolecules, enzymes and membranes structures.

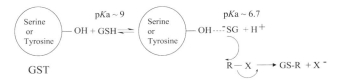

Fig. 31.43 Reaction mechanism of GST.[22]

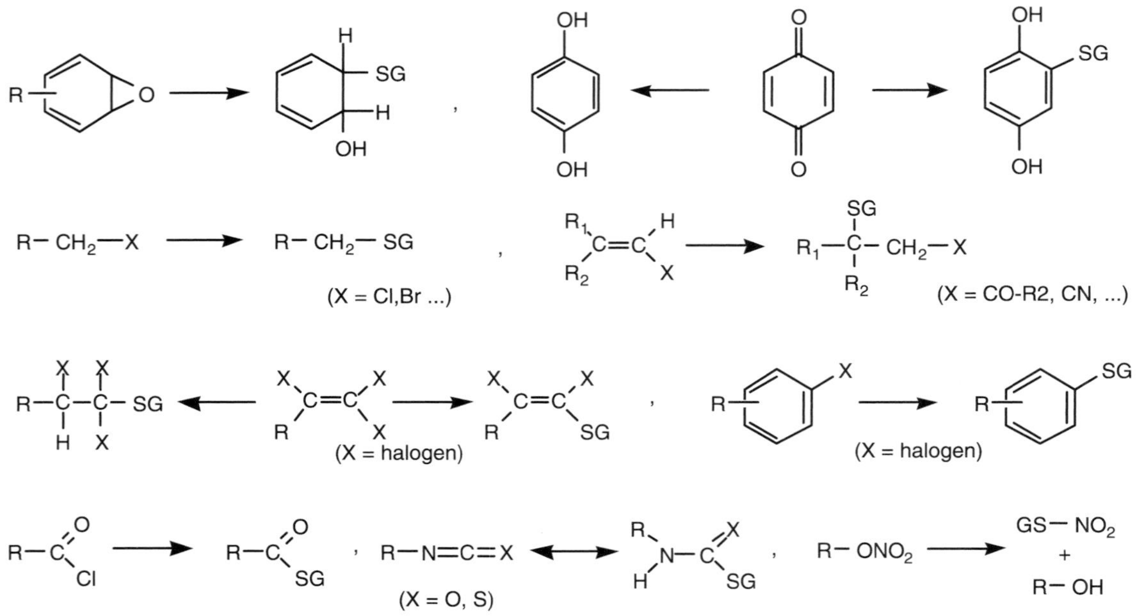

Fig. 31.44 Major reactions of conjugation with glutathione.

Organic nitrate esters such as nitroglycerine are vaso-dilators whose action results from their reduction to nitric oxide (NO). Glutathione and other thiols play an important role in this activation, a thionitrate being formed in the first step. *N*-Reduction may then proceed by various routes having nitrite (NO_2^-) or *S*-nitrosoglutathione (GS-NO) as an intermediate.

The diversity of all these reactions can be explained by the three-dimensional organization of the GST into a two-domain structure. The first domain contains the conserved active site responsible for catalysis, whereas the variable second domain, which interacts with the xenobiotic substrate of the reaction, provides substrate specificity and versatility.

H   Other reactions of conjugation

A number of other routes of xenobiotic conjugation have been reported, but their importance is restricted to a few exogenous substrates. While phosphorylation is of great significance in the processing of endogenous compounds and macromolecules, relatively few xenobiotics form phosphate esters. The enzymes involved are various phosphotransferases. Thus, a number of antiviral nucleoside analogues yield the mono-, di- and triphosphates *in vitro* and *in vivo*, e.g. zidovudine (AZT). However the pharmacological efficacy of the drug is greatly impaired through extensive glucuronidation on the terminal 5'hydroxy mediated mainly by the UGT2B7 isoform in human.

The reaction of hydrazines with endogenous carbonyls occurs nonenzymatically and involves a variety of carbonyl compounds, namely aldehydes (mainly acetaldehyde) and ketones (e.g. acetone, pyruvic acid and α-ketoglutaric acid). The products thus formed are hydrazones which may be excreted as such or undergo further transformation.

IV   CONCLUSION

The drug discovery process integrates early in the development of lead compounds, the data accumulated on the biotransformation of drugs and on the reaction mechanisms involved. Such information is essential for predicting the chemical pathways and the possible formation of active/reactive metabolites. It allows the selection of the drug candidates among thousands of compounds generated through combinatorial chemistry. The use of new *in vitro* models for drug biotransform-ation with recombinant human drug metabolizing enzymes or 'humanized animals' and the development of powerful molecular modelling methods are invaluable tools to predict the *in vivo* human situation better and avoid animal/human interspecies differences. Finally, the introduction of powerful techniques (X-ray structure resolution, protein and genetic engineering), the combination of sophisticated approaches (genomics, proteomics and bioinformatics) used in the framework of the Human Genome Project will contribute to remarkable advances in identifying novel therapeutic targets and new reaction mechanisms.

REFERENCES

1. Silverman, R.B. (1992) *The Organic Chemistry of Drug Design and Drug Action*. Academic Press, San Diego.
2. Testa, B. and Jenner, P. (1976) *Drug Metabolism. Chemical and Biochemical Aspects*. Marcel Dekker, New York.
3. Stenlake, J.B. (1979) *Foundations of Molecular Pharmacology, The Chemical Basis of Drug Action*, **2**. The Athlone Press-University of London, London.
4. Yamada, T., Morisseau, C., Maxwell, J.E., Argiriadi, M.A., Christianson, D.W. and Hammock, B.D. (2000) Biochemical evidence for the involvement of tyrosine in epoxide activation during the catalytic cycle of epoxide hydrolase. *J. Biol. Chem.* **275**: 23082–23088.
5. Nebert, D.W. (1994) Drug-metabolizing enzymes in ligand-modulated transcription. *Biochem. Pharmacol.* **47**: 25–37.
6. Jaggi, R., Addison, R.S., Suther, B.D. and Dickinson, R.G. (2002) Conjugation of desmethylnaproxen in the rat, a novel acylglucuronide-sulfate diconjugate as a major biliary metabolite. *Drug Metab. Dispos.* **30**: 161–166.
7. Lau, E.Y. and Bruice, T.C. (2000) Comparison of the dynamics for ground-state and transition-state structures of the active site of catechol-*O*-methyltransferase. *J. Am. Chem. Soc.* **122**: 7165–7171.
8. Männisto, P.T. and Kaakkola, S. (1999) Catechol-*O*-methyltransferase (COMT): biochemistry, molecular biology, pharmacology, and clinical efficacy of the new selective COMT inhibitors. *Pharmacol. Rev.* **51**: 593–628.
9. Rodrigues-Lima, F., Deloménie, C., Goodfellow, G.H., Grant, D.M. and Dupret, J.M. (2001) Homology modelling and structural analysis of human arylamine *N*-acetyltransferase NAT1: evidence for the conservation of a cysteine protease catalytic domain and an active loop. *Biochem. J.* **356**: 327–334.
10. Vessey, D.A. and Kelley, M. (2001) Characterization of the reaction mechanism for the XL-I form of bovine liver xenobiotic/medium chain fatty acid:CoA ligase. *Biochem. J.* **357**: 283–288.
11. Ouzzine, M., Magdalou, J., Burchell, B. and Fournel-Gigleux, S. (1999) An internal signal sequence mediates the targeting and retention of the human UDP-glucuronosyltransferase 1A6 to the endoplasmic reticulum. *J. Biol. Chem.* **274**: 31401–31409.
12. Ritter, J.K. (2000) Roles of glucuronidation and UDP-glucuronosyltransferases in xenobiotic bioactivation reactions. *Chem. Biol. Interactions* **129**: 171–193.
13. Georges, H., Jarecki, I., Netter, P., Magdalou, J. and Lapicque, F. (1999) Glycation of human serum albumin by acylglucuronides of nonsteroidal anti-inflammatory drugs of the series of phenylpropionates. *Life Sci.* **65**: 151–156.
14. Ouzzine, M., Antonio, L., Burchell, B., Netter, P., Fournel-Gigleux, S. and Magdalou, J. (2001) Importance of histidine residues for the function of the human liver UDP-glucuronosyltransferase UGT1A6: evidence for the catalytic role of histidine 370. *Mol. Pharm.* **58**: 1609–1615.
15. Said, M., Battaglia, E., Elass, A., Cano, V., Ziegler, J.C., *et al.* (1998) Mechanism of inhibition of rat liver bilirubin UDP-glucuronosyltransferase by triphenylalkyl derivatives. *J. Biochem. Mol. Toxicol.* **12**: 19–27.
16. Battaglia, E., Elass, A., Drake, R.R., Paul, P., Treat, S., *et al.* (1995) Characterization of a new class of inhibitors of the recombinant human liver UDP-glucuronosyltransferase, UGT1A6. *Biochim. Biophys. Acta* **1243**: 9–14.
17. Coughtrie, M.W., Sharp, S., Maxwell, K. and Innes, N.P. (1998) Biology and function of the reversible sulfation pathway catalysed by human sulfotransferases and sulfatases. *Chem. Biol. Interactions* **109**: 3–27.
18. Glatt, H. (2000) Sulfotransferases in the bioactivation of xenobiotics. *Chem. Biol. Interactions* **129**: 141–170.
19. Negishi, M., Pedersen, L.G., Petrotchenko, E., Shevtsov, S., Gorokhov, A., Kakuta, Y. and Pedersen, L.C. (2001) Structure and function of sulfotransferases. *Arch. Biochem. Biophys.* **390**: 149–157.
20. Duffel, M.W., Marshal, A.D., McPhie, P., Sharma, V. and Jakoby, W.B. (2001) Enzymatic aspects of the phenol (aryl) sulfotransferases. *Drug Metabab. Rev.* **33**: 369–395.
21. Dajani, R., Cleasby, A., Neu, M., Wonacott, A.J., Jhoti, H., *et al.* (1999) X-ray crystal structure of human dopamine sulfotransferase. SULT1A3. Molecular modeling and quantitative structure–activity relationship analysis demonstrate a molecular basis for sulfotransferase substrate specificity. *J. Biol. Chem.* **274**: 37862–37868.
22. Armstrong, R.N. (1998) Mechanistic imperatives for the evolution of glutathione transferase. *Curr. Opin. Chem. Biol.* **2**: 618–623.

32

CHEMICAL MECHANISMS OF TOXICITY: BASIC KNOWLEDGE FOR DESIGNING SAFER DRUGS

Anne-Christine Macherey and Patrick M. Dansette

La matière demeure et la forme se perd.
The matter remains and the form is lost.   **Ronsard**

I INTRODUCTION

Toxicity is the result of the more or less harmful action of chemicals on a living organism. Toxicology, the study of toxicity, is situated at the border of chemistry, biology and in some cases, physics. Molecular toxicology tries to elucidate the mechanisms by which chemicals exert their toxic effects. Because many foreign chemicals enter the body in inert but unexcretable forms, biotransformations are an important aspect of the fate of xenobiotics.[1,2] In the case of drugs, metabolic conversions may be required for therapeutic effect (prodrugs). In other cases, metabolism results in a loss of the biological activity. Sometimes, biotransformations produce

toxic metabolites. The last process is called toxication or bioactivation. It should be emphasized that the general principles of pharmacology embrace the occurrence of toxic events: although biotransformation processes are often referred to as detoxication, the metabolic products are, in a number of cases, more toxic than the parent compounds. For drugs, whether biotransformations lead to the formation of toxic metabolites or to variations in therapeutic effects depends on intrinsic (such as the genetic polymorphism of some metabolism pathways) and extrinsic (such as the dose, the route or the duration) factors. The biochemical conversions are usually of an enzymatic nature and yield reactive intermediates which may be implicated in the toxicity as far as the final metabolites. The primary events which constitute the beginning of the toxic effect may result, after metabolism, from an inhibition of a specific (and in most cases enzymatic) cellular function, an alkylating attack or an oxidative stress.

With regard to the toxicity arising from metabolites (indirect toxicity), three cases may be distinguished (Fig. 32.1):

(A) Biotransformation begins with the transient formation of a reactive intermediate, whose lifetime is long enough to allow an attack on cellular components. This occurs when a reactive intermediate (such as a radical or a carbenium ion) is formed and reacts rapidly with nucleophilic functions in cellular macromolecules (such as unsaturated lipids, proteins, nucleic acids), thus leading to their degradation and finally to cellular necrosis.

(B) The first step of the metabolic process yields a primary metabolite which can, in some cases, accumulate in

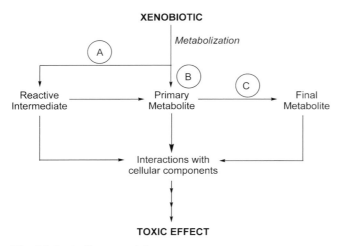

Fig. 32.1   Indirect toxicity.

the cell and reacts with cellular components before being transformed.

(C)   The final metabolites, when in excess, may accumulate and react with cellular macromolecules.

Usually, metabolic conversions are divided into two major types of reactions. Phase I reactions, or functionalization reactions, involve the introduction of a polar functionality such as a hydroxyl group into the xenobiotic structure. During phase II reactions, this group is subsequently coupled (or conjugated) with an endogenous cofactor which contains a functional group that is usually ionized at physiological pH. This ionic functional group facilitates active excretion into the urinary and/or hepatobiliary system.

Because bioactivation is mainly an activation of xenobiotics to electrophilic forms which are entities capable of reacting irreversibly with tissue nucleophiles, biotransformations leading to toxic metabolites are in most cases phase I reactions. However, phase II reactions may also give rise to toxic phenomena, e.g. when conjugation produces a toxic metabolite, or when it is responsible for a specific target organ toxicity by acting as delivery form to particular sites in the body where it is hydrolysed and exerts a localized effect. Also, the final toxic metabolite may be formed by combinations of several phase I and phase II reactions.

II   REACTIONS INVOLVED IN THE BIOACTIVATION PROCESSES

During the biotransformations affecting xenobiotics, four major kinds of chemical reactions may occur: oxidations (by far the most important), reductions, substitutions and eliminations. As phase I and II reactions are parts of this classification, these four classes of reactions can to give rise to toxic metabolites.

A   Oxidation

Several enzymatic systems are involved during the oxidative transformations of xenobiotics. Whether substances acted upon one enzyme rather than another depends not only on its specific function, but also on the electromolecular environment. The most important is the microsomal drug metabolizing system, known as cytochrome P-450 monooxygenase, which is localized mainly in the liver and which is involved in virtually all biological oxidations of xenobiotics.[3] Those include *C*-, *N*- and *S*-oxidations, *N*-, *O*- and *S*-dealkylation, deaminations and certain dehalogenations. Under anaerobic conditions, it can also catalyse reductive reactions. The cytochrome P-450 monooxygenase system is a multienzymatic complex constituted by the cytochrome P-450 haemoprotein, the flavoprotein enzyme NADPH cytochrome P-450 reductase, and the unsaturated phospholipid phosphatidylcholine. The catalytic mechanism of cytochrome P-450 involves a formal $(FeO)^{3+}$ complex formed by the elimination of H_2O from the iron site after two electrons have been added (Fig. 32.2).

Another oxidative enzyme is the FAD-containing monooxygenase, which is capable of oxidizing nucleophilic nitrogen, sulfur and organophosphorus compounds. The flavoprotein binds NADPH, oxygen and then the substrate. The oxidized metabolite is released, followed by NADPH. Alcohol dehydrogenase and aldehyde dehydrogenase catalyse the oxidation of a variety of alcohols and aldehydes into aldehydes and acids in the liver. Xanthine oxidase oxidizes several purine derivatives, such as theophylline. The monoamine oxidase (MAO) and diamine oxidase convert amines into alkyl or aryl aldehydes by means of

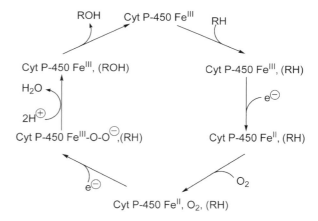

Fig. 32.2   Catalytic cycle of cytochrome P-450 monooxygenase.

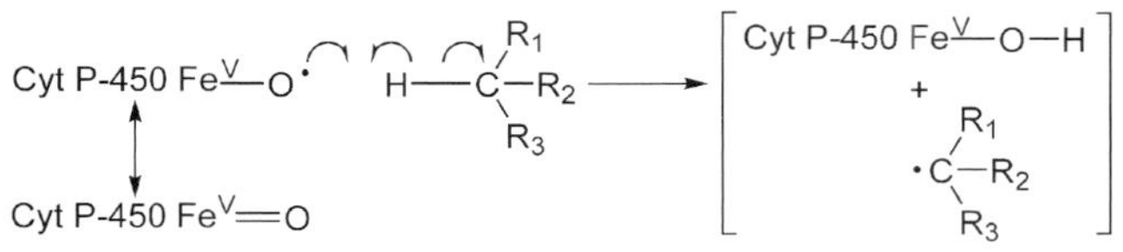

Fig. 32.3 Cytochrome P-450 oxidation process.

the abstraction of two hydrogen atoms, one from the nitrogen and the other from the α carbon, and a subsequent hydrolysis. Peroxidases are oxidative enzymes which couple the reduction of hydrogen peroxide and lipid hydroperoxides to the oxidation of other substrates. This co-oxidation is responsible for the production of reactive electrophiles from aromatic amines (e.g. the highly carcinogenic benzidine), phenols, hydroquinones, polycyclic aromatic hydrocarbons, etc.

The oxidation reactions can then be described in terms of a rather common chemistry that involves the abstraction of either a hydrogen atom or a nonbonded (or π) electron by the iron–oxo porphyrin complex (Fig. 32.3). The high-valent complex electronic configuration is unknown, but is usually written as $Fe^V = 0$. The one-electron oxidation yields transient radicals (Fig. 32.4) which are transformed into more stable forms. They can incorporate an oxygen atom by a radical recombination with dioxygen. This yields an oxidized derivative that may sometimes be more toxic than the parent compound or susceptible to further metabolic conversions. Free radicals may also bind to the site of their formation, thus leading to inhibition or inactivation of the enzyme. When the radical is not efficiently controlled by the iron, it may leave the active site. The subsequent 'free' radical is able to produce damage to unsaturated fatty acids, thus leading to lipid peroxidation and destruction of the cellular structure. Another mode of the radical stabilization is a second one-electron oxidation, which consists of the loss of another electron. The fate of free radicals is now extensively studied because of their great capacities for forming covalent bonds with cellular macromolecules.[4−6]

C–H bond oxidations

These oxidations, which are usually catalysed by cytochrome P-450 monooxygenases, produce hydroxylated

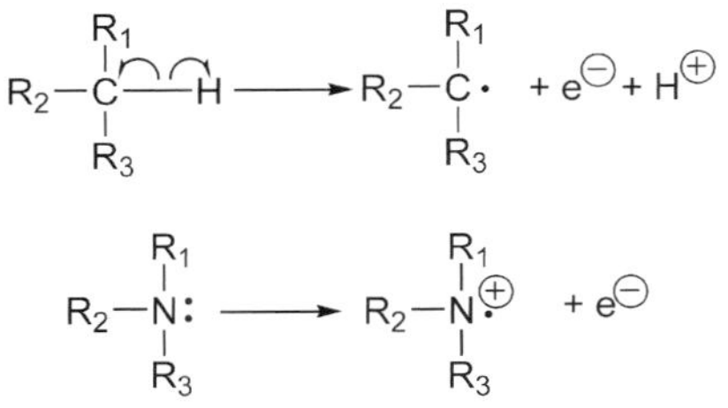

Fig. 32.4 One-electron oxidation.

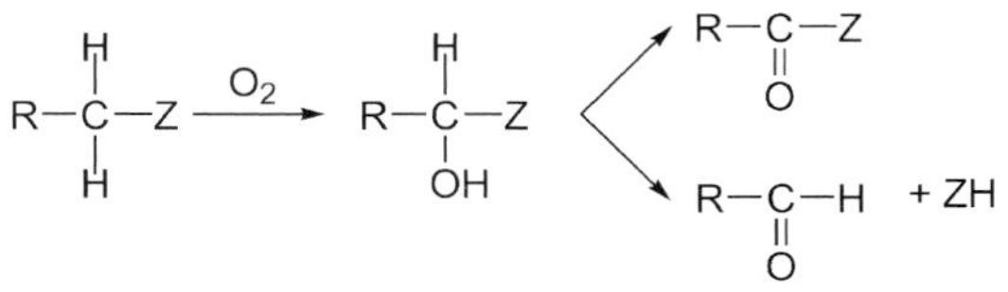

Fig. 32.5 C–H bond oxidation in the α-position to a heteroatom.

derivatives.[7] When the C–H bond is located in the α position to a heteroatom (such as O, S, N, halogen), the α hydroxylated derivative obtained is often unstable and may be further oxidized or cleaved (Fig. 32.5).

The antibiotic chloramphenicol is oxidized by cytochrome P-450 monooxygenase to chloramphenicol oxamyl chloride formed by the oxidative dechlorination of the dichloromethyl moiety of chloramphenicol.[8] The reactive metabolite binds to the ε-amino group of a lysine residue in the cytochrome P-450 (Fig. 32.6). This yields an adduct that blocks the electron transport from NADPH cytochrome P-450 reductase.[9] This type of mechanism is termed a suicide-substrate mechanism.

In the case of chloroform, the unstable trichloromethanol loses hydrochloric acid and forms phosgene, which is very reactive (Fig. 32.7).[10]

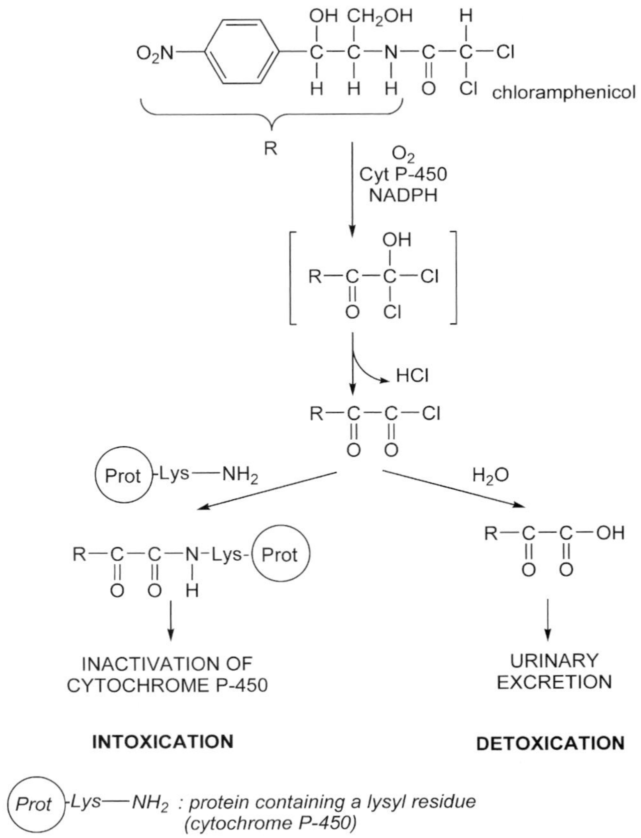

Fig. 32.6 Metabolic activation of chloramphenicol.

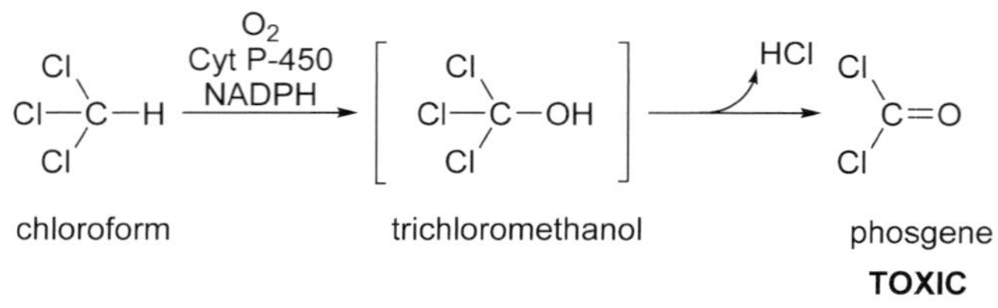

Fig. 32.7 Oxidation of chloroform.

Tertiary amines containing at least one hydrogen on the α carbon may either be *N*-oxidized (leading to an *N*-oxide in the case of tertiary amines), or *C*-oxidized, thus leading to a carbinolamine. The latter, often being unstable, usually splits into a secondary amine and an aldehyde moiety (Fig. 32.8). Several electron transfer mechanisms are proposed.[3]

During the oxidation of nitrosamines, the hydroxylated derivative formed cleaves spontaneously into highly reactive metabolites capable of alkylating nucleophilic sites in the cellular components.

Unsaturated bond oxidations

Double bonds are oxidized by cytochrome P-450 mono-oxygenases into epoxides, which are generally very reactive. Epoxides are considered responsible for the toxicity of the unsaturated compounds.

The hepatocarcinogenicity of aflatoxin B_1 (AFB$_1$) is known to be due to the epoxide (AFB$_1$-oxide) formed, which binds directly with the N-7 atom of a guanine molecule in DNA (Fig. 32.9).[11]

Aromatic chemicals are metabolized into unstable arene-oxides, which, as epoxides, are comparable to potentially equivalent electrophilic carbocations. These metabolites react easily with thiol groups derived from proteins, leading, for example, to hepatotoxicity. Bromobenzene (Fig. 32.10) is oxidized into a 3-4 epoxide, which does not exhibit mutagenic or carcinogenic activity, but reacts nonenzymatically with liver proteins and produces hepatic necrosis. A secondary P450-catalysed oxidation to hydroquinone

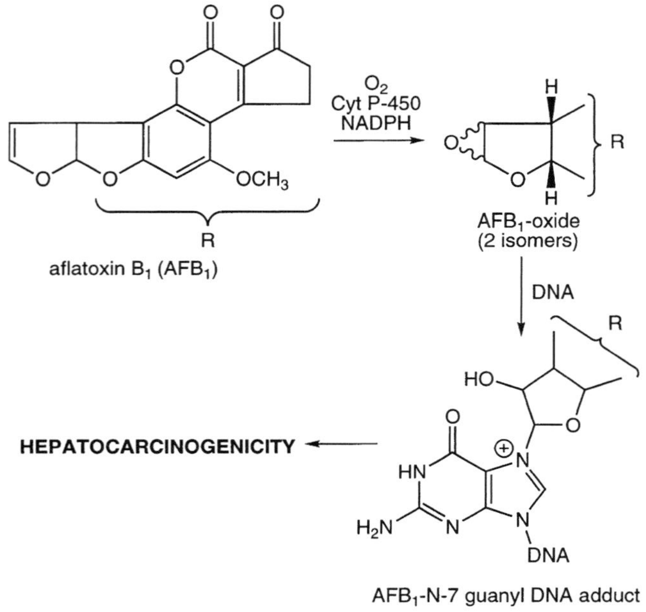

Fig. 32.9 Oxidation of aflatoxin B_1.

and benzoquinone is now evocated. In this alternative pathway, conjugation with glutathione can lead to the formation of products which may elicit their toxicity elsewhere from the liver and especially in the kidney.[12]

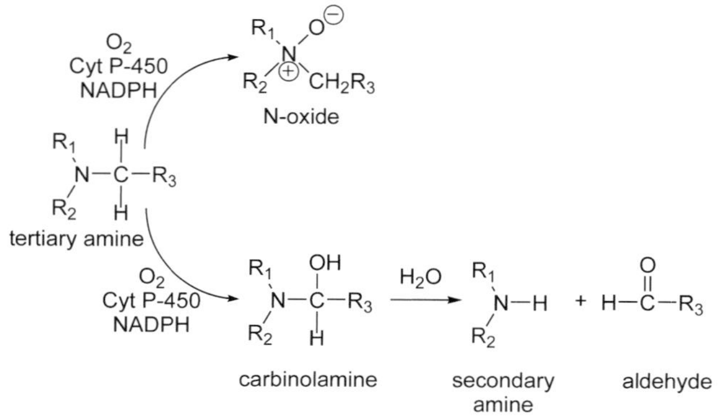

N-oxidations

Tertiary amines are transformed into *N*-oxides (generally less toxic), but primary and secondary amines are oxidized into hydroxylated derivatives (hydroxylamines). This

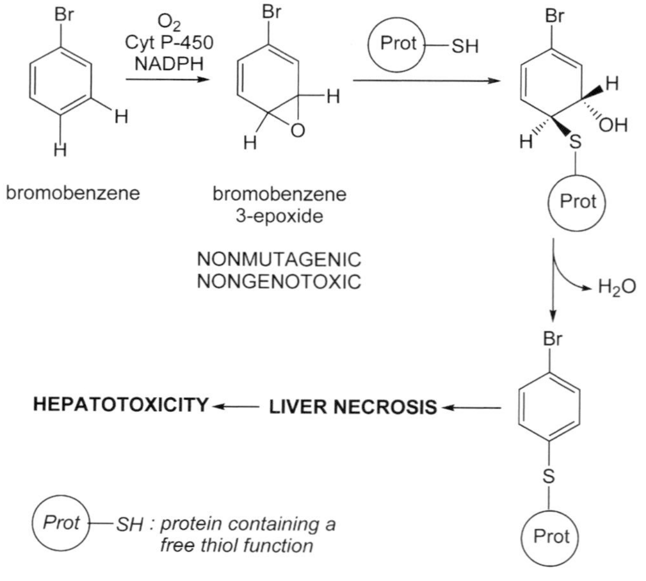

Fig. 32.8 Oxidation of a tertiary amine.

Fig. 32.10 Metabolism of bromobenzene.

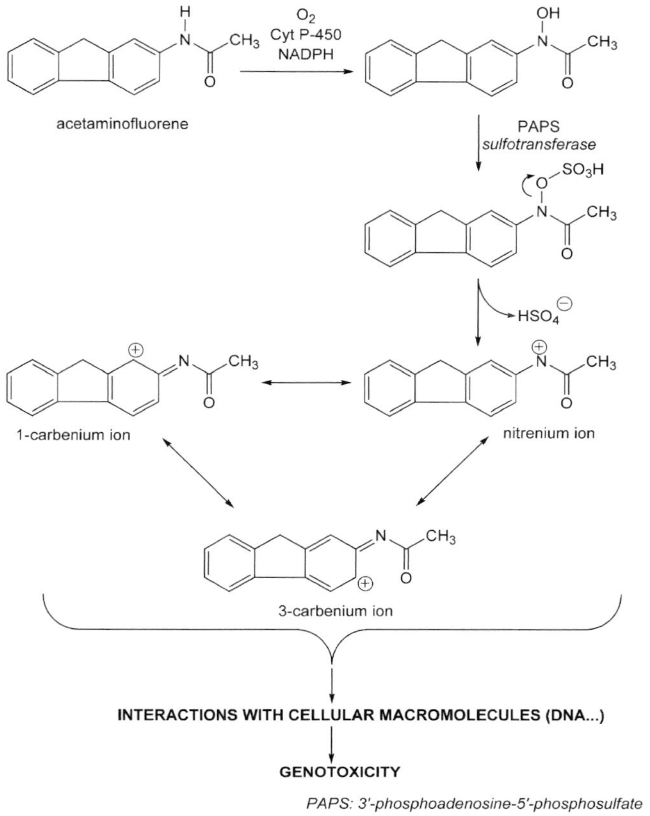

Fig. 32.11 *N*-oxidation of acetaminofluorene.

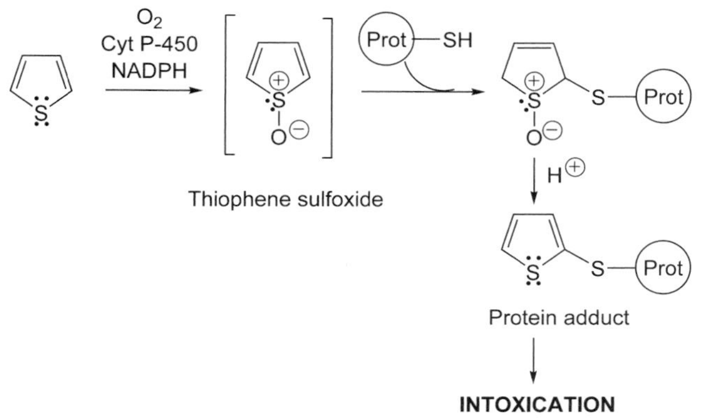

Fig. 32.12 Oxidation of thiophene.

oxidation is responsible for the hepatotoxicity of acetamino-2-fluorene (Fig. 32.11).[13]

Nitrenium ions may occur during bioactivation of aromatic amines and amides, which are usually *N*-oxidized into *N*-hydroxylated derivatives. By direct elimination or esterification followed by elimination, the latter may be transformed into highly reactive nitrenium ions. Nitrenium ions are of great importance because of the equilibrium with their mesomeric carbocationic forms and the subsequent reactivity with cellular nucleophilic macromolecules (nucleic acids, etc.).

Heteroatom oxidations

Heteroatoms such as nitrogen or sulfur are oxidized on their free peripheric electrons (Fig. 32.12) as described for thiophene.[14] Halogenated aromatic compounds, may also be oxidized by cytochrome P-450 monooxygenases, yielding hypervalent halogenated compounds.

B Oxidative stress

Oxidative stress has been defined as a disturbance in the pro-oxidant–antioxidant balance in favour of the pro-oxidant state resulting from alterations in the redox state of the cell.

The stepwise reduction of oxygen into superoxide anion, hydrogen peroxide, hydroxyl radical and finally water, which accounts for about 5% of the normal oxygen reduction (versus 95% by means of the mitochondrial electron-transport chain), may be increased by the redox cycling of some xenobiotics such as quinones or nitro-aromatic derivatives. These compounds are susceptible to one-electron reduction which yields radical structures which may be back-oxidized to the parent compound. During this reoxidation, oxygen is reduced into superoxide anion. The oxygen reduction products are highly reactive entities that attack all the cellular components, especially when their normal degradation systems (superoxide dismutase, glutathione peroxidase, catalase) are overburdened. The poly-unsaturated lipids are especially sensitive to these attacks because they are susceptible to a membrane-degrading peroxidation.

C Reduction

Reductive biotransformations of several compounds such as polyhalogenated, keto, nitro and azo derivatives, are catalysed by a variety of enzymes which differ according to the substrates and the species. The liver cytochrome P-450-dependent drug metabolizing system is capable of reducing *N*-oxide, nitro and azo bonds, whereas the cytosolic nitrobenzene reductase activity is mainly due to cytochrome P-450 reductase, which transforms nitrobenzene into its hydroxylamino derivative. NADPH cytochrome *c* reductase is also able to catalyse the reduction of nitro compounds. These metabolic conversions may also be brought about by gastrointestinal anaerobic bacteria.

Reductive processes that occur during the metabolism of xenobiotics involve either one-electron reduction or a two-electron transfer. Ionic reduction using a hydride occurs *in vivo* during the reduction catalysed by NADH or NADPH enzyme, whereas one-electron reduction releases a radical structure which may contribute to the toxic effect.

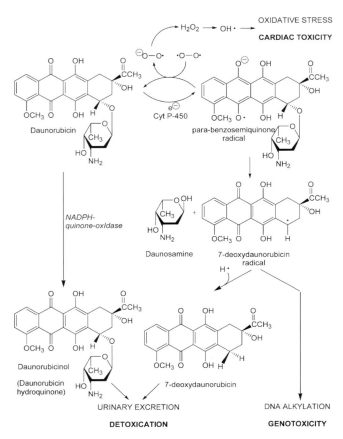

Fig. 32.13 Biotransformations of daunorubicin.

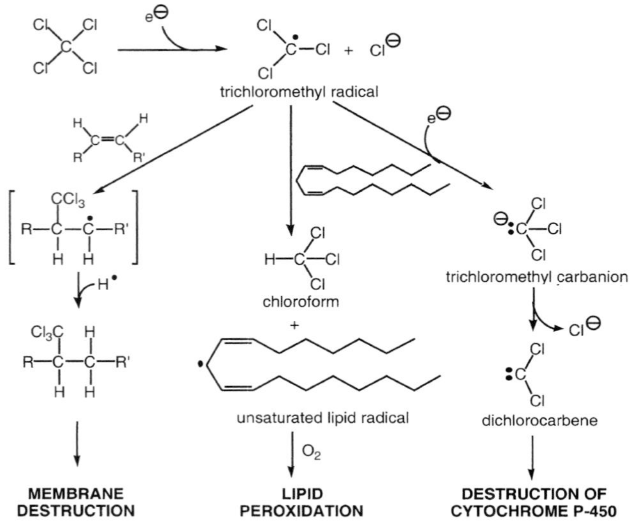

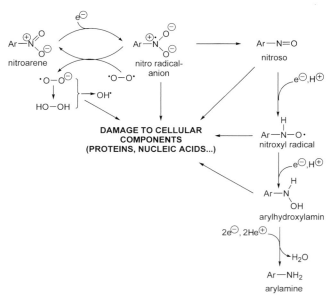

Fig. 32.14 Reduction of polyhalogenated compounds.

Figure 32.13 illustrates the biotransformations affecting the anthracycline antitumour drug daunomycin.[15] Recent studies suggest that nitric oxide synthases may contribute to the cardiotoxicity, probably because of their structural similarities with P450 reductase.[16]

Polyhalogenated compound reduction

Some polyhalogenated compounds such as CCl_4, $BrCCl_3$, halothane (CF_3—$CHBrCl$), when in the presence of the reduced form of cytochrome P-450, may undergo a reductolysis[17,18] (Fig. 32.14), which leads to a radical that may be transformed by different pathways.

The radical formed may add directly on the unsaturated lipid bonds or initiate an unsaturated lipid peroxidation or undergo another one-electron reduction. The last reaction yields a carbene that can form a carbenic complex with the iron of the reductive form of cytochrome P-450. Polyhalogenated compound reductions give rise to several reactive intermediates: radicals, carbenes and peroxides, whose participation in the toxic effect varies greatly.

Nitro compound reduction

The different steps of the biotransformations that produce a primary amine from an aromatic nitro compound involve the nitro radical-anion, the nitroso derivative, the nitroxyl radical, the hydroxylamine and then the primary amine (Fig. 32.15). Each of these different intermediates may contribute to the toxicity. Hydroxylamines are often responsible for methaemoglobinaemia, whereas mutagenic and carcinogenic activity may be due to the combination of nitro radical-anion, nitroso derivatives or esterified hydroxylamine (such as sulfate derivatives) with cellular macromolecules.

Carcinogenicity may also be the result of the oxidative stress subsequent to the formation of oxygen-reduction

Fig. 32.15 Reductive biotransformation of nitro arene compounds.

products (superoxide anion, hydrogen peroxide, hydroxyl radical) during the redox cycling of the nitro radical-anion, which restores the parent nitro compound.

Azo compound reduction

Azo compounds are susceptible to reduction, first to hydrazo intermediates which are reductively cleaved into the appropriate amines. It has been proposed that the first step, as with nitro compounds, is the formation of an azo-anion radical.

D   Substitutions: hydrolysis and conjugation

Among substitution reactions, ester and amide hydrolyses are of common occurrence, and often operate during detoxication processes. In addition to specific enzymes, the stomach and the kidney are areas where acid-catalysed hydrolyses may occur, whereas base-catalysed reactions may be assisted by the alkaline pH of intestine.

Phase II, or conjugation reactions are also substitution reactions, which proceed by means of an endogenous and generally activated nucleophile. In mammals, six major conjugation reactions of xenobiotics exist and are mediated by transferase enzymes. The specificity for the endogenous agent is high, but the specificity for the xenobiotic is broader. To a great extent, conjugation produces excretable and nontoxic metabolites and thus is referred to as detoxication, but exceptions exist in each class of conjugation reaction.

Glucuronic acid conjugation

This substitution involves the transfer of a glucuronic acid from uridine diphosphate glucuronic acid (UDPGA) to a functional group in the xenobiotic substrate. The group may be a hydroxyl, carboxylic acid, amino or sulfur functions. Glucuronides are never directly implicated in toxicity, but are sometimes responsible for target-organ toxicity. Aromatic amines may be converted in the liver into *N*-glucuronides, which are excreted in the urine and broken down in the bladder (because of the acidic pH) to liberate the proximate hydroxylamine carcinogen.

Sulfation

Sulfate conjugation gives a polar and ionized conjugate by means of the esterification of a hydroxyl group with sulfate ion (in the form of 3′-phosphoadenosine-5′-phosphosulfate or PAPS). The reaction is catalysed by a hydrosoluble sulfotransferase. Sulfation sometimes gives rise to reactive intermediates that may undergo further reactions to yield electrophilic metabolites. In the case of 2-acetaminofluorene, the O-sulfate moiety is a facile leaving group, and this cleavage produces nitrenium ions which act as alkylating agents for DNA (Fig. 32.11).

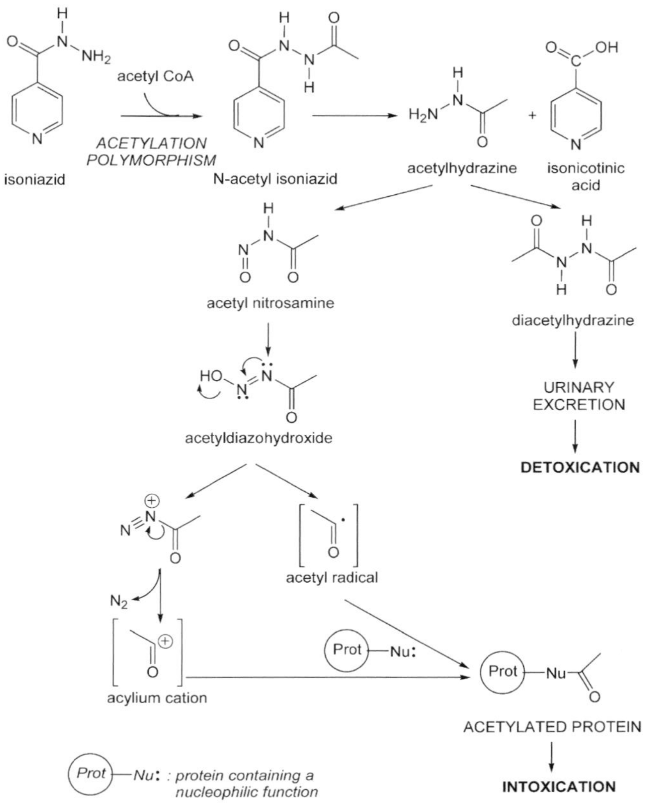

Fig. 32.16   Bioactivation of isoniazid.

Acetylation

Acetylation is a very common metabolic reaction which occurs with amino, hydroxyl or sulfhydryl groups. The acetyl group is transferred from acetyl-coenzyme A and the reaction is catalysed by acetyltransferases. An important aspect of this kind of substitution is the genetic polymorphism of one acetyltransferase in humans, who are divided into fast and slow acetylators. In a few cases, the conjugates are further metabolized to toxic compounds, as is seen with isoniazid. Some evidence exists that acetylation of the antitubercular isoniazid leads to enhanced hepatotoxicity of the drug.[19,20] Acetylation followed by hydrolysis and cytochrome P-450-dependent oxidation yields free acetyl radicals[21] or acylium cations which may acetylate the nucleophilic macromolecule functions (Fig. 32.16).

Glutathione conjugation

Substitution reactions of xenobiotics with glutathione are the most important and contribute efficiently to detoxication. Nevertheless, in some cases such as vicinal dihalogenated compounds, glutathione conjugation produces monosubstituted derivatives which may cycle into a highly electrophilic sulfonium ion (Fig. 32.17).[22]

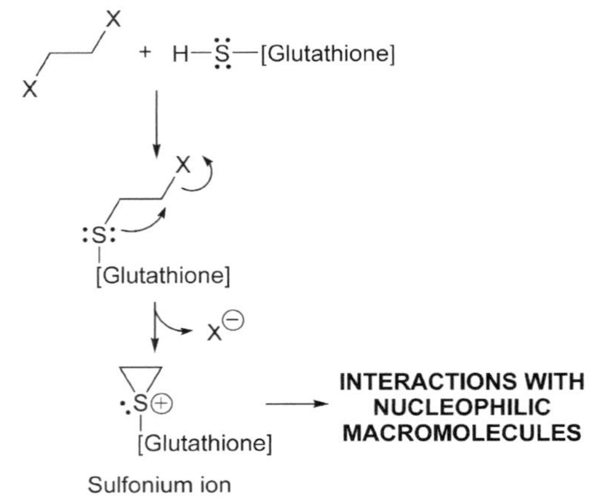

Fig. 32.17   Bioactivation to sulfonium ion.

Methylation

Methylation is rarely of quantitative importance in the metabolization of xenobiotics. The methyl group is transferred from the nucleotide *S*-adenosyl-L-methionine (SAM) by means of a methyl transferase. The functional groups include primary, secondary and tertiary amines, pyridines, phenols, catechols, thiophenols, etc. The azaheterocycle pyridine is metabolized to the *N*-methylpyridinium ion, which is more toxic than pyridine itself[23] (Fig. 32.18). The binding properties of the ionized metabolite are disturbed by the loss of its hydrophobic feature, resulting from the polarity inversion.

E   Eliminations

Eliminations of hydracide or halogen sometimes occur during the metabolism of halogenated xenobiotics and lead to an alkene. The double bond may be oxidized into an epoxide by means of oxidative enzyme systems as discussed above. Dehydrogenation, dehydrochloration and dechloration are (with oxidation) the different metabolic pathways of the γ-isomer of the insecticide hexachlorocyclohexane (lindane).

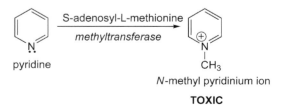

Fig. 32.18   Bioactivation of pyridine.

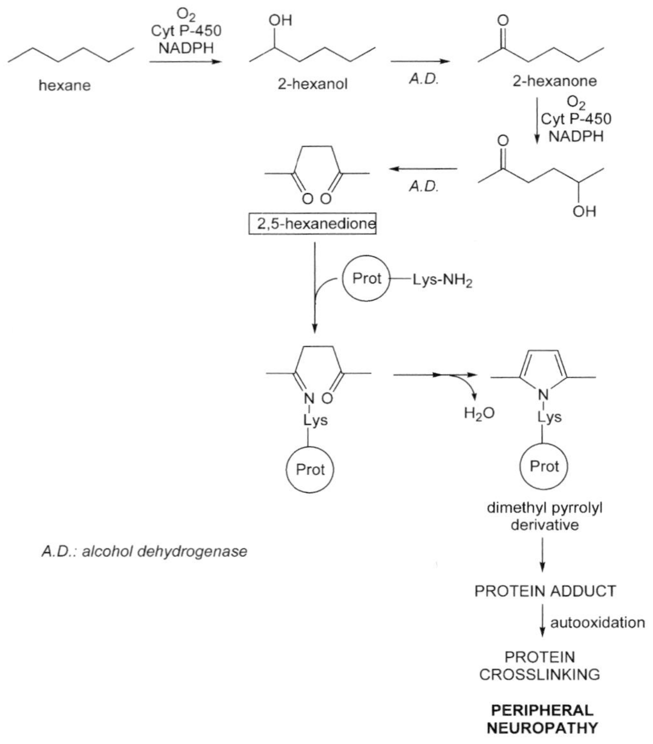

Fig. 32.19   Bioactivation of hexane.

F   Further biotransformations leading to the ultimate toxicant

Other reactions must be mentioned beside the major reactions described above. These reactions may be responsible for the transformation of a toxic metabolite into the ultimate toxicant.[24] Rearrangements and cyclizations are examples of reactions involved in these processes. In the case of the solvent hexane (Fig. 32.19), the toxic metabolite 2,5-hexanedione is formed by four successive oxidations of the molecule. The condensation of the γ-dicetone with the lysyl amino group of a neurofilament protein is followed by a cyclization according to a Paal–Knorr-type reaction. This is the initial process that explains the hexane-induced neurotoxicity.[25] A further auto-oxidation of the *N*-pyrrolyl derivatives leads to the cross-linking of the axonal intermediate filament proteins and the subsequent occurrence of the peripheral neurotoxicity.[26]

III   EXAMPLES OF METABOLIC CONVERSIONS LEADING TO TOXIC METABOLITES

The formation of toxic metabolites and/or intermediates during the metabolization of drugs may occur by a considerable variety of pathways which are mediated by several enzyme systems. The following five examples do not represent an exhaustive list of the bioactivation processes, but are samples of original, significant and/or well-known drugs whose biotransformations lead to toxic compounds by the four main types of reactions discussed above. Two of them (acetaminophen, tienilic acid) are cytochrome P-450-mediated oxidations. Halothane acts through both oxidative and reductive biotransformations. Valproic acid is toxic through its elimination product. The toxicity of troglitazone seems to involve two distinct metabolic pathways, leading to both alkylating and oxidative stress.

A   Acetaminophen

The analgesic acetaminophen (4-hydroxyacetanilide, paracetamol) exhibits hepatotoxicity when administered in very high doses (approximately 250 mg kg^{-1}).[27] The metabolite responsible is known to be the *N*-acetyl-*p*-benzo-quinoneimine (NAPQI) (Fig. 32.20).[28] The formation of NAPQI may proceed via the isoform 2E1 of cytochrome P-450 monooxygenase,[29] but also via peroxidases such as prostaglandine hydroperoxidase.

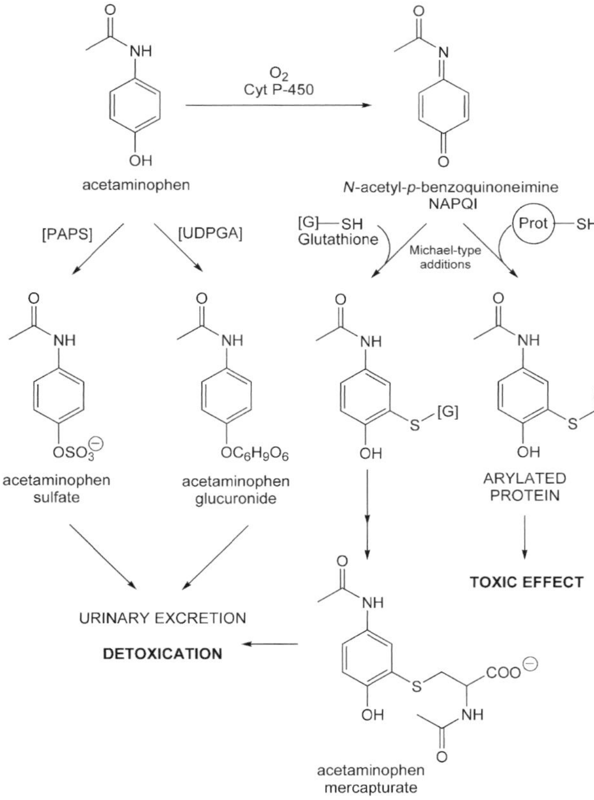

Fig. 32.20 Biotransformation pathway of acetaminophen.

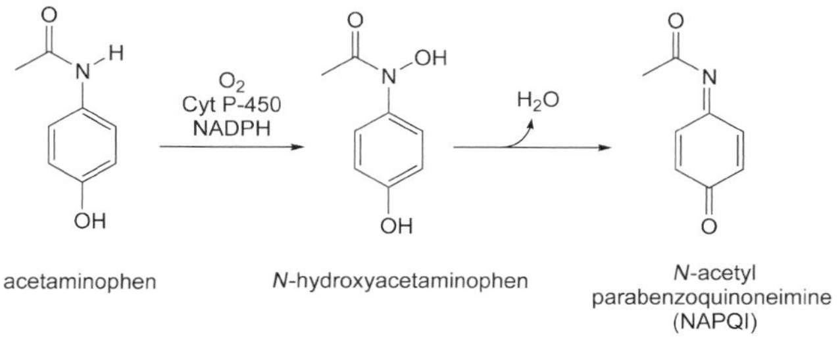

Fig. 32.21 Oxidation of acetaminophen according to the '*N*-hydroxyacetaminophen pathway'.

The most commonly described mechanism proposes that metabolic activation occurs through *N*-oxidation of acetaminophen to *N*-hydroxyacetaminophen followed by dehydratation to NAPQI (Fig. 32.21).[30]

However, it seems that *N*-hydroxyacetaminophen is not a major intermediate in the oxidation of acetaminophen. The formation of *N*-acetyl-*p*-benzo-quinoneimine probably proceeds by two successive one-electron oxidations[31] (Fig. 32.22).

During the first step, a one-electron oxidation yields a phenoxy radical (Ar-O$^{\bullet}$).[32] The presence of the radical was supported by fast flow ESR spectroscopy in the presence of horseradish peroxidase. In the second one-electron oxidation, the phenoxy radical is oxidized to NAPQI. As described in Fig. 32.20, the highly electrophilic NAPQI may easily react with glutathione or protein thiol groups according to a Michael-type addition. The attack of liver protein thiol groups and the subsequent adduct formation is frequently mentioned in the mechanism of acetaminophen hepatotoxicity. Such proteins were recently identified in mice: NAPQI forms covalent adducts with glyceraldehyde-3-phosphate dehydrogenase,[33] calreticulin and the thiol: protein disulfide reductases Q1 and Q5.[34]

Another hypothesis for the mechanism of toxicity is supported by the oxidative potency of NAPQI, but still suffers from lack of evidence.[35] NAPQI is a good oxidant for thiol functions of cellular components and pyridine nucleotides. Moreover, it may undergo a redox cycling with

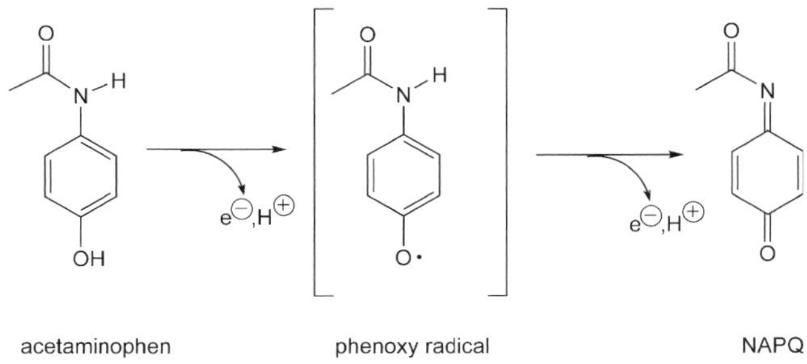

Fig. 32.22 Oxidation of acetaminophen by means of the phenoxy radical.

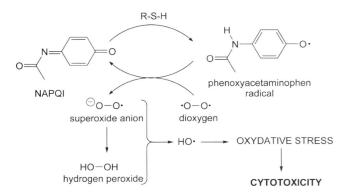

Fig. 32.23 Redox cycling of *N*-acetylparabenzoquinone imine.

formation of superoxide anions by means of an oxygen one-electron reduction (Fig. 32.23).

The stepwise reduction of oxygen produces hydrogen peroxide and finally a hydroxyl radical, which is a strong oxidant implicated in the cellular oxidative stress. This oxidative stress causes a glutathione depletion, a disruption of the cellular calcium regulation and modifications of cellular proteins, thus leading to cell death. Some biochemical parameters related to necrotic and apoptotic process are affected in acteaminophen-exposed PC12 cells transfected with CYP 2E1.[36,37]

It therefore appears that both covalent (e.g. alkylation) and non-covalent (e.g. oxidative stress) interactions play major roles in the pathogenesis of acute lethal cell injury caused by NAPQI.[38] At present, it is not possible to identify which of these two interactions is the critical event in initiating acetaminophen hepatotoxicity.

B   Tienilic acid

Tienilic acid is a uricosuric diuretic drug that may cause immunoallergic hepatitis in 1 in 10 000 patients, a side-effect that resulted in its withdrawal from circulation. The immunoallergic hepatitis was associated with the appearance of circulating anti-reticulum antibodies, called anti-LKM$_2$ antibodies, which are directed towards a liver endoplasmic reticulum protein.[39,40] From these observations, the mechanism of the immunotoxicity associated with the prolonged use of tienilic acid was elucidated by the Mansuy group.[41,42]

Tienilic acid is oxidized in the liver by the cytochrome P-450 monooxygenase to 5-hydroxytienilic acid, which is the major urinary metabolite (about 50% in humans). This oxidation occurs through an electrophilic intermediate capable of alkylating very specifically the cytochrome P-450.[43] This suicide-substrate inactivation is also observed with many xenobiotics such as alkenes with terminal

unsaturation, alkynes, strained cycloalkylamines, 4-alkyldi-hydropyridines, benzodioxoles and some tertiary amines. The irreversible binding of the compound with cytochrome P-450 leads to the appearance of antibodies against both the modified protein and its native form, and the subsequent destruction of hepatocytes.

In humans, the bioactivation of tienilic acid as its reactive intermediate depends on cytochrome P-450 2C9. This isoform is one of the major forms of cytochrome P-450 in the human liver. It has recently been demonstrated that in the presence of cytochrome P-450 thiophene is oxidized *in vivo* to yield thiophene sulfoxide (Fig. 32.12). This unusual function is a very electrophilic species capable of reacting with thiol group nucleophiles such as glutathione (detoxication) or proteins. This interaction with free proteins that contain thiol groups may give rise to an adduct and the potential associated toxicity.

In the case of tienilic acid, a sulfoxide is probably formed,[44] and this electrophile would be especially strong because of the mesomeric effect caused by the keto function. Addition of water to the sulfoxide, according to the Michael reaction, may occur at the activated position on the thiophene ring, thus yielding a 5-hydroxydihydrothiophene sulfoxide which, after dehydration, should give 5-hydroxy tienilic acid. Similarly, an amino acid nucleophilic function of the active site of cytochrome P-450 monooxygenase apoprotein may react with the same electrophilic centre, thus giving rise to an adduct between the activated tienilic acid and cytochrome P-450 2C9. Tienilic acid adducts on CYP 2C9 were recently identified directly by mass spectroscopy.[45]

The inactivation of cytochrome P-450 2C9 by covalent binding with the active metabolite of tienilic acid seems correlated with the appearance of anti-LKM$_2$ antibodies in patients showing an immunoallergic hepatitis (Fig. 32.24).

C   Halothane

Halothane is a widely used anaesthetic drug that occasionally results in severe hepatitis. About 60 to 80% of the dose is eliminated in unmetabolized form during the 24 hours following administration to patients. This compound is metabolized in the presence of cytochrome P-450 mono-oxygenase CYP 2E1 according to the two main pathways[7] depicted in Fig. 32.25.

The major biotransformation pathway involves an oxidative step with introduction of an oxygen atom and the subsequent formation of halohydrin. The unstable halohydrin loses hydrobromic acid to yield trifluoroacetyl chloride, which in turn is hydrolysed to trifluoroacetic acid. This final metabolite is found in the urine.[46]

In conditions of low levels of oxygen, a reductive pathway (10%) is enhanced and yields a free radical

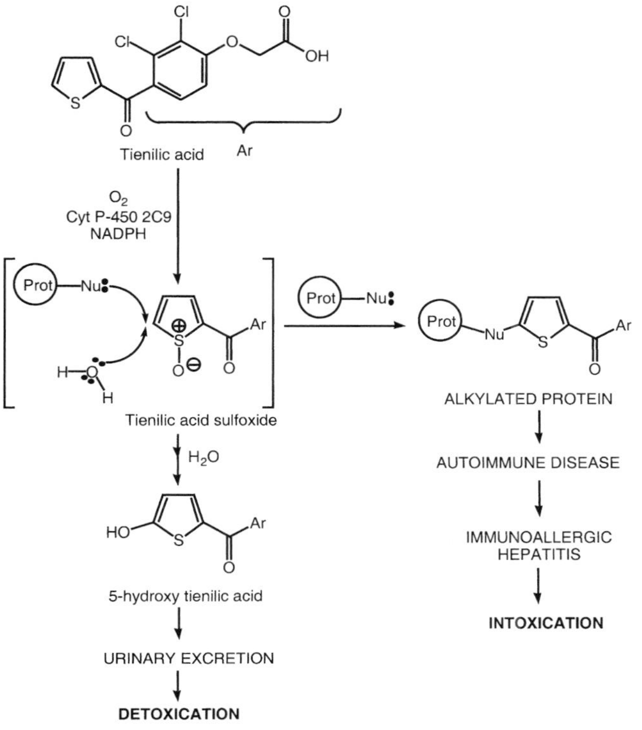

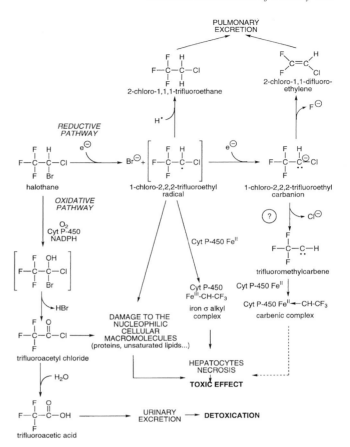

Fig. 32.24 Tienilic acid biotransformation to tienilic acid sulfoxide.

Fig. 32.25 The major metabolic pathways of halothane.

intermediate characterized as 1-chloro-2,2,2-trifluoroethyl radical. Another one-electron reduction produces the 1-chloro-2,2,2-trifluoroethyl carbanion, which may undergo two possible kinds of eliminations. One is the abstraction of a fluoride ion according to a E1Bc elimination, which yields 2-chloro-1,1-difluoroethylene. This metabolite is eliminated by exhalation. Early studies suggested that a second elimination process might be a α-elimination of a chloride ion, which produces trifluoromethylcarbene,[47] but this was later reconsidered.[48] It was hypothesized that a carbenic complex with the Fe[II] in the active site might lead to inactivation of the cytochrome P-450, but this inactivation is now thought to be due to the formation of an iron-σ-alkyl complex derived from the 1-chloro-2,2,2-trifluoroethyl radical.

The initially formed 1-chloro-2,2,2-trifluoroethyl radical may also cause a radical attack of the polyunsaturated lipids which produces 2-chloro-1,1,1-trifluoroethane. This mechanism is identical to the pathway described with the trichloromethyl radical formed during the one-electron reduction of carbon tetrachloride (Fig. 32.14). The trichloromethyl radical may initiate a peroxidation of the unsaturated lipids from the membrane and the subsequent liberation of chloroform.

Several studies have demonstrated that halothane hepatotoxicity is mainly due to an immune reaction towards modified proteins of the liver. In fact, these proteins are trifluoroacetylated on their ε-NH$_2$-lysyl residue by the trifluoroacetyl chloride formed during the oxidative metabolization of halothane.[49,50] The product of the reaction can act as a foreign epitope and the drug–protein conjugate, called neoantigen, elicits an immune response toward the liver[51] (Fig. 32.26).

A related fluorocarbon used in air conditioning systems, HCFC 1,2,3 is metabolized to the same acyl halide and was recently implicated in an epidemic of liver disease in nine workers of a Belgian factory.[52] All patients had serum antibodies to trifluoroacetylated proteins.

D Valproic acid

Valproic acid is an anticonvulsant agent used for the therapy of epilepsy, which occasionally results in hepatotoxicity in young children. The toxicity is characterized by mitochondrial damage, impairment of fatty acid β-oxidation and lipid accumulation.

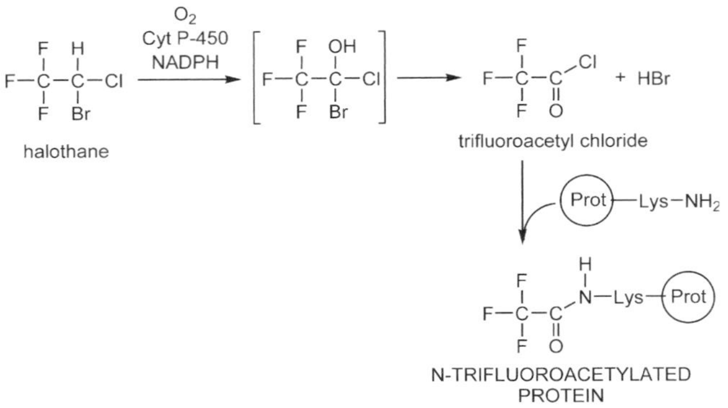

Fig. 32.26 Biotransformation of halothane to trifluoro-acetyl chloride and the subsequent binding to protein.

It has been proposed that hepatotoxicity is a consequence of the further biotransformation of the valproic acid metabolite 2-propyl-4-pentenoic acid (also called Δ^4 VPA).[53]

As depicted in Fig. 32.27, Δ^4 VPA is not formed by dehydration of 4- and 5-hydroxy valproic acids, which are, with the glucuronide conjugate, the major metabolites of valproic acid.[54] The mechanism is proposed to involve an initial hydrogen abstraction to generate a transient free radical intermediate. It has been demonstrated that the carbon-centred radical was localized at the C4 position. The radical undergoes both recombination (which yields 4-hydroxy valproic acid) and elimination (which produces the unsaturated derivative Δ^4 VPA). The formation of these metabolites is catalysed by the cytochrome P-450 mixed-function oxidase CYP 4B1.[55] Δ^4 VPA is a hepatotoxic and strong teratogenic compound in animal models. In addition to this metabolic pathway, valproic acid undergoes

biotransformation leading to (E)-Δ^2-VPA which is devoided of embryotoxic effect in rodents.[56]

Further biotransformations of Δ^4 VPA involve both the liver microsomal cytochrome P-450 enzymes and the fatty acid β-oxidation pathway (Fig. 32.28). The mixed-function oxidase system metabolizes the unsaturated metabolite to a γ-butyrolactone[57] derivative through a chemically reactive entity that is a suicide-substrate inhibitor of cytochrome P-450. The alkylation of the prosthetic haem by means of the radical occurs prior to the formation of the epoxide.[58] Thus the epoxide is not involved in the cytochrome P-450 inhibition.

The β-oxidation cycle activates Δ^4 VPA to its coenzyme A derivative and, through sequential steps of β-oxidation, yields 3-oxo 2-propyl-4-pentenoic acid.[59] This final metabolite is believed to be a reactive electrophilic species that alkylates 3-ketoacyl-CoA thiolase (the terminal enzyme of β-oxidation) by means of a Michael-type addition through nucleophilic attack at the olefinic terminus.[60]

E Troglitazone

Troglitazone ((±)-5-[4-(6-hydroxy-2,5,7,8-tetramethyl-chroman-2-ylmethoxy) benzyl]-2,4-thiazolidinedione) is an oral insulin sensitizer belonging to the thiazolidinedione class of compounds used for the treatment of type II diabetes. Its withdrawal from the US market was the consequence of the recent occurrence of hepatic failure leading sometimes to death.

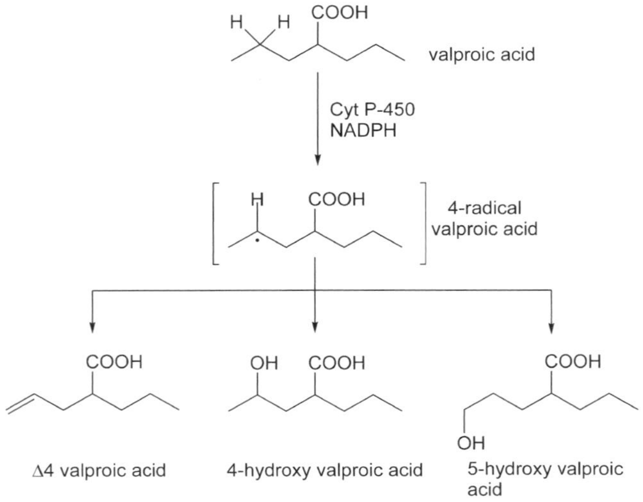

Fig. 32.27 Bioactivation of valproic acid to Δ^4 VPA.

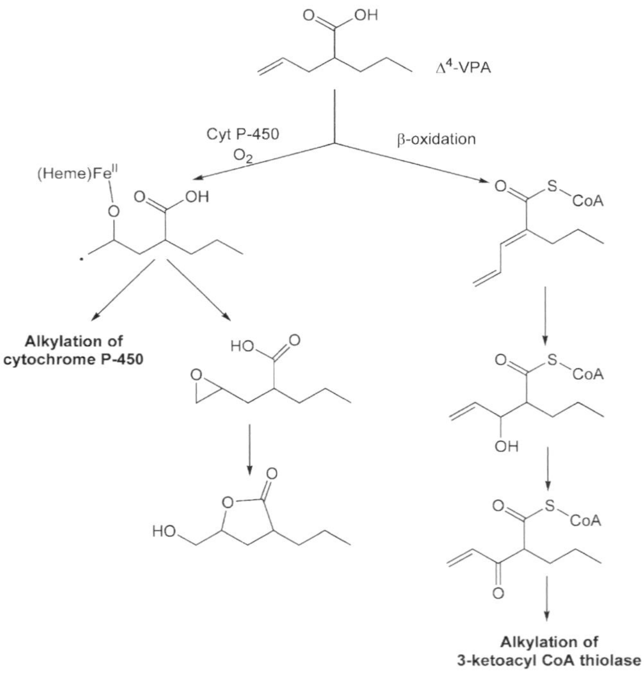

Fig. 32.28 Bioactivation of Δ^4 VPA.

It was first demonstrated that troglitazone is metabolized mainly to sulfate and glucuronide conjugates.[61] Also troglitazone is an inducer of CYP3A.[62] The mechanism of toxicity is still unclear, but seems to proceed according to two distinct pathways. This is suspected by the demonstration that incubation of troglitazone with P450 isoforms in the presence of glutathione give rise to at least five GSH conjugates.[63] Identification of these adducts permitted proposal of the two pathways which are described in Figs 32.29 and 32.30.

As described in Fig. 32.29, oxidative cleavage of the thiazolidinedione ring probably generates highly electrophilic α-ketoisocyanate and sulfenic acid intermediates. This P450 3A-mediated oxidation would afford a reactive sulfoxide intermediate which undergoes spontaneous ring opening. The second pathway (Fig. 32.30) consists of a

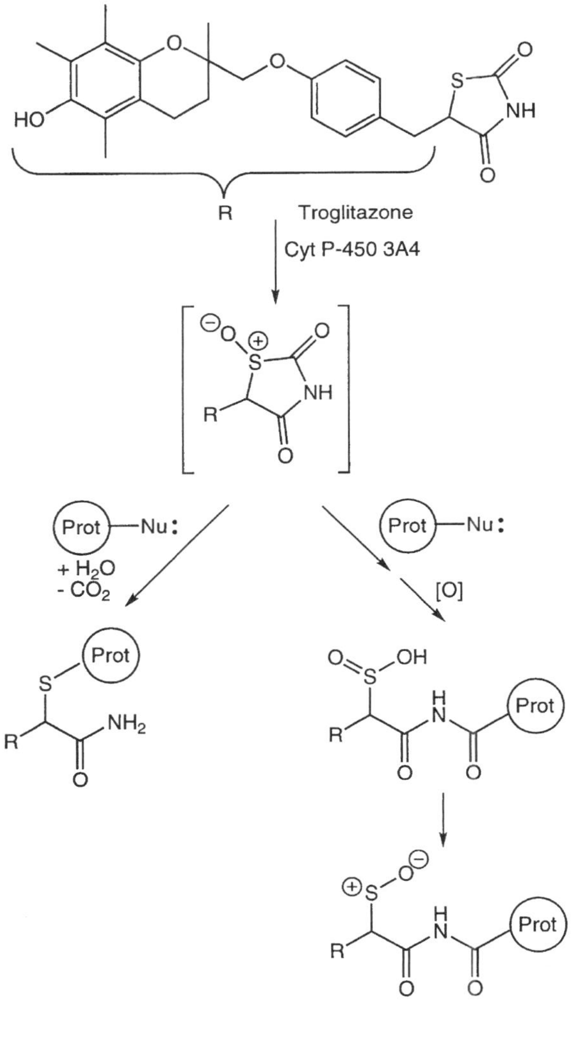

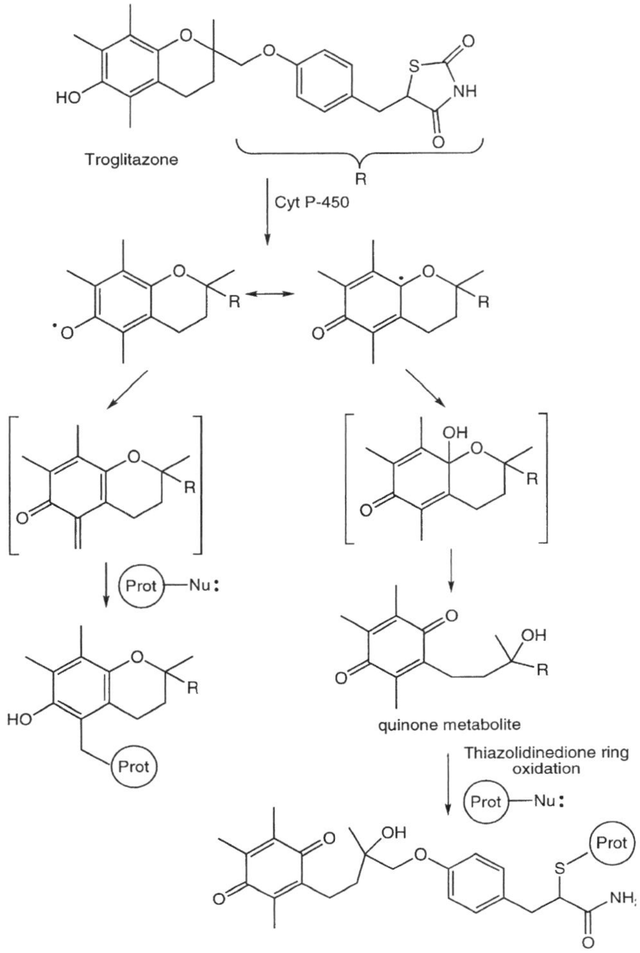

Fig. 32.30 Oxidation of the chromane ring of troglitazone.

Fig. 32.29 Oxidation of the thiazolidinedione ring of troglitazone.

P450 3A-mediated,[64] one-electron oxidation of the phenolic hydroxyl group leading to an unstable hemiaketal which opens spontaneously to form the quinone metabolite. This undergoes thiazolidinedione ring oxidation according to the pathways initially described. Alternatively, a P450-mediated hydrogen abstraction may occur on the phenoxy radical, leading to a *o*-quinone methide derivative.

It is now well established that troglitazone undergoes several metabolic transformation mediated by the 3A isoform of P450, leading to numerous electrophilic species. Thus toxicity acts probably both by covalent binding to hepatic proteins and oxidative stress through redox cycling process. The implication of the thiazolidinedione moiety is less likely since the more recent drugs of this series seem devoid of toxicity.

IV   CONCLUSION

In the foregoing it has been emphasized that almost all metabolic reactions are capable of producing reactive metabolites. This bioactivation yields toxic compounds which may act directly or indirectly[38] (Fig. 32.31). The emergence of toxicity may be the outcome of the interactions of metabolites or reactive intermediates with biological targets such as cellular macromolecules. Often, covalent bonds are formed during a phenomenon which may be referred to as 'alkylating stress'. The specific inhibition of an enzyme by its own substrate (suicide-substrate) is a peculiar feature of alkylating stress. Other compounds exhibit their toxicity by inducing the generation of reactive oxygen species, thus producing alterations in the redox state of the cell. Such a variety makes it difficult to point at molecular functions susceptible to produce toxic effects through bioactivation. However, some major toxophoric groups may be highlighted (Table 32.1). They may be implicated in acute or chronic toxicity. These patterns must be of particular concern in drug design.

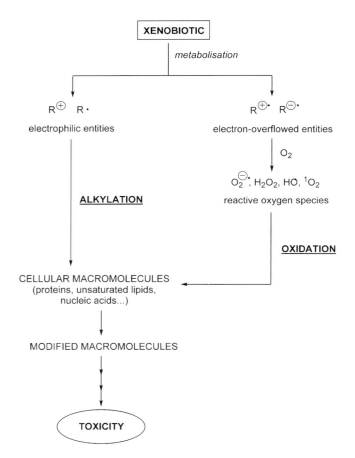

Fig. 32.31   Alkylating and oxidative stresses.

Table 32.1   Some major toxophoric groups and their bioactivation mechanisms

Toxophoric group	Bioactivation mechanism
Azocompounds	Nitrenium ions, tautomeric carbonium ions
Acetamides	
Aromatic/Heterocyclic amines	
Nitro compounds	
Nitroaromatic compounds	Radical formation/oxidative stress
Bromoarenes	Arene oxide formation
Ethinyl	Ketene formation/haem destruction
Furanes	Furane oxide formation
Pyrroles	Pyrrole oxide
Nitrogen mustard	Aziridium ions
Nitroso compounds	Diazonium ions/haem adduct/radical formation
Hydrazines	
Nitrosamines	Carbenium ions/DNA alkylation
Polyhalogenated compounds	Radical and carbene formation/episulfonium with GSH
Quinone	Semiquinone radical formation/oxidative stress/thiol trapping
Thioamides	Thiourea formation
Thiophene	Thiophene sulfoxide formation
Vinyl	Epoxidation/haem destruction

Generally, the formation of toxic metabolites is not the only pathway of biotransformation, and the overall metabolism is constituted of detoxication and bioactivation processes. The toxic metabolites are themselves often further detoxified. The duality between a beneficial detoxication phenomenon (metabolism, drug resistance) and the occurrence of a toxic effect represents the cost for a rapid adaptation to the transformation of any xenobiotic.

REFERENCES

1. Sipes, G. and Gandolfi, A.J. (1991) Biotransformation of toxicants. In Amdur, M.O., Doull, J. and Klaassen, C.D. (eds). *Casarett and Doull's Toxicology: The Basic Science of Poisons*, 4th edn, pp. 88–126. Pergamon Press, New York.
2. Alvares, A.P. and Pratt, W.B. (1990) Pathways of drug metabolism. In Pratt, W.B. and Taylor, P. (eds). *Principles of Drug Action: The Basis of Pharmacology*, 3rd edn, pp. 227–300. Churchill Livingstone, New York.
3. Guenguerich, F.P. (2001) Common and uncommon cytochrome P450 reactions related to metabolism and chemical toxicity. *Chem. Res. Toxicol.* **14**: 611–650.
4. Mason, R.P. and Chignell, C.F. (1982) Free radicals in pharmacology and toxicology — selected topics. *Pharmacol. Rev.* **33**: 189–211.

5. Aust, S.D., Chignell, C.F., Bray, T.M., Kalyanaraman, B. and Mason, R.P. (1993) Free radicals in toxicology. *Toxicol. Appl. Pharmacol.* **120**: 168–178.

6. Singal, P.K., Petkau, A., Gerrard, J.M., Hrushovetz, S. and Foerster, J. (1988) Free radicals in health and disease. *Mol. Cell. Biochem.* **84**: 121–122.

7. Anders, M.W. and Pohl, L.R. (1985) Halogenated alkanes. In Anders, M.W. (ed.). *Bioactivation of Foreign Compounds*, pp. 284–315.

8. Pohl, L.R., Nelson, S.D. and Krishna, G. (1978) Investigation of the mechanism of metabolic activation of chloramphenicol by rat liver microsomes: identification of a new metabolite. *Biochem. Pharmacol.* **27**: 491–496.

9. Halpert, J.R., Miller, N.E. and Gorsky, L.D. (1985) On the mechanism of the inactivation of the major phenobarbital-inducible isozyme of rat liver cytochrome P-450 by chloramphenicol. *J. Biol. Chem.* **260**: 8397–8403.

10. Cresteil, T., Beaune, P., Leroux, J.P., Lange, M. and Mansuy, D. (1979) Biotransformation of chloroform by rat and human liver microsomes; *in vitro* effect of some enzyme activities and mechanism of irreversible binding to macromolecules. *Chem. Biol. Interact.* **24**: 153–165.

11. Benasutti, M., Ejadi, S., Whitlow, M.D. and Loechler, E.L. (1988) Mapping the binding site of aflatoxin B1 in DNA: systematic analysis of the reactivity of aflatoxin B1 with guanines in different DNA sequences. *Biochem.* **27**: 472–481.

12. Rietjens, I.M., den Besten, C., Hanzlik, R.P. and van Bladeren, P.J. (1997) Cytochrome P450-catalysed oxidation of halobenzene derivatives. *Chem. Res. Toxicol.* **10**: 629–635.

13. Verna, L., Whysner, J. and Williams, G.M. (1996) 2-Acetylaminofluorene mechanistic data and risk assessment: DNA reactivity, enhanced cell proliferation and tumor initiation. *Pharmacol. Ther.* **71**: 83–105.

14. Dansette, P.M., Do Cao Thang, El Amri, H. and Mansuy, D. (1992) Evidence for thiophene-S-oxide as a primary reactive metabolite of thiophene *in vivo*: formation of a dihydrothiophene sulfoxide mercapturic acid. *Biochem. Biophys. Res. Commun.* **186**: 1624–2163.

15. Gaudiano, G. and Koch, T.H. (1991) Redox chemistry of anthracycline antitumor drugs and use of captodative radicals as tools for its elucidation and control. *Chem. Res. Toxicol.* **4**: 2–16.

16. Garner, A.P., Paine, M.J.I., Rodriguez-Crespo, I., Chinje, E.C., Ortiz de Montellano, P., Stratford, I.J., Tew, D.G. and Wolf, C.R. (1999) Nitric oxide synthases catalyse the activation of redox cycling and bioreductive anticancer agents. *Cancer Res.* **59**: 1929–1934.

17. Butler, T.S. (1961) Reduction of carbon tetrachloride *in vivo* and reduction of carbon tetrachloride and chloroform *in vitro* by tissues and tissue constituents. *J. Pharmacol. Exp. Ther.* **134**: 311–315.

18. Mico, B.A., Branchflower, R.V. and Pohl, L.R. (1983) Formation of electrophilic chlorine from carbon tetrachloride—Involvment of cytochrome P450. *Biochem. Pharmacol.* **32**: 2357–2359.

19. Grant, D.M., Hugues, N.C., Janezic, S.A., Goodfellow, G.H., Chen, H.J., *et al.* (1997) Human acetyltransferase polymorphisms. *Mutat. Res.* **376**: 61–70.

20. Timbrell, J.A., Mitchell, J.R., Snodgrass, W.R. and Nelson, S.D. (1980) Isoniazid hepatotoxicity: The relationship between covalent binding and metabolism *in vivo*. *J. Pharmacol. Exp. Ther.* **213**: 364–369.

21. Sinha, B.K. (1987) Activation of hydrazine derivatives to free radicals in the perfused rat liver: a spin-trapping study. *Biochim. Biophys. Acta.* **924**: 261–269.

22. Weber, G.L., Steewyk, R.C., Nelson, S.D. and Pearson, P.G. (1995) Identification of *N*-acetylcysteine conjugates of 1,2-dibromo-3-chloropropane: evidence for cytochrome P450 and glutathione mediated bioactivation pathways. *Chem. Res. Toxicol.* **8**: 560–573.

23. D'Souza, J., Caldwell, J. and Smith, R.L. (1980) Species variations in the *N*-methylation and quaternization of [^{14}C]pyridine. *Xenobiotica* **10**: 151–157.

24. Miller, E.C. and Miller, J.A. (1981) Mechanisms of chemical carcinogenesis. *Cancer* **47**: 1055–1064.

25. De Caprio, A.P., Strominger, L.N. and Weber, P. (1983) Neurotoxicity and protein binding of 2,5-hexanedione in the hen. *Toxicol. Appl. Pharmacol.* **68**: 297–307.

26. Genter St Clair, M.B., Amarnath, V., Moody, M.A., Anthony, D.C., Anderson, C.W. and Graham, D.G. (1988) Pyrrole oxidation and protein cross-linking as necessary steps in the development of gamma-diketone neuropathy. *Chem. Res. Toxicol.* **1**: 179–185.

27. Thomas, S.H.L. (1993) Paracetamol (acetaminophen) poisoning. *Pharm. Ther.* **60**: 91–120.

28. Dahlin, D.C., Miwa, G.T., Lu, A.Y. and Nelson, S.D. (1984) *N*-Acetyl-*p*-benzoquinone imine: a cytochrome P-450-mediated oxidation product of acteaminophen. *Proc. Natl Acad. Sci. USA* **81**: 1327–1331.

29. Chen, W., Koenig, L.L., Thompson, S.J., Peter, R.M., Rettie, A.E., Trager, W.F. and Nelson, S.D. (1998) Oxidation of acetaminophen to its toxic quinone imine and nontoxic catechol metabolites by baculovirus-expressed and purified human cytochromes P450 2E1 and 2A6. *Chem. Res. Toxicol.* **11**: 295–301.

30. Mitchell, J.R., Jollow, D.J., Gillette, J.R. and Brodie, B.B. (1973) Drug metabolism as a cause of drug toxicity. *Drug Metab. Dispos.* **1**: 418–423.

31. Rao, D.N.R., Fischer, V. and Mason, R.P. (1990) Glutathione and ascorbate reduction of the acetaminophen radical formed by peroxidase. Detection of the glutathione disulfide radical formed by the ascorbyl radical. *J. Biol. Chem.* **265**: 844–847.

32. Fischer, V., West, P.R., Harman, L.S. and Mason, R.P. (1985) Free-radical metabolites of acetaminophen and a dimethylated derivative. *Environ. Health Perspect.* **64**: 127–137.

33. Dietze, E.C., Schafer, A., Omichinski, J.G. and Nelson, S.D. (1997) Inactivation of glyceraldehyde-3-phosphate dehydrogenase by a reactive metabolite of acetaminophen and mass spectral characterization of an arylated active site peptide. *Chem. Res. Toxicol.* **10**: 1097–1103.

34. Zhou, L., McKenzie, B.A., Eccleston, E.D., Jr., Srivastava, S.P., Chen, N., *et al.* (1996) The covalent binding of [14C]acetaminophen to mouse hepatic microsomal proteins: the specific binding to calreticulin and the two forms of the thiol:protein disulfide oxidoreductases. *Chem Res Toxicol.* **9**: 1176–1182.

35. Rosen, G.M., Singletary, W.V., Jr., Rauckman, E.J. and Killenberg, P.G. (1983) Acetaminophen hepatotoxicity. An alternative mechanism. *Biochem. Pharmacol.* **32**: 2053–2059.

36. Holownia, A., Mapoles, J., Menez, J.F. and Braszko, J.J. (1997) Acetaminophen metabolism and cytotoxicity in PC12 cells transfected with cytochrome P4502E1. *J. Mol. Med.* **75**: 522–527.

37. Dai, Y. and Cederbaum, A.I. (1995) Cytotoxicity of acetaminophen in human cytochrome P4502E1-transfected HepG2 cells. *J. Pharmacol. Exp. Ther.* **273**: 1497–1505.

38. Nelson, S.D. and Pearson, P.G. (1990) Covalent and noncovalent interactions in acute lethal cell injury caused by chemicals. *Annu. Rev. Pharmacol. Toxicol.* **30**: 169–195.

39. Homberg, J.C., André, C. and Abuaf, N. (1984) A new anti-liver-kidney microsome antibody (anti-LKM$_2$) in tienilic acid-induced hepatitis. *Clin. Exp. Immunol.* **55**: 561–570.

40. Dansette, P.M., Bonierbale, E., Minoletti, C., Beaune, P.H., Pessayre, D. and Mansuy, D. (1998) Drug-induced immunotoxicity. *Eur. J. Drug Metab. Pharmacokinet.* **23**: 443–451.

41. Lopez-Garcia, M.P., Dansette, P. and Mansuy, D. (1994) Thiophene derivatives as new mechanism-based inhibitors of cytochromes P-450: inactivation of yeast-expressed human liver cytochrome P-450 2C9 by tienilic acid. *Biochem.* **33**: 166–175.

42. Lecoeur, S., Bonierbale, E., Challine, D., Gautier, J.-C., Valadon, P., *et al.* (1994) Specificity of *in vitro* binding of tienilic acid metabolites

to human live microsomes in relationship to the type of hepatotoxicity: comparison with two directly hepatotoxic drugs. *Chem. Res. Toxicol.* **7**: 434–442.

43. Mansuy, D. (1997) Molecular structure and hepatotoxicity: compared data about two closely related thiophene compounds. *J. Hepatol.* **26**: 22–25.

44. Lopez-Garcia, P.M., Dansette, P., Valadon, P., Amar, C., Beaune, P.H., *et al.* (1993) Human liver P-450 expressed in yeast as tools for reactive-metabolite formation studies. Oxidative activation of tienilic acid by P-450 2C9 and P-450 2C10. *Eur. J. Biochem.* **213**: 223–232.

45. Koenigs, L.L., Peter, R.M., Hunter, A.P., Haining, R.L., Rettie, A.E., *et al.* (1999) Electrospray ionization mass spectrometric analysis of intact cytochrome P450: identification of tienilic acid adducts to P450 2C9. *Biochem.* **38**: 2312–2319.

46. Harris, J.W., Pohl, L.R., Martin, J.L. and Anders, M.W. (1991) Tissue acylation by the chlorofluorocarbon substitute 2,2-dichloro-1,1,1-trifluoroethane. *Proc. Natl Acad. Sci. USA* **88**: 1407–1410.

47. Mansuy, D., Nastainczyk, W. and Ullrich, V. (1974) The mechanism of halothane binding to microsomal cytochrome P-450. *Naunyn-Schmiedeberg's Arch. Pharmacol.* **285**: 315–324.

48. Ahr, H.J., King, L.J., Nastainczyk, W. and Ullrich, V. (1982) The mechanism of reductive dehalogenation of halothane by liver cytochrome P-450. *Biochem. Pharmacol.* **31**: 383–390.

49. Pohl, L.R. (1993) An immunochemical approach of identifying and characterizing protein targets of toxic reactive metabolites. *Chem. Res. Toxicol.* **6**: 786–793.

50. Kenna, J.G., Neuberger, J. and Williams, R. (1988) Evidence for expression in human liver of halothane-induced neoantigens recognized by antibodies in sera from patients with halothane hepatitis. *Hepatology* **8**: 1635–1641.

51. Pohl, L.R., Kenna, J.G., Satoh, H. and Christ, D. (1989) Neoantigens associated with halothane hepatitis. *Drug Metab. Rev.* **20**: 203–217.

52. Hoet, P., Graf, M.L., Bourdi, M., Pohl, L.R., Duray, P.H., *et al.* (1997) Epidemic of liver disease caused by hydrochlorofluorocarbons used as ozone-sparing substitutes of chlorofluorocarbons. *Lancet* **350**: 556–559.

53. Baillie, T.A. (1988) Metabolic activation of valproic acid and drug-mediated hepatotoxicity. Role of the terminal olefin 2-n-propyl-4-pentenoic acid. *Chem. Res. Toxicol.* **1**: 195–199.

54. Rettie, A.E., Rettenmeier, A.W., Howald, W.N. and Baillie, T.A. (1987) Cytochrome P-450-catalyzed formation of Δ^4 VPA, a toxic metabolite of valproic acid. *Science* **235**: 890–893.

55. Rettie, A.E., Sheffels, P.R., Korzekwa, K.R., Gonzalez, F.J., Philpot, R.M. and Baillie, T.A. (1995) CYP4 isozyme specificity and the relationship between omega-hydroxylation and terminal desaturation of valproic acid. *Biochem.* **34**: 7889–7895.

56. Kassahun, K. and Baillie, T.A. (1993) Cytochrome P-450-mediated dehydrogenation of 2-n-propyl-2(E)-pentenoic acid, a pharmacologically-active metabolite of valproic acid, in rat liver microsomal preparations. *Drug Metab. Dispos.* **21**: 242–248.

57. Prickett, K.S. and Baillie, T.A. (1986) Metabolism of unsaturated derivative of valproic acid in rat liver microsomes and destruction of cytochrome P-450. *Drug Metab. Dispos.* **14**: 221–229.

58. Ortiz de Montellano, P.R., Yost, G.S., Mico, B.A., Dinizo, S.E., Correia, M.A. and Kambara, H. (1979) Destruction of cytochrome P-450 by isopropyl-4-pentenamide and methyl-2-isopropyl-4-pentenoate: mass spectrometric characterization of prosthetic heme adducts and nonparticipation of epoxide metabolites. *Biophys.* **197**: 524–533.

59. Rettenmeier, A.W., Gordon, W.P., Prickett, K.S., Levy, R.H. and Baillie, T.A. (1986) Biotransformation and pharmacokinetics in the rhesus monkey of 2-n-propyl-4-pentenoic acid, a toxic metabolite of valproic acid. *Drug Metab. Dispos.* **14**: 454–464.

60. Rettenmeier, A.W., Prickett, K.S., Gordon, W.P., Bjorge, S.M., Chang, S.-L., *et al.* (1985) Studies on the biotransformation in the perfused rat liver of 2-n-propyl-4-pentenoic acid, a metabolite of the antiepileptic drug valproic acid. Evidence for the formation of chemically reactive intermediates. *Drug Metab. Dispos.* **13**: 81–96.

61. Kawai, K., Kawasaki-Tokui, Y., Odaka, T., Tsuruta, F., Kazui, M., *et al.* (1997) Disposition and metabolism of the new oral antidiabetic drug troglitazone in rats, mice and dogs. *Arzneim. Forsch.* **47**: 356–368.

62. Ramachandran, V., Kostrubsky, V.E., Komoroski, B.J., Zhang, S., Dorko, K., *et al.* (1999) Troglitazone increases cytochrome P-450 3A protein and activity in primary cultures of human hepatocytes. *Drug Metab. Dispos.* **27**: 1194–1199.

63. Kassahun, K., Pearson, P.G., Tang, W., McIntosh, I., Leung, K., *et al.* (2001) Studies on the metabolism of troglitazone to reactive intermediates *in vitro* and *in vivo*. Evidence for novel biotransformation pathways involving quinone methide formation and thiazolidinedione ring scission. *Chem. Res. Toxicol.* **14**: 62–70.

64. Yamazaki, H., Shibata, A., Suzuki, M., Nakajima, M., Shimada, N., *et al.* (1999) Oxidation of troglitazone to a quinone-type metabolite catalyzed by cytochrome P-450 2C8 and P-450 3A4 in human liver microsomes. *Drug Metab Dispos.* **27**: 1260–1266.

33

DESIGNING PRODRUGS AND BIOPRECURSORS

Camille G. Wermuth

La façon de donner vaut mieux que ce que l'on donne. The manner of giving counts more than what one gives

Piere Corneille, *Le Menteur*, Acte 1, Scene 1.

I GENERAL INTRODUCTION

Therapeutic approaches based on molecular pharmacology mostly use *in vitro* models (membrane or enzyme preparations, cell or microorganism cultures, isolated organs, etc.). In the last two decades they have led to the discovery of numerous potent and quite selective agents. As examples, we can mention, the H_2 histamine antagonists burimamide and cimetidine,[1] the $GABA_A$ receptor antagonist gabazine,[2,3] the hydroxymethyl-glutaryl-CoA reductase inhibitor mevastatin,[4,5] the cholecystokinin antagonist asperlicin,[6] the anticancer drugs taxol[7,8] and neocarzinostatine[9] and the corticotropin-releasing factor (CRF) receptor modulators.[10,11]

However, the bioavailability of molecules exclusively screened through *in vitro* assays, can be low. Owing to the polarity of the functional groups present in the molecule, they may be poorly absorbed or incorrectly distributed. They may also, as a result of their vulnerability, be the subject of early metabolic destruction such as first-pass effects or any other kind of degradation leading to a short biological half-life. For such molecules *in vivo* administration is limited to the parenteral route and their clinical usefulness is thus restricted. Sometimes an adequate pharmaceutical formulation (micro-encapsulation, sustained-release or enterosoluble preparations) can overcome these drawbacks, but often the galenic formulation is inoperant and a chemical modification of the active molecule is necessary to correct its pharmacokinetic insufficiences. This *chemical formulation* process, whose objective is to convert an interesting active molecule into a clinically acceptable drug, often involves the so-called 'prodrug design'.

Initially, the term prodrug was introduced by Albert to describe 'any compound that undergoes biotransformation prior to exhibiting its pharmacological effects'.[12] Such a broad definition includes accidental historic prodrugs (aspirin and salicylic acid), active metabolites (imipramine and desmethylimipramine) and compounds intentionally prepared to improve the pharmacokinetic profile of an active molecule. From this point of view the term 'drug latentiation' proposed by Harper[13] is more appropriate for prodrug design as it indicates that there is an intention. Drug latentiation is defined as 'the chemical modification of a biologically active compound to form a new compound that, upon *in vivo* enzymatic attack, will liberate the parent compound'. Even this definition is too broad and a careful survey of the specialized literature led us to divide the prodrugs into two classes: the carrier prodrugs, and the bioprecursors.[14,15]

The *carrier prodrugs* result from a temporary linkage of the active molecule with a transport moiety that is frequently of lipophilic nature. A simple hydrolytic reaction cleaves this transport moiety at the correct moment (e.g. pivampicillin, bacampicillin). Such prodrugs are *per se* less active than the parent compounds or even inactive. The transport moiety (carrier group) will be chosen for its nontoxicity and its ability to ensure the release of the active principle with efficient kinetics.

The *bioprecursors* do not imply a temporary linkage between the active principle and a carrier group, but result

from a molecular modification of the active principle itself. This modification generates a new compound, able to be a substrate for the metabolizing enzymes, the metabolite being the expected active principle. This approach exemplifies the active metabolite concept in the prospective application (sulindac, fenbufen, losartan).

II THE CARRIER-PRODRUG PRINCIPLE

The carrier-prodrug principle (Fig. 33.1) consists of 'the attachment of a carrier group to the active drug to alter its physicochemical properties and then the subsequent enzyme attack to release the active drug moiety'.[13] 'Prodrugs can thus be viewed as drugs containing specialized nontoxic protective groups used in a transient manner to alter or eliminate undesirable properties in the parent molecule'.[16]

A well-designed carrier-prodrug satisfies the following criteria:[17,18]

(1) The linkage between the drug substance and the transport moiety is usually a covalent bond.
(2) As a rule the prodrug is inactive or less active than the parent compound.
(3) The linkage between the parent compound and the transport moiety must be broken *in vivo*.
(4) The prodrug, as well as the *in vivo* released transport moiety, must be nontoxic.
(5) The generation of the active form must take place with rapid kinetics to ensure effective drug levels at the site of action and to minimize either direct prodrug metabolization or gradual drug inactivation.

An example of prodrug design taking into account these criteria is found in orally active ampicillin derivatives.[20–22] Ampicillin is one of the main β-lactam antibiotics. It is widely used as a broad-spectrum antibiotic, but it suffers from poor absorption when administered orally; only about 40% of the drug is absorbed. In other words, to achieve the same clinical efficiency and the same blood level one must give two to three times more ampicillin by mouth than by intramuscular injection. The clinical tolerance of orally given ampicillin may be affected, the nonabsorbed part of the drug destroying the intestinal flora. Therefore, numerous

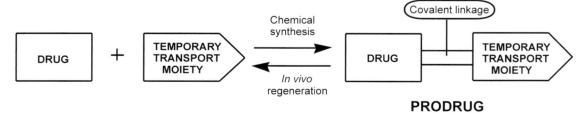

Fig. 33.1 The carrier-prodrug principle.[19]

attempts have been made to improve these poor absorption properties.

Figure 33.2 represents two prodrugs of ampicillin: pivampicillin and bacampicillin. They both result from the esterification of the polar carboxylic group with a lipophilic, enzymatically labile ester. The main properties of these prodrugs can be summarized as follows:

(1) The absorption of these compounds is nearly quantitative (98–99%).
(2) The generation of free ampicillin in the blood stream is rapid (less than 15 minutes).
(3) The released carrier molecules are formaldehyde and pivalic acid (trimethyl acetic acid) for pivampicillin and acetaldehyde, ethanol and carbon dioxide in the case of bacampicillin. These three latter compounds are natural metabolites in the human body. This may explain the better tolerance of bacampicillin compared with pivampicillin.
(4) The serum levels attained following oral administration of bacampicillin are similar to those obtained after intramuscular injection of an equimolecular amount of free ampicillin.
(5) The clinical trials confirm the efficiency and the safety of the prodrugs. Owing to their good absorption, the drugs are given at lower dosage than ampicillin: 0.8–1.0 g daily is sufficient in common infections as compared with 2.0 g daily for ampicillin.
(6) It has been shown, and this seems to be a rule for prodrugs, that pivampicillin and bacampicillin are inactive *per se*, the antibiotic potency appearing only *in vivo* after the release of free ampicillin.

III THE BIOPRECURSOR-PRODRUG PRINCIPLE

As already mentioned above, bioprecursor prodrugs result from a molecular modification of the active principle generating a new compound, able to be a substrate for the metabolizing enzymes, the metabolite being the expected active principle. The bioprecursor-prodrug approach exemplifies the active metabolite concept in the prospective application. A survey of a great number of examples of active metabolites shows that they belong exclusively to the phase I products and result from one of the reactions mentioned in Table 33.1. As such reactions follow some general rules, they can often be predicted. Taking into account the common metabolic pathways, one can imagine the design of a given molecule so that it will be converted *in vivo* into the desired compound by one or more of the phase I reactions. In other words, the active metabolite concept can be used in a forward-looking way ('metabolic synthesis'). By analogy to the retrosynthetic reasoning usual in organic chemistry we can imagine retrometabolic reasoning in prodrug design. Such reasoning can lead to a particular group of prodrugs for which we proposed the name *bioprecursors* or *bioprecursor prodrugs*.[14,15]

A typical example of an effective bioprecursor prodrug is given by the anti-inflammatory drug sulindac. Sulindac,

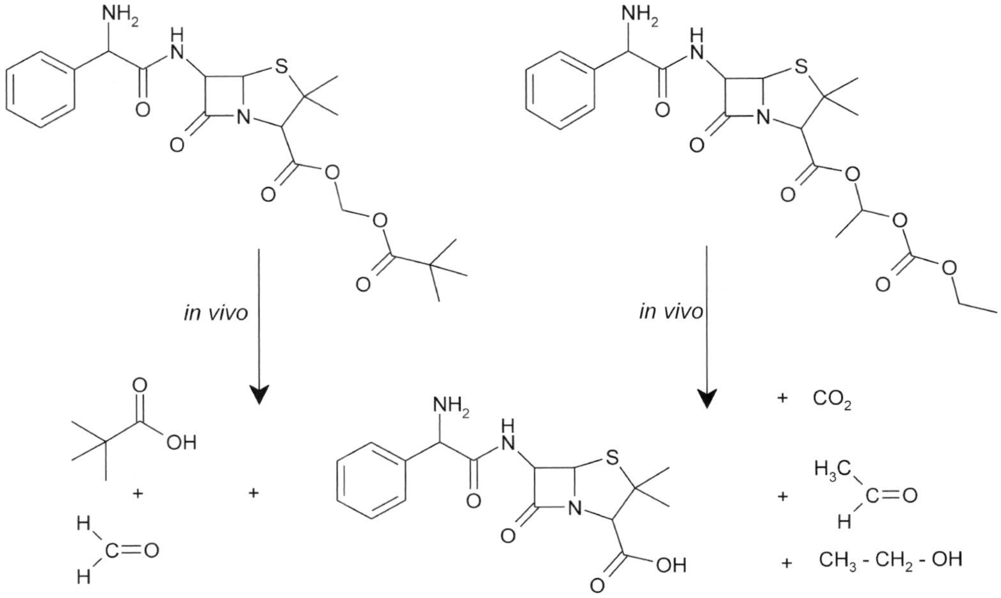

Fig. 33.2 Prodrugs derived from ampicillin.[20–22]

Table 33.1 *Phase I reactions*[23]

Oxidative reactions	Reductive reactions	Reactions without change in the state of oxidation
Oxidation of alcohol, carbonyl and acid functions, hydroxylation of aliphatic carbon atoms, hydroxylation of alicyclic carbon atoms, oxidation of aromatic carbon atoms, oxidation of carbon—carbon double bonds, oxidation of nitrogen-containing functional groups, oxidation of silicon, phosphorus, arsenic, and sulfur, oxidative *N*-dealkylation, oxidative *O*- and *S*-dealkylation, oxidative deamination, other oxidative reactions	Reduction of carbonyl groups, reduction of alcoholic groups and carbon—carbon double bonds, reduction of nitrogen-containing functional groups, other reductive reactions	Hydrolysis of esters and ethers, hydrolytic cleavage of C—N single bonds, hydrolytic cleavage of nonaromatic heterocycles, hydration and dehydration at multiple bonds, new atomic linkages resulting from dehydration reactions, hydrolytic dehalogenation: removal of hydrogen halide molecules, various reactions

cis-5-fluoro-2-methyl-1-[*p*-(methylsulphinyl) benzylidene] indene-3 acetic acid[24] is a nonsteroidal anti-inflammatory agent having a broad spectrum of activity in animal models and in humans. The two quantitatively significant biotransformations undergone by sulindac in laboratory species[25] and in humans[26,27] involve only changes in the oxidation state of the sulphinyl substituent, namely, irreversible oxidation of the parent (sulindac) to sulphone and reversible reduction to sulphide (Fig. 33.3), the latter being the active species.[28] In two *in vitro* models of inflammation, prostaglandin synthetase inhibition and inhibition of platelet aggregation, the sulphide has activities comparable to those of indomethacin, whereas sulindac itself is devoid of activity.

Nevertheless, sulindac is the preferred compound for clinical applications: an oral dosage of this inactive bioprecursor will circumvent initial exposure of gastric and intestinal mucosa to the active drug, and might thus provide a therapeutic advantage in comparison with the sulphide dosing.

IV PRACTICAL APPLICATIONS OF PRODRUG DESIGN

The domain of application of the prodrug approach is illustrated in Fig. 33.4. In practice, prodrugs usually achieve

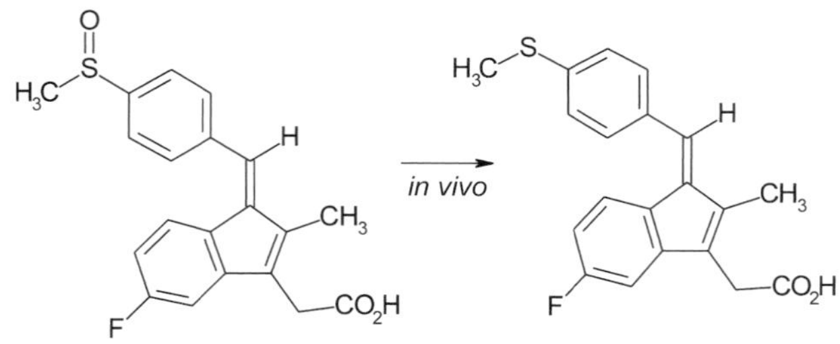

Fig. 33.3 Reductive bioactivation of sulindac.[28]

one of the five following goals: increased lipophilicity, increased duration of pharmacological effects, increased site specificity, decreased toxicity and adverse reactions, and improvement in drug formulation (stability, water solubility, suppression of an undesirable organoleptic or physicochemical property). For applications in the field of insecticides, see Drabek and Neumann.[29]

V CARRIER PRODRUGS: APPLICATION EXAMPLES

Bioactive compounds and drugs usually bear a limited number of polar functional groups suitable for carrier-prodrug synthesis. Among these, the most frequent are the alcoholic and the phenolic hydroxyls, the amino group, and the carboxylic function. The aim of the following sections is to illustrate how such groups can be used to prepare prodrugs with improved pharmacokinetic properties.

A Improvement of the bioavailability and the biomembrane passage

The biomembrane passage of a drug depends primarily on its physicochemical properties and especially on its partition coefficient (Chapters 21 and 22). Thus the transient attachment of a lipophilic carrier group to an active principle can provide better bioavailability, mostly by facilitating cell-membrane crossing by passive diffusion. Not only peroral absorption, but also rectal absorption, ocular drug delivery and dermal drug delivery, are dependent on passive diffusion. Finally, lipophilic carriers can sometimes be useful to reduce first-pass metabolism.[31]

Derivatization of drugs containing alcoholic or phenolic hydroxy groups

Starting from hydroxylic derivatives, high lipophilicity can simply be obtained by esterification with lipophilic

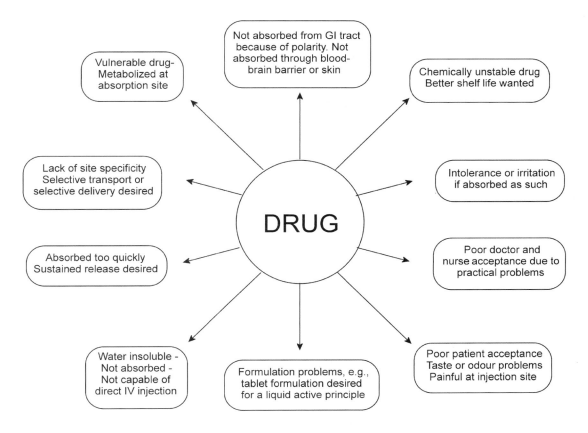

Fig. 33.4 Shortcomings that may be overcome through chemical formulation.[17,19,30]

carboxylic acids. *Dipivaloyl-epinephrine* for example (Fig. 33.5) crosses the cornea and is used in the treatment of glaucoma.[32] The β-blocker timolol contains a secondary amino group with a pK_a of 9.2 and since this group is highly protonated at pH 7.4, the compound shows a low lipophilicity at physiological pH (log P = −0.04) which in turn is unfavourable for corneal penetration. The corresponding butyryl ester has an increased lipophilicity

(log P = 2.08) and causes a four- to six-fold increase in the corneal absorption of timolol following topical administration to rabbits.[31]

In a similar manner, *dibenzoyl-2-amino-6,7-dihydroxy-tetrahydronaphthalene* (DB-ADTN) reaches the CNS, whereas the parent dopamine agonist ADTN does not.[33,34] For dipivaloyl-epinephrine and dibenzoyl-ADTN, the selective acylation of the phenolic hydroxyl groups was

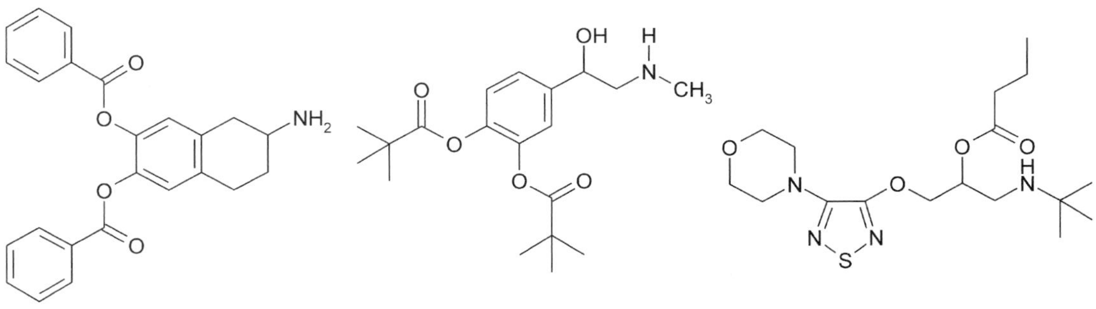

dibenzoyl-ADTN   dipivaloyl-epinephrine   butyryl-timolol

Fig. 33.5 Lipophilic prodrugs of hydroxy compounds with facilitated membrane penetration.[31–34]

achieved in a strong acidic medium, the amino function being protected by protonation.[33,35] Acylated thymidine analogues such as *3'-O-hexyl-5'-amino-2'-deoxythymidine* are prodrugs for topical application against herpes simplex type 1 (HSV-1) viruses.[36] Diacetyl and dipropionyl guanine derivatives, given orally to mice, provided concentrations of the parent drug that were more than 15-fold higher than those observed after dosing with the nonacylated parent drug.[37] In augmenting the lipophilicity and simultaneously destroying the crystal lattice energy, the 2',3'-diacetate of the antiviral agent 6-methoxypurine arabinoside allowed a five-fold increase in bioavailability and a three-fold increase in water solubility in comparison to the nonacetylated drug.[38] As a consequence, an intravenous formulation could be developed.

Esterification of hydroxylic functions with poly-acids (e.g. succinic acid, phosphoric acid) represents an excellent way to prepare water-soluble prodrugs (see Chapter 36).

Derivatization of drugs containing a carbonyl function: aldehydes and ketones

The ethylene ketal derivative of prostaglandin E_2 (dinoprostone) possesses much improved solid-state stability (see Chapter 39). Functionalized spirothiazolidines of hydrocortisone and hydrocortisone 21-acetate (Fig. 33.6), prepared with cysteine esters or related β-aminothiols, have improved topical anti-inflammatory activity. It is speculated that the Schiff base intermediate formed upon ring-opening may accumulate in the skin by binding (through its SH function) to thiol groups in the skin.[39]

Simple and substituted oximes are biostable unless intramolecular assistance is provided. This is the case for the oximes derived from oxyamino acetic acid, that are possible water-soluble prodrugs of ketones and aldehydes (see Chapter 36).

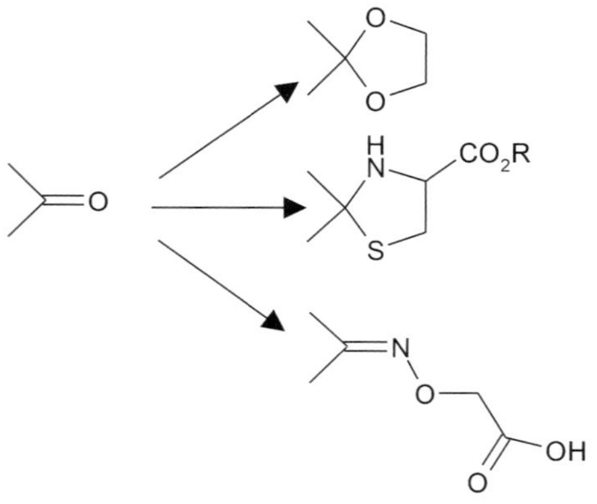

Fig. 33.6 Prodrug possibilities starting from aldehydes or ketones.

Derivatization of drugs containing a carboxylic acid function

Lipophilic prodrugs can also be derived from a carboxylic function, the most commonly used derivatives being *carboxylic esters*. Simple esters of aliphatic alcohols are attractive as they are cheap to prepare, chemically stable, and yield harmless hydrolysis products.[40] Typical representatives of such prodrugs are tyrosine methyl ester,[41] levodopa ethyl ester,[42] nipecotic acid ethyl ester,[43] enalaprilat ethyl ester,[44,45] trandolapril,[46] γ-aminobutyric acid cetyl ester[47,48] and methotrexate cetyl ester.[49]

Lipoidal prodrugs, in which the carboxyl function esterifies the free alcoholic hydroxyl of 1,2- or 1,3-diglyceride, are well absorbed and show high lymphotropism.[50,51] Owing to their greasy aspect and to their difficult purification these compounds have no industrial application.

The widespread use of *acyloxymethyl esters* in antibiotic chemistry, as illustrated above for bacampicillin, was initiated by Jansen and Russel[52] at Wyeth Laboratories and successfully applied to pivampicillin,[20] talampicillin[22] and cephalosporins.[53] In each of these cases, the oral absorption of the antibiotic was improved by some two- to three-fold over that of the parent compound. The acyloxymethyl derivatization was also extended to amino acids such as α-methyl-DOPA,[54] isoguvacine[55] and tranexamic acid,[56] anti-inflammatory drugs such as niflumic acid[57] and indomethacin,[58] and quinolone antibacterials such as norfloxacin.[59]

Mixed anhydrides represent original attempts for preparing prodrugs of carboxylic or phosphonic acid. *Clodronic acid dianhydrides* (Fig. 33.7), for example, were shown to be novel bioreversible prodrugs of clodronate. They are more lipophilic than the parent clodronate, stable against chemical hydrolysis, and hydrolyse enzymatically to clodronate in human serum.[60]

Methoxyimino bioisosteres of carboxylic acid anhydrides, which can be prepared by *O*-acylation of the corresponding hydroxamates, represent another method of preparing prodrugs of carboxylic acids (Fig. 33.8). One of this type of derivatives, compound FOX 988, was designed as a prodrug of 4-(2,2-dimethyl-1-hydroxypropyl) benzoic acid. In the liver this prodrug is metabolized at a rate sufficient enough to possess hypoglycemic potency (an ED_{50} of 65 μmol kg^{-1}, 28 mg kg^{-1} day^{-1}, for glucose lowering).[61]

However, by avoiding significant escape of the released active metabolite to the systemic circulation, it avoids testicular toxicity at doses up to 1500 μmol kg^{-1} day^{-1}. The mechanism of action of 4-(2,2-dimethyl-1-hydroxypropyl) benzoic acid consists of the sequestration of coenzyme A (CoA) in the mitochondria, thus inhibiting medium-chain acyltransferase and, as a consequence, hepatic neoglucogenesis.[62]

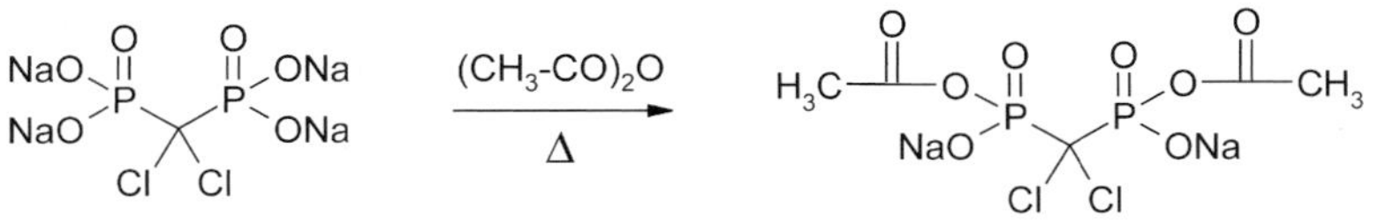

Fig. 33.7 Use of anhydrides as lipophilic prodrugs of phosphonates.

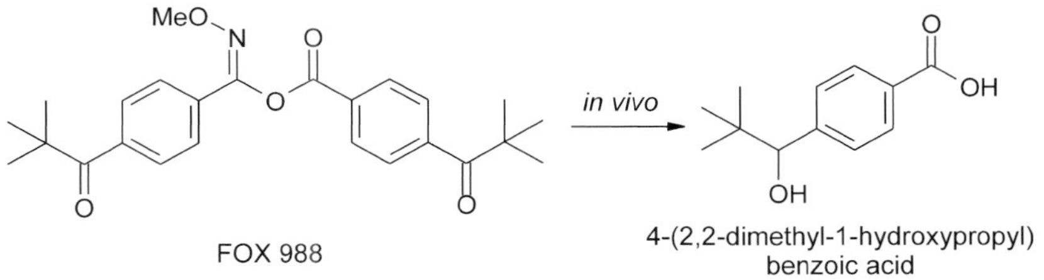

FOX 988

4-(2,2-dimethyl-1-hydroxypropyl) benzoic acid

Fig. 33.8 Methoxyimino bioisosteres of carboxylic acid anhydrides as prodrugs for carboxylic acids.

Primary amides of carboxylic acids are easily converted to the corresponding acid (e.g. depamide, progabide) in humans and can thus be used in prodrug design. Amides of ketoprofen-derived arylacetic acids possess a therapeutic index one order of magnitude greater than that of indomethacin.[63]

Derivatization of amines

Owing to the slow *in vivo* cleavage rate of the *N*-substituted amides, acylation of amines is generally not recommended. Better possibilities are offered by activated amides, peptides, imines and soft quaternary ammonium salts. However, the use of simple *N*-acyl derivatives must not systematically be discarded. The *N*-benzoyl- or *N*-pivaloyl derivatives of the inhibitory neurotransmitter GABA are examples of compounds able to penetrate the blood–brain barrier and to abolish pentetrazole- and bicuculline-induced convulsions. It was also demonstrated in rats that, following subcutaneous injection, rat-brain homogenates liberate free GABA from these amides.[64] There is even some biochemical and pharmacological evidence suggesting that N-pivaloyltaurine crosses the blood–brain barrier.[65]

Imines[66] and enamines,[67,68] stabilized through hydrogen bonds, can also be effective prodrugs of primary amines (Fig. 33.9).

Small peptides constitute an alternative way of derivatizing amines. The hypotensive drug milodrine, for example (Fig. 33.9), is the well-absorbed transport from

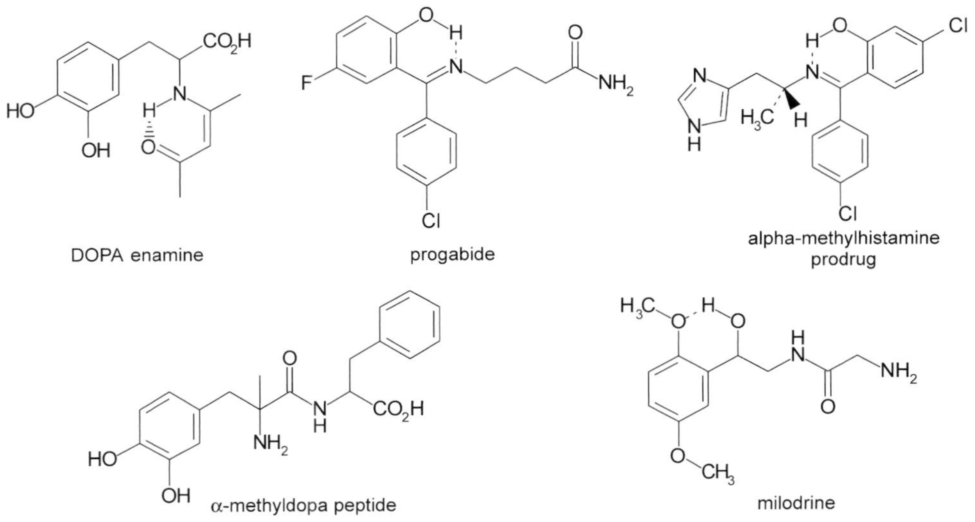

DOPA enamine

progabide

alpha-methylhistamine prodrug

α-methyldopa peptide

milodrine

Fig. 33.9 Imine, enamine and peptide prodrugs derived from amino functions.

which 2-(2,5-dimethoxyphenyl)-2-hydroxyethylamine (ST-1059) is liberated by enzymic cleavage of the glycine residue.[69] Orally given to fasting Wistar rats, *N*(*Z*-alanyl) amide of the hypoglycemic sulfonylurea carbutamide demonstrated a four to six times higher potency than the parent sulfonamide. The compound is well tolerated and is metabolized to the parent drug, the amino acid moiety just modifying the bioavailability.[70] The L-serine amide of the tubulin polymerization inhibitor analogue AC-7739 is a potent combretastatin A4 analogue with clearly improved bioavailability and reduced toxicity.[71]

Among the numerous variations made around the α-methyl-DOPA molecule, acylation with a glycyl–glycyl residue was claimed to improve oral bioavailability.[72] For a series of anticandidal di- and tripeptides containing *m*-fluorophenylalanine (m-FPhe), competitive antagonism studies supported peptide transport-mediated entry of the warhead m-FPhe inside the cell.[73] Dipeptides derived from α-methyldopa (Fig. 33.9) show a 10 to 20-fold better penetration of the intestinal wall than α-methyldopa itself.[74]

Derivatization of amidines

Improved oral activity for amidines resulted from the formation of carbamates of 2,5-bis(4-amidinophenyl)-furan, an effective drug against *Pneumocystis carinii* pneumonia (Fig. 33.10).[75]

Prodrugs for compounds with acidic NH functions

Prodrugs obtained by N-alkoxycarbonyloxymethylation of 5-fluorouracil show improved delivery properties. Both 1- and 3-alkoxycarbonyloxymethyl derivatives are hydrolysed quantitatively to 5-fluorouracil but the 3-substituted derivatives show greater promise as prodrugs since they combine adequate stability in aqueous solution with a high susceptibility to undergo hydrolysis in plasma.[76] Sulfonamides, but also carboxamides, carbamates and other NH-acidic compounds (Fig. 33.11) can be acylated with various groups[77] or converted into phtalidyl derivatives.[78]

Hetacillin[79,80] and droxicam[81] are examples of simultaneous cyclic protections of an acidic NH function and an amino or a hydroxylic group located in the vicinity (Fig. 33.12).

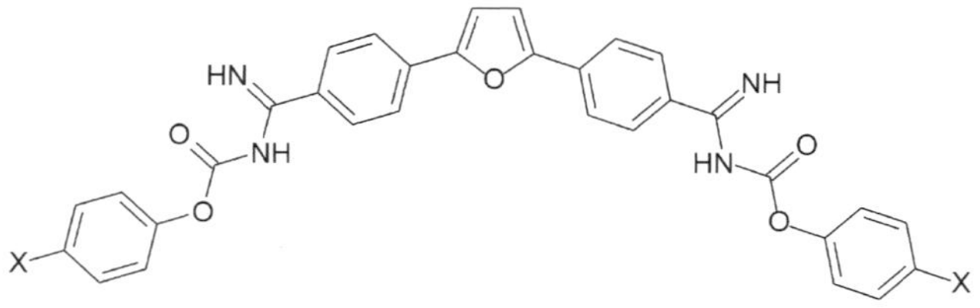

Fig. 33.10 Carbamates as amidine prodrugs.

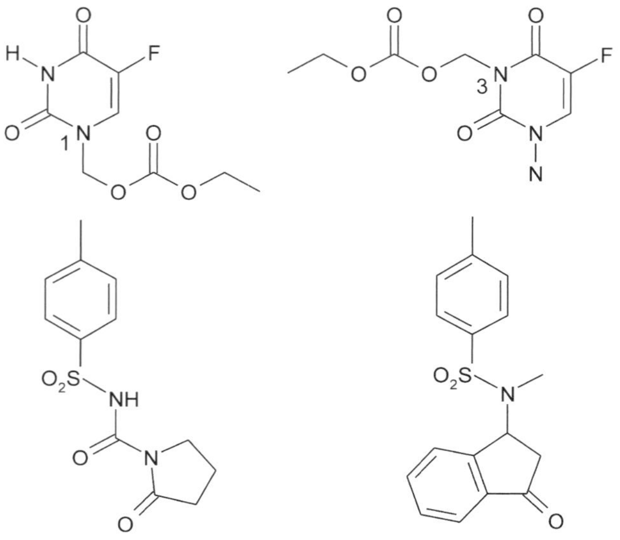

Fig. 33.11 Prodrugs of acidic NH functions.

B Site-specific delivery

Many hopes rested on the prodrug approach as a means to achieve the targeting of drugs for a specific site in the body. In fact, only a few convincing examples are found in the literature and we are becoming somewhat disillusioned about the real possibilities of the approach. In principle, two targeting possibilities can be considered:[31] First, one can design a prodrug that affords an increased or selective transport of the parent drug to the site of action (*site-directed* drug delivery). Second, one can design a derivative that goes everywhere, but undergoes bioactivation only inside the target organ (*site-specific* drug release).

Site-directed drug delivery

Most of the successes in achieving site-directed drug delivery through prodrugs have been through localized delivery of lipophilic prodrugs (eyes, skin) with increased permeability characteristics. Systemic site-directed delivery, that is, to a specific internal site or organ through a selective transport, is very difficult to achieve. Nevertheless, some possibilities of local enrichment or of privileged entry to a given organ or into the central nervous system are found in the literature. Thus the L-glutamic analogue of iproniazide presents a preferential monoamine oxidase inhibition in the brain,[82] whereas the palmitoyl isopropylhydrazide demonstrates a clear cardiac selectivity (Table 33.2).

The propensity of fatty chains to concentrate in cardiac tissue is also illustrated by the findings made in the field of myocardial imaging agents. An iodine and tellurium-containing fatty acid (Fig. 33.13) has a high

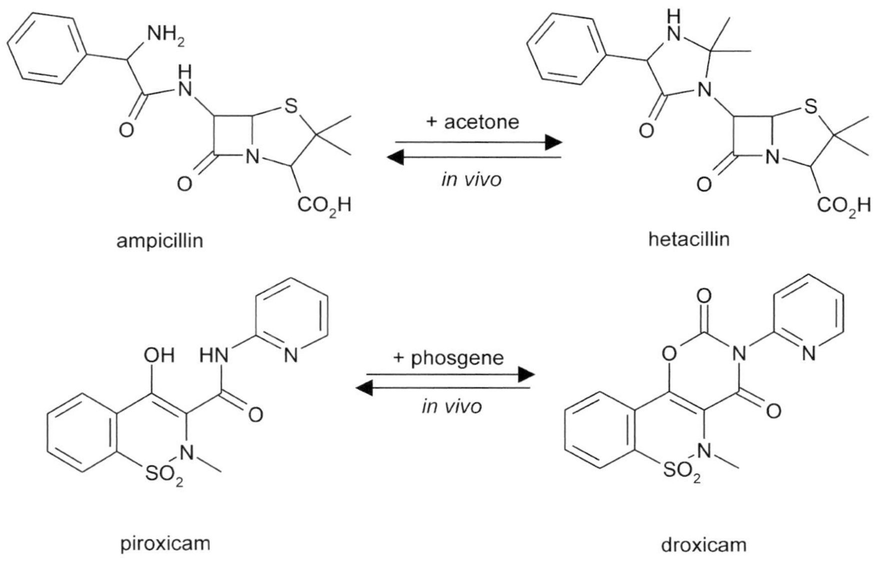

Fig. 33.12 Cyclic protections of two neighbouring functions.

heart uptake, and heart/blood ratios remained high for several hours: 13:1 after 1 hour and 9:1 after 4 hours.[83]

Coupling of drugs to modified bile acids was recently proposed for liver-specific targeting.[84] The rationale is based on the recognition of bile acid-linked drugs by the endogenous bile acid transport system. Chlorambucil, an alkylating cytostatic agent, HR-780, an inhibitor of HMG-CoA reductase, and an oxaproline peptide, an inhibitor of prolyl-4-hydroxylase, were chosen for conjugation to bile acids (Fig. 33.14).

Cytotoxic radicals derived from daunosamine, linked through a glutaric acid spacer to the amino terminal D-Phe residue of somatostatin, are claimed to selectively target prostatic, renal, ovarian and pancreatic cancers.[85]

The 2,3-dichlorophenoxyacetic moiety of ethacrynic acid was claimed to have a high affinity for renal tissue[86] and 2-thiouracil and 6-propylthiouracil exhibit marked affinities for melatonin-producing tissues.[87,88] They were therefore (unsuccessfully) tested for the treatment of malignant melanoma.[89] Many other examples of selective conducting moieties are found in X-ray contrast media and radioisotope imaging agents.[90] Similar efforts were made to find selective cancer chemotherapeutics, and a variety

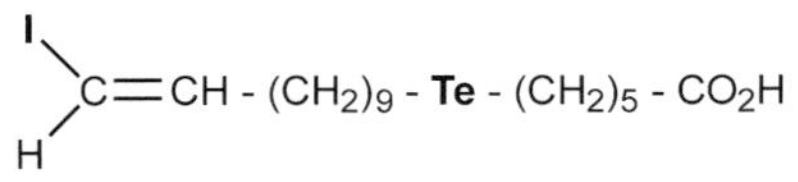

Table 33.2 *Effect of the acyl group in isopropyl hydrazide on selective transport*[82]

Isopropyl hydrazide	% Monoamine increase		Ratios
	Cardiac catecholamines	Cerebral 5-HTP	
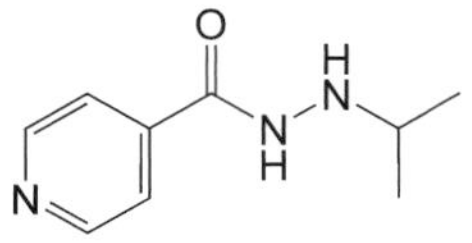	100	100	1.0
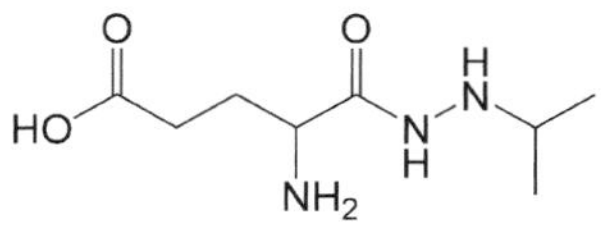	75	250	3.3
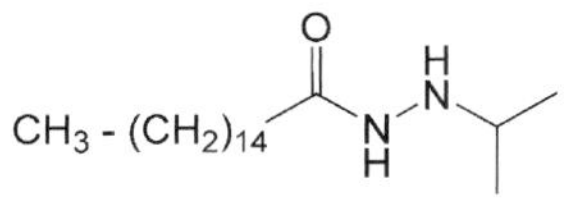	145	60	0.4

Fig. 33.13 Myocardial imaging agent.

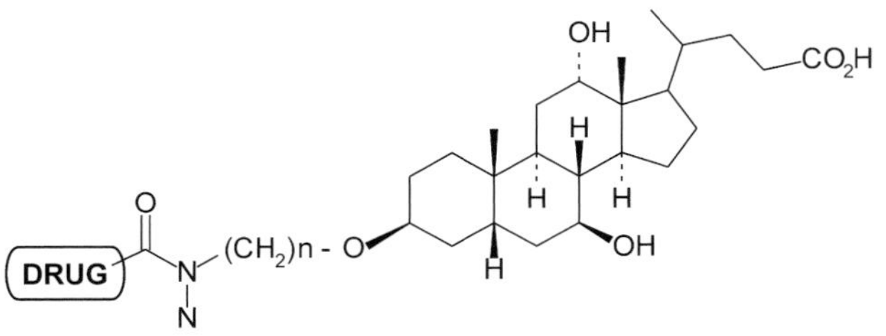

Fig. 33.14 Bile acids for liver-specific targeting.[84]

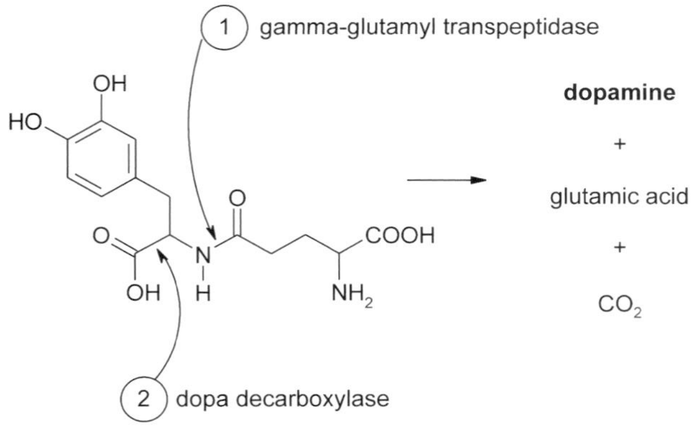

Fig. 33.15 Selective renal vasodilatation with γ-gluta-myl-DOPA.[94]

of drugs that is known to accumulate selectively in particular tissues have also been tried. Mustard derivatives of amino acids, steroid hormones, tetracyclines, quinacrine and uracil are examples.[91] Site-directed cancer chemotherapy can involve drugs bound to specific antibodies. Daunomycin, conjugated via an oxidized dextran bridge with anti-B-cell lymphoma 38C-13 cell-surface IgM antibodies, given to 38C-tumour-bearing mice, resulted in increased life span and even complete cures.[92] Some other attempts to achieve delivery to the CNS can be found in Chapters 38 and 39.

Site-specific drug release

The whole strategy of site-specific release of a given drug lies in the discovery of an enzyme present in high concentrations in the target organ and quasi-absent elsewhere. An appropriate prodrug can then be designed using the selective cleavage possibility offered by the enzyme.

A selective renal vasodilatation, for example, is produced by administration of γ-glutamyldopa. It is well known that L-dopa is a precursor of the neurotransmitter dopamine, which plays an important role in the CNS and in the kidneys. The association of L-dopa with a peripheral dopa-decarboxylase inhibitor allows preferential dopamine production in the brain and can be considered at present the best therapeutic possibility for Parkinson's disease.

On the renal side, a prodrug of L-dopa, γ-glutamyl-L-3,4-dihydroxyphenylalanine (γ-glutamyldopa), produces a specific vasodilatation of renal tissue. Indeed, the γ-glutamyl derivatives of amino acids and peptides accumulate in the kidneys where they undergo a selective metabolic process (for a review see Magnan *et al.*[93]). The successive actions of two enzymes present in high concentration in the kidney, γ-glutamyl transpeptidase and L-aromatic aminoacid decarboxylase, release dopamine locally from γ-glutamyldopa (Fig. 33.15).

In mice, the renal levels of dopamine, after γ-glutamyldopa, are five times higher than after an equimolar administration of L-dopa. A perfusion of $10 \, \mu M \, g^{-1}$ 30 minutes of γ-glutamyldopa in rats produces a 60% increase of the renal plasmatic flux.[94] The same dose of L-dopa induces no vasodilatation. Massive administration of γ-glutamyldopa (20 times the preceding dose) produces only a weak pressor effect, demonstrating that the systemic effects of the prodrug are low. The same principle was used for the synthesis of γ-glutamyl derivatives of dopamine itself and diacyldopamines.[95,96]

Similarly, it is possible to obtain a kidney-selective accumulation of sulfamethoxazole by administering the drug in the form of *N*-acetyl-γ-glutamate.[97] The regeneration of the free sulfamide requires the initial deacylation of the glutamic moiety thanks to an *N*-acylamino acid deacylase which is also present in the kidney in high concentrations (Fig. 33.16). The γ-glutamyl strategy for confining drug

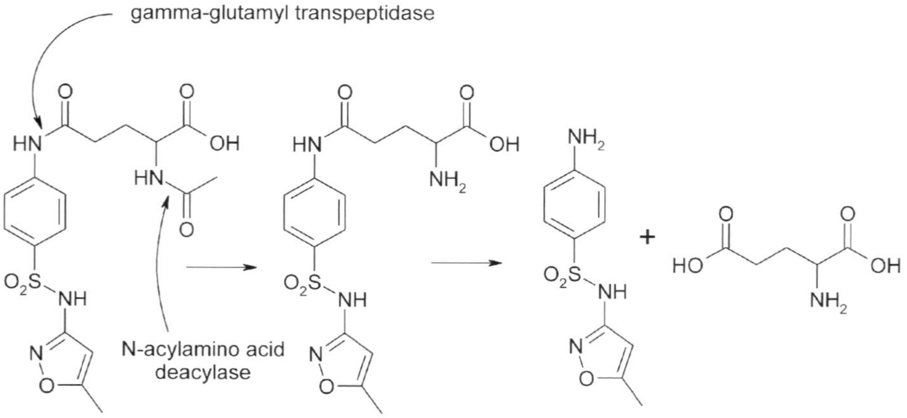

Fig. 33.16 Kidney-selective release of sulfamethoxazole.[97]

action to the kidney and the urinary tract implies that the prodrug under consideration can function as a substrate for γ-glutamyl transpeptidase and, eventually, for N-acylamino acid deacylase.

The unique glucosidase activity of the colonic microflora has been utilized to deliver selectively steroid prodrugs useful in treating inflammatory bowel disease.[98] Dexamethasone 21-β-D-glucoside appeared to be a good candidate as nearly 60% of an oral dose of the prodrug reached the caecum in the form of free steroid. Given orally, the parent dexamethasone is absorbed almost exclusively from the small intestine and less than 1% reaches the caecum.[99] In various tumour tissues, the activity of the enzyme uridine phosphorylase is markedly higher than in the surrounding normal tissues. This observation prompted the synthesis of 5-fluorouracil prodrugs. Among them, 5′-deoxy-5-fluorouracil shows high antitumour activity and less host toxicity compared with fluorouracil. This favourable therapeutic index is attributed to a preferential bioactivation by uridine phosphorylase in the tumour cells.[100,101]

Paclitaxel 2′-carbamates and 2′-carbonates were prepared in order to fulfil the requirements of an ideal prodrug, i.e. low cytotoxicity for healthy tissue, high stability against unspecific enzymes, and fast hydrolysis by the tumour-associated enzyme. These water-soluble prodrugs are specifically activated by plasmin in the tumour cells and show on average a decrease in cytotoxicity of more than 8000-fold in comparison with the parent drug.[102]

C Prolonged duration of action

Unless they are accumulated in the fatty tissues, orally administrated drugs are not expected to act for much longer than their transit period in the gastrointestinal tract (12–48 hours). For drugs that are rapidly cleared from the body, the duration of activity is even shorter, and a frequent dosing within the 24-hour period is required to maintain adequate plasma concentrations. This frequent dosing of short half-life drugs results in sharp peak–valley plasma concentration–time profiles, and consequently patient compliance is often poor.[103] However, for therapeutic, epidemiological, sociological and political reasons, durations of action prolonged over weeks or months are desired. The easiest administration route is represented by intramuscular injection of depot preparations. The most successful applications are found in the domain of hormonal steroids, of antipsychotic drugs and, to a lesser extent, of antibiotics. The general strategy consists of preparing lipophilic prodrugs, dissolved or suspended in oily vehicles, and administering them by deep intramuscular injection.

Contraceptive steroids

Progestagens such as norethysterone enanthate and medroxyprogesterone acetate (MPA) have long durations of

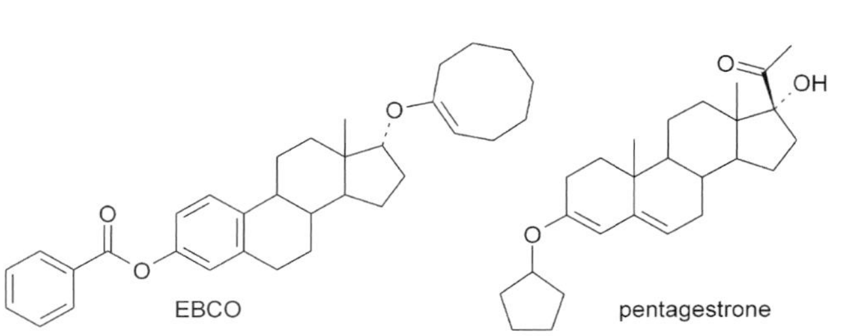

Fig. 33.17 Enol ethers as long-lasting steroid prodrugs.[106,107]

activity (3 months), primarily due to slow release from the injection site and storage in the fatty tissues.[104] Progestagen–estrogen combinations such as dihydroxyprogesterone acetophenide/estradiol enanthate or MPA/estradiol cypionate (= cyclopentylpropionate) are administered on a monthly basis.

Treatment of menopause

The symptomatic treatment of menopause (sweating, hot flushes, depression) has been successfully accomplished by the use of 200 mg of dehydroepiandrosterone-3-heptanoate and 4 mg of estradiol-17-β-valerate in a suspension of a castor oil–benzyl benzoate vehicle.[105]

In the estradiol prodrug estradiol 3-benzoate 17-β-cyclooctenyl ether (EBCO), the phenolic hydroxyl group is masked as a benzoyl ester and the alcoholic 17-β-hydroxyl, as an enol ether derived from cyclooctanone (Fig. 33.17). Given orally to rats as a suspension in sesame oil, this derivative was active for 1 to 2 weeks because it was stored in body fat.[106]

Enol ethers were also used for the synthesis of other long-lasting steroidal drugs such as penmestrol or pentagestrone.[107]

Antipsychotics

Clinically, depot neuroleptics possess several advantages over the short-acting oral forms. Among these, the main advantages are: (a) ease of administration, (b) reliable therapeutic effect with no increase in tolerance, (c) enhanced patient compliance, (d) reduced relapse and rehospitalization rate, and (f) enhanced rate and incidence of 'normal life' reintegration and resocialization.[104]

V USE OF CASCADE PRODRUGS

Classical carrier-linked prodrugs may sometimes be ineffective because the prodrug linkage is too stable (amides, nonactivated esters). In such cases a β-assistance, provided by an easily *in vivo*-generated nucleophile can represent an interesting solution. The release of the active

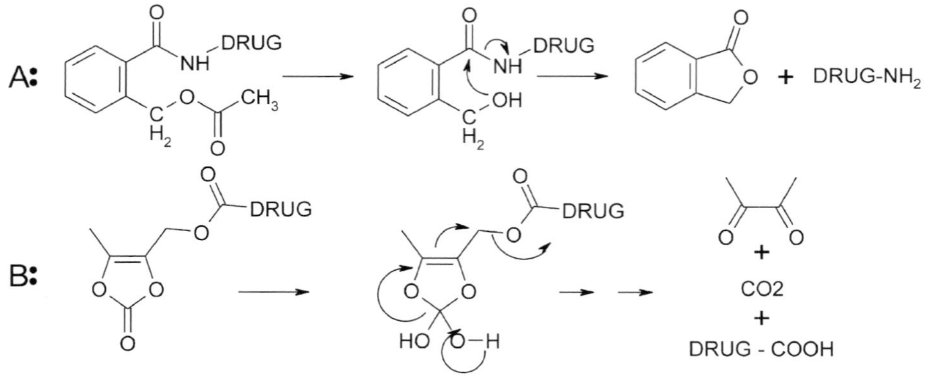

Fig. 33.18 Cascade latentiation. A: 2-Acyloxymethylbenzoic acids provide amides the lability of esters.[108,109] B: Substituted vinyl esters as lipophilic cascade carrier for carboxylic acid-containing drugs.[54,110]

molecule from the prodrug proceeds through a two-step trigger mechanism for which the name 'cascade latentiation' was coined by Cain in 1975.[108,109] The concept, also called distal hydrolysis[40] or the double prodrug concept[31,111] is illustrated by the use of 2-acyloxymethylbenzoic acids as amine protective functions providing amides with the lability of esters (Fig. 33.18A) and by the use of substituted vinyl esters [= (2-oxo-1,3-dioxol-yl)methyl esters] as lipophilic cascade carriers for carboxylic acid-containing drugs such as ampicillin[110] or α-methyldopa[54] or various cephalosporins[12,21,36,112] (Fig. 33.18B).

A Water-soluble taxol prodrugs

Taxol is a potent microtubule-stabilizing agent which has been approved for cancer treatment. Despite taxol's therapeutic promise, its aqueous insolubility (<0.004 mg ml^{-1}) hampers its clinical application. Nicolaou *et al.*[113,114] report the design, synthesis and biological

activity of prodrugs designed to improve water solubility and which can also be considered as cascade prodrugs (Fig. 33.19).

The mechanistic rationale of the design of the first two protaxols lies in the spontaneous decomposition of the carbonate ester after the abstraction of one of the activated protons or of an acidic proton (Fig. 33.20). The release of taxol from the pyridinium prodrug (taxol-2'-methylpyridinium acetate; taxol-2'-MPA) is presumed to be the result of a nucleophilic attack by water or another nucleophile at the 2' position of the pyridinium moiety[114] (Fig. 33.20).

B Bioactivation of an antibacterial prodrug

Although the amino acid (1) is a potent inhibitor of CMP-KDO synthetase (Fig. 33.21), a key enzyme in the biosynthesis of the lipopolysaccharide of Gram-negative bacteria, it is unable to reach its cytoplasmic target and is therefore inactive as an antibacterial agent. Simple

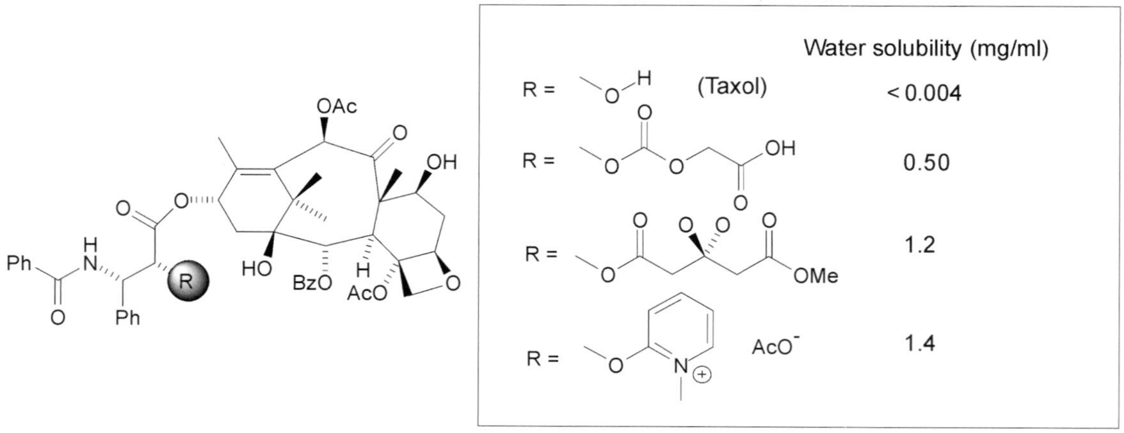

Fig. 33.19 Water-soluble protaxols.[113,114]

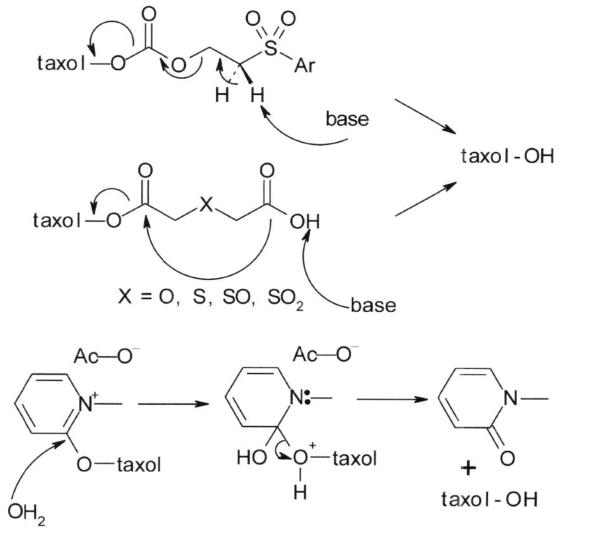

Fig. 33.20 Taxol release mechanisms from protaxols.[113,114]

lipophilic esters are of no use in enhancing the delivery of the amino acid (**1**) since they are not cleaved by the bacteria. The double prodrug (**3**) has, on the other hand, recently been found to solve the problem.[115]

Upon entry into bacterial cells, the disulphide bond in compound (**3**) is reduced by sulphydryl compounds present in the intracellular milieu, resulting in the formation of the

thiol (**2**). This is highly unstable and the active amino acid (**1**) is formed by a rapid, intramolecular displacement.

C Double prodrugs derived from pilocarpine

Monoesters of pilocarpic acid are potentially useful prodrug forms for ocular administration and enable an efficient penetration through the corneal membrane. Unfortunately, they suffer from poor solution stability as in aqueous solution they cyclize spontaneously to pilocarpine.[116] However, double esters derived from pilocarpic acid (Fig. 33.22) possess a high stability in aqueous solution (shelf lives of more than 5 years at 20°C were estimated). At the same time, they are readily converted to pilocarpine under conditions simulating those occurring *in vivo* through a sequential process involving enzymatic hydrolysis of the *O*-acyl bond followed by spontaneous lactonization of the intermediate pilocarpic monoester.[117]

D Double prodrugs for peptides

Amsberry and Borchardt[118,119] have applied Cain's cascade concept to prepare lipophilic polypeptide prodrugs. The amine functionality of the polypeptide is coupled to 2′-acylated derivatives of 3-(2′,5′-dihydroxy-4′,6′-dimethyl-phenyl)-3,3-dimethylpropionic acid (Fig. 33.23). Under simulated physiological conditions the parent amine is regenerated in a two-step process: enzymatic hydrolysis

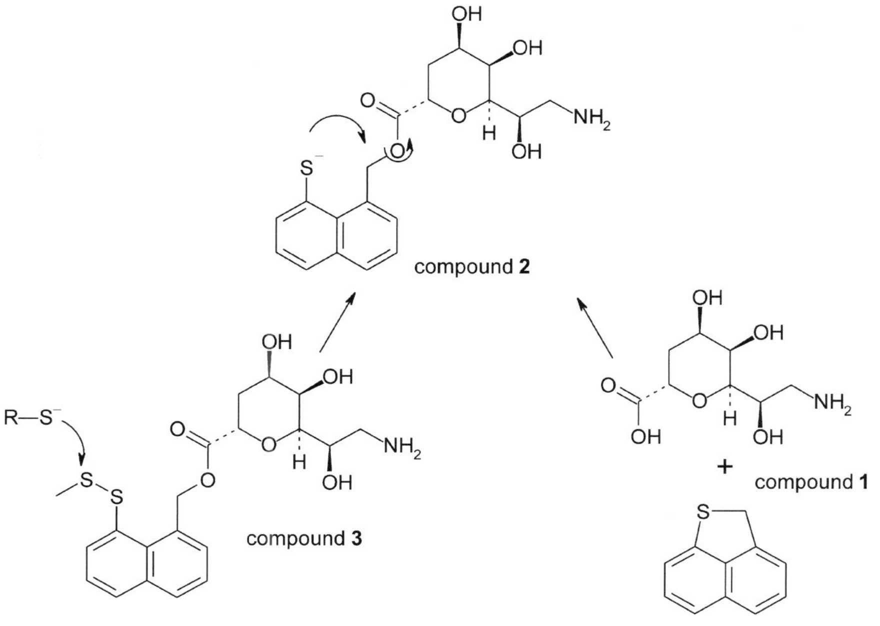

Fig. 33.21 Bioactivation of the antibacterial prodrug of an impermeant inhibitor of 3-deoxy-D-manno-2-octulosonate cytidylyl-transferase.[115]

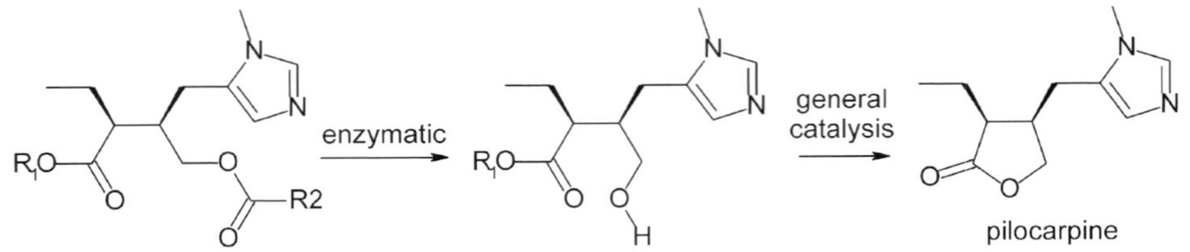

Fig. 33.22 Double esters derived from pilocarpic acid are readily converted to pilocarpine under conditions simulating those occurring *in vivo*.[117]

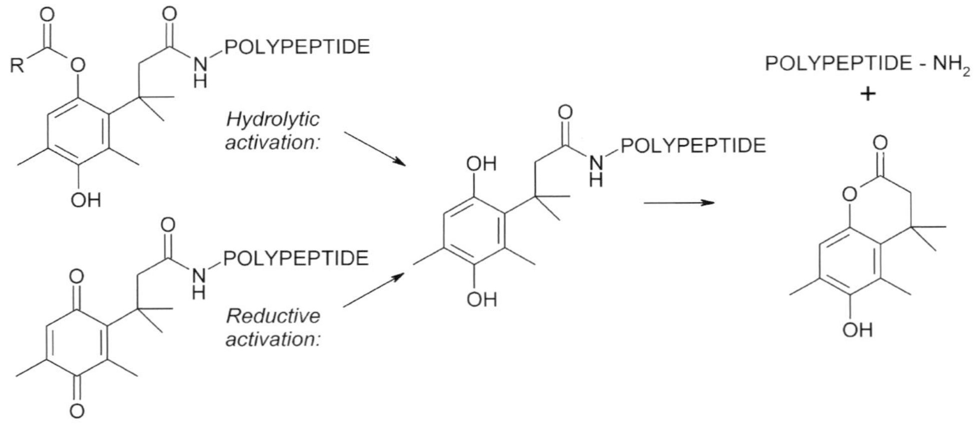

Fig. 33.23 Proposed conversion of esterase-sensitive and redox-sensitive double prodrugs of peptides.[118,119]

of the phenolic ester, followed by a nonenzymatic intramolecular cyclization leading to the release of the free amine (polypeptide) and a lactone.

The lactonization step is highly favoured because of the steric pressure created by the three methyl groups ('trimethyl lock' concept). An alternative to the hydrolytic first step involves a bioreductive generation of the intermediate phenolic amide (Fig. 33.23).

VI SOFT DRUGS

The 'soft' quaternary ammonium salts developed by Bodor *et al.*[120–123] are vulnerable derivatives of their 'hard'

analogues. In general, they show the same type of activity but with a much shorter half-life, as is the case for the soft analogue of cetylpyridinium chloride (Fig. 33.24). Both compounds have the same hydrophobic chain length, and thus similar surface-active and antimicrobial properties. However, the soft analogue is about 40 times less toxic than its hard analogue in terms of LD_{50}.[117] This is because the soft analogue undergoes a fast and easy hydrolytic deactivation resulting in the simultaneous destruction of the positive quaternary head and the surface-active properties.

In a similar way, the tetradecyloxymethyl quaternary salt of pilocarpine allows an enhanced penetration through

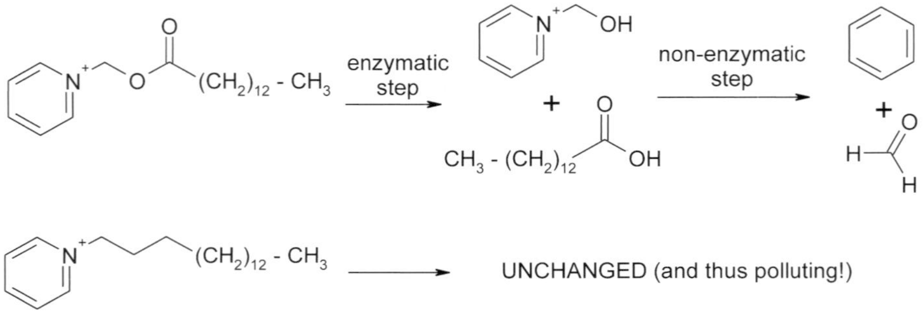

Fig. 33.24 The soft analogue of cetylpyridinium chloride.[121]

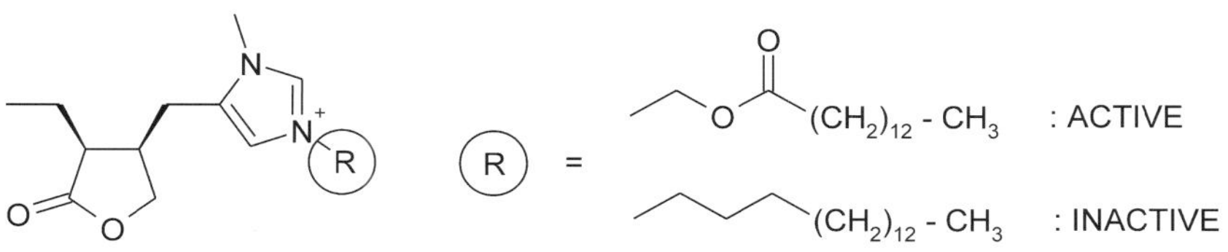

Fig. 33.25 The soft quaternary derivative of pilocarpine allows an enhanced penetration through the cornea.[124]

the cornea followed by a facile hydrolytic cleavage to pilocarpine (Fig. 33.25). The corresponding hard analogue (*N*-cetyl-pilocarpine) is unable to regenerate the parent drug and lacks any activity.[124]

Conclusion

The carrier prodrug approach is particularly successful in the field of antibiotics and in the improvement of some pharmacokinetic parameters. Other prodrug examples are less convincing; they have, nevertheless, been included in this chapter to illustrate the 'state of the art'.[125] Probably most of them have never been tested in humans or the laboratory. The design of carrier prodrugs represents in medicinal chemistry the counterpart of the design of protective groups in organic chemistry. Both approaches have much in common; in both of them imagination has no limits and reigns as master. However, among the enormous number of candidates, only very few attain real success and celebrity.

VII BIOPRECURSOR PRODRUGS: APPLICATION EXAMPLES

The following examples illustrate the bioprecursor-prodrug approach although the intentional use of bioprecursor design is relatively recent and in some cases there are doubts about the prospective or the retrospective character of the design. The first examples relate to oxidative bioactivations; they are followed by examples of reductive

bioactivations and finally by nonredox reactions. Often, however, the active species results from a cascade of metabolic reactions involving oxidative as well as reductive processes, complicated by hydrolytic reactions or hydration–dehydration sequences.

VIII OXIDATIVE BIOACTIVATIONS

A Dexpanthenol and 3-pyridine-methanol as provitamins

A simple example of bioprecursor prodrug is found in dexpanthenol and 3-pyridine-methanol (Fig. 33.26). These primary alcohols are the reduced forms of the vitaminic factors pantothenic acid and nicotinic acid, respectively. Dexpanthenol has the advantage over the parent drug of being more stable, especially towards racemization.

B Oxidative bioactivation of losartan

Losartan is a nonpeptide angiotensin II receptor antagonist used as antihypertensive medication.[126] It can also be considered as a bioprecursor prodrug in so far as, *in vivo*, the primary alcoholic function is oxidized into a carboxylic function (Fig. 33.27) and the obtained acid represents the actual active principle.[127]

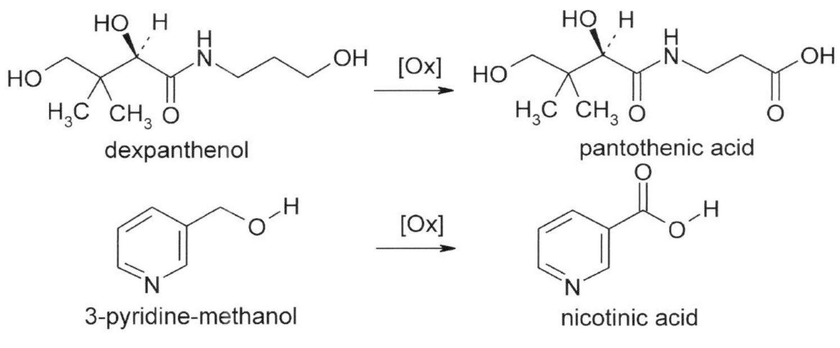

Fig. 33.26 Dexpanthenol and 3-pyridine-methanol are provitamins yielding again the parent molecules after *in vivo* oxydation.

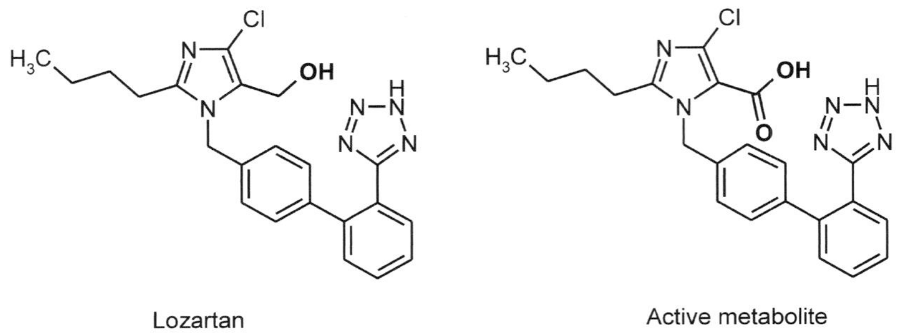

Fig. 33.27 Oxidative bioactivation of losartan.

C Methylenedioxy derivatives as bioprecursors of catechols

Various substituted and unsubstituted methylenedioxy derivatives of apomorphine and *N-n*-propylnorapomorphine have been studied by Baldessarini *et al.*[128] and one of these, 10,11-methylenedioxy-*N-n*-propylnorapomorphine, was found to be both a long-acting and an orally efficient prodrug (Fig. 33.28). The oral activity of the compound can be ascribed to the protection of the catechol system from the first-pass effects by the methylenedioxy group. The conversion to the free catechol is possible thanks to the hepatic microsomal enzymes (see Chapters 28 and 29 on drug metabolism).

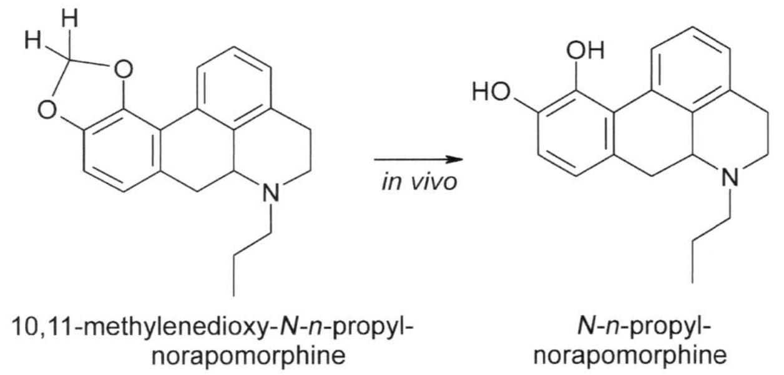

Fig. 33.28 Methylenedioxy derivatives as bioprecursors of catechols.[128]

D Site-specific delivery of the acetylcholinesterase reactivator 2-PAM to the brain

N-Methylpyridinium-2-carbaldoxime (2-PAM, (a); Fig. 33.29) constitutes the most potent reactivator of acetylcholinesterase poisoned through organophosphorus acylation. However, due to its quaternary nitrogen, 2-PAM penetrates the biological membranes poorly and does not appreciably cross the blood–brain barrier. For this compound Bodor and his colleagues[129] designed an ingenious dihydropyridine-pyridinium salt type of redox delivery system. The active drug is administered as its 5,6-dihydropyridine derivative (Pro-2-PAM, (b)), which exists as a stable immonium salt (c). The lipoidal (b) ($pK_a = 6.32$) easily penetrates the blood–brain barrier where it is oxidized to the active (a).

A dramatic increase in the brain delivery of 2-PAM by the use of Pro-2-PAM is thus achieved, resulting in a reactivation of phosphorylated brain acetylcholinesterase *in vivo*.[130,131]

E 6-Deoxyacyclovir as a bioprecursor of acyclovir

The antiherpetic agent acyclovir suffers from a poor oral bioavailability, only 10–20% of an oral dose being absorbed in humans. This can be essentially ascribed to a low water solubility due to strong interaction forces in the crystal lattice. The corresponding deoxo derivative (6-deoxyacyclovir) was shown by Krenitsky *et al.*[132] to be

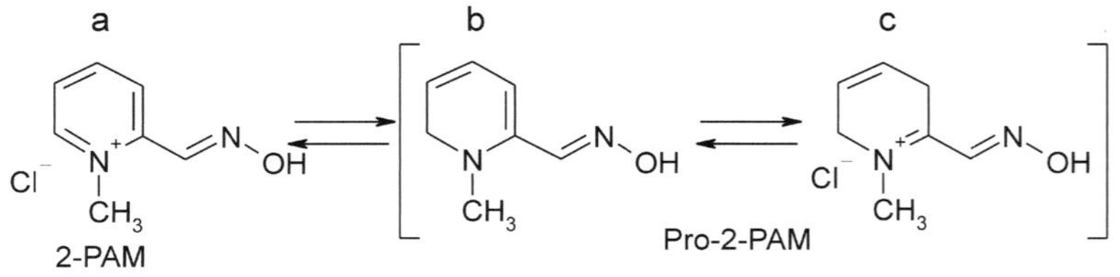

Fig. 33.29 Dihydro derivatives of 2-PAM.[129]

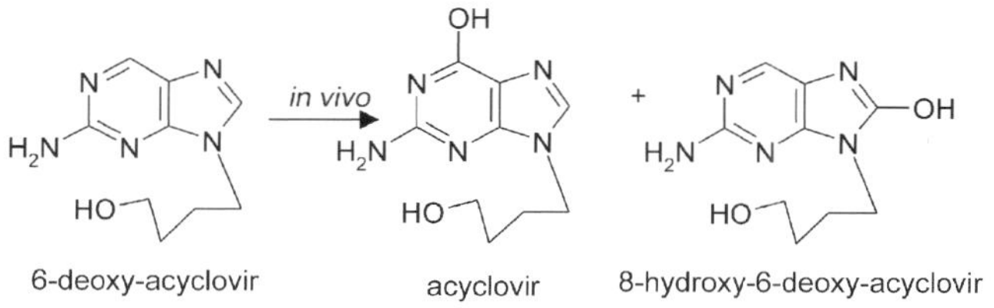

Fig. 33.30 6-Deoxyacyclovir as a bioprecursor of acyclovir.[132,133]

18 times more water soluble and to be rapidly oxidized *in vivo* by xanthine oxidase to the parent drug (Fig. 33.30). Studies in rats and in human volunteers showed that orally administered 6-deoxyacyclovir has a five to six times greater bioavailability than has acyclovir.[132,133]

F Bioactivation of cyclophosphamide

Cyclophosphamide is a cytotoxic (cytostatic), cell cycle nonspecific, antiproliferative agent which is used in such diverse medical problems as neoplasia, tissue transplantation and inflammatory diseases.[134] Chemically, it is an inert bioprecursor for a potent nitrogen mustard alkylation agent (Fig. 33.31).

Cyclophosphamide was synthesized by Arnold and coworkers[137-139] in the hope that it would be inert until activated by an enzyme present in the body, especially in the tumour. The activation mechanism is believed to require an initial oxidative dealkylation, followed by a spontaneous or phosphoramidase-catalysed hydrolysis to the parent nitrogen mustard.[135,136]

IX REDUCTIVE BIOACTIVATIONS

A Reductive bioactivation of nitrogen mustards

Many conventional anticancer drugs display relatively poor selectivity for neoplastic cells, and solid tumours are

particularly resistant to both radiation and chemotherapy. However, in solid tumours there are a few unique and important microenvironmental properties such as localized hypoxia, nutrient deprivation and low pH.[140] On the other hand, as shown above for sulindac, sulphoxides can undergo two major biotransformations: reversible reduction to the sulphide and irreversible oxidation to the sulfone. The oxidation to the sulphone is the dominant process under normal physiological conditions, but the reduction to the sulphide becomes significant under anaerobiotic conditions.[141] Taking advantage of these findings, Kwon *et al.*[142] devised a hypoxia-selective alkylating bioprecursor prodrug (Fig. 33.32).

B Reductive bioactivation of omeprazole

Omeprazole effectively inhibits gastric secretion by inhibiting the gastric H^+, K^+-ATPase.[143] This enzyme is responsible for gastric acid production, and is located in the secretory membranes of parietal cells. Thus, omeprazole is proposed as an anti-ulcerative drug, specially in the treatment of Zollinger–Ellison syndrome.[144]

In vivo omeprazole is transformed into the active inhibitor, a cyclic sulphenamide (Fig. 33.33), which forms disulphide bridges with the thiol groups of the enzyme and thus inactivates it.[145,146] The high specificity in the action of omeprazole ($pK_a = 4.0$) is due to its preferential concentration in the rather acidic parietal cells where it is activated. In neutral regions of the body, omeprazole is rather stable and only partially converted to the active species.

X MIXED BIOACTIVATION MECHANISMS

A Mixed oxidative/reductive bioactivation of dioxolanes

The prodrug SAH 51-641 (1) (Fig. 33.34) is a potent hypoglycemic agent, which acts by inhibiting hepatic

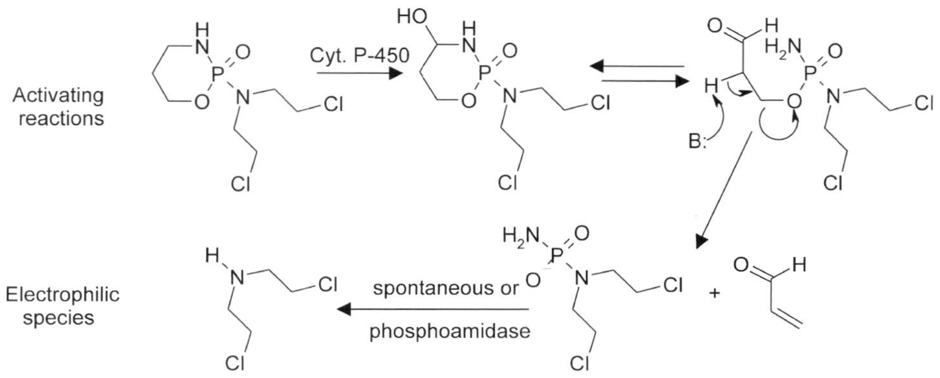

Fig. 33.31 Bioactivation of cyclophosphamide.[135,136]

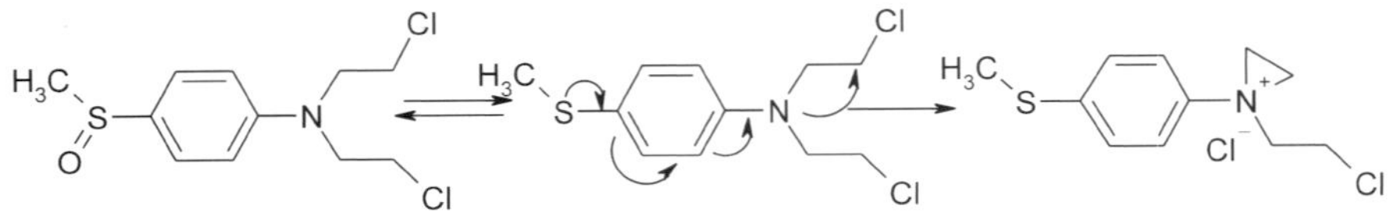

Fig. 33.32 Hypoxia-selective nitrogen mustard.[142]

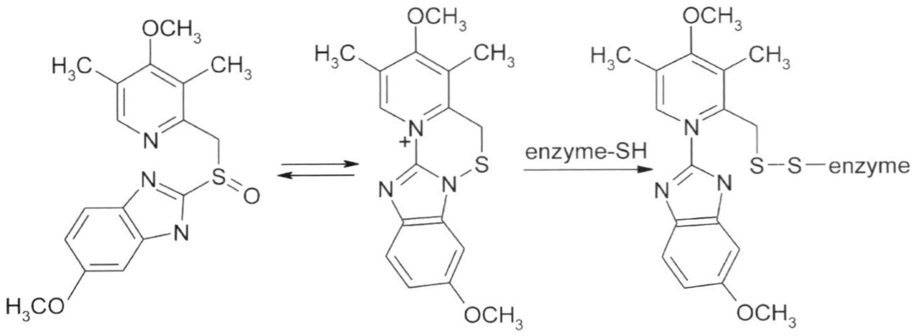

Fig. 33.33 Reductive bioactivation of omeprazole.

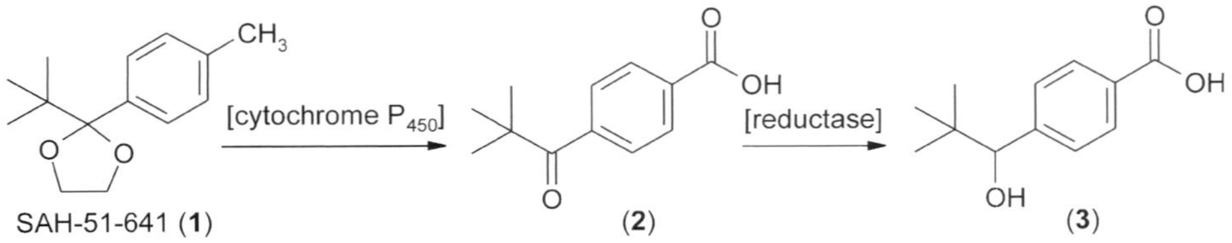

Fig. 33.34 Mixed oxidative/reductive bioactivation of dioxolanes.

gluconeogenesis via an inhibition of fatty acid oxidation.[62] This compound is metabolized by a sequential oxidation/ reduction to the corresponding keto-acid (2) and the hydroxy-acid (3). Compound (3) is a substrate for the medium-chain fatty acyl CoA ligase and represents the actual active agent.[61]

B Arylacetic acids from aroylpropionic precursors

The metabolic pathways involved in the biotransformation of nicotine and haloperidol involve the initial formation of an aroylpropionic acid (3-nicotinoyl-propionic and 3-(*p*-fluorobenzoyl)-propionic acid, respectively). These aroyl-propionic acids subsequently undergo a progressive degradation of the oxobutyric side-chain and finally yield arylacetic acids[23] (Fig. 33.35).

This information was used to design bucloxic acid,[147,148] fenbufene[149,150] and furobufene,[151] which are all bioprecursor forms of anti-inflammatory arylace-tic acids (Fig. 33.36). For all these compounds the bioactivation takes place through a multistep process implying reductive, oxidative and hydration–dehydration sequences.

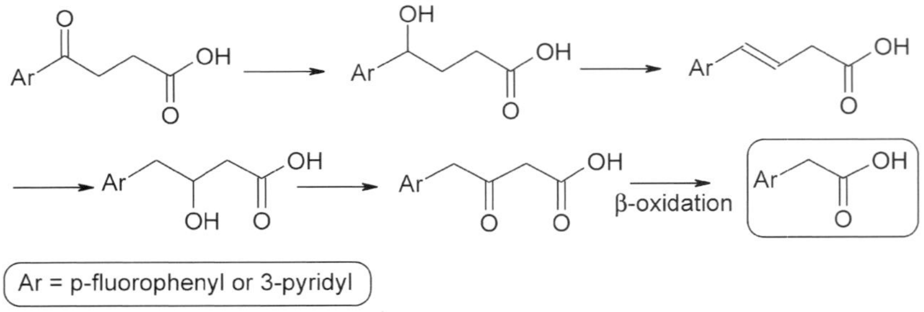

Ar = p-fluorophenyl or 3-pyridyl

Fig. 33.35 Progressive metabolic degradation of β-aroylpropionic acids into arylacetic acids.[23]

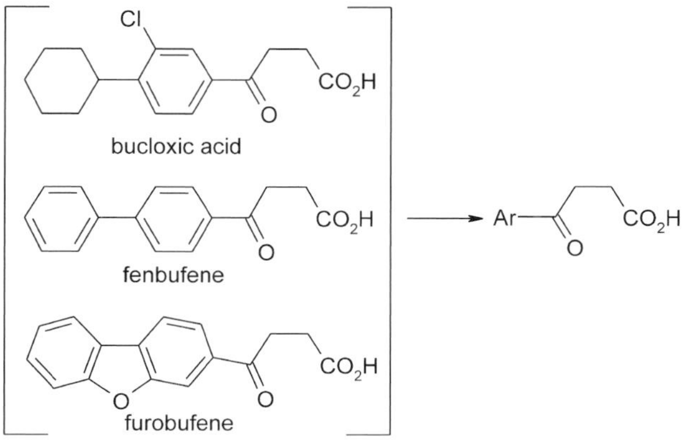

Fig. 33.36 Anti-inflammatory agents presenting the aroylpropionic structure.[147,149,151]

More recently, arylhexenoic acids were shown to undergo a similar metabolic degradation to arylacetic acids.[152] The hexenoic analogue of indomethacin (Fig. 33.37) acts as a prodrug of indomethacin and provides sustained analgesia at oral dosings of 30 mg kg^{-1} to mice (phenylquinone writhing test) or to rats (yeast-induced hyperalgesia test).

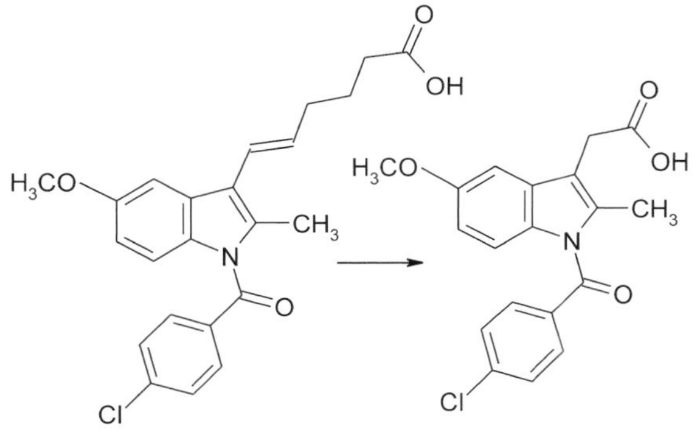

Fig. 33.37 The hexenoic analogue of indomethacin as bioprecursor prodrug of indomethacin.

XI REACTIONS WITHOUT CHANGE IN THE STATE OF OXIDATION

As a rule, within the bioprecursor category of prodrugs, nonredox reactions are infrequent. An example is found in the *in vivo* generation of L-cysteine from its cyclic thiocarbamate.

A L-2-Oxothiazolidine-4-carboxylate: a cysteine delivery system

The enzyme 5-oxo-L-prolinase which catalyses the conversion of 5-oxo-L-proline to L-glutamate coupled to the consumption of ATP (Fig. 33.38), was shown by Williamson and Meister[153] to also act on a synthetic substrate, L-2-oxothiazolidine-4-carboxylate, which is an analogue of 5-oxoproline with the 4-methylene group replaced by sulphur.

The enzyme, which exhibits a similar affinity for the analogue and the natural substrate, is inhibited by the analogue *in vitro* and *in vivo*. L-3-Oxothiazolidine-4-carboxylate thus serves as a potent inhibitor of the γ-glutamyl cycle at the 5-oxoprolinase step. Administration of L-2-oxothiazolidine-4-carboxylate to mice deprived of hepatic glutathione led to restoration of normal hepatic glutathione levels. Since L-2-oxothiazoline-4-carboxylate is an excellent substrate of the enzyme, it may serve as an intracellular delivery system for cysteine and thus has potential as a therapeutic agent for conditions in which there is depletion of hepatic glutathione.

XII DISCUSSION

A Bioprecursors versus carrier prodrugs

A comparative balance-sheet established for the two prodrug approaches led us to the following conclusions (Table 33.3):

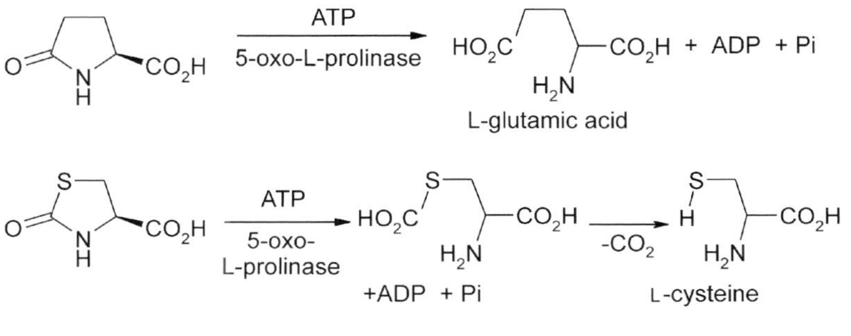

Fig. 33.38 L-2-Oxothiazolidine-4-carboxylate: an intracellular cysteine delivery system.[153]

Table 33.3 *Bioprecursors versus carrier prodrugs*

	Prodrugs	
	Carrier prodrugs	Bioprecursors
Constitution	Active principle + carrier group	No carrier group
Lipophilicity	Strongly modified	Slightly modified
Bioactivation	Hydrolytic	Oxidative or reductive
Catalysis	Chemical or enzymic	Only enzymic

- The *bioavailability* of carrier prodrugs is modulated by using a transient transport moiety; such a linkage is not implied for bioprecursors which result from a molecular modification of the active principle itself.
- The *lipophilicity* is generally the subject of a profound alteration of the parent molecule in the case of carrier prodrugs, whereas it remains practically unchanged for bioprecursors.
- The *bioactivation* process is exclusively hydrolytic for carrier prodrugs; it involves mostly redox systems for bioprecursors.
- The *catalysis* leading to the active principle is hydrolytic (either through general catalysis or through extra-hepatic enzymes) for carrier prodrugs. For bioprecursors, it seems largely restricted to phase I metabolizing enzymes.

B   Existence of mixed-type prodrugs

In some cases the design of mixed-type prodrugs can be advantageous as illustrated in the following three examples.

Disulphide thiamine prodrugs

The thiamin (vitamin B_1) molecule contains a quaternary ammonium functionality and is thus badly absorbed. In healthy patients the necessary amounts of thiamin are absorbed thanks to an active transport mechanism coupled with ATP consumption. However, these mechanisms are rapidly saturable and easily inhibited, especially by chronic alcoholic consumption. As a consequence of the insufficient absorption of thiamin, alcoholism often entails Wernicke's encephalopathy (neurological disorders such as nystagmus, ocular motor nerve paralysis, memory losses, disorientation). The design of lipophilic prodrugs, able to reach the CNS by passive diffusion was then undertaken: compounds like (a) and (b) result from lipophilic disulphide derivation of the open ring thiolate anion corresponding to thiamine (Fig. 33.39).[154]

Such compounds can also be considered as carrier prodrugs, in so far as the thiolate is linked to an *n*-propylthio (a) or a tetrahydrofuranylmethylenethio (b) transport moiety, or as bioprecursors, in so far as a bioreductive

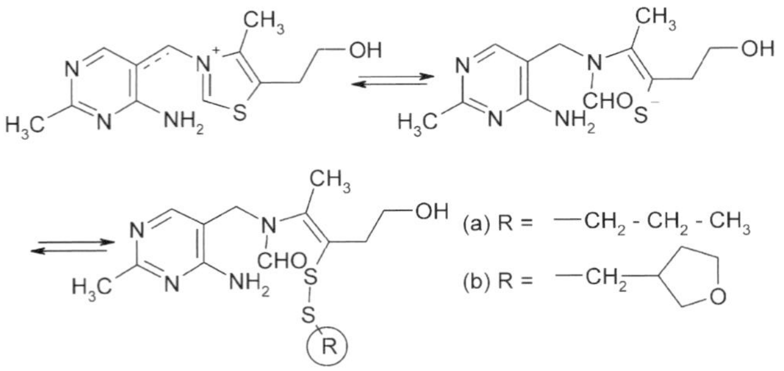

Fig. 33.39 Disulphide thiamine prodrugs as example of mixed-type prodrugs.[154]

cleavage in the thiolate anion is needed to generate the active thiamine; the thiolate anion then functions as a less polar (no quaternary ammonium function) precursor form of thiamine. After oral administration higher thiamine blood levels were observed, in healthy volunteers as well as in cirrhotic patients with the prodrug, than with thiamine hydrochloride.[155]

Trigonelline esters and amines

Generalizing the dihydropyridine ↔ pyridinium salt redox delivery system, successfully applied to 2-PAM, Bodor and his coworkers proposed an astute sustained release methodology for brain delivery, based on the mixed-type prodrugs.[156] The biologically active compound is linked to a lipoidal dihydropyridine carrier that easily penetrates the blood–brain barrier (Fig. 33.40).

Enzymatic oxidation *in vivo* by the NAD ↔ NADH system of the carrier part to the ionic pyridinium salt prevents its elimination from the brain, while elimination from the general circulation is accelerated. Subsequent cleavage of the quaternary carrier-drug species results in sustained delivery of the drug in the brain and a facile elimination of the nontoxic carrier part (trigonelline or its *N*-benzyl analogue).

XIII   DIFFICULTIES AND LIMITATIONS

The introduction of prodrugs in human therapy has given successful results in overcoming undesirable properties such as poor absorption, too rapid biodegradation and formulation problems. It can be expected that an increasing number of medicinal chemists will be tempted by this approach. However, they must keep in mind that prodrug design can also give rise to a large number of new difficulties, especially in the assessment of pharmacological, pharmacokinetic, toxicological and clinical properties.

At the *pharmacological level*, for example, because bioactivation is necessary to create the active species, these compounds cannot be submitted to preliminary *in vitro*

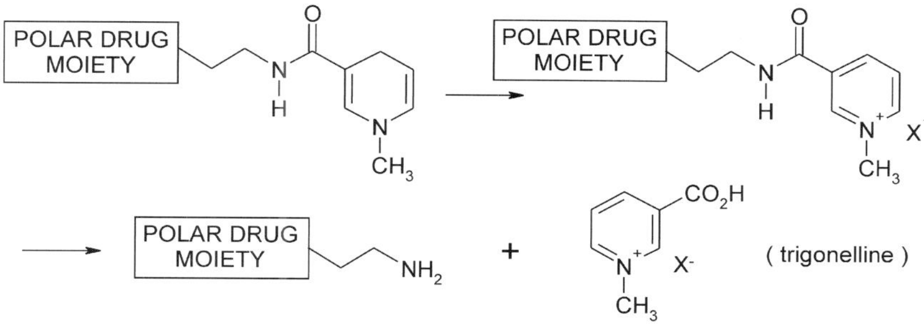

Fig. 33.40 Trigonelline amides (or esters) as examples of mixed-type prodrugs. [156]

screening tests, namely, binding studies, neurotransmitter re-uptake, measurements of enzymatic inhibition and activity on isolated organs.

The measurements of *pharmacokinetic parameters* can lead to numerous misinterpretations. Thus, pivampicilin has a half-life of 103 minutes in a buffered aqueous solution at 37°C, but it falls to less than 1 minute after addition of only 1% of mouse or rat serum. In the presence of human serum (10%), however, it is 50 minutes, whereas in whole human blood it is only 5 minutes. These results exemplify the care required to avoid incorrect conclusions. In addition, when a prodrug and the parent molecule are compared, one must take into account the differences in their respective time-courses of action. The maximum activity can appear later for the prodrug than for the parent compound, and often the comparison of the area under the curve could constitute a better criterion.

At the *toxicological level*, even when prodrugs derive from well-known active principles, they have to be regarded as new entities. Undesirable side-effects can appear that are directly related to the prodrug (allergy to bucloxic acid) or derived from the bioactivation process (formation of unwanted or unexpected metabolites) or which can be attributed to the temporary transport moiety (digestive intolerance to pivampicillin, antivitamin-PP activity of nicaphenine). This latter case is particularly illustrative: an apparently innocent carrier groups such as *N*-hydroxyethyl-nicotinamide appeared as a promising candidate for improving the absorption of acidic anti-inflammatory drugs or clofibric acid.[157–159] However, during the clinical studies, side-effects similar to vitamin PP deficiency appeared, suggesting that *N*-hydroxyethylnicotinamide could function as a nicotinamide antimetabolite. The compounds had then to be withdrawn (H. Cousse, Pierre Fabre & Co, personal communication).

In a review of potential hazards of the prodrug approach, Gorrod[160] cites four toxicity mechanisms:

(1) Formation of a toxic metabolite of the total prodrug, which is not produced by the parent drug.
(2) Consumption of a vital constituent (for example glutathione) during the prodrug activation process. As L-cysteine is needed for the biosynthesis of glutathione, a supply with L-cysteine prodrugs can eventually confer some protection of the hepatic cells.[161]
(3) Generation of a toxic derivative from a transport moiety supposed to be 'inert'.
(4) Release of a pharmacokinetic modifier (causing enzymatic induction, displacing protein-bound molecules, altering drug excretion, etc.)

At the *clinical stage* eventually, the predictive value of animal experiments is also questionable. Thus, for two prodrugs derived from α-methyldopa, the active doses in the rat were identical; nonetheless, they turned out to be very different during the clinical investigations. One compound was just as active as α-methyldopa, whereas the other one was three to four times more active.[162,163]

An application file for a new prodrug should take into account all these aspects and can in no way be regarded just as a complement to the main file.

XIV CONCLUSION

In the future it would be preferable to distinguish the carrier-prodrug and the bioprecursor approaches. The first one, consisting of the attachment of a temporary carrier group to an active principle, has largely proved its utility in the design of orally active antibiotics and more generally wherever high bioavailability in plasma or peripheral organs is required. The CNS delivery of drugs using carrier prodrugs, is less convincing in so far as usually high dosages are needed to ascertain clinical efficiency (1−2 g progabide per day, for example).

The design of bioprecursors, which represents a creative application of the active metabolite concept in the forward-looking way, seems a priori more adequate for CNS delivery, but it still has to prove the reality of its clinical usefulness.

Mixed-type approaches gave good results for thiamine and appear to be an interesting alternative when each individual approach fails.

REFERENCES

1. Ganellin, C.R. (1993) Discovery of cimetidine, ranitidine and other H_2-receptor Histamine Antagonists. In Ganellin, C.R. and Roberts, S.M. (eds). *Medicinal Chemistry. The Role of Organic Chemistry in Drug Research*, pp. 228–255. Academic Press, London.

2. Wermuth, C.G. and Bizière, K. (1986) Pyridazinyl-GABA derivatives: a new class of synthetic $GABA_A$ antagonists. *Trends Pharmacol. Sci.* **7**: 421–424.

3. Rognan, D., Boulanger, T., Hoffmann, R., *et al.* (1992) Structure and molecular modeling of $GABA_A$ receptor antagonists. *J. Med. Chem.* **35**: 1969–1977.

4. Endo, A. (1985) Compactin (ML-236B) and related compounds as potential cholesterol-lowering agents that inhibit HMG-Co A reductase. *J. Med. Chem.* **28**: 401–405.

5. Lee, T.J. (1987) Synthesis, SARs and therapeutic potential of HMG-Co A reductase inhibitors. *Trends Pharmacol. Sci.* **8**: 442–446.

6. Chang, R.S.L. and Lotti, V.Y. (1986) Biochemical and pharmacological characterization of an extremely potent and selective nonpeptide CCK antagonist. *Proc. Natl Acad. Sci. USA* **83**: 4923–4926.

7. Wani, M.C., Taylor, H.L., Wall, M.E., *et al.* (1971) Plant antitumor agents. IV. The isolation and structure of taxol, a novel antileukemic and antitumor agent from *Taxus brevifolia*. *J. Amer. Chem. Soc.* **93**: 2325–2327.

8. Schiff, P.B., Fant, J. and Horwitz, S.B. (1979) Promotion of microtubule assembly *in vitro* by taxol. *Nature* **277**: 665–669.

9. Ishida, M., Miyazaki, K., Kumagai, K. and Rikimaru, M. (1965) Neocarzinostatine, an antitumor antibiotic of high molecular weight isolation, physical properties and biological activities. *J. Antibiot.* **18**: 68.

10. Gully, D., Roger, P. and Wermuth, C.G. (1999) Substituted 4-phenylaminothiazoles, their process of preparation and the pharmaceutical compositions containiing them. US Patent 5880135.

11. Gilligan, P.J., Robertson, D.W. and Zaczek, R. (2000) Corticotropin releasing factor modulators: progress and opportunities for new therapeutic agents. *J.Med.Chem.* **43**: 1641–1658.

12. Albert, A. (1958) Chemical aspects of selective toxicity. *Nature (London)* **182**: 421–423.

13. Harper, N.J. (1959) Drug latentiation. *J. Med. Pharm. Chem.* **1**: 467–500.

14. Wermuth, C.G. (1983) Bioprécurseurs contre prodrogues. *Drug Metabolism and Drug Design: Quo Vadis?*, pp. 253–271. Sanofi-Clin-Midy, Montpellier.

15. Wermuth, C.G. (1984) Designing prodrugs and bioprecursors. *Drug Design: Fact or Fantasy?*, pp. 47–72. Academic Press, London.

16. Sinkula, A.A. (1977) Prodrugs, protective groups and the medicinal chemist. *Medicinal Chemistry*, pp. 125–133. Elsevier, Amsterdam.

17. Stella, V. (1975) Pro-drugs: an overview and definition. *Pro-drugs as a Novel Drug Delivery System*, pp. 1–115. American Chemical Society, Washington DC.

18. Wermuth, C.G. (1980) Les prodrogues, des médicaments plus sûrs et plus maniables. *Bull. Soc. Pharm. Bordeaux* **119**: 107–129.

19. Wermuth, C.G. (1981) Modulation of natural substances in order to improve their pharmacokinetic properties. *Natural Products as Medicinal Agents*, pp. 185–216. Hippokrates Verlag, Stuttgart.

20. Daehne, W.V., Frederiksen, E., Gundersen, E., *et al.* (1970) Acyloxymethyl esters of ampicillin. *J. Med. Chem.* **13**: 607–612.

21. Bodin, N.O., Ekström, B., Forsgren, U., *et al.* (1975) Bacampicilline: a new orally well-absorbed derivative of ampicillin. *Antimicrob. Agents Chemother.* **8**: 518–525.

22. Clayton, J.P., Cole, M., Elson, S.W., *et al.* (1976) Preparation, hydrolysis, and oral absorption of lactonyl esters of penicillins. *J. Med. Chem.* **19**: 1385–1391.

23. Testa, B. and Jenner, P. (1976) *Drug Metabolism, Chemical and Biochemical Aspects.* Marcel Dekker, New York.

24. Shen, T.I., Witzel, B.E., Jones, H., *et al.* (1972) Synthesis of a new anti-inflammatory agent, *cis*-5-fluoro-2-methyl-1-[*p*-(methylsulfinyl) benzylidenyl]-indene-3-acetic acid. *Fed. Proc.* **31**: 577.

25. Hucker, H.B., Stauffer, S.C., White, S.D., *et al.* (1973) Physiological disposition and metabolic fate of a new anti-inflammatory agent, *cis*-5-fluoro-2-methyl-1-[*p*-(methylsulfinyl) benzylidenyl]-indene-3-acetic acid in the rat, dog, rhesus monkey, and man. *Drug Metab. Dispos.* **1**: 721–736.

26. Duggan, D.E., Hare, L.E., Ditzler, C.A., *et al.* (1977) The disposition of sulindac. *Clin. Pharmacol. Ther.* **21**: 326–335.

27. Duggan, D.E., Hooke, K.F., Noll, R.M., *et al.* (1978) Comparative biodisposition of sulindac and metabolites in five species. *Biochem. Pharmacol.* **27**: 2311–2320.

28. Duggan, D.E., Hooke, K.F., Risley, E.A., *et al.* (1977) Identification of the biologically active form of sulindac. *J. Pharmacol. Exp. Ther.* **201**: 8–13.

29. Drabek, J. and Neumann, R. (1985) Proinsecticides. *Progress in Pesticide Biochemistry and Toxicology*, pp. 35–86. John Wiley & Sons, Chichester.

30. Higuchi, T. and Stella, V. (1975) Pro-drugs as novel drug delivery systems. In Society, A.C. (ed.). *ACS Symposium Series.* American Chemical Society, Washington DC.

31. Bundgaard, H. (1991) Design and application of prodrugs. *A Textbook of Drug Design and Development*, 1st edn, pp. 113–191. Harwood Academic Publisher, Chur.

32. McClure, D. (1975) The effect of a pro-drug of epinephrine (dipivaloyl-epinephrine) in glaucoma — general pharmacology, toxicology and clinical experience. *Pro-Drugs as Novel Drug Delivery Systems*, pp. 224–235. American Chemical Society, Washington DC.

33. Horn, A.S. (1980) Pro-drugs of dopaminergic agonists. *Chem. Ind.* 441–444.

34. Westerink, B.H.C., Dijkstra, D., Feenstra, M.G.P., *et al.* (1980) Dopaminergic prodrugs: brain concentration and neurochemical effects of 5,6- and 6,7-ADTN after administration as dibenzoyl esters. *Eur. J. Pharmacol.* **61**: 7–15.

35. Tullar, B.F., Minatoya, H. and Lorenz, R.R. (1976) Esters of *N*-tert-butylarterenol. Long-acting new bronchodilators with reduced cardiac effects. *J. Med. Chem.* **19**: 834–838.

36. Lin, T.S. (1984) Synthesis and *in vitro* antiviral activity of $3'$-O-acyl derivatives of $5'$-amino-$2'$-deoxy thymidine: potential prodrugs for topical application. *J. Pharm. Sci.* **73**: 1568–1571.

37. Harnden, M.R., Jarvest, R.L., Boyd, M.R., *et al.* (1989) Prodrugs of the selective antiherpesvirus agent 9-[4-hydroxy-3-(hydroxymethyl)but-1-yl]guanine (BRL 39123) with improved gastrointestinal absorption properties. *J. Med. Chem.* **32**: 1738–1743.

38. Jones, L.A., Moorman, A.R., Chamberlain, S.D., *et al.* (1992) Di- and triester prodrugs of the Varicella-Zoster antiviral agent 6-methoxy-purine arabinoside. *J. Med. Chem.* **35**: 56–63.

39. Bodor, N., Sloan, K.B., Little, R.J., *et al.* (1982) Soft drugs 4. 3-Spirothiazolidines of hydrocortisone and its derivatives. *Int. J. Pharm.* **10**: 307–321.

40. Collis, A.J. (1993) Drug access and prodrugs. *Medicinal Chemistry — The Role of Organic Chemistry in Drug Research*, 2nd edn, pp. 61–82. Academic Press, London.

41. Anden, N.E., Corrodi, H., Dahlström, A., Fuxe, K. and Högfelt, T. (1966) Effects of tyrosine hydroxylase inhibition on the amine levels of central monoamine neurons. *Life Sci.* **5**: 561–568.

42. Anonymous (2001) Etilevodopa. *Drugs of the Future* **26**: 219–223.

43. Frey, H.H., Popp, C. and Löscher, W. (1979) Influence of inhibitors of the high affinity GABA uptake on seizure threshold in mice. *Neuropharmacology* **18**: 581–590.

44. Ulm, E.H., Hichens, M., Gomez, H.J., *et al.* (1982) Enalapril maleate and a lysine analogue (MK-521): disposition in man. *Br. J. Pharmacol.* **14**: 357–362.

45. Swanson, B.L., Vlasses, P.H., Ferguson, R.K., *et al.* (1984) Influence of food on the bioavailability of enalapril. *J. Pharm. Sci.* **73**: 1655–1657.

46. Zannad, F. (1993) Trandolapril, how does it differ from other angiotensin converting enzyme inhibitors? *Drugs* **46**: 172–183.

47. Tsybina, N.M., Ostrovskaya, R.U., Protopova, T.V., *et al.* (1974) Synthesis and pharmacological activity of gamma-aminobutyric acid derivatives. *Khim. Pharm. Zh.* **17**: 10–13.

48. Ostrovskaya, R.U., Parin, V.V. and Tsybina, N.M. (1972) The comparative neurotropic potency of gamma-aminobutyric acid and its cetyl ester. *Byul. Eksp. Biol. Med.* **73**: 51–55.

49. Beardsley, G.P. and Rosowsky, A. (1980) Effect of methotrexate γ-monohexadecyl ester (γ-MHxMTX) on nucleoside uptake by human leukemic cells. *Proc. Am. Assoc. Cancer Res.* **21**: (71st Meeting)264.

50. Wermuth, C.G., Gaignault, J.-C. and Marchandeau, C. (1996) Designing prodrugs and bioprecursors I: carrier prodrugs. *The Practice of Medicinal Chemistry*, pp. 671–696. Academic Press, London.

51. Jones, G. (1980) Lipoidal pro-drug analogues of various anti-inflammatory agents. *Chem. Ind (London)* 452–456.

52. Jansen, A.B.A. and Russel, T.J. (1965) Some novel penicillin derivatives. *J. Chem. Soc.* 2127–2132.

53. Binderup, E., Godtfredsen, W.O. and Roholt, K. (1971) Orally active cephaloglycin esters. *J. Antibiot.* **24**: 767–773.

54. Saari, W.S., Halczenko, W., Cochran, D.W., *et al.* (1984) 3-Hydroxy-α-methyltyrosine progenitors: synthesis and evaluation of some (2-oxo-1,3-dioxol-4-yl)methyl esters. *J. Med. Chem.* **27**: 713–717.

55. Falch, E., Krogsgaard-Larsen, P. and Christensen, A.V. (1981) Esters of isoguvacine as potential prodrugs. *J. Med. Chem.* **24**: 285–289.

56. Svahn, C.M., Merenyi, F. and Karlson, L. (1986) Tranexamic acid derivatives with enhanced absorption. *J. Med. Chem.* **29**: 448–453.

57. Torriani, H. (1979) Talniflumate. *Drugs Fut.* **4**: 448–450.

58. Torriani, H. (1982) Talmetacin. *Drugs Fut.* **7**: 823–824.

59. Alexander, J., Fromtling, R.A., Bland, J.A., *et al.* (1991) (Acyloxy) alkyl carbamate prodrugs of norfloxacin. *J. Med. Chem.* **34**: 78–81.

60. Ahlmark, M., Vepsaelaeinen, J., Taipale, H., *et al.* (1999) Bisphosphonate prodrugs: synthesis and *in vitro* evaluation of novel clodronic acid dianhydrides as bioreversible prodrugs of clodronate. *J.Med.Chem.* **42**: 1473–1476.

61. Aicher, T.D., Bebernitz, G.R., Bell, P.A., *et al.* (1999) Hypoglycemic Prodrugs of 4-(2,2-Dimethyl-1-oxopropyl)benzoic Acid. *J.Med.Chem.* **42**: 153–163.

62. Young, D.A., Ho, R.S., Bell, P.A., *et al.* (1990) Inhibition of hepatic glucose production by SDZ 51641. *Diabetes* **39**: 1408–1413.

63. Walsh, D.A., Moran, H.W., Shamblee, D.A., *et al.* (1990) Anti-inflammatory agents. 4. Syntheses and biological evaluation of potential prodrugs of 2-amino-3-benzoylbenzeneacetic acid and 2-amino-3-(4-chlorobenzoyl)benzeneacetic acid. *J. Med. Chem.* **33**: 2296–2304.

64. Galzinga, L., Garbin, L., Bianchi, M. and Marzotto, A. (1978) Properties of two derivatives of γ-aminobutyric acid (GABA) capable of abolishing cardiazol- and bicuculline-induced convulsions in the rat. *Arch. int. Pharmacodyn.* **235**: 73–85.

65. Ahtee, L., Halmekoski, J., Heinonen, H. and Koskimies, A. (1979) Comparison of the central nervous system actions of taurine and N-pivaloyltaurine. *Br. J. Pharmacol* **66**: 480P.

66. Kaplan, J.P., Raizon, B., Desarmenien, M., *et al.* (1980) New anticonvulsants: Schiff bases of γ-aminobutyric acid and γ-aminobutyramide. *J. Med. Chem.* **23**: 702–704.

67. Jensen, N.P., Friedman, J.J., Kropp, H. and Kahan, F.M. (1980) Use of acetylacetone to prepare a prodrug of cycloserine. *J. Med. Chem.* **23**: 6–8.

68. Bodor, N.S., Sloan, K.B. and Hussain, A.A (1975) Novel Transient Pro-drug Forms of L-DOPA. US Patent 3 891 696 (24 June 1975; Inter'X Res. Corp.).

69. Koch, H. (1981) ST-1059. *Drugs Future* **6**: 244–246.

70. Vicentini, C.B., Guarneri, M. and Sarto, G. (1983) Hypoglycemic compounds. Sulfonylurea derivatives containing amino acids and dipeptides. *Farmaco, Ed Sci.* **38**: 595–608.

71. Rubenstein, S.M., Baichwal, V., Beckmann, H., *et al.* (2001) Hydrophilic, pro-drug analogues of T138067 are efficacious in controlling tumor growth *in vivo* and show a decreased ability to cross the blood–brain barrier. *J. Med. Chem.* **44**: 3599–3605.

72. Boehringer (1976) Dérivés de la L-(3,4-dihydroxy-phényl)-2-méthyl alanine et leur préparation. Belgium Patent 839 362 (9 March 1976; Boehringer Mannheim GMBH).

73. Kingsbury, W.D., Boehm, J.C., Mehta, R.J. and Grappel, S.F. (1983) Transport of antimicrobial agents using peptide carrier systems: anticandidal activity of *m*-fluorophenylalanine peptide conjugates. *J. Med. Chem.* **26**: 1725–1729.

74. Hu, M., Subramanian, P., Mosberg, H.I. and Amidon, G.L. (1989) Use of the peptide carrier system to improve the intestinal absorption of L-α-methyldopa: carrier kinetics, intestinal permeabilities, and *in vitro* hydrolysis of dipeptidyl derivatives of L-α-methyldopa. *Pharm. Res.* **6**: 66–70.

75. Rahmathullah, S.M., Hall, J.E., Bender, B.C., *et al.* (1994) Prodrugs for amidines: synthesis and anti-*Pneumocystis carinii* activity of carbamates of 2,5-bis(4-amidinophenyl)furan. *J.Med.Chem.* **42**: 3994–4000.

76. Buur, A., Bundgaard, H. and Falch, E. (1986) Prodrugs of 5-fluorouracil. VII. Hydrolysis kinetics and physicochemical properties of *N*-ethoxy- and *N*-phenoxycarbonylmethyl derivatives of 5-fluorouracil. *Acta Pharm. Suec.* **23**: 205–216.

77. Larsen, J.D. and Bundgaard, H. (1987) Prodrug forms for the sulfonamide group. I. Evaluation of *N*-acyl derivatives, *N*-sulfonylamidines, *N*-sulfonyl-sulfilimines and sulfonylureas as possible prodrug derivatives. *Int. J. Pharm.* **37**: 87–95.

78. Bundgaard, H., Buur, A., Hansen, K.T., *et al.* (1988) Prodrugs as drug delivery systems. 77. Phtalidyl derivatives as prodrug forms for amides, sulfonamides, carbamates and other NH-acidic compounds. *Int. J. Pharm.* **45**: 47–57.

79. Hardcastle, G.A., Johnson, D.A., Panetta, C.A., *et al.* (1966) The preparation and structure of hetacillin. *J. Org. Chem.* **31**: 897–899.

80. Bundgaard, H. (ed.). (1985) Design of prodrugs: bioreversible derivatives for various functional groups and chemical entities. *Design of Prodrugs*, pp. 1–92. Elsevier, Amsterdam.

81. Anonymous (1986) Droxicam. *Drugs Fut.* **11**: 835–836.

82. Zeller, P., Pletscher, A., Gey, K.F., *et al.* (1959) Amino acid and fatty acid hydrazides: chemistry and action on monoamine oxidase. *Ann. N.Y. Acad. Sci.* **80**: 555–567.

83. Knapp, F.F.J., Goodman, M.M., Callahan, A.P., *et al.* (1983) New myocardial imaging agents: stabilization of radioiodine as a terminal vinyl iodide moiety on tellurium fatty acids. *J. Med. Chem.* **26**: 1293–1300.

84. Wess, G., Kramer, W., Schubert, G. *et al.* (1993) Coupling of Drugs to Modified Bile Acids for Liver Specific Targeting. *Abst. Pap. 205th Meet. Am. Chem. Soc.* Pt. 1, MEDI 152.

85. Nagy, A. and Schally, A.V. (2001) Targeted cytotoxic somatostatin analogs: a modern approach to the therapy of various cancers. *Drugs of the Future* **26**: 261–270.

86. Biel, J.H. and Martin, Y.C. (1971) Organic synthesis as a source of new drugs. *Drug Discovery — Science and Development in a Changing Society*, pp. 81–111. American Chemical Society, Washington DC.

87. Whittaker, J.R. (1971) Biosynthesis of a thiouracil pheomelanin in embryonic pigment cells exposed to thiouracil. *J. Biol. Chem.* **246**: 6217–6226.

88. Dencker, L., Larsson, B., Olander, K., Ullberg, S. and Yokota, M. (1979) False precursors of melanin as selective melanoma seekers. *Br. J. Cancer* **39**: 449–452.

89. Wätjen, F., Buchardt, O. and Langvad, E. (1982) Affinity therapeutics. 1. Selective incorporation of 2-thiouracil derivatives in murine melanomas. Cytostatic activity of 2-thiouracil arotinoids, 2-thiouracil retinoids, arotinoids and retinoids. *J. Med. Chem.* **25**: 956–960.

90. Ariëns, E.J. (1971) Modulation of pharmacokinetics by molecular manipulation. *Drug Design*, pp. 1–127. Academic Press, New York.

91. Ariëns, E.J. (1975) Pharmacological basis of cancer therapy. In *Twenty-Seventh Annual Symposium on Fundamental Cancer Research*, 1974. The University of Texas M.D. Anderson Hospital and Tumor Institute at Houston: The Williams and Wilkins Company, Baltimore.

92. Hurwitx, E., Kashi, R., Burowsky, D., *et al.* (1983) Site-directed chemotherapy with a drug bound to anti-idiotypic antibody to a lymphoma cell-surface IgM. *Int. J. Cancer* **31**: 745–748.

93. Magnan, S.D.J., Shirota, F.N. and Nagasawa, H.T. (1982) Drug latentiation by γ-glutamyl transpeptidase. *J. Med. Chem.* **25**: 1018–1021.

94. Wilk, S., Mizoguchi, H. and Orlowski, M. (1978) Gamma-glutamyl-DOPA: a kidney specific dopamine precursor. *J. Pharmacol. Exp. Ther.* **206**: 227–232.

95. Kyncl, J.J., Minard, F.N. and Jones, P.H. (1979) Peripheral dopamine receptors. *Symposium on Peripheral Dopaminergic Receptors—Strasbourg*, July 1978, pp. 369–380. Pergamon Press, Oxford.

96. Jones, P.H., Kyncl, J., Ours, C.W. and Somani, P. (1977) Esters of γ-Glutamyl Amide of Dopamine. US Patent 4 017 636 (12 April 1977; Abott Laboratories).

97. Orlowski, M., Mizoguchi, H. and Wilk, S. (1979) *N*-acyl-γ-glutamyl derivatives of sulfamethoxazole as models of kidney-selective prodrugs. *J. Pharmacol. Exp. Ther.* **212**: 167–172.

98. Friend, D.R. and Chang, G.W. (1985) Drug glycosides: potential prodrugs for colon-specific drug delivery. *J. Med. Chem.* **28**: 51–57.

99. Friend, D.R. and Chang, G.W. (1984) A colon-specific drug-delivery system based on drug glycosides and the glycosidases of colonic bacteria. *J. Med. Chem.* **27**: 261–266.

100. Cook, A.F., Holman, M.J., Kramer, M.J. and Trown, P.W. (1979) Florinated pyrimidine nucleosides. 3. Synthesis and antitumor activity of a series of 5′-deoxy-5-fluoropyrimidine nucleosides. *J. Med. Chem.* **22**: 1330–1335.

101. Au, J.L.-S., Walker, J.S. and Rustum, Y. (1983) Pharmacokinetic studies of 5-fluorouracil and 5′-deoxy-5-fluorouridine in rats. *J. Pharmacol. Exptl. Ther.* **227**: 174–180.

102. de Groot, F.M.H., van Berkom, L.W.A. and Scheeren, H.W. (2000) Synthesis and biological evaluation of 2′-carbamate-linked and 3′-carbonate-linked prodrugs of paclitaxel: selective activation by the tumor-associated protease plasmin. *J.Med.Chem.* **43**: 3093–3102.

103. Stella, V.J., Charman, W.N.A. and Naringrekar, V.H. (1985) Prodrugs, do they have advantages in clinical practice? *Drugs* **29**: 455–473.

104. Sinkula, A.A. (1985) Sustained drug action accomplished by the prodrug approach. In Bundgaard, H. (ed.). *Design of Prodrugs*, pp. 157–176. Elsevier, Amsterdam.

105. Dusterberg, B. and Wendt, H. (1983) Plasma levels of dehydroepiandrosterone and 17β-estradiol after intramuscular administration of Gynodian-Depot® in three women. *Hormone Res.* **17**: 84–89.

106. Falconi, G., Galetti, F., Celasco, G. and Gardi, R. (1972) Oral long-lasting estrogenic activity of estradiol 3-benzoate 17-cyclooctenyl ether. *Steroids* **20**: 627–632.

107. Ercoli, A. and Gardi, R. (1960) Δ⁴-3-Keto steroidal ethers. Paradoxical dependency of their effectiveness on the administration route. *J. Am. Chem. Soc.* **82**: 746–748.

108. Cain, B.F. (1975) The role of structure–activity studies in the design of antitumor agents. *Cancer Chemother. Rep.* **59**: 679–683.

109. Cain, B.F. (1976) 2-Acyloxymethylbenzoic acids. Novel amine protective functions providing amides with the lability of esters. *J. Org. Chem.* **41**: 2029–2031.

110. Sakamoto, F., Ikeda, S. and Tsukamoto, G. (1984) Studies on Prodrugs. II. Preparation and characterization of (5-substituted 2-oxo-1,3-dioxolen-4-yl)methyl esters of ampicillin. *Chem. Pharm. Bull.* **32**: 2241–2248.

111. Bundgaard, H. (1991) Novel chemical approaches in prodrug design. *Drugs Fut.* **16**: 443–458.

112. Borgman, R.J., Smith, R.V. and Keiser, J.E. (1975) The acetylation of apomorphine. An improved method for the selective preparation of diacetylapomorphine utilizing trifluoroacetic acid/acetyl bromide. *Synthesis* 249–250.

113. Nicolaou, K.C., Riemer, C., Kerr, M.A., *et al.* (1993) Design, synthesis and biological activity of protaxols. *Nature (London)* **364**: 464–466.

114. Nicolaou, K.C., Guy, R.K., Pitsinos, E.N. and Wrasidlo, W. (1994) A water-soluble prodrug of Taxol with self-assembling properties. *Angew. Chem. Internat. Ed. Engl.* **33**: 1583–1586.

115. Norbeck, D.W., Rosenbrook, W., Kramer, J.B., *et al.* (1989) A novel prodrug of an impermeant inhibitor of 3-deoxy-D-manno-2-octulosonate cytidylyl-transferase has antibacterial activity. *J. Med. Chem.* **32**: 625–629.

116. Bundgaard, H., Falch, E., Larsen, C. and Mikkelson, T.J. (1986) Pilocarpine prodrugs I. Synthesis, stability, bioconversion, and physicochemical properties of sequentially labile pilocarpine acid diesters. *J. Pharm. Sci.* **75**: 36–44.

117. Bundgaard, H., Falch, E., Larsen, C., *et al.* (1986) Pilocarpine prodrugs II. Synthesis, stability, bioconversion and physicochemical properties of sequentially labile pilocarpine acid diesters. *J. Pharm. Sci.* **75**: 775–783.

118. Amsberry, K.L. and Borchardt, R.T. (1991) Amine prodrugs which utilize hydroxy-amide lactonization. I. A potential redox-sensitive amide prodrug. *Pharm. Res.* **8**: 323–330.

119. Amsberry, K.L., Gerstenberger, A.E. and Borchardt, R.T. (1991) Amine prodrugs which utilize hydroxy-amide lactonization. II. A potential esterase-sensitive prodrug. *Pharm. Res.* **8**: 455–461.

120. Bodor, N.S. (1977) Novel approaches for the design of membrane transport properties of drugs. In Roche, E.B. (ed.). *Design of Biopharmaceutical Properties through Prodrugs and Analogs*, pp. 98–135. American Pharmaceutical Association, Washington DC.

121. Bodor, N.S., Kaminski, J.J. and Selk, S. (1980) Soft drugs. 1. Labile quaternary ammonium salts as soft antimicrobials. *J. Med. Chem.* **23**: 469–474.

122. Bodor, N.S. and Kaminski, J.J. (1980) Soft drugs. 2. Soft alkylating compounds as potential antitumor agents. *J. Med. Chem.* **232**: 566–569.

123. Bodor, N.S., Woods, R., Raper, C., *et al.* (1980) Soft drugs. 3. A new class of anticholinergic agents. *J. Med. Chem.* **23**: 474–480.

124. Bodor, N. (1985) Prodrugs versus soft drugs. In Bundgaard, H. (ed.). *Design of Prodrugs*, pp. 333–354. Elsevier, Amsterdam.

125. Pitman, I.H. (1981) Pro-drugs of amides, imides, and amines. In *Medicinal Research Reviews*, pp. 189–214. John Wiley & Sons, New York.

126. Carini, D.J., Duncia, J.V., Aldrich, P.E., Chiu, A.T., Johnson, A.L., Piera, M.E., Price, W.A., Santella, J.B., 3rd, Wells, G.J., Wexler,

R.R., *et al.* (1991) Nonpeptide angiotensin II receptor antagonists: the discovery of a series of *N*-(biphenylymethyl) imidazoles as potent, orally active antihypertensives. *J.Med.Chem.* **34**: 2525–2547.

127. Wong, P.C., Price, W.A., Jr, Chiu, A.T., Duncia, J.V., Corini, D.J., Wexler, R.R., Johnson, A.L. and Timmermans, P.B. (1990) Nonpeptide angiotension II receptor antagonists. XI. Pharmacology of Exp 3174: an active metabolite of DuP 753, an orally active antihypertensive agent. *J. Pharmacol. Exp. Ther.* **255**: 211–217.

128. Baldessarini, R.J., Neumeyer, J.L., Campbell, A., *et al.* (1982) An orally effective, long-acting dopaminergic prodrug: (-)-10,11-methylenedioxy-*N*-propylnorapomorphine. *Eur. J. Pharmacol.* **77**: 87–88.

129. Bodor, N., Shek, E. and Higuchi, T. (1976) Improved delivery through biological membranes. 1. Synthesis and properties of 1-methyl-1,6-dihydropyridine-2-carbaldoxime, a pro-drug of N-methyl-pyridinium-2-carbaldoxime chloride. *J.Med.Chem.* **19**: 102–107.

130. Shek, E., Higuchi, T. and Bodor, N. (1976) Improved delivery through biological membranes. 2. Distribution, excretion, and metabolism of *N*-methyl-1,6-dihydropyridine-2-carbaldoxime hydrochloride, a pro-drug of *N*-methylpyridinium-2-carbaldoxime chloride. *J. Med. Chem.* **19**: 108–112.

131. Shek, E. and Higuchi, T. (1976) Improved delivery through biological membranes. 3. Delivery of improved delivery through biological membranes. 3. Delivery of *N*-methylpyridinium-2-carbaldoxime chloride through the blood-brain-barrier in its dihydropyridine pro-drug form. *J. Med. Chem.* **19**: 113–117.

132. Krenitsky, T.A., Hall, W.W., de Miranda, P., *et al.* (1984) Deoxyacyclovir: a xanthine oxidase-activated prodrug of acyclovir. *Proc. Natl Acad. Sci. USA* **31**: 3209–3213.

133. Whiteman, P.D., Bye, A., Fowle, A.S.E., *et al.* (1984) Tolerance and pharmacokinetics of A515U, an acyclovir analogue in healthy volunteers. *Eur. J. Clin. Pharmacol.* **27**: 471–475.

134. Gershwin, M.E., Goetzl, E.J. and Steinberg, A.D. (1974) Cyclophosphamide: use in practice. *Ann. Intern. Med.* **80**: 531–540.

135. Silverman, R.B. (1992) The Organic Chemistry of Drug Design and Drug Action, p. 380. Academic Press, San Diego.

136. Zon, G. (1982) Cyclophosphamide analogues. In Ellis, G.P. and West, G.B. (eds). *Progress in Medicinal Chemistry*, pp. 205–246. Elsevier, Amsterdam.

137. Arnold, H., Bourseaux, F. and Brock, N. (1958) NeuartigeKrebs-Chemotherapeutika aus der Gruppe der zyklischen *N*-Lost-phosphamidester. *Naturwissenschaften* **45**: 64–66.

138. Arnold, H., Bourseaux, F. and Brock, N. (1958) Chemotherapeutic action of a cyclic nitrogen mustard phosphamide ester (B518-ASTA) in experimental tumours of the rat. *Nature* **181**: 931.

139. Arnold, H. (1967) Ueber die Chemie neuer zytostatisch wirksamer *N*-Chloroaethyl-phosphorsäureesterdiamide. In Spitzy, K. and Haschet, H. (eds). *Proceedings of the Fifth International Congress on Chemotherapy*, pp. 751–754. Verlag der Wiener Medizinischen Akademie, Wien.

140. Kennedy, K.A., Teicher, B.A., Rockwell, S. and Sartorelli, A.C. (1980) The hypoxic tumor cell: a target for selective cancer chemotherapy. *Biochem. Pharmacol.* **29**: 1–8.

141. Davis, P.J. and Guenthner, L.E. (1985) Sulindac oxidation/reduction by microbial cultures; microbial models for mammalian metabolism. *Xenobiotica* **15**: 845–857.

142. Kwon, C.-H., Blanco, D.R. and Baturay, N. (1992) *p*-(Methylsulfinyl)phenyl nitrogen mustard as a novel bioreductive prodrug selective against hypoxic tumours. *J. Med. Chem.* **35**: 2137–2139.

143. Wallmark, B., Brändström, A. and Larsson, H. (1984) Evidence for acid-induced transformation of omeprazole into an active inhibitor of (H⁺ + K⁺)-ATPase within the parietal cell. *Biochim. Biophys. Acta* **778**: 549–558.

144. Lamers, C.B.H.W., Lind, T., Moberg, S., *et al.* (1984) Omeprazole in Zollinger-Ellison syndrome. *N. Engl. J. Med.* **310**: 758–761.

145. Im, W.I., Shi, J.C., Blakeman, D.P. and McGrath, J.P. (1985) Omeprazole, a specific inhibitor of gastric (H⁺-K⁺)-ATPase, is a H⁺-activated oxidizing agent of sulfhydryl groups. *J. Biol. Chem.* **260**: 4591–4597.

146. Lindberg, P., Nordberg, P., Alminger, T., *et al.* (1986) The mechanism of action of the gastric acid secretion inhibitor omeprazole. *J. Med. Chem.* **33**: 1327–1329.

147. Krausz, F., Demarne, H., Vaillant, J., *et al.* (1974) Anti-inflammatoires non stéroidiques: Dérivés de l'acide phényl-4-butyrique et phényl-4, oxo-4, butyrique. *Arzneim.-Forsch (Drug Res.)* **24**: 1360–1364.

148. Gros, P.M., Davi, H.J., Chasseaud, L.F. and Hawkins, D.R. (1974) Metabolic and pharmacokinetic study of bucloxic acid. *Arzneim.-Forsch (Drug Res.)* **24**: 1385–1390.

149. Kohler, C., Tolman, E., Wooding, W. and Ellenbogen, L. (1980) A review of the effects of Fenbufen and a metabolite, biphenylacetic acid, on platelet biochemistry and function. *Arzneim.-Forsch (Drug Res.)* **30**: 702–707.

150. Chicarelli, F.S., Eisner, H.J. and Van Lear, G.E. (1980) Disposition and metabolism of Fenbufen in several laboratory animals. *Arzneim.-Forsch (Drug Res.)* **30**: 707–715.

151. Martel, R.R., Rochefort, J.G., Klicius, J. and Dobson, T.A. (1974) Anti-inflammatory properties of Furobufen. *Can. J. Physiol. Pharmacol.* **52**: 669–673.

152. Gilard, J.W. and Belanger, P. (1987) Metabolic synthesis of arylacetic acid antiinflammatory drugs from arylhexenoic acids. 2. Indomethacin. *J. Med. Chem.* **30**: 2051–2058.

153. Williamson, J.M. and Meister, A. (1981) Stimulation of hepatic glutathione formation by administration of L-2-oxothiazolidine-4-carboxylate, a 5-oxo-l-prolinase substrate. *Proc. Natl Acad. Sci. USA* **78**: 936–939.

154. Matsukawa, T., Yuruki, S. and Oka, Y. (1962) The synthesis of S-acylthiamine derivatives and their stability. *Ann. NY Acad. Sci.* **98**: 430–444.

155. Thomson, A.D., Frank, O., Baker, H. and Leevy, C.M. (1971) Thiamine propyl disulphide: absorption and utilization. *Ann. Intern. Med.* **74**: 529–534.

156. Bodor, N., Farag, H.H. and Brewster, III, M.E. (1981) Site-specific, sustained release of drugs to the brain. *Science* **214**: 1370–1372.

157. Cousse, H., Casadio, S. and Mouzin, G. (1978) L'hydroxy éthyl nicotinamide vecteur d'acides thérapeutiquement actifs. *Trav. Soc. Pharm. Montpellier* **38**: 71–76.

158. Casadio, S., Cousse, H. and Mouzin, G. (1978) Nouvelles formes modulées de médicaments utilisant les N-hydroxy alcoyl pyridine carboxamides comme vecteur. French Patent No. 77 13478 (May 2, 1977; P. Fabre S.A.).

159. Vezin, J.C., Mouzin, G., Cousse, H. and Casadio, S. (1979) Nicafenine, a new analgesic. *Arzneimit.-Forsch.* **29**: 1659–1661.

160. Gorrod, J.W. (1980) Potential hazards of the pro-drug approach. *Chem. Ind.* 457–461.

161. Roberts, J.C., Nagasawa, H.T., Zera, R.T., *et al.* (1987) Prodrugs of L-cysteine as protective agents against acetaminophen-induced hepatotoxicity. 2-(Polyhydroxyalkyl)- and 2-(Polyacetoxyalkyl)thiazolidine-4(R)-carboxylic Acids. *J. Med. Chem.* **30**: 1891–1896.

162. Saari, W.S., Freedman, M.B., Hartman, R.D., *et al.* (1978) Synthesis and antihypertensive activity of some ester progenitors of methyldopa. *J. Med. Chem.* **21**: 746–753.

163. Vickers, S., Duncan, C.A., White, S.D., *et al.* (1978) Evaluation of succinimidoethyl and pivaloyloxyethyl esters as progenitors of methyldopa in man, rhesus monkey, dog, and rat. *Drug Metab. Dispos.* **6**: 640–646.

34

MACROMOLECULAR CARRIERS FOR DRUG TARGETING

Etienne H. Schacht, Katleen De Winne, Katty Hoste and Stefan Vansteenkiste

It is a capital mistake to theorize before one has data. Insensibly one begins to twist facts to suit theories instead of theories to suit facts.
Conan Doyle

I INTRODUCTION

One of the major problems in chemotherapy is the limited selectivity of most common drugs. Following administration, the active agent is distributed all over the body. Only a certain fraction reaches the target cells. The undesired interaction with healthy cells can lead to severe side-effects. This is a major problem in cancer chemotherapy. Over the past three decades intensive efforts have been made to design novel systems able to deliver the drug more efficiently to the target site.

Most approaches are based on combinations of the drug with a polymer. The latter serves as a carrier system wherein the drug is dispersed or dissolved, or to which it is covalently linked. Cells, microspheres, nanospheres, liposomes, proteins, antibodies, hormones, natural and synthetic polymers, and other systems have been used as carriers. For a detailed discussion of these systems the reader is referred to some recent reviews and books.[1-8]

The objective of this chapter is to demonstrate the possibility of achieving drug targeting by linking drugs with soluble polymers. The aim is to discuss concepts and structure–property relationships and to illustrate applications with some selected examples. The term 'carrier' in this paper will be restricted to water-soluble polymers to which a drug is covalently linked. Polymer–drug complexes will not be included.

II CONCEPT

From the mid-1950s onwards, a large variety of drugs has been covalently attached to many natural or synthetic polymers. Initially, the only rationale was that converting the drug into a macromolecular prodrug might reduce renal excretion and hence increase the duration of activity. There was at that time little concern about the cellular uptake and processing of these conjugates. In 1975 Ringsdorf proposed a more rationalized model for macromolecular prodrugs[9] (Fig. 34.1). He upgraded enormously the potential of this concept by calling attention to the possibility of introducing on to the polymeric carrier not only the drug moiety but also groups that can influence the solubility properties as well as groups that can alter the body distribution and promote cell selectivity.

In this model the polymeric carrier can be either an inert or a biodegradable polymer. The drug can be fixed directly or via a spacer group onto the polymer backbone. If the polymeric conjugate is pharmacologically active as a whole, the product can be regarded as a polymer drug. Conjugates which are active after release of the parent drug are termed

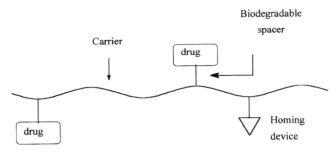

Fig. 34.1 Ringsdorf model.

macromolecular prodrugs. The hydrophilic–lipophilic balance of the polymer conjugate can be varied by proper selection of the backbone and side-group constituents. The proper selection of this spacer provides the possibility of controlling the site and the rate of release of the active drug from the conjugate by hydrolytic or enzymatic cleavage. The most challenging aspect of this model is the possibility of altering the body distribution and cell uptake by attaching cell-specific or nonspecific uptake enhancers (homing devices). This model, although still oversimplified, has been an important milestone in the history of polymeric prodrug design. It made clear that a more rational design was needed based on information arising from biological work. This remarkable paper has also catalysed the interest of biologists and pharmacists in synthetic polymers. As more information becomes available from cell biology and molecular biology, polymer chemists are trying to design tailor-made polymeric carriers that better fulfil the specified requirements.

III TYPE OF CARRIERS USED

The selection of polymers as candidate carriers is made on the basis of the following criteria:

- chemical composition: availability of suitable functional groups for covalent coupling with drugs;
- biocompatibility: preferably nontoxic, nonimmunogenic;
- biodegradability;
- availability.

A number of reviews cover what has been achieved over the past 20 years in the field of soluble polymers as potential drug carriers.[10–15] The polymers selected for preparing macromolecular prodrugs can be categorized according to: (1) the chemical nature (vinylic or acrylic polymers, polysaccharides, poly(α-amino acids)); (2) the back bone stability (biodegradable polymers, stable polymers); (3) the origin (natural polymers, synthetic polymers); and (4) the

Table 34.1 *Overview of macromolecular prodrug conjugates*

Drug carrier	Drug	Ref.
Vinyl polymers		
N-(2-hydroxypropyl)methacryl-amide copolymers	Adriamycin	16–18
	Various others	
Poly(1-vinyl-2-pyrrolidone-co-maleic anhydride)	Quinidine	19
	6-Purinethiol	19
	5-Aminosalicyl acid	20
Poly(1-vinyl-2-pyrrolidone-co-vinylamine)	Chlorambucil	21
Poly(styrene-co-maleic anhydride)	Neocarzinostatin	22
Poly(divinylether-co-maleic anhydride)	Cyclophosphamide	23
	Methotrexate	24
Polysaccharides		
Dextran	Procainamide	25, 26
	Daunomycin	27
	Interferon	28
	Mitomycin C	29, 30
Inulin	Procainamide	26
	Antibiotics	31
Biozan R	Ampiciline	32
Carboxymethyldextran	Daunomycin	33
Chitosan	5-Fluorouracil	34
Synthetic poly(α-amino acids)		
Poly(L-lysine)	Methotrexate	35
Poly(L-aspartic acid)	Daunorubricin	36
Poly(2-hydroxyethyl)-D,L-aspartamide	Naproxen and others	37–39
Poly(2-hydroxyethyl)-L-glutamine	Mitomycin	40
Proteins		
Human serum albumin	Primaquine	41
Bovine serum albumin	Methotrexate	42
Poly(ethylene glycol)	Enzymes/drugs	43, 44
	Asparaginase	45
	Insulin	46
	Ibuprofen	47

molecular weight (oligomers, polymers). Examples of frequently used carriers are listed in Table 34.1.

Vinyl polymers can be easily prepared by radical polymerization of the corresponding vinyl monomer. Enormous variability in the polymer composition and properties can be achieved by copolymerization of selected monomers. This makes it possible to tailor-make the structure to meet the requirements of any system under consideration. For that reason, vinyl-type polymers are interesting drug carrier candidates. However, vinyl polymers are not biodegradable. Hence, in order to avoid undesirable storage, the molecular weight should at least be

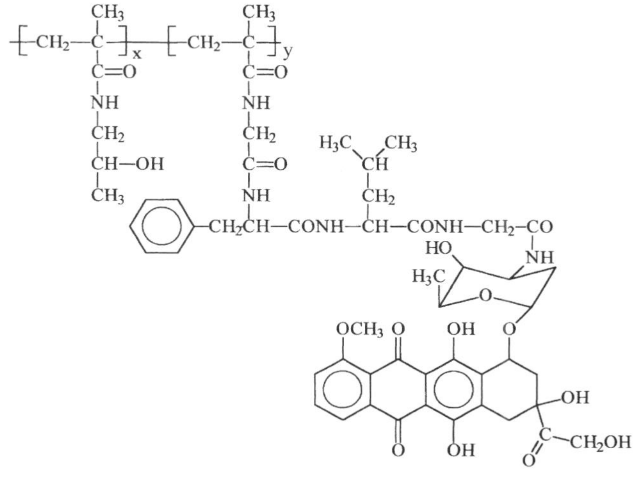

Fig. 34.2 Structure of PHMPA copolymer containing adriamycin.

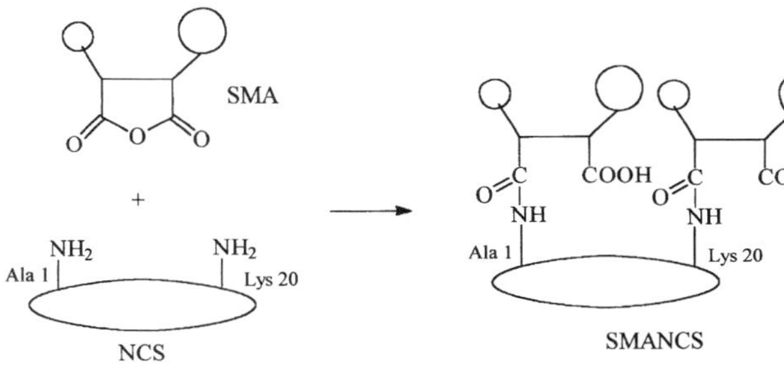

Fig. 34.3 Diagrammatic representation of the reaction between SMA and NCS to produce the conjugate SMANCS.

below the renal filtration limit (40–50 kDa) and for this reason the future of nondegradable polymeric carriers is questionable. The answer has to come from the ongoing clinical evaluations of some polyvinylic-type macromolecular prodrugs. At present, the most intensively studied vinyl polymers are copolymers of *N*-(2-hydroxypropyl)-methacrylamide (PHPMA).[48–53] These polymers have been used as carriers for several drug molecules. A vast amount of information is available about the biodistribution, immunogenicity and biological activity. PHPMA proved to be nontoxic and nonimmunogenic *in vivo* in animals. It was demonstrated that PHPMA–adriamycin conjugates are remarkably less toxic than the free drug. The HPMA copolymer with adriamycin as an antitumour drug, PK1 (Fig. 34.2), was developed by Duncan and Kopecek and reached phase II clinical trials for the treatment of breast, colon and non-small-cell lung cancer.[54] The only polymeric prodrug with targeting moiety which entered early clinical trials for the treatment of primary and secondary liver cancer is PK2.[55] This conjugate is a PHPMA copolymer with adriamycin as the antitumour agent and *N*-acylated galactosamine as the targeting group. It was developed for targeting to the liver by facilitating the interaction with hepatocyte asiaglycoprotein receptors. Preliminary clinical studies on paclitaxel and camptothecin PHPMA analogues

demonstrated greater toxicity than PK1.[56] This is probably due to the ester bond between drug and spacer.

Another polyvinylic prodrug, SMANCS (Fig. 34.3), is a conjugate of a low molecular weight styrene maleic anhydride copolymer (SMA, 1.6 kDa) and neocarzinostatin (NCS). It is already marketed in Japan for the treatment of hepatocellular carcinoma.

Synthetic poly(α-amino acids) like poly(L-lysine), poly(L-glutamic acid), poly[(*N*-hydroxyalkyl) glutamines] can be made by ring-opening polymerization of the *N*-carboxyanhydride monomers (Fig. 34.4). These polymers have functionalities in their side-groups (amine, hydroxyl, carboxyl) that allow covalent coupling with drug molecules. Generally, poly(L-amino acids) are biodegradable, whereas their *D*-enantiomers are not. The cationic nature of poly(L-lysine) in plasma causes it to interact easily with most cell membranes. Unfortunately the polymer is toxic.[57] Succinylation of the polymer converts it into a less toxic polyacid that still can be used as drug carrier[58] (Fig. 34.5).

Polysaccharides are another interesting class of drug carriers. Much attention has been directed to the use of dextran. Sezaki and his group prepared dextran–mitomycin conjugates by coupling mitomycin C with dextran modified with either aminocaproic acid or 6-bromohexanoic acid[29,30] (Fig. 34.6). The pharmacokinetics of these conjugates proved to be dependent on the molecular weight and the electrical charge of the polymer derivative.

The selection of dextran as drug carrier has mostly been based on its clinical use as a plasma expander and its claimed biodegradability. However, it was demonstrated by

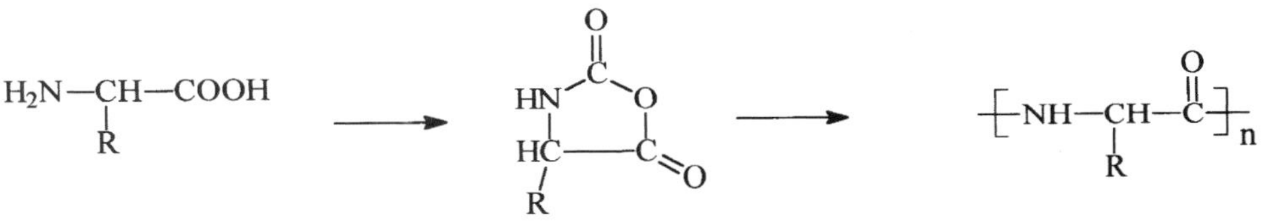

Fig. 34.4 Ring-opening polymerization of the *N*-carboxyanhydride monomers.

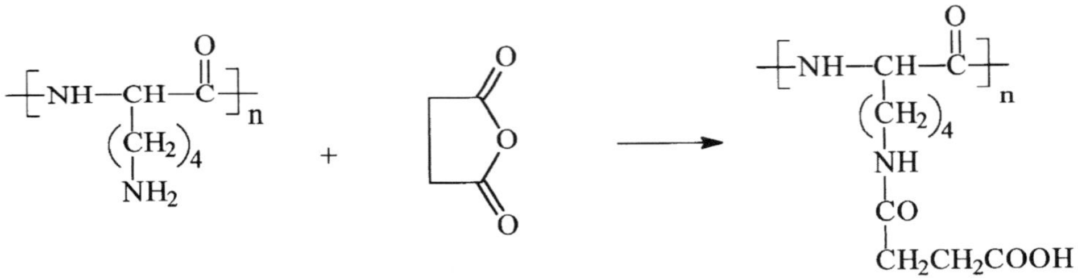

Fig. 34.5 Succinylation of poly(L-lysine).

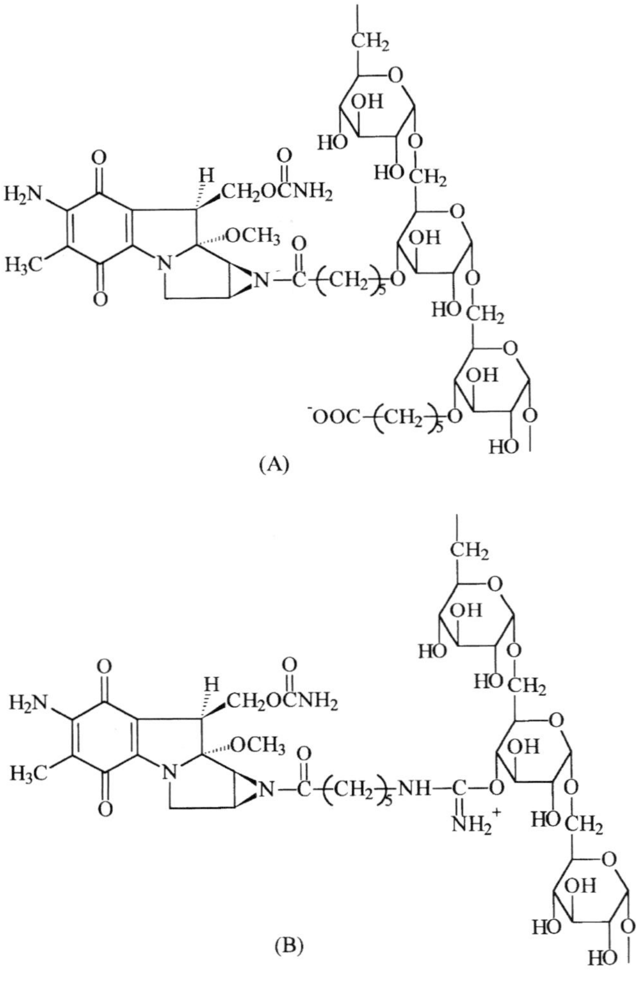

(A)

(B)

Fig. 34.6 Chemical structure of (A) anionic and (B) cationic mitomycin C–dextran conjugates.

Vercauteren *et al* that the *in vitro* degradation of dextrans in the presence of lysosomal glucosidases or endodextranases is rather slow. Moreover, it was shown that chemical modification of the dextran further reduces its biodegradability.[59]

Proteins such as serum albumin have also been frequently used for preparing polymeric prodrugs. An interesting example is the work of Meyer and coworkers who used mannosylated serum albumin as carrier for antiviral drugs.[60,61] A disadvantage of proteins is their complexity in chemical composition, which complicates the identification of the final conjugates.

Poly(ethylene glycol) (PEG) has been used to modify a number of therapeutically interesting proteins. It has been clearly demonstrated by Abuchowski *et al.* [44,45] that grafting of PEG onto proteins reduces their immunogenicity, improves their resistance to proteolytic degradation and improves their thermostability.

Micelle-forming block copolymers were introduced by Yokoyama and colleagues.[62–64] Conjugates of adriamycin with poly(ethylene glycol)–poly(aspartamide) block copolymers tend to form micelles (Fig. 34.7). It was demonstrated that these systems have a very high *in vivo* antitumour activity.

IV METHODS OF DRUG RELEASE

Drug molecules are generally linked to the polymeric carrier via a spacer group. The drug can be released during plasma circulation or after cell uptake. Low molecular weight

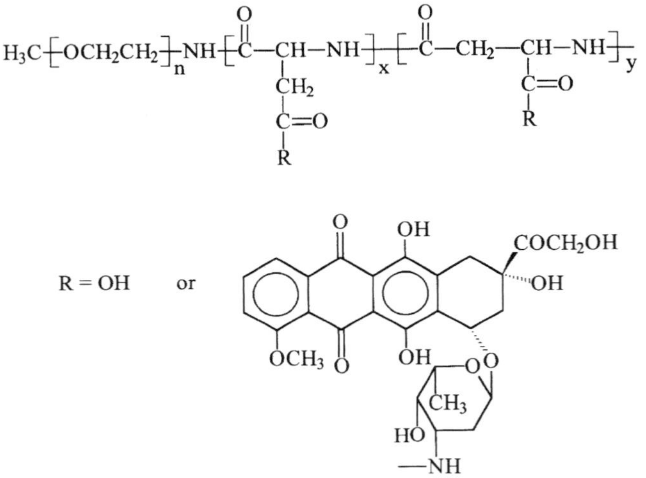

Fig. 34.7 Adriamycin-conjugated poly(ethylene glycol)-poly(aspartic acid) block copolymer.

molecules can enter cells by diffusion. Macromolecules normally do not pass across plasma membranes and their capture by cells is restricted to passive or active endocytosis. The endocytic capture of macromolecules and its significance for drug delivery have been discussed in depth by Duncan.[65] In an endocytic process, polymers enter the cell in pinocytic vesicles which rapidly join the endosomal compartment before moving on to fuse with lysosomes containing a variety of enzymes, including peptidases. If the spacer is a good substrate for these lysosomal enzymes, lysosomotropic drug delivery is feasible. Lysosomotropic drug delivery depends on the choice of drug, carrier and linkage between drug and carrier. The *in vivo* release can be due to passive hydrolysis or can be caused by a more specific mode of cleavage, such as enzymatic release or pH-controlled release.

Passive hydrolysis. Esters, carbonates, amides and urethanes are susceptible to hydrolysis. Drugs linked with the spacer via such bonds will be released in aqueous media. The rate of release will decrease in the order ester > carbonate > urethane > amide. Cleavage may also occur at the level of the spacer–backbone bond so that spacer–drug moieties can be released as well.

Enzyme-assisted hydrolysis. This is likely to occur for conjugates having oligopeptide spacers. The rate and site of the cleavage will depend on the composition of the peptide. Trouet and coworkers prepared albumin–adriamycin conjugates with oligopeptide spacers between the drug and the carrier.[66] As shown in Table 34.2, *in vitro* release of adriamycin increased with increasing length of the peptidic spacer. Only conjugates with tri- or tetrapeptide spacers expressed high *in vivo* anticancer activity, observed by prolonged lifespan of treated mice. Since the conjugates with tri- or tetrapeptide spacers released minor amounts of free drug in serum, the significant antitumour effect was attributed to lysosomal drug release. From these data it follows that site-specific drug release by proper molecular design of the conjugate is feasible. This is further substantiated by the excellent work of Kopecek and Duncan who evaluated a series of PHPMA–adriamycin conjugates

with different oligopeptide spacers.[65,67] *In vitro* release studies in media containing lysosomal enzymes clearly demonstrated that drug release can be tailored by the length and composition of the peptide spacer (Table 34.2). *In vivo* experiments on animals confirmed that the pharmacological activity of the polymer–drug conjugate depends on the nature of the spacer.

The importance of the spacer composition on the drug release was also demonstrated by De Marre *et al.* for poly[(2-hydroxyethyl)-L-glutamine] PHEG-peptide-mito-mycin C conjugates.[40,68] Conjugates with a spacer having glycine as *C*-terminal amino acid were more susceptible to hydrolytic release in aqueous buffer or serum than those having a more hydrophobic-terminal amino acid. Tetrapeptide spacers were more susceptible to cleavage by lysosomal enzymes than tripeptides (Table 34.3).

pH-controlled drug release. Polymers entering the endosomal or lysosomal compartment are exposed to an acidic medium (pH 4.5–5.5). Using an acid-cleavable spacer between the drug and carrier can subsequently result in a pH-controlled intracellular drug release. In the literature two types of acid-sensitive spacer are frequently studied and reviewed:[69] the hydrazon linkage and the *N-cis*-aconityl spacer (Fig. 34.8). In order to study the relationship between the acid-sensitivity and cytotoxicity of adriamycin–immunoconjugates, Kaneko and coworkers synthetized a series of conjugates with hydrazon spacers.[70] The immunoconjugate with propanoyl-hydra-zonspacer showed the highest *in vitro* and *in vivo* antitumour activity. The research group of Kratz developed and evaluated a series of transferrin and albumin conjugates with anthracyclines.[71] Coessens and coworkers linked the antibiotic streptomycin to dextran and to poly-[*N*-(2-hydroxyethyl)-L-glutamine] via a carboxylic hydra-zon linkage.[72] Release of streptomycin was demonstrated at lysosomal pH.

Shen and Ryser developed daunorubicin-linked amino-ethyl polyacrylamide beads (Affi-gel 701) or poly(D-lysine) via an *N-cis*-aconityl spacer.[73] The *cis*-aconityl linkage between drug and carrier was readily hydrolysed at pH 4 but not appreciably at pH 6. Poly(D-lysine) caused 90%

Table 34.2 *Degradation of HPMA copolymers by rat liver lysosomal enzymes, effect of peptidyl side-chain*

Spacer	Percentage of doxorubicin released after 24 h
P-Gly-Phe-Gly-Dox	64
P-Gly-Leu-Gly-Dox	18.5
P-Gly-Phe-Leu-Gly-Dox	59.3
P-Gly-Leu- Phe-Gly-Dox	70.7

Table 34.3 *Release of mitomycin C (MMC) by hydrolysis of PHEG-tripeptide or tetrapeptide-MMC conjugates by tritosomes at pH 5.5 after 3 hours*

Tripeptide spacers	Percentage MMC release	Tetrapeptide spacer	Percentage MMC release
Gly-Phe-Leu	2.4	Gly-Gly-Phe-Leu	3.1
Gly-Gly-Leu	2.5	Gly-Phe-Leu-Gly	57.7
Gly-Phe-Phe	2.7	Gly-Phe-Ala-Leu	74.6
Gly-Phe-Gly	7.1	Ala-Leu-Ala-Leu	81.0

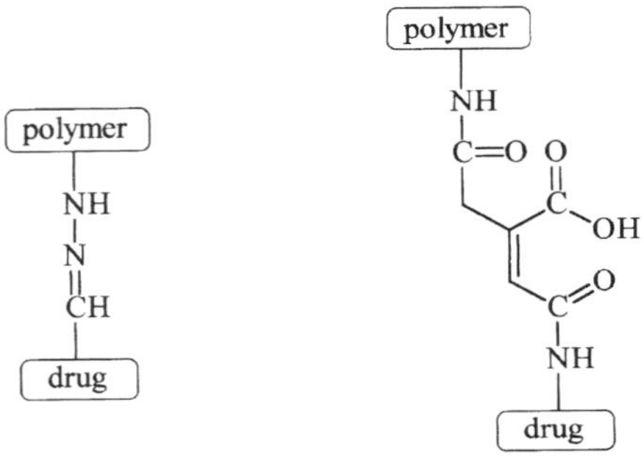

Fig. 34.8 The hydrazon and *N-cis*-aconityl spacer.

inhibition of growth of WEH 1-5 cells cultured *in vitro*. The conjugate was able to enter the cells by pinocytosis and, on reaching the lysosomal compartment, liberate daunorubicin. Drug release must be due to pH sensitivity of the *cis*-aconityl linkage, since the poly(D-lysine) is not biodegradable.

Reduction-sensitive spacers. Shen and Ryser coupled methotrexate via a disulphide-containing spacer with poly(D-lysine).[74] The conjugate was able to enter methotrexate-resistant cell lines cultured *in vitro*. There was evidence for a reductive cleavage of the spacer in the cytosol compartment. This example indicates another possibility for achieving intracellular release of drugs from polymeric conjugates by proper selection of the spacer.

V MODIFICATION OF PHARMACOKINETICS AND IMMUNOGENICITY BY POLYMER CONJUGATION

The profile of plasma concentration of drugs is an important determinant of their quantitative access to peripheral targets. The plasma profile is usually measured as the area under the curve (AUC). In general, slow renal elimination and metabolic inactivation promote better access of drugs to remote targets, although this can also cause elevated toxicity. Many drugs in routine use are membrane permeable because their sites of action are intracellular, and such drugs typically exhibit high volumes of distribution and rapid plasma clearance. Conjugation to hydrophilic macromolecular carriers can prevent rapid renal excretion and restrict drug entry into cells to pinocytic mechanisms, markedly prolonging plasma circulation time.[75,76] It is important that polymers used for this type of pharmacokinetic modification are well tolerated, showing good biocompatibility with blood and tissues. To date, the greatest success has been gained using neutral or slightly negatively charged polymers, since these materials have limited interaction with the negatively charged chondroitin sulphates and heparan sulphates of the endothelial wall and remain in circulation for a prolonged time.[77,78]

Molecular size is important in determining rates of glomerular filtration of large water-soluble molecules circulating in the plasma. For example, small proteins (< 40 kDa) are rapidly excreted in the urine with half-lives of only a few minutes.[53,79] Conjugation of small peptides or proteins to hydrophillic soluble polymers can prevent rapid glomerular filtration and lead to much greater AUC. In most cases the resulting polymer–protein conjugate is thought to take the form of a colloid, with the protein core protected from interaction with other macromolecular plasma components by a hydrophillic polymeric shield. Hence, this procedure also decreases the immunogenicity of the protein component and permits repeated administration of foreign proteins.[44,80] One disadvantage of the approach is that the derivatized protein usually has decreased access to macromolecular substrates and cell surfaces. Hence, modification of murine antibodies for drug targeting has been largely unsuccessful.[51] However, the approach is particularly useful where the protein is an enzyme with a low molecular weight substrate found in the plasma. For example, L-asparaginase is an enzyme which hydrolyses L-asparagine to yield L-aspartic acid and ammonia. The depletion of L-asparagine from blood plasma can be used to inhibit growth of certain tumours. In clinical trials, however, the enzyme displayed a very short plasma half-life ($T^{1/2} =$ 1.5 hours) and an anti- L-asparaginase antibody was soon produced in patients, which nullified the pharmacological activity. These problems were solved simultaneously by conjugating the enzyme with poly(ethylene glycol) (PEG). The immunogenic potential of the conjugated enzyme was decreased by 99% compared with the native form, and plasma circulation was greatly extended.[81] In the early 1990s PEG-L-asparaginase (ONCASPAR®) was one of the first to be on the market.[54]

VI ACTIVE TARGETING

A Intra-arterial injection and embolization

An important physical means of achieving tumour-selective delivery of drugs is by direct injection into the artery feeding the tumour. Although it is usually impossible to gain access to arteries supplying exclusively tumour tissue, the regional selectivity of exposure to cytotoxic drugs that can be achieved in this way is much better than for intravenous injection. In addition, physiological differences between the vascular supply to tumour and normal tissues can sometimes be exploited to give additional selectivity of action; for example, the use of vasoconstricting agents to decrease the relative blood supply through normal vasculature, which is reported to be more responsive than tumour vasculature.[82-84] The artery most widely used for this type of clinical treatment is the hepatic artery, which supplies most of the blood to advanced hepatic metastases of colorectal carcinomas as well as primary liver tumours. Recently, great effort has been expended in developing techniques for injection into arteries supplying breast and renal tumours.

Apart from direct injection of drugs into tumour-feeding arteries, various attempts have been made to embolize the tumour capillary bed using microparticulate formulations of drugs. A range of drugs and delivery vehicles (simple aqueous solutions, oils or particulate formulations) have undergone evaluation.[85] One approach, of particular note, involving soluble macromolecular drug carriers is SMANCS. In the clinical formulation, neocarzinostatin (NCS) (molecular weight 10 700) is conjugated to two chains of a styrene–maleic anhydride copolymer (SMA) (molecular weight average 1500, polydispersity < 1.2) through primary amino functions at NCS positions 1 (alanine) and 20 (lysine).[75,86] The SMA copolymer is itself derivatized with an alkyl group (usually butyl) which determines the overall hydrophobicity of the conjugate.[87] Although the resulting conjugate is sometimes applied intravenously, when it binds to albumin and shows prolonged circulation compared with parent NCS, the more successful use of SMANCS has followed its dissolution in lipiodol and hepatic arterial injection to treat primary hepatocellular carcinoma.

Primary hepatoma usually exhibits a leaky, sinusoidal, endothelial layer, permitting relatively easy interstitial access of oils administered intra-arterially.[88] The lack of organized lymphatic drainage from the tumour (see later) results in long-term retention of oils that can be exploited for angiographic imaging of primary hepatic tumour masses. When SMANCS is dissolved in lipiodol, it remains selectively associated with the oily phase, diffusing out slowly into adjacent tumour tissues over a period of many weeks. One obvious strength of the SMANCS approach is that cytotoxicity of the conjugate depends on the rate of diffusion of SMANCS out of the lipiodol; variations in alkyl derivatization of the polymeric SMA component can influence the oil–plasma partition coefficient of SMANCS, permitting development of drugs with precisely optimized release rates. This approach has been widely investigated in Japan for treatment of primary hepatoma, with remarkable response rates.[89,90] SMANCS is now clinically used for treatment of primary hepatoma.

B Antibody conjugates

Antibodies represent the most universally applicable active targeting agent, having exquisite selectivity for recognition of small antigenic epitopes. The antibody–antigen interaction can be so strong that antitumour effects of drug–antibody conjugates can be improved many times compared with the free drug.[91] Frequently, however, the quantity of drug that can be selectively targeted is limited by the number of antigens available. Hence, in cancer therapy the targeted-drug approach has been most successful for extremely potent agents such as plant toxins, which in conjugation with antibodies have been termed the 'immunotoxins'.[92] Unfortunately, selectivity of immunotoxins delivery in human subjects, particularly where the target is a solid tumour, has been inadequate to mediate reproducible therapeutic responses.[93] One factor limiting this approach is the poor access of antibodies to tumour cells deep within the tumour interstitium, and recently greater attention has shifted to identifying antigens associated specifically with the vasculature serving the tumour.[94] Tumour-associated endothelial cells represent a target that is accessible, vulnerable and crucial to tumour survival; initial reports on selective destruction have been encouraging.[95]

One approach of particular note in cancer therapy has been the use of antibodies for tumour-targeted delivery of enzymes capable of activating innocuous prodrugs to highly cytotoxic species[96,97] (Fig. 34.9). This approach is known by the acronym ADEPT (antibody-directed-enzyme-prodrug therapy) and one major advantage over conventional antibody-targeting is the inherent amplification stage, meaning that for every successful enzyme-targeting event a very large number of prodrug molecules can be activated.[98,99] The novelty of this approach is such that it has attracted great interest and certain versions have already received clinical appraisal. Initial results have been promising, although dogged with such problems as poor water solubility of prodrugs, and the approach is currently being refined for further development.

C Macromolecular glycoconjugates as carrier systems

Sugar-specific receptors are plasma membrane components (either glycoproteins and glycolipids, called lectins[100]) of

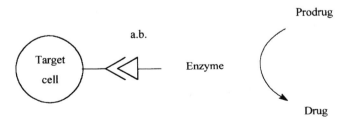

Fig. 34.9 Schematic representation of the ADEPT concept.

Table 34.4 *Membrane lectins from various sources*[78,79]

Origin	Sugar specificity
Murine liver macrophages	D-Mannose, L-fucose, D-galactose and N-acetyl-D-glucosamine
Rat and human hepatocytes	D-Galactose
Mouse spleen	D-Galactose/N-acetyl-D-glucosamine
Human fibroblasts	D-Mannose-6-phosphate
Mouse L1210 leukaemia cells	L-Fucose

many mammalian cells. The first membrane lectin was characterized on hepatocytes by Ashwell and Hanford.[101] Endogenous lectins, generally multivalent in their binding and recognition capacities, are found on numerous normal and malignant cells (Table 34.4). Biologically they are vital elements in a complex network of 'biosignalling', acting as sugar-specific receptors and/or mediating endocytosis of specific glycoconjugates.[102,103] Therefore, glycoconjugates may be used as carriers to specifically deliver biological active agents such as metabolite inhibitors, toxic drugs, biological response modifiers or genetic material to intracellular compartments (e.g. lysosomotropic delivery[104]).

The possibility of using hepatic lectins recognizing galactose as targets for drug delivery is particularly attractive and has been investigated exhaustively. A well-known target is the asialoglycoprotein receptor (ASGP-R),[105] which is easily accessible to the vascular circulation, being situated predominantly on the sinusoidal surfaces of hepatocytes. Moreover, it is present in relatively large numbers, allowing the delivery of therapeutic doses of bioactive agents. In addition, the hepatocyte constitutes a valuable target for treatments of various diseases including lysosomal storage diseases, metabolic deficiencies, hepatitis, parasitic infections (e.g. malaria) and cancer treatment (e.g. liver metastasis).

The feasibility of liver targeting is well documented. As an example, Vansteenkiste and colleagues[106] prepared a range of copolymers based on the carbohydrate dextran substituted with pendent side-chains terminated in 1-O-linked monosaccharides, including mannose and galactose (mono-gal) as well as clusters of three galactose molecules (tris-gal).[107] Following intravenous injection (50 μg) to rats, both galactosylated dextrans were cleared rapidly with plasma half-lives of <2 minutes, mainly into the liver. However, a significant difference in liver disposition was demonstrated between the tris-gal-substituted dextran (71% of the injected dose) and its mono-gal analogue (43%). Moreover, subcellular distribution experiments[108] indicated that mono-gal dextran is accumulated within the lysosomal compartment of liver

hepatocytes. In contrast, the tris-gal-substituted polymer shows a greater affinity for the galactose-specific receptor *in vivo* and also shows a high level of association with the cell surface of hepatocytes (a variation of the ADEPT concept).

Drug delivery to macrophages (e.g. Kupffer cells) offers a second potentially attractive goal in the development of targeted treatment of various malfunctions, notably parasitic disorders such as Leishmaniasis or enzyme deficiencies such as Gaucher's syndrome. Moreover, since macrophages are part of the immune system, they can be activated and rendered tumoricidal by immunostimulating agents (e.g. N-acetylmuramyldipeptide, MDP).

Mannosylated carriers can also fulfil an important role not only in active drug targeting but also in receptor blocking. It was demonstrated that mannosylated dextrans were useful as transient receptor blockers *in vivo* for a 791T/36-ricin toxin A immunotoxin.[107] The circulation half-life of the immunotoxin was prolonged by a factor of 3–4 up to 40 minutes following co-injection of an excess of mannosylated dextran. The liver disposition of the immunotoxin was markedly reduced from 43% to 1% of the recovered dose. The influence of the molecular size as well as the sugar loading of the competing polysaccharide was demonstrated to be small.

D Targeting to angiogenic vessels

Angiogenesis is an important process for tumour growth and metastasis of solid tumours.[109] Endothelial cells in angiogenic vessels of solid tumours show an increased expression of several cell surface proteins that stimulate cell invasion and proliferation.[110,111] These proteins include receptors for different angiogenic growth factors[112] such as the vascular endothelial cell growth factor (VEGF) and they also include the $\alpha_v\beta_3$ integrin receptor.[113] One type of integrin receptor binding peptide is the RGD (arginine-glycine-aspartic acid)-containing peptides. These peptides bind to the integrin receptors with high affinity and can therefore be used as targeting moieties for drug delivery. Moreover, it is known that peptides containing the RGD sequence inhibit experimental metastasis.[114] Arap and co-workers showed that the coupling of cyclic RGD and NGR peptides (CDCRGDCFC and CNGRCVSGCAGRC) to the anticancer drug adriamycin resulted in an increased efficacy of the drug against human breast cancer xenografts in mice.[115] The research group of Mayumi prepared conjugates of RGD-peptides (RGD and RGDS) and poly(ethylene glycol) (PEG).[116] The inhibitory effect of these conjugates, examined on experimental metastasis in mice, was demonstrated to be superior to the free RGD peptides.

Another mediator of angiogenesis is endothelium-derived nitric oxide (NO).[117] VEGF stimulates the release of NO from cultured human umbilical venous endothelial cells. Xanthine oxidase (XO) is an enzyme with anticancer activity that generates cytotoxic reactive oxygen species, including superoxide anion radical and hydrogen peroxide.[118] It was anticipated by Maeda and coworkers that the binding of XO to blood vessels can cause reaction of the formed superoxide anion (O_2^-) with the endogenously formed NO and thus cause severe vascular damage.[119] Chemical conjugation of XO to PEG resulted in an enhancement of the antitumour activity of XO. It was also demonstrated that administration of hypoxanthine, a substrate of XO, after administration of PEG-XO resulted in a significant suppression of tumour growth and with no tumour growth after 52 days.

VII PASSIVE TARGETING

Numerous studies over the past 40 years have reported an apparent passive targeting of soluble macromolecules to solid tumours. Various causes have been suggested, including rapid pinocytosis by tumour cells *in vivo*, although the most likely explanation was proposed by Matsumura and Maeda in 1986[120] and results from the absence of organized lymphatic drainage of solid tumours. Slow or absent convection of fluid through solid tumour masses results in the formation of hypoxic regions and the elevation of interstitial hydrostatic pressure.[121] In response to poor nutritional supply, the tumours produce angiogenic and capillary-permeabilizing factors which result in nonspecific leakiness of the tumour vasculature and the extravasation and subsequent extravascular retention of macromolecules (and soluble drug conjugates) from the bloodstream. This effect, termed the 'enhanced permeability and retention effect' (EPR effect) is thought to be important for nutrition and for the laying down of new tumour stroma. Similar phenomena are observed in processes of wound healing.[122] Intensive studies are underway to elucidate factors controlling tumour vascular hyperpermeability, although the effect is already being examined clinically using anthracyclines conjugated to hydrophillic macromolecular carriers.

A number of macromolecular drug conjugates have taken advantage, sometimes inadvertently, of the EPR effect.[123–127] One of the best-characterized is the conjugate of doxorubicin (DOX) with copolymers based on *N*-(2-hydroxypropyl)methacrylamide (HPMA) which is now in phase II clinical trials in the UK.[128] The form under clinical evaluation has weight average molecular weight of approximately 25 000 and the anthracycline is attached via a tetrapeptide side-chain designed as a substrate for lysosomal thiol-proteases. In preclinical studies the conjugate has shown a passive tumoritropism to subcutaneous tumours, together with sustained activation of the drug conjugate *in situ*. Following intravenous administration to mice bearing solid subcutaneous B16F10 melanomas, free DOX (5 mg DOX per kg body weight) produced tumour levels of only 0.55 μg DOX per g tumour, while the same dose of DOX administered as polymer conjugate achieved levels up to 7.5 μg per g tumour. The decreased peripheral toxicity of polymer-conjugated DOX permits the use of greater doses, and administration of doses of 18 mg DOX per kg results in tumour levels of drug up to 22 μg per g.[129] Hence the impressive anticancer activity observed is thought to result from a combination of passive tumoritropism and decreased peripheral toxicity.[130]

Some of the parameters influencing the tumoritropic behaviour of soluble macromolecules have now been elucidated,[75,130] permitting development of more sophisticated approaches using biodegradable carriers and linkages carefully designed for selective cleavage by tumour-associated enzymes.

VIII OVERCOMING 'MULTIDRUG' RESISTANCE

A major problem of using antitumour agents in cancer chemotherapy is multidrug resistance. Such resistance is mainly due to the overexpression of the P-glycoprotein (Pgp). This transmembrane glycoprotein, encoded by the *mdr1* gene, functions as an energy-dependent efflux pump reducing the intracellular accumulation of anticancer drug molecules such as anthracyclines and vinca alkaloids.[131]

By conjugation of those cytostatics to a macromolecular carrier, the drug under the form of the polymeric prodrug is taken up by endocytosis and subsequently, the efflux pumps are circumvented.[132] When the free drugs are administered, the drugs enter the cell by diffusion through the plasma membrane and are recognized by the Pgp pumps. When the drugs are endocytosed by the cell as polymer–drug conjugates or physically entrapped in nanospheres, the drug is released in the lysosomes and enters the cytoplasm close to the nucleus. There it can interact with the DNA or with the topoisomerases before being removed by the efflux pumps.

Several research groups have demonstrated that polymeric drug delivery systems can bypass tumour cell multidrug resistance. Cuvier and coworkers[133] demonstrated that *in vitro* resistance of tumour cells to adriamycin can be overcome by using adriamycin-loaded polyalkylcyanoacrylate nanospheres. *In vivo*, a prolonged survival of mice with P388–adriamycin resistant tumour was demonstrated using adriamycin-loaded nanospheres, while the treatment with free adriamycin was ineffective. The possibility of overcoming multidrug resistance is also demonstrated with liposomes, bovine serum albumin–adriamycin conjugates and antibody-targeted drugs conjugated to PHMPA copolymer.[134–136] Kopecek and coworkers have studied the exposure of human ovarian carcinoma cells to free drug and PHMPA copolymer-bound adriamycin and mesochlorin e(6).[137,138] Neither PHMPA conjugates induce Pgp-mediated multidrug resistance.

IX IMMUNOPROTECTIVE THERAPY WITH POLYMERIC PRODRUGS

Recently, it was found that the use of polymeric antitumour drug derivatives may play an important role in the protection of the cancer patient's immune system. One of the mechanisms that induces the programmed cell death or apoptosis of cancer cells, is the Fas–Fas ligand interaction.[139] Fas and Fas ligand (FasL) are both transmembrane proteins.[140] Both receptor and ligand are expressed either constitutively or after activation on most of the cells of the immune system.[141,142] Fas is also expressed on cancer cells. Interaction between Fas and FasL triggers a cascade of signals, that eventually results in apoptosis.[143,144] It has been reported, however, that several tumour cell lines can express FasL.[145–152] Hence, they are able to kill cells of the immune system expressing Fas. This mechanism is called the Fas counterattack. The counterattack of the tumour cells not only prevents the eradication of the cancer cells, but also participates in the destruction of the immune system.

The counterattack mechanism is often favoured by nonfunctioning,[149] down-regulation or loss[153] of the cancer cell Fas receptors. Moreover, it has been reported that treatment with antitumour drugs promotes the induction of Fas ligands on the cancer cells.[154]

There is a strong indication that treatment with macromolecular drug derivatives can overcome this. Rihova and coworkers found a strong expression of FasL on the SW620 human metastatic colorectal cancer cell line when it was exposed to doxorubicin or mitomycin C (MMC).[155] However, when the cell line was exposed to polymeric derivatives of these drugs, no increase of the FasL was noticed on the SW620, not even when higher concentrations were used. The drug derivatives used in this experiment were MMC bound via a GFAL spacer on to PEG-grafted poly[N^5-(2-hydroxyethyl-L-glutamine)] (PHEG) and doxorubicin coupled via a GFLG spacer onto poly-[N-(2-hydroxypropyl)methacrylamide] (PHPMA) with or without anti-CD71 mAbs as targeting group. These results suggest that the expression of Fas ligands on cancer cells is different when they are exposed to free antitumour drugs or to their macromolecular derivatives. This is an important outcome that might indicate that polymeric prodrugs are able to protect the patient's immune system.

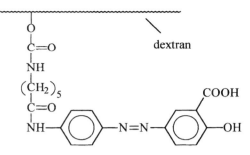

Fig. 34.10 Structure of azo-coupled dextran–5-ASA prodrug.

X ORAL DRUG DELIVERY

The preferred mode for drug administration is undoubtedly the oral route, but the efficiency of oral administration is often limited by premature uptake or degradation. For drugs that need to enter systemic circulation, the adsorption window is situated in the upper intestine. For treatment of inflammations in the lower part of the GI tract, uptake in the small intestine is to be avoided. A typical example is the treatment of ulcerative colitis and Crohn's disease with 5-aminosalicylic acid (5-ASA). The parent drug is not efficient because of premature uptake in the upper intestine. One possible solution is the use of polymeric conjugates of 5-ASA linked to a polymeric carrier via an azo bond. It is well known that the colon is a reductive medium that can split azo bonds with formation of amino constituents. It is anticipated that polymer–5-ASA conjugate will pass intact through the upper part of the GI tract and reach the colon, where 5-ASA will be released. We have prepared in our laboratory a series of azo-coupled polymer–5-ASA conjugates[156] (Fig. 34.10).

In vitro experiments carried out in a bioreactor simulating the human intestinal microbial environment demonstrated that the parent drug is readily cleaved from the carrier in the colon-simulating part of the reactor.[157] Independently of our work, Kopecek and his group also prepared polymeric prodrugs of 5-ASA using PHPMA as carrier.[158]

REFERENCES

1. Robinson, J.R. and Lee, H.L. (eds). (1987) *Controlled Drug Delivery: Fundamentals and Applications*, 2nd edn. Marcel Dekker, New York.
2. Duncan, R. and Seymour, L.W. (1989) *Controlled Release Technologies: A Survey of Research and Commercial Applications*. Elsevier Advanced Technology, Oxford.
3. Kim, S.W., Kopecek, J. and Knutsen, K. (1992) In Anderson, J.M. (ed.). *Advances in Drug Delivery Systems*, Part 5. Elsevier, Amsterdam.
4. Kim, S.W., Kopecek, J. and Knutsen, K. (1994) In Anderson, J.M. (ed.). *Advances in Drug Delivery Systems*, Part 6. Elsevier, Amsterdam.
5. Baker, R. (1987) *Controlled Release of Biologically Active Agents*. Wiley, New York.

6. Vert, M. (1986) Polyvalent polymeric drug carriers. *CRC Crit. Rev. Ther. Drug Carrier Syst.* **2**: 291–327.

7. Okano, T., Yui, N., Yokoyama, M. and Yoshida, R. (1994) *Advances in Polymeric Systems for Drug Delivery.* Gordon and Breach Science Publisher, Gordon and Breach Science Publisher, Tokyo.

8. Barry, B.W. (1983) Drug delivery systems. *CHEMTECH* 38–44.

9. Ringsdorf, H. (1975) Structure and properties of pharmacologically active polymers. *J. Polym. Sci. Symp.* **51**: 135–153.

10. Donaruma, L.G. (1974) Synthetic biologically active polymers. *Prog. Polym. Sci.* **4**: 1–25.

11. Batz, H.G. (1977) Polymeric drugs. *Adv. Polym. Sci.* **23**: 25–53.

12. Ottenbrite, R.M. (1980) Introduction to biology and medicine. In Donaruma, L.G., Ottenbrite, R.M. and Vogl, O. (eds). *Anionic Polymeric Drugs.* Wiley, New York.

13. Duncan, R. and Kopecek, J. (1984) Soluble synthetic polymers as potential drug carriers. *Adv. Polym. Sci.* **57**: 51–101.

14. Ferruti, P. and Tanzi, M.C. (1986) New polymeric and oligomeric matrices as drug carriers. *CRC Crit. Rev. Ther. Drug Carrier Syst.* **2**: 175–244.

15. Kim, S.W., Peterson, R.V. and Feijen, J. (1980) Polymeric drug delivery systems. In Arlens, E. (ed.). *Drug Design*, vol. X, pp. 193–250. Academic Press, New York.

16. Kopecek, J., Rejmanova, P., Duncan, R. and Lloyd, J.B. (1985) Controlled release of drug model from N-(2-hydroxypropyl)methacrylamide copolymers. *Ann. NY Acad. Sci.* **446**: 93–104.

17. Rihova, B., Ulbrich, K., Strohalm, J., Vetvicka, V., Bilej, M., Duncan, R. and Kopecek, J. (1989) Biocompatibility of N-(2-hydroxypropyl)-methacrylamide copolymers containing adriamycin. Immunogenicity, effect on hematopoietic stem cells in bone marrow *in vivo* and effect on mouse splenocytes and human peripheral blood lymphocytes *in vitro*. *Biomaterials* **10**: 335–342.

18. Duncan, R. (1992) Drug–polymer conjugates: potential for improved chemotherapy. *Anticancer Drugs* **3**: 175–210.

19. Pato, J., Azori, M. and Tudos, F. (1987) Quinidine convalently bound to a dextran carrier. *J. Bioact. Biocomp. Polymers* **2**: 142–147.

20. Mora, M., Pato, J. and Tudos, F. (1989) Polymeric prodrugs, 6. Synthesis and examination of 6-purinethiol bound to poly(1-vinyl-2-pyrrolidone-co-maleic acid). *Makromol. Chem.* **190**: 1967–1974.

21. Soutif, J.C., Mouity-Moussounda, F. and Brosse, J.C. (1983) Polymeric carriers of glycerol derivatives, 2. Chlorambucil derivatives. *Makromol. Chem. Rapid Commun.* **4**: 61–64.

22. Yasuhiro, M. and Maeda, H. (1986) A new concept for macromolecular therapeutics in cancer chemotherapy: mechanism of tumourotropic accumulation of proteins and the antitumour agent smancs. *Cancer Res.* **46**: 6387–6392.

23. Hirano, T., Ringsdorf, H. and Zaharko, D.Z. (1980) Antitumour activity of monomeric and polymeric cyclophosphamide derivatives compared with *in vitro* hydrolysis. *Cancer Res.* **40**: 2263–2267.

24. Przybylski, E., Fell, E., Ringsdorf, H. and Zaharko, D.Z. (1978) Pharmacologically active polymers. 17. Synthesis and characterization of polymeric derivatives of the antitumour agent methotrexate. *Makromol. Chem.* **179**: 1719–1733.

25. Remon, J.P., Duncan, R. and Schacht, E. (1984) Polymer-drug combinations: pinocytic uptake of modified polysaccharides containing procaimide moieties by rat visceral yolk sacs cultured *in vitro*. *J. Controlled Rel.* **1**: 47–56.

26. Schacht, E. (1985) Use of polysaccharides as drug carriers. *Ann. Natl Acad. Sci. USA* **446**: 199–212.

27. Bernstein, A., Hurwitz, E., Maron, R., Arnon, R., Sela, M. and Wilchek, M. (1978) Higher antitumour efficacy of daunomycin when linked to dextran, *in vivo* and *in vitro* studies. *J. Natl Cancer Inst.* **60**: 379–384.

28. Konieezny, M., Charytonowicz, D. and Inglot, A.D. (1982) Search for carriers for non-covalent binding of interferon among 1,3,5-triazine derivatives of dextran. *Arch. Immunol. Ther. Exp.* **30**: 1–10.

29. Sezaki, H. and Hashida, M. (1984) Macromolecule-drug conjugates in targeted cancer chemotherapy. *CRC Crit. Rev. Therapeut. Drug Carr. Syst.* **1**: 1–38.

30. Matsumoto, S., Yamamoto, A., Takakura, Y., Hashida, M. and Sezaki, H. (1986) Cellular interaction and *in vitro* antitumour activity of mitomycin C-dextran conjugate. *Cancer Res.* **46**: 4463–4468.

31. Molteni, L. (1979) Dextrans as drug carriers. In Gregoriadis, G. (ed.). *Drug Carriers in Biology and Medicine*, pp. 107–125. Academic Press, London.

32. Simonescu, C.R., Popa, M.I. and Dimitriu, S. (1984) Bioactive polymers. XIV. Immobilization of ampicillin on biozan R. *Z. Naturforsch.* **39C**: 397–401.

33. Hurwitz, E., Wilchek, M. and Pitha, J. (1980) Soluble macromolecules as carriers of daunomycin. *J. Appl. Biochem.* **2**: 25–36.

34. Ouchi, Y., Banba, T., Matsumoto, T., Suzuki, S. and Suzuki, M. (1989) Antitumour activity of chitosan and chitin immobilized 5-fluorouracils through hexamethylene spacers via carbamoyl bonds. *J. Bioact. Biocomp. Pol.* **4**: 362–371.

35. Ryser, H. and Shen, W.C. (1978) Conjugation of methotrexate to poly(l-lysine) increases drug transport and overcomes drug resistance in cultured cells. *Proc. Natl Acad. Sci. USA* **75**: 3867–3870.

36. Zunino, F., Giulliani, F., Savi, G., Dasdia, T. and Gambetta, R. (1982) Anti-tumour activity of daunorubicin linked to poly-L-aspartic acid. *J. Pharmocol. Exp. Ther.* **30**: 465–469.

37. Giammona, G., Puglisi, G., Carlisi, B., Pignatello, R., Spadaro, A. and Caruso, A. (1989) Polymeric prodrugs: α-β-poly(N-hydroxyethyl)-DL-aspartamide as macromolecular carrier for some non-steroidal anti-inflammatory agents. *Int. J. Pharm.* **57**: 55–62.

38. Friedmann, G., Aichaoui, H. and Brini, M. (1981) Polymères à propriétés pharmacologiques potentiels: greffage par liaison amide et (ou) ester d'un hypoglycémiant. *Makromol. Chem.* **182**: 337–347.

39. De Machado, M., Neuse, E.W. and Perlwitz, A.G. (1992) Water-soluble polyamides as potential drug carriers. V. Carboxy-functionalized polyaspartamides and copolyaspartamides. *Angew. Makromol. Chem.* **195**: 35–56.

40. De Marre, A., Soyez, H. and Schacht, E. (1994) Synthesis of macromolecular mitomycin C derivatives. *J. Controlled Release* **32**: 129–137.

41. Trouet, A., Baurain, R., Deprez-De Campaneere, D. and Pirson, P. (1982) Targeting of antitumoral antiprotozoal drugs by covalent linkage to protein carriers. In Gregoriadis, G., Senior, J. and Trouet, A. (eds). *Targeting of Drugs*, pp. 19–30. Plenum Press, New York.

42. Chu, B.C.F. and Whiteley, J.M. (1979) Control of solid tumour metastases with a high molecular weight derivative of methotrexate. *J. Natl Cancer Inst.* **62**: 79–82.

43. Harris, J.M. (1992) Introduction to biotechnical and biomedical applications of poly(ethylene glycol). In Harris, J.M. (ed.). *Poly (ethylene glycol) Chemistry: Biotechnical and Biomedical Applications*, pp. 1–14. Plenum Press, New York.

44. Abuchowski, A., Es, T.V., Palczuk, N.C. and Davis, F.F. (1977) Alteration of immunological properties of bovine serum albumin by covalent attachment of polyethylene glycol. *J. Biol. Chem.* **252**: 3578–3581.

45. Abuchowski, A., Kazo, G.M., Verhoest, C.R., Jr., Van Es, T., *et al.* (1984) Cancer Therapy with chemically modified enzymes. 1. A property of polyethylene glycol asparaginase conjugates. *Cancer Bio. Chem. Biophys.* **7**: 175–186.

46. Abuchowski, A. and Davis, F.F. (1981) In Hosenberg, J. and Roberts, J. (eds). *Enzymes as Drugs.* Wiley, New York.

47. Cecchi, R., Rusconi, L., Tanzi, M.C., Danusso, F. and Ferruti, P. (1981) Synthesis and Pharmacological evaluation of 4-isobutylphenyl-2-propionic acid (ibuprofen). *J. Med. Chem.* **24**: 622–625.

48. Duncan, R., Kopecek, J. and Lloyd, J.B. (1983) Development of N-(2-hydroxypropyl)-methacrylamide copolymers as carriers of therapeutic agents. In Chiellini, E. and Guisti, P. (eds). *Polymers in Medicine: Biomedical and Pharmacological Applications*, pp. 97–114. Plenum Press, New York.

49. Lloyd, J.B., Duncan, R. and Pratten, M.K. (1983) Soluble synthetic polymers as targetable agents for intracellular drug release. *Br. Polym. J.* **15**: 158–159.

50. Duncan, R., Kopecek, J., Rejmanova, P. and Lloyd, J.B. (1983) Targeting of N-(2-hydroxypropyl)-methacrylamide copolymers to liver by incorporation of galactose residues. *Biochem. Biophys. Acta* **755**: 518–521.

51. Seymour, L.W., Flanagan, P.A., Al-Shamkhani, A., Subr, V., Ulbrich, K., Cassidy, J. and Duncan, R. (1991) Synthetic polymers conjugated to monoclonal antibodies: vehicles for tumour-targeted drug delivery. *Select. Cancer Ther.* **7**: 59–73.

52. Flanagan, P.A., Kopeckova, P., Kopecek, J. and Duncan, R. (1989) Evaluation of antibody-N-(2-hydroxypropyl)-methacrylamide copolymer conjugates as targetable drug-carriers. 1. Binding, pinocytic uptake and intracellular distribution of transferrin and anti-transferrin receptor antibody-conjugates. *Biochim. Biophys. Acta* **993**: 83–91.

53. Seymour, L.W., Duncan, R., Strohalm, J. and Kopecek, J. (1987) Effect of molecular weight of N-(2-hydroxypropyl)methacrylamide copolymers on body distribution and rate of excretion after subcutaneous, intraperitoneal and intravenous administration to rats. *J. Biomed. Mater. Res.* **21**: 1341–1358.

54. Duncan, R (2001) Polymer therapeutics. In: Business Briefing, PharmaTech 2001, pp. 178–184.

55. Seymour, L.W., Ferry, D.R., Anderson, D., Hesslewood, S., *et al.* (2002) Hepatic drug targeting: phase I evaluation of polymer-bound doxorubicin. *J. Clin. Oncol.* **20**: 1668–1676.

56. Terwogt, J.M.M., Huinink, W.W.T., Schellens, J.H.M., Schot, M., *et al.* (2001) Phase I clinical and pharmacokinetic study of PNU166945, a novel water-soluble polymer-conjugated prodrug of paclitaxel. *Anticancer Drugs* **12**: 315–323.

57. Sela, M. and Katchalski, E. (1959) Biological properties of poly-α-amino acids. *Adv. Protein Chem.* **14**: 391–478.

58. Monsigny, M., Roche, A.C., Midoux, P. and Mayer, R. (1994) Glycoconjugates as carriers for specific delivery of therapeutic drugs and genes. *Adv. Drug Del. Rev.* **14**: 1–24.

59. Vercauteren, R., Schacht, E. and Duncan, R. (1992) Effect of the chemical modification of dextran on the degradation by rat liver lysosomal enzymes. *J. Bioact. Biocomp. Polym.* **7**: 346–357.

60. Franssen, E.J.F., Moolenaar, F., De Zeeuw, D. and Meijer, D.K.F. (1994) Drug Targeting to the kidney with low-molecular weight proteins. *Adv. Drug Del. Rev.* **14**: 67–88.

61. Seymour, L.W. (1994) Soluble polymers for lectin-mediated drug targeting. *Adv. Drug Del. Rev.* **14**: 89–112.

62. Yokoyama, M., Miyauchi, M., Yamada, N., Okano, T., Sakurai, Y., Kataoka, K. and Inoue, S. (1990) Characterization and anticancer activity of the micelle forming polymeric anti-cancer drug adriamycin-conjugated poly(ethylene glycol)-poly(aspartic acid) block copolymer. *Cancer Res.* **50**: 1693–1700.

63. Yokoyama, M., Okano, T., Sakurai, Y., Ekimoto, H., Shibazaki, C. and Kataoka, K. (1991) Toxicity and anti-tumour activity against solid tumours of micelle-forming polymeric anti-cancer drug and its extremely long circulation in blood. *Cancer Res.* **51**: 3229–3236.

64. Yokoyama, M. (1992) Block copolymers as drug carriers. *Crit. Rev. Ther. Drug Carrier Syst.* **9**: 213–248.

65. Duncan, R. (1987) Selective endocytosis of macromolecular drug carriers. In Robinson, J.R. and Lee, V.H. (eds). *Controlled Drug Delivery: Fundamentals and Applications*, 2nd edn, pp. 581–607. Marcel Dekker, New York.

66. Trouet, A., Masquelier, M., Baurain, R. and Deprez-De Campeneere, D. (1982) A covalent linkage between daunorubicin and protein that is stable in serum and reversible by lysosomal hydrolases, as required for a lysosomotropic drug-carrier conjugate: *in vitro* and *in vivo* studies. *Proc. Natl Acad. Sci. USA* **79**: 626–629.

67. Subr, V., Strohalm, J., Ulbrich, K., Duncan, R. and Hume, Z. (1992) Polymers containing enzymatically degradable bonds, XII. Effect of spacer structure on the rate of release of daunomycin and adriamycin from poly[N-(2-hydroxypropyl)-methacrylamide] copolymer drug carriers *in vitro* and antitumour activity measured *in vivo*. *J. Controlled Rel.* **18**: 123–132.

68. De Marre, A., Seymour, L.W. and Schacht, E.H. (1994) Evaluation of the hydrolytic and enzymatic stability of macromolecular mitomycin C derivatives. *J. Controlled Res.* **31**: 89–97.

69. Kratz, F., Beyer, U. and Schütte, M.T. (1999) Drug–polymer conjugates containing acid-cleavable bonds carrier systems. *Crit. Rev. Ther. Drug Carrier Syst.* **16**: 245–288.

70. Kaneko, T., Willner, D., Monkovic, I., Knipe, J.O., *et al.* (1991) New hydrazone derivatives of adriamycin and their immunoconjugates. A correlation between acid stability and cytotoxicity. *Bioconjugate Chem.* **2**: 133–141.

71. Kratz, F., Beyer, U., Collery, P., Lechenault, F., *et al.* (1998) Preparation, characterization and *in vitro* efficacy of albumin conjugates of doxorubicin. *Biol. Pharm. Bull.* **21**: 56–61.

72. Coessens, V., Schacht, E. and Domurado, D. (1996) Synthesis of polyglutamine and dextran conjugates of streptomycin with an acid-sensitive drug-carrier linkage. *J. Controlled Rel.* **38**: 141–150.

73. Shen, W.C. and Ryser, H.J.P. (1981) Cis-aconityl spacer between daunomycin and macromolecular carriers: a model of pH-sensitive linkage releasing drug from a lysosomotropic conjugate. *Biochem. Biophys. Res. Commun.* **102**: 1048–1054.

74. Shen, W.C., Ryser, H.J.P. and La Manna, L. (1985) Disulfide spacer between methotrexate and poly(D-lysine). *J. Biol. Chem.* **260**: 10905–10908.

75. Maeda, H., Seymour, L.W. and Miyamoto, Y. (1992) Conjugates of anticancer agents and polymer: advantages of macromolecular therapeutics *in vivo*. *Bioconj. Chem.* **3**: 351–362.

76. Seymour, L.W., Ulbrich, K., Strohalm, J. and Duncan, R. (1990) The pharmacokinetics of polymer-bound adriamycin. *Biochem. Pharmacol.* **39**: 1125–1131.

77. Takakura, Y., Kitajima, M., Matsumoto, S., Hashida, M. and Sezaki, H. (1987) Development of a novel polymeric prodrug of mitomycin C, mitomycin C-dextran conjugate with anionic charge. 1. Physicochemical characteristics and *in vitro* and *in vivo* antitumour activities. *Int. J. Pharm.* **37**: 135–143.

78. Sezaki, H. and Hashida, M. (1984) Macromolecule-drug conjugates in targeted cancer chemotherapy. *CRC Crit. Rev. Ther. Drug Carrier Syst.* **1**: 1–38.

79. Ogino, T., Inoue, M., Ando, Y., Arai, H. and Morino, Y. (1988) Chemical modification of superoxide dismutase. Extension of plasma half-life of the enzyme through its reversible binding to circulating albumin. *Int. J. Peptide Protein Res.* **32**: 153–159.

80. Abuchowski, A., McCoy, J.R., Palczuk, N.C., Es, T.V. and Davis, F.F. (1977) Effect of covalent attachment of polyethylene glycol on immunogenicity and circulating life of bovine liver catalase. *J. Biol. Chem.* **252**: 3582–3586.

81. Kamisaki, Y., Wada, H., Yagura, H., Matsushima, A. and Inada, Y. (1981) Reduction in immunogenicity and clearance rate of *Escherichia coli* L-asparaginase by modification with monomethoxypoly(ethylene glycol). *J. Pharmacol. Exp. Ther.* **216**: 410–414.

82. Li, C.J., Miyamoto, Y., Kojima, Y. and Maeda, H. (1993) Augmentation of tumour delivery of macromolecular drugs with reduced bone marrow delivery by elevating blood pressure. *Br. J. Cancer* **67**: 975–980.

83. Suzuki, M., Hori, K., Abe, I., Saito, S. and Sato, H. (1981) A new approach to cancer chemotherapy: a selective enhancement of tumour blood flow with angiotensin II. *J. Natl. Cancer Inst.* **67**: 663–669.

84. Hori, K., Suzuki, M., Tanda, S., Saito, S., Shiozaki, S. and Zhang, Q.H. (1991) Fluctuation in tumour blood flow under normotension and the effect of angiotensin II induced hypertension. *Jpn. J. Cancer Res.* **82**: 1309–1316.

85. Willmott, N. (1987) Chemoembolisation in regional cancer chemotherapy: a rationale. *Cancer Treat. Rev.* **14**: 143–156.

86. Maeda, H. (1991) SMANCS and polymer-conjugated macromolecular drugs: advantages in cancer chemotherapy. *Drug Delivery Rev.* **6**: 181–202.

87. Hirayama, S., Sato, F., Oda, T. and Maeda, H. (1986) Stability of high molecular weight anticancer agent SMANCS and its transfer from oil-phase to water-phase. *Jpn J. Antibiot.* **39**: 815–822.

88. Maeda, H. (1992) The tumour blood vessel as an ideal target for macromolecular anticancer agents. *J. Controlled Rel.* **19**: 315–324.

89. Konno, T., Maeda, H., Iwai, K., Tashiro, S., Mati, S., Morinaga, T., Mochinaga, M., Hiraoka, T. and Yokoyama, I. (1983) Effect of arterial administration of high-molecular-weight anticancer agent SMANCS with lipid lymphographic agent on hepatoma: a preliminary report. *Eur. J. Cancer Clin. Oncol.* **19**: 1053–1065.

90. Konno, T. and Maeda, H. (1987) Targeting chemotherapy of hepatocellular carcinoma: arterial administration of smancs/lipiodol. In Okuda, K. and Ishak, K.G. (eds). *Neoplasms of the Liver*, pp. 343–352, chap. 27. Springer-Verlag, New York.

91. Trail, P.A., Willner, D., Lasch, S.J., Henderson, A.J., Hofstead, S., Casazza, A.M., Firestone, R.A. and Hellstrom, K.E. (1993) Cure of xenografted human carcinomas by BR96-doxorubicin immunoconjugates. *Science* **261**: 212–215.

92. Wawrzynczak, E.J. and Derbyshire, E.J. (1992) Immunotoxins: the power and the glory. *Immunology Today* **13**: 381–383.

93. Vitetta, E.S., Stone, M., Amlot, P., Fay, J., *et al.* (1991) Phase I immunotoxin trial in patients with B-cell lymphoma. *Cancer Res.* **51**: 4052–4058.

94. Wang, J.M., Kumar, S., Pye, D., Van Agthoven, A.J., Krupinski, J. and Hunter, R.D. (1993) A monoclonal antibody detects heterogenicity in vascular endothelium of tumours and normal tissues. *Int. J. Cancer* **54**: 363–370.

95. Burrows, F.J. and Thorpe, P.E. (1993) Eradication of large solid tumours in mice with an immunotoxin directed against tumour vasculature. *Proc. Natl. Acad. Sci.* **90**: 8996–9000.

96. Bagshawe, K.D., Springer, C.J., Searle, F., Antoniw, P., Sharma, S.K., Melton, R.G. and Sherwood, R.F. (1988) A cytotoxic agent can be generated selectively at cancer sites. *Br. J. Cancer* **58**: 700–703.

97. Springer, C.J., Bagshawe, K.D., Sharma, S.K., Searle, F., *et al.* (1991) Ablation of human choriocarcinoma xenografts in nude mice by antibody-directed enzyme prodrug therapy (ADEPT) with three novel compounds. *Eur. J. Cancer* **27**: 1361–1366.

98. Senter, P.D., Su, P.D.D., Katsuragi, T., Sakai, T., *et al.* (1991) Generation of 5-fluorouracil from 5-fluorocytosine by monoclonal antibody-cytosine deaminase conjugates. *Bioconjugate Chem.* **2**: 447–451.

99. Sharma, S.K., Bagshawe, K.D., Springer, C.J., Burge, P.J., *et al.* Antibody directed enzyme prodrug therapy (ADEPT): a three-phase system. *Disease Markers* **9**: 225–231.

100. Goldstein, I.J., Hugues, R.C., Monsigny, M., Osawa, T. and Sharon, N. (1980) What should be called a lectin? *Nature* **285**: 66.

101. Ashwell, G. and Harford, J. (1982) Carbohydrate-specific receptors of the liver. *Annu. Rev. Biochem.* **51**: 531–554.

102. Sharon, N. and Lis, H. (1989) Lectins as cell recognition molecules. *Science* **246**: 227–234.

103. Monsigny, M., Roche, A.C. and Midoux, P. (1988) Endogenous lectins and drug targeting. *Ann. NY Acad. Sci.* **551**: 399–414.

104. De Duve, C., De Barsy, T., Poole, B., Trouet, A., *et al.* (1974) Lysosomotropic agents. *Biochem. Pharmacol.* **23**: 2495–2531.

105. Schwartz, A.L. (1984) The hepatic asialoglycoprotein receptor. *CRC Crit. Rev. Biochem.* **16**: 207–233.

106. Vansteenkiste, S., Schacht, E., Duncan, R., Seymour, L., Pawluczyk, I. and Baldwin, R. (1991) Fate of glycosylated dextrans after *in vivo* administration. *J. Controlled Rel.* **16**: 91–100.

107. Vansteenkiste, S., De Marre, A. and Schacht, E. (1992) Synthesis of glycosylated dextrans. *J. Bioact. Compat. Polymers* **7**: 4–14.

108. Anderson, D., Vansteenkiste, S., Schacht, E.H., Sen, S.V. and Seymour, L.W. (1994) *In vitro* binding specificity of glycosylated dextrans to the asialoglycoprotein receptor of primary hepatocytes. *Eur. J. Pharm.* **3**: 339–345.

109. Weidner, N., Semple, J.P., Welch, W.R. and Folkman, J. (1991) Tumor angiogenesis and metastasis — correlation in invasive breast carcinoma. *N. Engl J Med.* **324**: 1–8.

110. Yancopoulos, G.D., Klagsbrun, M. and Folkman, J. (1998) Vasculogenesis, angiogenesis, and growth factors: ephrins enter the fray at the border. *Cell* **93**: 661–664.

111. Eliceiri, B.P. and Cheresh, D.A. (2001) Adhesion events in angiogenesis. *Curr Opin. Cell Biol.* **13**: 563–568.

112. Martiny-Baron, G. and Marme, D. (1995) VEGF-mediated tumor angiogenesis – a new target for cancer-therapy. *Curr. Opin. Biotechnol.* **6**: 675–680.

113. Brooks, P.C., Montgomery, A.M.P., Rosenfeld, M., Reisfeld, R.A., *et al.* (1994) Integrin $\alpha_v\beta_3$ antagonist promote tumor-regression by inducing apoptosis of angiogenic blood vessels. *Cell* **79**: 1157–1164.

114. Humphries, M.J., Olden, K. and Yamada, K.M. (1986) A synthetic peptide from fibronectin inhibits experimental metastasis of muring melanoma cells. *Science* **233**: 467–470.

115. Arap, W., Pasqualini, R. and Ruoslahti, E. (1998) Cancer treatment by targeted drug delivery to tumor vasculature in a mouse model. *Science* **279**: 377–380.

116. Maeda, M., Izuno, Y., Kawasaki, K., Kaneda, Y., *et al.* (1997) Amino acids and peptides XXX. Preparation of Arg-Gly-Asp (RGD) hybrids with poly(ethylene glycol) analogs and their antimetastatic effect. *Chem. Pharm. Bull.* **45**: 1788–1792.

117. Cooke, J.P. and Losordo, D.W. (2002) Nitric oxide and angiogenesis. *Circulation* **105**: 2133–2135.

118. Fridovich, L. (1970) Quantitative aspects of the production of superoxide anion radical by milk xanthine oxidase. *J. Biol. Chem.* **245**: 4053–4057.

119. Sawa, T., Wu, J., Akaike, T. and Maeda, H. (2000) Tumor-targeting chemotherapy by a xanthine oxidase-polymer conjugate that generates oxygen-free radicals in tumor tissue. *Cancer Res.* **60**: 666–671.

120. Matsumura, Y. and Maeda, H. (1986) A new concept for macromolecular therapeutics in cancer chemotherapy: mechanism of tumouritropic accumulation of proteins and antitumour agent SMANCS. *Cancer Res.* **46**: 6387–6392.

121. Jain, R.K. (1991) Vascular and interstitial barriers to the delivery of therapeutic agents in tumours. *Cancer Metast. Rev.* **9**: 253–266.

122. Senger, D.R., Van de Water, L., Brown, I.F., Nagy, J.A., Yeo, K.T., *et al.* (1993) Vascular permeability factor in tumour biology. *Cancer Metast. Rev.* **12**: 303–324.

123. Berstein, A., Hurwitz, E., Maron, R., Arnon, R., *et al.* (1978) Higher antitumour efficacy of daunomycin when linked to dextran: *in vivo* and *in vitro* studies. *J. Natl Cancer Inst.* **60**: 379–384.

124. Zunino, F., Pratesi, G. and Pezzoni, G. (1987) Increased therapeutic efficacy and reduced toxicity of doxorubicin linked to pyran copolymer via the side chain of the drug. *Cancer Treat. Reps.* **71**: 367–373.

125. Trouet, A. and Jolles, G. (1984) Targeting of daunorubicin by association with DNA or proteins: a review. *Sem. Oncol.* **11**: 64–73.

126. Seymour, L.W. (1992) Passive tumour targeting of soluble macro-molecules and drug conjugates. *Crit. Rev. Therapeut. Drug Carr. Syst.* **9**: 135–187.

127. Cassidy, J., Duncan, R., Morrison, G.J., Strohalm, J., *et al.* (1989) Activity of N-(2-hydroxypropyl)methacrylamide copolymers containing daunomycin against a rat tumour model. *Biochem. Pharmacol.* **38**: 875–880.

128. Duncan, R., Seymour, L.W., O'Hare, K.B., Flanagan, P.A., *et al.* (1992) Preclinical evaluation of polymer-bound doxorubicin. *J. Controlled Rel.* **18**: 123–132.

129. Seymour, L.W., Ulbrich, K., Steyger, P.S., Brereton, M., *et al.* (1994) Tumourtropism and anticancer afficacy of polymer-based doxorubicin prodrugs in the treatment of subcutaneous murine B16F10 melanoma. *Br. J. Cancer* **70**: 636–641.

130. Seymour, L.W., Miyamoto, Y., Maeda, H., Brereton, M., *et al.* (1995) Influence of molecular weight on passive tumour accumulation of a soluble macromolecular drug carrier. *Eur. J. Cancer* **31A**: 766–770.

131. Moscow, J.A. and Cowan, K.H. (1988) Multidrug resistance. *J. Natl Cancer* **80**: 14–20.

132. Kopecek, J., Kopeckova, P., Minko, T. and Zheng-Rong, L. (2000) HPMA copolymer-anticancer drug conjugates: design, activity and mechanism of action. *Eur. J. Pharma. Biopharm.* **50**: 61–81.

133. Cuvier, C., Roblot-Treupel, L., Millot, J.M., Lizard, G., *et al.* (1992) Doxorubicin-loaded nanospheres bypass tumor cell multidrug resistance. *Biochem. Pharmacol.* **44**: 509–517.

134. Warren, L., Jardilier, J.C., Malarska, A. and Akeli, M.G. (1992) Increased accumulation of drugs in multidrug resistant cell induced by liposomes. *Cancer Res.* **52**: 3241–3245.

135. Ohkawa, K., Hatano, T., Yamada, K., Joh, K., *et al.* (1993) Bovine serum albumin-doxorubicin conjugate overcomes multidrug resistance in a rat hepatoma. *Cancer Res.* **53**: 4238–4242.

136. Stastny, M., Strohalm, J., Plocova, D., Ulbrich, K. and Rihova, B. (1999) A possibility to overcome P-glycoprotein (PGP)-mediated multidrug resistance by antibody-targeted drugs conjugated to N-(2-hydroxypropyl)methacrylamide (HPMA) copolymer carrier. *Eur. J. Cancer* **35**: 459–466.

137. Minko, T., Kopeckova, P., Pozharov, V. and Kopecek, J. (1998) HPMA copolymer bound adriamycin overcomes MDR1 gene encoded resistance in human ovarian carcinoma cell line. *J. Controlled Rel.* **54**: 223–233.

138. Tijerina, M., Fowers, K.D., Kopeckova, P. and Kopecek, J. (2000) Chronic exposure of human ovarian carcinoma cells to free or HPMA copolymer-bound mesochlorin e(6) does not induce P-glycoprotein-mediated multidrug resistance. *Biomaterials* **21**: 2203–2210.

139. Maher, S., Toomey, D., Condron, C. and Bouchier-Hayes, D. (2002) Activation induced cell death: the controversial role of Fas and Fas ligand in immune privilege and tumour counterattack. *Immunol. Cell Biol.* **80**: 131–137.

140. Nagata, S. (1997) Apoptosis by death factor. *Cell* **88**: 355–365.

141. Daniel, P.T., Scholz, C., Westermann, J., Dorken, B. and Pezzutto, A. (1998) Dendritic cells prevent CD95-mediated T lymphocyte death through costimulatory signals. *Gene Ther. Cancer* **28**: 173–177.

142. Restifo, N. (2000) Not so Fas: re-evaluating the mechanism of immune privilege and tumor escape. *Nat. Med.* **6**: 493–495.

143. Krammer, P.H., Dhein, J., Walczak, H., Behermann, I., *et al.* (1994) The role of APO-1-mediated apoptosis in the immune system. *Immunol. Rev.* **142**: 175–191.

144. O'Connell, J., Bennett, M.W., O'Sullivan, G.C., Collins, J.K. and Shanahan, F. (1999) The Fas counterattack: cancer as a site of immune privilege. *Immunol. Today* **20**: 46–52.

145. O'Connell, J., O'Sullivan, G.C., Collins, J.K. and Shanahan, F. (1996) The Fas counterattack: Fas-mediated T cell killing by colon cancer cells expressing Fas ligand. *J. Exp. Med.* **184**: 1075–1082.

146. Hahne, M., Rinoldi, D., Schroter, M., Romero, P., *et al.* (1996) Melanoma cell expression of Fas(Apo-1/CD95) ligand: implications for tumor immune escape. *Science* **274**: 1363–1366.

147. Strand, S., Hofmann, W.J., Hug, H., Muller, M., *et al.* (1996) Lymphocyte apoptosis induced by CD95 (APO-1/Fas) ligand-expressing tumor cells — a mechanism of immune evasion? *Nat. Med.* **2**: 1361–1366.

148. Niehans, G.A., Brunner, T., Frizelle, S.P., Liston, J.C., *et al.* (1997) Human lung carcinomas express Fas ligand. *Cancer Res.* **57**: 1007–1012.

149. Von Bernstorff, W., Spanjaard, R.A., Chan, A.K., Lockhart, D.C., *et al.* (1999) Pancreatic cancer cells can evade immune surveillance via nonfunctional Fas(APO-1/CD95) receptors and aberrant expression of functional Fas ligand. *Surgery* **125**: 73–84.

150. Bennett, M.W., O'Connell, J., O'Sullivan, G.C., Brady, C., *et al.* (1998) The Fas counterattack *in vivo*: apoptotic depletion of tumor-infiltrating lymphocytes associated with Fas ligand expression by human esophageal carcinoma. *J. Immunol.* **160**: 5669–5675.

151. Mitsiades, N., Poulaki, V., Kotoula, V., Leone, A. and Tsokos, M. (1998) Fas ligand is present in tumors of the Ewing's sarcoma family and is cleaved into a soluble form by a metalloprotease. *Am. J. Pathol.* **153**: 1947–1956.

152. Friesen, C., Herr, I., Krammer, P.H. and Debatin, K.-M. (1996) Involvement of the CD95 (Apo-1/Fas) receptor/ligand system in drug-induced apoptosis in leukemia cells. *Nat. Med.* **2**: 574–577.

153. Walker, P.R., Saas, P. and Dietrich, P.-Y. (1997) Role of Fas ligand (CD95L) in immune escape: the tumor cell strikes back – commentary. *J. Immunol.* **15**: 4521–4524.

154. Friesen, C., Fulda, S. and Debatin, K.-M. (1999) Cytotoxic drugs and the CD95 pathways. *Leukemia* **13**: 1854–1858.

155. Rihova, B., Strohalm, J., Hoste, K., Jelinkova, M., *et al.* (2001) Immunoprotective therapy with targeted anticancer drugs. *Macromol. Symp.* **172**: 21–28.

156. Callant, D. and Schacht, E. (1990) Macromolecular prodrugs of 5-aminosalicylic acid, 1. Azo-conjugates. *Makromol. Chem.* **191**: 529–536.

157. Schacht, E., Gevaert, A., Kenawy, E.R., Molly, K., Verstraete, W., *et al.* (1996) Polymers for colon specific drug delivery. *J. Controlled Rel.* **39**: 327–338.

158. Kopeckova, P. and Kopecek, J. (1990) Release of 5-aminosalicylic acid from bioadhesive N-(2-hydroxypropyl)methacrylamide co-polymers by azoreductases *in vitro*. *Makromol. Chem.* **191**: 2037–2045.

35

PREPARATION OF WATER-SOLUBLE COMPOUNDS THROUGH SALT FORMATION

P. Heinrich Stahl

There is nothing in the universe but alkali and acid, from which nature composes all things.

Ott. Tachenius (1671)

I INTRODUCTION

An optimal drug candidate is not only a highly potent ligand, selective to the receptor species to which it has been designed to bind, has more favourable kinetics of binding than its naturally competing ligands, and has adequate chemical and metabolic stability. Moreover, it should satisfy a series of requirements set by areas remote from the immediate environment of the receptors. Most important, there is the need for substantial transport from outside the organism to the sites of action. Biopharmaceutically relevant substance properties can enable or limit the crossing of absorption barriers and for administering a drug in dosed amounts or suitable concentration. Preparations need to be designed and developed for timely and quantitative delivery of the active substance. Although

pharmaceutical formulation techniques offer numerous possibilities, there are limitations in mastering problems incurred by substances with suboptimal inherent properties, and the least of the consequences is that they slow down the drug development process.

An NCE (new chemical entity) rarely meets all the requirements of an ideal drug substance. As lead optimization concentrates towards improvement of biological activity as evaluated by *in vitro* tests, other desirable properties often fall behind, although the awareness is increasing that acceptable pharmacokinetic properties should be built into a drug molecule as well.

Functional groups that turn a molecule into an electrolyte provide the opportunity for the formation of salts. Salt formation is a means to change the properties while the chemical structure of the active entity is left intact. Yet, a new chemical entity is created by salt formation with all the consequences.

About half of all present-day drugs are used as salts, and for the chemist it is obvious to try improving the properties of a drug candidate by the formation of a suitable salt, in particular with the intention to enhance solubility and in consequence, the rate of dissolution as prerequisite parameters for absorption. However, one should be aware, that high solubility alone does not guarantee good absorption.

In this chapter not only the topic of water-soluble salts but also some general points to consider when the final form of a drug substances will be decided upon. Although it is generally accepted that solubility in aqueous media is the most crucial physicochemical property of a drug substance, it is not appropriate that only a single parameter should dominate such an important decision. Since here only the essentials of the complex field of salt formation and selection can be dealt with, the reader may refer to a recent monograph on the subject.[1]

II THE SOLUBILITY OF COMPOUNDS IN WATER

A Prediction of solubility

Solubility is understood as the concentration in a solution found in equilibrium with an excess of the solid solute at a given temperature. It is generally accepted that solubility in aqueous media is the most crucial physicochemical property of a drug substance. In order to get an estimate of the solubility of a nonelectrolyte at a very early stage, many computational approaches have been made. Among them, a surprisingly simple and yet effective tool is available, the General Solubility Equation (GSE) as developed and refined during the last two decades by Yalkowsky and his coworkers[2,3] using a thermodynamically sound approach for establishing a semi-empirical correlation:

$$\log S = 0.5 - 0.01(Mp - 25) - \log K_{ow} \quad (1)$$

where S is the molar solubility of the solute, K_{ow} the octanol/water partition coefficient, which can be calculated from the structural formula (ClogP),[4] and Mp is the melting point (in centigrade, °C) as the only experimental data, representing the easiest accessible descriptor of the strength of a solid's crystal lattice. Based on the assumption, that the nonionized form of an electrolyte may be regarded as a nonelectrolyte, an extension of the GSE has recently been proposed. By simply combining equation (1) with the solubility–ionization relationships (equations in Table 35.2), it is possible to construct pH solubility profiles for acids, bases and zwitterions.[5]

B Ionization of weak electrolytes

In physicochemical terms, about two-thirds of all existing drug entities belong to the class of weak electrolytes, i.e. substances which in aqueous solution are at least partly present as ions. They are formed by releasing protons (acids) into, or by accepting protons (bases) from, an aqueous environment. Ionized species are easily hydrated and hence are generally more soluble in an aqueous phase than their nonionized source. The aqueous solubility of weak electrolytes is influenced and can be controlled by adjusting the pH of the solution via the equilibrium between the nonionized and the ionized species.

The dissociation of a monoprotic acid HA is described by the equilibrium between the concentrations of the nonionized acid molecule (= [HA]) and of its anion (= [A$^-$]):

$$[HA] \overset{K_a}{\rightleftharpoons} [H^+] + [A^-] \quad (2)$$

The equilibrium constant K_a of this reaction is the ionization constant, which is defined by

$$K_a = \frac{[H^+] \cdot [A^-]}{[HA]} \quad (3)$$

Because ionization constants are small and inconvenient figures, they are expressed as their negative decadic logarithms:

$$pK_a = -\left(^{10}\log K_a\right) \quad (4)$$

As an example, the K_a of acetic acid converts from 1.738×10^{-5} to $pK_a = 4.76$. Similarly, for a monobasic compound the dissociation equilibrium is expressed as follows:

$$[BH^+] \overset{K_a}{\rightleftharpoons} [B + H^+] \quad (5)$$

with the dissociation constant

$$K_a = \frac{[H^+] \cdot [B]}{[HB^+]} \quad (6)$$

Since the protonated base can be considered as an acid corresponding to the free base, it is possible to characterize both, acids and bases with the same parameter, i.e. the acid ionization constant. The K_a or the pK_a, respectively, is the key parameter indicating the strength of an electrolyte (Table 35.1).

Table 35.1 *Examples of acids and bases aligned according to strength*

	pK_a	Acids	Bases	pk_a	
Strong	1.3	Oxalic acid, $pK_{a,1}$	Caffeine	0.6	Weak
	1.6	Saccharin	Quinine, $pK_{a,2}$	4.1	
	2.8	Benzylpenicillin	Pyridine	5.2	
	3.0	Salicylic acid, $pK_{a,1}$	Papaverine	5.9	
	4.1	Diclofenac	Apomorphine	7.0	
	6.5	Sulfadiazine	Benzoctamine	7.6	
	7.1	Sulfathiazole	Quinine, $pK_{a,1}$	8.0	
	7.4	Phenobarbital, $pK_{a,1}$	Cocaine	8.4	
	8.4	Phenytoin	Imipramine	9.5	
	8.8	Theophyllin	Atropine	9.7	
	10.0	Phenol	Amantadine	10.3	
	11.8	Phenobarbital, $pK_{a,2}$	Piperidine	11.2	
Weak	13.8	Salicylic acid, $pK_{a,2}$	Arginine, $pK_{a,1}$	13.2	Strong

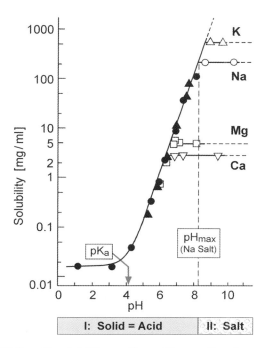

Fig. 35.1 pH solubility profiles of four salts of *naproxen* (pK_a = 4.15) in water at 25°C. Starting material was naproxen acid (full symbols) or the respective salt (open symbols). (Redrawn from Cowhan.[6])

C Solubility of electrolytes

General features of pH solubility profiles

Monoprotic acids and bases. An ideal pH solubility profile for a weakly monoprotic acidic compound is exemplified by *naproxen* **1**, as shown in Fig. 35.1. It is obtained by increasing the pH, e.g. by stepwise adding amounts of a base, for example sodium hydroxide, to a suspension of the acid and determining the concentration of dissolved acid after equilibration. The pH range of the diagram is divided into two regions marked as region I and II, respectively, according to the nature of the excess solid phase in equilibrium with the saturated solution. In region I (pH < 8.3 in Fig. 35.1) the excess solid phase is the free acid, whereas in region II (pH > 8.3), it is the sodium salt. In region I, the total solubility is described by the sum of concentrations of the neutral and the dissociated fractions of the acid:

Naproxen

1

Dexoxadrol

2

Valsartan

3

Baclofen

4

Flubendazole

5

Chlordiazepoxide

6

Terfenadine

7

Diclofenac

8

Flurbiprofen

9

Cyclopentamine hydrochloride

10

Amfenac sodium

11

Seproxetine

12

13

Salmeterol xinafoate

$$S = [HA] + [A^-] \quad (7)$$

$$S = S_0\left(1 + \frac{K_a}{[H^+]}\right) \quad (8)$$

where S is the total solubility at any given pH, S_0 is the intrinsic solubility of the free acid, [HA] and [A$^-$] represent concentrations in solution of the undissociated and dissociated forms, respectively, and K_a is the acid dissociation constant defined in equation (3).

Region I may be further subdivided into a range where the solubility is essentially independent of the pH (pH < ca.3) and the range adjacent to region II, between pH approximately 3 and 7.3 characterized by an exponential dependence of solubility on pH.

The total solubility in region II is described by:

$$S = \left(1 + \frac{[H^+]}{K_a}\right)\sqrt{K_{sp}} \quad (9)$$

$$K_{sp} = [Na^+] \cdot [A^-] \quad (10)$$

where K_{sp} is the solubility product of the salt, and $\sqrt{K_{sp}} = S_s$ is the intrinsic solubility of the salt.

Thus, the two equations (8) and (9) are sufficient to describe the entire pH solubility profile of a monoprotic acid, and the point of intersection is defined by equation (11):

$$pH_{max} = pK_a + \log\frac{S_0}{\sqrt{K_{sp}}} = pK_a + \log S_0 - \tfrac{1}{2}\log K_{sp} \quad (11)$$

For a weak base, the pH solubility profile is essentially the mirror image of that of a weak acid along a vertical axis, as illustrated in Fig. 35.2 by the central stimulant and analgesic *dexoxadrol* **2**.

As in the case of the acid, there is a region in equilibrium with excess undissolved free base, here at high pH values and a region with high and constant solubility at low pH values. An analogous set of equations applies for the two legs of the theoretical pH solubility profile (equations (11) and (12)), whereas for the position of pH_{max}, equation (10) is equally valid for bases.

High pH region (II) : $S = [B] + [BH^+]$

$$= S_0\left(1 + \frac{[H^+]}{K_a}\right) \quad (12)$$

Low pH region (I) : $S = \left(1 + \frac{K_a}{[H^+]}\right)\sqrt{K_{sp}} \quad (13)$

pH$_{max}$. At pH_{max}, both the nonionized species and the respective salt coexist in the undissolved solid. Bogardus

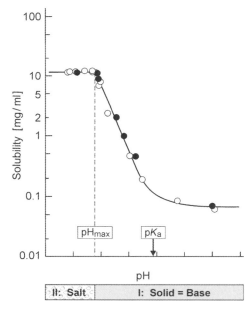

Fig. 35.2 pH solubility profile of *dexoxadrol* in water at approximately 23°C. Open and full circles indicate different analytical methods. (Redrawn from Kramer and Flynn.[7])

et al.[8] reported that indeed the excess solid in equilibrium with the saturated solution of doxycycline at pH_{max} contained both the free base and the hydrochloride salt phases.

pH_{max} constitutes a further descriptor of the solubility profile specific for the particular salt, as may be recognized from the profiles of the four salts of naproxen presented in Fig. 35.1. The different counterions limit the solubility at different levels, resulting in differences as high as two orders of magnitude. A counterion can gain influence on solubility only at pH values where the drug substance is ionized, and salt formation cannot improve solubility in pH regions where the nonionized species prevails. It should be noted that the solubilization capacity by pH control for weak electrolytes is limited by the solubility product K_{sp} of the salt formed. Once the solubility of the salt is reached, pH control cannot further solubilize the compound.

Diprotic acids and bases. Many drug substances bear more than one ionizable function: acidic or basic functional groups or, in case of amphoteric substances, even both types in the same molecule. In order to evaluate their influence on solubility, the key parameter is again the pK_a. In Table 35.2, the equations describing the pH profiles for mono- and diprotic acids and bases are given in their exponential form. As an example, the graph in Fig. 35.3 shows the profile of a diprotic base with pK_a values relatively close to each other. As upon addition of an acid to a suspension of the base the pH decreases, the solubility starts to increase by one order of

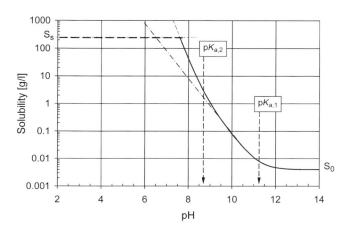

Fig. 35.3 Theoretical pH solubility profile of a diprotic base with the pK_a values $pK_{a,1} = 11.25$ and $pK_{a,2} = 8.7$; $S_0 = 0.004$. The point-dashed line shows the course of solubility if there were only the first basic function.

magnitude as the pH decreases by one unit once it gets below the first protonation step, $pK_{a,1}$. However, as the pH drops below the second ionization step, $pK_{a,2}$, the solubility increases by two orders of magnitude per one pH unit. Similar relationships apply for diprotic acids. A further aspect of diprotic electrolytes may be illustrated by the speciation diagram (Fig. 35.4) of the diprotic acid *valsartan* **3** ($pK_{a,1} = 3.90$, carboxylic acid; $pK_{a,2} = 4.73$, tetrazole). This graph shows the relative abundance of the ionic and

Table 35.2 *Solubility equations for mono- and diprotic acids and bases*

$pH \leq pH_{max}$		$pH \geq pH_{max}$	
Monoprotic acid			
$S = S_0 \cdot \left(1 + 10^{pH - pK_a}\right)$	(17)	$S = S_S \cdot \left(1 + 10^{pK_a - pH}\right)$	(18)
Diprotic acid			
$S = S_0 \cdot \left(1 + 10^{pH - pK_{a,1}}\right) \cdot \left(1 + 10^{pH - pK_{a,2}}\right)$	(19)	$S = S_{S,2} \cdot \left(1 + 10^{pK_{a,2} - pH}\right)$	(20)
Monoprotic base			
$S = S_S \cdot \left(1 + 10^{pH - pK_a}\right)$	(21)	$S = S_0 \cdot \left(1 + 10^{pK_a - pH}\right)$	(22)
Diprotic base			
$S = S_{S,2} \cdot \left(1 + 10^{pH - pK_a}\right)$	(23)	$S = S_0 \cdot \left(1 + 10^{pK_{a,1} - pH}\right) \cdot \left(1 + 10^{pK_{a,2} - pH}\right)$	(24)

$S_{S,2}$ is the solubility of the neutral salt of the diprotic acid or base, respectively. Other symbols are as used in the text.

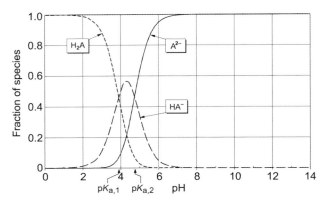

Fig. 35.4 pH species distribution diagram of *valsartan* (see text).

nonionic species of the acid in solution as a function of pH. The maximum fraction that the mono-anion can reach is at pH 4.2, and amounts to a percentage of 56% of the dissolved acid, while at the same time 22% each of the nonionized acid and the di-anion are simultaneously present. A consequence from such relations is that it will depend on the relative solubilities of the respective ion pairs, which of them would form the excess solid in equilibrium with a saturated aqueous solution in the intermediate pH range, and it is possible that from aqueous solutions a pure acidic salt cannot be isolated.

Special types of amphoteric substances are the zwitterionic compounds. Every molecule that has an acidic pK_a lower than the pK_a of its basic function is a zwitterion, because such a substance has acidic and basic groups strong enough to neutralize one another; hence, they are 'inner salts'. As shown in Fig. 35.5 the solubility profile of *baclofen* **4** ($pK_{a,1} = 3.85$, proton gained; $pK_{a,2} = 9.25$,

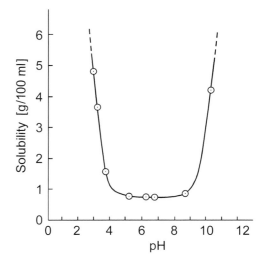

Fig. 35.5 Solubility pH profile of *baclofen* at 25°C in aqueous buffers.

proton lost), a typical example of this class, exhibits a solubility minimum in the pH range where the zwitterion is the predominant species.

Practical evaluation of solubility

The principle of solubility determination was already briefly described above. The suspension system should be held at a constant and controlled temperature. Before sampling it should be given sufficient time for equilibration, which is accelerated by using finely ground material. Ideally, equilibrium is assured by repeated sampling. The concentration in the filtered samples is determined by a suitable analytical method, most frequently by direct UV or by HPLC assay; the latter would help to detect any degradation during the equilibration process. The frequently applied 'quick' method of heating a sample in water to achieve complete dissolution and cooling to room temperature leads more often than not to erroneous results due to supersaturation effects and also due to changes of the excess solid like polymorphic transformation or change of the state of hydration induced by higher temperature. Sometimes solid samples are obtained by lyophilization and are therefore amorphous. An amorphous substance is both, much more soluble and more hygroscopic, than the same material is in a crystalline state.

It is important to note all relevant parameters of the data points in a solubility experiment: temperature, pH, concentration, nature of the excess solid, i.e. whether salt or free acid (base) is present, and whether changes with reference to the initial sample material concerning the polymorphic state or the degree of hydration have taken place. While such careful studies are appropriate at a later stage, i.e. prior to the final candidate selection, other techniques are applied for screening large numbers of compounds at an earlier stage. As for screening the biological activity, high throughput techniques are also applied for solubility screening. A typical nonspecific microtechnique is based on the turbidimetric indication of the transition between complete and incomplete dissolution when stock solutions of substances in DMSO are diluted with water or aqueous buffers.[9] This technique can be performed on 96-well microtiter plates and adapted for a first screening of salt formers and establishing pH solubility profiles.

III ACIDS AND BASES FOR SALT FORMATION

Numerous acids and bases are in use for providing the counterions to form salts. Figures 35.6 and 35.7 show acids and bases, indicating the frequency of their use in drug salts of prescription drug products. However, salt forming agents are material, which are not expressively approved by the health authorities for that particular use, and hence there are

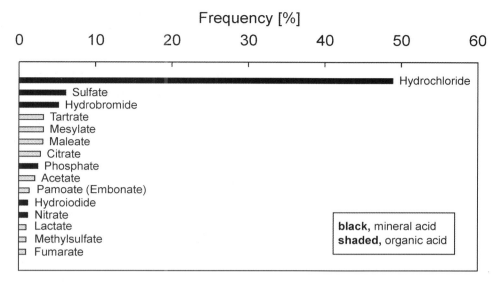

Fig. 35.6 Percentage of the 15 most frequently used acids for salt formation, resulting from a survey of the 1998 edition of *Index Nominum*.

no official lists. A drug is approved not as an isolated active entity but only in the context of the drug product as a whole, i.e. as the dosage form intended for the market, containing the active ingredient in its complete chemical form. The counter-ion is not a separate item of application. The use of such 'inactive' moieties is rather justified by their prior occurrence as salt formers in established drugs, and their further use as safe 'inactive' moiety in drug salts can be largely derived from their presence in marketed drug products. A periodically updated register of approved drug products is the *Orange Book*[10] which lists drug products

whose safety and efficacy has been evaluated, so it may be searched also for salt forms of drug substances.

The high frequency of hydrochlorides and sodium salts has not only the background of convenience but can be derived from the fact that both ions are the most abundant electrolytes in the body and hence are expected to cause no alterations of physiological functions. Other salt-forming acids are naturally occurring components of food or natural substrates of the intermediary metabolism. The number of organic bases usable as salt formers is much smaller, because generally, amines and other nitrogen bases have

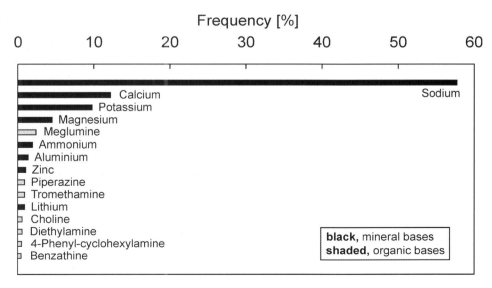

Fig. 35.7 Percentage of the 15 most frequently used bases for salt formation resulting from a survey of the 1998 edition of *Index Nominum*.

their own pharmacodynamic activity, unless they are very rapidly metabolized. Exceptions are the essential basic amino acids, e.g. arginine and lysine.

A Selection of salt formers

pK_a

Which acids or bases should be considered for forming salts with a drug substance? This must chiefly be decided with regard of physicochemical parameters, in the first place based on the pK_a values involved. When forming a salt the acid transfers a proton to the conjugate base, which in turn must be selected to be ready for accepting the proton. This is generally the case if the pK_a of the acid is at least two units lower than the pK_a of the base. This corresponds to a situation where in water both components are ionized to a degree of at least 99%. Strong mineral acids such as HCl ($pK_a = -6$) or H_2SO_4 ($pK_a = -3$) could form solid salts with the anthelmintic *flubendazole* **5** having a pK_a value as low as 4.1, whereas with acetic acid or with benzoic acid ($pK_a = 4.2$) an attempt to prepare a salt would not be successful with such a very weak base. For basic drugs, Gould[11] has published detailed physicochemical relationships along a decision analysis procedure including useful tables of salt-forming acids.

Resulting pH

The next parameter to consider is the pH resulting in the aqueous solution of a salt. The pH may be estimated even before the salt has been prepared, since equations (14)–(16) for the three general cases of salt types of weak electrolytes sallow a theoretical approach:

weak acid × strong base: pH

$$= \tfrac{1}{2}(pK_{a,acid} + pK_w + \log c) \tag{14}$$

weak base × strong acid: pH

$$= \tfrac{1}{2}(pK_{a,base} - \log c) \tag{15}$$

weak acid × weak base: pH

$$= \tfrac{1}{2}(pK_{a,acid} + pK_{a,base}) \tag{16}$$

where, pK_w is the negative ionization exponent of water with the value of 14 at 25°C, and c is the concentration of the salt. How the solubility and pH can be estimated, using equations (12) and (15), respectively, is demonstrated for the weakly basic tranquillizer *chlordiazepoxide* **6** ($pK_a = 4.6$). For a solution containing 50 mg ml^{-1} (0.167 M) as the hydrochloride salt, the calculated pH is 2.7. With the known solubility of the nonionized base (2 mg ml^{-1}), equation (12) predicts a solubility of 200 mg ml^{-1} at this

pH (in fact, the solubility of the hydrochloride in water is 100 mg ml$^{-1} = 0.33$ M). Trying to prepare an acetate of chlordiazepoxide would make little sense for three reasons: (1) the pK_a of acetic acid is just about the same as that of the drug base; (2) an equimolar solution of base and acetic acid would have a calculated pH of 4.78, at which pH in 1 ml a mere 4 mg of the base are soluble, a solubility value not really worthwhile for a salt; (3) if a solid acetate were feasible it would, on contact with water or humid air, release the volatile acetic acid, and the solid drug base would be left behind.

The common-ion effect

While the preference for preparing hydrochlorides and sodium salts is favoured by the physiological abundance of Cl^- and Na^+ ions, these can also bring about negative effects. One of them is the suppression of solubility, which becomes particularly apparent with hydrochlorides and sodium salts of moderate to low intrinsic solubility. The presence of additional ions of the same kind, e.g. in physiological saline, in the stomach and in blood, causes a reduction of solubility by the law of mass action. This 'common-ion effect' depresses the solubility of *terfenadine* (**7**) hydrochloride in water (2 mg ml^{-1} at pH of approximately 5) down to a tenth when 0.05 M NaCl is added. The solubility of *diclofenac* **8** sodium in water (25°C) is 21.3 mg ml^{-1}, but 6.7 mg ml^{-1} in physiological saline. Under the same conditions, the corresponding figures for 4-(5,6-dimethyl-2-benzofuranyl)-piperidine hydrochloride, an experimental antidepressant, are 3.8 and 0.44 mg ml^{-1}, respectively. While hydrochlorides of aliphatic amines are highly soluble (oxprenolol: 720 mg g^{-1} solution at 25°C), those of cycloaliphatic and heteroaromatic nitrogen bases are often sparingly soluble. Numerous examples have been reported where hydrochloride salts exhibited lower solubility than other salts.[12–16] Solubility and dissolution rate of hydrochlorides administered orally may be further suppressed by the common-ion effect,[17–19] since the chloride ion is present at concentrations between 100 and 140 mval l^{-1} in gastric fluid.[20] There are examples of pyrimidine derivatives whose hydrochlorides are even less soluble than the free bases. Bogardus and Blackwood[8,21] compared the intrinsic dissolution rates of doxycycline monohydrate and hydrochloride dihydrate in 0.1 M HCl. Although the hydrochloride is better soluble in water, it dissolved six times slower than the free base form due to the common-ion effect.

Due to their strength, mineral acids (and the alkali hydroxides) are appreciated for forming stable salts with very weak bases (and acids, respectively), they will be considered in first place, provided the above pH estimations would not exclude them in individual cases. Once several alternative salt formers have been selected, experimental screening will identify those rendering high solubility.

Tong and Whitesell[22] have described a substance-sparing procedure. Using less than 50 mg of investigational material, they exemplified their *in situ* salt screening method with GW1818X, a basic drug candidate (a piperidine base, $pK_a = 8.02$; relative molar mass 571.60) having a solubility of 4.4 mg l^{-1} in water. They investigated the solubilities of six salts using 0.1 M aqueous solutions of HCl, methanesulfonic, phosphoric, citric, succinic and tartaric acids and simultaneously characterized the excess materials that were crystalline except for the oily citrate and tartrate residues. The rank order of solubility in water (mg ml^{-1}) found with the salt samples subsequently prepared in gram amounts, was as follows: mesylate (9.2) > phosphate (6.8) > succinate (2.9) > hydrochloride (1.2). However, the initial screening had already shown that the intrinsic solubility of the phosphate salt, which was reached at a lower pH (i.e. in region II), had indeed the highest solubility (13.5 mg ml^{-1}) in the series, as was already found by *in situ* screening.

Predictability of salt solubility

Given an acidic or basic drug, the straightforward determination of the salt forming partner leading to a salt with the desired water solubility would be highly desirable. However, so far the predictive tools for estimating solubility based on the structure of the constituting ions are not at hand. Although some empirical structure–solubility relationships have been studied within closely related drug series, or with the salts of a given drug with a series of salt-forming ions, the pharmaceutical scientist is left with empirical tendencies or rank orders. So often, the highest solubilities are achieved with mesylates, lactates and salts of polyhydroxy acids of basic drugs, and with the potassium, tromethamine, meglumine and arginine salts of acidic drugs. Figure 35.1 reflects a typical rank order. As an additional observation, the series of *diclofenac* salts presented in Table 35.3 demonstrates the nonuniform temperature dependence of solubility even in a series of closely related members. In Fig. 35.8, linear solubility profiles of *terfenadine* 7 are shown, obtained with four different acids; the few experimental points are supplemented with theoretical curves. The frequently encountered rank order: lactate > methanesulfonate > hydrochloride > phosphate is found in this example. Here also the common-ion effect becomes dominant towards the low pH end of the profile because (1) for reaching the low pH values, the acid concentration of the respective acid has to be increased, which also raises the concentration of the respective anion; and (2) the overall level of solubility of this drug is rather low. Nevertheless, sometimes surprising changes of rank orders may be found, as in the example of GW1818X mentioned

Table 35.3 *Solubility of diclofenac salts in water*

Salt	25°C	pH	37°C	pH
Diclofenac (acid)	0.0071	5.8	0.0120	6.9
Calcium	0.58	6.8	0.62	6.8
Sodium	19.08	7.8	22.5	7.8
Potassium	47.05	7.8	120.5	8.2
Ethylenediamine	1.76	7.2	2.86	7.2
Diethylamine	13.95	7.6	18.3	7.6
Meglumine[a]	14.5	8.0	15.9	7.6
HEP[b]	44.7	8.3	484	7.8

Solubility (mg ml^{-1}) is expressed as the free acid; the pH of the saturated solution is given.
[a]1-deoxy-1-(methylamino)-D-glucitol.
[b]2-pyrrolidino-2-ethanol.

above with the highly soluble phosphate salt, whereas most phosphates are rather sparingly soluble.

The solubility of salts is determined by (1) the interactions of the drug ions among themselves; and (2) between themselves and their counter-ions in the crystal lattice of the solid salt, represented by the lattice energy $\Delta G_{lattice}$; (3) between each sort of ions and water, expressed as the solvation energies of the cation ΔG_{cation}, and of the anion, ΔG_{anion}. The molar free energy of solution, $\Delta G_{solution}$, is the balance between those interactions:

$$\Delta G_{solution} = \Delta G_{cation} + \Delta G_{anion} - \Delta G_{lattice} \quad (17)$$

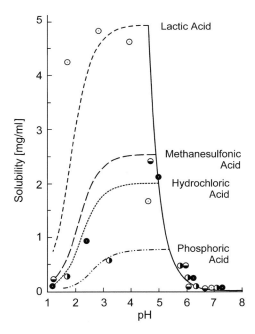

Fig. 35.8 pH profiles of the solubility of *terfenadine* with different acids. (Redrawn from Streng *et al.*[15])

In principle both the lattice energy and the hydration energies increase with an increase in cation or anion charge and decrease with an increase in ionic radius. They are also expected to increase with the polarity or with the hydrogen-bonding nature of the counterion. The overall effect of a given structural change on water solubility will depend on which terms, the lattice energy or the hydration energies, are more sensitive to the change in structure. Due to the complexity of drug molecules and in consequence, the multitude of the number and nature of interactions, the contribution of each term cannot yet be reliably estimated.

Anderson and Conradi[23] studied the effect of six ammonium salts of *flurbiprofen* **9**, differing in the hydrophobicity of the cation, on the water solubility. The solubility correlated well with the melting point of the salts, but the tromethamine salt with the most hydrophilic cation ranked only third in solubility. Chowhan[6] in his investigation of the sodium, potassium, magnesium and calcium salts of four carboxylic acids found that even the rank order of the salts of this cation series is not uniform. The result of a study on the sodium salts of 11 drugs, Rubino[24] confirms the correlation of melting point and salt solubility. How the melting point and other characteristics can be used to guide the process leading towards a salt of desired properties is presented in Table 35.4.

Other criteria

Not only physicochemical but also safety aspects can play a role when the decision is made as to which salt formers to include in a search for the optimum salt. Acids and bases that are traditionally in use for preparing salts are regarded as biologically 'inert'. However, not all of them are entirely indifferent. Some of them are indeed inert, such as the chloride, phosphate, or sodium ions. As an example, sodium ions are well tolerated, whereas elevated concentrations of potassium ions can cause local necrosis of the mucosa of the gastrointestinal tract. Salicylic acid, though formerly sometimes used as counter-ion, has its own pharmacological activity and can therefore not be regarded as an inert salt former. The route of administration, the prospective doses or concentrations, the therapeutic area of the drug candidate and the expected treatment regimen (acute or chronic) need to be taken into account even at this early stage. A comprehensive compilation of salt-forming acids and bases and their characteristics is found in Stahl and Wermuth.[1]

IV OBTAINING SOLID SALTS

A Some principles and practical considerations of preparation

Based on pH profiles of solubility, aqueous solutions can be prepared by adjusting the pH with a suitable acid or base in concentrations just necessary to achieve a solution, be it for immediate use for tests by the biologist, as well as for developing injectable formulations by the pharmacist. This is even done in production scale in pharmaceutical manufacturing if solutions are the final product, or if for some reason it is preferred to produce, isolate and store the free drug acid (or base). However, while in solubility screening experiments certain salt-forming acids or bases, respectively, may prove suitable for increasing solubility in water by pH adjustment, this is not the only requirement to be met for salt formation. In addition, conditions must be found for preparing and isolating a salt. After all the resulting solid is expected to be a physically stable and stoichiometrically reproducible product.

To this end, the process of salt formation is performed preferably in nonaqueous media. No drug salt can be isolated under conditions where it is fully ionized. Ion pairs constituting the salt need an environment that suppresses dissociation as much as possible. Drug substance and salt-forming agent are dissolved separately in solvents of moderate polarity, e.g. alcohols, esters, ethers and ketones. If one of the components is available in aqueous solution only (e.g. phosphorous acid), the other solvent should be water-miscible. While the solutions of the components are gradually mixed, the salt frequently starts to precipitate. If not, crystallization may be induced by raising the degree of supersaturation, which can be effected by cooling, by evaporating the solvent or by slow addition of a miscible anti-solvent (nonsolvent), or by a combination of these measures.

The preparation of salts is illustrated by two example procedures for a hydrochloride salt and for a sodium salt.

Cyclopentamine hydrochloride **10**: 141 g (1 mol) 1-cyclopentyl-2-methylaminopropane (= *cyclopentamine*) are dissolved in 500 ml of dry ether. Dry hydrogen chloride is passed into the solution until the weight of the mixture has increased by 36 g. During the addition of the hydrogen chloride, the hydrochloride salt of cyclopentamine precipitates as a white powder. The salt is filtered off and washed with dry ether. Cyclopentamine hydrochloride thus prepared melts at about 113–115°C. The yield is practically quantitative.

Amfenac sodium **11**: A stirred solution of 111 g (0.43 mol) of 2-amino-3-benzoyl-phenylacetic acid (= *amfenac*) in 777 ml of tetrahydrofuran is treated with 31.3 g (0.39 mol) of 50% NaOH. After cooling the solution at 0°C for 3 h, the solid which precipitates is collected by filtration to yield 64 g (56%) of the expected sodium salt, m.p. 245–252°C. An analytical sample is obtained by dissolving 1.0 g of the crude salt in 10 ml of 95% EtOH and treating the solution with 5 ml of isopropyl ether. The pure sodium salt precipitates slowly to yield 0.9 g of yellow solid, m.p. 254–255.5°C.

Table 35.4 Manipulating characteristics of basic drugs by change of salt form

Characteristic	Direction of change		Direction of change	
Melting point	*Decrease*		*Increase*	
Action to achieve change	1	Use more flexible aliphatic acids with aromatic bases	1	Use small counter ions, e.g. Cl$^-$
	2	Move to more highly substituted acids that destroy crystal symmetry	2	If aromatic base use aromatic anions
			3	If base has hydrogen bonding properties use small hydroxy carboxylic acids
Consequences of change	1	Increases solubility	1	Decreases solubility
	2	May form noncrystallizing liquid (rarely possible, rarely desired!)	2	Reduces processing problems
Solubility, dissolution rate	*Increase*		*Decrease*	
Action to achieve change	1	Decrease melting point	1	Increase melting point
	2	Decrease pK_a and increase soly. of acid	2	Increase hydrophobicity of acid
	3	Increase hydroxylation of acid		
	4	Move to small organic acid in case of common ion effect		
Consequences of change	1	Increases bioavailability	1	Suspensions physically more stable (reduced rate of particle growth)
	2	Enables possibly liquid formulation	2	Solid dosage forms: retarded release
Stability	*Decrease*		*Increase*	
Action to achieve change	—		1	Increase hydrophobicity of acid
			2	Use carboxylic rather than sulfonic or mineral acids
			3	(in case of acid-labile a.s. base:) Use acid of higher pK_a to reduce pH of adsorbed water
			4	Decrease solubility
			5	Increase crystallinity
Consequences of change	—		1	Reduces hygroscopicity
			2	Improves resistance to attack by environmental agents
Wettability	*Decrease*		*Increase*	
Action to achieve change	1	Increase hydrophobicity of acid	1	Increase polarity of counterion
			2	Use acid with high degree of hydroxylation
			3	Use acid of lower pK_a
			4	Attempt recrystallization from different solvents to alter crystal habit
Consequences of change	1	Reduces degree of hygroscopicity/deliquescence	1	Improves dissolution and bioavailability

Adapted from Gould.[11]

In the crystallization of salts, water may assume different roles. At times, the presence of traces up to stoichiometrical amounts may favour crystallization, in particular if preferably a hydrated salt form has a strong tendency to crystallize. In other cases, strictly anhydrous conditions may be mandatory which should be observed with highly water-soluble salts. Also, if salt formation of very weakly acidic (basic) substances with strong bases (acids) is intended the presence of water would cause partly protolytic disproportionation of the salt, then the result would be a salt contaminated with the free acid (base) form of the drug substance.

A series of typical preparative procedures for salt formation can be found in Wermuth and Stahl.[25]

B Solid-state properties

Polymorphism and hydrate formation

A salt, once isolated, may maintain its physical properties for long periods of storage, and repeated observations would render the same characteristics, e.g. melting point, or solubility. However, many salts are not only able to crystallize in different crystal forms but also may transform from one form to a different one even in the solid state: they exhibit the phenomenon of polymorphism. In addition, more frequently than with nonelectrolytes, pseudopoly-morphism is encountered: the stoichiometric inclusion of solvent molecules in the crystal lattice producing solvates, and in particular hydrates, if the solvent is water. Such variations of a drug's solid state are highly important for the manufacturing of solid and semisolid dosage forms since the different forms differ in all their substance parameters and properties, including solubility and dissolution. Once they are encountered, both, polymorphs and hydrates, can become disturbing, especially if one form transforms to a different one during production. A consequence may then be that the final drug substance or salt cannot be delivered reproducibly in the same form. Transformation of poly-morphs can also take place during storage of the substance, or during processing, manufacturing or storage of dosage forms. Therefore, the thermodynamically most stable form is preferred. This also includes stable hydrates or stable anhydrates, where 'stable' means that within acceptable ranges of storage and processing temperatures and humid-ity, no change will take place with respect to the crystal polymorphic state or the degree of hydration. Anhydrous material can form hydrates on contact with water or when exposed to humid air. Generally, hydrates are less soluble in water than the corresponding anhydrates. (One of the very rare exceptions to this rule is reported in Kristl *et al.*[26]) Although salt hydrates usually form rapidly, the rate of their formation can vary widely, taking from minutes to months. Thus, the solubility and dissolution rate of an anhydrous salt can be initially high, leading to supersaturation with respect to a hydrate which might appear with delay, whereupon the concentration in a solution drops as the hydrate crystallizes, or a dissolution process would slow down due to the transformation of the undissolved anhydrous solid to a hydrate. Equilibria of hydrate formation in humid air are graphically presented as water sorption isotherms as shown in Fig. 35.9 for the potassium salt of *diclofenac* **8**. When exposed to stepwise increasing humidity, this salt takes up 4 mol of water, once 60% relative humidity is exceeded. Whereas in other cases, a hydrate would revert to the anhydrate if the humidity drops below the same minimum value when it has been formed (in this case, below 60%), this sorption isotherm exhibits hysteresis: the salt hydrate retains water even at lower humidity. Only below 40% is the hydrate water lost completely. The question must be asked

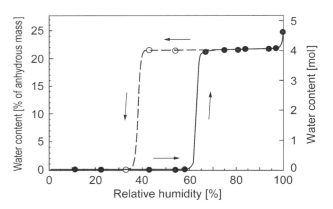

Fig. 35.9 Water sorption isotherm (23°C) of diclofenac potassium at atmospheric pressure. Full circles, adsorp-tion; empty circles, desorption.

whether the propensity to form a hydrate might interfere with, for example, tablet production. For this particular case the following consideration leads to the conclusion that no interference would occur. If a wet granulation process is applied to form a homogeneous granular mass for producing tablets, an aqueous binder solution is worked into the powder mixture consisting of the drug salt and excipients. Clearly, the tetrahydrate is intermediatedly formed during this procedure. Checking the sorption isotherm against the temperature and humidity conditions of the subsequent drying of the wet granulation at a temperature above 60°C it can be concluded that the drying process not only removes the excess water introduced by the binder solution, but also the hydrate water completely. However, when the intermediate tablet mass is further processed, care must be taken that the humidity in the manufacturing facility is controlled to values below 60% so that the anhydrous state will be maintained, and the final tablet product will be sealed in blister packs. This example can make clear that hydrates are not generally prohibitive in drug product development and manufacture, if the climatic stability ranges of the different forms of the particular substance are sufficiently known and such climatic con-ditions are applied as to avoid transformations.

Chemical stability

The salt form can have an effect on the chemical stability of the drug. If a potentially reactive salt-forming agent has been chosen unwittingly the resulting salt may decompose under certain conditions. Such cases have been reported for fumarate and maleate salts. A pH-dependent adduct for-mation with maximum reactivity at pH $\approx$ 5 took place with the dimaleate of the development compound PGE-7762928 (two basic nitrogens; $pK_{a,1} = 4.3$; $pK_{a,2} = 10.6$).[27] Further, with the selective serotonin reuptake inhibitor *seproxetine* (= *S*-norfluoxetine; **12**) maleate hemihydrate, the inter-action of the primary amine group with the anion in aqueous solution (optimum pH range 5.5–8.5) resulted in adduct formation.[28]

The stability in the solid state of the free acid and three salts of a prostaglandin derivative were compared on samples stored at 33°C, protected from light. After 2 months, > 10% of the acid, > 50% of the sodium salt, < 2% of the potassium salt and 2–3% of the Tris (= tromethamine) salt were decomposed. The surpris-ingly large differences could be correlated neither with the water solubility nor with the melting point.[29]

Powell[30] found a dramatic difference in stability between the phosphate and the sulfate salts of codeine in solution at room temperature. Whereas the phosphate solution had a shelf life of 1.1 years, the extrapolated shelf life of the sulfate was 44 years. The low stability of the former was ascribed to a catalytic effect of the phosphate anion on the degradation of codeine.

These few examples underline the importance of the stability aspect of salt selection.

V CRITERIA FOR SALT SELECTION

The selection of the final salt form must take into account that each one of the different salts of a drug candidate has its own profile of properties, and an ideal profile would match the biopharmaceutical and technological require-ments for a particular dosage form and route of adminis-tration. Rarely high solubility can be the only criterion of decision. In addition, the requirements of the envisaged dosage form need to be respected. For example, a salt with the highest solubility, but a pH of 2.8 for the saturated solution in water, might be corrosive to tableting machin-ery,[31] which would preclude manufacturing of a 200-mg tablet product. So the decision could vote for a less soluble, but also less aggressive salt. In contrast, for preparing an injectable solution such a low pH could be tolerated if the solution is designed for administration by addition to an intravenous infusion. A step-wise route to salt selection has been proposed by Morris and coworkers.[32] Often the result of the decision process is a compromise between several competing property profiles, or if one salt form is intended to be used in different dosage forms.

A Essential parameters and properties

Here are summarized the essential factors that constitute the pharmaceutically relevant profile of a drug candidate and hence for any salt in consideration.

The *solubility* (pH profile) should be as high as possible for injectables, moderate for peroral dosage forms.

Hygroscopicity, the adsorptive uptake and release of water or the deliquescence of highly soluble salts, can be decisive for technical problems. As an example, sodium valproate, used as an anticonvulsive, deliquesces at humidity ≥ 43% relative humidity and therefore requires special precautions during processing for solid dosage forms.

The propensity to form hydrates in water or in humid air must be known. If hydrates form, the conditions of formation and the climatic range (temperature, humidity) of stability must be known, and the compatibility of both for the form to be chosen must be assessed.

Morphic state: It must be clear whether the salt is monomorphic or polymorphic. If it is polymorphic, the relationship between the polymorphic forms (monotropic, enantiotropic), their relative thermodynamic stability (stable or metastable at room temperature) and the temperature range of their physical stability need to be clarified.

Crystallinity: The degree of crystallinity must be known. Amorphous materials are thermodynamically not stable, and stabilization is likely to take place under a broad range of conditions and half-life times. The morphology and crystal habit is of interest for mechanical processing procedures such as milling, mixing, powder flow and compression.

Melting point: A melting point lower than approximately 100°C can cause problems during mechanical handling and processing. In particular, melting or lumping can occur when comminution by milling is attempted.

Chemical stability under a variety of macro- and micro-environmental conditions is a highly critical property determining the expected shelf life of products. The stability of the bulk solid and in solution must be studied, whereby the influence of environmental factors (pH, temperature, humidity, oxygen, light) and including the presence of various excipients on drug stability must be established.

Corrosiveness: salts of weak bases with strong acids need to be tested whether or not they are corrosive on tableting tools.

Such profiles are studied in preformulation programs,[33] but some of the properties must already be known before the decision for the final salt form is made. A change of the salt form of a drug substance at later stages of product development must be avoided. Otherwise, many essential studies would then have to be repeated with the new salt at a cost of additional time and resources. Besides the essential parameters and properties of immediate pharma-ceutical-technological and biopharmaceutical interest men-tioned above, further issues may have to be considered in the final decision concerning other areas: (1) chemical processing, handling and economy: the yield and quality of crystallization, of milling; in addition, the equivalent mass ratio *active entity:salt* can play a role, if a drug is likely to be used in high doses, because a large counter-ion could cause unwieldy large doses to be administered; (2) patent aspects: physical and chemical variations of a drug substance can be employed for extending the proprietary substance protection by a suitable strategy of patent application for salt forms or crystal modifications that are characterized by particularly advantageous (pharmaceutical, biopharmaceutical, medical or technical) properties and can open up new areas of application. It should be mentioned that salt formation is not only used to improve solubility, but also for reducing solubility. Thus, sparingly soluble salts can be used for developing peroral slow-release drug products. Typical applications are the pamoates (= embonates) of imipramine, pyrantel, pyrvinium and others; also the resinates, in which the drug substance is loaded on an ion exchanger.[34,35] *Salmeterol xinafoate* **13** (= 1-hydroxy-2-naphthoate) is an example of tailoring the properties of a bronchodilating beta-stimulant by means of salt formation to an optimum performance in inhalation. The sparingly soluble xinafoate salt dissolves slowly thereby contributing to the long-acting properties of the drug substance. At the same time, mucosal irritation by osmotic effects is avoided.

REFERENCES

1. Stahl, P.H. and Wermuth, C.G., (eds) (2002) *Handbook of Pharmaceutical Salts — Preparation, Selection and Use*. Verlag Helvetica Chimica Acta, Zürich.
2. Yalkowsky, S.H. (1985) Solubility and solubilization of nonelectrolytes. In Yalkowsky, S.H. (ed.). *Techniques of Solubilization of Drugs*, pp. 1–14. Marcel Dekker, New York.
3. Jain, N. and Yalkowsky, S.H. (2001) Estimation of the aqueous solubility. I: Application to organic nonelectrolytes. *J. Pharm. Sci.* **90**: 234–252.
4. CLOGP (2001) BioByte Corp. and Pomona College: Claremont, CA, USA.
5. Yang, G., Jain, N. and Yalkowsky, S.H. (2001) Estimation of the aqueous solubility. III: Application to weak-electrolyte drugs in buffered solutions. *AAPS: 2001 Annual Meeting & Exposition*. Denver, CO, USA, *AAPS PharmSci.* **3** Number 3 (Suppl.), poster No. W4446.
6. Chowhan, Z.T. (1978) pH-Solubility profiles of organic carboxylic acids and their salts. *J. Pharm. Sci.* **67**: 1257–1260.
7. Kramer, S.F. and Flynn, G.L. (1972) Solubility of organic hydrochlorides. *J. Pharm. Sci.* **61**: 1896–1904.
8. Bogardus, J.B. and Blackwood, R. K. J. (1979) Dissolution rates of doxycycline free base and hydrochloride salts. *J. Pharm. Sci.* **68**: 1183–1184.
9. Bevan, C.D. and Lloyd, R.S. (2000) A high-throughput screening method for the determination of aqueous drug solubility using laser nephelometry in microtiter plates. *Anal. Chem.* **72**: 1781–1787.
10. *The Electronic Orange Book: Approved Drug Products with Therapeutic Equivalence Evaluations (Internet version of the Orange Book)*. US Department of Health and Human Services, Public Health Service, Food and Drug Administration, Center for Drug Evaluation and Research (CDER), Office of Information Technology, Division of Data Management and Services. http://www.fda.gov/cder/ob/default.htm.
11. Gould, P.L. (1986) Salt selection for basic drugs. *Int. J. Pharm.* **33**: 201–217.
12. Senior, N. (1973) Some observations on the formulation and properties of chlorhexidine. *J. Soc. Cosmet. Chem.* **24**: 259–278.
13. Nudelman, A., McCaully, R.J. and Bell, S.C. (1974) Water-soluble derivatives of 3-oxy-substituted 1,4-benzodiazepines. *J. Pharm. Sci.* **63**: 1880–1885.
14. Agharkar, S., Lindenbaum, S. and Higuchi, T. (1976) Enhancement of solubility of drug salts by hydrophilic counterions: properties of organic salts of an antimalarial drug. *J. Pharm. Sci.* **65**: 747–749.
15. Streng, W.H., Hsi, S.K., Helms, P.E. and Tan, H. G. H. (1984) General treatment of pH-solubility profiles of weak acids and bases and the effects of different acids on the solubility of a weak base. *J. Pharm. Sci.* **73**: 1679–1684.
16. Hussain, M.A., Wu, L.S., Koval, C. and Hurwitz, A.R. (1992) Parenteral formulation of the *kappa* agonist analgesic, DuP 747, via micellar solubilization. *Pharm. Res.* **9**: 750–752.
17. Miyazaki, S., Nakano, M. and Arita, T. (1975) A comparison of solubility characteristics of free bases and hydrochloride salts of tetracycline antibiotics in hydrochloric acid solutions. *Chem. Pharm. Bull.* **23**: 1197–1204.
18. Miyazaki, S., Oshiba, M. and Nadai, T. (1980) Unusual solubility and dissolution behaviour of pharmaceutical hydrochloride salts in chloride-containing media. *Int. J. Pharm.* **6**: 77–85.
19. Miyazaki, S., Oshiba, M. and Nadai, T. (1981) Precaution on use of hydrochloride salts in pharmaceutical formulation. *J. Pharm. Sci.* **70**: 594–596.
20. Lentner, C. (ed.). (1981) *Geigy Scientific Tables*, 8th edn, vol. 1, p. 145. Ciba-Geigy AG, Basel.
21. Bogardus, J.B. and Blackwood, R.K.J. (1979) Solubility of doxycycline in aqueous solution. *J. Pharm. Sci.* **68**: 188–194.
22. Tong, W.Q. and Whitesell, G. (1998) *In situ* salt screening — a useful technique for discovery support and preformulation studies. *Pharm. Dev. Technol.* **3**: 215–223.
23. Anderson, B.D. and Conradi, R.A. (1985) Predictive relationships in the water solubility of salts of a nonsteroidal anti-inflammatory drug. *J. Pharm. Sci.* **74**: 815–820.
24. Rubino, J.T. (1989) Solubilities and solid state properties of the sodium salts of drugs. *J. Pharm. Sci.* **78**: 485–489.
25. Wermuth, C.G. and Stahl, P.H. (2002) Selected procedures for the preparation of pharmaceutically acceptable salts. In Stahl, P.H. and Wermuth, C.G. (eds). *Handbook of Pharmaceutical Salts — Preparation, Selection and Use*, pp. 249–263. Verlag Helvetica Chimica Acta, Zürich.
26. Kristl, A., Srčič, S., Vrečer, F., Šuštar, B. and Vojnovic, D. (1996) Polymorphism and pseudopolymorphism: influencing the dissolution properties of the guanine derivative acyclovir. *Int. J. Pharm.* **139**: 231–235.
27. Gazda, M., Dansereau, R.J., Crail, D.J., Sakkab, D.H. and Virgin, M.M. (1998) Maleate adduct formation during solid-oral dose form screening of PGE-7762928. *Pharm. Sci.* **1**: S403.
28. Schildcrout, S.A., Risley, D.S. and Kleemann, R.L. (1993) Drug–excipient interactions of seproxetine maleate hemihydrate: Isothermal stress methods. *Drug Dev. Ind. Pharm.* **19**: 1113–1130.
29. Anderson, B.D. (1985) Prodrugs for improved formulation properties. In Bundgaard, H. (ed.). *Design of Prodrugs*, pp. 243–269. Elsevier Science, Biomedical Division, Amsterdam.

30. Powell, M.F. (1986) Enhanced stability of codeine sulfate: effect of pH, buffer, and temperature on the degradation of codeine in aqueous solution. *J. Pharm. Sci.* **75**: 901–903.

31. Stahl, P.H. (1993) Characterization and improvement of the stability behaviour of drug substances. In Grimm, W. and Krummen, K. (eds). *Stability Testing in the EC, Japan and the USA*, pp. 15–32. Wissenschaftliche Verlagsgesellschaft, Stuttgart.

32. Morris, K.R., Fakes, M.G., Thakur, A.B., *et al.* (1994) An integrated approach to the selection of optimal salt form for a new drug candidate. *Int. J. Pharm.* **105**: 209–217.

33. Wells, J.I. (1988) *Pharmaceutical Preformulation*. Ellis Horwood, Chichester.

34. Borodkin, S.S. (1993) Ion exchange resins and sustained release. In Swarbrick, J. and Boylan, J.C. (eds). *Encyclopedia of Pharmaceutical Technology*, pp. 203–216. Marcel Dekker, New York.

35. Borodkin, S.S. (1991) Ion-exchange resin delivery systems. In Tarcha, P.J. (ed.). *Polymers for Controlled Drug Delivery*, pp. 215–230. CRC Press, Boca Raton.

36

PREPARATION OF WATER-SOLUBLE COMPOUNDS BY COVALENT ATTACHMENT OF SOLUBILIZING MOIETIES

Camille G. Wermuth

Ajouter à sa queue, ôter à ses oreilles
Add to her tail, remove from her ears

Jean de La Fontaine[1]

I INTRODUCTION

The strategy described here aims to convert a water-insoluble drug into a water-soluble one by attaching *covalently* an appropriate solubilizing side chain. Surprisingly few reviews covering this subject are found in the literature.[2–5] Seen from the chemical side, the solubilizing moiety can be a neutral hydrophilic group or an ionizable organic base or acid. With the exception of possible crystallization problems, no major difficulties are expected in the synthesis of such compounds. One problematic aspect of the solubilization approach to be taken is to decide if the solubilizing moiety has to be fixed in a reversible manner, generating a prodrug, or in an irreversible manner, yielding a new chemical entity. In the latter case the solubilizing procedure may exact a cost in terms of the recognition mechanisms, the soluble analogue being less potent or even showing a different pharmacological profile. In some instances changes in one part of the molecule have to be compensated by changes at another part. As Jean de La Fontaine said of the elephant, 'Add to her tail, remove from her ears'. In addition, the solubilized analogue of an already approved drug is considered by the governmental drug agencies as a totally new chemical entity, demanding a completely new development process. The financial investment that is then necessary can only be justified if enough sales of the solubilized form are expected. As a consequence the attachment of solubilizing moieties has to be considered very early in the drug discovery process or else limited to drugs with sizeable markets and undertaken when all other solubilizing stratagems fail. Despite these difficulties, many examples are found in therapy of successful drug solubilization by means of a chemical transformation of a parent drug (Table 36.1).

Chemically solubilized active principles render possible the preparation of parenteral, and especially intravenous forms appreciated in clinical practice. But even at the preclinical level, the use of water-soluble molecules is recommended as they are effectively much easier to study

The Practice of Medicinal Chemistry
ISBN 0-12-744481-5

Table 36.1 *Successful examples of drug solubilization by a chemical means*

Solubilizing side chain	Therapeutic class	Compound
Phosphoric ester	Steroids	Betamethazone
Phosphoric ester	Vitamins	Menadione
Hemisuccinate	Cardiotonics	Benfurodil
Hemisuccinate	Antibiotics	Chloramphenicol
Hemisuccinate	Steroides	Prednisolone
Hemisuccinate	Benzodiazepines	Oxazepam
Acidic side chain	Theophylline	Etaphylline
Acidic side chain	Antisyphilitic	Solusalvarsan
Neutral side chain	Analgesic	Glafenine
Neutral side chain	Bronchodilator	Dyphylline
Neutral side chain	Antibacterial	Sulfapyridine *N*-glucoside
Basic side chain	Antibiotics	Rolitetracycline
Basic side chain	Flavonoids	Solurutine
Basic side chain	Morphine	Pholcodine

by *in vitro* tests, in cell or microorganism cultures and on isolated organs. The inconveniences are that chemically modified structures may show modified pharmacological, pharmacokinetic and toxicological properties.

II SOLUBILIZATION STRATEGIES

Three points are decisive in terms of the solubilizing strategies: How will the solubilizing moiety be grafted? Where will it be grafted? What kind of side chain will be utilized?

A How will the solubilizing moiety be grafted?

The solubilizing chain can be *reversibly* or *irreversibly* grafted to the parent molecule. In the case of reversible linkages we are dealing in fact with prodrugs. Reversible linkages are usually provided by esters, peptides or glucosides.

Irreversible attachment of side chains is achieved by *O*- and *N*-alkylation and creation of C—C bonds. The grafted side chains can be basic (dimethylaminoethyl or morpholinoethyl chains), acidic (carboxylic, sulfonic, etc.) or neutral (glyceryl).

Intermediate situations are found for enol and phenol phosphates, as well as for some amides. For these compounds only partial reversibility is observed *in vivo*.

B Where will it be grafted?

First of all a careful examination of the parent molecule must be undertaken in order to identify the parts of the molecule that present adequate chemical reactivity and are suitable as attachment points for the solubilizing chain. Functions such as OH, SH, NH, acidic CH or CO_2H are reactive sites that furnish nucleophilic or basic entities. Conversely aromatic double bonds are sensitive to electrophilic attack, whereas carbonyl groups and conjugated carbon–carbon double bonds are sensitive to nucleophilic attack. The second criterion that has to be considered is of a biological nature: the solubilizing chain can only be grafted to those parts of the molecule that are not involved in the drug–receptor interaction. Fixed at the wrong place, the solubilizing chain can totally inactivate the molecule.

C What kind of solubilizing chain will be utilized?

The size of the solubilizing chain is one of the selection parameters. The chains can be limited to the strict minimum and simply represent functional groups, or they can be made from larger residues containing several atoms (Table 36.2). The nature of the side chains is the second selection parameter. It has to be decided if they can be ionizable (acidic or basic) or non ionizable.

Acidic ionizable moieties (e.g. carboxylic acids) yield readily crystallizable compounds and often do not alter the pharmacologic profile of the parent molecule. However, owing to their amphiphilic nature, they can show hemolytic properties. In addition only a limited number of cations can be used to neutralize them. Traditional inorganic cations such as sodium, potassium or magnesium can induce mineral surcharges and are no longer recommended. They are avantageously replaced by organic bases such as tromethamine, lysine or *N*-methylglucamine (see Chapter 35).

Basic ionizable moieties (e.g. substituted amines) can be neutralized by a very large number of organic and inorganic acids (see Chapter 35). The salts obtained are also readily crystallized and usually show less surface-active properties

Table 36.2 *Small and large solubilizing moieties*

Small groups or simple functionalities	Larger solubilizing moieties
—CO_2H	R—OH → R—O—CH_2—CH_2—CO_2H
—SO_3H, —OSO_3H	R—NH_2 → R—NH—CH_2—CH_2—CH_2—SO_3H
—PO_3H_2, —OPO_3H_2	$(R)_2C = O$ → $(R)_2C = N$—O—CH_2—CO_2H
—NH_2, —NHR, —NR_2	R—OH → O-morpholinylethyl
N-oxides	R—OH → O-glucoside
S-oxides	R—OH → O—CO—CH_2—CH_2—CO_2H
Sulfones	R—OH → *m*-O—C_6H_4—SO_3H

than salts from acidic chains. Their main disadvantage, which somewhat limits their utility, is their tendency to interfere with biogenic amines and neurotransmitters. In other words, attaching a basic amine functional group can seriously modify the pharmacological activity with regard to the parent drug.

Drugs with acidic side chains cannot be mixed with drugs having basic side chains, as it is likely that a salt formed between the two drugs might precipitate.

Nonionizable moieties (e.g. polyhydroxylated chains) do not present this disadvantage and are compatible with other drug preparations. As they can be delivered at pH values close to 7, they do not produce painful injections. The main problems encountered with nonionizable solubilizing moities is their lesser propensity to crystallize. In addition increased cost can be expected from the necessity of added protection/deprotection steps during their synthesis.

III ACIDIC SOLUBILIZING CHAINS

When planning solubilization by means of a carboxylic acid side chain, one has to take into account the therapeutic properties peculiar to the carboxylic group. Thus all arylacetic acids show more or less potent anti-inflammatory activities and many α-functionalized carboxylic acids are chelating agents. Among them we find the chelating α-amino acids[6] and antibacterial nalidixic acid-derived quinolones (for references see Hammond[7]) and probably kynurenic acid analogues acting as antagonists at the glycine site.[8]

A Direct introduction of acidic functions

Direct introduction of a solubilizing function can be achieved by carboxylation and by sulfonation. The historical example of carboxylation is the Kolbe synthesis of salicylic acid. Sulfonation was employed to solubilize guaiacol, camphor and 7-chloro-8-hydroxyquinoline (Fig. 36.1).

In a recent example, aryl-carboxylic solubilization was used to solubilize core-modified porphyrins.[9] Solubilizing aryl carboxylic functions were also replaced by their tetrazole or 1,2,4-oxadiazolone bioisosteres in the design

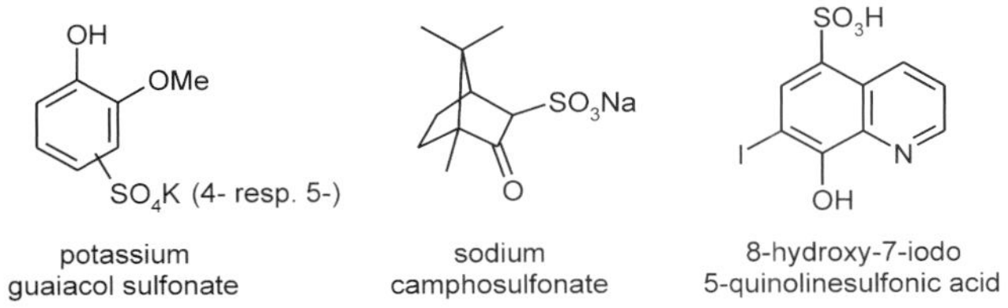

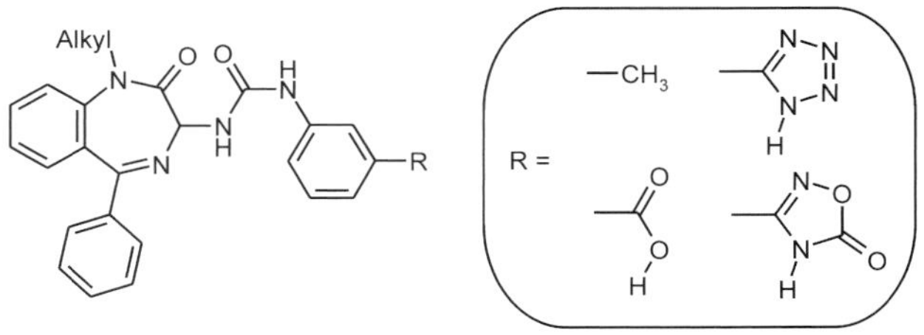

Fig. 36.2 Carboxylic acid and heterocyclic bioisosteres as solubilizing groups.

of second generation, benzodiazepine-derived, CCK-B antagonists (Fig. 36.2).[10]

B Alkylation of OH and NH functions with acidic chains

This procedure alkylates the hydroxy and the amino groups already present on the molecule with reactive intermediates bearing acidic functional residues (Table 36.3).

These compounds are prepared starting from chloroacetic acid or its ethyl ester. For chains longer than acetic, cyanoethylation and hydrolysis of the nitrile obtained leads to the propionic chain, alkylation with ethyl 4-bromobutyrate and saponification leads to the butyric chain. The propanesulfonic chains are particularly accessible by means of ring opening of propane-sultone.

Dihydroartemisinin ethers. A water-soluble derivative of artemisinin, the sodium salt of artesunic acid (the succinic half-ester derivative of dihydroartemisinin; Fig. 35.2), can be administered by intravenous injection, a property that makes it especially useful in the treatment of advanced and potentially lethal cases of *Plasmodium falciparum*. Sodium artesunate is capable of rapidly reversing parasitaemia and causing the restoration to consciousness of the comatose cerebral malaria patient. The utility of sodium artesunate, however, is impaired by its poor stability due to the facile hydrolysis of the ester linkage. To overcome the ease of hydrolysis of the ester function in sodium artesunate, Lin *et al.*[15] prepared a series of analogues in which the solubilizing moiety is joined to dihydroartemisinin by an

Table 36.3 *Alkylation of OH and NH functions with acidic chains*

Starting derivative	Solubilized analogue	Example	Reference
Ar — OH	Ar—O—CH$_2$—CO$_2$H	Solusalvarsan	3
Ar — OH	Ar —O— CH$_2$ — CO$_2$H	Flavodic acid	11
Ar — NH$_2$	Ar — NH — CH$_2$ — CO$_2$H	Acediasulfone	12
Ar — NH$_2$	Ar — NH — CH$_2$ — CO$_2$H	Iodopyracet	13
Ar — NH$_2$	Ar — NH — CH$_2$ — SO$_2$H	Sulfoxone sodium	14

potassium guaiacol sulfonate

sodium camphosulfonate

8-hydroxy-7-iodo 5-quinolinesulfonic acid

Fig. 36.1 Sulfonic acid solubilization.

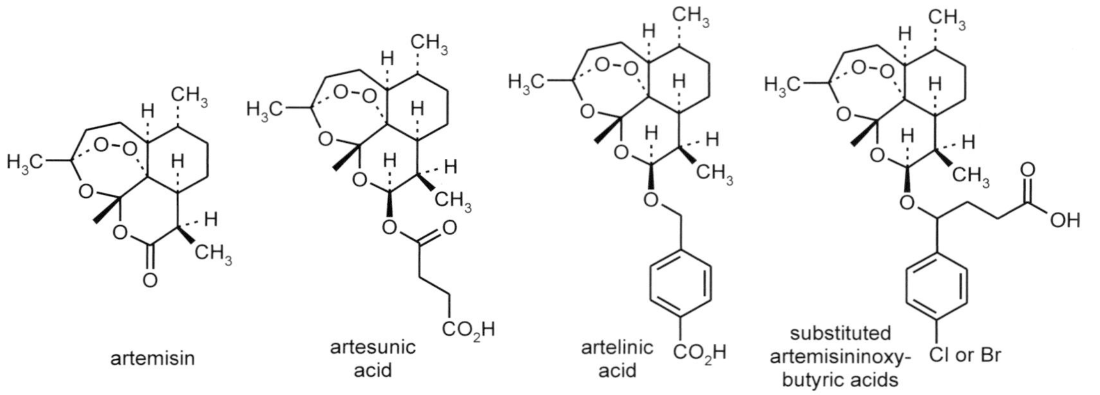

Fig. 36.3 Solubilized forms of artemisinin.[15,16]

ether rather than an ester linkage. One of the compounds prepared, artelinic acid (Fig. 36.3) is both soluble and stable in 2.5% K_2CO_3 solution and possesses superior *in vivo* activity against *Plasmodium berghei* in comparison to artermisinin or artesunic acid.[15]

Continuing the search for water-soluble dihydroartemisin derivatives with higher efficacy and longer plasma half-life than artesunic or artelinic acid, Lin *et al.* prepared a series of dihydroartemisinoxy-butyric acids bearing an aryl substituent at the 4 position of the butyric side chain (Fig. 36.3). The *p*-chlorophenyl and the *p*-bromophenyl derivatives showed a 5–8-fold increase in *in vitro* antimalarial activity against D-6 and W-2 clones of *Plasmodium falciparum* than artemisin or artelinic acid. They also showed *in vivo* oral antimalarial activity superior to that of artelinic acid.[16] Other ether-linked artemisinin-solubilizing chains containing asymmetric centres did not show activities superior of that of artelinic acid.[17]

C Acylation of OH and NH functions with acidic chains

The acylation of OH and NH functions with acidic chains is probably the most popular mode of acidic solubilization. Alcohols and phenols are converted into half-esters such as hemisuccinates, hemiglutarates, hemiphtalates,[18] meta-benzenesulfonates[19] but also into phosphates or even sulfates (Fig. 36.4). All these derivatives can give water-soluble sodium or amine salts. Similar acylation possibilities exist for amines, but peptide-like derivatives are often preferred because the enzymatic regeneration of the parent molecule *in vivo* is easier.

Carboxylic half-esters (e.g. hemisuccinates) of phenols are easily hydrolysed in aqueous solution and are therefore not recommended for the solubilization of phenolic compounds. Even hemisuccinates of alcohols suffer somewhat from stability problems and must be supplied as

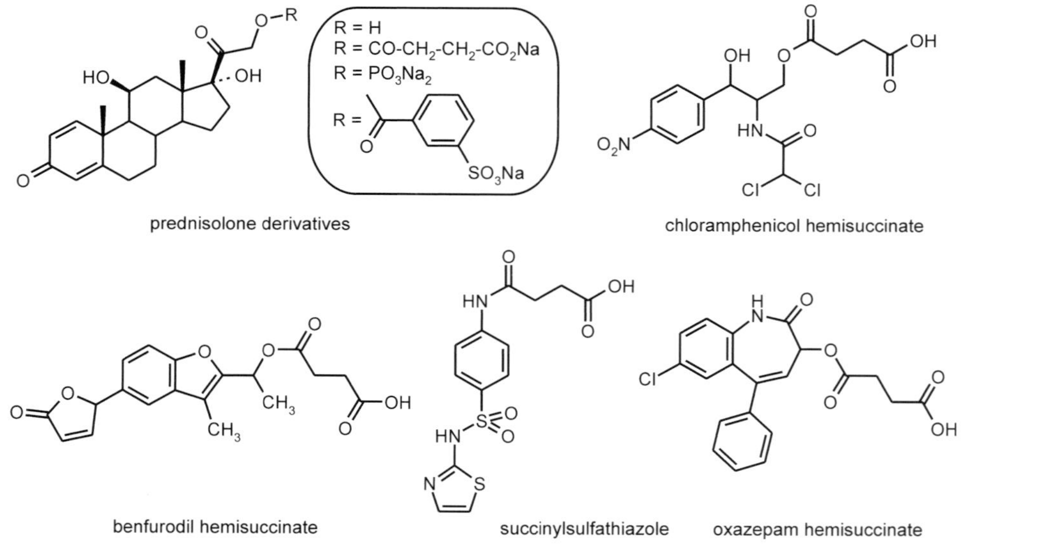

Fig. 36.4 Acylation of OH and NH functions with acidic chains.

lyophilized (freeze-dried) powders for reconstitution in water and used within 48 h (see, for example, the monograph Chloramphenicol Sodium Succinate or Hydrocortisone Sodium Succinate in *The Handbook of Injectable Drugs*,[20] see also Anderson *et al.*[21,22]).

An additional difficulty occurring with hemisuccinates was discovered by Sandman *et al.*[23] These authors, in studying the stability of chloramphenicol succinate, found an unusual partial acyl transfer reaction of the succinyl group to give a cyclic hemi-orthoester (Fig. 36.5).

Apparently hemiglutarates of phenolic drugs are more stable than hemisuccinates, an example is provided by the water-soluble diglutaryl-probucol which prevents cell-induced LDL oxidation (Fig. 36.6).[24]

In the search for an improvement of solution stability, i.e. in minimizing the ester hydrolysis and decreasing the acyl migration, Anderson and coworkers[25] synthesized a series of more stable water-soluble methylprednisolone esters. Several of the analogues were shown to have shelf-lives in solution of greater than 2 years at room temperature. Ester hydrolysis studies of these compounds in human and monkey serum indicated that derivatives having anionic solubilizing residues, such as carboxylate or sulfonate, are more slowly hydrolysed by serum esterases than compounds with a cationic solubilizing moiety (tertiary amine).[26]

Phosphoric esters (Fig. 36.7) are generally more stable. They have been used in the steroid[27,28] and in the vitamin field (vitamin C,[29] vitamin B$_1$,[30] benfotiamine,[31,32] riboflavine,[33] dihydrovitamin K$_1$[34]). Riboflavine-5′-phosphoric

acid dihydrate for example, has a solubility in water of 112 g l^{-1} at pH 6.9, compared with 0.06–0.33 g l^{-1} for riboflavine itself. Phosphoric esters of trichloroethanol, diphenylhydantoin (open form) and clindamycin are discussed in Chapter 39.

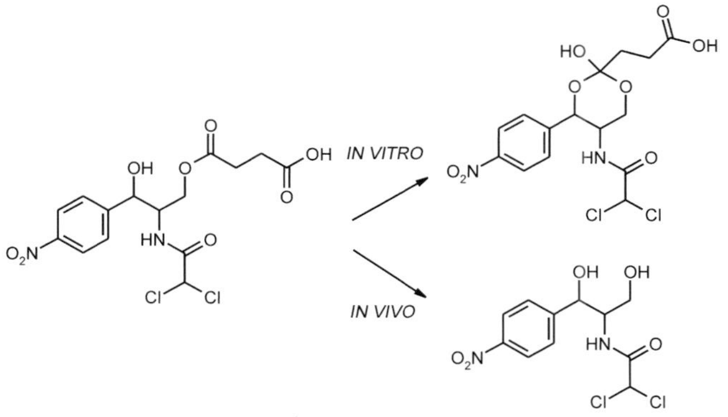

A large number of reported peptidomimetic compounds possess very low aqueous solubility at physiological pH owing to the high lipophilicity inherent in these structures. Phosphorylation can yield improved biological activities for such compounds. This is at least the case for the phosphorylated neurokinin-1 receptor antagonist[35] and the HIV protease inhibitor[36] of Fig. 36.7 described by scientists from Merck and Upjohn, respectively. Clean phosphorylation methods are now available: some of them are shown in Fig. 36.8.

Formation of sulfate esters is one of the metabolic conjugation reactions (phase II reactions, see Chapter 31). Sulfates of estradiol,[47] glucose,[48] menadiol,[49] and hydroxyethyl-theophylline[50] have been prepared (Fig. 36.9). As a rule sulfuric acid esters, compared with their phosphoric analogues, are resistant to enzymatic hydrolysis *in vivo*[51,52] and their conversion to the parent drug is questionable.

Sulfonic acids can be prepared by direct sulfonation (see above, Fig. 36.1). Compounds containing conjugated double bonds have been solubilized by addition of sodium bisulfite. Treatment of menadione (vitamin K$_3$) with sodium bisulfite leads to two addition compounds (Fig. 36.10). Mild warming of the reactants for a short time predominantly affords adduct (a) which arises from attack of bisulfite ion at carbon 2. Heating at reflux for an extended period yields adduct (b) from addition of bisulfite ion to carbon in position 3.[53]

In a similar way the treatment of N^4-cinnamylidenesulfanilamide (prepared from cinnamic aldehyde and sulfanylamide) with sodium bisulfite affords noprylsulfamide (Fig. 36.10), according to the 'soluseptazine principle' (noprylsulfamide is also called soluseptazine).[54] Noprylsulfamide is freely soluble in water (200 g l^{-1}), and breaks down in the body with the liberation of free sulfanilamide. Treatment of 6-chloropurine riboside with *p*-aminobenzenesulfonic acid leads to the highly water-soluble N^6-(*p*-sulfophenyl)adenosine (solubility > 1.5 g ml^{-1}, ≈ 3 M).

This compound (Fig. 36.11) is a potent A$_1$ adenosine agonist in receptor binding, in inhibitory electrophysiological effects in hippocampal slices, and in inhibition of lipolysis *in vivo*.[55]

Treatment of primary and secondary amines with formaldehyde and sodium bisulfite generates stable methanesulfonates that can also act as solubilizing groups. The first example of this reaction found in the literature is the conversion of *p*-phenetidine into the corresponding methanesulfonate.[56] The compound obtained (Fig. 36.12) still possesses antipyretic properties and is much less toxic than *p*-phenetidine. Nevertheless it did not break into a new market.[3]

Fig. 36.5 Formation of cyclic hemi-orthoesters from a hemisuccinate.[4,23]

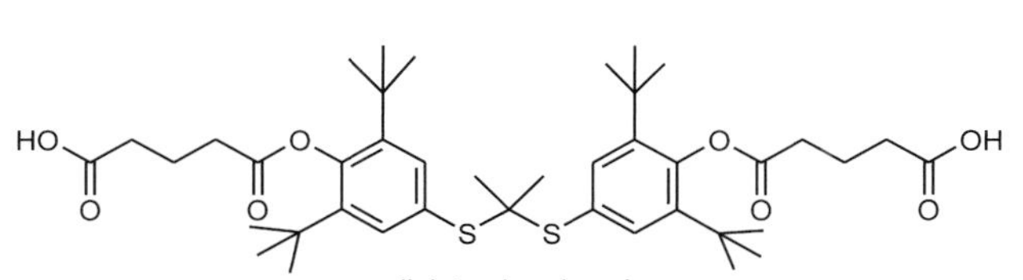

diglutaryl-probucol

Fig. 36.6 Structure of the diglutaryl ester of probucol.[24]

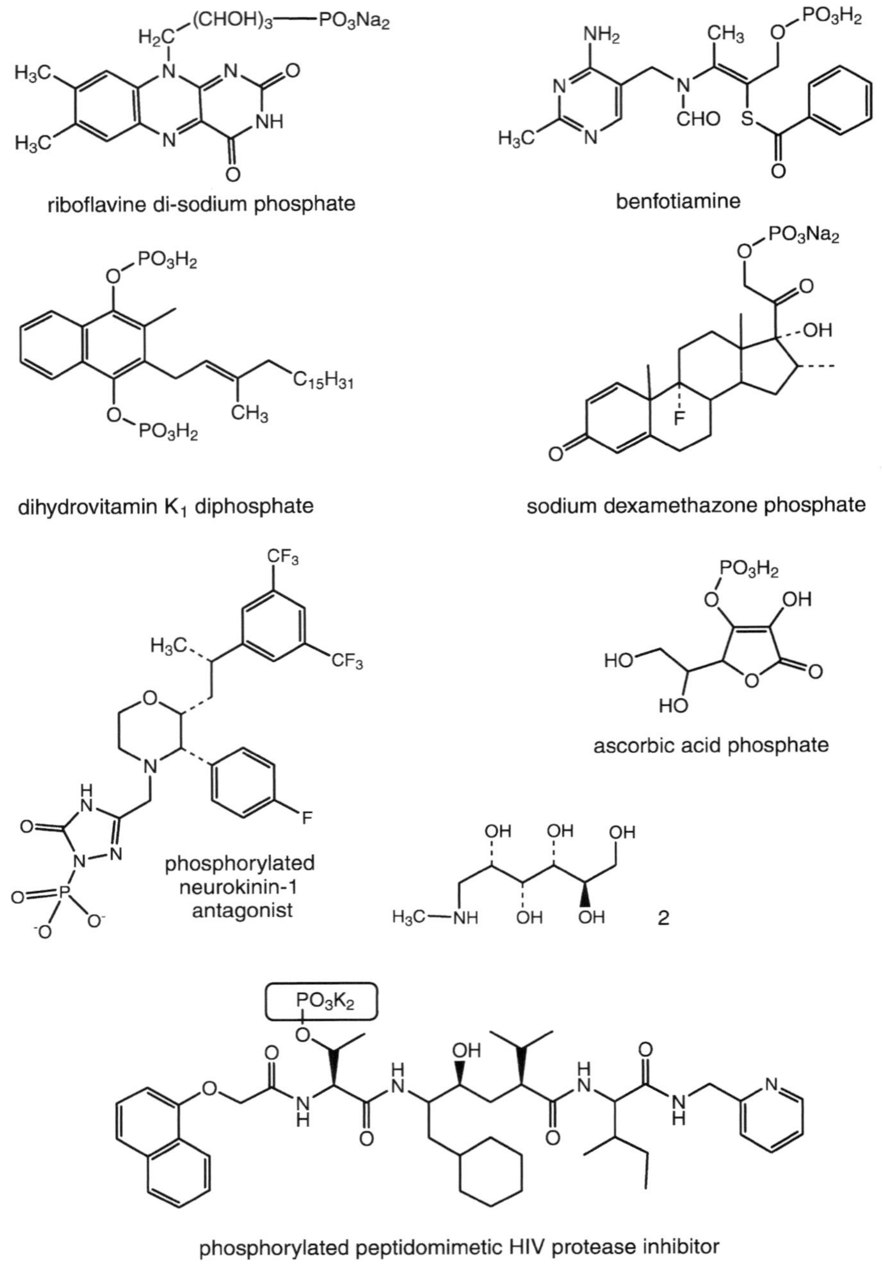

riboflavine di-sodium phosphate

benfotiamine

dihydrovitamin K₁ diphosphate

sodium dexamethazone phosphate

phosphorylated
neurokinin-1
antagonist

ascorbic acid phosphate

phosphorylated peptidomimetic HIV protease inhibitor

Fig. 36.7 Phosphate esters.

Applied to noraminopyrine, the same solubilization strategy led to dipyrone,[57,58] a water-soluble (1 g 1.5 ml⁻¹), injectable form of aminopyrine (Pyramidon) used worldwide. Replacement of formaldehyde by acetaldehyde to yield ethanesulfonates has been claimed to lead to compounds with faster hydrolysis kinetics *in vivo*.[59] Replacement of formaldehyde by glucose afforded glucosulfone sodium, a soluble preparation of the leprostatic 4,4′-diaminodiphenylsulfone.[60,61] Replacement of formal-

dehyde-bisulfite by formaldehydesulfoxylate is also claimed in some references.[3,62]

Ketones can be solubilized as carboxymethoximes (Fig. 36.13). Menadoxime is freely water-soluble, can be sterilized by autoclaving and, like menadione itself, shows high antihaemorrhagic activity.[63] The carboxymethoxime of griseofulvin, on the other hand, is devoid of activity *in vitro*, as well as *in vivo*.[64] This may be due either to an absence of conversion to the parent molecule, or to rapid

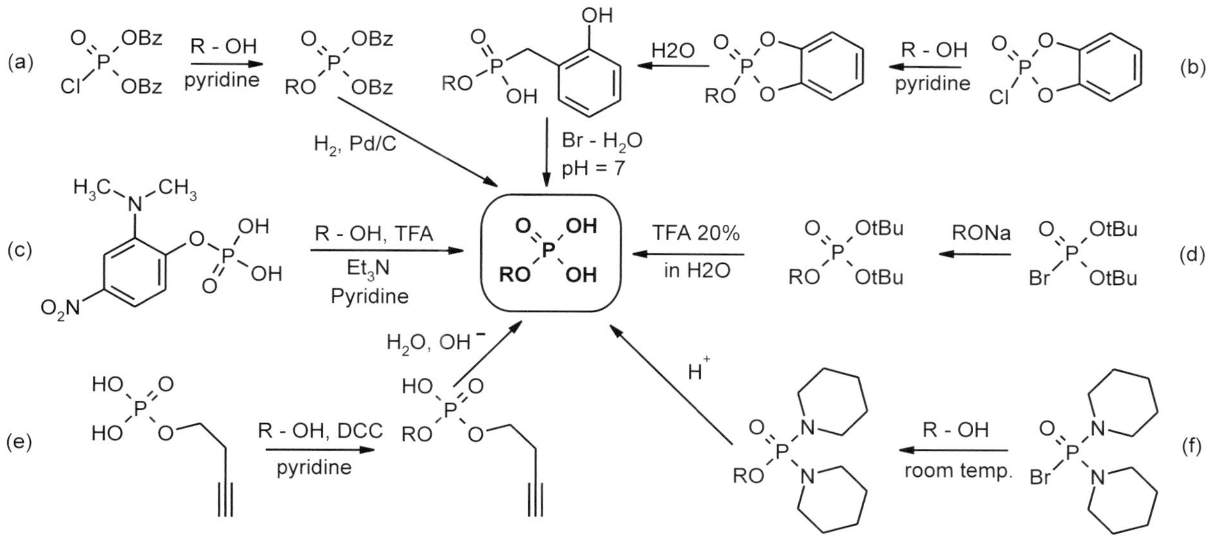

Fig. 36.8 Useful syntheses of monophosphate esters: (a) ref. 37; (b) refs 38,39; (c) ref. 40; (d) ref 41,42; (e) ref. 43–45; (f) ref. 46.

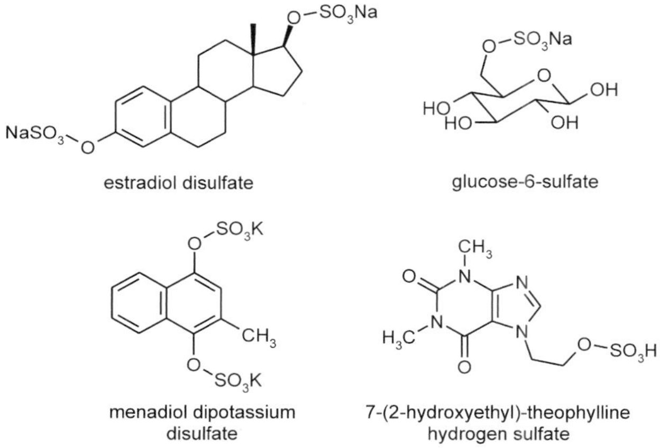

Fig. 36.9 Sulfuric esters.

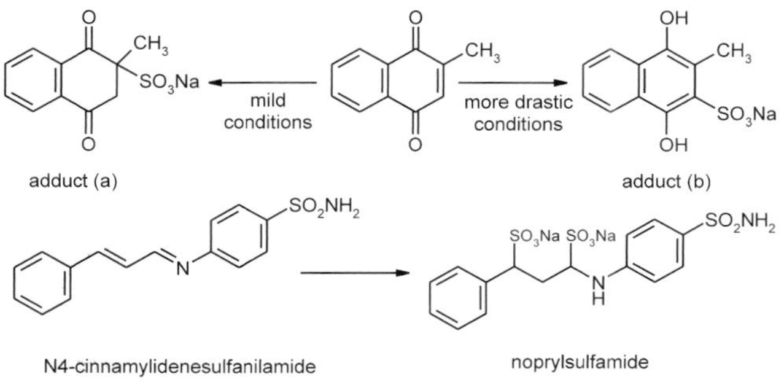

estradiol disulfate

glucose-6-sulfate

menadiol dipotassium disulfate

7-(2-hydroxyethyl)-theophylline hydrogen sulfate

N^6-(*p*-sulfophenyl)adenosine

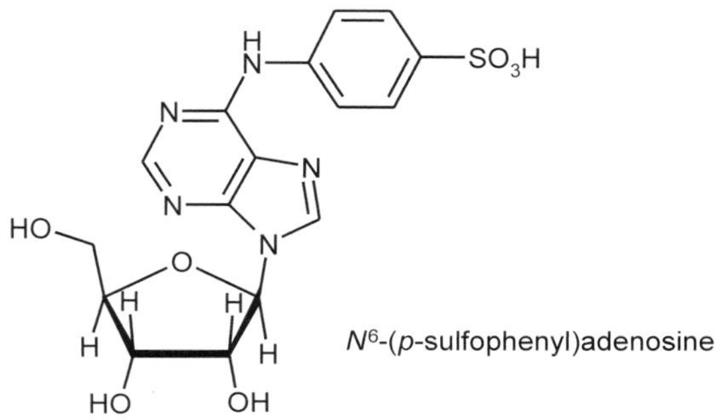

Fig. 36.11 N^6-(*p*-sulfophenyl)adenosine, a freely water-soluble adenosine A1 receptor agonist.[55]

adduct (a)

adduct (b)

N4-cinnamylidenesulfanilamide

noprylsulfamide

Fig. 36.10 Bisulfite adducts.

renal elimination. Unsubstituted oximes can be converted enzymatically to their corresponding ketones.[65] For carboxymethoximes intramolecular assistance should even facilitate the hydrolysis to the initial carbonyl function. This was shown to be the case for the carboxymethoxime of naloxone.[66] Other substituted oximes seem to be rather resistant as apparent from the metabolic stability of noxiptylin.[67]

IV BASIC SOLUBILIZING CHAINS

Solubilization with basic side chains involves two essential strategies: either direct binding of the amine function on a carbon atom of the parent molecule, or linking it to a function already present: alcoholic or phenolic hydroxyl, carboxylic acid, amine or amide.

A Direct attachment of a basic residue

Simple tertiary amines can be grafted to a carbon skeleton, either by exchange reactions or by Mannich reactions. The hydrochloride salt of the camptothecin derivative (Fig. 36.14) in (a) is soluble in water at concentrations up

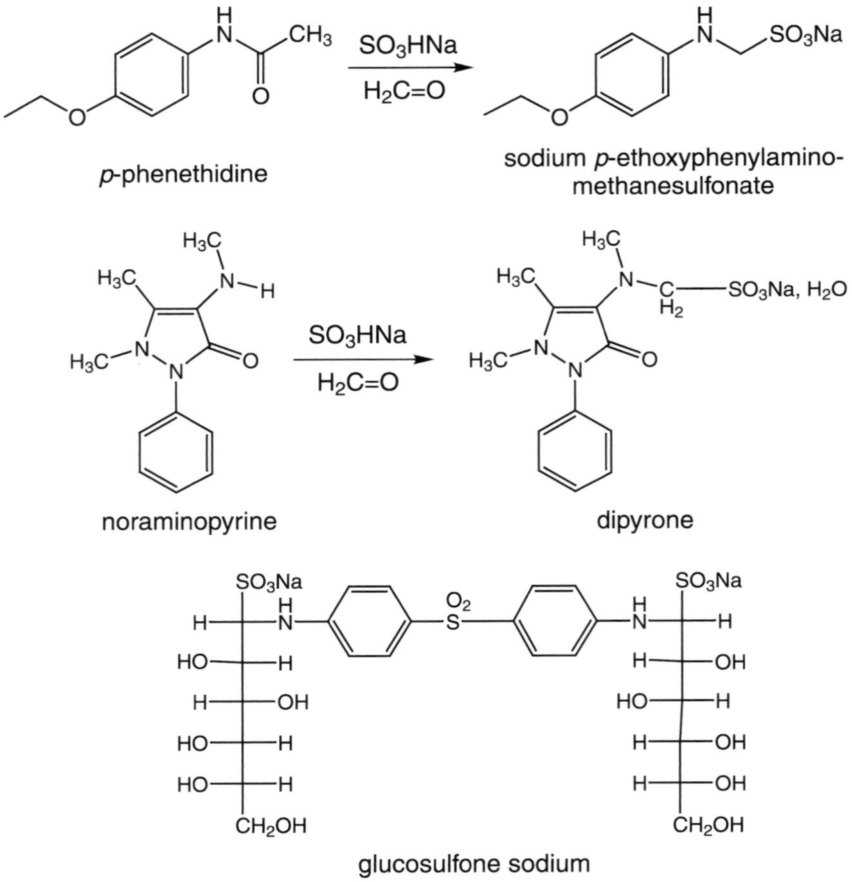

Fig. 36.12 Sulfonates.

to 1 mg ml^{-1}; the comparable value for camptothecin itself is 0.0025 mg ml^{-1}.[68] Similar results were obtained in solubilizing the benzodiazepine (b)[69] and the quinazolinone (c).[70] The adenosine A_1 antagonist KW-3902 (d) was solubilized in an original manner by converting it to an amidinic and cyclized bioisostere (e).[71]

B Attachment of the solubilizing moiety to an alcoholic hydroxyl

Esterification of an alcoholic hydroxyl with dialkylglycine or its analogues is a very popular mode of solubilization of alcohols. It is illustrated (Fig. 36.15) by soluble esters of forskolin,[72] of allopurinol[73] and of metronidazole.[74] This mode of solubilization can also be applied to phenols such as paracetamol (see Chapter 31).

Many α-amino acid esters or related short-chained aliphatic amino esters show satisfying hydrolysis kinetics in plasma, but exhibit poor stability in aqueous solution.[74] This poor stability is predominantly due to electron withdrawal by the positively charged amino group, but

may also involve intramolecular catalysis or assistance of the ester hydrolysis by the neighbouring amino group. The replacement of the glycine unit by its benzologue, as shown for the metronidazole derivative of Fig. 36.14, prevents the hydrolysis-facilitating effect of the amino group. Alternative solutions place the amino group more distant from the ester linkage in using 6-aminocaproic acid esters and sebacic acid-derived spacer groups[25] or render the ester function more resistant by replacing it by a carbonate or a carbamate function.[75]

C Attachment of the solubilizing moiety to an acidic NH function

The NH group in theophylline is reactive and can easily be alkylated, yielding among others, etamiphyllin (Fig. 36.16) which is rendered water-soluble as hydrochloride or as camphosulfonate.[76]

Rolitetracycline, the Mannich base derived from the carboxamide function of tetracycline, formaldehyde and pyrrolidine,[77] is surprisingly stable and is used as an injectable form of tetracycline.

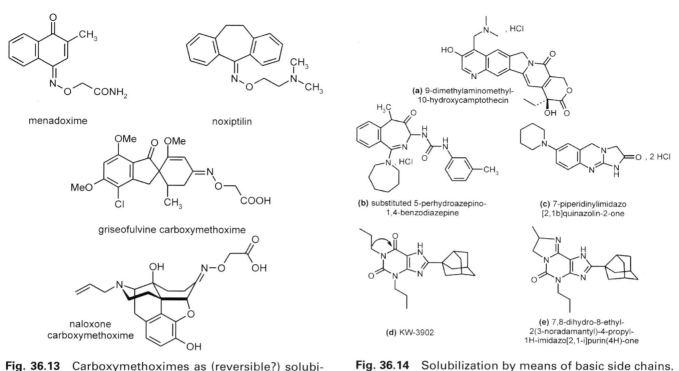

menadoxime

noxiptilin

griseofulvine carboxymethoxime

naloxone
carboxymethoxime

Fig. 36.13 Carboxymethoximes as (reversible?) solubilizing chains for carbonyl-containing molecules.

(a) 9-dimethylaminomethyl-10-hydroxycamptothecin

(b) substituted 5-perhydroazepino-1,4-benzodiazepine

(c) 7-piperidinylimidazo[2,1b]quinazolin-2-one

(d) KW-3902

(e) 7,8-dihydro-8-ethyl-2(3-noradamantyl)-4-propyl-1H-imidazo[2,1-i]purin(4H)-one

Fig. 36.14 Solubilization by means of basic side chains.

D Attachment of the solubilizing moiety to a basic NH₂ function

An appropriate manner of solubilizing basic NH_2 functions is to form peptides with common amino acids. In Chapter 33, the L-lysine peptide of the ring-opened form of diazepam was described. Similarly acylation of the 3-amino group of the pyrrolidine ring in a series of quinolone antibacterial agents yielded interesting compounds (Fig. 36.17).

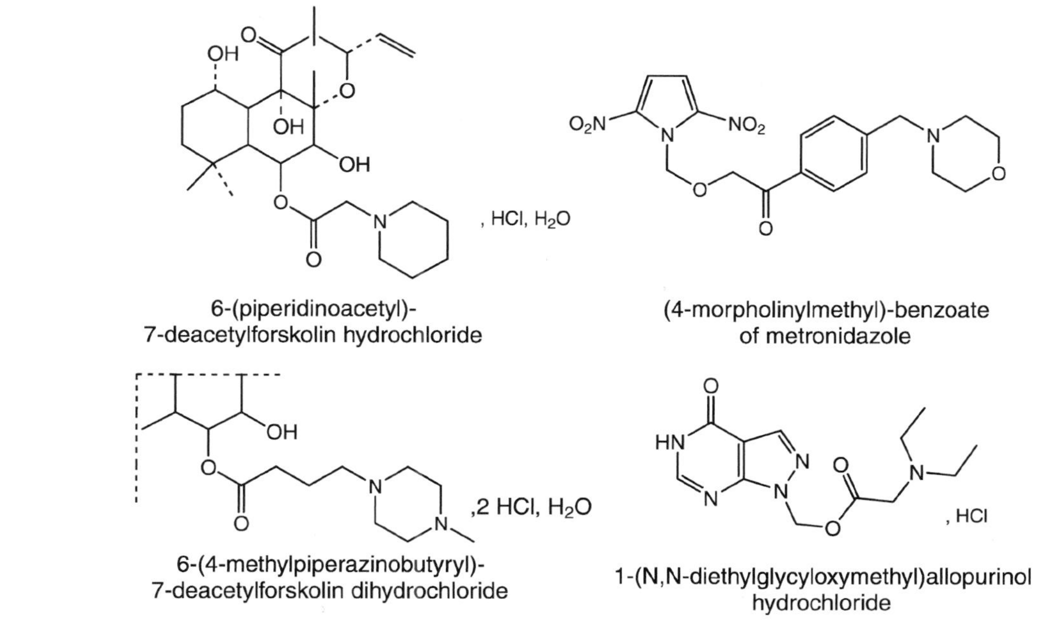

6-(piperidinoacetyl)-7-deacetylforskolin hydrochloride

(4-morpholinylmethyl)-benzoate of metronidazole

6-(4-methylpiperazinobutyryl)-7-deacetylforskolin dihydrochloride

1-(N,N-diethylglycyloxymethyl)allopurinol hydrochloride

Fig. 36.15 Dialkylglycines and related aminoesters.

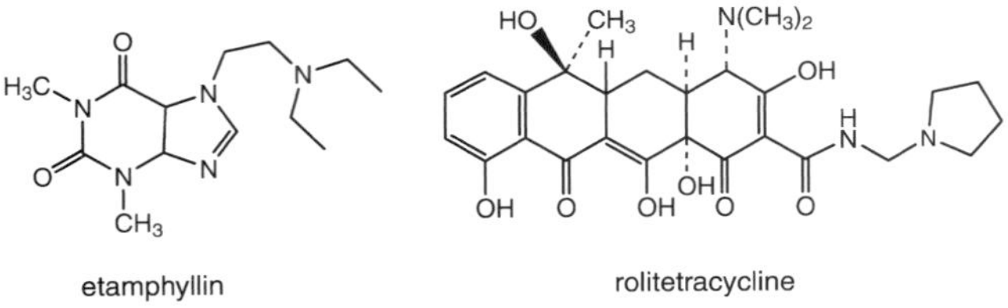

Fig. 36.16 Basic chains on acidic NH functions.

The amino acid analogues were less active *in vitro*, but had equal or increased efficacy *in vivo*. Indeed, it was shown that these compounds, which were stable to acid and base under the reaction conditions for their preparation, were rapidly cleaved in serum to give the parent quinolones. The amino acid derivatives showed a 3–70 times improved solubility when compared with the parent compounds.[78]

E Attachment of the solubilizing moiety to carboxylic acid functionalities

As carboxylic esters of aminoalcohols are usually too sensitive to hydrolysis, amides with aminoalkylamines are preferred (Fig. 36.18).

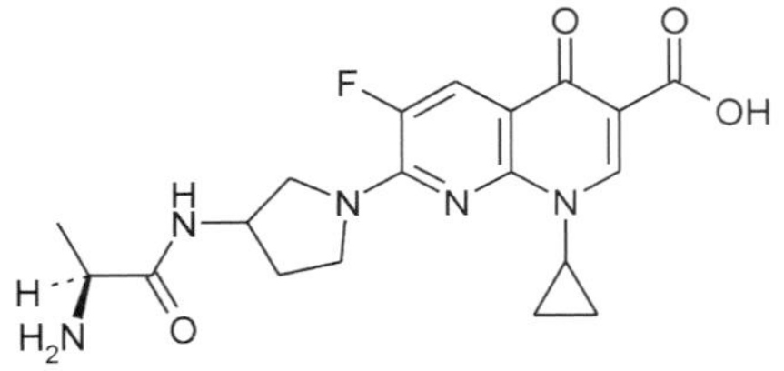

compound PD131112

Fig. 36.17 *S*-Alanine-derived peptide of a quinolone antibacterial agent confers high water-solubility and is sensitive to enzymatic cleavage *in vivo*.[78]

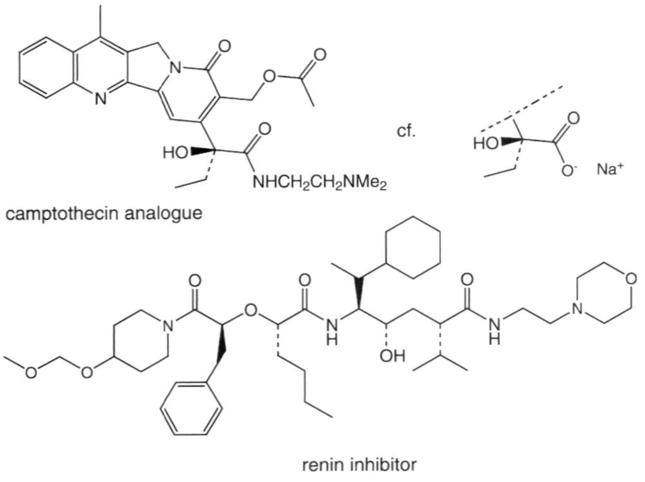

Fig. 36.18 Basic amides of carboxylic acids.

The water-soluble E-lactone ring modified 7-ethylcamptothecin analogue bearing a dimethylaminoethyl amidic chain compares favourably with the sodium salt resulting simply from the lactone ring opening.[79] Introduction of basic substituents into modified hydroxyethylene dipeptide isosteres gave inhibitors with improved solubility, as well as improved potency against human plasma rennin.[80]

V NONIONIZABLE SIDE CHAINS

The most frequently employed solubilization approaches using nonionizable moieties involve hydroxylated and polyoxymethylenic side chains or glucosides and their analogues.

A Glycolyl and glyceryl side chains

These chains are present in some classical drugs such as the muscle relaxant mephenesin[81] and the bronchodilator dyphylline (diprophylline)[82] (Fig. 36.19). More recent applications are found in the venotropic troxerutin[83] and in the analgesic-anti-inflammatory drugs glafenin[19,84] and etofenamate.[85]

B Polyethylene glycol derivatives

Only a few examples of solubilization involving esters or ethers of poly(ethylene glycol) are found in the literature and it is not always clear whether the main purpose of the polyoxyethylenic chain grafting was to increase the aqueous solubility or to produce another improvement such as sustained release. One of the most representative examples is the antitussive benzonatate[86] (Fig. 36.20). In a similar way Nagakawa and colleagues prepared soluble forms of vitamins A and E, as well as of various steroids (prednisolone, testosterone, hydrocortisone, gitoxine).[87,88]

The symmetrical attachment of the local anaesthetic procaine to poly(ethylene glycol) increases the duration of action[89] and probably improves water-solubility. The hypnotic etodroxizine bears a three-ethylene-oxy-unit

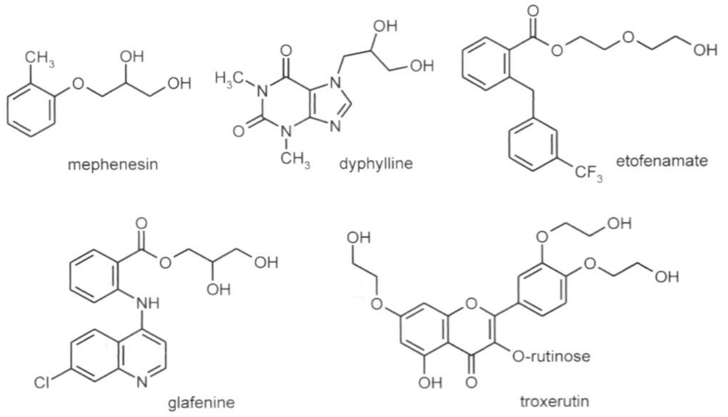

Fig. 36.19 Glycolyl and glyceryl side chains.

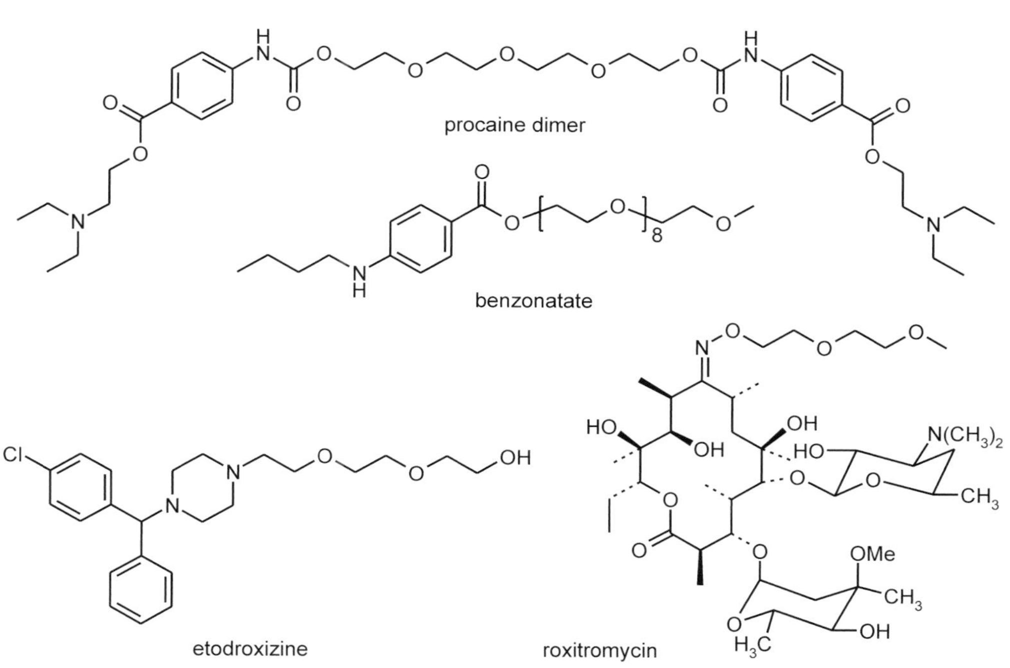

Fig. 36.20 Polyethylene glycol derivatives.

chain, but is nevertheless used as a dimaleate.[90] In roxithromycin the oxygenated side chain is attached to the oxygen atom of the oxime of the antibiotic erythromycin.[91]

C Glucosides and related compounds

Despite the ubiquitous distribution in plants, man-made glucosidic derivatives of alcohols or phenols are rarely prepared in medicinal chemistry. An old-timer is the sedative-hypnotic α-chloralose (Fig. 36.21) which is presently used only as a surgical anaesthetic for laboratory animals.

Other examples are deoxycorticosterone β-maltoside[92,93] and menthol β-glucoside. Deoxycorticosterone glycosides show various solubilities depending on the sugar conjugate (Table 36.4).[94]

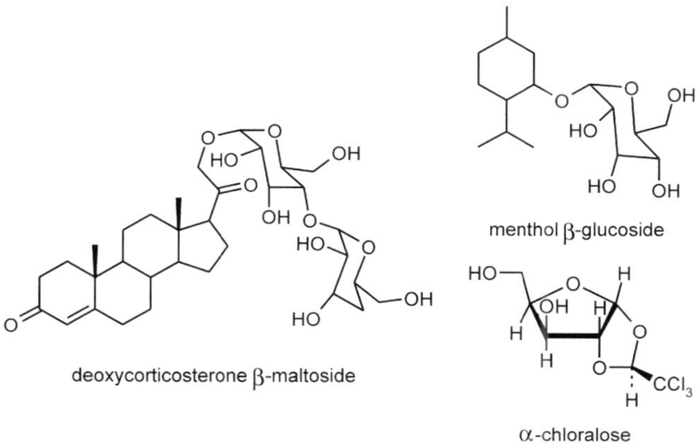

Table 36.4 *Solubilities of deoxycorticosterone glycosides*[94]

Deoxycorticosterone glycoside	Solubility in water
Glucoside	1.2‰
Galactoside	2.2‰
Lactoside	3.4‰
Lactosidoglucoside	Unlimited

Menthol β-glucoside is a water-soluble, nonirritating prodrug of menthol that can be used, like glucovanillin, the β-D-glucoside of vanillin, as a pharmaceutic flavour adjuvant.[95] The use of sugar moieties as drug carriers has been reviewed by Chavis and Imbach.[96]

Attachment of the sugar moiety to nitrogen atoms of amine, amide or hydrazine functions is much more frequently encountered (Fig. 36.22). Prontoglucal is the N^4-β-D-glucoside of sulfanylamide,[97] the tuberculostatic glyconiazide is the isonicotinoylhydrazone of D-glucuronic acid lactone,[98] and glucometacin results from the amidification of indomethacin with D-glucosamine.[99]

Many highly water-soluble radiological contrast agents are solubilized as sugar conjugates. This is the case for metrizamide[100] and compound P 297.[101]

VI CONCLUDING REMARKS

The different chemical solutions to solubilizing problems discussed in this chapter reveal that in many cases

Fig. 36.21 Glycosidic derivatives of alcoholic functions.

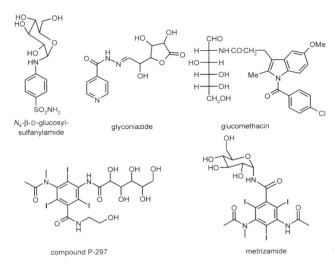

Fig. 36.22 Glycosidic derivatives of amine, amide and hydrazine functions.

the chemical transformation used also improves the activity profile of the parent molecule. This can be due to purely *pharmacokinetic* factors such as a better resorption from the organism and faster transport and diffusion. These factors also explain why solubilized drugs are generally faster eliminated and therefore show fewer symptoms of toxicity. However, the *pharmacological* profile also can be affected. Chlorpromazine for example (Fig. 36.23) has neuroleptic properties, whereas the parent phenothiazine possesses anthelminthic properties. In this example the attachment of the basic moiety has totally modified the pharmacological profile. However, the replacement of the basic moiety by its

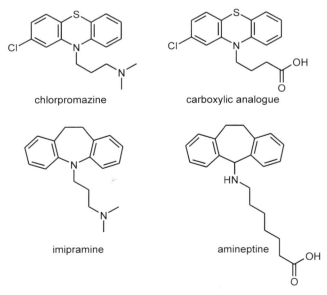

Fig. 36.23 Acidic and basic side chains on tricyclic skeletons.

carboxylic counterpart yielded a compound totally inactive as neuroleptic (C.G. Wermuth, unpublished result).

Conversely, in the tricyclic antidepressant series, the passage from the basic imipramine to the acidic amineptine conserved the antidepressant properties. In other words, there are no general rules available for the selection of the most appropriate solubilizing moiety. It is therefore recommended, that for each new solubilization problem, acidic, basic and neutral solubilized versions of the parent molecule be prepared.

REFERENCES

1. La Fontaine, J. de La Besace. (1993) *Fables Livre I* Fable VII, pp. 42–43. Florilège, Paris.
2. Marini-Bettolo, G.B. (1948) Metodos modernos de solubilizacion de medicamentos organicos. *Ann. Asoc. Quim. Pharm. Urugay* **50**: 3–17.
3. Büchi, J. (1963) *Grundlagen der Arzneimittelforschung und der Synthetischen Arzneimittel*, pp. 220–235. Birkhaüser Verlag, Basel.
4. Stella, V. (1963) Prodrugs: an overview and definition. In Higuchi, T. and Stella, V. (eds). *Prodrugs as Novel Drug Delivery Systems*, pp. 1–115. American Chemical Society, Washington, DC.
5. Silverman, R.B. (1992) *The Organic Chemistry of Drug Design and Drug Action*, pp. 358–360. Academic Press, San Diego.
6. Albert, A. (1979) *Selective Toxicity*, pp. 403–440. Chapman and Hall, London.
7. Hammond, M.L. (1993) Recent advances in anti-infective agents. In Bristol, J.A. (ed.). *Ann. Rep. Med. Chem.* pp. 119–130. Academic Press, San Diego.
8. Leeson, P.D., Carling, R.W., Kulagowski, J.J., *et al.* (1993) Drugs interacting with the glycine binding site. In Testa, B., Kyburz, E., Fuhrer, W. and Giger, R. (eds). *Perspectives in Medicinal Chemistry*, pp. 239–257. Verlag Helvetica Chemica Acta/VHC, Basel.
9. Stilts, C.E., Nelen, M.I., Hilmey, D.G., *et al.* (2000) Water-soluble, core-modified porphyrins as novel, longer-wavelength-absorbing sensitizers for photodynamic therapy. *J. Med. Chem.* **43**: 2403–2410.
10. Bock, M.G., DiPardo, R.M., Mellin, E.C., *et al.* (1994) Second-generation benzodiazepine CCK-B antagonists. Development of subnanomolar analogs with selectivity and water solubility. *J. Med. Chem.* **37**: 722–724.
11. Blaise, R. Derivatives of flavone. French patent 80122, 22 March 1963, to Société Anonyme pour L'Industrie clinique.
12. Jackson, E.L. (1948) Certain N-Alkyl, N-carboxyalkyl and N-hydroxyalkyl derivatives of 4,4′-diaminodiphenyl sulfone. *J. Am. Chem. Soc.* **70**: 680–684.
13. Reitmann, J. (1935) Aliphatic amine salts of halogenated pyridones containing an acid group. US Patent 1 993 039 (5 March 1935; to Winthrop Chemical Company, Inc., New York).
14. Bauer, H. (1939) Organic compounds in chemotherapy. II. The Preparation of formaldehyde sulfoxylate derivatives of sulfanilamide and of amino compounds. *J. Am. Chem. Soc.* **61**: 617–618.
15. Lin, A.J., Klayman, D.L. and Milhous, W. (1987) Antimalarial activity of new water-soluble dihydroartemisin derivatives. *J. Med. Chem.* **30**: 2147–2150.
16. Lin, A.J., Sikry, A.B. and Kyle, D.E. (1997) Antimalarial activity of new dihydroartemisin derivatives. 7. 4-(*p*-Substituted phenyl)-4(R or S)-[10 (α or β)-dihydroartemisinoxy]butyric acids. *J. Med. Chem.* **40**: 1396–1400.

17. Lin, A.J., Lee, M. and Klayman, D.L. (1989) Antimalarial activity of new water-soluble dihydroartemisin derivatives. 2. Stereospecificity of the ether side chain. *J. Med. Chem.* **32**: 1249–1252.

18. Coker, J.D., Elks, J., May, P.J. *et al.* (1965) Action of some steroids on the central nervous system of the mouse. I. Synthetic Methods. *J. Med. Chem.* **8**: 417–425.

19. Allais, A., Rousseau, G., Girault, P. *et al.* (1966) Sur l'activité analgésique et antiinflammatoire des 4-(2'-alcoxycarbonyl phényl-amino) quniléïnes. *Eur. J. Med. Chem.* 65–70.

20. Trissel, L.A. (1986) *Handbook on Injectable Drugs*. American Society of Hospital Pharmacists, Bethesda.

21. Anderson, B.D. and Taphouse, V. (1981) Initial rate studies of hydrolysis and acyl migration in methylprednisolone 21-hemisuccinate and 17-hemisuccinate. *J. Pharm. Sci.* **70**: 181–186.

22. Anderson, B.D., Conradi, R.A. and Lambert, J.W. (1984) Carboxyl group catalysis of acyl transfer reactions in corticosteroid 17- and 21-monoesters. *J. Pharm. Sci.* **73**: 604–611.

23. Sandman, B., Szulczewski, D., Winheuser, J. and Higuchi, T. (1970) Rearrangement of chloramphenicol-3-monosuccinate. *J. Pharm. Sci.* **59**: 427–429.

24. Parthasarathy, S. (1992) Diglutarylprobucol. *J. Clin. Invest.* **89**: 1618.

25. Anderson, B.D., Conradi, R.A. and Knuth, K.E. (1985) Strategies in the design of solution-stable, water-soluble prodrugs: I: A physical-organic approach to pro-moiety selection for 21-esters of corticosteroids. *J. Pharm. Sci.* **74**: 365–374.

26. Anderson, B.D., Conradi, R.A., Spilman, C.H. and Forbes, A.D. (1985) Strategies in the design of solution-stable, water-soluble prodrugs. III: Influence of the pro-moiety on the bioconversion of 21-esters of corticosteroids. *J. Pharm. Sci.* **74**: 382–387.

27. Flynn, G.L. and Lamb, D.J. (1970) Factors influencing solvolysis of corticosteroid-21-phosphate esters. *J. Pharm. Sci.* **59**: 1433–1438.

28. Melby, J.C. and St Cyr, M. (1961) Comparative studies on absorption and metabolic disposal of water-soluble corticosteroid esters. *Metabolism* **10**: 75–82.

29. Cutolo, E. and Larizza, A. (1961) Synthesis of 3-phosphoric ester of L-ascorbic acid. *Gazz. Chim. Ital.* **91**: 964–972.

30. Wenz, A., Göttmann, G. and Koop, H. (1961) Verfahren zur Herstellung der freien Base und der Salze des Aneurin-orthophos-phor-säureesters. Ger. Patent 1 110 646 (13 July 1961; E. Merck A. G. Darmstadt).

31. Ito, A., Hamanaka, W., Takagi, H., *et al.* (1960) Verfahren zur Herstellung eines Acylierungsproduktes von Vitamin-B₁-orthophos-phorsäureester und Salzen davon. Ger. Patent 1 130 811 (14 April 1960 to Sankyo Kabushiki Kaisha, Tokyo).

32. Wada, T., Tagaki, H., Minakami, H., *et al.* (1961) A new thiamine derivative, S-benzoylthiamine O-monophosphate. *Science* **134**: 195–196.

33. Viscontini, M., Ebnoether, C. and Karrer, P. (1952) Einfache synthese kristallisierter lactoflavin-5'-phosphorsäure (Coferment des Flavin-enzyms). *Helv. Chim. Acta* **35**: 457–459.

34. Fieser, L.F. (1946) Antihemorrhagic esters and methods for producing the same. US Patent 2,407,823 (17 September 1946 to Research Corporation, New York).

35. Hale, J.J., Mills, S.G., MacCoss, M., *et al.* (2000) Phosphorylated morpholine acetal human neurokinin-1 receptor antagonists as water-soluble prodrugs. *J. Med. Chem.* **43**: 1234–1241.

36. Chong, K.-T., Ruwart, M.J., Hinshaw, R.R., *et al.* (1993) Peptidomi-metic HIV protease inhibitors: phosphate prodrugs with improved biological activities. *J. Med. Chem.* **36**: 2575–2577.

37. Fieser, L.F. and Fieser, M. (1967) *Reagents for Organic Synthesis*, p. 198. John Wiley, New York.

38. Khawaja, T.A. and Reese, C. (1966) o-Phenylene phosphochloridate. A convenient phosphorylating agent. *J. Am. Chem. Soc.* **88**: 3446–3447.

39. Khawaja, T.A., Reese, C.B. and Stewart, J.C.M. (1970) A convenient general procedure for the conversion of alcohols into their monophos-phate esters. *J. Chem. Soc. (C)* 2090–2100.

40. Taguchi, Y. and Mushika, Y. (1975) 2-(N,N-Diethylamino)-4-nitro-phenyl phosphate and its use in the selective phosphorylation of unprotected nucleosides. *Tetrahedron Lett.* **24**: 1913–1916.

41. Gajda, T. and Zwierzak, A. (1976) Phase-transfer-catalysed halogena-tion of Di-t-butyl phosphite. Preparation of Di-t-butyl phosphorohali-dates. *Synthesis* 243–244.

42. Gajda, T. and Zwierzak, A. (1977) Di-t-butyl phosphorobromidate. A new selective phosphorylating agent containing acid-labile protect-ing groups. *Synthesis* 623–625.

43. Tener, G.M. (1961) 2-Cyanoethyl phosphate and its use in the synthesis of phosphate esters. *J. Am. Chem. Soc.* **83**: 159–168.

44. Moffatt, J.G. (1963) The synthesis of orotidine-5' phosphate. *J. Am. Chem. Soc.* **85**: 1118–1123.

45. Brownfield, R.B. and Shultz, W. (1963) A direct method for the preparation of steroid-21-phosphates. *Steroids* **2**: 597–603.

46. Montgomery, H.A.C. and Turnbull, J.H. (1958) Phosphoramidic halides; phosphorylating agents derived from morpholine. *J. Chem. Soc.* 1963–1967.

47. Fex, H., Lundvall, K.E. and Olsson, A. (1968) Hydrogen sulfates of natural estrogens. *Acta Chem. Scand.* **22**: 254–264.

48. Guiseley, K.B. and Ruoff, P.M. (1961) Monosaccharide sulfates. I. Glucose-6-sulfate. Preparations, characterization of the crystalline potassium salt, and kinetic studies. *J. Org. Chem.* **26**: 1248–1254.

49. Fieser, L.F. and Fry, E.M. (1940) Water-soluble antihemorrhagic esters. *J. Am. Chem. Soc.* **62**: 228–229.

50. Stieglitz, E. and Matz, M. (1958) Verfahren zur Herstellung von Xanthinalkylschwefelsäuren oder deren Salze. *Ger. Auslegeschrift* 1,090,669 (*11 January 1958 to Arzneimittelfabrik Krewel-Leuffen G.m.b. H., Eitorf/Sieg*).

51. Miyabo, S., Nakamura, T., Kuwazima, S. and Kishida, S. (1981) A comparison of the bioavailability and potency of dexamethazone phosphate and sulphate in man. *Eur. J. Clin. Pharmacol.* **20**: 277–282.

52. Williams, D.B., Varia, S.A., Stella, V.J. and Pitman, I.H. (1983) Evaluation of the prodrug potential of the sulfate ester of acetamino-phen and 3-hydroxymethyl-phenytoin. *Int. J. Pharm.* **14**: 113–120.

53. Greenberg, F.H., Leung, K.K. and Leung, M. (1917) The reaction of vitamin K₃ with sodium bisulfite. *J. Chem. Ed.* **48**: 632–634.

54. Despois, R.L. (1938) Procédé de préparation de composés aminés aromatiques solubles possédant une valeur thérapeutique. French Patent 831 366 (7 June 1938; Rhône-Poulenc).

55. Jacobson, K.A., Nikodijevic, O., Ji, X.-D., *et al.* (1992) Synthesis and biological activity of N⁶-(p-sulfophenyl)alkyl and N⁶-sulfoalkyl derivatives of adenosine: water-soluble and peripherally selective adenosine agonists. *J. Med. Chem.* **35**: 4143–4149.

56. Anonymous (1907) Verfahren zur Darstellung von p-äthoxyphenyla-minomethylschwefligsauren Salzen. German patent 209 695 (8 May 1907, to Roberto Lepetit, Garessio, Italy).

57. Anonymous (1911) Verfahren zur Darstellung von ω-methylschwe-fligsauren Salzen aminosubstituierter Arylpyrazolone. German patent 254 711 (21 July 1911, to Farbwerke vorm. Meister Lucius and Brüning in Hoechst am Main).

58. Anonymous (1911) Verfahren zur Darstellung von ω-methylschwe-fligsauren Salzen aminosubstituierter Arylpyrazolone. German patent 259 503 (17 September 1911, to Farbwerke vorm. Meister Lucius & Brüning in Hoechst am Main).

59. Mutch, N. (1941) A new sulphonamide (sulfonamide E.O.S.). *Br. Med. J.* **2**: 503–507.

60. Jain, B.C., Iyer, B.H. and Guha, P.C (1946) The preparation of promin. *Science and Culture* **11**: 568–569 (C.A. 1946, **40**: 4687⁷).

61. Anonymous (1949) Soluble Diphenylsulfone. Swiss Patent 234 108 (1 Dec. 1944, Aktien-Gesellschaft vorm. B. Siegfried).

62. Bockmühl, M., Krohs, W., Racke, F. and Windisch, K (1940) *N*-Methylsulphites and *N*-methanesulphinic acid salts of 1-aryl-2,3-dialkyl-4-alkylaminopyrazolones. US Patent 2 193 788 (19 March 1940, Winthrop Chemical Company, Inc.).

63. Holland, D.O. (1949) Preparation of 2-alkyl-1,4-naphtaquinone-4-carboxy-alkoximes. British Patent 621 934 (22 April 1949, to Glaxo Laboratories Ltd).

64. Fischer, L.J. and Riegelman, S. (1967) Absorption and activity of some derivatives of griseofulvin. *J. Pharm. Sci.* **56**: 469–476.

65. Tatsumi, K. and Ishigai, M. (1987) Oxime-metabolizing activity of liver aldehyde oxidase. *Arch. Biochem. Biophys.* **253**: 413–418.

66. Negwer, M. (1987) *Organic-Chemical Drugs and their Synonyms. Entry No. 5760: Codoxime.* Akademie Verlag, Berlin.

67. Aichinger, G., Behner, O., Hoffmeister, F. and Schütz, S. (1969) Basische tricyclische Oximinoäther und ihre pharmakologischen Eigenschaften. *Arzneim.-Forsch.* **19**: 838–845.

68. Kingsbury, W.D., Boehm, J.C., Jakas, D.R., *et al.* (1991) Synthesis of water-soluble (aminoalkyl)camptothecin analogues: inhibition of topoisomerase I and antitumor activity. *J. Med. Chem.* **34**: 98–107.

69. Showell, G.A., Bourrain, S., Neduvelil, J.G., *et al.* (1994) High-affinity and potent, water-soluble 5-amino-1,4-benzodiazepine CCKB/gastrin receptor antagonists containing a cationic solubilizing group. *J. Med. Chem.* **37**: 719–721.

70. Ishikawa, F., Saegusa, J., Inamura, K., *et al.* (1985) Cyclic guanidines. 17. Novel (*N*-substituted amino)imidazo[2,1-b]quinazolin-2-ones: water-soluble platelet aggregation inhibitors. *J. Med. Chem.* **28**: 1387–1393.

71. Suzuki, F., Shimada, J., Nonaka, H., *et al.* (1992) 7,8-Dihydro-8-ethyl-2-(3-noradamantyl)-4-propyl-1H-imidazo[2,1-i]purin-5(4H)-one: a potent and water-soluble adenosine A1 antagonist. *J. Med. Chem.* **35**: 3572–3581.

72. Khandelwal, Y., Rajeshwari, K., Rajagopalan, R., *et al.* (1988) Cardiovascular effects of new water-soluble derivatives of forskolin. *J. Med. Chem.* **31**: 1872–1879.

73. Bundgaard, H. and Falch, E. (1985) Improved rectal and parenteral delivery of allopurinol using the prodrug approach. *Arch. Pharm. Chem. Sci. Ed.* **13**: 39–48.

74. Bundgaard, H., Falch, E. and Jensen, E. (1989) A novel solution-stable, water-soluble prodrug type for drugs containing a hydroxyl or an NH-acidic group. *J. Med. Chem.* **32**: 2503–2507.

75. Balkovec, J.M., Black, R.M., Hammond, M.L., *et al.* (1992) Synthesis, stability, and biological evaluation of water-soluble prodrugs of a new echinocandin lipopeptide. Discovery of a potential clinical agent for the treatment of systemic candidiasis and *Pneumocystis carinii* pneumonia (PCP). *J. Med. Chem.* **35**: 194–198.

76. Klosa, J. (1955) Beitrag zur Reaktionsfähigkeit der 7-Stellung des Theophyllins. 2.Mitt. Über Synthesen in der Theophyllinreihe. *Arch. Pharm.* **288**: 301–303.

77. Gottstein, W.J., Minor, W.F. and Cheney, L.C. (1959) Carboxamido derivatives of tetracyclines. *J. Am. Chem. Soc.* **81**: 1198–1201.

78. Sanchez, J.P., Domagala, J.M., Heifetz, C.L., *et al.* (1992) Quinolone antibacterial agents. Synthesis and structure-activity relationships of a series of amino acid prodrugs of racemic and chiral 7-(3-amino-1-pyrrolidinyl)quinolones. Highly soluble quinolone prodrugs with *in vivo* Pseudomonas activity. *J. Med. Chem.* **35**: 1764–1773.

79. Sawada, S., Yaegashi, T. and Furuta, T., *et al.* (1993) Chemical modification of an antitumor alkaloid, 20(S)-camptothecin: E-lactone ring modified water-soluble derivatives of 7-ethylcamptothecin. *Chem. Pharm. Bull.* **41**: 310–313.

80. Boyd, S.A., Fung, A.K.L., Baker, W.R., *et al.* (1992) C-Terminal modifications of nonpeptide renin inhibitors: improved oral bioavail-ability via modification of physicochemical properties. *J. Med. Chem.* **35**: 1735–1746.

'81. Morch, P. (1947) Glykresin. *Arch. Pharm. Chem.* **54**: 327–332.

82. Roth, H.J. (1959) Zur Darstellung von β-oxyalkyl-dimethyl-purinen; Umsetzung von theophyllin und theobromin mit 1,2-epoxyden. *Arch. Pharm.* **292**: 234–238.

83. Courbat, P.J. (1969) Qercetin and quercetin gycoside. US Patent 3 420 815 (7 January 1969, to Zyma S.A.).

84. Mouzin, G., Cousse, H. and Autin, J.M. (1980) A new, convenient synthesis of Glafenine and Floctafenine. *Synthesis* 54–55.

85. Boltze, K.H. and Kreisfeld, H. (1977) Zur chemie von etofenamat, einem antiphlogisticum aus der klasse der N-arylanthranylsäurederi-vate. *Arzneimit.-Forsch.-Drug Design* **27**: 1300–1312.

86. Matter, M. (1955) Polyethoxy ethers of isocyclic organic carboxylic acids. US Patent 2 714 608 (2 August 1955, to Ciba Pharmaceutical Products, Inc., Summit, N.J.).

87. Nagakawa, T., Muneyuki, T. and Mori, Y. (1961) Water solubilization by means of polyethylene glycol derivatives. Japanese Patent 7455 CA (1961) 55, 5879b.

88. Nagakawa, T., Muneyuki, T. and Mori, Y. (1960) Water solubilization of drugs. Japanese Patent 17006 (17 November 1960, to Shionogi & Co, Ltd), CA (1961) 55, 21494h.

89. Weiner, B.-Z. and Zilkha, A. (1973) Polyethylene glycol derivatives of procaine. *J. Med. Chem.* **16**: 573–574.

90. Morren, H. (1959) New derivatives of *N*-mono-benzhydril-piperazine and process for the preparation thereof. British Patent 817 231 (21 July 1959).

91. Gouin d'Ambrières, S., Lutz, A. and Gasc, J.C. (1982) Novel erythromycin A derivatives. US Patent 4 349 545 (14 September 1982, to Roussel-UCLAF).

92. Miescher, K., Fischer, W.H. and Meystre, C. (1942) 6. Über steroide (33. Mitteilung). Über glukoside des desoxy-corticosterons. *Helv. Chim. Acta* **25**: 40–42.

93. Miescher, K. and Meystre, C. (1943) 26. Über steroide. 34. Mitteilung. Über sacharide des desoxy-corticosterons II. *Helv. Chim. Acta* **26**: 224–233.

94. Meystre, C. and Miescher, K. (1944) Über steroide (35. Mitteilung). Zur darstellung von saccharidderivaten der steroide. *Helv. Chim. Acta* **27**: 231–236.

95. Higashiyama, T. and Sakata, I. (1972) Mentholglykoside, verfahren zur ihrer herstellung, ihre verwendung zur entwicklung des pfefferminzgeschmack sowie diese verbindungen enthaltende arznei-mittel. German Offen. 2 242 237 (28 August 1972, to Toyo Hakka Kogyo Kabushiki Kaishy).

96. Chavis, C. and Imbach, J.-L. (1977) Sur une méthode de pharmacomodulation:l'utilisation du vecteur sucre. In Combet-Farnoux, C. (ed.), *Actualités de Chimie Thérapeutique*, pp. 3–28. Société de Chimie Thérapeutique, Châtenay-Malabry.

97. Bognar, R. and Nanasi, P. (1953) *N*-Substituted glycosylamines derived from sulphanilamide and p-aminosalicylic acid. *J. Chem. Soc.* 1703–1708.

98. Sah, P. (1953) D-Glucuronol lactone isonicotinyl hydrazone. *J. Am. Chem. Soc.* **75**: 2512–2513.

99. Demetrio, A., Ganzina, F., Magi, M., *et al.* (1974) Anti-inflammatory di-(+)glucosamide of 1-(4-chlorobenzoyl)-2-methyl-5-methoxyindo-le-3-acetic acid. German Patent 2 223 051 (13 December 1973, to SIR Lab. Chimico Biologici SpA), CA 80, 83529e.

100. Almen, T.H.O., Haavaldsen, J. and Nordal, V. (1972) *N*-(2,4,6-Triiodobenzoyl)-sugar amines. US Patent 3 701 771 (31 October 1972, to Nyegaard & Co, A/S).

101. Hilae, S.K., Dauth, G.W., Hess, K.H. and Gilman, S. (1978) Development and evaluation of a new water-soluble iodinated myelographic contrast medium with markedly reduced convulsive effects. *Radiology* **126**: 417–422.

37

DRUG SOLUBILIZATION WITH ORGANIC SOLVENTS, SURFACTANTS AND LIPIDS

P. Heinrich Stahl

I INTRODUCTION

Drug design does not primarily include concern for a particular level neither of solubility nor of other aspects of drug delivery. In the quest for drugs with high biological activity, low solubility is rarely regarded to be an obstacle since *in vitro* testing aims for active concentrations as low as possible, at least in the micromolar but still better, in the nanomolar range. Thus, the problems emerge in full size at the time of *in vivo* studies, and in particular when safety studies in animals require high quantities to be administered. Then the formulator is left with the challenging task of optimizing the solubility for dosage forms of a new drug. For resolving this task there are two principal areas in focus: for the parenteral route of administration an injectable form, preferably as a solution, has to be provided, whereas for the peroral route provisions need to be made for appropriately fast dissolution and absorption of the drug dose. However, many drug candidates lack the desired degree of solubility in water, for which reason various techniques of solubilization are employed.

In principle there are three quite different routes to improve the solubility of sparingly soluble drug candidates.

On *Route 1*, the deficient solubility of a drug candidate is overcome by physical means ranging from solubilization by cosolvents to complex pharmaceutical–technological systems. This kind of problem-solving is handled in the realm of the pharmacist during formulation development, and is discussed in this chapter.

Route 2 is a physicochemical approach usually applied in the chemist's laboratory. It is the preparation of salts that are better soluble than the parent compound. The structure of the parent compound remains unaltered; nevertheless, new chemical entities are created by salt formation. Obviously, this option is open only to candidates bearing ionizable groups. Solubilization by salt formation is the subject of Chapter 35. Complexation is a further means of solubilization with features similar to salt formation (see Chapter 38).

Route 3 changes the compound by synthetic means. Introducing into a molecule solubilizing functional groups can form the part of a lead-optimizing program. In a prodrug approach the compound's structure is modified by covalent attachment of suitable functional groups. However, this is done in such a way that the bond is easily cleaved hydrolytically or enzymatically, and thus the parent compound is reconstituted in the body ideally after the prodrug has passed the absorption barriers. Such chemical approaches are described in Chapter 33.

The ideal degree of dispersity of a drug substance in an aqueous phase is the molecular level of dispersion, both for providing and applying drug dosage forms and for the absorption in the body. If the substance properties do not allow complete dissolution of a drug dose in the aqueous body fluids or if aqueous solutions of a required concentration have to be prepared, other degrees of dispersity and heterogeneous dispersion systems need to be utilized. An overview of the physical systems is given in Fig. 37.1.

The Practice of Medicinal Chemistry
ISBN 0-12-744481-5

631

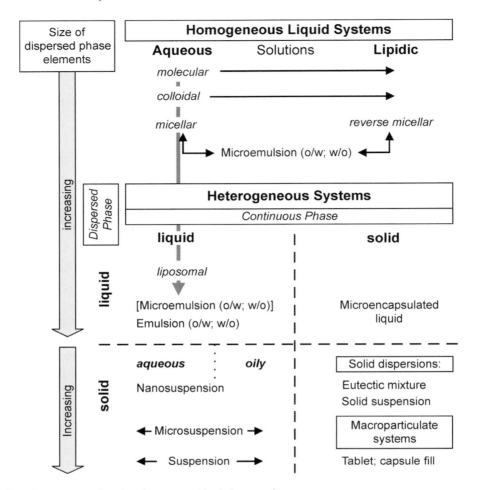

Fig. 37.1 Physical systems occurring in pharmaceutical dosage forms.

II FACTORS CONTROLLING SOLUBILITY

A The nature of drug substances

Before dealing with solubilization in the later sections, we must be aware of some qualitative characteristics, as it is necessary to know the nature of the compound before deciding on possible formulation approaches. From a pragmatic point of view, any one of the wide varieties of drug substances may be positioned in a field according to the prevailing physicochemical nature, linking qualitatively chemical structure and solubility behaviour. Figure 37.2 visualizes this perception.

Ideally, a 'well-balanced' drug molecule (position 1) as represented by, for example, antipyrine or caffeine, is hydrophilic enough to be soluble in the aqueous body fluids as well as in the pharmaceutically preferred aqueous buffers, but is also sufficiently lipophilic for permeating the biological membranes, thus having no absorption problems. A polar substance is soluble but lacks the lipophilicity for membrane permeation (position 2; examples: sorbitol and

other sugar alcohols). Likewise, the presence of an ionic charge (position 3; examples: quarternary ammonium bases, zwitterionic substances) causes poor absorption in spite of good dissolution. In contrast, truly lipophilic compounds

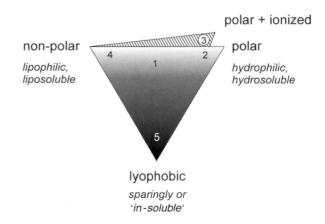

Fig. 37.2 Polarity and solubility of drug substances.

(position 4) such as the retinoids, have severe limitations in dissolving in aqueous media but do not have problems in penetrating absorption barriers. Now, current drug research turns out increasing numbers of drug candidates, which lack sufficient solubility, in both aqueous or lipid solvents. Their structures typically contain H-bond acceptors and donors responsible for highly cohesive properties of the solid state. Griseofulvin and uric acid are typical representatives of this type of lyophobic compounds.

At this point mention should be made of the Biopharmaceutics Classification System (BCS) for drug substances, suggested by Amidon and coworkers[1] and adopted by the F.D.A. in a guidance paper[2] concerning the evaluation of bioequivalence of immediate-release solid oral dosage forms. The BCS defines four classes according to the performance of drug substances after oral administration. These are highly soluble and highly permeable (Class I), low soluble and highly permeable (Class II), highly soluble and low permeable (Class III), and low soluble and low permeable (Class IV) drugs. It is easy to identify the position that the four classes would take in the triangular diagram with the position numbers discussed above.

Several parameters are suitable as measures of polarity: the dielectric constant ε, the Hildebrand solubility parameter δ and its extension, the three-component (or three-dimensional) Hansen solubility parameters, which cannot be discussed here. However, the most practical value and significance has gained $P_{octanol}$ because it is the most easily accessible parameter, and many useful correlations for its utilization in a number of relationships have been worked out in the past decades.

It is also common to view log $P_{octanol}$ as a measure of lipophilicity. This perception leads to the erroneous

expectation that a compound with a high log P value should dissolve easily in lipids. However, this is not necessarily the case as the example Table 37.1 demonstrates. Substance 2, which is the methyl homologue of substance 1, has a higher partition coefficient. Nevertheless, the calculated solubility in octanol is just about half of the lower homologue with its significantly lower partition coefficient. This is due to the much lower solubility of substance 2 in water (only a fourth of that of substance 1). Hence it is more appropriate to regard log $P_{octanol}$ as a measure of hydrophobicity, i.e. the tendency to escape an aqueous environment.

III WATER-COSOLVENT SYSTEMS

For a chemist, it is obvious to resort to water-miscible organic solvents if desired concentrations cannot be achieved with purely aqueous solvents. Cosolvency, the addition of water-miscible solvents to an aqueous system, is one of the oldest, most powerful, and most popular means of solubilization, rendering thermodynamically stable homogeneous solutions. Cosolvents are used in 13% of FDA-approved parenteral products.[3] Pharmaceutically acceptable solvents are listed in Table 37.2. Fig. 37.3 shows the structures of the solvents. Of these, the four bold face ones are used in approximately 66% of those products. They are also used in peroral dosage forms like syrups and drop solutions. The objective is to use, in a mixture with water, the minimum proportion of cosolvent necessary for increasing the solubility of a substance sufficient to accommodate the dose to be administered. Thereby a solution volume of 1 ml being the ideal for an intramuscular injection. For estimating the solubility of a given solute in cosolvent–water mixtures a relationship between suitable parameters of the solute and the cosolvent would be helpful.

A straightforward and reliable approach for selecting cosolvents and predicting their solubilization effects on drugs that requires little or no experimental data, and thus minimal time and drug, is the log-linear model proposed by Yalkowsky and coworkers.[4-6] This model describes an exponential increase in a nonpolar drug's solubility with a linear increase in cosolvent concentration. The relationship is described by:

$$\log S_{tot} = \log S_w + \sigma \cdot f_c \qquad (1)$$

where S_{tot} is the total solute solubility in the cosolvent–water mixture, S_w is its water solubility, σ is the cosolvent solubilization power for the particular cosolvent-solute system, and f_c is the volume fraction of the cosolvent in the aqueous mixture.

The σ term can be obtained from the slope of the $\log(S_{tot}/S_w)$ versus cosolvent volume fraction (f_c) profile of each selected drug and cosolvent. This implies the need for

Table 37.1 *Example of the property 'lipophilicity'*

Substance 1: R = H

Substance 2: R = -CH₃

Parameter	Substance 1 R = H	Substance 2 R = −CH₃
Solubility in water S_w (mg 100 ml^{-1}) at 25°C	210	40
log $P_{octanol}$ (pH = 7.4)	1.81	2.26
$P_{octanol}$	64.4	182
Calculated solubility in octanol (mg 100 ml^{-1}) $S_{octanol} = P_{octanol} \cdot S_w$	13.5	7.28
Ratio $S_{octanol}$ (Subst. 1 / Subst. 2)	**1**	0.54

Table 37.2 *Pharmaceutically used hydrophilic solvents*

Solvent	Remarks on use
Ethanol 96%	Moderately hemolytic cosolvent in injectables, up to 25%; in oral dosage forms.
Propylene glycol (1,2-propanediol)	In injectables up to 70%, low hemolytic activity. In oral dosage forms.
Glycerol (glycerine)	In injectables up to 30%; in oral dosage forms.
Polyethylene glycol (PEG 200, PEG 300, **PEG 400**, PEG 600)	In injectable and in oral dosage forms. Can be used undiluted even for subcutaneous administration without tissue damage.
Isopropanol	For topical (external) use only
Butylene glycol (1,3-butanediol)	Miscible with water or ethanol.
Benzyl alcohol (phenylmethanol)	In injectables up to 4%; used for its preservative and local anesthetic activity. Limited miscibility with water.
Tetrahydrofurfuryl alcohol	Moderately hemolytic cosolvent for injectables.
Glycofurol	In injectables up to 50%; moderately hemolytic.
Solketal (glycerol dimethylketal)	Moderately hemolytic cosolvent for injectables.
Glycerol formal	Moderate hemolytic activity. Mixture of isomers.
Dimethyl-isosorbide	Low hemolytic activity.
Diglyme (diethylene glycol dimethyl ether)	Low hemolytic activity.
N,N-dimethylacetamide	For injectables.
N-methyl-2-pyrrolidone	For experimental purpose and as an intermediary solvent only.
2-pyrrolidone	Occasionally in experimental use.
Dimethylsulfoxide (DMSO)	Acceptable for investigational purpose only. Although listed in U.S.P. XX for topical use, definitively not recommended for formulation of dosage forms intended for use in humans.

The most frequently used solvents are in bold print.

experimental data for each cosolvent–solute system, a time- and material-consuming prospect. However, it was demonstrated that a linear relationship also exists between σ and the logarithm of the solute's log $P_{octanol}$ (log P_{ow} in the equations).[7] Therefore σ can be related to the solute and the cosolvent by the following simple relationship:

$$\sigma = S \cdot \log P_{ow} + T \qquad (2)$$

where S and T are constants dependent only on the particular cosolvent and log P_{ow} is the solute's partition coefficient. In other words, the only *ab initio* data required to predict the solubility of a solute is the compound's octanol–water partition coefficient. Fortunately, this value can be determined experimentally or accurately predicted by a number of computational methods (ClogP® for example). A useful form of the log-linear equation is obtained by substituting σ from equation (2) into equation (1) to give:

$$\log S_{tot} = \log S_w + f_c(S \cdot \log P_{ow} + T) \qquad (3)$$

which expresses the total solubility in a mixed solvent system solely in terms of the pure components: water,

1
Tioconazole

2
Caffeine

3
Oxfenicine

cosolvent, and solute—obviating the need for any individual solute–cosolvent experiments. In Fig. 37.4, the solubility is plotted against the composition (per cent of volume) of the water–cosolvent mixtures for three drug substances representative for different polarities: tioconazole **1** (log $P_{octanol}$ = 4.52), caffeine **2** (log $P_{octanol}$ = 0.57), and oxfenicine **3** (log $P_{octanol}$ = −2.5). Several features can be derived from the graphs. (1) As the polarity of a compound increases, the solubilization by cosolvents decreases, eventually turning to *de*solvation in case of polar substances. The nonpolar tioconazole exhibits a logarithmic increase of solubility as the proportion of cosolvent increases, whereas a polar compound's solubility decreases, following

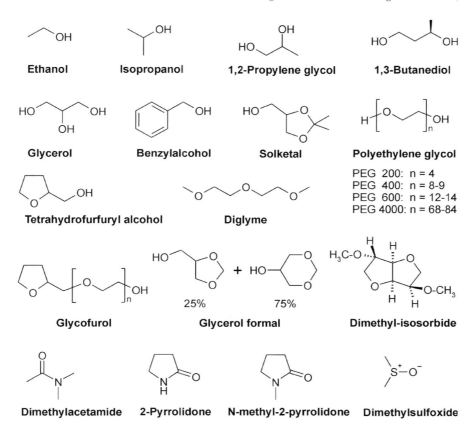

Fig. 37.3 Structures of hydrophilic solvents.

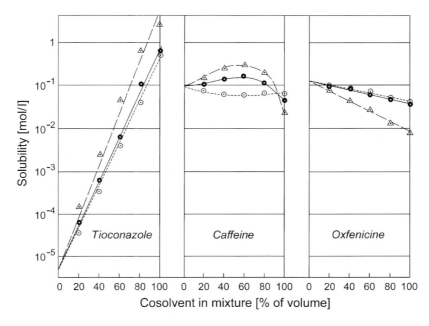

Fig. 37.4 Solubility of tioconazole, caffeine, and oxfenicine in cosolvent-water mixtures. Cosolvents are ethanol (triangles), propylene glycol (black circles) and polyethylene glycol (PEG 400; hollow circles). Redrawn from Gould *et al.*[8]

the model represented by equation (3); (2) for nonpolar solutes, the solubilization power of the cosolvents increases in the series glycerol < PEG 400 < propylene glycol < ethanol. The lower the polarity of the cosolvent, the higher is its solubilization power for nonpolar solutes. As can be seen above, this correlates well with the log P_{ow} of the cosolvents; and (3) the intermediate case of caffeine does not follow this log-linear relationship, instead generally, a parabolic course is observed with semipolar compounds. The deviation from linearity is related to the nonideality of water–cosolvent interactions. To account for such deviations, equation (3) can be empirically modified by the addition of a quadratic term to provide a more accurate description of the data, although no theoretical foundation can be given:

$$\log S_{tot} = \log S_w + \sigma' \cdot f_c - \tau \cdot f_c \qquad (4)$$

where σ' and τ are constants for each solvent. For the system propylene glycol–water the coefficients $\sigma' = \log P_{ow}$ of the solute and $\tau = 0.5$ work reasonably well. In consequence, for semipolar substances the maximum of solubilization must be sought experimentally. But at least the rank order of the degree of solubilization is reflected by the relative magnitude of σ values that may be calculated from the drug's log $P_{octanol}$ value and the solvent coefficients in Table 37.3.

As exemplified by the case of oxfenidine, the solubilization approach by cosolvent mixtures appears unattractive for polar substances simply because they are less soluble in nonpolar solvents than in water. Electrolytes are at least partly ionized, and therefore they exhibit pH-dependent solubility in aqueous media, while the fraction of the less water-soluble nonionized species may benefit from cosolvent solubilization as shown in Fig. 37.5. Although the solubility of dexoxadrol hydrochloride (see Chapter 35) in water exceeds that of the free base by more than two orders of magnitude, still an increase can be achieved by additional cosolvent (PEG 300), although the effect is much less pronounced than with the free base.

Finally, it should be noted that the log-linear model holds for mixed cosolvent systems in the following form:

$$\log S_{mix} = \log S_w + \sum (\sigma \cdot f_c) \qquad (5)$$

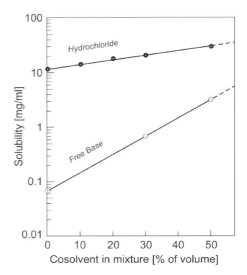

Fig. 37.5 Solubility of the free base and the hydrochloride of dexoxadrol at 23°C in a water-cosolvent system, as a function of the percentage of the cosolvent, PEG 300. Redrawn from Kramer and Flynn.[9]

where the individual solubilization powers and volume fractions of each cosolvent are linearly summed up.

Cosolvent-solubilized drug products are favoured in products for parenteral administration since they are homogeneous solutions. The compositions of three commercial injectable products are given in Table 37.4. The first product (1) has a low percentage of cosolvent in the separate solvent ampule. The drug substances are provided as a dry powder because of limited stability in solution. The second one (2) is solubilized with two cosolvents amounting to 50% of the total volume, whereas in the third product the drug dose is dissolved in a water-free mixture of 'cosolvents'. This draws attention to a further point to consider when cosolvency is employed as the formulation principle. The composition has to be adjusted in such a way that the effect of dilution of the drug solution by the aqueous body fluids (blood, tissue interstitial fluid) during administration is anticipated. In order to avoid precipitation upon injection, solubilization needs to be maintained until the drug substance is diluted to its final concentration in the circulating bloodstream or in an infusion fluid, if the product is added to an intravenous infusion.[10] Otherwise, there is the risk that precipitated drug particles could mechanically or chemically irritate tissue, or even clog blood vessels.

IV SOLUBILIZATION MEDIATED BY SURFACTANTS

Substances with polar and nonpolar moieties in their molecules, imparting surface-active properties, are

Table 37.3 *Coefficients of cosolvents for calculating the solubilization power σ according to equation (3)*

Cosolvent	Cosolvent log $P_{octanol}$	S	T
Ethanol	−0.31	0.93	0.38
Propylene glycol	−0.92	0.76	0.57
Polyethylene glycol 400	−0.88	0.73	1.19
Glycerol	−1.96	0.34	0.29

Taken from reference 35.

Table 37.4 *Composition of cosolvent-solubilized drug products*

	Component	Amount
(1) Librium ® for i.m. injection		in 2 ml
In 5-ml dry filled ampoule:	chlordiazepoxide · HCl	100 mg
In 2-ml solvent ampoule:	maleic acid	1.6 %
buffering agent	NaOH	to adjust pH
solvents		
– solvent and surfactant	polysorbate 80	1.5 %
– solvent and preservative	benzyl alcohol	4 %
– solvents	propylene glycol	20 %
	water	to volume
(2) Bactrim ® solution for i.v. infusion		in 5 ml
Active substances	trimethoprim	80 mg
	sulfamethoxazole	400 mg
Preservative	benzyl alcohol	1 %
Preservative	ethanolamine	0.3 %
Antioxidant	sodium metabisulfite	0.1 %
Solvents	propylene glycol	40 %
	ethanol	10 %
	water	to volume
(3) Ativan ® for i.m. injection		in 1 ml
Active substance	lorazepam	2 or 4 mg
solvent	PEG 400	0.18 ml
solvent and preservative	benzyl alcohol	2%
solvent	propylene glycol	to volume

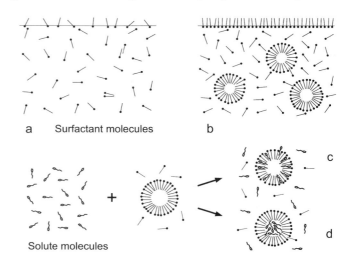

Fig. 37.6 Formation of surfactant micelles and micellar solubilization.

amphiphilic (or amphipatic). In aqueous solution they are not randomly distributed. Instead, they adopt a specific orientation at interfaces according to the nature of the polarity of the molecules and that of the interface. At the air–water interface their polar region is oriented toward the aqueous surface, while the nonpolar region is pointing outwards into the low-polar gas phase (Fig. 37.6a). The nonpolar region attaches to the surface of suspended hydrophobic particles while the polar group mediates the intimate contact to the aqueous bulk phase. At low concentrations, surfactants are therefore used as wetting agents. As the concentration in aqueous solution exceeds a critical value, the critical micelle concentration ('CMC') surfactants assemble cooperatively to form soluble structured aggregates (Fig. 37.6b). The CMC is a characteristic of a particular surfactant and depends on environmental factors like temperature, pH, ionic strength and others. Surfactant micelles are in a dynamic equilibrium with the surfactant monomers. Therefore the number of molecules constituting a micelle is distributed around an average value over time and over a population of micelles. For example, the average aggregation number $\bar{n}$ in micelles of sodium laurylsulfate is 64 at 25°C; for other surfactants this number varies over orders of magnitude from 2–8 (bile salts) to many thousands. With $\bar{n}$ as that of sodium laurylsulfate the shape of micelles is likely to be globular; as $\bar{n}$ increases the shapes will tend to become disks, rods or even vesicles. Surfactant micelles can accommodate solute molecules in a number of orientations, arranged according to their polarity and hydrophobicity: (1) at the interface between the micelle surface and the aqueous bulk phase; (2) between the polar hydrophilic head groups; (3) within the hydrophobic part (the palisades layer) of the micelle shell (Fig. 37.6c); or (4) in the core (Fig. 37.6d). This is the mechanism of micellar solubilization.

The totality of micelles represents a colloidal phase, into which a substance present in the aqueous phase partitions. The capacity of the micellar phase to solubilize a solute can therefore be expressed as a partition coefficient K_m. Hence a linear relationship can be expected between the concentration S_{mic} of substance solubilized by micelles and the concentration of the surfactant C_{sfc} in the system. Because only micelles contribute to the solubilizing effect but not the monomeric surfactant molecules, the critical micelle concentration C_{cmc} must be subtracted from the total of surfactant concentration. The resulting total solute concentration of solute in the micellar solution is then:

$$S_{tot} = S_w + S_{mic} \qquad (6)$$

where

$$S_{mic} = K_m \cdot (C_{sfc} - C_{cmc}) \qquad (7)$$

and

$$K_m = S_{mic}/S_w \qquad (8)$$

The solubilization of chloramphenicol with one member each of four classes of surfactants is shown in Fig. 37.7, while Table 37.5 gives the respective parameters. The typically linear relationship in micellar solubilization is obvious in the graph Fig. 37.7. In Table 37.5, the respective data reflecting the solubilization power of the surfactants are presented.

For the characterization of surfactants the hydrophile/lipophile balance system has been devised by Griffin,[12] originally meant for selecting nonionic surfactants as emulsifiers for stabilizing oil-and-water emulsions. According to the ratio of polar and nonpolar groups in the molecule, a surfactant is assigned a number, the HLB value, on a scale originally ranging from 1 to 20 with the most lipophilic surfactant at the low end. The more hydrophilic a surfactant is the higher its HLB value. Figure 37.8 gives an overview of the use-assignment of the HLB classification of surfactants and also the HLB-related visual impression of surfactant suspensions in water. For inclusion of ionic surfactants the scale was later extended beyond 20. The HLB values correlate with polarity-indicating parameters like the $\log P_{octanol}$ and the dielectric constant. Of the countless surfactants available, the pharmaceutically most frequently used ones are listed in Table 37.6 by their HLB values, and their structures shown in Fig. 37.9.

Two compositions of the liposoluble vitamin K compound phytomenadione are listed in Table 37.7. In (4) a nonionic surfactant is used, in (5) lecithin-bile acid mixed micelles are the solubilizing principle.

Due to their strong interfacial activity, surfactants can elicit adverse reactions in biological systems. Cell membrane constituents (phospholipids, cholesterol, etc.) can be solubilized at supermicellar concentrations. This impairs the integrity of membranes whereby their permeability for drugs but also for other substances can be enhanced. At low concentrations such alterations are reversible and membranes recover rapidly, but higher concentrations can cause irreversible damage to mucosal and other tissue. Some of the polyglycolized surfactants used as solubilizers in parenteral and oral applications, in

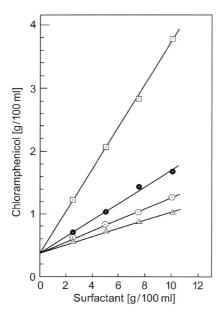

Fig. 37.7 Solubilization of chloramphenicol by surfactants. Squares, cetrimide; black circles, sodium laurylsulfate; empty circles, polysorbate 80; polyethylene glycol 40 monostearate, triangles. Compiled and redrawn from Aboutaleb and Abdelzaher.[11]

Table 37.5 *Solubilization of chloramphenicol by surfactants*

Surfactant	Solubility [g per g solubilizer]		K_m (g g^{-1})	
	25°C	35°C	25°C	35°C
Cetrimide	0.3386	0.393	96	82
SDS	0.135	0.2263	40	50
Polysorbate 80	0.0875	0.1224	24	25
Myrji 52	0.0634	0.082	19	19

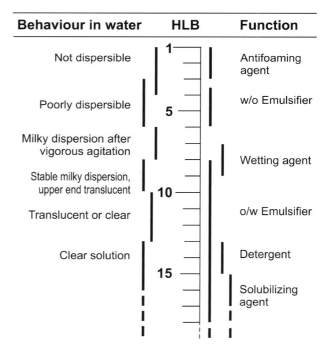

Fig. 37.8 Use and behaviour in water of surfactants, according to HLB ranges.

Table 37.6 *Selected surfactants used in formulations*

Surfactant[a]	Chemical name or other description	Type	HLB	CMC	Use
Glycerol monostearate		N	3.8		po, par
Span 80	sorbitan mono-oleate	N	4.3		po, ex
Lecithin	from soybeans or egg yolk	Z	7–10		po, par, ex
DPPC	dipalmitoyl phosphatidylcholine			0.47 nM	
Cremophor EL	polyethylene glycol 35 castor oil	N	12–14	0.02 wt.%	po, par
Cremophor RH 40	polyethylene glycol 40 hydrogenated castor oil	N	14–16		po, par
Solutol HS 15	polyethylene glycol 660 hydroxystearate, mono- and di-esters	N	14–16	0.005–0.02 wt.%	po, par
Vitamin E TPGS	D-α-tocopheryl polyethylene glycol 1000 succinate	N	15–19	0.02 wt.%	po, par
Tween 80[b]	polysorbate 80, polyethylene glycol 20 sorbitan mono-oleate	N	15	20 mM	po, par
Myrji 52[b]	polyethylene glycol 40 monostearate	N	16.9		po, ext
Brji 35[b]	polyethylene glycol 23 lauryl ether	N	16.9	0.09 mM[c]	po, ext
Bile salts	Deoxycholate Na	A	20–25	1.5 mM[c]	po, par
	Taurocholate Na			3.3 mM[c]	
	Taurodeoxycholate Na			2.7 mM[c]	
	Glycocholate			7.1 mM[c]	
Pluronic F-68	polyethylenglycol-polypropylenglycol block copolymer	N	29	190 mg l^{-1}	po, par
Sodium laurylsulfate	Dodecylsulfate sodium	A	40	8.2 mM	po, ext
Docusate sodium	Dioctylsulfosuccinate sodium	A	40	2.5 mM	po, ext
Cetrimonium bromide	Cetyltrimethylammonium bromide	C	40	1 mM[c]	ext

[a]Structural formula in Fig. 37.9.

[b]Typical members of large families of surfactants with the number of polyoxyethylene units and the ester acyl moieties varied.

[c]50 mM Na$^+$ A, anionic; C, cationic; N, non-ionic; Z, zwitterionic; ext, topical; par, parenteral; po, oral applications.

particular Cremophor EL, were reported to occasionally elicit anaphylactic reactions.[13–15] Cationic surfactants must never be used in parenterals; they have their own pharmacological activity and are bactericidal, for which reason they are used as antiseptics and disinfectants in topical and other external preparations only. Lecithins as the constituents of all biological membranes are the surfactants and solubilizers tolerated best in parenteral and peroral applications.

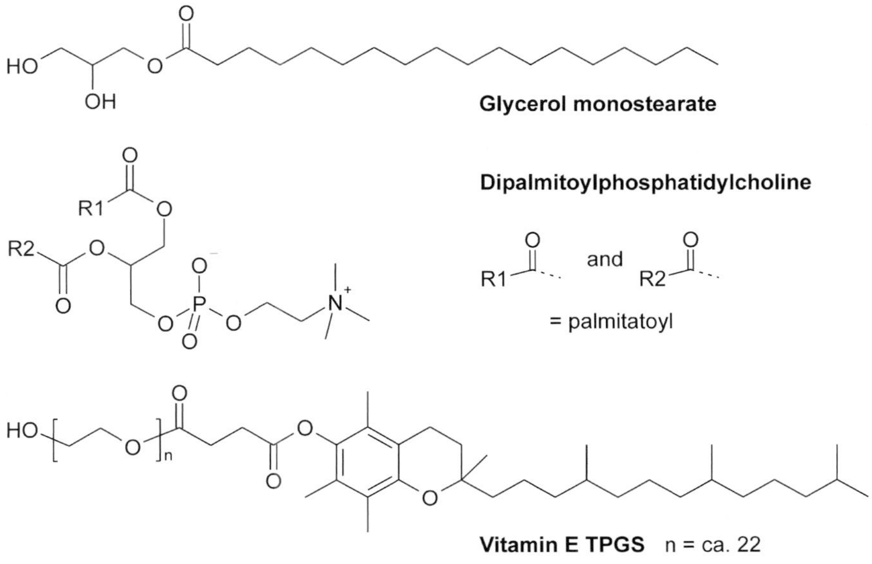

Fig. 37.9 Structures of surfactants. Note: The 'Tweens' are derivatives of the same sorbitol anhydride ring structures including isosorbide (not shown) as shown for Span 80 (*Continued over the page*).

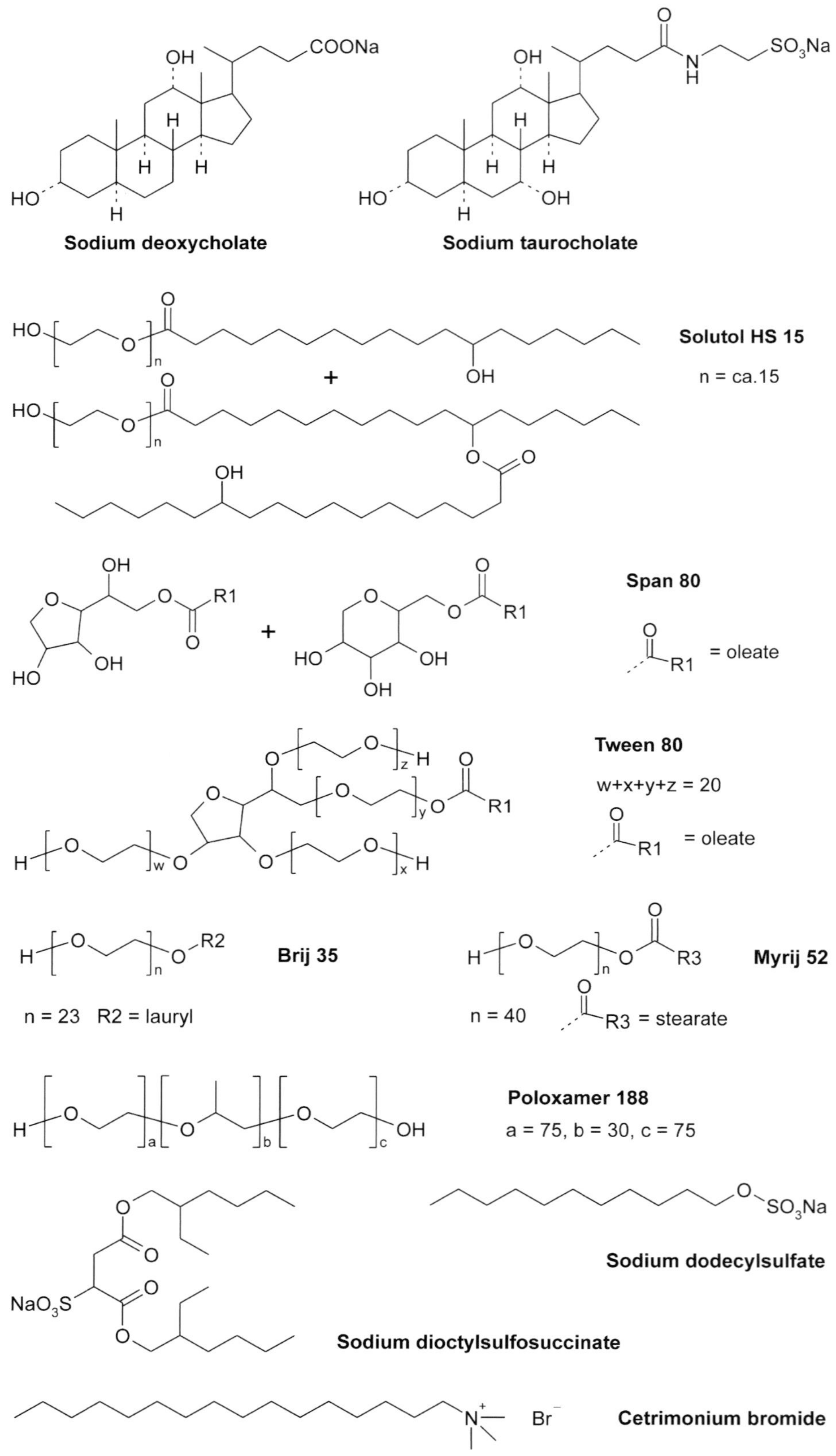

Sodium deoxycholate

Sodium taurocholate

Solutol HS 15

n = ca.15

Span 80

$\overset{O}{\underset{\cdot\cdot}{\parallel}}R1$ = oleate

Tween 80

w+x+y+z = 20

$\overset{O}{\underset{\cdot\cdot}{\parallel}}R1$ = oleate

Brij 35

n = 23   R2 = lauryl

Myrij 52

n = 40 $\overset{O}{\underset{\cdot\cdot}{\parallel}}R3$ = stearate

Poloxamer 188

a = 75, b = 30, c = 75

Sodium dodecylsulfate

Sodium dioctylsulfosuccinate

Cetrimonium bromide

Fig. 37.9   (*continued*)

Table 37.7 *Compositions of a surfactant-solubilized drug product*

(4) AquaMephyton® solution for s.c./i.m. injection		in 1 ml
Active substance	phytomenadione	2 or 10 mg
Solubilizer	polyoxyethylated fatty acid derivative	70 mg
Preservative	benzyl alcohol	0.9%
To adjust isotonicity	dextrose	37.5 mg
Solvent	water	to volume
(5) Konakion® for i.m. injection		**in 1 ml**
Active substance	phytomenadione	2 mg
Solubilizer	soy lecithin glycocholic acid	
NaOH/HCl		to adjust pH
Solvent	water	to volume

In order to keep the use concentrations low, the solubilizing power of one surfactant is rarely utilized alone. Thus, combinations of two or more surfactants are used to optimize the solubilization effect by the formation of mixed micelles, especially with lecithin and bile acids. For injectable solutions mostly several solubilization principles are combined as can be seen in the examples. By employing pH adjustment and 5% sodium dodecyl sulfate, the solubility of the investigational cancer drug pyroxamide could thus be increased by a factor of nearly half a million over its intrinsic solubility.[16]

Although the micellar solubilization can be demonstrated *in vitro* there is no warrant for a corresponding enhancement of bioavailability by the oral route in any case. With particular combinations of drug substance and surfactant, even reduced instead of improved bioavailability has been observed. Such effects depend on the concentration of the surfactant and may have various reasons such as shift of equilibrium in favour of micellar inclusion or formation of other types of drug–surfactant aggregates.

The presence of hydrophilic and hydrophobic regions in the same molecule is not a monopoly of surfactants: most drug molecules feature hydrophilic and hydrophobic regions as well, and this is indeed the structural requirement for their ability to permeate cell membranes and pass absorption barriers. The amphiphilic properties are obvious in some drug classes, e.g. the local anaesthetics.[17] On some occasions a drug substance may exceed the CMC, whereupon micellar 'self-solubilization' can occur. A recent example was described by Hussain and coworkers[18] who found that the hydrochloride of the basic analgesic DuP 747 is surface active; the saturated solution (3 mg ml^{-1} at 22°C) lowered the surface tension of water to 50 dyn cm^{-1}. It was hoped

that a different salt might be soluble enough to reach higher concentrations exceeding a CMC. This was indeed the case: the CMC of the methanesulfonate was found to be in the region of 4 mg ml^{-1} and a micellar solubility of 60 mg ml^{-1} could be achieved.

V SOLUBILIZATION BY LIPID VEHICLES

Liposoluble drug substances like the fat soluble vitamins and steroids can be dissolved in lipid vehicles like natural triglycerides (i.e. purified fatty oils), semi-synthetic triglycerides like the mean-chain triglycerides (e.g. Miglyol 812 consisting of the $C_8 — C_{12}$ fatty acid triglycerides) and synthetic fatty acid esters (e.g. isopropyl myristate) for oral and parenteral administration, though restricted to subcutaneous and intramuscular injections. When given intravenously, oils would form large droplets leading to fatal fat embolism, therefore homogeneous lipid phases must not be injected intravasally, i.e. into veins and arteries. However, this problem can be circumvented if the oil is finely dispersed to a droplet size that can pass through narrow capillaries.

A Emulsions

Emulsions are heterogeneous dispersions of immiscible liquids; pharmaceutically both types, oil in water (o/w) and water in oil (w/o), are of interest. Mechanical work is required to break up the liquid to be dispersed to small droplets. Thermodynamically emulsions are unstable systems, because the interfacial tension between the two liquids causes droplets of the disperse phase to coalesce, approaching the state of complete phase separation. To counteract this tendency an emulsifying agent must be added that occupies the interfaces between dispersed droplets and bulk liquid thereby lowering the interfacial tension.

Whereas emulsion systems play their most prominent role in products for topical applications (creams, ointments), they are also prepared for peroral and even parenteral administration. The typical composition of a nutritional fat emulsion for intravenous administration is 200 g soybean oil or safflower oil, 12 g egg phospholipids and 25 g glycerol per litre emulsion with the pH adjusted in the range from 5.5 to 8. Most important is the fat particle size that must not exceed 0.5 μm. For experimental purpose, such fat emulsions (e.g. Abbolipid®, Intralipid®, Lipofundin®) may be used as vehicles for lipophilic drug candidates. However, depending on the drug properties, emulsion formulations require optimization by tailoring the oily phase and emulsifiers. As an example, for the intravenous administration of the poorly soluble antimitotic agent taxol, large volumes of a micellar and cosolvent-mediated dispersion consisting of Cremophor EL + ethanol + isotonic saline (5 + 5 + 90) have to be infused to reach the therapeutic dose of 30 mg because the solubility of the drug is only 0.6 mg ml^{-1} in that vehicle. Intralipid as an infusion vehicle was ruled out (solubility of taxol in soybean oil is 0.3 mg ml^{-1}). Therefore, an emulsion was developed based on triacetin as the oil phase (solubility of taxol in triacetin is 75 mg ml^{-1}). An optimized taxol emulsion consists of 1.5% soy lecithin, 1.5% poloxamer 188, 2% ethyl oleate, 50% triacetin, and 10 mg taxol ml^{-1}.[19]

It has been demonstrated that the gastrointestinal absorption of highly lipophilic substances can be enhanced impressively when given in lipid vehicles by entering 'piggy-back' the lymphatic route of fat absorption where bile acids are the potent natural solubilizers. Charman and Stella[20,21] have suggested that drugs require a log $P_{octanol}$ in excess of 5 and a triglyceride solubility of at least 50 mg ml^{-1} before mesenteric lymphatic transport is likely to become a major contributor to bioavailability. Simple solutions of the antimalarial halofantrine hydrochloride (solubility in water: approximately ~1 μg ml^{-1}, calculated log P = 8.5) in triglycerides increased the lymphatic uptake of the drug compared with an aqueous suspension. The long chain glycerides (peanut oil) were most effective and surpassed by far the medium-chain triglycerides (fractionated coconut oil) and the still less effective tributyrin.[22]

B   Microemulsions

While surfactants in submicellar concentrations, are used to promote the dissolution of solid drug particles by wetting surfaces and thus making them accessible for the aqueous solvent phase, they act as micellar carriers in concentrations above the CMC as described. Moreover under certain conditions, in still higher concentrations they are essential constituents of a particular type of liquid system, the microemulsions. However, other than the term 'microemulsion' would suggest, such systems have little in common with ordinary (or macro-) emulsions. Some of the characteristics and essential differences are listed in Table 37.8. Preparing a microemulsion requires four components: (1) an oil phase; (2) an aqueous phase; (3) a surfactant with a high HLB; and (4) a co-surfactant with a lower HLB. As the microemulsion state is existent only within certain limits of mixing ratios of the constituents, it is necessary to search for those domains of existence. For this purpose, the observations made when the ratios of the four components are systematically varied, would have to be entered into a three-dimensional space within a tetrahedron (Fig. 37.10). However, this would be difficult to handle. Instead, the system is reduced to three components by keeping the percentage of one component constant, or setting a fixed ratio of two of the components, e.g. the surfactant/co-surfactant ratio, a procedure that also reflects the experimental search strategy. For each surfactant/co-surfactant ratio a so-called pseudoternary diagram is then established, and with a series of such reduced diagrams the whole mixing space can be described. Such diagrams reveal the extreme complexity of quarternary systems with areas of o/w emulsions, lamellar phases, liquid crystalline, micellar, reverse micellar phases; phases with two (o/w emulsions; w/o emulsions) and with three coexisting phases, and isotropic phases with o/w microemulsions and with w/o microemulsions. The elaboration of microemulsions is not easy, but can be quite rewarding. Figure 37.11 shows the relevant fraction of a pseudoternary diagram. It describes the microemulsion domain of an experimental system composed of polyoxyethylene-7 lauryl ether as the surfactant, decanol as the co-surfactant, dodecane and water, and tetracycline as the model drug. The three different systems prepared from

Table 37.8   *Characteristics of emulsions and microemulsions*

Property	Emulsions	Microemulsions
Thermodynamic stability	Not stable; can be made sufficiently stable in practical terms	Stable
Effect of temperature changes; freeze–thaw cycles	Will break the emulsion irreversibly	System may undergo changes of state, but reverts to initial state after return to the outset condition
Appearance	Opaque, milky, translucent	Clear, opalescent
Formation	Require input of mechanical work for size reduction of the phase to be dispersed (homogenization)	Form spontaneously
Particle size	> 300 nm	3–300 nm

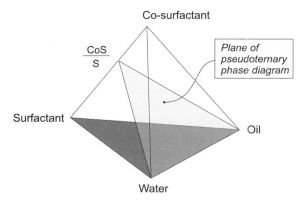

Fig. 37.10 Three-dimensional space within a tetrahedron for representing 4-component compositions.

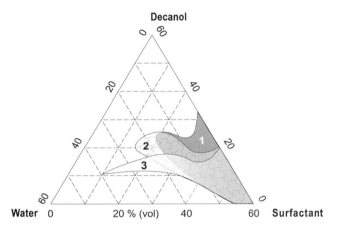

Fig. 37.11 Pseudoternary phase diagram of the system water + dodecane + polyoxyethylene-(7)-lauryl ether + decanol, where the dodecane fraction is maintained at a constant 40% (v/v), for which reason the scales run from zero to 60% instead of 100% of a full diagram. The areas labelled 1, 2 and 3 delineate the domain of the oil-continuous isotropic microemulsion at 20, 25 and 35°C, respectively. Redrawn from Ziegenmeyer and Führer.[23]

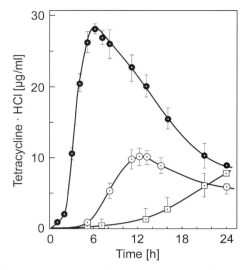

Fig. 37.12 *In vitro* skin penetration of tetracycline hydrochloride from three preparations based on the same content of dodecane and water but differing in the proportion of decanol. Black circles, microemulsion, hollow circles, gel phase, squares, cream-like phase. Redrawn from Ziegenmeyer and Führer.[23]

Drug-carrying microemulsions so far have been used in the topical and transdermal field, but with the natural phospholipids the principle can be applied even to parenterally administered products. The breakthrough in peroral application was achieved with the successful attempts of Kissel and Posanski[24] to increase the bioavailability of cyclosporin A by the microemulsion principle. A solution formulation for an oral microemulsion with a drug content of 10% cyclosporin A is composed of corn oil mono-, di-, and triglycerides, polyethylene glycol-40 hydrogenated castor oil, propylene glycol and anhydrous ethanol. This product, replacing the former SEDDS (see below), is actually a concentrate which upon dilution with an aqueous phase generates the effective microemulsion.

This principle of anhydrous concentrates was named 'SMEDDS', self micro-emulsifying drug delivery systems. Such formulations lack the aqueous phase. On dilution, a SMEDDS spontaneously converts to an optically clear, thermodynamically stable microemulsion, which contains the drug in molecular dispersion. The same principle of a water-free concentrate which leads to a 'macro'-emulsion is called SEDDS, a self-emulsifying drug delivery system. A recent review on self-dispersing lipid based systems was published by Gershanik and Benita.[25]

C Liposomal systems

A class of systems that can be differentiated by their particular morphology from other lipid-containing drug

the same components but in different physical states, the microemulsion produced the highest skin penetration *in vitro*, as can be seen in Fig. 37.12.

The pharmaceutical significance of microemulsions lies in the fact that they are not merely solvents that solubilize drug substances hard to dissolve in other solvents, but serve as systems that impart high penetration power to a drug product. For this reason they have gained increasing interest. Limiting factors for their use on a broad scale is the need for well-tolerated components, particularly surfactants and cosurfactants whose concentrations are higher than in ordinary emulsions. Furthermore, thermodynamic stability must be maintained over a temperature range between 4°C and 40°C.

carrier systems are the liposomes: globular in shape, they consist of one (unilamellar vesicles) or several (multilamellar vesicles) lipid double layers of natural or synthetic phospholipids. Because the phospholipids are amphipathic in aqueous media, their thermodynamic phase properties and self-assembling characteristics evoke entropically driven assembly of their hydrophobic regions into spherical bilayers. Those layers are referred to as *lamellae*. As in natural vesicles, cholesterol may contribute to the rigidity of liposomes. The size and morphology of liposomes varies depending on the lipids used and the technique applied for producing them. So the terms *large unilamellar vesicles* (LUV), *small unilamellar vesicles* (SUV), and *multilamellar vesicles* (MLV) are the terms addressing the morphology in brief. The size range of MLVs is $0.1-5.0$ μm, of SUVs $0.02-0.05$ μm, and LUVs range up from 0.06 μm.

Liposomes provide a number of important advantages over other dispersed systems. They can encapsulate drugs more firmly than micelles do, as their structure is more persisting. Therefore they must be regarded as drug delivery systems, rather than simple vehicles. Liposomal preparations can replace some commercial products containing less well-tolerated solubilizing agents, thus providing useful alternative dosage forms for intravenous administration. Due to the lipid nature of the liposomes the body distribution of the drugs can be quite different from the non-encapsulated drug. In fact, the tolerability of certain drugs can be improved.

There are numerous options for combining phospholipids with an aqueous phase, but two major methods are used to make liposomal systems for drug delivery. The first is simple hydration (swelling) of the phospholipid. This is followed by high-intensity agitation using sonication or a high-shear impeller. Liposomes are then sized by filtration, extrusion or elutriation. The second method is an emulsion technique. Phospholipid is first dissolved in an organic solvent (typically methylene chloride). The solution is added to an aqueous medium with vigorous agitation. Then the organic solvent is removed under reduced pressure. The resulting liposomal dispersion is also sized by filtration or extrusion. In general, the first method yields multilamellar products, while the second yields products with few lamellae. Although the techniques are simple in principle, the development of a drug product requires many steps from the selection of the lipid composition for the drug substance, studies on the amount of drug included into the liposomes, separation of non-encapsulated drug, studies on the rate of drug leaking out of the liposomes. These steps have to be repeated until the optimal composition, size and size distribution of the liposomes and the tight control over the many parameters of the production technique are established.

The surprisingly simple compositions of liposomal products do not reflect the enormous difficulties of their development and production. Examples of drugs available as injectable liposomal products are amphotericin B and doxorubicin. The composition of a lyophilized liposomal amphotericin B product (7) is given in Table 37.9 next to a product (6) which upon reconstitution results in a micellar dispersion. A third liposomal product of the same polyene antifungal antibiotic is provided as an opaque solution with two phospholipids in a 1:1 drug-to-lipid molar ratio. The two phospholipids, L-α-dimyristoylphosphatidylcholine (DMPC) and L-α-dimyristoylphosphatidylglycerol (DMPG) are used in a 7:3 molar ratio.

The potential of liposomal systems but at the same time the compexity of such a development and production process leading to a useful product is illustrated by the solution of the challenging task to convert an insoluble dye pigment to a clear liposomal solution, suitable as a photosensitizer for subcutaneous injection in the photodynamic therapy of skin

Table 37.9 *Composition of a micellar-solubilized (6) and a liposomal (7) amphotericin-B injectable product*

(6) Constituents	Amphotec ® (Sequus)	(7) Constituents	Ambisome ® (Fujisawa)
Amphotericin B	50 mg	Amphotericin B	50 mg
Sodium cholesteryl sulfate	26.4 mg	Hydrogenated soy phosphatidylcholine	213 mg
Tromethamine	5.64 mg	Cholesterol	52 mg
		Distearylphosphatidylglycerol	84 mg
Di-sodium edetate dihydrate	0.372 mg	α-Tocopherol	0.64 mg
Lactose	950 mg	Sucrose	900 mg
HCl	to adjust pH	Disodium succinate hexahydrate	27 mg
To be reconstituted with sterile water for injection	10 ml	To be reconstituted with sterile water for injection	12 ml
		pH (reconstituted)	5.0–6.0

neoplasms. The flow sheet (Fig. 37.13) shows the route of preparation.[26] An additional problem was to keep control of the state of aggregation of the dye which has a high tendency to form molecular aggregates. It requires monomeric zinc phthalocyanine to yield the maximal dye concentration in the tumour tissue and to reduce the load in the liver. Therefore the lipid composition and the process had to be tuned for encapsulation of the dye as monomers and to avoid the formation of aggregates.[27]

In 1975, Papahadjopoulos *et al.*[28] reported on their observation of the formation of large cylindrical or cigar-shaped structures (200–1000 nm), when Ca^{2+} was added to suspensions of unilamellar vesicles (liposomes, 20–50 nm) of phosphatidylserine. These multilamellar structures, named cochleates, consist of many layers of sheets of phosphatidylserine bilayers rolled in spiral configuration, the bilayers being cross-linked by the bivalent calcium ions bridging the phosphate groups of the adjacent layers. Recently, cochleates have been used to trap drug substances for very stable formulations of antifungal antibiotics and vaccines offering distinct advantages over liposomal delivery systems.[29,30]

Being the best-tolerated surfactants, the purified natural and semi-synthetic phospholipids turn out as the most versatile materials for solubilizing poorly soluble drug substances and to promote their absorption. Their use ranges from micellar solubilization agents over emulsifiers to the self-organized liposomal systems and microemulsions. Recently a technology has emerged that utilizes specific phospholipid mixtures, derived from membrane lipids, as matrix-forming carriers tailored for the particular drug substance. Such formulations can combine high drug loads with simple technology (co-evaporation) and result in a high bioavailability of the respective drug. In a study in dogs with cyclosporin-A dissolved in two semisolid lipid matrices (lipid:drug weight ratio = 5:1, in contrast to a liposomal formulation where this ratio would be 20:1), the bioavailability relative to a SMEEDS was 122 and 131%, respectively.[31]

VI OTHER TECHNIQUES TO IMPROVE SPARINGLY SOLUBLE DRUG SUBSTANCES

A Nanosuspensions

An alternative to any type of solution is to keep the drug substance in a particulate solid state. This is the normal formulation approach taken for solid dosage forms for oral administration. As long as the rate of dissolution is sufficient to release and dissolve the drug dose in the gastric and intestinal fluids, no problems will arise. The rate of dissolution J of a solid depends mainly on two parameters, namely the solubility C_s in the aqueous phase and on the specific surface area A of the drug dose that is exposed to the solvent according to the Nernst–Brunner modified Noyes–Whitney equation:

$$J = -\frac{dM}{dt} = k \cdot A(C_s - C_t) \qquad (10)$$

where dM/dt is the mass dissolved per unit time, k the permeability of the diffusion boundary layer covering the particle surface, and C_t the concentration in the bulk solution at time t. Comminution of a substance increases the specific surface area, for which reason sparingly soluble drugs are milled or even micronized, i.e. ground by means of an air jet mill to a median of particle size distribution of 5–10 μm.

Measures need to be taken so that agglomeration does not take place and that a solid formulation (tablet mass or capsule fill) will become fully wetted when it is exposed to aqueous fluids.

For milling solids in air by mechanical grinding or by air-jet mills (used for 'micronization') a lower limit of about 2 μm is given by re-agglomeration counteracting the comminution. A further degree of comminution can be achieved only if the solid is suspended in a liquid wherein,

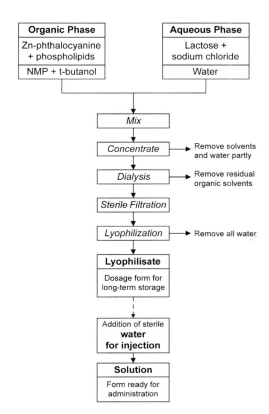

Fig. 37.13 Preparation of injectable liposomal zinc phthalocyanine.

after any mechanical impact, fragmented particles are immediately separated by the liquid. Wet-milling can be performed in pearl-mills where small beads made from such different materials as glass, zirconium oxide or hard polystyrene are agitated by high-speed stirring blades. While the beads collide, they hit and shatter suspended drug particles. The degree of comminution depends on the duration of the action, which is measured in hours. Thereafter the milling beads are separated from the suspension by filtration. For experimental purposes, the pearl-mill technique can be scaled down to a volume of about 10 ml.

A different technique is the piston-and-gap homogenization in the so-called 'French press', where the particle suspension is squeezed under high pressure (100 to 1500 bar) through a narrow gap of about 25 μm. The acceleration of the liquid to extremely high velocity causes cavitation and creates shear forces high enough to disintegrate solid particles to fragments of submicron size. In order to arrive at a desired size and size distribution of particles a batch is run through several homogenization cycles.

A third technique is the 'Microfluidizer' system. Again driven by high pressures up to 2750 bar, the particle suspension passes through an interaction chamber without any moving parts. The interaction chamber has bored channels whose geometry causes the flow of fluid to pass zones with high shear alternating with zones of high impact. Various interaction chambers with the flow geometry adapted to different tasks of comminution or emulsion homogenization are available.[32]

These techniques of size reduction require intensive cooling to dissipate the heat that is created by the high input of mechanical energy. Suspensions with particle sizes between 200 and 800 nm can be achieved. Agglomeration of particles is suppressed by the addition of surfactants such as polysorbate 80 or lecithin. Furthermore, the additives and a narrow size distribution help in minimizing particle growth, the so-called Ostwald ripening, whose driving force is the higher solubility of submicron particles as described below.

Unlike ordinary suspensions with particle sizes down to a few μm, nanosuspensions deserve attention not only for the enormous increase of the specific surface area, which brings about a high dissolution rate upon dilution. Moreover, the extremely small particles have a solubility exceeding the common solubility C_s in equilibrium with macroscopic particles. As the diameter of particles falls below 1 μm, the solubility increases with shrinking particle size (although normally perceived as a substance characteristic, for a given solvent composition being dependent only on temperature) as described by the Kelvin equation:

$$C_{s,particle} = C_s \cdot e^{\left(\frac{2\gamma M}{r\rho RT} \right)} \tag{11}$$

where $C_{s,particle}$ is the solubility of the small drug particle with the radius r, the density ρ and the molecular mass M, and γ is the interfacial tension between drug particle and the solvent surrounding the particle. The gain in solubility adds to the enlarged specific surface area for accelerated drug dissolution, which then produces a higher bioavailability of the sparingly soluble drug substance.

Nanoparticles can be used for oral products, for the various parenteral routes of administration, for topical use and for pulmonal inhalation as aerosol after nebulizing.

In vivo studies proved the excellent performance of drug nanoparticles. For example, in humans the oral administration of the analgesic naproxene as drug nanoparticles led to an AUC (0–2 h) of 97.5 mg h l^{-1} compared with just 44.7 mg h l^{-1} for a commercial suspension and of 32.7 mg h l^{-1} for the same drug as a tablet product. The corresponding t_{max} values were 1.96 h for the nanoparticles, 3.33 h and 3.20 h for the two commercial products. In a canine study, oral administration of the gonadotropin inhibitor danazol as a suspension of nanoparticles led to an absolute bioavailability of 82.3%, whereas with the conventional dispersion only 5.2% was achieved.

B Hydrosols

A different approach to suspensions of colloidal particles of poorly water-soluble drugs in the size range 1 nm–1 μm can be taken by rapid precipitation from homogeneous solution. Organic solutions of substance with very low water solubility are very rapidly introduced into an excess of water. The resulting hydrosols are colloidal aqueous dispersions, which require additives for stabilization, e.g. poloxamer 188 or modified gelatin products as used in plasma volume expanders. The solvents must be removed by freeze drying or spray drying.[33]

C Solid dispersions

A further means to generate very small particles of a sparingly soluble substance are the solid dispersions. Their field of application are dosage forms for oral administration. Such systems contain a drug substance in a solid carrier and can represent different physical states in a solid matrix: eutectic mixtures, solid solutions (i.e., solid solute in a solid solvent), glass solutions, glass suspensions, or they contain amorphous drug in crystalline matrix. Solid dispersions are prepared by either melting the components or dissolving the drug substance in the melt of the carrier, or by evaporating a solution of the combined components to dryness ('co-evaporates'). The following have been used as carriers: polyols (pentaerythrol), sugars (lactose, glucose, sucrose, trehalose), acids (adipic, ascorbic, citric, succinic and bile acids), urea, nicotinamide, acetamide; polymeric hydrophilic materials (polyethylene glycols, polyvinylpyrrolidone), lipids, and blends of such materials. When exposed to an aqueous phase, the matrix of a hydrophilic solid dispersion dissolves and releases the poorly soluble drug as a suspension of very small particles that are finer than can be achieved by micronization. From a real solid solution, or if solubilizing matrix components have been chosen, there is the chance to achieve supersaturation during dissolution in aqueous medium, and if it persists over a sufficient time period then a substantially higher absorption of the drug may result. As an example, the phase diagram of the system griseofulvin-succinic acid is presented in Fig. 37.14. Two solid dispersions have been prepared by melting, one of them was the eutectic, the other was meant to be a solid solution containing 20 weight-% griseofulvin. The dissolution rates of the solid dispersions were much higher than those of physical mixtures of the same composition (Fig. 37.15). Moreover, it can be seen that high concentrations are achieved exceeding the solubility of griseofulvin (12.5 μg ml^{-1} at 37 °C). Both effects can promote absorption *in vivo*; the drug substance is made available faster and at higher concentration.

Lipophilic drugs are preferentially dissolved in melts of, for example, the waxy Gelucire 44/14 (lauroyl macrogol-glycerides), or in semisolid mixtures with lauroglycol (propylene glycol laurate). Such drug-containing lipids

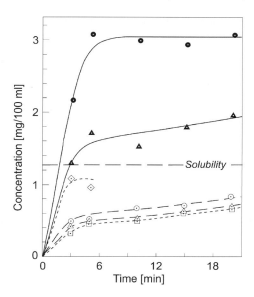

Fig. 37.15 Dissolution of griseofulvin-succinic acid preparations in water at 37°C. Black symbols denote melt-products, hollow symbols: plain griseofulvin or physical mixtures, respectively. Circles, solid dispersion and physical mixture, composition S (see phase diagram Fig. 37.9); triangles, eutectic solid dispersion and physical mixture, composition E; squares, crystalline griseofulvin; diamonds, micronized griseofulvin. Redrawn from Goldberg *et al.*[33]

swell in the aqueous gastro-intestinal fluids to form emulsion suspensions as a vehicle that promotes absorption of the finely dispersed drug.

VII CONCLUSION

Of the many materials and techniques available for improving the performance of drug candidates and substances with the 'birth defect' poor solubility, the basic and essential ones have been described. They can be used or adapted for first *in vivo* formulations, but serve likewise as the basis for developing dosage forms for use in humans. Beyond those easy and straightforward tools, which are based on well-understood physicochemical relations, there are numerous sophisticated and specialized systems and technologies available that are based on lipids, on amphiphilic surfactants and in particular, on the dazzling self-structuring tendency of the phospholipids. However, the expenses for probing advanced delivery techniques should be wisely weighed against the medicinal chemist's engagement in optimizing the candidate design by synthetic means.

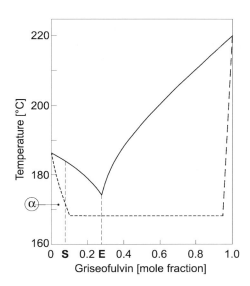

Fig. 37.14 Phase diagram of the system griseofulvin-succinic acid. E indicates the composition of the eutectic corresponding to 66 weight-% griseofulvin, S is the composition of a sample with 20 weight-% griseofulvin. Redrawn from Goldberg *et al.*[34]

REFERENCES

1. Amidon, G. L., Lennernäs, H., Shah, V. P. and Crison, J. R. (1995) A theoretical basis for a biopharmaceutic drug classification: the correlation of *in vitro* drug product dissolution and *in vivo* bioavailability. *Pharm. Res.* **12**: 413–420.

2. U.S. Department of Health and Human Services, F.D.A., Center for Drug Evaluation and Research (CDER) (2000, August 31), *Guidance for Industry. Waiver of In Vivo Bioavailability and Bioequivalence Studies for Immediate-Release Solid Oral Dosage Forms Based on a Biopharmaceutics Classification System.* Office of Training and Communications, Division of Communications Management, Drug Information Branch, HFD-210, 5600 Fishers Lane, Rockville, MD.

3. Nema, S., Washkuhn, R. and Brendel, R. (1997) Excipients and their use in injectable products. *PDA J. Pharm. Sci. Technol.* **51**: 161–171.

4. Yalkowsky, S. H. and Roseman, T. J. (1981) Solubilization of drugs by co-solvents. In Yalkowsky, S. H. (ed.), *Techniques of Solubilization of Drugs*, pp. 91–134. Marcel Dekker, New York.

5. Yalkowsky, S. H., Flynn, G. and Amidon, G. L. (1972) Solubility of nonelectrolytes in polar solvents. *J. Pharm. Sci.* **61**: 983–984.

6. Yalkowsky, S. H., Valvani, S. C. and Amidon, G. L. (1976) Solubility of nonelectrolytes in polar solvents IV: Nonpolar drugs in mixed solvents. *J. Pharm. Sci.* **65**: 1488–1494.

7. Valvani, S. C., Yalkowsky, S. H. and Roseman, T. J. (1981) Solubility and partitioning IV: Aqueous solubility and octanol-water partition coefficients of liquid nonelectrolytes. *J. Pharm. Sci.* **70**: 502–507.

8. Gould, P. L., Goodman, M. and Hanson, P. A. (1984) Investigation of the solubility relationships of polar, semi-polar and non-polar drugs in mixed co-solvent systems. *Int. J. Pharm.* **19**: 149–159.

9. Kramer, S. F. and Flynn, G. L. (1972) Solubility of organic hydrochlorides. *J. Pharm. Sci.* **61**: 1896–1904.

10. Yalkowsky, S. H. and Valvani, S. C. (1977) Precipitation of solubilized drugs due to injection or dilution. *Drug Intell. Clin. Pharm.* **11**: 417–419.

11. Aboutaleb, A. E. and Abdelzaher, A. (1981) A study on the solubilization of chloramphenicol. *Pharmazie* **36**: 102–106.

12. Griffin, W. C. (1949) *J. Soc. Cosmet. Chem.* **1**: 311.

13. Monteil, A., Navarrot, P., Kienlen, J. and DuCailar, J. (1976) [2 cases of anaphylactic-type reaction to alphadione (Alfatesine, Althesin, CT 13.41)]. *Ann. Anesthesiol. Fr.* **17**: 71–76.

14. Volcheck, G. W. and Van Dellen, R. G. (1998) Anaphylaxis to intravenous cyclosporine and tolerance to oral cyclosporine: case report and review. *Ann. Allergy Asthma Immunol.* **80**: 159–163.

15. Ebo, D., Piel, G. C., Conraads, V. and Stevens, W. J. (2001) IgE-mediated anaphylaxis after first intravenous infusion of cyclosporin. *Ann. Allergy Asthma Immunol.* **87**: 243–245.

16. Jain, N., Yang, G. and Yalkowsky, S. H. (2001) Solubilization of NSC-639829 by combined effects of pH with cosolvents, surfactants, and complexants. *Int. J. Pharm.* **225**: 41–47.

17. Thoma, K. and Albert, K. (1983) Amphiphilic drugs. 2. Relation between colloidal properties and pharmaceutic-technological, pharmacokinetic and pharmacodynamic behavior. *Pharmazie* **38**: 807–817.

18. Hussain, M. A., Wu, L. S., Koval, C. and Hurwitz, A. R. (1992) Parenteral formulation of the kappa agonist analgesic, DuP 747, via micellar solubilization. *Pharm. Res.* **9**: 750–752.

19. Tarr, B. D., Sambandan, T. G. and Yalkowsky, S. H. (1987) A new parenteral emulsion for the administration of taxol. *Pharm. Res.* **4**: 162–165.

20. Charman, W. N. (1992) Lipid vehicle and formulation effects on intestinal lymphatic drug transport. In Charman, W. N. and Stella, V. L. (eds). *Lymphatic Transport of Drugs*, pp. 113–179. CRC Press, Boca Raton.

21. Charman, W. N. (2000) Lipids, lipophilic drugs, and oral drug delivery — Some emerging concepts (Mini-Review). *J. Pharm. Sci.* **89**: 967–978.

22. Charman, W. N. (1997) Lipids, lymph and lipidic formulations. *Bull. Tech. Gattefossé* **90**: 27–33.

23. Ziegenmeyer, J. and Führer, C. (1980) Mikroemulsionen als topische Arzneiform. *Acta Pharm. Tech.* **26**: 273–275.

24. Drewe, J., Meier, R., Vonderscher, J. D., *et al.* (1992) Enhancement of the oral absorption of cyclosporin in man. *Br. J. Clin. Pharm.* **34**: 60–64.

25. Gershanik, T. and Benita, S. (2000) Self-dispersing lipid formulations for improving oral absorption of lipophilic drugs. *Eur. J. Pharm. Biopharm.* **50**: 179–188.

26. Isele, U., Van Hoogevest, P., Hilfiker, R., *et al.* (1994) Large-scale production of liposomes containing monomeric zinc phthalocyanine by controlled dilution of organic solvents. *J. Pharm. Sci.* **83**: 1608–1616.

27. Isele, U., Schieweck, K., Kessler, R., *et al.* (1995) Pharmacokinetics and body distribution of liposomal zinc phthalocyanine in tumor-bearing mice: influence of aggregation state, particle size, and composition. *J. Pharm. Sci.* **84**: 166–173.

28. Papahadjopoulos, D., Vail, W. J., Jacobson, K. and Poste, G. (1975) Cochleate lipid cylinders: formation by fusion of unilamellar lipid vesicles. *Biochim. Biophys. Acta* **394**: 483–491.

29. (a) Zarif, L., Graybill, J., Perlin, D. and Mannino, R. (2000) Cochleates: New lipid-based drug delivery system. *J. Liposome Res.* **10**: 523–538; (b) Zarif, L. (2002) Elongated supramolecular assemblies in drug delivery. Review. *J. Control Rel.* **80**: 7–23.

30. Santangelo, R., Paderu, P., Delmas, G. *et al.* (2000) Efficacy of oral cochleate-amphotericin B in a mouse model of systemic candidiasis. *Antimicrob. Agents Chemother.* **44**: 2356–2360.

31. Leigh, M., van Hoogevest, P. and Tiemessen, H. (2001) Optimising the oral bioavailability of the poorly water-soluble drug cyclosporin A using membrane lipid technology. *Drug Del. Syst. Sci.* **1**: 73–77.

32. Illig, K., Mueller, R.L., Ostrander, K.D. and Swanson, J.R. (1996) Use of Microfluidizer® processing for preparation of pharmaceutical suspensions. *Pharm. Technol.* 78.

33. Gassmann, P., List, M., Schweitzer, A. and Sucker, H. (1994) Hydrosols — alternatives for the parenteral application of poorly water soluble drugs. *Eur. J. Pharm. Biopharm.* **40**: 64–72.

34. Goldberg, A. H., Gibaldi, M. and Kanig, J. L. (1966) Increasing dissolution rates and gastrointestinal absorption of drugs via solid solutions and eutectic mixtures III. Experimental evaluation of griseofulvin-succinic acid solid solution. *J. Pharm. Sci.* **55**: 487–492.

35. Alvarez-Nunez, F.A., Pinsuwan, S., Lerkpulsawad, S. and Yalkowsky, S.H. (1998) Effect of the most common co-solvents upon the extent of solubilization of some drugs: log-linear model. AAPS Annual Meeting, San Francisco, CA, USA. Volume **1** (4), Poster 2433.

38

IMPROVEMENT OF DRUG PROPERTIES BY CYCLODEXTRINS

Kaneto Uekama and Fumitoshi Hirayama

I INTRODUCTION

Cyclodextrins (CyDs) were first isolated in 1891 as degradation products of starch and were characterized as cyclic oligosaccharides (Fig. 38.1).[1] The α-, β- and γ-CyDs are the most common natural CyDs, consisting of six, seven and eight glucose units, respectively (Table 38.1). Because of their different internal cavity diameters, each CyD shows a different degree of molecular encapsulation with different-sized guest molecules.[2,3] These CyDs have therefore been utilized for the modification of physical, chemical or biological properties of guest molecules. In the pharmaceutical field, CyDs have been recognized as potent candidates to overcome the undesirable properties of drug molecules through the formation of inclusion complexes.[4-6] Recently, a number of new dosage forms have been developed as drug

delivery systems (DDS). For the design of such advanced dosage forms various kinds of CyD derivatives have been prepared to extend the physicochemical properties and inclusion capacity of natural CyDs as novel drug carriers.[7,8] This chapter deals with some aspects of the utilization of chemically modified CyDs in pharmaceutical formulations, and will discuss some fundamental characteristics of CyDs which should be considered in the development of advanced dosage forms.

II PHARMACEUTICALLY USEFUL CYDS

To obtain drug carrier properties better than those of natural CyDs, the hydroxyl groups of Ds are available as starting points for structural modification, and various functional groups have been incorporated in the CyD molecule (Table 38.2). These chemically modified CyDs can be classified into three types: hydrophilic, hydrophobic and ionizable derivatives.[8] Hydrophilic derivatives such as methylated CyDs,[9,10] hydroxyalkylated CyDs,[11-15] and branched CyDs[16,17] deserve special attention because their solubility in water is high, suggesting their use as solubilizers for poorly water-soluble drugs rather than the use of surface-active agents. In contrast, hydrophobic CyDs are useful as sustained-release drug carriers of water-soluble drugs[18,19] and of peptides,[20] since they have the ability to decrease the solubility of guest molecules. On the other hand, the ionizable CyDs can modify the release rate of drug, depending on the pH of the solution[21] and bind to the surface membranes of cells, which may alter the function of biological barriers.[22] Sulfates and sulfoalkyl ethers of CyDs have been evaluated as a new class of parenteral drug carriers with heparin-mimetic biological activities. However, new CyD derivatives must be thoroughly characterized before practical use in pharmaceutical formulations.[23] In this section, some physicochemical and biological profiles of CyD derivatives are briefly described, comparing them with those of natural CyDs.

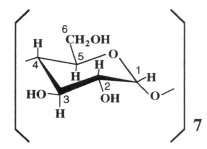

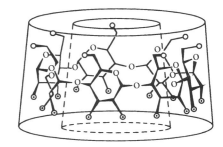

Fig. 38.1 Structure of β-cyclodextrin.

A Physicochemical profiles of CyDs

Widespread use of natural CyDs as hosts for drugs is restricted by their low aqueous solubility, particularly that of β-CyD.[24] Methylation or hydroxyalkylation of the hydroxyl groups of β-CyD has been used to obviate this problem. For example, hydroxyalkylated CyDs are amorphous mixtures of chemically related components with different degrees of substitution.[11,12] This multicomponent character prevents crystallization, and thus the hydroxyalkylated CyDs have higher solubility (> 50%) in both water and ethanol. The solubility of natural β-CyD in water increases with increase in temperature. On the other hand, 2,6-dimethyl-β-CyD (DM-β-CyD) shows exothermic dissolution in water, so that the solubility decreases with increase in temperature, because of dehydration at elevated temperatures. Thus, methylated CyDs have clouding points, a behaviour similar to that of nonionic surfactants. The solubilities of maltosyl-β-CyD (G2-β-CyD) and hydroxypropyl-β-CyD (HP-β-CyD) show a little temperature dependence.[25,26] Such information is particularly useful for the design of aqueous injectable CyD solutions which are to be heat-sterilized. The aqueous solubility of ionizable CyDs depends on the pH of the solution. *O*-Carboxymethyl-*O*-ethyl-β-CyD (CME-β-CyD), in which the hydroxyl groups of ethylated β-CyD are substituted by carboxymethyl groups, is slightly soluble in low pH regions but freely soluble in neutral and alkaline regions owing to the ionization of the carboxyl group (pK_a 3–4).[21] Thus, CME-β-CyD can serve as an enteric-type drug carrier similar to

carboxylmethylethylcellulose, but may be of greater advantage than the cellulose derivatives for the stabilization of labile drugs owing to the inclusion ability.[27]

The glycosidic bonds of CyDs are fairly stable in alkaline solution, whereas they are hydrolytically cleaved by strong acids to give linear oligosaccharides.[28] The ring-opening rate of CyDs increases with increasing cavity size, and is accelerated when the ring is distorted. For example, permethylated γ-CyD (TM-γ-CyD) having a distorted ring conformation is most susceptible to acid-catalysed hydrolysis.[6] It should be noted that the ring-opening rate of β-CyD is decreased by the addition of guest molecules, the decrease being marked for guests with a close fit to the β-CyD cavity.[29] This reduction of rate can be attributed to the inhibition of access of catalytic oxonium ions to the glycosidic bond because the CyD cavity is occupied by guests. The glycosidic bonds of CyDs are cleaved by some starch-degrading enzymes with the proper substrate specificity, although the reaction rate is much slower than that of linear sugars.[17] Generally, the introduction of substituents on the hydroxyl groups slows enzymatic hydrolysis of CyDs by lowering the affinity of CyDs for enzymes or owing to the intrinsic reactivity of enzymes.

B Biological profiles of CyDs

α- and β-CyDs are resistant to metabolism in the body, whereas γ-CyD, having a large cavity, is hydrolysed even by human salivary α-amylase.[30] On the other hand, β-CyD is hardly hydrolysed at all in whole blood of rats, rabbits, dogs and humans, and also in rat liver homogenates.[31] In the case of G2-β-CyD, the α-1,4 glycosidic bonds in the CyD ring and the α-1,6 bond at the junction between the CyD ring and the substituents are hydrolytically stable in human body fluids. Subacute or subchronic intravenous administration of HP-β-CyD to rats and monkeys showed no significant alteration in the morphological and clinical pathology parameters.[32] When G2-β-CyD and HP-β-CyD were administered intravenously rats, they disappeared rapidly from the plasma[31,33] and were recovered almost completely as a form of water-soluble G1-β-CyD and intact

Table 38.1 *Some characteristics of natural CyDs*

CyD	Number of glucose units	Molecular weight	Cavity[a] diameter(Å)	Solubility[b] (g dl^{-1})
α-CyD	6	973	5	15
β- CyD	7	1135	6	1.85
γ- CyD	8	1297	8	23

[a] Estimated by the Corey–Pauling–Koltun model.
[b] In water at 25°C.

Table 38.2 *Pharmaceutically useful β-cyclodextrin derivatives*

Derivative	Characteristic	Possible use (dosage form)
Hydrophilic derivatives		
Methylated β-CyD	Soluble in cold water and in organic solvents, surface active, haemolytic	Oral, dermal, mucosal[a]
DM-β-CyD		
TM-β-CyD		
DMA-β-CyD	Soluble in water, low haemolytic	Parenteral
Hydroxyalkylated β-CyD		
2-HE-β-CyD	Amorphous mixture with different degrees of substitution, highly water- soluble (> 50%), low toxicity	Oral, dermal, mucosal, parenteral (intravenous)
2-HP-β-CyD		
3-HP-β-CyD		
2,3-DHP-β-CyD		
Branched β-CyD		
G$_1$-β-CyD	Highly water-soluble (> 50%)	Oral, mucosal, parenteral (intravenous)
G$_2$-β-CyD	low toxicity	
GUG-β-CyD		
Hydrophobic derivatives		
Alkylated β-CyD		
DE-β-CyD	Water insoluble, soluble in organic solvents, surface active	Oral, parenteral (subcutaneous) (slow release)
TE-β-CyD		
Acylated β-CyD		
TAcyl-β-CyD	Water insoluble, soluble in organic solvents	Oral, dermal (slow release)
TValeryl-β-CyD	Film formation	
Ionizable derivatives		
Anionic β-CyD		
CME-β-CyD	pK$_a$ = 3 to 4, soluble at pH > 4	Oral, dermal, mucosal (delayed release)
β-CyD·sulphate	pK$_a$ < 1, water soluble	Oral, mucosal
SBE-β-CyD		Parenteral (intravenous)
β-CyD·phosphate		
Al β-CyD·sulphate	Water insoluble	(slow release)

Abbreviations: DM, 2,6-di-*O*-methyl; TM, 2,3,6-tri-*O*-methyl; DMA, acetylated DM-β-CyD; 2-HE, 2-hydroxyethyl; 2-HP, 2-hydroxypropyl; 3-HP, 3-hydroxypropyl; 2,3-DHP, 2,3-dihydroxypropyl; G$_1$, glycosyl; G$_2$, maltosyl; GUG, Glucuronyl-glucosyl; DE, 2,6-di-*O*-ethyl;TE, 2,3,6-tri-*O*-ethyl; CME, *O*-carboxymethyl-*O*-ethyl; TAcyl, 2,3,6-tri-*O*-acyl (C$_2$ ~ C$_{18}$); TValeryl, 2,3,6-tri-*O*-valeryl; SBE, sulfobutyl ether.

[a] Mucosal: nasal, sublingual, ophthalmic, pulmonary, rectal, vaginal, etc.

HP-β-CyD, respectively. Since the nephrotoxicity of natural β-CyD at higher doses was ascribed to the crystallization of less-soluble β-CyD or its cholesterol complex in renal tissues,[34] the metabolic fates of G2-β-CyD and HP-β-CyD are suggestive of lower renal toxicity compared with the parent β-CyD. To assess tolerance via the parenteral administration route, various blood chemistry parameters in rats and rabbits after the multiple intravenous adminis-trations of hydrophilic β-CyDs were compared with those of the parent β-CyD.[31] Multiple injections of β-CyD or DM-β-CyD at a total dose of 900 mg kg^{-1} in rats and 1200 mg kg^{-1} in rabbits produced some kidney and liver failure, while those for HP-β-CyD, G2-β-CyD and β-CyD sulfate at the same doses failed to induce any kidney or liver failure,

suggesting that these hydrophilic β-CyDs can be safely used in parenteral formulations.

The haemolytic activities of natural CyDs are reported to be in the order β- > α- > γ-CyD.[35,36] These differences are ascribed to the differential solubilization of membrane components by each CyD. When the CyD cavity is modified by chemical derivatization, its effects on cell membranes can be changed dramatically from those of parent CyDS.[13,17] When the muscle tissue damage due to the injection of hydrophilic CyDs was compared with that of mannitol and nonionic surfactants, following a single injection (100 mg ml^{-1}) of the compounds into *M. vastus lateralis* of rabbits (Fig. 38.2), α-CyD and DM-β-CyD showed a relatively high irritation reaction, the degree of

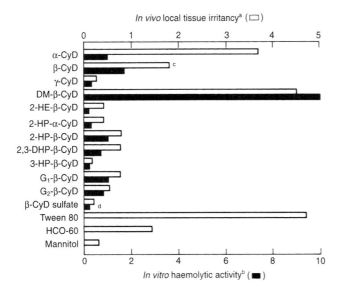

In vivo local tissue irritancy[a] (□)

In vitro haemolytic activity[b] (■)

Fig. 38.2 *In vivo* local tissue irritancy and *in vitro* haemolytic activity of CyDs. [a]Dose (100 mg ml^{-1}) injected into *M. vastus lateralis* of rabbits. [b]The reciprocal of the concentration (w/v%) of CyDs to induce 50% lysis of human erythrocytes. [c]Used as suspension due to the limited solubility. [d]Less than 0.2 w/v%.

which corresponded to that of Tween 80®. On the other hand, G$_2$-β-CyD, HP-β-CyD, SBE-β-CyD, DMA-β-CyD and CyD sulfates showed no or only slight irritation reaction, the degree of which was comparable to those of γ-CyD, mannitol and HCO-60®.

It is generally recognized that the gastrointestinal (GI) absorption of CyDs in intact form is limited because of their bulky and hydrophilic nature. Only an insignificant amount of intact β-CyD was absorbed from the GI tract in rats.[37,38]

In the case of the rectal route, when oleaginous suppositories containing β-CyD derivatives were administered to the rat rectum, large amounts of intact β-CyDs were excreted into the urine up to 24 h after administration.[39] The relatively high absorption observed for β-CyDs was ascribed to a change in the permeability of the rectal mucosa and/or an interaction between the surface-active β-CyDs and glycerides, which are principle components of the suppository bases. Similarly, the hydrophilic CyDs are supposed to be hardly absorbed through the skin at all. DM-β-CyD and HP-β-CyD, however, do permeate into the skin, particularly when they are applied under occlusive-dressing conditions and/or using vehicles containing absorption-promoting agents.[40,41] HP-β-CyD, even when it is applied as an aqueous solution under nonocclusive condition, penetrates into the skin of rats and distributes homogeneously over all epidermal and dermal structures without irritation.

III IMPROVEMENT OF DRUG PROPERTIES

An important characteristic of CyDs is the formation of inclusion complexes in both the solution and solid states, in which each guest molecule is surrounded by the hydrophobic environment of the CyDs cavity. This can lead to the alteration of physicochemical properties of guest molecules, and can eventually have considerable pharmaceutical potential (Table 38.3).[42] In the practical application of CyDs, attention should be directed towards the dissociation equilibrium and stoichiometry of the inclusion complex. When a CyD complex is dissolved in water or introduced into body fluids, it dissociates rapidly to free components in equilibrium with the complex. The stability constant (K_c) is

Table 38.3 *Pharmaceutical products containing CyDs*

Component	Dosage form	Efficacy	Trade name
Prostaglandin E$_1$-α-CyD			
20 μg/ampoule	Injection	Chronic arteriosclerotic obstruction	Prostandin
500 μg/vial	Injection	Hypertension during surgery	Prostandin 500
Prostaglandin E$_2$-β-CyD	Tablet	Labour induction	Prostarmon E
Limaprost-α-CyD	Tablet	Buerger's disease	Opalmon
Benexate-β-CyD	Capsule	Antiulcer	Ulgut, Lonmiel
Nitroglycerin-β-CyD	Sublingual tablet	Angina pectoris	Nitropen
Cefotiam hexetil hydrochloride-α-CyD	Tablet	Antibiotic	Pansporin T
Piroxicam-β-CyD	Tablet and suppository	Analgesic, antiinflammatory	Brexin, Cicladol
Iodine-β-CyD	Gargling	Throat disinfection	Mena-Gargle
New oral cephalosporin α-CyD	Tablet	Antibiotic	Meiact
Hydrocortisone-HP-β-CyD	Liquid	Mouth washing	Dexocort
Itraconazole-HP-β-CyD	Liquid	Esophageal candidiosis	Sporanox
Ziprasidone-SBE-β-CyD	Intramuscular injection	Antipsychotic	Geodon

Table 38.4 *Effect of temperature on stability constants (K_c) of inclusion complexes of ethyl 4-biphenylylacetate with various β-CyDs, and their thermodynamic parameters in water*

Complex	K_c (1 mol^{-1})				ΔH (kJ mol^{-1})	ΔS (J K^{-1} mol^{-1})
	15°C	25°C	37°C	45°C		
β-CyD	3330	3050	2850	1630	−18.2	5.7
HP-β-CyD	7950	4200	3000	1650	−37.2	−55.7
DM-β-CyD	31500	12500	10000	5700	−40.4	−57.1

a useful index for estimating the binding strength of the complex and changes in the physicochemical properties of a guest in the complex. The degree of dissociation is dependent on the magnitude of K_c, and various environmental factors such as dilution, temperature, pH and additives, will affect the K_c value. Since the formation of the inclusion complex is usually exothermic (Table 38.4), the dissociation is facilitated by raising the temperature.[40] In the case of ionizable guests, change of the pH of solutions shifts the equilibrium to such an extent and in such a direction that the unionized form is predominantly included in the cavity. This is due to the favouring of the inclusion of hydrophobic guests compared with hydrophilic guests. The obvious requirement for inclusion complexation is that the hydrophobic moiety of a guest molecule must fit entirely or partially into the CyD cavity. In the case of prostaglandin E$_1$ (PGE$_1$) (Fig. 38.3), α-CyD, with a smaller cavity, preferentially includes an aliphatic chain of the PGE$_1$ molecule, while β-CyD accommodates the five-membered ring of PGE$_1$.[6] In contrast, the larger γ-CyD cavity accommodates PGE$_1$ in such a manner that the whole PGE$_1$ molecule penetrates the cavity. When a guest molecule is too large to be included in one CyD cavity or the host cavity is too small to include a whole guest, more than one CyD is available for complete inclusion. However, the inclusion complexation often shows a different stoichiometry depending on the guest/host concentration employed, which consequently affects the physicochemical properties of the guest molecules.

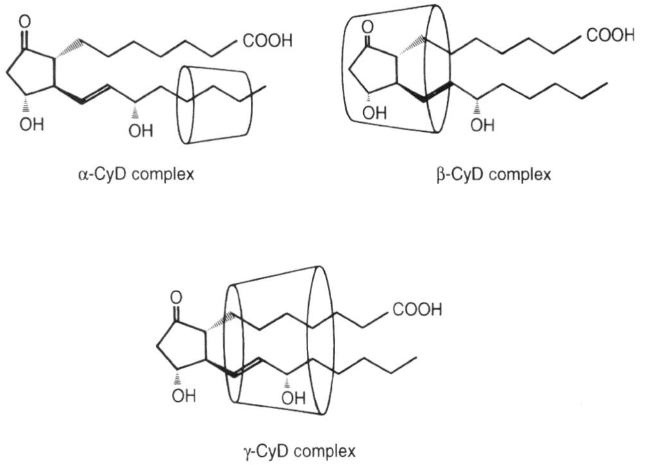

A Solubilization

The solubilizing effect of CyDs is related both to their ability to form inclusion complexes and to the intrinsic solubility of the host molecule in water. The former factor is related to the K_c values. Among β-CyD and its derivatives, DM-β-CyD shows the highest solubilizing effect for poorly water-soluble drugs, probable owing to the elongation of the cavity.[7,9,10] The steric effects of substituents also play a role in the solubilizing ability of the host molecules. For example, HP-β-CyD, similarly to DM-β-CyD, has higher complexing ability, but its solubilizing ability is lowered at very high degrees of substitution.[13] Glucose and maltose units in the branched β-CyD also hinder the inclusion of drug molecules.[17]

HP-β-CyD can be recommended as a parenteral drug carrier because of its low toxicity, high toleration and excellent solubilizing abilities.[43] In the preparation of an HP-β-CyD-based pharmaceutical formulation, the solubility of some lipophilic drugs in water was synergistically increased with increasing temperature, suggesting that heating or heat-sterilization may be useful steps.[44] A detailed evaluation of HP-β-CyD as an excellent solubilizer and stabilizer for brain-targeting chemical delivery systems such as estradiol-dihydropyridine conjugate has been described.[45,46] The hydrophilic CyDs can solubilize some specific lipids from biological membranes through the rapid and reversible formation of inclusion complexes, leading to an increase in the membrane permeability.[35] This may allow the extended use of CyDs as adjuvants to improve the transmucosal absorption of poorly absorbable peptide and protein drugs,[47] which will be discussed in later sections.

The solubility of guests changes according to the stoichiometry of the complexes: for example, solubilities

Fig. 38.3 Cavity-size dependency of inclusion complexation of CyDs with prostaglandin E$_1$.

in water of nocloprost, a derivative of PGE_1[48] and sofalcon, an anti-ulcer agent,[49] increased on 1:1 complexation with β- and γ-CyDs, respectively, owing to the partial inclusion. On the other hand, the solubilities were decreased by 1:2 complexation at higher CyD concentrations because of the complete inclusion of the whole guest molecules within the CyD dimers, which are less hydrated.

B Stabilization in solution

The drug must remain sufficiently stable not only during storage but also in the GI fluids, since reactions which result in a product that is pharmacologically inactive or less active will reduce the therapeutic effectiveness. CyDs are known to accelerate or decelerate various kinds of reactions depending on the nature of the complex formed.[2,4] Generally, when a drug's active centre is included in the CyD cavity there is a deceleration effect, and the reaction rate is dependent primarily on the amount of free drug concentration resulting from the dissociation of the complex. On the other hand, if the drug does not totally fit into the cavity or is only partially included, leaving the active centre sterically fixed in close proximity to the catalysts, it experiences an acceleration effect.[28]

Prostanoids

Prostaglandins (PGs) are essentially long-chain unsaturated fatty acids containing a substituted cyclopentane ring system. The β-hydroxyketo moiety of E-type prostaglandins (PGEs) is extremely susceptible to dehydration under acidic or alkaline conditions to give A-type prostaglandins (PGAs), which are isomerized subsequently to form B-type prostaglandins (PGBs) under alkaline conditions. The biological activities of PGEs decrease with progress of these reactions. The chemical instability and the low aqueous solubilities of PGEs have limited dosage form design and presented a substantial challenge to pharmaceu-

tical scientists. Natural CyDs have successfully been applied to PGEs, and stable and soluble complexes were first marketed in Japan as a PGE_2-β-CyD tablet and PGE_1-α-CyD injection (see Table 38.3). In aqueous solution, however, attempts to stabilize PGEs were rather disappointing because of the positive catalytic effect of narural CyDs (Fig. 38.4). In most cases, the stabilizing effect of TM-β-CyD is smaller than that of DM-β-CyD, preventing deep penetration of the bulky PGE_2 molecules into the narrow TM-β-CyD cavity.[50] G_2-β-CyD has been shown to improve the undesirable properties of PGE_1 in lyophilized preparations.[51] The decomposition of lyophilized PGE_1 is significantly retarded by both β-CyD and G_2-β-CyD. The rapid-dissolution property of lyophilized PGE_1 with G_2-β-CyD is maintained during storage, while it tends to decrease with β-CyD, depending on the moisture-adsorbing and wetting properties and crystallinity changes of the additives.

Prostacyclin (PGI_2) is a potent therapeutic agent in the treatment of thrombosis and relaxation of vascular smooth muscle. However, this compound undergoes an extremely facile hydrolysis of the vinyl-ether moiety to yield 6-keto-$PGF_1\alpha$ in aqueous solution, losing its activity within a few minutes. Upon binding to CyDs, the rate of hydrolysis of PGI_2 is retarded in the order β- > α- > γ-CyD. The retardation effect of CyDs seems to be at least in part a result of the inhibition of intramolecular carboxylate ion catalysis due to the decrease in acidity of the terminal carboxyl group. Methylated β-CyDs in which the hydroxyl groups are blocked show a much greater retardation effect in the hydrolysis of PGI_2 in comparison with that of parent β-CyD.[52]

Cardiac glycosides

Digoxin, one of the potent cardiac glycosides, is susceptible to hydrolysis in an acidic medium, and therapeutic efficiency, as well as oral bioavailability, may decrease as a result. In the degradation pathways of digoxin, prevention

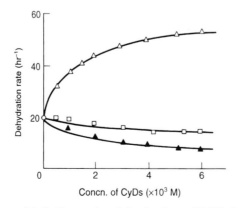
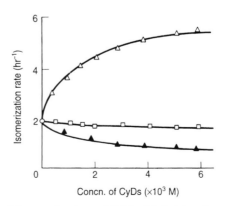

Fig. 38.4 Effects of β-CyDs on the dehydration of PGE_2 (left) and isomerization of PGA_2 (right) in phosphate buffer (pH 11.0, $\mu = 0.2$) at 60°C. ○, PG alone; △, β-CyD; ▲, DM-β-CyD; □, TM-β-CyD.

of the appearance of digoxigenin might be clinically important because the cardioactivity of digoxigenin is about one-tenth of that of digoxin, but other digoxosides (mono- and bisdigoxosides) possess approximately the same activity. In the presence of CyDs, acid hydrolysis of digoxin is suppressed in the order β- > α- > γ-CyD, where β-CyD inhibits the conversion from digoxosides to digoxigenin almost completely. ^{1}H NMR data reveal that the A-ring of digoxin is located at the entrance to the α-CyD cavity, that it could penetrate further into the β-CyD cavity, and that it is loosely bound to γ-CyD.[53] This indicates that either a smaller (α-CyD) or a larger (γ-CyD) cavity is unfavourable for preventing the hydrolysis of digoxin. Moreover, the methylated CyDs suppressed the hydrolysis of digitoxin in the order DM-β- > α- > TM-β-CyD, where DM-β-CyD completely inhibited the conversion from digitoxosides to digitoxigenin.[54] In a dissolution study of digoxin tablets, increase in dissolution rate and decrease in acid hydrolysis were achieved by β- and γ-CyDs, where the appearance of digoxigenin was almost negligible even 60 min after the initiation of the dissolution test.[55] When tablets containing a γ-CyD complex of digoxin were administered sublingually to human volunteers, the serum levels of drug were significantly increased compared with digoxin alone, owing to the prevention of acid hydrolysis in the stomach.[56] The stoichiometric ratio is in practice responsible for the dosage form design of a drug molecule. In the case of digoxin, for example, one digoxin molecule needs four molecules of γ-CyD to form a 1:4 stable crystalline complex, resulting in an eight-fold increase in the molecular weight of digoxin.[53] This may facilitate the preparation of tablets containing small amounts of drug, resulting in a better uniformity of content. In the case of high-dosage drugs, the increase in molecular weight becomes a disadvantage, since the relationship between the required dose and molecular weight determines the feasibility of oral administration in CyD complexes.

C   Control of solid properties

Many solid compounds are known to exist in different crystalline modifications such as amorphous, crystalline or glassy states, affecting solubility, dissolution rate, stability and bioavailability. In order to improve the pharmaceutical potential, therefore, it is important to control the crystallization, the polymorphic transition, and whisker generation of solid drugs.[5,6]

Modification of crystalline states of nifedipine

The oral bioavailability of crystalline nifedipine, a potent calcium-channel antagonist, is very low because of its poor solubility and its slow dissolution rate in water. Various hydrophilic macromolecules such as poly(vinylpyrrolidone) (PVP) and poly(ethylene glycols) have been used to improve the dissolution characteristics of nifedipine.[57] However, amorphous nifedipine in these matrices gradually crystallizes during storage at high temperature and humidity, deteriorating its rapid dissolution characteristics. We have reported that crystalline nifedipine is converted to an amorphous state by spray-drying with HP-β-CyD, and the oral bioavailability is improved significantly.[58] HP-β-CyD was useful in preventing the crystal growth of amorphous nifedipine, maintaining a relatively fine and uniform size of crystals even under adverse storage conditions (60°C, 75% RH). Moreover, the rapidly dissolving form of the metastable state (Form B; m.p. 163°C) transiently formed at an early stage of storage of amorphous nifedipine, and the glassy state (transition temperature at 48°C) of nifedipine could be obtained by cooling melts of stable form of nifedipine (Form A: m.p. 171°C). In the presence of HP-β-CyD, the crystallization of the glassy nifedipine and the polymorphic transition of Form B to Form A were significantly suppressed, indicating that Form B could be prepared in high yield (> 75%) by heating in the amorphous HP-β-CyD matrix. Although the initial dissolution rate of nifedipine increased in the order of glassy state > Form B > Form A, the glassy state of nifedipine was readily converted to Form A of larger crystal size at higher humidity and temperature. These facts suggest that HP-β-CyD is particularly useful for the selective preparation of the fast-dissolving Form B, and will provide a rational basis for the design of formulation and storage conditions in solid dosage forms of nifedipine.

Stabilization of carmofur

Carmofur (1-hexylcarbamoyl-5-fluorouracil, HCFU) is one of the masked forms of 5-fluorouracil (5-FU) and has been widely used to treat carcinomas of breast and the GI tract. It is expected that the highly hygroscopic character of natural CyDs may significantly influence degradation of HCFU, because HCFU is extremely susceptible to base- and water-catalysed hydrolysis to give 5-FU. In fact, the degradation rate of HCFU in the solid CyD complex was very fast under accelerated conditions (70°C, 75% RH). This problem can be solved by adding an organic acid, such as citric or tartaric acid, as a pH-controlling agent.[59] The organic acids may provide an acidic environment around the CyD complex after moisture sorption, since HCFU is chemically stable under acidic conditions. The methylated CyDs are highly effective in preventing the decomposition of HCFU in the solid state for long periods under accelerated storage conditions (40°C, 75% RH), because they are less hygroscopic than natural CyDs. Since the stabilizing effect of CyDs is responsible for the bioavailability of HCFU,[60,61] the acidic property of CME-β-CyD is particularly effective in improving the oral bioavailability of HCFU, preventing the degradation of HCFU into 5-FU, which irritates the GI mucosa.[27]

D Release control

For the design of advanced oral formulations, control of drug release rate from dosage forms is of critical importance in realizing their therapeutic efficacy. Most of the slow-release preparations have been aimed at achieving zero-order release of drugs to provide a constant blood level for extended periods. For this purpose, wax-type matrices or water-soluble cellulose derivatives are generally used as slow-release carriers for water-soluble drugs. When hydroxyl groups of CyDs are substituted by ethyl, acetyl or longer alkyl groups, the solubility of these CyDs in water decreases proportionally to their degree of substitution or the length of the alkyl chains.[18,19] We have demonstrated that the ethylated β-CyDs, such as DE-β-CyD and perethylated β-CyD (TE-β-CyD), are useful as slow-release carriers of isosorbide dinitrate[62] and diltiazem hydrochloride.[63] New series of peracylated β-CyDs with different alkyl chains (acetyl to octanoyl) also were prepared by acylating all hydroxyl groups of β-CyD, and their physical properties were evaluated with anticipation of more effective slow-release carriers for water-soluble drugs.[64] The concentrated solutions of peracylated β-CyDs in organic solvents were highly viscous and sticky and gelation took place upon evaporation of the solvents. Since these properties were thought to be particularly useful for a slow-release carrier, the solid complexes of peracylated β-CyDs with water-soluble drugs were prepared and their *in vitro* and *in vivo* release behaviours were examined. Drug release was markedly retarded by the complexation with peracylated β-CyDs in the decreasing order of the solubility of the host molecules. Since the peracylated derivatives with substituents longer than the hexanoyl moiety hardly released the water-soluble drugs, peracetyl- (TA-), perbutanoyl- (TB-) and perhexanoyl- (TH-)β-CyDs were used in the *in vivo* studies following oral administration of the hydrophobic complexes to dogs. As shown in Fig. 38.5, TB-β-CyD suppressed a peak plasma level of molsidomine, a peripheral vasodilator, and maintained a sufficient drug level for long periods, while other peracylated β-CyDs with shorter or longer chains were ineffective in controlling the *in vivo* release behaviour of molsidomine. It is noteworthy that the TB-β-CyD complex significantly increased the AUC and prolonged the mean residence time (MRT) of the drug. The prominent retarding effect of TB-β-CyD was ascribable to its mucoadhesive property and hydrophobicity compared with other peracylated β-CyDs.[64] These facts suggest that peracylated β-CyDs are useful as novel multifunctional carriers to modify the release rate of water-soluble drugs.

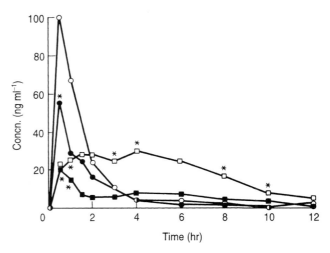

Fig. 38.5 Plasma levels of molsidomine after oral administration of capsules containing molsidomine or its peracylated β-CyD complexes (equivalent to molsidomine at 10 mg body^{-1}) in dogs. ○, Molsidomine alone (diluent: starch); ●, TA-β-CyD complex; □, TB-β-CyD complex; ◆, TH-β-CyD complex. Each point represents the mean of 3–6 dogs. $^*p < 0.05$ versus molsidomine alone.

CME-β-CyD is effective in modifying the release rate of water-soluble drugs. For example, the CME-β-CyD complex of diltiazem, a potent calcium-channel antagonist, releases the drug very slowly in the stomach but rapidly in the intestine, the main absorption site. The *in vitro* release behaviour was clearly reflected in the plasma levels of diltiazem after oral administration to dogs, where the complex produced a two-fold increase in bioavailability compared with the drug alone.[21] Such enhancement may arise from the reduction of the first pass effect — the metabolism in the liver immediately after absorption is saturated owing to the high local concentration of the drug at the intestinal tract. To investigate the delayed-release characteristics of the CME-β-CyD complex, the *in vivo* release of diltiazem was evaluated using dogs whose gastric acidity had been controlled by pretreatment with tetragastrin or omeprazole.[65] Following oral administration of the CME-β-CyD complex tablets in dogs with high controlled gastric acidity, the absorption of diltiazem was significantly retarded, since the complex dissolved only in the intestinal fluid after passing through the stomach. In dogs with low gastric acidity, rapid absorption of diltiazem was observed owing to the dissolution of the complex in the gastric fluid. In this delayed-release formulation, a good correlation between the *in vitro* and *in vivo* release rates of diltiazem was obtained in both groups of dogs with controlled gastric acidity.[66]

A suitable combination of various CyD derivatives is effective in sustaining the release rate of drugs. For example, the release rate of theophylline, which has a narrow therapeutic range in the blood level, can be controlled by hybridizing its hydrophilic, hydrophobic and ionizable CyD complexes.[66] This formulation has many advantages in reducing the frequency of dosing, prolonging the drug efficacy and avoiding the toxicity associated with the administration of a simple plain tablet of theophylline.

The oily injection of buserelin acetate (BLA), a luteinizing hormone-releasing hormone (LHRH) agonist, with a sustained-release feature can be achieved using hydrophobic CyDs.[67] The release of BLA from the peanut-oil suspension into the aqueous phase was significantly retarded by the complexation with TA-CyDs. A single subcutaneous injection of the oily suspension of BLA containing TA-β-CyD and TA-γ-CyD in rats provided the retardation of plasma BLA levels, giving 25 and 30 times longer mean residence times, respectively, than that with BLA alone. Simultaneously, the suppression of plasma testosterone levels to induce castration, the pharmacological effect of BLA, continued for 1 to 2 weeks and significant weight reduction in genital organs was observed due to the antigonadal effect. Since TA-β-CyD and TA-γ-CyD were degraded enzymatically in rat skin homogenates, both TA-CyDs can be useful as bioabsorbable sustained-release carriers for the water-soluble peptides following subcutaneous injection of oily suspensions.[68]

E Enhancement of drug absorption

Many factors affect drug absorption and they are largely dependent on the pharmaceutical formulation and administration route. In order to enhance drug absorption from CyD complexes, the dissociation equilibrium can be controlled by adjusting environmental factors.[69] Fig. 38.6 shows the overall process of drug absorption from the complex in the presence of competing agent, where k_d is the dissolution rate constant, K_1 is the stability constant of the drug-CyD complex, K_2 is the stability constant of the competing agent–CyD complex, and k_a is the absorption rate constant of the drug. When the solid complex is administered orally, the drug absorbed after the complex is dissolved and dissociated, since only the free form of the drug in solution is capable of penetrating the lipid barrier of the GI tract. Therefore, the absorption of the drug from the complex is mainly dependent on the magnitude of the dissolution rate (k_d) and the stability constants (K_1 and K_2) of the complexes; high dissolution rates and the relative ability of the complexes ($K_2 > K_1$) favour a free drug which is readily available for absorption. Since the displacement of drug from the cavity by exogenous and endogenous substances in the formulation and GI tract is responsible for acceleration of the drug absorption, the equilibrium can be controlled by adding an appropriate competitor and by adjusting its amount.[70] Such competition may occur not only in the body fluids containing various biological components, such as lipids and sterols, but also in pharmaceutical dosage forms containing various excipients.

Gastrointestinal absorption

The hydrophilic CyDs are useful for improving oral bioavailability of poorly water-soluble drugs, including steroids, cardiac glycosides, nonsteroidal anti-inflammatory drugs, barbiturates, antiepileptics, benzodiazepines, anti-diabetics, vasodilators, etc.[1-6] These improvements are mainly due to the increase in solubility, dissolution rate and wettability of the drugs through formation of inclusion complexes. As described in Fig. 38.6, CyDs are supposed to act only as carrier materials and help to transport the drug through an aqueous medium to the lipophilic absorption site in the GI tract. Results fully confirming this view were obtained in the enhancement of the absorption of spironolactone,[71] proscillaridin,[72] prednisolone,[73] benzodiazepines,[74] cardiac glycosides,[53] acetohexamide,[75] and anti-inflammatory drugs[76] using natural CyDs. Additionally, CyDs cause some modification of the GI mucosa at high concentrations owing to the removal of membrane components such as cholesterol, phospholipids, or proteins.[77,78] Therefore, free CyDs after dissociation of the complex may alter the lipid barrier of the absorption site, which consequently facilitates drug absorption. However, despite the larger stability constant, DM-β-CyD significantly enhances both the extent and rate of absorption of drugs compared with parental β-CyD, including fat-soluble vitamins,[3,79,80] anti-inflammatory drugs,[81,82] and antitumour drugs.[83] For example, the rapidly dissolving DM-β-CyD complex of α-tocopheryl nicotinate, following oral administration to fasting dogs, resulted in the area under the plasma concentration-time curve (AUC) being about 70 times as great as that of the drug alone.[84] To explain this

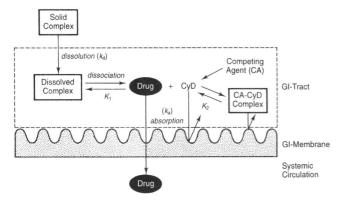

Fig. 38.6 Effect of competing agent on drug absorption from inclusion complex following dissolution and dissociation in gastrointestinal tract.

anomalous enhancing effect of DM-β-CyD on fatty drugs, additional factors influencing drug absorption were considered. Since highly surface-active DM-β-CyD is supposed to substitute bile in GI tract,[3] the lipophilic drugs can be finely emulsified by DM-β-CyD, which eventually provides the marked GI absorption in the fasting condition.

Transdermal absorption

Recent studies on human volunteers demonstrated that CyDs have a significant safety margin in dermal application.[85] Optimized release of the drug from the topical preparation containing its CyD complex may be obtained by using a vehicle in which the complex is barely dissociated and maintains a high thermodynamic activity. For example, the *in vitro* release rate of corticosteroids from water-containing ointments (hydrophilic, absorptive or polyacrylic base) was markedly increased by the hydrophilic CyDs, whereas in other ointments (a fatty alcohol propylene glycol (FAPG) or macrogol base) CyDs retarded the drug release.[86] The enhancement drug release can be ascribed to the increase in solubility, diffusibility and concentration of the drug in the aqueous phase of the ointment through water-soluble complex formation. Moreover, the drug in the CyD complex may be displaced by some components of the treatment, depending on the stability constant of the complex.[87,88] CyDs also interact with some components of skin. For example, DM-β-CyD markedly extracts cholesterol from rabbit skin *in vitro* (Fig. 38.7), a process which may reduce the function of skin as a barrier and may eventually contribute in part to the enhancement of drug absorption.

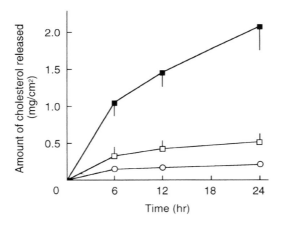

Fig. 38.7 Release profiles of cholesterol from rabbit skin treated with β-CyDs (1.2 mM) in isotonic phosphate buffer (pH 7.4) at 37°C. ○, Control; □, β-CyD; ■, DM-β-CyD. Each point represents the mean ± SE of 12 experiments.

When a hydrophilic ointment containing anti-inflammatory ethyl 4-biphenylylacetate (EBA) or its CyD complexes was applied to the skin of rats, the release of EBA from the ointment into the skin was assisted by DM-β-CyD or HP-β-CyD, while β-CyD had no appreciable effect.[41] Although the entry of CyDs into the skin made the stratum corneum more resistant to the permeation of the drug, the greater release of EBA from the vehicle compensated for the negative effect on the skin permeation of the drug and delivered the drug more effectively to the site of action. The application of EBA in the complexed form resulted in a transient rise in the total amount of EBA and its active metabolite, biphenylylacetic acid (BPAA) in the stratum corneum and in the viable skin. Interestingly, the fraction of active BPAA in the viable skin was increased when EBA was applied in the complexed form, indicating that β-CyDs assist the bioconversion of EBA to BPAA in the skin and consequently facilitate the delivery of active BPAA to subcutaneous tissues, where its action is most desired. In the model of carrageenan-induced acute oedema in rat paw, the inflammation was inhibited by pretreatment with ointments containing EBA-CyD complexes of DM-β-CyD or HP-β-CyD.[41,89]

Rectal absorption

The release of drug from suppository bases is one of the important factors in the transmucosal absorption of drugs, since the rectal fluid is small in volume and is viscous compared with GI fluid. Generally, the hydrophilic CyDs enhance the release of poorly water-soluble drugs from fatty bases because of the lesser interaction of the resultant complex with vehicles,[90,91] which eventually improves the rectal absorption. For example, the methylated CyDs significantly enhance the rectal absorption of hydrophobic drugs such as flurbiprofen,[40] HCFU,[61] and BPAA.[26] The superior effect of methylated CyDs can be explained by the faster release of the drug together with the lowering of the affinity of the complexed drug for the oleaginous suppository base. HP-β-CyD is particularly useful for improving the rectal absorption of hydrophobic drugs from fatty bases. The most illuminating effect of HP-β-CyD was obtained for the enhancement of rectal absorption of EBA, a lipophilic prodrug of BPAA.[26,91] The relative potency of β-CyD analogues in enhancing the dissolution rate of EBA in water and reducing the binding affinity of drugs to the fatty base was DM-β-CyD complex > HP-β-CyD complex > β-CyD complex; this order was consistent with that of the magnitudes of the stability constants of the complexes. However, *in vivo* absorption of EBA was enhanced in the order HP-β-CyD complex > DM-β-CyD complex > β-CyD complex in rats after single and multiple administrations of suppositories containing the complexes. The enhancement of rectal absorption of EBA *in vivo* can be explained by the facts that HP-β-CyD increases the release rate of EBA from the vehicle and stabilizes it in the rectal lumen and that the drug is partly absorbed in the form of the complex. The rather small enhancing effect of DM-β-CyD was ascribable to the considerable dissociation of the complex in the vehicle together with increased viscosity of the suppository base, since DM-β-CyD is extremely surface-active and oil-soluble compared with HP-β-CyD. Consequently, HP-β-CyD had the highest potential to improve rectal absorption of EBA among the three β-CyDs tested.

Nasal absorption

Highly water-soluble CyD complexes of steroid hormones are well suited to nasal administration.[92,93] This kind of formulation may provide a rapid rise of drug levels in systemic circulation and avoid intestinal and hepatic first pass metabolism of the drugs. Inherently, the blood level of endogenous hormones rises a few times a day in episodes lasting approximately one hour. Such pulsatile release of steroids can be imitated by the nasal administration of water-soluble CyD complexes, which may provide some desirable pharmacological profiles as demonstrated in sublingual administration of a rapidly dissolving complex of steroids with HP-β-CyD.[94] The effects of CyDs on the nasal epithelial membranes seem to be of minor importance for the absorption enhancement, because CyDs will lose their ability to interact with the membranes when their cavities are occupied by the steroids. Nasal preparations must be evaluated critically for their possible effect on the nasal mucociliary functions, which are known to defend the respiratory tract against dust, allergens and bacteria. When compared with other absorption-promoting agents and preservatives used in nasal formulations, DM-β-CyD exerted only a minor effect on the ciliary beat frequency of human nasal adenoid tissue *in vitro*.[95] This may represent an advantage of DM-β-CyD over the enhancers in promoting the nasal absorption of drugs, especially for short-term therapy. A nasal spray containing estradiol solubilized by DM-β-CyD was effective in the treatment of symptoms of estrogen deficiency in bilateral oophorectomized women, and the twice daily administration of this formulation over a period of 6 months was well-tolerated by the patients.[96]

F   Reduction of side-effects

The molecular entrapment of a drug into the CyD cavity may prevent direct contact of the drug with biological surfaces, and both the drug's entry into the cells of nontargeted tissues and local irritation are thus decreased. Since the CyD complex eventually dissociates into its components, there is no drastic loss of the therapeutic benefits of the drug. Therefore, CyDs act as wafer-like carriers which decrease the drug-induced local tissue damage at the administration site and then deliver the drug close to the site of its action.

Reduction of local irritancy

CyDs alleviate muscular tissue damage following the intramuscular injection of drugs.[97,98] The protective effects of CyDs may be attributed mainly to the poor affinity of the hydrophilic complexes of drugs for the sarcolemmal membranes of muscle fibres, a situation expected from the results of *in vitro* haemolysis studies. Similarly, CyDs diminished the ulcerogenic potency of several acidic anti-inflammatory drugs when these were administered orally.[99] In addition, HP-β-CyD significantly reduced the irritation of rectal mucosa in rats caused by BPAA, both for single and multiple administrations of EBA-β-CyD complexes in oleaginous suppositories (Fig. 38.8).[91]

Chlorpromazine (CPZ), a typical antipsychotic agent, frequently causes cutaneous phototoxic and photoallergic responses in patients being treated with prolonged and high doses. These adverse effects may be mainly attributable to the toxic photoproducts of CPZ. We have reported that DM-β-CyD significantly reduced the photosensitized skin irritation caused by CPZ in guinea pigs according to gross and histological examination.[100] The inhibitory effect of DM-β-CyD was ascribed to the alteration of the photochemical reactivity of CPZ, rather than the direct interaction of the photoproducts with DM-β-CyD. When CPZ was photoirradiated with DM-β-CyD, promazine, which is less

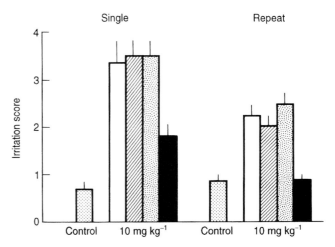

Fig. 38.8 Irritation effects of single or repeated administration (four times, 12-h intervals) of EBA or its β-CyD complexes as suppositories (10 mg kg^{-1} as EBA) on rectal mucosa in rats. Open bars, EBA alone; hatched bars, β-CyD complex; tinted bars, DM-β-CyD complex; solid bars, HP-β-CyD complex. Each value represents the mean ± SE of five rats.

toxic than CPZ, was produced in high yield.[101] In addition, DM-β-CyD suppresses the formation of numerous oxidation and polymerization photoproducts which are responsible for CPZ-photosensitized skin irritation. β-CyD derivatives also decreased the photoinduced free-radical production from CPZ,[102] the photodecarboxylation of benoxaprofen,[103] and the photodimerization of protriptyline,[104] resulting in the reduction of phototoxicity.

Systemic detoxication

The addition of β-CyD to dialysis fluids accelerated the removal of phenobarbital by peritoneal dialysis, thereby proving effective in the treatment of drug overdose.[105] HP-β-CyD is useful not only in the administration of drugs but also in redistributing other endogeneous lipophiles in the body. When HP-β-CyD was infused intravenously into a patient with familial hypervitaminosis A, retinyl ester overloading in the liver was reduced.[106] The possibility of HP-β-CyD redistributing cholesterol deposited in the vascular systems of hereditary hyperlipidaemic rabbits was reported.[107] A single intravenous administration of HP-β-CyD to rabbits slightly and temporarily decreased the level of total cholesterol in the serum. Similar results were obtained using normal rats.[108] Repeated administration of HP-β-CyD to the rabbits led to a gradual increase in total cholesterol in the circulation and eventually to a slight relief of atherosclerotic lesions in the thoracic aorta. HP-β-CyD may serve as an artificial lipid carrier to catalytically augment the establishment of equilibria in lipid distribution.[109]

Gentamicin, an aminoglycoside antibiotic, is widely used in the clinical treatment of Gram-negative infections, but its use is sometimes complicated by the development of drug-induced acute renal failure. Gentamicin is thought to interact with negatively charged phospholipids of lysosomal membranes in the proximal tubular cells, the interaction of which may lead eventually to lysosomal dysfunction, resulting in necrosis of the cells. Since some polyanions such as dextran sulphates are able to interact electrostatically with gentamicin and to reduce the drug's entry into the renal cortex, the effects of CyD sulphates on development of rat renal dysfunction induced with gentamicin were studied.[110] Daily subcutaneous infection of gentamicin (100 mg kg^{-1}, 14 days) developed nephrotoxicity in the rat, as assessed by an increase in serum urea nitrogen (Fig. 38.9) and histopathological changes in the renal cortex. When CyD sulphates were given intraperitoneally at 300 mg kg^{-1} at 6-h intervals after gentamicin administration, they protected the rats against the drug-induced renal impairment, while parent CyDs were ineffective. Since the postadministration of CyD sulphates did not reduce the total amount of gentamicin accumulated in the kidney, the protection may occur through interference with

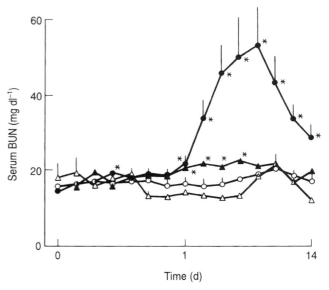

Fig. 38.9 Effect of β-CyD sulphate (300 mg kg^{-1} per day, i.p., 6h post-administration) on serum levels of urea nitrogen in rats treated with gentamicin (100 mg kg^{-1} per day, s.c.) for 14 days. ○, Saline control; △, β-CyD sulphate alone; ●, gentamicin alone; ▲, gentamicin with β-CyD sulphate. Each value represents the mean ± SE of 3–12 rats. $^{*}p < 0.05$ versus saline control.

intracellular events leading from the drug accumulation to nephrotoxicity. Since CyD sulphates have anti-inflammatory activity and have high affinity for growth factors and stabilize them against proteolysis, the effect of CyD sulphates on the renal regeneration processes should be considered as another possible protective mechanism. In any case, CyD sulphates may serve as potent antidotes against renal failure associated with aminoglycoside treatment.

G   Use in peptide and protein drugs

There are considerable hurdles in the practical use of biologically active peptides and proteins because of chemical and biological instability, poor absorption through biological membranes, rapid plasma clearance, peculiar dose–response curves, and immunogenicity. Many attempts have addressed these problems by chemical modification or by co-administration of adjuvants to promote absorption of peptides and proteins and protect them from proteolytic enzymes. The absorption-enhancing effects of CyDs mimic those of bile salts in regard to increased membrane permeability accompanied by inhibition of proteolysis, although they may differ somewhat in their manner of action on membranes. Thus, CyD complexation seems to be an attractive alternative to these approaches, but the field is still in its infancy.[69]

Absorption enhancement

Internasal delivery of peptide and protein drugs is severely restricted by presystemic elimination due to enzymatic degradation or mucocillary clearance and by the limited extent of mucosal membrane permeability. α-CyD has been shown to remove some fatty acids from nasal mucosa and to enhance the nasal absorption of leuprolide acetate in rats and dogs.[111] The utility of chemically modified CyDs as absorption enhancers for peptide drugs in rats has been demonstrated.[69] For example, DM-β-CyD was shown to be a potent enhancer of insulin absorption in rats,[112] and a minimal effective concentration of DM-β-CyD for absorption enhancement exerted only a mild effect on the *in vitro* ciliary movement.[113] The scope of interaction of insulin with CyDs is limited, because CyDs can only partially include the hydrophobic amino acid residues in peptides with small stability constants.[112] Under *in vivo* conditions, these complexes will readily dissociate into separate components, and hence the displacement by membrane lipids may further destabilize the complexes. The direct interaction of peptides with CyDs is therefore of minor importance in the enhancement of nasal absorption. Of the hydrophilic CyDs tested, DM-β-CyD had the most prominent inhibitory effect on the enzymatic degradation of both BLA and insulin in rat nasal tissue homogenates. Because of the limited interaction between peptides and CyDs, they may reduce the proteolytic activities of enzymes by preventing the formation of the enzyme–substrate complexes. This view was supported by the following observations. Leucine aminopeptidase in the nasal mucosa is known to cleave the B-chain of insulin from the N-terminal end. CyDs, especially DM-β-CyD and HP-β-CyD, reduce the activity of leucine aminopeptidase in a concentration-dependent manner.[114] The inhibition of proteolysis by these CyDs may participate in the absorption enhancement of peptides. Another potential barrier to the nasal absorption of peptide and protein drugs is the limitation in the size of hydrophilic pores through which they are thought to pass. The methylated CyDs significantly extracted membrane lipids, depending on the size and hydrophobicity of the CyD cavity in which lipids were included.[36] Therefore, lipid solubilization mediated by CyDs may result in transcellular processes, and these changes could be transmitted to the paracellular region, which is the most likely route for the transport of polypeptides.

The combined effect of β-CyD with absorption enhancers such as sodium glycocholate or Azone® on the nasal absorption of human fibroblast interferon-β in powder form in rabbits has been described.[115] HP-β-CyD was useful as a biocompatible solubilizer for lipophilic absorption enhancers involved in the nasal preparations of peptides.[116] When insulin was administered nasally to rats, simultaneous use of an oily penetration enhancer, HPE-101, (1-[2-(decylthio)-ethyl]azacyclopentane-2-one) or oleic acid solubilized in HP-β-CyD showed a marked increase in serum immunoreactive insulin levels and marked hypoglycaemia (Fig. 38.10). The potentiation of the enhancing effect of HPE-101 by HP-β-CyD can be explained by the facilitated transfer of HPE-101 into the nasal mucosa. Studies on the release of membrane proteins and scanning electron-microscopic observations of rat nasal mucosa indicated that the local mucosal damage due to the combination with HP-β-CyD may not be serious obstacles to their safe use.

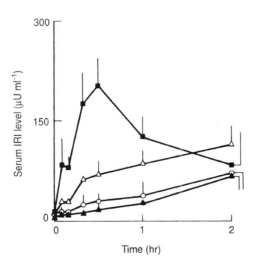 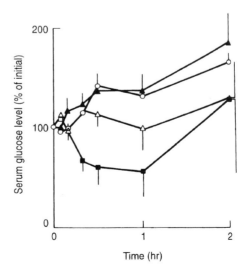

Fig. 38.10 Serum levels of immunoreactive insulin (IRI) and glucose after nasal administration of insulin (21 U/body) with HPE-101 (1% w/v) and/or HP-β-CyD (10% w/v) to rats. ○, Insulin alone; ▲, with HP-β-CyD; △, with HPE-101; ■, with HPE-101 and HP-β-CyD. Each point represents the mean ± SE of four rats.

Stabilization

Initial use of HP-β-CyD in peptide formulations has been reviewed.[117] For example, HP-β-CyD prevented the heat-induced charge alterations and loss of antigen-binding capacity of lyophilized monoclonal antibody (MN12) during storage.[118] Also, G_2-β-CyD has been shown to be an effective stabilizer in preventing the degradation of lyophilized tumour necrosis factor (TNF) during storage and freeze-drying processes.[119]

Basic fibroblast growth factor (bFGF) is a potent mitgen that stimulates the proliferation of a wide variety of cells and could play a crucial role in wound healing processes. The therapeutic potential of bFGF as not been fully realized, however, because of its susceptibility to proteolytic inactivation and short duration of retention at the site of action. Sulphated oligosaccharides, including a sodium salt of β-CyD sulphate (Na·β-CyD sul), have high affinity for bFGF and protect it from heat, acid and proteolytic degradation. Unfortunately, the highly hydrophilic nature of Na·β-CyD sul is not suited to the design of bFGF formulations with controlled-release features. A water-insoluble aluminium salt of β-CyD sulfate (Al·β-CyD sul) was prepared, and its possible utility as a stabilizer and sustained-release carrier for recombinant human bFGF was evaluated. An adsorbate of bFGF with Al·β-CyD sul was prepared by incubating the protein with a suspension of Al CyD sul in water.[120] The mitogenetic activity of bFGF released from the adsorbate, as indicated by the proliferation of kidney cells of baby hamster (BHK-21), was almost comparable with that of the intact bFGF. Interestingly, Al·β-CyD sul significantly protected bFGF from proteolytic degradation by pepsin and α-chymotrypsin compared with their sodium salts and other oligosaccharides (Table 38.5). The *in vivo* release of bFGF from the adsorbate was proportional to the increase in the Al·β-CyD sul/protein ratio in the adsorbate. Of the bFGF preparations evaluated, the adsorbate of bFGF with Al·β-CyD sul, when given subcutaneously to rats, showed the most prominent increase in the formation of granulation tissues, owing to the stabilization and slow release of the mitogen. These results suggest that the adsorbate of bFGF with Al·β-CyD sul has potent therapeutic efficacy for wound healing, and may be applicable to oral protein formulations for the treatment of intestinal mucosal erosions.

Prevention of self-association

Self-association of the insulin molecule into oligomers and macromolecular aggregates leads to complications in the development of long-term insulin therapeutic systems and limits the rate of subcutaneous absorptions, a process which is too slow to mimic the physiological plasma insulin profile at the time of meal consumption. These problems are further complicated by the tendency for insulin to adsorb on to the

Table 38.5 Effects of β-CyD suls and additives (25 mg ml^{-1}) on stability of bFGF(250 μg ml^{-1}) in the presence of pepsin(50 μg ml^{-1},pH 1.6) for 3 h at 37°C

CyDs	pH	Remaining bFGF(%)[a]
Without additives	1.67	ND
α-CyD	1.67	ND
β-CyD	1.60	ND
γ-CyD	1.66	ND
HP-β-CyD	1.65	ND
NA·α-CyD·Sul	1.74	1.6 ± 1.3
NA·β-CyD·Sul	1.74	3.3 ± 1.4
NA·γ-CyD·Sul	1.73	5.1 ± 1.0
NA·HP-β-CyD·Sul	1.63	4.5 ± 3.2
Al·α-CyD·Sul	3.81	96.4 ± 1.2
Al·β-CyD·Sul	3.88	99.2 ± 2.6
Al·γ-CyD·Sul	3.82	96.5 ± 3.1
Al·HP-β-CyD·Sul	3.86	93.6 ± 2.1
Dextran sulfate	1.84	20.2 ± 9.6
Sucralfate	3.78	59.5 ± 6.1

[a] Each value represents the mean ± SD of three determinations. ND, not detectable.

surfaces of containers and devices, perhaps by mechanisms similar to those inducing aggregation. Thus, many attempts have been made to prevent aggregation and surface adsorption in parenteral formulations.[119] Among the hydrophilic CyDs tested, HP-β-CyD and G_2-β-CyD significantly reduced the adsorption to containers and the self-association of insulin at neutral pH. Both hydrophilic β-CyDs facilitated the permeation of insulin through ultrafiltration membranes (Fig. 38.11). The increased

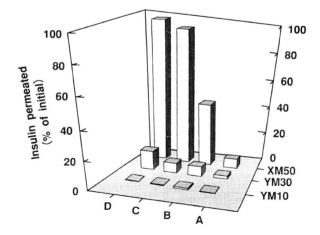

Fig. 38.11 Effects of β-CyDs (0.1 M) and EDTA (0.2 mM) on permeation of insulin (0.1 mM) through ultrafiltration membranes in phosphate buffer (pH 6.8, μ = 0.2) at 25°C. (A) Insulin alone, (B) with EDTA, (C) with HP-β-CyD, (D) with G_2-β-CyD. The membranes of YM-10, YM-30 and XM-50 were rated at pore sizes 10 000, 30 000, and 50 000 nominal molecular weight cut-off, respectively.

permeation of insulin was much greater than that of EDTA which is known to prevent the self-association of insulin by sequestering zinc ions from insulin oligomers. By the addition of β-CyDs, the surface tension of insulin solutions was increased, whereas it was decreased by EDTA. In the circular dichroism (CD) spectra of insulin, β-CyDs increased the negative CD intensity around 208 nm assigned to the α-helix structure of insulin, while it decreased that around 275 nm assigned to the antiparallel β-structure of insulin oligomers. These spectral changes were in good agreement with those observed when insulin aggregates are dissociated to monomer or lower-order aggregates. In ^{1}H NMR measurements, G_2-β-CyD significantly altered the aromatic regions of insulin, particularly B24(Phe) and B26(Tyr), which are involved in the association of insulin to form an antiparallel β-sheet around the carboxyl terminal of the B-chain. The inhibition mechanism of β-CyDs seems to be different from that of EDTA: EDTA sequesters the binding zinc ions from insulin and dissociates the hexamer or higher-order aggregates to the dimer, whereas β-CyDs may interact with hydrophobic amino acid residues to prevent the direct contact of insulin molecules.

H Combined use of CyDs with additives

In the previous sections, the inclusion ability of CyDs has mainly been focused on modification of the pharmaceutical properties of drug molecules. However, practical formulations usually contain considerable amounts of pharmaceutical excipients and/or additives to maintain the efficacy and safety of the drug molecules. To extend their functions, attention should be given to the improvement of the properties of pharmaceutical additives by means of complexation with CyDs. Since CyDs are capable of modifying the physicochemical properties of hydrophobic guest molecules, a suitable combination of CyDs with additives will be particularly useful in the potentiation of drug efficacy.[69] In this section, therefore, a new strategy of CyDs applications will be described for the design of advanced dosage forms.

Modified release of nifedipine

Conventional formulations of nifedipine must be dosed either twice or three times daily because of the short elimination half-life, due to considerable first pass metabolism, producing significant fluctuations in plasma drug concentrations. To attain a prolonged therapeutic effect and a reduced incidence of side-effects, many attempts have been made to maintain a suitable plasma level of nifedipine for long periods with much reduced frequency of administration. In the design of a modified-release formulation of nifedipine the following desirable attributes were sought: (a) the release should be at least 90% within 6–8 h, which is

the average passage time of a tablet in the human GI tract after oral administration; (b) adequate release of nifedipine in the early stage is necessary to reduce the significant first pass metabolism in the intestine and liver, which may offer a more balanced bioavailability; (c) nifedipine should be released according to zero-order kinetics over an extended period, to take advantage of such therapeutic advantages as duration of pharmacological effect and reduction of side-effects. With these goals, a series of formulations of double-layer tablets containing a fast-release portion (FRP) and a slow-release portion (SRP) were designed.[121,122] The *in vitro* release rate of nifedipine from tablets containing nifedipine-HP-β-CyD (molar ratio of 1:1) and small amount of HCO-60®, non-ionic surfactant, was much greater than those of drug or the complex alone, and this fast-release property was maintained for long periods of storage. In contrast, the release of nifedipine from hydroxypropylcellulose (HPC) solid dispersions was retarded with increasing viscosity of the HPC (L, M, H) or amount of each HPC. An optimal formulation of the double-layer tablet was tested by changing the ratios of the components (Fig. 38.12).[123]A double-layer tablet consisting of HP-β-CyD with 3% HCO-60®/(HPC-L:HPC-M) in a weight ratio of 1/(1.5:1.5) was found to be an appropriate modified-release formulation because it showed zero-order release from the SRP after the burst release from the FRP. This preparation exhibited prolonged plasma nifedipine levels without decrease of AUC in dogs, and the retarding behaviour was superior to that of a commercially available slow-release nifedipine product (Adalat L-20®). It is noteworthy that the release rate of nifedipine from this formulation was hardly affected by the pH of the medium and the rotation speed of the paddle in the dissolution test during accelerated storage (60°C, 75% RH). These findings suggest that a combination of HP-β-

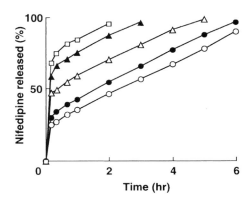

Fig. 38.12 Release profiles of nifedipine from double-layer tablets (25 mg as 5 mg nifedipine) consisting of nifedipine-HP-β-CyD with 3% HCO-60 as fast-release portion (FRP) and nifedipine-HPC-M as slow-release portion (SRP) in various weight ratios (FRP:SRP) in water at 37°C. ○, 1:3; ●, 1:2; △, 1:1; ▲, 2:1; □, 3:1.

CyD, HCO-60® and HPCs can serve as a modified-release carrier of nifedipine and can be applied to other poorly water-soluble drugs with short elimination half-lives.

Transdermal delivery of prostaglandin

Transdermal delivery of PGE_1 as an alternative to parenteral injection has engendered much recent interest in the treatment of peripheral vascular disorders. Since PGE_1 is chemically unstable and permeates poorly into the skin, a topical preparation capable of overcoming these problems needs to be devised to realise the full therapeutic potential of PGE_1. CME-β-CyD markedly improved the chemical stability of PGE_1 in aqueous solution and ointments.[124] Topical application of an ointment containing PGE_1-CME-β-CyD complex supplemented with HPE-101, an oily penetration enhancer, on to the skin of hairless mice resulted in a significant increase in the cutaneous blood flow owing to the vasodilating action of PGE_1.[125] The combination of CME-β-CyD and HPE-101 enhanced the percutaneous penetration of PGE_1 in a synergistic manner; CME-β-CyD assisted the release of HPE-101 from the ointment base and its entry into the skin, which may facilitate the percutaneous penetration of PGE_1.[126] Furthermore, this formulation suppressed the bioconversion of PGE_1 to give less pharmacologically active metabolites during the passage through the skin, which delivers intact PGE_1 more effectively to the site of action. Figure 38.13 shows the effects of topically applied PGE_1 or its CyD complexes in FAPG ointments supplemented with HPE-101 on the

peripheral vascular occlusive sequelae induced by sodium laurate in the ears of rabbits.[127] After the intra-arterial injection of sodium laurate, vascular occlusive lesions became progressively aggravated, finally resulting in degeneration or necrosis over the entire area of the ear. PGE_1 and its CyD complexes, when supplemented with HPE-101, significantly protected against the vascular occlusive sequelae. In particular, the ointment containing PGE_1–CME-β-CyD complex supplemented with HPE-101 showed the most marked inhibitory effect on the progress of the lesions; only regional discoloration was observed on day 7 after the injection of laurate. The inhibitory effect of PGE_1 ointments was consistent with the sequence of magnitude of their vasodilating actions. Skin compatibility tests of PGE_1 ointments suggested no serious obstacles to their safe use. Accordingly, topical application of PGE_1 may be a rational choice for minimizing systemic side-effects and patient compliance for long-term therapy.

Transmucosal delivery of morphine

The potent opioid morphine has been used for preoperative sedation and anxiolysis in paediatric patients and for the management of postoperative pain. In advanced cancer patients there is a great clinical need for the development of long-acting rectal preparations of morphine in the treatment of intractable chronic pain. However, the use of prolonged release of drugs from suppositories is limited, showing a wide interindividual variation of rectal bioavailability. We have demonstrated that some hydrophilic CyDs enhanced the rectal absorption of morphine in rabbits, and xanthan gum, a polysaccharide-type polymer with high swelling capacity, sustained the *in vitro* release of morphine from the suppository.[128] An attempt was made to design a more effective rectal delivery system for morphine using α-CyD as an absorption enhancer and xanthan gum as a swelling hydrogel in Witepsol H-15 hollow-type suppositories, and this was tested in rabbits.[129] With the combination of α-CyD and xanthan gum, the plasma morphine level was significantly sustained, with an improved bioavailability. Observation of the distribution behaviour of suppositories in rabbit rectum and colon after rectal administration showed that the xanthan gum prevented the upward spread of the drug in the rectum. This suggests that the absorption site for morphine should be limited to the lower part of the rectum so as to eliminate the first pass metabolism of morphine in the liver. In addition, gross and microscopic observations indicated that this preparation was less irritating to the rectal mucosa. From the viewpoints of safety and efficacy, therefore, this retentive opioid preparation may promise excellent therapeutic potential for the treatment of severe malignant cancer pain, offering an improvement in quality of life.

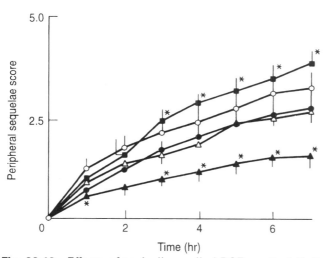

Fig. 38.13 Effects of topically applied PGE_1 or its β-CyD complexes in FAPG ointments (500 mg, 0.01% PGE_1) supplemented with HPE-101 (3 w/w%) on laurate-induced peripheral occlusive sequelae in ears of rabbits. ■, FAPG base; ○, PGE_1 alone; ●, PGE_1 with HPE-101; Δ, β-CyD complex with HPE-101; ▲, CME-β-CyD complex with HPE-101. Each point represents the mean ± SE of 3–6 rabbits. *$p < 0.05$ vs PGE_1 with HPE-101.

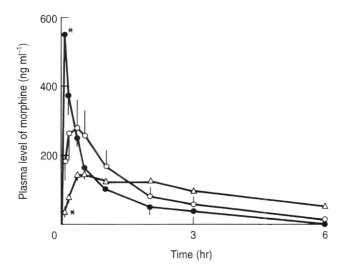

Fig. 38.14 Plasma levels of morphine after nasal administration of morphine hydrochloride (MH; 1 mg kg^{-1}) with DM-β-CyD or HP-γ-CyD in rats. ○, MH alone; ●, with DM-β-CyD (50 mM); Δ, with HP-γ-CyD (200 mM). Each point represents the mean ± SE of 2–8 rats. $^{*}p < 0.05$ versus drug alone.

Noninvasive transmucosal routes of administration for morphine have received much current attention because these routes could bypass the gastrointestinal degradation and hepatic first pass metabolism of morphine, as well as providing ease of administration. We have found that the nasal bioavailability of morphine in solution was greater than that of the rectal suppository in rats. Of the CyD derivatives tested, DM-β-CyD increased the rate of nasal absorption of morphine, when its plasma drug level–time profile was similar to that obtained after intravenous administration. Furthermore, DM-β-CyD increased the cerebrospin fluid (CSF) level of morphine after nasal administration, suggesting that DM-β-CyD may facilitate the transport of morphine not only to the systemic circulation but also to CSF. Interestingly, HP-γ-CyD sustained plasma levels of morphine (Fig. 38.14), probably owing to the formation of less membrane permeable complex and/or the increase in the viscosity of the solution, since a viscosity-enhancing polymer, HPC-H, sustained plasma levels of nasally administered morphine. These findings suggest that the combination of CyDs and viscous polymers may be useful for optimizing the transmucosal delivery of morphine and potentiating its analgesic action.

IV CYD-BASED SITE-SPECIFIC DRUG DELIVERY

CyD complexes are in equilibrium with guest and host molecules in aqueous solution, with the degree of the dissociation being dependent on the magnitude of the stability constant of the complex.[2,28] This property of CyD complexes is a desirable quality, because the complex dissociates to give free CyDs and drug at the absorption site, and only the drug in a free form enters into the systemic circulation.[6] However, the inclusion equilibrium is sometimes disadvantageous when drug targeting is to be attempted, because the complex dissociates before it reaches the organs or tissues to which it is to be delivered. One of the methods to prevent the dissociation is to bind a drug covalently to CyDs. A number of CyD derivatives have been synthesized for many purposes such as the construction of sophisticated enzyme models and reagents or catalysts for regio- and stereo-selective synthesis or chiral recognition.[130,131] CyD derivatives coupled to drugs such as enkephalin, doxorubicin, methicillin, ibuprofen and nucleobase were synthesized for modification of biological potencies.[132–136] In this section, recent results on site-specific delivery using CyDs are described.

A Colon targeting

CyDs are known to be barely capable of being hydrolysed and only slightly absorbed in passage through the stomach and small intestine. However, they are fermented by colonic microflora into small saccharides and thus absorbed as maltose or glucose in the large intestine.[137] Most bacteroid strains isolated from the human colon are capable of degrading CyDs, as shown by their ability to grow on CyDs using them as the sole carbon source and by the stimulation of cyclodextranase activity by exposure to CyDs.[138] This biodegradation property of CyDs is useful as a colon-targeting carrier, and thus CyD prodrugs may serve as a source of site-specific delivery of drugs to colon. Taking these factors into account, we have designed the CyD conjugates of a nonsteroidal anti-inflammatory drug, biphenylylacetic acid (BPAA),[139–141] and a steroidal drug, prednisolone,[142–144] and a short-chain fatty acid, *n*-butyric acid,[145] anticipating new candidates for colon-specific delivery prodrugs.

The amide- and ester-type conjugates of BPAA with three CyDs (α-, β- and γ-CyDs) are shown in Fig. 38.15. The release profiles of BPAA after incubation of the ester conjugates in rat gastrointestinal tract contents, intestine

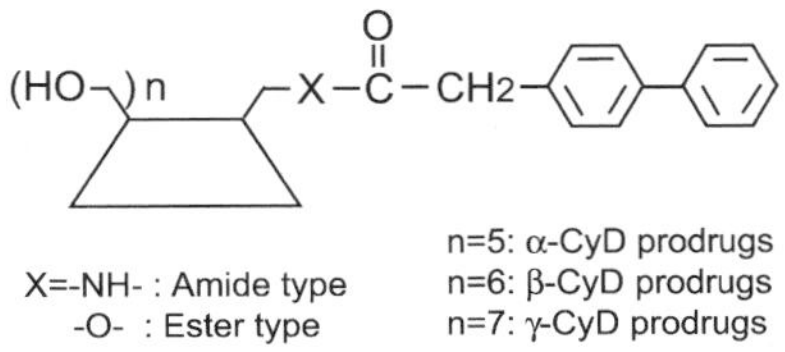

Fig. 38.15 Structures of BPAA/CyD amide and ester conjugates.

and liver homogenates, and blood in isotonic buffer solutions were compared with those of ethyl ester biphenyl-ylacetate, a simple ethyl ester of BPAA. Ethyl biphenyly-lacetate was easily hydrolysed in liver and gastrointestinal homogenates and also in blood. However, it was stable enough in the cecal and colonic contents. In sharp contrast, the α- and γ-CyD ester conjugates release BPAA significantly in cecal and colonic contents, while no appreciable drug release from the conjugates was observed on incubation with other contents of fluids. When the ester conjugates were incubated with rat cecal contents, the α- and γ-CyD conjugates produced BPAA quantitatively, while the β-CyD conjugate released the drug in only small amounts, probably because of its low solubility in water, compared with the α- and γ-CyD conjugates. Although the drug release patterns of the three CyD conjugates are different from each other, the *in vitro* data clearly suggest that the ester conjugates are firstly subject to the ring-opening of CyDs by bacterial enzymes to give the triose and maltose conjugates through longer linear oligosaccharide conjugates. Thus, the ester linkage of the small saccharide/BPAA conjugates could be highly suscep-tible to hydrolysis. On the other hand, the amide conjugates hardly released BPAA at all, despite fermentation to small saccharide/BPAA conjugates, because the amide linkage is resistant to the second hydrolysis. These results suggest the involvement of two type enzymes in the drug release of the CyD conjugates, i.e. firstly sugar-degrading enzymes and secondly ester-hydrolysing enzymes. This consecutive hydrolysis mechanism was demonstrated in the hydrolysis of the *n*-butyric acid/β-CyD ester conjugate catalysed by α-amylase (*Aspergillus oryzae*) and carboxylic esterase (porcine liver).[145] Figure 38.16 shows the serum levels of BPAA after oral administration of BPAA/α-, β- and γ-CyD

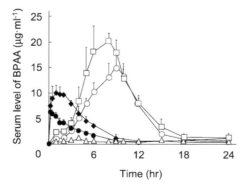

Fig. 38.16 Serum levels of BPAA after oral adminis-tration of the BPAA ester conjugates with α-CyD ($\bigcirc$), β-CyD ($\triangle$) and γ-CyD ($\blacklozenge$), BPAA alone ($\square$) or BPAA/β-CyD complex ($\bullet$) (equivalent to 10 mg kg^{-1} BPAA) to rats. Each point represents the mean $\pm$ S.E. of three experiments.

ester conjugates in rats, compared with the drug alone and β-CyD complex (a noncovalent inclusion complex prepared by the kneading method in a molar ratio of 1:1). The fast dissolving form of the β-CyD complex shows a rapid increase and decrease in the serum drug levels, compared with the drug alone. In the case of the β-CyD conjugate, little increase in serum level was observed, probably due to the low aqueous solubility. However, the serum drug levels of the α- and γ-CyD conjugates increased after a lag time of about 3 h and reached maximum levels at about 9 and 8 h, respectively, accompanying a significant increase in the extent of bioavailability. The extents of bioavailability for the α- and γ-CyD conjugates were about four and five times larger than that of BPAA alone, respectively. The anti-inflammatory effect of the BPAA/γ-CyD ester conjugate after oral administration was evaluated using the model of carageenan-induced acute edema in rat paw. The β-CyD complex, a fast dissolving form of BPAA, showed the rapid, but short anti-inflammatory response (about 3 h), because the drug is absorbed mainly from the upper small intestine. On the other hand, the response of the CyD conjugates was highest 12 h after the administration. These *in vivo* results indicate that the CyD conjugates need a lag time to exhibit the anti-inflammatory activity, because BPAA is produced after the conjugate had reached the cecum and colon.

Glucocorticoids are known to be effective in the treatment of various inflammatory and allergic disorders, and have been used for ulcerative colitis patients. However, oral and intravenous administrations of glucocorticoids to patients with severe ulcerative colitis are restricted, because of the undesirable systemic side-effects such as adrenosup-pression, immunosuppression, hypertension and osteoporo-sis, etc. In order to reduce such side-effects, rectal application has been used for therapy of ulcerative colitis. However, the chronic use of glucocorticoids in high doses often causes systemic side-effects even under topical application. One of the methods to overcome this problem, is to develop a prodrug which can be poorly absorbed from intestinal tracts and slowly releases the active drug at the site of action. The colon-specific drug delivery system may be particularly useful for the treatment of ulcerative colitis. Therefore, two kinds of prednisolone-appended CyD conjugates were prepared, i.e. prednisolone is covalently bound to CyD by ester and amide linkages through a spacer of succinic acid, as shown in Fig. 38.17.[142-144] The amide conjugate was prepared by binding prednisolone 21-hemisuccinate covalently to the amino group of mono(6-deoxy-6-amino)-β-CyD. This amide conjugate rapidly released prednisolone, due to the involvement of an effective intramolecular catalysis of the amide group, giving prednisolone and mono(6-deoxy-6-succimino)-β-CyD (half-life of the release at pH 7.0, 25°C = 6.5 min). There-fore, the amide conjugate can work as a fast-releasing prodrug of prednisolone, but is not suitable for the colon

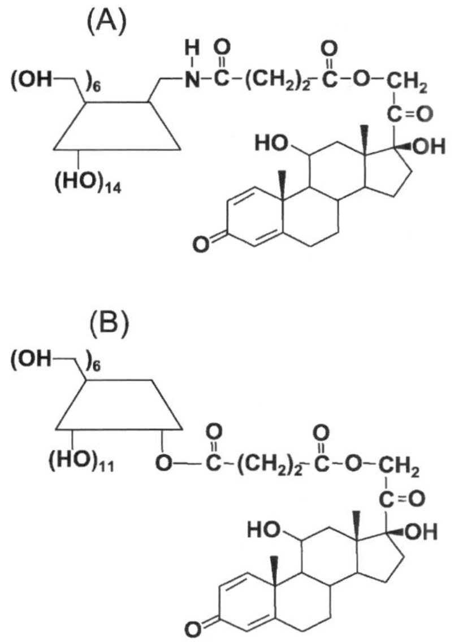

Fig. 38.17 Structures of prednisolone amide (A) and ester (B) conjugates with CyDs.

delivery system, because the amide conjugate spontaneously releases the drug before it reaches the colon. On the other hand, the ester conjugate was prepared by coupling the carboxyl group of prednisolone 21-hemisuccinate to one of the secondary hydroxyl groups of CyDs. The drug release rate of the ester conjugate was much slower than the amide conjugate, thus working as a colon delivery prodrug. Interestingly, the ester conjugates are very soluble (> 50% w/v) in water, the solubility being > 1000 times those of prednisolone and its succinate. This is in contrast to the decreased solubility of CyD derivatives conjugated at the primary hydroxyl groups. For example, the aqueous solubility of the BPAA and *n*-butyric acid conjugated at the primary hydroxyl group of β-CyD was about 1/10 that of the parent drugs. The decreased solubility is attributable to strong intermolecular association between the drug moieties and the neighbouring CyD cavity, forming a stable columnar packing structure in the solid state. On the other hand, the increased solubility of the prednisolone conjugated at the secondary hydroxyl group of β-CyD is due to the formation of the self-inclusion complex, which inhibits the intermolecular stacking association of the conjugate. The prednisolone-appended α-CyD ester conjugate was intracolonically administered to 2,4,6-trinitrobenzenesulfonic acid (TNBS)-induced colitis rats, and its anti-inflammatory and systemic adverse effects were compared with those of prednisolone alone and prednisolone/HP-β-CyD complex (a noncovalent inclusion complex in a molar ratio of 1:1). The colonic damage score (CDS), the ratio of distal colon wet weight to body weight (C/B) and the myeloperoxidase (MPO) activity were evaluated as measures of the

therapeutic effect of prednisolone, whereas the ratio of thymus wet weight to body weight (T/B) was evaluated as a measure of the side-effect of the drug. Healthy rats gave the CDS of 0, the C/B ratio of 0.0018, the MPO activity of 0.00016 units per mg tissues, and the T/B ratio of 0.0016. In the case of the control experiment where the saline solution without drugs was administered after the TNBS treatment, the CDS value, the C/B ratio and the MPO activity were 7.9, 0.0069 and 0.0145 units per mg tissue, respectively. The HP-β-CyD (3.1% w/v) and α-CyD (2.2% w/v) solutions without drugs gave the therapeutic indices the same as those of the control experiments, indicating no anti-inflammatory effects of the CyDs on the TNBS-induced colitis under experimental conditions. On the other hand, the intracolonic administration of prednisolone alone, the prednisolone/HP-β-CyD complex and the prednisolone/α-CyD conjugate (equivalent doses of 5 and 10 mg kg^{-1} of prednisolone) significantly decreased the CDS, the C/B ratio and the MPO activity. The CDS value and the MPO activity were in the order of the complex < the conjugate ≈ prednisolone alone, indicating the higher anti-inflammatory activity of the HP-β-CyD complex followed by the conjugate and the drug alone. The thymus atrophy is known to be a typical systemic adverse effect in steroid therapy. The prednisolone/HP-β-CyD complex gave the smallest T/B ratio (5.2−5.6 × 10^{-4}), indicating the significantly higher adverse effect. The decrease of the T/B ratio was also observed in the administration of prednisolone alone. On the other hand, the prednisolone/α-CyD conjugate gave the larger T/B ratio which was the same as that (1.16 × 10^{-3}) of the control experiment, indicating the small systemic adverse effect. These results suggest that the conjugate can alleviate the adverse effect of prednisolone without reducing its therapeutic effect. The *in vitro* and *in vivo* hydrolysis studies indicated that the conjugate maintains the local concentration in the colon at a low but constant level for a long period, due to the slow ester hydrolysis. The high anti-inflammatory effect with a low adverse effect was also observed in the oral administration of the ester conjugate to the TNBS-induced colitis rats.[144b] Therefore, CyDs can serve promoieties for colon-specific targeting prodrugs, and the CyD prodrug approach can provide a versatile means for construction of colon-specific delivery systems of certain drugs.

B Cell targeting

β-Lactam antibiotics show their lethal effect by inhibiting the synthesis of bacterial peptidoglycan, thereby disrupting the cell morphology and eventually resulting in cell lysis and death. Because the antibacterial effect of β-lactam is affected by the permeability of the outer membrane of bacteria, it is important in the evaluation of drug efficiency to measure the diffusion rate of drugs across the membrane. However, the β-lactamase enzyme expressed at the cell surface interferes with the measurement. In order to solve this problem, a β-lactamase inhibitor that is water-soluble and could reach the cell membrane, but not penetrate it, must be prepared. Nedra Kurunarate et al.[134] chose β-CyD as a bulky moiety for the prevention of drug entry across channels in the bacterial outer membrane; thus, they prepared β-CyD conjugates linked to an antibiotic, methicilline, at the molecular terminus through spacer of different chain lengths. The extent of inhibition of the surface β-lactamase of *P. aeruginosa* in the presence of 10 mM of each of the conjugates was as follows: 6% using the conjugate with the spacer length of 12.0 Å, 20% using the 14.0 Å conjugate, 43% using the 16.8 Å conjugate, 77% using the 20.0 Å conjugate, 91% using the 22.6 Å conjugate, and 100% using 23.7 Å conjugate. They concluded that the length of the spacer should be greater than 16 Å for optimum inhibition of β-lactamase in the outer membrane. Kim et al.[146] reported that the α-CyD conjugate with an antitumour sulfonylurea selectively blocks NADH oxidase activity at the external plasma membrane surface of HeLa cells.

Intercellular recognition events are fundamental to many biological processes in which oligosaccharides on cell surface glycoproteins or lectins have been responsible for cell–cell recognition and adhesion. Therefore, CyD derivatives bearing small saccharides may be useful as a carrier for transporting active drugs to sugar receptors, such as lectins located on the cell surface. In accordance with this concept, several CyD derivatives with mono- and di-saccharides, such glucose, galactose, mannose, fucose, etc., have been prepared and investigated for binding characteristics to sugar-specific receptors.[147–149] The β-CyD conjugates with galactose exhibited higher recognition by the galactose-specific *K. bulgaricus* cell wall lectin.[150] It is reported that some galactose and fucose conjugates have a significant cytotoxic effect on the human rectal adenocarcinoma cell line, with P-glycoprotein-positive multidrug resistance.[151] The sugar-substituted CyD derivatives may offer a new way of delivering drugs selectively to specific cell surfaces of organs such as liver and colon, although the uptake of drugs into cells may decrease due to the presence of a bulky, hydrophilic CyD moiety.

Péan et al.[152,153] reported that β- and γ-CyD derivatives coupled with the neuropeptide substance P bind to the neurokinin receptor of rat brain and cultured Chinese hamster ovary (CHO) cells transfected with the human neurokinin receptor gene, *in vitro*. The *in vivo* intracerebral injection of substance P/γ-CyD conjugate in the rat striatum induced a massive internalization of the receptor, and the conjugate recognized the neurokinin receptor-bearing neurons. Schaschke et al.[154] prepared the β-CyD conjugates linked covalently to tetra- and hepta-peptide analogues of

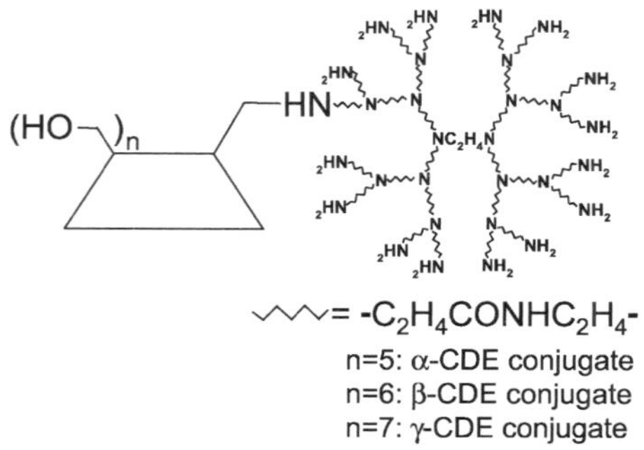

$\sim\!\sim\!\sim$ = -C₂H₄CONHC₂H₄-
n=5: α-CDE conjugate
n=6: β-CDE conjugate
n=7: γ-CDE conjugate

Fig. 38.18 Structures of CDE conjugates.

gastrin through succinyl amide bond and investigated their binding characteristics of G-protein-coupled CCK-B/gastrin receptor expressed in CHO cells. The receptor affinity of the heptapeptide/β-CyD conjugate was identical to that of the unconjugated tetrapeptide, although it was weaker than that of corresponding unconjugated heptapeptide. However, the functional assay monitored by inositol phosphate production was potent as the corresponding parent peptides. They concluded by using docking experiments of the conjugates into the receptor model that the high hormonal potency of the heptapeptide conjugate is ascribed to a synergetic effect of the specific intermolecular interaction of the peptide with the receptor and the unspecific interaction of the β-CyD moiety with the receptor surface.

Gene therapy requires carriers that can efficiently and safely transfer the gene into the nucleus of the desired cells. Nonviral vectors have many advantages over viral vectors, such as ease of manufacture, safety, low immunogenicity, and molecular attachment of targeting ligand. Among these nonviral vectors, starburst polyamidoamine dendrimers have received attention as cationic vector for gene transfection, because of their unique molecular shape and the condensed positive charges. To improve the transfection efficiency, we prepared the α-, β- and γ-CyD conjugates with the starburst dendrimer (CDE conjugates, Fig. 38.18), and investigated their transfection ability in NIH3T3 and RAW264.7 cells.[155] The CDE conjugates interacted electrostatically with plasmid DNA (pDNA) and protected the degradation of pDNA by DNAase I. The CDE conjugates showed a potent luciferase gene expression, especially in the α-CDE conjugate which provided the greatest transfection activity (approximately 100 times higher than those of the dendrimer alone and of the physical mixture of the dendrimer and α-CyD) among the three conjugates. These findings suggest that the α-CDE conjugate could be a new and preferable nonviral vector of pDNA.

C   Brain targeting

When a drug is covalently coupled to 1-methyl-1,4-dihydronicotinic acid through an enzymatically labile linkage, its lipophilicity increases and allows selective delivery of drugs into the brain across the blood–brain barrier, which is characterized by the endothelial cells of cerebral capillaries that have tight continuous circumferential junctions, thus restricting the passage of polar drugs to the brain.[156] After entry into the brain, the dihydropyridine moiety is oxidized by oxidoreductase to 1-methylpyridinium cation. This polar drug/1-methylpyridinium cation is trapped in the brain, and releases the drug by action of other enzymes. This is an essential concept of Bodor's chemical delivery system, and is applied to brain targeting of drugs such as steroids, antitumour agents and calcium-channel antagonists.[157] The problem, however, is that the prodrugs of the chemical delivery system are poorly water-soluble due to the presence of the lipophilic moiety. HP-β-CyD solved this solubility problem by means of soluble complex formation, together with enhancing the chemical stability of dihydronicotinic acid in water. For example, the i.v. administration of estradiol chemical delivery systems (5 mg kg^{-1}) solubilized in 20% HP-β-CyD produced a higher concentration of the prodrug in rat brain, which was almost the same as that produced by the administration of the prodrug (15 mg kg^{-1}) solution in dimethylsulfoxide.[45]

Unfortunately, application of CyD conjugates to brain targeting are few. One of the examples is the β-CyD conjugate with δ-opioid receptor peptide, *N*-leucine-enkephalin and its cyclic analogue [*p*-I-Phe^4, D-Pen^2, D-Pen^5]enkephalin, where the carbonyl group of the C terminal leucine was coupled with 6-amino-6-deoxy β-CyD and in the latter conjugate, all hydroxyl groups of CyDs were further methylated to increase the lipophilicity.[158,159] Although the potency of the latter conjugate decreased in receptor binding assay and in the *in vitro* guinea pig intestine and mouse spermatic duct bioassay systems, it showed potent antinociceptive properties when given i.c.v. and i.v. in the mouse tail flick test and exhibited no toxicity. Furthermore, it showed prolonged action in the bioassay. Although the detailed transport mechanism of these conjugates to the brain has not been fully elucidated, this methodology can be applied to other neuropharmaceuticals, such as morphine. The β-CyD conjugate with *N*-leucine-enkephalin is of interest, because it has a vacant cavity and can include a neurotropic drug, dothiepin. This inclusion property of conjugates may be useful from the viewpoint of drug formulation, because two different drugs can be incorporated into the CyD molecule.

V CONCLUSION

The natural CyDs and their synthetic derivatives have successfully been applied to improve undesirable drug properties such as solubility, stability or bioavailability. In addition, the enhancement of drug activity and selective transfer or the reduction of side-effects has been achieved by means of inclusion complex formation. The bioadaptable CyDs can serve as potent absorption enhancers or coenhancers, and the release rate of water-soluble drugs, including peptides, can be appropriately modified by combination of hydrophobic, hydrophilic and ionizable CyDs. Since CyDs are also useful in extending the functions of pharmaceutical additives, the combination of molecular encapsulation with other carrier materials will be an effective and a valuable tool in the improvement of drug properties. Although the various CyDs described here have many advantages as novel drug carriers for development of advanced dosage forms, the toxicological issues together with their biological fates need to be investigated in detail.

REFERENCES

1. Szejtli, J. (1982) *Cyclodextrins and Their Inclusion Complexes.* Akademiai Kiadó, Budapest.
2. Saenger, W. (1980) Cyclodextrin inclusion compounds in research and industry. *Angew. Chem. Int. Ed Engl.* **19**: 344–362.
3. Szejtli, J. (1988) *Cyclodextrin Technology.* Kluwer Academic Publishers, Dordrecht, The Netherlands.
4. Uekama, K. (1981) Pharmaceutical applications of cyclodextrin complexations. *Yakugaku Zasshi* **101**: 857–873.
5. Duchêne, D. (1987) *Cyclodextrins and Their Industrial Uses.* Editions de Sante, Paris.
6. Uekama, K. and Otagiri, M. (1987) Cyclodextrins in drug carrier systems. *CRC Crit. Rev. Ther. Drug Carrier Systems* **3**: 1–40.
7. Duchêne, D. (1991) *New trends in Cyclodextrins and Their Derivatives.* Editions de Sante, Paris.
8. Uekama, K., Hirayama, F. and Irie, T. (1998) Cyclodextrin drug carrier systems. *Chem. Rev.* **98**: 2045–2076.
9. Szejtli, J. (1983) Dimethyl-β-cyclodextrin as parente drug carrier. *J. Incl. Phenom.* **1**: 135–150.
10. Uekama, K. (1985) Pharmaceutical applications of methylated cyclodextrins. *Pharm. Int.* **6**: 61–65.
11. Pitha, J. and Pitha, J. (1985) Amorphous water-soluble derivatives of cyclodextrins, non-toxic dissolution enhancing excipients. *J. Pharm. Sci.* **74**: 987–990.
12. Müller, B. W. and Brauns, U. (1985) Solubilization of drugs by modified β-cyclodextrins. *Int. J. Pharm.* **26**: 77–88.
13. Yoshida, A., Arima, H., Uekama, K. and Pitha, J. (1988) Pharmaceutical evaluation of hydroxyalkyl ethers of β-cyclodextrins. *Int. J. Pharm.* **46**: 217–222.
14. Yoshida, A., Yamamoto, M., Irie, T., *et al.* (1989) Some pharmaceutical properties of 3-hydroxypropyl- and 2,3-dihydroxypropyl-β-cyclodextrins and their solubilizing and stabilizing abilities. *Chem. Pharm. Bull* **37**: 1059–1063.
15. Brewster, M. E., Simpkins, J. W., Hora, M. S., Stern, W. C. and Bodor, N. (1989) The potential use of cyclodextrins in parenteral formulations. *J. Parent. Sci. Technol.* **43**: 231–240.
16. Koizumi, K., Utamura, T., Sato, M. and Yagi, Y. (1986) Isolation and characterization of branched cyclodextrins. *Carbohydr. Res.* **153**: 55–67.

17. Yamamoto, M., Yoshida, A., Hirayama, F. and Uekama, K. (1989) Some physicochemical properties of branched β-cyclodextrins and their inclusion characteristics. *Int. J. Pharm.* **49**: 163–171.

18. Hirayama, F. (1993) Development and pharmaceutical evaluation of hydrophobic derivatives as modified-release drug carrier. *Yakugaku Zasshi* **113**: 425–437.

19. Hirayama, F., Kurihara, M., Horichi, Y., *et al.* (1993) Preparation of heptakis(2,6-di-*O*-ethyl)-[3-cyclodextrin and its nuclear magnetic resonance spectroscopic characterization. *Pharm. Res.* **10**: 208–213.

20. Uekama, K., Arima, H., Irie, T., *et al.* (1989) Sustained release of buserelin acetate, a luteinizing hormone-releasing hormone agonist, from the injectable oily preparation utilizing ethylated β-cyclodextrin. *J. Pharm. Pharmacol.* **41**: 874–876.

21. Uekama, K., Horiuchi, Y., Irie, T. and Hirayama, F. (1989) *O*-Carboxymethyl-*O*-ethyl-cyclomaltoheptaose as a delayed-release type drug carrier. Improvement of the oral bioavailability of diltiazem. *Carbohydr. Res.* **192**: 323–330.

22. Folkman, J., Weisz, P. B., Joullie, M. M., *et al.* (1989) Control of angiogenesis with synthetic heparin substitutes. *Science* **243**: 1490–1493.

23. Uekama, K. (1987) Cyclodextrin inclusion compounds: effects on stability and bio-pharmaceutical properties. In Breimer, D. D. and Speiser, P. (eds). *Topics in Pharmaceutical Sciences*, pp. 181–194. Elsevier, Amsterdam.

24. Coleman, A. W., Nicolis, I., Keller, N. and Dalbiez, J. P. (1992) Aggregation of cyclodextrins; an explanation of the abnormal solubility of β-cyclodextrin. *J. Incl. Phenom.* **13**: 139–143.

25. Okada, Y., Kubota, Y., Koizumi, K., *et al.* (1988) Some properties and inclusion behaviour of branched cyclodextrins. *Chem. Pharm. Bull.* **36**: 2176–2185.

26. Arima, H., Kondo, T. and Irie, T. (1992) Use of water-soluble β-cyclodextrin derivatives as carriers of anti-inflammatory drug biphenylylacetic acid in rectal delivery. *Yakugaku Zasshi* **112**: 65–72.

27. Horiuchi, Y., Hirayama, F. and Uekama, K. (1991) Improvement of stability and bioavailability of 1-hexylcarbamoyl-5-fluorouracil (HCFU) by *O*-carboxymethyl-*O*-ethyl-β-cyclodextrin. *Yakugaku Zasshi* **111**: 592–599.

28. Bender, M. L. and Komiyama, M. (1978) *Cyclodextrin Chemistry*. Springer-Verlag, Berlin.

29. Hirayama, F., Kurihara, M., Utsuki, T. and Uekama, K. (1993) Inhibitory effect of guest molecules on acid-catalyzed ring-opening of β-cyclodextrin. *J. Chern. Soc., Chem. Commun.* 1578–1580.

30. Marshall, J. J. and Miwa, I. (1981) Kinetic difference between hydrolyses of γ-cyclodextrin by human salivary and pancreatic α-amylases. *Biochim. Biophys. Acta.* **661**: 142–147.

31. Yamamoto, M., Aritomi, H., Irie, T., *et al.* (1991) Biopharmaceutical evaluation of maltosyl-β-cyclodextrin as a parenteral drug carrier. *S.T.P. Pharm. Sci.* **1**: 397–402.

32. Brewster, M. E., Estes, K. S. and Bodor, N. (1990) An intravenous toxicity of study of 2-hydroxypropyl-β-cyclodexttin, a useful drug solubilizer, in rats and monkeys. *Int. J. Pharm.* **59**: 231–243.

33. Frijlink, H. W., Visser, J., Hefting, N. R., *et al.* (1990) The pharmacokinetics of β-cyclodextrin and hydroxypropyl-β-cyclodextrin in the rat. *Pharm. Res.* **7**: 1248–1252.

34. Frank, D. W., Gray, J. E. and Weaver, R. N. (1976) Cyclodextrin nephrosis in the rat. *Am. J. Pathol.* **83**: 367–382.

35. Irie, T. and Uekama, K. (1997) Pharmaceutical application of cyclodextrins III. Toxicological issues and safety. *J. Pharm. Sci.* **86**: 147–162.

36. Ohtani, Y., Irie, T., Uekama, K., *et al.* (1989) Differential effects of α-, β-,and γ-cyclodextrins on human erythrocytes. *Eur. J. Biochem.* **186**: 17–22.

37. Gerloczy, A., Fonagy, A., Keresztes, P., *et al.* (1985) Absorption, distribution, excretion and metabolism of orally administered ^{14}C-β-cyclodextrin in rats. *Arzneimittel-Forsch.* **35**: 1042–1047.

38. Poelma, F. G. J., Tukker, J. J., Hilbers, H. W. and Jansen, A. C. A. (1989) Intestinal absorption of drugs II. The effects of inclusion in cyclodextrins on the absorption of dantrolene. *J. Incl. Phenom.* **7**: 423–430.

39. Arima, H., Kondo, T., Irie, T. and Uekama, K. (1992) Enhanced rectal absorption and reduced local irritation of anti-inflammatory drug ethyl 4-biphenylyl acetate in rats by complexation with water-soluble β-cyclodextrin derivatives and formulations as oleaginous suppository. *J. Pharm. Sci.* **81**: 1119–1125.

40. Uekama, K., Imai, T., Maeda, T., *et al.* (1985) Improvement of dissolution and suppository release characteristics of flurbiprofen by inclusion complexation with heptakis(2,6-di-*O*-methyl)-β-cyclodextrin. *J. Pharm. Sci.* **74**: 841–845.

41. Arima, H., Adachi, H., Irie, T., *et al.* (1990) Enhancement of the anti-inflammatory effect of ethyl 4-biphenylylacetate in ointment by β-cyclodextrin derivatives: increased absorption and localized activation of the prodrug in rats. *Pharm. Res.* **7**: 1152–1156.

42. Zhang, M.-Q. and Rees, D.C. (1999) A review of recent application of cyclodextrins for drug discovery. *Exp. Opin. Ther. Patents* **9**: 1697–1717.

43. Dietzel, K., Estes, K.S., Brewster, M.E., *et al.* (1990) The use of 2-hydroxypropyl-β-cyclodextrin as a vehicle for intravenous administration of dexamethasone in dogs. *Int. J. Pharm.* **59**: 225–230.

44. Hoshino, T., Uekama, K. and Pitha, J. (1993) Increase in temperature enhances solubility of drugs in aqueous solutions of hydroxypropylcyclodextrins. *Int. J. Pharm.* **98**: 239–242.

45. Brewster, M.E., Estes, K.S., Loftsson, T., *et al.* (1988) Improved delivery through biological membranes. XXXI: Solubilization and stabilization of an estradiol chemical delivery system by modified β-cyclodextrins. *J. Pharm. Sci.* **77**: 981–985.

46. Millard, W.J., Romano, T.M., Bodor, N. and Simpkins, J.W. (1990) Growth hormome (GH) secretory dynamics in animals administered estradiol utilizing a chemical delivery system. *Pharm. Res.* **7**: 1011–1018.

47. Merkus, F. W. H. M., Verhoef, J. C., Romeijn, S. G. and Schipper, N. G. M. (1991) Absorption enhancing effect of cyclodextrins on intranasally administered insulin in rats. *Pharm. Res.* **8**: 588–592.

48. Kurihara, M., Hirayama, F., Uekama, K. and Yamasaki, M. (1990) Improvement of some pharmaceutical properties of nocloprost by β- and β-cyclodextrin complexations. *J. Incl. Phenom.* **8**: 363–373.

49. Utsuki, T., Imamura, K., Hirayama, F. and Uekama, K. (1993) Stoichiometry-dependent changes of solubility and photo reactivity of an antiulcer agent, 2′-carboxymethoxy-4,4′-bis(3-methyl-2-butenyloxy)chalcone, in cyclodextrin inclusion complexes. *Eur. J. Pharm. Sci.* **1**: 81–87.

50. Hirayama, F., Kurihara, M. and Uekama, K. (1984) Improving the aqueous stability of prostaglandin E_2 and prostaglandin A_2 by inclusion complexation with methylated β-cyclodextrins. *Chem. Pharm. Bull.* **32**: 4237–4240.

51. Yamamoto, M., Hirayama, F. and Uekama, K. (1992) Improvement of stability and dissolution of prostaglandin E_1 by maltosyl-β-cyclodextrin in lyophilized formulation. *Chem. Pharm. Bull.* **40**: 747–751.

52. Hirayama, F., Kurihara, M. and Uekama, K. (1987) Improvement of chemical instability of prostacyclin in aqueous solution by complexation with methylated cyclodextrins. *Int. J. Pharm.* **35**: 193–199.

53. Uekama, K., Fujinaga, T., Hirayama, F., *et al.* (1983) Improvement of the oral bioavailability of digitalis glycosides by cyclodextrin complexation. *J. Pharm. Sci.* **72**: 1338–1341.

54. Yoshida, A., Yamamoto, M., Hirayama, F. and Uekama, K. (1988) Improvement of chemical instability of digitoxin in aqueous solution by complexation with β-cyclodextrin derivatives. *Chem. Pharm. Bull.* **36**: 4075–4080.

55. Uekama, K., Fujinaga, T., Hirayama, F., *et al.* (1982) Effects of cyclodextrins on the acid hydrolysis of digoxin. *J. Pharm. Pharmacol* **34**: 627–630.

56. Seo, H. and Uekama, K. (1989) Enhanced bioavailability of digoxin by γ-cyclodextrin complexation: evaluation for sublingual and oral administrations in human. *Yakugaku Zasshi* **109**: 778–782.

57. Sugimoto, I., Kuchiki, K. and Nakagawa, H. (1981) Stability of nifedipine-polyvinylpyrrolidone coprecipitate. *Chem. Pharm. Bull.* **29**: 1715–1723.

58. Uekama, K., Ikegami, K., Wang, Z., *et al.* (1992) Inhibitory effect of 2-hydroxypropyl-β-cyclodextrin on crystal-growth of nifedipine during storage: superior dissolution and oral bioavailability compared with polyvinylpyrrtolidone K-30. *J. Pharm. Pharmacol.* **44**: 73–78.

59. Kikuchi, M., Hirayama, F. and Uekama, K. (1987) Improvement of chemical instability of carmofur in β-cyclodextrin solid complex by utilizing some organic acids. *Chem. Pharm. Bull* **35**: 315–319.

60. Kikuchi, M., Uemura, Y., Hirayama, F., *et al.* (1984) Improvement of some pharmaceutical properties of carmofur by cyclodextrin complexation. *J. Incl. Phenom.* **2**: 623–630.

61. Kikuchi, M., Hirayama, F. and Uekama, K. (1987) Improvement of oral and rectal bioavailability of carmofur by methylated β-cyclodextrin complexations. *Int. J. Pharm.* **38**: 191–198.

62. Hirayama, F., Hirashima, N., Abe, K., *et al.* (1988) Utilization of diethyl-β-cyclodextrin as a sustained-release carrier for isosorbide dinitrate. *J. Pharm. Sci.* **77**: 233–236.

63. Uekama, K., Hirashima, N., Horiuchi, Y., *et al.* (1987) Ethylated β-cyclodextrins as hydrophobic drug carriers: sustained release of diltiazem in the rat. *J. Pharm. Sci.* **76**: 660–661.

64. Uekama, K., Horikawa, T., Yamanaka, M. and Hirayama, F. (1994) Peracylated β-cyclodextrin as novel sustained-release carrier for water-soluble drug, molsiodomine. *J. Pharm. Pharmacol.* **46**: 714–717.

65. Uekama, K., Horikawa, T., Horiuchi, Y. and Hirayama, F. (1993) *In vitro* and *in vivo* evaluation of delayed-release behaviour of diltiazem from its *O*-carboxymethyl-*O*-ethyl-β-cyclodextrin complex. *J. Control. Rel.* **25**: 99–106.

66. Horiuchi, Y., Abe, K., Hirayama, F. and Uekama, K. (1991) Control of theophylline by β-cyclodextrin derivatives: hybridizing of hydrophilic and ionizable β-cyclodextrin complexes. *J. Control. Rel.* **15**: 177–182.

67. Matsubara, K., Kuriki, T., Arima, H., *et al.* (1990) Possible use of diethyl-β-cyclodextrin in preparation of sustained release oily injection of busenelin acetate (LHRH agonist). *Drug Del. Syst.* **5**: 95–99.

68. Matsubara, K., Irie, T. and Uekama, K. (1994) Controlled-release of the LHRH agonist buserelin acetate from injectable suspensions containing triacetylated cyclodextrins in the oil vehicle. *J. Control. Rel.* **31**: 173–180.

69. Uekama, K., Hirayama, F. and Irie, T. (1994) Application of cyclodexterin. In Boer, A. G. (ed.). *Absorption Enhancement: Concept, Possibilities and Limitations*, vol. 3, pp. 411–456. Harwood Publishers, Amsterdam.

70. Tokumura, T., Nanba, M., Tsushima, Y., *et al.* (1986) Enhancement of bioavailability of cinnarizine from its β-cyclodextrin complex on oral administration with DL-phenylalanine as a competing agent. *J. Pharm. Sci.* **75**: 391–394.

71. Seo, H., Tsuruoka, M., Hashimoto, T., *et al.* (1983) Enhancement of oral bioavailability of spironolactone by β- and γ-cyclodextrin complexations. *Chem. Pharm. Bull.* **31**: 286–291.

72. Uekama, K., Fujinaga, T., Otagiri, M., *et al.* (1983) Improvement of dissolution and chemical stability of proscillaridin by cyclodextrin complexations. *Acta Pharm. Suec.* **20**: 287–294.

73. Uekama, K., Otagiri, M., Uemura, Y., *et al.* (1983) Improvement of oral bioavailability of prednisolone by β-cyclodextrin complexation in humans. *J. Pharmacobio-Dyn.* **6**: 124–127.

74. Uekama, K., Narisawa, S., Hirayama, F. and Otagiri, M. (1983) Improvement of dissolution and absorption characteristics of benzodiazepines by cyclodextrin complexation. *Int. J. Pharm.* **16**: 327–338.

75. Uekama, K., Matsuo, N., Hirayama, F., *et al.* (1980) Enhanced bioavailability of acetohexamide by β-cyclodextrin complexation. *Yakugaku Zasshi* **100**: 903–909.

76. Miyaji, T., Inoue, Y., Acarrurk, F., *et al.* (1992) Improvement of oral bioavailability of fenbufen by cyclodextrin complexations. *Acta Pharm. Nord* **4**: 17–22.

77. Irie, T., Tsunenari, Y., Uekama, K. and Pitha, J. (1988) Effect of bile on the intestinal absorption of α-cyclodextrin in rats. *Int. J Pharm.* **43**: 41–44.

78. Nakanishi, K., Nadai, T., Masada, M. and Miyajima, K. (1992) Effect of cyclodextrins on biological membrane II. Mechanism of enhancement of the intestinal absorption of non-absorbable drug by cyclodextrins. *Chem. Pharm. Bull.* **40**: 1252–1256.

79. Horiuchi, Y., Kikuchi, M., Hirayama, F., *et al.* (1988) Improvement of bioavailability of menaquinone-4 by dimethyl-β-cyclodextrin complexation following oral administration. *Yakugaku Zasshi.* **108**: 1093–1100.

80. Ueno, M., Ijitsu, T., Horiuchi, Y., *et al.* (1989) Improvement of dissolution and absorption characteristics of ubidecarenon by dimethyl-β-cyclodextrin complexation. *Acta Pharm. Nord.* **2**: 99–104.

81. Otagiri, M., Imai, T. and Uekama, K. (1982) Enhanced oral bioavailability of anti-inflammatory drug flurbiprofen in rabbits by tri-*O*-methyl-β-cyclodextrin complexation. *J. Pharmacobio-Dyn.* **5**: 1027–1029.

82. Nakai, Y., Yamamoto, K., Terada, K., *et al.* (1983) Interaction of tri-*O*-methyl-β-cyclodextrin with drugs, enhanced bioavailability of ketoprofen in rats when administered with tri-*O*-methyl-β-cyclodextrin. *Chem. Pharm. Bull* **31**: 3745–3747.

83. Kikuchi, M. and Uekama, K. (1988) Enhancement of antitumor activity of carmofur (HCFU) by dimethyl-β-cyclodextrin complexation in P-388 leukemia-bearing mice. *Xenobiotic Metab. Dispos.* **3**: 267–273.

84. Uekama, K., Horiuchi, Y., Kikuchi, M., *et al.* (1988) Enhanced dissolution and oral bioavailability of α-tocopheryl esters by dimethyl-β-cyclodextrin complexations. *J. Incl. Phenom.* **6**: 167–174.

85. Duchêne, D., Wouessidjewe, D. and Poelman, M.-C. (1991) Dermal uses of cyclodextrin and derivatives. In Duchene, D. (ed.). *New Trends in Cyclodextrins and Their Derivatives*, pp. 447–481. Editions de Sante, Paris.

86. Otagiri, M., Fujinaga, T., Sakai, A. and Uekama, K. (1984) Effects of β- and γ-cyclodextrins on release of betamethasone from ointment bases. *Chem. Pharm. Bull.* **32**: 2401–2405.

87. Orienti, I., Zecchi, V., Bettasi, V. and Fini, A. (1991) Release of ketoprofen from dermal bases in presence of cyclodextrins: effect of the affinity constant determined in semisolid vehicles. *Arch. Pharm. (Weinheim)* **324**: 943–947.

88. Uekama, K., Otagiri, M., Sakai, A., *et al.* (1985) Improvement in the percutaneous absorption of beclomethasone dipropionate by γ-cyclodextrin complexation. *J. Pharm. Pharmacol.* **37**: 532–535.

89. Arima, H., Adachi, H., Irie, T. and Uekama, K. (1990) Improved drug delivery through the skin by hydrophilic β-cyclodextrins: enhancement of anti-inflammatory effect of 4-biphenylylacetic acid in rats. *Drug. Invest.* **2**: 155–161.

90. Uekama, K., Maeda, T. and Arima, H., *et al.* (1986) Possible utility of β-cyclodextrin complexation in the preparation of biphenylylacetic acid suppository. *Yakugaku Zasshi* **106**: 1126–1130.

91. Arima, H., Irie, T. and Uekama, K. (1989) Differences in the enhancing effects of water-soluble β-cyclodextrins on the release of ethyl 4-biphenylyl acetate, an anti-inflammatory agent from an oleaginous suppository base. *Int. J. Pharm.* **57**: 107–115.

92. Hermens, W. A. J. J., Deurloo, M. J. M., Romeyn, S. G., *et al.* (1990) Nasal absorption enhancement of 17β-estradiol by dimethyl-β-cyclodextrin in rabbits and rats. *Pharm. Res.* **7**: 500–503.

93. Schipper, N. G. M., Hermens, W. A. J. J., Romeyn, S. G., *et al.* (1990) Nasal absorption of 17β-estradiol and progesterone from a dimethyl-cyclodextrin inclusion formulation in rats. *Int. J. Pharm.* **64**: 61–66.

94. Stuenkel, C. A., Dudley, R. E. and Yen, S. S. C. (1991) Sublingual administration of testosterone-hydroxypropyl-β-cyclodextrin inclusion complex simulates episodic androgen release in hypogonadal men. *J. Clin. Endocrinol. Metab.* **72**: 1054–1059.

95. Schipper, N. G. M., Verhoef, J. C. and Merkus, F. W. H. M. (1991) The nasal mucociliary clearance: relevance to nasal drug delivery. *Pharm. Res.* **8**: 807–814.

96. Hermens, W. A. J. J., Belder, C. W. J., Merkus, J. M. W. M., *et al.* (1991) Intranasal estradiol administration to oophorectomized women. *Eur. J. Obstet. Gynecol. Reprod. Biol.* **40**: 35–41.

97. Yoshida, A., Yamamoto, M., Itoh, T., *et al.* (1990) Utility of 2-hydroxypropyl-β-cyclodextrin in an intramuscular injectable preparation of nimodipine. *Chem. Pharm. Bull.* **38**: 176–179.

98. Irie, T., Otagiri, M., Uekama, K., *et al.* (1984) Alleviation of the chlorpromazine-induced muscular tissue damage by β-cyclodextrin complexation. *J. Incl. Phenom.* **2**: 637–644.

99. Imai, T., Maeda, T., Otagiri, M. and Uekama, K. (1987) Improvement of absorption characteristics and reduction of irritation on stomach of flurbiprofen by complexation with various cyclodextrins. *Xenobiotic Metab. Dispos.* **2**: 657–664.

100. Ishida, K., Hoshino, T., Irie, T. and Uekama, K. (1988) Alleviation of chlorpromazine-photosensitized contact dermatitis by β-cyclodextrin derivatives and their possible mechanisms. *Xenobiotic Metab. Dispos.* **3**: 377–386.

101. Hoshino, T., Ishida, K., Irie, T., *et al.* (1989) An attempt to reduce the photosensitizing potential of chlorpromazine with the simultaneous use of β- and dimethyl-β-cyclodextrins in guinea pigs. *Arch. Dermatol Res.* **281**: 60–65.

102. Uekama, K., Irie, T. and Hirayama, F. (1978) Participation of cyclodextrin inclusion catalysis in photolysis of chlorpromazine to give promazine in aqueous solution. *Chem. Lett.* 1109–1112.

103. Hoshino, T., Ishida, K., Irie, T., *et al.* (1988) Reduction of photohemolytic activity of benoxaprofen by β-cyclodextrin complexation. *J. Incl. Phenom.* **6**: 415–423.

104. Hoshino, T., Ishida, K., Irie, T., *et al.* (1987) Alleviation in protriptyline-photosensitized skin irritation by di-*O*-methyl-β-cyclodextrin complexation. *Int. J. Pharm.* **38**: 265–267.

105. Perrin, J. H., Field, F. P., Hansen, D. A., *et al.* (1978) β-Cyclodextrin as an aid to peritoneal dialysis. Renal toxicity of β-cyclodextrin in the rat. *Res. Commun. Chem. Pathol. Pharmacol.* **19**: 373–376.

106. Carpenter, T. O., Pettifor, J. M., Russell, R. M., *et al.* (1987) Severe hypervitaminosis A in siblings: evidence of variable tolerance to rerinol intake. *J. Pediatr.* **111**: 507–512.

107. Irie, T., Fukunaga, K., Garwood, M. K., *et al.* (1992) Hydroxypropylcyclodextrins in parenteral use. II. Effects on transport and disposition of lipids in rabbits and humans. *J. Pharm. Sci.* **81**: 524–528.

108. Frijlink, H. W., Eissen, A. C., Hefting, N. R., *et al.* (1991) The effect of parenterally administered cyclodextrins on cholesterol levels in the rat. *Pharm. Res.* **8**: 9–16.

109. Irie, T., Fukunaga, K. and Pitha, J. (1992) Hydroxypropylcyclodextrins in parenteral use. I. Lipid dissolution and effects on lipid transfers *in vitro*. *J. Pharm. Sci.* **81**: 521–523.

110. Uekama, K., Shiotani, K., Irie, T., *et al.* (1993) Protective effects of cyclodextrin sulphates against gentamicin-induced nephrotoxicity in the rat. *J. Pharm. Pharmacol.* **45**: 745–747.

111. Shimamoto, T. (1987) Pharmaceutical aspects: nasal and depot formulations of leuprolide. *J. Androl.* **8**: S14–S16.

112. Irie, T., Wakamatsu, K., Arima, H., *et al.* (1992) Enhancing effects of cyclodextrins on nasal absorption of insulin. *Int. J. Pharm.* **84**: 129–139.

113. Schipper, N. G. M., Verhoef, J., Romeijn, S. G. and Merkus, F. W. H. M. (1992) Absorption enhancers in nasal insulin delivery and their influence on nasal ciliary functioning. *J. Control. Rel.* **21**: 173–186.

114. Irie, T., Arima, H., Abe, K., *et al.* (1992) Possible mechanisms of cyclodextrin-enhanced nasal absorption of peptide and protein drugs. In Hedges, A.R. (ed.). *Minutes of the 6th International Symposium on Cyclodextrins*, pp. 503–508. Editions de Sante, Paris.

115. Maitani, Y., Igawa, T., Machida, Y. and Nagai, T. (1986) Intranasal administration of β-interferon in rabbits. *Drug Des. Del.* **1**: 65–70.

116. Irie, T., Abe, K., Adachi, H., *et al.* (1992) Potential use of 2-hydroxypropyl-β-cyclodextrin in designing nasal preparations of insulin involving lipophilic absorption enhancer HPE-101. *Drug Del. Syst.* **7**: 91–95.

117. Brewster, M. E., Hora, M. S., Simpkins, J. W. and Bodor, N. (1991) Use of 2-hydroxypropyl-β-cyclodextrin as a solubilizing and stabilizing excipient for protein drugs. *Pharm. Res.* **8**: 792–795.

118. Ressing, M. E., Jiskoot, W., Talsma, H., *et al.* (1992) The influence of sucrose, dextran, and hydroxypropyl-β-cyclodextrin as lyoprotectants for a freeze-dried mouse IgG$_{2a}$ monoclonal antibody (MN12). *Pharm. Res.* **9**: 266–270.

119. Uekama, K., Yamamoto, M., Irie, T. and Hirayama, F. (1992) Pharmaceutical evaluation of maltosyl-β-cyclodextrin as a drug carrier in parenteral formulation. In Hedges, A.R. (ed.). *Minutes of the 6th International Symposium on Cyclodextrins*, pp. 491–496. Editions de Sante, Paris.

120. Fukunaga, K., Hijikata, S., Ishimura, K., *et al.* (1994) Aluminium β-cyclodextrin sulphate as a stabilizer and sustained-release carrier for basic fibroblast growth factor. *J. Pharm. Pharmacol.* **46**: 168–171.

121. Wang, Z., Ikegami, K., Hirayama, F. and Uekama, K. (1993) Release characteristics of nifedipine from 2-hydroxypropyl-β-cyclodextrin complex during storage and its modification of hybridizing polyvinylpyrrolidone K-30. *Chem. Pharm. Bul.* **41**: 1822–1826.

122. Wang, Z., Hirayama, F. and Uekama, K. (1993) Design and *in vitro* evaluation of a modified-release oral dosage form of nifedipine by hybridization of hydroxypropyl-β-cyclodextrin and hydroxypropyl-cellulose. *J. Pharm. Pharmacol.* **45**: 942–946.

123. Wang, Z., Hirayama, F. and Uekama, K. (1994) *In vivo* and *in vitro* evaluation of modified-release oral dosage form of nifedipine by hybridization of hydroxypropyl-β-cyclodextrin and hydroxypropyl-celluloses in dog. *J. Pharm. Pharmacol.* **46**: 505–507.

124. Adachi, H., Irie, T., Hirayama, F. and Uekama, K. (1992) Stabilization of prostaglandin E$_1$ in fatty alcohol propylene glycol ointment by acidic cyclodextrin derivative, *O*-carboxymethyl-*O*-ethyl-β-cyclodextrin. *Chem. Pharm. Bull.* **40**: 1586–1591.

125. Uekama, K., Adachi, H., Irie, T., *et al.* (1992) Improved transdermal delivery of prostaglandin E$_1$ through hairless mouse skin: combined use of carboxymethyl-ethyl-β-cyclodextrin and penetration enhancers. *J. Pharm. Pharmacol.* **44**: 119–121.

126. Adachi, H., Irie, T., Uekama, K., *et al.* (1993) Combination effects of *O*-carboxymethyl-*O*-ethyl-β-cyclodextrin and penetration enhancer HPE-101 on transdermal delivery of prostaglandin E$_1$ in hairless mice. *Eur. J. Pharm. Sci.* **1**: 117–123.

127. Adachi, H., Irie, T., Uekama, K., *et al.* (1992) Inhibitory effect of prostaglandin E$_1$ on laurate-induced peripheral vascular occlusive

sequelae in rabbits: optimized topical formulation with β-cyclodextrin derivative and penetration enhancer HPE-101. *J. Pharm. Pharmacol.* **44**: 1033–1035.

128. Tobino, Y., Torii, H., Ikeda, K., *et al.* (1991) Release control of morphine hydrochloride in suppository. *Kyushu Yakugaku-Kai Kaiho* **45**: 15–20.

129. Uekama, K., Kondo, T., Nakamura, K., *et al.* (1995) Modification of rectal absorption of morphine from hollow-type suppositories with a combination of α-cyclodextrins and viscosity-enhancing polysaccharide. *J. Pharm. Sci.* **84**: 15–20.

130. König, W. A., Lutz, S. and Wenz, G. (1988) Modified cyclodextrins-novel, highly enantioselective stationary phases for gas chromatography. *Angew. Chem., Int. Ed. Engl.* **27**: 979–980.

131. Breslow, R. (1995) Biomimetic chemistry and artificial enzymes: catalysis by design. *Acc. Chem. Res.* **28**: 146–153.

132. Parrot-Lopez, H., Djedaini, F., Perly, B., *et al.* (1990) An approach to vectorization of pharmacologically active molecules: the covalent binding of Leu-enkephalin to a modified β-cyclodextrin. *Tetrahedron Lett.* **31**: 1999–2002.

133. Tanaka, H., Kominato, K., Yamamoto, R., *et al.* (1994) Synthesis of doxorubicin-cyclodextrin conjugates. *J. Antibiot.* **47**: 1025–1029.

134. Nedra Karunaratne, D., Farmer, S. and Hancock, R. E. W. (1993) Synthesis of bulky β-lactams for inhibition of cell surface β-lactamase activity. *Bioconjugate Chem.* **4**: 434–439.

135. Coates, J. H., Easton, C. J., van Eyk, S. J., *et al.* (1991) Chiral differentiation in the deacylation of 6^A-O-{2-[4-(2-methylpropyl)-phenyl] propanoyl}-β-cyclodextrin. *J. Chem. Soc., Chem. Commun.* 759–760.

136. Djedaini-Pilard, F., Perly, B., Dupas, S., *et al.* (1993) Specific interaction and stabilization between host and guest: complexation of ellipticine in a nucleobase functionalized cyclodextrin. *Tetrahedron Lett.* **34**: 1145–1148.

137. Flourie, B., Molis, C., Achour, L., *et al.* (1993) Fate of β-cyclodextrin in the human intestine. *J. Nutr.* **123**: 676–680.

138. Antenucci, R. and Palmer, J. K. (1984) Enzymatic degradation of α- and β-cyclodextrins by bacteroides of the human colon. *J. Agric. Food Chem.* **32**: 1316–1321.

139. Hirayama, F., Minami, K. and Uekama, K. (1996) *In vitro* evaluation of biphenylylacetic acid–β-cyclodextrin conjugate as colon-targeting prodrug: drug release behavior in rat biological fluids. *J. Pharm. Pharmacol.* **48**: 27–31.

140. Uekama, K., Minami, K. and Hirayama, F. (1997) 6^A-O-[(4-biphenylyl)acetyl]-α-β-, and -γ-cyclodextrins and 6^A-deoxy-6^A [[(4-biphenylyl)acetyl]amino]-α-, -β-, and -γ-cyclodextrins: potential prodrugs for colon-specific delivery. *J. Med. Chem.* **40**: 2755–2761.

141. Minami, K., Hirayama, F. and Uekama, K. (1998) Colon-specific drug delivery based on a cyclodextrin prodrug: release behavior of biphenylylacetic acid from its cyclodextrin conjugates in rat intestinal tracts after oral administration. *J. Pharm. Sci.* **87**: 715–720.

142. Yano, H., Hirayama, F., Arima, H. and Uekama, K. (2000) Hydrolysis behavior of prednisolone 21-hemisuccinate/β-cyclodextrin amide conjugate: involvement of intramolecular catalysis of amide group in drug release. *Chem. Pharm. Bull.* **48**: 1125–1128.

143. Yano, H., Hirayama, F., Arima, H. and Uekama, K. (2001) Preparation of prednisolone-appended α-, β- and γ-cyclodextrins: substitution at secondary hydroxyl groups and *in vitro* hydrolysis behavior. *J. Pharm. Sci.* **90**: 493–503.

144. (a) Yano, H., Hirayama, F., Arima, H. and Uekama, K. (2001) Prednisolone-appended α-cyclodextrin: alleviation of systemic adverse effect of prednisolone after intracolonic administration in 2, 4,6-trinitrobenzenesulfonic acid-induced colitis rats. *J. Pharm. Sci.* **90**: 2103–2112; (b) Yano., H, Hirayama, F., Kamada, M. *et al.* (2002)

Colon-specific delivery of prednisolone-appended α-cyclodextrin conjugate: alleviation of systemic side effect after oral administration. *J. Control. Rel.* **79**: 103–112.

145. Hirayama, F., Ogata, T., Yano, H., *et al.* (2000) Release characteristics of a short-chain fatty acid, *n*-butyric acid, from its β-cyclodextrin ester conjugate in rat biological media. *J. Pharm. Sci.* **89**: 1486–1495.

146. Kim, C., MacKellar, W. C., Cho, N., *et al.* (1997) Impermeant antitumor sulfonylurea conjugates that inhibit plasma membrane NADH oxidase and growth of HeLa cells in culture: identification of binding proteins from sera of cancer patients. *Biochim. Biophys. Acta* **1324**: 171–181.

147. Parrot-Lopez, H., Leray, E. and Coleman, A. W. (1993) New β-cyclodextrin derivatives possessing biologically active saccharide antennae. *Supramol. Chem.* **3**: 37–42.

148. Robertis, L., Lancelon-Pin, C., Driguez, H., *et al.* (1994) Synthesis of new oligosaccharidyl-thio-β-cyclodextrins (CDS): a novel family of potent drug-targeting vectors. *Bioorg. Med. Chem. Lett.* **4**: 1127–1130.

149. Leray, E., Parrot-Lopez, H., Auge, C., *et al.* (1995) Chemical-enzymatic synthesis and bioactivity of mono-6-[gal-β-1,4-glcNAc-β-(1,6′)-hexyl]amido-6-deoxy-cycloheptaamylose. *J. Chem. Soc., Chem. Commun.* 1019–1020.

150. Imata, H., Kubota, K., Hattori, K., *et al.* (1997) The specificity of association between concanavalin A and oligosaccharide-branched cyclodextrins with an optical biosensor. *Bioorg. Med. Chem. Lett.* **7**: 109–112.

151. Attioui, F., Al-Omar, A., Leray, E., *et al.* (1994) Recognition ability and cytotoxicity of some oligosaccharidyl substituted β-cyclodextrins. *Biol. Cell* **82**: 161–167.

152. Péan, C., Wijkhuisen, A., Djedaini-Pilard, F., *et al.* (1998) Pharmacological investigations of new peptide-cyclodextrins. In Torres Labandeira, J.J. and Vila-Jato, J.L. (eds). *Proceedings of the 9th International Symposium on Cyclodextrins*, pp. 387–390. Kluwer Academic Press, Dordrecht.

153. Péan, C., Fischer, J., Doly, A., *et al.* (1998) *In vitro* and *in vivo* investigations of the specific binding of substance P-γ-cyclodextrin adducts (SB-γ-CD) to rat brain NK1 receptors. In Torres Labandeira, J.J. and Vila-Jato, J.L. (eds). *Proceedings of the 9th International Symposium on Cyclodextrins*, pp. 403–406. Kluwer Academic Press, Dordrecht.

154. Schaschke, N., Fiori, S., Weyher, E., *et al.* (1998) Cyclodextrin as carrier of peptide hormones: conformational and biological properties of β-cyclodextrin/gastrin constructs. *J. Am. Chem. Soc.* **120**: 7030–7038.

155. Arima, H., Kihara, F., Hirayama, F. and Uekama, K. (2001) Enhancement of gene expression by polyamidoamine dendrimer conjugates with α-, β-, and γ-cyclodextrin. *Bioconjugate. Chem.* **12**: 476–484; Kihara, F., Arima, H., Tsutsumi, T. *et al.* (2002) Effects of structure of polyamidoamine dendrimer of gene transfer efficiency of the dendrimer conjugate with α-cyclodextrin. *ibid.* **13**: 1211–1219.

156. Begley, D. J. (1996) The blood–brain barrier: principles for targeting peptides and drugs to the central nervous system. *J. Pharm. Pharmacol.* **48**: 136–146.

157. Bodor, N. (1985) Prodrugs versus soft drugs. In Bundgaard, H. (ed.). *Design of Prodrugs*, pp. 333–354. Elsevier, Amsterdam.

158. Djedaini-Pilard, F., Désalos, J. and Perly, B. (1993) Synthesis of a new carrier: *N* (Leu-enkephalin)yl-6-amino-6-deoxy-cyclomaltoheptaose. *Tetrahedron Lett.* **34**: 2457–2460.

159. Hristova-Kazmierski, M. K., Horan, P., Davis, P., *et al.* (1993) A new approach to enhance bioavailability of biologically active peptides: conjugation of a δ opioid agonist to β-cyclodextrin. *Bioorg. Med. Chem. Lett.* **3**: 831–834.

39

CHEMICAL AND PHYSICOCHEMICAL SOLUTIONS TO FORMULATION PROBLEMS

Camille G. Wermuth

You may readily infer that such substances as agreeably titillate the sense (of smell) are composed of smooth round atoms. Those that seem bitter and harsh are more tightly compacted of hooked particles and accordingly tear their way into our sense and rend our bodies by their inroads. **Lucretius, 47 BC**[1]

I INTRODUCTION

Medicinal chemists aim to discover new lead compounds and to optimize them to highly potent, selective, and nontoxic new drug candidates. However, the selected drug candidates can still present some drawbacks of physicochemical nature (insufficient stability, inappropriate aggregation state) or which might affect the patient's compliance (causticity, irritation, painful injections, undesirable organoleptic properties). To render the compound marketable, the pharmacists must then invent an adequate pharmaceutical formulation, stable on keeping and perfectly adapted to its clinical use. In leaving this task to the formulation pharmacists, the chemists are not always aware of their own role in the formulation phase and the final presentation of a new drug. They also probably overestimate the intervention possibilities of the formulation pharmacists. Of course, new pharmaceutical technologies such as microencapsulation or cyclodextrin complexation can overcome various drawbacks, but they are costly and often replaceable by simple chemical derivations. Medicinal chemists should therefore also feel concerned with these problems and tackle them in the early research stage, before the investment in a moderately satisfying molecule has become too large for an alternative to be sought. Besides the pharmaceutical approach to formulation problems there is a place for chemical solutions.

II INCREASING CHEMICAL STABILITY

The shelf-life of an organic chemical is the time taken for its pharmacological activity to fall by an acceptable amount. While this cannot be universally defined, about 10% decomposition may be considered acceptable, unless the decomposition products are toxic. Among possible decomposition paths, the following are the most frequent: hydrolysis, oxidation, photochemical degradations, racemization, thermal decomposition, chemical interactions, and microbial degradations.[2] Usually, selected pharmaceutical prevention procedures and manufacturing practices can overcome decomposition problems. For example, hydrolyses are avoided in dissolving the active principle in an anhydrous solvent, oxidations are prevented in replacing atmospheric oxygen by an inert gas or by adding antioxidants, and so on. However, it can happen that chemical derivatizations are needed to achieve satisfactory drug formulations. Some of them are illustrated by the examples described below. More detailed

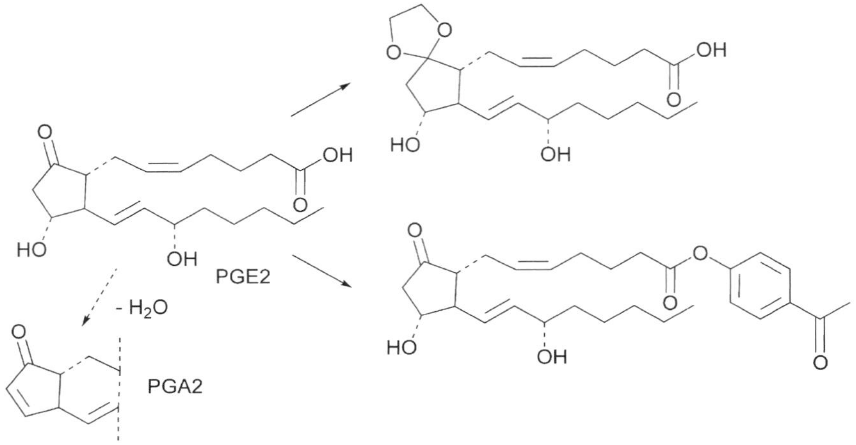

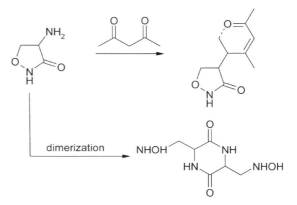

Fig. 39.1 Stabilized derivatives of prostaglandin E_2: ethylene ketal and crystalline ester with *p*-hydroxyacetophenone.[5,6]

reviews on the basic physical and chemical principles that determine the stability of drugs are available in specialized monographs.[3,4]

Pressing problems with prostaglandin E_2 (dinoprostone) and many other prostaglandins are chemical instability and difficulties in handling a liquid compound. Effectively, prostaglandin E_2 is a crystalline solid (m.p. 63°C), stable at room temperature for short periods, but it liquefies and decomposes rapidly after a few months. The instability of dinoprostone is due to the 9,11β-ketol system in which the activated C_{11} hydroxyl undergoes facile elimination to give prostaglandin A_2. The corresponding ethylene ketal derivative (Fig. 39.1) possesses much improved solid-state stability. It may be an orally useful prodrug form, since it readily undergoes an acid-catalysed hydrolysis back to the parent prostaglandin under conditions similar to those prevailing in the stomach.[5]

The crystalline ester with *p*-hydroxyacetophenone (Fig. 39.1), or with some other phenolic compounds, represents another means of overcoming these difficulties.[6] Storage of the crystalline esters at room temperature for 22–30 months resulted in no detectable degradation as shown by silica TLC; the esters remained as white solids. The free acid, on the other hand, underwent 44–59% degradation in 12 months at room temperature.

Cycloserine forms a bioreversible condensation product with acetylacetone (Fig. 39.2). This greatly depresses the dimerizing tendency of the parent drug, which occurs particularly rapidly in concentrated aqueous media and can even take place in the solid state.[7]

Pilocarpine is widely used as a topical miotic for controlling the elevated intraocular pressure associated with glaucoma. Besides its low lipophilicity, which stimulated the search for prodrugs,[8] pilocarpine has a short duration of action, its lactonic ring being rapidly opened to yield pilocarpic acid. The synthesis of its isosteric

Fig. 39.2 Derivatization of the primary amino function of cycloserine. A bioreversible enamine, probably stabilized by an intramolecular hydrogen bond, is formed by reaction with acetylacetone.[7]

carbamate (Fig. 39.3), which is as effective as pilocarpine, has greatly improved the stability of the former lactonic ring.[9] Although the phthalimido-bearing pyridinone (1) (Fig. 39.4) is a potent reverse transcriptase inhibitor ($IC_{50} = 30$ nM), hydrolytic instability under physiological conditions precludes the demonstration of its antiviral activity in cell culture. This instability was recognized to be due to the aminal structure of the exocyclic NH. Subsequent efforts led to the substitution of the NH with a methylene unit and the replacement of the phthalimido moiety by a benzoxazole. One of the prepared compounds, 3-[(benzoxazol-2-yl)ethyl]-5-ethyl-6-methylpyridin-2(1*H*)-one (2) (Fig. 39.4), is as potent ($IC_{50} = 23$ nM) as the phthalimido derivative but is 10 times more stable towards acid hydrolysis and shows good oral bioavailability.[10]

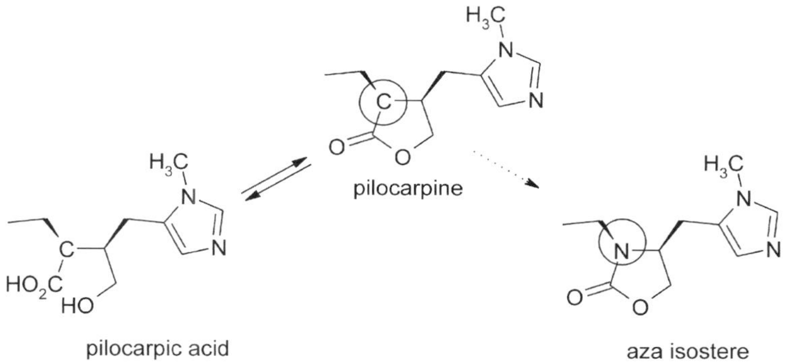

Fig. 39.3 Stabilization of the lactonic ring of pilocarpine. In preparing the carbamic analogue of pilocarpine, the sensitivity of the carbonyl group towards nucleophilic attacks is greatly decreased.[9]

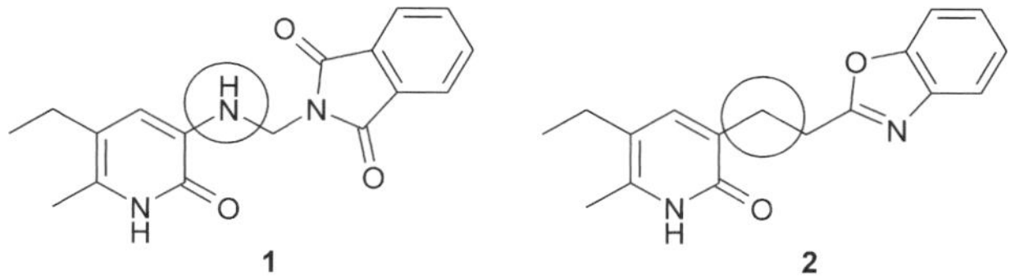

Fig. 39.4 Improved chemical stability through suppression of an aminal function.[10]

Cefoxitin is a broad-spectrum, semisynthetic cephamycin antibiotic. The free acid form is a white, crystalline, practically insoluble solid and therefore the decision was made to use the sodium salt to provide a sterile solution for intravenous administration. It soon became evident that the sodium salt of cefoxitin would have limited stability in solution and would not lend itself to the formulation of a marketable solution product. Stability studies highlighted the advantages of sterile crystalline solid over an amorphous freeze-dried product, and also the need for rubber stopper screening studies to eliminate interactions between sodium cefoxitin powder and rubber, and finally the profound effect of oxygen on the coloration rate of the product.[11]

The *N*-nitrosoureas such as lomustine, carmustine and tauromustine are important alkylating antineoplastic agents which have demonstrated activity against a wide spectrum of tumours. Although some *N*-nitrosoureas (e.g. lomustine) are sufficiently stable to be administered orally, generally they are very unstable in aqueous solutions. Addition of an equimolar amount of Tris (tris-hydroxymethyl aminomethane) forms a complex which is stable in aqueous solution and which increases shelf-life of the nitrosoureas. The rate of degradation of the drugs in the complex is 1.5–2.5 times slower than outside the complex.[12] In a similar way the antiulcerative drug omeprazole is claimed to be more stable if prepared as a magnesium salt.[13] Probably the compound forms a 2:1 complex (Fig. 39.5).

Cyclodextrin complexation (see Chapter 38) can also represent a way of improving the stability and solubility of

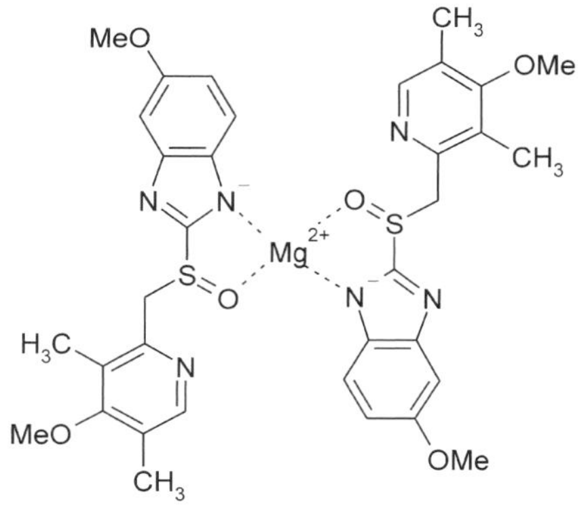

Fig. 39.5 Formation of a magnesium salt stabilizes omeprazole.

sensitive drugs such as thalidomide. Thalidomide is currently in clinical use for the treatment and prevention of graft-versus-host disease in leukaemia patients after bone marrow transplantation. However, this drug is sparingly soluble in aqueous solutions (50 μg ml^{-1}) and is readily hydrolysed. Complexation with hydroxypropyl β-cyclodexrin increases the solubility to 1700 μg ml^{-1} and extends the half-life of a dilute solution from 2.1 to 4.1 hours.[14] Other vulnerable and sparingly soluble drugs stabilized by means of cyclodextrin complexation are the nonsteroidal anti-inflammatory drugs diclofenac, piroxicam and indomethacin[15] and the anthracycline antibiotic daunorubicin.[16]

The presence of an alkyl substituent at N3 in the oxazaphosphorine ring stabilizes *N*-substituted 4-(alkylthio)cyclophosphamides from 'spontaneous' decomposition. On the basis of this finding, several *N*-methyl-4-thiocyclophosphamide derivatives were synthesized and examined as prodrugs of 4-hydroxycyclophosphamide, the activated species of cyclophosphamide.[17]

All prodrugs are stable in aqueous buffer but undergo *N*-demethylation when incubated with rat hepatic

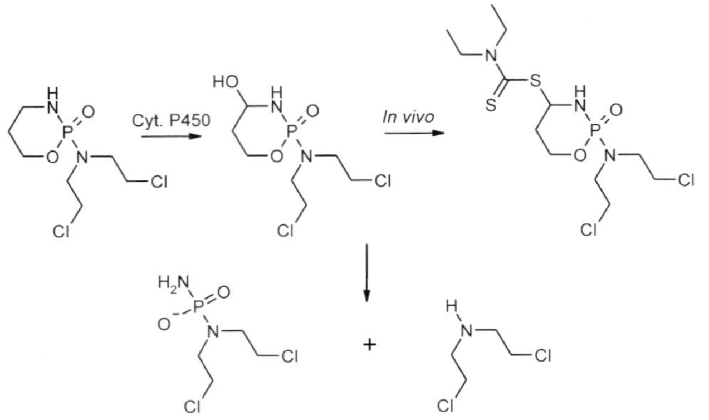

Fig. 39.6 Stable precursor of 4-hydroxycyclophosphamide.[17]

microsomes, forming alkylating species. *N*-Methyl-4-(diethyldithiocarbamyl) cyclophosphamide particularly (Fig. 39.6) shows notable *in vitro* toxicity against mouse 3T3 cells and human tumour cells.

The half-life of the antitumour antibiotic leinamycin in aqueous solution at pH 7 at 37°C is about 6 hours and the hydrolysis of the reactive dithiolanone moiety could be reasons for instability. However, this dithiolanone moiety of leinamycin has been shown to be essential for its DNA-cleaving activity and antiproliferative activity. Accordingly, Kanda *et al.*[18] devised more stable thioester derivatives, resulting from various alkylations in the presence of potassium carbonate (Fig. 39.7).

These derivatives have a unique 3-isothiazolidinone-1-oxide moiety and can act as prodrugs that provide dithiolanone compounds in biological media. Particularly compound '3c' was found to show potent antitumour activity against human tumour xenografts, such as lung, liver, ovary, prostate, and colon carcinomas.[18]

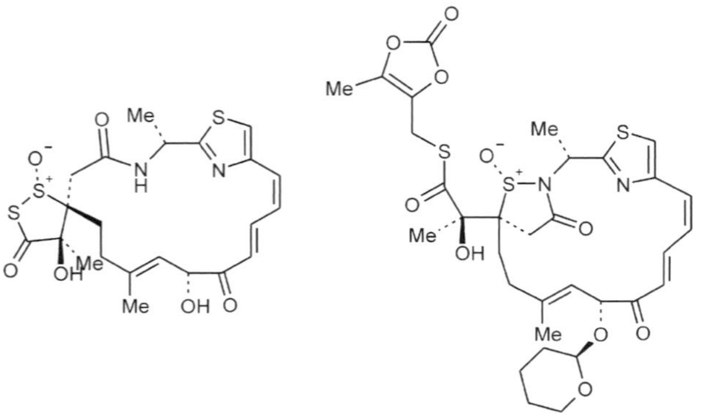

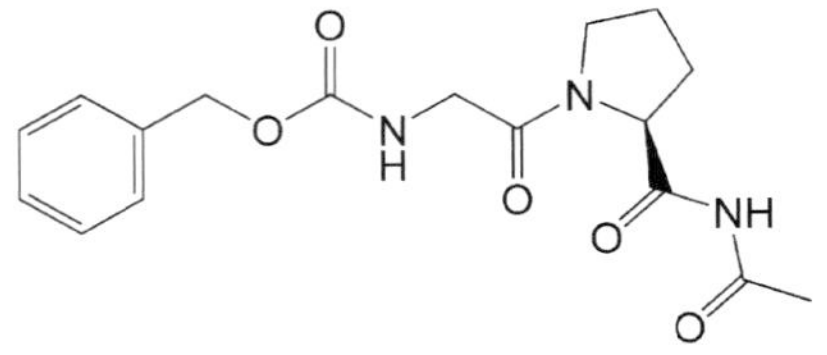

Fig. 39.8 *N*-Acetylation of *Z*-Gly-ProNH$_2$ protects against the hydrolytic activity of prolyl endopeptidase.[20]

III IMPROVED FORMULATION OF PEPTIDES AND PROTEINS

Enhancement of proteolytic stability of peptides can be achieved by well-established procedures such as synthesis of retropeptides, isosteric replacements of the peptidic bond, *N*-methylation and the use of non-natural D-amino acids.[19,20] Prodrugs of peptides can also be helpful, see Chapter 30.

However, sometimes very simple derivatives can afford protection against enzymatic degradation. Thus the *N*-acetyl derivative of *Z*-glycylprolylamide (Fig. 39.8) is six times more resistant to gut prolyl endopeptidase (which normally hydrolyses the terminal primary amide) than the nonacetylated parent molecule.[20]

In an analogous way, Leu-enkephalin and Met-enkephalin were rendered resistant to aminopeptidases[21] by condensation with various aldehydes and ketones to form 4-imidazolidine derivatives (Fig. 39.9).

Interesting case histories about formulation, characterization and stability of protein drugs are discussed in the book edited by Pearlman and Wang.[22]

IV DEALING WITH MESOMORPHIC CRYSTALLINE FORMS

Polymorphism is a keyword of considerable importance in the life-sciences and especially in the pharmaceutical

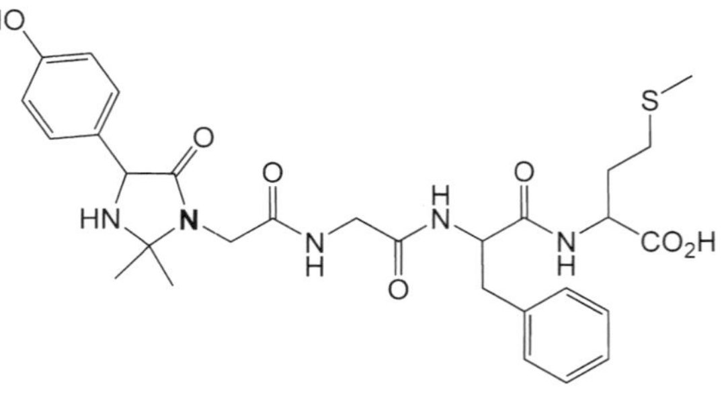

leinamycin   bi-protected leinamycin prodrug "3c"

Fig. 39.7 Thioester derivatives of leinamycin are prodrugs with improved stability.[18]

Fig. 39.9 Condensation of acetone with Met-enkephalin yields a 4-imidazolidinone with increased stability towards aminopeptidases.[21]

industry. It abridges the fact that a solid compound can exist in different crystalline forms which can have different physical chemical properties. To ensure no variations in the product are due to different solid-state properties, care must be taken in selecting the most appropriate solid-state form for the substance and in ensuring a reproducible production of this form.

With regard to the pharmaceutical industry, it has been shown that more than half of the drug substances described in monographs crystallize in more than one solid-state form, being either polymorphs, solvates, or both.[23] A list of about 600 polymorphic drug substances was published and discussed by Giron.[24] The solid-state form of a drug substance can influence a variety of properties, namely the solubility and rate of dissolution or the chemical stability or stability against excipients. Thus, the regulatory bodies require an exhaustive search for polymorphic forms of a drug substance.[25,26] The manufacturer is required to make a substantiated choice for one of the forms, or a defined mixture of forms. Changes in the polymorphic form of the batches produced are seen as indicative of changes in the production process, also requiring the reproducible crystallization of a certain solid state form. The choice of the solid-state form of a new drug substance is up to the applicant and should be made by considering all aspects, such as chemical stability and stability against excipients, dissolution behaviour and bioavailability, and last but not least, thermodynamic stability of the solid-state form and ease and reproducibility of production.

Some of the possible crystalline states are metastable and can be converted into more stable forms with different physicochemical properties. Two types of transformation are possible: the reversible enantiotropic transformation in which the polymorphic forms can change from one into the other and vice versa, and the irreversible monotropic transformation, which is now most often seen and which involves the change from a thermodynamically unstable form to a more stable one. The different crystalline forms of a given compound can be distinguished by their melting point and solubility, by scanning calorimetry, by thermogravimetric analysis, by infrared spectrometry, by X-ray powder diffraction, and by scanning electron microscopy.

As a rule, metastable polymorphs will tend to have increased solubility and a faster dissolution rate than a stable polymorph. Although four-fold increases in solubility are sometimes observed between the two forms,[27,28] a 50–100% increase in dissolution rate represents the most usual situation.[29] Riboflavin is an exception. It has three polymorphs with solubilities of 60 mg l^{-1}, 810 mg l^{-1} and 1200 mg ml^{-1}.[30] When a metastable form is placed in contact with solvents, it can change progressively to the most stable but least soluble form. An example is provided by the acidic and amorphous form of novobiocine, which can hardly be administered as a suspension because of its tendency to turn into a much less soluble crystalline form.[31] Spray-drying of pure drugs often results in amorphous products with increased rates of dissolution, and eventually an auxiliary ingredient can be added to yield solid spray-dispersions of drugs.[32]

The transformation from one crystalline form to the other can also occur during the manufacturing process. Chloroquine diphosphate crystals, for example, can be obtained anhydrous by storing the hydrate at elevated temperatures. Dehydration is favoured by grinding of the raw material. The transition of anhydrous chloroquine diphosphate into a second hydrated form is possible by storing the drug at a high relative humidity. Compression of the raw material resulted in the formation of another crystal form.[33] In a similar way, grinding the polymorphic form A of chloramphenicol stearate in the presence of colloidal silica converts it into form B.[34] Taken together, these findings emphasize the necessity for standardization of the manufacturing processes and closer examination of the solid drug as part of the quality control.

V INCREASING THE MELTING POINT

A less frequent but none the less interesting problem arises in the chemical modification of liquid, or low-melting, active principles into solid prodrugs, suitable for tablet or capsule preparation. Indeed, solid dosage forms are still the most widely used for the administration of medicines, for reasons of patient acceptability and convenience in terms of product stability and ease of manufacture. Their preparation implies that the active principle can itself be handled as a stable solid, an objective that is usually attained by one of the following strategies: formation of a salt or a molecular complex, formation of a crystalline covalent derivative, introduction of symmetry.

Salt or complex formation

Many basic active compounds, belonging to various therapeutic classes (antihistaminics, neuroleptics, local anaesthetics, etc.) are oily liquids. Salification with an appropriate acid (hydrochloric, phosphoric, tartaric, etc.) represents a good means of converting them into the desired aggregation state (see Chapter 34). Neutral liquids can sometimes be converted into solids through formation of molecular complexes. The central and respiratory stimulant nikethamide is a slightly viscous oil or a low-melting (m.p. 24–26°C) crystalline solid. With calcium thiocyanate it forms a solid combination made of two molecules of *N,N*-diethylnicotinamide and one molecule of $Ca(SCN)_2$ that can be used for tableting.[30] Chloral hydrate, another low-melting solid, yields a crystalline and odourless 2:1 complex with phenazone;[35,36] see below. Chloral

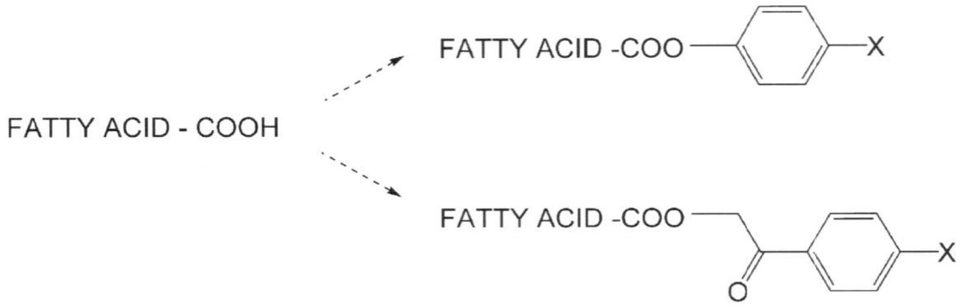

Fig. 39.10 Crystalline derivatives of fatty acids.

formamide (Merck Index) and Chloral *N*-acetylglycinamide[37] are other solid complexes.

(2) Covalent derivatives

The methodology consists of linking the liquid active principle, by means of a covalent bond, to a nontoxic moiety having a strong tendency to crystallize. Liquid or greasy fatty acids can be easily converted into phenacyl esters[38] and, as shown above, substances related to prostaglandins give crystalline esters with *p*-hydroxyacetophenone (Figs 39.1 and 39.10) or with some other phenolic compounds.[6] Chloral yields a crystalline hemiacetal (Fig. 39.17) with the phenolic hydroxyl of paracetamol.[39]

Trichloroethanol can be converted to its phosphate ester, which gives a crystalline monosodium salt,[40] the same possibility exists for the low-melting (m.p. 28°C) guaiacol (Guaiacol phosphate: Merck Index) (Fig. 39.11). Phenols can be acylated (guaiacol benzoate, phenylacetate or valerate) or alkylated (guaifenesin, guaiapate) (Fig. 39.11).

Introduction of symmetry

It is well known that symmetric molecules have higher melting points than nonsymmetric ones, e.g. *p*-nitrotoluene is a solid, *o*- and *m*-nitrotoluenes are liquids. For chemical formulation of pharmaceuticals, symmetry can be obtained by preparing twin drugs (see Chapter 16). Low molecular weight duplication reagents are formaldehyde or glyoxylic acid (methylene-bis type derivatives), phosgene

or phosphorus oxychloride (carbonate or phosphate diesters), and small bifunctionalized molecules (ethylene glycol, ethylenediamine esters or amides).

Symmetric carbonates of guaiacol or ethyl salicylate are crystalline solids (Fig. 39.12), pentaerythritol tetranitrate is the solid counterpart of the liquid trinitrine, and petrichloral is again a stabilized chloral derivative. Methenamine, formulated as an enteric-coated tablet, is used as a urinary tract antibacterial. Following absorption, the compound is eliminated in the urine, where formaldehyde is generated in the acidic environment.

VI GASTROINTESTINAL IRRITABILITY AND PAINFUL INJECTIONS

A Gastrointestinal irritability

Gastrointestinal disturbancies can occur for several reasons: (1) direct contact of the drug with the gastric mucosa, producing a localized irritating and necrotizing effect, (2) irritation of the gastric mucosa through an indirect mechanism, e.g. stimulation of the gastric secretion, (3) irritation of the intestinal mucosa, (4) inhibition of the mucopolysaccharide biosynthesis, (5) destruction of the intestinal flora.

Chemical formulations mainly deal with the direct irritating properties of drugs containing phenolic or acidic groups. For example, to eliminate gastric irritation produced

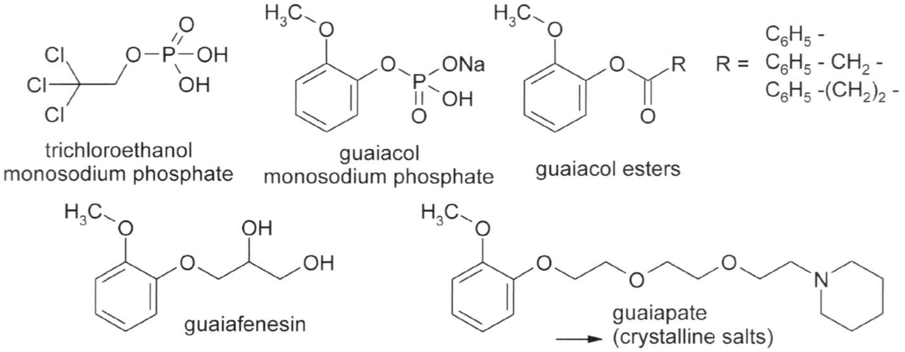

Fig. 39.11 Covalent solid derivatives of liquid active principles.

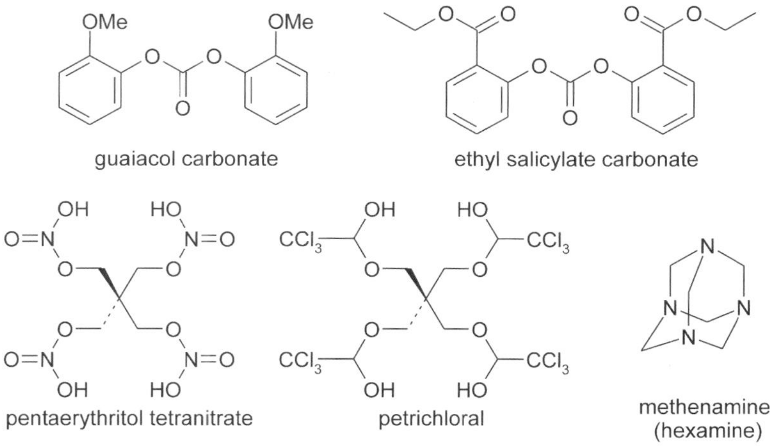

guaiacol carbonate   ethyl salicylate carbonate

pentaerythritol tetranitrate   petrichloral   methenamine (hexamine)

Fig. 39.12 Compounds for which the solid state results from molecular symmetry.

by salicylic acid, aspirin was developed. Another possibility is given by carbonate esters (*n*-hexyl carbonate), which are rapidly hydrolysed *in vivo* and might have hydrophobic properties adequate to allow absorption and distribution over a greater area of the GI tract, thus reducing local irritation.[41]

For nonsteroidal anti-inflammatory carboxylic acids such as mefenamic acid[42] or *N*-(7-chloro-4-quinolyl)anthranilic acid,[43] glyceryl esters are claimed to be less irritating. Alternatively, the ulcerogenicity of indomethacin derivatives was reduced by formation of the ester with glycolic acid[44,45] or the peptide with serine.[41,46] Another indomethacin-related anti-inflammatory drug, sulindac, is an inactive sulfoxide and becomes only activated after absorption and reduction into the corresponding sulfide. Thus the initial exposure of gastric and intestinal mucosa to the active drug is circumvented (see Chapter 33).

B   Avoidance of painful injections

When an initially painful intravenous or intramuscular injection must be administered repetitively, patient reluctance develops. Injection pains are usually accompanied by haemorrhage, oedema, inflammation and tissue necrosis.[47] Among the factors responsible for painful injections, the most important are the drug's solubility in aqueous medium, the viscosity, the pH and the hypo- or hyperosmotic character of the injected drug solution, the amount of the injected volume, the site of injection, the pain tolerance of the patient, and the technique of administration. Other factors include precipitation of the drug at the injection site, and localized cell lysis.[47]

Excessive concentrations of the active compound at the injection site (initial peak concentrations) are avoided by masking the irritating agent through complexation. The free concentrations of iron or of calcium are reduced by using calcium gluconate or laevulinate for intravenous injections and iron–sorbitol citrate for intramuscular injections.[48]

The low aqueous solubility of the antibiotic clindamycin hydrochloride is responsible for the pain experienced on intramuscular injection. Phosphorylation improves the aqueous solubility from 3 mg ml^{-1} to > 150 mg ml^{-1} and avoids the pain resulting from injection of the parent clindamycin.[49,50] The phosphorylated drug (Fig. 39.13) possesses little or no intrinsic antibacterial activity,[51] however, owing to the enzymatic action of phosphatases, it is rapidly converted to the parent drug. The half-life for hydrolysis *in vivo* is approximately 10 min and only 1–2% of an intravenous dose is eliminated in the urine as unchanged prodrug.[52]

The anticonvulsant drug phenytoin (5,5-diphenylhydantoin) is sparingly soluble in water (0.02 mg ml^{-1}) and is therefore formulated as sodium enolate for intravenous or intramuscular injections. The solvent is a mixture of 40% propylene glycol, 10% ethanol and 50% water and the pH of the solution is alkaline (pH $\sim$ 12), owing to the weak acidity ($pK_a = 8.3$) of the drug.[53] Such a formulation allows concentrations as high as 50 mg ml^{-1}, but there is a serious

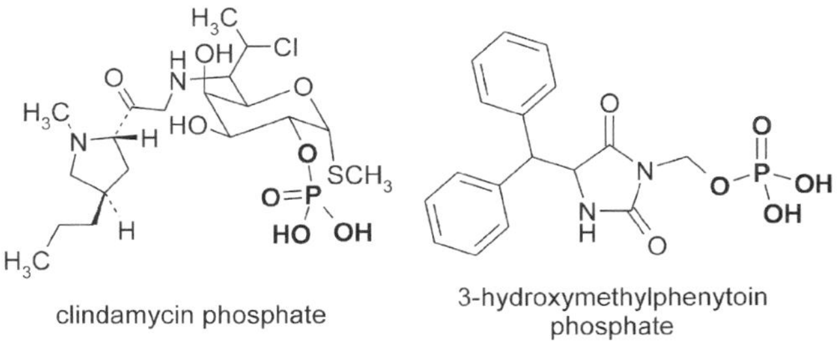

clindamycin phosphate   3-hydroxymethylphenytoin phosphate

Fig. 39.13 Phosphate esters of clindamycin and of 3-hydroxymethylphenytoin (as sodium salts) are freely soluble in water and allow painless injections.

risk of phenytoin precipitation at the physiological pH (7.4) of the injection site. On the other hand, propylene glycol is a cardiac depressant (see Chapter 36). The disodium salt of the phosphate ester of 3-hydroxymethylphenytoin was found to be extremely soluble (4500 times phenytoin), pharmacologically inert and able to regenerate free phenytoin rapidly and quantitatively with no apparent irritation.[54]

VII SUPPRESSION OF UNDESIRABLE ORGANOLEPTIC PROPERTIES

The use of flavours and flavour modifiers represents the first attempt to improve pharmaceuticals in masking undesirable organoleptic properties such as taste, odour and 'feel' factors (for a review see Adjei *et al.*[55]). When this approach is ineffective, chemical modifications have to be considered.

A Odour

Despite its interesting antiseptic properties, iodoform was rejected by the medical community because of its disagreeable odour. A first historical attempt towards an odourless substitute was bismuth iodosubgallate (Airoform®) which is practically odourless (Fig. 39.14).

Bismuth iodosubgallate was followed by the introduction of iodochlorhydroxyquin (Vioform®). Unfortunately, this latter compound has been linked with the occurrence of subacute myelo-optic neuropathy (SMON syndrome) in Japan. Decisive progress came with the iodine-free chlorquinaldol (Sterosan®).

Thiamine chloride has a slight, but disagreeable and penetrating odour and often causes vomiting when incorporated into paediatric polyvitaminic preparations. The corresponding monophosphoric ester (Fig. 39.15) is totally odourless and shows increased stability.[56]

The development of diethyl dithioisophtalate (Etisul®) as an antituberculosis and antileprosy drug originated from the observation that ethyl mercaptan, an evil smelling, low-boiling, inflammable liquid was active as an antitubercular agent. Among a number of derivatives prepared and which would be expected to be metabolized to ethyl mercaptan

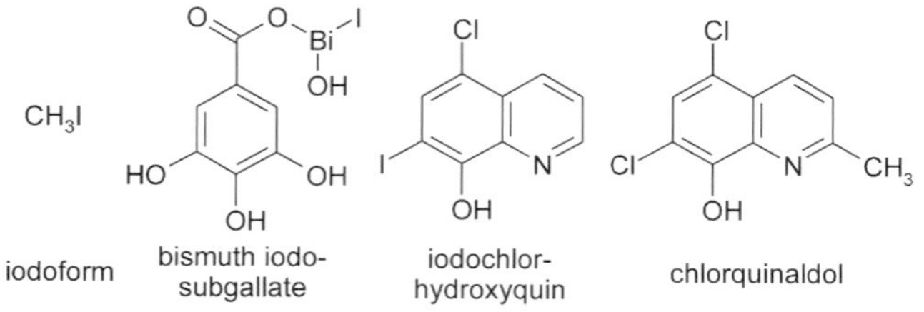

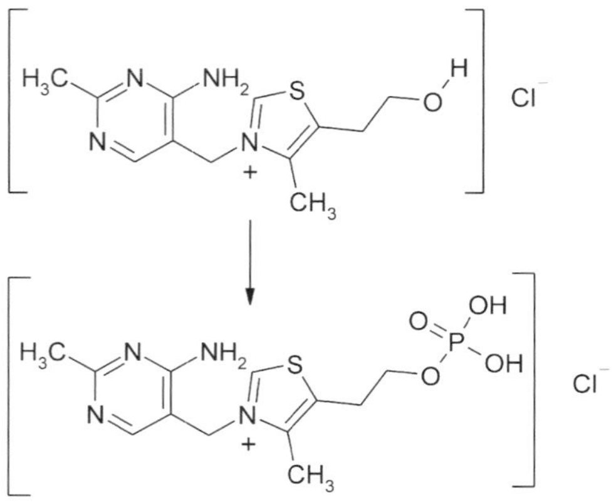

Fig. 39.15 Thiamine chloride and its odourless monophosphate ester.[56]

in vivo, the most successful was diethyl dithioisophtalate (Fig. 39.16), an almost bland odourless oil, which is effective in experimental tuberculosis parenterally.[57,58]

Besides its bad taste and to its gastric and intestinal irritating properties chloral hydrate has an unpleasant odour. All these disadvantages have been overcome by complexing two molecules of chloral hydrate with one molecule of phenazone to give dichloralphenazone.[35,36] Dichloralphenazone (Fig. 39.17) possesses all the hypnotic and sedative properties of chloral hydrate, and is less toxic and free from the unpleasant odour and taste of chloral. Another masked form of chloral, cloracetadol, results from a hemiacetalic bonding between chloral hydrate and paracetamol (Fig. 39.17). As such, this combination represents three achievements: attenuation of the bad odour, obtaining of a solid state, and associative synthesis.[39] Trichloroethanol, the active metabolite of chloral, is deodorized as a symmetric ester with carbonic acid. The ester is prepared from trichloroethanol and phosgene. It is a tasteless, crystalline solid with limited solubility in water. Pharmacological comparisons show its equivalence with trichloroethanol.[59]

B Taste

Taste is a complex combination of sensations including gustation, olfaction, tactility and responses to heat and cold.

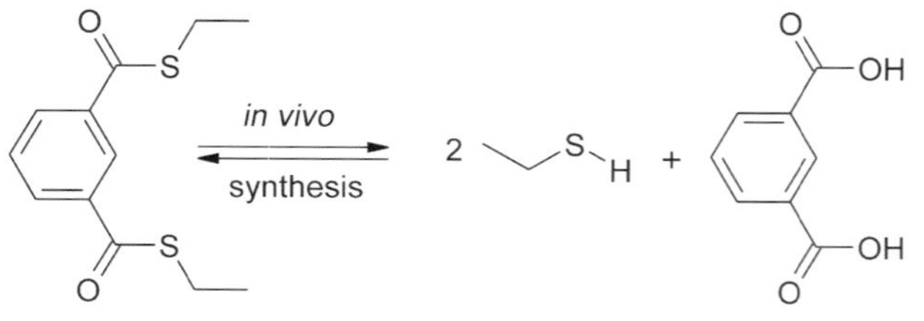

Fig. 39.14 Change from iodoform to odourless antiseptics.

Fig. 39.16 Diethyl dithioisophtalate an odourless administration form of ethyl mercaptan.[57,58]

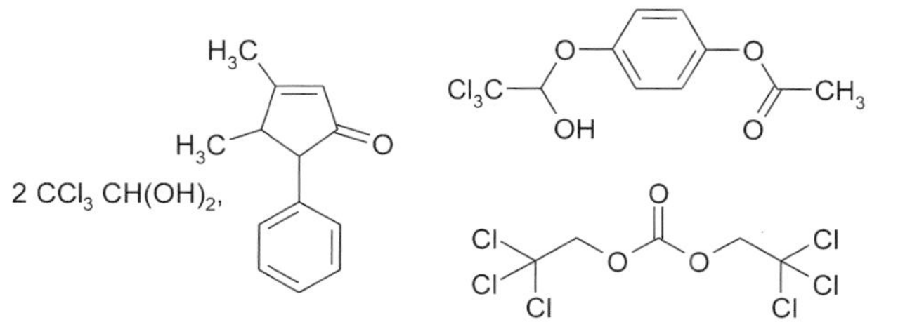

Fig. 39.17 Chloral hydrate yields an odourless 2:1 complex with phenazone[35,36] and a hemiacetal with paracetamol.[39] Trichloroethanol is deodorized as carbonate ester.[59]

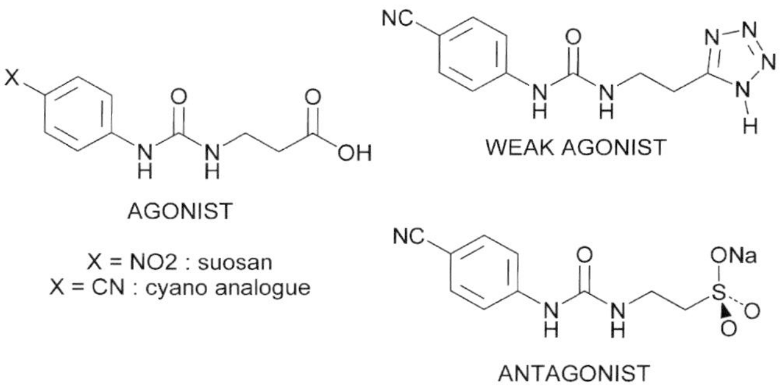

Fig. 39.18 Suosan analogues.[68,69]

In humans the sensation of taste is confined primarily to the dorsal face of the tongue, the soft palate, the epiglottis and parts of the gullet. In children, however, the taste receptors are distributed over larger areas of the mouth.[47] Basically there are four so-called primary tastes: sour, bitter, salty and sweet. Others include astringent, metallic and alkaline (soapy). For pharmaceuticals the main problem resides in masking the bitterness of some orally administrated pediatric formulations, such as antibiotic-containing suspensions and syrups. Two strategies are possible: decreasing the drug solubility, sufficient to reduce it below the threshold value for taste receptor activation, or modifying the shape and electronic potential of the molecule. The first strategy is well illustrated by the formation of very sparingly soluble salts or esters (Table 39.1). In the gastrointestinal tract or following absorption the derivatives are cleaved with formation of the parent drugs. However, insolubility in water is not always a prerequisite and in some cases water-solubilizing groups achieve the suppression of bitterness. Examples are *N*-arylanthranilic glyceryl esters[42] or lincomycin phosphate.[66,67]

An illustration of the second strategy is provided by the *p*-cyano analogues of the nonnutritive sweetener suosan (Fig. 39.18). The replacement of the planar carboxylic group by the bioisosteric tetrazolyl group yields less potent,

but still sweet compounds.[68] However, replacement of the carboxylic group by the tetrahedral sulfonic group yields an *antagonist* of the sweet taste response which inhibits the sweetness perception of a variety of sweeteners without having any effect on the sour or salty taste response. Surprisingly, the sulfonic antagonist also antagonized the bitter taste response to reference compounds such as caffeine, quinine and naringine.[69]

Tinti and Nofre[70,71] observed that the combination of the cyano analogue of suosan with the elements of asparatame yields 'superaspartame' which has a potency 14 000 times that of sucrose (aspartame has a potency of about 180 times that of sucrose, Fig. 39.19). Subsequently, these authors discovered sweet guanidines such as the compound SC-45647 which has a potency 28 000 times that of sucrose.[71,72] More recently, they reported *N*-alkyl-substituted aspartame derivatives such as neotame which has a potency 10 000 times that of sucrose.[73] For these three high-potency sweeteners, Walters *et al.* were able to identify a five-point pharmacophore model in which the carboxylate, the two hydrophobic groups and two NH groups match well

Table 39.1 *Sparingly soluble salts or esters of bitter drugs*

Parent molecule	Derivative	Reference
Dextropropoxyphene	Napsylate salt	Gruber *et al.*[60]
Tetracycline	3,4,5-Trimethoxybenzoate salt	Rocca and Rusconi[61]
Chloramphenicol	Palmitic ester	Glazko *et al.*[62]
Sulphafurazole (sulfisoxazole)	*N'*-acetyl	McEvoy[63]
Erythromycin	Ethylsuccinate	Murphy[64]
Clindamycin	Dialkylcarbonate esters	Sinkula *et al.*[65]

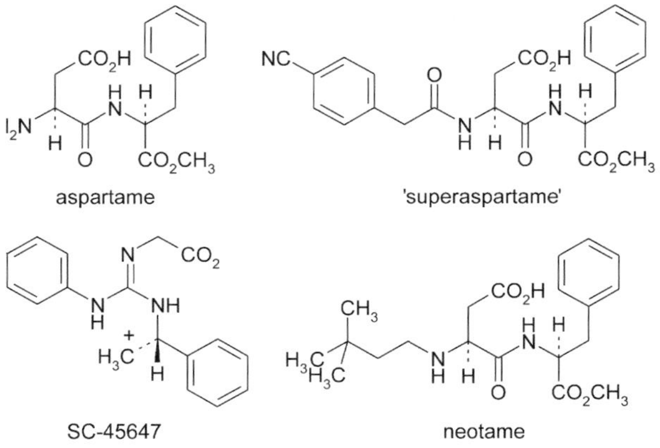

Fig. 39.19 Aspartame and aspartame–suosan hybrids.[70–73]

sterically and with respect to the orientation of hydrogen-bond donors and acceptors.

A series of qualitative and empirical rules was collected by Sinkula.[74] These may provide some clues to the medicinal chemist for improving the taste quality of objectionable tasting drugs (Table 39.2).

Table 39.2 *Empirical and qualitative structure-taste relationships*[74]

1.	Compounds containing several hydroxyl groups are usually *sweet* (glycols, sugars and other carbohydrates)
2.	An increase in molecular weight, i.e. extending a homologous series, often changes from *sweet to bitter*. Further increases in homologation reduce drug solubility below taste threshold values
3.	Nitro groups present in a molecule are usually indicative of *bitter* taste (chloramphenicol, picric acid)
4.	Alkylation of an amine or amide usually produces a *sweet* tasting substance
5.	Alkylation of a hydroxyl soup (etherification) *destroys sweet* taste
6.	Alkylation of an imide *eliminates sweet taste*; *N*-alkyl saccharins are tasteless
7.	Addition of a phenyl group to a drug molecule causes or increases bitter taste and may be due to enhanced lipophilicity and/or preferential binding with bitter taste receptors
8.	An increase in chain branching decreases sweet taste and increases *bitter* taste
9.	Esterification *enhances sweetness*, e.g. ethyl butyrate, aspartyl dipeptide esters
10.	Introduction of unsaturation into a molecule *increases bitterness and pungency*
11.	Primary amines *enhance sweet taste* especially if in close proximity to an electronegative group (—COOH), i.e. in certain D- and L-amino acids
12.	Secondary, tertiary amines and quaternary ammonium salts are *bitter*, especially alkaloids, antibiotics and quaternary ammonium drugs
13.	Monosubstituted ureas are usually *sweet* although some are tasteless. Urea itself is *bitter*. Symmetrical molecules are often *bitter*
14.	Aliphatic polyhalogenated compounds are sweet (chloroform, 1,2-dichloroethane, dichloromethane)
15.	Aromatic halogen substitution *increases bitterness* and is a function of the atomic weight of the halogen (see, for example, halogenated saccharins)
16.	The presence of sulfur in an aliphatic molecule is usually associated with *bitterness* (—SH, —S—, —S—S—, C = S)
17.	Sulfonic acids are usually *bitter or tart*
18.	Aldehydes are *sweet*. Aldehydo-semicarbazones are *not as sweet* as the parent aldehydes. Aldehydophenylhydrazones *are not sweet*
19.	Oximes are usually *sweet or tasteless*

REFERENCES

1. Lucretius, T.C. (1951) *The Nature of the Universe*, Translated by Latham. Penguin Books, London.
2. Shotton, E. and Ridgway, K. (1974) *Physical Pharmaceutics*, pp. 313–331. Clarendon Press, Oxford.
3. Connors, K.A., Amidon, G.L. and Stella, V.J. (1986) *Chemical Stability of Pharmaceutics: a Handbook for Pharmacists*, 2nd edn. John Wiley & Sons, New York.
4. Essig, D., Hofer, J., Schmidt, P.C. and Stumpf, H. (1986) *Stabilisierungstechnologie — Wege zur haltbaren Arzneiform*. Wissenschaftliche Verlagsgesellschaft GmbH, Stuttgart.
5. Cho, M.J., Bundy, G.L. and Biermacher, J.J. (1977) Prostaglandin Prodrugs. 5. Prostaglandin E2 ethylene Ketal. *J.Med. Chem.* **20**: 1525–1527.
6. Morozowich, W., Oesterling, T.O., Miller, W.L., *et al.* (1979) Prostaglandine prodrugs I: Stabilization of dinoprostone (prostaglandin E2) in solid state through formation of crystalline C1-phenyl esters. *J. Pharm. Sci.* **68**: 833–836.
7. Jensen, N.P., Friedman, J.J., Kropp, H. and Kahan, F.M. (1980) Use of acetylacetone to prepare a prodrug of cycloserine. *J. Med. Chem.* **23**: 6–8.
8. Bundgaard, H. (1985) *Design of Prodrugs*, pp. 55–61. Elsevier, Amsterdam.
9. Sauerberg, P., Chen, J., WoldeMussie, E. and Rapoport, H. (1989) Cyclic carbamate analogues of pilocarpine. *J. Med. Chem.* **32**: 1322–1326.
10. Hoffman, J.M., Smith, A.M., Rooney, C.S., *et al.* (1993) Synthesis and evaluation of 2-pyridone derivatives as HIV-1-specific reverse transcriptase inhibitors. 4. 3-[(Benzoxazol-2-yl) ethyl]-5-ethyl-6-methylpyridin-2(1*H*)-one and analogues. *J. Med. Chem.* **36**: 953–966.
11. Portnoff, J.B., Henley, M.W. and Restaino, F.A. (1983) The development of sodium cefoxitin as a dosage form. *J. Parenteral Sci. Technol.* **37**: 180–185.
12. Loftsson, T. and Fridriksdottir, H. (1992) Stabilizing effect of Tris(hydroxymethyl) aminomethane on *N*-nitrosoureas in aqueous solutions. *J. Pharm. Sci.* **81**: 197–198.
13. Brändström, A.E. (1988) Base addition salts of omeprazole. United States patent 4,738,974 Aktiebolaget Hassle, Sweden.
14. Krenn, M., Gamcsik, M.P., Vogelsang, G.B., *et al.* (1992) Improvements in solubility and stability of thalidomide upon complexation with hydroxypropyl-β-cyclodextrin. *J. Pharm. Sci.* **81**: 685–689.
15. Backensfeld, T., Müller, B.W. and Kolter, K. (1991) Interaction of NSA with cyclodextrins and hydroxypropyl-cyclodextrin derivatives. *Int. J. Pharm.* **74**: 65–93.
16. Suenaga, A., Bekers, O., Beijnen, J.H., *et al.* (1992) Stabilization of danorubicin and 4-demethoxydaunorubicin on complexation with octakis(2,6-di-*O*-methyl)-γ-cyclodextrin in acidic aqueous solution. *Int. J. Pharm.* **82**: 29–37.
17. Moon, K.Y., Kwon, C.H., Shirota, F.N. and Baturay, N.Z (1993) Design, synthesis, and evaluation of *N*-methyl-4-thiocyclophosphamide derivatives as chemically stable, alternative prodrugs of 4-hydroxycyclophosphamide. In *Proc. Amer. Assoc. Cancer Res.*, 84 Meet., pp. 267.
18. Kanda, Y., Ashizawa, T., Kakita, S., *et al.* (1999) Synthesis and antitumor activity of novel thioester derivatives of leinamycin. *J. Med. Chem.* **42**: 1330–1332.
19. Plattner, J.J. and Norbeck, D.W. (1990) Obstacles to drug development from peptide leads. *Drug Discovery Technologies*, pp. 92–126. Ellis Horwood Limited, Chichester.
20. Møss, J. and Bundgaard, H. (1992) Prodrugs of peptides. 17. Bioreversible derivatization of the C-terminal prolineamide residue in peptides to afford protection against prolyl endopeptidase. *Int. J. Pharm.* **82**: 91–97.

21. Rasmussen, G.J. and Bundgaard, H. (1991) Prodrugs of Peptides. 15. 4-Imidazolidinone prodrug derivatives of enkephalins to prevent aminopeptidase-catalyzed metabolism in plasma and absorptive mucosae. *Int. J. Pharm.* **76**: 113–122.

22. Pearlman, R. and Wang, Y.J. (1996) Formulation, characterization, and stability of protein drugs. Case Histories. In Bortchardt, R.T. (ed.). *Pharmaceutical Biotechnology*. Plenum Press, New York.

23. Henck, J.O., Griesser, U.J. and Burger, A. (1997) Polymorphil von Arzneistoffen: eine wirtschaftliche Herausforderung? *Pharm. Ind.* **59**: 165–169.

24. Giron, D. (1995) Thermal analysis and calorimetric methods in the characterization of polymorphs and solvates. *Thermochimica Acta* **248**: 1–59.

25. Byrn, S., Pfeiffer, R., Ganey, M., *et al.* (1995) Pharmaceutical solids: a strategic approach to regulatory considerations. *Pharm. Res.* **2**: 945–954.

26. Beckmann, W. (2000) Seeding the desired polymorph: background possibilities, limitations, and case studies. *Org. Process Res. Dev.* **4**: 372–383.

27. Lin, S.L. (1972) Preformulation investigation II: dissolution kinetics and thermodynamic parameters of polymorphs of an experimental antihypertensive. *J. Pharm. Sci.* **61**: 1423–1430.

28. Aguiar, A.J. and Zelmer, J.E. (1969) Dissolution behavior of polymorphs of chloramphenicol palmitate and mefenamic acid. *J. Pharm. Sci.* **58**: 983–987.

29. Shefter, E. (1981) Solubilization by solid-state manipulation. *Techniques of Solubilization of Drugs*, pp. 159–182. Marcel Dekker, Inc, New York.

30. Shotton, E. and Ridgway, K. (1974) *Physical Pharmaceutics*, p. 337. Clarendon Press, Oxford.

31. Mullins, J.D. and Macek, T.J. (1960) Some pharmaceutical properties of Novobiocin. *J. Amer. Pharm. Assoc., Sci.* **49**: 245–248.

32. Nürnberg, E. (1980) Darstellung und Eigenschaften pharmazeutisch relevanter Sprühtrocknungsprodukte, eine Übersicht. *Acta Pharm. Technol.* **26**: 39–67.

33. Bjerga Bjaen, A.K., Nord, K., Furuseth, S., *et al.* (1993) Polymorphism of chloroquine diphosphate. *Int. J. Pharm.* **92**: 183–189.

34. Forni, R., Coppi, G., Iannucelli, V., *et al.* (1988) The grinding of the polymorphic forms of chloramphenicol stearic ester in the presence of colloidal silica. *Acta Pharm. Suec.* **25**: 173–180.

35. Willis, G.C. and Arendt, E.C. (1954) Relative efficacy of sodium amytal, chloral hydrate and dichloralphenazone. *Canad. Med. Assoc. J.* **71**: 126–128.

36. Rice, W.B. and McColl, J.W. (1956) A comparison of dichloralphenazone and chloral hydrate. *J. Amer. Pharm. Assoc., Sci.* **45**: 137–141.

37. Bruce, W. F. (1957) Chloral derivatives and methods for their preparation. US Patent 2 784 237 (5 March 1957; American Home Products).

38. Vogel, A.I. (1989) *Vogel's Textbook of Practical Organic Chemistry*, 5th edn., pp. 1261–1265. Longman Scientific & Technical, London.

39. Anonymous (1963) Médicaments à base de paraacétylaminophénoxy-1 trichloro-2 éthanol. French Pat. M 2031 (May 24, 1962; Brevets Pharmaceutiques et Cosmétologiques S.A., Switzerland). *Chem. Abstr.* **60**: 9201a.

40. Hems, B.A., Atkinson, R.M., Early, M. and Tomich, E.G. (1962) Trichloroethyl phosphate. *Br. Med. J.* **1**: 1834–1835.

41. Dittert, L.W., Caldwell, H.C., Ellison, T., *et al.* (1968) Carbonate ester prodrugs of salicylic acid. *J. Pharm. Sci.* **57**: 828–831.

42. Anonymous (1973) Glyceryl-*N*-(substituted phenyl) anthranilates in the treatment of inflammation. US Patent 3 767 811 (23, October 1973; Schering Corporation).

43. Allais, A., Rousseau, G., Girault, P., *et al.* (1966) Sur l'activité analgésique et antiinflammatoire des 4-(2′-alcoxycarbonyl phénylamino) quinoléïnes. *Eur. J. Med Chem.* **1**: 65–70.

44. Boltze, K.-H., Brendler, O., Dell, H.-D. and Jacobi, H. (1975) Novel substituted indole compound, process for the preparation and therapeutic compositions Containing it. German Patent 3 910 952 (7 October 1975, Troponwerke Dinklage & Co., Cologne).

45. Boltze, K.-H., Brendler, O., Jacobi, H., *et al.* (1980) Chemische Struktur und antiphlogistische Wirkung in der Reihe der substituierten Indol-3-essigsäuren. *Arzneim.-Forsch. (Drug Res.)* **30**: 1314–1325.

46. Anonymous (1975) Dérivés d'indolylacétylamino-acide, leur procédé de préparation et les compositions pharmaceutiques comprenant ces composés. Belgium Patent 826 711 (14 March 1975; Schering Aktiengesellschaft, Berlin und Bergkamen).

47. Sinkula, A.A. and Yalkowsky, S.H. (1975) Rationale for design of biologically reversible drug derivatives: prodrugs. *J. Pharm. Sci.* **64**: 181–210.

48. Berger, F.M. (1952) The anticonvulsant activity of carbamate esters of certain 2,2-disubstituted-1,3-propanediols. *J. Pharmacol. Exp. Ther.* **104**: 229–233.

49. Edmondson, H.T. (1973) Parenteral and oral clindamycin therapy in surgical infections: a preliminary report. *Ann. Surg.* **168**: 637–642.

50. Gray, J.E., Weaver, R.N., Moran, J. and Feenstra, E.S. (1974) The parenteral toxicity of clindamycin 2-phosphate in laboratory animals. *Toxicol. Appl. Pharmacol.* **27**: 308–321.

51. Brodasky, T.F. and Lewis, C. (1972) An *in vivo–in vitro* comparison of the 2- and 3-phosphate esters of clindamycin. *J. Antibiot.* **25**: 230–238.

52. DeHaan, R.M., Metzler, C.M., Schellenberg, D. and Van Denbosch, W.D. (1973) Pharmacokinetic studies of clindamycin phosphate. *J. Clin. Pharmacol.* **13**: 190–209.

53. Stella, V.J., Charman, W.N.A. and Naringrekar, V.H. (1985) Prodrugs, do they have advantages in clinical practice? *Drugs* **29**: 455–473.

54. Varia, A.A. and Stella, V.J. (1984) Phenytoin prodrugs. VI: *In vivo* evaluation of a phosphate ester prodrug of phenytoin after parenteral administration to rats. *J. Pharm. Sci.* **73**: 1087–1090.

55. Adjei, A.L., Doyle, R. and Reiland, T. (1988) Flavors and flavor modifiers. In Swarbuck, J. and Boylan, J.C. (eds). *Encyclopedia of Pharmaceutical Technology*, pp. 101–139. Marcel Dekker, Inc, New York.

56. Wenz, A., Göttmann, G. and Koop, H. (1961) Verfahren zur Herstellung der freien Base und der Salze des Aneurin-orthophosphor-säureesters. German Patent 1 110 646 (13 July 1961; E. Merck A.G. Darmstadt).

57. Davies, G.E. and Driver, G.W. (1957) The antituberculous activity of ethylthiolesters with particular reference to diethyldithiol isophtalate. *Br. J. Pharmacol.* **12**: 434–437.

58. Davies, G.E. and Driver, G.W. (1958) Inhibitory action of ethyl mercaptan on intracellular tubercle bacilli. *Nature (London)* **182**: 664–665.

59. Caldwell, H.C., Adams, H.J., Rivard, D.E. and Swintosky, J.V. (1967) Trichloroethyl carbonate I. Synthesis, physical properties, and pharmacology. *J. Pharm. Sci.* **56**: 920–921.

60. Gruber, J.C.M., Stephens, V.C. and Terrill, P.M. (1971) Propoxyphene napsylate: chemistry and experimental design. *Toxicol. Appl. Pharmacol.* **19**: 423–426.

61. Rocca, G. and Rusconi, L. (1972) Procédé pour la préparation de triméthoxybenzoate de tétracycline et les médicaments qui en dérivent. French Patent 2 099 449 (17 March 1972: Officina Therapeutica Italiana, SRL).

62. Glazko, A.J., Edgerton, W.H., Dill, W.A. and Lenz, W.R. (1952) Chloromycetin palmitate: a synthetic ester of chloromycetin. *Antibiot. Chemother.* **2**: 234–242.

63. McEvoy, J.P. (1974) *American Hospital Formulary Service*, p. 24. American Society of Hospital Pharmacists, Washington DC.

64. Murphy, H.W. (1954) Esters of erythromycin. *Antibiotics Annual* 500–521.

65. Sinkula, A.A., Morozowich, W. and Rowe, E.L. (1973) Chemical modifications of lincomycin: synthesis and bioactivity of selected 2,7-dialkylcarbonate esters. *J. Pharm. Sci.* **62**: 1106–1111.

66. Morozowich, W., Sinkula, A.A., Karnes, H.A., *et al.* (1969) Synthesis and bioactivity of lincomycin-2-phosphate. *J. Pharm. Sci.* **58**: 1485–1489.

67. Morozowich, W., Sinkula, A.A., MacKellar, F.A. and Lewis, C.J. (1973) Synthesis and bioactivity of lincomycin-2-monoesters. *J. Pharm. Sci.* **62**: 1102–1105.

68. Owens, W.H. (1990) Tetrazoles as carboxylic acid surrogates in the suosan sweetener series. *J. Pharm. Sci.* **79**: 826–828.

69. Muller, G.W., Culberson, J.C., Roy, G., *et al.* (1992) Carboxylic acid replacement structure–activity relationships in suosan type sweeteners. A sweet taste antagonist. 1. *J. Med. Chem.* **35**: 1747–1751.

70. Tinti, J.-M. and Nofre, C. (1984) Synthetic sweeteners. French Demande FR 2 533 210; *Chem. Abstr.* **101:** 152354k.

71. Tinti, J.-M. and Nofre, C. (1991) Design of sweeteners. A rational approach. *Sweeteners: Discovery, Molecular Design, and Chemo-reception*, pp. 88–99. American Chemical Society, Washington DC.

72. Nofre, C., Tinti, J.-M. and Chatzopoulos-Ouar, F (1987) Preparation of (phenylguanidino)- and [[(1-phenylamino)ethyl]amino]acetic acids as sweeteners. European Patent Appl. EP 241 395; *Chem. Abstr.* **109:** 190047k.

73. Nofre, C. and Tinti, J.-M. (1996) N-substituted derivatives of aspartame useful as sweetening agents. U.S. Patent 5 480 668; *Chem. Abstr.* **112**: 1996 235846.

74. Sinkula, A.A. (1977) Design of improved taste properties through structural modification. *Design of Biopharmaceutical Properties through Prodrugs and Analogs*, pp. 422–445. American Pharmaceutical Association–Academy of Pharmaceutical Sciences, Washington.

40

DISCOVER A DRUG SUBSTANCE, FORMULATE AND DEVELOP IT TO A PRODUCT

Bruno Galli and Bernard Faller

I INTRODUCTION

Today's drug development companies are increasingly pressed to improve efficiency and profitability to sustain growth. The ability to keep a technical development fast, qualitatively high and consistent has become essential. Formulating and adapting the future drug product concomitantly with the available information provided by the different involved disciplines is on one hand mandatory and almost continuous, but also difficult to bring to an end. Basically there is always something to improve. Probably formulation development is the most exposed part in the development process, which needs constant adaption. Too often, this continuous optimization makes this part of the process time-critical. Besides some general basic information on the discovery and the development process, this chapter aims to draw attention to some strategic points at the technical level which are very important to keep the overall speed of the project high and to avoid repetition. Beside raising the awareness of hidden pitfalls and providing sufficient background information regarding formulation development, the information also aims to provide the background to ask the right questions at the right time. Links to the repository of the essential regulatory documents of the European Regulatory Agency, the FDA and of the ICH organization, which rule every development, are also provided. The development of a standard oral dosage was envisaged as the model to focus the information.

II DISCOVER THE DRUG SUBSTANCE

Knowledge of drug substance properties evolves as one moves through the drug discovery process. Usually lead compounds are obtained from high-throughput screening or other sources, such as literature or competitor compounds. Initially, very little is known about drug substance characteristics as one mainly focuses on potency and selectivity. Later, properties influencing drug disposition become important and need to be optimized in parallel to potency. Finally, some substance characteristics critical for development are determined in early development or at an interface between Research and Development.

Drug discovery activities can be divided into 3 steps: (1) exploratory research; (2) early discovery program; and (3) mature discovery program.

A Exploratory research (target finding)

The aim of the exploratory phase is to identify new therapeutic targets or mechanisms of action. Historically, this phase was mainly driven by comparison of phenotypes and genotypes, elucidation of biochemical pathways and luck, as in the past, therapeutic breakthrough often emerged from unexpected experimental results, in apparent contrast with rational drug discovery. Nowadays, molecular genetics and proteomics are emerging new technologies which are thought to help identify larger numbers of new targets of interest.

Once a target of interest has been identified one needs to evaluate its drug worthiness. In other words, how difficult would it be to translate the new concept into a commercial drug? Many factors influence worthiness:

* What is the amount of structural information available on the target?
* Does one need to go for an agonist or an antagonist?
* Where in the body is the target located (extracellular versus intracellular, free versus membrane bound)?
* Are predictive animal models available?
* Is the disease of interest acute or chronic?
* How clinically relevant might the target be?

The other question one needs to address early relates to the market value of the new approach. Is it an improvement of an existing medicine? What are the potential advantages over the existing treatment? Is it likely to be a life-saving drug (i.e. cancer) or is it designed to improve quality of life (i.e. obesity)? What is the size of the patient population? What is the targeted drug delivery pathway?

B Early discovery program (lead finding)

Once a target has been identified one can start an early discovery program. In this time period one essentially tries to validate the concept (or the target) in order to decide whether it is worth investing more resources in that direction.

This includes assay(s) development, synthesis of reference or competitor compounds and setting up the basic animal models needed in pharmacology. Here it is worth mentioning that validation of the target can often not be done with full predictive power because: (1) animals behave differently from humans; and (2) the models are always a simplification of the real pathophysiological condition encountered in patients. Once the assay(s) are in place one can initiate lead-finding activities in order to identify lead compounds in the compound collection of the company. At this point the size, quality and chemical diversity of the accessible compound collection plays a major role. Common problems in high-throughput screening are stability/purity of the solution archived and solubility. Poor solubility can significantly alter the results as much lower concentrations than expected are actually in solution. Once hits are obtained from high-throughput screening they need to be validated or confirmed. The assay is repeated with fresh stock solutions, which also undergo structure confirmation. On some occasions re-synthesis is necessary.

C Mature discovery program (lead optimization)

Once the concept has been validated using reference compounds and suitable animal models, a number of other activities can be initiated. The main goal of this phase of discovery is to optimize the lead compound(s) with respect to potency, selectivity, pharmacokinetic properties and toxicity. The difficulty is to optimize all the above in parallel, particularly because one needs to find the appropriate balance between properties that can be uneasy to reconcile. For example, potency and permeability are often in conflict due to the chemical constraints involved: binding is often gained through addition of hydrogen bond donor or acceptor atoms, while the same features usually hinder permeability (and thus oral absorption). Another example lies in the balance between binding and solubility: binding is usually gained during lead optimization at the expense of an increased lipophilicity which in turn is unfavourable for solubility, and therefore to drug disposition. In the past, the initial focus was mainly on optimization of potency and selectivity while the pharmacokinetic and toxicity aspects were addressed at a later stage. The major drawback of this approach is that at the time one focuses on ADME properties it is usually too late to introduce changes in the molecule. Today one tries to optimize ADME properties and potency in parallel. This new strategy calls for the development of tools, which allow high-throughput profiling and have substance requirements compatible with drug discovery. At that point in time, it is usually helpful to set up a flow chart to optimize the resources and streamline decisions (i.e. what assay is done when, what is done sequentially versus in parallel). Due to the relatively low solubility of many drug candidates, preformulation is often necessary even in the early phase of drug discovery. This is important to get good quality dose–response (and exposure) data in animal models particularly when the therapeutic dose is relatively high (>5 mg kg^{-1}). Preformulation ingredients should ideally not affect the model itself and contain ingredients compatible with repeated administration.

One usually applies for a patent protection at this early stage to protect the invention and/or its use. At what point one wants to file the patent is a subtle balance between too early (the patent protection clock starts) and too late (competition).

D Research–development interface

Initially one synthesizes only a few milligrams of substance, which is used for *in vitro* assays. The most interesting compounds are then re-synthesized on a slightly larger scale (100–500 mg) for testing in animal models. Finally only a few compounds will be selected for further evaluation and larger (1 to 100 g) quantities made. At that point in time one starts to characterize the drug substance in greater detail, which eventually leads to some disappointment. For example, when one looks at the crystallinity and polymorphism of different batches it is not unusual that some physicochemical properties, such as solubility change substantially (decrease by more than 1 log unit, for example) compared with earlier determinations done with batches which were not characterized with respect to polymorphism. One drug substance can lead to many polymorphs (and pseudo-polymorphs) which need to be characterized with respect to their physicochemical properties (solubility, stability, hygroscopicity, dissolution rate). Formation of hydrates and solvates is another problem encountered in early development. Other properties which are not too critical in the early phase start to become important, like compound 'stickiness' which can lead to substantial loss of material through passive adsorption to walls of vials, tubes or pipettes. Particularly highly potent lipophilic compounds are subject to such phenomena, which may lead to erratic dosing.

The current bottleneck is that many of the drug substance properties important for development are only obtained in early development, at a time where it is late to introduce chemical changes in the molecule itself. In the future one hopes to move some of the determinations of the drug substance characteristics to an earlier stage, where medicinal chemistry alternatives are still possible.

III DEFINING EXPERIMENTAL FORMULATIONS, THE CREATIVE PHASE

A Basic thoughts on oral formulation

Various major and minor factors determine the bioavailability of a compound, to what extent and for how long a drug substance is present in the blood circulation and in the actual target tissue.

'Pfizer's rule of five' (see Lipinsky *et al.*[1] and the previous chapter) primarily allows positioning of the compound with respect to potential problems of oral absorption. It might be of interest at this point to also know what substance properties the customer or the person charged with formulation would like to ideally obtain:

- LogP of <3 and >0
- Permeability in the CaCo-2 model better than mannitol as the reference substance
- Compound withstanding exposure to water and oxygen at room temperature
- Anticipated dose in humans of <500 mg
- Dose number of <20 (according to Amidon *et al.*[2])

What is the purpose of a formulation?

A formulation is mainly intended for accurate dosing and preservation of the physical and chemical integrity of the compound. Additionally it aims to preserve the degree of dispersion of the molecule or of the crystals embedded in excipients ensuring the dissolution properties which are defined.

Figure 40.1 shows the main interrelated factors, which lead individual substances to the formulation and thereafter to the dosage form. In the early stages efforts will primarily consist of trying to understand and combine Basic Factors so as to readily set up a useful, versatile and reliable formulation principle. The Derived Factors will predominantly have to be investigated, understood and validated at later development stages, possibly in parallel or shortly after having fixed the quality of the drug substance (see also Fig. 40.3).

Only to a minor extent is the task of a formulation to improve the absorption process which is normally mainly driven by the passive diffusion (physicochemistry) of the compound. This is somehow in contrast with many cases where special formulation efforts allowed compounds to be rescued and to compensate for their poor physicochemical properties. A good and integrated formulation approach should therefore always consider physicochemistry of the compound first relating it to the key information for any formulation activity, namely the envisaged dose.

However, there are also biological factors such as metabolism, for example, in liver, GIT-tissue and skin, as well as active molecule transport systems like P-gP or MRP, for example, in the gastrointestinal tract, brain or cancer tissue, which influence the bioavailability and mislead the prediction of the dose. The influence of such active transports is particularly noted if the main driving forces of the absorption process, physicochemical parameters like solubility, dissolution rate, free concentration and permeability are low.

The anticipated dose triggers the selection of a route of administration as well, simply because of limitations with respect to mass transport. Table 40.1 provides a rough idea about the achievable doses related to the administration route.

B Suggested sequence of activities prior to starting formulation

Prior to starting multiple formulation efforts and labour-intensive *in vivo* formulation optimization experiments, an assessment should be made as to whether drug metabolism and transport are relevant for the substance to be formulated. This can be done by implementing stepwise the following procedures:

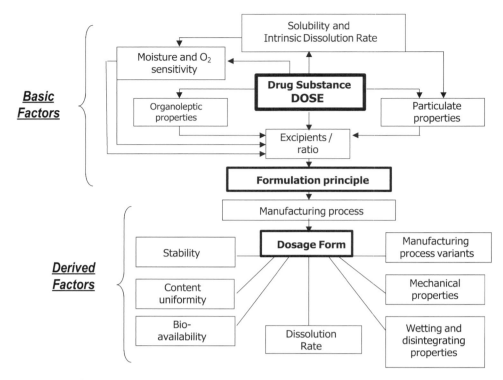

Fig. 40.1 Main interrelated factors of an oral formulation/dosage form.

(1) *In silico* estimation of the physicochemical properties, such as calculated LogP and polar surface area (see Palm *et al.*[3]).

(2) Experimental determination of physicochemical properties, such as pK_a, solubilities, chemical stability or even special amphiphilic properties (for the latter see Seelig *et al.*[4]).

(3) Assessment of the ability of the compound to pass through membranes using *in vitro* models, such as CaCo-2 cell layers together with a well-documented reference compound such as, for example, mannitol (see Arthusson *et al.*[5]).

Table 40.1 *Approximate dosages in humans depending on the administration route chosen*

Administration route	'Realistic' dosages related to drug substance (humans)
Oral	0.25–1000 mg
Passive transdermal	0.05–5 mg
Inhalation	0.01–5 mg
Nasal, buccal, sublingual	0.05–2 mg
Injection	Mainly depending on volume that can be administered and on physicochemical properties

See reference 6.

(4) Assessment *in vitro* of the metabolic pattern, biological stability, protein binding and partitioning *in vitro* and on a need basis followed by a verification *in vivo*.

C Biopharmaceutical classification of compounds

Provided that metabolism and active transport systems are not the dominating factors for the drug absorption process, there are, in general terms, two factors that can fairly well describe the absorption by passive diffusion: solubility together with permeability through membranes. Consequently, four classes of substances can be defined according to Amidon *et al.*[2]:

(1) *High solubility–high permeability* (e.g. Metoprolol, l-Dopa): Formulation principles for class I compounds are normally straightforward. Aqueous solutions and *in situ* suspensions cover the needs for an early formulation.

(2) *Low solubility–high permeability* (e.g. Carbamazepine, Nifedipine): Generations of pharmacists have struggled to overcome the difficulties related to this class of compound. Scientific progress in the field of formulation of class II compounds have delivered many interesting formulation principles over the last two decades. Thus, handling of this class is relatively well understood. Success is mainly due to improvements in controlling

the solubility and dissolution rate by measures, such as for example, the selection of the appropriate salt, the use of the appropriate crystal form, micronization, controlling of the wettability, the addition of surfactants and the use of microemulsions or solid dispersions, just to give a few examples. These measures have become the basis for adequate formulations of this category of substances which is handled (within a certain frame and limits) much better than in the past.

(3) *High solubility–low permeability* (e.g. Atenolol, Terbutaline): The bioavailability of class III compounds can in practice not be influenced by a formulation. Rarely, bioavailability can be improved by increasing the dose or by modifying the permeability of the gastrointestinal membrane. However, modification of the gastrointestinal membrane structure by penetration or absorption enhancers always bears the risk of improving the uptake of all kinds of toxins from the intestinal lumen and also of causing severe irritation.

(4) *Low solubility–low permeability:* Real class IV compounds are not likely to be formulated. For highly potent drugs there might be the opportunity to raise the dose but in general one starts looking for suitable back-ups and to switch to a different class of compounds. Changing the route of administration might be viable (e.g. parenteral), but this has consequently a strong impact on the attractiveness of the product or limitations in dosing but it may be helpful for products in therapeutic areas such as oncology.

How do we proceed at a practical level?

Solubility always has to be considered together with the envisaged dose (anticipated needed dose and reference volume of a stomach where it has to dissolve). This leads simplistically to define:

$$\text{Dose number} = \frac{M_o/V_o}{C_s},$$

where M_o = dose to be administered; V_o = 250 ml (by definition the reference stomach volume); C_s = solubility.

Dose numbers of >20 are by definition related to 'low soluble and dissolution rate-limited' substances (see Amidon *et al.*[2]). In such cases special formulations are to be considered from scratch in the formulation strategy.

A permeability assessment at an early stage has to be done using, for example, the CaCo-2 cell model and mannitol as a reference substance. Low permeable substances can be defined as to be less permeable than mannitol.

Low permeable compounds always create serious formulation problems whether it is due to the physicochemistry or due to active transport systems. For both the CaCo-2 model provides useful information. However, the crucial question at this stage is to sort out, in view of the evaluation of alternative routes, whether the observed low bioavailability is due to metabolism or low permeability or due to both. Thus, permeability is always complex and not straightforward to assess.

In summary (see also Lipper[6]):

- Poor biopharmaceutical properties may sometimes be corrected by formulation, but at the expense of time and resources.
- Poor solubility and stability may be amenable to being fixed by formulation.
- Poor permeability is very difficult to correct by formulation.
- First-pass metabolism problems cannot be fixed by oral formulations.

D Which formulation principles are used?

At early stages stable solutions are preferred because of their defined physical state assumed as to be monomolecularly dispersed systems.

Simple micellar dispersed solutions are also directly usable by the experimenter.

Complex micellar systems like microemulsions or special solid systems (suspensions or solid dispersions) usually require a basic pharmaceutical development strategy and sufficient understanding of the technical concept to ensure appropriate quality and reproducibility of the formulation. These systems can usually be properly elaborated only at later development stages because of a lack of sufficient drug substance at early stages!

In early preclinical drug testing, it is preferable to use excipients that can also be used in later development phases, including clinical development. Whenever possible, the same formulation principle should be used throughout preclinical testing. This applies to all studies in experimental animals including pharmacological, toxicological and pharmacokinetic studies.

To switch without a rationale to different formulation principles may significantly affect the consistency of *in vivo* results such as, for example, bioavailability or in a worst case the pharmacodynamics. Not adhering to this simple rule has resulted many times in false interpretations of *in vivo* results and delays in compound development at later, more expensive stages (Fig. 40.2).

With the complexity and the multitude of open biological and physicochemical questions at early stages, it becomes imperative to work closely and to interact using a multidisciplinary approach. Information from iterative learning loops for a formulation is on one hand derived from physical stress (time, temperature and humidity) and on the other hand information is provided sequentially from biological tests as pharmacokinetic results. Of particular interest is the wide range of dosing encountered: from

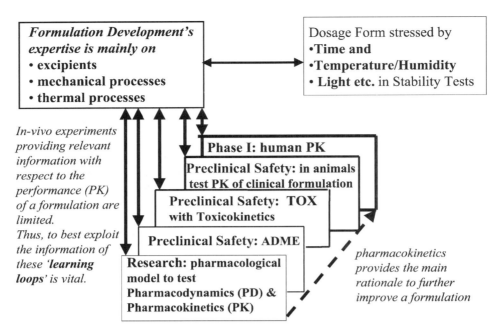

Fig. 40.2 Iterative learning loops to support the optimization of a formulation.

the pharmacological doses up to toxicological doses. The main difficulties consist in properly assessing and distilling useful information out of these experiments as early as possible and integrating it into the dosage form concept. Boundless information flow is of particular importance as by tradition the different contributing units belong to different line functions and departments not always with the same objectives. The final result of these efforts is to have from the start a good estimate of dose (even if only in animals at this point), safety margin, route of administration and a consistent formulation approach for the subsequent expensive clinical trials (see point 1 in Fig. 40.4).

IV PHARMACEUTICAL DEVELOPMENT IN INDUSTRY

Over the last decade the free flow of information, pressure from health authorities and considerable efforts in science have driven drug development in three main directions: towards 'evidence-based medicine', cost savings and speed. An excellent review on the state of the science in the field of drug development and its different approaches, together with critical assessments are provided by Lesko *et al.*[7]

Figure 40.3 provides a general overview of the main drug development stages, an approximate overall average duration and the likelihood of success for a drug as a function of its development stage.

With 'creativity at early stage' it means that besides setting up an innovative therapeutic concept, there is also much information generated to feed the main product development strategy. The concentrated information should at this stage also be used to proactively set up ideas and concepts for the future: life cycle management and its related drug product strategy. However, it is very important to proceed with balanced risk management, always considering the respective chances of success. In this situation it is worth constantly monitoring the competitive situation on the drug product side, and to act accordingly. At a later stage, once the major clinical trials have been started, it will be much more difficult to make changes to the drug product and its quality, given the efforts to be undertaken at this stage. All departments will be involved in narrowing specifications for process parameters submission and forthcoming inspections by the health authorities. Normally at the beginning of the phase IIb clinical studies the market form of an oral drug should be defined, in order to consistently support the wider clinical trials and to avoid the necessity of bridging studies.

Because of the huge costs incurred during the large clinical phase III trials many companies have started in the recent past to implement rigorous selection processes allowing selection of the most suitable and most promising projects. An excellent review on the different current philosophies is provided by Shillingford and Vose.[8] To be able to decide on the relative importance of the selected criteria, the envisaged target product profile for therapeutic and commercial success needs to be decided early on. This triggers beneficial early interactions between marketing/commercial and R&D functions leading to criteria which

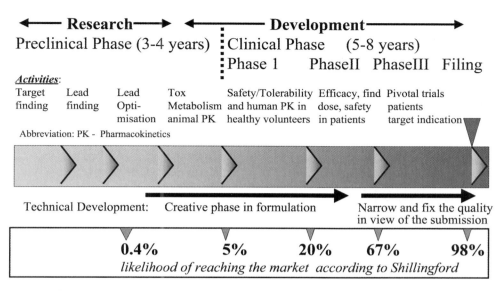

Fig. 40.3 Main stages of drug development.

allow constant challenge of the project regarding its viability from different points of view. From the cited authors, a table of go/no go criteria is taken and slightly modified. It reflects the thoughts resulting from the collaboration at an early stage of research, development and marketing/business functions (Table 40.2).

With the challenging goal of becoming faster and better estimating the likelihood of success, proof-of-concept approaches based on monitoring surrogate markers, bio-markers, functional imaging and early clinical results are investigated in studies involving small numbers of volunteers to support the fast decision-making process. However, there are also risks associated with these fast investigations made within the tight frameworth of resources. Prior to starting investigations in man at preclinical stage the risks might be:

• Profile of the compound could suffer from a lack of an optimal formulation and thus show low bioavailability.

Table 40.2 *Criteria and data supporting go or no-go decision-making*

Criterion	Go!	No-Go!
PD activity at tolerable doses	Reproducible and related to clinically relevant dose	Absent, variable, poorly dose-related
PD duration	Allows dosing regimen acceptable to patients	Requires inconvenient complex dosing regimen
PK characteristics	Linear with dose and time, low inter/intra subject variability	Nonlinear with dose and time, high inter/intra subject variability
PK/PD relationship	Well-defined, predictable	Inadequate, unpredictable
Safety, tolerability profile	Predictable, wide therapeutic range	Unpredictable, narrow therapeutic range, potential drug–drug interactions
Bioavailability	Acceptable, allowing therapeutic levels, predictable, low inter/intra subject variability	Unpredictable, therapeutic levels not reached, high inter/intra subject variability
Physicochemical properties	Allows adequate exposure in human studies in an appropriate formulation Reasonable stability	Adequate exposure in human studies in an appropriate formulation not reached Instability
Commercial viability	Adequate market size and/or price profile to achieve return of investment	Market too small and/or low priced to ensure return of investment

PK, pharmacokinetics; PD, pharmacodynamics.
See reference 8.

- Validation of clinical relevance is not possible in the animal model.
- Metabolism in man differs from that observed in animals.
- Indication chosen may be uncertain and this can affect risk or benefit assessment.
- Forecast regarding the market condition's competitive situation and projected return of investment assessment may be wrong. In the case of an innovative new approach it may not be assessable and therefore results in low interest.

Once human data are available at the clinical phase II stage, the problems may derive from other topics such as:

- Lack of understanding of the relationship dose–pharmacokinetics–pharmacodynamics in the disease state.
- Relevance of any biological markers used to define clinical outcomes.
- Uncertainty of the therapeutic range.
- Uncertainty as to what consists a clinically relevant effect.
- Relevance of unexpected findings.
- Relative return of investment with other competing approaches or with a more attractive back-up compound.

Point 2 in Fig. 40.4 shows at which stage such decisions often occur. It is evident that it may occur in the middle of the synthesis of campaign 3 of the drug substance, or of the development of the market form and of long-term toxicity studies. It illustrates as well that to start everything in parallel increases the risk of making a wrong investment at a stage with still a relatively low chance of success (see Fig. 40.3). It goes without saying that the alternative of minimizing the risk by starting all activities in sequence may on the other hand save resources, but at the price of becoming extremely slow. The right balance has to be struck.

V FIXING THE QUALITY AND DEVELOPING THE PRODUCT IN A REGULATED ENVIRONMENT

As already mentioned it is vital that after the positive outcome of the clinical phase II studies and once the whole organization really wants to develop the product the full development process starts as soon as possible. The whole process now will tend to gradually fix the variables by narrowing the specifications, fixing the analytical methods, defining the launch sites, the equipment, the production

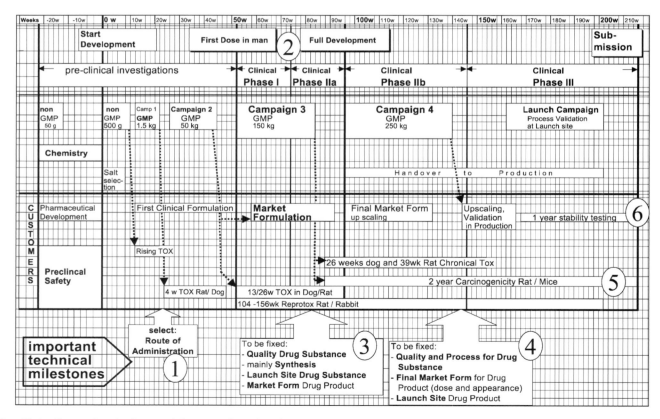

Fig. 40.4 Scenario of a fast oral drug product development showing the interdependence of some involved selected key disciplines from a technical point of view.

batch sizes, the final market product and by generating stability data. In parallel, optimization of the profitability is run concomitantly with assuring the supply chain. It is very important to quickly come to a controlled, robust process in order to timely validate and to consistently meet the three topics which are relevant to the health authorities during the filing and inspection process: efficacy, safety, quality.

Point 3 in Fig. 40.4 relates to the start of fixing the quality of the drug substance, the main part of the synthesis, the definition of the launch site to the outcome of a mid-size drug substance campaign. The experience gained during the campaign and perhaps first hints of the human dose (phase I results) allow anticipation of the target production volume and based on this information, decisions on the production site must be taken. The latter is particularly important as it needs to be filed and is inspected prior to the approval. The quality of the drug substance at this stage is also very important for the toxicity studies to be initiated. It would be detrimental to have to repeat at a later time-point extensive toxicity studies because of new by-products, which were not present in previous drug substance qualities. This is particularly important for the carcinogenicity studies, which are long and usually time-critical (point 5 in Fig. 40.4). As a general rule one tries to become gradually 'cleaner' with forthcoming batches. However, to become 'cleaner' may have a strong impact on the properties relevant to the drug product and it is very important to involve both disciplines (chemical and pharmaceutical development) in this decision-making process and mutually agree on the properties and constraints.

It is important that after campaign 4 when chemistry and pharmaceutical development are ready to initiate their activities of process validations that the variables of drug substance and product are definitively fixed (point 4 in Fig. 40.4). Validation according to the definition means to provide documented evidence that the processes consistently works. It is logical for the validation processes that the analytical methods need to be validated first and all other activities subsequently.

Last but not least, point 6 in Fig. 40.4 shows the importance of long-term stability tests of the drug product. Often together with the carcinogenicity studies their availability trigger the submission process and therefore they are project-wise on the critical path.

Links to the repository of the essential regulatory documents of the European Regulatory Agency (EMEA), the US Food and Drug Administration (FDA) and of the tripartite International Committee of Harmonization (Europe, USA and Japan) ICH organization are provided here:

EMEA: http://www.emea.eu.int/index/indexh1.htm
FDA: http://www.fda.gov/cder/guidance/index.htm
ICH: http://www.ifpma.org/ich5q.html

Regulatory information is readily and freely accessible today. As the data are updated, we highly recommend periodically checking the latest information provided and also to dedicate some time to reading documents which are in the review process. They are also accessible and should be mandatory reading for all those involved in the developmental process.

REFERENCES

1. Lipinsky, C.A., Lombardo, F., Dominy, B.W. and Feeney, P.J. (1997) Experimental and computational approaches to estimate solubility and permeability in drug discovery and development setting. *Adv. Drug Deliv. Rev.* **23**: 3–25.
2. Amidon, G., Lennernas, H., Shah, V.P. and Crison, J.R. (1995) Theoretical basis for a biopharmaceutical drug classification: correction of *in vitro* drug product dissolution and *in vivo* bioavailability. *Pharm. Res.* **12**: 413–420.
3. Palm, K., Stenberg, P., Luthman, K. and Artursson, P. (1997) Polar molecular surface properties predict intestinal absorption in humans. *Pharm. Res.* **14**: 568–571.
4. Seelig, A., Gottschlich, R. and Devant, R.M. (1994) A method to determine the ability of drugs to diffuse through the blood–brain barrier. *Proc. Natl Acad. Sci. USA* **91**: 68–72.
5. Arthusson, P., *et al.* (1991) Correlation between oral absorption in humans and apparent drug permeability coefficient in human intestinal epithelial (CaCo-2) cells. *Biophys. Res. Commun.* **157**: 880–885.
6. Lipper, R. (1999) E pluribus product. *Mod. Drug Disc.* **2**: 55–60.
7. Lesko, L.J., Rowland, M., Peck, C.C. and Blaschke, T.F. (2000) Optimizing the science of drug development: opportunities for better candidate selection and accelerated evaluation in humans. *Pharm. Res.* **17**: 1335–1344.
8. Shillingford, C.A. and Vose, C.W. (2001) Effective decision-making: progressing compounds through clinical development. *Drug Disc. Today* **6**: 941–946.

41

DRUG NOMENCLATURE

Sabine Kopp-Kubel

Ordnung is die Verbindung des Vielen nach einer Regel
Order is the combination of amounts according to a rule.

Emmanuel Kant

I TRADE NAMES AND NONPROPRIETARY NAMES

Most products available on the market are nowadays identified by a trade name. This is also true in the pharmaceutical field. In many countries trade names, also called trademarks or brand names, are used when prescribing, dispensing, selling, promoting or buying a medicament. Trade names are usually selected by the owner of the product and registered in national trademark or patent offices. They are private property and can be used only with the consent of the owner of the trademark.[1, 2]

In most cases brand names are chosen for a finished pharmaceutical product, i.e. for one or various active drug substances in a defined dosage form and formulation. Therefore, pharmaceutical preparations containing the same active drug substance are frequently sold under different brand/trade names, not only in different countries, but even within the same country (see Fig. 41.1). In practice, this means that the number of trade names in one country is usually much higher than the number of active drug substances marketed and used.

Nonproprietary names, also called generic or common names, are intended to be used as public property without restraint, i.e. nobody should own any rights on their usage. These names are usually designated by national or international nomenclature commissions.

Both trade names and nonproprietary names are normally published first in the form of proposals. Comments may be made and objections raised for a certain time period before final publication.

Although both nonproprietary names and trade names may appear similar in form to an outsider, there is, in fact, a big difference. Firstly, nonproprietary names are designations to identify the active pharmaceutical drug substance rather than the final product. Secondly, the selection of a nonproprietary name follows established rules so that the name itself communicates to the medical and pharmaceutical health professional to which therapeutic or pharmacological group the active drug substance belongs.

II DRUG NOMENCLATURE

A International Nonproprietary Names for pharmaceutical substances (INN)

History

During the twentieth century, the rapid development of pharmaceutical chemistry has brought with it the need to identify large numbers of active drug substances by unique, universally available and accepted names. The systematic chemical name, codified by international bodies, including the International Union for Pure and Applied Chemistry (IUPAC) and International Union of Biochemistry (IUB) has the advantage of unambiguously defining a specific

Fig. 41.1 Various trade names for one substance, example *paracetamol*.

Fig. 41.2 Various common names for one substance, example *paracetamol*.

chemical substance, but it is often very long, difficult to memorize and practically incomprehensible for the non-chemist. Moreover, it gives no indication as to the therapeutic action of the substance.

In order to avoid citation of difficult chemical names, generic names came into being. However, in the beginning different names were independently assigned to the same substance in different countries. For example, not everybody would know that acetaminophen, *N*-(4-hydroxyphenyl)acetamide, 4'-hydroxyacetanilide, *p*-acetamidophenol, *N*-acetyl-*p*-aminophenol, acetomenophen and *paracetamol* are the same substance (see Fig. 41.2).

The World Health Organization's (WHO) international nomenclature programme was established in 1950 when member countries passed a resolution at the World Health Assembly officially initiating the programme on International Nonproprietary Names for pharmaceutical substances (INN; in French: Dénominations Communes Internationales (DCI); in Spanish: Denominaciones Comunes Internacionales (DCI)). Subsequently an international panel of experts developed a *Procedure for the Selection of Recommended International Nonproprietary Names for Pharmaceutical Substances (INNs)* (see Fig. 41.3) and *General Principles for Guidance in Devising INNs for Pharmaceutical Substances* (see Fig. 41.4).

When WHO's programme first started, the experts had to coordinate the activities of existing national nomenclature programmes, which were especially active in France, the Nordic countries, the UK and the USA. As a result of the national programmes' activities, many substances already

had different well-established national names. Members of the newly established international nomenclature programme were faced with the difficulty of choosing a single name in these instances — *paracetamol* in the example given above (see Fig. 41.5).

Since then, the activities of national nomenclature commissions have been coordinated in order to achieve international standardization in nomenclature under the auspices of WHO according to article 2a and 2u of its constitution:[3]

> *In order to achieve its objective, the functions of the World Health Organization shall be: (a) to act as the directing and co-ordinating authority on international health work;... (u) to develop, establish and promote international standards with respect to food, biological, pharmaceutical and similar products ... procedure (application, detailed steps)*

Requests for recommended international nonproprietary names are submitted on a form to the World Health Organization (INN programme, World Health Organization, 20, avenue Appia, 1211 Geneva-27, Switzerland).[4] These proposals are then submitted by the WHO Secretariat on behalf of the Director-General to the members of the *WHO Expert Advisory Panel on the International Pharmacopoeia and Pharmaceutical Preparations* designated for this purpose. The following information has to be provided on the form:

- Name and address of manufacturer and/or originator, name of responsible person

PROCEDURE FOR THE SELECTION OF RECOMMENDED INTERNATIONAL NONPROPRIETARY NAMES FOR PHARMACEUTICAL SUBSTANCES

The following procedure shall be followed by the World Health Organization in the selection of recommended international nonproprietary names for pharmaceutical substances, in accordance with the World Health Assembly resolution WHA3.11:

1.   Proposals for recommended international nonproprietary names shall be submitted to the World Health Organization on the form provided therefore.

2.   Such proposals shall be submitted by the Director-General of the World Health Organization to the members of the Expert Advisory Panel on the International Pharmacopoeia and Pharmaceutical Preparations designated for this purpose, for consideration in accordance with the "General principles for guidance in devising International Nonproprietary Names", appended to this procedure. The name used by the person discovering or first developing and marketing a pharmaceutical substance shall be accepted, unless there are compelling reasons to the contrary.

3.   Subsequent to the examination provided for in article 2, the Director-General of the World Health Organization shall give notice that a proposed international nonproprietary name is being considered.

   A.   Such notice shall be given by publication in the *Chronicle of the World Health Organization*[1] and by letter to Member States and to national pharmacopoeia commissions or other bodies designated by Member States.

4.   Notice may also be sent to specific persons known to be concerned with a name under consideration.

   A.   Such notice shall:

      (i)   set forth the name under consideration;

      (ii)   identify the person who submitted a proposal for naming the substance, if so requested by such person;

      (iii) identify the substance for which a name is being considered;

      (iv) set forth the time within which comments and objections will be received and the person and place to whom they should be directed;

   B.   state the authority under which the World Health Organization is acting and refer to these rules of procedure.

   C.   In forwarding the notice, the Director-General of the World Health Organization shall request that Member States take such steps as are necessary to prevent the acquisition of proprietary rights in the proposed name during the period it is under consideration by the World Health Organization.

5.   Comments on the proposed name may be forwarded by any person to the World Health Organization within four months of the date of publication, under article 3, of the name in the *Chronicle of the World Health Organization [1959-1986 in WHO Chronicle, since 1987 in WHO Drug Information]*.

6.   A formal objection to a proposed name may be filed by any interested person within four months of the date of publication, under article 3, of the name in the *Chronicle of the World Health Organization*.[1]

   A.   Such objection shall:

      (i)   identify the person objecting;

      (ii)   state his interest in the name;

      (iii) set forth the reasons for his objection to the name proposed.

7.   Where there is a formal objection under article 5, the World Health Organization may either reconsider the proposed name or use its good offices to attempt to obtain withdrawal of the objection. Without prejudice to the consideration by the World Health Organization of a substitute name or names, a name shall not be selected by the World Health Organization as a recommended international nonproprietary name while there exists a formal objection thereto filed under article 5 which has not been withdrawn.

8.   Where no objection has been filed under article 5, or all objections previously filed have been withdrawn, the Director-General of the World Health Organization shall give notice in accordance with subsection A of article 3 that the name has been selected by the World Health Organization as a recommended international nonproprietary name.

9.   In forwarding a recommended international nonproprietary name to Member States under article 7, the Director-General of the World Health Organization shall:

   A.   request that it be recognized as the nonproprietary name for the substance; and

   B.   request that Member States take such steps as are necessary to prevent the acquisition of proprietary rights in the name, including prohibiting registration of the name as a trade-mark or trade-name.

Fig. 41.3 Procedure for the selection of recommended international nonproprietary names for pharmaceutical substances.

GENERAL PRINCIPLES FOR GUIDANCE IN DEVISING INTERNATIONAL NONPROPRIETARY NAMES FOR PHARMACEUTICAL SUBSTANCES

1. International Nonproprietary Names (INN) should be distinctive in sound and spelling. They should not be inconveniently long and should not be liable to confusion with names in common use.

2. The INN for a substance belonging to a group of pharmacologically related substances should, where appropriate, show this relationship. Names that are likely to convey to a patient an anatomical, physiological, pathological or therapeutic suggestion should be avoided.

These primary principles are to be implemented by using the following secondary principles:

3. In devising the INN of the first substance in a new pharmacological group, consideration should be given to the possibility of devising suitable INN for related substances, belonging to the new group.

4. In devising INN for acids, one-word names are preferred; their salts should be named without modifying the acid name, e.g. 'oxacillin' and 'oxacillin sodium', 'ibufenac' and 'ibufenac sodium'.

5. INN for substances which are used as salts should in general apply to the active base or the active acid. Names for different salts or esters of the same active substance should differ only in respect of the name of the inactive acid or the inactive base.
   For quaternary ammonium substances, the cation and anion should be named appropriately as separate components of a quaternary substance and not in the amine-salt style.

6. The use of an isolated letter or number should be avoided; hyphenated construction is also undesirable.

7. To facilitate the translation and pronunciation of INN, 'f' should be used instead of 'ph', 't' instead of 'th', 'e' instead of 'ae' or 'oe', and 'i' instead of 'y'; the use of the letters 'h' and 'k' should be avoided.

8. Provided that the names suggested are in accordance with these principles, names proposed by the person discovering or first developing and marketing a pharmaceutical preparation, or names already officially in use in any country, should receive preferential consideration.

9. Group relationship in INN (see Guiding Principle 2) should if possible be shown by using a common stem. The following list contains examples of stems for groups of substances, particularly for new groups [list see text]. There are many other stems in active use. Where a stem is shown without any hyphens it may be used anywhere in the name.

Fig. 41.4 General principles for guidance in devising International Nonproprietary Names (INN) for pharmaceutical substances.

Fig. 41.5 One international name for one substance, example *paracetamol*.

- Suggested nonproprietary name(s) (various proposals possible)
- Chemical name (following IUPAC rules)
- Molecular formula
- Graphic formula
- Stereochemical information
- Therapeutic use and pharmacological mode of action
- Code, trademark (known or contemplated)
- Date of commencement in human subjects of clinical trials
- Agreement of the applicant to establish a new CAS registry number, if necessary.

Each name proposed by the originator of such a request is then examined to determine whether it complies with the rules and guiding principles for selection of a name. When all members of the *WHO Expert Advisory Panel on the International Pharmacopoeia and Pharmaceutical Preparations* designated to select nonproprietary names agree to a name, it is published first as a proposed INN. For a 4-month period, any person can forward comments or a formal objection, e.g. on the grounds of similarity to existing trademarks. When there is no objection, the name is published a second time and WHO gives notice to member states that the name has been selected by the World Health Organization as a recommended international nonproprietary name.

Selection process and selection criteria

General rules were established at the beginning of the INN programme in order to allow health professionals to understand the rationale for a number of new names for medicaments. At first some countries used shortened chemical names as generic names, but this system was found to be very limited, since many molecules contain similar elements and groups, such as phenol, chlor, methyl or benzyl-rings, in their chemical structures. In addition, in most cases a name that indicates relationship to a group of pharmacologically similarly acting substances is more meaningful to users.

The following principles should in general be applied when selecting an INN. The name should: (1) be distinctive in sound and spelling; (2) not be too long; and (3) show relationship to substances with the same pharmacological action. In addition the new name should not conflict with any existing common names or trademarks, and patients should not be confronted with nonproprietary names that are likely to have anatomical, physiological or pathological connotations: e.g. a name starting *cancer-* would not be acceptable.

In principle, INNs are given only to the active base or the active acid. Names for different salts or esters of the same active substance should differ only in respect of the name of the inactive moiety of the molecule. For example, *oxacillin* and *ibufenac* are INNs and their salts are named *oxacillin sodium* and *ibufenac sodium*. The latter are also called *modified INNs (INNM)*. [Note that before the existence of this rule, some INNs were published for salts; the term modified INN may, therefore, sometimes be used for a base or acid; e.g. *levothyroxine sodium* was published as INN and *levothyroxine* may thus be referred to as INNM.]

To facilitate the transliteration and pronunciation of International Nonproprietary Names for pharmaceutical substances certain letters, such as 'h' and 'k', should be avoided. Preference is given to 'f' instead of 'ph', 't' instead of 'th', 'e' instead of 'ae' or 'oe' and 'i' instead of 'y'. The INN for *amphetamine* is, therefore, spelt *amfetamine*.

When devising an INN it is important to be aware of possible language problems. Since the name is used worldwide, not only should certain letters be avoided, but experts need to be aware of unsuitable connotations in the major languages spoken in the world. A name may appear excellent for an English speaker, but unacceptable in another language. For example the name 'inglicretin' could remind a French speaker of the term 'crétin anglais' (stupid Englishman) and might therefore not be the best choice for a medicament.

As INNs should show relationship to other substances of similar pharmacological action, common stems have been created. A large number of such common stems are in use, and new stems are created whenever necessary.[5] Some examples are given in Table 41.1.

Examples of INNs published for calcium channel blockers, *nifedipine* derivatives are: *amlodipine, barnidipine, benidipine, cilnidipine, cronidipine, darodipine, efonidipine, elgodipine, felodipine, flordipine, furnidipine, isradipine, lacidipine, levniguldipine, manidipine, mesudipine, nicardipine, nifedipine, niludipine, nilvadipine,*

Table 41.1 *Common stems in INN formation*

Stem	Pharmacotherapeutic group
-ac	Anti-inflammatory agents of the ibufenac group
-actide	Synthetic polypeptides with a corticotropin-like action
-adol/-adol-	Analgesics
-ast	Anti-asthmatic, anti-allergic substances not acting primarily as antihistaminics
-astine	Antihistaminics
-azepam	Diazepam derivatives
-bactam	β-Lactamase inhibitors
bol	Steroids, anabolic
-buzone	Anti-inflammatory analgesics, phenylbutazone derivatives
-cain-	Antifibrillant substances with local anaesthetic activity
-caine	Local anaesthetics
cef-	Antibiotics, cefalosporanic acid derivatives
-cillin	Antibiotics, derivatives of 6-aminopenicillanic acid
-conazole	Systemic antifungal agents, miconazole derivatives
cort	Corticosteroids, except prednisolone derivatives
-dipine	Calcium channel blockers, nifedipine derivatives
-fibrate	Clofibrate derivatives
gest	Steroids, progestogens
gli-	Sulfonamide hypoglycaemics
io	Iodine-containing contrast media
-ium	Quaternary ammonium compounds
-metacin	Anti-inflammatory substances, indometacin derivatives
-mycin	Antibiotics, produced by *Streptomyces* strains
-nidazole	Antiprotozoal substances, metronidazole derivatives
-olol	β-adrenoreceptor antagonists
-oxacin	Antibacterial agents, nalidixic acid derivatives
-pride	Sulpiride derivatives
pril(at)	Angiotens-converting enzyme inhibitors
-profen	Anti-inflammatory substances, ibuprofen derivatives
prost	Prostaglandins
-relin	Hypophyseal hormone release-stimulating peptides
-terol	Bronchodilators, phenethylamine derivatives
-tidine	H_2-histamine receptor antagonists
-trexate	Folic acid antagonists
-verine	Spasmolytics with a papaverine-like action

nimodipine, nisoldipine, nitrendipine, oxodipine, palonidipine, pranidipine, riodipine, sagandipine, sornidipine, teludipine, etc.

Recent developments

When requesting selection of an INN the manufacturer has often not yet finalized the precise indications for the therapeutic use of the compound. A name is usually requested during the development phase of a new compound, which means that the request is generally submitted to WHO during the clinical trials phase. A name is, however, needed as soon as an application for registration of a product is forwarded to the national authorities. This means that the naming process is close to all new scientific developments in the pharmaceutical field. External expertise is often needed for specific questions concerning new therapeutic groups and new types of products.

During the last few years the selection process has become more complex. New receptors and pharmacological actions are discovered more and more frequently. This means in many cases that new stems have to be created. However, there is sometimes a structural relationship to existing molecules and experts have to decide whether an existing stem may be used or whether a new one must be established. Fibrinogen receptor antagonists are an example. These substances act as platelet aggregation inhibitors for which the stem *-grel* existed for several years. The nomenclature experts have to decide whether the same stem should be used for the fibrinogen receptor antagonists or whether the group of new molecules is so important that a new stem needs to be established.

On the other hand, a new mode of action is sometimes discovered for an existing substance. If further substances are developed with a similar mode of action, the question arises whether a new stem is needed, which would mean modifying the 'old' name for the first compound in the series. For example *albifylline* and *pentoxifylline* are *N*-methylxanthine derivatives and the stem *-fylline* was therefore chosen for their names. These substances have been found to also suppress tumour necrosis factor-α.[6] The experts decided to retain the stem *-fylline* in this case, since the 'new' action was nevertheless based on the typical xanthine-mediated inhibition of phosphodiesterase.

New approaches to naming pharmaceutical substances may be needed in the near future because of the increasing research using molecular design. 'Simple' derivatives of known compounds are becoming more and more rare. Chemistry based on receptor structure and molecular design focuses more on synthesizing compounds to fit receptor binding sites. This will mean that nomenclature will have to move in the same direction. Chemical relationship will need to be looked at from a different standpoint, and the pharmacological activity might have to be considered in

almost all cases as a basis for assigning a given substance to a group.

Substances produced by biotechnology are another challenge for the nomenclature committee.[7] New schemes and concepts need to be developed on a worldwide basis. One example is the scheme for common stems for naming monoclonal antibodies as given below.

(1) **General stem**
    -mab

(2) **Substems for source of product**
human	*-u-*
rat	*-a-*
hamster	*-e-*
primate	*-i-*
mouse	*-o-*
chimeras	*-xi-*

(3) **Substems for disease or target group**
bacterial	*-ba(c)-*
cardiovascular	*-ci(r)-*
immunomodulator	*-li(m)-*
viral	*-vi(r)-*

Substems for tumours
colon	*-co(l)-*
testis	*-go(t)-*
ovary	*-go(v)-*
mammary	*-ma(r)-*
melanoma	*-me(l)-*
prostate	*-pr(o)-*
miscellaneous	*-tu(m)-*

Whenever there is a problem in pronunciation, the final letter of the substems for diseases or targets may be deleted, e.g. -co(l)-, vi(r), li(m), etc.

(4) **Prefix** The prefix should be random, e.g. the only requirement is to contribute to a euphonious and distinctive name.

(5) **Second word** If the product is radiolabelled or conjugated to another chemical, such as a toxin, identification of this conjugate is accomplished by use of a separate, second word or acceptable chemical designation. For monoclonals conjugated to a toxin, the *tox-* stem must be included as part of the name selected for the toxin.

Examples of INNs are: *altumomab, balimomab, biciromab, dorlimomab aritox, imciromab, maslimomab, nebacumab, satumomab, sevirumab, telimomab aritox, tuvirumab*, etc.

Publication

Newly selected proposed INNs are published, after the originator of the request for a name has been informed, in a list in *WHO Drug Information,* a journal published by WHO.[8] For example:

propranololum 1-isopropylamino-3-(1-naphthyloxy)-2-propranol
propranolol $C_{16}H_{21}NO_2$ 525-66-6   *β-adrenoreceptor antagonist*

Two lists of proposed INNs are published per year. If no objection has been raised during a 4-month period, the proposed name is published a second time as a recommended INN as the following example:

propranololum 1-isopropylamino-3-(1-naphthyloxy)-2-propranol
propranolol $C_{16}H_{21}NO_2$

Lists of both proposed and recommended lists of INNs, are sent together with a circular letter to WHO Member States (at present 192), to national pharmacopoeia commissions and to other bodies designated by Member States. In this letter, the World Health Organization requests that Member States should take such steps as are necessary to prevent the acquisition of proprietary rights in the name, including prohibiting registration of the name as a tradename.

Up to now more than 8000 INNs have been selected. All names ever selected are published in a cumulative list of INNs, which is updated periodically.[9] The generic names are presented in alphabetical order by Latin name. Each entry includes:

- Equivalent nonproprietary names: in Latin, English, French, Russian, Spanish and now also Arabic and Chinese;
- A reference to the INN list in which the name was originally proposed or recommended;
- Reference to substances that have been abandoned or never been marketed;
- Reference to national nonproprietary names;
- Reference to pharmacopoeial monographs or similar official references;
- Names issued by the International Organization for Standardization (ISO)
- The molecular formula;
- Its Chemical Abstracts Service (CAS) number.

All INNs are now available on the web. Information on INNs is available via Internet as shown here.

INN request form:
http://www.who.int/medicines/organization/qsm/activities/qualityassurance/inn/innform2000.pdf

Procedure for the Selection of Recommended International Nonproprietary Names for Pharmaceutical Substances:
http://www.who.int/medicines/organization/qsm/activities/qualityassurance/inn/innproc.html

General Principles for Guidance in Devising International Nonproprietary Names for Pharmaceutical Substances:
http://www.who.int/medicines/organization/qsm/activities/qualityassurance/inn/inngen.html

On-line access to published INNs:
http://mednet.who.int

B Common names selected by the International Standards Organization (ISO)

Selection principles

The International Organization for Standardization (ISO) has laid down principles for selecting common names for pesticides and other agrochemicals.[10] These principles are comparable to the guiding principles for selecting INNs and have a similar purpose: to provide short, distinctive and easily pronounced names for substances whose full chemical names are too complex for convenient use. The names chosen should not be permitted to become privately owned trademarks. ISO names are also given for salts and complex esters, as well as mixtures of isomers. The work of the INN and ISO committees sometimes overlaps, especially in the field of veterinary medicine. The two committees collaborate to avoid different names used for the same compound.

National nomenclature

Since the INN programme came into existence, WHO has coordinated the activities of national nomenclature commissions. Several INN experts are secretaries to national nomenclature commissions and in most cases the WHO secretariat also acts as corresponding member of these commissions. Differences between national and international nomenclature has become rare.

In most countries national nomenclature commissions are part of or closely linked to the national pharmacopoeia. Some countries no longer have a commission and publish the INNs directly as national names in their legal publications, e.g. in Germany and the Nordic countries; the latter published previously Nordic Pharmacopoeia Names (NFN). In others, the national nomenclature commission adopts INNs in the language of the country as national names.

All European directives include INNs as 'usual terminology' (point 3 of Article 4 (2) of Directive 65/65/EEC). The European Pharmacopoeia uses INNs in the main titles of monographs.

National nomenclature commissions select and publish the following national nonproprietary names (address of responsible authority given in brackets):

- **BAN** (British Approved Names) (The Secretary, British Pharmacopoeia Commission, Market Towers, 1 Nine Elms Lane, London SW8 5NQ, UK).
- **DCF** (Dénominations Communes Françaises) (Secretariat of the French Pharmacopoeia Commission at the Drug Agency, Direction des Laboratoires et des Contrôles, Unité Pharmacopée, 145–147, Boulevard Anatole France, 93200 Saint-Denis, France).
- **JAN** (Japanese Accepted Names) (Japanese Ministry of Health and Welfare, New Drugs Division, Pharmaceuticals Affairs Bureau, 1-2-2, Kasumigaseki, Chiyoda-ku, Tokyo 100, Japan).
- **USAN** (United States Adopted Names) (United States Adopted Names Council, American Medical Association, P.O. Box 10970, Chicago, Illinois 60610, USA).

Other names:
- **PEN** (Pharmacy Equivalent Name). A Pharmacy Equivalent Name is a short and simple name that is offered by the United States Pharmacopeia (USP) as a standardized term. The PEN name for a dosage form containing two or more therapeutic drug substances may be devised by combining portions of the official names of the component drug substances. A 'Co-' prefix indicates that the product is a combination dosage form. PEN names are not official titles in the USP pharmacopeia.

III USE AND PROTECTION OF NONPROPRIETARY NAMES

WHO's INN programme has been actively providing nonproprietary names since 1953. During this period more than 8000 names have been published. New pharmaceutical substances are continually being developed and some 150 new names are published every year.

In order to avoid confusion, it is strongly recommended by the World Health Organization that new drug substances are identified by codes rather than arbitrary names until nonproprietary names have been designated.

Use of nonproprietary names

Nonproprietary names are intended to be used in pharmacopoeias, labelling, advertising, drug regulation and scientific literature, and as product names, e.g. for generics.

Some countries have defined the minimum size of characters in which the generic name must be printed under the trademark labelling and advertising. In Canada, the USA and Uruguay, the generic name must appear prominently in type at least half as large as that used for the proprietary or brand name. Certain countries, for example Mexico, have even gone as far as abolishing trademarks for the public sector.

Protection of nonproprietary names

The consequence of introducing INN common stems into trademarks, which seems to be increasingly popular, hampers the selection of new nonproprietary names within the established system. Given that all new INNs should be distinctive from existing INNs, without similarity to trademarks, this practice causes confusion to the health professional, may be the source of serious errors in prescribing and dispensing and hinders the selection of future names for compounds in the same group of substances. Based on recommendations made by the WHO Expert Committee on the Use of Essential Drugs, a resolution[11] was adopted during the 46th World Health Assembly requesting Member States to:

> *enact rules or regulations, as necessary, to ensure that international nonproprietary names ... are always displayed prominently; to encourage manufacturers to rely on their corporate name and the international nonproprietary names, rather than on trademarks, to promote and market multisource products introduced after patent expiration; to develop policy guidelines on the use and protection of international nonproprietary names, and to discourage the use of names derived from INNs, and particularly names including INN stems in trademarks.*

IV SUMMARY

The existence of an international nomenclature for pharmaceutical substances, in the form of INNs, has proved since 1950 to be important for the safe prescription and dispensing of medicines to patients, and for communication and exchange of information among health professionals worldwide. INNs identify pharmaceutical substances by unique names that are globally recognized and are public property. They are also known as generic names.

Common stems are developed for the selection of INNs to communicate to the health professionals the type of pharmaceutical product in question. National and international nomenclature commissions collaborate closely to select a single name of worldwide acceptability for each active substance that is to be marketed as a pharmaceutical.

To avoid confusion, which could jeopardize the safety of patients, nonproprietary names and their common stems should not be used in trademarks. The selection of further names within a series should not be hindered by the use of a common stem in a brand name.

The author alone is responsible for the views expressed in this article.

REFERENCES

1. Wehrli, A. (1986) Pharmaceuticals: Trademarks versus generic names. *Trademark World Journal*, 31–35.
2. WHO (1987) Trademarks versus generic names for pharmaceuticals. A conflict that requires resolution. *WHO Drug Information* **1**: 39–40.
3. WHO (2001) *Basic Documents*, forty-first edition. World Health Organization, Geneva.
4. WHO (1990) *INN Application Form*. World Health Organization, Geneva.
5. WHO (2001) The use of common stems in the selection of INN for pharmaceutical substances. World Health Organization, Geneva.
6. Semmler, J., Gebert, U., Eisenhut, T., *et al.* (1993) Xanthine derivatives: Comparison between suppression of tumor necrosis factor-α production and inhibition of cAMP phosphodiesterase activity. *Immunology* **78**: 520–525.
7. Wehrli, A. (1992) Generic names for biotechnology-derived products. *Drug News Persp.* **5**: 55–58.
8. Examples of published INN lists: List 86 of proposed INNs: *WHO Drug Information*, 2002, 15(2); List 47 of recommended INNs: *WHO Drug Information*, 2002, 15(3). World Health Organization, Geneva.
9. *Cumulative List No. 10 of International Nonproprietary Names (INN) for Pharmaceutical Substances*. World Health Organization, Geneva, Switzerland.
10. ISO Standard Pesticides and Other Agrochemicals (1988) Principles for the selection of common names (ISO/DIS 257, 1988), International Organization for Standardization, Geneva.
11. Forty-sixth World Health Assembly Resolution on Nonproprietary Names for Pharmaceutical Substances (WHA46.19) (1993) World Health Organization, Geneva.

42

LEGAL ASPECTS OF PRODUCT PROTECTION — WHAT A MEDICINAL CHEMIST SHOULD KNOW ABOUT PATENT PROTECTION

Maria Souleau

Εν οιδα οτl ονδεν οιδα ΠΛΑΤΩΝ
The only thing that I know is that I don't know anything

Plato

I INTRODUCTION

A patent is a grant to an inventor of the exclusive right to use and profit from his invention for a limited term. In exchange, the inventor discloses the invention in such a way that an expert could follow it without doing research. Patents play an important role in the progress of national economies and in progress, in general.

During the last two centuries, the USA and some western European countries have undergone phenomenal industrial and economic development. In this process, research and innovation played a predominant role and patents were instrumental, inasmuch as inventions were no longer kept secret but were made known to the public, thus inducing new research and innovation. The post-war development of Japan's industrial potential is an example which demonstrates the role and the importance of the patent system.

A A history of the patent system prior to 1883

Originally patents were intended to encourage the introduction of new technologies into a country so as to develop native industry, rather than to provide a just reward to the inventor. In order to encourage and reward the publication of new knowledge, privileges were granted giving the owner of the technology the exclusive right to exploit that knowledge in the relevant territory for a limited period. Until the fifteenth century, the various states were scarcely concerned with furthering the development of industry and generally speaking manufacturing secrecy remained the rule. The statute for inventors decreed by the Council of the Venetian Republic on 19 March 1474, can be considered as the first law on patents.[1] This statute for inventors stipulated that:

whoever will make in this City [Republic of Venice] any new and ingenious artifice, not made previously in our state, will be obliged to register it at the office of our proveditors of the Commune, as soon as it will be reduced to perfection, so that it will be possible to use and apply it. It shall be forbidden to anyone else in any our land and place to make any other artifice to the image and similarity of that one without consent and licence of the author during the term of ten years.

Studies of the detailed records of early grants in other countries suggest that the idea of a patent system spread over Europe from Italy with emigrating Venetian glassworkers.

In England, as from 1315, the grant of monopolies by the Crown had been widely used as a means of promoting and regulating trade and as a source of revenue. As a means of attracting new industries from overseas, letters of protection were also granted to groups of foreign workers. From the beginning, an essential condition of the validity of English patents for invention rights was that the invention should actually be worked in the country. The practice of granting patent monopolies became firmly established in England in the reign of Elisabeth I. Between 1561 and 1590 more than 50 monopolies were granted to encourage English industry and to make the country self-sufficient. Some of these patents were patents of invention for new industries, but the Queen's need for money also led her to grant other monopolies giving the right to control established industries. Such patents were clearly against the public interest and provoked vigorous protest in the House of Commons. So, in 1601, the Queen issued a proclamation revoking the objectionable patents and enforced the monopoly rights grants for invention. In 1623, under the reign of James I, the 'Statute of Monopolies' prescribed the cases for which there could be a monopoly. According to this statute an invention had to be novel and nondetrimental to the general interest. This law established also a time limit for the 'duration of the monopoly'. After the passage of the Statute of Monopolies, there was no further legislation in England on patents of invention for over 200 years.

The Statute of Monopolies was followed by similar legislation in the American colonies, where settlers began to petition the legislature or governor for monopoly patents similar to those they had been familiar with in England. In 1641 the General Court of Massachusetts adopted a 'Body of Liberties' prohibiting the granting of monopolies except for 'such new inventions as are profitable to the country and that for a short time'. After the American War of Independence, the old concept of a patent granted as an act of grace and favour by the Crown began to be replaced by the idea that an inventor had a natural right to his property. In South Carolina, an act of 1784 provided for 14 years monopoly rights to inventors and most of the other new states continued and extended the colonial practice of rewarding inventors with the grant of patent rights. When the delegates of the states met to draft the Constitution of the United States of America it was recognized that it would be more effective to replace the separate state patents by a single form of protection for the whole country. In 1790 Thomas Jefferson drafted the first American law of patents. This law stipulated a monopoly to inventors of 17 years from the grant for novel inventions useful to the country.

In France, the early practice of the royal grant of patents (from 1543) for inventions was confirmed by an edict of Louis XV in 1762 which prohibited permanent privileges, but provided for inventors' patents to be limited to 15 years. In 1789, with the revolution, the privileges of inventors were abolished. Nevertheless on 7 January 1791, the Constitutional Assembly passed the first law on patents. In the words of the preamble of the new French law,

every novel idea whose realisation or development can become useful to Society belongs primarily to the man who conceived it, and it would be a violation of the rights of man in the very essence if an industrial invention were not regarded as the property of its creator.

According to this French law for a disclosure of the invention, the state granted to the author a 'manufacturing monopoly' for a limited period. Under the law of 1791, French patents were granted for 5, 10 or 15 years. Patents became null and void if not worked within 2 years or if the patentee sought to take out a patent abroad. Examination of applications was still required, but in 1844 this was abolished. According to the 1844 law, the inventor was not obliged to delimit his monopoly by claims, but the inventor was obliged to describe his invention to the best of his ability.[2]

In the seventeenth and eighteenth centuries, with the development of commerce and the beginnings of industry, a certain number of states thus adopted patent laws which gave their inventors protection against the infringement of their inventions. In Europe subsequent developments in the nineteenth century were greatly influenced by competing economic interest. Industrialists who recognized the value of patents in protecting their businesses increasingly pressed for the patent system to be strengthened and expanded. A uniform patent law for the whole of Germany was adopted in 1877 with strict examination for 'novelty' and 'level of invention'.

Before enumerating the main conventions and treaties, still in force, it must be noted that almost every country has now adopted a patent law.

B Main conventions and treaties

In the middle of the nineteenth century it was already apparent, that industrial property rights should extend beyond national boundaries. In 1883, in order to facilitate the transfer of technology, a large number of states created the *Paris Convention*, an international convention. The Paris Convention established major basic rules concerning industrial property and was a remarkable piece of legislation for its time. The convention was revised in subsequent international conferences in Brussels in 1900, Washington in 1911, The Hague in 1925, London in 1934, Lisbon in 1958 and Stockholm in 1967. The main provisions which apply between the member countries are now:[3]

(1) Foreigners receive in each country the same treatment as nationals of that country.

(2) An applicant for a patent in one country receives the benefit of his original filing date for any subsequent application for that invention, filed in another country within one year (this is the priority clause — *Priority Right*).

(3) Patents in different countries for the same invention are independent, so that they are not affected by refusal, revocation or expire in any other country.

(4) Importation by the patentee of goods produced in one country is not to entail forfeiture of patent protection for those goods.

(5) Each country may take steps to prevent abuses resulting from the exclusive rights arising from patents, e.g. failure to use the invention, but the patent may be revoked only if compulsory licensing is not a sufficient remedy.

Concerning the Priority Right, no extension of the 12-month period can be allowed. After 12 months, the Priority Right is lost.

In 1978, two treaties came into force which had a great impact on the world patent system unequalled since the Paris Convention of 1883.

These two treaties were the Convention of the Grant of European Patents, named the *European Patent Convention (EPC)*, signed in Munich in 1973, set up by the Council of Europe, and the *Patent Cooperation Treaty (PCT)*, signed in Washington in 1970, established by the World Intellectual Property Organization. Furthermore, the Council of the European Economic Community has, for some time, been endeavouring to bring into force a Convention for the European Patent for the Common Market (the Community Patent Convention) for which the basic text was signed in Luxembourg in December 1975 by the EEC countries at that time. This has been supplemented by a Community Patent Agreement, initiated by the 12 EEC countries in December 1985.

The European Patent Convention is in addition to the national patent laws which exist in each EPC-contracting state. Even though the contracting states had to modify their laws in order to harmonize it with the European Patent Convention.

The European Patent Convention and the Patent Cooperation Treaty each establishes a system for filing a single patent application to obtain, ultimately, national patents in countries designated in the application.

As concerns the European Patent Convention (EPC), the possibility of filing a European Patent Application is open to countries both inside and outside the EPC-member states. According to the European Patent Convention, by means of a direct single filing at the European Patent Office (EPO) or indirect single filing through certain EPC-member states national patent offices, if the law of a contracting state permits, it is possible to obtain national patents in the contracting states.

In July 2002, the contracting states were: Austria, Belgium, Bulgaria, Cyprus, Czech Republic, Denmark, Estonia, Finland, France, Germany, Greece, Ireland, Italy, Luxembourg, Monaco, the Netherlands, Portugal, Spain, Slovakia, Sweden, Switzerland and Liechtenstein, Turkey and United Kingdom.

The European Patent Organization has also concluded agreements on cooperation in the field of patents and on extending the protection conferred by European patents (Extension Agreements) with a number of states which are not party to the EPC. These agreements form the basis of an extension system providing European patent applicants with a simple and cost-effective way of obtaining patent protection in these countries. At the applicant's request patents can be extended to these countries where they will have the same effects as national applications and patents and will enjoy substantially the same protection as patents granted by the EPO for the member states of the European Patent Organization. At present, extension to the following states may be requested: Slovenia (as from 1 March 1994), Lithuania (as from 5 July 1994), Latvia (as from 1 May 1995), Albania (as from 1 February 1996), Romania (as from 15 October 1996), former Yugoslav Republic of Macedonia (as from 1 November 1997).

According to the European Patent Convention, a European Patent Application is submitted to a single examination system. An applicant files a patent application in one of the official languages of the European Patent Convention, namely German, English and French. This application is examined at the EPO for formalities and then a prior art — the invention must not have been disclosed in or by prior publication or prior patents, or must not have been involved in public use or public sale, or must not have been a part of or suggested of prior knowledge — search and a search report are performed. The patent application and the search report are then published.

After publication of the application and the search report, the applicant has to request substantive examination for patentability to be performed at the EPO. The examination of the application is performed by the Examining Division of the EPO. On completion of the examination, which includes an examination for inventive step as well as for novelty (novelty and inventive step are basic requirements for patentability; see the section on basic and formal requirements below), either the application is rejected or the European Patent is granted. The specification as granted is published and the grant of the European patent leads to national patents being registered in the designated EPC countries.

Opposition can be filed at the EPO after the European Patent has been granted. The results of the opposition, i.e. amendment or revocation of the European Patent, apply equally to all the national patents resulting from the European Patent. Furthermore, each national patent can be amended or revoked by national courts under the relevant national law.

The EPO has set out to establish a practice which is a compromise between various traditions and practices (German, British etc.). The officials in charge of the development procedure have attempted to avoid the worst features of the existing national systems. The EPO regards itself as having the function of granting patents rather than refusing them, and with the proviso that the claimed invention is patentable, an application will be allowed without prolonged arguments and numerous official actions.[4,5]

The *Patent Cooperation Treaty* (PCT) is open to all countries of the world, which have acceded to the Convention of Paris. A great number of countries has ratified and acceded to the Patent Cooperation Treaty. The procedures of the Patent Cooperation Treaty are under the control of World Intellectual Property Organization (WIPO, Geneva). Under the provisions of the Patent Cooperation Treaty an applicant files a single patent application named an 'international application' at the national patent office in any country which has acceded to the Patent Cooperation Treaty and in a language acceptable by that country. The national office at which the international application is filed is known as the 'receiving office'. The application must designate the countries for which national patents are required and must request a search to be performed, named an 'international search', by a competent Patent Office. It is also possible to designate a European application the so-called 'regional designation'.

The international application is examined with the required formalities. A prior art search is performed and a search report is sent to the applicant. The application and the international search report are published. This is known as the 'international phase' of the procedure.

Articles 31 to 42 PCT provide for an 'international preliminary examination' to be performed during the international phase (Chapter II of the Patent Cooperation Treaty). Countries having acceded to the Patent Cooperation Treaty can make a declaration that they will not be bound by Chapter II of the Patent Cooperation Treaty. An applicant resident in a country which is bound by Chapter II of the Patent Cooperation Treaty can make a request for an 'international preliminary examination' to be performed for selected patent offices, known as 'elected offices'. After the issue of the search report (and the international preliminary examination, if applicable and requested) the application passes to the national patent office of each of the designated countries and this is the start of the 'national phase' of the procedure. On request of the applicant, a substantive examination takes place in each of the designated countries, (and the EPO, if designated), under the laws of that country.

After examination in each designated country, the application can be rejected or a patent can be granted. Opposition to the issued national patents can be filed individually, if such procedure is provided for under national laws or conventions and the issued national patents can be amended or revoked, if these are provided for under national laws.

Compared with direct filing at the national offices or at the EPO, the PCT route affords the advantage of postponing for at least 18 months the payment of the national or regional filing fees, and search, designation and claims fees if any, although the total costs is eventually greater.[4,6,7]

To conclude, currently there are a number of choices available for filing a patent. An applicant can file national applications. If an applicant wishes to obtain a national patent in one or more countries which are EPC contracting states or PCT contracting states, it is possible to file a PCT application designating all the selected countries which are contracting states (including a European application) or an application at the EPO designating all the selected countries which are EPC contracting states.

II DEFINITION OF A PATENT—PATENT RIGHTS

A patent is 'piece of property' issued by a state or a state organization which grants to the owner the exclusive rights of the claimed invention.[8] In fact, in exchange for the disclosure of the invention by its owner (author), the state grants to the latter an exclusive right.

In the majority of countries, a patent bestows a right of preventing any third party, from manufacturing, offering, selling, using, importing or stocking the product which is the subject of the invention without the consent of the owner of the patent.

The exclusive rights given by a patent do not extend, in general, to acts carried out in private and for noncommercial purposes, nor usually experimentation with a product when the final aim is not the marketing of the said product. Experimentation in order to make an improvement had not been considered as an act of infringement.[9] However, experimentation with a pharmaceutical product with the aim of marketing it just after the expiry of the patent should be regarded in some countries, in particular European countries, as an act of infringement.

The patent can be considered as a right to 'exclude'. On the other hand, a patent taken out in any country has several important limitations. The most significant are:

- *The territorial limitation.* The patent provides a legal protection only within the frontiers of the state in which it is granted.

- *The time limitation.* A patent has a limited life which in a great number of countries is further restricted by being subjected to the payment of maintenance fees or the like, so that failure to pay such fees automatically invalidates the patent.
- *The exploitation limitation.* The claimed invention according to the patent specification may only be made, used, exercised and sold.

It must be noted that once a patent has been granted, this does not mean that the patentee is free to go ahead and make, use or sell the subject of the invention. If the invention is an improvement on an earlier product, process or piece of apparatus which is the subject of an earlier non-expired patent having a claim which would be infringed by the improvement, it would be necessary to get a licence from the earlier patentee.

III KIND OF INVENTIONS

Inventions can be divided into four main categories:

(1) Inventions concerning products or machines
(2) Inventions concerning processes
(3) Inventions concerning the use of products already known for other uses (in the pharmaceutical file primary therapeutic use and secondary therapeutic uses)
(4) Inventions concerning compositions or combinations.

IV SUBJECTS OF PATENTS: BASIC AND FORMAL REQUIREMENTS FOR FILING A PATENT

Since a patent is a creation of a statute in every country, it is essential to examine the statutes for which a patent may be granted. To be able to file a patent application a certain number of basic and formal requirements have to be met. Basic requirements are quite similar in the main industrial countries. Formal requirements vary a good deal from one country to another. Inventors, for example chemists, are mainly concerned with basic requirements and some formal requirements like disclosure of the invention, i.e. description (see Section IV.B).

A Basic requirements

Depending on the country, there are two or three such requirements. Concerning the European Patent Convention, the criteria for patentability of an invention to be considered are laid down in Articles 52 to 57 EPC.[4,5]

Article 52(1) EPC defines three basic criteria for an invention which must be satisfied if a patent is to be granted for the invention:

- that the invention must be susceptible of industrial application
- must be new
- must involve an inventive step.

Article 52(2) EPC defines certain items which are not regarded as inventions. Discoveries, scientific theories and mathematical methods, aesthetic creations, schemes, rules and methods for performing mental acts, playing games or doing business and programs for computers and presentation of information are not considered as inventions. It must be noted that according to the wording of Article 52(2) EPC, this list is not exhaustive.

Article 52(3) is important when considering what is excluded from patentability, since it explicitly states that the exclusion relates only '*to the extent to which a European patent application... relates to such subject matter or activities as such*'. Thus under certain circumstances other principles will apply if the subject matter or activities listed in Article 52(2) are associated with a technical use. For example, the discovery of X-rays as such is not patentable, but the use of X-rays in a process could be protected by means of a patent. A number of EPO decisions follow the general principles of the European Patent Convention in that invention are excluded from patentability only to the extent that application relates to the excluded subject matter of Article 52(2) as such. There is no problem in patenting an invention which includes technical and nontechnical features.

Article 52(4) EPC defines certain items which shall not be regarded as inventions susceptible of industrial application. Article 52(4) of the EPC, which defines patentable inventions, uses the conditions of industrial application to reject a category of inventions. This Article states that methods for treatment of the human or animal body by surgery or therapy and diagnostic methods practised on the human or animal body are not regarded as inventions susceptible of industrial application. A very important Decision of the Enlarged Board of Appeal of EPO is Decision GR 01/83. With this decision the Enlarged Board of Appeal has confirmed that a claim to 'the use of a substance or composition of the treatment of the human or animal body by therapy' is not allowable, but a claim to 'the use of a substance or composition for the manufacture of a medicament for a specified new and inventive use' is allowable. Article 52(4) EPC does not exclude from being as inventions susceptible of industrial applications substances or composition for use in methods of treatment of the human or animal body by surgery or therapy. Apparatus for such treatment are also patentable.

Article 53 EPC defines certain inventions for which patents will not be granted. Two types of invention for which patents cannot be granted can be identified:

(1) Inventions the publication or exploitation of which would be contrary to public order or morality. Exploitation is not deemed to be contrary to public order or morality, merely because it is prohibited by law or regulation in some or all of the EPC contracting states. This prohibition is only likely to be invoked if it is probable that the invention is regarded as so abhorrent that the grant of patent rights would be socially unacceptable.

(2) Inventions of plants or animal varieties or essentially biological processes for the production of plants or animals. However, microbiological processes and the products thereof are patentable.

It must be mentioned that the patentability of animals has given rise to much discussion. The United States Patent and Trademarks Office (USPTO) following the decision in 'Ex parte Allen' 2 USPQ 2nd 1425 (1987), regards unnaturally accruing living organisms including animals, as patentable.

The Examining Division of the EPO rejected the application for the 'Harvard mouse' *inter alia* as excluded from patentability by Article 53(b) EPC. This decision was appealed and the Board of Appeal disagreed with the reasoning of the Examining Division. The board agreed that the process claims for the production of transgenic non-human mammals through chromosomal incorporation of an activated oncogene sequence into the genome of the nonhuman mammal did not involve an 'essentially biological process' within the meaning of Art. 53(b).

Article 54 EPC indicates that an invention shall be considered to be new if it does not form part of the state of the art.

Article 55 EPC defines certain disclosures of an invention which shall not be taken into consideration when determining its novelty under Article 54.

Article 56 EPC defines inventions involving an inventive step.

Article 57 EPC specifies the criteria for an invention to be considered as susceptible of industrial application.

Concerning the United States patent law (Patent — Act US Code Title 35 — Patents), Section 101 of the Patent Act states:

whoever invents or discovers any new and useful process, machine, manufacture or composition of matter, or any new and useful improvement thereof, may obtain a patent therefor, subject to the conditions and requirements of this title.

Section 101 of the Patent Act expresses the novelty requirement in general terms. Section 102 attempts to spell out just what is not to be considered novel in the patent law sense. With regard to the utility, this is only considered in the Section 101 and in the first paragraph of Section 112 which states:

The specification shall contain a written description of the invention, and the manner and process of making and using, it …

The foregoing two provisos are the only ones of the Patent Act which make mention of utility. Nevertheless, courts have almost uniformly regarded an exhibition of utility — along with novelty and unobviousness — to be the affirmative requisites of every valid patent.[10]

Section 103 of the Patent Act makes patentability depend upon, in addition to novelty and utility, the nonobvious nature of the subject matter sought to be patented.

Japanese law also defines novelty, inventive step and industrial application as the basic criteria required for an invention.

Novelty

Novelty is the *sine qua non* condition of every invention. The requirement for novelty is very severe since it has an absolute character and since any prior public knowledge is destructive of novelty.[11] An invention may be protected if it is new and thus enriches technology. However, within this broad definition the term may be subjected to different interpretations. It appears simple enough, but has developed quite differently under the various patents systems.

Article 54(1) and (2) of the European Patent Convention stipulate:[5]

An invention shall be considered to be new if it does not form part of the art.

The state of the art shall be held to comprise everything made available to the public by means of a written or oral description, by use, or in any other way, before the date of filing of the European patent application.

The definition of *the state of the art* according to Article 54(2) encompasses all conceivable forms of disclosure, and expressly includes written as well as oral description as well as use. In order to cover all other possible forms of disclosure this definition also includes the making available of information.

Under European law and under the law of most countries in the world, except the United States, a disclosure of the invention by the inventor before an application is filed (first application allowing the Priority Right) will destroy novelty, i.e. patentability.

For a document 'to be made available to the public' it does not matter whether any member of the public actually read it. The point at issue is whether, on the balance of probabilities, any member of the public had access to it. A document available to the public in a library[4,12] or in a patent office file[4,13] would represent a disclosure even if it could be proved that no member of the public had actually read the document. A document is not made available to the public upon posting but on delivery.[4]

The chemical composition of a product is state of the art when the product as such is available to the public and can be analysed and reproduced by a skilled person, irrespective of whether or not particular reasons can be identified for analysing the composition.[14]

In order to ascertain if there is prior knowledge, three criteria are taken into account: public disclosure, sufficient disclosure and date of disclosure. It will thus be investigated whether at the filing date of the patent application it could be known about the invention. It will be also taken into consideration whether the disclosure is sufficient for an expert to be able to perform the invention without an inventive act. Finally the date of the disclosure will be considered, which must be certain and not in doubt. In order to destroy novelty, it must of course be prior to the filing date of the patent application, or prior of the priority date, if any.

As noted above, according to Article 54 EPC, the definition of the state of the art is very wide. There is no restriction as to where and when a disclosure was made available to the public, nor are there any restrictions as to the language and manner of the disclosure. There are however, certain exceptions relating to disclosure in breach of confidence and displays at certain international exhibitions defined in Article 55 EPC.[4] Since the obligation to secrecy is particularly important during licensing negotiations and demonstrations sometimes prior to patent application filing, various principles have been adopted and are taken into consideration by the EPO (this is also the case in the law of many countries). According to the EPO's practice the subject matter of a patent is not made available to the public if there is an express or tacit agreement to secrecy. If a person under the obligation to secrecy does make the information available to the public in abuse of this obligation, then from that moment this information forms part of the state of the art and thus destroys novelty. However, such information is not prejudicial to the invention.

A prior disclosure in a document or by use destroys the novelty of any claimed invention that is directly and unambiguously derivable therefrom.[15] The disclosure not only destroys the novelty of specifically mentioned features, but also destroys the novelty of features that are implicit to those skilled in the art (an expert). A specific disclosure of a chemical compound destroys the novelty of a generic claim embracing it.[16] For example, a disclosure of the use of rubber in circumstances where its elastic properties are used, would destroy novelty of the use of an elastic material in general.[17]

On the other hand, a generic disclosure does not usually destroy the novelty of a specific example failing within it or even of a subgeneric claim. Thus the disclosure of a genus does not disclose its species and the disclosure of a range does not disclose elements within that range. In this case, the claimed species or subgenus must not include, and must show an unexpected advantage over, prior disclosed species of the genus. This is similar to the situation in many countries, and concerning EPO, it was confirmed by a very first decision of Technical Board Appeal.[18]

Consideration must be given not only to specifically described features but also to whether by following the prior disclosure there would be an inevitable result.[4]

Concerning enantiomers, the EPO's Technical Board of Appeal confirmed that the disclosure of a racemate did not destroy the novelty of the enantiomers, even though they would be understood as being present in the racemate (unseparated form). According the EPO's decision, the enantiomers had not been 'individualized' in the prior art.[4,19]

To attack novelty it is not permitted to combine two or more prior disclosures.

It may in some cases be difficult to prove the date relating to oral disclosure. A written confirmation, even updated, of the oral disclosure enforces it.

The concept of absolute novelty at the filing date of the patent application is now to be found in practically all patent laws, with the exception however, as said before, of the law of the United States. Article 102 of the American law on patents[20] which establishes the criteria of novelty in the United States in effect is as follows:

A person shall be entitled to a patent unless:

(a) the invention was known or used by others in this country, or patented or described in a printed publication in this or a foreign country, before the invention thereof by the applicant for patent, or,

(b) the invention was patented or described in a printed publication in this or a foreign country or in public use or on sale in this country, more than one year prior to the date of the application for patent in the United States, or,

(c) he has abandoned the invention, or

(d) the invention was first patented or caused to be patented, or was the subject of an inventor's certificate, by the applicant or his legal representatives or assigns in a foreign country prior to the date of the application for patent in this country on an application for patent or inventor's certificate field more than twelve months before the filing of the application in the United Sates, or

(e) the invention was described in a patent granted on an application for patent by another filed in the United States before the invention thereof by the applicant for patent, or on an international application by another who has fulfilled the requirements of paragraphs (1), (2) and (4) of section 371(c) of this title before the invention thereof by the applicant for patent, or

(f) he did not himself invent the subject matter sought to be patented, or

(g) before the applicant's invention thereof the invention was made in this country by another who had not abandoned, suppressed, or concealed it. In determining priority of invention there shall be considered not only the respective dates of conception and reduction to practice of the invention, but also the reasonable diligence of one who was first to conceive and last to reduce to practice, from a time prior to conception by the other.

Paragraphs (b) and (d) of this article provide a grace period of one year which enables an inventor in the USA to test an invention without such testing being regarded as destroying novelty. So the invention can therefore be freely developed without the possibility of personal disclosures being counted against the inventor. During this grace period of one year counting from the first disclosure, the inventor can then file patent application.

It must be emphasized here that all other countries in the world have opted for the criterion of absolute novelty at the filing date of the patent application and disclosures of the inventors before filing applications destroy the novelty of their own invention. Thus, an inventor in the United States taking advantage of this possibility offered by Article 102 of the American law will no longer have the chance to file in foreign countries since there is no longer novelty at the filing date of the patent application.

Article 102 of the American law refers moreover to the invention and not to the filing date of the patent application. This article thus created the so-called '*first inventor system*' as contrasted with the so-called '*first depositor system*' used in almost all other countries.

Paragraph (g) of Article 102 of the US law lays down a ruling on the question of 'double patenting' different from that of other countries and the corresponding Article of EPC [Article 54(3)]. According to paragraph (g) of Article 102 of the US law, in the case of 'double patenting' the patent will be granted to the inventor who not only has the earliest date for conception and for putting the invention into practice, but who also has displayed diligence in putting the invention to use. Under 54(3) EPC, the content of the art includes the content of other European patent application field earlier than, but published on or after the filing date of a European application. Concerning co-pending application

with the same priority date, there is no express provision in the EPC. The EPO will allow both to proceed to the grant. But if applicants are the same, the EPO will require amendment in order to avoid double patenting.

Effect of priority right. According to the Paris Convention, a person who has filed an earlier application in any state which is party to the Paris Convention may claim for a later filed application for the same invention priority from the date of filing of the earlier application. As a result, the date of filing of the earlier application becomes the effective date of filing of the later application, if this later application is filed in a state party to the Paris Convention. This is also stipulated in Article 87(1) EPC.

Thus, a document, or oral disclosure or co-pending European patent application will be cited by the EPO as part of the state of the art if such a document or oral disclosure is believed to have been publicly available or such an application is believed to disclose matter having a priority date before the filing date of the application.[15] The test for determining whether subject matter is entitled to priority earlier than that of the date on which the application containing that matter was filed is the same as the test novelty, i.e. the disclosure in a single earlier application from which priority is claimed must be sufficient to destroy the novelty of a claim to that subject matter in the later application, for which priority is claimed.

Inventive step

A patentable invention requires the exercise of the inventive faculty or the exercise of more than the expected skill of the art. In other words, the invention must have been nonobvious to a person skilled in the art when it was made.

From the inspection of the first patent laws, the criterion of novelty had been regarded as insufficient. German patent law therefore demanded that for a patent to be granted, it had to have a certain 'inventive height' (or 'inventive level') or a certain 'technical advance'. The United States patent laws speaks of 'unobviousness'.

Article 52(1) EPC requires an invention to involve an inventive step if it is to be patentable, and Article 56 EPC states that:

An invention will be considered as involving an inventive step if, having regard to the state of the art, it is nonobvious to a person skilled in the art ...

The state of the art for this purpose is defined as not including documents under Article 54(3) EPC, i.e. patent applications which were not published at the filing date of the patent applications under consideration. In fact, it is normal that it cannot really be expected that an invention

'B' should display an inventive step as compared with another invention 'A', which was totally unknown at the filing date of the application 'B' being considered. (As to novelty, the contents of European patent applications still unpublished at the filing date are retained, to avoid granting two patents for the same invention, i.e. double patenting.)

Since the criteria for an inventive step is expressed in a negative manner, no positive virtues are required in order to satisfy Article 56. In other words, there is no requirement for a technical advance. However a technical advance may be useful in arguing for the existence of an inventive step.[4] As noted by Dr R. Singer[21] (Chairman of the Legal Board of Appeal of the European Patent Office):

in order to understand the term inventive step it is useful to divide technical knowledge into three levels. The first level includes all that is known, the mentioned state of the art, i.e. nothing new. The second level contains everything that is new but does not involve an inventive step, i.e. everyday developments of the state of the art. This is the so-called patent-free area. Finally, the third level includes everything that is new and which is also inventive and can be protected by patents.

As with any assessment based upon a qualitative judgement, often it is a difference of opinion between the different people involved in the assessment of the inventive step (the applicant, the representative if any, the examiner and the opponent, if any).

During the examination by the EPO an inventive step may arise from devising a solution to a known problem. It may also arise from the identification of a problem to be solved and its solution, even though the solution is obvious once the problem has been identified. Each case should be carefully analysed on its own merits.

To reach a conclusion as to whether an invention includes an inventive step it is necessary to determine the difference between it and the whole of the state of the art. As cited above, the state of the art comprises everything made available to the public before the filing of the application or its priority date.

When considering whether there is inventive step, the question must always be asked: Who is the person skilled in the art? Who is presumed to know all the relevant prior art? *A person skilled in the art* is presumed to have access to the entire state of the art and in particular, is presumed to possess common general knowledge applicable to his field of interest. Furthermore, it is part of the normal activities of the skilled person to select from the material known to him the most appropriate one. The skilled person is not regarded as having any inventive ingenuity. So, the teaching of a document may have narrower implications for a skilled person and broader implications for a potential inventor who

first perceives the problem which the future invention is intended to solve.

The skilled person is a 'practical person'. According to a EPO Decision,[22] if the problem prompts the skilled person to seek a solution in another technical field, the person of ordinary skill in that other field is the appropriate person to solve the problem. The assessment of inventive step must therefore be based on the skill of that latter person. EPO's Directives[15] point out that the skilled person need not be an individual. In certain circumstances depending of the nature of the invention, it may be more appropriate to think as a group of specialists pooling their knowledge.

The assessment of an inventive step is achieved by starting from the state of the art viewed objectively.[23] In an evaluation of the state of the art to determine an inventive step it is permissible to combine the disclosures of two or more documents, but only where to do so would have been obvious to the skilled person. Thus, documents should not be combined if it is unlikely that a skilled person would combine them to solve his problem, especially where there is an inherent incompatibility.

If an invention differs from the state of the art by the substitution of an alleged equivalent for one of its elements, it is not necessary to combine a secondary reference of the prior art in a related subject matter to establish lack of inventive step, with the proviso, that the type of substitution is generally known and a skilled person would expect the general effects given by the invention. However, if the substitution gives an unexpectedly greater effect or a nonobvious additional effect, inventive step exists. Additionally if the inventor was confronted with many alternative possibilities as potential solutions to his problem and he had no particular reason to select the one leading to the invention, inventive step should be recognized.

A situation encountered in chemical cases is where two distinct substances, each comprising two or more structural elements, have a common property and a claim is presented to a new substance combining structural elements from each of the known substances and having the same property. If the common property of the two known substances is indicated as being associated with each substance as a whole and not with the structural elements selected for the new substance, the new substance may not be obvious from the two known substances.[4,24]

In a combination of known features, an inventive step should be recognized if the combined features mutually support each other in their effect and so a new technical result is achieved. According the EPO's Guidelines[15] for examination of patent applications there is an inventive step 'if in a mixture of medicines consisting of a analgesic and a sedative (tranquillizer) it was found that through the addition of the sedative — which intrinsically appeared to have no analgesic effect — the analgesic effect of the analgesic was intensified in a way which could not have been predicted from the known properties of the active substances'.

An unexpected effect is one of the arguments for the existence of an inventive step. The unexpected effect must be established by evidence. Very often it is necessary to present appropriate comparative tests with materials from the closest prior art. If a particular effect or property would be expected qualitatively, applicants are often asked to supply evidence to support an inventive step by demonstrating a significant quantitative improvement.

In many cases a technical advantage may be foreseeable and therefore not unexpected. In these cases there is a lack of inventive step.

The EPO Technical Board of Appeal have held that if the prior art very strongly and clearly directs the skilled person in a particular direction, there is no inventive step involved in following that direction, irrespective of the results. This is known as the 'one-way street' situation and has led to considerable discussion amongst commentators. 'Obvious to try' is not a standard for denying the existence of an inventive step. In accordance with recent practice in national courts in Europe and in the United States, the EPO is reluctant to maintain an 'obvious to try' objection in the face of unexpected results.

Unexpected effect in one species or subgenus selected from a known group is sufficient to render the species or subgenus patentable, because an inventive step can be acknowledged. Such inventions are called 'selection inventions'. The principle of selection invention has been upheld by the EPO in many decisions.

A chemical intermediate which has no inherent useful properties other than as a starting material for a useful end-product claimed in the same application may depend on the useful end-product for its inventive step because it is part of the same invention.[4] On the other hand, according to a decision of EPO's Technical Board of Appeal, a new chemical intermediate may be inventive if it is prepared in the course of a multistage process, providing that the intermediate itself makes a contribution to the quantitative effect of the process.[25]

According to the American law of patents, the criterion of inventive step is replaced by the criterion of unobviousness. Nonobvious subject matter and conditions for patentability are set forth in Section 103 of the US Patent Law which states that:

A patent may not be obtained though the invention is not identically disclosed or described as set forth in Section 102 of this title, if the differences between the subject matter sought to be patented and the prior art are such that the subject matter as a whole would have been obvious at the time the invention was made to a person having ordinary skill in the art to which said subject matter pertains. Patentability shall not be negatived by the manner in which the invention was made.

In general, in determining whether a patentable invention has been made or not, the American Patent Office (USPTO) and a court ascertain, among other points, whether the invention[1] produced new, improved or unexpected results; satisfied a long-felt want; solved an outstanding problem; was contraindicated in the prior art; succeeded over unsuccessful efforts of others; went into extensive use and enjoyed public acquiescence; and had commercial success. On the other hand, inventions are ordinarily held to be unpatentable when they involve nothing more than a change of form, size or location; a substitution of materials or equivalents; a change of proportions or ingredients of a composition without a surprising result; superior or excellent workmanship.

The invention must be capable of industrial application

If novelty is the *sine qua non* condition of invention, then industrial application or utility is its *raison d'être*. The laws of almost every country required that the invention should be capable of industrial application.

Industrial application is defined by Article 57 of the EPC as follows:[5]

An invention shall be considered as susceptible of industrial application if it can be made or used in any kind of industry including agriculture.

It would seem that the term 'industrial' will be interpreted broadly so as to include most commercial activities. For example, according to EPO decisions,[26,27] professional use of a cosmetic process in a beauty parlour is found to fall within the meaning of 'industrial application'.

The criterion of 'utility' found in American law is quite close to the criterion of 'industrial application' which is found in the laws of a good number of countries and in particular in the EPC. A chemical product for which no industrial application nor any pharmacological property had been found would not be patentable under the United States patent law, because it would not have any utility.

Of interest is the decision of the US Supreme Court in the Manson case (383 US 519) which overruled the utility doctrine enunciated by the Court of Customs and Patent Appeals in the 'Nelson case' (47 CCPA 1031) and which adjudicated the question of what constitutes utility in a chemical process. The invention involved a process for making certain steroids.[1] The court held that the utility of the product must be established and that merely providing a product to be available for 'use testing' by the trade was insufficient. In the decision, it was pointed out that

'It was never intended that a patent be granted upon a product, or a process producing a product, unless such product be useful'.

The court said that it did not mean to disparage 'the importance of contributions to the fund of scientific information short of the invention of something useful, or that we are blind to the prospect that what now seems without 'use' may tomorrow command the grateful attention of the public. But a patent is not a hunting license. It is not a reward for the search but compensation for its successful conclusion. A patent system must be related to the world of commerce rather than to the realm of philosophy'.

In view of the Supreme Court decision, it is clear that as many affirmative statements and positive results should be included in a patent specification as possible.

B Formal requirements

The patent application must have a single subject. This principe is known as *unity of invention*. According to EPC's Article 82:

> *The European patent application shall relate to one invention only or to a group of inventions so linked as to form a single general inventive concept.*

Unity of invention is required for documentary research reasons. In the United States, the requirement of unity of invention (single subject per patent application) is investigated with the greatest strictness.

If the patent application relates to more than one invention and the applicant wishes to protect two or more inventions, the 'invention or inventions' can be the subject of one or more *divisional applications*. The divisional application may be filed at any time before the grant of the 'parent application'. A divisional application cannot be filed if the parent application has been withdrawn or is deemed to have been withdrawn. Lack of unity is not a cause of invalidity of a granted patent.

The patent application must contain a *description* explaining the invention.

The description shall:

(a) specify the technical field to which the invention relates;
(b) indicate the background art which as far as is known to the applicant, can be regarded as useful for understanding the invention and preferably cite the documents reflecting such art;
(c) disclose the invention as claimed and state any advantageous effect of the invention with reference to the background art;
(d) describe in detail at least one way of carrying out the invention claimed using examples where appropriate and referring to the drawings if any;

(e) indicate explicitly, when it is not obvious from the nature of the invention the way in which the invention is capable of exploitation in industry; and
(f) briefly describe the figures in the drawings if any.

In fact, the patent application must explain the invention in a sufficiently clear and complete way for a person skilled in the art to be able to reproduce the invention without having to carry out research or lengthy and delicate operations.

Application to be filed in the United States must in addition give the best method of realizing the invention.

In general, it is not necessary to give detailed results of trials of use of the invention, for example, biological pharmacological or clinical results for pharmaceuticals. But sometimes, it can be useful to present comparative results between the products of the application and the products of the prior art, in order to demonstrate the advantage of the invention.

The patent application must contain *claims*. According to Article 84 of EPC:

> *The claims shall define the matter for which protection is sought. They shall be clear and concise and be supported by the description.*

In fact, the claims shall define the matter for which protection is sought in terms of the technical features of the invention.

The European Patent Convention enumerates four kinds of claims corresponding to the four main categories of invention. But in fact there exist two basic types of claims: claims for products or machines and claims for activities (processes, utilization).[15]

Any claim stating the essential features of an invention may be followed by one or more claims concerning particular embodiments of that invention.

It is possible to have several independent claims in a European patent application. This is in particular permitted in order to claim certain permutations of what are termed different categories.

A patent application to be filed at the EPO can have different types of claims. The different categories permissible are:

- A product, a process adapted for the manufacture of the product and a use of the product
- A process and an apparatus or means specifically designed for carrying out the process
- A product, a process specially adapted for making the product and an apparatus or means specially designed for carrying out the process.

In American practice, in an application for invention concerning compounds only products and methods of use claims are generally admitted by the examiners of the Patent Office. For the process for the manufacture of the products, a divisional application is often required. In all cases, the claims cannot extend more widely than the description, completed if appropriate with drawings.

Products claims bestow a protection that can be regarded as absolute. Until the beginning of 1990s in most of Eastern Europe patents systems it was not possible to claim new compounds.

This was also the case especially for pharmaceuticals in Canada, Austria, Spain, Greece, in most Scandinavian countries, except Sweden, and in South American countries. Pharmaceutical were protected only by process claims. Now the patent laws have been changed and pharmaceutical compounds can be claimed in most countries. A product or a family of products can be defined equally well by a general chemical formula (Markush formula), as by specific chemical names.

In chemistry, a *process claim* describes a number of series of reactions leading to the production of a given chemical product. Process claims have the effect of preventing a third party from using the same process. With regard to the product, a process claim is ineffective against a third party who might use a different process to prepare the same product.

Sometimes, a patentable product cannot be defined by reference to its structure, composition or other reproducible property. Such products are protected (claimed) by '*product by process*' claims. A product by process claim covers a product independently of the process used to obtain it.

Concerning the *first therapeutic use*, for a known compound which is shown to have a specific therapeutic use, it is possible according to the EPO's practice, to have a broad claim in the form such as 'substance or composition X for use in medicine'. Formulation of claims such as 'the use of a substance X for the therapeutic treatment of man or animals' are not permitted, because they are regarded as methods of medicinal treatment.

With regard to the *second therapeutic use*, as noted above, the EPO's Enlarged Board of Appeal approved a wording such as 'the use of a substance X for the manufacture of a medicine for therapeutic application'.[15] In the USA, the patentability of 'second therapeutic uses' has been recognized for a long time and many methods of use patents may be granted for the same product. It is now possible to protect second therapeutic uses in many countries. Claims for second nonmedical use are also admitted by the Patent Laws of many countries. A claim covering a novel use of a known product will only cover the claimed use. Other uses of the compound and the compound as such are not protected.

Compositions or combinations if they yield an unexpected result of their own or a technical effect of their own are patentable, but the mere juxtaposition of two means is not by itself patentable. *Composition claims* are admitted in the patents laws of most countries. A patent covering a particular composition, or a given composition will cover the composition or combination effectively claimed and not each constituent of the composition or the combination.

Another formal requirement is that the patent application must *designate the true inventors*. The true inventors are the persons who have actually participated in achieving the invention. People who have only carried out routine tests (biological tests relating to screening, for example) should not be regarded as true inventors. A failure to designate any inventor in the request of a grant of a European patent creates a formal deficiency. According to the EPC, there is no penalty if an incorrect designation of inventors exists, apart from procedures for correction of the designation. So, the resulting patent cannot be opposed or revoked merely on the grounds of wrong inventor designation.

Identifying the true inventor is very important in the United States. According to US patent law, any failure to designate the true inventor is a cause for invalidation of a patent.

Patent applications must be filed in the *official language* or one of the official languages of the country in which the application is filed. As mentioned above European Patent Applications must be filed in one of the three official languages of the European Patent Office, i.e. German, English or French.

Filing a Patent Application also gives rise to *payment of fees*, i.e. filing fees, examination fees (if any) and to payment of annuities. Payment of annuities or periodical fees must be paid throughout the whole life of the patent. Fees must be paid in due time. Failure to pay fees automatically invalidates the patent or the patent application.

For filing a patent application in foreign countries, a *legal representative* is required. For European patent applications, parties to proceedings may act in relation to EPO in the following alternative ways:

(a) An individual may act for himself.
(b) A real person may act through an authorized representative who may be a professional representative, a legal practitioner or any employee of the real person.
(c) A legal person (i.e. a body corporate) may act for itself that is to say by the actions of any responsible officer of the legal person.
(d) A legal person may act through an authorized representative who may be a professional representative, a legal practitioner or any employee of the legal person.

Individual legal persons having neither residence nor principal place of business in any EPC contracting state must act through a representative. This representative must be a professional representative or a legal practitioner.

Patenting is a complicated task. It is fully recommended to applicants without experience of patenting to prepare patent applications and to act through professionals (patent attorney, patent lawyers).

V LIFETIME OF PATENTS

At the beginning of the 1900s, the lifetime of patents varied enormously from one country to another. With the introduction of the European Patent System, the countries ratifying the EPC opted for a lifetime of 20 years from the filing date. Since then, other countries not participating in the EPC have also opted for the lifetime of 20 years. Before 8 June 1995, a US patent had a term of 17 years from the issue date of the patent. Under the new law, enacted 8 December 1994, a patent issuing from an application filed on or after 8 June 1995 has a term that begins when the patent issues but that ends 20 years after the earliest effective US filing date. As for patents that are in force on, or issue from US applications filed before, 8 June 1995, such a patent has a term that is the greater of 17 years from the patent's issue date or 20 years from the application's earliest effective US filing date, whichever is longer. Canada used to have the same system until 1989. From January 1989, the lifetime of a Canadian patent is 20 years from filing. Concerning Japan, the lifetime of a patent was 15 years from the publication of the application, but could not exceed 20 years from filing. Under the new law, which will is enforced as from 1 July 1995, the lifetime of a patent is 20 years from the filing date of the patent application.

Information concerning patent law and lifetime of patents for all countries throughout the world is given in the *Manual for the Handling of Applications for Patents, Designs and Trade Marks Throughout the World* (Octroi-bureau Losen Stiger, Amsterdam, 2001).

In 1984, the USA adopted a law which enabled the lifetime of certain American patents to be extended. According to this law, it is possible to obtain a maximum extension of 5 years to holders of pharmaceutical patents when the marketing of a pharmaceutical product has been delayed in its registration by the FDA. In 1988, Japan has adopted a law which enables the life of a pharmaceutical patent to be extended for a period of 2 to 5 years. France and Italy adopted a Certificate of Complementary Protection of pharmaceutical patents in 1990.

The laws concerning the French and Italian Certificate of Complementary Protection were abrogated in January 1993. Since this date the *Community Supplementary Protection Certificate* has been in force. According to the Community Certificate, it is possible to obtain a maximum extension of 5 years to holders of pharmaceutical patents. This certificate concerns all the EEC and EPC contracting states.

VI OWNERSHIP OF PATENTS

A patent is a grant from the State and constitutes a property. As such, it is subject to all of the laws and regulations to property. The grant of a patent gives to the patentee a title to an intangible and incorporate right in the nature of a franchise. The title of a patent constitutes a right to the patentee until it is divested from him or by voluntary grant or by other legal means. A patent can be conveyed like any other piece of property. It can be assigned and transferred to an other party.

A patent may be licensed under the terms which give to the licensee the right to make, use or sell the patented invention. It is in this respect similar to a piece of property which may be leased and may be also sold. A license may be exclusive or nonexclusive. It may be restricted as to a certain geographical territory, or to certain industrial fields, or may even be restricted to a certain specific factory or other place of commerce and industry.

In the USA, the titular of a patent can be only the inventor who may assignee his patent application to a company or to an organization, whereas in the other countries the titular may be the inventor, a company or an organization.

VII INFRINGEMENT OF A PATENT

An act of infringement is an attack of the rights of the patentee in the territory where a patent is in force and effect. A person may be guilty of 'direct infringement' when he either makes, uses or sells the entire invention as defined by the claims of a patent. A person may be guilty of 'contributory infringement' if he cooperates with another, in infringing a patent. For example, a chemical company which sells substances to another for infringing a patent, can be sued just like a direct counterfeiter.

An action for infringement can only be launched as from the day the patent application is published. The reason for this rule is to protect third parties who might have infringed a patent or patent application of which they are unaware. According to the laws of certain countries, it is possible to notify a suspected counterfeiter of the existence of a patent application. Concerning infringements, the judges rule according to the claims of the patent as granted.

Suits for the infringement of patents, must be instituted in national tribunals. Infringement of a national patent issued from a European patent is dealt with under national law of the contracting states and may be heard by the national courts. Generally, proof of infringement must be provided by the owner of the patent. Such proof can be witnesses, documents, purchase of products and so on.

In every case where a patent has been held to be valid and infringed, the patent owner is entitled to have an injunction issued against the counterfeiter. The injunction will restrict the counterfeiter from making, using or selling the invention. Additionally the patent owner may obtain the payment of compensation to make up for the loss of profit or for the loss suffered. According to the law of some countries, patented articles may be withdrawn and delivered up to the proprietor of the patent. In some countries relief for the infringing action may be provided by criminal penalties, including a prison sentence.

In order to determine infringement, the subject matter protected by a patent is compared with the act alleged to constitute an infringement. Slavish reproduction of the subject matter protected by a patent constitutes an act of infringement. The reproduction of a patented invention by varying only details or by using equivalents (variants of execution) also contitutes infringement.

VIII PATENTS AS A SOURCE OF INFORMATION

Patent literature is a useful source of technical, legal and commercial information. It is extremely rich in technical information. In a patent application, applicants must disclose the invention in a sufficiently clear and complete way for a person skilled in the art to be able to reproduce the invention. New and inventive concepts give rise to a considerable number of refinements, improvements and modifications. Often, the importance of an invention is related to the number of patents and patent applications which protect it.

Complete information about the patents of the competitors is useful before the beginning of a new research program. Liberty of exploitation (freedom to use) has to be evaluated before the commercialization and preferably, before the development of a new product, process or machine. In this case, patent applications, patents and the corresponding patent statutes of competitors must be studied in detail.

The negotiating and conclusion of licence agreement have become a specialized profession. Information obtained by patents can help companies to find new products or processes for expansion or diversification of their manufacturing activities.

IX PATENTING IN THE PHARMACEUTICAL INDUSTRIES

In certain trades and industries and in particular in the pharmaceutical industries it is generally not possible to inhibit competition effectively by keeping key formulae, products and techniques secret. Pharmaceuticals is a classic case of an industry dependent on patents. In fact, patent protection is of great importance to the pharmaceutical research and development process. Practically all new drugs which were developed after the Second World War were developed by private pharmaceutical firms in the United States, Federal Republic of Germany, Switzerland, France, Great Britain, the Nordic European countries and Japan. These countries provide the strongest known patent protection for pharmaceuticals. Countries with similar industrial standards and possibilities, but not providing patent protection for pharmaceuticals cannot show any drug innovations of importance.

In any case, having obtained a patent, the patentee should not think he is heir to a fortune. Many worthwhile inventions have failed commercially for lack of interest or funds to promote them. Additionally, the grant of a patent is no way a guarantee that the invention is significant or that the patent is valid. It should therefore be of no surprise that one of the factors considered by pharmaceutical management in deciding whether to develop a new drug, in addition to its medical and commercial potential and cost of manufacturing, is the strength and duration of patent protection.

Concerning strength, the best kind of protection, and really the only truly effective kind, is that which protects the product itself. A patent covering a product, as opposed to one covering only a process for making the product, or a use for the product, can prevent others from manufacturing or selling the product for any purpose, no matter how it is made. Since chemists can often find new processes for making a compound, a patent on a process can be circumvented by use of the new processes and the owner of a patent no longer has an exclusive position with respect to the drug. The same is true concerning a patent covering only a particular use. A new use may permit a copier to sell the product free of the original use limited patent, even though the product may be prescribed and used for the original use. Patents of new use are only strong if the new use of a compound requires a special and new formulation.

Another aspect of protection, in addition to the type of protection available, is the term of the patent.

For competitive reasons, a patent application must be filed as soon as possible. In fact, a patent application for a new compound or series of compounds must be filed as soon as the new compound or series of compounds have been made and found to have significant and interesting activity. This is necessary because others may have made the same invention, and failure to file promptly may result in a total loss of patent rights (in most countries, except the United States, patents are granted to the first party to file a patent application on the subject matter). There are many cases in which the same invention is independently made by more than one party within as short a period as 2 or 3 weeks.

However, in filing an application as soon as possible many problems can be generated. The term of a patent starts to run even before the product is marketed. As a period of at least 7–10 years is required for the development of a new drug after a compound and its analogues have been prepared and biologically tested, the patent terms can expire before the product begins to produce any real profits. In the pharmaceutical industry, many candidate products are not developed because of the short patent term that would remain after development.

Under the Paris Convention, initial filing can be followed up in other countries where protection is desired. The benefit of the initial early filing date is obtained if applications are filed within a year of this date. Thus, the choice of countries must be decided before knowing the fully importance of an invention. Commercial interest and effective protection against infringers or competitors offered by the 'local' patent law must be taken into consideration in the choice of countries.

Often, at the time a patent application is filed for a new compound, research and development staff do not know which compound in a serial will turn out to be the product to be developed. In addition, the chosen compound or compounds may well be chemically related to compounds patented earlier, and this fact complicates the question of 'patentability'. Even before the full importance of an invention is known, a number of needs must be met at this early stage when a patent application is filed. In order to adequately protect an invention, the applicant must describe how to realize and how to use it. Normally, a chemist will invent a series of compounds, all having the same type of activity to a greater or lesser degree. Inventors cannot afford to simply patent for the best compound, even if they know which compound is the best. Competitors would read the patent and make the next closest relative of the compound, thereby depriving the inventor of part of his invention. Thus examples must be described of the group of the invented compounds. Moreover a company may change its interest to another compound in the series as it learns more about the compounds. Potency, stability, side-effect profile, bioavailability and difficulty or cost of manufacturing may indicate that later prepared compounds in a series are preferable for development. For this reason, it is also necessary for a variety of compounds in the series to be described.

The other significant problem in obtaining valid and effective patent protection relates to requirements of the patent law variously referred to in different countries as the requirement for nonobviousness, or attainment of a sufficient inventive level. Normally the requirement that an invention be useful or capable of industrial application is not a problem. The novelty requirement is also not usually a problem. However, even in the usual case where compounds are novel, few compounds are of such distinctiveness that no compound of similar structure has ever previously been published. Thus, it must be established, either by arguments of chemical difference or of biological properties, that the new compounds merit patent protection. Inventors' previous patents and publications often complicate the situation and it is most distressing when inventors have difficulty patenting their chosen compounds because of their earlier public disclosure of compounds long since discarded.

The opportunity for filing a patent application has to be well evaluated. Sometimes, it is preferable to keep an invention secret, if the infringement of the future patent related to this invention will be difficult to demonstrate. This concerns especially process patents for preparing commercially successful compounds not protected by a 'compound patent'. Infringement of such patents can generally be demonstrated by identifying impurities due to the process (if any) or by witness. If the demonstration of infringement is not possible, competitors may use the process they are interested in, after the publication of the patent application, without any jeopardy or sanction.

X CONCLUSION

Patent systems throughout the world are today based on the principle of territorial rights. As an invention will be protected in that state in which a patent has been applied for and granted, inventors and applicants are therefore forced to apply for patents for an invention in several states. This involves the use of different laws and rules and a considerable financial outlay. Inventors and applicants are often confronted with the legalistic problems and phraseology of the world of patent protection.

Even if the phraseology can be easily understood and adopted, legalistic problems require a good knowledge of the law of the countries and the corresponding practice and jurisprudence.

There is a correlation between research and development and industrialization. Patents provide a stimulus for investment in research and development. Investment in the type of development which has yielded great progress in medicinal chemistry requires a healthy patent strategy.

REFERENCES

1. Greenstrat, C.H. (1972) In Liebesny, F. (ed.). *Mainly on Patents*. The Use of Industrial Property and Its Literature. Butterworth, London.
2. Mathély, P. (1992) *Le droit français des brevets d'invention*. Journal des notaires et des avocats, Paris.
3. Bodeyhausey, G.H.C. (1969) *Guide d'application de la Convention de Paris pour la Protection de la Propriété Industrielle*. BIRPI, Genève.
4. *European Patents Handbook* (1991) Release 10. Longman, Harlow.

5. Druck, R. (2002) *European Patent Convention*. European Patent Office.

6. *Traité de coopération en matière de brevets (PCT) et règlement d'exécution du PCT* (1998) Organisation Mondiale de Propriété Industrielle, Geneva.

7. *PCT — Guide du déposant* (2000) No. 6. Organisation Mondiale de la Propriété Industrielle (OMPI), Genève.

8. Chavanne, A. and Burst, J.J. (1990) *Droit de la Propriété Industrielle*, 3rd edn. Dalloz, Paris.

9. *PIBD*, III 2.3.5., Tribunal de Grande Instance de Paris. Juillet. 8 1982.

10. Rosenberg, P. (1975) *Patent Law Fundamentals*. Clark Boardman, New York.

11. Decision of the Technical Boar of Appeal 3.3.1, T 31/84, 4 May 1985. *Official Journal of the European Patent Office* **11/86**: 369–372.

12. Decision of the Technical Board of Appeal 3.3.1, T 381/87, 10 November 1988. *Official Journal of the European Patent Office* **5/90**: 213–223.

13. Decision of the Technical Board of Appeal T 444/88 (1990) 12 IPD Decision IPD 13222.

14. Opinion of the Enlarged Board of Appeal G1/92, 18 December 1992. *Official Journal of the European Patent Office* **5/93**: 277.

15. *Directives relatives à l'examen pratique à l'Office Européen des brevets* (2001) Publication Office Européen des brevets. Pindar Infotek, York.

16. Decision of the Technical Board of Appeal 3.3.1 T 188/83, 30 July 1984. *Official Journal of the European Patent Office* **11/84**: 555–562.

17. Decision of the Technical Board of Appeal 3.3.1 T 181/82, 28 February 1984. *Official Journal of the European Patent Office* **9/84**: 401–414.

18. Decision of the Technical Board of Appeal 3.3.1 T. 01/80, 6 April 1981. *Official Journal of the European Patent Office* **9/81**: 349.

19. Decision of the Technical Board of Appeal 3.4.1. T, 296/87, 30 August 1988. *Official Journal of the European Patent Office* **5/90**: 195–212.

20. *Patents and Trademark Laws* (1989) The Bureau of National Affairs, Washington, DC.

21. Singer, R. (1981) *The New European Patent System*. In Devons, D.J. (ed.). Seminar Services International.

22. Decision of the Technical Board of Appeal 3.2.1 T 195/84, 10 October 1985. *Official Journal of the European Patent Office* **5/86**: 121–128.

23. Decision of the Technical Board of Appeal 3.3.1 T 24/81, 13 October 1982. *Official Journal of the European Patent Office* **4/83**: 133–142.

24. Decision of the Technical Board of Appeal 3.3.1 T 20/83, 17 March 1983. *Official Journal of the European Patent Office* **9/86**: 295–301.

25. Decision of the Technical Board of Appeal 3.3.2 T 163/84, 21 August 1986. *Official Journal of the European Patent Office* **7/87**: 301–308.

26. Decision of the Technical Board of Appeal 3.3.1 T 36/83, 14 May 1985. *Official Journal of the European Patent Office* **9/86**: 295–301.

27. Decision of the Technical Board of Appeal 3.3.1 T 144/83, 27 March 1986. *Official Journal of the European Patent Office* **9/86**: 301–305.

43

THE CONSUMPTION AND PRODUCTION OF PHARMACEUTICALS

Bryan G. Reuben

The end of the medical art is health and that of economics is wealth   **Aristotle (*Ethics*)**

Medicinal chemistry is of great academic and intellectual interest. The elucidation of the arachidonic acid cascade, for example, is one of the most fascinating pieces of chemistry of our generation. The study of the chemistry of the brain, so that we are able to understand how it is that we can understand, is one of the great frontiers of science.

To the author, at least, it is of far greater significance than the problem of the origins of the universe, most of which is cold, inhospitable and distant, which came into being an indecently long time ago and which will continue long after any of us is able to take an interest in it.

Unlike astrophysics, however, medicinal chemistry is an applied science. It is lavishly funded not because of its philosophical centrality, but because it provides the hope that human disease can be cured or alleviated. Its paymasters intend that mankind, or at least those sections of it with access to advanced medical care, live longer and more comfortably. Its practitioners are judged by this criterion. There are few Nobel prizes for those who discover, say, the biochemical origins of a rodent-specific dermatitis.

As the test of success is pragmatic, serendipity plays an important role in medicinal chemistry. The discoverers of sulphonamides thought that dyestuffs might prove efficacious because they bonded specifically to certain tissues, as in Ehrlich's classic experiment. In the end, Prontosil worked not because it was a dye but because it cleaved in the gut to p-aminobenzenesulphonic acid. We still honour Domagk, just as we honour Fleming, whose chance discovery of penicillin would have been meaningless had not Florey and Chain solved the problem of its purification and Coghill and his coworkers (who did not get Nobel prizes) the problem of its large-scale production.

Medicines are thus judged by their results. A successful drug can be manufactured reasonably easily, has negligible side-effects, is widely prescribed, makes a lot of money and is perceived as making a major contribution to health care. In this chapter, we shall discuss which drugs are important on which criteria and say something about the technical, social and economic problems of the pharmaceutical industry.

The Practice of Medicinal Chemistry
ISBN 0-12-744481-5

I 'IMPORTANT' DRUGS

To the person whose life is saved by it, any drug is important. Nonetheless, some are more significant than others either because they provide better returns for their manufacturers or because they are more frequently prescribed.

A The top-earning drugs

From the point of view of the manufacturer, the money side is the most important. A research-oriented drug company needs adequate cash flow to maintain its organization, fund its research and keep its shareholders happy. The world market for drugs in 2000 was $369 billion. Table 43.1 shows the 50 best-sellers. Most of them are still under patent or their patents have only recently expired; thus their price and profitability remain high.

Three of the top five drugs are for heart disease, reflecting that this is the major cause of death in the developed world. Two of the top six, including the first on the list, are for ulcers. Until the mid-1970s, ulcers could be treated only by surgery and were generally uncomfortable rather than fatal. The importance of these drugs is that they improve the quality of life rather than saving it. Seven of the top 50 drugs are antibiotics reflecting man's conquest of infectious disease. Conventional antibiotics, however, are fairly cheap and the ones listed here all have special properties to justify their prices.

Two drugs, celecoxib and rofecoxib, are listed for arthritis, an illness that is painful and disabling, but not life-threatening. The human hormone erythropoietin occurs twice made by different manufacturers and is used mainly to counter anaemia arising from renal dialysis. Together with human insulin and interferon-alpha, it is made using recombinant DNA technology. Three antidepressants (fluoxetine, sertraline and venlaxafine) but only one anxiolytic (alprazolam) are listed, reflecting the swing in medical opinion from benzodiazepine anxiolytics to antidepressants for treatment of poorly defined mental illness. Finally, there are three drugs — fluticasone, salmeterol and montelukast — for asthma, the only disease for which developed world mortality is rising.

The above compounds are the glamorous pharmaceuticals, the results of fairly recent research and mostly based on a clear idea of the enzyme systems they are supposed to influence. The most widely consumed pharmaceuticals, however, are aspirin, paracetamol (called acetaminophen in the USA) and vitamin C. These can all be bought over-the-counter. Annual world consumption is of the order of 50 000 tonnes each, an order of magnitude higher than consumption of, say, penicillin V. At the other extreme come materials such as the anti-cancer drug, methotrexate, with annual

production of about 150 kg. The prices of the large tonnage drugs are so low, however, that sales of aspirin and paracetamol together are probably less than $0.5 billion, and even that figure is achieved by producers' combining them with other drugs, dressing up the products and selling them to the consumer by extensive advertising. Nonetheless, the number of people who take these drugs to alleviate the symptoms of colds, reduce pain and so on is astronomical. In the UK, some years ago, it appeared that consumption of aspirin-containing tablets averaged 200 per year per man, woman and child.

B The most widely prescribed drugs

More important in terms of public health than the top-earning drugs are the drugs that are most widely prescribed, although there is overlap between the two. Table 43.2 shows the 50 most frequently prescribed drugs in the United States. Table 43.3 classifies the top 50 drugs in both tables by therapeutic class and also lists world sales of each therapeutic class and the proportion of drugs in each therapeutic class in the UK.

On a world basis, cheap antibiotics are the most important drugs, and amoxicillin is the second most frequently prescribed drug in the USA, although this is not clear from the table because of the way prescription data are compiled. In the developed world, heart drugs as a whole are more frequently prescribed, partly because antibiotics are prescribed for a single bout of infection, whereas heart drugs are generally taken for the remainder of a sufferer's life. After heart drugs come drugs for the central nervous system. Non-narcotic analgesics (aspirin and paracetamol) are usually included in this category, which remains extremely important even though prescriptions for anxiolytics and appetite suppressants have been dropping since the mid-1980s as doctors have recognized the problems of tolerance and addiction. The so-called SSRIs (selective serotonin reuptake inhibitors) have to some extent replaced them. Drugs against asthma are increasingly prescribed, reflecting greater incidence of the disease and better diagnosis. In the respiratory drugs group, these are lumped together with drugs to ease the symptoms of coughs and colds. Prescriptions for gastrointestinal drugs have also increased in recent years, reflecting the wide range of new drugs available. Twenty years ago, only anticholinergic drugs of the atropine group were used against ulcers. Now there are H_2-antagonists and proton pump inhibitors. Also the role of *Heliobacter pylori* in ulcers has been recognized by the simultaneous use of antibiotic treatment, clarithromycin being the favoured compound.

Other widely prescribed drugs include the nonsteroid anti-inflammatories, used to counter the pain and inflammation of arthritis, hypoglycaemics for diabetes sufferers,

Table 43.1 *The top 50 drugs by world sales 2000*

	Product	Generic name	Company	Therapeutic category	Pharmacological class	2000 sales ($US million)
1	Losec	omeprazole	AstraZeneca	GI	Proton pump inhibitor	6260
2	Zocor	simvastatin	Merck	H	HMG CoA Reductase inhibitor	5280
3	Lipitor	atorvastatin	Pfizer	H	HMG CoA Reductase inhibitor	5031
4	Norvasc	amlodipine	Pfizer	H	Calcium antagonist	3362
5	Prevacid	lansoprazole	TAP Holdings	GI	Proton pump inhibitor	2740
6	Procrit	erythropoietin	Johnson & Johnson	B	Red blood cell growth factor	2709
7	Celebrex	celecoxib	Pharmacia	RH	COX-2 inhibitor	2614
8	Prozac	fluoxetine	Eli Lilly	CNS	SSRI	2585
9	Zyprexa	olanzapine	Eli Lilly	CNS	Atypical anti-psychotic	2366
10	Seroxat/Paxil	paroxetine	GlaxoSmithKline	CNS	SSRI	2350
11	Claritin	loratadine	Schering-Plough	R	Anti-histamine	2194
12	Vioxx	rofecoxib	Merck	RH	COX-2 inhibitor	2160
13	Zoloft	sertraline	Pfizer	CNS	SSRI	2140
14	Epogen	erythropoietin	Amgen	B	Red blood cell growth factor	1963
15	Premarin	oestrogen	American Home Products	GI	Conjugated oestrogens	1870
16	Augmentin	amoxicillin & clavulanate	GlaxoSmithKline	INF	Penicillin	1848
17	Pravachol	pravastatin	Bristol-Myers Squibb	H	HMG CoA Reductase inhibitor	1817
18	Vasotec	enalapril	Merck	H	ACE inhibitor	1790
19	Glucophage	metformin	Bristol-Myers Squibb	END	Biguanide	1732
20	Cozaar	losartan	Merck	H	Angiotensin II antagonist	1715
21	Tylenol	acetaminophen & codeine	Johnson & Johnson	RH	NSAID	1680
22	Ciprobay	ciprofloxacin	Bayer AG	INF	Quinolone	1648
23	Risperdal	risperidone	Johnson & Johnson	CNS	5HT2 & D2 antagonist	1603
24	Mevalotin	pravastatin	Sankyo	H	HMG CoA Reductase inhibitor	1602
25	Taxol	paclitaxel	Bristol-Myers Squibb	C	Taxoid	1592

(continued on next page)

Table 43.1 (continued)

	Product	Generic name	Company	Therapeutic category	Pharmacological class	2000 sales ($US million)
26	Zithromax	azithromycin	Pfizer	INF	Macrolide	1382
27	Intron A	interferon alpha-2b	Schering-Plough	C	Alpha interferon	1360
28	Viagra	sildenafil	Pfizer	OBS	Phosphodiesterase V inhibitor	1344
29	Flixotide/Flovent	fluticasone proprionate	GlaxoSmithKline	R	Corticosteroid	1334
30	Neurontin	gabapentin	Pfizer	CNS	GABA agonist	1334
31	Fosamax	alendronate	Merck	RH	Bisphosphonate	1275
32	Klacid/Biaxin	clarithromycin	Abbott Laboratories	INF	Macrolide	1241
33	Neupogen	filgrastim	Amgen	C	Colony stimulating factor	1224
34	Neoral	cyclosporin A	Novartis	C	Immunosuppressant	1215
35	Zestril	lisinopril	AstraZeneca	H	ACE inhibitor	1188
36	Effexor	venlafaxine	American Home Products	CNS	SNRI	1159
37	Humulin R	insulin	Eli Lilly	END	Insulin	1137
38	Levaquin	levofloxacin	Johnson & Johnson	INF	Quinolone	1089
39	Allegra/Telfast	fexofenadine	Aventis	R	Anti-histamine	1077
40	Prinivil	lisinopril	Merck & Co	H	ACE inhibitor	1075
41	Imigran/Imitrex	sumatriptan	GlaxoSmithKline	CNS	5HT1B & 1D agonist	1069
42	Adalat	nifedipine	Bayer AG	H	Calcium antagonist	1067
43	Diflucan	fluconazole	Pfizer	INF	Imidazole	1014
44	Rocephin	ceftriaxone	Roche Holdings	INF	Cephalosporin	1013
45	Lovenox	enoxaparin	Aventis	B	Low molecular weight heparin	962
46	Ortho-Novum	norethisterone & ethinyloestradiol	Johnson & Johnson	OBS	Oestrogen agonist	956
R	Serevent	salmeterol xinofoate	GlaxoSmithKline	R	Beta 2 agonist	943
48	Plavix	clopidogrel	Bristol-Myers Squibb	B	Platelet ADP antagonist	903
49	Zantac	ranitidine	GlaxoSmithKline	GI	H2 antagonist	872
50	Singulair	montelukast	Merck	R	Leukotriene antagonist	860

NSAID, Nonsteroid anti-inflammatory drug; SNRI, Selective serotonin and nonadrenaline reuptake inhibitor; SSRI, Selective serotonin reuptake inhibitor

Table 43.2 Top 50 most frequently prescribed drugs (United States, 2000)

	Generic name	Brand name	Manufacturer	Therapeutic category[c]	Therapeutic class
1	Hydrocodone + acetaminophen[a,b]	Hydrocodone w/APAP	Various	CNS	Analgesic
2	Atorvastatin	Lipitor	Parke-Davis	H	HMG CoA reductase inhibitor
3	Conjugated Estrogens	Premarin	Wyeth-Ayerst	OBS	Hormone replacement therapy
4	Levothyroxine[a]	Synthroid	Knoll	END	Thyroid
5	Atenolol	Atenolol	Various	H	Beta-blocker
6	Furosemide (oral)	Furosemide	Various	H	Diuretic
7	Omeprazole	Prilosec	Astra	GI	Proton pump inhibitor
8	Albuterol[b]	Albuterol	Various	R	Asthma
9	Amlodipine	Norvasc	Pfizer	H	Calcium antagonist
10	Alprazolam	Alprazolam	Various	H	Anxiolytic
11	Propoxyphene napsylate + acetaminophen[a,b]	Propoxyphene N/APAP	Various	CNS	Analgesic
12	Metformin	Glucophage	B-M Squibb	END	Diabetes
13	Cephalexin	Cephalexin	Various	INF	Cephalosporin antibiotic
14	Amoxicillin[a]	Amoxicillin	Various	INF	Beta-lactam antibiotic
15	Loratadine	Claritin	Schering	R	Non-sedating antihistamine
16	Amoxicillin[a]	Trimox	Apothecon	INF	Beta-lactam antibiotic
17	Hydrochlorothiazide[a]	Hydrochlorothiazide	Various	H	Diuretic
18	Sertraline	Zoloft	Pfizer	CNS	Selective serotonin reuptake inhibitor
19	Azithromycin	Zithromax (Z-Pack)	Pfizer	INF	Macrolide antibiotic
20	Fluoxetine	Prozac	Lilly	CNS	Selective serotonin reuptake inhibitor
21	Ibuprofen	Ibuprofen	Various	RH	Nonsteroid anti-inflammatory
22	Paroxetine	Paxil	SK Beecham	CNS	Selective serotonin reuptake inhibitor
23	Triamterene/hydrochlorothiazide[a]	Triamterene/HCTZ	Various	H	K-sparing diuretic
24	Celecoxib	Celebrex	Searle	RH	COX-2 inhibitor
25	Acetaminophen[a,b]/Codeine	Acetaminophen/Codeine	Various	CNS	Analgesic
26	Lansoprazole	Prevacid	Tap Pharm	GI	Proton pump inhibitor
27	Lisinopril	Zestril	Zeneca	H	ACE inhibitor

(continued on next page)

Table 43.2 *(continued)*

	Generic name	Brand name	Manufacturer	Therapeutic category[c]	Therapeutic class
28	Amoxicillin[a]/Clavulanate	Augmentin	SK Beecham	INF	Beta-lactam antibiotic
29	Conj Estrogens/ Medroxyprogesterone	Prempro	Wyeth-Ayerst	OBS	Hormone replacement therapy
30	Prednisone	Prednisone (oral)	Various	END	Steroid
31	Simvastatin	Zocor	Merck	H	HMG CoA reductase inhibitor
32	Rofecoxib	Vioxx	Merck	RH	COX-2 inhibitor
33	Norgestimate/Ethinyl Estradiol	Ortho Tri-Cyclen	Ortho Pharm	OBS	Oral contraceptive
34	Lorazepam	Lorazepam	Various	CNS	Anxiolytic
35	Trimethoprim/Sulfamethoxazole	Trimethoprim/Sulfamethoxazole	Various	INF	Antibacterial
36	Digoxin	Lanoxin	Glaxo-Wellcome	H	Cardiac stimulant
37	Metoprolol[a]	Metoprolol Tartrate	Various	H	Beta-blocker
38	Amitriptyline	Amitriptyline	Various	CNS	Tricyclic antidepressant
39	Ranitidine	Ranitidine	Various	GI	H2 receptor antagonist
40	Levothyroxine[a]	Levoxyl	Jones Medical Ind	END	Thyroid
41	Fexofenadine	Allegra	Hoechst-Marion-Roussel	R	Non-sedating antihistamine
42	Amoxicillin[a]	Amoxil	SK-Beecham	INF	Beta-lactam antibiotic
43	Ciprofloxacin	Cipro	Bayer Pharm	INF	Quinolone antibiotic
44	Zolpidem	Ambien	Searle	CNS	Hypnotic
45	Cetirizine	Zyrtec	Pfizer	R	Non-sedating antihistamine
46	Naproxen	Naproxen	Various	RH	Nonsteroid anti-inflammatory
47	Warfarin	Coumadin	Dupont	H	Anticoagulant
48	Quinapril	Accupril	Parke-Davis	H	ACE inhibitor
49	Pravastatin	Pravachol	B-M Squibb	H	HMG CoA reductase inhibitor
50	Sildenafil Citrate	Viagra	Pfizer	OBS	Erectile dysfunction

*Source:*http://www.rxlist.com/top200.htm. Numbers of prescriptions were not divulged. For estimates for 1998, see Reuben B.G. (2001) Pharmaceuticals. In *Encyclopedia of Physical Science and Technology*, Academic Press, San Diego (Table 1).[17] In 2000, there would have been about 100 m acetaminophen-containing prescriptions, 75 m amoxicillin, 60 m conjugated estrogen, 47 m levothyroxine, 45 m hydrochlorothiazide in various mixtures. By No. 10 in the table, numbers are down to about 25 m, No. 20 about 20 m, No. 30 about 18 m, No. 40 about 13 m and No. 50 about 12 m. These compounds are listed two or more times in this list for reasons that are not clear. In most cases, all the products from different manufacturers are bundled together, but in these cases exceptions have been made. Amoxicillin, for example, occurs in positions 24, 16 and 42 and in combination with potassium clavulanate at position 28. If added together, amoxicillin would move up to No. 2 in the list.

[a] These compounds are listed two or more times in this list for reasons that are not clear. In most cases, all the products from different manufacturers are bundled together, but in these cases exceptions have been made. Amoxicillin, for example, occurs in positions 24, 16 and 42 and in combination with potassium clavulanate at position 28. If added together, amoxicillin would move up to No. 2 in the list.

[b] Acetaminophen is known as paracetamol in Europe; albuterol is known as salbutamol.

[c] See Table 43.3.

Table 43.3 *Drug market by therapeutic class: World sales, and percentage of top-50 drugs by value of world sales and by US prescriptions, and by percentage of UK prescriptions*

Therapeutic class	World market 2000 ($ billion)	World market 2000 (%)	Prescriptions (UK % 1999)	% top 50 by world sales 2000	% top 50 by US prescriptions 2000
Gastrointestinal system (GI)	a	a	7.3	6	6
Cardiovascular system (H)	72.1	19.5	19.9	20	26
Respiratory system (R)	34.6	9.4	8.3	10	8
Central nervous system (CNS)	57.3	15.5	17.2	16	20
Infections (INF)	37.7	10.2	7.3	14	16
Endocrine system (END)	a	a	6.3	8	14
Obstetrics, gynaecology and urinary tract disorders (OBS)	20	5.4	2.2	2	2
Malignant disease and immunosuppression (C)	20	5.4	0.5	8	0
Nutrition and blood (B)	11.7	3.2	2.5	8	0
Musculoskeletal and joint diseases (RH)	21.3	5.8	4.6	8	8
Other (MISC)	71.9	19.5	12.0	0	0

a Included in the 'other' category.

Sources: Columns 2 and 3 Medical Healthcare Marketplace Guide 2000–2001, Dorland Healthcare Information; column 3 Statistical Bulletin 2000/20, Department of Health, UK Office of National Statistics; columns 4 and 5, see Tables 43.1 and 43.2.

L-thyroxine and thyroid hormone for thyroid sufferers, and antihistamines for hay fever sufferers. The chemical class of steroids plays an unusual role in the list, being the active component of oral contraceptives, some anti-asthma drugs and some anti-inflammatories. In particular, they are the active ingredients of conjugated oestrogenic hormone, taken as hormone replacement therapy by menopausal women. This has been among the three most widely prescribed drugs for many years. Few chemical classes resemble steroids in being therapeutically active in a number of areas, and the most important of these are shown in Table 43.4.

C National differences in prescribing

National differences in prescribing can be identified and are related to national wealth, national culture and attitudes towards the role of drugs in medical treatment. Expenditure per head per year in different countries varies dramatically as shown in Table 43.5. Total expenditure and a drug price index, where available, are also shown. Drugs are cheap in France and Italy, expensive in the USA and Japan. Comparable prescription numbers are difficult to establish, but in the UK and USA the annual number of prescriptions per person is about 10, in France it is between 25 and 35.

The inability of third world countries to pay for drugs is evident, and typical countries, whether rich or poor, spend of the order 1% of GDP on drugs. Thus Bangladeshi spending of $3.3/head is symptomatic of poverty rather than national culture.

Culture needs to be invoked, however, to explain differences between developed countries. A belief in the prophylactic benefits of taking medicines is widespread in Belgium, France, Greece, Italy and Spain but less prevalent in Britain, Ireland, the Netherlands and the Scandinavian countries. Germany, Austria and the United States have a culture with elements in common with both groups.[1]

Table 43.4 *Chemical groups with a range of pharmaceutical applications*

Chemical class	Example	Therapeutic use
Phenylpiperidines	Meperidine (Pethedine)	Opioid analgesic
	Loperamide	Antidiarrhoeal
	Haloperidol	Neuroleptic
	Terfenadine	Antihistamine
Sulfonamides/sulphones	Sulphamethoxazole	Antibacterial
	Acetazolamide	Carbonic anhydrase inhibitor used in glaucoma and (infrequently) as a diuretic and in epilepsy
	Glyburide	Hypoglycaemic
	Dapsone	Antileprotic/antimalarial
Steroids	Ethynyloestradiol, Norethindrone, Medroxyprogesterone	Sex hormones used in oral contraceptives and for menstrual and menopausal disorders.
	Prednisone	Anti-inflammatories in arthritis; immunosuppressives in transplant operations
	Beclomethasone dipropionate	Anti-asthma
	Estramustine phosphate	Prostate cancer
	Testosterone esters	Anabolic hormones used for body building after surgery
Prostaglandins	Prostaglandin $F_{2\alpha}$ and E_2	For abortions and to induce labour. $F_{2\alpha}$ reduces intraocular pressure and may be of use in glaucoma
	Misoprostol	For gastric ulcers
	Prostaglandin E_1	Vasodilator used to treat vascular disease of the leg. Also for ductal-dependent congenital heart disease in new-born babies
	Prostacyclin	Antithrombotic and vasodilator. Inhibits platelet aggregation. Used experimentally to inhibit clotting in operations where blood circulates outside the body

Table 43.5 *Total expenditure on pharmaceuticals, expenditure/head and price indexes for selected countries, 2000*

Country	US $ billion	$/head	Price index
USA	144	501	140
Canada	8.8	281	
France	17.2	289	70
Germany	17.2	209	115
Italy	11.1	190	80
Spain	7.5	182	90
UK	11.3	188	110
Other EU	14.2	195	
Japan	51.5	406	
Argentina	4.3	118	
Brazil	6.3	36	
Mexico	5.7	56	
Bangladesh	0.4	3	
China	12.5	10	
India	3.8	4	
Pakistan	1.1	8	
World total	327.5	53	

Sources: M.L. Burstall and B.G. Reuben, Will the European pharmaceutical industry survive? *Spectrum*, Decision Resources, 5 September 2001, 11-1–11-16; M.L. Burstall and B.G. Reuben, Challenges of drug pricing and availability in developing countries, *Spectrum*, Decision Resources, 12 April 2002, 4-1–4-18.

Precisely what constitutes the cultural differences is unclear. It has been suggested that it is related to the proportion of Roman Catholics in the population, but that scarcely explains the gap between Italy and Ireland. There is a loose correlation with climate, people from more temperate climates requiring fewer drugs, but that does not explain the difference between Belgium and the Netherlands. The perception of doctor and patient of their roles, however, is at the root of the discrepancies. A patient visiting an Italian doctor will emerge with a prescription in 95% of cases. In the UK, this drops to below 70%. The giving of a prescription is seen to a greater or lesser extent in high- and low-prescribing countries as a way of signalling the conclusion of a consultation. To hand out a prescription reinforces the doctor's image of him or herself as a doctor (that's what doctors are supposed to do) and the patient's image of him or herself as a patient (patients are given medicine; by giving me a prescription, the doctor is confirming that I am ill, have not troubled him or her unnecessarily and am justified in taking the day off work to visit the surgery).

The differences between countries are less clear-cut, nonetheless, than the above figures imply. The drugs discussed so far are those that must be prescribed by a physician. Some drugs are classified as 'generally regarded as safe', however, which means that they can be sold direct to the public, that is over the counter (OTC) rather than on prescription. Diet supplements are apparently much more popular in Germany than in the United States but, in fact, the Americans buy them over the counter instead of having them prescribed. This reflects both culture and systems of health insurance reimbursement.

Apart from differences in quantity, there are disparities between countries in the types of drugs prescribed. In the USA and UK, almost all the leading drugs are therapeutically active; in Germany, France and Italy, many prescriptions are for 'comfort' drugs that are generally agreed to have little therapeutic value. Estimates from 1992 are given in Table 43.6. The Germans show enthusiasm for ointments based on nonsteroid anti-inflammatories and salicylic acid derivatives. They also take peripheral vasodilators and antihypotensives, drugs that are rarely used by the British and Americans. They are concerned, as are the French, with problems of the digestive system and take drugs that supposedly reinvigorate the gall bladder. The French also take peripheral vasodilators and numerous drugs for the *crise de foie*.

In comparison, the British and Americans take large numbers of antibiotics. These are indeed therapeutically active, but the number of prescriptions compared with

Table 43.6 *The 'useless' drugs league*

	Top 25 Products			Top 50 Products			'Useless' drugs as % sales
	A	B	C	A	B	C	
Italy	11	7	7	25	15	10	21.2
France	16	4	5	26	14	10	20.5
Germany	19	5	1	35	9	6	11.9
UK	24	1	0	46	4	0	NA

Class A: Products agreed internationally to be therapeutically effective. Class B: Second-line therapy, open to misuse, more expensive than similar products, or combination products with no advantage over monosubstances. Class C: Drugs with no evidence of efficacy. *Source*: Health Economics Centre, Cesav, Italy, reported in *SCRIP* (1993) **1860**: 4. Reproduced with permission from PJB Publications Ltd, 2002.

Germany suggests that in many cases they too are prescribed as 'comfort' drugs for conditions that are either self-limiting or viral infections insensitive to antibiotics or both. The problem has not been ignored by the authorities. In Germany prescriptions for products 'of disputed efficacy' dropped by 13.4% between 1998 and 1999. Since 1993, the Germans have saved DM4.8 billion (about 2% of the drug budget) by bringing pressure on doctors.[2] In general, patterns of prescribing in the developed world are drawing closer together.

As noted, the above tables exclude OTC drugs. OTC sales vary between countries, Switzerland having the highest level recorded in the literature at 28.8% of drug sales and Spain coming lowest at 5.1%. The OTC market in 1997 amounted to $15 billion and in Europe $13 billion.

In most developed countries, the therapeutically active constituents of OTC drugs are largely confined to non-narcotic analgesics, antihistamines, antacids and anti-inflammatory steroid ointments. In some countries, antibiotics are available without prescription and the narcotic analgesic, codeine, is sometimes permitted mixed with aspirin or paracetamol. As governments try to increase the proportion of drug costs paid by patients, the range of OTC drugs increases. Compounds that have moved from prescription only to over the counter in the UK in recent years include the antiviral, acyclovir; the H_2-antagonists, ranitidine and famotidine; the steroid ointment, clobetasone; the hair-growing compound, minoxidil, and various nicotine replacement therapies. There are moves in the United States to extend the basis for OTC classification from drugs for mild conditions to those for chronic illnesses — high blood pressure, high cholesterol and osteoporosis.[3]

The other major group of drugs omitted from the prescriptions list is those given in hospitals. Typically for developed countries they amount to about 10% of drug sales although in the United States it rises as high as 22%. Drugs that are largely confined to hospitals include numerous antibiotics (e.g. third generation cephalosporins, lincomycin and vancomycin) used to counter potentially resistant bacteria found in hospitals, together with blood products (e.g. albumin), intravenous products (e.g. total parenteral nutrition, dextrose) and the anticoagulant heparin.[4]

II SOURCES OF DRUGS

Drugs are obtained from five sources: animal and vegetable extracts, biological sources, fermentation and chemical synthesis. Our ancestors, lacking synthetic skill, drew primarily on animal and vegetable sources. Macbeth's witches brewed up 'eye of newt, toe of frog and liver of blaspheming Jew', which must have had a largely psychological effect, not to mention encouraging anti-

Semitism, but they remembered to add pharmacological activity with 'root of hemlock picked in the dark'.

A Vegetable sources

Plants continue to this day to provide a range of medicinal alkaloids (e.g. papaverine, atropine, codeine, quinidine), glycosides (e.g. digoxin), steroids (e.g. diosgenin, stigmasterol, sitosterol, the raw materials for medicinal steroids) and vitamins (e.g. vitamin E from soya bean oil footstocks). This is discussed in Chapter 6. A current problem is the anticancer drug paclitaxel (Taxol) a diterpenoid ester produced in 0.02% yield from the stem bark of the Pacific yew tree *Taxis brevifolia*. This can never be an adequate source. Semi-synthetic routes starting with taxol precursors also found in yew trees have eased the problem somewhat, but a practicable synthetic route is currently sought.

B Animal sources

Animal sources are less convenient than vegetable sources for production of commercial quantities of drugs, but some pharmaceuticals are obtained from slaughterhouse wastes. Heparin and bile acids are examples. Cow and pig insulin have largely been displaced by human insulin generated using recombinant DNA. Conjugated oestrogenic hormone is obtained from the urine of pregnant mares, with the bonus that the animal does not have to be slaughtered. Gelatine is used to encapsulate active materials.

Human blood, given by blood donors, is a source not only of cellular material (red cells, white cells, platelets), but of blood plasma, which is subjected to protein fractionation to give albumin, anti-haemophilia factors and immunoglobulins. The possibility of transmission of CJD (prions are not really understood) and AIDS, if sterilization procedures fail, combined with the feasibility of manufacturing proteins by recombinant DNA technology mean that there is uncertainty about the long-term future of the blood products industry. Indeed, the problems with bovine spongiform encephalitis (BSE) and the related fatal human brain disorder, Creutzfeld–Jakob disease (CJD), have led to unease about cattle-derived substances.

C Biological sources

Biological sources are primarily used for vaccines. Vaccines are suspensions of living or killed micro-organisms, components or products thereof. They are produced in living systems. Eggs are the most widely used medium at present, but there is increasing use of animal cell cultures. Bacterial vaccines, for example the diphtheria vaccine, can be cultured. There is an overlap here with fermentation processes (Section II.D).

Recombinant DNA processes can be used to make subunit vaccines. These can be classified either as biological or fermentation processes. Hepatitis B vaccines are made by genetic manipulation of yeast or animal cells. The cells then express the hepatitis B virus outer coat protein, which is harmless to the host because there is no DNA present.

There are two types of immunization, active and passive. In active immunization, the vaccine is administered to the patient, whose immune system responds by producing antibodies and cells sensitized to the vaccine and hence conferring immunity to the disease. In passive immunization, a donor who has had a specific disease donates blood, and an immunoglobulin is isolated from the blood plasma. This is then injected into the patient.

Some immunoglobulins cannot be obtained from normal, healthy blood donors and are obtained instead by inoculation of volunteers, so that they develop the required antibody. Such procedures are used for immunoglobulins against vaccinia, tetanus, measles, hepatitis A, Rhesus incompatibility and rabies.[5] The biological 'factory' for the immune globulin, in such cases, is the body of the volunteer.

In the future, monoclonal antibodies will have a major role in the development of new diagnostic and therapeutic products.

D Fermentation

Fermentation in its simplest forms, such as the production of bread or wine, is very old technology. In its application with recombinant DNA to produce pharmacologically active proteins (erythropoetin, human growth hormone, human insulin) it is central to the modern biotechnological revolution. In between, it is the method for production of antibiotics and vitamin B_{12} and provides a step in the production of vitamin C and the synthesis of cortisone from diosgenin.

Plant cell culture might be classified as fermentation or production from a biological or vegetable source. It is not yet widely applied. There were reports that a digoxin producer modified the useless glycosides from *Digitalis lanata* by plant cell culture and thus converted them to digoxin and increased his yield per plant. The Japanese are certainly using plant cell culture to produce shikonin, a red pigment with anti-inflammatory properties.

E Chemical synthesis

In spite of the importance of the above methods, chemical synthesis is the most important method of drug manufacture. It is responsible for almost all heart drugs, drugs for the central nervous system, anti-ulcer drugs, analgesics and antihistamines. Many pharmaceuticals are made simply from readily available bulk chemicals. For example,

the nonsteroid anti-inflammatory agent, ibuprofen, is made from toluene and propylene, two of the seven basic building blocks of the petrochemical industry.[6] Chemical synthesis is also used to modify materials made from other sources, and chemical methods of extraction are used in the downstream processing of products from other sources.

The combination of methods illustrates the eclecticism of the pharmaceutical industry. The use of fermentation in the production of vitamin C and cortisone was mentioned above. In the production of the semisynthetic penicillins, penicillin G or V is first made by fermentation. It is then cleaved to 6-aminopenicillanic acid by an immobilized enzyme, and a new side chain is added by chemical means. With the thrombolytic drug, Eminase, a plasminogen-streptokinase activator is made by a cell-cloning biotechnological technique. This is then chemically modified by addition of a *p*-anisoyl group to give a longer-lived product that can be administered by injection rather than through a drip.

III MANUFACTURE OF DRUGS

The largest tonnage organic chemical is ethylene, made in the United States and in Western Europe on a scale of about 26 and 21 million tonnes/year, respectively. It sells for about $0.5 per kg. The largest medicinal chemicals are aspirin, paracetamol (acetaminophen) and vitamin C, made worldwide on a scale of 30 000–50 000 tonnes per year. A total of about 149 000 tonnes of medicinal chemicals was manufactured in the United States in 1992 (more recent figures are not available). Ninety-two thousand tonnes of these were sold for $2.4 billion at an average price of $25.90 per kg.[7] These medicinal chemicals were formulated into products, however, which sold for about $35 billion, amounting to almost $200 per kg of raw material. A Merck executive remarked, 'We buy chemicals at the price of silver and sell them at that of gold'.

The manufacture of pharmaceutical chemicals in some ways resembles a scaled-up version of the organic chemical syntheses carried out in the laboratory and, in others, a scaled down version of the processes used in the heavy chemicals industry. The heavy chemicals industry manufactures commodity chemicals that are largely undifferentiated and have to be sold at the ruling market price. The individual company, having little control over prices, is therefore preoccupied with reducing costs. A major source of cost reduction is economies of scale, and those in turn are related to the use of continuous rather than batch processes. The former are well-nigh universal in the heavy chemicals sector.

The pharmaceutical industry, in contrast, manufactures a large number of chemicals in small quantities to a high level of purity. Economies of scale are relatively inaccessible.

The product, especially if it is under patent, is highly differentiated and can be sold, like other speciality chemicals, at a price that reflects its value-in-use, rather than its manufacturing cost. The industry is preoccupied with quality control, and this is more easily achieved in batch equipment. Purity is of greater importance than high yield. The process must always be operated in such a way that the quality of the final product is maintained, and this leads to certain design features not normally found in chemical plants. An analytical department intervenes at every stage to monitor the progress of a drug through the system. There is a code of good manufacturing practice (GMP) for the pharmaceutical industry, which lays down guidelines within the European Community,[8] and this is matched by regulations of the Food and Drug Administration (FDA) in the United States. These dominate the organization of pharmaceutical manufacturing sites. Uniform standards are desirable, and a common technical document for submissions for new drug approval and an agreement on common GMP guidelines were agreed at San Diego in November 2000. The implementation of these, so that different countries would mutually recognize each others' plant inspection procedures, is proving more difficult.

A Good manufacturing practice

Medicinal products must be fit for their intended use, comply with the requirements of market authorization and not place patients at risk because of inadequate safety, quality or efficacy. These objectives are easy to state, but to achieve them requires a comprehensively designed and correctly implemented system of quality assurance. It is important that this is documented and its effectiveness monitored. Records must be kept that are open to inspection by validating bodies, and there must also be procedures for self-inspection and quality audit that allow appraisal of the quality assurance system. Management must be adequately trained and their responsibilities minutely defined. Two key posts are the heads of production and of quality control. These posts are required by the guidelines to be independent of one another.

Quality control is part of GMP. It is concerned with sampling, specification and testing and with the organization, documentation and release procedures that ensure that tests have indeed been carried out. No materials must be released for use or sale until their quality has been judged satisfactory. For example, the purity of most drugs is greater than 99.9% and the content of the remaining 0.1% is usually specified in the Pharmacopoeia and in the quality control specifications. Absorbent cotton fibres may present a problem and ointments are specified to contain less than 1 fibre cm^{-3}, the fibre being less than 1 mm long. Contamination by the ubiquitous *Penicillium* micro-organisms are a particular problem (see below).

B Plant design

A large pharmaceuticals manufacturer will operate several different plants. The plant for the manufacture of active ingredients by chemical synthesis will be separate from the plant that formulates the active ingredients into finished medicines. Fermentation plant will usually be on a totally different site geographically from a chemical synthesis plant. The reason for this is to minimize contamination by the *Penicillium* micro-organism, among others. Its concentration in finished products must be kept below 1 ppm to avoid allergic reaction in penicillin-sensitive patients. Bacteria insensitive to an antibiotic may grow in a medium containing it, and most countries demand the absence of harmful varieties (e.g. salmonella) and specify a limit for total bacterial count. Thus the U.S. Pharmacopoeia quotes 5000 cm^{-3} for a gelatine base. The cautious manufacturer will prefer to build his chemical plant far away from the micro-organism-rich environment of a fermentation plant. There is also likely to be a separation from plant making biological medicinal products, and these have separate regulations.

As far as equipment is concerned, the pharmaceutical industry rarely follows the chemical industry pattern of purpose-built plant, carefully optimized and with the larger sections fabricated on site. In general, equipment is bought 'off the shelf' and assembled like a child's erector set.

Optimization is sacrificed to availability of plant items. A secondary benefit to emerge from this is that lead times (i.e. the time taken to build plant) are lower than in the chemical industry. Because kilogram quantities of a drug are required for testing before it is known whether a drug will be a success or not, the initial manufacturing process usually involves a batch preparation on general purpose equipment. There may be a need for changes to the initial method subsequently on economic or safety grounds (e.g. one of the reagents may be pyrophoric), but the need to register a validated process means that there will be a conservative attitude to process change.

The equipment supplier will therefore aim to offer versatile, multi-purpose equipment that can be used in a range of processes. Most of it will be stainless steel or glass-lined to avoid problems with contamination or corrosion. In spite of its versatility, however, the plant, once built, will rarely be changed. Any new process or important modification of an existing process must be validated, and critical phases of an existing manufacturing process must be regularly revalidated. If the manufacturer wants to add a recycle stream, for example, clearance must first be obtained.

Wherever possible, a single reaction vessel is dedicated to one particular process, and usually each preparation is performed in a self-contained area. This situation has eased somewhat in recent years, and more multi-use equipment is being installed. There is still not the flexibility in equipment use that one might expect in comparison, say, with the dyestuffs industry, but there is a trend in that direction.

Effluent disposal is a central problem. Yields of pharmaceuticals are low compared with the heavy chemicals industry, and waste streams, perhaps containing pharmacologically active materials, must be incinerated or scrupulously purified before discharge.

An exception to the rule of small, general purpose equipment arises with fermentation processes. A reactor producing penicillin G will certainly be dedicated and may have a capacity of 250 tonnes and a 30 000 h.p. stirrer. Because yields are so low, the weight of penicillin produced per run will be two orders of magnitude smaller. One of the problems of downstream processing is to reduce the volume of the penicillin-containing stream to more manageable proportions and to dispose of the effluent fermentation liquid without imposing an unacceptable biochemical oxygen demand on the aqueous environment.

The availability of on-line microcomputers and microprocessor-controlled equipment is extending many of the benefits of continuous processing to the batch processor. The pharmaceutical industry has been a leader in the employment of these new techniques.[9,10]

C Formulation and packaging — sterile products

Plant for the production of active ingredients resembles plant for the production of other small tonnage chemicals. Formulation and packaging plant, in which the active ingredient is converted into saleable form, is less familiar to the industrial chemist. The reason is the need for a controlled environment. Some straightforward chemical products to be taken by mouth do not need to be tableted in a sterile environment. For others, such as injectable antibiotics, it is essential. Although the regulations do not demand sterility for all products, most companies would carry out all their formulating in a controlled environment, the severity of which would depend on the nature of the product. Walls and floors are rounded off to allow efficient cleaning. Sinks and drains are avoided as far as possible. Packaging is carried out in individual cubicles to avoid cross-contamination if there is spillage. Production of sterile medicines is carried out in clean areas whose entry is through airlocks both for people and goods. The areas are supplied with suitably filtered air and are classified A, B, C or D according to the required characteristic of the air. These are shown in Table 43.7. The actual regulations are more detailed and the table merely summarizes a few of the points. In particular, it illustrates the need for personnel to wear suitable clothing that will prevent contamination by viruses, bacteria or particulate matter.

Manufacturing operations for sterile products are divided into two groups, one where the product is sterilized terminally, that is at the end of processing after it has been sealed in its container, and the other where some of the processing must be conducted aseptically. For terminally sterilized products, a grade C environment is usually adequate. Aseptic preparation can be performed in a grade C environment provided the product is sterile filtered later in the process. If not, a grade A zone with a grade B background is required.

Table 43.7 *Air classification system for the manufacture of sterile products*

Grade	Maximum permitted numbers of particles per m^3		Maximum permitted number of viable micro-organisms per m^3	Approximate equivalent US Federal Standard 209C	Clothing for each grade
	$\geq 0.5 \mu M$	$\geq 5 \mu M$			
A[a]	3500	None	$<1^b$	100	
B	3500	None	5^b	100	Clean sterilized protective garments provided at least once a day. Gloves disinfected during operations. Masks and gloves changed every working session. Disposable clothing a possibility.
C	350 000	2000	100	10 000	Hair covered. Two-piece nonfibre-shedding trouser suit gathered at wrists and with high neck required.
D	3 500 000	20 000	500	100 000	Hair covered. Protective clothing and shoes required.

The number of air changes should generally be higher than 20 per hour in a room with good air-flow pattern and appropriate HEPA filters. *Source*: reference 8.
[a] Laminar air-flow work station. Systems should provide homogeneous air speed of 0.3 m s^{-1} vertical and 0.45 m s^{-1} horizontal flow.
[b] Only reliable when a large number of air samples is taken.

The provision of sterile areas in a packaging and formulation plant means that such a plant may cost as much as a conventional manufacturing plant for the active ingredient. The tableting and packaging equipment are relatively inexpensive, and most of the money is spent on equipment to produce a controlled environment.

The formulation of active ingredients into a satisfactory dosage form is a science in itself. It is beyond the scope of this chapter, but a brief discussion is given in reference 6.

D Choice of reagents

The special characteristics of the pharmaceutical industry also affect choice of reagents. In general, processing costs are higher than in the chemical industry as a whole, and capital costs are lower. Because of the high unit value of pharmaceuticals, expensive chemicals are practicable. For example, lithium aluminium hydride might be used as a reducing agent while, in the petrochemical industry, hydrogen is the only economic reagent.

Pharmaceutical companies are involved in long, multistep synthetic processes. A modest way in which they try to keep costs within bounds is by the use of standard intermediates that can be used for a number of products. For example, all the different chemically modified penicillins are based on 6-aminopenicillanic acid, which can consequently be manufactured on a much larger scale than any of the individual penicillins. Such intermediates are known as synthons.

In general, companies prefer to reduce the number of synthetic steps they have to carry out themselves by purchase of synthons — high quality raw materials, reagents and key intermediates — from fine chemical manufacturers. These manufacturers are often prepared to develop reagents and intermediates specifically for one customer in the expectation that they will have further uses in the future. Reagents that increase reaction selectivities are often identified in academic laboratories, but cannot be used on a tonnage scale because of lack of availability or cost. Fine chemical companies can find niche markets by developing methods of making these reagents cheaply available. Examples are Hünig's base, N,N-diisopropyl-N-ethylamine[11] (i-C_3H_7)$_2$NCH$_2$CH$_3$, and oxalyl chloride[12] (COCl$_2$)$_2$. An improvement in selectivity is economically attractive not only because it gives higher yields but also because it produces less waste; the reduction of waste is currently a key environmental issue.

Companies may contract out some intermediate stages in a drug's synthesis (see Section IIIF) or conduct them in an isolated part of the plant if they offer particular hazards. Nitrations would probably be carried out in a special nitrating plant with equipment designed to prevent runaway reactions; phosgene would be handled only under carefully controlled conditions because of its toxicity. Experience is similarly needed with organofluorine compounds, reactions at cryogenic temperatures and fermentation routines for steroids and antibiotics. A firm that has been engaged in such operations for many years accumulates know-how, and this gives them a great advantage over new entrants to the industry.

There is currently a question mark over the importance of chirality. In May 1992 the FDA published a policy statement 'strongly urging companies to evaluate racemates and enantiomers for new drugs' and declaring that drugs be sold only in their enantiomerically effective form. This hard line has softened and the marketing of single enantiomers is not yet mandatory. Since Monsanto's pioneering asymmetric synthesis in the early 1980s of L-dopa, a drug for Parkinson's disease, chiral syntheses and separations have been of intense interest. This was accentuated by experiments with mice showing that it was the D-isomer of thalidomide that was responsible for birth defects, while the L-isomer was the hypnotic. With rabbits, however, both forms were active. Meanwhile, thalidomide has returned to the market with applications in leprosy, Hansen's disease, tuberculosis, lupus and cachexia, the weight loss and metabolic wasting associated with AIDS.[13] It has orphan drug status (Section IV.C) in some of these applications.

Sometimes both enantiomers of a drug are pharmacologically active and, in such cases, there is no reason to favour one over the other. In other cases, a chiral drug, when administered, racemises in the body anyway. Nonetheless, there are numerous examples of drugs where the administration of a racemic mixture means that patients are needlessly exposed to pharmacologically inactive chemicals. Even if the racemic mixture is synthesized but separated in the production stage, half of the production may be wasted. It might in the end be cheaper to make the correct enantiomer to start with. Hence the interest in chiral intermediates and chiral catalysts. Many specialist firms are springing up to try to find niche markets in this area.

E Downstream processing

The reagents used by the pharmaceutical industry are more complex and sophisticated than those employed by the heavy organic chemical industry. The same is true of the purification and isolation of the products. The small simple molecules of the petrochemical industry can usually be separated from impurities by distillation. In the pharmaceutical industry, this remains true for solvents, less true for intermediates and rarely true for end products. In the case of medicinal products from biological sources or made by fermentation, it is never true.

Much downstream processing is performed by the well-documented, but less widely used unit processes of chemical engineering — solvent extraction, filtration, leaching, adsorption and crystallization. Freeze drying, centrifugation, ion exchange and preparative chromatography are valuable options. Cross-flow filtration has become important especially since the development of anisotropic membranes. These permit adequate flow rates, but are still sufficiently robust to withstand high pressures and the wear and tear of industrial scale operations. Cross-flow techniques may be used for microfiltration, reverse osmosis and ultrafiltration.

Sterilization is a unit process not found in the heavy chemical industry. The aim of sterilization is to kill and/or remove bacteria and pyrogens. Pyrogens are fragments of bacterial cell walls, which may produce fever as a reaction to foreign proteins if they enter the bloodstream of a sensitive person.

For heat-stable materials, sterilization by steam for 15 min at 121°C kills all bacteria and viruses. Pasteurization for a longer period at a lower temperature is also effective. Dry heat at a temperature above 250°C destroys pyrogens.

Heat-sensitive materials may be sterilized by radiation, providing the product and its container are not radiation-sensitive. It is particularly used with plastics and more widely used with containers than with products. An alternative is ethylene oxide gas. This presents problems in that the gas must be brought into contact with the cell walls of contaminating bacteria. Ethylene oxide is highly reactive and can only be used when it can be proved not to react with the product.

Filtration through a 0.22 μm filter removes bacteria and moulds, but not all viruses or mycoplasma. It is not considered adequate when terminal sterilization is practicable.

F Outsourcing

Outsourcing, that is contracting out of all or part of one's business was a major feature of the pharmaceutical industry's policies in the 1990s. Contract manufacturing alone accounted for some $12 billion of business. Initially companies outsourced low technology activities such as cleaning and security. Then came stages of the manufacturing process that the outsourcing company preferred not to have to deal with, then parts of the testing. Testing drugs on humans had always been done in hospitals, away from company premises, but animal testing was also outsourced. More complex chemicals for use as raw materials were purchased, so that the early stages of synthesis were accomplished by specialist firms. Companies making patent-protected drugs usually preferred to perform the final synthetic stages and the formulation in-house. In addition to outsourcing, there was also a trend towards collaborative projects, and initial ideas were purchased from small 'ideas' companies before being developed in-house.

The main reasons given for outsourcing were, firstly, that it enabled projects to go ahead where in-house resources were lacking. No money was saved, but the project was speeded up. Secondly, the company might gain access to specialized skills, for example, experience with particular synthetic reactions or with the formulation of such things as transdermal patches. Thirdly, by outsourcing some activities, the company could focus on its core competencies. Finally, a company could avoid setting up a specialist department to undertake a particular function, which might not be needed once a project was concluded.

IV SOCIAL AND ECONOMIC FACTORS

The economics of the research-based pharmaceutical industry is as remote from classical economics as its technology is from that of the heavy chemical industry. Neither the manufacturers nor the consumers conform to the concept of a free market.

Producers of pharmaceuticals are monopolists to the extent that they have patents on their products. Instead of conventional competition on price and quality, there is competition in innovation between rival producers, so that a number of manufacturers may offer, for example, different H_2-antagonists for the treatment of gastric ulcers. Insofar as the patient requires a particular product when it is under patent, however, he is faced by a monopoly producer.

At the consumer's end, the situation is even more confused. Drugs are taken by the patient, prescribed by the physician and paid for (to a varying extent in different countries) by the state or by insurance companies. Thus the usual market constraints are lacking. On the other hand, the state, insurance companies and hospitals sometimes have much of the power of monopoly purchasers — so-called monopsonists. In many countries, the state licenses individual pharmaceuticals and sets their prices. In the negotiations with drug companies, the usual price determinants of supply and demand are replaced by political considerations.

The above factors mean that the market for pharmaceuticals is subject to severe distortions. Certain aspects will be discussed in this section, notably the pattern and cost of innovation, the role of patents, orphan drugs, the rise of generic drugs, parallel trade and the attempts by governments at cost containment.

A Pattern and cost of innovation

The modern pharmaceutical industry dates back to the discovery of the sulphonamide drugs in 1935. The industry has grown since 1950 typically at 7–15% per year, and over 95% of drugs available originated after that date. An astonishing series of innovations has increased life expectancy and improved the quality of life in developed and to a lesser extent in developing countries. The industry has moved from a 'molecular roulette' system, by which new drugs were developed on the basis of trial and error to a sophisticated system hinging on an understanding of the enzyme systems the drugs are intended to influence.

The industry was surprisingly unregulated in its early days but, since the thalidomide disaster in 1960, regulation has become increasingly strict and the cost of developing new drugs has escalated. Figure 43.1 shows the number of new products launched world-wide each year since 1980, together with total spending on research and development by American and European companies at 1998 prices.

The cost of introducing a new molecular entity (NME) to the drug market in the mid-1980s was estimated at $114 million, measured in 1987 dollars.[14] An NME is a truly novel product not a line extension of a drug already on the market. The figure includes the cost of failures, that is drugs that were eliminated during testing. This high cost is accentuated by the increased time required for drug discovery and testing. The average new chemical entity in the above survey took 11.8 years to reach launch. This had two consequences. Firstly, the patent life remaining to the drug, during which its development costs could be recovered, was much shortened. Secondly, the return on the R&D outlay was delayed. To get a true cost for the development of a new drug, one must discount the expenditure to the moment of launch. If a discount rate of 9% is assumed, then a figure of $231 million is obtained. The above calculation was seminal and the various

uncertainties and assumptions in it have been discussed and the figures updated. The estimates for new drugs in the 1990s were of the order of $500 million.[15]

This huge figure explains why only large multinationals with generous cash flow from other projects can be serious players in the game of pharmaceutical innovation. Smaller companies may come up with bright ideas that they sell to the multinationals, but only the giants can undertake the testing and registration procedures. Indeed, not only is pharmaceutical innovation confined to large companies, but it is also confined to countries that can provide the resources and manpower for such high technology operations.

The regions where drug discovery takes place are shown in Table 43.8. The big players are the USA, UK, Germany, Japan, Switzerland, France and Belgium, although the whole of the European industry is showing signs of weakness.[16] Meanwhile, in terms of foreign sales of pharmaceutical products,[17] everyone outside the European Community shows a balance of trade deficit whereas within the Community, only Spain and Italy show a deficit and that is small. The reason for the small United States deficit is that United States companies are happy to manufacture abroad and in particular see Ireland as their gateway to Europe. The Japanese, although innovative, concentrate on their large home market and prefer to license abroad products that they might otherwise export.

The governments of the leading drug-producing countries are in a curious position. On the one hand, they are anxious to limit the social security costs of drugs at home. On the other hand, they are anxious to promote

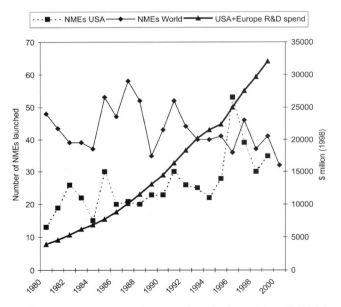

Fig. 43.1 Numbers of new chemical entities (NMEs) launched worldwide and in the United States 1980–2000 and United States plus Europe R&D pharmaceutical R&D expenditure at 1998 prices.

Table 43.8 *NMEs 1975–2000 by nationality of company*

	Europe	USA	Japan	Other	Joint	Total
1975–90	454	223	171	NA	NA	810
1991	22	13	11	2	4	52
1992	17	9	10	0	0	36
1993	14	10	15	1	0	40
1994	15	11	13	1	0	40
1995	19	11	10	1	0	41
1996	20	10	5	1	0	36
1997	21	14	5	2	4	46
1998	12	18	2	5	0	37
1999	17	13	4	1	6	41
2000	12	13	6	1	0	32
1991–2000	169	122	81	15	14	401
NMEs/year						
1975–90	28.4	13.9	10.7	NA	NA	50.6
1991–2000	16.9	12.2	8.1	1.5	1.4	40.1

Source: Scrip Magazine, *SCRIP*, Annual Review Issues, Reproduced with permission from PJB Publications Ltd, 2002. European Commission 1997: Single Market Review, Impact on Manufacturing, Pharmaceutical Products, Table 4.17, 63.

foreign trade in pharmaceuticals and to ensure that their industry remains prosperous. The countries with less drug-producing capability may have ambitions to build up their own industry, but otherwise are more determined to limit drug expenditure. Until about 1990, the drug companies were successful in fighting off cost containment measures but since then governments are collaborating on global measures which, they hope, will save money without wrecking the industry.[18,19] Whether this will be successful remains to be seen.

B Patents

'I knew that a country without a patent office … was just a crab,' said Mark Twain, 'and couldn't travel any way but sideways or backwards'. Patent protection is recognized as a cornerstone of the success of the industrialized market economies of Western Europe and North America in the nineteenth and twentieth centuries. Without it, individuals would have been discouraged from investing in innovation fearing that, even if their research was a success, they would never recover development costs because of cut-price competition from imitators.

As described in Chapter 42, a patent provides the discoverer of a 'non-obvious' invention with a monopoly to exploit it for a specified period after its registration in return for disclosure of details of the invention. The period in the United States is 17 years from the granting of the patent, while the European Community has adopted a European Patent Convention, which grants 20 years from the filing of the patent application. This also applies in some countries outside the Community, for example Switzerland. In Japan, a patent lasts for 20 years from filing or 15 years from publication, whichever is shorter. As there is an interval between the filing and granting of a patent, all these patent lives are similar. Some countries, especially in the Third World, do not have patent laws and their citizens are not bound by foreign patents. Such countries have difficulty gaining access to First World technology and are unable to export to countries that subscribe to patent conventions. Trade sanctions may be imposed on offenders, for example the United States took action against Brazil in 1988. Meanwhile, the Trade-Related Intellectual Property (TRIP) agreement, which came into force in 1995, declared that all World Trade Organization countries had to recognize patents, 50 of them not having done so previously.[20] While the agreement was a landmark, its full implementation will take some time.[21]

The pharmaceutical industry invariably seeks patent protection for new drugs. The problem is that the time taken to develop and test a new drug has lengthened. In 1962, the process of invention and testing took on average only 2 years. By the late 1980s, this had lengthened to 11.8 years.

(This figure dropped below 11 years briefly in the mid-1990s, but has risen again to more than 13 years in 1999.[22]) The effective patent life of a drug varied from country to country and drug to drug. Lis and Walker[23,24] concluded that effective patent life had shortened on average to between 6 and 8 years. Thus the time for a company to recoup its investment had been more than halved, while the cost of development had risen exponentially.

The industry campaigned with some success for patent term restoration. The United States Federal Food, Drug and Cosmetic Act was amended in 1984. It extended the patent term of pharmaceuticals and biologicals by half the time from which the first investigational new drug application (INDA) became effective until the date of filing of the new drug application (NDA) plus all the time the FDA takes to review the new drug application, up to a maximum of 5 years. A limitation is that a company cannot have more than a 14-year patent life by this method, and time can also be subtracted if the FDA feels the company is not pursuing the project with sufficient diligence.

A Japanese patent law came into effect in 1988, which allows patent term restoration up to a maximum of 5 years. Restoration is applicable only to products that have taken more than 2 years to obtain marketing authorization after patent grant. The effect of this is similar to the United States legislation.

The European Community lagged behind in legislation, partly because of the legal difficulties in amending the European Patent Convention. Agreement was reached on a maximum of 15 years protection from the date of first marketing. After patent expiry, companies can apply for a supplementary protection certificate for up to 5 years. A series of court rulings on data protection make it likely that some innovators will gain an extra 3 to 4 years marketing exclusivity on top of this, but the total protection must not exceed 15 years.

C Orphan drugs

The spiralling cost of drug development also created problems with drugs for rare diseases. About 4000 rare diseases are described in the literature. Many are known by the names of those who first described them — Huntingdon's chorea, Paget's disease, Tourette's syndrome and so on. Many might or do respond to chemotherapy. It is in the public interest that such diseases should be treated, but the problem is who should pay.

The cost of developing a new drug in the 1990s was estimated at $500 million. If a 15-year patent life and a 9% discount rate are assumed then, to cover the development costs alone, a world-wide income of about $65 million per year is required, about a quarter of this in the European Community. There are about 20 000 sufferers from

Huntingdon's chorea in the Community, which means $3250 per person. Add in the other costs of drug production and the patient is being asked to pay at least $10 000 annually.

Such a sum is lower than the cost of the anti-haemophilia Factor VIII in the massive doses that need to be administered to patients who have developed antibodies to Factor VIII or the cost of β-interferon for multiple sclerosis. It is of the same order of cost as a week in hospital. One might argue that social security funds should be willing to pay large sums for the relief of a tiny minority of the population, but there is a counter-argument that limited funds should be spread where they can do most good to most people. Whatever view is taken, research on rare diseases — and there are many that are rarer than Huntingdon's chorea — must appear uneconomic to pharmaceutical companies planning their research expenditure. Even if they find a drug to cure one, the disease may be so rare that they cannot hope to recoup their expenditure. Alternatively, a drug may be available but they do not want to produce it, either because it is not patentable or because the cost of testing is too high. For a truly rare disease, there are problems assembling enough patients for statistical analysis of test results. One can add to these factors that the companies fear litigation, damages and loss of their good name if things go wrong.

Compounds that no one wants to develop are known as orphan drugs.[25] Because of the difficulties surrounding their development, there was a vigorous campaign in the United States to alter the law. One of the principal spokesmen was David Abelow, who suffered from neurofibromatosis, which became widely known as a result of the film *The Elephant Man*. John Merrick, the subject of the film, had it in a severe form.

As a result of the campaign, the Orphan Drugs Act was passed in 1983, which allows for grants, provides tax credits, eases regulatory pressures and permits 7 years' exclusive marketing rights for non-patentable drugs to companies making them. It is possible for the US Secretary of State to hold the product licence and for the supplier not to be responsible for the safety of the product. In the UK, drugs can be given a clinical trials' certificate, which allows patients under close supervision to be given a drug that has not been fully tested. An orphan drug is defined in the United States as one used for the treatment of a disease or condition affecting fewer than 200 000 people there. If more than 200 000 people are affected, it could alternatively be shown that the cost of developing the drug and making it available would not be covered by sales in the United States.

By January 1991, 50 orphan products had been approved for the treatment of 58 rare conditions.[26] Companies had received $12 million in tax credits for orphan research. The 'classic' orphan drug cases had been sorted out. For example, trientine hydrochloride, $H_2N(CH_2)_2NH(CH_2)_2NH(CH_2)_2NH_2 \cdot HCl$, a chelating agent used to remove excess copper in Wilson's disease for patients who did not respond to penicillamine, and which had been a major concern of Scheinberg and Walshe[27] in Cambridge, was licensed in 1985.

The US Congress became concerned in 1990 about what they saw as abuse by some companies of the Orphan Drug Act, or at any rate that the act was not always working as intended. In some cases, compounds were registered as orphan drugs when a disease was rare, but then it became an epidemic and the number of sufferers rose well above the 200 000 mark. For example, the AIDS virus was identified in 1981. By 1986 there were only 23 000 cases in the United States and the drug azidothymidine (AZT) was genuinely orphan. Since then, many millions of cases have been diagnosed world-wide. Sales of AZT in 1992 amounted to a little under $200 million. A similar rise in demand applied to drugs for the more common side-effects of AIDS. Epoetin alfa, one of the best-selling drugs listed in Table 43.1, obtained orphan drug status for treatment of anaemia associated with AZT therapy, prematurity and end-stage renal disease in both dialysed and non-dialysed patients.

Legislation to amend the Orphan Drug Act has been slow. Early in 1994 a compromise proposal was introduced, which would reduce the period of marketing exclusivity from 7 to 4 years, allow more than one version of an orphan drug if they were developed simultaneously, and allow withdrawal of exclusivity for orphan products when the patient population exceeds 200 000. Orphan drugs 'of limited commercial potential' could be granted an additional 3 years exclusivity if sales information and related data justified it.[28]

It was 1999 before the European Community followed suit on orphan drugs. They agreed full or partial exemption from registration fees, 10 years exclusive marketing and help with chemical R&D for drugs for life-threatening or chronically debilitating illnesses with a patient population less than five per 10 000.[19]

An interesting scientific question raised by the new regulations was the point at which two drugs should be regarded as 'identical'. The definition affects not only whether a novel drug could be patented as a new molecular entity but also whether an out-of-patent generic drug (see Section IV.D) could be approved on the basis of an ANDA.

With small molecules, there was no problem. If structures were different, apart from salts and esters, then the drug was a new chemical entity. With macromolecules, however, small and insignificant differences might occur which, it was felt, did not constitute genuine innovation. For example, two protein drugs would be considered the same if the only difference between them was due to post-transitional events; infidelity of transcription or translation; or minor differences in amino acid sequence. Different glycosylation patterns or tertiary structures would only be considered significant if the drug was shown to be clinically superior.

Even allowing for government grants and tax concessions, the cost of developing orphan drugs is high and the price is bound to be substantial. The original price set for azidothymidine drew widespread protests and was, in fact, reduced on the grounds that sales were likely to be much higher than had been anticipated.

There are other diseases whose incidence is likely to remain static. Cystic fibrosis is the most common genetic disorder in Caucasian populations. It afflicts between one in 1600 and one in 8000. In the UK, about 400 infants per year are born with the disorder and less than 25% would be expected to live until their thirties. The population of sufferers depends on survival rates but, with a United States incidence of one in 3800,[29] a figure of the order of 10 000 is probable, well within the orphan drug limit. Recombinant DNase, Pulmozyme (dornase alpha) is an orphan drug that breaks down the thick mucous secretions associated with the disease. It gained regulatory approval in 1994 but the cost per patient is about $10 000 per year.[30]

The orphan drug that caused the greatest furore was alglucerase (Ceredase), a treatment for Gaucher's disease. Gaucher's disease is characterized by the presence of enlarged lipid-containing histiocytes (Gaucher cells) in the bone marrow. The cells cause bone pain and necrosis of some bones and make the patient liable to bone fractures. The disease (to which Ashkenazi Jews are genetically disposed) is due to a deficiency of the enzyme glucocerebrosidase. There are about 2000–3000 sufferers in Europe and a similar number in the United States who can benefit from enzyme replacement therapy. Ceredase is a modified preparation of glucocerebrosidase derived from placentas. It was approved in the United States in April 1991, in Israel in 1992 and in the European Union in 1994.[31] Supplies are limited and, depending on body weight, response and so on, a year's supply in the UK costs about $500 000. It was described as 'the orphan drug that broke the camel's back',[32] yet is none the less available on the UK National Health Service.

Thus the Orphan Drug Act has been a success in that companies have been encouraged to develop drugs for rare diseases. On the other hand, the US government is currently paying the bill and in certain cases feels that it is being asked to pay excessively for drugs that could legitimately be asked to stand on their own. Whether it proves possible to 'fine-tune' the legislation to make it more cost-effective remains to be seen.

D Generic pharmaceuticals

Innovative drugs, particularly orphan drugs, present economic problems for health administrators. The same is true at the other end of the spectrum, for drugs that have been on the market for many years and are out of patent. The original

manufacturer is then open to imitation and competition. Out-of-patent pharmaceuticals fall into four categories:[33]

(1) Those manufactured by their inventors under their own brand names. For example, Hoechst still sells Lasix® even though furosemide is out of patent.
(2) Those marketed by non-originating companies under their own brand names. For example, amoxycillin is sold as Larotid®, Polymox®, Trimox®, Utimox® and Wymox® as well as under the originator's name, Amoxil®. These are known as branded generics.
(3) Out-of-patent pharmaceuticals marketed by non-originating companies under a generic name plus a company name or prefix or suffix. Furosemide is sold under its own name by Geneva, Lederle and Rugby, and the company name features on the label. The companies undertake limited promotional activity. In some statistics these are included in branded generics.
(4) Out-of-patent pharmaceuticals marketed by non-originating companies under a generic name with minimum mention of the company's name. The bottle may be labelled furosemide BP or furosemide USP and the producer appears only in small letters or not at all.

Categories 1 to 4 are known as multisource drugs, while categories 2 to 4 are the true generics. Group 4 (minimum name generics) are illegal in most European countries since the doctor must specify the source, hence the generics category is limited to groups 2 and 3.

Although some generics companies try to develop new dosage forms, calendar packs and so on, generic drugs in general sell entirely on price. This is resented by the research-based companies who see their profit margins being eroded after what they regard as an inadequate period of patent protection. They present the public with a picture of hole-in-the-wall generics companies flooding the market with inferior, inadequately tested drugs made on the cheap. In 1988, three multinationals sent key legislators in the United States a 'memorandum in opposition' that featured photographs of modern pharmaceutical manufacturing plant compared with the garbage-strewn vacant lot of a 'New York generics manufacturer'.[34] Grotesque as the attack was, it is only fair to note that shortly afterwards the generics industry was rocked by a series of revelations that indeed showed serious failure among some firms to comply with testing standards. Among other infringements, various producers had apparently submitted samples for testing of drugs not made by them but by the originators.

In fact, most generics are made by the same companies that manufacture ethical pharmaceuticals. About 80% of multisource drugs in the United States are made by the approximately 60 member companies of the Pharmaceutical Manufacturers' Association. If one excludes category 4 then the proportion drops to 70%. The figure is anyway rising as research-based companies take over generics companies as an insurance against government measures to contain pharmaceutical prices.

Furthermore, it is the originating companies who can manufacture out-of-patent products on the cheap. They have already amortized their plant, they have personnel already available, they have optimized the process and the product can share analytical and distribution facilities with in-patent products. The reason that the research based companies appear to have higher costs is an accounting convention that loads the costs of research and development of new products on to products already being sold. This is legitimate in that the cash flow for research and development must come out of current income, but it is an illusion to see it as a higher cost. Rather it is a question of cross-subsidy.

Until the 1980s, the issue of generics was not significant. Doctors in general prescribed drugs by the names they knew and loved, which were usually the brand names given by the originators.

In the United States, the situation was revolutionized by the Waxman-Hatch Act of 1984. Generics companies were granted the right to market out-of-patent drugs under their generic names without having to repeat all the tests performed by the originator. Instead, they were able to submit an abbreviated new drug application (ANDA) in which they had to prove that the chemical composition and bioavailability of the drug matched the standards previously agreed by the FDA and the originator. Products regarded as equivalent are listed in the 'Orange Book'.[35] The proportion of new prescriptions written generically in the United States (probably minimum-name generics are meant here) has remained at approximately 14% for many years. Another 44% are being filled generically so that between 50 and 60% (~33% by value) of all prescriptions are generics. The reason is that laws in all 50 states permit the pharmacist to substitute a less expensive generic even when the doctor prescribes a brand name. In some states, doctors must write 'prescribe as written' if they want the patient to receive the branded product and in other states they must sign on one of two lines. Always they must specify that the branded product be supplied if that is what they wish.

In United Kingdom medical schools, doctors are trained to write generic names rather than brand names. By 2000, the proportion of prescriptions written generically had risen to 71% and the number dispensed generically to 52%, the 19% difference being where the prescription was written with a generic name, but the product was still under patent. Because it is the older, cheaper products that are available as generics, the 52% of prescriptions written generically were responsible for 22% of expenditure.

In France, the generics market has been largely confined to hospitals. Perhaps 2–3% of prescriptions are filled generically. Even in-patent drugs in France are cheap, and French doctors have hitherto been hostile to generics, although this is changing slowly.

Generics established themselves from the early 1980s in Germany and by 1989 they accounted for 21.9% of all prescription. Multisource products accounted for 50.4%. The extensive and complicated health service reforms since then have boosted the generics sector to 46.4% by number and 31.8% by value in 1999.[36,37]

From a simplistic viewpoint, the distinction between originators' drugs and generics looks like the difference between, say, Kellogg's corn flakes and a supermarket's own brand. Two factors modify this picture. Firstly, the consumer is able to tell if one brand of corn flakes tastes as good as another and how much the difference is worth financially. Secondly, if the consumer does not pay directly for the product, then there is no reason to choose a cheaper one with an unpromoted brand name. As in all economics problems, it is the organization that pays that complains, and governments are anxious to promote cheaper drugs whether generic or otherwise.

E   Parallel trade

In the past, barriers to international trade were erected in almost every country. Imports of drugs were restricted by tariff or non-tariff barriers to economize on imports, keep down the national health care bill and protect the local pharmaceutical industry. This is no longer the situation. The GATT agreement of 1994 established the World Trade Organization and provided a framework for freer trade in drugs (and other goods). With the United States acting as the bulldozer, other traditional barriers may eventually be demolished. However, a different problem remains: the parallel trade in drugs. This situation occurs when the same product has significantly different prices in two countries. As noted in Section I.C, prices of a given pharmaceutical vary widely between countries. Companies tend to set higher prices in richer countries, although Government intervention may distort the pattern. Traders can then buy the drug in the low-priced market and sell it in the high-priced one; this form of arbitrage forces prices to converge downward. The threat to the incomes of international pharmaceutical companies is obvious.

A long series of court cases in Europe has established, however, that parallel trade is legal within the EU, even when the price differences that make it worthwhile result from the actions of governments rather than companies. Priority was given to Article 30 of the Treaty of Rome, which forbids all measures that establish quantitative restrictions on imports, rather than to Article 36 of the treaty, which permits such restrictions on the grounds of public health or the infringement of intellectual property rights. Accordingly, parallel trade remains a major cause of anxiety for the research-based firms. Although it accounts for no more than 2–3% of the total EU pharmaceutical market, the proportion is much higher in certain countries. In the UK it is about 10%. Furthermore, most of the products affected are popular and profitable patent-protected medicines, such as omeprazole and fluoxetine.

The real worry for pharmaceutical companies has always been the possible extension of parallel trade. The most immediate threat is in Europe, where supply problems have in practice limited the trade. European wholesalers are mostly oriented to their national markets; they are therefore often unable to provide traders with the volume of products desired. As pan-European wholesalers emerge, this constraint may disappear; indeed, such wholesalers may enter the trade themselves.

Could parallel trade develop outside Europe? The TRIP agreement specifically excludes this type of situation from the treaty's scope. Drug prices are high in the United States, lower in Canada, and very much lower in Mexico, so parallel importing from Mexico to the United States would be attractive. At present, the US Prescription Drug Marketing Act prevents the reimport of a patented product by a company other than the US patent holder. As neither the World Trade Agreement nor the North American Free Trade Agreement (NAFTA) deals with this issue, United States law prevails for the time being. Whether it will continue to do so is a moot point.

There is also a question about the supply of drugs cheaply to developing countries, which might then be re-exported to more attractive markets. The South African Government decided in 2001 to import cheap 'pirate' AIDS drugs. An Indian generics company offered to supply combination therapy at $350 per year compared with the United States cost of $10 400 per year.[38] The World Trade Organization (WTO) meeting at Doha in November 2001 agreed that least developed countries (LDCs) had the right to grant compulsory licences; to determine the grounds on which it did so; and to decide what constituted a public health crisis which would justify a country to issue a compulsory licence under a fast-track procedure. The period of transition during which members of the WTO must introduce 20-year patent protection for pharmaceutical products was extended from 2006 to 2016 and possibly beyond. During the transitional period countries can import generic forms of any drug still in patent without the permission of the original patent holder. In effect, this legalizes imports of, for example, medicines for AIDS from, say, India up to 1 January 2005, when the transitional period for India expires. It would also permit a company in an LDC to make what is needed for the local market up to 2016.

The snag is that, in much of Africa, even the lower priced therapy costs several times the per capita income, and the medical infrastructure for diagnosis of illness and delivery of drugs is lacking. The major pharmaceutical companies decided not to sue. The long-term consequences are still unclear. On the one hand, if the decision allows Africa's many AIDS sufferers to receive treatment, it cannot be a bad thing. On the other hand, particularly if parallel trade develops and the patent system is undermined, there is a much reduced incentive for companies to invest in treatments for tropical diseases.

F Pharmacoeconomics

The organizations that actually pay for medicines — governments, insurers and so on — have only recently started to use the power conferred by their expenditure. A consequence is that a drug company has not only to show that its new product is safe and works, but also that it is cost-effective. In Australia, this has been spelled out in legislation. Since 1993, any drug submitted for approval there must be accompanied not only by the results of clinical trials but also by an economic impact analysis. In 1999 the UK set up a National Institute for Clinical Excellence (NICE) to advise the National Health Service on the cost-effectiveness of health care technologies. Other countries ask formally or informally for pharmacoeconomic analysis.[39] Economic impacts can be measured in a variety of ways,[40] for example, cost–effectiveness, cost–utility or full cost–benefit studies.

Pharmacoeconomic analysis must be based on comparisons with existing treatments. It must cover not only the spending on the drug but possible savings on other medical services, for example cost of a period in hospital or of providing social services for an elderly patient at home.

The exercise is relatively simple if a drug is only a marginal improvement on its competitors. Small improvements in hospital bed occupancy and so on can readily be computed. On the other hand, drugs that offer only marginal improvements will rarely be able to command a high enough price to justify development costs. Thus the exercise is mainly of importance to companies trying to justify very high prices for 'breakthrough' products.

Cost–effectiveness studies measure the cost in relation to outcome. The classic cost–effectiveness study was carried out on the pioneering antiulcer drug, cimetidine. The figures are shown in Table 43.9. The major saving in cimetidine-treated patients was in hospital costs. It was also shown that among non-hospitalized patients who were on older drugs, 2 days of work per week were missed on average, while among the cimetidine-treated group only 1 day was missed. A similar study showed the benefits of the anticancer drug carboplatin in spite of its costing 10 times as much as the older drug cisplatin. The point is that the latter has to be given to inpatients, while the former can be used as outpatient therapy. When hospital costs are taken into account, the more expensive drug offers higher cost-effectiveness.[41]

Table 43.9 *Costs and benefits of cimetidine. Average annual Michigan Medicaid expenditures per patient with duodenal ulcer*

	Control group ($)	Cimetidine treated group ($)
Physician	109	57
Hospital	602	97
Drugs	10[a]	66
Total	721	220

Source: Patterson, M.L. (1983) *Management and Decision Economics* **4**: 50.
[a] Does not include antacids that are excluded from Michigan Medicaid.

A third example compared the cost in Sweden of treating patients who had had myocardial infarction with the β-blocker metoprolol. The direct health service costs for treatment were reduced from 17 120 to 12 310 Swedish Kröner, a saving of 28%.[42] A fourth example was used to justify hepatitis B vaccination in Japan. A net saving of 16 billion yen was estimated as a result of vaccination costing 1.3 billion yen.[43]

The problem for hospital administrators is that the largest savings come from a reduction in time spent in hospital. In countries with long hospital waiting lists such as the UK, however, beds do not remain empty. They are reoccupied at once. During a patient's stay in hospital, the first few days are the most expensive, and subsequently only minimal care and nursing are required. Consequently, by hastening the turnover of patients, a hospital will have more patients in the early expensive stage of their stay. Although the average cost per patient treated will decrease, the overall hospital costs will increase and so will the cost per patient per day. Governments might complain and the efficient hospitals will be the ones that are criticized. With 'inefficient' care one patient is treated at a cost of x. With efficient care three patients are treated at a cost of $2x$. More patients benefit but at higher total cost. (In a country with too many hospital beds such as France, the unoccupied bed might remain empty in which case, although the direct costs of hospitalization are saved, the overheads and depreciation will be spread among fewer patients and the cost per patient per day will again appear to rise.)

The Health Service reforms in the UK in 1991 avoided this trap and used instead the cost–utility concept. Cost-effectiveness takes into account only the cost of a single standardized outcome; cost–utility looks at the costs of a range of different outcomes, such as extra years of survival and improved quality of life. A Dutch study of two cholesterol-lowering drugs, for example, showed that cholestyramine cost 131 000 Dutch Guilders per year of life saved compared with simivastatin which cost 31 500 Dutch Guilders.[44]

The above example is attractive in that it compares two forms of therapy. What it does not do is place a value on an extra year of life gained or indeed on the quality of life. Economists have worked hard at producing measures to quantify these two concepts. A number of indexes have been produced, the meaning of which is uncertain and discussion of which is beyond the scope of this chapter. They include the Nottingham Health Profile, the American 'Sickness Impact Profile' and the Rosser Health Index.[45] The Rosser Index has been used to produce a 'quality adjusted life year' or QALY. Each year of life is discounted by a factor reflecting the disability and distress suffered during the year. The factor is obtained by asking people to estimate what number of years of perfect health would be equivalent to 10 years of life in a particular state of disability.

The technique is clearly subjective and there is room for much development and validation. Nonetheless, it makes possible some worthwhile comparisons. Table 43.10 shows the treatment costs per QALY produced by different types of

Table 43.10 *Treatment costs at 1990 prices and technology of various interventions*

Treatment	Cost/QALY (£)
Cholesterol testing and diet only (all adults aged 40–69)	200
Neurosurgical intervention for head injury	240
GP advice to stop smoking	270
Neurosurgical intervention for subarachnoid haemorrhage	490
Anti-hypertensive therapy to prevent stroke (ages 40–69)	940
Pacemaker implantation	1100
Hip replacement[a]	1180
Valve replacement for aortic stenosis	1410
Coronary artery bypass graft (left main disease, severe angina)	2090
Kidney transplant	4710
Breast cancer screening	5780
Heart transplant	7840
Cholesterol testing and treatment (incrementally, all adults aged 25–39)	14 150
Home haemodialysis	17 260
Coronary artery bypass graft (one vessel disease, moderate angina)	18 830
Hospital haemodialysis	21 970
Erythropoetin treatment for anaemia in dialysis patients (assuming 10% mortality reduction)[b]	54 380
Neurosurgical intervention for malignant intercranial tumours	107 780

[a] Does not alter survival, but improves QALY.
[b] Before epoetin alfa became available.
Source: Maynard, A. (1990) The Upjohn Lecture, quoted in *SCRIP* **1568**: 8. Reproduced with permission from PJB Publications Ltd, 2002.

health intervention. The figures are somewhat dated, but illustrate the principle. Simple preventive measures such as changing diet or giving up smoking produce the cheapest benefits. Reduction of blood pressure to avert strokes and selective testing for cholesterol are also cost-effective. Hospital haemodialysis for renal failure is much less cost-effective than kidney transplants. That, together with erythropoietin treatment for dialysis patients and brain surgery for malignant tumours, comes at the most expensive end of the list.

The final type of analysis is the full cost–benefit study. The term is often used to cover all methods described above, but strictly it applies only to studies where every cost and every benefit are taken into account. This includes such things as the contribution to the national economy of people whose lives are saved or prolonged. Such a study would be immensely complicated, but a partial study of hypertension and stroke in Britain has been performed.[46] Between the mid-1950s and mid-1980s, new cases of stroke dropped from 2.4 to 1.75 per 1000 population. The cost of stroke to the Health Service in 1984 was £550 million. Had the incidence been at the 1950s level, it would have cost $550 \times 2.4/1.75 = £754$ million. In addition there was a contribution to the national economy of £322 million through a reduction in premature mortality. A total gain of £526 million can be set against the cost of anti-hypertensive drugs (at manufacturers' prices) of £185 million. The social benefit of control of hypertension is evident.

Economics is not an exact science and the question of the cash value of a year of life or a quality adjusted year of life (whose? where? under what conditions?) is one that no-one would find easy to answer, if there is an answer, and many would find it offensive even to ask the question. The problem is that with limited resources a community has to channel cash to where it will do the most good.

A second approach, related to pharmacoeconomics but leading to fewer moral problems, is outcomes analysis. Physicians, notably those collaborating with Health Management Organizations (HMOs) in the United States but also elsewhere, are required to report their diagnoses and prescriptions to the HMO, which analyses the data and is then in a position to recommend the most effective treatment for a particular diagnosis. The benefit of this is that it should show up marginal differences between treatments and provide solid statistical advice for participating physicians. The drawback, if it is a drawback, is that it reduces a physician's autonomy. The physician might say, 'I have known the patient for 30 years and he or she is suffering from W and should be prescribed X', while the database says that he or she is suffering from Y and should be prescribed Z. Skilled physicians would wish to back their own judgement but, given the litigation currently surrounding medical intervention, they might very well prefer to practise defensive medicine and take the computer's advice.

The question then arises as to why to bother with a physician in the first place. Diagnostic packages are now readily available and uninsured people (and over 50 million Americans lack medical insurance) might well opt for a visit to their local Internet café where they could obtain diagnosis and prescription, also via the Internet, for perhaps a tenth of what a physician would charge. Perhaps the pharmaceutical would come from a 'pirate' source, but there is still a question as to whether this low-grade treatment is better than no treatment at all. The above example is an extreme one, but outcome analysis is here to stay and medical autonomy is certain to decrease as physicians are squeezed between the empowerment of their clients and the demands of their paymasters.

V THE FUTURE OF THE PHARMACEUTICAL INDUSTRY

The years since 1935 have seen the rise of the pharmaceutical industry to its present eminence. Despite the thalidomide disaster and a handful of other tragic episodes, the industry has flourished. A formulary of life-saving and life-improving drugs has been discovered. Two questions are now paramount. Firstly, can society go on spending ever more of its national product on health care and, secondly, can the momentum be maintained? Have all the easy illnesses been dealt with so that only the difficult ones are left?

A Cost-containment measures

Governments are increasingly worried about the rise in health care expenditure. The pharmaceutical industry would say that drugs provide the best possible value within that system and permit treatment at home of patients who would otherwise be costing much more in hospital. Drugs in the United States and Europe average about 10% and 13% of health care spending, respectively, so that savings on drugs have to be huge to make much impact on overall health care costs. Much smaller cuts in hospital costs in percentage terms would have a greater effect. On the other hand, medical and paramedical personnel enjoy high status and public support. The pharmaceutical industry does not: its reputation is equivocal and its very success in commercial terms make it more vulnerable to attack. Official controls on pharmaceutical expenditure are found everywhere in Europe and North America.[47]

Most European countries control the permitted prices of individual prescription drugs. A price is agreed at the time a new product is introduced and it may not be increased even to compensate for inflation without official permission. The exceptions to this system are Denmark, which permits free pricing and Ireland, which permits free pricing in principle, but sets an upper limit based on average prices in other countries. Germany and the Netherlands permit free pricing but have reference prices (see below). The UK regulates prices indirectly via limits on profitability. The United States has free pricing, but stimulates generic competition. In all countries, OTC drugs and prescription drugs that are not reimbursed by the national health care system are free from controls.

The UK Department of Health negotiates with companies a maximum rate of return on capital employed in their sales to the National Health Service. An overall rate is fixed for the industry as a whole and within this figure an individual firm is awarded a particular rate 'having regard to the scale and nature of the company's relevant investments and activities and the associated long-term risks'. The system may perhaps have to be changed because of EC legislation.

The German and Dutch reference pricing systems set reference prices for three groups of pharmaceutical products:

(1) Containing the same active ingredients (i.e. multi-source products)
(2) Containing active ingredients that are comparable therapeutically and pharmacologically (e.g. benzodiazepine anxiolytics)
(3) That have comparable effects, especially if combination products (e.g. antihypertensive–diuretic combinations).

The reference price is related to the average price of the various products on the market. A physician may prescribe as he or she pleases, but if the product costs more than the reference price, the patient has to pay the difference. The result of reference pricing was that virtually all producers dropped their price to the reference price and, where they were in a position to do so, raised the price of their in-patent (therefore not reference-priced) products. This did not necessarily work to the advantage of the generics producers because the cut in the price of the branded product undermined their own major selling point.

An alternative to price controls is to stimulate competition from generics producers, and this is the path that has been followed in the United States and Canada. This was discussed in Section IV.D. In Europe, Germany, Denmark and the Netherlands permit generic substitution.

Pharmaceutical spending may also be controlled by regulation of the drugs that may be prescribed. Denmark, France, Germany and Japan have positive lists. The United Kingdom, the Netherlands and Germany have negative lists. Both reduce the proportion of prescriptions that are reimbursed. In principle, a positive list should indicate a more considered approach to prescribing than a negative list, but in practise other factors seem much more important.

It is difficult to deny reimbursement to popular drugs of doubtful efficacy.

A way for governments to save money is for part of the burden to be shifted to the patient, that is patient copayments. These are more or less universal. Two systems operate: a flat-rate copayment unconnected with the price of the drug as in Ireland, the Netherlands and the UK, or a proportion of the price of the drug as in Spain, Denmark, Belgium and France. Exemptions are common. For example, in the UK 85% of prescriptions were exempt in 2000. Encouragement of OTC drugs is another way in which patients can be induced to contribute. Self-diagnosis and prescription worries doctors, but there has been an increase in the scope of drugs licensed for OTC sale and this may be expected to increase.

The final method for discouraging lavish prescribing is drug budgets for general practitioners. These have been introduced in Germany and the UK, but were discontinued in the latter with a change of Government in 1997. In Germany the effects were immediate and considerable. In the first quarter of 1993, drug volume fell by 17% and spending by 24% compared with 1992. There was a substantial drop in sales volume by the large research-based companies and massive increases by the bottom of the market generics producers.[48] There was also a serious effect on the pharmacists, whose turnover in 1993 was reported to be down by 13%.

In the UK, not all general practitioners were budget holders. Furthermore, the level of generic prescribing in the UK has always been relatively high and there was smaller opportunity for switching. Hence any pattern was swamped by other factors.

The research-based pharmaceutical industry has fought against cost-containment measures. To some extent, the patent term restoration legislation was intended as a sweetener to diminish industry anxiety about the other initiatives. Financial pressures have persuaded a number of research-based companies to take over generics companies to provide them with some sort of insurance against a swing in that direction. The irony of this is that the added respectability given to generics by such policies has increased their rate of penetration of the market.[49] In addition to take-overs, there have been numerous mergers that will enable companies to spread the research risk, although there is evidence that mergers result in a market share lower than the joint market shares of the participants.[21] In the end, however, the industry will continue to flourish only if it continues to produce a stream of innovative drugs. What are the possibilities?

B Trends in pharmaceuticals

Technological forecasting of pharmaceutical progress is as risky, if less expensive, than drug research itself. There have been bitter disappointments in the past 15 years in areas where success was confidently expected. For example, elucidation of the role of endorphins and enkephalins in the perception of pain by the brain gave the hope that effective but non-addictive substitutes for morphine would be available by the mid-1980s. In the event, no drugs based on such structures were developed. Instead, slow delivery forms of morphine have become available so that pain can be 'titrated'. The success of the technique has meant the UK alone now consumes as much morphine as did the world 25 years ago.

Prostaglandins have been another disappointment. They have orphan drug status in a number of applications, but developments have been slow. Our understanding of the arachidonic acid cascade, the biosynthetic route to prostaglandins, has led to advances in the development of prostaglandin inhibitors (e.g. the nonsteroid anti-inflammatories). Prostaglandin analogues, on the other hand, have at present only a handful of indications (Table 43.4).

Statistics on the compounds in research at present are shown in Table 43.11. These give a picture of how the major players see the future. The first column lists them by therapeutic category. Drugs against cancer take the first two

places. Cancer is second only to heart disease as a cause of death in the developed world. Cancer drugs are registered more readily than other drugs because they are the only hope for many sufferers. Progress has been made to the extent that the 5-year survival rate increases steadily. New cancer drugs will continue to reach the market, but it is unlikely that one drug will be the answer to more than a handful of cancers.

Gene therapy comes third in the list, reflecting the 'blue skies' research related to the unravelling of the human genome. Cardiovascular drugs come fourth because heart disease is the major cause of death in the developed world. Anti-asthma drugs appear in sixth place, asthma being on the increase.

Antiviral drugs come eleventh and seventeenth. They are an important research area, but progress has been slow. Acyclovir and related drugs for herpes simplex and varicella-zoster are now well established. AIDS cannot be cured but its progression can be dramatically slowed by combination therapy of two reverse transcriptase inhibitors (e.g. azidothymidine and lamivudine) plus a protease inhibitor (e.g. saquinavir). Unfortunately, viruses have the ability to modify themselves in unpredictable ways. An

Table 43.11 *Drugs in research by therapeutic category and mode of action, 2001*

Rank	Therapeutic category	Number of compounds	Mode of action[a]	Number of compounds
1	Anticancer, other (K6Z) [b]	827	Immunostimulant (IM +)	637
2	Anticancer, immunological (K3)	381	Angiogenesis inhibitor (ANGG-)	143
3	Gene therapy (T4A)	325	Apoptosis stimulant (APOP+)	114
4	Cardiovascular (C9Z)	301	Immunosuppressant (IM-)	97
5	Prophylactic vaccines (J7A1)	292	DNA antagonist (DNA-)	92
6	Antiasthma (R8A)	261	Topoisomerase 2 inhibitor (TO-2-)	79
7	Neuroprotective (N7C)	258	Estrogen agonist (ESTR+)	72
8	Alimentary/metabolic (A16)	234	Cell wall synthesis inhibitor (SY-CW-)	63
9	Antidiabetic (A10B)	233	Protein synthesis antagonist (PRT-SY-)	61
10	Anti-inflammatory (M1A1)	219	Cyclo-oxygenase inhibitor (OX-C-)	60
11	Antiviral, other (J5Z)	214	Bone formation stimulant (BONE+)	56
12	Analgesic, other (N2Z)	209	Arachidonic acid antagonist (ARACH+)	52
13	Biotechnology, other (T5Z)	207	Calcium channel antagonist (CHA-CH-)	48
14	Immunosuppressant (I5)	203	Prostaglandin synthase inhibitor (SY-PG-)	48
15	Memory enhancer (N6D)	195	Insulin agonist (INS+)	46
16	Genomics technology (T4C)	194	Thrombin inhibitor (PR-SE-T-)	45
17	Antiviral HIV (J5A)	173	Sodium channel antagonist (CHA-NA-)	44
18	Recombinant vaccine (T2B)	170	Lipocortins synthesis agonist (LIPCOR+)	43
19	Recombinants, other (T2Z)	159	Opioid μ-receptor agonist (OPI-M+)	43
20	Formulation, neurological (F1N)	156	RNA-directed DNA polymerase inhibitor (PO-RDNA-)	43

[a] The number of compounds listed by mode of action is smaller than that listed by therapeutic category partly because this is a top-20 not a comprehensive list, but mainly because of the omission of the 'unidentified pharmacological activity' category.

[b] The therapy classifications have been adapted from the system used by the European Pharmaceutical Market Research Association (EPhMRA). The EPhMRA system was designed to classify marketed products, and modifications were made to produce a system better suited to drugs in research and development.

Source: Pharmaprojects Update Analysis, May 2001, p.3. Reproduced with permission from Pharmaprojects, PJB Publications Ltd, 2002. For further information please refer to www.pharmaprojects.co.uk

antiviral that would cure influenza or the common cold would be a best seller.

Antibiotics are registered relatively quickly because they are taken for short periods, and long-term toxicity tests are not as important. They did not appear in the top-20 drugs in research in 1993 and there were repeated articles in the press bemoaning the proliferation of antibiotic-resistant superbugs, notably methicillin-resistant staphyloccus aureus (MRSA) and vancomycin resistant enterococci (VRE). It was against the run of play, therefore, that Synercid, a synergistic mixture of the streptogramins, quinupristin and dalfopristin, appeared in 1999. They are active against MRSA and some strains of VRE. A totally new group of antibiotics, the oxazolidinones is also active against resistant bacteria, and linezolid was licensed in 2000. The emergence of these two drugs emphasizes that one should be careful extrapolating from the past into the future — there had been no new group of antibiotics for several decades — and that the most highly featured areas of research are not necessarily the ones where breakthroughs occur.

Ranking of compounds in research by mode of action are shown in the second column of Table 43.11. Discussion of these mechanisms appears elsewhere in this book, but it is worth noting that the list is headed by two anticancer drug development strategies whose validity has yet to be established.

Examination of Table 43.11, however, does not indicate the changes in technology and modes of thinking within the pharmaceutical industry in the 1990s. Combinatorial chemistry and high throughput screening, which are methods of synthesizing and screening drug molecules with the aid of computers, have been developed. Computer-based databases have been developed for assessment of data from clinical trials and evaluation of performance and cost-effectiveness of drugs. Molecular graphics and analytical methods have, of course, progressed and become part of the technological furniture. There has also been progress in biotechnology, with two areas of research — genomics and proteomics — becoming significant. The human genome has been deciphered. New sciences of genomics and proteomics have been developed. If these techniques are as successful as the drug companies hope, then they will add up to a chemotherapeutic revolution.[50]

There are many obstacles to this revolution. Innovation is not a linear process, and the science simply may not work. It may work but take too long and be too expensive for the people who fund it. Political pressure on the drug companies to set lower prices may mean that resources are not available. In most areas of the pharmaceutical industry, there is progress towards harmonization, and the prospects for reasonable policies are encouraging. In the area of pricing and reimbursement, there is conflict. On almost any scenario, there is a mismatch between what the drug companies are hoping to deliver and what the paymasters —

governments, insurance companies and so on — are prepared to pay for. How this conflict will be resolved remains in doubt.

C Conclusion

There is still a large number of interesting and important problems to be solved by the pharmaceutical industry. Many of them can be solved, but at a cost. All drugs are expensive to develop and drugs for rare diseases are prohibitively expensive on a *per capita* basis unless there is a government subsidy. In the 1990s, for the first time as an overt policy issue, society has questioned the amount of money spent on health care. There is general agreement that some sort of rationing is inevitable, but none on who is to do the rationing or how. Will a balance be struck on the basis of ability to pay or on some sort of cost–benefit analysis? Consider the following situation. It appears that the consumption of health care increases exponentially with age and is five to ten times higher at 80–84 than at 60–64.[18,51] The number of old people is increasing rapidly throughout the developed world. Demand for additional health care comes and will come increasingly in the future from the elderly. On a cost–utility analysis, however, treatment of the elderly is not an attractive option. The elderly are less likely to recover from their illnesses. If they do, the fact that they are elderly to start with means that the years of life gained will be smaller. The elderly will have already retired so that their contribution to national income is zero or (because of their pensions) negative.

If health care is to be rationed by cost, then doctors will have the task of turning away patients who cannot pay (and if the doctors are soft-hearted, then the hospitals will turn them away instead). If care is to be allocated on a cost–utility basis, then the doctor will have to find out the patient's age before deciding whether to offer a coronary by-pass, a packet of anti-angina pills or a loaded revolver.

There is thus an ethical challenge to governments, to economists and, worst of all because they are the ones to see the patients, the medical profession. Doctors do not want to be the people to do the rationing, particularly if it is to be done overtly against a series of guidelines, so that they can be criticized afterwards. They have taken the Hippocratic oath and are committed to 'carry out regimen for the benefit of the sick and keep them from harm and wrong'. But, if a doctor does not take the decision, will it be taken by an economist or a politician? Will a patient abandoned by a doctor be permitted to seek a second opinion, and at whose expense? Will there be a health care sector for people whose treatment under the social security system would not be cost-effective, but who are willing and able to pay?

At the beginning of this chapter, I quoted Aristotle's distinction between health and wealth. My late father, Jacob Victor Reuben, to whom this chapter is dedicated, qualified as a doctor in 1920. For his first 15 years in practice, there was scarcely a drug he could prescribe that would cure disease. The most he could do in many cases was to advise patients to stay in bed, take some aspirin and plenty of fluids, and allow nature to run its course. If the patients died, he could say in all honesty that he had done everything in his power to save them. Is his successor 80 years later going to have to say to patients, at any rate the elderly ones, that there are many things he or she *could* do to save them, but they are all too expensive, and why not stay in bed, take some aspirin and plenty of fluids, and allow nature to run its course?

REFERENCES

1. (a) O'Brien, B. (1984) *Patterns of European Diagnosis and Prescribin.* Office of Health Economics, London. (b) Payer, L. (1990) *Medicine and Culture.* Gollancz, London. (c) Teeling-Smith, G. (1991) *Patterns of Prescribing.* Office of Health Economics, London.
2. Arzneiverordnungs Report 2000. (2001) Springer Verlag, Berlin.
3. *FDA re-examines basis for OTC*, New York Times, 28 June 2000.
4. *SCRIP* **1310**: 18.
5. Myllylä, G. (1991) Whole blood and plasma procurement and the impact of plasmapheresis. In Harris, J.R. (ed.). *Blood Separation and Plasma Fractionation*, p. 38. Wiley-Liss, New York.
6. Reuben, B.G. and Wittcoff, H.A. (1990) *Pharmaceutical Chemicals in Perspective*, pp. 337–341. Wiley, New York.
7. US International Trade Commission (1994) *Synthetic Organic Chemicals, US Production and Sales 1992.* Washington DC.
8. Commission of the European Communities (1992) *Rules Governing Medicinal Products in the European Community, Volume IV. Good Manufacturing Practice for Medicinal Products.* Office for Official Publications of the European Community, Luxembourg.
9. Last, P.E. (1987) *Chem. Br.* **23**: 1073.
10. Nathan, S. (2001) *Pharmaceutical Visions*, pp. 54–57. Spring.
11. Brathwaite, M.J. and Ketterman, C.L. (1993) *Chem. Ind.* 1042–1044.
12. Jackson, A. and Angoh, G. (1993) *Chem. Ind.* 1046–1048.
13. Fox, J. (1993) *Chem. Ind.* 270.
14. DiMasi, J.A., Hansen, R.W., Grabowski, H.G. and Lasagna, L. (1991) *J. Health Econ.* **10**: 107.
15. Kettler, H.E. (1999) *Updating the Cost of a New Chemical Entity.* Office of Health Economics, London.
16. Burstall, M.L. and Reuben, B.G. (2001) SPECTRUM Pharmaceutical Industry Dynamics, 5 September 2001, 11-1–11-17.
17. *International Trade Statistics Year Book*, United Nations, New York, published annually. See also Reuben, B.G. (1991) Pharmaceuticals. In Meyers, R.A. (ed.). *Encyclopedia of Physical Science and Technology.* Academic Press, New York.
18. Burstall, M.L., Reuben, B.G. and Reuben, A.J. (1999) Pricing and reimbursement regulation in Europe: the industry perspective—an update. *DIA Journal* **33**: 669–688.
19. See EC No. 141/2000 and 847/2000.
20. UNCTAD (1996) The TRIPS Agreement and developing countries. United Nations, Geneva.
21. Burstall, M.L. and Reuben, B.G. (1999) Outlook for the World Pharmaceutical Industry to 2010. Decision Resources, Waltham MA.
22. CMR International News, 2000, **18:2**, 1.
23. Lis, Y. and Walker, S.R. (1989) *Pharmaceutical Patents.* Centre for Medicines Research: Carshalton, Surrey.
24. Lis, Y. and Walker, S.R. (1989) In Griffith, J.P. (ed.). *Medicines: Regulation, Research and Risk.* Queen's University, Belfast.
25. For general background, see (a) Asbury, C.H. (1985) *Orphan Drugs: Medical versus Market Value.* Lexington Books, Lexington, MA. (b) Asbury, C.H. (1991) The Orphan Drug Act: the first seven years. *J. Am. Med. Assoc.* **265**: 893.
26. A full list of orphan drugs up to 1991 appears in *Drug Evaluations Annual 1992*, American Medical Association, Chicago, IL, 1992.
27. Scheinberg, I.H. and Walshe, J.M. (eds). (1985) *Orphan Diseases and Orphan Drugs.* Manchester University Press, London.
28. *SCRIP* Magazine (1 April 1994) **1910**: 15.
29. *Cystic Fibrosis.* Office of Health Economics, London, 1986.
30. *SCRIP* Magazine (January 1994) 48.
31. *SCRIP* (29 March 1994) *1909*: 21.
32. Tobias, G. and Faigen, N. (1992) *SCRIP* Review: 12.
33. Reuben, B.G. and Burstall, M.L. (1989) *Generic Pharmaceuticals—The Threat.* Economists Advisory Group, London.
34. *Consumers Reports* (1987) **8**: 80 (Magazine published in USA).
35. *Approved Drug products with Therapeutic Equivalence Evaluations.* US Department of Health and Human Services, Rockville, MD, annual updated by monthly supplements.
36. Burstall, M.L. and Reuben, B.G. (1992) *Cost Containment in the European Pharmaceutical Market.* MarketLetter, London.
37. Arzneiverordnugs-Report 2000, Springer Verlag, Heidelberg, 2001.
38. New York Times, 7 February 2001.
39. CMR/OHE R&D Briefing, No. 33, p. 3, 2001, http://www.cmr.org
40. Teeling Smith, G. (ed.)., (1990) *Measuring the Benefits of Medicines: the Future Agenda.* Office of Health Economics, London.
41. Tighe, M. and Goodman, S. (1998) Carboplatin versus cisplatin. *The Lancet* **2**: 1372.
42. Olsson, G., Levin, L.A. and Rehnquist, N. (1987) Economic consequences of post-infarction prophylaxis with β-blockers. *Br. Med. J.* **294**: 339.
43. Institute of Statistical Studies (1998) Technological Assessment of Drug Therapy—Analysis of Cost–Benefit Studies on Pharmaceuticals, Tokyo.
44. Martens, L.L. and Finn, P. (1991) In Teeling Smith, G. (ed.). *Patterns of Prescribing. Office of Health Economics*, London.
45. Rosser, R.M. (1984) A history of the development of health indicators. In Teeling Smith, G. (ed.). *Measuring the Social Benefits of Medicine.* Office of Health Economics, London.
46. Teeling Smith, G. (1989) The economics of hypertension and stroke. *Am. Heart J.* **119** (3) pt 2: 725.
47. Burstall, M.L. and Reuben, B.G. (1990) *Critics of the Pharmceutical Industry*, especially pp. 19–69. REMIT Consultants, London.
48. Munnich, F. (1993) IIR's *Pharma Pricing Meeting '93. SCRIP* **1861**: 6.
49. *SCRIP* (1994) **1916**: 15.
50. Reuben, B.G. (2000) A third chemotherapeutic revolution? *Chem. Ind.* 214–217.
51. (a) Olsson, G.J. (1988) *R. Soc. Med.* **81**: 242. (b) Goldman, L., Sia, B., Cook, F., *et al.* (1988) *N. Engl. J. Med.* **319**: 152.

INDEX

(**Bold** page numbers indicate main entries; *Italic* page numbers indicate illustrations/tables)